1997

UNIFORM BUILDING CODE™

VOLUME 2

STRUCTURAL ENGINEERING DESIGN PROVISIONS

Fourth Printing

Publication Date: April 1997

ISSN 0896-9655
ISBN 1-884590-89-6 (soft cover edition)
ISBN 1-884590-90-X (loose leaf edition)
ISBN 1-884590-93-4 (3-vol. set—soft cover)
ISBN 1-884590-94-2 (3-vol. set—loose leaf)

by

International Conference of Building Officials

5360 WORKMAN MILL ROAD
WHITTIER, CALIFORNIA 90601-2298
(800) 284-4406 • (562) 699-0541

PRINTED IN THE U.S.A.

Preface

The *Uniform Building Code*™ is dedicated to the development of better building construction and greater safety to the public by uniformity in building laws. The code is founded on broad-based principles that make possible the use of new materials and new construction systems.

The *Uniform Building Code* was first enacted by the International Conference of Building Officials at the Sixth Annual Business Meeting held in Phoenix, Arizona, October 18-21, 1927. Revised editions of this code have been published since that time at approximate three-year intervals. New editions incorporate changes approved since the last edition.

The *Uniform Building Code* is designed to be compatible with related publications to provide a complete set of documents for regulatory use. See the publications list following this preface for a listing of the complete family of Uniform Codes and related publications.

Code Changes. The ICBO code development process has been suspended by the Board of Directors and, because of this action, changes to the *Uniform Building Code* will not be processed. For more information, write to the International Conference of Building Officials, 5360 Workman Mill Road, Whittier, California 90601-2298. An analysis of changes between editions is published in the *Analysis of Revisions to the Uniform Codes*.

Marginal Markings. Solid vertical lines in the margins within the body of the code indicate a change from the requirements of the 1994 edition except where an entire chapter was revised, a new chapter was added or a change was minor. Where an entire chapter was revised or a new chapter was added, a notation appears at the beginning of that chapter. The letter **F** repeating in line vertically in the margin indicates that the provision is maintained under the code change procedures of the International Fire Code Institute. Deletion indicators (◗) are provided in the margin where a paragraph or item listing has been deleted if the deletion resulted in a change of requirements.

Three-Volume Set. Provisions of the *Uniform Building Code* have been divided into a three-volume set. Volume 1 accommodates administrative, fire- and life-safety, and field inspection provisions. Chapters 1 through 15 and Chapters 24 through 35 are printed in Volume 1 in their entirety. Any appendix chapters associated with these chapters are printed in their entirety at the end of Volume 1. Excerpts of certain chapters from Volume 2 are reprinted in Volume 1 to provide greater usability.

Volume 2 accommodates structural engineering design provisions, and specifically contains Chapters 16 through 23 printed in their entirety. Included in this volume are design standards that have been added to their respective chapters as divisions of the chapters. Any appendix chapters associated with these chapters are printed in their entirety at the end of Volume 2. Excerpts of certain chapters from Volume 1 are reprinted in Volume 2 to provide greater usability.

Volume 3 contains material, testing and installation standards.

Metrication. The *Uniform Building Code* was metricated in the 1994 edition. The metric conversions are provided in parenthesis following the English units. Where industry has made metric conversions available, the conversions conform to current industry standards.

Formulas are also provided with metric equivalents. Metric equivalent formulas immediately follow the English formula and are denoted by "For **SI:**" preceding the metric equivalent. Some formulas do not use dimensions and, thus, are not provided with a metric equivalent. Multiplying conversion factors have been provided for formulas where metric forms were unavailable. Tables are provided with multiplying conversion factors in subheadings for each tabulated unit of measurement.

CODES AND RELATED PUBLICATIONS

The International Conference of Building Officials (ICBO) publishes a family of codes, each correlated with the *Uniform Building Code*™ to provide jurisdictions with a complete set of building-related regulations for adoption. Some of these codes are published in affiliation with other organizations such as the International Fire Code Institute (IFCI) and the International Code Council (ICC). Reference materials and related codes also are available to improve knowledge of code enforcement and administration of building inspection programs. Publications and products are continually being added, so inquiries should be directed to Conference headquarters for a listing of available products. Many codes and references are also available on CD-ROM or floppy disk. These are denoted by (*). The following publications and products are available from ICBO:

CODES

*Uniform Building Code, Volumes 1, 2 and 3. The most widely adopted model building code in the United States, the performance-based *Uniform Building Code* is a proven document, meeting the needs of government units charged with the enforcement of building regulations. Volume 1 contains administrative, fire- and life-safety and field inspection provisions; Volume 2 contains structural engineering design provisions; and Volume 3 contains material, testing and installation standards.

*Uniform Mechanical Code™. Provides a complete set of requirements for the design, construction, installation and maintenance of heating, ventilating, cooling and refrigeration systems; incinerators and other heat-producing appliances.

International Plumbing Code™. Provides consistent and technically advanced requirements that can be used across the country to provide comprehensive regulations of modern plumbing systems. Setting minimum regulations for plumbing facilities in terms of performance objectives, the IPC provides for the acceptance of new and innovative products, materials and systems.

International Private Sewage Disposal Code™. Provides flexibility in the development of safety and sanitary individual sewage disposal systems and includes detailed provisions for all aspects of design, installation and inspection of private sewage disposal systems.

International Mechanical Code™. Establishes minimum regulations for mechanical systems using prescriptive and performance-related provisions. It is founded on broad-based principles that make possible the use of new materials and new mechanical designs.

Uniform Zoning Code™. This code is dedicated to intelligent community development and to the benefit of the public welfare by providing a means of promoting uniformity in zoning laws and enforcement.

*Uniform Fire Code™, Volumes 1 and 2. The premier model fire code in the United States, the *Uniform Fire Code* sets forth provisions necessary for fire prevention and fire protection. Published by the International Fire Code Institute, the *Uniform Fire Code* is endorsed by the Western Fire Chiefs Association, the International Association of Fire Chiefs and ICBO. Volume 1 contains code provisions compatible with the *Uniform Building Code,* and Volume 2 contains standards referenced from the code provisions.

*Urban-Wildland Interface Code™. Promulgated by IFCI, this code regulates both land use and the built environment in designated urban-wildland interface areas. This newly developed code is the only model code that bases construction requirements on the fire-hazard severity exposed to the structure. Developed under a grant from the Federal Emergency Management Agency, this code is the direct result of hazard mitigation meetings held after devastating wildfires.

Uniform Housing Code™. Provides complete requirements affecting conservation and rehabilitation of housing. Its regulations are compatible with the *Uniform Building Code.*

Uniform Code for the Abatement of Dangerous Buildings™. A code compatible with the *Uniform Building Code* and the *Uniform Housing Code* which provides equitable remedies consistent with other laws for the repair, vacation or demolition of dangerous buildings.

Uniform Sign Code™. Dedicated to the development of better sign regulation, its requirements pertain to all signs and sign construction attached to buildings.

Uniform Administrative Code™. This code covers administrative areas in connection with adoption of the *Uniform Building Code,*

Uniform Mechanical Code and related codes. It contains provisions which relate to site preparation, construction, alteration, moving, repair and use and occupancies of buildings or structures and building service equipment, including plumbing, electrical and mechanical regulations. The code is compatible with the administrative provisions of all codes published by the Conference.

Uniform Building Security Code™. This code establishes minimum standards to make dwelling units resistant to unlawful entry. It regulates swinging doors, sliding doors, windows and hardware in connection with dwelling units of apartment houses or one- and two-family dwellings. The code gives consideration to the concerns of police, fire and building officials in establishing requirements for resistance to burglary which are compatible with fire and life safety.

Uniform Code for Building Conservation™. A building conservation guideline presented in code format which will provide a community with the means to preserve its existing buildings while achieving appropriate levels of safety. It is formatted in the same manner as the *Uniform Building Code,* is compatible with other Uniform Codes, and may be adopted as a code or used as a guideline.

Dwelling Construction under the Uniform Building Code™. Designed primarily for use in home building and apprentice training, this book contains requirements applicable to the construction of one- and two-story dwellings based on the requirements of the *Uniform Building Code.* Available in English or Spanish.

Dwelling Construction under the Uniform Mechanical Code™. This publication is for the convenience of the homeowner or contractor interested in installing mechanical equipment in a one- or two-family dwelling in conformance with the *Uniform Mechanical Code.*

Supplements to UBC and related codes. Published in the years between editions, the Supplements contain all approved changes, plus an analysis of those changes.

Uniform Building Code—1927 Edition. A special 60th anniversary printing of the first published *Uniform Building Code.*

One and Two Family Dwelling Code. Promulgated by ICC, this code eliminates conflicts and duplications among the model codes to achieve national uniformity. Covers mechanical and plumbing requirements as well as construction and occupancy.

Application and Commentary on the One and Two Family Dwelling Code. An interpretative commentary on the *One and Two Family Dwelling Code* intended to enhance uniformity of interpretation and application of the code nationwide. Developed by the three model code organizations, this document includes numerous illustrations of code requirements and the rationale for individual provisions.

Model Energy Code. This code includes minimum requirements for effective use of energy in the design of new buildings and structures and additions to existing buildings. It is based on American Society of Heating, Refrigeration and Air-conditioning Engineers Standard 90A-1980 and was originally developed jointly by ICBO, BOCA, SBCCI and the National Conference of States on Building Codes and Standards under a contract funded by the United States Department of Energy. The code is now maintained by ICC and is adopted by reference in the *Uniform Building Code.*

National Electrical Code®. The electrical code used throughout the United States. Published by the National Fire Protection Association, it is an indispensable aid to every electrician, contractor, architect, builder, inspector and anyone who must specify or certify electrical installations.

TECHNICAL REFERENCES AND EDUCATIONAL MATERIALS

Analysis of Revisions to the Uniform Codes™. An analysis of changes between the previous and new editions of the Uniform Codes is provided. Changes between code editions are noted either at the beginning of chapters or in the margins of the code text.

*Handbook to the Uniform Building Code. The handbook is a completely detailed and illustrated commentary on the *Uniform Building Code,* tracing historical background and rationale of the codes through the current edition. Also included are numerous drawings and figures clarifying the application and intent of the code provisions. Also available in electronic format.

*Handbook to the Uniform Mechanical Code. An indispensable tool for understanding the provisions of the current UMC, the handbook traces the historical background and rationale behind the UMC provisions, includes 160 figures which clarify the intent and application of the code, and provides a chapter-by-chapter analysis of the UMC.

*Uniform Building Code Application Manual. This manual discusses sections of the *Uniform Building Code* with a question-and-answer format, providing a comprehensive analysis of the intent of the code sections. Most sections include illustrative examples. The manual is in loose-leaf format so that code applications published in *Building Standards* magazine may be inserted. Also available in electronic format.

*Uniform Mechanical Code Application Manual. As a companion document to the *Uniform Mechanical Code,* this manual provides a comprehensive analysis of the intent of a number of code sections in an easy-to-use question-and-answer format. The manual is available in a loose-leaf format and includes illustrative examples for many code sections.

*Uniform Fire Code Applications Manual. This newly developed manual provides questions and answers regarding UFC provisions. A comprehensive analysis of the intent of numerous code sections, the manual is in a loose-leaf format for easy insertion of code applications published in IFCI's *Fire Code Journal.*

Quick-Reference Guide to the Occupancy Requirements of the 1997 UBC. Code requirements are compiled in this publication by occupancy groups for quick access. These tabulations assemble requirements for each occupancy classification in the code. Provisions, such as fire-resistive ratings for occupancy separations in Table 3-B, exterior wall and opening protection requirements in Table 5-A-1, and fire-resistive ratings for types of construction in Table 6-A, are tabulated for quick reference and comparison.

Plan Review Manual. A practical text that will assist and guide both the field inspector and plan reviewer in applying the code requirements. This manual covers the nonstructural and basic structural aspects of plan review.

Field Inspection Manual. An important fundamental text for courses of study at the community college and trade or technical school level. It is an effective text for those studying building construction or architecture and includes sample forms and checklists for use in the field.

Building Department Administration. An excellent guide for improvement of skills in departmental management and in the enforcement and application of the Building Code and other regulations administered by a building inspection department. This textbook will also be a valuable aid to instructors, students and those in related professional fields.

Building Department Guide to Disaster Mitigation. This new, expanded guide is designed to assist building departments in developing or updating disaster mitigation plans. Subjects covered include guidelines for damage mitigation, disaster-response management, immediate response, mutual aid and inspections, working with the media, repair and recovery policies, and public information bulletins. This publication is a must for those involved in preparing for and responding to disaster.

Building Official Management Manual. This manual addresses the unique nature of code administration and the managerial duties of the building official. A supplementary insert addresses the budgetary and financial aspects of a building department. It is also an ideal resource for those preparing for the management module of the CABO Building Official Certification Examination.

Legal Aspects of Code Administration. A manual developed by the three model code organizations to inform the building official on the legal aspects of the profession. The text is written in a logical sequence with explanation of legal terminology. It is designed to serve as a refresher for those preparing to take the legal module of the CABO Building Official Certification Examination.

Illustrated Guide to Conventional Construction Provisions of the UBC. This comprehensive guide and commentary provides detailed explanations of the conventional construction provisions in the UBC, including descriptive discussions and illustrated drawings to convey the prescriptive provisions related to wood-frame construction.

Introduction to the Uniform Building Code. A workbook that provides an overview of the basics of the UBC.

Uniform Building Code Update Workbook. This manual addresses many of the changes to the administrative, fire- and life-safety, and inspection provisions appearing in the UBC.

UMC Workbook. Designed for independent study or use with instructor-led programs based on the *Uniform Mechanical Code,* this comprehensive study guide consists of 16 learning sessions, with the first two sessions reviewing the purpose, scope, definitions and administrative provisions and the remaining 14 sessions progressively exploring the requirements for installing, inspecting and maintaining heating, ventilating, cooling and refrigeration systems.

UBC Field Inspection Workbook. A comprehensive workbook for studying the provisions of the UBC. Divided into 12 sessions, this workbook focuses on the UBC combustible construction requirements for the inspection of wood-framed construction.

Concrete Manual. A publication for individuals seeking an understanding of the fundamentals of concrete field technology and inspection practices. Of particular interest to concrete construction inspectors, it will also benefit employees of concrete producers, contractors, testing and inspection laboratories and material suppliers.

Reinforced Concrete Masonry Construction Inspector's Handbook. A comprehensive information source written especially for masonry inspection covering terminology, technology, materials, quality control, inspection and standards. Published jointly by ICBO and the Masonry Institute of America.

You Can Build It! Sponsored by ICBO in cooperation with CABO, this booklet contains information and advice to aid "do-it-yourselfers" with building projects. Provides guidance in necessary procedures such as permit requirements, codes, plans, cost estimation, etc.

Guidelines for Manufactured Housing Installations. A guideline in code form implementing the *Uniform Building Code* and its companion code documents to regulate the permanent installation of a manufactured home on a privately owned, nonrental site. A commentary is included to explain specific provisions, and codes applying to each component part are defined.

Accessibility Reference Guide. This guide is a valuable resource for architects, interior designers, plan reviewers and others who design and enforce accessibility provisions. Features include accessibility requirements, along with detailed commentary and graphics to clarify the provisions; cross-references to other applicable sections of the UBC and the Americans with Disabilities Act Accessibility Guidelines; a checklist of UBC provisions on access and usability requirements; and many other useful references.

Educational and Technical Reference Materials. The Conference has been a leader in the development of texts and course material to assist in the educational process. These materials include vital information necessary for the building official and subordinates in carrying out their responsibilities and have proven to be excellent references in connection with community college curricula and higher-level courses in the field of building construction technology and inspection and in the administration of building departments. Included are plan review checklists for structural, nonstructural, mechanical and fire-safety provisions and a full line of videotapes and automated products.

Table of Contents—Volume 1
Administrative, Fire- and Life-Safety, and Field Inspection Provisions

Table of Contents—Volume 2
Structural Engineering Design Provisions

Table of Contents—Volume 3
Material, Testing and Installation Standards

EFFECTIVE USE OF THE
UNIFORM BUILDING CODE

The following procedure may be helpful in using the *Uniform Building Code:*

1. Classify the building:

    A. **OCCUPANCY CLASSIFICATION:** Compute the floor area and occupant load of the building or portion thereof. See Sections 207 and 1002 and Table 10-A. Determine the occupancy group which the use of the building or portion thereof most nearly resembles. See Sections 301, 303.1.1, 304.1, 305.1, 306.1, 307.1, 308.1, 309.1, 310.1, 311.1 and 312.1. See Section 302 for buildings with mixed occupancies.

    B. **TYPE OF CONSTRUCTION:** Determine the type of construction of the building by the building materials used and the fire resistance of the parts of the building. See Chapter 6.

    C. **LOCATION ON PROPERTY:** Determine the location of the building on the site and clearances to property lines and other buildings from the plot plan. See Table 5-A and Sections 602.3, 603.3, 604.3, 605.3 and 606.3 for fire resistance of exterior walls and wall opening requirements based on proximity to property lines. See Section 503.

    D. **ALLOWABLE FLOOR AREA:** Determine the allowable floor area of the building. See Table 5-B for basic allowable floor area based on occupancy group and type of construction. See Section 505 for allowable increases based on location on property and installation of an approved automatic fire sprinkler system. See Section 504.2 for allowable floor area of multistory buildings.

    E. **HEIGHT AND NUMBER OF STORIES:** Compute the height of the building, Section 209, and determine the number of stories, Section 220. See Table 5-B for the maximum height and number of stories permitted based on occupancy group and type of construction. See Section 506 for allowable story increase based on the installation of an approved automatic fire-sprinkler system.

2. Review the building for conformity with the occupancy requirements in Sections 303 through 312.

3. Review the building for conformity with the type of construction requirements in Chapter 6.

4. Review the building for conformity with the exiting requirements in Chapter 10.

5. Review the building for other detailed code regulations in Chapters 4, 7 through 11, 14, 15, 24 through 26, and 30 through 33, and the appendix.

6. Review the building for conformity with structural engineering regulations and requirements for materials of construction. See Chapters 16 through 23.

Volume 2

Chapters 1 through 15 are printed in Volume 1 of the *Uniform Building Code*.

Chapter 16

STRUCTURAL DESIGN REQUIREMENTS

NOTE: This chapter has been revised in its entirety.

Division I—GENERAL DESIGN REQUIREMENTS

SECTION 1601 — SCOPE

This chapter prescribes general design requirements applicable to all structures regulated by this code.

SECTION 1602 — DEFINITIONS

The following terms are defined for use in this code:

ALLOWABLE STRESS DESIGN is a method of proportioning structural elements such that computed stresses produced in the elements by the allowable stress load combinations do not exceed specified allowable stress (also called working stress design).

BALCONY, EXTERIOR, is an exterior floor system projecting from a structure and supported by that structure, with no additional independent supports.

DEAD LOADS consist of the weight of all materials and fixed equipment incorporated into the building or other structure.

DECK is an exterior floor system supported on at least two opposing sides by an adjoining structure and/or posts, piers, or other independent supports.

FACTORED LOAD is the product of a load specified in Sections 1606 through 1611 and a load factor. See Section 1612.2 for combinations of factored loads.

LIMIT STATE is a condition in which a structure or component is judged either to be no longer useful for its intended function (serviceability limit state) or to be unsafe (strength limit state).

LIVE LOADS are those loads produced by the use and occupancy of the building or other structure and do not include dead load, construction load, or environmental loads such as wind load, snow load, rain load, earthquake load or flood load.

LOAD AND RESISTANCE FACTOR DESIGN (LRFD) is a method of proportioning structural elements using load and resistance factors such that no applicable limit state is reached when the structure is subjected to all appropriate load combinations. The term "LRFD" is used in the design of steel and wood structures.

STRENGTH DESIGN is a method of proportioning structural elements such that the computed forces produced in the elements by the factored load combinations do not exceed the factored element strength. The term "strength design" is used in the design of concrete and masonry structures.

SECTION 1603 — NOTATIONS

D = dead load.

E = earthquake load set forth in Section 1630.1.

E_m = estimated maximum earthquake force that can be developed in the structure as set forth in Section 1630.1.1.

F = load due to fluids.

H = load due to lateral pressure of soil and water in soil.

L = live load, except roof live load, including any permitted live load reduction.

L_r = roof live load, including any permitted live load reduction.

P = ponding load.

S = snow load.

T = self-straining force and effects arising from contraction or expansion resulting from temperature change, shrinkage, moisture change, creep in component materials, movement due to differential settlement, or combinations thereof.

W = load due to wind pressure.

SECTION 1604 — STANDARDS

The standards listed below are recognized standards (see Section 3504).

1. Wind Design.

  1.1 ASCE 7, Chapter 6, Minimum Design Loads for Buildings and Other Structures

  1.2 ANSI EIA/TIA 222-E, Structural Standards for Steel Antenna Towers and Antenna Supporting Structures

  1.3 ANSI/NAAMM FP1001, Guide Specifications for the Design Loads of Metal Flagpoles

SECTION 1605 — DESIGN

1605.1 General. Buildings and other structures and all portions thereof shall be designed and constructed to sustain, within the limitations specified in this code, all loads set forth in Chapter 16 and elsewhere in this code, combined in accordance with Section 1612. Design shall be in accordance with Strength Design, Load and Resistance Factor Design or Allowable Stress Design methods, as permitted by the applicable materials chapters.

> **EXCEPTION:** Unless otherwise required by the building official, buildings or portions thereof that are constructed in accordance with the conventional light-framing requirements specified in Chapter 23 of this code shall be deemed to meet the requirements of this section.

1605.2 Rationality. Any system or method of construction to be used shall be based on a rational analysis in accordance with well-established principles of mechanics. Such analysis shall result in a system that provides a complete load path capable of transferring all loads and forces from their point of origin to the load-resisting elements. The analysis shall include, but not be limited to, the provisions of Sections 1605.2.1 through 1605.2.3.

1605.2.1 Distribution of horizontal shear. The total lateral force shall be distributed to the various vertical elements of the lateral-force-resisting system in proportion to their rigidities considering the rigidity of the horizontal bracing system or diaphragm. Rigid elements that are assumed not to be part of the lateral-force-resisting system may be incorporated into buildings, provided that their effect on the action of the system is considered and provided for in the design.

Provision shall be made for the increased forces induced on resisting elements of the structural system resulting from torsion due to eccentricity between the center of application of the lateral forces and the center of rigidity of the lateral-force-resisting system. For accidental torsion requirements for seismic design, see Section 1630.6.

1605.2.2 Stability against overturning. Every structure shall be designed to resist the overturning effects caused by the lateral forces specified in this chapter. See Section 1611.6 for retaining walls, Section 1615 for wind and Section 1626 for seismic.

1605.2.3 Anchorage. Anchorage of the roof to walls and columns, and of walls and columns to foundations, shall be provided to resist the uplift and sliding forces that result from the application of the prescribed forces.

Concrete and masonry walls shall be anchored to all floors, roofs and other structural elements that provide lateral support for the wall. Such anchorage shall provide a positive direct connection capable of resisting the horizontal forces specified in this chapter but not less than the minimum forces in Section 1611.4. In addition, in Seismic Zones 3 and 4, diaphragm to wall anchorage using embedded straps shall have the straps attached to or hooked around the reinforcing steel or otherwise terminated so as to effectively transfer forces to the reinforcing steel. Walls shall be designed to resist bending between anchors where the anchor spacing exceeds 4 feet (1219 mm). Required anchors in masonry walls of hollow units or cavity walls shall be embedded in a reinforced grouted structural element of the wall. See Sections 1632, 1633.2.8 and 1633.2.9 for earthquake design requirements.

1605.3 Erection of Structural Framing. Walls and structural framing shall be erected true and plumb in accordance with the design.

SECTION 1606 — DEAD LOADS

1606.1 General. Dead loads shall be as defined in Section 1602 and this section.

1606.2 Partition Loads. Floors in office buildings and other buildings where partition locations are subject to change shall be designed to support, in addition to all other loads, a uniformly distributed dead load equal to 20 pounds per square foot (psf) (0.96 kN/m^2) of floor area.

> **EXCEPTION:** Access floor systems shall be designed to support, in addition to all other loads, a uniformly distributed dead load not less than 10 psf (0.48 kN/m^2) of floor area.

SECTION 1607 — LIVE LOADS

1607.1 General. Live loads shall be the maximum loads expected by the intended use or occupancy but in no case shall be less than the loads required by this section.

1607.2 Critical Distribution of Live Loads. Where structural members are arranged to create continuity, members shall be designed using the loading conditions, which would cause maximum shear and bending moments. This requirement may be satisfied in accordance with the provisions of Section 1607.3.2 or 1607.4.2, where applicable.

1607.3 Floor Live Loads.

1607.3.1 General. Floors shall be designed for the unit live loads as set forth in Table 16-A. These loads shall be taken as the minimum live loads in pounds per square foot of horizontal projection to be used in the design of buildings for the occupancies

listed, and loads at least equal shall be assumed for uses not listed in this section but that create or accommodate similar loadings.

Where it can be determined in designing floors that the actual live load will be greater than the value shown in Table 16-A, the actual live load shall be used in the design of such buildings or portions thereof. Special provisions shall be made for machine and apparatus loads.

1607.3.2 Distribution of uniform floor loads. Where uniform floor loads are involved, consideration may be limited to full dead load on all spans in combination with full live load on adjacent spans and alternate spans.

1607.3.3 Concentrated loads. Provision shall be made in designing floors for a concentrated load, L, as set forth in Table 16-A placed upon any space $2^1/_2$ feet (762 mm) square, wherever this load upon an otherwise unloaded floor would produce stresses greater than those caused by the uniform load required therefor.

Provision shall be made in areas where vehicles are used or stored for concentrated loads, L, consisting of two or more loads spaced 5 feet (1524 mm) nominally on center without uniform live loads. Each load shall be 40 percent of the gross weight of the maximum-size vehicle to be accommodated. Parking garages for the storage of private or pleasure-type motor vehicles with no repair or refueling shall have a floor system designed for a concentrated load of not less than 2,000 pounds (8.9 kN) acting on an area of 20 square inches (12 903 mm^2) without uniform live loads. The condition of concentrated or uniform live load, combined in accordance with Section 1612.2 or 1612.3 as appropriate, producing the greatest stresses shall govern.

1607.3.4 Special loads. Provision shall be made for the special vertical and lateral loads as set forth in Table 16-B.

1607.3.5 Live loads posted. The live loads for which each floor or portion thereof of a commercial or industrial building is or has been designed shall have such design live loads conspicuously posted by the owner in that part of each story in which they apply, using durable metal signs, and it shall be unlawful to remove or deface such notices. The occupant of the building shall be responsible for keeping the actual load below the allowable limits.

1607.4 Roof Live Loads.

1607.4.1 General. Roofs shall be designed for the unit live loads, L_r, set forth in Table 16-C. The live loads shall be assumed to act vertically upon the area projected on a horizontal plane.

1607.4.2 Distribution of loads. Where uniform roof loads are involved in the design of structural members arranged to create continuity, consideration may be limited to full dead loads on all spans in combination with full roof live loads on adjacent spans and on alternate spans.

> **EXCEPTION:** Alternate span loading need not be considered where the uniform roof live load is 20 psf (0.96 kN/m^2) or more or where load combinations, including snow load, result in larger members or connections.

For those conditions where light-gage metal preformed structural sheets serve as the support and finish of roofs, roof structural members arranged to create continuity shall be considered adequate if designed for full dead loads on all spans in combination with the most critical one of the following superimposed loads:

1. Snow load in accordance with Section 1614.

2. The uniform roof live load, L_r, set forth in Table 16-C on all spans.

3. A concentrated gravity load, L_r, of 2,000 pounds (8.9 kN) placed on any span supporting a tributary area greater than 200 square feet (18.58 m^2) to create maximum stresses in the member,

whenever this loading creates greater stresses than those caused by the uniform live load. The concentrated load shall be placed on the member over a length of $2^1/_2$ feet (762 mm) along the span. The concentrated load need not be applied to more than one span simultaneously.

4. Water accumulation as prescribed in Section 1611.7.

1607.4.3 Unbalanced loading. Unbalanced loads shall be used where such loading will result in larger members or connections. Trusses and arches shall be designed to resist the stresses caused by unit live loads on one half of the span if such loading results in reverse stresses, or stresses greater in any portion than the stresses produced by the required unit live load on the entire span. For roofs whose structures are composed of a stressed shell, framed or solid, wherein stresses caused by any point loading are distributed throughout the area of the shell, the requirements for unbalanced unit live load design may be reduced 50 percent.

1607.4.4 Special roof loads. Roofs to be used for special purposes shall be designed for appropriate loads as approved by the building official.

Greenhouse roof bars, purlins and rafters shall be designed to carry a 100-pound-minimum (444.8 N) concentrated load, L_r, in addition to the uniform live load.

1607.5 Reduction of Live Loads. The design live load determined using the unit live loads as set forth in Table 16-A for floors and Table 16-C, Method 2, for roofs may be reduced on any member supporting more than 150 square feet (13.94 m^2), including flat slabs, except for floors in places of public assembly and for live loads greater than 100 psf (4.79 kN/m^2), in accordance with the following formula:

$$R = r (A - 150) \qquad (7\text{-}1)$$

For **SI:**

$$R = r (A - 13.94)$$

The reduction shall not exceed 40 percent for members receiving load from one level only, 60 percent for other members or R, as determined by the following formula:

$$R = 23.1 (1 + D/L) \qquad (7\text{-}2)$$

WHERE:

A = area of floor or roof supported by the member, square feet (m^2).

D = dead load per square foot (m^2) of area supported by the member.

L = unit live load per square foot (m^2) of area supported by the member.

R = reduction in percentage.

r = rate of reduction equal to 0.08 percent for floors. See Table 16-C for roofs.

For storage loads exceeding 100 psf (4.79 kN/m^2), no reduction shall be made, except that design live loads on columns may be reduced 20 percent.

The live load reduction shall not exceed 40 percent in garages for the storage of private pleasure cars having a capacity of not more than nine passengers per vehicle.

1607.6 Alternate Floor Live Load Reduction. As an alternate to Formula (7-1), the unit live loads set forth in Table 16-A may be reduced in accordance with Formula (7-3) on any member, including flat slabs, having an influence area of 400 square feet (37.2 m^2) or more.

$$L = L_o \left(0.25 + \frac{15}{\sqrt{A_I}} \right) \qquad (7\text{-}3)$$

For **SI:**

$$L = L_o \left[0.25 + 4.57 \left(\frac{1}{\sqrt{A_I}} \right) \right]$$

WHERE:

A_I = influence area, in square feet (m^2). The influence area A_I is four times the tributary area for a column, two times the tributary area for a beam, equal to the panel area for a two-way slab, and equal to the product of the span and the full flange width for a precast T-beam.

L = reduced design live load per square foot (m^2) of area supported by the member.

L_o = unreduced design live load per square foot (m^2) of area supported by the member (Table 16-A).

The reduced live load shall not be less than 50 percent of the unit live load L_o for members receiving load from one level only, nor less than 40 percent of the unit live load L_o for other members.

SECTION 1608 — SNOW LOADS

Snow loads shall be determined in accordance with Chapter 16, Division II.

SECTION 1609 — WIND LOADS

Wind loads shall be determined in accordance with Chapter 16, Division III.

SECTION 1610 — EARTHQUAKE LOADS

Earthquake loads shall be determined in accordance with Chapter 16, Division IV.

SECTION 1611 — OTHER MINIMUM LOADS

1611.1 General. In addition to the other design loads specified in this chapter, structures shall be designed to resist the loads specified in this section and the special loads set forth in Table 16-B.

1611.2 Other Loads. Buildings and other structures and portions thereof shall be designed to resist all loads due to applicable fluid pressures, F, lateral soil pressures, H, ponding loads, P, and self-straining forces, T. See Section 1611.7 for ponding loads for roofs.

1611.3 Impact Loads. Impact loads shall be included in the design of any structure where impact loads occur.

1611.4 Anchorage of Concrete and Masonry Walls. Concrete and masonry walls shall be anchored as required by Section 1605.2.3. Such anchorage shall be capable of resisting the load combinations of Section 1612.2 or 1612.3 using the greater of the wind or earthquake loads required by this chapter or a minimum horizontal force of 280 pounds per linear foot (4.09 kN/m) of wall, substituted for E.

1611.5 Interior Wall Loads. Interior walls, permanent partitions and temporary partitions that exceed 6 feet (1829 mm) in height shall be designed to resist all loads to which they are subjected but not less than a load, L, of 5 psf (0.24 kN/m^2) applied perpendicular to the walls. The 5 psf (0.24 kN/m^2) load need not be applied simultaneously with wind or seismic loads. The deflection of such walls under a load of 5 psf (0.24 kN/m^2) shall not exceed $^1/_{240}$ of the span for walls with brittle finishes and $^1/_{120}$ of the span for walls with flexible finishes. See Table 16-O for earthquake design requirements where such requirements are more restrictive.

> **EXCEPTION:** Flexible, folding or portable partitions are not required to meet the load and deflection criteria but must be anchored to the supporting structure to meet the provisions of this code.

1611.6 Retaining Walls. Retaining walls shall be designed to resist loads due to the lateral pressure of retained material in accordance with accepted engineering practice. Walls retaining drained soil, where the surface of the retained soil is level, shall be designed for a load, H, equivalent to that exerted by a fluid weighing not less than 30 psf per foot of depth (4.71 kN/m^2/m) and having a depth equal to that of the retained soil. Any surcharge shall be in addition to the equivalent fluid pressure.

Retaining walls shall be designed to resist sliding by at least 1.5 times the lateral force and overturning by at least 1.5 times the overturning moment, using allowable stress design loads.

1611.7 Water Accumulation. All roofs shall be designed with sufficient slope or camber to ensure adequate drainage after the long-term deflection from dead load or shall be designed to resist ponding load, P, combined in accordance with Section 1612.2 or 1612.3. Ponding load shall include water accumulation from any source, including snow, due to deflection. See Section 1506 and Table 16-C, Footnote 3, for drainage slope. See Section 1615 for deflection criteria.

1611.8 Hydrostatic Uplift. All foundations, slabs and other footings subjected to water pressure shall be designed to resist a uniformly distributed uplift load, F, equal to the full hydrostatic pressure.

1611.9 Flood-resistant Construction. For flood-resistant construction requirements, where specifically adopted, see Appendix Chapter 31, Division I.

1611.10 Heliport and Helistop Landing Areas. In addition to other design requirements of this chapter, heliport and helistop landing or touchdown areas shall be designed for the following loads, combined in accordance with Section 1612.2 or 1612.3:

1. Dead load plus actual weight of the helicopter.

2. Dead load plus a single concentrated impact load, L, covering 1 square foot (0.093 m^2) of 0.75 times the fully loaded weight of the helicopter if it is equipped with hydraulic-type shock absorbers, or 1.5 times the fully loaded weight of the helicopter if it is equipped with a rigid or skid-type landing gear.

3. The dead load plus a uniform live load, L, of 100 psf (4.8 kN/m^2). The required live load may be reduced in accordance with Section 1607.5 or 1607.6.

1611.11 Prefabricated Construction.

1611.11.1 Connections. Every device used to connect prefabricated assemblies shall be designed as required by this code and shall be capable of developing the strength of the members connected, except in the case of members forming part of a structural frame designed as specified in this chapter. Connections shall be capable of withstanding uplift forces as specified in this chapter.

1611.11.2 Pipes and conduit. In structural design, due allowance shall be made for any material to be removed for the installation of pipes, conduits or other equipment.

1611.11.3 Tests and inspections. See Section 1704 for requirements for tests and inspections of prefabricated construction.

SECTION 1612 — COMBINATIONS OF LOADS

1612.1 General. Buildings and other structures and all portions thereof shall be designed to resist the load combinations specified in Section 1612.2 or 1612.3 and, where required by Chapter 16, Division IV, or Chapters 18 through 23, the special seismic load combinations of Section 1612.4.

The most critical effect can occur when one or more of the contributing loads are not acting. All applicable loads shall be considered, including both earthquake and wind, in accordance with the specified load combinations.

1612.2 Load Combinations Using Strength Design or Load and Resistance Factor Design.

1612.2.1 Basic load combinations. Where Load and Resistance Factor Design (Strength Design) is used, structures and all portions thereof shall resist the most critical effects from the following combinations of factored loads:

$$1.4D \tag{12-1}$$
$$1.2D + 1.6L + 0.5 \, (L_r \text{ or } S) \tag{12-2}$$
$$1.2D + 1.6 \, (L_r \text{ or } S) + (f_1 L \text{ or } 0.8W) \tag{12-3}$$
$$1.2D + 1.3W + f_1 L + 0.5 \, (L_r \text{ or } S) \tag{12-4}$$
$$1.2D + 1.0E + (f_1 L + f_2 S) \tag{12-5}$$
$$0.9D \pm (1.0E \text{ or } 1.3W) \tag{12-6}$$

WHERE:

f_1 = 1.0 for floors in places of public assembly, for live loads in excess of 100 psf (4.9 kN/m^2), and for garage live load.

= 0.5 for other live loads.

f_2 = 0.7 for roof configurations (such as saw tooth) that do not shed snow off the structure.

= 0.2 for other roof configurations.

> **EXCEPTIONS:** 1. Factored load combinations for concrete per Section 1909.2 where load combinations do not include seismic forces.
>
> 2. Factored load combinations of this section multiplied by 1.1 for concrete and masonry where load combinations include seismic forces.
>
> 3. Where other factored load combinations are specifically required by the provisions of this code.

1612.2.2 Other loads. Where F, H, P or T are to be considered in design, each applicable load shall be added to the above combinations factored as follows: 1.3F, 1.6H, 1.2P and 1.2T.

1612.3 Load Combinations Using Allowable Stress Design.

1612.3.1 Basic load combinations. Where allowable stress design (working stress design) is used, structures and all portions thereof shall resist the most critical effects resulting from the following combinations of loads:

$$D \tag{12-7}$$
$$D + L + (L_r \text{ or } S) \tag{12-8}$$
$$D + \left(W \text{ or } \frac{E}{1.4} \right) \tag{12-9}$$
$$0.9D \pm \frac{E}{1.4} \tag{12-10}$$
$$D + 0.75 \left[L + (L_r \text{ or } S) + \left(W \text{ or } \frac{E}{1.4} \right) \right] \tag{12-11}$$

No increase in allowable stresses shall be used with these load combinations except as specifically permitted elsewhere in this code.

1612.3.2 Alternate basic load combinations. In lieu of the basic load combinations specified in Section 1612.3.1, structures and portions thereof shall be permitted to be designed for the most critical effects resulting from the following load combinations. When using these alternate basic load combinations, a one-third increase shall be permitted in allowable stresses for all combinations including W or E.

$$D + L + (L_r \text{ or } S) \qquad (12\text{-}12)$$

$$D + L + \left(W \text{ or } \frac{E}{1.4} \right) \qquad (12\text{-}13)$$

$$D + L + W + \frac{S}{2} \qquad (12\text{-}14)$$

$$D + L + S + \frac{W}{2} \qquad (12\text{-}15)$$

$$D + L + S + \frac{E}{1.4} \qquad (12\text{-}16)$$

$$0.9D \pm \frac{E}{1.4} \qquad (12\text{-}16\text{-}1)$$

EXCEPTIONS: 1. Crane hook loads need not be combined with roof live load or with more than three fourths of the snow load or one half of the wind load.

2. Design snow loads of 30 psf (1.44 kN/m^2) or less need not be combined with seismic loads. Where design snow loads exceed 30 psf (1.44 kN/m^2), the design snow load shall be included with seismic loads, but may be reduced up to 75 percent where consideration of siting, configuration and load duration warrant when approved by the building official.

1612.3.3 Other loads. Where F, H, P or T are to be considered in design, each applicable load shall be added to the combinations specified in Sections 1612.3.1 and 1612.3.2. When using the alternate load combinations specified in Section 1612.3.2, a one-third increase shall be permitted in allowable stresses for all combinations including W or E.

1612.4 Special Seismic Load Combinations. For both Allowable Stress Design and Strength Design, the following special load combinations for seismic design shall be used as specifically required by Chapter 16, Division IV, or by Chapters 18 through 23:

$$1.2D + f_1L + 1.0E_m \qquad (12\text{-}17)$$

$$0.9D \pm 1.0E_m \qquad (12\text{-}18)$$

WHERE:

f_1 = 1.0 for floors in places of public assembly, for live loads in excess of 100 psf (4.79 kN/m^2), and for garage live load.

= 0.5 for other live loads.

SECTION 1613 — DEFLECTION

The deflection of any structural member shall not exceed the values set forth in Table 16-D, based on the factors set forth in Table 16-E. The deflection criteria representing the most restrictive condition shall apply. Deflection criteria for materials not specified shall be developed in a manner consistent with the provisions of this section. See Section 1611.7 for camber requirements. Span tables for light wood-frame construction as specified in Chapter 23, Division VII, shall conform to the design criteria contained therein. For concrete, see Section 1909.5.2.6; for aluminum, see Section 2003; for glazing framing, see Section 2404.2.

Division II—SNOW LOADS

SECTION 1614 — SNOW LOADS

Buildings and other structures and all portions thereof that are subject to snow loading shall be designed to resist the snow loads, as determined by the building official, in accordance with the load combinations set forth in Section 1612.2 or 1612.3.

Potential unbalanced accumulation of snow at valleys, parapets, roof structures and offsets in roofs of uneven configuration shall be considered.

Snow loads in excess of 20 psf (0.96 kN/m^2) may be reduced for each degree of pitch over 20 degrees by R_s as determined by the formula:

$$R_s = \frac{S}{40} - \frac{1}{2} \qquad (14\text{-}1)$$

For **SI:**
$$R_s = \frac{S}{40} - 0.024$$

WHERE:

R_s = snow load reduction in pounds per square foot (kN/m^2) per degree of pitch over 20 degrees.

S = total snow load in pounds per square foot (kN/m^2).

For alternate design procedure, where specifically adopted, see Appendix Chapter 16, Division I.

Division III—WIND DESIGN

SECTION 1615 — GENERAL

Every building or structure and every portion thereof shall be designed and constructed to resist the wind effects determined in accordance with the requirements of this division. Wind shall be assumed to come from any horizontal direction. No reduction in wind pressure shall be taken for the shielding effect of adjacent structures.

Structures sensitive to dynamic effects, such as buildings with a height-to-width ratio greater than five, structures sensitive to wind-excited oscillations, such as vortex shedding or icing, and buildings over 400 feet (121.9 m) in height, shall be, and any structure may be, designed in accordance with approved national standards.

The provisions of this section do not apply to building and foundation systems in those areas subject to scour and water pressure by wind and wave action. Buildings and foundations subject to such loads shall be designed in accordance with approved national standards.

SECTION 1616 — DEFINITIONS

The following definitions apply only to this division:

BASIC WIND SPEED is the fastest-mile wind speed associated with an annual probability of 0.02 measured at a point 33 feet (10 000 mm) above the ground for an area having exposure category C.

EXPOSURE B has terrain with buildings, forest or surface irregularities, covering at least 20 percent of the ground level area extending 1 mile (1.61 km) or more from the site.

EXPOSURE C has terrain that is flat and generally open, extending $1/2$ mile (0.81 km) or more from the site in any full quadrant.

EXPOSURE D represents the most severe exposure in areas with basic wind speeds of 80 miles per hour (mph) (129 km/h) or greater and has terrain that is flat and unobstructed facing large bodies of water over 1 mile (1.61 km) or more in width relative to any quadrant of the building site. Exposure D extends inland from the shoreline $1/4$ mile (0.40 km) or 10 times the building height, whichever is greater.

FASTEST-MILE WIND SPEED is the wind speed obtained from wind velocity maps prepared by the National Oceanographic and Atmospheric Administration and is the highest sustained average wind speed based on the time required for a mile-long sample of air to pass a fixed point.

OPENINGS are apertures or holes in the exterior wall boundary of the structure. All windows or doors or other openings shall be considered as openings unless such openings and their frames are specifically detailed and designed to resist the loads on elements and components in accordance with the provisions of this section.

PARTIALLY ENCLOSED STRUCTURE OR STORY is a structure or story that has more than 15 percent of any windward projected area open and the area of opening on all other projected areas is less than half of that on the windward projection.

SPECIAL WIND REGION is an area where local records and terrain features indicate 50-year fastest-mile basic wind speed is higher than shown in Figure 16-1.

UNENCLOSED STRUCTURE OR STORY is a structure that has 85 percent or more openings on all sides.

SECTION 1617 — SYMBOLS AND NOTATIONS

The following symbols and notations apply to the provisions of this division:

C_e = combined height, exposure and gust factor coefficient as given in Table 16-G.

C_q = pressure coefficient for the structure or portion of structure under consideration as given in Table 16-H.

I_w = importance factor as set forth in Table 16-K.

P = design wind pressure.

q_s = wind stagnation pressure at the standard height of 33 feet (10 000 mm) as set forth in Table 16-F.

SECTION 1618 — BASIC WIND SPEED

The minimum basic wind speed at any site shall not be less than that shown in Figure 16-1. For those areas designated in Figure 16-1 as special wind regions and other areas where local records or terrain indicate higher 50-year (mean recurrence interval) fastest-mile wind speeds, these higher values shall be the minimum basic wind speeds.

SECTION 1619 — EXPOSURE

An exposure shall be assigned at each site for which a building or structure is to be designed.

SECTION 1620 — DESIGN WIND PRESSURES

Design wind pressures for buildings and structures and elements therein shall be determined for any height in accordance with the following formula:

$$P = C_e \, C_q \, q_s \, I_w \qquad (20\text{-}1)$$

SECTION 1621 — PRIMARY FRAMES AND SYSTEMS

1621.1 General. The primary frames or load-resisting system of every structure shall be designed for the pressures calculated using Formula (20-1) and the pressure coefficients, C_q, of either Method 1 or Method 2. In addition, design of the overall structure and its primary load-resisting system shall conform to Section 1605.

The base overturning moment for the entire structure, or for any one of its individual primary lateral-resisting elements, shall not exceed two thirds of the dead-load-resisting moment. For an entire structure with a height-to-width ratio of 0.5 or less in the wind direction and a maximum height of 60 feet (18 290 mm), the combination of the effects of uplift and overturning may be reduced by one third. The weight of earth superimposed over footings may be used to calculate the dead-load-resisting moment.

1621.2 Method 1 (Normal Force Method). Method 1 shall be used for the design of gabled rigid frames and may be used for any structure. In the Normal Force Method, the wind pressures shall be assumed to act simultaneously normal to all exterior surfaces. For pressures on roofs and leeward walls, C_e shall be evaluated at the mean roof height.

1621.3 Method 2 (Projected Area Method). Method 2 may be used for any structure less than 200 feet (60 960 mm) in height except those using gabled rigid frames. This method may be used in stability determinations for any structure less than 200 feet (60 960 mm) high. In the Projected Area Method, horizontal pressures shall be assumed to act upon the full vertical projected area of the structure, and the vertical pressures shall be assumed to act simultaneously upon the full horizontal projected area.

SECTION 1622 — ELEMENTS AND COMPONENTS OF STRUCTURES

Design wind pressures for each element or component of a structure shall be determined from Formula (20-1) and C_q values from Table 16-H, and shall be applied perpendicular to the surface. For outward acting forces the value of C_e shall be obtained from Table 16-G based on the mean roof height and applied for the entire height of the structure. Each element or component shall be designed for the more severe of the following loadings:

1. The pressures determined using C_q values for elements and components acting over the entire tributary area of the element.

2. The pressures determined using C_q values for local areas at discontinuities such as corners, ridges and eaves. These local pressures shall be applied over a distance from a discontinuity of 10 feet (3048 mm) or 0.1 times the least width of the structure, whichever is less.

The wind pressures from Sections 1621 and 1622 need not be combined.

SECTION 1623 — OPEN-FRAME TOWERS

Radio towers and other towers of trussed construction shall be designed and constructed to withstand wind pressures specified in this section, multiplied by the shape factors set forth in Table 16-H.

SECTION 1624 — MISCELLANEOUS STRUCTURES

Greenhouses, lath houses, agricultural buildings or fences 12 feet (3658 mm) or less in height shall be designed in accordance with Chapter 16, Division III. However, three fourths of q_s, but not less than 10 psf (0.48 kN/m^2), may be substituted for q_s in Formula (20-1). Pressures on local areas at discontinuities need not be considered.

SECTION 1625 — OCCUPANCY CATEGORIES

For the purpose of wind-resistant design, each structure shall be placed in one of the occupancy categories listed in Table 16-K. Table 16-K lists importance factors, I_w, for each category.

Division IV—EARTHQUAKE DESIGN

SECTION 1626 — GENERAL

1626.1 Purpose. The purpose of the earthquake provisions herein is primarily to safeguard against major structural failures and loss of life, not to limit damage or maintain function.

1626.2 Minimum Seismic Design. Structures and portions thereof shall, as a minimum, be designed and constructed to resist the effects of seismic ground motions as provided in this division.

1626.3 Seismic and Wind Design. When the code-prescribed wind design produces greater effects, the wind design shall govern, but detailing requirements and limitations prescribed in this section and referenced sections shall be followed.

SECTION 1627 — DEFINITIONS

For the purposes of this division, certain terms are defined as follows:

BASE is the level at which the earthquake motions are considered to be imparted to the structure or the level at which the structure as a dynamic vibrator is supported.

BASE SHEAR, *V,* is the total design lateral force or shear at the base of a structure.

BEARING WALL SYSTEM is a structural system without a complete vertical load-carrying space frame. See Section 1629.6.2.

BOUNDARY ELEMENT is an element at edges of openings or at perimeters of shear walls or diaphragms.

BRACED FRAME is an essentially vertical truss system of the concentric or eccentric type that is provided to resist lateral forces.

BUILDING FRAME SYSTEM is an essentially complete space frame that provides support for gravity loads. See Section 1629.6.3.

CANTILEVERED COLUMN ELEMENT is a column element in a lateral-force-resisting system that cantilevers from a fixed base and has minimal moment capacity at the top, with lateral forces applied essentially at the top.

COLLECTOR is a member or element provided to transfer lateral forces from a portion of a structure to vertical elements of the lateral-force-resisting system.

COMPONENT is a part or element of an architectural, electrical, mechanical or structural system.

COMPONENT, EQUIPMENT, is a mechanical or electrical component or element that is part of a mechanical and/or electrical system.

COMPONENT, FLEXIBLE, is a component, including its attachments, having a fundamental period greater than 0.06 second.

COMPONENT, RIGID, is a component, including its attachments, having a fundamental period less than or equal to 0.06 second.

CONCENTRICALLY BRACED FRAME is a braced frame in which the members are subjected primarily to axial forces.

DESIGN BASIS GROUND MOTION is that ground motion that has a 10 percent chance of being exceeded in 50 years as determined by a site-specific hazard analysis or may be determined from a hazard map. A suite of ground motion time histories with dynamic properties representative of the site characteristics shall be used to represent this ground motion. The dynamic effects of the Design Basis Ground Motion may be represented by the Design Response Spectrum. See Section 1631.2.

DESIGN RESPONSE SPECTRUM is an elastic response spectrum for 5 percent equivalent viscous damping used to represent the dynamic effects of the Design Basis Ground Motion for the design of structures in accordance with Sections 1630 and 1631. This response spectrum may be either a site-specific spectrum based on geologic, tectonic, seismological and soil characteristics associated with a specific site or may be a spectrum constructed in accordance with the spectral shape in Figure 16-3 using the site-specific values of C_a and C_v and multiplied by the acceleration of gravity, 386.4 in./sec.^{2} (9.815 m/sec.^{2}). See Section 1631.2.

DESIGN SEISMIC FORCE is the minimum total strength design base shear, factored and distributed in accordance with Section 1630.

DIAPHRAGM is a horizontal or nearly horizontal system acting to transmit lateral forces to the vertical-resisting elements. The term "diaphragm" includes horizontal bracing systems.

DIAPHRAGM or SHEAR WALL CHORD is the boundary element of a diaphragm or shear wall that is assumed to take axial stresses analogous to the flanges of a beam.

DIAPHRAGM STRUT (drag strut, tie, collector) is the element of a diaphragm parallel to the applied load that collects and transfers diaphragm shear to the vertical-resisting elements or distributes loads within the diaphragm. Such members may take axial tension or compression.

DRIFT. See "story drift."

DUAL SYSTEM is a combination of moment-resisting frames and shear walls or braced frames designed in accordance with the criteria of Section 1629.6.5.

ECCENTRICALLY BRACED FRAME (EBF) is a steel-braced frame designed in conformance with Section 2213.10.

ELASTIC RESPONSE PARAMETERS are forces and deformations determined from an elastic dynamic analysis using an unreduced ground motion representation, in accordance with Section 1630.

ESSENTIAL FACILITIES are those structures that are necessary for emergency operations subsequent to a natural disaster.

FLEXIBLE ELEMENT or system is one whose deformation under lateral load is significantly larger than adjoining parts of the system. Limiting ratios for defining specific flexible elements are set forth in Section 1630.6.

HORIZONTAL BRACING SYSTEM is a horizontal truss system that serves the same function as a diaphragm.

INTERMEDIATE MOMENT-RESISTING FRAME (IMRF) is a concrete frame designed in accordance with Section 1921.8.

LATERAL-FORCE-RESISTING SYSTEM is that part of the structural system designed to resist the Design Seismic Forces.

MOMENT-RESISTING FRAME is a frame in which members and joints are capable of resisting forces primarily by flexure.

MOMENT-RESISTING WALL FRAME (MRWF) is a masonry wall frame especially detailed to provide ductile behavior and designed in conformance with Section 2108.2.5.

ORDINARY BRACED FRAME (OBF) is a steel-braced frame designed in accordance with the provisions of Section

2213.8 or 2214.6, or concrete-braced frame designed in accordance with Section 1921.

ORDINARY MOMENT-RESISTING FRAME (OMRF) is a moment-resisting frame not meeting special detailing requirements for ductile behavior.

ORTHOGONAL EFFECTS are the earthquake load effects on structural elements common to the lateral-force-resisting systems along two orthogonal axes.

OVERSTRENGTH is a characteristic of structures where the actual strength is larger than the design strength. The degree of overstrength is material- and system-dependent.

$P\Delta$ EFFECT is the secondary effect on shears, axial forces and moments of frame members induced by the vertical loads acting on the laterally displaced building system.

SHEAR WALL is a wall designed to resist lateral forces parallel to the plane of the wall (sometimes referred to as vertical diaphragm or structural wall).

SHEAR WALL-FRAME INTERACTIVE SYSTEM uses combinations of shear walls and frames designed to resist lateral forces in proportion to their relative rigidities, considering interaction between shear walls and frames on all levels.

SOFT STORY is one in which the lateral stiffness is less than 70 percent of the stiffness of the story above. See Table 16-L.

SPACE FRAME is a three-dimensional structural system, without bearing walls, composed of members interconnected so as to function as a complete self-contained unit with or without the aid of horizontal diaphragms or floor-bracing systems.

SPECIAL CONCENTRICALLY BRACED FRAME (SCBF) is a steel-braced frame designed in conformance with the provisions of Section 2213.9.

SPECIAL MOMENT-RESISTING FRAME (SMRF) is a moment-resisting frame specially detailed to provide ductile behavior and comply with the requirements given in Chapter 19 or 22.

SPECIAL TRUSS MOMENT FRAME (STMF) is a moment-resisting frame specially detailed to provide ductile behavior and comply with the provisions of Section 2213.11.

STORY is the space between levels. Story x is the story below Level x.

STORY DRIFT is the lateral displacement of one level relative to the level above or below.

STORY DRIFT RATIO is the story drift divided by the story height.

STORY SHEAR, V_x, is the summation of design lateral forces above the story under consideration.

STRENGTH is the capacity of an element or a member to resist factored load as specified in Chapters 16, 18, 19, 21 and 22.

STRUCTURE is an assemblage of framing members designed to support gravity loads and resist lateral forces. Structures may be categorized as building structures or nonbuilding structures.

SUBDIAPHRAGM is a portion of a larger wood diaphragm designed to anchor and transfer local forces to primary diaphragm struts and the main diaphragm.

VERTICAL LOAD CARRYING FRAME is a space frame designed to carry vertical gravity loads.

WALL ANCHORAGE SYSTEM is the system of elements anchoring the wall to the diaphragm and those elements within the diaphragm required to develop the anchorage forces, including subdiaphragms and continuous ties, as specified in Sections 1633.2.8 and 1633.2.9.

WEAK STORY is one in which the story strength is less than 80 percent of the story above. See Table 16-L.

SECTION 1628 — SYMBOLS AND NOTATIONS

The following symbols and notations apply to the provisions of this division:

A_B = ground floor area of structure in square feet (m²) to include area covered by all overhangs and projections.

A_c = the combined effective area, in square feet (m²), of the shear walls in the first story of the structure.

A_e = the minimum cross-sectional area in any horizontal plane in the first story, in square feet (m²) of a shear wall.

A_x = the torsional amplification factor at Level x.

a_p = numerical coefficient specified in Section 1632 and set forth in Table 16-O.

C_a = seismic coefficient, as set forth in Table 16-Q.

C_t = numerical coefficient given in Section 1630.2.2.

C_v = seismic coefficient, as set forth in Table 16-R.

D = dead load on a structural element.

D_e = the length, in feet (m), of a shear wall in the first story in the direction parallel to the applied forces.

$E, E_h,$
E_m, E_v = earthquake loads set forth in Section 1630.1.

$F_i, F_n,$
F_x = Design Seismic Force applied to Level i, n or x, respectively.

F_p = Design Seismic Forces on a part of the structure.

F_{px} = Design Seismic Force on a diaphragm.

F_t = that portion of the base shear, V, considered concentrated at the top of the structure in addition to F_n.

f_i = lateral force at Level i for use in Formula (30-10).

g = acceleration due to gravity.

$h_i, h_n,$
h_x = height in feet (m) above the base to Level i, n or x, respectively.

I = importance factor given in Table 16-K.

I_p = importance factor specified in Table 16-K.

L = live load on a structural element.

Level i = level of the structure referred to by the subscript i. "$i = 1$" designates the first level above the base.

Level n = that level that is uppermost in the main portion of the structure.

Level x = that level that is under design consideration. "$x = 1$" designates the first level above the base.

M = maximum moment magnitude.

N_a = near-source factor used in the determination of C_a in Seismic Zone 4 related to both the proximity of the building or structure to known faults with magnitudes and slip rates as set forth in Tables 16-S and 16-U.

N_v = near-source factor used in the determination of C_v in Seismic Zone 4 related to both the proximity of the building or structure to known faults with magnitudes and slip rates as set forth in Tables 16-T and 16-U.

PI = plasticity index of soil determined in accordance with approved national standards.

R = numerical coefficient representative of the inherent overstrength and global ductility capacity of lateral-force-resisting systems, as set forth in Table 16-N or 16-P.

r = a ratio used in determining ρ. See Section 1630.1.

$S_A, S_B,$
$S_C, S_D,$
S_E, S_F = soil profile types as set forth in Table 16-J.

T = elastic fundamental period of vibration, in seconds, of the structure in the direction under consideration.

V = the total design lateral force or shear at the base given by Formula (30-5), (30-6), (30-7) or (30-11).

V_x = the design story shear in Story x.

W = the total seismic dead load defined in Section 1630.1.1.

w_i, w_x = that portion of W located at or assigned to Level i or x, respectively.

W_p = the weight of an element or component.

w_{px} = the weight of the diaphragm and the element tributary thereto at Level x, including applicable portions of other loads defined in Section 1630.1.1.

Z = seismic zone factor as given in Table 16-I.

Δ_M = Maximum Inelastic Response Displacement, which is the total drift or total story drift that occurs when the structure is subjected to the Design Basis Ground Motion, including estimated elastic and inelastic contributions to the total deformation defined in Section 1630.9.

Δ_S = Design Level Response Displacement, which is the total drift or total story drift that occurs when the structure is subjected to the design seismic forces.

δ_i = horizontal displacement at Level i relative to the base due to applied lateral forces, f, for use in Formula (30-10).

ρ = Redundancy/Reliability Factor given by Formula (30-3).

Ω_o = Seismic Force Amplification Factor, which is required to account for structural overstrength and set forth in Table 16-N.

SECTION 1629 — CRITERIA SELECTION

1629.1 Basis for Design. The procedures and the limitations for the design of structures shall be determined considering seismic zoning, site characteristics, occupancy, configuration, structural system and height in accordance with this section. Structures shall be designed with adequate strength to withstand the lateral displacements induced by the Design Basis Ground Motion, considering the inelastic response of the structure and the inherent redundancy, overstrength and ductility of the lateral-force-resisting system. The minimum design strength shall be based on the Design Seismic Forces determined in accordance with the static lateral force procedure of Section 1630, except as modified by Section 1631.5.4. Where strength design is used, the load combinations of Section 1612.2 shall apply. Where Allowable Stress Design is used, the load combinations of Section 1612.3 shall apply. Allowable Stress Design may be used to evaluate sliding or overturning at the soil-structure interface regardless of the design approach used in the design of the structure, provided load com-

binations of Section 1612.3 are utilized. One- and two-family dwellings in Seismic Zone 1 need not conform to the provisions of this section.

1629.2 Occupancy Categories. For purposes of earthquake-resistant design, each structure shall be placed in one of the occupancy categories listed in Table 16-K. Table 16-K assigns importance factors, I and I_p, and structural observation requirements for each category.

1629.3 Site Geology and Soil Characteristics. Each site shall be assigned a soil profile type based on properly substantiated geotechnical data using the site categorization procedure set forth in Division V, Section 1636 and Table 16-J.

> **EXCEPTION:** When the soil properties are not known in sufficient detail to determine the soil profile type, Type S_D shall be used. Soil Profile Type S_E or S_F need not be assumed unless the building official determines that Type S_E or S_F may be present at the site or in the event that Type S_E or S_F is established by geotechnical data.

1629.3.1 Soil profile type. Soil Profile Types S_A, S_B, S_C, S_D and S_E are defined in Table 16-J and Soil Profile Type S_F is defined as soils requiring site-specific evaluation as follows:

1. Soils vulnerable to potential failure or collapse under seismic loading, such as liquefiable soils, quick and highly sensitive clays, and collapsible weakly cemented soils.

2. Peats and/or highly organic clays, where the thickness of peat or highly organic clay exceeds 10 feet (3048 mm).

3. Very high plasticity clays with a plasticity index, $PI > 75$, where the depth of clay exceeds 25 feet (7620 mm).

4. Very thick soft/medium stiff clays, where the depth of clay exceeds 120 feet (36 576 mm).

1629.4 Site Seismic Hazard Characteristics. Seismic hazard characteristics for the site shall be established based on the seismic zone and proximity of the site to active seismic sources, site soil profile characteristics and the structure's importance factor.

1629.4.1 Seismic zone. Each site shall be assigned a seismic zone in accordance with Figure 16-2. Each structure shall be assigned a seismic zone factor Z, in accordance with Table 16-I.

1629.4.2 Seismic Zone 4 near-source factor. In Seismic Zone 4, each site shall be assigned a near-source factor in accordance with Table 16-S and the Seismic Source Type set forth in Table 16-U. The value of N_a used to determine C_a need not exceed 1.1 for structures complying with all the following conditions:

1. The soil profile type is S_A, S_B, S_C or S_D.

2. $\rho = 1.0$.

3. Except in single-story structures, Group R, Division 3 and Group U, Division 1 Occupancies, moment frame systems designated as part of the lateral-force-resisting system shall be special moment-resisting frames.

4. The exceptions to Section 2213.7.5 shall not apply, except for columns in one-story buildings or columns at the top story of multistory buildings.

5. None of the following structural irregularities is present: Type 1, 4 or 5 of Table 16-L, and Type 1 or 4 of Table 16-M.

1629.4.3 Seismic response coefficients. Each structure shall be assigned a seismic coefficient, C_a, in accordance with Table 16-Q and a seismic coefficient, C_v, in accordance with Table 16-R.

1629.5 Configuration Requirements.

1629.5.1 General. Each structure shall be designated as being structurally regular or irregular in accordance with Sections 1629.5.2 and 1629.5.3.

1629.5.2 Regular structures. Regular structures have no significant physical discontinuities in plan or vertical configuration or in their lateral-force-resisting systems such as the irregular features described in Section 1629.5.3.

1629.5.3 Irregular structures.

1. Irregular structures have significant physical discontinuities in configuration or in their lateral-force-resisting systems. Irregular features include, but are not limited to, those described in Tables 16-L and 16-M. All structures in Seismic Zone 1 and Occupancy Categories 4 and 5 in Seismic Zone 2 need to be evaluated only for vertical irregularities of Type 5 (Table 16-L) and horizontal irregularities of Type 1 (Table 16-M).

2. Structures having any of the features listed in Table 16-L shall be designated as if having a vertical irregularity.

> **EXCEPTION:** Where no story drift ratio under design lateral forces is greater than 1.3 times the story drift ratio of the story above, the structure may be deemed to not have the structural irregularities of Type 1 or 2 in Table 16-L. The story drift ratio for the top two stories need not be considered. The story drifts for this determination may be calculated neglecting torsional effects.

3. Structures having any of the features listed in Table 16-M shall be designated as having a plan irregularity.

1629.6 Structural Systems.

1629.6.1 General. Structural systems shall be classified as one of the types listed in Table 16-N and defined in this section.

1629.6.2 Bearing wall system. A structural system without a complete vertical load-carrying space frame. Bearing walls or bracing systems provide support for all or most gravity loads. Resistance to lateral load is provided by shear walls or braced frames.

1629.6.3 Building frame system. A structural system with an essentially complete space frame providing support for gravity loads. Resistance to lateral load is provided by shear walls or braced frames.

1629.6.4 Moment-resisting frame system. A structural system with an essentially complete space frame providing support for gravity loads. Moment-resisting frames provide resistance to lateral load primarily by flexural action of members.

1629.6.5 Dual system. A structural system with the following features:

1. An essentially complete space frame that provides support for gravity loads.

2. Resistance to lateral load is provided by shear walls or braced frames and moment-resisting frames (SMRF, IMRF, MMRWF or steel OMRF). The moment-resisting frames shall be designed to independently resist at least 25 percent of the design base shear.

3. The two systems shall be designed to resist the total design base shear in proportion to their relative rigidities considering the interaction of the dual system at all levels.

1629.6.6 Cantilevered column system. A structural system relying on cantilevered column elements for lateral resistance.

1629.6.7 Undefined structural system. A structural system not listed in Table 16-N.

1629.6.8 Nonbuilding structural system. A structural system conforming to Section 1634.

1629.7 Height Limits. Height limits for the various structural systems in Seismic Zones 3 and 4 are given in Table 16-N.

> **EXCEPTION:** Regular structures may exceed these limits by not more than 50 percent for unoccupied structures, which are not accessible to the general public.

1629.8 Selection of Lateral-force Procedure.

1629.8.1 General. Any structure may be, and certain structures defined below shall be, designed using the dynamic lateral-force procedures of Section 1631.

1629.8.2 Simplified static. The simplified static lateral-force procedure set forth in Section 1630.2.3 may be used for the following structures of Occupancy Category 4 or 5:

1. Buildings of any occupancy (including single-family dwellings) not more than three stories in height excluding basements, that use light-frame construction.

2. Other buildings not more than two stories in height excluding basements.

1629.8.3 Static. The static lateral force procedure of Section 1630 may be used for the following structures:

1. All structures, regular or irregular, in Seismic Zone 1 and in Occupancy Categories 4 and 5 in Seismic Zone 2.

2. Regular structures under 240 feet (73 152 mm) in height with lateral force resistance provided by systems listed in Table 16-N, except where Section 1629.8.4, Item 4, applies.

3. Irregular structures not more than five stories or 65 feet (19 812 mm) in height.

4. Structures having a flexible upper portion supported on a rigid lower portion where both portions of the structure considered separately can be classified as being regular, the average story stiffness of the lower portion is at least 10 times the average story stiffness of the upper portion and the period of the entire structure is not greater than 1.1 times the period of the upper portion considered as a separate structure fixed at the base.

1629.8.4 Dynamic. The dynamic lateral-force procedure of Section 1631 shall be used for all other structures, including the following:

1. Structures 240 feet (73 152 mm) or more in height, except as permitted by Section 1629.8.3, Item 1.

2. Structures having a stiffness, weight or geometric vertical irregularity of Type 1, 2 or 3, as defined in Table 16-L, or structures having irregular features not described in Table 16-L or 16-M, except as permitted by Section 1630.4.2.

3. Structures over five stories or 65 feet (19 812 mm) in height in Seismic Zones 3 and 4 not having the same structural system throughout their height except as permitted by Section 1630.4.2.

4. Structures, regular or irregular, located on Soil Profile Type S_F, that have a period greater than 0.7 second. The analysis shall include the effects of the soils at the site and shall conform to Section 1631.2, Item 4.

1629.9 System Limitations.

1629.9.1 Discontinuity. Structures with a discontinuity in capacity, vertical irregularity Type 5 as defined in Table 16-L, shall not be over two stories or 30 feet (9144 mm) in height where the weak story has a calculated strength of less than 65 percent of the story above.

> **EXCEPTION:** Where the weak story is capable of resisting a total lateral seismic force of Ω_o times the design force prescribed in Section 1630.

1629.9.2 Undefined structural systems. For undefined structural systems not listed in Table 16-N, the coefficient R shall be substantiated by approved cyclic test data and analyses. The following items shall be addressed when establishing R:

1. Dynamic response characteristics,

2. Lateral force resistance,

3. Overstrength and strain hardening or softening,

4. Strength and stiffness degradation,

5. Energy dissipation characteristics,

6. System ductility, and

7. Redundancy.

1629.9.3 Irregular features. All structures having irregular features described in Table 16-L or 16-M shall be designed to meet the additional requirements of those sections referenced in the tables.

1629.10 Alternative Procedures.

1629.10.1 General. Alternative lateral-force procedures using rational analyses based on well-established principles of mechanics may be used in lieu of those prescribed in these provisions.

1629.10.2 Seismic isolation. Seismic isolation, energy dissipation and damping systems may be used in the design of structures when approved by the building official and when special detailing is used to provide results equivalent to those obtained by the use of conventional structural systems. For alternate design procedures on seismic isolation systems, refer to Appendix Chapter 16, Division III, Earthquake Regulations for Seismic-isolated Structures.

SECTION 1630 — MINIMUM DESIGN LATERAL FORCES AND RELATED EFFECTS

1630.1 Earthquake Loads and Modeling Requirements.

1630.1.1 Earthquake loads. Structures shall be designed for ground motion producing structural response and seismic forces in any horizontal direction. The following earthquake loads shall be used in the load combinations set forth in Section 1612:

$$E = \rho E_h + E_v \qquad (30\text{-}1)$$

$$E_m = \Omega_o E_h \qquad (30\text{-}2)$$

WHERE:

E = the earthquake load on an element of the structure resulting from the combination of the horizontal component, E_h, and the vertical component, E_v.

E_h = the earthquake load due to the base shear, V, as set forth in Section 1630.2 or the design lateral force, F_p, as set forth in Section 1632.

E_m = the estimated maximum earthquake force that can be developed in the structure as set forth in Section 1630.1.1.

E_v = the load effect resulting from the vertical component of the earthquake ground motion and is equal to an addition of $0.5C_aID$ to the dead load effect, D, for Strength Design, and may be taken as zero for Allowable Stress Design.

Ω_o = the seismic force amplification factor that is required to account for structural overstrength, as set forth in Section 1630.3.1.

ρ = Reliability/Redundancy Factor as given by the following formula:

$$\rho = 2 - \frac{20}{r_{max} \sqrt{A_B}} \qquad (30\text{-}3)$$

For **SI:**

$$\rho = 2 - \frac{6.1}{r_{max} \sqrt{A_B}}$$

WHERE:

r_{max} = the maximum element-story shear ratio. For a given direction of loading, the element-story shear ratio is the ratio of the design story shear in the most heavily loaded single element divided by the total design story shear. For any given Story Level i, the element-story shear ratio is denoted as r_i. The maximum element-story shear ratio r_{max} is defined as the largest of the element story shear ratios, r_i, which occurs in any of the story levels at or below the two-thirds height level of the building.

For braced frames, the value of r_i is equal to the maximum horizontal force component in a single brace element divided by the total story shear.

For moment frames, r_i shall be taken as the maximum of the sum of the shears in any two adjacent columns in a moment frame bay divided by the story shear. For columns common to two bays with moment-resisting connections on opposite sides at Level i in the direction under consideration, 70 percent of the shear in that column may be used in the column shear summation.

For shear walls, r_i shall be taken as the maximum value of the product of the wall shear multiplied by $10/l_w$ (For **SI:** $3.05/l_w$) and divided by the total story shear, where l_w is the length of the wall in feet (m).

For dual systems, r_i shall be taken as the maximum value of r_i as defined above considering all lateral-load-resisting elements. The lateral loads shall be distributed to elements based on relative rigidities considering the interaction of the dual system. For dual systems, the value of ρ need not exceed 80 percent of the value calculated above.

ρ shall not be taken less than 1.0 and need not be greater than 1.5, and A_B is the ground floor area of the structure in square feet (m²). For special moment-resisting frames, except when used in dual systems, ρ shall not exceed 1.25. The number of bays of special moment-resisting frames shall be increased to reduce r, such that ρ is less than or equal to 1.25.

> **EXCEPTION:** A_B may be taken as the average floor area in the upper setback portion of the building where a larger base area exists at the ground floor.

When calculating drift, or when the structure is located in Seismic Zone 0, 1 or 2, ρ shall be taken equal to 1.

The ground motion producing lateral response and design seismic forces may be assumed to act nonconcurrently in the direction of each principal axis of the structure, except as required by Section 1633.1.

Seismic dead load, W, is the total dead load and applicable portions of other loads listed below.

1. In storage and warehouse occupancies, a minimum of 25 percent of the floor live load shall be applicable.

2. Where a partition load is used in the floor design, a load of not less than 10 psf (0.48 kN/m²) shall be included.

3. Design snow loads of 30 psf (1.44 kN/m²) or less need not be included. Where design snow loads exceed 30 psf (1.44 kN/m²), the design snow load shall be included, but may be reduced up to 75 percent where consideration of siting, configuration and load duration warrant when approved by the building official.

4. Total weight of permanent equipment shall be included.

1630.1.2 Modeling requirements. The mathematical model of the physical structure shall include all elements of the lateral-force-resisting system. The model shall also include the stiffness and strength of elements, which are significant to the distribution of forces, and shall represent the spatial distribution of the mass and stiffness of the structure. In addition, the model shall comply with the following:

1. Stiffness properties of reinforced concrete and masonry elements shall consider the effects of cracked sections.

2. For steel moment frame systems, the contribution of panel zone deformations to overall story drift shall be included.

1630.1.3 $P\Delta$ effects. The resulting member forces and moments and the story drifts induced by $P\Delta$ effects shall be considered in the evaluation of overall structural frame stability and shall be evaluated using the forces producing the displacements of Δ_S. $P\Delta$ need not be considered when the ratio of secondary moment to primary moment does not exceed 0.10; the ratio may be evaluated for any story as the product of the total dead, floor live and snow load, as required in Section 1612, above the story times the seismic drift in that story divided by the product of the seismic shear in that story times the height of that story. In Seismic Zones 3 and 4, $P\Delta$ need not be considered when the story drift ratio does not exceed $0.02/R$.

1630.2 Static Force Procedure.

1630.2.1 Design base shear. The total design base shear in a given direction shall be determined from the following formula:

$$V = \frac{C_v I}{R T} W \qquad (30\text{-}4)$$

The total design base shear need not exceed the following:

$$V = \frac{2.5 C_a I}{R} W \qquad (30\text{-}5)$$

The total design base shear shall not be less than the following:

$$V = 0.11 C_a I W \qquad (30\text{-}6)$$

In addition, for Seismic Zone 4, the total base shear shall also not be less than the following:

$$V = \frac{0.8 Z N_v I}{R} W \qquad (30\text{-}7)$$

1630.2.2 Structure period. The value of T shall be determined from one of the following methods:

1. **Method A:** For all buildings, the value T may be approximated from the following formula:

$$T = C_t (h_n)^{3/4} \qquad (30\text{-}8)$$

WHERE:

C_t = 0.035 (0.0853) for steel moment-resisting frames.

C_t = 0.030 (0.0731) for reinforced concrete moment-resisting frames and eccentrically braced frames.

C_t = 0.020 (0.0488) for all other buildings.

Alternatively, the value of C_t for structures with concrete or masonry shear walls may be taken as $0.1/\sqrt{A_c}$ (For **SI:** $0.0743/\sqrt{A_c}$ for A_c in m^2).

The value of A_c shall be determined from the following formula:

$$A_c = \Sigma A_e \left[0.2 + (D_e/h_n)^2 \right] \qquad (30\text{-}9)$$

The value of D_e/h_n used in Formula (30-9) shall not exceed 0.9.

2. **Method B:** The fundamental period T may be calculated using the structural properties and deformational characteristics of the resisting elements in a properly substantiated analysis. The analysis shall be in accordance with the requirements of Section 1630.1.2. The value of T from Method B shall not exceed a value 30 percent greater than the value of T obtained from Method A in Seismic Zone 4, and 40 percent in Seismic Zones 1, 2 and 3.

The fundamental period T may be computed by using the following formula:

$$T = 2\pi \sqrt{\left(\sum_{i=1}^{n} w_i \delta_i{}^2 \right) \div \left(g \sum_{i=1}^{n} f_i \delta_i \right)} \qquad (30\text{-}10)$$

The values of f_i represent any lateral force distributed approximately in accordance with the principles of Formulas (30-13), (30-14) and (30-15) or any other rational distribution. The elastic deflections, δ_i, shall be calculated using the applied lateral forces, f_i.

1630.2.3 Simplified design base shear.

1630.2.3.1 General. Structures conforming to the requirements of Section 1629.8.2 may be designed using this procedure.

1630.2.3.2 Base shear. The total design base shear in a given direction shall be determined from the following formula:

$$V = \frac{3.0 C_a}{R} W \qquad (30\text{-}11)$$

where the value of C_a shall be based on Table 16-Q for the soil profile type. When the soil properties are not known in sufficient detail to determine the soil profile type, Type S_D shall be used in Seismic Zones 3 and 4, and Type S_E shall be used in Seismic Zones 1, 2A and 2B. In Seismic Zone 4, the Near-Source Factor, N_a, need not be greater than 1.3 if none of the following structural irregularities are present: Type 1, 4 or 5 of Table 16-L, or Type 1 or 4 of Table 16-M.

1630.2.3.3 Vertical distribution. The forces at each level shall be calculated using the following formula:

$$F_x = \frac{3.0 C_a}{R} w_i \qquad (30\text{-}12)$$

where the value of C_a shall be determined in Section 1630.2.3.2.

1630.2.3.4 Applicability. Sections 1630.1.2, 1630.1.3, 1630.2.1, 1630.2.2, 1630.5, 1630.9, 1630.10 and 1631 shall not apply when using the simplified procedure.

> **EXCEPTION:** For buildings with relatively flexible structural systems, the building official may require consideration of $P\Delta$ effects and drift in accordance with Sections 1630.1.3, 1630.9 and 1630.10. Δ_s shall be prepared using design seismic forces from Section 1630.2.3.2.

Where used, Δ_M shall be taken equal to 0.01 times the story height of all stories. In Section 1633.2.9, Formula (33-1) shall read $F_{px} = \dfrac{3.0 C_a}{R} w_{px}$ and need not exceed $1.0 C_a w_{px}$, but shall not be less than $0.5 C_a w_{px}$. R and Ω_o shall be taken from Table 16-N.

1630.3 Determination of Seismic Factors.

1630.3.1 Determination of Ω_o. For specific elements of the structure, as specifically identified in this code, the minimum design strength shall be the product of the seismic force overstrength factor Ω_o and the design seismic forces set forth in Section 1630. For both Allowable Stress Design and Strength Design, the Seismic Force Overstrength Factor, Ω_o, shall be taken from Table 16-N.

1630.3.2 Determination of R. The notation R shall be taken from Table 16-N.

1630.4 Combinations of Structural Systems.

1630.4.1 General. Where combinations of structural systems are incorporated into the same structure, the requirements of this section shall be satisfied.

1630.4.2 Vertical combinations. The value of R used in the design of any story shall be less than or equal to the value of R used in the given direction for the story above.

> **EXCEPTION:** This requirement need not be applied to a story where the dead weight above that story is less than 10 percent of the total dead weight of the structure.

Structures may be designed using the procedures of this section under the following conditions:

1. The entire structure is designed using the lowest R of the lateral-force-resisting systems used, or

2. The following two-stage static analysis procedures may be used for structures conforming to Section 1629.8.3, Item 4.

> 2.1 The flexible upper portion shall be designed as a separate structure, supported laterally by the rigid lower portion, using the appropriate values of R and ρ.

> 2.2 The rigid lower portion shall be designed as a separate structure using the appropriate values of R and ρ. The reactions from the upper portion shall be those determined from the analysis of the upper portion amplified by the ratio of the (R/ρ) of the upper portion over (R/ρ) of the lower portion.

1630.4.3 Combinations along different axes. In Seismic Zones 3 and 4 where a structure has a bearing wall system in only one direction, the value of R used for design in the orthogonal direction shall not be greater than that used for the bearing wall system.

Any combination of bearing wall systems, building frame systems, dual systems or moment-resisting frame systems may be used to resist seismic forces in structures less than 160 feet (48 768 mm) in height. Only combinations of dual systems and special moment-resisting frames shall be used to resist seismic forces in structures exceeding 160 feet (48 768 mm) in height in Seismic Zones 3 and 4.

1630.4.4 Combinations along the same axis. For other than dual systems and shear wall-frame interactive systems in Seismic Zones 0 and 1, where a combination of different structural systems is utilized to resist lateral forces in the same direction, the value of R used for design in that direction shall not be greater than the least value for any of the systems utilized in that same direction.

1630.5 Vertical Distribution of Force. The total force shall be distributed over the height of the structure in conformance with Formulas (30-13), (30-14) and (30-15) in the absence of a more rigorous procedure.

$$V = F_t + \sum_{i=1}^{n} F_i \qquad (30\text{-}13)$$

The concentrated force F_t at the top, which is in addition to F_n, shall be determined from the formula:

$$F_t = 0.07\, T\, V \qquad (30\text{-}14)$$

The value of T used for the purpose of calculating F_t shall be the period that corresponds with the design base shear as computed using Formula (30-4). F_t need not exceed $0.25V$ and may be considered as zero where T is 0.7 second or less. The remaining portion of the base shear shall be distributed over the height of the structure, including Level n, according to the following formula:

$$F_x = \frac{(V - F_t)\, w_x\, h_x}{\sum_{i=1}^{n} w_i\, h_i} \qquad (30\text{-}15)$$

At each level designated as x, the force F_x shall be applied over the area of the building in accordance with the mass distribution at that level. Structural displacements and design seismic forces shall be calculated as the effect of forces F_x and F_t applied at the appropriate levels above the base.

1630.6 Horizontal Distribution of Shear. The design story shear, V_x, in any story is the sum of the forces F_t and F_x above that story. V_x shall be distributed to the various elements of the vertical lateral-force-resisting system in proportion to their rigidities, considering the rigidity of the diaphragm. See Section 1633.2.4 for rigid elements that are not intended to be part of the lateral-force-resisting systems.

Where diaphragms are not flexible, the mass at each level shall be assumed to be displaced from the calculated center of mass in each direction a distance equal to 5 percent of the building dimension at that level perpendicular to the direction of the force under consideration. The effect of this displacement on the story shear distribution shall be considered.

Diaphragms shall be considered flexible for the purposes of distribution of story shear and torsional moment when the maximum lateral deformation of the diaphragm is more than two times the average story drift of the associated story. This may be determined by comparing the computed midpoint in-plane deflection of the diaphragm itself under lateral load with the story drift of adjoining vertical-resisting elements under equivalent tributary lateral load.

1630.7 Horizontal Torsional Moments. Provisions shall be made for the increased shears resulting from horizontal torsion where diaphragms are not flexible. The most severe load combination for each element shall be considered for design.

The torsional design moment at a given story shall be the moment resulting from eccentricities between applied design lateral forces at levels above that story and the vertical-resisting elements in that story plus an accidental torsion.

The accidental torsional moment shall be determined by assuming the mass is displaced as required by Section 1630.6.

Where torsional irregularity exists, as defined in Table 16-M, the effects shall be accounted for by increasing the accidental torsion at each level by an amplification factor, A_x, determined from the following formula:

$$A_x = \left[\frac{\delta_{max}}{1.2\, \delta_{avg}} \right]^2 \qquad (30\text{-}16)$$

WHERE:

δ_{avg} = the average of the displacements at the extreme points of the structure at Level x.

δ_{max} = the maximum displacement at Level x.

The value of A_x need not exceed 3.0.

1630.8 Overturning.

1630.8.1 General. Every structure shall be designed to resist the overturning effects caused by earthquake forces specified in Section 1630.5. At any level, the overturning moments to be resisted shall be determined using those seismic forces (F_t and F_x) that act on levels above the level under consideration. At any level, the incremental changes of the design overturning moment shall be distributed to the various resisting elements in the manner prescribed in Section 1630.6. Overturning effects on every element shall be carried down to the foundation. See Sections 1612 and 1633 for combining gravity and seismic forces.

1630.8.2 Elements supporting discontinuous systems.

1630.8.2.1 General. Where any portion of the lateral-load-resisting system is discontinuous, such as for vertical irregularity Type 4 in Table 16-L or plan irregularity Type 4 in Table 16-M, concrete, masonry, steel and wood elements supporting such discontinuous systems shall have the design strength to resist the combination loads resulting from the special seismic load combinations of Section 1612.4.

> **EXCEPTIONS:** 1. The quantity E_m in Section 1612.4 need not exceed the maximum force that can be transferred to the element by the lateral-force-resisting system.
>
> 2. Concrete slabs supporting light-frame wood shear wall systems or light-frame steel and wood structural panel shear wall systems.

For Allowable Stress Design, the design strength may be determined using an allowable stress increase of 1.7 and a resistance factor, ϕ, of 1.0. This increase shall not be combined with the one-third stress increase permitted by Section 1612.3, but may be combined with the duration of load increase permitted in Chapter 23, Division III.

1630.8.2.2 Detailing requirements in Seismic Zones 3 and 4. In Seismic Zones 3 and 4, elements supporting discontinuous systems shall meet the following detailing or member limitations:

1. Reinforced concrete elements designed primarily as axial-load members shall comply with Section 1921.4.4.5.

2. Reinforced concrete elements designed primarily as flexural members and supporting other than light-frame wood shear wall systems or light-frame steel and wood structural panel shear wall systems shall comply with Sections 1921.3.2 and 1921.3.3. Strength computations for portions of slabs designed as supporting elements shall include only those portions of the slab that comply with the requirements of these sections.

3. Masonry elements designed primarily as axial-load carrying members shall comply with Sections 2106.1.12.4, Item 1, and 2108.2.6.2.6.

4. Masonry elements designed primarily as flexural members shall comply with Section 2108.2.6.2.5.

5. Steel elements designed primarily as axial-load members shall comply with Sections 2213.5.2 and 2213.5.3.

6. Steel elements designed primarily as flexural members or trusses shall have bracing for both top and bottom beam flanges or chords at the location of the support of the discontinuous system and shall comply with the requirements of Section 2213.7.1.3.

7. Wood elements designed primarily as flexural members shall be provided with lateral bracing or solid blocking at each end of the element and at the connection location(s) of the discontinuous system.

1630.8.3 At foundation.
See Sections 1629.1 and 1809.4 for overturning moments to be resisted at the foundation soil interface.

1630.9 Drift.
Drift or horizontal displacements of the structure shall be computed where required by this code. For both Allowable Stress Design and Strength Design, the Maximum Inelastic Response Displacement, Δ_M, of the structure caused by the Design Basis Ground Motion shall be determined in accordance with this section. The drifts corresponding to the design seismic forces of Section 1630.2.1, Δ_S, shall be determined in accordance with Section 1630.9.1. To determine Δ_M, these drifts shall be amplified in accordance with Section 1630.9.2.

1630.9.1 Determination of Δ_S. A static, elastic analysis of the lateral force-resisting system shall be prepared using the design seismic forces from Section 1630.2.1. Alternatively, dynamic analysis may be performed in accordance with Section 1631. Where Allowable Stress Design is used and where drift is being computed, the load combinations of Section 1612.2 shall be used. The mathematical model shall comply with Section 1630.1.2. The resulting deformations, denoted as Δ_S, shall be determined at all critical locations in the structure. Calculated drift shall include translational and torsional deflections.

1630.9.2 Determination of Δ_M. The Maximum Inelastic Response Displacement, Δ_M, shall be computed as follows:

$$\Delta_M = 0.7 R\Delta_S \qquad (30\text{-}17)$$

> **EXCEPTION:** Alternatively, Δ_M may be computed by nonlinear time history analysis in accordance with Section 1631.6.

The analysis used to determine the Maximum Inelastic Response Displacement Δ_M shall consider $P\Delta$ effects.

1630.10 Story Drift Limitation.

1630.10.1 General. Story drifts shall be computed using the Maximum Inelastic Response Displacement, Δ_M.

1630.10.2 Calculated. Calculated story drift using Δ_M shall not exceed 0.025 times the story height for structures having a fundamental period of less than 0.7 second. For structures having a fundamental period of 0.7 second or greater, the calculated story drift shall not exceed 0.020 times the story height.

> **EXCEPTIONS:** 1. These drift limits may be exceeded when it is demonstrated that greater drift can be tolerated by both structural elements and nonstructural elements that could affect life safety. The drift used in this assessment shall be based upon the Maximum Inelastic Response Displacement, Δ_M.
>
> 2. There shall be no drift limit in single-story steel-framed structures classified as Groups B, F and S Occupancies or Group H, Division 4 or 5 Occupancies. In Groups B, F and S Occupancies, the primary use shall be limited to storage, factories or workshops. Minor accessory uses shall be allowed in accordance with the provisions of Section 302. Structures on which this exception is used shall not have equipment attached to the structural frame or shall have such equipment detailed to accommodate the additional drift. Walls that are laterally supported by the steel frame shall be designed to accommodate the drift in accordance with Section 1633.2.4.

1630.10.3 Limitations. The design lateral forces used to determine the calculated drift may disregard the limitations of Formula (30-6) and may be based on the period determined from Formula (30-10) neglecting the 30 or 40 percent limitations of Section 1630.2.2, Item 2.

1630.11 Vertical Component.
The following requirements apply in Seismic Zones 3 and 4 only. Horizontal cantilever components shall be designed for a net upward force of $0.7C_aIW_p$.

In addition to all other applicable load combinations, horizontal prestressed components shall be designed using not more than 50 percent of the dead load for the gravity load, alone or in combination with the lateral force effects.

SECTION 1631 — DYNAMIC ANALYSIS PROCEDURES

1631.1 General. Dynamic analyses procedures, when used, shall conform to the criteria established in this section. The analysis shall be based on an appropriate ground motion representation and shall be performed using accepted principles of dynamics.

Structures that are designed in accordance with this section shall comply with all other applicable requirements of these provisions.

1631.2 Ground Motion. The ground motion representation shall, as a minimum, be one having a 10-percent probability of being exceeded in 50 years, shall not be reduced by the quantity R and may be one of the following:

1. An elastic design response spectrum constructed in accordance with Figure 16-3, using the values of C_a and C_v consistent with the specific site. The design acceleration ordinates shall be multiplied by the acceleration of gravity, 386.4 in./sec.^{2} (9.815 m/sec.^{2}).

2. A site-specific elastic design response spectrum based on the geologic, tectonic, seismologic and soil characteristics associated with the specific site. The spectrum shall be developed for a damping ratio of 0.05, unless a different value is shown to be consistent with the anticipated structural behavior at the intensity of shaking established for the site.

3. Ground motion time histories developed for the specific site shall be representative of actual earthquake motions. Response spectra from time histories, either individually or in combination, shall approximate the site design spectrum conforming to Section 1631.2, Item 2.

4. For structures on Soil Profile Type S_F, the following requirements shall apply when required by Section 1629.8.4, Item 4:

    4.1 The ground motion representation shall be developed in accordance with Items 2 and 3.

    4.2 Possible amplification of building response due to the effects of soil-structure interaction and lengthening of building period caused by inelastic behavior shall be considered.

5. The vertical component of ground motion may be defined by scaling corresponding horizontal accelerations by a factor of two-thirds. Alternative factors may be used when substantiated by site-specific data. Where the Near Source Factor, N_a, is greater than 1.0, site-specific vertical response spectra shall be used in lieu of the factor of two-thirds.

1631.3 Mathematical Model. A mathematical model of the physical structure shall represent the spatial distribution of the mass and stiffness of the structure to an extent that is adequate for the calculation of the significant features of its dynamic response. A three-dimensional model shall be used for the dynamic analysis of structures with highly irregular plan configurations such as those having a plan irregularity defined in Table 16-M and having a rigid or semirigid diaphragm. The stiffness properties used in the analysis and general mathematical modeling shall be in accordance with Section 1630.1.2.

1631.4 Description of Analysis Procedures.

1631.4.1 Response spectrum analysis. An elastic dynamic analysis of a structure utilizing the peak dynamic response of all modes having a significant contribution to total structural response. Peak modal responses are calculated using the ordinates of the appropriate response spectrum curve which correspond to the modal periods. Maximum modal contributions are combined in a statistical manner to obtain an approximate total structural response.

1631.4.2 Time-history analysis. An analysis of the dynamic response of a structure at each increment of time when the base is subjected to a specific ground motion time history.

1631.5 Response Spectrum Analysis.

1631.5.1 Response spectrum representation and interpretation of results. The ground motion representation shall be in accordance with Section 1631.2. The corresponding response parameters, including forces, moments and displacements, shall be denoted as Elastic Response Parameters. Elastic Response Parameters may be reduced in accordance with Section 1631.5.4.

1631.5.2 Number of modes. The requirement of Section 1631.4.1 that all significant modes be included may be satisfied by demonstrating that for the modes considered, at least 90 percent of the participating mass of the structure is included in the calculation of response for each principal horizontal direction.

1631.5.3 Combining modes. The peak member forces, displacements, story forces, story shears and base reactions for each mode shall be combined by recognized methods. When three-dimensional models are used for analysis, modal interaction effects shall be considered when combining modal maxima.

1631.5.4 Reduction of Elastic Response Parameters for design. Elastic Response Parameters may be reduced for purposes of design in accordance with the following items, with the limitation that in no case shall the Elastic Response Parameters be reduced such that the corresponding design base shear is less than the Elastic Response Base Shear divided by the value of R.

1. For all regular structures where the ground motion representation complies with Section 1631.2, Item 1, Elastic Response Parameters may be reduced such that the corresponding design base shear is not less than 90 percent of the base shear determined in accordance with Section 1630.2.

2. For all regular structures where the ground motion representation complies with Section 1631.2, Item 2, Elastic Response Parameters may be reduced such that the corresponding design base shear is not less than 80 percent of the base shear determined in accordance with Section 1630.2.

3. For all irregular structures, regardless of the ground motion representation, Elastic Response Parameters may be reduced such that the corresponding design base shear is not less than 100 percent of the base shear determined in accordance with Section 1630.2.

The corresponding reduced design seismic forces shall be used for design in accordance with Section 1612.

1631.5.5 Directional effects. Directional effects for horizontal ground motion shall conform to the requirements of Section 1630.1. The effects of vertical ground motions on horizontal cantilevers and prestressed elements shall be considered in accordance with Section 1630.11. Alternately, vertical seismic response may be determined by dynamic response methods; in no case shall the response used for design be less than that obtained by the static method.

1631.5.6 Torsion. The analysis shall account for torsional effects, including accidental torsional effects as prescribed in Section 1630.7. Where three-dimensional models are used for analysis, effects of accidental torsion shall be accounted for by appropriate adjustments in the model such as adjustment of mass locations, or by equivalent static procedures such as provided in Section 1630.6.

1631.5.7 Dual systems. Where the lateral forces are resisted by a dual system as defined in Section 1629.6.5, the combined system shall be capable of resisting the base shear determined in accordance with this section. The moment-resisting frame shall conform to Section 1629.6.5, Item 2, and may be analyzed using either the procedures of Section 1630.5 or those of Section 1631.5.

by wind, the calculated story drift based on Δ_M or $^1/_2$ inch (12.7 mm), whichever is greater.

2. Connections to permit movement in the plane of the panel for story drift shall be sliding connections using slotted or oversize holes, connections that permit movement by bending of steel, or other connections providing equivalent sliding and ductility capacity.

3. Bodies of connections shall have sufficient ductility and rotation capacity to preclude fracture of the concrete or brittle failures at or near welds.

4. The body of the connection shall be designed for the force determined by Formula (32-2), where $R_p = 3.0$ and $a_p = 1.0$.

5. All fasteners in the connecting system, such as bolts, inserts, welds and dowels, shall be designed for the forces determined by Formula (32-2), where $R_p = 1.0$ and $a_p = 1.0$.

6. Fasteners embedded in concrete shall be attached to, or hooked around, reinforcing steel or otherwise terminated to effectively transfer forces to the reinforcing steel.

1633.2.5 Ties and continuity. All parts of a structure shall be interconnected and the connections shall be capable of transmitting the seismic force induced by the parts being connected. As a minimum, any smaller portion of the building shall be tied to the remainder of the building with elements having at least a strength to resist $0.5 \, C_a I$ times the weight of the smaller portion.

A positive connection for resisting a horizontal force acting parallel to the member shall be provided for each beam, girder or truss. This force shall not be less than $0.5 \, C_a I$ times the dead plus live load.

1633.2.6 Collector elements. Collector elements shall be provided that are capable of transferring the seismic forces originating in other portions of the structure to the element providing the resistance to those forces.

Collector elements, splices and their connections to resisting elements shall resist the forces determined in accordance with Formula (33-1). In addition, collector elements, splices, and their connections to resisting elements shall have the design strength to resist the combined loads resulting from the special seismic load of Section 1612.4.

> **EXCEPTION:** In structures, or portions thereof, braced entirely by light-frame wood shear walls or light-frame steel and wood structural panel shear wall systems, collector elements, splices and connections to resisting elements need only be designed to resist forces in accordance with Formula (33-1).

The quantity E_M need not exceed the maximum force that can be transferred to the collector by the diaphragm and other elements of the lateral-force-resisting system. For Allowable Stress Design, the design strength may be determined using an allowable stress increase of 1.7 and a resistance factor, ϕ, of 1.0. This increase shall not be combined with the one-third stress increase permitted by Section 1612.3, but may be combined with the duration of load increase permitted in Division III of Chapter 23.

1633.2.7 Concrete frames. Concrete frames required by design to be part of the lateral-force-resisting system shall conform to the following:

1. In Seismic Zones 3 and 4 they shall be special moment-resisting frames.

2. In Seismic Zone 2 they shall, as a minimum, be intermediate moment-resisting frames.

1633.2.8 Anchorage of concrete or masonry walls. Concrete or masonry walls shall be anchored to all floors and roofs that provide out-of-plane lateral support of the wall. The anchorage shall provide a positive direct connection between the wall and floor or roof construction capable of resisting the larger of the horizontal forces specified in this section and Sections 1611.4 and 1632. In addition, in Seismic Zones 3 and 4, diaphragm to wall anchorage using embedded straps shall have the straps attached to or hooked around the reinforcing steel or otherwise terminated to effectively transfer forces to the reinforcing steel. Requirements for developing anchorage forces in diaphragms are given in Section 1633.2.9. Diaphragm deformation shall be considered in the design of the supported walls.

1633.2.8.1 Out-of-plane wall anchorage to flexible diaphragms. This section shall apply in Seismic Zones 3 and 4 where flexible diaphragms, as defined in Section 1630.6, provide lateral support for walls.

1. Elements of the wall anchorage system shall be designed for the forces specified in Section 1632 where $R_p = 3.0$ and $a_p = 1.5$.

In Seismic Zone 4, the value of F_p used for the design of the elements of the wall anchorage system shall not be less than 420 pounds per lineal foot (6.1 kN per lineal meter) of wall substituted for E.

See Section 1611.4 for minimum design forces in other seismic zones.

2. When elements of the wall anchorage system are not loaded concentrically or are not perpendicular to the wall, the system shall be designed to resist all components of the forces induced by the eccentricity.

3. When pilasters are present in the wall, the anchorage force at the pilasters shall be calculated considering the additional load transferred from the wall panels to the pilasters. However, the minimum anchorage force at a floor or roof shall be that specified in Section 1633.2.8.1, Item 1.

4. The strength design forces for steel elements of the wall anchorage system shall be 1.4 times the forces otherwise required by this section.

5. The strength design forces for wood elements of the wall anchorage system shall be 0.85 times the force otherwise required by this section and these wood elements shall have a minimum actual net thickness of $2^1/_2$ inches (63.5 mm).

1633.2.9 Diaphragms.

1. The deflection in the plane of the diaphragm shall not exceed the permissible deflection of the attached elements. Permissible deflection shall be that deflection that will permit the attached element to maintain its structural integrity under the individual loading and continue to support the prescribed loads.

2. Floor and roof diaphragms shall be designed to resist the forces determined in accordance with the following formula:

$$F_{px} = \frac{F_t + \sum\limits_{i=x}^{n} F_i}{\sum\limits_{i=x}^{n} w_i} \, w_{px} \qquad (33\text{-}1)$$

The force F_{px} determined from Formula (33-1) need not exceed $1.0 C_a I w_{px}$, but shall not be less than $0.5 C_a I w_{px}$.

When the diaphragm is required to transfer design seismic forces from the vertical-resisting elements above the diaphragm to other vertical-resisting elements below the diaphragm due to offset in the placement of the elements or to changes in stiffness in the vertical elements, these forces shall be added to those determined from Formula (33-1).

3. Design seismic forces for flexible diaphragms providing lateral supports for walls or frames of masonry or concrete shall be

determined using Formula (33-1) based on the load determined in accordance with Section 1630.2 using a R not exceeding 4.

4. Diaphragms supporting concrete or masonry walls shall have continuous ties or struts between diaphragm chords to distribute the anchorage forces specified in Section 1633.2.8. Added chords of subdiaphragms may be used to form subdiaphragms to transmit the anchorage forces to the main continuous crossties. The maximum length-to-width ratio of the wood structural subdiaphragm shall be $2^1/_2$:1.

5. Where wood diaphragms are used to laterally support concrete or masonry walls, the anchorage shall conform to Section 1633.2.8. In Seismic Zones 2, 3 and 4, anchorage shall not be accomplished by use of toenails or nails subject to withdrawal, wood ledgers or framing shall not be used in cross-grain bending or cross-grain tension, and the continuous ties required by Item 4 shall be in addition to the diaphragm sheathing.

6. Connections of diaphragms to the vertical elements in structures in Seismic Zones 3 and 4, having a plan irregularity of Type 1, 2, 3 or 4 in Table 16-M, shall be designed without considering either the one-third increase or the duration of load increase considered in allowable stresses for elements resisting earthquake forces.

7. In structures in Seismic Zones 3 and 4 having a plan irregularity of Type 2 in Table 16-M, diaphragm chords and drag members shall be designed considering independent movement of the projecting wings of the structure. Each of these diaphragm elements shall be designed for the more severe of the following two assumptions:

Motion of the projecting wings in the same direction.

Motion of the projecting wings in opposing directions.

> **EXCEPTION:** This requirement may be deemed satisfied if the procedures of Section 1631 in conjunction with a three-dimensional model have been used to determine the lateral seismic forces for design.

1633.2.10 Framing below the base. The strength and stiffness of the framing between the base and the foundation shall not be less than that of the superstructure. The special detailing requirements of Chapters 19 and 22, as appropriate, shall apply to columns supporting discontinuous lateral-force-resisting elements and to SMRF, IMRF, EBF, STMF and MMRWF system elements below the base, which are required to transmit the forces resulting from lateral loads to the foundation.

1633.2.11 Building separations. All structures shall be separated from adjoining structures. Separations shall allow for the displacement Δ_M. Adjacent buildings on the same property shall be separated by at least Δ_{MT} where

$$\Delta_{MT} = \sqrt{(\Delta_{M1})^2 + (\Delta_{M2})^2} \qquad (33\text{-}2)$$

and Δ_{M1} and Δ_{M2} are the displacements of the adjacent buildings.

When a structure adjoins a property line not common to a public way, that structure shall also be set back from the property line by at least the displacement Δ_M of that structure.

> **EXCEPTION:** Smaller separations or property line setbacks may be permitted when justified by rational analyses based on maximum expected ground motions.

SECTION 1634 — NONBUILDING STRUCTURES

1634.1 General.

1634.1.1 Scope. Nonbuilding structures include all self-supporting structures other than buildings that carry gravity loads and resist the effects of earthquakes. Nonbuilding structures shall

be designed to provide the strength required to resist the displacements induced by the minimum lateral forces specified in this section. Design shall conform to the applicable provisions of other sections as modified by the provisions contained in Section 1634.

1634.1.2 Criteria. The minimum design seismic forces prescribed in this section are at a level that produce displacements in a fixed base, elastic model of the structure, comparable to those expected of the real structure when responding to the Design Basis Ground Motion. Reductions in these forces using the coefficient R is permitted where the design of nonbuilding structures provides sufficient strength and ductility, consistent with the provisions specified herein for buildings, to resist the effects of seismic ground motions as represented by these design forces.

When applicable, design strengths and other detailed design criteria shall be obtained from other sections or their referenced standards. The design of nonbuilding structures shall use the load combinations or factors specified in Section 1612.2 or 1612.3. For nonbuilding structures designed using Section 1634.3, 1634.4 or 1634.5, the Reliability/Redundancy Factor, ρ, may be taken as 1.0.

When applicable design strengths and other design criteria are not contained in or referenced by this code, such criteria shall be obtained from approved national standards.

1634.1.3 Weight W. The weight, W, for nonbuilding structures shall include all dead loads as defined for buildings in Section 1630.1.1. For purposes of calculating design seismic forces in nonbuilding structures, W shall also include all normal operating contents for items such as tanks, vessels, bins and piping.

1634.1.4 Period. The fundamental period of the structure shall be determined by rational methods such as by using Method B in Section 1630.2.2.

1634.1.5 Drift. The drift limitations of Section 1630.10 need not apply to nonbuilding structures. Drift limitations shall be established for structural or nonstructural elements whose failure would cause life hazards. $P\Delta$ effects shall be considered for structures whose calculated drifts exceed the values in Section 1630.1.3.

1634.1.6 Interaction effects. In Seismic Zones 3 and 4, structures that support flexible nonstructural elements whose combined weight exceeds 25 percent of the weight of the structure shall be designed considering interaction effects between the structure and the supported elements.

1634.2 Lateral Force. Lateral-force procedures for nonbuilding structures with structural systems similar to buildings (those with structural systems which are listed in Table 16-N) shall be selected in accordance with the provisions of Section 1629.

> **EXCEPTION:** Intermediate moment-resisting frames (IMRF) may be used in Seismic Zones 3 and 4 for nonbuilding structures in Occupancy Categories 3 and 4 if (1) the structure is less than 50 feet (15 240 mm) in height and (2) the value R used in reducing calculated member forces and moments does not exceed 2.8.

1634.3 Rigid Structures. Rigid structures (those with period T less than 0.06 second) and their anchorages shall be designed for the lateral force obtained from Formula (34-1).

$$V = 0.7C_a IW \qquad (34\text{-}1)$$

The force V shall be distributed according to the distribution of mass and shall be assumed to act in any horizontal direction.

1634.4 Tanks with Supported Bottoms. Flat bottom tanks or other tanks with supported bottoms, founded at or below grade, shall be designed to resist the seismic forces calculated using the procedures in Section 1634 for rigid structures considering the entire weight of the tank and its contents. Alternatively, such tanks may be designed using one of the two procedures described below:

1. A response spectrum analysis that includes consideration of the actual ground motion anticipated at the site and the inertial effects of the contained fluid.

2. A design basis prescribed for the particular type of tank by an approved national standard, provided that the seismic zones and occupancy categories shall be in conformance with the provisions of Sections 1629.4 and 1629.2, respectively.

1634.5 Other Nonbuilding Structures. Nonbuilding structures that are not covered by Sections 1634.3 and 1634.4 shall be designed to resist design seismic forces not less than those determined in accordance with the provisions in Section 1630 with the following additions and exceptions:

1. The factors R and Ω_o shall be as set forth in Table 16-P. The total design base shear determined in accordance with Section 1630.2 shall not be less than the following:

$$V = 0.56C_a IW \qquad (34\text{-}2)$$

Additionally, for Seismic Zone 4, the total base shear shall also not be less than the following:

$$V = \frac{1.6\, ZN_v\, I}{R}\, W \qquad (34\text{-}3)$$

2. The vertical distribution of the design seismic forces in structures covered by this section may be determined by using the provisions of Section 1630.5 or by using the procedures of Section 1631.

> **EXCEPTION:** For irregular structures assigned to Occupancy Categories 1 and 2 that cannot be modeled as a single mass, the procedures of Section 1631 shall be used.

3. Where an approved national standard provides a basis for the earthquake-resistant design of a particular type of nonbuilding structure covered by this section, such a standard may be used, subject to the limitations in this section:

The seismic zones and occupancy categories shall be in conformance with the provisions of Sections 1629.4 and 1629.2, respectively.

The values for total lateral force and total base overturning moment used in design shall not be less than 80 percent of the values that would be obtained using these provisions.

SECTION 1635 — EARTHQUAKE-RECORDING INSTRUMENTATIONS

For earthquake-recording instrumentations, see Appendix Chapter 16, Division II.

Division V—SOIL PROFILE TYPES

SECTION 1636 — SITE CATEGORIZATION PROCEDURE

1636.1 Scope. This division describes the procedure for determining Soil Profile Types S_A through S_F in accordance with Table 16-J.

1636.2 Definitions. Soil profile types are defined as follows:

S_A Hard rock with measured shear wave velocity, $\bar{v}_s > 5{,}000$ ft./sec. (1500 m/s).

S_B Rock with 2,500 ft./sec. $< \bar{v}_s \le 5{,}000$ ft./sec. (760 m/s $< \bar{v}_s \le 1500$ m/s).

S_C Very dense soil and soft rock with 1,200 ft./sec. $< \bar{v}_s \le 2{,}500$ ft./sec. (360 m/s $\bar{v}_s \le 760$ m/s) or with either $\bar{N} > 50$ or $\bar{s}_u \ge 2{,}000$ psf (100 kPa).

S_D Stiff soil with 600 ft./sec. $\le \bar{v}_s \le 1{,}200$ ft./sec. (180 m/s $\le \bar{v}_s \le 360$ m/s) or with $15 \le \bar{N} \le 50$ or 1,000 psf $\le \bar{s}_u \le 2{,}000$ psf (50 kPa $\le \bar{s}_u \le 100$ kPa).

S_E A soil profile with $\bar{v}_s < 600$ ft./sec. (180 m/s) or any profile with more than 10 ft. (3048 mm) of soft clay defined as soil with $PI > 20$, $w_{mc} \ge 40$ percent and $s_u < 500$ psf (25 kPa).

S_F Soils requiring site-specific evaluation:

1. Soils vulnerable to potential failure or collapse under seismic loading such as liquefiable soils, quick and highly sensitive clays, collapsible weakly cemented soils.

2. Peats and/or highly organic clays [$H > 10$ ft. (3048 mm) of peat and/or highly organic clay where H = thickness of soil].

3. Very high plasticity clays [$H > 25$ ft. (7620 mm) with $PI > 75$].

4. Very thick soft/medium stiff clays [$H > 120$ ft. (36 580 mm)].

> **EXCEPTION:** When the soil properties are not known in sufficient detail to determine the soil profile type, Type S_D shall be used. Soil Profile Type S_E need not be assumed unless the building official determines that Soil Profile Type S_E may be present at the site or in the event that Type S_E is established by geotechnical data.

The criteria set forth in the definition for Soil Profile Type S_F requiring site-specific evaluation shall be considered. If the site corresponds to this criteria, the site shall be classified as Soil Profile Type S_F and a site-specific evaluation shall be conducted.

1636.2.1 $\bar{v}_s$, Average shear wave velocity. $\bar{v}_s$ shall be determined in accordance with the following formula:

$$\bar{v}_s = \frac{\sum\limits_{i=1}^{n} d_i}{\sum\limits_{i=1}^{n} \frac{d_i}{v_{si}}} \qquad (36\text{-}1)$$

WHERE:

d_i = thickness of Layer i in feet (m).

v_{si} = shear wave velocity in Layer i in ft./sec. (m/sec).

1636.2.2 $\bar{N}$, average field standard penetration resistance and $\bar{N}_{CH}$, average standard penetration resistance for cohesionless soil layers. $\bar{N}$ and $\bar{N}_{CH}$ shall be determined in accordance with the following formula:

$$\bar{N} = \frac{\sum\limits_{i=1}^{n} d_i}{\sum\limits_{i=1}^{n} \frac{d_i}{N_i}} \qquad (36\text{-}2)$$

and

$$\bar{N}_{CH} = \frac{d_s}{\sum\limits_{i=1}^{n} \frac{d_i}{N_i}} \qquad (36\text{-}3)$$

WHERE:

d_i = thickness of Layer i in feet (mm).

d_s = the total thickness of cohesionless soil layers in the top 100 feet (30 480 mm).

N_i = the standard penetration resistance of soil layer in accordance with approved nationally recognized standards.

1636.2.3 $\bar{s}_u$, Average undrained shear strength. $\bar{s}_u$ shall be determined in accordance with the following formula:

$$\bar{s}_u = \frac{d_c}{\sum\limits_{i=1}^{n} \frac{d_i}{s_{ui}}} \qquad (36\text{-}4)$$

WHERE:

d_c = the total thickness $(100 - d_s)$ of cohesive soil layers in the top 100 feet (30 480 mm).

s_{ui} = the undrained shear strength in accordance with approved nationally recognized standards, not to exceed 5,000 psf (250 kPa).

1636.2.4 Soft clay profile, S_E. The existence of a total thickness of soft clay greater than 10 feet (3048 mm) shall be investigated where a soft clay layer is defined by $s_u < 500$ psf (24 kPa), $w_{mc} \ge 40$ percent and $PI > 20$. If these criteria are met, the site shall be classified as Soil Profile Type S_E.

1636.2.5 Soil profiles S_C, S_D and S_E. Sites with Soil Profile Types S_C, S_D and S_E shall be classified by using one of the following three methods with $\bar{v}_s$, $\bar{N}$ and $\bar{s}_u$ computed in all cases as specified in Section 1636.2.

1. $\bar{v}_s$ for the top 100 feet (30 480 mm) ($\bar{v}_s$ method).

2. $\bar{N}$ for the top 100 feet (30 480 mm) ($\bar{N}$ method).

3. $\bar{N}_{CH}$ for cohesionless soil layers ($PI < 20$) in the top 100 feet (30 480 mm) and average $\bar{s}_u$ for cohesive soil layers ($PI > 20$) in the top 100 feet (30 480 mm) ($\bar{s}_u$ method).

1636.2.6 Rock profiles, S_A and S_B. The shear wave velocity for rock, Soil Profile Type S_B, shall be either measured on site or estimated by a geotechnical engineer, engineering geologist or seismologist for competent rock with moderate fracturing and weathering. Softer and more highly fractured and weathered rock shall either be measured on site for shear wave velocity or classified as Soil Profile Type S_C.

The hard rock, Soil Profile Type S_A, category shall be supported by shear wave velocity measurement either on site or on profiles of the same rock type in the same formation with an equal or greater degree of weathering and fracturing. Where hard rock conditions are known to be continuous to a depth of 100 feet (30 480 mm), surficial shear wave velocity measurements may be extrapolated to assess $\bar{v}_s$. The rock categories, Soil Profile Types S_A and

S_B, shall not be used if there is more than 10 feet (3048 mm) of soil between the rock surface and the bottom of the spread footing or mat foundation.

The definitions presented herein shall apply to the upper 100 feet (30 480 mm) of the site profile. Profiles containing distinctly different soil layers shall be subdivided into those layers designated by a number from 1 to n at the bottom, where there are a total of n distinct layers in the upper 100 feet (30 480 mm). The symbol i then refers to any one of the layers between 1 and n.

TABLE 16-A—UNIFORM AND CONCENTRATED LOADS

USE OR OCCUPANCY		UNIFORM LOAD[1] (psf)	CONCENTRATED LOAD (pounds)
Category	Description	× 0.0479 for kN/m²	× 0.004 48 for kN
1. Access floor systems	Office use	50	2,000[2]
	Computer use	100	2,000[2]
2. Armories		150	0
3. Assembly areas[3] and auditoriums and balconies therewith	Fixed seating areas	50	0
	Movable seating and other areas	100	0
	Stage areas and enclosed platforms	125	0
4. Cornices and marquees		60[4]	0
5. Exit facilities[5]		100	0[6]
6. Garages	General storage and/or repair	100	7
	Private or pleasure-type motor vehicle storage	50	7
7. Hospitals	Wards and rooms	40	1,000[2]
8. Libraries	Reading rooms	60	1,000[2]
	Stack rooms	125	1,500[2]
9. Manufacturing	Light	75	2,000[2]
	Heavy	125	3,000[2]
10. Offices		50	2,000[2]
11. Printing plants	Press rooms	150	2,500[2]
	Composing and linotype rooms	100	2,000[2]
12. Residential[8]	Basic floor area	40	0[6]
	Exterior balconies	60[4]	0
	Decks	40[4]	0
	Storage	40	0
13. Restrooms[9]			
14. Reviewing stands, grandstands, bleachers, and folding and telescoping seating		100	0
15. Roof decks	Same as area served or for the type of occupancy accommodated		
16. Schools	Classrooms	40	1,000[2]
17. Sidewalks and driveways	Public access	250	7
18. Storage	Light	125	
	Heavy	250	
19. Stores		100	3,000[2]
20. Pedestrian bridges and walkways		100	

[1]See Section 1607 for live load reductions.
[2]See Section 1607.3.3, first paragraph, for area of load application.
[3]Assembly areas include such occupancies as dance halls, drill rooms, gymnasiums, playgrounds, plazas, terraces and similar occupancies that are generally accessible to the public.
[4]When snow loads occur that are in excess of the design conditions, the structure shall be designed to support the loads due to the increased loads caused by drift buildup or a greater snow design as determined by the building official. See Section 1614. For special-purpose roofs, see Section 1607.4.4.
[5]Exit facilities shall include such uses as corridors serving an occupant load of 10 or more persons, exterior exit balconies, stairways, fire escapes and similar uses.
[6]Individual stair treads shall be designed to support a 300-pound (1.33 kN) concentrated load placed in a position that would cause maximum stress. Stair stringers may be designed for the uniform load set forth in the table.
[7]See Section 1607.3.3, second paragraph, for concentrated loads. See Table 16-B for vehicle barriers.
[8]Residential occupancies include private dwellings, apartments and hotel guest rooms.
[9]Restroom loads shall not be less than the load for the occupancy with which they are associated, but need not exceed 50 pounds per square foot (2.4 kN/m²).

TABLE 16-B

1997 UNIFORM BUILDING CODE

TABLE 16-B—SPECIAL LOADS[1]

USE		VERTICAL LOAD	LATERAL LOAD
Category	Description	(pounds per square foot unless otherwise noted)	
		$\times$ 0.0479 for kN/m^2	
1. Construction, public access at site (live load)	Walkway, see Section 3303.6	150	
	Canopy, see Section 3303.7	150	
2. Grandstands, reviewing stands, bleachers, and folding and telescoping seating (live load)	Seats and footboards	120[2]	See Footnote 3
3. Stage accessories (live load)	Catwalks	40	
	Followspot, projection and control rooms	50	
4. Ceiling framing (live load)	Over stages	20	
	All uses except over stages	10[4]	
5. Partitions and interior walls, see Sec. 1611.5 (live load)			5
6. Elevators and dumbwaiters (dead and live loads)		2 $\times$ total loads[5]	
7. Mechanical and electrical equipment (dead load)		Total loads	
8. Cranes (dead and live loads)	Total load including impact increase	1.25 $\times$ total load[6]	0.10 $\times$ total load[7]
9. Balcony railings and guardrails	Exit facilities serving an occupant load greater than 50		50[8]
	Other than exit facilities		20[8]
	Components		25[9]
10. Vehicle barriers	See Section 311.2.3.5		6,000[10]
11. Handrails		See Footnote 11	See Footnote 11
12. Storage racks	Over 8 feet (2438 mm) high	Total loads[12]	See Table 16-O
13. Fire sprinkler structural support		250 pounds (1112 N) plus weight of water-filled pipe[13]	See Table 16-O
14. Explosion exposure	Hazardous occupancies, see Section 307.10		

[1]The tabulated loads are minimum loads. Where other vertical loads required by this code or required by the design would cause greater stresses, they shall be used.
[2]Pounds per lineal foot ($\times$ 14.6 for N/m).
[3]Lateral sway bracing loads of 24 pounds per foot (350 N/m) parallel and 10 pounds per foot (145.9 N/m) perpendicular to seat and footboards.
[4]Does not apply to ceilings that have sufficient total access from below, such that access is not required within the space above the ceiling. Does not apply to ceilings if the attic areas above the ceiling are not provided with access. This live load need not be considered as acting simultaneously with other live loads imposed upon the ceiling framing or its supporting structure.
[5]Where Appendix Chapter 30 has been adopted, see reference standard cited therein for additional design requirements.
[6]The impact factors included are for cranes with steel wheels riding on steel rails. They may be modified if substantiating technical data acceptable to the building official is submitted. Live loads on crane support girders and their connections shall be taken as the maximum crane wheel loads. For pendant-operated traveling crane support girders and their connections, the impact factors shall be 1.10.
[7]This applies in the direction parallel to the runway rails (longitudinal). The factor for forces perpendicular to the rail is 0.20 $\times$ the transverse traveling loads (trolley, cab, hooks and lifted loads). Forces shall be applied at top of rail and may be distributed among rails of multiple rail cranes and shall be distributed with due regard for lateral stiffness of the structures supporting these rails.
[8]A load per lineal foot ($\times$ 14.6 for N/m) to be applied horizontally at right angles to the top rail.
[9]Intermediate rails, panel fillers and their connections shall be capable of withstanding a load of 25 pounds per square foot (1.2 kN/m^2) applied horizontally at right angles over the entire tributary area, including openings and spaces between rails. Reactions due to this loading need not be combined with those of Footnote 8.
[10]A horizontal load in pounds (N) applied at right angles to the vehicle barrier at a height of 18 inches (457 mm) above the parking surface. The force may be distributed over a 1-foot-square (304.8-millimeter-square) area.
[11]The mounting of handrails shall be such that the completed handrail and supporting structure are capable of withstanding a load of at least 200 pounds (890 N) applied in any direction at any point on the rail. These loads shall not be assumed to act cumulatively with Item 9.
[12]Vertical members of storage racks shall be protected from impact forces of operating equipment, or racks shall be designed so that failure of one vertical member will not cause collapse of more than the bay or bays directly supported by that member.
[13]The 250-pound (1.11 kN) load is to be applied to any single fire sprinkler support point but not simultaneously to all support joints.

TABLE 16-C—MINIMUM ROOF LIVE LOADS[1]

	METHOD 1			METHOD 2		
	Tributary Loaded Area in Square Feet for Any Structural Member					
	× 0.0929 for m²					
	0 to 200	201 to 600	Over 600			
	Uniform Load (psf)			Uniform Load[2] (psf)	Rate of Reduction *r* (percentage)	Maximum Reduction *R* (percentage)
ROOF SLOPE	× 0.0479 for kN/m²					
1. Flat[3] or rise less than 4 units vertical in 12 units horizontal (33.3% slope). Arch or dome with rise less than one eighth of span	20	16	12	20	.08	40
2. Rise 4 units vertical to less than 12 units vertical in 12 units horizontal (33% to less than 100% slope). Arch or dome with rise one eighth of span to less than three eighths of span	16	14	12	16	.06	25
3. Rise 12 units vertical in 12 units horizontal (100% slope) and greater. Arch or dome with rise three eighths of span or greater	12	12	12	12	No reductions permitted	
4. Awnings except cloth covered[4]	5	5	5	5		
5. Greenhouses, lath houses and agricultural buildings[5]	10	10	10	10		

[1]Where snow loads occur, the roof structure shall be designed for such loads as determined by the building official. See Section 1614. For special-purpose roofs, see Section 1607.4.4.

[2]See Sections 1607.5 and 1607.6 for live load reductions. The rate of reduction *r* in Section 1607.5 Formula (7-1) shall be as indicated in the table. The maximum reduction *R* shall not exceed the value indicated in the table.

[3]A flat roof is any roof with a slope of less than $1/4$ unit vertical in 12 units horizontal (2% slope). The live load for flat roofs is in addition to the ponding load required by Section 1611.7.

[4]As defined in Section 3206.

[5]See Section 1607.4.4 for concentrated load requirements for greenhouse roof members.

TABLE 16-D—MAXIMUM ALLOWABLE DEFLECTION FOR STRUCTURAL MEMBERS[1]

TYPE OF MEMBER	MEMBER LOADED WITH LIVE LOAD ONLY (*L.*)	MEMBER LOADED WITH LIVE LOAD PLUS DEAD LOAD (*L. + K.D.*)
Roof member supporting plaster or floor member	*l*/360	*l*/240

[1]Sufficient slope or camber shall be provided for flat roofs in accordance with Section 1611.7.

L.—live load.

D.—dead load.

K.—factor as determined by Table 16-E.

l—length of member in same units as deflection.

TABLE 16-E—VALUE OF "K"

WOOD		REINFORCED CONCRETE[2]	STEEL
Unseasoned	Seasoned[1]		
1.0	0.5	$T/(1+50\rho')$	0

[1]Seasoned lumber is lumber having a moisture content of less than 16 percent at time of installation and used under dry conditions of use such as in covered structures.

[2]See also Section 1909 for definitions and other requirements.

ρ' shall be the value at midspan for simple and continuous spans, and at support for cantilevers. Time-dependent factor *T* for sustained loads may be taken equal to:

five years or more	2.0
twelve months	1.2
six months	1.4
three months	1.0

TABLE 16-F
TABLE 16-G

1997 UNIFORM BUILDING CODE

TABLE 16-F—WIND STAGNATION PRESSURE (q_s) AT STANDARD HEIGHT OF 33 FEET (10 058 mm)

Basic wind speed (mph)[1] ($\times$ 1.61 for km/h)	70	80	90	100	110	120	130
Pressure q_s (psf) ($\times$ 0.0479 for kN/m^2)	12.6	16.4	20.8	25.6	31.0	36.9	43.3

[1]Wind speed from Section 1618.

TABLE 16-G—COMBINED HEIGHT, EXPOSURE AND GUST FACTOR COEFFICIENT (C_e)[1]

HEIGHT ABOVE AVERAGE LEVEL OF ADJOINING GROUND (feet) $\times$ 304.8 for mm	EXPOSURE D	EXPOSURE C	EXPOSURE B
0-15	1.39	1.06	0.62
20	1.45	1.13	0.67
25	1.50	1.19	0.72
30	1.54	1.23	0.76
40	1.62	1.31	0.84
60	1.73	1.43	0.95
80	1.81	1.53	1.04
100	1.88	1.61	1.13
120	1.93	1.67	1.20
160	2.02	1.79	1.31
200	2.10	1.87	1.42
300	2.23	2.05	1.63
400	2.34	2.19	1.80

[1]Values for intermediate heights above 15 feet (4572 mm) may be interpolated.

TABLE 16-H—PRESSURE COEFFICIENTS (C_q)

STRUCTURE OR PART THEREOF	DESCRIPTION	C_q FACTOR

1. Primary frames and systems	**Method 1** (Normal force method)
	Walls:
	    Windward wall
	    Leeward wall
	Roofs[1]:
	Wind perpendicular to ridge
	Leeward roof or flat roof
	Windward roof
	    less than 2:12 (16.7%)
	    Slope 2:12 (16.7%) to less than 9:12 (75%)
	    Slope 9:12 (75%) to 12:12 (100%)
	    Slope > 12:12 (100%)
	Wind parallel to ridge and flat roofs
		
		0.8 inward
		0.5 outward
		
		0.7 outward
		
		0.7 outward
		0.9 outward or 0.3 inward
		0.4 inward
		0.7 inward
		0.7 outward

	Method 2 (Projected area method)
	On vertical projected area
	    Structures 40 feet (12 192 mm) or less in height
	    Structures over 40 feet (12 192 mm) in height
	On horizontal projected area[1]
		
		1.3 horizontal any direction
		1.4 horizontal any direction
		0.7 upward

2. Elements and components not in areas of discontinuity[2]	Wall elements
	    All structures
	    Enclosed and unenclosed structures
	    Partially enclosed structures
	    Parapets walls
		1.2 inward
		1.2 outward
		1.6 outward
		1.3 inward or outward

	Roof elements[3]
	    Enclosed and unenclosed structures
	    Slope < 7:12 (58.3%)
	    Slope 7:12 (58.3%) to 12:12 (100%)
	
	    Partially enclosed structures
	    Slope < 2:12 (16.7%)
	    Slope 2:12 (16.7%) to 7:12 (58.3%)
	    Slope > 7:12 (58.3%) to 12:12 (100%)
		
		1.3 outward
		1.3 outward or inward
		
		1.7 outward
		1.6 outward or 0.8 inward
		1.7 outward or inward

3. Elements and components in areas of discontinuities[2,4,5]	Wall corners[6]
	
	Roof eaves, rakes or ridges without overhangs[6]
	    Slope < 2:12 (16.7%)
	    Slope 2:12 (16.7%) to 7:12 (58.3%)
	    Slope > 7:12 (58.3%) to 12:12 (100%)
	For slopes less than 2:12 (16.7%)
	    Overhangs at roof eaves, rakes or ridges, and canopies
		1.5 outward or 1.2 inward
		
		2.3 upward
		2.6 outward
		1.6 outward
		
		0.5 added to values above

4. Chimneys, tanks and solid towers	Square or rectangular Hexagonal or octagonal Round or elliptical	1.4 any direction 1.1 any direction 0.8 any direction

5. Open-frame towers[7,8]	Square and rectangular
	    Diagonal
	    Normal
	Triangular
		
		4.0
		3.6
		3.2

6. Tower accessories (such as ladders, conduit, lights and elevators)	Cylindrical members
	    2 inches (51 mm) or less in diameter
	    Over 2 inches (51 mm) in diameter
	Flat or angular members
		
		1.0
		0.8
		1.3

7. Signs, flagpoles, lightpoles, minor structures[8]		1.4 any direction

[1]For one story or the top story of multistory partially enclosed structures, an additional value of 0.5 shall be added to the outward C_q. The most critical combination shall be used for design. For definition of partially enclosed structures, see Section 1616.

[2]C_q values listed are for 10-square-foot (0.93 m²) tributary areas. For tributary areas of 100 square feet (9.29 m²), the value of 0.3 may be subtracted from C_q, except for areas at discontinuities with slopes less than 7 units vertical in 12 units horizontal (58.3% slope) where the value of 0.8 may be subtracted from C_q. Interpolation may be used for tributary areas between 10 and 100 square feet (0.93 m² and 9.29 m²). For tributary areas greater than 1,000 square feet (92.9 m²), use primary frame values.

[3]For slopes greater than 12 units vertical in 12 units horizontal (100% slope), use wall element values.

[4]Local pressures shall apply over a distance from the discontinuity of 10 feet (3048 mm) or 0.1 times the least width of the structure, whichever is smaller.

[5]Discontinuities at wall corners or roof ridges are defined as discontinuous breaks in the surface where the included interior angle measures 170 degrees or less.

[6]Load is to be applied on either side of discontinuity but not simultaneously on both sides.

[7]Wind pressures shall be applied to the total normal projected area of all elements on one face. The forces shall be assumed to act parallel to the wind direction.

[8]Factors for cylindrical elements are two thirds of those for flat or angular elements.

TABLE 16-I
TABLE 16-K

1997 UNIFORM BUILDING CODE

TABLE 16-I—SEISMIC ZONE FACTOR Z

ZONE	1	2A	2B	3	4
Z	0.075	0.15	0.20	0.30	0.40

NOTE: The zone shall be determined from the seismic zone map in Figure 16-2.

TABLE 16-J—SOIL PROFILE TYPES

SOIL PROFILE TYPE	SOIL PROFILE NAME/GENERIC DESCRIPTION	AVERAGE SOIL PROPERTIES FOR TOP 100 FEET (30 480 mm) OF SOIL PROFILE		
		Shear Wave Velocity, $\bar{v}_s$ feet/second (m/s)	Standard Penetration Test, $\bar{N}$ [or $\bar{N}_{CH}$ for cohesionless soil layers] (blows/foot)	Undrained Shear Strength, $\bar{s}_u$ psf (kPa)
S_A	Hard Rock	> 5,000 (1,500)	—	—
S_B	Rock	2,500 to 5,000 (760 to 1,500)		
S_C	Very Dense Soil and Soft Rock	1,200 to 2,500 (360 to 760)	> 50	> 2,000 (100)
S_D	Stiff Soil Profile	600 to 1,200 (180 to 360)	15 to 50	1,000 to 2,000 (50 to 100)
$S_E{}^1$	Soft Soil Profile	< 600 (180)	< 15	< 1,000 (50)
S_F	Soil Requiring Site-specific Evaluation. See Section 1629.3.1.			

[1]Soil Profile Type S_E also includes any soil profile with more than 10 feet (3048 mm) of soft clay defined as a soil with a plasticity index, $PI > 20$, $w_{mc} \geq 40$ percent and $s_u < 500$ psf (24 kPa). The Plasticity Index, PI, and the moisture content, w_{mc}, shall be determined in accordance with approved national standards.

TABLE 16-K—OCCUPANCY CATEGORY

OCCUPANCY CATEGORY	OCCUPANCY OR FUNCTIONS OF STRUCTURE	SEISMIC IMPORTANCE FACTOR, I	SEISMIC IMPORTANCE[1] FACTOR, I_p	WIND IMPORTANCE FACTOR, I_w
1. Essential facilities[2]	Group I, Division 1 Occupancies having surgery and emergency treatment areas Fire and police stations Garages and shelters for emergency vehicles and emergency aircraft Structures and shelters in emergency-preparedness centers Aviation control towers Structures and equipment in government communication centers and other facilities required for emergency response Standby power-generating equipment for Category 1 facilities Tanks or other structures containing housing or supporting water or other fire-suppression material or equipment required for the protection of Category 1, 2 or 3 structures	1.25	1.50	1.15
2. Hazardous facilities	Group H, Divisions 1, 2, 6 and 7 Occupancies and structures therein housing or supporting toxic or explosive chemicals or substances Nonbuilding structures housing, supporting or containing quantities of toxic or explosive substances that, if contained within a building, would cause that building to be classified as a Group H, Division 1, 2 or 7 Occupancy	1.25	1.50	1.15
3. Special occupancy structures[3]	Group A, Divisions 1, 2 and 2.1 Occupancies Buildings housing Group E, Divisions 1 and 3 Occupancies with a capacity greater than 300 students Buildings housing Group B Occupancies used for college or adult education with a capacity greater than 500 students Group I, Divisions 1 and 2 Occupancies with 50 or more resident incapacitated patients, but not included in Category 1 Group I, Division 3 Occupancies All structures with an occupancy greater than 5,000 persons Structures and equipment in power-generating stations, and other public utility facilities not included in Category 1 or Category 2 above, and required for continued operation	1.00	1.00	1.00
4. Standard occupancy structures[3]	All structures housing occupancies or having functions not listed in Category 1, 2 or 3 and Group U Occupancy towers	1.00	1.00	1.00
5. Miscellaneous structures	Group U Occupancies except for towers	1.00	1.00	1.00

[1]The limitation of I_p for panel connections in Section 1633.2.4 shall be 1.0 for the entire connector.
[2]Structural observation requirements are given in Section 1702.
[3]For anchorage of machinery and equipment required for life-safety systems, the value of I_p shall be taken as 1.5.

TABLE 16-L—VERTICAL STRUCTURAL IRREGULARITIES

IRREGULARITY TYPE AND DEFINITION	REFERENCE SECTION
1. **Stiffness irregularity—soft story** A soft story is one in which the lateral stiffness is less than 70 percent of that in the story above or less than 80 percent of the average stiffness of the three stories above.	1629.8.4, Item 2
2. **Weight (mass) irregularity** Mass irregularity shall be considered to exist where the effective mass of any story is more than 150 percent of the effective mass of an adjacent story. A roof that is lighter than the floor below need not be considered.	1629.8.4, Item 2
3. **Vertical geometric irregularity** Vertical geometric irregularity shall be considered to exist where the horizontal dimension of the lateral-force-resisting system in any story is more than 130 percent of that in an adjacent story. One-story penthouses need not be considered.	1629.8.4, Item 2
4. **In-plane discontinuity in vertical lateral-force-resisting element** An in-plane offset of the lateral-load-resisting elements greater than the length of those elements.	1630.8.2
5. **Discontinuity in capacity—weak story** A weak story is one in which the story strength is less than 80 percent of that in the story above. The story strength is the total strength of all seismic-resisting elements sharing the story shear for the direction under consideration.	1629.9.1

TABLE 16-M—PLAN STRUCTURAL IRREGULARITIES

IRREGULARITY TYPE AND DEFINITION	REFERENCE SECTION
1. **Torsional irregularity—to be considered when diaphragms are not flexible** Torsional irregularity shall be considered to exist when the maximum story drift, computed including accidental torsion, at one end of the structure transverse to an axis is more than 1.2 times the average of the story drifts of the two ends of the structure.	1633.1, 1633.2.9, Item 6
2. **Re-entrant corners** Plan configurations of a structure and its lateral-force-resisting system contain re-entrant corners, where both projections of the structure beyond a re-entrant corner are greater than 15 percent of the plan dimension of the structure in the given direction.	1633.2.9, Items 6 and 7
3. **Diaphragm discontinuity** Diaphragms with abrupt discontinuities or variations in stiffness, including those having cutout or open areas greater than 50 percent of the gross enclosed area of the diaphragm, or changes in effective diaphragm stiffness of more than 50 percent from one story to the next.	1633.2.9, Item 6
4. **Out-of-plane offsets** Discontinuities in a lateral force path, such as out-of-plane offsets of the vertical elements.	1630.8.2; 1633.2.9, Item 6; 2213.9.1
5. **Nonparallel systems** The vertical lateral-load-resisting elements are not parallel to or symmetric about the major orthogonal axes of the lateral-force-resisting system.	1633.1

TABLE 16-N

1997 UNIFORM BUILDING CODE

TABLE 16-N—STRUCTURAL SYSTEMS[1]

BASIC STRUCTURAL SYSTEM[2]	LATERAL-FORCE-RESISTING SYSTEM DESCRIPTION	R	Ω_o	HEIGHT LIMIT FOR SEISMIC ZONES 3 AND 4 (feet) × 304.8 for mm
1. Bearing wall system	1. Light-framed walls with shear panels			
	a. Wood structural panel walls for structures three stories or less	5.5	2.8	65
	b. All other light-framed walls	4.5	2.8	65
	2. Shear walls			
	a. Concrete	4.5	2.8	160
	b. Masonry	4.5	2.8	160
	3. Light steel-framed bearing walls with tension-only bracing	2.8	2.2	65
	4. Braced frames where bracing carries gravity load			
	a. Steel	4.4	2.2	160
	b. Concrete[3]	2.8	2.2	—
	c. Heavy timber	2.8	2.2	65
2. Building frame system	1. Steel eccentrically braced frame (EBF)	7.0	2.8	240
	2. Light-framed walls with shear panels			
	a. Wood structural panel walls for structures three stories or less	6.5	2.8	65
	b. All other light-framed walls	5.0	2.8	65
	3. Shear walls			
	a. Concrete	5.5	2.8	240
	b. Masonry	5.5	2.8	160
	4. Ordinary braced frames			
	a. Steel	5.6	2.2	160
	b. Concrete[3]	5.6	2.2	—
	c. Heavy timber	5.6	2.2	65
	5. Special concentrically braced frames			
	a. Steel	6.4	2.2	240
3. Moment-resisting frame system	1. Special moment-resisting frame (SMRF)			
	a. Steel	8.5	2.8	N.L.
	b. Concrete[4]	8.5	2.8	N.L.
	2. Masonry moment-resisting wall frame (MMRWF)	6.5	2.8	160
	3. Concrete intermediate moment-resisting frame (IMRF)[5]	5.5	2.8	—
	4. Ordinary moment-resisting frame (OMRF)			
	a. Steel[6]	4.5	2.8	160
	b. Concrete[7]	3.5	2.8	—
	5. Special truss moment frames of steel (STMF)	6.5	2.8	240
4. Dual systems	1. Shear walls			
	a. Concrete with SMRF	8.5	2.8	N.L.
	b. Concrete with steel OMRF	4.2	2.8	160
	c. Concrete with concrete IMRF[5]	6.5	2.8	160
	d. Masonry with SMRF	5.5	2.8	160
	e. Masonry with steel OMRF	4.2	2.8	160
	f. Masonry with concrete IMRF[3]	4.2	2.8	—
	g. Masonry with masonry MMRWF	6.0	2.8	160
	2. Steel EBF			
	a. With steel SMRF	8.5	2.8	N.L.
	b. With steel OMRF	4.2	2.8	160
	3. Ordinary braced frames			
	a. Steel with steel SMRF	6.5	2.8	N.L.
	b. Steel with steel OMRF	4.2	2.8	160
	c. Concrete with concrete SMRF[3]	6.5	2.8	—
	d. Concrete with concrete IMRF[3]	4.2	2.8	—
	4. Special concentrically braced frames			
	a. Steel with steel SMRF	7.5	2.8	N.L.
	b. Steel with steel OMRF	4.2	2.8	160
5. Cantilevered column building systems	1. Cantilevered column elements	2.2	2.0	35[7]
6. Shear wall-frame interaction systems	1. Concrete[8]	5.5	2.8	160
7. Undefined systems	See Sections 1629.6.7 and 1629.9.2	—	—	—

N.L.—no limit

[1]See Section 1630.4 for combination of structural systems.

[2]Basic structural systems are defined in Section 1629.6.

[3]Prohibited in Seismic Zones 3 and 4.

[4]Includes precast concrete conforming to Section 1921.2.7.

[5]Prohibited in Seismic Zones 3 and 4, except as permitted in Section 1634.2.

[6]Ordinary moment-resisting frames in Seismic Zone 1 meeting the requirements of Section 2211.6 may use a R value of 8.

[7]Total height of the building including cantilevered columns.

[8]Prohibited in Seismic Zones 2A, 2B, 3 and 4. See Section 1633.2.7.

TABLE 16-O—HORIZONTAL FORCE FACTORS, a_p AND R_p

ELEMENTS OF STRUCTURES AND NONSTRUCTURAL COMPONENTS AND EQUIPMENT[1]	a_p	R_p	FOOTNOTE
1. Elements of Structures			
A. Walls including the following:			
(1) Unbraced (cantilevered) parapets.	2.5	3.0	
(2) Exterior walls at or above the ground floor and parapets braced above their centers of gravity.	1.0	3.0	2
(3) All interior-bearing and nonbearing walls.	1.0	3.0	2
B. Penthouse (except when framed by an extension of the structural frame).	2.5	4.0	
C. Connections for prefabricated structural elements other than walls. See also Section 1632.2.	1.0	3.0	3
2. Nonstructural Components			
A. Exterior and interior ornamentations and appendages.	2.5	3.0	
B. Chimneys, stacks and trussed towers supported on or projecting above the roof:			
(1) Laterally braced or anchored to the structural frame at a point below their centers of mass.	2.5	3.0	
(2) Laterally braced or anchored to the structural frame at or above their centers of mass.	1.0	3.0	
C. Signs and billboards.	2.5	3.0	
D. Storage racks (include contents) over 6 feet (1829 mm) tall.	2.5	4.0	4
E. Permanent floor-supported cabinets and book stacks more than 6 feet (1829 mm) in height (include contents).	1.0	3.0	5
F. Anchorage and lateral bracing for suspended ceilings and light fixtures.	1.0	3.0	3, 6, 7, 8
G. Access floor systems.	1.0	3.0	4, 5, 9
H. Masonry or concrete fences over 6 feet (1829 mm) high.	1.0	3.0	
I. Partitions.	1.0	3.0	
3. Equipment			
A. Tanks and vessels (include contents), including support systems.	1.0	3.0	
B. Electrical, mechanical and plumbing equipment and associated conduit and ductwork and piping.	1.0	3.0	5, 10, 11, 12, 13, 14, 15, 16
C. Any flexible equipment laterally braced or anchored to the structural frame at a point below their center of mass.	2.5	3.0	5, 10, 14, 15, 16
D. Anchorage of emergency power supply systems and essential communications equipment. Anchorage and support systems for battery racks and fuel tanks necessary for operation of emergency equipment. See also Section 1632.2.	1.0	3.0	17, 18
E. Temporary containers with flammable or hazardous materials.	1.0	3.0	19
4. Other Components			
A. Rigid components with ductile material and attachments.	1.0	3.0	1
B. Rigid components with nonductile material or attachments.	1.0	1.5	1
C. Flexible components with ductile material and attachments.	2.5	3.0	1
D. Flexible components with nonductile material or attachments.	2.5	1.5	1

[1]See Section 1627 for definitions of flexible components and rigid components.

[2]See Sections 1633.2.4 and 1633.2.8 for concrete and masonry walls and Section 1632.2 for connections for panel connectors for panels.

[3]Applies to Seismic Zones 2, 3 and 4 only.

[4]Ground supported steel storage racks may be designed using the provisions of Section 1634. Chapter 22, Division VI, may be used for design, provided seismic design forces are equal to or greater than those specified in Section 1632.2 or 1634.2, as appropriate.

[5]Only attachments, anchorage or restraints need be designed.

[6]Ceiling weight shall include all light fixtures and other equipment or partitions that are laterally supported by the ceiling. For purposes of determining the seismic force, a ceiling weight of not less than 4 psf (0.19 kN/m²) shall be used.

[7]Ceilings constructed of lath and plaster or gypsum board screw or nail attached to suspended members that support a ceiling at one level extending from wall to wall need not be analyzed, provided the walls are not over 50 feet (15 240 mm) apart.

[8]Light fixtures and mechanical services installed in metal suspension systems for acoustical tile and lay-in panel ceilings shall be independently supported from the structure above as specified in UBC Standard 25-2, Part III.

[9]W_p for access floor systems shall be the dead load of the access floor system plus 25 percent of the floor live load plus a 10-psf (0.48 kN/m²) partition load allowance.

[10]Equipment includes, but is not limited to, boilers, chillers, heat exchangers, pumps, air-handling units, cooling towers, control panels, motors, switchgear, transformers and life-safety equipment. It shall include major conduit, ducting and piping, which services such machinery and equipment and fire sprinkler systems. See Section 1632.2 for additional requirements for determining a_p for nonrigid or flexibly mounted equipment.

[11]Seismic restraints may be omitted from piping and duct supports if all the following conditions are satisfied:

  [11.1]Lateral motion of the piping or duct will not cause damaging impact with other systems.

  [11.2]The piping or duct is made of ductile material with ductile connections.

  [11.3]Lateral motion of the piping or duct does not cause impact of fragile appurtenances (e.g., sprinkler heads) with any other equipment, piping or structural member.

  [11.4]Lateral motion of the piping or duct does not cause loss of system vertical support.

  [11.5]Rod-hung supports of less than 12 inches (305 mm) in length have top connections that cannot develop moments.

  [11.6]Support members cantilevered up from the floor are checked for stability.

(Continued)

TABLE 16-O
TABLE 16-Q

1997 UNIFORM BUILDING CODE

FOOTNOTES TO TABLE 16-O—(Continued)

[12]Seismic restraints may be omitted from electrical raceways, such as cable trays, conduit and bus ducts, if all the following conditions are satisfied:

[12.1]Lateral motion of the raceway will not cause damaging impact with other systems.

[12.2]Lateral motion of the raceway does not cause loss of system vertical support.

[12.3]Rod-hung supports of less than 12 inches (305 mm) in length have top connections that cannot develop moments.

[12.4]Support members cantilevered up from the floor are checked for stability.

[13]Piping, ducts and electrical raceways, which must be functional following an earthquake, spanning between different buildings or structural systems shall be sufficiently flexible to withstand relative motion of support points assuming out-of-phase motions.

[14]Vibration isolators supporting equipment shall be designed for lateral loads or restrained from displacing laterally by other means. Restraint shall also be provided, which limits vertical displacement, such that lateral restraints do not become disengaged. a_p and R_p for equipment supported on vibration isolators shall be taken as 2.5 and 1.5, respectively, except that if the isolation mounting frame is supported by shallow or expansion anchors, the design forces for the anchors calculated by Formula (32-1), (32-2) or (32-3) shall be additionally multiplied by a factor of 2.0.

[15]Equipment anchorage shall not be designed such that lateral loads are resisted by gravity friction (e.g., friction clips).

[16]Expansion anchors, which are required to resist seismic loads in tension, shall not be used where operational vibrating loads are present.

[17]Movement of components within electrical cabinets, rack- and skid-mounted equipment and portions of skid-mounted electromechanical equipment that may cause damage to other components by displacing, shall be restricted by attachment to anchored equipment or support frames.

[18]Batteries on racks shall be restrained against movement in all directions due to earthquake forces.

[19]Seismic restraints may include straps, chains, bolts, barriers or other mechanisms that prevent sliding, falling and breach of containment of flammable and toxic materials. Friction forces may not be used to resist lateral loads in these restraints unless positive uplift restraint is provided which ensures that the friction forces act continuously.

TABLE 16-P—R AND Ω_o FACTORS FOR NONBUILDING STRUCTURES

STRUCTURE TYPE	R	Ω_o
1. Vessels, including tanks and pressurized spheres, on braced or unbraced legs.	2.2	2.0
2. Cast-in-place concrete silos and chimneys having walls continuous to the foundations.	3.6	2.0
3. Distributed mass cantilever structures such as stacks, chimneys, silos and skirt-supported vertical vessels.	2.9	2.0
4. Trussed towers (freestanding or guyed), guyed stacks and chimneys.	2.9	2.0
5. Cantilevered column-type structures.	2.2	2.0
6. Cooling towers.	3.6	2.0
7. Bins and hoppers on braced or unbraced legs.	2.9	2.0
8. Storage racks.	3.6	2.0
9. Signs and billboards.	3.6	2.0
10. Amusement structures and monuments.	2.2	2.0
11. All other self-supporting structures not otherwise covered.	2.9	2.0

TABLE 16-Q—SEISMIC COEFFICIENT C_a

SOIL PROFILE TYPE	SEISMIC ZONE FACTOR, Z				
	$Z = 0.075$	$Z = 0.15$	$Z = 0.2$	$Z = 0.3$	$Z = 0.4$
S_A	0.06	0.12	0.16	0.24	$0.32N_a$
S_B	0.08	0.15	0.20	0.30	$0.40N_a$
S_C	0.09	0.18	0.24	0.33	$0.40N_a$
S_D	0.12	0.22	0.28	0.36	$0.44N_a$
S_E	0.19	0.30	0.34	0.36	$0.36N_a$
S_F	See Footnote 1				

[1]Site-specific geotechnical investigation and dynamic site response analysis shall be performed to determine seismic coefficients for Soil Profile Type S_F.

TABLE 16-R—SEISMIC COEFFICIENT C_v

SOIL PROFILE TYPE	SEISMIC ZONE FACTOR, Z				
	$Z = 0.075$	$Z = 0.15$	$Z = 0.2$	$Z = 0.3$	$Z = 0.4$
S_A	0.06	0.12	0.16	0.24	$0.32N_v$
S_B	0.08	0.15	0.20	0.30	$0.40N_v$
S_C	0.13	0.25	0.32	0.45	$0.56N_v$
S_D	0.18	0.32	0.40	0.54	$0.64N_v$
S_E	0.26	0.50	0.64	0.84	$0.96N_v$
S_F	See Footnote 1				

[1]Site-specific geotechnical investigation and dynamic site response analysis shall be performed to determine seismic coefficients for Soil Profile Type S_F.

TABLE 16-S—NEAR-SOURCE FACTOR N_a[1]

SEISMIC SOURCE TYPE	CLOSEST DISTANCE TO KNOWN SEISMIC SOURCE[2,3]		
	≤ 2 km	5 km	≥ 10 km
A	1.5	1.2	1.0
B	1.3	1.0	1.0
C	1.0	1.0	1.0

[1]The Near-Source Factor may be based on the linear interpolation of values for distances other than those shown in the table.

[2]The location and type of seismic sources to be used for design shall be established based on approved geotechnical data (e.g., most recent mapping of active faults by the United States Geological Survey or the California Division of Mines and Geology).

[3]The closest distance to seismic source shall be taken as the minimum distance between the site and the area described by the vertical projection of the source on the surface (i.e., surface projection of fault plane). The surface projection need not include portions of the source at depths of 10 km or greater. The largest value of the Near-Source Factor considering all sources shall be used for design.

TABLE 16-T—NEAR-SOURCE FACTOR N_v[1]

SEISMIC SOURCE TYPE	CLOSEST DISTANCE TO KNOWN SEISMIC SOURCE[2,3]			
	≤ 2 km	5 km	10 km	≥ 15 km
A	2.0	1.6	1.2	1.0
B	1.6	1.2	1.0	1.0
C	1.0	1.0	1.0	1.0

[1]The Near-Source Factor may be based on the linear interpolation of values for distances other than those shown in the table.

[2]The location and type of seismic sources to be used for design shall be established based on approved geotechnical data (e.g., most recent mapping of active faults by the United States Geological Survey or the California Division of Mines and Geology).

[3]The closest distance to seismic source shall be taken as the minimum distance between the site and the area described by the vertical projection of the source on the surface (i.e., surface projection of fault plane). The surface projection need not include portions of the source at depths of 10 km or greater. The largest value of the Near-Source Factor considering all sources shall be used for design.

TABLE 16-U—SEISMIC SOURCE TYPE[1]

SEISMIC SOURCE TYPE	SEISMIC SOURCE DESCRIPTION	SEISMIC SOURCE DEFINITION[2]	
		Maximum Moment Magnitude, M	Slip Rate, SR (mm/year)
A	Faults that are capable of producing large magnitude events and that have a high rate of seismic activity	$M \geq 7.0$	$SR \geq 5$
B	All faults other than Types A and C	$M \geq 7.0$ $M < 7.0$ $M \geq 6.5$	$SR < 5$ $SR > 2$ $SR < 2$
C	Faults that are not capable of producing large magnitude earthquakes and that have a relatively low rate of seismic activity	$M < 6.5$	$SR \leq 2$

[1]Subduction sources shall be evaluated on a site-specific basis.

[2]Both maximum moment magnitude and slip rate conditions must be satisfied concurrently when determining the seismic source type.

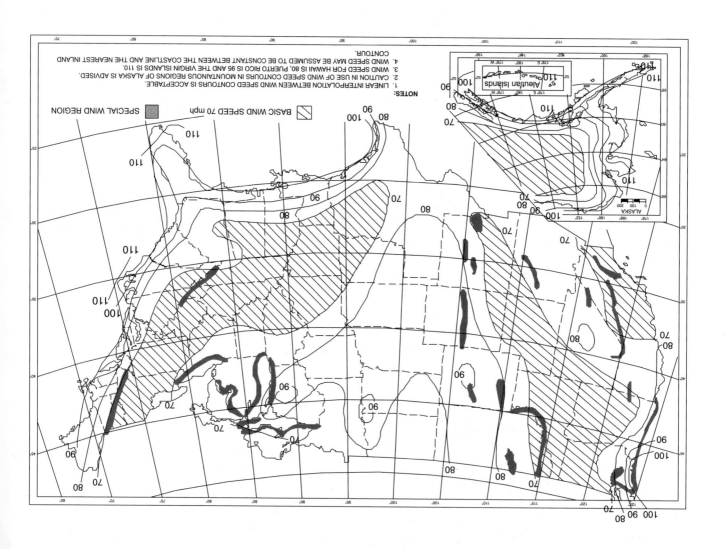

FIGURE 16-1—MINIMUM BASIC WIND SPEEDS IN MILES PER HOUR (× 1.61 for km/h)

FIGURE 16-1

1997 UNIFORM BUILDING CODE

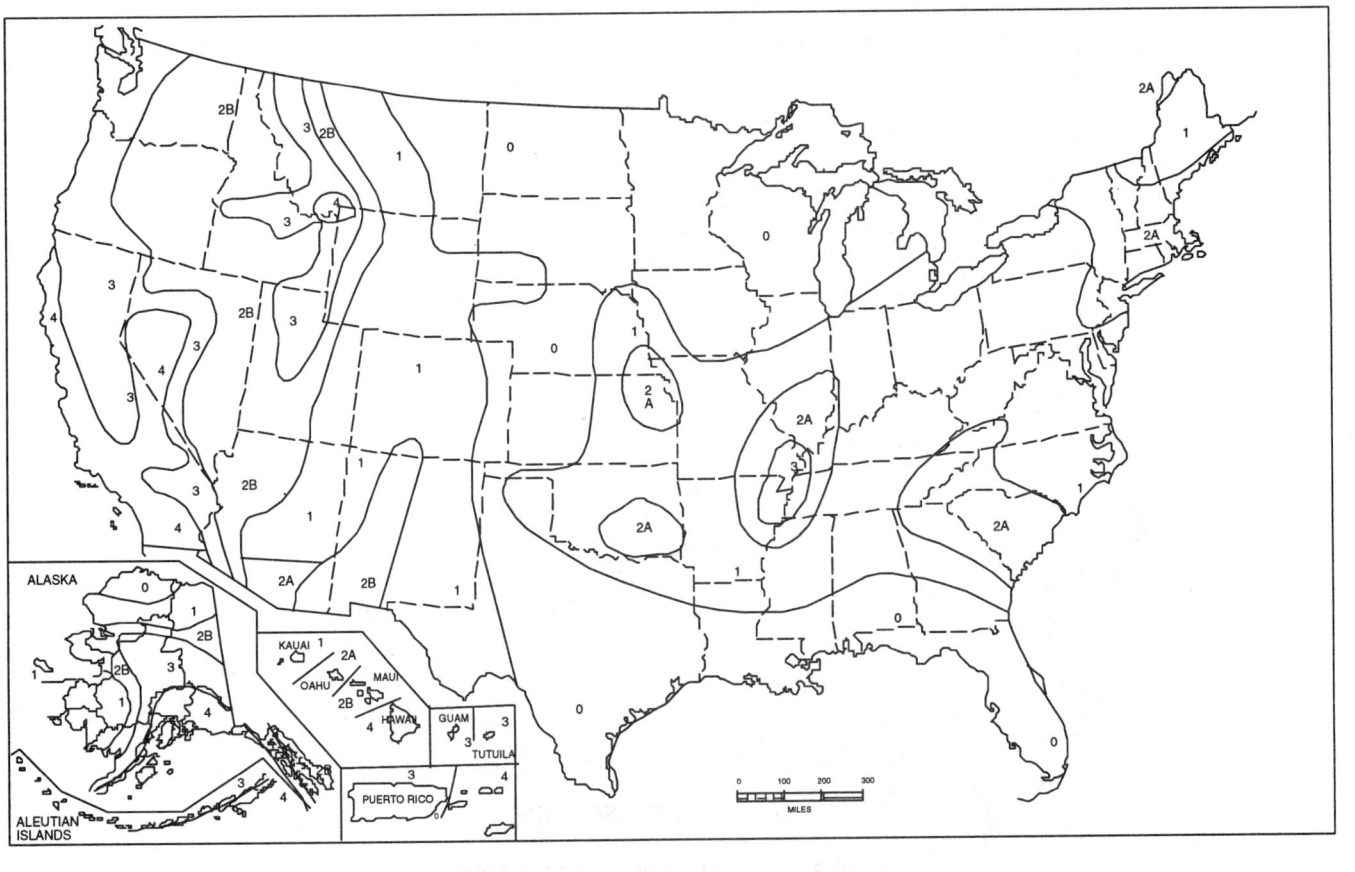

FIGURE 16-2—SEISMIC ZONE MAP OF THE UNITED STATES
For areas outside of the United States, see Appendix Chapter 16.

FIGURE 16-3

1997 UNIFORM BUILDING CODE

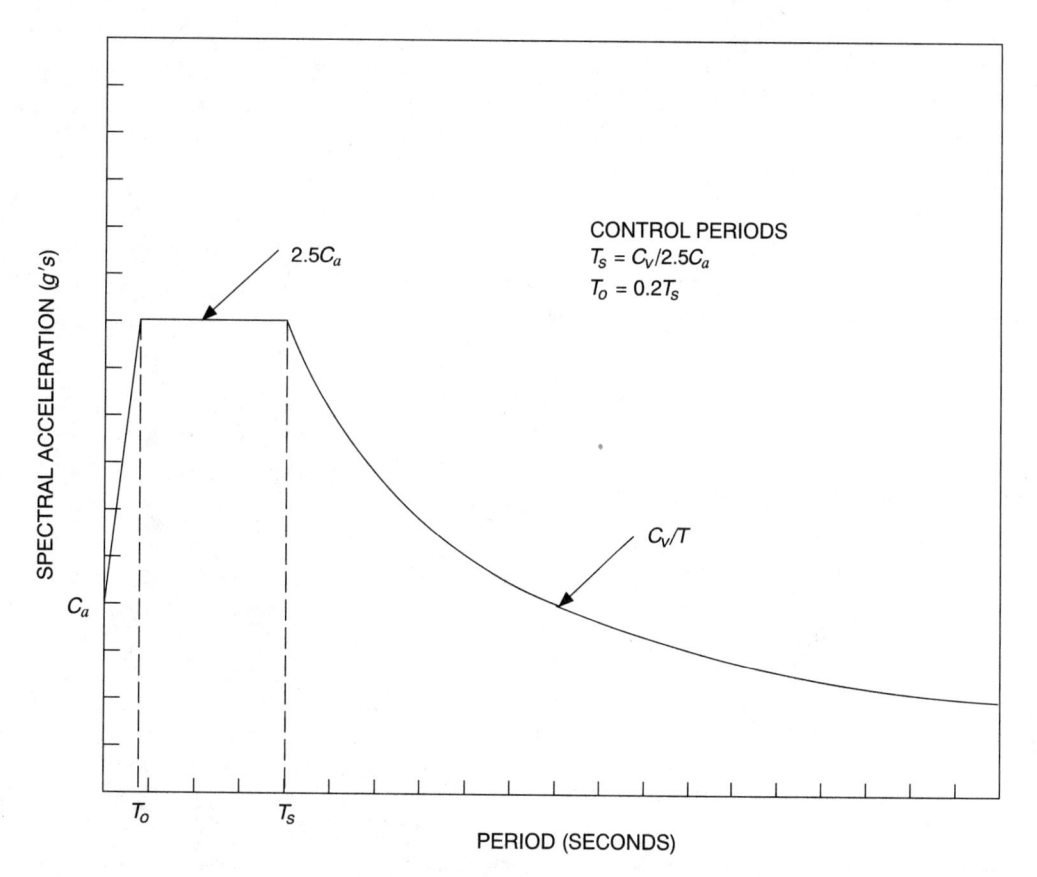

CONTROL PERIODS
$T_s = C_v/2.5C_a$
$T_o = 0.2T_s$

FIGURE 16-3—DESIGN RESPONSE SPECTRA

Chapter 17
STRUCTURAL TESTS AND INSPECTIONS

SECTION 1701 — SPECIAL INSPECTIONS

1701.1 General. In addition to the inspections required by Section 108, the owner or the engineer or architect of record acting as the owner's agent shall employ one or more special inspectors who shall provide inspections during construction on the types of work listed under Section 1701.5.

> **EXCEPTION:** The building official may waive the requirement for the employment of a special inspector if the construction is of a minor nature.

1701.2 Special Inspector. The special inspector shall be a qualified person who shall demonstrate competence, to the satisfaction of the building official, for inspection of the particular type of construction or operation requiring special inspection.

1701.3 Duties and Responsibilities of the Special Inspector. The special inspector shall observe the work assigned for conformance to the approved design drawings and specifications.

The special inspector shall furnish inspection reports to the building official, the engineer or architect of record, and other designated persons. All discrepancies shall be brought to the immediate attention of the contractor for correction, then, if uncorrected, to the proper design authority and to the building official.

The special inspector shall submit a final signed report stating whether the work requiring special inspection was, to the best of the inspector's knowledge, in conformance to the approved plans and specifications and the applicable workmanship provisions of this code.

1701.4 Standards of Quality. The standards listed below labeled a "UBC Standard" are also listed in Chapter 35, Part II, and are part of this code. The other standards listed below are recognized standards. (See Sections 3503 and 3504.)

1. Concrete.

ASTM C 94, Ready-mixed Concrete

2. Connections.

Specification for Structural Joints Using ASTM A 325 or A 490 Bolts-Load and Resistance Factor Design, Research Council of Structural Connections, Section 1701.5, Item 6.

Specification for Structural Joints Using ASTM A 325 or A 490 Bolts-Allowable Stress Design, Research Council of Structural Connections, Section 1701.5, Item 6.

3. Spray-applied Fire-resistive Materials.

UBC Standard 7-6, Thickness and Density Determination for Spray-applied Fire-resistive Materials

1701.5 Types of Work. Except as provided in Section 1701.1, the types of work listed below shall be inspected by a special inspector.

1. Concrete. During the taking of test specimens and placing of reinforced concrete. See Item 12 for shotcrete.

> **EXCEPTIONS:** 1. Concrete for foundations conforming to minimum requirements of Table 18-I-C or for Group R, Division 3 or Group U, Division 1 Occupancies, provided the building official finds that a special hazard does not exist.
>
> 2. For foundation concrete, other than cast-in-place drilled piles or caissons, where the structural design is based on an f'_c no greater than 2,500 pounds per square inch (psi) (17.2 MPa).

> 3. Nonstructural slabs on grade, including prestressed slabs on grade when effective prestress in concrete is less than 150 psi (1.03 MPa).
>
> 4. Site work concrete fully supported on earth and concrete where no special hazard exists.

2. Bolts installed in concrete. Prior to and during the placement of concrete around bolts when stress increases permitted by Footnote 5 of Table 19-D or Section 1923 are utilized.

3. Special moment-resisting concrete frame. For moment frames resisting design seismic load in structures within Seismic Zones 3 and 4, the special inspector shall provide reports to the person responsible for the structural design and shall provide continuous inspection of the placement of the reinforcement and concrete.

4. Reinforcing steel and prestressing steel tendons.

4.1 During all stressing and grouting of tendons in prestressed concrete.

4.2 During placing of reinforcing steel and prestressing tendons for all concrete required to have special inspection by Item 1.

> **EXCEPTION:** The special inspector need not be present continuously during placing of reinforcing steel and prestressing tendons, provided the special inspector has inspected for conformance to the approved plans prior to the closing of forms or the delivery of concrete to the jobsite.

5. Structural welding.

5.1 **General.** During the welding of any member or connection that is designed to resist loads and forces required by this code.

> **EXCEPTIONS:** 1. Welding done in an approved fabricator's shop in accordance with Section 1701.7.
>
> 2. The special inspector need not be continuously present during welding of the following items, provided the materials, qualifications of welding procedures and welders are verified prior to the start of work; periodic inspections are made of work in progress; and a visual inspection of all welds is made prior to completion or prior to shipment of shop welding:
>
> > 2.1 Single-pass fillet welds not exceeding $^5/_{16}$ inch (7.9 mm) in size.
> >
> > 2.2 Floor and roof deck welding.
> >
> > 2.3 Welded studs when used for structural diaphragm or composite systems.
> >
> > 2.4 Welded sheet steel for cold-formed steel framing members such as studs and joists.
> >
> > 2.5 Welding of stairs and railing systems.

5.2 **Special moment-resisting steel frames.** During the welding of special moment-resisting steel frames. In addition to Item 5.1 requirements, nondestructive testing as required by Section 1703 of this code.

5.3 **Welding of reinforcing steel.** During the welding of reinforcing steel.

> **EXCEPTION:** The special inspector need not be continuously present during the welding of ASTM A 706 reinforcing steel not larger than No. 5 bars used for embedments, provided the materials, qualifications of welding procedures and welders are verified prior to the start of work; periodic inspections are made of work in progress; and a visual inspection of all welds is made prior to completion or prior to shipment of shop welding.

6. High-strength bolting. The inspection of high-strength A 325 and A 490 bolts shall be in accordance with approved nationally recognized standards and the requirements of this section.

While the work is in progress, the special inspector shall determine that the requirements for bolts, nuts, washers and paint; bolted parts; and installation and tightening in such standards are met. Such inspections may be performed on a periodic basis in accordance with the requirements of Section 1701.6. The special inspector shall observe the calibration procedures when such procedures are required by the plans or specifications and shall monitor the installation of bolts to determine that all plies of connected materials have been drawn together and that the selected procedure is properly used to tighten all bolts.

7. **Structural masonry.**

7.1 For masonry, other than fully grouted open-end hollow-unit masonry, during preparation and taking of any required prisms or test specimens, placing of all masonry units, placement of reinforcement, inspection of grout space, immediately prior to closing of cleanouts, and during all grouting operations.

EXCEPTION: For hollow-unit masonry where the f'_m is no more than 1,500 psi (10.34 MPa) for concrete units or 2,600 psi (17.93 MPa) for clay units, special inspection may be performed as required for fully grouted open-end hollow-unit masonry specified in Item 7.2.

7.2 For fully grouted open-end hollow-unit masonry during preparation and taking of any required prisms or test specimens, at the start of laying units, after the placement of reinforcing steel, grout space prior to each grouting operation, and during all grouting operations.

EXCEPTION: Special inspection as required in Items 7.1 and 7.2 need not be provided when design stresses have been adjusted as specified in Chapter 21 to permit noncontinuous inspection.

8. **Reinforced gypsum concrete.** When cast-in-place Class B gypsum concrete is being mixed and placed.

9. **Insulating concrete fill.** During the application of insulating concrete fill when used as part of a structural system.

EXCEPTION: The special inspections may be limited to an initial inspection to check the deck surface and placement of reinforcing. The special inspector shall supervise the preparation of compression test specimens during this initial inspection.

10. **Spray-applied fire-resistive materials.** As required by UBC Standard 7-6.

11. **Piling, drilled piers and caissons.** During driving and testing of piles and construction of cast-in-place drilled piles or caissons. See Items 1 and 4 for concrete and reinforcing steel inspection.

12. **Shotcrete.** During the taking of test specimens and placing of all shotcrete and as required by Sections 1924.10 and 1924.11.

EXCEPTION: Shotcrete work fully supported on earth, minor repairs and when, in the opinion of the building official, no special hazard exists.

13. **Special grading, excavation and filling.** During earth-work excavations, grading and filling operations inspection to satisfy requirements of Chapter 18 and Appendix Chapter 33.

14. **Smoke-control system.**

14.1 During erection of ductwork and prior to concealment for the purposes of leakage testing and recording of device location.

14.2 Prior to occupancy and after sufficient completion for the purposes of pressure difference testing, flow measurements, and detection and control verification.

15. **Special cases.** Work that, in the opinion of the building official, involves unusual hazards or conditions.

1701.6 Continuous and Periodic Special Inspection.

1701.6.1 Continuous special inspection. Continuous special inspection means that the special inspector is on the site at all times observing the work requiring special inspection.

1701.6.2 Periodic special inspection. Some inspections may be made on a periodic basis and satisfy the requirements of continuous inspection, provided this periodic scheduled inspection is performed as outlined in the project plans and specifications and approved by the building official.

1701.7 Approved Fabricators. Special inspections required by this section and elsewhere in this code are not required where the work is done on the premises of a fabricator registered and approved by the building official to perform such work without special inspection. The certificate of registration shall be subject to revocation by the building official if it is found that any work done pursuant to the approval is in violation of this code. The approved fabricator shall submit a certificate of compliance that the work was performed in accordance with the approved plans and specifications to the building official and to the engineer or architect of record. The approved fabricator's qualifications shall be contingent on compliance with the following:

1. The fabricator has developed and submitted a detailed fabrication procedural manual reflecting key quality control procedures that will provide a basis for inspection control of workmanship and the fabricator plant.

2. Verification of the fabricator's quality control capabilities, plant and personnel as outlined in the fabrication procedural manual shall be by an approved inspection or quality control agency.

3. Periodic plant inspections shall be conducted by an approved inspection or quality control agency to monitor the effectiveness of the quality control program.

4. It shall be the responsibility of the inspection or quality control agency to notify the approving authority in writing of any change to the procedural manual. Any fabricator approval may be revoked for just cause. Reapproval of the fabricator shall be contingent on compliance with quality control procedures during the past year.

SECTION 1702 — STRUCTURAL OBSERVATION

Structural observation shall be provided in Seismic Zone 3 or 4 when one of the following conditions exists:

1. The structure is defined in Table 16-K as Occupancy Category 1, 2 or 3,

2. The structure is required to comply with Section 403,

3. The structure is in Seismic Zone 4, N_a as set forth in Table 16-S is greater than one, and a lateral design is required for the entire structure,

EXCEPTION: One- and two-story Group R, Division 3 and Group U Occupancies and one- and two-story Groups B, F, M and S Occupancies.

4. When so designated by the architect or engineer of record, or

5. When such observation is specifically required by the building official.

The owner shall employ the engineer or architect responsible for the structural design, or another engineer or architect designated by the engineer or architect responsible for the structural design, to perform structural observation as defined in Section 220. Observed deficiencies shall be reported in writing to the owner's representative, special inspector, contractor and the building official. The structural observer shall submit to the building official a written statement that the site visits have been made and identifying any reported deficiencies that, to the best of the structural observer's knowledge, have not been resolved.

SECTION 1703 — NONDESTRUCTIVE TESTING

In Seismic Zones 3 and 4, welded, fully restrained connections between the primary members of ordinary moment frames and special moment-resisting frames shall be tested by nondestructive methods for compliance with approved standards and job specifications. This testing shall be a part of the special inspection requirements of Section 1701.5. A program for this testing shall be established by the person responsible for structural design and as shown on plans and specifications.

As a minimum, this program shall include the following:

1. All complete penetration groove welds contained in joints and splices shall be tested 100 percent either by ultrasonic testing or by radiography.

> **EXCEPTIONS: 1.** When approved, the nondestructive testing rate for an individual welder or welding operator may be reduced to 25 percent, provided the reject rate is demonstrated to be 5 percent or less of the welds tested for the welder or welding operator. A sampling of at least 40 completed welds for a job shall be made for such reduction evaluation. Reject rate is defined as the number of welds containing rejectable defects divided by the number of welds completed. For evaluating the reject rate of continuous welds over 3 feet (914 mm) in length where the effective throat thickness is 1 inch (25 mm) or less, each 12-inch increment (305 mm) or fraction thereof shall be considered as one weld. For evaluating the reject rate on continuous welds over 3 feet (914 mm) in length where the effective throat thickness is greater than 1 inch (25 mm), each 6 inches (152 mm) of length or fraction thereof shall be considered one weld.
>
> **2.** For complete penetration groove welds on materials less than $^5/_{16}$ inch (7.9 mm) thick, nondestructive testing is not required; for this welding, continuous inspection is required.
>
> **3.** When approved by the building official and outlined in the project plans and specifications, this nondestructive ultrasonic testing may be performed in the shop of an approved fabricator utilizing qualified test techniques in the employment of the fabricator.

2. Partial penetration groove welds when used in column splices shall be tested either by ultrasonic testing or radiography when required by the plans and specifications. For partial penetration groove welds when used in column splices, with an effective throat less than $^3/_4$ inch (19.1 mm) thick, nondestructive testing is not required; for this welding, continuous special inspection is required.

3. Base metal thicker than $1^1/_2$ inches (38 mm), when subjected to through-thickness weld shrinkage strains, shall be ultrasonically inspected for discontinuities directly behind such welds after joint completion.

Any material discontinuities shall be accepted or rejected on the basis of the defect rating in accordance with the (larger reflector) criteria of approved national standards.

SECTION 1704 — PREFABRICATED CONSTRUCTION

1704.1 General.

1704.1.1 Purpose. The purpose of this section is to regulate materials and establish methods of safe construction where any structure or portion thereof is wholly or partially prefabricated.

1704.1.2 Scope. Unless otherwise specifically stated in this section, all prefabricated construction and all materials used therein shall conform to all the requirements of this code. (See Section 104.2.8.)

1704.1.3 Definition.

PREFABRICATED ASSEMBLY is a structural unit, the integral parts of which have been built up or assembled prior to incorporation in the building.

1704.2 Tests of Materials. Every approval of a material not specifically mentioned in this code shall incorporate as a proviso the kind and number of tests to be made during prefabrication.

1704.3 Tests of Assemblies. The building official may require special tests to be made on assemblies to determine their durability and weather resistance.

1704.4 Connections. See Section 1611.11.1 for design requirements of connections for prefabricated assemblies.

1704.5 Pipes and Conduits. See Section 1611.11.2 for design requirements for removal of material for pipes, conduit and other equipment.

1704.6 Certificate and Inspection.

1704.6.1 Materials. Materials and the assembly thereof shall be inspected to determine compliance with this code. Every material shall be graded, marked or labeled where required elsewhere in this code.

1704.6.2 Certificate. A certificate of approval shall be furnished with every prefabricated assembly, except where the assembly is readily accessible to inspection at the site. The certificate of approval shall certify that the assembly in question has been inspected and meets all the requirements of this code. When mechanical equipment is installed so that it cannot be inspected at the site, the certificate of approval shall certify that such equipment complies with the laws applying thereto.

1704.6.3 Certifying agency. To be acceptable under this code, every certificate of approval shall be made by an approved agency.

1704.6.4 Field erection. Placement of prefabricated assemblies at the building site shall be inspected by the building official to determine compliance with this code.

1704.6.5 Continuous inspection. If continuous inspection is required for certain materials where construction takes place on the site, it shall also be required where the same materials are used in prefabricated construction.

> **EXCEPTION:** Continuous inspection will not be required during prefabrication if the approved agency certifies to the construction and furnishes evidence of compliance.

Chapter 18

FOUNDATIONS AND RETAINING WALLS

Division I—GENERAL

SECTION 1801 — SCOPE

1801.1 General. This chapter sets forth requirements for excavation and fills for any building or structure and for foundations and retaining structures.

Reference is made to Appendix Chapter 33 for requirements governing excavation, grading and earthwork construction, including fills and embankments.

1801.2 Standards of Quality. The standards listed below labeled a "UBC Standard" are also listed in Chapter 35, Part II, and are part of this code.

1. **Testing.**

    1.1 UBC Standard 18-1, Soils Classification

    1.2 UBC Standard 18-2, Expansion Index Test

SECTION 1802 — QUALITY AND DESIGN

The quality and design of materials used structurally in excavations, footings and foundations shall conform to the requirements specified in Chapters 16, 19, 21, 22 and 23.

Excavations and fills shall comply with Chapter 33.

Allowable bearing pressures, allowable stresses and design formulas provided in this chapter shall be used with the allowable stress design load combinations specified in Section 1612.3.

SECTION 1803 — SOIL CLASSIFICATION—EXPANSIVE SOIL

1803.1 General. For the purposes of this chapter, the definition and classification of soil materials for use in Table 18-I-A shall be according to UBC Standard 18-1.

1803.2 Expansive Soil. When the expansive characteristics of a soil are to be determined, the procedures shall be in accordance with UBC Standard 18-2 and the soil shall be classified according to Table 18-I-B. Foundations for structures resting on soils with an expansion index greater than 20, as determined by UBC Standard 18-2, shall require special design consideration. If the soil expansion index varies with depth, the variation is to be included in the engineering analysis of the expansive soil effect upon the structure.

SECTION 1804 — FOUNDATION INVESTIGATION

1804.1 General. The classification of the soil at each building site shall be determined when required by the building official. The building official may require that this determination be made by an engineer or architect licensed by the state to practice as such.

1804.2 Investigation. The classification shall be based on observation and any necessary tests of the materials disclosed by borings or excavations made in appropriate locations. Additional studies may be necessary to evaluate soil strength, the effect of moisture variation on soil-bearing capacity, compressibility, liquefaction and expansiveness.

In Seismic Zones 3 and 4, when required by the building official, the potential for seismically induced soil liquefaction and soil instability shall be evaluated as described in Section 1804.5.

> **EXCEPTIONS:** 1. The building official may waive this evaluation upon receipt of written opinion of a qualified geotechnical engineer or geologist that liquefaction is not probable.
>
> 2. A detached, single-story dwelling of Group R, Division 3 Occupancy with or without attached garages.
>
> 3. Group U, Division 1 Occupancies.
>
> 4. Fences.

1804.3 Reports. The soil classification and design-bearing capacity shall be shown on the plans, unless the foundation conforms to Table 18-I-C. The building official may require submission of a written report of the investigation, which shall include, but need not be limited to, the following information:

1. A plot showing the location of all test borings and/or excavations.

2. Descriptions and classifications of the materials encountered.

3. Elevation of the water table, if encountered.

4. Recommendations for foundation type and design criteria, including bearing capacity, provisions to mitigate the effects of expansive soils, provisions to mitigate the effects of liquefaction and soil strength, and the effects of adjacent loads.

5. Expected total and differential settlement.

1804.4 Expansive Soils. When expansive soils are present, the building official may require that special provisions be made in the foundation design and construction to safeguard against damage due to this expansiveness. The building official may require a special investigation and report to provide these design and construction criteria.

1804.5 Liquefaction Potential and Soil Strength Loss. When required by Section 1804.2, the potential for soil liquefaction and soil strength loss during earthquakes shall be evaluated during the geotechnical investigation. The geotechnical report shall assess potential consequences of any liquefaction and soil strength loss, including estimation of differential settlement, lateral movement or reduction in foundation soil-bearing capacity, and discuss mitigating measures. Such measures shall be given consideration in the design of the building and may include, but are not limited to, ground stabilization, selection of appropriate foundation type and depths, selection of appropriate structural systems to accommodate anticipated displacements, or any combination of these measures.

The potential for liquefaction and soil strength loss shall be evaluated for a site peak ground acceleration that, as a minimum, conforms to the probability of exceedance specified in Section 1631.2. Peak ground acceleration may be determined based on a site-specific study taking into account soil amplification effects. In the absence of such a study, peak ground acceleration may be assumed equal to the seismic zone factor in Table 16-I.

1804.6 Adjacent Loads. Where footings are placed at varying elevations, the effect of adjacent loads shall be included in the foundation design.

1804.7 Drainage. Provisions shall be made for the control and drainage of surface water around buildings. (See also Section 1806.5.5.)

SECTION 1805 — ALLOWABLE FOUNDATION AND LATERAL PRESSURES

The allowable foundation and lateral pressures shall not exceed the values set forth in Table 18-I-A unless data to substantiate the use of higher values are submitted. Table 18-I-A may be used for design of foundations on rock or nonexpansive soil for Type II One-hour, Type II-N and Type V buildings that do not exceed three stories in height or for structures that have continuous footings having a load of less than 2,000 pounds per lineal foot (29.2 kN/m) and isolated footings with loads of less than 50,000 pounds (222.4 kN).

Allowable bearing pressures provided in Table 18-I-A shall be used with the allowable stress design load combinations specified in Section 1612.3.

SECTION 1806 — FOOTINGS

1806.1 General. Footings and foundations shall be constructed of masonry, concrete or treated wood in conformance with Division II and shall extend below the frost line. Footings of concrete and masonry shall be of solid material. Foundations supporting wood shall extend at least 6 inches (152 mm) above the adjacent finish grade. Footings shall have a minimum depth as indicated in Table 18-I-C, unless another depth is recommended by a foundation investigation.

The provisions of this section do not apply to building and foundation systems in those areas subject to scour and water pressure by wind and wave action. Buildings and foundations subject to such loads shall be designed in accordance with approved national standards. See Section 3302 for subsoil preparation and wood form removal.

1806.2 Footing Design. Except for special provisions of Section 1808 covering the design of piles, all portions of footings shall be designed in accordance with the structural provisions of this code and shall be designed to minimize differential settlement when necessary and the effects of expansive soils when present.

Slab-on-grade and mat-type footings for buildings located on expansive soils may be designed in accordance with the provisions of Division III or such other engineering design based on geotechnical recommendation as approved by the building official.

1806.3 Bearing Walls. Bearing walls shall be supported on masonry or concrete foundations or piles or other approved foundation system that shall be of sufficient size to support all loads. Where a design is not provided, the minimum foundation requirements for stud bearing walls shall be as set forth in Table 18-I-C, unless expansive soils of a severity to cause differential movement are known to exist.

> **EXCEPTIONS:** 1. A one-story wood- or metal-frame building not used for human occupancy and not over 400 square feet (37.2 m²) in floor area may be constructed with walls supported on a wood foundation plate when approved by the building official.
>
> 2. The support of buildings by posts embedded in earth shall be designed as specified in Section 1806.8. Wood posts or poles embedded in earth shall be pressure treated with an approved preservative. Steel posts or poles shall be protected as specified in Section 1807.9.

1806.4 Stepped Foundations. Foundations for all buildings where the surface of the ground slopes more than 1 unit vertical in 10 units horizontal (10% slope) shall be level or shall be stepped so that both top and bottom of such foundation are level.

1806.5 Footings on or Adjacent to Slopes.

1806.5.1 Scope. The placement of buildings and structures on or adjacent to slopes steeper than 1 unit vertical in 3 units horizontal (33.3% slope) shall be in accordance with this section.

1806.5.2 Building clearance from ascending slopes. In general, buildings below slopes shall be set a sufficient distance from the slope to provide protection from slope drainage, erosion and shallow failures. Except as provided for in Section 1806.5.6 and Figure 18-I-1, the following criteria will be assumed to provide this protection. Where the existing slope is steeper than 1 unit vertical in 1 unit horizontal (100% slope), the toe of the slope shall be assumed to be at the intersection of a horizontal plane drawn from the top of the foundation and a plane drawn tangent to the slope at an angle of 45 degrees to the horizontal. Where a retaining wall is constructed at the toe of the slope, the height of the slope shall be measured from the top of the wall to the top of the slope.

1806.5.3 Footing setback from descending slope surface. Footing on or adjacent to slope surfaces shall be founded in firm material with an embedment and setback from the slope surface sufficient to provide vertical and lateral support for the footing without detrimental settlement. Except as provided for in Section 1806.5.6 and Figure 18-I-1, the following setback is deemed adequate to meet the criteria. Where the slope is steeper than 1 unit vertical in 1 unit horizontal (100% slope), the required setback shall be measured from an imaginary plane 45 degrees to the horizontal, projected upward from the toe of the slope.

1806.5.4 Pools. The setback between pools regulated by this code and slopes shall be equal to one half the building footing setback distance required by this section. That portion of the pool wall within a horizontal distance of 7 feet (2134 mm) from the top of the slope shall be capable of supporting the water in the pool without soil support.

1806.5.5 Foundation elevation. On graded sites, the top of any exterior foundation shall extend above the elevation of the street gutter at point of discharge or the inlet of an approved drainage device a minimum of 12 inches (305 mm) plus 2 percent. The building official may approve alternate elevations, provided it can be demonstrated that required drainage to the point of discharge and away from the structure is provided at all locations on the site.

1806.5.6 Alternate setback and clearance. The building official may approve alternate setbacks and clearances. The building official may require an investigation and recommendation of a qualified engineer to demonstrate that the intent of this section has been satisfied. Such an investigation shall include consideration of material, height of slope, slope gradient, load intensity and erosion characteristics of slope material.

1806.6 Foundation Plates or Sills. Wood plates or sills shall be bolted to the foundation or foundation wall. Steel bolts with a minimum nominal diameter of $1/2$ inch (12.7 mm) shall be used in Seismic Zones 0 through 3. Steel bolts with a minimum nominal diameter of $5/8$ inch (16 mm) shall be used in Seismic Zone 4. Bolts shall be embedded at least 7 inches (178 mm) into the concrete or masonry and shall be spaced not more than 6 feet (1829 mm) apart. There shall be a minimum of two bolts per piece with one bolt located not more than 12 inches (305 mm) or less than seven bolt diameters from each end of the piece. A properly sized nut and washer shall be tightened on each bolt to the plate. Foundation plates and sills shall be the kind of wood specified in Section 2306.4.

1806.6.1 Additional requirements in Seismic Zones 3 and 4. The following additional requirements shall apply in Seismic Zones 3 and 4.

1. Sill bolt diameter and spacing for three-story raised wood floor buildings shall be specifically designed.

2. Plate washers a minimum of 2 inch by 2 inch by $^3/_{16}$ inch (51 mm by 51 mm by 4.8 mm) thick shall be used on each bolt.

1806.7 Seismic Zones 3 and 4. In Seismic Zones 3 and 4, horizontal reinforcement in accordance with Sections 1806.7.1 and 1806.7.2 shall be placed in continuous foundations to minimize differential settlement. Foundation reinforcement shall be provided with cover in accordance with Section 1907.7.1.

1806.7.1 Foundations with stemwalls. Foundations with stemwalls shall be provided with a minimum of one No. 4 bar at the top of the wall and one No. 4 bar at the bottom of the footing.

1806.7.2 Slabs–on–ground with turned–down footings. Slabs–on–ground with turned-down footings shall have a minimum of one No. 4 bar at the top and bottom.

> **EXCEPTION:** For slabs-on-ground cast monolithically with a footing, one No. 5 bar may be located at either the top or bottom.

1806.8 Designs Employing Lateral Bearing.

1806.8.1 General. Construction employing posts or poles as columns embedded in earth or embedded in concrete footings in the earth may be used to resist both axial and lateral loads. The depth to resist lateral loads shall be determined by means of the design criteria established herein or other methods approved by the building official.

1806.8.2 Design criteria.

1806.8.2.1 Nonconstrained. The following formula may be used in determining the depth of embedment required to resist lateral loads where no constraint is provided at the ground surface, such as rigid floor or rigid ground surface pavement.

$$d = \frac{A}{2}\left(1 + \sqrt{1 + \frac{4.36h}{A}}\right) \qquad (6\text{-}1)$$

WHERE:

$A = \dfrac{2.34P}{S_1 b}$

b = diameter of round post or footing or diagonal dimension of square post or footing, feet (m).

d = depth of embedment in earth in feet (m) but not over 12 feet (3658 mm) for purpose of computing lateral pressure.

h = distance in feet (m) from ground surface to point of application of "P."

P = applied lateral force in pounds (kN).

S_1 = allowable lateral soil-bearing pressure as set forth in Table 18-I-A based on a depth of one third the depth of embedment (kPa).

S_3 = allowable lateral soil-bearing pressure as set forth in Table 18-I-A based on a depth equal to the depth of embedment (kPa).

1806.8.2.2 Constrained. The following formula may be used to determine the depth of embedment required to resist lateral loads where constraint is provided at the ground surface, such as a rigid floor or pavement.

$$d^2 = 4.25\frac{Ph}{S_3 b} \qquad (6\text{-}2)$$

1806.8.2.3 Vertical load. The resistance to vertical loads is determined by the allowable soil-bearing pressure set forth in Table 18-I-A.

1806.8.3 Backfill. The backfill in the annular space around columns not embedded in poured footings shall be by one of the following methods:

1. Backfill shall be of concrete with an ultimate strength of 2,000 pounds per square inch (13.79 MPa) at 28 days. The hole shall not be less than 4 inches (102 mm) larger than the diameter of the column at its bottom or 4 inches (102 mm) larger than the diagonal dimension of a square or rectangular column.

2. Backfill shall be of clean sand. The sand shall be thoroughly compacted by tamping in layers not more than 8 inches (203 mm) in depth.

1806.8.4 Limitations. The design procedure outlined in this section shall be subject to the following limitations:

The frictional resistance for retaining walls and slabs on silts and clays shall be limited to one half of the normal force imposed on the soil by the weight of the footing or slab.

Posts embedded in earth shall not be used to provide lateral support for structural or nonstructural materials such as plaster, masonry or concrete unless bracing is provided that develops the limited deflection required.

1806.9 Grillage Footings. When grillage footings of structural steel shapes are used on soils, they shall be completely embedded in concrete with at least 6 inches (152 mm) on the bottom and at least 4 inches (102 mm) at all other points.

1806.10 Bleacher Footings. Footings for open-air seating facilities shall comply with Chapter 18.

> **EXCEPTIONS:** Temporary open-air portable bleachers as defined in Section 1008.2 may be supported upon wood sills or steel plates placed directly upon the ground surface, provided soil pressure does not exceed 1,200 pounds per square foot (57.5 kPa).

SECTION 1807 — PILES — GENERAL REQUIREMENTS

1807.1 General. Pile foundations shall be designed and installed on the basis of a foundation investigation as defined in Section 1804 where required by the building official.

The investigation and report provisions of Section 1804 shall be expanded to include, but not be limited to, the following:

1. Recommended pile types and installed capacities.

2. Driving criteria.

3. Installation procedures.

4. Field inspection and reporting procedures (to include procedures for verification of the installed bearing capacity where required).

5. Pile load test requirements.

The use of piles not specifically mentioned in this chapter shall be permitted, subject to the approval of the building official upon submission of acceptable test data, calculations or other information relating to the properties and load-carrying capacities of such piles.

1807.2 Interconnection. Individual pile caps and caissons of every structure subjected to seismic forces shall be interconnected by ties. Such ties shall be capable of resisting, in tension or compression, a minimum horizontal force equal to 10 percent of the larger column vertical load.

EXCEPTION: Other approved methods may be used where it can be demonstrated that equivalent restraint can be provided.

1807.3 Determination of Allowable Loads. The allowable axial and lateral loads on piles shall be determined by an approved formula, by load tests or by a foundation investigation.

1807.4 Static Load Tests. When the allowable axial load of a single pile is determined by a load test, one of the following methods shall be used:

Method 1. It shall not exceed 50 percent of the yield point under test load. The yield point shall be defined as that point at which an increase in load produces a disproportionate increase in settlement.

Method 2. It shall not exceed one half of the load which causes a net settlement, after deducting rebound, of 0.01 inch per ton (0.000565 mm/N) of test load which has been applied for a period of at least 24 hours.

Method 3. It shall not exceed one half of that load under which, during a 40-hour period of continuous load application, no additional settlement takes place.

1807.5 Column Action. All piles standing unbraced in air, water or material not capable of lateral support, shall conform with the applicable column formula as specified in this code. Such piles driven into firm ground may be considered fixed and laterally supported at 5 feet (1524 mm) below the ground surface and in soft material at 10 feet (3048 mm) below the ground surface unless otherwise prescribed by the building official after a foundation investigation by an approved agency.

1807.6 Group Action. Consideration shall be given to the reduction of allowable pile load when piles are placed in groups. Where soil conditions make such load reductions advisable or necessary, the allowable axial load determined for a single pile shall be reduced by any rational method or formula approved by the building official.

1807.7 Piles in Subsiding Areas. Where piles are driven through subsiding fills or other subsiding strata and derive support from underlying firmer materials, consideration shall be given to the downward frictional forces which may be imposed on the piles by the subsiding upper strata.

Where the influence of subsiding fills is considered as imposing loads on the pile, the allowable stresses specified in this chapter may be increased if satisfactory substantiating data are submitted.

1807.8 Jetting. Jetting shall not be used except where and as specifically permitted by the building official. When used, jetting shall be carried out in such a manner that the carrying capacity of existing piles and structures shall not be impaired. After withdrawal of the jet, piles shall be driven down until the required resistance is obtained.

1807.9 Protection of Pile Materials. Where the boring records of site conditions indicate possible deleterious action on pile materials because of soil constituents, changing water levels or other factors, such materials shall be adequately protected by methods or processes approved by the building official. The effectiveness of such methods or processes for the particular purpose shall have been thoroughly established by satisfactory service records or other evidence which demonstrates the effectiveness of such protective measures.

1807.10 Allowable Loads. The allowable loads based on soil conditions shall be established in accordance with Section 1807.

EXCEPTION: Any uncased cast-in-place pile may be assumed to develop a frictional resistance equal to one sixth of the bearing value

of the soil material at minimum depth as set forth in Table 18-I-A but not to exceed 500 pounds per square foot (24 kPa) unless a greater value is allowed by the building official after a soil investigation as specified in Section 1804 is submitted. Frictional resistance and bearing resistance shall not be assumed to act simultaneously unless recommended after a foundation investigation as specified in Section 1804.

1807.11 Use of Higher Allowable Pile Stresses. Allowable compressive stresses greater than those specified in Section 1808 shall be permitted when substantiating data justifying such higher stresses are submitted to and approved by the building official. Such substantiating data shall include a foundation investigation including a report in accordance with Section 1807.1 by a soils engineer defined as a civil engineer experienced and knowledgeable in the practice of soils engineering.

SECTION 1808 — SPECIFIC PILE REQUIREMENTS

1808.1 Round Wood Piles.

1808.1.1 Material. Except where untreated piles are permitted, wood piles shall be pressure treated. Untreated piles may be used only when it has been established that the cutoff will be below lowest groundwater level assumed to exist during the life of the structure.

1808.1.2 Allowable stresses. The allowable unit stresses for round wood piles shall not exceed those set forth in Chapter 23, Division III, Part I.

The allowable values listed in Chapter 23, Division III, Part I, for compression parallel to the grain at extreme fiber in bending are based on load sharing as occurs in a pile cluster. For piles which support their own specific load, a safety factor of 1.25 shall be applied to compression parallel to the grain values and 1.30 to extreme fiber in bending values.

1808.2 Uncased Cast-in-place Concrete Piles.

1808.2.1 Material. Concrete piles cast in place against earth in drilled or bored holes shall be made in such a manner as to ensure the exclusion of any foreign matter and to secure a full-sized shaft. The length of such pile shall be limited to not more than 30 times the average diameter. Concrete shall have a specified compressive strength f'_c of not less than 2,500 psi (17.24 MPa).

EXCEPTION: The length of pile may exceed 30 times the diameter provided the design and installation of the pile foundation is in accordance with an approved investigation report.

1808.2.2 Allowable stresses. The allowable compressive stress in the concrete shall not exceed $0.33f'_c$. The allowable compressive stress of reinforcement shall not exceed 34 percent of the yield strength of the steel or 25,500 psi (175.7 MPa).

1808.3 Metal-cased Concrete Piles.

1808.3.1 Material. Concrete used in metal-cased concrete piles shall have a specified compressive strength f'_c of not less than 2,500 psi (17.24 MPa).

1808.3.2 Installation. Every metal casing for a concrete pile shall have a sealed tip with a diameter of not less than 8 inches (203 mm).

Concrete piles cast in place in metal shells shall have shells driven for their full length in contact with the surrounding soil and left permanently in place. The shells shall be sufficiently strong to resist collapse and sufficiently watertight to exclude water and foreign material during the placing of concrete.

Piles shall be driven in such order and with such spacing as to ensure against distortion of or injury to piles already in place. No pile shall be driven within four and one-half average pile diameters of a pile filled with concrete less than 24 hours old unless approved by the building official.

1808.3.3 Allowable stresses. Allowable stresses shall not exceed the values specified in Section 1808.2.2, except that the allowable concrete stress may be increased to a maximum value of $0.40f'_c$ for that portion of the pile meeting the following conditions:

1. The thickness of the metal casing is not less than 0.068 inch (1.73 mm) (No. 14 carbon sheet steel gage).

2. The casing is seamless or is provided with seams of equal strength and is of a configuration that will provide confinement to the cast-in-place concrete.

3. The specified compressive strength f'_c shall not exceed 5,000 psi (34.47 MPa) and the ratio of steel minimum specified yield strength f_y to concrete specified compressive strength f'_c shall not be less than 6.

4. The pile diameter is not greater than 16 inches (406 mm).

1808.4 Precast Concrete Piles.

1808.4.1 Materials. Precast concrete piles shall have a specified compressive strength f'_c of not less than 3,000 psi (20.68 MPa), and shall develop a compressive strength of not less than 3,000 psi (20.68 MPa) before driving.

1808.4.2 Reinforcement ties. The longitudinal reinforcement in driven precast concrete piles shall be laterally tied with steel ties or wire spirals. Ties and spirals shall not be spaced more than 3 inches (76 mm) apart, center to center, for a distance of 2 feet (610 mm) from the ends and not more than 8 inches (203 mm) elsewhere. The gage of ties and spirals shall be as follows:

For piles having a diameter of 16 inches (406 mm) or less, wire shall not be smaller than 0.22 inch (5.6 mm) (No. 5 B.W. gage).

For piles having a diameter of more than 16 inches (406 mm) and less than 20 inches (508 mm), wire shall not be smaller than 0.238 inch (6.0 mm) (No. 4 B.W. gage).

For piles having a diameter of 20 inches (508 mm) and larger, wire shall not be smaller than $1/4$ inch (6.4 mm) round or 0.259 inch (6.6 mm) (No. 3 B.W. gage).

1808.4.3 Allowable stresses. Precast concrete piling shall be designed to resist stresses induced by handling and driving as well as by loads. The allowable stresses shall not exceed the values specified in Section 1808.2.2.

1808.5 Precast Prestressed Concrete Piles (Pretensioned).

1808.5.1 Materials. Precast prestressed concrete piles shall have a specified compressive strength f'_c of not less than 5,000 psi (34.48 MPa) and shall develop a compressive strength of not less than 4,000 psi (27.58 MPa) before driving.

1808.5.2 Reinforcement. The longitudinal reinforcement shall be high-tensile seven-wire strand. Longitudinal reinforcement shall be laterally tied with steel ties or wire spirals.

Ties or spiral reinforcement shall not be spaced more than 3 inches (76 mm) apart, center to center, for a distance of 2 feet (610 mm) from the ends and not more than 8 inches (203 mm) elsewhere.

At each end of the pile, the first five ties or spirals shall be spaced 1 inch (25 mm) center to center.

For piles having a diameter of 24 inches (610 mm) or less, wire shall not be smaller than 0.22 inch (5.6 mm) (No. 5 B.W. gage). For piles having a diameter greater than 24 inches (610 mm) but

less than 36 inches (914 mm), wire shall not be smaller than 0.238 inch (6.0 mm) (No. 4 B.W. gage). For piles having a diameter greater than 36 inches (914 mm), wire shall not be smaller than $1/4$ inch (6.4 mm) round or 0.259 inch (6.6 mm) (No. 3 B.W. gage).

1808.5.3 Allowable stresses. Precast prestressed piling shall be designed to resist stresses induced by handling and driving as well as by loads. The effective prestress in the pile shall not be less than 400 psi (2.76 MPa) for piles up to 30 feet (9144 mm) in length, 550 psi (3.79 MPa) for piles up to 50 feet (15 240 mm) in length, and 700 psi (4.83 MPa) for piles greater than 50 feet (15 240 mm) in length.

The compressive stress in the concrete due to externally applied load shall not exceed:

$$f_c = 0.33f'_c - 0.27fp_c$$

WHERE:

fp_c = effective prestress stress on the gross section.

Effective prestress shall be based on an assumed loss of 30,000 psi (206.85 MPa) in the prestressing steel. The allowable stress in the prestressing steel shall not exceed the values specified in Section 1918.

1808.6 Structural Steel Piles.

1808.6.1 Material. Structural steel piles, steel pipe piles and fully welded steel piles fabricated from plates shall conform to UBC Standard 22-1 and be identified in accordance with Section 2202.2.

1808.6.2 Allowable stresses. The allowable axial stresses shall not exceed 0.35 of the minimum specified yield strength F_y or 12,600 psi (86.88 MPa), whichever is less.

> **EXCEPTION:** When justified in accordance with Section 1807.11, the allowable axial stress may be increased above 12,600 psi (86.88 MPa) and $0.35F_y$, but shall not exceed $0.5F_y$.

1808.6.3 Minimum dimensions. Sections of driven H-piles shall comply with the following:

1. The flange projection shall not exceed 14 times the minimum thickness of metal in either the flange or the web, and the flange widths shall not be less than 80 percent of the depth of the section.

2. The nominal depth in the direction of the web shall not be less than 8 inches (203 mm).

3. Flanges and webs shall have a minimum nominal thickness of $3/8$ inch (9.5 mm).

Sections of driven pipe piles shall have an outside diameter of not less than 10 inches (254 mm) and a minimum thickness of not less than $1/4$ inch (6.4 mm).

1808.7 Concrete-filled Steel Pipe Piles.

1808.7.1 Material. The concrete-filled steel pipe piles shall conform to UBC Standard 22-1 and shall be identified in accordance with Section 2202.2. The concrete-filled steel pipe piles shall have a specified compressive strength f'_c of not less than 2,500 psi (17.24 MPa).

1808.7.2 Allowable stresses. The allowable axial stresses shall not exceed 0.35 of the minimum specified yield strength F_y of the steel plus 0.33 of the specified compressive strength f'_c of concrete, provided F_y shall not be assumed greater than 36,000 psi (248.22 MPa) for computational purposes.

> **EXCEPTION:** When justified in accordance with Section 2807.11, the allowable stresses may be increased to $0.50 F_y$.

1808.7.3 Minimum dimensions. Driven piles of uniform section shall have a nominal outside diameter of not less than 8 inches (203 mm).

SECTION 1809 — FOUNDATION CONSTRUCTION— SEISMIC ZONES 3 AND 4

1809.1 General. In Seismic Zones 3 and 4 the further requirements of this section shall apply to the design and construction of foundations, foundation components and the connection of superstructure elements thereto.

1809.2 Soil Capacity. The foundation shall be capable of transmitting the design base shear and overturning forces prescribed in Section 1630 from the structure into the supporting soil. The short-term dynamic nature of the loads may be taken into account in establishing the soil properties.

1809.3 Superstructure-to-Foundation Connection. The connection of superstructure elements to the foundation shall be adequate to transmit to the foundation the forces for which the elements were required to be designed.

1809.4 Foundation-Soil Interface. For regular buildings, the force F_t as provided in Section 1630.5 may be omitted when determining the overturning moment to be resisted at the foundation-soil interface.

1809.5 Special Requirements for Piles and Caissons.

1809.5.1 General. Piles, caissons and caps shall be designed according to the provisions of Section 1603, including the effects of lateral displacements. Special detailing requirements as described in Section 1809.5.2 shall apply for a length of piles equal to 120 percent of the flexural length. Flexural length shall be considered as a length of pile from the first point of zero lateral deflection to the underside of the pile cap or grade beam.

1809.5.2 Steel piles, nonprestressed concrete piles and prestressed concrete piles.

1809.5.2.1 Steel piles. Piles shall conform to width-thickness ratios of stiffened, unstiffened and tubular compression elements as shown in Chapter 22, Division VIII.

1809.5.2.2 Nonprestressed concrete piles. Piles shall have transverse reinforcement meeting the requirements of Section 1921.4.

> **EXCEPTION:** Transverse reinforcement need not exceed the amount determined by Formula (21-2) in Section 1921.4.4.1 for spiral or circular hoop reinforcement or by Formula (21-4) in Section 1921.4.4.1 for rectangular hoop reinforcement.

1809.5.2.3 Prestressed concrete piles. Piles shall have a minimum volumetric ratio of spiral reinforcement no less than 0.021 for 14-inch (356 mm) square and smaller piles, and 0.012 for 24-inch (610 mm) square and larger piles unless a smaller value can be justified by rational analysis. Interpolation may be used between the specified ratios for intermediate sizes.

TABLE 18-I-A—ALLOWABLE FOUNDATION AND LATERAL PRESSURE

CLASS OF MATERIALS[1]	ALLOWABLE FOUNDATION PRESSURE (psf)[2] $\times$ 0.0479 for kPa	LATERAL BEARING LBS./SQ./FT./FT. OF DEPTH BELOW NATURAL GRADE[3] $\times$ 0.157 for kPa per meter	LATERAL SLIDING[4] Coefficient[5]	LATERAL SLIDING[4] Resistance (psf)[6] $\times$ 0.0479 for kPa
1. Massive crystalline bedrock	4,000	1,200	0.70	
2. Sedimentary and foliated rock	2,000	400	0.35	
3. Sandy gravel and/or gravel (GW and GP)	2,000	200	0.35	
4. Sand, silty sand, clayey sand, silty gravel and clayey gravel (SW, SP, SM, SC, GM and GC)	1,500	150	0.25	
5. Clay, sandy clay, silty clay and clayey silt (CL, ML, MH and CH)	1,000[7]	100		130

[1]For soil classifications OL, OH and PT (i.e., organic clays and peat), a foundation investigation shall be required.
[2]All values of allowable foundation pressure are for footings having a minimum width of 12 inches (305 mm) and a minimum depth of 12 inches (305 mm) into natural grade. Except as in Footnote 7, an increase of 20 percent shall be allowed for each additional foot (305 mm) of width or depth to a maximum value of three times the designated value. Additionally, an increase of one third shall be permitted when considering load combinations, including wind or earthquake loads, as permitted by Section 1612.3.2.
[3]May be increased the amount of the designated value for each additional foot (305 mm) of depth to a maximum of 15 times the designated value. Isolated poles for uses such as flagpoles or signs and poles used to support buildings that are not adversely affected by a $^1/_2$-inch (12.7 mm) motion at ground surface due to short-term lateral loads may be designed using lateral bearing values equal to two times the tabulated values.
[4]Lateral bearing and lateral sliding resistance may be combined.
[5]Coefficient to be multiplied by the dead load.
[6]Lateral sliding resistance value to be multiplied by the contact area. In no case shall the lateral sliding resistance exceed one half the dead load.
[7]No increase for width is allowed.

TABLE 18-I-B—CLASSIFICATION OF EXPANSIVE SOIL

EXPANSION INDEX	POTENTIAL EXPANSION
0-20	Very low
21-50	Low
51-90	Medium
91-130	High
Above 130	Very high

TABLE 18-I-C—FOUNDATIONS FOR STUD BEARING WALLS—MINIMUM REQUIREMENTS[1,2,3]

NUMBER OF FLOORS SUPPORTED BY THE FOUNDATION[4]	THICKNESS OF FOUNDATION WALL (inches) $\times$ 25.4 for mm Concrete	THICKNESS OF FOUNDATION WALL (inches) $\times$ 25.4 for mm Unit Masonry	WIDTH OF FOOTING (inches)	THICKNESS OF FOOTING (inches)	DEPTH BELOW UNDISTURBED GROUND SURFACE (inches)
			$\times$ 25.4 for mm		
1	6	6	12	6	12
2	8	8	15	7	18
3	10	10	18	8	24

[1]Where unusual conditions or frost conditions are found, footings and foundations shall be as required in Section 1806.1.
[2]The ground under the floor may be excavated to the elevation of the top of the footing.
[3]Interior stud bearing walls may be supported by isolated footings. The footing width and length shall be twice the width shown in this table and the footings shall be spaced not more than 6 feet (1829 mm) on center.
[4]Foundations may support a roof in addition to the stipulated number of floors. Foundations supporting roofs only shall be as required for supporting one floor.

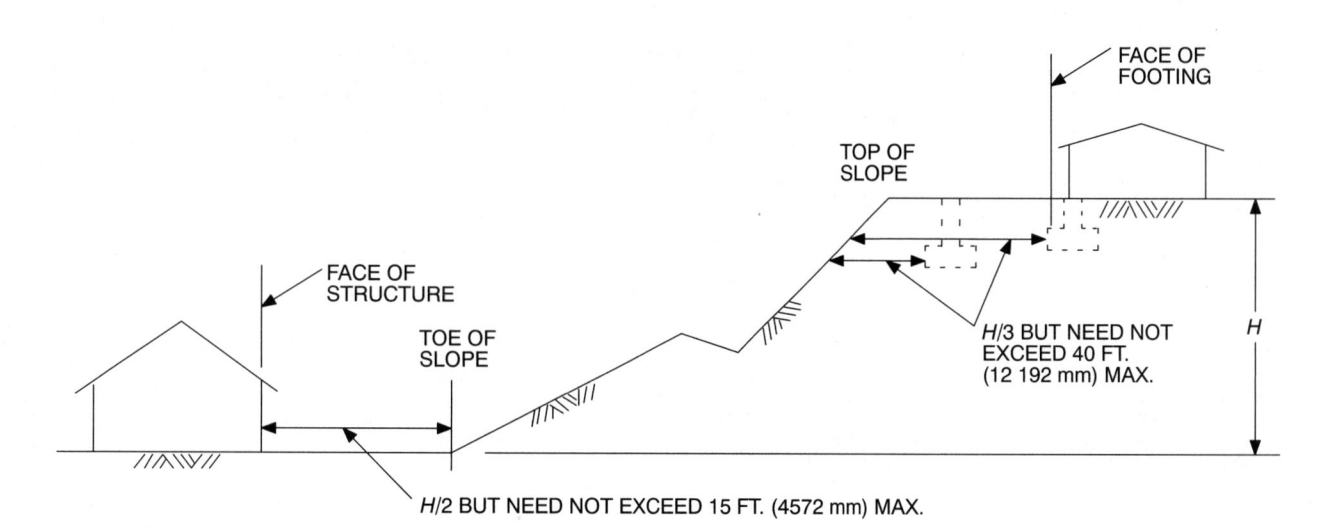

FIGURE 18-I-1—SETBACK DIMENSIONS

Division II—DESIGN STANDARD FOR TREATED WOOD FOUNDATION SYSTEM

Based on National Forest Products Association, Technical Report No. 7

SECTION 1810 — SCOPE

The basic design and construction requirements for treated wood foundation systems are set forth in this division. Included are criteria for materials, preservative treatment, soil characteristics, environmental control, design loads and structural design.

SECTION 1811 — MATERIALS

1811.1 Lumber. Lumber shall be of a species and grade for which allowable unit stresses are set forth in Chapter 23, Division III, Part I, and shall bear the grade mark of, or have a certificate of inspection issued by, an approved lumber grading or inspection bureau or agency.

1811.2 Plywood. All plywood shall be bonded with exterior glue and be grade marked indicating conformance with UBC Standard 23-2 and shall bear the grade mark of an approved plywood inspection agency.

1811.3 Fasteners in Preservative-treated Wood. Fasteners in preservative-treated wood shall be approved silicon bronze or copper, stainless steel or hot-dipped zinc-coated steel. Silicon bronze, copper and stainless steel fasteners are acceptable for all ground contact and moisture situations. Hot-dipped zinc-coated nails may be used for basement and crawl space wall construction where polyethylene sheeting is applied to the below-grade portion of the exterior wall and for wood basement floor construction, provided the polyethylene sheeting is placed in accordance with Section 1812.4. In addition, crawl space construction shall be located in soils having good drainage, such as GW, GP, SW, SP, GM and SM types. Other types of steel or metal fasteners shall be permitted only if adequate comparative tests for corrosion resistance, including the effects associated with the wood treating chemicals, indicate an equal or better performance. Zinc-coated fasteners shall be coated after manufacture to their final form, including pointing, heating, threading or twisting, as applicable. Electrogalvanized nails or staples and hot-dipped zinc-coated staples shall not be used.

Framing anchors shall be of hot-dipped zinc-coated A-446 Grade A sheet steel conforming to UBC Standard 22-1.

1811.4 Gravel, Sand or Crushed Stone for Footings Fill. Gravel shall be washed and well graded. The maximum size stone shall not exceed $^3/_4$ inch (19 mm). Gravel shall be free from organic, clayey or silty soils.

Sand shall be coarse, not smaller than $^1/_{16}$-inch (1.6 mm) grains and shall be free from organic, clayey or silty soils.

Crushed stone shall have a maximum size of $^1/_2$ inch (12.7 mm).

1811.5 Polyethylene Sheeting. Polyethylene sheeting shall conform to requirements approved by the building official.

1811.6 Sealants. The materials used to attach the polyethylene sheets to each other or to the plywood shall be capable of adhering to those materials to form a continuous seal.

The material used for caulking joints in plywood sheathing shall be capable of adhering to the wood to provide a moisture seal under the conditions of temperature and moisture content at which it will be applied and used.

1811.7 Preservative Treatment. All lumber and plywood required to be preservative treated shall be pressure treated and bear the FDN grade mark. After treatment, each piece of lumber and plywood shall be dried to a moisture content not exceeding 19 percent. Each piece of treated lumber and plywood shall bear an approved quality mark or that of an approved inspection agency which maintains continuing supervision, testing and inspection over the quality of the product, and shall be identified.

Where FDN lumber is cut or drilled after treatment, the cut surface shall be field treated with the following preservatives by repeated brushing, dipping or soaking until the wood absorbs no more preservative: ammoniacal copper arsenate (ACA), chromated copper arsenate (CCA), fluor chrome arsenate phenol (FCAP), acid copper chromate (ACC), or copper napthenate.

Copper napthenate shall be prepared with a solvent conforming to AWPA Standard P5. The preservative concentration shall contain a minimum of 2 percent copper metal. Preparations made by manufacturers of preservatives can also be used.

Waterborne preservatives ACA and CCA, Types A, B and C, shall have a minimum concentration of 3 percent in solution. Waterborne preservatives FCAP and ACC may be used for field treatment of material originally treated with CCA and ACA waterborne preservatives and the concentration of FCAP or ACC shall be a minimum of 5 percent in solution.

All lumber and plywood used in exterior foundation walls (except the upper top plate), all interior-bearing wall framing and sheathing posts or other wood supports used in crawl spaces; all sleepers, joists, blocking and plywood subflooring used in basement floors; and all other plates, framing and sheathing in the ground or in direct contact with concrete shall be preservative treated. Where a significant portion of a bottom story wall is above adjacent ground level, such as when a building is situated on sloping terrain, the portion of wall to be considered as foundation wall shall be based on good engineering practice. Some members in such a wall may not require preservative treatment, such as window or door headers or the top plate. As a minimum, all exterior wall framing lumber and plywood sheathing less than 6 inches (152 mm) above finished grade shall be preservative treated.

1811.8 Soil Characteristics. Soils are defined herein in accordance with the Unified Soil Classification System (see UBC Standard 18-1). Design properties are provided in Table 18-I-A or by a qualified soils engineer who, by approval of the building official, may assign other values based on soil tests or local experience.

Backfill of CH type (inorganic clays of high plasticity) or other types of expansive soils shall not be compacted dry. Backfill with MH soil types (inorganic silts, micaceous or diatomaceous fine sandy or silty soils, elastic silts) shall be well compacted to prevent surface water infiltration.

Organic soils, OL, OH and P_t are unsatisfactory for foundations unless specifically approved by the building official after a qualified soils engineer advises on the design of the entire soil-structural system.

SECTION 1812 — DRAINAGE AND MOISTURE CONTROL

1812.1 General. The following sections present requirements to achieve dry and energy-efficient below-grade habitable space that is located above the permanent water table. Floors located below the permanent water table are not permitted unless special moisture control measures are designed by persons qualified in accordance with the authority having jurisdiction. (See Section 1804.7.)

1812.2 Area Drainage. Adjacent ground surface shall be sloped away from the structure with a gradient of at least $^1/_2$ inch (12.7 mm) per foot for a distance of 6 feet (1829 mm) or more. Provisions shall be made for drainage to prevent accumulation of surface water.

1812.3 Subgrade Drainage. A porous layer of gravel, crushed stone or sand shall be placed to a minimum thickness of 4 inches (102 mm) under basement floor slabs and all wall footings. For basement construction in MH and CH type soils, the porous layer under footings and slab shall be at least 6 inches (152 mm) thick.

Where there is basement space below grade, a sump shall be provided to drain the porous layer unless the foundation is installed in GW-, GP-, SW-, SP-, GM- and SM-type soils. The sump shall be at least 24-inch (610 mm) diameter or 20-inch (508 mm) square, shall extend at least 24 inches (610 mm) below the bottom of the basement floor slab and shall be capable of positive gravity or mechanical drainage to remove any accumulated water.

1812.4 Sheeting and Caulking. Polyethylene sheeting of 6-mil (0.153 mm) thickness shall be applied over the porous layer. A concrete slab shall be poured over the sheeting or a wood basement floor system shall be constructed on the sheeting. Where wood floors are used, the polyethylene sheeting shall be placed over wood sleepers supporting the floor joists. Sheeting should not extend beneath the wood footing plate.

In basement construction, joints between plywood panels in the foundation walls shall be sealed full length with caulking compound. Any unbacked panel joints shall be caulked at the time the panels are fastened to the framing.

Six-mil-thick (0.153 mm) polyethylene sheeting shall be applied over the below-grade portion of exterior basement walls prior to backfilling, except in GW-, GP-, SW-, SP-, GM- and SM-type soils. Joints in the polyethylene sheeting shall be lapped 6 inches (152 mm) and bonded with a sealant. The top edge of the polyethylene sheeting shall be bonded with a sealant to the plywood sheeting. A treated lumber or plywood strip shall be attached to the wall to cover the top edge of the polyethylene sheeting. The wood strip shall extend at least 2 inches (51 mm) above and 5 inches (127 mm) below finish grade level to protect the polyethylene from exposure to light and from mechanical damage at or near grade. The joint between the strip and the wall shall be caulked full length prior to fastening the strip to the wall. Alternatively, asbestos-cement board, brick, stucco or other covering appropriate to the architectural treatment may be used in place of the wood strip. The polyethylene sheeting shall extend down to the bottom of the wood footing plate, but shall not overlap or extend into the gravel footing.

1812.5 Perimeter Drainage Control. The space between the side of a basement excavation and the exterior of a basement wall shall be backfilled for half the height of the excavation with the same material used for footings, except that for basements located in GW-, GP-, SW-, SP-, GM- and SM-type soils, or other sites that are well drained and acceptable to the authority having jurisdiction, the granular fill need not exceed a height of 1 foot (305 mm) above the footing. The top of this granular fill outside basement foundation walls and footings shall be covered with strips of 6-mil-thick (0.153 mm) polyethylene sheeting or Type 30 felt, with adjacent strips lapped to provide for water seepage while preventing excessive infiltration of fine soils. Perforated sheeting or other filter membrane may also be used to control infiltration of fines.

1812.6 Alternate Drainage System. If a continuous concrete footing rather than a composite wood and gravel footing is used

with the wood foundation in basement construction, the concrete shall be placed over a 4-inch-thick (102 mm) layer of gravel, crushed stone or sand that is arranged to allow drainage of water from the granular backfill outside the footing to the porous layer under the slab. Alternately, drainage across the concrete footing shall be provided by transverse pipes or drain tiles embedded in the concrete every 6 linear feet (1829 mm) around the foundation.

1812.7 Insulation. Where insulation is applied between studs in exterior basement walls but the insulation is not flush with the exterior wall sheathing and does not extend down to the bottom plate, blocking shall be installed between the studs at the lower end of the insulation to prevent convection currents.

SECTION 1813 — DESIGN LOADS

1813.1 General. All parts of the wood foundation system shall be designed and constructed to provide safe support for all anticipated loads within the stress limits specified by this code. Design loads shall not be less than those specified in Chapter 16.

Design loads shall include downward forces acting on the wall from dead loads and roof and floor live loads, plus the lateral pressure from soil. Where applicable, the foundation also shall be designed to resist wind, earthquake and other static or dynamic forces. The foundation system shall be designed for the most severe distribution, concentration or combination of design loads deemed proper to act on the structure simultaneously.

1813.2 Soil Loads. Lateral pressure of the soil on the wall shall be considered in accordance with Section 1611.6.

SECTION 1814 — STRUCTURAL DESIGN

1814.1 General. Structural design of wood foundations shall be in accordance with established structural engineering and wood design practices as set forth in Chapter 23.

1814.2 Allowable Stresses. Allowable unit stresses for lumber and plywood shall be as provided in Section 2304. Design stresses for framing lumber shall be based on use under dry conditions (19 percent maximum moisture content), except that stresses for footing plates and crawl space framing shall be based on use under wet conditions. Design stresses for plywood sheathing shall be based on use under damp (moisture content 16 percent or more) conditions.

1814.3 Allowable Loads on Fastenings. Allowable loads for steel nails and framing anchors shall be in accordance with Section 2318. Allowable loads for stainless steel Type 304 or 316, silicon bronze or copper nails shall be developed on a comparable basis to loads allowed for common steel nails. Allowable loads for stainless steel Type 304 or 316, silicon bronze or copper staples or other fasteners shall be in accordance with good engineering practice.

1814.4 Footing Design. The treated wood foundation systems incorporate a composite footing consisting of a wood footing plate and a layer of gravel, coarse sand or crushed stone. The wood footing plate distributes the axial design load from the framed wall to the gravel layer which in turn distributes it to the supporting soil.

Soil-bearing pressure under the gravel, sand or crushed stone footings shall not exceed the allowable soil bearing values from Table 18-I-A except as permitted by Section 1805.

Footing plate width shall be determined by allowable bearing pressure between the footing plate and the granular part of the footing. Gravel, sand or crushed stone under the footing plate shall be compacted to provide an allowable bearing capacity of 3,000 psf (144 kPa) when required by the design, otherwise an allowable bearing capacity of 2,000 psf (96 kPa) shall be assumed.

When the footing plate is wider than the bottom wall plate, the tension stress perpendicular to grain induced in the bottom face of the footing plate shall not exceed one third the allowable unit shear stress for the footing plate. Use of plywood strips to reinforce the lumber footing plate is acceptable.

Thickness and width of the granular footing shall be determined by allowable bearing pressure between the gravel, sand or crushed stone and the supporting soil, assuming the downward load from the wood footing plate is distributed outward through the gravel, sand or crushed stone footing at an angle of 30 degrees from vertical at each edge of the footing plate. Additionally, the gravel, sand or crushed stone footing shall have a width not less than twice the width and a thickness not less than three-quarters the width of the wood footing plate and shall be confined laterally by backfill, granular fill, undisturbed soil, the foundation wall or other equivalent means.

The bottom of the wood footing plate shall not be above the maximum depth of frost penetration unless the gravel, sand or crushed stone footing extends to the maximum depth of frost penetration and is either connected to positive mechanical or gravity drainage, at or below the frost line, or is installed in GW-, GP-, SW-, SP-, GM- and SM-type soils where the permanent water table is below the frost line. A granular footing connected to a positively drained sump (see Section 1812.3) by a trench filled with gravel, sand or crushed stone, or by an acceptable pipe connection, shall be considered to be drained to the level of the bottom of the sump or the bottom of the connecting trench or pipe, whichever is higher.

Where the bottom of the wood footing plate of a crawl space wall is not below the frost line, the top of the gravel, sand or crushed stone outside the wall shall be covered as required in Section 1812.5 for basement construction to prevent excessive infiltration of fine soils.

Where a wood footing plate is close to finished grade, such as when a deep granular footing is used to reach the frost line, the granular footing shall be protected against surface erosion or mechanical disturbance.

Posts and piers and their footings in basements or crawl spaces shall be in accordance with Sections 1806 and 2306.

Footings under posts or piers may be of treated wood, treated wood and gravel, precast concrete or other approved material.

1814.5 Foundation Wall Design. Foundation wall studs shall be designed for stresses due to combined bending moment and axial loading resulting from lateral soil pressure and downward live and dead loads on the foundation wall, and for shear stresses due to lateral soil pressure. Top and bottom wall plates shall be designed for bearing of the studs on the plates. Joints in footing plate and upper top plate shall be staggered at least one stud space from joints in the adjacent plate to provide continuity between wall panels. Framing at openings in wall and floor systems and at other points of concentrated loads shall be designed with adequate capacity for the concentrated loads.

Plywood wall sheathing shall be designed for the shear and bending moment between studs due to soil pressures.

Joints, fastenings and connections in the wood foundation system shall be adequate to transfer all vertical and horizontal forces to the footing or to the applicable floor system. Connections at the top of the foundation wall shall be designed to transfer lateral soil load into the floor assembly. Lateral load at the bottom of a basement wall shall be transferred to the basement floor through bearing of the studs against the floor. Lateral load at the bottom of a crawl space wall shall be resisted by the soil inside the footing.

Foundation walls subject to racking loads due to earthquake, wind or differential soil pressure forces shall be designed with adequate shear strength to resist the most severe racking load or combination of loads, but earthquake and wind forces shall not be assumed to act simultaneously. Where a bottom wall plate of 1-inch (25 mm) nominal thickness has been used, the bottom of the wall shall be considered an unsupported panel edge when determining shear resistance of the wall.

1814.6 Interior Load-bearing Walls. Interior load-bearing walls in basements or crawl spaces shall be designed to carry the applicable dead and live loads in accordance with standard engineering practice and the requirements of this code.

1814.7 Basement Floor Design. Concrete slab basement floors shall be designed in accordance with requirements of this code but shall not be less than $3^1/_2$ inches (89 mm) in thickness.

Wood basement floors shall be designed to withstand axial forces and bending moments resulting from lateral soil pressures at the base of the exterior foundation walls and floor and live and dead loads. Floor framing shall be designed to meet joist deflection requirements of this code.

Unless special provision is made to resist sliding caused by unbalanced lateral soil loads, wood basement floors shall be limited to applications where the differential depth of fill on opposing exterior foundation walls is 2 feet (610 mm) or less.

Joists in wood basement floors shall bear tightly against the narrow face of studs in the foundation wall or directly against a band joist that bears on the studs. Plywood subfloor shall be continuous overlapped joists or over butt joints between in-line joists. Where joists are parallel to the wall, sufficient blocking shall be provided between joists to transfer lateral forces from the base of the wall into the floor system.

Where required, resistance to uplift or restraint against buckling shall be provided by interior-bearing walls or appropriately designed stub walls anchored in the supporting soil below.

Sleepers, joists, blocking and plywood subflooring used in basement floors shall meet the treatment requirements of Section 1811.7.

1814.8 Uplift or Overturning. Design of the structure for uplift or overturning shall be in accordance with the requirements of this code.

Division III—DESIGN STANDARD FOR DESIGN OF SLAB-ON-GROUND FOUNDATIONS TO RESIST THE EFFECTS OF EXPANSIVE SOILS AND COMPRESSIBLE SOILS

SECTION 1815 — DESIGN OF SLAB-ON-GROUND FOUNDATIONS [BASED ON DESIGN OF SLAB-ON-GROUND FOUNDATIONS OF THE WIRE REINFORCEMENT INSTITUTE, INC. (AUGUST, 1981)]

1815.1 Scope. This section covers a procedure for the design of slab-on-ground foundations to resist the effects of expansive soils in accordance with Division I. Use of this section shall be limited to buildings three stories or less in height in which gravity loads are transmitted to the foundation primarily by means of bearing walls constructed of masonry, wood or steel studs, and with or without masonry veneer.

1815.2 Symbols and Notations.

$1\text{-}c$ = soil/climatic rating factor. See Figure 18-III-8.

A_s = area of steel reinforcing (square inch per foot) (mm^2 per m) in slab. See Figure 18-III-1.

C_o = overconsolidation coefficient. See Figure 18-III-2.

C_s = soil slope coefficient. See Figure 18-III-3.

C_w = climatic rating. See Figure 18-III-4.

E_c = creep modulus of elasticity of concrete.

f_y = yield strength of reinforcing.

I_c = cracked moment of inertia of cross section.

k_l = length modification factor-long direction. See Figure 18-III-5.

k_s = length modification factor-short direction. See Figure 18-III-5.

L = total length of slab in prime direction.

L' = total length of slab (width) perpendicular to L.

L_c = design cantilever length $(l_c k)$—See Figures 18-III-5 and 18-III-6.

l_c = cantilever length as soil function.

M_l = design moment in long direction.

M_s = design moment in short direction.

PI = plasticity index.

S = maximum spacing of beams. See Figure 18-III-7.

V = design shear force (total).

w = weight per square foot (N/m^2) of building and slab.

q_u = unconfined compressive strength of soil.

Δ = deflection of slab, inch (mm).

1815.3 Foundation Investigation. A foundation investigation of the site shall be conducted in accordance with the provisions of Section 1804.

1815.4 Design Procedure.

1815.4.1 Loads. The foundation shall be designed for a uniformly distributed load which shall be determined by dividing the actual dead and live loads for which the superstructure is designed, plus the dead and live loads contributed by the foundation, by the area of the foundation.

> **EXCEPTIONS:** 1. For one-story metal and wood stud buildings, with or without masonry veneer, and when the design floor live load is 50 pounds per square foot (2.4 kN/m^2) or less, a uniformly distributed load of 200 pounds per square foot (9.6 kN/m^2) may be assumed in lieu of calculating the effects of specific dead and live loads.

2. Those conditions where concentrated loads are of such magnitude that they must be considered are not covered by this section.

1815.4.2 Determining the effective plasticity index. The effective plasticity index to be used in the design shall be determined in accordance with the following procedures:

1. The plasticity index shall be determined for the upper 15 feet (4572 mm) of the soil layers and where the plasticity index varies between layers shall be weighted in accordance with the procedures outlined in Figure 18-III-9.

2. Where the natural ground slopes, the plasticity index shall be increased by the factor C_s determined in accordance with Figure 18-III-3.

3. Where the unconfined compressive strength of the foundation materials exceeds 6,000 pounds per square foot (287.4 kPa), the plasticity index shall be modified by the factor C_o determined in accordance with Figure 18-III-2. Where the unconfined compressive strength of the foundation materials is less than 6,000 pounds per square foot (287.4 kPa), the plasticity index may be modified by the factor C_o determined in accordance with Figure 18-III-2.

The value of the effective plasticity index is that determined from the following equation:

Effective plasticity index =
weighted plasticity index $\times C_s \times C_o$

Other factors that are capable of modifying the plasticity index such as fineness of soil particles and the moisture condition at the time of construction shall be considered.

1815.5 Beam Spacing and Location. Reinforced concrete beams shall be provided around the perimeter of the slab, and interior beams shall be placed at spacings not to exceed that determined from Figure 18-III-7. Slabs of irregular shape shall be divided into rectangles (which may overlap) so that the resulting overall boundary of the rectangles is coincident with that of the slab perimeter. See Figure 18-III-10.

1815.6 Beam Design. The following formulas shall be used to calculate the moment, shear and deflections, and are based on the assumption that the zone of seasonal moisture changes under the perimeter of the slab is such that the beams resist loads as a cantilever of length L_c:

$$M = \frac{w L' (L_c)^2}{2}$$

$$V = wL'L_c$$

$$\Delta = \frac{w L' (L_c)^4}{4 E_c I_c}$$

The calculations shall be performed for both the long and short directions. Deflection shall not exceed $L_c/480$.

1815.7 Slab Reinforcing. The minimum slab thickness shall be 4 inches (102 mm), and the maximum spacing of reinforcing bars shall be 18 inches (457 mm). The amount of reinforcing shall be determined in accordance with Figure 18-III-1. Slab reinforcing shall be placed in both directions at the specified amounts and spacing.

SECTION 1816 — DESIGN OF POSTTENSIONED SLABS ON GROUND (BASED ON DESIGN SPECIFICATION OF THE POSTTENSIONING INSTITUTE)

1816.1 Scope. This section covers a procedure for the design of slab-on-ground foundations to resist the effects of expansive soils in accordance with Division I. Use of this section shall be limited to buildings three stories or less in height in which gravity loads are transmitted to the foundation primarily by means of bearing walls constructed of masonry, wood or steel studs, and with or without masonry veneer.

1816.2 List of Symbols and Notations.

A = area of gross concrete cross section, in.^{2} (mm^2).

A_b = bearing area beneath a tendon anchor, in.^{2} (mm^2).

A'_b = maximum area of the portion of the supporting surface that is geometrically similar to and concentric with the loaded area, in.^{2} (mm^2).

A_{bm} = total area of beam concrete, in.^{2} (mm^2).

AC = activity ratio of clay.

A_o = coefficient in Formula (16-13-1).

A_{ps} = area of prestressing steel in.^{2} (mm^2).

A_{sl} = total area of slab concrete, in.^{2} (mm^2).

B = constant used in Formula (16-13-1).

b = width of an individual stiffening beam, in. (mm).

B_w = assumed slab width (used in Section 1816.4.12), in. (mm).

C = constant used in Formula (16-13-1).

c = distance between CGC and extreme cross-section fibers, in. (mm).

$CEAC$ = cation exchange activity.

CGC = geometric centroid of gross concrete section.

CGS = center of gravity of prestressing force.

C_p = coefficient in Formula (16-35) for slab stress due to partition load—function of k_s.

C_Δ = coefficient used to establish allowable differential deflection (see Table 18-III-GG).

e = eccentricity of posttensioning force (perpendicular distance between the CGS and the CGC), in. (mm).

E_c = long-term or creep modulus of elasticity of concrete, psi (MPa).

EI = expansion index (see Table 18-I-B and UBC Standard 18-2).

e_m = edge moisture variation distance, ft. (m).

e_n = base of natural (Naperian) logarithms.

E_s = modulus of elasticity of the soil, psi (MPa).

f = applied flexural concrete stress (tension or compression), psi (MPa).

f_B = section modulus factor for bottom fiber.

f_{bp} = allowable bearing stress under tendon anchorages, psi (MPa).

f_c = allowable concrete compressive stress, psi (MPa).

f'_c = 28-day concrete compressive strength, psi (MPa).

f'_{ci} = concrete compressive strength at time of stressing tendons, psi (MPa).

f_{cr} = concrete modulus of rupture, flexural tension stress which produces first cracking, psi (MPa).

f_e = effective prestress force, lbs. (N).

f_p = minimum average residual prestress compressive stress, psi (MPa).

f_{pi} = allowable tendon stress immediately after stressing, psi (MPa).

f_{pj} = allowable tendon stress due to tendon jacking force, psi (MPa).

f_{pu} = specified maximum tendon tensile stress, psi (MPa).

f_T = section modulus factor for top fiber.

f_t = allowable concrete flexural tension stress, psi (MPa).

g = moment of inertia factor.

H = thickness of a uniform thickness foundation, in. (mm).

h = total depth of stiffening beam, measured from top surface of slab to bottom of beam (formerly d, changed for consistency with ACI-318), in. (mm).

I = gross concrete moment of inertia, in.^{4} (mm^4).

k = depth-to-neutral axis ratio; also abbreviation for "kips" (kN).

k_s = soil subgrade modulus, pci (N/mm^3).

L = total slab length (or total length of design rectangle) in the direction being considered (short or long), perpendicular to W, ft. (m).

L_L = long length of the design rectangle, ft. (m).

L_S = short length of the design rectangle, ft. (m).

M_L = maximum applied service load moment in the long direction (causing bending stresses on the short cross section) from either center lift or edge lift swelling condition, ft.-kips/ft. (kN·m/m).

M_{max} = maximum moment in slab under load-bearing partition, ft.-kips/ft. (kN·m/m).

M_S = maximum applied service load moment in the short direction (causing bending stresses on the long cross section) from either center lift or edge lift swelling condition, ft.-kips/ft. (kN·m/m).

n = number of stiffening beams in a cross section of width W.

P = a uniform unfactored service line load (P) acting along the entire length of the perimeter stiffening beams representing the weight of the exterior building material and that portion of the superstructure dead and live loads that frame into the exterior wall. P does not include any portion of the foundation concrete, lbs./ft. (N/m).

P_e = effective prestress force after losses due to elastic shortening, creep and shrinkage of concrete, and steel relaxation, kips (kN).

PI = plasticity index.

P_i = prestress force immediately *after* stressing and anchoring tendons, kips (kN).

P_r = resultant prestress force after *all* losses (including those due to subgrade friction), kips (kN).

q_{allow} = allowable soil-bearing pressure, psf (N/m^2).

q_u = unconfined compressive strength of the soil, psf (N/m^2).

S = interior stiffening beam spacing, ft. If beam spacings vary, the average spacing may be used if the ratio between the largest and smallest spacing does not exceed 1.5. If the ratio between the largest and smallest spacing exceeds 1.5, use $S = 0.85x$ (largest spacing).

S_b = section modulus with respect to bottom fiber, in.^{3} (mm^3).

SG = prestress loss due to subgrade friction, kips (kN).

S_T = section modulus with respect to top fiber, in.^{3} (mm^3).

t = slab thickness in a ribbed (stiffened) foundation, in. (mm).

V = controlling service load shear force, larger of V_S or V_L, kips/ft. (kN/m).

v = service load shear stress, psi (MPa).

v_c = allowable concrete shear stress, psi (MPa).

V_L = maximum service load shear force in the long direction from either center lift or edge lift swelling condition, kips/ft. (kN/m).

V_S = maximum service load shear force in the short direction from either center lift or edge lift swelling condition, kips/ft. (kN/m).

W = foundation width (or width of design rectangle) in the direction being considered (short or long), perpendicular to L, ft. (m).

W_{slab} = foundation weight, lbs. (kg).

y_m = maximum differential soil movement or swell, in. (mm).

α = slope of tangent to tendon, radians.

β = relative stiffness length, approximate distance from edge of slab to point of maximum moment, ft. (m).

Δ = expected service load differential deflection of slab, including correction for prestressing, in. (mm) ($\Delta = \Delta_o \pm \Delta_p$).

Δ_{allow} = allowable differential deflection of slab, in. (mm).

Δ_o = expected service load differential deflection of slab (without deflection caused by prestressing), in. (mm).

Δ_p = deflection caused by prestressing, in. (mm).

μ = coefficient of friction between slab and subgrade.

1816.3 Foundation Investigation.

A foundation investigation of the site shall be conducted in accordance with the provisions of Section 1804.

1816.4 Structural Design Procedure for Slabs on Expansive Soils.

1816.4.1 General. This procedure can be used for slabs with stiffening beams (ribbed foundations) or uniform thickness foundations. To design a uniform thickness foundation, the designer must first design a ribbed foundation that satisfies all requirements of the design procedure for ribbed foundations. The fully conformant ribbed foundation is then converted to an equivalent uniform thickness foundation.

The design procedure for posttensioned foundations constructed over expansive clays should include the following steps, with the pertinent sections shown in parentheses:

1. Assemble all the known design data (Section 1816.4.2).

2. Divide an irregular foundation plan into overlapping rectangles and design each rectangular section separately (Figure 18-III-11).

3. Assume a trial section for a ribbed foundation in both the long and short directions of the design rectangle (Section 1816.4.3).

4. Calculate the applied service moment the section will be expected to experience in each direction for either the center lift or edge lift condition (Section 1816.4.7).

5. Determine the flexural concrete stresses caused by the applied service moments and compare to the allowable flexural concrete stresses (Sections 1816.4.4 and 1816.4.7).

6. Determine the expected differential deflections and compare with the allowable differential deflections (Section 1816.4.9).

7. Calculate the applied service shear force and shear stress in the assumed sections and compare the applied shear stress with the allowable shear stress (Section 1816.4.10).

8. Convert the ribbed foundation to an equivalent uniform thickness foundation, if desired (Section 1816.4.11).

9. Repeat Steps 4 through 8 for the opposite swelling condition.

10. Check the design for the first swelling condition to ascertain if adjustments are necessary to compensate for any design changes resulting from the second design swelling condition addressed in Step 9.

11. Check the effect of slab-subgrade friction to ensure a residual compressive stress of 50 psi (0.35 MPa) at the center of each design rectangle in both directions. Adjust posttensioning force, if necessary (Section 1816.4.6).

12. Calculate stresses due to any heavy concentrated loads on the slab and provide special load transfer details when necessary (Section 1816.4.12).

1816.4.2 Required design data. The soils and structural properties needed for design are as follows:

1. **Soils properties.**

    1.1   Allowable soil-bearing pressure, q_{allow}, in pounds per square foot (newtons per square meter).

    1.2   Edge moisture variation distance, e_m, in feet (meters).

    1.3   Differential soil movement, y_m, in inches (millimeters).

    1.4   Slab-subgrade friction coefficient, μ.

2. **Structural data and materials properties.**

    2.1   Slab length, L, in feet (meters) (both directions).

    2.2   Perimeter loading, P, in pounds per foot (newtons per meter).

    2.3   Average stiffening beam spacing, S, in feet (meters) (both directions).

    2.4   Beam depth, h, in inches (millimeters).

    2.5   Compressive strength of the concrete, f'_c, in pounds per square inch (MPa).

    2.6   Allowable flexural tensile stress in the concrete, f_t, in pounds per square inch (MPa).

    2.7   Allowable compressive stress in the concrete, f_c, in pounds per square inch (MPa).

    2.8   Type, grade and strength of the prestressing steel.

    2.9   Prestress losses in kips per inch (kN per mm).

1816.4.3 Trial section assumptions.

1816.4.3.1 Assume beam depth and spacing. An initial estimate of the depth of the stiffening beam can be obtained from solving either Formula (16-21) or Formula (16-22) for the beam depth yielding the maximum allowable differential deflection. The procedure is as follows:

1. Determine the maximum distance over which the allowable differential deflection will occur, L or 6β, whichever is smaller. As a first approximation, use $\beta = 8$ feet (2.44 m).

2. Select the allowable differential deflection Δ_{allow}:

    2.1   Center lift (assume $C_\Delta = 360$):

$$\Delta_{allow} = \frac{12(L \text{ or } 6\beta)}{C_\Delta} = \frac{12(L \text{ or } 6\beta)}{360} \quad (16\text{-}1)$$

For **SI:** 1 inch = 25.4 mm.

2.2   Edge lift (assume $C_\Delta = 720$):

$$\Delta_{allow} = \frac{12(L \text{ or } 6\beta)}{C_\Delta} = \frac{12(L \text{ or } 6\beta)}{720} \quad (16\text{-}2)$$

For SI: 1 inch = 25.4 mm.

Alternatively, C_Δ may be selected from Table 18-III-GG, which presents sample C_Δ values for various types of superstructures.

3. Assume a beam spacing, S, and solve for beam depth, h:

3.1   Center lift (from Formula 16-20):

$$h^{1.214} = \frac{(y_m L)^{0.205}(S)^{1.059}(P)^{0.523}(e_m)^{1.296}}{380\Delta_{allow}} \quad (16\text{-}3\text{-}1)$$

$$h = \left(\frac{(y_m L)^{0.205}(S)^{1.059}(P)^{0.523}(e_m)^{1.296}}{380\Delta_{allow}}\right)^{0.824} \quad (16\text{-}3\text{-}2)$$

For SI: 1 inch = 25.4 mm.

3.2   Edge lift (from Formula 16-21):

$$h^{0.85} = \frac{(L)^{0.35}(S)^{0.88}(e_m)^{0.74}(y_m)^{0.76}}{15.9\Delta_{allow}(P)^{0.01}} \quad (16\text{-}4\text{-}1)$$

$$h = \left(\frac{(L)^{0.35}(S)^{0.88}(e_m)^{0.74}(y_m)^{0.76}}{15.9\Delta_{allow}(P)^{0.01}}\right)^{1.176} \quad (16\text{-}4\text{-}2)$$

For SI: 1 inch = 25.4 mm.

Select the larger h from Formula (16-3-2) or (16-4-2). In the analysis procedure, the beam depth h must be the same for all beams in both directions. If different beam depths are selected for the *actual* structure (such as a deeper edge beam), the analysis shall be based on the smallest beam depth actually used.

1816.4.3.2 Determine section properties. The moment of inertia, section modulus, and cross-sectional area of the slabs and beams, and eccentricity of the prestressing force shall be calculated for the trial beam depth determined above in accordance with normal structural engineering procedures.

1816.4.4 Allowable stresses.

The following allowable stresses are recommended:

1. **Allowable concrete flexural tensile stress:**

$$f_t = 6\sqrt{f'_c} \quad (16\text{-}5)$$

For SI: $\qquad f_t = 0.5\sqrt{f'_c}$

2. **Allowable concrete flexural compressive stress:**

$$f_c = 0.45f'_c \quad (16\text{-}6)$$

3. **Allowable concrete bearing stress at anchorages.**

3.1   At service load:

$$f_{bp} = 0.6f'_c\sqrt{\frac{A'_b}{A_b}} \leq f'_c \quad (16\text{-}7)$$

3.2   At transfer:

$$f_{bp} = 0.8f'_{ci}\sqrt{\frac{A'_b}{A_b} - 0.2} \leq 1.25f'_{ci} \quad (16\text{-}8)$$

4. **Allowable concrete shear stress:**

$$v_c = 1.7\sqrt{f'_c} + 0.2f_p \quad (16\text{-}9)$$

For SI: $\qquad v_c = 0.14\sqrt{f'_c} + 0.2f_p$

5. **Allowable stresses in prestressing steel.**

5.1   Allowable stress due to tendon jacking force:

$$f_{pj} = 0.8f_{pu} \leq 0.94f_{py} \quad (16\text{-}10)$$

5.2   Allowable stress immediately after prestress transfer:

$$f_{pi} = 0.7f_{pu} \quad (16\text{-}11)$$

1816.4.5 Prestress losses. Loss of prestress due to friction, elastic shortening, creep and shrinkage of the concrete, and steel relaxation shall be calculated in accordance with Section 1918.6.

1816.4.6 Slab-subgrade friction. The effective prestressing force in posttensioned slabs-on-ground is further reduced by the frictional resistance to movement of the slab on the subgrade during stressing as well as the frictional resistance to dimensional changes due to concrete shrinkage, creep and temperature variations. The resultant prestress force, P_r, is the difference between the effective prestress force and the losses due to subgrade friction:

$$P_r = P_e - SG \quad (16\text{-}12\text{-}1)$$

For SI: 1 pound = 4.45 kN.

where SG can be conservatively taken as:

$$SG = \frac{W_{slab}}{2000}\mu \quad (16\text{-}12\text{-}2)$$

For SI: 1 pound = 4.45 kN.

The largest amount of prestress loss due to slab-subgrade friction occurs in the center regions of the slab. The greatest structural requirement for prestress force, however, is at the location of the maximum moment, which occurs at approximately one β-length inward from the edge of the slab. For normal construction practices, the value of the coefficient of friction μ should be taken as 0.75 for slabs on polyethylene and 1.00 for slabs cast directly on a sand base.

The maximum spacing of tendons shall not exceed that which would produce a minimum average effective prestress compression of 50 psi (0.35 MPa) after allowance for slab-subgrade friction.

1816.4.7 Maximum applied service moments. The maximum moment will vary, depending on the swelling mode and the slab direction being designed. For design rectangles with a ratio of long side to short side less than 1.1, the formulas for M_L [Formulas (16-13-1) and (16-15)] shall be used for moments in both directions.

1. **Center lift moment.**

1.1   Long direction:

$$M_L = A_o[B(e_m)^{1.238} + C] \quad (16\text{-}13\text{-}1)$$

For SI: 1 ft. kips/ft. = 4.45 kN·m/m.

WHERE:

$$A_o = \frac{1}{727}\left[(L)^{0.013}(S)^{0.306}(h)^{0.688}(P)^{0.534}(y_m)^{0.193}\right] \quad (16\text{-}13\text{-}2)$$

and for:

$$0 \leq e_m \leq 5 \quad B = 1, \quad C = 0 \qquad (16\text{-}13\text{-}3)$$

$$e_m > 5 \quad B = \left(\frac{y_m - 1}{3}\right) \leq 1.0 \qquad (16\text{-}13\text{-}4)$$

$$C = \left[8 - \frac{P - 613}{255}\right]\left[\frac{4 - y_m}{3}\right] \geq 0 \qquad (16\text{-}13\text{-}5)$$

 1.2 Short direction.

For $L_L/L_S \geq 1.1$:

$$M_S = \left[\frac{58 + e_m}{60}\right] M_L \qquad (16\text{-}14)$$

For SI: 1 ft.·kips/ft. = 4.45 kN·m/m.

For $L_L/L_S < 1.1$:

$$M_S = M_L$$

2. **Edge lift moment.**

 2.1 Long direction:

$$M_L = \frac{(S)^{0.10}(he_m)^{0.78}(y_m)^{0.66}}{7.2(L)^{0.0065}(P)^{0.04}} \qquad (16\text{-}15)$$

For SI: 1 ft.·kips/ft. = 4.45 kN·m/m.

 2.2 Short direction.

For $L_L/L_S \geq 1.1$:

$$M_S = h^{0.35}\left[\frac{19 + e_m}{57.75}\right] M_L \qquad (16\text{-}16)$$

For SI: 1 ft.·kips/ft. = 4.45 kN·m/m.

For $L_L/L_S < 1.1$:

$$M_S = M_L$$

Concrete flexural stresses produced by the applied service moments shall be calculated with the following formula:

$$f = \frac{P_r e}{A} \pm \frac{M_{L,S}}{S_{t,b}} \pm \frac{P_r e}{S_{t,b}} \qquad (16\text{-}17)$$

For SI: 1 pound per square inch = 0.0069 MPa.

The applied concrete flexural stresses f shall not exceed f_t in tension and f_c in compression.

1816.4.8 Cracked section considerations. This design method limits concrete flexural tensile stresses to $6\sqrt{f'_c}$ (For **SI:** $0.5\sqrt{f'_c}$). Since the modulus of rupture of concrete is commonly taken as $f_{cr} = 7.5\sqrt{f'_c}$ (For **SI:** $f_{cr} = 0.625\sqrt{f'_c}$), slabs designed with this method will theoretically have no *flexural* cracking. Some cracking from restraint to slab shortening is inevitable in posttensioned slabs on ground, as it is in elevated posttensioned concrete members. Nevertheless, the limitation of flexural tensile stresses to a value less than the modulus of rupture justifies the use of the gross concrete cross section for calculating all section properties. This is consistent with standard practices in elevated post tensioned concrete members.

1816.4.9 Differential deflections. Allowable and expected differential deflections may be calculated from the formulas presented in the following sections.

1816.4.9.1 Relative stiffness length. β may be calculated as follows:

$$\beta = \frac{1}{12}\sqrt[4]{\frac{E_c I}{E_s}} \qquad (16\text{-}18)$$

For SI:

$$\beta = \frac{1}{1000}\sqrt[4]{\frac{E_c I}{E_s}}$$

If the creep modulus of elasticity of the concrete E_c is not known, it can be closely approximated by using half of the normal or early life concrete modulus of elasticity. If the modulus of elasticity of the clay soil E_s is not known, use 1,000 psi (6.89 MPa). I in Formula (16-18) is the gross moment of inertia for the entire slab cross section of width W, in the appropriate direction (short or long).

1816.4.9.2 Differential deflection distance. The differential deflection may not occur over the entire length of the slab, particularly if the slab is longer than approximately 50 feet (15.24 m). Thus, the effective distance for determining the allowable differential deflection is the smaller of the two distances, L or 6β, both expressed in feet (meters).

1816.4.9.3 Allowable differential deflection, Δ_{allow} (in inches) (mm).

 1. Center lift or edge lift:

$$\Delta_{allow} = \frac{12(L \text{ or } 6\beta)}{C_\Delta} \qquad (16\text{-}19)$$

For SI:

$$\Delta_{allow} = \frac{1000(L \text{ or } 6\beta)}{C_\Delta}$$

The coefficient C_Δ is a function of the type of superstructure material and the swelling condition (center or edge lift). Sample values of C_Δ for both swelling conditions and various superstructure materials are shown in Table 18-III-GG.

1816.4.9.4 Expected differential deflection without prestressing, Δ_o (in inches) (mm):

 1. Center lift:

$$\Delta_o = \frac{(y_m L)^{0.205}(S)^{1.059}(P)^{0.523}(e_m)^{1.296}}{380(h)^{1.214}} \qquad (16\text{-}20)$$

For SI: 1 inch = 25.4 mm.

 2. Edge lift:

$$\Delta_o = \frac{(L)^{0.35}(S)^{0.88}(e_m)^{0.74}(y_m)^{0.76}}{15.9(h)^{0.85}(P)^{0.01}} \qquad (16\text{-}21)$$

For SI: 1 inch = 25.4 mm.

1816.4.9.5 Deflection caused by prestressing, Δ_p (in inches) (mm). Additional slab deflection is produced by prestressing if the prestressing force at the slab edge is applied at any point other than the *CGC*. The deflection caused by prestressing can be approximated with reasonable accuracy by assuming it is produced by a concentrated moment of $P_e e$ applied at the end of a cantilever with a span length of β. The deflection is:

$$\Delta_p = \frac{P_c e \beta^2}{2E_c I} \qquad (16\text{-}22)$$

For SI: 1 inch = 25.4 mm.

If the tendon *CGS* is higher than the concrete *CGC* (a typical condition), Δ_p increases the edge lift deflection and decreases the center lift deflection. Deflection caused by prestressing is normally small and can justifiably be ignored in the design of most posttensioned slabs on ground.

1816.4.9.6 Compare expected to allowable differential deflection. If the expected differential deflection as calculated by either Formula (16-20) or (16-21), adjusted for the effect of prestressing, exceeds that determined from Formula (16-19) for the appropriate swelling condition, the assumed section must be stiffened.

1816.4.10 Shear.

1816.4.10.1 Applied service load shear. Expected values of service shear forces in kips per foot (kN per meter) of width of slab and stresses in kips per square inch (kN per square millimeter) shall be calculated from the following formulas:

1. Center lift.

   1.2 Long direction shear:

$$V_L = \frac{1}{1940}\left[(L)^{0.09}(S)^{0.71}(h)^{0.43}(P)^{0.44}(y_m)^{0.16}(e_m)^{0.93}\right] \quad (16\text{-}23)$$

For **SI:** 1 kips/ft. = 14.59 kN/m.

   1.1 Short direction shear:

$$V_s = \frac{1}{1350}\left[(L)^{0.19}(S)^{0.45}(h)^{0.20}(P)^{0.54}(y_m)^{0.04}(e_m)^{0.97}\right] \quad (16\text{-}24)$$

For **SI:** 1 kips/ft. = 14.59 kN/m.

2. Edge lift (for both directions):

$$V_S \text{ or } V_L = \frac{(L)^{0.07}(h)^{0.4}(P)^{0.03}(e_m)^{0.16}(y_m)^{0.67}}{3.0(S)^{0.015}} \quad (16\text{-}25)$$

For **SI:** 1 kips/ft. = 14.59 kN/m.

1816.4.10.2 Applied service load shear stress, v. Only the beams are considered in calculating the cross-sectional area resisting shear force in a ribbed slab:

1. Ribbed foundations:

$$v = \frac{VW}{nhb} \quad (16\text{-}26)$$

For **SI:** 1 pound per square inch = 0.0069 MPa.

2. Uniform thickness foundations:

$$v = \frac{V}{12H} \quad (16\text{-}27)$$

For **SI:** 1 pound per square inch = 0.0069 MPa.

1816.4.10.3 Compare v to v_c. If v exceeds v_c, shear reinforcement in accordance with ACI 318-95 shall be provided. Possible alternatives to shear reinforcement include:

1. Increasing the beam depth,

2. Increasing the beam width,

3. Increasing the number of beams (decrease the beam spacing).

1816.4.11 Uniform thickness conversion. Once the ribbed foundation has been designed to satisfy moment, shear and differential deflection requirements, it may be converted to an equivalent uniform thickness foundation with thickness H, if desired. To convert a ribbed slab of width, W (ft.) (m) and moment of inertia, I (in.^{4}) (mm^4) to a uniform thickness foundation of width, W (ft.) (m) and depth, H (ft.) (m), use the following formula:

$$I = \frac{(12W)H^3}{12} \quad (16\text{-}28)$$

Solve for H:

$$H = \sqrt[3]{\frac{I}{W}} \quad (16\text{-}29)$$

For **SI:**

$$H = \sqrt[3]{\frac{12I}{1000W}}$$

1816.4.12 Calculation of stress in slabs due to load-bearing partitions. The formula for the allowable tensile stress in a slab beneath a bearing partition may be derived from beam-on-elastic foundation theory. The maximum moment directly under a point load P in such a beam is:

$$M_{\max} = -\frac{P\beta}{4} \quad (16\text{-}30)$$

For **SI:** 1 ft.·kips/ft. = 4.45 kN·m/m.

WHERE:

$$\beta = \left[\frac{4E_cI}{k_sB_w}\right]^{0.25} \leq S_{t,b} \quad (16\text{-}31)$$

For **SI:** 1 ft.·kips/ft. = 4.45 kN·m/m.

with $E_c = 1,500,000$ psi (10 341 MPa) and $k_s = 4$ pci (0.001 N/mm^3):

$$\frac{I}{B_w} = \frac{B_wt^3}{12B_w} = \frac{t^3}{12}$$

$$\beta = \left[\frac{4(1,500,000)t^3}{4(12)}\right]^{0.25} = 18.8t^{0.75} \quad (16\text{-}32)$$

therefore:

$$M_{\max} = -\frac{18.8Pt^{0.75}}{4} = -4.7Pt^{0.75} \quad (16\text{-}33)$$

For **SI:** 1 ft.·kips/ft. = 4.45 kN·m/m.

The formula for applied tensile stress f_t is:

$$f_t = \frac{P_r}{A} - \frac{M_{\max}c}{I} \quad (16\text{-}34)$$

For **SI:** 1 pound per square inch = 0.0069 MPa.

and since:

$$\frac{I}{c} = \frac{B_wt^3}{12}\left(\frac{2}{t}\right) = \frac{B_wt^2}{6} = \frac{12t^2}{6} = 2t^2$$

the applied tensile stress is:

$$f = \frac{P_r}{A} - \frac{4.7Pt^{0.75}}{2t^2} = \frac{P_r}{A} - C_p\frac{P}{t^{1.25}} \quad (16\text{-}35)$$

For **SI:** 1 pound per square inch = 0.0069 MPa.

For uniform thickness foundations substitute H for t in Formulas (16-32), (16-33) and (16-35). The value of C_p depends on the assumed value of the subgrade modulus k_s. The following table illustrates the variation in C_p for different values of k_s:

TYPE OF SUBGRADE	k_s, lb./in.^{3} (0.00027 for N/mm^3)	C_p
Lightly compacted, high plastic compressible soil	4	2.35
Compacted, low plastic soil	40	1.34
Stiff, compacted, select granular or stabilized fill	400	0.74

If the allowable tensile stress is exceeded by the results of the above analysis, a thicker slab section should be used under the loaded area, or a stiffening beam should be placed directly beneath the concentrated line load.

SECTION 1817 — APPENDIX A (A PROCEDURE FOR ESTIMATION OF THE AMOUNT OF CLIMATE CONTROLLED DIFFERENTIAL MOVEMENT OF EXPANSIVE SOILS)

In general, the amount of differential movement to be expected in a given expansive soil should be based on recommendations supplied by a registered geotechnical engineer. The geotechnical engineer may use various soil testing procedures to provide a basis for these recommendations. A procedure developed in part through the PTI-sponsored research project at Texas A & M University that may be used by geotechnical engineers (in conjunction with accumulated experience with local soils conditions) as an aid for estimation of expected differential movements of expansive soils is presented in this appendix. This procedure is applicable only in those cases where site conditions have been corrected so that soil moisture conditions are controlled by the climate alone.

The information necessary to determine the differential movement using the procedure in this appendix is the type and amount of clay, the depth to constant or equilibrium suction, the edge moisture variation distance, the magnitude of the equilibrium suction, and the field moisture velocity. With this information either known or estimated, differential movements may be selected from Tables 18-III-A to 18-III-O for the center lift condition, or Tables 18-III-P to 18-III-DD for the edge lift condition.

Procedures for determining or estimating the necessary items of soil information are as follows:

1. Select a Thornthwaite Moisture Index from Figure 18-III-14 or 18-III-15. Alternatively, extreme annual values of the Thornthwaite Index may be calculated for a given site using Thornthwaite's procedures.

2. Obtain an estimate of the edge moisture variation distance, e_m, for both edge lift and center lift loading conditions from Figure 18-III-14.

3. Determine the percent of clay in the soil and the predominant clay mineral. The predominant type of clay can be determined by performing the following tests and calculations and by using Figure 18-III-15.

    3.1 Determine the plastic limit (PL) and the plasticity index (PI) of the soil.

    3.2 Determine the percentage of clay sizes in the material passing the U.S. No. 200 (75 µm) sieve (Hydrometer Test).

    3.3 Calculate the activity ratio of the soil:

$$A_c = \frac{PI}{\text{(Percent passing U.S. No. 200}\ [75\ \mu\text{m]\ sieve} \leq 0.002\text{mm)}} \quad (1)$$

    3.4 Calculate the Cation Exchange Activity. A discussion of procedures for determining Cation Exchange Capacity for use in calculating Cation Exchange Activity is presented in Appendix B.

$$CEAC = \frac{PI^{1.17}}{\text{(Percent passing U.S. No. 200}\ [75\ \mu\text{m]\ sieve} \leq 0.002\text{mm)}} \quad (2)$$

    3.5 Enter Figure 18-III-15 with the A_c and $CEAC$. The soil type is determined by the intersection of the two entries. Note that the same mineral type is obtained from Figure 18-III-15 for a significant range of values of A_c and $CEAC$. This indicates that the determination of the

mineral type is relatively insensitive to the precision by which the Atterberg Limits and other soil parameters have been determined. In the case of doubt as to the predominant mineral type, the clay may be conservatively classified as montmorillonite.

4. Depth to constant soil suction can be estimated as the depth below which the ratio of water content to plastic limit is constant. At times it will be the depth to an inert material, an unweathered shale, or to a high water table. Constant soil suction can be estimated with reasonable accuracy from Figure 18-III-16 if it is not actually determined in the laboratory; however, for most practical applications, the design soil suction value will seldom exceed a magnitude of pF 3.6.

5. Moisture velocity can be approximated by using a velocity equal to one half of the Thornthwaite Moisture Index [expressed in inches/year (mm/year)] for the construction site, converted to inches/month. To allow for extreme local variations in moisture velocity, this value shall not be assumed to be less than 0.5 in./month (12.7 mm/month), and the maximum moisture velocity shall be 0.7 in./month (17.8 mm/month).

6. Using values of edge moisture distance variation, e_m, percent clay, predominant clay mineral (kaolinite, illite, or montmorillonite), depth to constant suction, soil suction, pF, and velocity of moisture flow determined in steps 1 through 5 above, enter the appropriate tables, Tables 18-III-A to 18-III-O for center lift and Tables 18-III-P to 18-III-DD for edge lift, and find the corresponding soil differential movements, y_m. The values of swell presented in the tables were obtained from a computer program based on the permeability of clays and the total potential of the soil water.

SECTION 1818 — APPENDIX B (SIMPLIFIED PROCEDURES FOR DETERMINING CATION EXCHANGE CAPACITY AND CATION EXCHANGE ACTIVITY)

Simplified Procedure for Determining Cation Exchange Capacity Using a Spectrophotometer

The Cation Exchange Capacity of soil samples may be determined by comparative means in the standard spectrophotometer device. This method of determining the Cation Exchange Capacity is used by the U.S. Soil Conservation Service. Data obtained by this method should be comparable with data for similar soils that have been measured by the U.S. Conservation Service. This simplified procedure is:

1. Place 10 grams of clay soil in a beaker and 100 ml of neutral 1 N ammonium acetate (NH$_4$AC) is added. This solution is allowed to stand overnight.

2. Filter the solution of Step 1 by washing through filter paper with 50 ml of NH$_4$AC.

3. Wash the material retained on the filter paper of Step 2 with two 150 ml washings of isopropyl alcohol, using suction. The isopropyl alcohol wash fluid should be added in increments of approximately 25 ml and the sample allowed to drain well between additions.

4. Transfer the soil and filter paper to a 800-ml flask. Add 50 ml M$_g$CI$_2$ solution and allow to set at least 30 minutes, but preferably 24 hours.

5. Under suction, filter the fluid resulting from Step 4.

6. Normally, the solution of M$_g$CI$_2$ must be diluted before it is placed in the spectrophotometer in Step 10. The dilution will vary from one piece of equipment to the next. The calculations given at the end of this section assume that 200 ml of distilled water have

been used to dilute 1 ml of the M_gCl_2 solution. The 200-to-1 dilution is fairly typical.

7. Prepare a standard curve by using 10 μg of nitrogen (in the NH_4 form) per ml of a standard solution in a 50 ml volumetric flask. Adjust the volume to approximately 25 ml, add 1 ml of 10 percent tartrate solution, and shake. Add 2 ml of Nessler's aliquot with rapid mixing. Add sufficient distilled water to bring the total volume to 50 ml. Allow color to develop for 30 minutes.

8. Repeat Step 7 for 1.0, 2.0, 4.0, and 8.0 ml aliquots of standard solution.

9. Insert the standard solution resulting from Steps 7 and 8 into the spectrophotometer. Record readings and plot the results to construct a standard curve. (The spectrophotometer is calibrated beforehand with distilled water.)

10. Extract 2.0 ml of sample aliquot from Step 6 and add 25 ml of distilled water in a 50 ml volumetric flask. Add 1 ml of 10 percent tartrate and shake. Add 2 ml of Nessler's aliquot with rapid mixing. Add sufficient distilled water to bring the total volume to 50 ml. Let the solution stand for 30 minutes and then insert into the spectrophotometer and record the transparency reading.

11. Typical calculations:

Weight of dry soil	= 10.64 grams
Spectrophotometer	= 81 percent
	= 24.5 μg/g from standard curve

Conversion:

$$\frac{24.5 \mu g}{2 \text{ ml/aliquot}} \times \frac{200 \text{ ml}}{1 \text{ ml}} \times \frac{50 \text{ ml}}{10.64 \mu g} \times \frac{1}{1,000 \mu/\text{mg}} \times$$

$$\frac{1}{14 \text{ mg/meq}} \times 100 \text{ g} = 82.2 \text{ meq/100g}$$

Equation for Cation Exchange Capacity

A 1979 study at Texas Tech University resulted in the following proposed modifications to the Pearring and Holt equations for Clay Activity, Cation Exchange Capacity, and Cation Exchange Activity:

Clay Activity $A_c = \dfrac{PI}{\% \text{ Clay}}$

Cation Exchange Capacity:

$CEC = (PL)^{1.17}$

Cation Exchange Activity:

$CEAC = \dfrac{(PL)^{1.17}}{\% \text{ Clay}}$

Symbols and Notations

PI = plasticity index.

PL = plastic limit.

% Clay = % Passing U.S. No. 200 sieve (75 μm) ≤ 0.002 mm.

Comparison of Methods of Determining Cation Exchange Capacity in Predominant Clay Mineral

A comparison of values of Cation Exchange Capacity using atomic absorption and spectrophotometer techniques is presented in Table 18-III-EE.

Comparison of clay mineral determination between atomic absorption of the correlation equations presented above is presented in Table 18-III-FF and Figure 18-III-17.

SECTION 1819 — DESIGN OF POSTTENSIONED SLABS ON COMPRESSIBLE SOILS (BASED ON DESIGN SPECIFICATIONS OF THE POSTTENSIONING INSTITUTE)

1819.1 General. The design procedure for foundations on compressible soils is similar to the structural design procedure in Section 1816.4, except that different equations are used and the primary bending deformation is usually similar to the edge lift loading case.

1819.2 List of Symbols and Notations.

M_{cs} = applied service moment in slab on compressible soil, ft.-kips/ft. (kN·m/m).

M_{ns} = moment occurring in the "no swell" condition, ft.-kips/ft. (kN·m/m).

V_{cs} = maximum service load shear force in slab on compressible soil, kips/ft. (kN/m).

V_{ns} = service load shear force in the "no swell" condition, kips/ft. (kN/m).

Δ_{cs} = differential deflection in a slab on compressible soil, in. (mm).

Δ_{ns} = differential deflection in the "no swell" condition, in. (mm).

δ = expected settlement, reported by the geotechnical engineer occurring in compressible soil due to the total load expressed as a uniform load, in. (mm).

1819.3 Slabs-on-ground Constructed on Compressible Soils. Design of slabs constructed on compressible soils can be done in a manner similar to that of the edge lift condition for slabs on expansive soils. Compressible soils are normally assumed to have allowable values of soil-bearing capacity, q_{allow}, equal to or less than 1,500 pounds per square foot (71.9 kN/m²). Special design equations are necessary for this problem due to the expected in situ elastic property differences between compressible soils and the stiffer expansive soils. These formulas are:

1. **Moment.**

    1.1 Long direction:

    $$M_{cs_L} = \left(\frac{\delta}{\Delta_{ns_L}}\right)^{0.5} M_{ns_L} \qquad (19\text{-}1)$$

    1.2 Short direction:

    $$M_{cs_S} = \left(\frac{970 - h}{880}\right) M_{cs_L} \qquad (19\text{-}2)$$

For **SI:** $\qquad M_{cs_S} = \left(\dfrac{24\,638 - h}{22\,352}\right) M_{cs_L}$

WHERE:

$$M_{ns_L} = \frac{(h)^{1.35}(S)^{0.36}}{80(L)^{0.12}(P)^{0.10}} \qquad (19\text{-}3)$$

For **SI:** $\qquad 1 \dfrac{\text{ft.·kip}}{\text{ft.}} = 4\,448\,031 \dfrac{\text{kN·m}}{\text{m}}$

$$\Delta_{ns_L} = \frac{(L)^{1.28}(S)^{0.80}}{133(h)^{0.28}(P)^{0.62}} \qquad (19\text{-}4)$$

For **SI:** 1 inch = 25.4 mm.

2. **Differential deflection:**

$$\Delta_{cs} = \delta e_n^{[1.78 - 0.103(h) - 1.65 \times 10^{-3}(P) + 3.95 \times 10^{-7}(P)^2]} \qquad (19\text{-}5)$$

3. **Shear.**

   3.1   Long direction:

$$V_{cs_L} = \left[\frac{\delta}{\Delta_{ns_L}}\right]^{0.30} V_{ns_L} \qquad (19\text{-}6)$$

WHERE:

$$V_{ns_L} = \frac{(h)^{0.90}(PS)^{0.30}}{550(L)^{0.10}} \qquad (19\text{-}7)$$

For **SI:** $\quad 1\,\dfrac{\text{kip}}{\text{ft.}} = 2100\,\dfrac{\text{kN}}{\text{m}}$

   3.2   Short direction:

$$V_{cs_S} = \left[\frac{116 - h}{94}\right] V_{cs_L} \qquad (19\text{-}8)$$

For **SI:** $\quad V_{cs_S} = \left[\dfrac{2946.4 - h}{2387.6}\right] V_{cs_L}$

TABLE 18-III-A—DIFFERENTIAL SWELL OCCURRING AT THE PERIMETER OF A SLAB FOR A CENTER LIFT SWELLING CONDITION IN PREDOMINANTLY KAOLINITE CLAY SOIL (30 PERCENT CLAY)

PERCENT CLAY (%)	DEPTH TO CONSTANT SUCTION (ft.) × 304.8 for mm	CONSTANT SUCTION (pF)	VELOCITY OF MOISTURE FLOW (inches/month) × 25.4 for mm/month	DIFFERENTIAL SWELL (inch) × 25.4 for mm — Edge Distance Penetration × 304.8 for mm							
				1 ft.	2 ft.	3 ft.	4 ft.	5 ft.	6 ft.	7 ft.	8 ft.
30	3	3.2	0.5	0.003	0.005	0.008	0.011	0.015	0.018	0.021	0.025
			0.7	0.004	0.008	0.012	0.017	0.021	0.026	0.032	0.038
		3.4	0.5	0.006	0.012	0.019	0.026	0.034	0.044	0.054	0.067
			0.7	0.008	0.017	0.027	0.038	0.052	0.069	0.091	0.124
		3.6	0.5	0.014	0.030	0.050	0.077	0.117	0.192	0.370	0.881
			0.7	0.018	0.042	0.074	0.125	0.226	0.487	1.252	3.530
	5	3.2	0.5	0.007	0.013	0.020	0.028	0.035	0.043	0.051	0.060
			0.7	0.009	0.019	0.029	0.040	0.051	0.062	0.075	0.089
		3.4	0.5	0.014	0.028	0.043	0.060	0.079	0.100	0.125	0.153
			0.7	0.018	0.039	0.062	0.088	0.119	0.157	0.207	0.279
		3.6	0.5	0.030	0.067	0.112	0.171	0.258	0.413	0.776	1.797
			0.7	0.042	0.096	0.166	0.276	0.486	1.009	2.499	6.879
	7	3.2	0.5	0.012	0.025	0.038	0.051	0.065	0.080	0.095	0.111
			0.7	0.017	0.035	0.063	0.073	0.093	0.115	0.139	0.164
		3.4	0.5	0.024	0.050	0.079	0.110	0.144	0.184	0.228	0.281
			0.7	0.034	0.071	0.113	0.616	0.218	0.287	0.379	0.514
		3.58	0.5	0.051	0.110	0.182	0.272	0.396	0.596	1.006	2.098
			0.7	0.071	0.157	0.269	0.431	0.712	1.346	3.081	8.129

TABLE 18-III-B—DIFFERENTIAL SWELL OCCURRING AT THE PERIMETER OF A SLAB FOR A CENTER LIFT SWELLING CONDITION IN PREDOMINANTLY KAOLINITE CLAY SOIL (40 PERCENT CLAY)

PERCENT CLAY (%)	DEPTH TO CONSTANT SUCTION (ft.) × 304.8 for mm	CONSTANT SUCTION (pF)	VELOCITY OF MOISTURE FLOW (inches/month) × 25.4 for mm/month	DIFFERENTIAL SWELL (inch) × 25.4 for mm — Edge Distance Penetration × 304.8 for mm							
				1 ft.	2 ft.	3 ft.	4 ft.	5 ft.	6 ft.	7 ft.	8 ft.
40	3	3.2	0.5	0.004	0.008	0.012	0.016	0.020	0.024	0.029	0.034
			0.7	0.005	0.010	0.016	0.022	0.029	0.035	0.043	0.050
		3.4	0.5	0.007	0.016	0.074	0.037	0.046	0.058	0.073	0.090
			0.7	0.011	0.023	0.036	0.051	0.070	0.092	0.122	0.166
		3.6	0.5	0.018	0.040	0.066	0.102	0.157	0.256	0.496	1.181
			0.7	0.025	0.056	0.100	0.168	0.303	0.653	1.677	4.728
	5	3.2	0.5	0.009	0.018	0.027	0.037	0.047	0.057	0.068	0.080
			0.7	0.012	0.025	0.038	0.053	0.067	0.083	0.100	0.118
		3.4	0.5	0.018	0.037	0.148	0.081	0.106	0.134	0.167	0.206
			0.7	0.025	0.052	0.083	0.118	0.159	0.210	0.277	0.374
		3.6	0.5	0.041	0.090	0.150	0.229	0.346	0.553	1.040	2.408
			0.7	0.057	0.128	0.224	0.371	0.652	1.353	3.349	9.215
	7	3.2	0.5	0.016	0.033	0.051	0.069	0.087	0.107	0.127	0.148
			0.7	0.023	0.046	0.071	0.098	0.125	0.155	0.186	0.220
		3.4	0.5	0.033	0.069	0.107	0.148	0.194	0.246	0.306	0.377
			0.7	0.045	0.095	0.152	0.216	0.292	0.385	0.507	0.689
		3.58	0.5	0.069	0.148	0.244	0.365	0.531	0.799	1.348	2.791
			0.7	0.095	0.210	0.360	0.577	0.953	1.803	4.126	—

TABLE 18-III-C
TABLE 18-III-D

1997 UNIFORM BUILDING CODE

TABLE 18-III-C—DIFFERENTIAL SWELL OCCURRING AT THE PERIMETER OF A SLAB FOR A CENTER LIFT SWELLING CONDITION IN PREDOMINANTLY KAOLINITE CLAY SOIL (50 PERCENT CLAY)

PERCENT CLAY (%)	DEPTH TO CONSTANT SUCTION (ft.) × 304.8 for mm	CONSTANT SUCTION (pF)	VELOCITY OF MOISTURE FLOW (inches/month) × 25.4 for mm/month	DIFFERENTIAL SWELL (inch) × 25.4 for mm — Edge Distance Penetration × 304.8 for mm							
				1 ft.	2 ft.	3 ft.	4 ft.	5 ft.	6 ft.	7 ft.	8 ft.
50	3	3.2	0.5	0.004	0.009	0.014	0.019	0.025	0.030	0.036	0.042
			0.7	0.006	0.013	0.020	0.028	0.036	0.044	0.053	0.063
		3.4	0.5	0.009	0.020	0.031	0.043	0.057	0.073	0.091	0.113
			0.7	0.013	0.028	0.044	0.064	0.086	0.115	0.153	0.207
		3.6	0.5	0.022	0.049	0.083	0.127	0.196	0.321	0.620	1.480
			0.7	0.031	0.070	0.124	0.210	0.380	0.818	2.103	5.926
	5	3.2	0.5	0.011	0.022	0.034	0.046	0.059	0.072	0.086	0.100
			0.7	0.015	0.031	0.048	0.066	0.084	0.104	0.125	0.148
		3.4	0.5	0.023	0.047	0.073	0.101	0.133	0.168	0.209	0.258
			0.7	0.031	0.066	0.105	0.148	0.200	0.264	0.347	0.469
		3.6	0.5	0.051	0.113	0.188	0.288	0.434	0.694	1.303	3.018
			0.7	0.071	0.160	0.281	0.465	0.817	1.696	4.196	—
	7	3.2	0.5	0.021	0.042	0.064	0.086	0.110	0.134	0.159	0.186
			0.7	0.028	0.058	0.090	0.122	0.157	0.194	0.233	0.276
		3.4	0.5	0.041	0.085	0.133	0.185	0.243	0.308	0.383	0.472
			0.7	0.057	0.120	0.191	0.272	0.366	0.483	0.636	0.864
		3.58	0.5	0.086	0.186	0.306	0.457	0.666	1.001	1.690	3.499
			0.7	0.119	0.263	0.452	0.723	1.194	2.260	5.172	—

TABLE 18-III-D—DIFFERENTIAL SWELL OCCURRING AT THE PERIMETER OF A SLAB FOR A CENTER LIFT SWELLING CONDITION IN PREDOMINANTLY KAOLINITE CLAY SOIL (60 PERCENT CLAY)

PERCENT CLAY (%)	DEPTH TO CONSTANT SUCTION (ft.) × 304.8 for mm	CONSTANT SUCTION (pF)	VELOCITY OF MOISTURE FLOW (inches/month) × 25.4 for mm/month	DIFFERENTIAL SWELL (inch) × 25.4 for mm — Edge Distance Penetration × 304.8 for mm							
				1 ft.	2 ft.	3 ft.	4 ft.	5 ft.	6 ft.	7 ft.	8 ft.
60	3	3.2	0.5	0.006	0.012	0.018	0.024	0.030	0.037	0.044	0.052
			0.7	0.008	0.017	0.025	0.034	0.044	0.054	0.065	0.077
		3.4	0.5	0.012	0.024	0.038	0.053	0.069	0.088	0.110	0.136
			0.7	0.016	0.033	0.053	0.077	0.104	0.138	0.184	0.249
		3.6	0.5	0.026	0.059	0.099	0.154	0.236	0.386	0.745	1.779
			0.7	0.036	0.083	0.149	0.252	0.455	0.983	2.527	7.124
	5	3.2	0.5	0.013	0.027	0.041	0.056	0.071	0.087	0.103	0.120
			0.7	0.019	0.038	0.058	0.080	0.102	0.126	0.151	0.178
		3.4	0.5	0.028	0.056	0.087	0.122	0.160	0.202	0.252	0.310
			0.7	0.037	0.078	0.125	0.177	0.240	0.316	0.417	0.564
		3.6	0.5	0.062	0.135	0.226	0.345	0.521	0.834	1.566	3.628
			0.7	0.086	0.193	0.337	0.559	0.982	2.039	5.049	—
	7	3.2	0.5	0.025	0.050	0.077	0.104	0.132	0.161	0.192	0.224
			0.7	0.034	0.070	0.108	0.147	0.189	0.233	0.281	0.332
		3.4	0.5	0.050	0.103	0.160	0.223	0.292	0.371	0.461	0.568
			0.7	0.069	0.144	0.229	0.326	0.440	0.580	0.765	1.038
		3.58	0.5	0.103	0.223	0.367	0.549	0.800	1.203	2.031	4.205
			0.7	0.142	0.316	0.543	0.870	1.436	2.717	6.217	—

TABLE 18-III-E—DIFFERENTIAL SWELL OCCURRING AT THE PERIMETER OF A SLAB FOR A CENTER LIFT SWELLING CONDITION IN PREDOMINANTLY KAOLINITE CLAY SOIL (70 PERCENT CLAY)

PERCENT CLAY (%)	DEPTH TO CONSTANT SUCTION (ft.) ×304.8 for mm	CONSTANT SUCTION (pF)	VELOCITY OF MOISTURE FLOW (inches/month) ×25.4 for mm/month	DIFFERENTIAL SWELL (inch) ×25.4 for mm — Edge Distance Penetration ×304.8 for mm							
				1 ft.	2 ft.	3 ft.	4 ft.	5 ft.	6 ft.	7 ft.	8 ft.
70	3	3.2	0.5	0.006	0.013	0.020	0.027	0.035	0.042	0.051	0.060
			0.7	0.010	0.019	0.029	0.040	0.051	0.063	0.076	0.089
		3.4	0.5	0.014	0.028	0.044	0.062	0.081	0.103	0.129	0.159
			0.7	0.019	0.039	0.063	0.090	0.122	0.162	0.215	0.292
		3.6	0.5	0.032	0.069	0.117	0.180	0.276	0.451	0.871	2.079
			0.7	0.043	0.098	0.174	0.294	0.532	1.148	2.952	8.322
	5	3.2	0.5	0.016	0.032	0.048	0.065	0.083	0.101	0.120	0.140
			0.7	0.022	0.044	0.068	0.093	0.119	0.146	0.176	0.208
		3.4	0.5	0.031	0.065	0.101	0.141	0.185	0.235	0.293	0.361
			0.7	0.043	0.082	0.146	0.207	0.280	0.370	0.487	0.659
		3.6	0.5	0.072	0.158	0.264	0.404	0.609	0.974	1.830	4.239
			0.7	0.100	0.225	0.394	0.653	1.147	2.381	5.893	—
	7	3.2	0.5	0.030	0.059	0.090	0.121	0.154	0.188	0.224	0.262
			0.7	0.040	0.082	0.126	0.172	0.221	0.273	0.328	0.388
		3.4	0.5	0.057	0.119	0.186	0.260	0.341	0.432	0.538	0.663
			0.7	0.080	0.168	0.267	0.381	0.514	0.678	0.893	1.213
		3.58	0.5	0.120	0.260	0.429	0.642	0.935	1.406	2.373	4.913
			0.7	0.166	0.369	0.634	1.016	1.677	3.175	7.263	—

TABLE 18-III-F—DIFFERENTIAL SWELL OCCURRING AT THE PERIMETER OF A SLAB FOR A CENTER LIFT SWELLING CONDITION IN PREDOMINANTLY ILLITE CLAY SOIL (30 PERCENT CLAY)

PERCENT CLAY (%)	DEPTH TO CONSTANT SUCTION (ft.) ×304.8 for mm	CONSTANT SUCTION (pF)	VELOCITY OF MOISTURE FLOW (inches/month) ×25.4 for mm/month	DIFFERENTIAL SWELL (inch) ×25.4 for mm — Edge Distance Penetration ×304.8 for mm							
				1 ft.	2 ft.	3 ft.	4 ft.	5 ft.	6 ft.	7 ft.	8 ft.
30	3	3.2	0.5	0.006	0.012	0.018	0.024	0.030	0.037	0.044	0.051
			0.7	0.008	0.016	0.024	0.033	0.043	0.053	0.064	0.075
		3.4	0.5	0.011	0.024	0.037	0.052	0.068	0.087	0.109	0.135
			0.7	0.016	0.034	0.054	0.077	0.104	0.138	0.182	0.248
		3.6	0.5	0.027	0.058	0.098	0.152	0.234	0.382	0.737	1.760
			0.7	0.036	0.083	0.147	0.249	0.451	0.973	2.500	7.049
	5	3.2	0.5	0.013	0.027	0.041	0.055	0.070	0.086	0.102	0.119
			0.7	0.018	0.037	0.057	0.078	0.100	0.124	0.149	0.176
		3.4	0.5	0.027	0.055	0.086	0.120	0.157	0.200	0.248	0.306
			0.7	0.037	0.078	0.124	0.176	0.238	0.319	0.413	0.558
		3.6	0.5	0.062	0.134	0.224	0.342	0.516	0.825	1.551	3.591
			0.7	0.084	0.190	0.333	0.553	0.971	2.016	4.991	—
	7	3.2	0.5	0.025	0.050	0.076	0.103	0.131	0.160	0.190	0.221
			0.7	0.034	0.070	0.107	0.146	0.187	0.231	0.278	0.329
		3.4	0.5	0.048	0.102	0.158	0.221	0.288	0.367	0.456	0.562
			0.7	0.068	0.143	0.227	0.323	0.436	0.574	0.757	1.028
		3.58	0.5	0.103	0.221	0.363	0.543	0.792	1.191	2.010	4.162
			0.7	0.141	0.313	0.537	0.861	1.421	2.689	6.153	—

TABLE 18-III-G
TABLE 18-III-H

1997 UNIFORM BUILDING CODE

**TABLE 18-III-G—DIFFERENTIAL SWELL OCCURRING AT THE PERIMETER OF A SLAB FOR
A CENTER LIFT SWELLING CONDITION IN PREDOMINANTLY ILLITE CLAY SOIL
(40 PERCENT CLAY)**

PERCENT CLAY (%)	DEPTH TO CONSTANT SUCTION (ft.) × 304.8 for mm	CONSTANT SUCTION (pF)	VELOCITY OF MOISTURE FLOW (inches/month) × 25.4 for mm/month	DIFFERENTIAL SWELL (inch) × 25.4 for mm Edge Distance Penetration × 304.8 for mm							
				1 ft.	2 ft.	3 ft.	4 ft.	5 ft.	6 ft.	7 ft.	8 ft.
40	3	3.2	0.5	0.008	0.016	0.025	0.034	0.043	0.052	0.063	0.073
			0.7	0.011	0.023	0.035	0.048	0.062	0.076	0.092	0.109
		3.4	0.5	0.017	0.034	0.054	0.075	0.099	0.126	0.157	0.194
			0.7	0.023	0.048	0.077	0.110	0.149	0.198	0.263	0.357
		3.6	0.5	0.039	0.085	0.142	0.220	0.338	0.552	1.065	2.542
			0.7	0.053	0.120	0.213	0.360	0.651	1.405	3.611	—
	5	3.2	0.5	0.019	0.039	0.059	0.080	0.101	0.124	0.147	0.172
			0.7	0.026	0.054	0.083	0.113	0.145	0.179	0.215	0.254
		3.4	0.5	0.039	0.080	0.124	0.173	0.227	0.288	0.358	0.442
			0.7	0.053	0.112	0.178	0.254	0.343	0.452	0.596	0.805
		3.6	0.5	0.089	0.194	0.323	0.494	0.745	1.192	2.239	5.184
			0.7	0.122	0.275	0.482	0.799	1.403	2.912	7.207	—
	7	3.2	0.5	0.035	0.072	0.109	0.148	0.188	0.230	0.274	0.320
			0.7	0.049	0.100	0.154	0.210	0.270	0.330	0.401	0.474
		3.4	0.5	0.070	0.146	0.228	0.318	0.417	0.528	0.657	0.811
			0.7	0.098	0.207	0.328	0.466	0.629	0.829	1.093	1.484
		3.58	0.5	0.147	0.318	0.524	0.784	1.143	1.719	2.902	6.008
			0.7	0.203	0.451	0.775	1.242	2.051	3.883	8.882	—

**TABLE 18-III-H—DIFFERENTIAL SWELL OCCURRING AT THE PERIMETER OF A SLAB FOR
A CENTER LIFT SWELLING CONDITION IN PREDOMINANTLY ILLITE CLAY SOIL
(50 PERCENT CLAY)**

PERCENT CLAY (%)	DEPTH TO CONSTANT SUCTION (ft.) × 304.8 for mm	CONSTANT SUCTION (pF)	VELOCITY OF MOISTURE FLOW (inches/month) × 25.4 for mm/month	DIFFERENTIAL SWELL (inch) × 25.4 for mm Edge Distance Penetration × 304.8 for mm							
				1 ft.	2 ft.	3 ft.	4 ft.	5 ft.	6 ft.	7 ft.	8 ft.
50	3	3.2	0.5	0.010	0.021	0.033	0.042	0.056	0.069	0.082	0.096
			0.7	0.014	0.030	0.046	0.063	0.081	0.099	0.120	0.142
		3.4	0.5	0.022	0.045	0.070	0.098	0.129	0.164	0.205	0.254
			0.7	0.030	0.063	0.101	0.144	0.196	0.260	0.344	0.467
		3.6	0.5	0.050	0.111	0.185	0.287	0.441	0.721	1.391	3.322
			0.7	0.069	0.156	0.278	0.470	0.851	1.836	4.720	—
	5	3.2	0.5	0.025	0.051	0.077	0.104	0.133	0.162	0.193	0.225
			0.7	0.035	0.071	0.110	0.148	0.190	0.235	0.282	0.333
		3.4	0.5	0.051	0.104	0.163	0.227	0.298	0.377	0.469	0.579
			0.7	0.070	0.147	0.233	0.332	0.449	0.592	0.779	1.054
		3.6	0.5	0.116	0.253	0.423	0.646	0.974	1.558	2.927	6.778
			0.7	0.159	0.359	0.629	1.043	1.834	3.807	9.422	—
	7	3.2	0.5	0.046	0.094	0.143	0.193	0.246	0.301	0.358	0.418
			0.7	0.064	0.131	0.201	0.275	0.353	0.436	0.524	0.620
		3.4	0.5	0.092	0.192	0.299	0.416	0.546	0.692	0.860	1.061
			0.7	0.129	0.271	0.429	0.610	0.823	1.085	1.429	1.940
		3.58	0.5	0.193	0.417	0.687	1.028	1.495	2.248	3.795	7.856
			0.7	0.265	0.590	1.013	1.624	2.682	5.075	—	—

TABLE 18-III-I—DIFFERENTIAL SWELL OCCURRING AT THE PERIMETER OF A SLAB FOR A CENTER LIFT SWELLING CONDITION IN PREDOMINANTLY ILLITE CLAY SOIL (60 PERCENT CLAY)

PERCENT CLAY (%)	DEPTH TO CONSTANT SUCTION (ft.) × 304.8 for mm	CONSTANT SUCTION (pF)	VELOCITY OF MOISTURE FLOW (inches/month) × 25.4 for mm/month	DIFFERENTIAL SWELL (inch) × 25.4 for mm / Edge Distance Penetration × 304.8 for mm							
				1 ft.	2 ft.	3 ft.	4 ft.	5 ft.	6 ft.	7 ft.	8 ft.
60	3	3.2	0.5	0.013	0.027	0.040	0.055	0.070	0.085	0.102	0.119
			0.7	0.018	0.037	0.057	0.078	0.100	0.123	0.148	0.176
		3.4	0.5	0.027	0.055	0.087	0.121	0.159	0.203	0.253	0.314
			0.7	0.037	0.078	0.124	0.178	0.241	0.320	0.424	0.576
		3.6	0.5	0.062	0.136	0.229	0.355	0.544	0.890	1.659	4.104
			0.7	0.085	0.194	0.344	0.582	1.052	2.268	5.831	—
	5	3.2	0.5	0.031	0.062	0.095	0.129	0.164	0.200	0.238	0.277
			0.7	0.043	0.088	0.134	0.183	0.235	0.290	0.348	0.411
		3.4	0.5	0.062	0.129	0.201	0.280	0.367	0.466	0.579	0.714
			0.7	0.086	0.181	0.288	0.410	0.554	0.730	0.962	1.301
		3.6	0.5	0.144	0.313	0.522	0.797	1.203	1.924	3.615	8.371
			0.7	0.197	0.444	0.778	1.289	2.265	4.702	—	—
	7	3.2	0.5	0.057	0.116	0.176	0.239	0.304	0.372	0.442	0.516
			0.7	0.079	0.162	0.249	0.340	0.436	0.538	0.647	0.766
		3.4	0.5	0.114	0.237	0.369	0.514	0.674	0.855	1.063	1.310
			0.7	0.156	0.333	0.528	0.752	1.015	1.340	1.764	2.395
		3.58	0.5	0.238	0.514	0.846	1.266	1.846	2.776	4.687	9.702
			0.7	0.328	0.730	1.252	2.006	3.312	6.268	—	—

TABLE 18-III-J—DIFFERENTIAL SWELL OCCURRING AT THE PERIMETER OF A SLAB FOR A CENTER LIFT SWELLING CONDITION IN PREDOMINANTLY ILLITE CLAY SOIL (70 PERCENT CLAY)

PERCENT CLAY (%)	DEPTH TO CONSTANT SUCTION (ft.) × 304.8 for mm	CONSTANT SUCTION (pF)	VELOCITY OF MOISTURE FLOW (inches/month) × 25.4 for mm/month	DIFFERENTIAL SWELL (inch) × 25.4 for mm / Edge Distance Penetration × 304.8 for mm							
				1 ft.	2 ft.	3 ft.	4 ft.	5 ft.	6 ft.	7 ft.	8 ft.
70	3	3.2	0.5	0.016	0.032	0.048	0.066	0.083	0.102	0.121	0.141
			0.7	0.021	0.044	0.068	0.092	0.119	0.147	0.177	0.209
		3.4	0.5	0.032	0.066	0.103	0.144	0.190	0.242	0.302	0.374
			0.7	0.044	0.093	0.148	0.212	0.287	0.381	0.510	0.686
		3.6	0.5	0.074	0.162	0.273	0.423	0.648	1.061	2.047	4.885
			0.7	0.101	0.231	0.409	0.692	1.251	2.700	6.940	—
	5	3.2	0.5	0.037	0.074	0.113	0.153	0.195	0.238	0.283	0.330
			0.7	0.051	0.104	0.159	0.218	0.279	0.345	0.414	0.489
		3.4	0.5	0.073	0.153	0.239	0.330	0.437	0.554	0.689	0.850
			0.7	0.102	0.215	0.342	0.487	0.659	0.869	1.145	1.548
		3.6	0.5	0.170	0.372	0.620	0.948	1.431	2.290	4.302	9.964
			0.7	0.234	0.528	0.925	1.534	2.696	5.597	—	—
	7	3.2	0.5	0.068	0.138	0.210	0.284	0.362	0.442	0.526	0.614
			0.7	0.094	0.193	0.296	0.404	0.519	0.640	0.771	0.911
		3.4	0.5	0.136	0.282	0.439	0.612	0.803	1.018	1.265	1.560
			0.7	0.188	0.396	0.629	0.895	1.209	1.594	2.100	2.851
		3.58	0.5	0.283	0.612	1.007	1.507	2.197	3.304	5.579	11.549
			0.7	0.391	0.869	1.490	2.388	3.943	7.462	—	—

TABLE 18-III-K
TABLE 18-III-L

1997 UNIFORM BUILDING CODE

TABLE 18-III-K—DIFFERENTIAL SWELL OCCURRING AT THE PERIMETER OF A SLAB FOR A CENTER LIFT SWELLING CONDITION IN PREDOMINANTLY MONTMORILLONITE CLAY SOIL (30 PERCENT CLAY)

PERCENT CLAY (%)	DEPTH TO CONSTANT SUCTION (ft.) × 304.8 for mm	CONSTANT SUCTION (pF)	VELOCITY OF MOISTURE FLOW (inches/month) × 25.4 for mm/month	DIFFERENTIAL SWELL (inch) × 25.4 for mm — Edge Distance Penetration × 304.8 for mm							
				1 ft.	2 ft.	3 ft.	4 ft.	5 ft.	6 ft.	7 ft.	8 ft.
30	3	3.2	0.5	0.006	0.013	0.020	0.027	0.035	0.043	0.051	0.060
			0.7	0.010	0.019	0.029	0.040	0.051	0.063	0.076	0.089
		3.4	0.5	0.014	0.028	0.044	0.062	0.081	0.102	0.128	0.159
			0.7	0.019	0.039	0.063	0.089	0.122	0.162	0.214	0.291
		3.6	0.5	0.031	0.069	0.116	0.179	0.275	0.450	0.868	2.073
			0.7	0.043	0.098	0.173	0.294	0.532	1.145	2.945	8.301
	5	3.2	0.5	0.016	0.032	0.048	0.065	0.083	0.101	0.120	0.140
			0.7	0.022	0.044	0.068	0.093	0.119	0.147	0.176	0.208
		3.4	0.5	0.032	0.065	0.102	0.141	0.186	0.235	0.293	0.361
			0.7	0.043	0.091	0.145	0.207	0.280	0.369	0.486	0.657
		3.6	0.5	0.072	0.158	0.264	0.403	0.607	0.972	1.826	2.514
			0.7	0.100	0.224	0.393	0.651	1.144	2.375	5.870	—
	7	3.2	0.5	0.029	0.059	0.089	0.121	0.154	0.188	0.223	0.261
			0.7	0.039	0.081	0.125	0.171	0.219	0.271	0.326	0.386
		3.4	0.5	0.058	0.120	0.187	0.261	0.342	0.433	0.537	0.662
			0.7	0.081	0.169	0.268	0.381	0.514	0.677	0.892	1.211
		3.58	0.5	0.120	0.259	0.427	0.639	0.932	1.402	2.367	4.900
			0.7	0.165	0.368	0.632	1.013	1.672	3.165	7.244	—

TABLE 18-III-L—DIFFERENTIAL SWELL OCCURRING AT THE PERIMETER OF A SLAB FOR A CENTER LIFT SWELLING CONDITION IN PREDOMINANTLY MONTMORILLONITE CLAY SOIL (40 PERCENT CLAY)

PERCENT CLAY (%)	DEPTH TO CONSTANT SUCTION (ft.) × 304.8 for mm	CONSTANT SUCTION (pF)	VELOCITY OF MOISTURE FLOW (inches/month) × 25.4 for mm/month	DIFFERENTIAL SWELL (inch) × 25.4 for mm — Edge Distance Penetration × 304.8 for mm							
				1 ft.	2 ft.	3 ft.	4 ft.	5 ft.	6 ft.	7 ft.	8 ft.
40	3	3.2	0.5	0.009	0.019	0.029	0.040	0.050	0.062	0.074	0.086
			0.7	0.014	0.027	0.042	0.057	0.074	0.091	0.109	0.129
		3.4	0.5	0.019	0.040	0.063	0.088	0.116	0.148	0.185	0.229
			0.7	0.027	0.057	0.091	0.130	0.177	0.234	0.311	0.422
		3.6	0.5	0.046	0.100	0.168	0.260	0.399	0.652	1.258	3.004
			0.7	0.062	0.142	0.251	0.425	0.769	1.660	4.267	7.762
	5	3.2	0.5	0.023	0.046	0.069	0.094	0.120	0.147	0.174	0.203
			0.7	0.032	0.064	0.098	0.134	0.172	0.212	0.255	0.301
		3.4	0.5	0.045	0.094	0.147	0.205	0.269	0.341	0.424	0.522
			0.7	0.063	0.133	0.211	0.300	0.405	0.535	0.704	0.952
		3.6	0.5	0.105	0.229	0.382	0.584	0.880	1.408	2.645	6.127
			0.7	0.144	0.325	0.569	0.944	1.658	3.441	8.517	—
	7	3.2	0.5	0.042	0.085	0.129	0.175	0.223	0.272	0.324	0.378
			0.7	0.059	0.119	0.183	0.249	0.320	0.394	0.475	0.561
		3.4	0.5	0.084	0.173	0.270	0.377	0.494	0.626	0.778	0.960
			0.7	0.116	0.244	0.387	0.550	0.743	0.980	1.291	1.753
		3.56	0.5	0.138	0.339	0.553	0.815	1.160	1.668	2.583	4.748
			0.7	0.218	0.480	0.813	1.271	2.004	3.504	7.412	—

TABLE 18-III-M—DIFFERENTIAL SWELL OCCURRING AT THE PERIMETER OF A SLAB FOR A CENTER LIFT SWELLING CONDITION IN PREDOMINANTLY MONTMORILLONITE CLAY SOIL (50 PERCENT CLAY)

PERCENT CLAY (%)	DEPTH TO CONSTANT SUCTION (ft.) ×304.8 for mm	CONSTANT SUCTION (pF)	VELOCITY OF MOISTURE FLOW (inches/month) ×25.4 for mm/month	DIFFERENTIAL SWELL (inch) ×25.4 for mm — Edge Distance Penetration ×304.8 for mm							
				1 ft.	2 ft.	3 ft.	4 ft.	5 ft.	6 ft.	7 ft.	8 ft.
50	3	3.2	0.5	0.012	0.026	0.038	0.052	0.066	0.081	0.097	0.113
			0.7	0.018	0.036	0.055	0.075	0.096	0.119	0.143	0.169
		3.4	0.5	0.026	0.053	0.083	0.116	0.153	0.195	0.243	0.301
			0.7	0.036	0.075	0.119	0.170	0.231	0.307	0.407	0.553
		3.6	0.5	0.059	0.131	0.220	0.340	0.522	0.854	1.648	3.935
			0.7	0.082	0.186	0.330	0.558	1.008	2.175	5.590	—
	5	3.2	0.5	0.030	0.060	0.091	0.124	0.157	0.192	0.228	0.266
			0.7	0.041	0.083	0.128	0.175	0.224	0.277	0.333	0.394
		3.4	0.5	0.059	0.123	0.193	0.268	0.352	0.446	0.555	0.684
			0.7	0.083	0.174	0.276	0.393	0.531	0.700	0.923	1.247
		3.6	0.5	0.137	0.299	0.500	0.764	1.152	1.844	3.464	8.025
			0.7	0.189	0.426	0.745	1.236	2.712	4.508	—	—
	7	3.2	0.5	0.055	0.111	0.169	0.229	0.291	0.356	0.423	0.495
			0.7	0.076	0.155	0.238	0.326	0.418	0.516	0.621	0.734
		3.4	0.5	0.106	0.226	0.354	0.493	0.646	0.819	1.019	1.256
			0.7	0.152	0.320	0.507	0.721	0.973	1.283	1.692	2.256
		3.56	0.5	0.207	0.444	0.724	1.068	1.519	2.185	3.382	6.219
			0.7	0.286	0.629	1.066	1.665	2.625	4.590	—	—

TABLE 18-III-N—DIFFERENTIAL SWELL OCCURRING AT THE PERIMETER OF A SLAB FOR A CENTER LIFT SWELLING CONDITION IN PREDOMINANTLY MONTMORILLONITE CLAY SOIL (60 PERCENT CLAY)

PERCENT CLAY (%)	DEPTH TO CONSTANT SUCTION (ft.) ×304.8 for mm	CONSTANT SUCTION (pF)	VELOCITY OF MOISTURE FLOW (inches/month) ×25.4 for mm/month	DIFFERENTIAL SWELL (inch) ×25.4 for mm — Edge Distance Penetration ×304.8 for mm							
				1 ft.	2 ft.	3 ft.	4 ft.	5 ft.	6 ft.	7 ft.	8 ft.
60	3	3.2	0.5	0.015	0.031	0.048	0.065	0.082	0.101	0.120	0.140
			0.7	0.022	0.044	0.068	0.092	0.119	0.147	0.176	0.209
		3.4	0.5	0.031	0.065	0.102	0.143	0.189	0.240	0.300	0.372
			0.7	0.044	0.093	0.147	0.211	0.286	0.380	0.503	0.683
		3.6	0.5	0.073	0.161	0.272	0.420	0.645	1.056	2.037	4.865
			0.7	0.101	0.229	0.407	0.689	1.246	2.689	6.912	—
	5	3.2	0.5	0.037	0.074	0.113	0.153	0.194	0.237	0.282	0.329
			0.7	0.050	0.103	0.158	0.217	0.278	0.343	0.412	0.487
		3.4	0.5	0.073	0.152	0.237	0.331	0.435	0.551	0.686	0.846
			0.7	0.102	0.214	0.341	0.485	0.655	0.865	1.140	1.541
		3.6	0.5	0.169	0.370	0.618	0.945	1.425	2.280	4.284	9.923
			0.7	0.234	0.526	0.922	1.528	2.686	5.574	—	—
	7	3.2	0.5	0.068	0.137	0.209	0.283	0.360	0.441	0.524	0.612
			0.7	0.093	0.191	0.294	0.402	0.516	0.637	0.767	0.907
		3.4	0.5	0.135	0.280	0.438	0.609	0.799	1.013	1.260	1.553
			0.7	0.188	0.395	0.627	0.892	1.204	1.587	2.092	2.840
		3.56	0.5	0.256	0.549	0.895	1.320	1.879	2.702	4.182	8.216
			0.7	0.354	0.779	1.317	2.059	3.247	5.677	—	—

TABLE 18-III-O
TABLE 18-III-P

1997 UNIFORM BUILDING CODE

TABLE 18-III-O—DIFFERENTIAL SWELL OCCURRING AT THE PERIMETER OF A SLAB FOR A CENTER LIFT SWELLING CONDITION IN PREDOMINANTLY MONTMORILLONITE CLAY SOIL (70 PERCENT CLAY)

PERCENT CLAY (%)	DEPTH TO CONSTANT SUCTION (ft.) × 304.8 for mm	CONSTANT SUCTION (pF)	VELOCITY OF MOISTURE FLOW (inches/month) × 25.4 for mm/month	DIFFERENTIAL SWELL (inch) × 25.4 for mm — Edge Distance Penetration × 304.8 for mm							
				1 ft.	2 ft.	3 ft.	4 ft.	5 ft.	6 ft.	7 ft.	8 ft.
70	3	3.2	0.5	0.018	0.037	0.057	0.077	0.098	0.120	0.143	0.167
			0.7	0.026	0.052	0.082	0.110	0.141	0.174	0.210	0.248
		3.4	0.5	0.038	0.078	0.122	0.171	0.225	0.287	0.358	0.443
			0.7	0.052	0.110	0.176	0.251	0.341	0.452	0.600	0.814
		3.6	0.5	0.088	0.192	0.324	0.502	0.769	1.258	2.428	5.796
			0.7	0.120	0.273	0.485	0.821	1.485	3.203	8.234	—
	5	3.2	0.5	0.044	0.088	0.134	0.182	0.231	0.283	0.336	0.392
			0.7	0.060	0.123	0.189	0.258	0.331	0.409	0.491	0.580
		3.4	0.5	0.088	0.182	0.284	0.395	0.519	0.658	0.818	1.008
			0.7	0.121	0.256	0.406	0.578	0.781	1.031	1.358	1.837
		3.6	0.5	0.202	0.441	0.737	1.126	1.698	2.717	5.104	11.822
			0.7	0.278	0.627	1.098	1.820	3.199	6.640	—	—
	7	3.2	0.5	0.081	0.163	0.249	0.338	0.429	0.525	0.624	0.729
			0.7	0.112	0.229	0.351	0.480	0.616	0.759	0.915	1.081
		3.4	0.5	0.162	0.334	0.522	0.727	0.952	1.207	1.501	1.851
			0.7	0.224	0.470	0.747	1.063	1.435	1.891	2.492	3.383
		3.56	0.5	0.305	0.655	1.067	1.573	2.239	3.219	4.983	9.162
			0.7	0.421	0.928	1.569	2.453	3.868	6.763	—	—

TABLE 18-III-P—DIFFERENTIAL SWELL OCCURRING AT THE PERIMETER OF A SLAB FOR AN EDGE LIFT SWELLING CONDITION IN PREDOMINANTLY KAOLINITE CLAY SOIL (30 PERCENT CLAY)

PERCENT CLAY (%)	DEPTH TO CONSTANT SUCTION (ft.) × 304.8 for mm	CONSTANT SUCTION (pF)	VELOCITY OF MOISTURE FLOW (inches/month) × 25.4 for mm/month	DIFFERENTIAL SWELL (inch) × 25.4 for mm — Edge Distance Penetration × 304.8 for mm							
				1 ft.	2 ft.	3 ft.	4 ft.	5 ft.	6 ft.	7 ft.	8 ft.
30	3	3.2	0.5	0.003	0.005	0.007	0.010	0.012	0.014	0.017	0.019
			0.7	0.004	0.007	0.010	0.013	0.017	0.020	0.023	0.026
		3.4	0.5	0.005	0.010	0.015	0.020	0.025	0.029	0.033	0.037
			0.7	0.007	0.014	0.021	0.027	0.033	0.039	0.045	0.050
		3.6	0.5	0.013	0.025	0.035	0.046	0.055	0.064	0.072	0.080
			0.7	0.018	0.034	0.048	0.061	0.073	0.084	0.094	0.104
		3.8	0.5	0.031	0.056	0.078	0.097	0.114	0.129	0.143	0.156
			0.7	0.042	0.075	0.102	0.125	0.145	0.164	0.180	0.195
	5	3.2	0.5	0.006	0.011	0.016	0.021	0.027	0.032	0.037	0.041
			0.7	0.008	0.015	0.022	0.030	0.037	0.044	0.050	0.057
		3.4	0.5	0.012	0.023	0.034	0.045	0.055	0.065	0.075	0.084
			0.7	0.016	0.032	0.047	0.061	0.075	0.089	0.101	0.114
		3.6	0.5	0.029	0.056	0.081	0.104	0.126	0.147	0.167	0.186
			0.7	0.041	0.077	0.110	0.141	0.169	0.195	0.220	0.243
		3.8	0.5	0.072	0.131	0.183	0.228	0.270	0.307	0.342	0.374
			0.7	0.100	0.178	0.244	0.300	0.350	0.395	0.436	0.474
	7	3.2	0.5	0.010	0.019	0.028	0.037	0.046	0.055	0.064	0.072
			0.7	0.013	0.026	0.039	0.052	0.064	0.076	0.088	0.099
		3.4	0.5	0.021	0.041	0.060	0.079	0.097	0.115	0.132	0.149
			0.7	0.029	0.056	0.083	0.108	0.133	0.157	0.180	0.202
		3.6	0.5	0.052	0.100	0.145	0.187	0.227	0.264	0.300	0.335
			0.7	0.073	0.138	0.198	0.254	0.305	0.353	0.399	0.441

TABLE 18-III-Q—DIFFERENTIAL SWELL OCCURRING AT THE PERIMETER OF A SLAB FOR AN EDGE LIFT SWELLING CONDITION IN PREDOMINANTLY KAOLINITE CLAY SOIL (40 PERCENT CLAY)

PERCENT CLAY (%)	DEPTH TO CONSTANT SUCTION (ft.) × 304.8 for mm	CONSTANT SUCTION (pF)	VELOCITY OF MOISTURE FLOW (inches/month) × 25.4 for mm/month	DIFFERENTIAL SWELL (inch) × 25.4 for mm — Edge Distance Penetration × 304.8 for mm							
				1 ft.	2 ft.	3 ft.	4 ft.	5 ft.	6 ft.	7 ft.	8 ft.
40	3	3.2	0.5	0.003	0.007	0.010	0.013	0.016	0.019	0.022	0.025
			0.7	0.006	0.009	0.014	0.018	0.022	0.026	0.030	0.034
		3.4	0.5	0.007	0.014	0.020	0.027	0.033	0.039	0.045	0.050
			0.7	0.010	0.019	0.028	0.037	0.045	0.052	0.060	0.067
		3.6	0.5	0.017	0.033	0.047	0.061	0.074	0.086	0.097	0.108
			0.7	0.024	0.045	0.064	0.081	0.098	0.112	0.126	0.139
		3.8	0.5	0.041	0.075	0.104	0.129	0.152	0.173	0.192	0.209
			0.7	0.056	0.100	0.136	0.167	0.195	0.219	0.241	0.261
	5	3.2	0.5	0.007	0.015	0.022	0.029	0.035	0.042	0.049	0.056
			0.7	0.010	0.020	0.030	0.040	0.049	0.058	0.067	0.076
		3.4	0.5	0.016	0.031	0.046	0.060	0.074	0.087	0.100	0.113
			0.7	0.022	0.043	0.063	0.082	0.101	0.119	0.136	0.152
		3.6	0.5	0.039	0.075	0.108	0.140	0.169	0.197	0.223	0.249
			0.7	0.064	0.103	0.147	0.188	0.226	0.261	0.295	0.326
		3.8	0.5	0.096	0.176	0.245	0.306	0.361	0.411	0.458	0.501
			0.7	0.134	0.239	0.326	0.402	0.469	0.529	0.584	0.634
	7	3.2	0.5	0.013	0.025	0.038	0.050	0.062	0.074	0.085	0.097
			0.7	0.018	0.035	0.052	0.069	0.085	0.102	0.117	0.133
		3.4	0.5	0.028	0.054	0.080	0.105	0.130	0.154	0.177	0.200
			0.7	0.039	0.076	0.111	0.145	0.178	0.210	0.241	0.271
		3.6	0.5	0.069	0.134	0.194	0.250	0.304	0.354	0.402	0.448
			0.7	0.098	0.185	0.266	0.340	0.409	0.473	0.534	0.591

TABLE 18-III-R—DIFFERENTIAL SWELL OCCURRING AT THE PERIMETER OF A SLAB FOR AN EDGE LIFT SWELLING CONDITION IN PREDOMINANTLY KAOLINITE CLAY SOIL (50 PERCENT CLAY)

PERCENT CLAY (%)	DEPTH TO CONSTANT SUCTION (ft.) × 304.8 for mm	CONSTANT SUCTION (pF)	VELOCITY OF MOISTURE FLOW (inches/month) × 25.4 for mm/month	DIFFERENTIAL SWELL (inch) × 25.4 for mm — Edge Distance Penetration × 304.8 for mm							
				1 ft.	2 ft.	3 ft.	4 ft.	5 ft.	6 ft.	7 ft.	8 ft.
50	3	3.2	0.5	0.004	0.008	0.012	0.016	0.020	0.024	0.028	0.032
			0.7	0.006	0.012	0.017	0.023	0.028	0.033	0.038	0.043
		3.4	0.5	0.009	0.017	0.026	0.034	0.041	0.049	0.056	0.063
			0.7	0.012	0.024	0.035	0.046	0.056	0.066	0.075	0.084
		3.6	0.5	0.022	0.042	0.059	0.076	0.092	0.107	0.121	0.135
			0.7	0.030	0.056	0.080	0.102	0.122	0.141	0.158	0.175
		3.8	0.5	0.052	0.094	0.130	0.162	0.191	0.217	0.240	0.262
			0.7	0.071	0.126	0.171	0.210	0.244	0.275	0.302	0.328
	5	3.2	0.5	0.009	0.018	0.027	0.036	0.044	0.053	0.061	0.070
			0.7	0.013	0.025	0.038	0.050	0.062	0.073	0.084	0.096
		3.4	0.5	0.020	0.039	0.057	0.075	0.092	0.109	0.126	0.142
			0.7	0.028	0.054	0.079	0.103	0.126	0.149	0.170	0.191
		3.6	0.5	0.049	0.094	0.136	0.175	0.212	0.247	0.280	0.312
			0.7	0.068	0.129	0.185	0.236	0.283	0.328	0.369	0.408
		3.8	0.5	0.120	0.220	0.307	0.384	0.453	0.516	0.574	0.628
			0.7	0.168	0.299	0.409	0.504	0.588	0.663	0.732	0.795
	7	3.2	0.5	0.016	0.032	0.047	0.062	0.077	0.092	0.107	0.121
			0.7	0.022	0.044	0.066	0.087	0.107	0.127	0.147	0.167
		3.4	0.5	0.035	0.068	0.100	0.132	0.163	0.193	0.222	0.250
			0.7	0.048	0.095	0.139	0.182	0.223	0.263	0.302	0.339
		3.6	0.5	0.087	0.168	0.243	0.314	0.381	0.444	0.504	0.562
			0.7	0.122	0.232	0.333	0.426	0.512	0.593	0.669	0.741

TABLE 18-III-S
TABLE 18-III-T

1997 UNIFORM BUILDING CODE

TABLE 18-III-S—DIFFERENTIAL SWELL OCCURRING AT THE PERIMETER OF A SLAB FOR AN EDGE LIFT SWELLING CONDITION IN PREDOMINANTLY KAOLINITE CLAY SOIL (60 PERCENT CLAY)

PERCENT CLAY (%)	DEPTH TO CONSTANT SUCTION (ft.) × 304.8 for mm	CONSTANT SUCTION (pF)	VELOCITY OF MOISTURE FLOW (inches/month) × 25.4 for mm/month	DIFFERENTIAL SWELL (inch) × 25.4 for mm / Edge Distance Penetration × 304.8 for mm							
				1 ft.	2 ft.	3 ft.	4 ft.	5 ft.	6 ft.	7 ft.	8 ft.
60	3	3.2	0.5	0.005	0.010	0.015	0.020	0.024	0.029	0.033	0.038
			0.7	0.007	0.014	0.021	0.027	0.033	0.040	0.046	0.052
		3.4	0.5	0.011	0.021	0.031	0.040	0.049	0.058	0.067	0.076
			0.7	0.015	0.029	0.042	0.055	0.067	0.079	0.090	0.101
		3.6	0.5	0.026	0.050	0.071	0.092	0.111	0.129	0.146	0.162
			0.7	0.036	0.068	0.097	0.123	0.147	0.169	0.190	0.210
		3.8	0.5	0.062	0.113	0.157	0.195	0.229	0.260	0.289	0.315
			0.7	0.085	0.151	0.205	0.252	0.293	0.330	0.363	0.394
	5	3.2	0.5	0.011	0.022	0.033	0.043	0.053	0.064	0.074	0.084
			0.7	0.015	0.031	0.045	0.060	0.074	0.088	0.101	0.115
		3.4	0.5	0.024	0.046	0.069	0.090	0.111	0.131	0.151	0.170
			0.7	0.033	0.065	0.095	0.124	0.152	0.179	0.205	0.230
		3.6	0.5	0.059	0.113	0.163	0.210	0.255	0.297	0.337	0.375
			0.7	0.082	0.155	0.222	0.284	0.341	0.394	0.444	0.491
		3.8	0.5	0.144	0.265	0.369	0.461	0.544	0.620	0.690	0.754
			0.7	0.201	0.360	0.492	0.606	0.707	0.798	0.880	0.956
	7	3.2	0.5	0.019	0.038	0.057	0.075	0.093	0.111	0.128	0.146
			0.7	0.027	0.053	0.079	0.104	0.129	0.153	0.177	0.201
		3.4	0.5	0.042	0.082	0.121	0.159	0.196	0.232	0.267	0.301
			0.7	0.058	0.114	0.167	0.219	0.268	0.316	0.363	0.408
		3.6	0.5	0.105	0.201	0.292	0.377	0.457	0.534	0.606	0.675
			0.7	0.147	0.279	0.400	0.512	0.616	0.713	0.804	0.891

TABLE 18-III-T—DIFFERENTIAL SWELL OCCURRING AT THE PERIMETER OF A SLAB FOR AN EDGE LIFT SWELLING CONDITION IN PREDOMINANTLY KAOLINITE CLAY SOIL (70 PERCENT CLAY)

PERCENT CLAY (%)	DEPTH TO CONSTANT SUCTION (ft.) × 304.8 for mm	CONSTANT SUCTION (pF)	VELOCITY OF MOISTURE FLOW (inches/month) × 25.4 for mm/month	DIFFERENTIAL SWELL (inch) × 25.4 for mm / Edge Distance Penetration × 304.8 for mm							
				1 ft.	2 ft.	3 ft.	4 ft.	5 ft.	6 ft.	7 ft.	8 ft.
70	3	3.2	0.5	0.006	0.012	0.017	0.023	0.028	0.034	0.039	0.044
			0.7	0.008	0.016	0.024	0.032	0.039	0.046	0.054	0.061
		3.4	0.5	0.012	0.024	0.036	0.047	0.058	0.068	0.078	0.088
			0.7	0.017	0.034	0.049	0.064	0.079	0.092	0.106	0.118
		3.6	0.5	0.030	0.058	0.084	0.107	0.130	0.151	0.170	0.189
			0.7	0.042	0.079	0.113	0.143	0.172	0.198	0.222	0.245
		3.8	0.5	0.072	0.132	0.183	0.228	0.268	0.304	0.338	0.369
			0.7	0.099	0.176	0.240	0.295	0.343	0.386	0.425	0.460
	5	3.2	0.5	0.013	0.026	0.038	0.050	0.062	0.074	0.086	0.098
			0.7	0.018	0.036	0.053	0.070	0.086	0.103	0.119	0.134
		3.4	0.5	0.028	0.054	0.080	0.105	0.130	0.153	0.176	0.199
			0.7	0.039	0.075	0.111	0.145	0.177	0.209	0.239	0.268
		3.6	0.5	0.068	0.132	0.191	0.246	0.298	0.347	0.393	0.438
			0.7	0.096	0.181	0.259	0.331	0.398	0.460	0.518	0.574
		3.8	0.5	0.169	0.309	0.431	0.539	0.636	0.724	0.806	0.881
			0.7	0.235	0.420	0.575	0.708	0.826	0.932	1.028	1.117
	7	3.2	0.5	0.022	0.044	0.066	0.087	0.109	0.129	0.150	0.170
			0.7	0.031	0.062	0.092	0.121	0.150	0.179	0.207	0.234
		3.4	0.5	0.048	0.095	0.141	0.185	0.228	0.270	0.311	0.351
			0.7	0.068	0.133	0.195	0.256	0.314	0.370	0.424	0.476
		3.6	0.5	0.122	0.235	0.341	0.441	0.534	0.623	0.708	0.789
			0.7	0.172	0.326	0.468	0.598	0.719	0.833	0.940	1.041

TABLE 18-III-U—DIFFERENTIAL SWELL OCCURRING AT THE PERIMETER OF A SLAB FOR AN EDGE LIFT SWELLING CONDITION IN PREDOMINANTLY ILLITE CLAY SOIL (30 PERCENT CLAY)

PERCENT CLAY (%)	DEPTH TO CONSTANT SUCTION (ft.) × 304.8 for mm	CONSTANT SUCTION (pF)	VELOCITY OF MOISTURE FLOW (inches/month) × 25.4 for mm/month	DIFFERENTIAL SWELL (inch) × 25.4 for mm — Edge Distance Penetration × 304.8 for mm							
				1 ft.	2 ft.	3 ft.	4 ft.	5 ft.	6 ft.	7 ft.	8 ft.
30	3	3.2	0.5	0.006	0.010	0.015	0.019	0.024	0.029	0.333	0.037
			0.7	0.007	0.014	0.020	0.027	0.033	0.039	0.045	0.051
		3.4	0.5	0.011	0.021	0.030	0.040	0.049	0.058	0.066	0.075
			0.7	0.015	0.029	0.042	0.054	0.067	0.078	0.089	0.100
		3.6	0.5	0.026	0.049	0.071	0.091	0.110	0.128	0.144	0.160
			0.7	0.036	0.067	0.096	0.121	0.145	0.168	0.186	0.208
		3.8	0.5	0.061	0.112	0.155	0.193	0.227	0.258	0.286	0.312
			0.7	0.084	0.149	0.203	0.250	0.290	0.327	0.360	0.390
	5	3.2	0.5	0.011	0.022	0.032	0.043	0.053	0.063	0.073	0.083
			0.7	0.015	0.030	0.045	0.059	0.073	0.087	0.100	0.114
		3.4	0.5	0.023	0.046	0.068	0.089	0.110	0.130	0.149	0.168
			0.7	0.033	0.064	0.094	0.123	0.150	0.177	0.202	0.227
		3.6	0.5	0.058	0.112	0.161	0.208	0.252	0.294	0.333	0.371
			0.7	0.081	0.154	0.220	0.281	0.337	0.390	0.439	0.486
		3.8	0.5	0.143	0.262	0.365	0.456	0.539	0.613	0.683	0.746
			0.7	0.199	0.356	0.487	0.600	0.699	0.789	0.871	0.946
	7	3.2	0.5	0.019	0.038	0.056	0.074	0.092	0.110	0.127	0.144
			0.7	0.027	0.052	0.078	0.103	0.127	0.152	0.176	0.198
		3.4	0.5	0.041	0.081	0.119	0.157	0.194	0.229	0.264	0.298
			0.7	0.058	0.113	0.166	0.217	0.266	0.313	0.359	0.404
		3.6	0.5	0.103	0.199	0.289	0.373	0.453	0.528	0.600	0.668
			0.7	0.145	0.276	0.396	0.507	0.609	0.706	0.796	0.882

TABLE 18-III-V—DIFFERENTIAL SWELL OCCURRING AT THE PERIMETER OF A SLAB FOR AN EDGE LIFT SWELLING CONDITION IN PREDOMINANTLY ILLITE CLAY SOIL (40 PERCENT CLAY)

PERCENT CLAY (%)	DEPTH TO CONSTANT SUCTION (ft.) × 304.8 for mm	CONSTANT SUCTION (pF)	VELOCITY OF MOISTURE FLOW (inches/month) × 25.4 for mm/month	DIFFERENTIAL SWELL (inch) × 25.4 for mm — Edge Distance Penetration × 304.8 for mm							
				1 ft.	2 ft.	3 ft.	4 ft.	5 ft.	6 ft.	7 ft.	8 ft.
40	3	3.2	0.5	0.007	0.014	0.021	0.028	0.035	0.041	0.048	0.054
			0.7	0.010	0.020	0.029	0.039	0.048	0.057	0.065	0.074
		3.4	0.5	0.015	0.030	0.044	0.058	0.071	0.083	0.096	0.108
			0.7	0.021	0.041	0.060	0.079	0.096	0.113	0.129	0.145
		3.6	0.5	0.037	0.071	0.102	0.131	0.158	0.184	0.208	0.231
			0.7	0.051	0.097	0.138	0.175	0.210	0.242	0.272	0.300
		3.8	0.5	0.088	0.161	0.224	0.278	0.328	0.372	0.413	0.451
			0.7	0.122	0.216	0.294	0.360	0.419	0.472	0.519	0.563
	5	3.2	0.5	0.016	0.031	0.047	0.062	0.076	0.091	0.105	0.120
			0.7	0.022	0.044	0.065	0.085	0.106	0.125	0.145	0.164
		3.4	0.5	0.034	0.066	0.098	0.129	0.158	0.187	0.216	0.243
			0.7	0.047	0.092	0.135	0.177	0.217	0.255	0.292	0.328
		3.6	0.5	0.084	0.161	0.233	0.300	0.364	0.424	0.481	0.535
			0.7	0.117	0.222	0.317	0.405	0.487	0.563	0.634	0.701
		3.8	0.5	0.206	0.378	0.527	0.659	0.777	0.886	0.985	1.078
			0.7	0.288	0.514	0.703	0.866	1.010	1.139	1.257	1.366
	7	3.2	0.5	0.027	0.054	0.081	0.107	0.133	0.158	0.183	0.208
			0.7	0.038	0.076	0.112	0.149	0.184	0.219	0.253	0.286
		3.4	0.5	0.059	0.117	0.172	0.227	0.279	0.331	0.381	0.430
			0.7	0.083	0.163	0.239	0.313	0.384	0.452	0.518	0.583
		3.6	0.5	0.149	0.288	0.417	0.539	0.654	0.762	0.866	0.965
			0.7	0.210	0.399	0.572	0.731	0.880	1.019	1.149	1.273

TABLE 18-III-W
TABLE 18-III-X

1997 UNIFORM BUILDING CODE

TABLE 18-III-W—DIFFERENTIAL SWELL OCCURRING AT THE PERIMETER OF A SLAB FOR AN EDGE LIFT SWELLING CONDITION IN PREDOMINANTLY ILLITE CLAY SOIL (50 PERCENT CLAY)

PERCENT CLAY (%)	DEPTH TO CONSTANT SUCTION (ft.) × 304.8 for mm	CONSTANT SUCTION (pF)	VELOCITY OF MOISTURE FLOW (inches/month) × 25.4 for mm/month	DIFFERENTIAL SWELL (inch) × 25.4 for mm Edge Distance Penetration × 304.8 for mm							
				1 ft.	2 ft.	3 ft.	4 ft.	5 ft.	6 ft.	7 ft.	8 ft.
50	3	3.2	0.5	0.009	0.019	0.028	0.037	0.045	0.054	0.062	0.071
			0.7	0.013	0.026	0.038	0.051	0.062	0.074	0.086	0.097
		3.4	0.5	0.020	0.039	0.057	0.075	0.092	0.109	0.125	0.141
			0.7	0.028	0.054	0.079	0.103	0.126	0.148	0.169	0.189
		3.6	0.5	0.048	0.093	0.134	0.172	0.207	0.241	0.272	0.303
			0.7	0.067	0.127	0.180	0.229	0.274	0.316	0.356	0.392
		3.8	0.5	0.116	0.211	0.292	0.364	0.428	0.486	0.540	0.589
			0.7	0.159	0.282	0.384	0.471	0.548	0.616	0.679	0.736
	5	3.2	0.5	0.021	0.041	0.061	0.080	0.100	0.119	0.138	0.156
			0.7	0.029	0.057	0.085	0.112	0.138	0.164	0.190	0.215
		3.4	0.5	0.044	0.087	0.128	0.168	0.207	0.245	0.282	0.318
			0.7	0.062	0.121	0.177	0.231	0.283	0.334	0.382	0.429
		3.6	0.5	0.109	0.211	0.305	0.393	0.476	0.554	0.629	0.700
			0.7	0.153	0.290	0.415	0.530	0.636	0.736	0.829	0.917
		3.8	0.5	0.269	0.494	0.689	0.861	1.016	1.158	1.288	1.409
			0.7	0.376	0.672	0.919	1.132	1.320	1.490	1.644	1.785
	7	3.2	0.5	0.036	0.071	0.106	0.140	0.173	0.207	0.240	0.272
			0.7	0.050	0.099	0.147	0.194	0.240	0.286	0.331	0.375
		3.4	0.5	0.077	0.153	0.225	0.296	0.365	0.432	0.498	0.562
			0.7	0.109	0.213	0.313	0.409	0.501	0.591	0.678	0.762
		3.6	0.5	0.195	0.376	0.545	0.704	0.854	0.997	1.132	1.261
			0.7	0.274	0.522	0.748	0.956	1.150	1.332	1.503	1.664

TABLE 18-III-X—DIFFERENTIAL SWELL OCCURRING AT THE PERIMETER OF A SLAB FOR AN EDGE LIFT SWELLING CONDITION IN PREDOMINANTLY ILLITE CLAY SOIL (60 PERCENT CLAY)

PERCENT CLAY (%)	DEPTH TO CONSTANT SUCTION (ft.) × 304.8 for mm	CONSTANT SUCTION (pF)	VELOCITY OF MOISTURE FLOW (inches/month) × 25.4 for mm/month	DIFFERENTIAL SWELL (inch) × 25.4 for mm Edge Distance Penetration × 304.8 for mm							
				1 ft.	2 ft.	3 ft.	4 ft.	5 ft.	6 ft.	7 ft.	8 ft.
60	3	3.2	0.5	0.012	0.023	0.034	0.045	0.056	0.067	0.077	0.087
			0.7	0.016	0.032	0.047	0.062	0.077	0.092	0.106	0.119
		3.4	0.5	0.025	0.048	0.071	0.093	0.114	0.135	0.155	0.174
			0.7	0.034	0.067	0.097	0.127	0.155	0.182	0.208	0.234
		3.6	0.5	0.060	0.114	0.165	0.212	0.256	0.297	0.337	0.374
			0.7	0.083	0.156	0.223	0.283	0.339	0.391	0.439	0.485
		3.8	0.5	0.143	0.260	0.361	0.450	0.529	0.601	0.667	0.728
			0.7	0.196	0.348	0.474	0.582	0.677	0.761	0.838	0.909
	5	3.2	0.5	0.025	0.050	0.075	0.099	0.123	0.147	0.170	0.193
			0.7	0.036	0.070	0.104	0.138	0.171	0.203	0.234	0.265
		3.4	0.5	0.055	0.107	0.158	0.208	0.256	0.303	0.348	0.393
			0.7	0.076	0.149	0.219	0.286	0.350	0.412	0.472	0.530
		3.6	0.5	0.135	0.260	0.376	0.485	0.588	0.685	0.777	0.864
			0.7	0.189	0.358	0.512	0.654	0.786	0.908	1.024	1.133
		3.8	0.5	0.333	0.611	0.851	1.064	1.255	1.430	1.591	1.740
			0.7	0.465	0.830	1.135	1.398	1.631	1.840	2.030	2.205
	7	3.2	0.5	0.044	0.088	0.130	0.173	0.214	0.255	0.296	0.336
			0.7	0.062	0.122	0.182	0.240	0.297	0.353	0.408	0.463
		3.4	0.5	0.096	0.188	0.278	0.366	0.451	0.534	0.615	0.694
			0.7	0.134	0.263	0.386	0.505	0.619	0.730	0.837	0.941
		3.6	0.5	0.241	0.465	0.674	0.870	1.055	1.231	1.398	0.558
			0.7	0.339	0.644	0.924	1.181	1.421	1.645	1.856	2.055

TABLE 18-III-Y—DIFFERENTIAL SWELL OCCURRING AT THE PERIMETER OF A SLAB FOR AN EDGE LIFT SWELLING CONDITION IN PREDOMINANTLY ILLITE CLAY SOIL (70 PERCENT CLAY)

PERCENT CLAY (%)	DEPTH TO CONSTANT SUCTION (ft.) × 304.8 for mm	CONSTANT SUCTION (pF)	VELOCITY OF MOISTURE FLOW (inches/month) × 25.4 for mm/month	DIFFERENTIAL SWELL (inch) × 25.4 for mm — Edge Distance Penetration × 304.8 for mm							
				1 ft.	2 ft.	3 ft.	4 ft.	5 ft.	6 ft.	7 ft.	8 ft.
70	3	3.2	0.5	0.014	0.027	0.041	0.054	0.067	0.079	0.092	0.104
			0.7	0.019	0.038	0.056	0.074	0.092	0.109	0.126	0.142
		3.4	0.5	0.029	0.057	0.084	0.111	0.136	0.160	0.184	0.207
			0.7	0.041	0.079	0.116	0.151	0.185	0.217	0.248	0.278
		3.6	0.5	0.071	0.136	0.196	0.252	0.305	0.354	0.401	0.445
			0.7	0.099	0.186	0.265	0.337	0.403	0.465	0.523	0.577
		3.8	0.5	0.170	0.310	0.430	0.535	0.630	0.715	0.794	0.866
			0.7	0.234	0.415	0.564	0.692	0.805	0.906	0.998	1.082
	5	3.2	0.5	0.030	0.060	0.089	0.118	0.147	0.175	0.202	0.230
			0.7	0.042	0.084	0.124	0.164	0.203	0.241	0.279	0.315
		3.4	0.5	0.065	0.128	0.188	0.247	0.305	0.360	0.414	0.467
			0.7	0.091	0.177	0.260	0.340	0.417	0.490	0.562	0.631
		3.6	0.5	0.161	0.309	0.448	0.577	0.699	0.815	0.925	1.029
			0.7	0.225	0.426	0.610	0.779	0.935	1.081	1.219	1.348
		3.8	0.5	0.396	0.727	1.013	1.266	1.494	1.702	1.894	2.072
			0.7	0.563	0.988	1.351	1.664	1.941	2.190	2.417	2.625
	7	3.2	0.5	0.062	0.104	0.155	0.205	0.255	0.304	0.352	0.400
			0.7	0.074	0.146	0.216	0.285	0.354	0.420	0.486	0.551
		3.4	0.5	0.114	0.224	0.331	0.436	0.537	0.636	0.732	0.826
			0.7	0.160	0.313	0.459	0.601	0.737	0.869	0.996	1.120
		3.6	0.5	0.287	0.553	0.802	1.036	1.256	1.465	1.664	1.854
			0.7	0.403	0.767	1.099	1.406	1.691	1.958	2.209	2.447

TABLE 18-III-Z—DIFFERENTIAL SWELL OCCURRING AT THE PERIMETER OF A SLAB FOR AN EDGE LIFT SWELLING CONDITION IN PREDOMINANTLY MONTMORILLONITE CLAY SOIL (30 PERCENT CLAY)

PERCENT CLAY (%)	DEPTH TO CONSTANT SUCTION (ft.) × 304.8 for mm	CONSTANT SUCTION (pF)	VELOCITY OF MOISTURE FLOW (inches/month) × 25.4 for mm/month	DIFFERENTIAL SWELL (inch) × 25.4 for mm — Edge Distance Penetration × 304.8 for mm							
				1 ft.	2 ft.	3 ft.	4 ft.	5 ft.	6 ft.	7 ft.	8 ft.
30	3	3.2	0.5	0.006	0.012	0.017	0.023	0.028	0.034	0.039	0.044
			0.7	0.008	0.016	0.024	0.032	0.039	0.046	0.053	0.060
		3.4	0.5	0.012	0.024	0.036	0.047	0.058	0.068	0.078	0.088
			0.7	0.017	0.034	0.049	0.064	0.078	0.092	0.105	0.118
		3.6	0.5	0.030	0.058	0.083	0.107	0.129	0.150	0.170	0.189
			0.7	0.042	0.079	0.112	0.143	0.171	0.197	0.222	0.245
		3.8	0.5	0.072	0.132	0.182	0.227	0.267	0.303	0.337	0.368
			0.7	0.099	0.176	0.239	0.294	0.342	0.385	0.423	0.459
	5	3.2	0.5	0.013	0.026	0.038	0.050	0.062	0.074	0.086	0.098
			0.7	0.018	0.036	0.053	0.070	0.086	0.102	0.118	0.134
		3.4	0.5	0.028	0.054	0.080	0.105	0.129	0.153	0.176	0.198
			0.7	0.039	0.075	0.110	0.144	0.177	0.208	0.238	0.268
		3.6	0.5	0.068	0.131	0.190	0.245	0.297	0.346	0.392	0.437
			0.7	0.095	0.181	0.259	0.330	0.397	0.459	0.517	0.572
		3.8	0.5	0.168	0.308	0.430	0.537	0.634	0.722	0.804	0.879
			0.7	0.235	0.419	0.573	0.706	0.824	0.929	1.025	1.114
	7	3.2	0.5	0.022	0.044	0.066	0.087	0.108	0.129	0.150	0.170
			0.7	0.031	0.062	0.092	0.121	0.150	0.178	0.206	0.234
		3.4	0.5	0.048	0.095	0.141	0.185	0.228	0.270	0.311	0.361
			0.7	0.068	0.133	0.195	0.255	0.313	0.369	0.423	0.475
		3.6	0.5	0.122	0.235	0.340	0.439	0.533	0.622	0.706	0.787
			0.7	0.171	0.326	0.466	0.597	0.718	0.831	0.937	1.038

TABLE 18-III-AA
TABLE 18-III-BB

1997 UNIFORM BUILDING CODE

TABLE 18-III-AA—DIFFERENTIAL SWELL OCCURRING AT THE PERIMETER OF A SLAB FOR AN EDGE LIFT SWELLING CONDITION IN PREDOMINANTLY MONTMORILLONITE CLAY SOIL (40 PERCENT CLAY)

PERCENT CLAY (%)	DEPTH TO CONSTANT SUCTION (ft.) × 304.8 for mm	CONSTANT SUCTION (pF)	VELOCITY OF MOISTURE FLOW (inches/month) × 25.4 for mm/month	DIFFERENTIAL SWELL (inch) × 25.4 for mm — Edge Distance Penetration × 304.8 for mm							
				1 ft.	2 ft.	3 ft.	4 ft.	5 ft.	6 ft.	7 ft.	8 ft.
40	3	3.2	0.5	0.009	0.017	0.025	0.033	0.041	0.049	0.056	0.064
			0.7	0.012	0.023	0.035	0.046	0.056	0.067	0.077	0.087
		3.4	0.5	0.018	0.035	0.052	0.068	0.084	0.099	0.113	0.128
			0.7	0.025	0.049	0.071	0.093	0.114	0.133	0.153	0.171
		3.6	0.5	0.044	0.084	0.121	0.155	0.187	0.218	0.246	0.274
			0.7	0.061	0.114	0.163	0.207	0.248	0.286	0.321	0.355
		3.8	0.5	0.105	0.191	0.264	0.329	0.387	0.440	0.488	0.533
			0.7	0.144	0.255	0.347	0.426	0.495	0.557	0.614	0.665
	5	3.2	0.5	0.019	0.037	0.055	0.073	0.090	0.107	0.125	0.141
			0.7	0.026	0.052	0.076	0.101	0.125	0.148	0.171	0.194
		3.4	0.5	0.040	0.078	0.116	0.152	0.187	0.221	0.255	0.287
			0.7	0.056	0.109	0.160	0.209	0.256	0.302	0.345	0.388
		3.6	0.5	0.099	0.190	0.275	0.355	0.430	0.501	0.568	0.633
			0.7	0.138	0.262	0.375	0.479	0.575	0.665	0.749	0.829
		3.8	0.5	0.244	0.447	0.623	0.779	0.919	1.047	1.164	1.274
			0.7	0.340	0.607	0.831	1.023	1.193	1.347	1.486	1.614
	7	3.2	0.5	0.032	0.064	0.095	0.126	0.157	0.187	0.217	0.246
			0.7	0.045	0.089	0.133	0.176	0.217	0.258	0.299	0.339
		3.4	0.5	0.070	0.138	0.204	0.268	0.330	0.391	0.450	0.508
			0.7	0.098	0.192	0.283	0.369	0.453	0.534	0.613	0.689
		3.6	0.5	0.176	0.340	0.493	0.637	0.772	0.901	1.023	1.140
			0.7	0.248	0.472	0.676	0.864	1.040	1.204	1.358	1.504

TABLE 18-III-BB—DIFFERENTIAL SWELL OCCURRING AT THE PERIMETER OF A SLAB FOR AN EDGE LIFT SWELLING CONDITION IN PREDOMINANTLY MONTMORILLONITE CLAY SOIL (50 PERCENT CLAY)

PERCENT CLAY (%)	DEPTH TO CONSTANT SUCTION (ft.) × 304.8 for mm	CONSTANT SUCTION (pF)	VELOCITY OF MOISTURE FLOW (inches/month) × 25.4 for mm/month	DIFFERENTIAL SWELL (inch) × 25.4 for mm — Edge Distance Penetration × 304.8 for mm							
				1 ft.	2 ft.	3 ft.	4 ft.	5 ft.	6 ft.	7 ft.	8 ft.
50	3	3.2	0.5	0.011	0.022	0.033	0.043	0.054	0.064	0.074	0.084
			0.7	0.016	0.031	0.045	0.060	0.074	0.088	0.101	0.115
		3.4	0.5	0.024	0.046	0.068	0.089	0.109	0.129	0.148	0.167
			0.7	0.033	0.064	0.093	0.122	0.149	0.175	0.200	0.224
		3.6	0.5	0.057	0.110	0.158	0.203	0.245	0.285	0.323	0.358
			0.7	0.079	0.150	0.213	0.271	0.325	0.375	0.421	0.465
		3.8	0.5	0.137	0.250	0.346	0.431	0.507	0.576	0.639	0.698
			0.7	0.188	0.334	0.454	0.558	0.649	0.730	0.804	0.871
	5	3.2	0.5	0.024	0.048	0.072	0.095	0.118	0.141	0.163	0.185
			0.7	0.034	0.067	0.100	0.132	0.163	0.194	0.224	0.254
		3.4	0.5	0.052	0.103	0.152	0.199	0.245	0.290	0.334	0.376
			0.7	0.073	0.143	0.210	0.274	0.335	0.395	0.452	0.508
		3.6	0.5	0.130	0.249	0.361	0.465	0.563	0.656	0.745	0.829
			0.7	0.181	0.343	0.491	0.627	0.753	0.871	0.981	1.086
		3.8	0.5	0.319	0.585	0.816	1.020	1.204	1.371	1.525	1.668
			0.7	0.445	0.796	1.088	1.340	1.563	1.764	1.946	2.114
	7	3.2	0.5	0.042	0.084	0.125	0.165	0.205	0.245	0.284	0.322
			0.7	0.059	0.117	0.174	0.230	0.285	0.339	0.391	0.443
		3.4	0.5	0.092	0.181	0.267	0.351	0.433	0.512	0.590	0.665
			0.7	0.129	0.252	0.370	0.484	0.594	0.700	0.802	0.902
		3.6	0.5	0.231	0.446	0.646	0.834	1.012	1.180	1.341	1.494
			0.7	0.325	0.618	0.885	1.132	1.362	1.577	1.779	1.970

TABLE 18-III-CC—DIFFERENTIAL SWELL OCCURRING AT THE PERIMETER OF A SLAB FOR AN EDGE LIFT SWELLING CONDITION IN PREDOMINANTLY MONTMORILLONITE CLAY SOIL (60 PERCENT CLAY)

PERCENT CLAY (%)	DEPTH TO CONSTANT SUCTION (ft.) × 304.8 for mm	CONSTANT SUCTION (pF)	VELOCITY OF MOISTURE FLOW (inches/month) × 25.4 for mm/month	DIFFERENTIAL SWELL (inch) × 25.4 for mm — Edge Distance Penetration × 304.8 for mm							
				1 ft.	2 ft.	3 ft.	4 ft.	5 ft.	6 ft.	7 ft.	8 ft.
60	3	3.2	0.5	0.014	0.027	0.041	0.054	0.066	0.079	0.091	0.104
			0.7	0.019	0.038	0.056	0.074	0.091	0.109	0.125	0.142
		3.4	0.5	0.029	0.057	0.084	0.110	0.135	0.160	0.183	0.206
			0.7	0.041	0.079	0.116	0.151	0.184	0.216	0.247	0.277
		3.6	0.5	0.071	0.136	0.195	0.251	0.303	0.352	0.399	0.433
			0.7	0.098	0.186	0.264	0.336	0.402	0.463	0.521	0.575
		3.8	0.5	0.169	0.309	0.428	0.533	0.627	0.712	0.790	0.863
			0.7	0.233	0.413	0.562	0.690	0.802	0.903	0.994	1.077
	5	3.2	0.5	0.030	0.060	0.090	0.118	0.146	0.174	0.202	0.229
			0.7	0.042	0.083	0.124	0.163	0.202	0.240	0.278	0.314
		3.4	0.5	0.065	0.127	0.188	0.246	0.303	0.359	0.413	0.465
			0.7	0.090	0.177	0.259	0.339	0.415	0.488	0.559	0.628
		3.6	0.5	0.160	0.308	0.446	0.575	0.697	0.812	0.921	1.025
			0.7	0.224	0.425	0.607	0.775	0.931	1.077	1.214	1.343
		3.8	0.5	0.395	0.724	1.009	1.261	1.488	1.695	1.886	2.063
			0.7	0.551	0.984	1.345	1.657	1.933	2.181	2.407	2.614
	7	3.2	0.5	0.052	0.104	0.155	0.205	0.254	0.303	0.351	0.398
			0.7	0.073	0.145	0.215	0.284	0.352	0.419	0.484	0.548
		3.4	0.5	0.113	0.223	0.330	0.434	0.535	0.633	0.729	0.823
			0.7	0.159	0.311	0.458	0.598	0.734	0.865	0.992	1.115
		3.6	0.5	0.286	0.551	0.799	1.031	1.251	1.459	1.658	1.847
			0.7	0.402	0.764	1.095	1.400	1.684	1.950	2.200	2.437

TABLE 18-III-DD—DIFFERENTIAL SWELL OCCURRING AT THE PERIMETER OF A SLAB FOR AN EDGE LIFT SWELLING CONDITION IN PREDOMINANTLY MONTMORILLONITE CLAY SOIL (70 PERCENT CLAY)

PERCENT CLAY (%)	DEPTH TO CONSTANT SUCTION (ft.) × 304.8 for mm	CONSTANT SUCTION (pF)	VELOCITY OF MOISTURE FLOW (inches/month) × 25.4 for mm/month	DIFFERENTIAL SWELL (inch) × 25.4 for mm — Edge Distance Penetration × 304.8 for mm							
				1 ft.	2 ft.	3 ft.	4 ft.	5 ft.	6 ft.	7 ft.	8 ft.
70	3	3.2	0.5	0.016	0.032	0.048	0.064	0.079	0.094	0.109	0.123
			0.7	0.023	0.045	0.067	0.088	0.109	0.129	0.149	0.169
		3.4	0.5	0.035	0.068	0.100	0.131	0.161	0.190	0.219	0.246
			0.7	0.048	0.094	0.138	0.179	0.219	0.258	0.294	0.330
		3.6	0.5	0.084	0.162	0.233	0.299	0.361	0.420	0.475	0.528
			0.7	0.117	0.221	0.314	0.400	0.479	0.552	0.620	0.684
		3.8	0.5	0.202	0.368	0.510	0.635	0.747	0.849	0.942	1.028
			0.7	0.277	0.492	0.669	0.822	0.955	1.075	1.184	1.284
	5	3.2	0.5	0.036	0.071	0.106	0.140	0.174	0.207	0.240	0.273
			0.7	0.050	0.099	0.147	0.195	0.241	0.286	0.331	0.374
		3.4	0.5	0.077	0.151	0.223	0.293	0.361	0.427	0.492	0.554
			0.7	0.108	0.210	0.309	0.403	0.494	0.582	0.666	0.748
		3.6	0.5	0.191	0.367	0.531	0.685	0.830	0.967	1.097	1.221
			0.7	0.267	0.506	0.724	0.924	1.110	1.283	1.446	1.600
		3.8	0.5	0.470	0.862	1.262	1.502	1.773	2.020	2.247	2.458
			0.7	0.656	1.172	1.603	1.974	2.303	2.598	2.867	3.114
	7	3.2	0.5	0.062	0.124	0.184	0.244	0.303	0.361	0.418	0.475
			0.7	0.087	0.173	0.256	0.339	0.419	0.499	0.577	0.653
		3.4	0.5	0.135	0.266	0.393	0.517	0.637	0.754	0.869	0.980
			0.7	0.189	0.371	0.545	0.713	0.875	1.031	1.182	1.329
		3.6	0.5	0.341	0.656	0.951	1.229	1.490	1.739	1.975	2.200
			0.7	0.479	0.910	1.304	1.668	2.006	2.323	2.621	2.903

TABLE 18-III-EE
TABLE 18-III-GG

1997 UNIFORM BUILDING CODE

TABLE 18-III-EE—COMPARISON OF METHODS OF DETERMINING CATION EXCHANGE CAPACITY

SOIL SAMPLE	CATION EXCHANGE CAPACITY (meq/100gm)	
	Atomic Absorption	Spectrophotometer[1]
01 - 01	21.1	20.2
31 - 02	28.2	26.2
53 - 05	14.7	7.0
72 - 06	71.4	72.8
73 - 06	21.6	18.9
86 - 08	45.0	50.0

[1]Bausch & Lomb "Spectronic-20."

TABLE 18-III-FF—COMPARISON OF CLAY MINERAL DETERMINATION METHODS

SOIL SAMPLE	PERCENT CLAY	ATTERBERG LIMITS		C.E.C. (meq./100 gm)		AC	CEAC		PREDOMINANT CLAY MINERAL		
		PL	PI	Flame Photometer	Correlation Equation		Flame Photometer	Correlation Equation	Flame Photometer	Correlation Equation	X-ray Defraction Analysis
31-02	33.5	16.5	26.6	28.2	26.6	0.79	0.84	0.80	Smectite	Smectite	Smectite
72-06	50.0	32.5	41.8	71.4	58.7	0.84	1.43	1.17	Smectite	Smectite	Smectite
86-08	47.0	25.1	36.4	45.0	43.4	0.77	0.96	0.92	Smectite	Smectite	Smectite

TABLE 18-III-GG—SAMPLE VALUES C_Δ

MATERIAL	CENTER LIFT	EDGE LIFT
Wood Frame	240	480
Stucco or Plaster	360	720
Brick Veneer	480	960
Concrete Masonry Units	960	1,920
Prefab Roof Trusses[1]	1,000	2,000

[1]Trusses that clearspan the full length or width of the foundation from edge to edge.

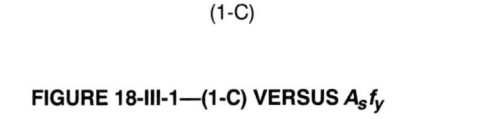

FIGURE 18-III-1—(1-C) VERSUS $A_s f_y$

FIGURE 18-III-2
FIGURE 18-III-3

1997 UNIFORM BUILDING CODE

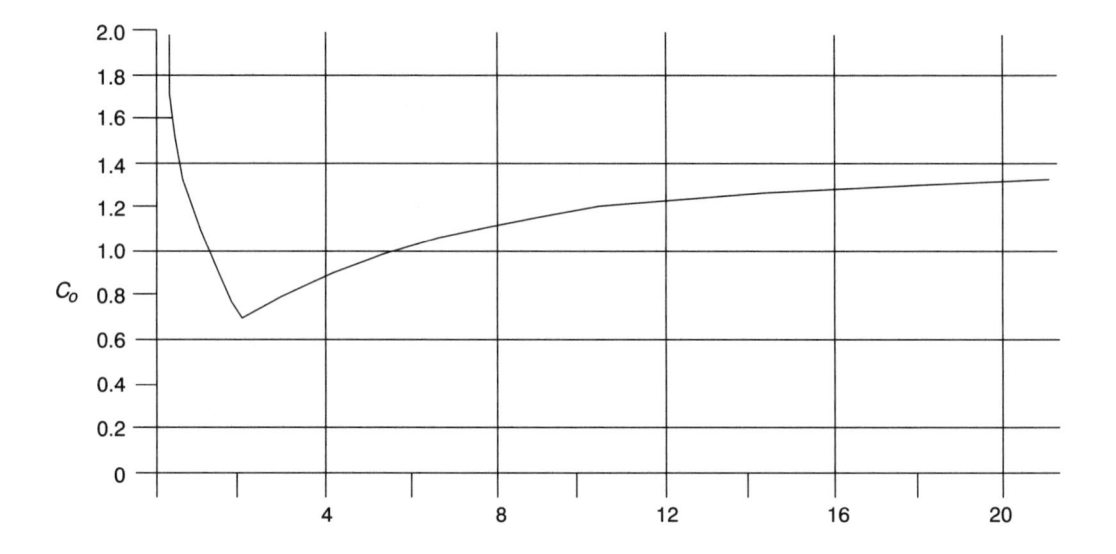

UNCONFINED COMPRESSIVE STRENGTH (q_u) KSF (x 47.9 for kPa)

**FIGURE 18-III-2—UNCONFINED COMPRESSIVE STRENGTH VERSUS
OVERCONSOLIDATED CORRECTION COEFFICIENT**

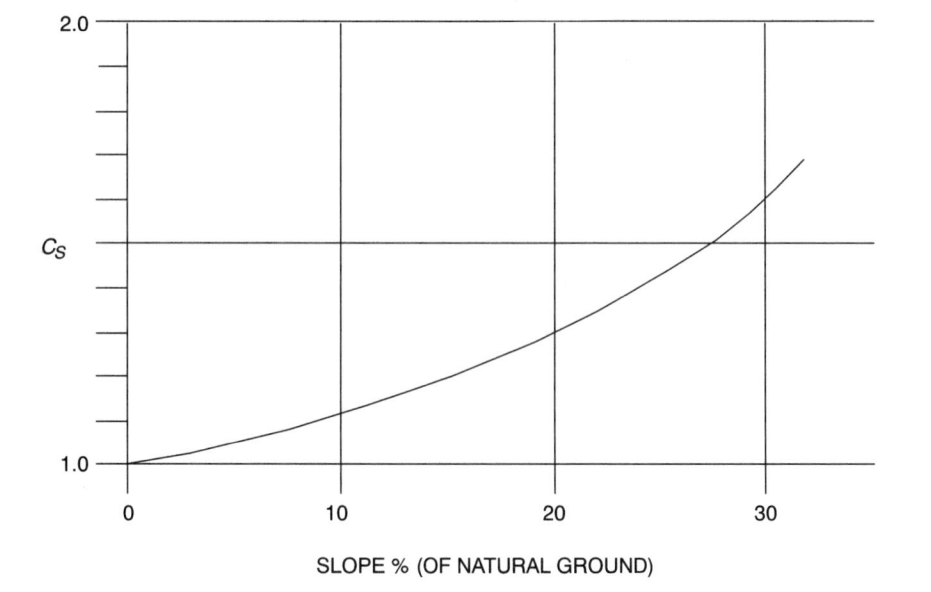

SLOPE % (OF NATURAL GROUND)

FIGURE 18-III-3—SLOPE OF NATURAL GROUND VERSUS SLOPE CORRECTION COEFFICIENT

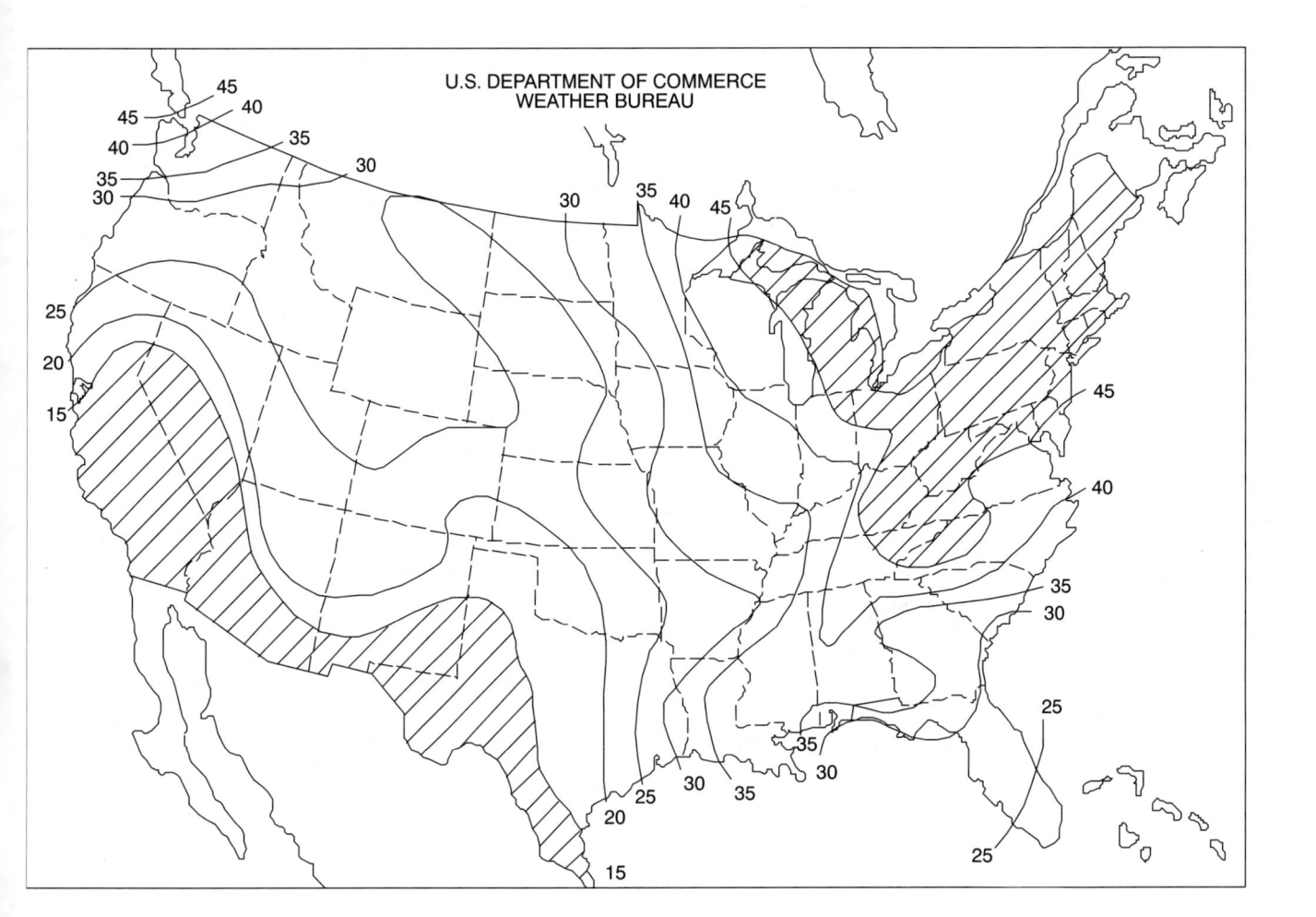

FIGURE 18-III-4—CLIMATIC RATING *(C_W)* CHART

FIGURE 18-III-5

1997 UNIFORM BUILDING CODE

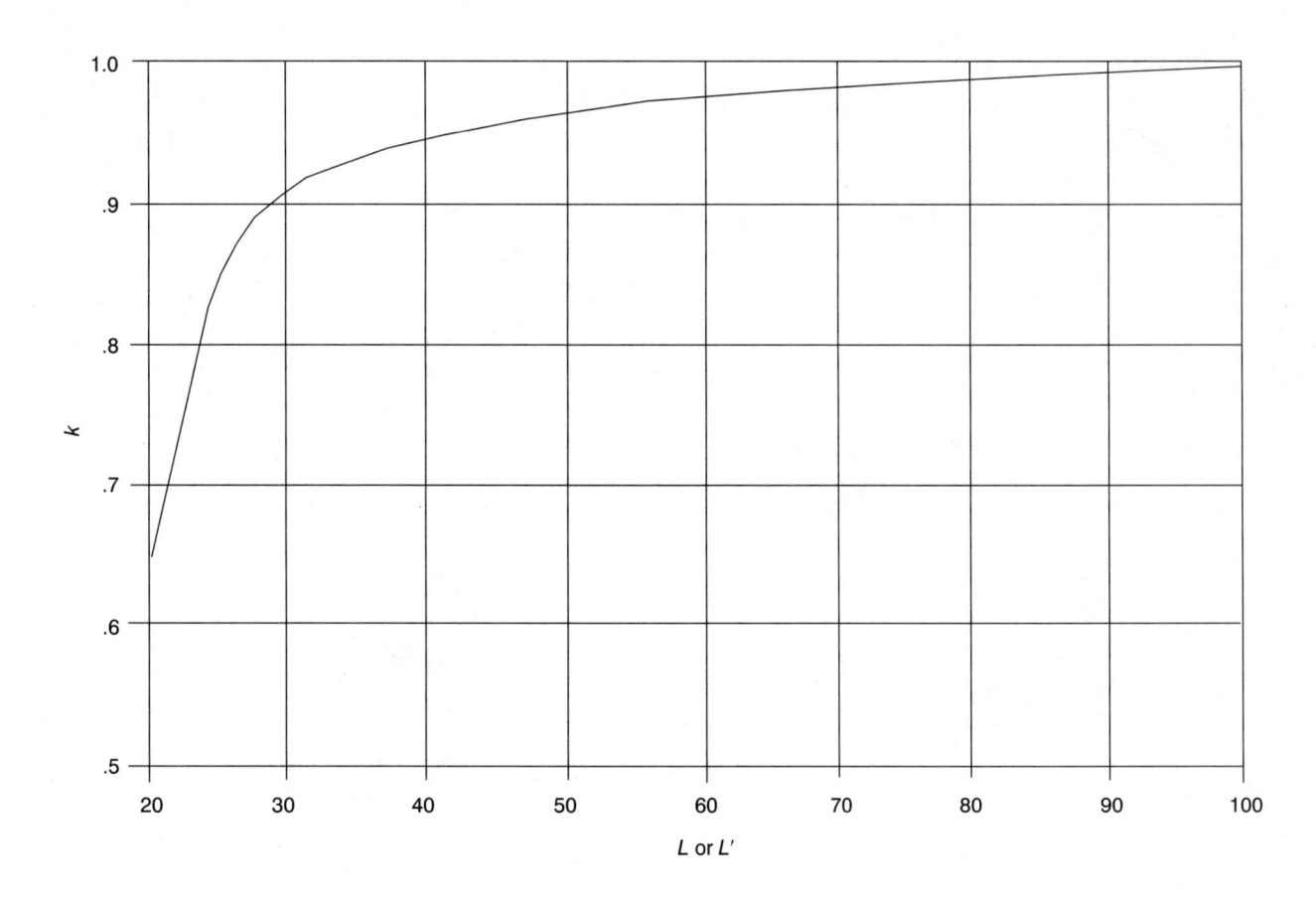

FIGURE 18-III-5—*L* or *L′* VERSUS *k*

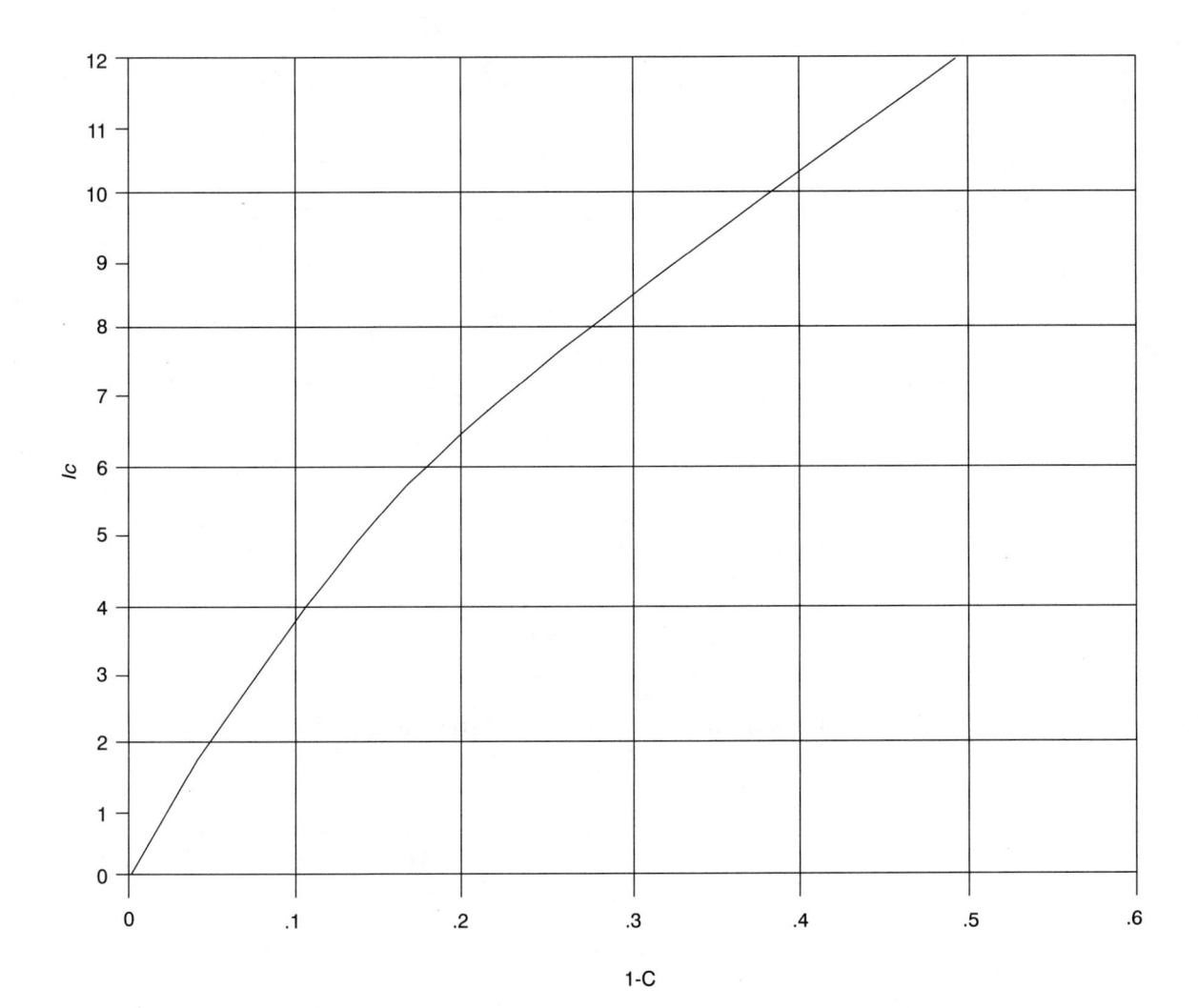

1-C

FIGURE 18-III-6—1-C VERSUS CANTILEVER LENGTH (*lc*)

FIGURE 18-III-7

1997 UNIFORM BUILDING CODE

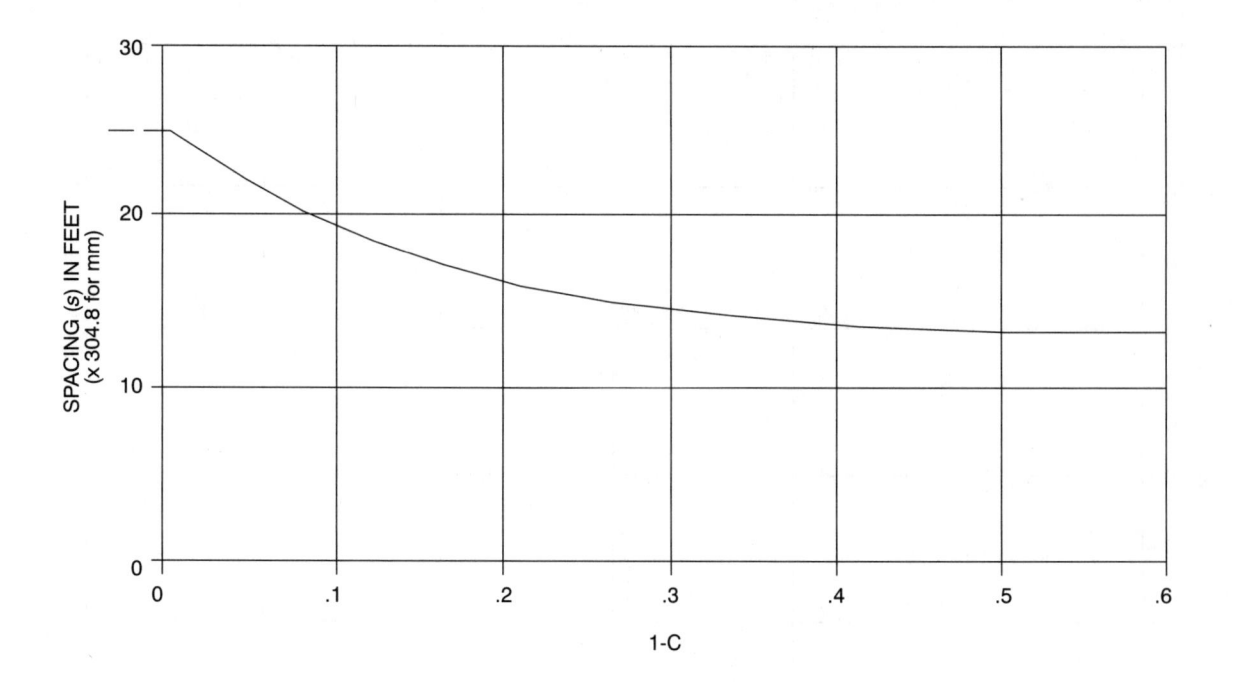

FIGURE 18-III-7—1-C VERSUS MAXIMUM BEAM SPACING

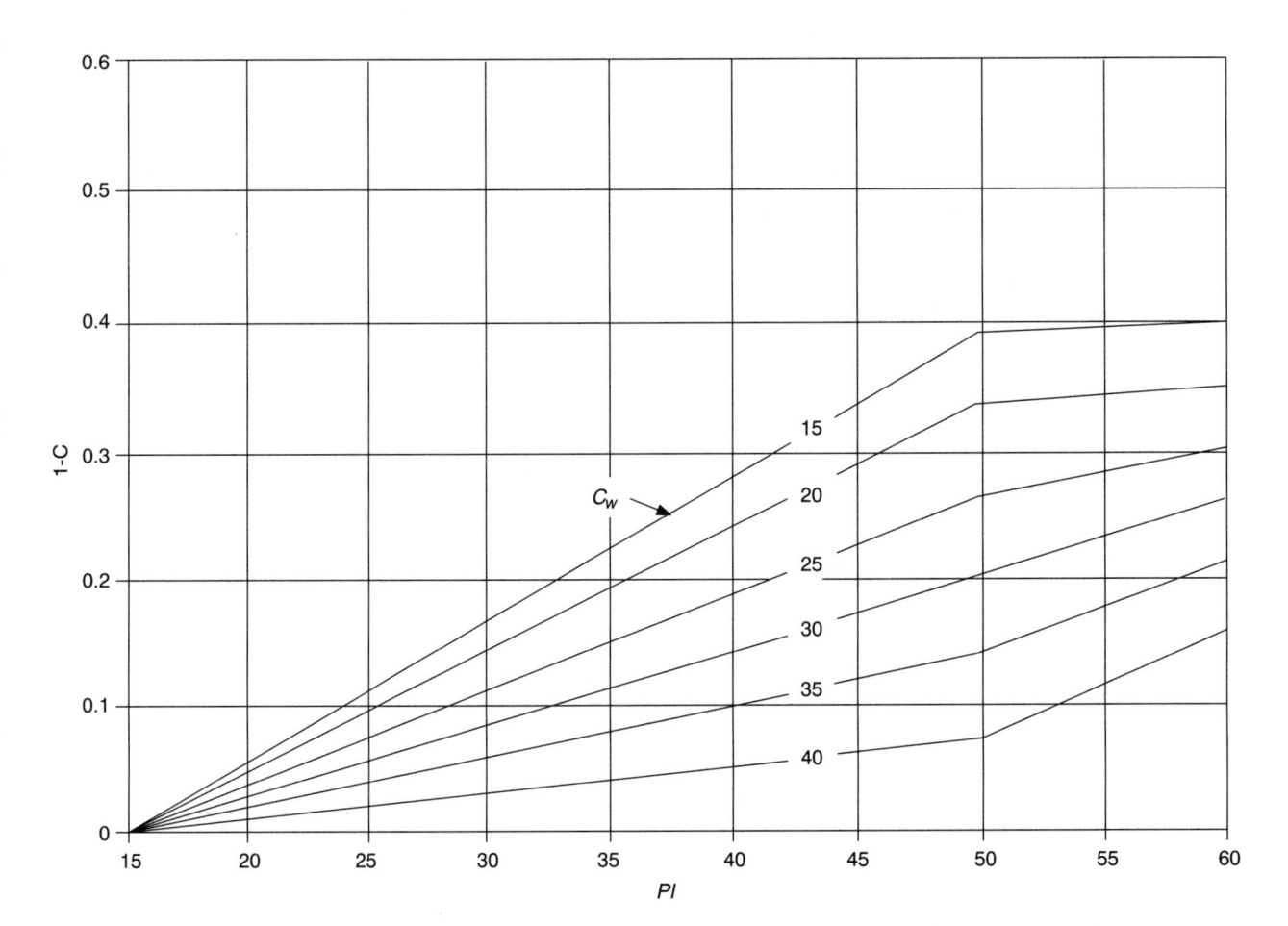

FIGURE 18-III-8—*PI* VERSUS (1-C)

FIGURE 18-III-9   **1997 UNIFORM BUILDING CODE**

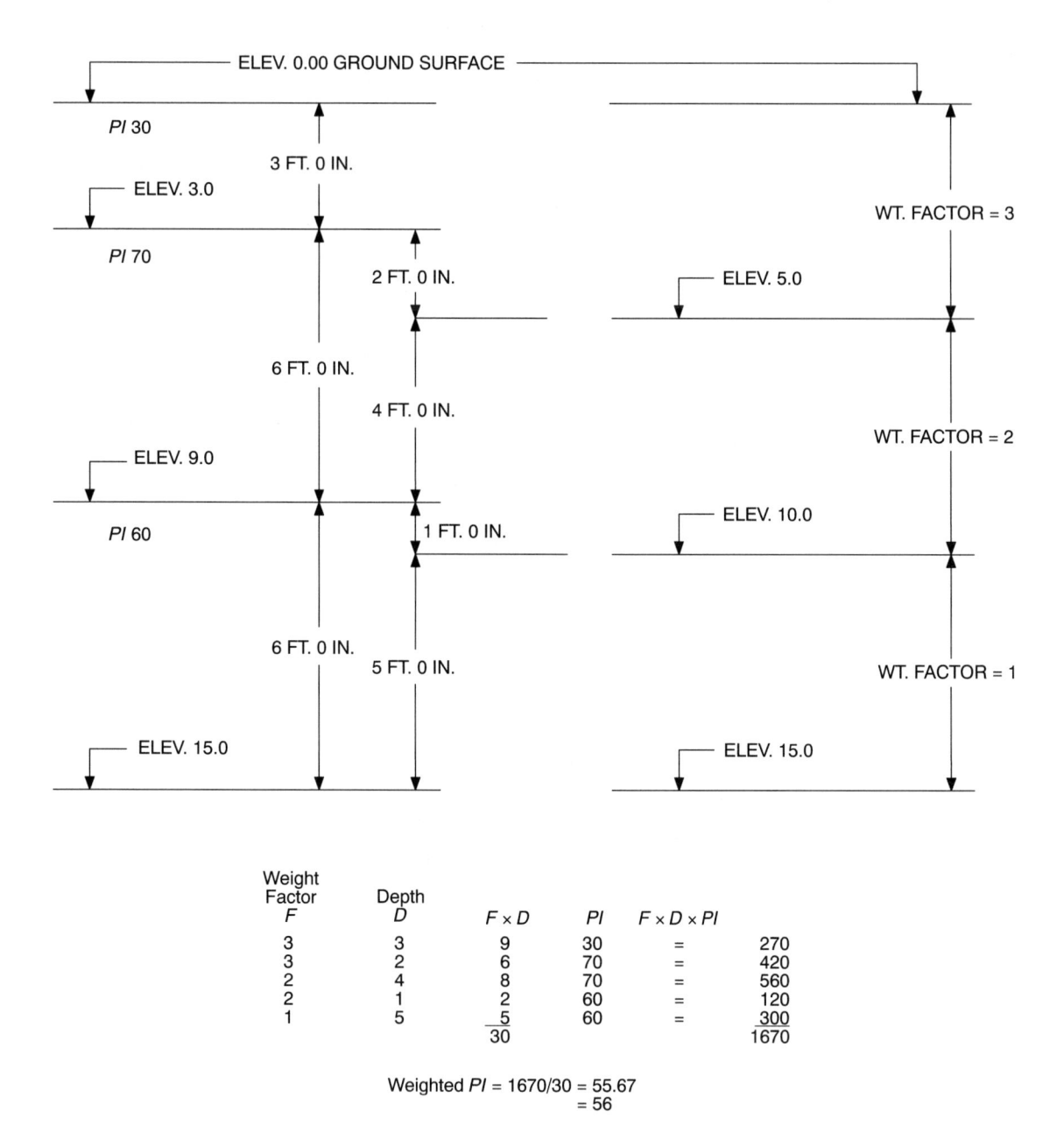

Weight Factor F	Depth D	$F \times D$	PI	$F \times D \times PI$	
3	3	9	30	=	270
3	2	6	70	=	420
2	4	8	70	=	560
2	1	2	60	=	120
1	5	5	60	=	300
		30			1670

Weighted PI = 1670/30 = 55.67
= 56

FIGURE 18-III-9—DETERMINING THE WEIGHTED PLASTICITY INDEX (PI)

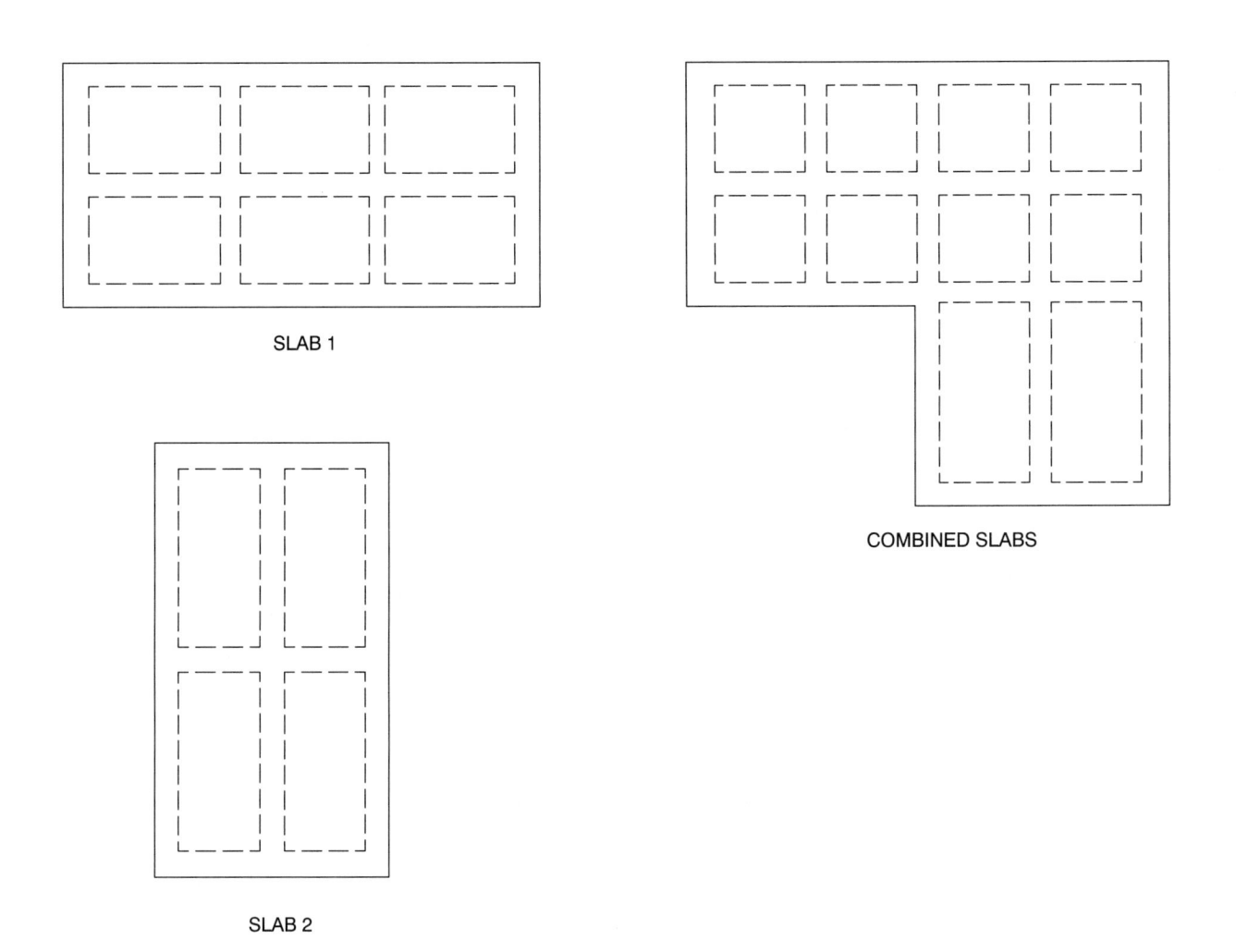

SLAB 1

SLAB 2

COMBINED SLABS

FIGURE 18-III-10—SLAB SEGMENTS AND COMBINED

FIGURE 18-III-11

1997 UNIFORM BUILDING CODE

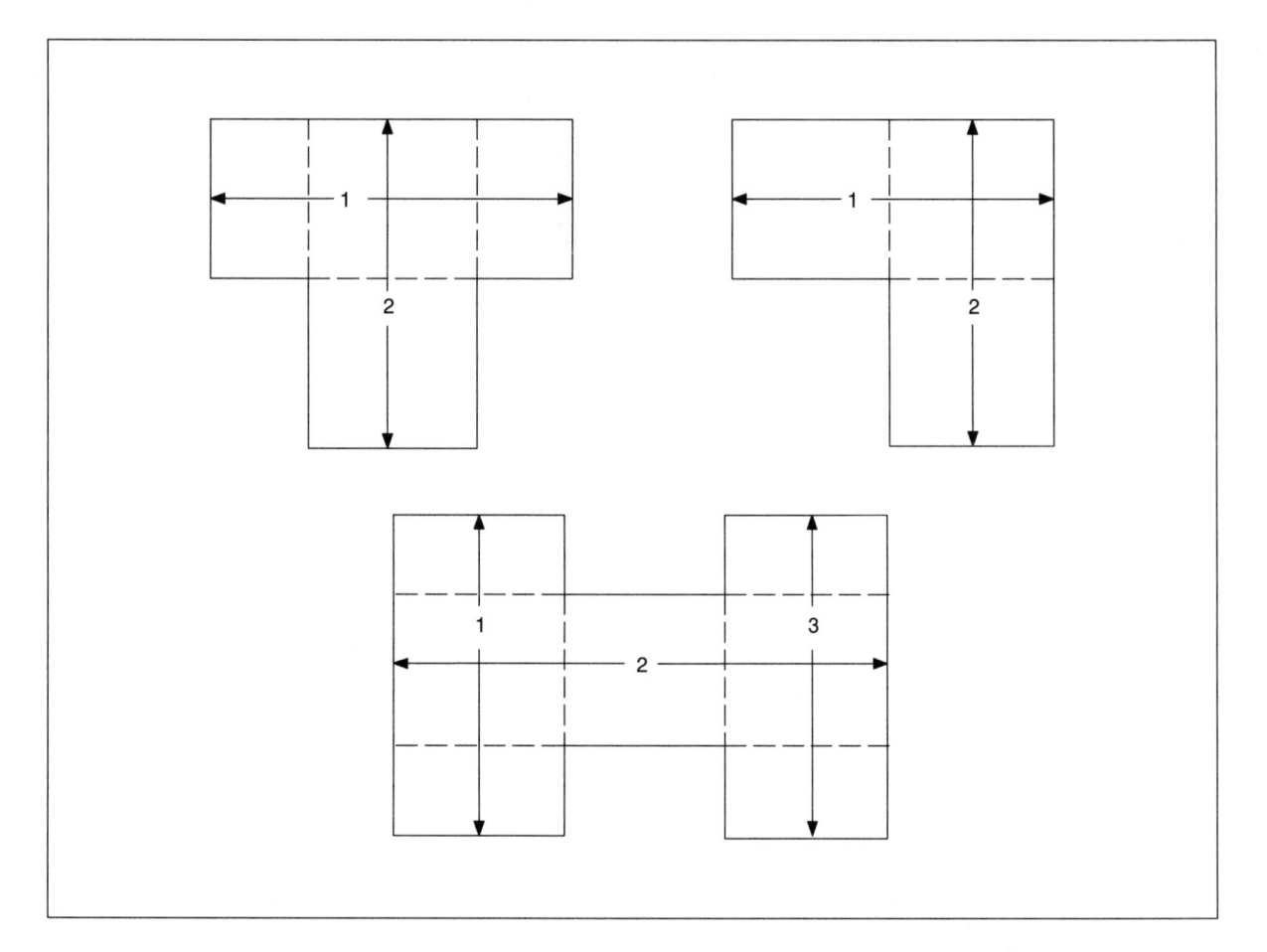

FIGURE 18-III-11—DESIGN RECTANGLES FOR SLABS OF IRREGULAR SHAPE

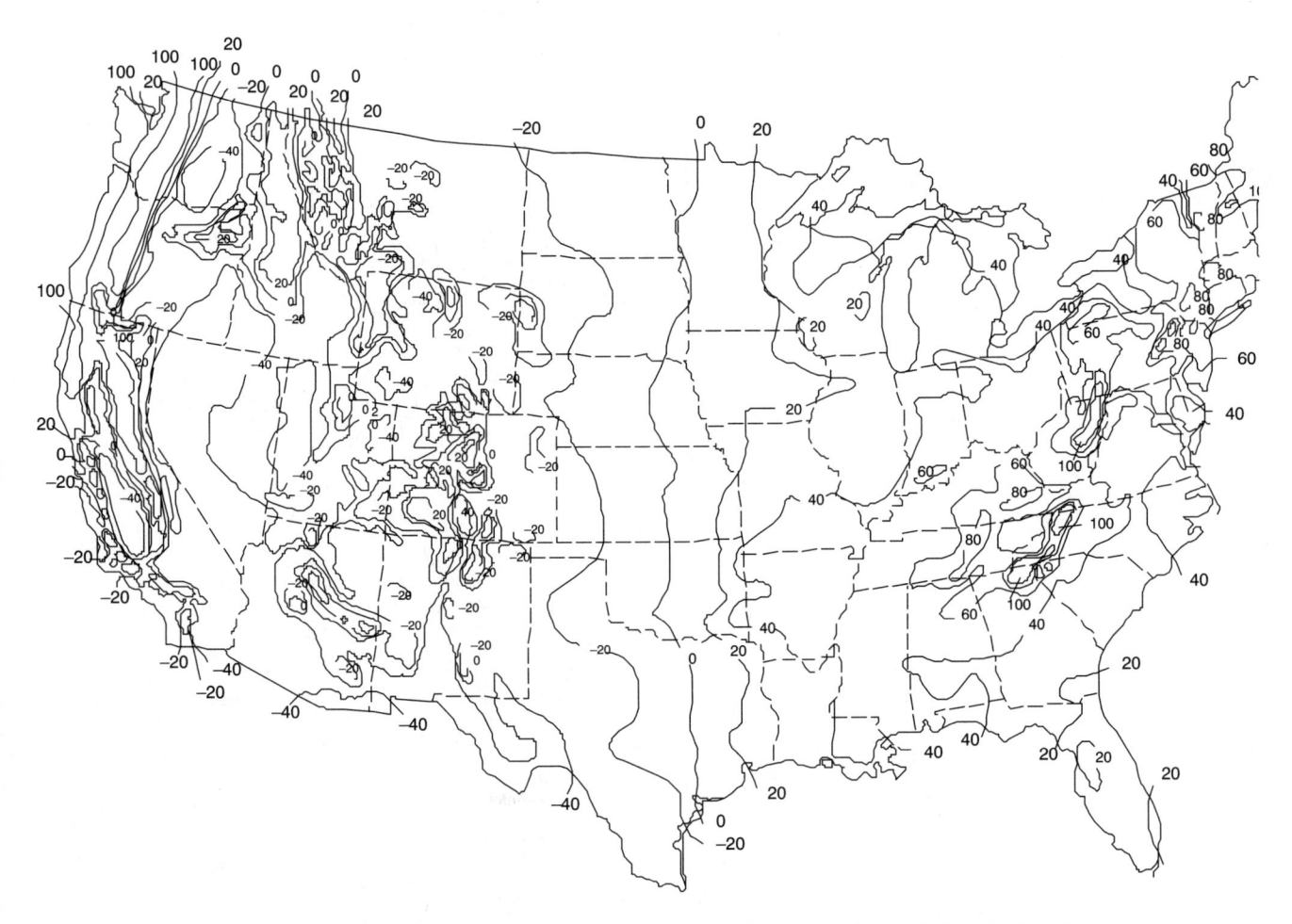

FIGURE 18-III-12—THORNTHWAITE MOISTURE INDEX DISTRIBUTION IN THE UNITED STATES

FIGURE 18-III-13-1

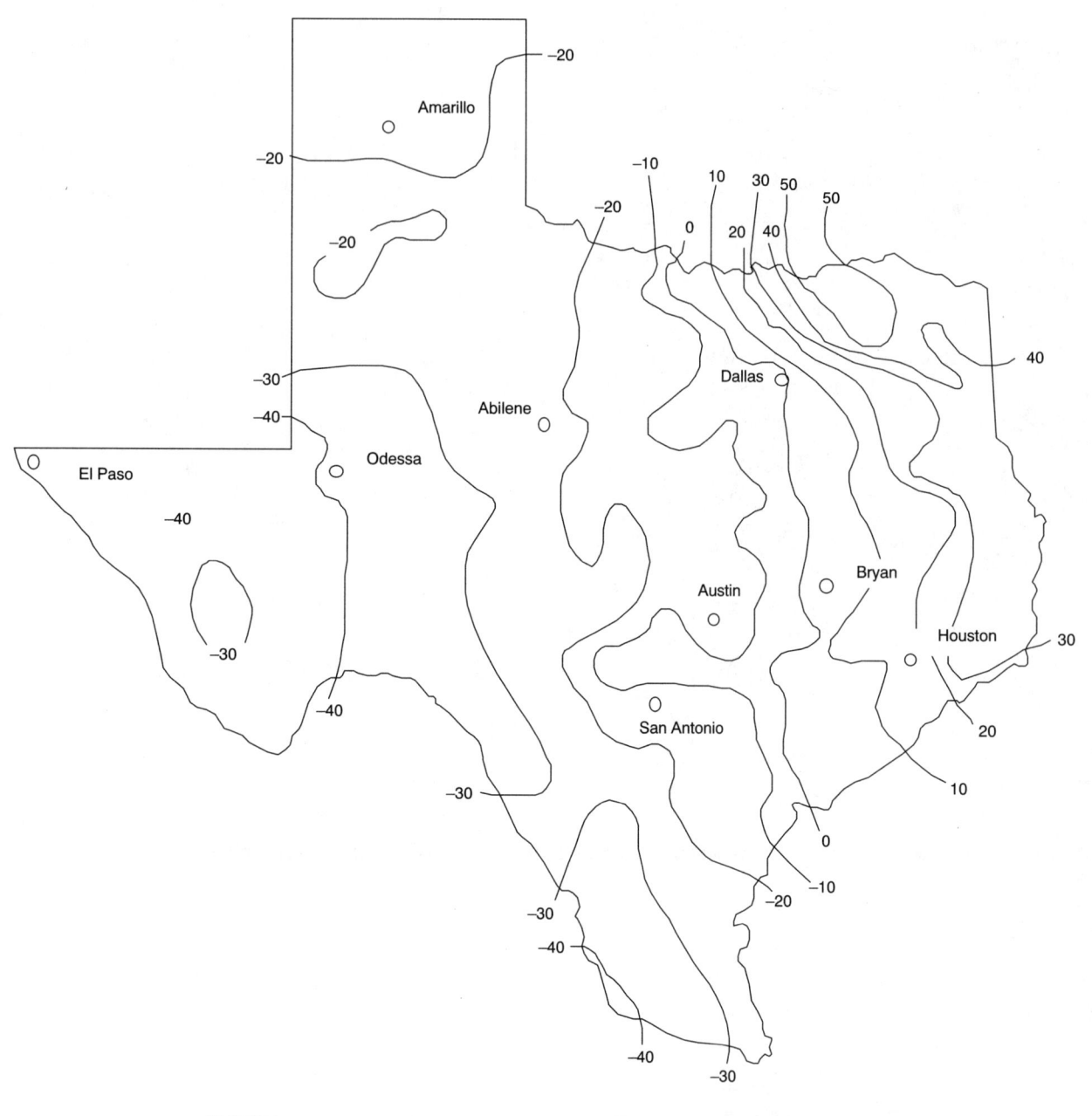

**FIGURE 18-III-13-1—THORNTHWAITE MOISTURE INDEX DISTRIBUTION FOR TEXAS
(20-YEAR AVERAGE, 1955-1974)**

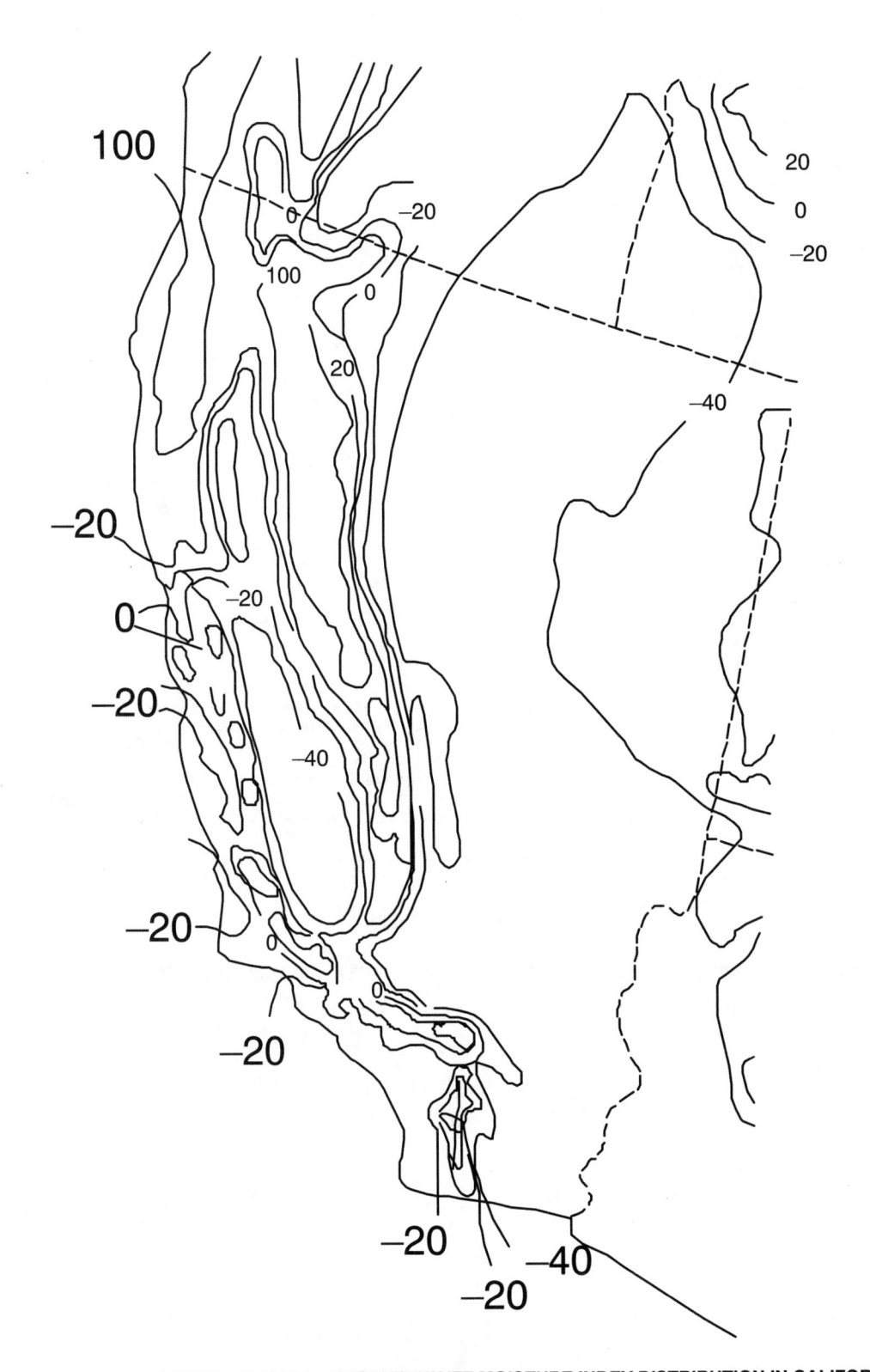

FIGURE 18-III-13-2—THORNTHWAITE MOISTURE INDEX DISTRIBUTION IN CALIFORNIA

FIGURE 18-III-14

1997 UNIFORM BUILDING CODE

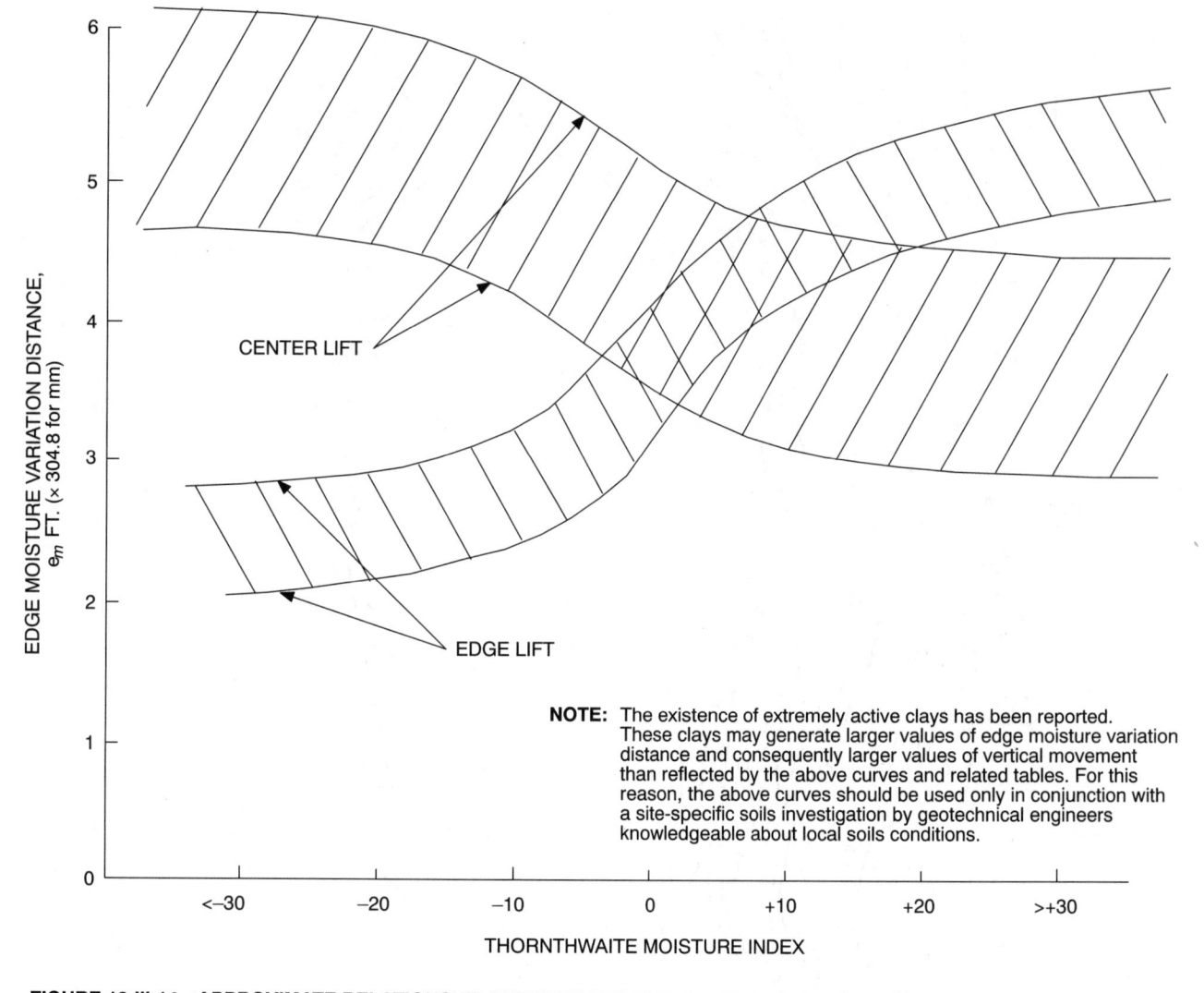

FIGURE 18-III-14—APPROXIMATE RELATIONSHIP BETWEEN THORNTHWAITE INDEX AND MOISTURE VARIATION DISTANCE

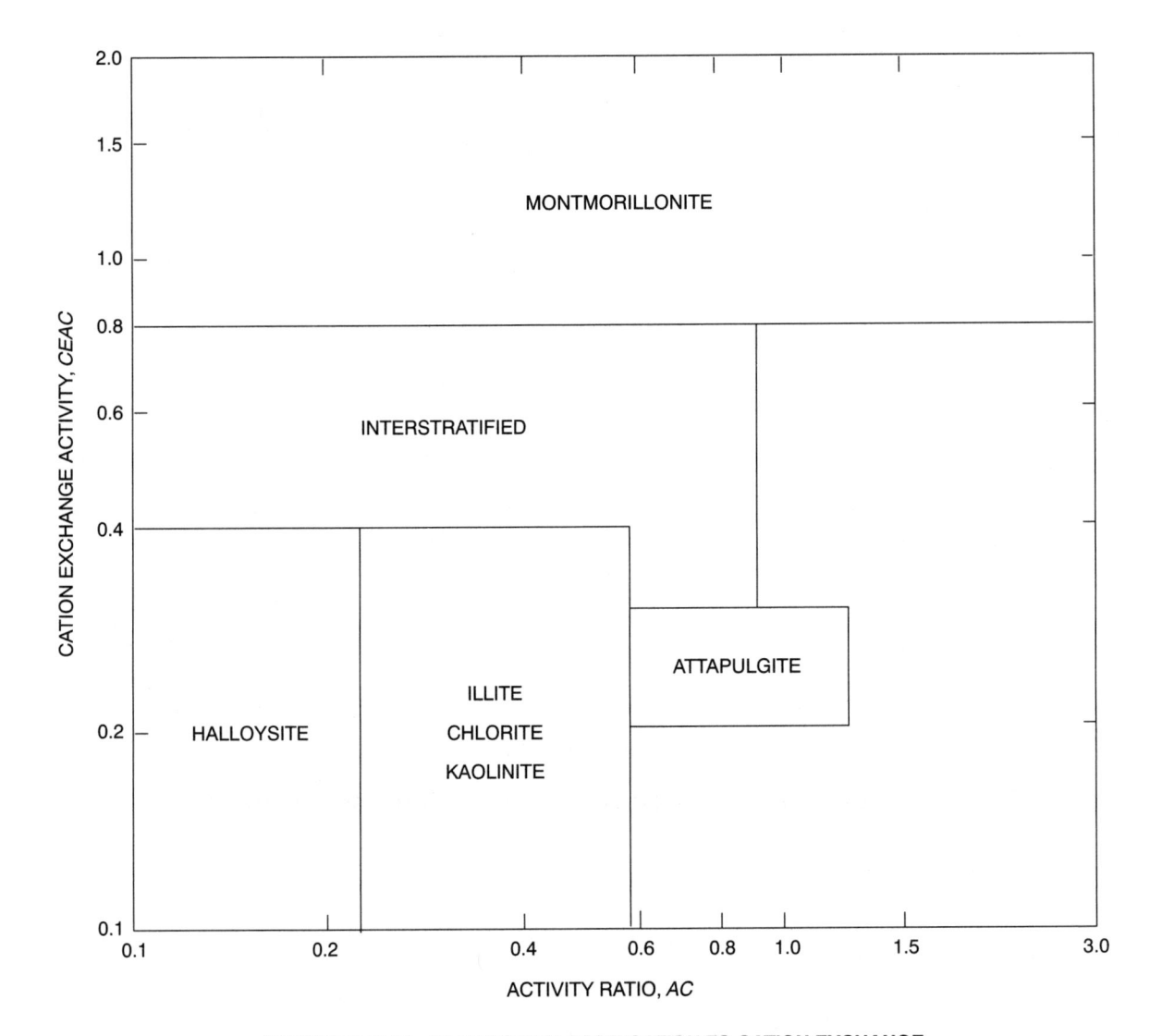

**FIGURE 18-III-15—CLAY TYPE CLASSIFICATION TO CATION EXCHANGE
AND CLAY ACTIVITY RATIO AFTER PEARRING AND HOLT**

FIGURE 18-III-16

1997 UNIFORM BUILDING CODE

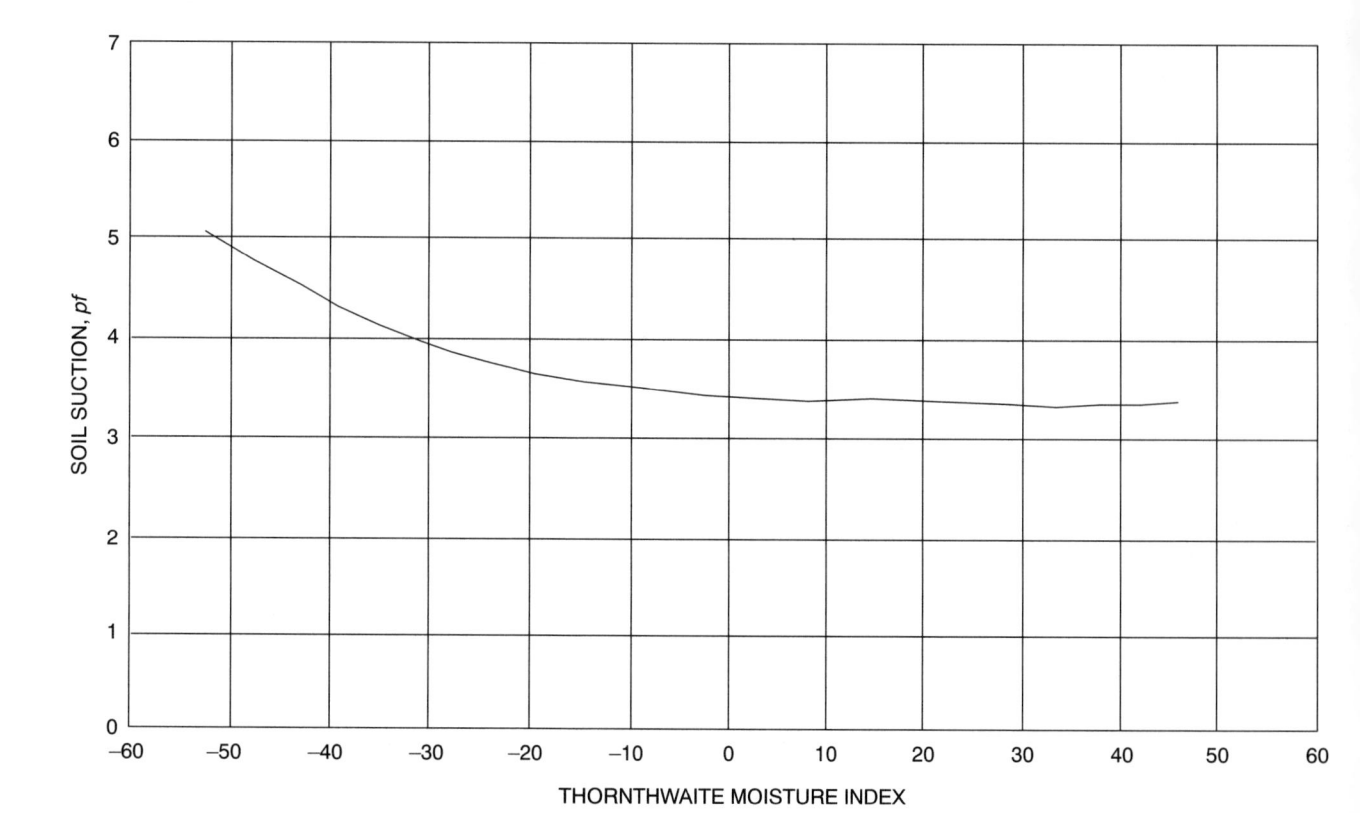

FIGURE 18-III-16—VARIATION OF CONSTANT SOIL SUCTION WITH THORNTHWAITE INDEX

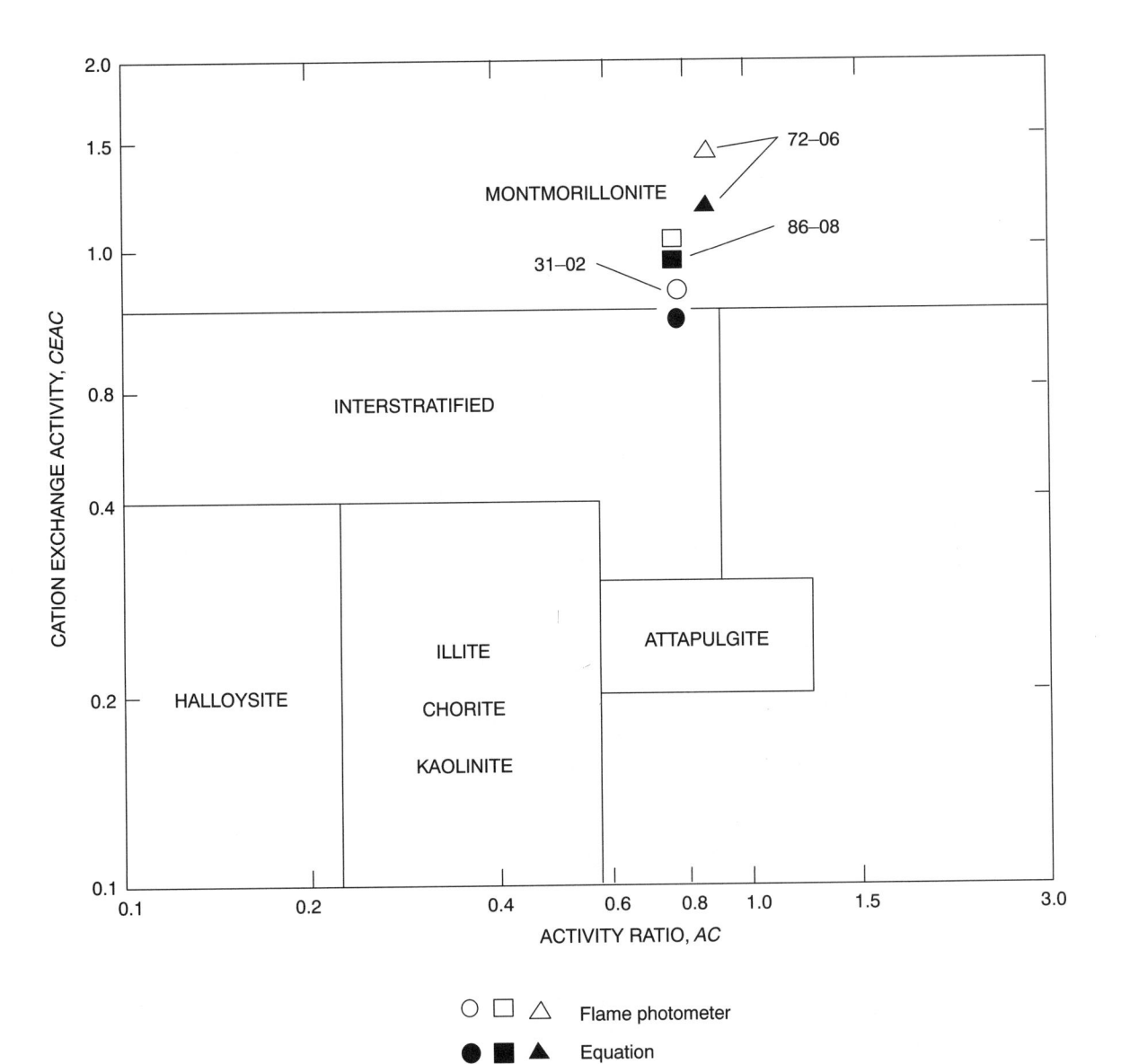

Chapter 19
CONCRETE

NOTE: This is a new division.

Division I — GENERAL

SECTION 1900 — GENERAL

1900.1 Scope. The design of concrete structures of cast-in-place or precast construction, plain, reinforced or prestressed shall conform to the rules and principles specified in this chapter.

1900.2 General Requirements. All concrete structures shall be designed and constructed in accordance with the requirements of Division II and the additional requirements contained in Section 1900.4 of this division.

1900.3 Design Methods. The design of concrete structures shall be in accordance with one of the following methods.

1900.3.1 Strength design (load and resistance factor design). The design of concrete structures using the strength design method shall be in accordance with the requirements of Division II.

1900.3.2 Allowable stress design. The design of concrete structures using the Allowable Stress Design Method shall be in accordance with the requirements of Division VI, Section 1926.

1900.4 Additional Design and Construction Requirements.

1900.4.1 Anchorage. Anchorage of bolts and headed stud anchors to concrete shall be in accordance with Division III.

1900.4.2 Shotcrete. In addition to the requirements of Division II, design and construction of shotcrete structures shall meet the requirements of Division IV.

1900.4.3 Reinforced gypsum concrete. Reinforced gypsum concrete shall be in accordance with Division V.

1900.4.4 Minimum slab thickness. The minimum thickness of concrete floor slabs supported directly on the ground shall not be less than $3^1/_2$ inches (89 mm).

1900.4.5 Unified design provisions for reinforced and prestressed concrete flexural and compression members. It shall be permitted to use the alternate flexural and axial load design provisions in accordance with Division VII, Section 1927.

1900.4.6 Alternative load-factor combination and strength-reduction factors. It shall be permitted to use the alternative load-factor and strength-reduction factors in accordance with Division VIII, Section 1928.

Division II

The contents of this division are patterned after, and in general conformity with, the provisions of Building Code Requirements for Reinforced Concrete (ACI 318-95) and commentary—ACI 318 R-95. For additional background information and research data, see the referenced American Concrete Institute (ACI) publication.

To make reference to the ACI commentary easier for users of the code, the section designations of this division have been made similar to those found in ACI 318. The first two digits of a section number indicates this chapter number and the balance matches the ACI chapter and section designation wherever possible. Italics are used in this chapter to indicate where the *Uniform Building Code* differs substantively from the ACI standard.

SECTION 1901 — SCOPE

The design of structures in concrete of cast-in-place or precast construction, plain, reinforced or prestressed, shall conform to the rules and principles specified in this chapter.

SECTION 1902 — DEFINITIONS

The following terms are defined for general use in this code. Specialized definitions appear in individual *sections*.

ADMIXTURE is material other than water, aggregate, or hydraulic cement used as an ingredient of concrete and added to concrete before or during its mixing to modify its properties.

AGGREGATE is granular material, such as sand, gravel, crushed stone and iron blast-furnace slag, and when used with a cementing medium forms a hydraulic cement concrete or mortar.

AGGREGATE, LIGHTWEIGHT, is aggregate with a dry, loose weight of 70 pounds per cubic foot (pcf) (1120 kg/m³) or less.

AIR-DRY WEIGHT is the unit weight of a lightweight concrete specimen cured for seven days with neither loss nor gain of moisture at 60°F to 80°F (15.6°C to 26.7°C) and dried for 21 days in 50 ± 7 percent relative humidity at 73.4°F ± 2°F (23.0°C ± 1.1°C).

ANCHORAGE in posttensioning is a device used to anchor tendons to concrete member; in pretensioning, a device used to anchor tendons during hardening of concrete.

BONDED TENDON is a prestressing tendon that is bonded to concrete either directly or through grouting.

CEMENTITIOUS MATERIALS are materials as specified in Section 1903 which have cementing value when used in concrete either by themselves, such as portland cement, blended hydraulic cements and expansive cement, or such materials in combination with fly ash, raw or other calcined natural pozzolans, silica fume, or ground granulated blast-furnace slag.

COLUMN is a member with a ratio of height-to-least-lateral dimension of 3 or greater used primarily to support axial compressive load.

COMPOSITE CONCRETE FLEXURAL MEMBERS are concrete flexural members of precast and cast-in-place concrete elements or both constructed in separate placements but so interconnected that all elements respond to loads as a unit.

COMPRESSION-CONTROLLED SECTION is a cross section in which the net tensile strain in the extreme tension steel at nominal strength is less than or equal to the compression-controlled strain limit.

COMPRESSION-CONTROLLED STRAIN LIMIT is the net tensile strain at balanced strain conditions. (See *Section B1910.3.2.*)

CONCRETE is a mixture of portland cement or any other hydraulic cement, fine aggregate, coarse aggregate and water, with or without admixtures.

CONCRETE, SPECIFIED COMPRESSIVE STRENGTH OF (f'_c), is the compressive strength of concrete used in design and evaluated in accordance with provisions of Section 1905, expressed in pounds per square inch (psi) (MPa). Whenever the quantity f'_c is under a radical sign, square root of numerical value only is intended, and result has units of psi (MPa).

CONCRETE, STRUCTURAL LIGHTWEIGHT, is concrete containing lightweight aggregate having an air-dry unit weight as determined by definition above, not exceeding 115 pcf (1840 kg/m³). In this code, a lightweight concrete without natural sand is termed "all-lightweight concrete" and lightweight concrete in which all fine aggregate consists of normal-weight sand is termed "sand-lightweight concrete."

CONTRACTION JOINT is a formed, sawed, or tooled groove in a concrete structure to create a weakened plane and regulate the location of cracking resulting from the dimensional change of different parts of the structure.

CURVATURE FRICTION is friction resulting from bends or curves in the specified prestressing tendon profile.

DEFORMED REINFORCEMENT is deformed reinforcing bars, bar and rod mats, deformed wire, welded smooth wire fabric and welded deformed wire fabric.

DEVELOPMENT LENGTH is the length of embedded reinforcement required to develop the design strength of reinforcement at a critical section. See Section 1909.3.3.

EFFECTIVE DEPTH OF SECTION *(d)* is the distance measured from extreme compression fiber to centroid of tension reinforcement.

EFFECTIVE PRESTRESS is the stress remaining in prestressing tendons after all losses have occurred, excluding effects of dead load and superimposed load.

EMBEDMENT LENGTH is the length of embedded reinforcement provided beyond a critical section.

EXTREME TENSION STEEL is the reinforcement (prestressed or nonprestressed) that is the farthest from the extreme compression fiber.

ISOLATION JOINT is a separation between adjoining parts of a concrete structure, usually a vertical plane, at a designed location such as to interfere least with performance of the structure, yet such as to allow relative movement in three directions and avoid formation of cracks elsewhere in the concrete and through which all or part of the bonded reinforcement is interrupted.

JACKING FORCE is the temporary force exerted by device that introduces tension into prestressing tendons in prestressed concrete.

LOAD, DEAD, is the dead weight supported by a member, as defined by *Section 1602* (without load factors).

LOAD, FACTORED, is the load, multiplied by appropriate load factors, used to proportion members by the strength design method of this code. See Sections 1908.1.1 and 1909.2.

LOAD, LIVE, *is the live load specified by Section 1602* (without load factors).

LOAD, SERVICE, *is the live and dead loads* (without load factors).

MODULUS OF ELASTICITY is the ratio of normal stress to corresponding strain for tensile or compressive stresses below proportional limit of material. See Section 1908.5.

NET TENSILE STRAIN is the tensile strain at nominal strength exclusive of strains due to effective prestress, creep, shrinkage and temperature.

PEDESTAL is an upright compression member with a ratio of unsupported height to average least lateral dimension of 3 or less.

PLAIN CONCRETE is structural concrete with no reinforcement or with less reinforcement than the minimum amount specified for reinforced concrete.

PLAIN REINFORCEMENT is reinforcement that does not conform to definition of deformed reinforcement.

POSTTENSIONING is a method of prestressing in which tendons are tensioned after concrete has hardened.

PRECAST CONCRETE is a structural concrete element cast in other than its final position in the structure.

PRESTRESSED CONCRETE is structural concrete in which internal stresses have been introduced to reduce potential tensile stresses in concrete resulting from loads.

PRETENSIONING is a method of prestressing in which tendons are tensioned before concrete is placed.

REINFORCED CONCRETE is structural concrete reinforced with no less than the minimum amounts of prestressing tendons or nonprestressed reinforcement specified in this code.

REINFORCEMENT is material that conforms to Section 1903.5.1, excluding prestressing tendons unless specifically included.

RESHORES are shores placed snugly under a concrete slab or other structural member after the original forms and shores have been removed from a larger area, thus requiring the new slab or structural member to deflect and support its own weight and existing construction loads applied prior to the installation of the reshores.

SHORES are vertical or inclined support members designed to carry the weight of the formwork, concrete and construction loads above.

SPAN LENGTH. See Section 1908.7.

SPIRAL REINFORCEMENT is continuously wound reinforcement in the form of a cylindrical helix.

SPLITTING TENSILE STRENGTH (f_{ct}) is the tensile strength of concrete. See Section 1905.1.4.

STIRRUP is reinforcement used to resist shear and torsion stresses in a structural member; typically bars, wires, or welded wire fabric (smooth or deformed) bent into L, U or rectangular shapes and located perpendicular to or at an angle to longitudinal reinforcement. (The term "stirrups" is usually applied to lateral reinforcement in flexural members and the term "ties" to those in compression members.) See "tie."

STRENGTH, DESIGN, is the nominal strength multiplied by a strength-reduction factor ϕ. See Section 1909.3.

STRENGTH, NOMINAL, is the strength of a member or cross section calculated in accordance with provisions and assumptions of the strength design method of this code before application of any strength-reduction factors. See Section 1909.3.1.

STRENGTH, REQUIRED, is the strength of a member or cross section required to resist factored loads or related internal moments and forces in such combinations as are stipulated in this code. See Section 1909.1.1.

STRESS is the intensity of force per unit area.

STRUCTURAL CONCRETE is all concrete used for structural purposes, including plain and reinforced concrete.

TENDON is a steel element such as wire, cable, bar, rod or strand, or a bundle of such elements, used to impart prestress to concrete.

TENSION-CONTROLLED SECTION is a cross section in which the net tensile strain in the extreme tension steel at nominal strength is greater than or equal to 0.005.

TIE is a loop of reinforcing bar or wire enclosing longitudinal reinforcement. A continuously wound bar or wire in the form of a circle, rectangle or other polygon shape without re-entrant corners is acceptable. See "stirrup."

TRANSFER is the act of transferring stress in prestressing tendons from jacks or pretensioning bed to concrete member.

WALL is a member, usually vertical, used to enclose or separate spaces.

WOBBLE FRICTION in prestressed concrete, is friction caused by unintended deviation of prestressing sheath or duct from its specified profile.

YIELD STRENGTH is the specified minimum yield strength or yield point of reinforcement in psi.

SECTION 1903 — SPECIFICATIONS FOR TESTS AND MATERIALS

1903.0 Notation.

f_y = specified yield strength of nonprestressed reinforcement, psi (MPa).

1903.1 Tests of Materials.

1903.1.1 The building official may require the testing of any materials used in concrete construction to determine if materials are of quality specified.

1903.1.2 Tests of materials and of concrete shall be made *by an approved agency and at no expense to the jurisdiction. Such tests shall be made in accordance with the standards listed in Section 1903.*

1903.1.3 A complete record of tests of materials and of concrete shall be available for inspection during progress of work and for two years after completion of the project, and shall be preserved by the inspecting engineer or architect for that purpose.

1903.1.4 Material and test standards. *The standards listed in this chapter labeled a "UBC Standard" are also listed in Chapter 35, Part II, and are part of this code. The other standards listed in this chapter are recognized standards. (See Sections 3503 and 3504.)*

1903.2 Cement.

1. ASTM C 845, Expansive Hydraulic Cement

2. ASTM C 150, Portland Cement

3. ASTM C 595 or ASTM C 1157, Blended Hydraulic Cements

1903.3 Aggregates.

1903.3.1 Recognized standards.

1. ASTM C 33, Concrete Aggregates

2. ASTM C 330, Lightweight Aggregates for Structural Concrete

3. ASTM C 332, Lightweight Aggregates for Insulating Concrete

4. ASTM C 144, Aggregate for Masonry Mortar

5. Aggregates failing to meet the above specifications but which have been shown by special test or actual service to produce concrete of adequate strength and durability may be used where authorized by the building official.

1903.3.2 The nominal maximum size of coarse aggregate shall not be larger than:

1. One fifth the narrowest dimension between sides of forms, or

2. One third the depth of slabs, or

3. Three fourths the minimum clear spacing between individual reinforcing bars or wires, bundles of bars, or prestressing tendons or ducts.

These limitations may be waived if, in the judgment of the *building official,* workability and methods of consolidation are such that concrete can be placed without honeycomb or voids.

1903.4 Water.

1903.4.1 Water used in mixing concrete shall be clean and free from injurious amounts of oils, acids, alkalis, salts, organic materials or other substances deleterious to concrete or reinforcement.

1903.4.2 Mixing water for prestressed concrete or for concrete that will contain aluminum embedments, including that portion of mixing water contributed in the form of free moisture on aggregates, shall not contain deleterious amounts of chloride ions. See Section 1904.4.1.

1903.4.3 Nonpotable water shall not be used in concrete unless the following are satisfied:

1903.4.3.1 Selection of concrete proportions shall be based on concrete mixes using water from the same source.

1903.4.3.2 Mortar test cubes made with nonpotable mixing water shall have seven-day and 28-day strengths equal to at least 90 percent of strengths of similar specimens made with potable water. Strength test comparison shall be made on mortars, identical except for the mixing water, prepared and tested in accordance with ASTM C 109 (Compressive Strength of Hydraulic Cement Mortars).

1903.5 Steel Reinforcement.

1903.5.1 Reinforcement shall be deformed reinforcement, except that plain reinforcement may be used for spirals or tendons, and reinforcement consisting of structural steel, steel pipe or steel tubing may be used as specified in this chapter.

1903.5.2 Welding of reinforcing bars shall conform to *approved nationally recognized standards.* Type and location of welded splices and other required welding of reinforcing bars shall be indicated on the design drawings or in the project specifications. ASTM reinforcing bar specifications, except for A 706, shall be supplemented to require a report of material properties necessary to conform to requirements in UBC Standard 19-1.

1903.5.3 Deformed reinforcements.

1903.5.3.1 ASTM A 615, A 616, A 617, A 706, A 767 and A 775, Reinforcing Bars for Concrete.

1903.5.3.2 Deformed reinforcing bars with a specified yield strength f_y exceeding 60,000 psi (413.7 MPa) may be used, provided f_y shall be the stress corresponding to a strain of 0.35 percent and the bars otherwise conform to approved national standards, see *ASTM A 615, A 616, A 617, A 706, A 767 and A 775.* See Section 1909.4.

1903.5.3.3 ASTM A 184, Fabricated Deformed Steel Bar Mats. For reinforced bars used in bar mats, see ASTM A 615, A 616, A 617, A 706, A 767 or A 775.

1903.5.3.4 ASTM A 496, Steel Wire, Deformed, for Concrete Reinforcement.

For deformed wire for concrete reinforcement, see *ASTM A 496,* except that wire shall not be smaller than size D4, and for wire with a specified yield strength f_y exceeding 60,000 psi (413.7 MPa), f_y shall be the stress corresponding to a strain of 0.35 percent, if the yield strength specified in design exceeds 60,000 psi (413.7 MPa).

1903.5.3.5 ASTM A 185, Steel Welded Wire, Fabric, Plain for Concrete Reinforcement.

For welded plain wire fabric for concrete reinforcement, see *ASTM 185,* except that for wire with a specified yield strength f_y exceeding 60,000 psi (413.7 MPa), f_y shall be the stress corresponding to a strain of 0.35 percent, if the yield strength specified in design exceeds 60,000 psi (413.7 MPa). Welded intersections shall not be spaced farther apart than 12 inches (305 mm) in direction of calculated stress, except for wire fabric used as stirrups in accordance with Section 1912.14.

1903.5.3.6 ASTM A 497, Welded Deformed Steel Wire Fabric for Concrete Reinforcement.

For welded deformed wire fabric for concrete reinforcement, see *ASTM A 497,* except that for wire with a specified yield strength f_y exceeding 60,000 psi (413.7 MPa), f_y shall be the stress corresponding to a strain of 0.35 percent, if the yield strength specified in design exceeds 60,000 psi (413.7 MPa). Welded intersections shall not be spaced farther apart than 16 inches (406 mm) in direction of calculated stress, except for wire fabric used as stirrups in accordance with Section 1912.13.2.

1903.5.3.7 Deformed reinforcing bars may be galvanized or epoxy coated. For zinc or epoxy-coated reinforcement, see ASTM A 615, A 616, A 617, A 706, A 767 and A 775 and ASTM A 934 *(Epoxy-Coated Steel Reinforcing Bars).*

1903.5.3.8 Epoxy-coated wires and welded wire fabric shall comply with ASTM A 884 *(Standard Specification for Epoxy-Coated Steel Wire and Welded Wire Fabric for Reinforcement).* Epoxy-coated wires shall conform to Section 1903.5.3.4 and epoxy-coated welded wire fabric shall conform to Section 1903.5.3.5 or 1903.5.3.6.

1903.5.4 Plain reinforcement.

1903.5.4.1 Plain bars for spiral reinforcement shall conform to approved national standards, see ASTM *A 615, A 616 and A 617.*

1903.5.4.2 For plain wire for spiral reinforcement, see *ASTM A 82* except that for wire with a specified yield strength f_y exceeding 60,000 psi (413.7 MPa), f_y shall be the stress corresponding to a strain of 0.35 percent, if the yield strength specified in design exceeds 60,000 psi (413.7 MPa).

1903.5.5 Prestressing tendons.

1903.5.5.1 1. ASTM A 416, Uncoated Seven-wire Stress-relieved Steel Strand for Prestressed Concrete

2. ASTM A 421, Uncoated Stress-relieved Wire for Prestressed Concrete

3. ASTM A 722, Uncoated High-strength Steel Bar for Prestressing Concrete

1903.5.5.2 Wire, strands and bars not specifically listed in *ASTM A 416, A 421 and A 722* may be used, provided they conform to minimum requirements of these specifications and do not have properties that make them less satisfactory than those listed.

1903.5.6 Structural steel, steel pipe or tubing.

1903.5.6.1 For structural steel used with reinforcing bars in composite compression members meeting requirements of Section 1910.16.7 or 1910.16.8, see *ASTM A 36, A 242, A 572 and A 588.*

1903.5.6.2 For steel pipe or tubing for composite compression members composed of a steel-encased concrete core meeting requirements of Section 1910.16.4, see *ASTM A 53, A 500 and A 501.*

1903.5.7 *UBC Standard 19-1, Welding Reinforcing Steel, Metal Inserts and Connections in Reinforced Concrete Construction*

1903.6 Admixtures.

1903.6.1 Admixtures to be used in concrete shall be subject to prior approval by the *building official.*

1903.6.2 An admixture shall be shown capable of maintaining essentially the same composition and performance throughout the work as the product used in establishing concrete proportions in accordance with Section 1905.2.

1903.6.3 Calcium chloride or admixtures containing chloride from other than impurities from admixture ingredients shall not be used in prestressed concrete, in concrete containing embedded aluminum, or in concrete cast against stay-in-place galvanized steel forms. See Sections 1904.3.2 and 1904.4.1.

1903.6.4 ASTM C 260, Air-entraining Admixtures for Concrete

1903.6.5 ASTM C 494 and C 1017, Chemical Admixtures for Concrete

1903.6.6 ASTM C 618, Fly Ash and Raw or Calcined Natural Pozzolans for Use as Admixtures in Portland Cement Concrete

1903.6.7 *ASTM C 989,* Ground-iron Blast-furnace Slag for Use in Concrete and Mortars

1903.6.8 Admixtures used in concrete containing ASTM C 845 expansive cements shall be compatible with the cement and produce no deleterious effects.

1903.6.9 Silica fume used as an admixture shall conform to ASTM C 1240 (*Silica Fume for Use in Hydraulic Cement Concrete and Mortar*).

1903.7 Storage of Materials.

1903.7.1 Cementitious materials and aggregate shall be stored in such manner as to prevent deterioration or intrusion of foreign matter.

1903.7.2 Any material that has deteriorated or has been contaminated shall not be used for concrete.

1903.8 Concrete Testing.

1. ASTM C 192, Making and Curing Concrete Test Specimens in the Laboratory

2. ASTM C 31, Making and Curing Concrete Test Specimens in the Field

3. ASTM C 42, Obtaining and Testing Drilled Cores and Sawed Beams of Concrete

4. ASTM C 39, Compressive Strength of Cylindrical Concrete Specimens

5. ASTM C 172, Sampling Freshly Mixed Concrete

6. ASTM C 496, Splitting Tensile Strength of Cylindrical Concrete Specimens

7. ASTM C 1218, Water-Soluble Chloride in Mortar and Concrete

1903.9 Concrete Mix.

1. *ASTM C 94, Ready-mixed Concrete*

2. ASTM C 685, Concrete Made by Volumetric Batching and Continuous Mixing

3. *UBC Standard 19-2, Mill-mixed Gypsum Concrete and Poured Gypsum Roof Diaphragms*

4. ASTM C 109, Compressive Strength of Hydraulic Cement Mortars

5. ASTM C 567, Unit Weight of Structural Lightweight Concrete

1903.10 Welding. *The welding of reinforcing steel, metal inserts and connections in reinforced concrete construction shall conform to UBC Standard 19-1.*

1903.11 Glass Fiber Reinforced Concrete. Recommended Practice for Glass Fiber Reinforced Concrete Panels, Manual 128.

SECTION 1904 — DURABILITY REQUIREMENTS

1904.0 Notation.

f'_c = specified compressive strength of concrete, psi (MPa).

1904.1 Water-Cementitious Materials Ratio.

1904.1.1 The water-cementitious materials ratios specified in Tables 19-A-2 and 19-A-4 shall be calculated using the weight of cement meeting ASTM C 150, C 595 or C 845 plus the weight of fly ash and other pozzolans meeting ASTM C 618, slag meeting ASTM C 989, and silica fume meeting ASTM C 1240, if any, except that when concrete is exposed to deicing chemicals, Section 1904.2.3 further limits the amount of fly ash, pozzolans, silica fume, slag or the combination of these materials.

1904.2 Freezing and Thawing Exposures.

1904.2.1 Normal-weight and lightweight concrete exposed to freezing and thawing or deicing chemicals shall be air entrained with air content indicated in Table 19-A-1. Tolerance on air content as delivered shall be ± 1.5 percent. For specified compressive strength f'_c greater than 5,000 psi (34.47 MPa), reduction of air content indicated in Table 19-A-1 by 1.0 percent shall be permitted.

1904.2.2 Concrete that will be subjected to the exposures given in Table 19-A-2 shall conform to the corresponding maximum water-cementitious materials ratios and minimum specified concrete compressive strength requirements of that table. In addition, concrete that will be exposed to deicing chemicals shall conform to the limitations of Section 1904.2.3.

1904.2.3 For concrete exposed to deicing chemicals, the maximum weight of fly ash, other pozzolans, silica fume or slag that is included in the concrete shall not exceed the percentages of the total weight of cementitious materials given in Table 19-A-3.

1904.3 Sulfate Exposure.

1904.3.1 Concrete to be exposed to sulfate-containing solutions or soils shall conform to the requirements of Table 19-A-4 or shall be concrete made with a cement that provides sulfate resistance and that has a maximum water-cementitious materials ratio and minimum compressive strength set forth in Table 19-A-4.

1904.3.2 Calcium chloride as an admixture shall not be used in concrete to be exposed to severe or very severe sulfate-containing solutions, as defined in Table 19-A-4.

1904.4 Corrosion Protection of Reinforcement.

1904.4.1 For corrosion protection of reinforcement in concrete, maximum water soluble chloride ion concentrations in hardened concrete at ages from 28 to 42 days contributed from the ingredients, including water, aggregates, cementitious materials and admixtures shall not exceed the limits of Table 19-A-5. When testing is performed to determine water soluble chloride ion content, test procedures shall conform to ASTM C 1218.

1904.4.2 If concrete with reinforcement will be exposed to chlorides from deicing chemicals, salt, salt water, brackish water, sea water or spray from these sources, requirements of Table 19-A-2 for water-cementitious materials ratio and concrete strength and the minimum concrete cover requirements of Section 1907.7 shall be satisfied. In addition, see Section 1918.14 for unbonded prestressed tendons.

SECTION 1905 — CONCRETE QUALITY, MIXING AND PLACING

1905.0 Notations.

f'_c = specified compressive strength of concrete, psi (MPa).

f'_{cr} = required average compressive strength of concrete used as the basis for selection of concrete proportions, psi (MPa).

f_{ct} = average splitting tensile strength of lightweight aggregate concrete, psi (MPa).

s = standard deviation, psi (MPa).

1905.1 General.

1905.1.1 Concrete shall be proportioned to provide an average compressive strength as prescribed in Section 1905.3.2, as well as satisfy the durability criteria of Section 1904. Concrete shall be produced to minimize frequency of strengths below f'_c as prescribed in Section 1905.6.2.3.

1905.1.2 Requirements for f'_c shall be based on tests of cylinders made and tested as prescribed in Section 1905.6.2.

1905.1.3 Unless otherwise specified, f'_c shall be based on 28-day tests. If other than 28 days, test age for f'_c shall be as indicated in design drawings or specifications.

Design drawings shall show specified compressive strength of concrete f'_c for which each part of structure is designed.

1905.1.4 Where design criteria in Sections 1909.5.2.3, 1911.2; and 1912.2.4, provide for use of a splitting tensile strength value of concrete, laboratory tests shall be made to establish value of f_{ct} corresponding to specified values of f'_c.

1905.1.5 Splitting tensile strength tests shall not be used as a basis for field acceptance of concrete.

1905.2 Selection of Concrete Proportions.

1905.2.1 Proportions of materials for concrete shall be established to provide:

1. Workability and consistency to permit concrete to be worked readily into forms and around reinforcement under conditions of placement to be employed without segregation or excessive bleeding.

2. Resistance to special exposures as required by Section 1904.

3. Conformance with strength test requirements of Section 1905.6.

1905.2.2 Where different materials are to be used for different portions of proposed work, each combination shall be evaluated.

1905.2.3 Concrete proportions, including water-cementitious materials ratio, shall be established on the basis of field experience and/or trial mixtures with materials to be employed (see Section 1905.3), except as permitted in Section 1905.4 or required by Section 1904.

1905.3 Proportioning on the Basis of Field Experience and Trial Mixtures.

1905.3.1 Standard deviation.

1905.3.1.1 Where a concrete production facility has test records, a standard deviation shall be established. Test records from which a standard deviation is calculated:

1. Must represent materials, quality control procedures and conditions similar to those expected, and changes in materials and proportions within the test records shall not have been more restricted than those for proposed work.

2. Must represent concrete produced to meet a specified strength or strengths f'_c within 1,000 psi (6.89 MPa) of that specified for proposed work.

3. Must consist of at least 30 consecutive tests or two groups of consecutive tests totaling at least 30 tests as defined in Section 1905.6.1.4, except as provided in Section 1905.3.1.2.

1905.3.1.2 Where a concrete production facility does not have test records meeting requirements of Section 1905.3.1.1, but does have a record based on 15 to 29 consecutive tests, a standard deviation may be established as the product of the calculated standard deviation and the modification factor of Table 19-A-6. To be acceptable, the test record must meet the requirements of Section 1905.3.1.1, Items 1 and 2, and represent only a single record of consecutive tests that span a period of not less than 45 calendar days.

1905.3.2 Required average strength.

1905.3.2.1 Required average compressive strength f'_{cr} used as the basis for selection of concrete proportions shall be the larger of Formula (5-1) or (5-2) using a standard deviation calculated in accordance with Section 1905.3.1.1 or 1905.3.1.2.

$$f'_{cr} = f'_c + 1.34s \qquad (5-1)$$

or

$$f'_{cr} = f'_c + 2.33s - 500 \qquad (5-2)$$

For SI: $\qquad f'_{cr} = f'_c + 2.33s - 3.45$

1905.3.2.2 When a concrete production facility does not have field strength test records for calculation of standard deviation meeting requirements of Section 1905.3.1.1 or 1905.3.1.2, required average strength f'_{cr} shall be determined from Table 19-B and documentation of average strength shall be in accordance with requirements of Section 1905.3.3.

1905.3.3 Documentation of average strength. Documentation that proposed concrete proportions will produce an average compressive strength equal to or greater than required average compressive strength (see Section 1905.3.2) shall consist of a field strength test record, several strength test records, or trial mixtures.

1905.3.3.1 When test records are used to demonstrate that proposed concrete proportions will produce the required average strength f'_{cr} (see Section 1905.3.2), such records shall represent materials and conditions similar to those expected. Changes in materials, conditions and proportions within the test records shall not have been more restricted than those for proposed work. For the purpose of documenting average strength potential, test records consisting of less than 30 but not less than 10 consecutive tests may be used, provided test records encompass a period of time not less than 45 days. Required concrete proportions may be established by interpolation between the strengths and proportions of two or more test records each of which meets other requirements of this section.

1905.3.3.2 When an acceptable record of field test results is not available, concrete proportions established from trial mixtures meeting the following restrictions shall be permitted:

1. Combination of materials shall be those for proposed work.

2. Trial mixtures having proportions and consistencies required for proposed work shall be made using at least three different water-cementitious materials ratios or cementitious materials contents that will produce a range of strengths encompassing the required average strength f'_{cr}.

3. Trial mixture shall be designed to produce a slump within ± 0.75 inch (± 19 mm) of maximum permitted, and for air-entrained concrete, within ± 0.5 percent of maximum allowable air content.

4. For each water-cementitious materials ratio or cementitious materials content, at least three test cylinders for each test age shall be made and cured. Cylinders shall be tested at 28 days or at test age designated for determination of f'_c.

5. From results of cylinder tests, a curve shall be plotted showing relationship between water-cementitious materials ratio or cementitious materials content and compressive strength at designated test age.

6. Maximum water-cementitious materials ratio or minimum cementitious materials content for concrete to be used in proposed work shall be that shown by the curve to produce the average strength required by Section 1905.3.2, unless a lower water-cementitious materials ratio or higher strength is required by Section 1904.

1905.4 Proportioning without Field Experience or Trial Mixtures.

1905.4.1 If data required by Section 1905.3 are not available, concrete proportions shall be based upon other experience or information, if approved by the building official. The required average compressive strength f'_{cr} of concrete produced with materials similar to those proposed for use shall be at least 1,200 psi (8.3 MPa) greater than the specified compressive strength, f'_c. This alternative shall not be used for specified compressive strength greater than 4,000 psi (27.58 MPa).

1905.4.2 Concrete proportioned by Section 1905.4 shall conform to the durability requirements of Section 1904 and to compressive strength test criteria of Section 1905.6.

1905.5 Average Strength Reduction. As data become available during construction, it shall be permitted to reduce the amount by which f'_{cr} must exceed the specified value of f'_c, provided:

1. Thirty or more test results are available and average of test results exceeds that required by Section 1905.3.2.1, using a standard deviation calculated in accordance with Section 1905.3.1.1, or

2. Fifteen to 29 test results are available and average of test results exceeds that required by Section 1905.3.2.1, using a standard deviation calculated in accordance with Section 1905.3.1.2, and

3. Special exposure requirements of Section 1904 are met.

1905.6 Evaluation and Acceptance of Concrete.

1905.6.1 Frequency of testing.

1905.6.1.1 Samples for strength tests of each class of concrete placed each day shall be taken not less than once a day, or not less than once for each 150 cubic yards (115 m^3) of concrete, or not less than once for each 5,000 square feet (465 m^2) of surface area for slabs or walls.

1905.6.1.2 On a given project, if the total volume of concrete is such that the frequency of testing required by Section 1905.6.1.1 would provide less than five strength tests for a given class of concrete, tests shall be made from at least five randomly selected batches or from each batch if fewer than five batches are used.

1905.6.1.3 When total quantity of a given class of concrete is less than 50 cubic yards (38 m^3), strength tests are not required when evidence of satisfactory strength is submitted to and approved by the building official.

1905.6.1.4 A strength test shall be the average of the strengths of two cylinders made from the same sample of concrete and tested at 28 days or at test age designated for determination of f'_c.

1905.6.2 Laboratory-cured specimens.

1905.6.2.1 Samples for strength tests shall be taken.

1905.6.2.2 Cylinders for strength tests shall be molded and laboratory cured and tested.

1905.6.2.3 Strength level of an individual class of concrete shall be considered satisfactory if both the following requirements are met:

1. Every arithmetic average of any three consecutive strength tests equals or exceeds f'_c.

2. No individual strength test (average of two cylinders) falls below f'_c by more than 500 psi (3.45 MPa).

1905.6.2.4 If either of the requirements of Section 1905.6.2.3 are not met, steps shall be taken to increase the average of subsequent strength test results. Requirements of Section 1905.6.4 shall be observed if the requirement of Item 2 of Section 1905.6.2.3 is not met.

1905.6.3 Field-cured specimens.

1905.6.3.1 If required by the building official, results of strength tests of cylinders cured under field conditions shall be provided.

1905.6.3.2 Field-cured cylinders shall be cured under field conditions, in accordance with Section 1903.8.

1905.6.3.3 Field-cured test cylinders shall be molded at the same time and from the same samples as laboratory cured test cylinders.

1905.6.3.4 Procedures for protecting and curing concrete shall be improved when strength of field-cured cylinders at test age designated for determination of f'_c is less than 85 percent of that of companion laboratory-cured cylinders. The 85 percent limitation shall not apply if field-cured strength exceeds f'_c by more than 500 psi (3.45 MPa).

1905.6.4 Investigation of low-strength test results.

1905.6.4.1 If any strength test (see Section 1905.6.1.4) of laboratory-cured cylinders falls below specified values of f'_c by more than 500 psi (3.45 MPa) (see Section 1905.6.2.3, Item 2) or if tests of field-cured cylinders indicate deficiencies in protection and curing (see Section 1905.6.3.4), steps shall be taken to ensure that load-carrying capacity of the structure is not jeopardized.

1905.6.4.2 If the likelihood of low-strength concrete is confirmed and calculations indicate that load-carrying capacity is significantly reduced, tests of cores drilled from the area in question shall be permitted. In such case, three cores shall be taken for each strength test more than 500 psi (3.45 MPa) below specified value of f'_c.

1905.6.4.3 If concrete in the structure will be dry under service conditions, cores shall be air dried [temperatures 60°F to 80°F (15.6°C to 26.7°C), relative humidity less than 60 percent] for seven days before test and shall be tested dry. If concrete in the structure will be more than superficially wet under service conditions, cores shall be immersed in water for at least 40 hours and be tested wet.

1905.6.4.4 Concrete in an area represented by core tests shall be considered structurally adequate if the average of three cores is equal to at least 85 percent of f'_c and if no single core is less than 75 percent of f'_c. Additional testing of cores extracted from locations represented by erratic core strength results shall be permitted.

1905.6.4.5 If criteria of Section 1905.6.4.4 are not met, and if structural adequacy remains in doubt, the responsible authority shall be permitted to order a strength evaluation in accordance with Section 1920 for the questionable portion of the structure, or take other appropriate action.

1905.7 Preparation of Equipment and Place of Deposit.

1905.7.1 Preparation before concrete placement shall include the following:

1. All equipment for mixing and transporting concrete shall be clean.

2. All debris and ice shall be removed from spaces to be occupied by concrete.

3. Forms shall be properly coated.

4. Masonry filler units that will be in contact with concrete shall be well drenched.

5. Reinforcement shall be thoroughly clean of ice or other deleterious coatings.

6. Water shall be removed from place of deposit before concrete is placed unless a tremie is to be used or unless otherwise permitted by the building official.

7. All laitance and other unsound material shall be removed before additional concrete is placed against hardened concrete.

1905.8 Mixing.

1905.8.1 All concrete shall be mixed until there is a uniform distribution of materials and shall be discharged completely before mixer is recharged.

1905.8.2 Ready-mixed concrete shall be mixed and delivered in accordance with requirements of *ASTM C 94 (Ready-Mixed Concrete)* or *ASTM C 685 (Concrete Made by Volumetric Batching and Continuous Mixing)*.

1905.8.3 Job-mixed concrete shall be mixed in accordance with the following:

1. Mixing shall be done in a batch mixer of an approved type.

2. Mixer shall be rotated at a speed recommended by the manufacturer.

3. Mixing shall be continued for at least $1^1/_2$ minutes after all materials are in the drum, unless a shorter time is shown to be satisfactory by the mixing uniformity tests of *ASTM C 94 (Ready-Mixed Concrete)*.

4. Materials handling, batching and mixing shall conform to applicable provisions of *ASTM C 94 (Ready-Mixed Concrete)*.

5. A detailed record shall be kept to identify:

    5.1 Number of batches produced;

    5.2 Proportions of materials used;

    5.3 Approximate location of final deposit in structure;

    5.4 Time and date of mixing and placing.

1905.9 Conveying.

1905.9.1 Concrete shall be conveyed from mixer to place of final deposit by methods that will prevent separation or loss of materials.

1905.9.2 Conveying equipment shall be capable of providing a supply of concrete at site of placement without separation of ingredients and without interruptions sufficient to permit loss of plasticity between successive increments.

1905.10 Depositing.

1905.10.1 Concrete shall be deposited as nearly as practicable in its final position to avoid segregation due to rehandling or flowing.

1905.10.2 Concreting shall be carried on at such a rate that concrete is at all times plastic and flows readily into spaces between reinforcement.

1905.10.3 Concrete that has partially hardened or been contaminated by foreign materials shall not be deposited in the structure.

1905.10.4 Retempered concrete or concrete that has been remixed after initial set shall not be used unless approved by the *building official.*

1905.10.5 After concreting is started, it shall be carried on as a continuous operation until placing of a panel or section, as defined by its boundaries or predetermined joints, is completed, except as permitted or prohibited by Section 1906.4.

1905.10.6 Top surfaces of vertically formed lifts shall be generally level.

1905.10.7 When construction joints are required, joints shall be made in accordance with Section 1906.4.

1905.10.8 All concrete shall be thoroughly consolidated by suitable means during placement and shall be thoroughly worked around reinforcement and embedded fixtures and into corners of forms.

1905.11 Curing.

1905.11.1 Concrete (other than high-early-strength) shall be maintained above 50°F (10.0°C) and in a moist condition for at least the first seven days after placement, except when cured in accordance with Section 1905.11.3.

1905.11.2 High-early-strength concrete shall be maintained above 50°F (10.0°C) and in a moist condition for at least the first three days, except when cured in accordance with Section 1905.11.3.

1905.11.3 Accelerated curing.

1905.11.3.1 Curing by high-pressure steam, steam at atmospheric pressure, heat and moisture or other accepted processes, may be employed to accelerate strength gain and reduce time of curing.

1905.11.3.2 Accelerated curing shall provide a compressive strength of the concrete at the load stage considered at least equal to required design strength at that load stage.

1905.11.3.3 Curing process shall be such as to produce concrete with a durability at least equivalent to the curing method of Section 1905.11.1 or 1905.11.2.

1905.11.3.4 When required by the building official, supplementary strength tests in accordance with Section 1905.6.3 shall be performed to assure that curing is satisfactory.

1905.12 Cold Weather Requirements.

1905.12.1 Adequate equipment shall be provided for heating concrete materials and protecting concrete during freezing or near-freezing weather.

1905.12.2 All concrete materials and all reinforcement, forms, fillers and ground with which concrete is to come in contact shall be free from frost.

1905.12.3 Frozen materials or materials containing ice shall not be used.

1905.13 Hot Weather Requirements. During hot weather, proper attention shall be given to ingredients, production methods, handling, placing, protection and curing to prevent excessive concrete temperatures or water evaporation that may impair required strength or serviceability of the member or structure.

SECTION 1906 — FORMWORK, EMBEDDED PIPES AND CONSTRUCTION JOINTS

1906.1 Design of Formwork.

1906.1.1 Forms shall result in a final structure that conforms to shapes, lines and dimensions of the members as required by the design drawings and specifications.

1906.1.2 Forms shall be substantial and sufficiently tight to prevent leakage of mortar.

1906.1.3 Forms shall be properly braced or tied together to maintain position and shape.

1906.1.4 Forms and their supports shall be designed so as not to damage previously placed structure.

1906.1.5 Design of formwork shall include consideration of the following factors:

1. Rate and method of placing concrete.

2. Construction loads, including vertical, horizontal and impact loads.

3. Special form requirements for construction of shells, folded plates, domes, architectural concrete or similar types of elements.

1906.1.6 Forms for prestressed concrete members shall be designed and constructed to permit movement of the member without damage during application of prestressing force.

1906.2 Removal of Forms, Shores and Reshoring.

1906.2.1 Removal of forms. Forms shall be removed in such a manner as not to impair safety and serviceability of the structure. Concrete to be exposed by form removal shall have sufficient strength not to be damaged by removal operation.

1906.2.2 Removal of shores and reshoring. The provisions of Section 1906.2.2.1 through 1906.2.2.3 shall apply to slabs and beams except where cast on the ground.

1906.2.2.1 Before starting construction, the contractor shall develop a procedure and schedule for removal of shores and installation of reshores and for calculating the loads transferred to the structure during the process.

1. The structural analysis and concrete strength data used in planning and implementing form removal and shoring shall be furnished by the contractor to the building official when so requested.

2. Construction loads shall *not* be supported on, or any shoring removed from, any part of the structure under construction except when that portion of the structure in combination with remaining forming and shoring system has sufficient strength to support safely its weight and loads placed thereon.

3. Sufficient strength shall be demonstrated by structural analysis considering proposed loads, strength of forming and shoring system and concrete strength data. Concrete strength data may be based on tests of field-cured cylinders or, when approved by the building official, on other procedures to evaluate concrete strength.

1906.2.2.2 Construction loads exceeding the combination of superimposed dead load plus specified live load shall *not* be supported on any unshored portion of the structure under construction, unless analysis indicates adequate strength to support such additional loads.

1906.2.2.3 Form supports for prestressed concrete members shall not be removed until sufficient prestressing has been applied to enable prestressed members to carry their dead load and anticipated construction loads.

1906.3 Conduits and Pipes Embedded in Concrete.

1906.3.1 Conduits, pipes and sleeves of any material not harmful to concrete and within limitations of this subsection may be embedded in concrete with approval of the *building official*, provided they are not considered to replace structurally the displaced concrete.

1906.3.2 Conduits and pipes of aluminum shall not be embedded in structural concrete unless effectively coated or covered to prevent aluminum-concrete reaction or electrolytic action between aluminum and steel.

1906.3.3 Conduits, pipes and sleeves passing through a slab, wall or beam shall not impair significantly the strength of the construction.

1906.3.4 Conduits and pipes, with their fittings, embedded within a column shall not displace more than 4 percent of the area of cross section on which strength is calculated or which is required for fire protection.

1906.3.5 Except when plans for conduits and pipes are approved by the *building official*, conduits and pipes embedded within a slab, wall or beam (other than those merely passing through) shall satisfy the following:

1906.3.5.1 They shall not be larger in outside dimension than one third the overall thickness of slab, wall or beam in which they are embedded.

1906.3.5.2 They shall be spaced not closer than three diameters or widths on center.

1906.3.5.3 They shall not impair significantly the strength of the construction.

1906.3.6 Conduits, pipes and sleeves may be considered as replacing structurally in compression the displaced concrete, provided:

1906.3.6.1 They are not exposed to rusting or other deterioration.

1906.3.6.2 They are of uncoated or galvanized iron or steel not thinner than standard Schedule 40 steel pipe.

1906.3.6.3 They have a nominal inside diameter not over 2 inches (51 mm) and are spaced not less than three diameters on centers.

1906.3.7 Pipes and fittings shall be designed to resist effects of the material, pressure and temperature to which they will be subjected.

1906.3.8 No liquid, gas or vapor, except water not exceeding 90°F (32.2°C) or 50 psi (0.34 MPa) pressure, shall be placed in the pipes until the concrete has attained its design strength.

1906.3.9 In solid slabs, piping, unless it is used for radiant heating or snow melting, shall be placed between top and bottom reinforcement.

1906.3.10 Concrete cover for pipes, conduit and fittings shall not be less than $1^1/_2$ inches (38 mm) for concrete exposed to earth or weather, or less than $^3/_4$ inch (19 mm) for concrete not exposed to weather or in contact with ground.

1906.3.11 Reinforcement with an area not less than 0.002 times the area of concrete section shall be provided normal to the piping.

1906.3.12 Piping and conduit shall be so fabricated and installed that cutting, bending or displacement of reinforcement from its proper location will not be required.

1906.4 Construction Joints.

1906.4.1 Surface of concrete construction joints shall be cleaned and laitance removed.

1906.4.2 Immediately before new concrete is placed, all construction joints shall be wetted and standing water removed.

1906.4.3 Construction joints shall be so made and located as not to impair the strength of the structure. Provision shall be made for transfer of shear and other forces through construction joints. See Section 1911.7.9.

1906.4.4 Construction joints in floors shall be located within the middle third of spans of slabs, beams and girders. Joints in girders shall be offset a minimum distance of two times the width of intersecting beams.

1906.4.5 Beams, girders or slabs supported by columns or walls shall not be cast or erected until concrete in the vertical support members is no longer plastic.

1906.4.6 Beams, girders, haunches, drop panels and capitals shall be placed monolithically as part of a slab system, unless otherwise shown in design drawings or specifications.

SECTION 1907 — DETAILS OF REINFORCEMENT

1907.0 Notations.

d = distance from extreme compression fiber to centroid of tension reinforcement, inches (mm).

d_b = nominal diameter of bar, wire or prestressing strand, inches (mm).

f_y = specified yield strength of nonprestressed reinforcement, psi (MPa).

l_d = development length, inches (mm). See Section 1912.

1907.1 Standard Hooks. "Standard hook" as used in this code is one of the following:

1907.1.1 One-hundred-eighty-degree bend plus $4d_b$ extension, but not less than $2^1/_2$ inches (64 mm) at free end of bar.

1907.1.2 Ninety-degree bend plus $12d_b$ extension at free end of bar.

1907.1.3 For stirrup and tie hooks:

1. No. 5 bar and smaller, 90-degree bend plus $6d_b$ extension at free end of bar, or

2. No. 6, No. 7 and No. 8 bar, 90-degree bend, plus $12d_b$ extension at free end of bar, or

3. No. 8 bar and smaller, 135-degree bend plus $6d_b$ extension at free end of bar.

4. *For stirrups and tie hooks in Seismic Zones 3 and 4, refer to the hoop and crosstie provisions of Section 1921.1.*

1907.2 Minimum Bend Diameters.

1907.2.1 Diameter of bend measured on the inside of the bar, other than for stirrups and ties in sizes No. 3 through No. 5, shall not be less than the values in Table 19-B.

1907.2.2 Inside diameter of bends for stirrups and ties shall not be less than $4d_b$ for No. 5 bar and smaller. For bars larger than No. 5, diameter of bend shall be in accordance with Table 19-B.

1907.2.3 Inside diameter of bends in welded wire fabric (plain or deformed) for stirrups and ties shall not be less than $4d_b$ for deformed wire larger than D6 and $2d_b$ for all other wires. Bends with inside diameter of less than $8d_b$ shall not be less than $4d_b$ from nearest welded intersection.

1907.3 Bending.

1907.3.1 All reinforcement shall be bent cold, unless otherwise permitted by the *building official.*

1907.3.2 Reinforcement partially embedded in concrete shall not be field bent, except as shown on the design drawings or permitted by the *building official.*

1907.4 Surface Conditions of Reinforcement.

1907.4.1 At the time concrete is placed, reinforcement shall be free from mud, oil or other nonmetallic coatings that decrease bond. Epoxy coatings of bars in accordance with Section 1903.5.3.7 shall be permitted.

1907.4.2 Reinforcement, except prestressing tendons, with rust, mill scale or a combination of both, shall be considered satisfactory, provided the minimum dimensions (including height of deformations) and weight of a hand-wire-brushed test specimen are not less than applicable specification requirements.

1907.4.3 Prestressing tendons shall be clean and free of oil, dirt, scale, pitting and excessive rust. A light oxide shall be permitted.

1907.5 Placing Reinforcement.

1907.5.1 Reinforcement, prestressing tendons and ducts shall be accurately placed and adequately supported before concrete is placed, and shall be secured against displacement within tolerances of this section.

1907.5.2 Unless otherwise *approved by the building official,* reinforcement, prestressing tendons and prestressing ducts shall be placed within the following tolerances:

1907.5.2.1 Tolerance for depth *d*, and minimum concrete cover in flexural members, walls and compression members shall be as follows:

	TOLERANCE ON d	TOLERANCE ON MINIMUM CONCRETE COVER
$d \leq 8$ in. (203 mm)	$\pm \, ^3/_4$ in. (9.5 mm)	$- \, ^3/_8$ in. (9.5 mm)
$d > 8$ in. (203 mm)	$\pm \, ^1/_2$ in. (12.7 mm)	$- \, ^1/_2$ in. (12.7 mm)

except that tolerance for the clear distance to formed soffits shall be minus $^1/_4$ inch (6.4 mm) and tolerance for cover shall not exceed minus one third the minimum concrete cover required by the approved plans or specifications.

1907.5.2.2 Tolerance for longitudinal location of bends and ends of reinforcement shall be ± 2 inches (± 51 mm) except at discontinuous ends of members where tolerance shall be $\pm \, ^1/_2$ inch (± 12.7 mm).

1907.5.3 Welded wire fabric (with wire size not greater than W5 or D5) used in slabs not exceeding 10 feet (3048 mm) in span shall be permitted to be curved from a point near the top of slab over the support to a point near the bottom of slab at midspan, provided such reinforcement is either continuous over, or securely anchored at, support.

1907.5.4 Welding of crossing bars shall not be permitted for assembly of reinforcement.

> **EXCEPTIONS:** *1. Reinforcing steel not required by design.*
>
> *2. When specifically approved by the building official, welding of crossing bars for assembly purposes in Seismic Zones 0, 1 and 2 may be permitted, provided that data are submitted to the building official to show that there is no detrimental effect on the action of the structural member as a result of welding of the crossing bars.*

1907.6 Spacing Limits for Reinforcement.

1907.6.1 The minimum clear spacing between parallel bars in a layer shall be d_b but not less than 1 inch (25 mm). See also Section 1903.3.2.

1907.6.2 Where parallel reinforcement is placed in two or more layers, bars in the upper layers shall be placed directly above bars in the bottom layer with clear distance between layers not less than 1 inch (25 mm).

1907.6.3 In spirally reinforced or tied reinforced compression members, clear distance between longitudinal bars shall not be less than $1.5d_b$ or less than $1^1/_2$ inches (38 mm). See also Section 1903.3.2.

1907.6.4 Clear distance limitation between bars shall apply also to the clear distance between a contact lap splice and adjacent splices or bars.

1907.6.5 In walls and slabs other than concrete joist construction, primary flexural reinforcement shall not be spaced farther apart than three times the wall or slab thickness, or 18 inches (457 mm).

1907.6.6 Bundled bars.

1907.6.6.1 Groups of parallel reinforcing bars bundled in contact to act as a unit shall be limited to four bars in one bundle.

1907.6.6.2 Bundled bars shall be enclosed within stirrups or ties.

1907.6.6.3 Bars larger than No. 11 shall not be bundled in beams.

1907.6.6.4 Individual bars within a bundle terminated within the span of flexural members shall terminate at different points with at least $40d_b$ stagger.

1907.6.6.5 Where spacing limitations and minimum concrete cover are based on bar diameter d_b, a unit of bundled bars shall be treated as a single bar of a diameter derived from the equivalent total area.

1907.6.7 Prestressing tendons and ducts.

1907.6.7.1 Clear distance between pretensioning tendons at each end of a member shall not be less than $4d_b$ for wire, or $3d_b$ for strands. See also Section 1903.3.2. Closer vertical spacing and bundling of tendons shall be permitted in the middle portion of a span.

1907.6.7.2 Bundling of posttensioning ducts shall be permitted if it is shown that concrete can be satisfactorily placed and if provision is made to prevent the tendons, when tensioned, from breaking through the duct.

1907.7 Concrete Protection for Reinforcement.

1907.7.1 Cast-in-place concrete (nonprestressed). The following minimum concrete cover shall be provided for reinforcement:

		MINIMUM COVER, inches (mm)
1.	Concrete cast against and permanently exposed to earth	3 (76)
2.	Concrete exposed to earth or weather:	
	No. 6 through No. 18 bar	2 (51)
	No. 5 bar, W31 or D31 wire, and smaller	$1^1/_2$ (38)
3.	Concrete not exposed to weather or in contact with ground:	
	Slabs, walls, joists:	
	No. 14 and No. 18 bar	$1^1/_2$ (38)
	No. 11 bar and smaller	$^3/_4$ (19)
	Beams, columns:	
	Primary reinforcement, ties, stirrups, spirals	$1^1/_2$ (38)
	Shells, folded plate members:	
	No. 6 bar and larger	$^3/_4$ (19)
	No. 5 bar, W31 or D31 wire, and smaller	$^1/_2$ (12.7)
4.	*Concrete tilt-up panels cast against a rigid horizontal surface, such as a concrete slab, exposed to the weather:*	
	No. 8 and smaller	*1 (25)*
	No. 9 through No. 18	*2 (51)*

1907.7.2 Precast concrete (manufactured under plant control conditions). The following minimum concrete cover shall be provided for reinforcement:

MINIMUM COVER,
inches (mm)

1. Concrete exposed to earth or
   weather:
      Wall panels:
        No. 14 and No. 18 bar ...... $1^1/_2$ (38)
        No. 11 bar and smaller ...... $^3/_4$ (19)
      Other members:
        No. 14 and No. 18 bar 2 (51)
        No. 6 through No. 11 bar .... $1^1/_2$ (38)
        No. 5 bar W31 or D31 wire,
         and smaller .... $1^1/_4$ (32)

2. Concrete not exposed to weather or
   in contact with ground:
      Slabs, walls, joists:
        No. 14 and No. 18 bar ...... $1^1/_4$ (32)
        No. 11 bar and smaller ...... $^5/_8$ (16)
      Beams, columns:
        Primary reinforcement ...... d_b but not less than $^5/_8$ (16) and need not exceed $1^1/_2$ (38)
        Ties, stirrups, spirals ...... $^3/_8$ (9.5)
      Shells, folded plate members:
        No. 6 bar and larger .... $^5/_8$ (16)
        No. 5 bar, W31 or D31 wire,
         and smaller .... $^3/_8$ (9.5)

1907.7.3 Prestressed concrete.

1907.7.3.1 The following minimum concrete cover shall be provided for prestressed and nonprestressed reinforcement, ducts and end fittings, except as provided in Sections 1907.7.3.2 and 1907.7.3.3.

MINIMUM COVER,
inches (mm)

1. Concrete cast against and
   permanently exposed to earth 3 (76)
2. Concrete exposed to earth or
   weather:
      Wall panels, slabs, joists 1 (25)
      Other members .... $1^1/_2$ (32)
3. Concrete not exposed to weather or
   in contact with ground:
      Slabs, walls, joists .... $^3/_4$ (19)
      Beams, columns:
        Primary reinforcement ...... $1^1/_2$ (38)
        Ties, stirrups, spirals 1 (25)
      Shells, folded plate members:
        No. 5 bars, W31 or D31 wire,
         and smaller .... $^3/_8$ (9.5)
        Other reinforcement .... d_b but not less than $^3/_4$ (19)

1907.7.3.2 For prestressed concrete members exposed to earth, weather or corrosive environments, and in which permissible tensile stress of Section 1918.4.2, Item 3, is exceeded, minimum cover shall be increased 50 percent.

1907.7.3.3 For prestressed concrete members manufactured under plant control conditions, minimum concrete cover for nonprestressed reinforcement shall be as required in Section 1907.7.2.

1907.7.4 Bundled bars. For bundled bars, minimum concrete cover shall be equal to the equivalent diameter of the bundle, but need not be greater than 2 inches (51 mm); except for concrete cast against and permanently exposed to earth, minimum cover shall be 3 inches (76 mm).

1907.7.5 Corrosive environments. In corrosive environments or other severe exposure conditions, amount of concrete protec-

tion shall be suitably increased, and denseness and nonporosity of protecting concrete shall be considered, or other protection shall be provided.

1907.7.6 Future extensions. Exposed reinforcement, inserts and plates intended for bonding with future extensions shall be protected from corrosion.

1907.7.7 Fire protection. When a thickness of cover for fire protection greater than the minimum concrete cover specified in Section 1907.7 is required, such greater thickness shall be used.

1907.8 Special Reinforcement Details for Columns.

1907.8.1 Offset bars. Offset bent longitudinal bars shall conform to the following:

1907.8.1.1 Slope of inclined portion of an offset bar with axis of column shall not exceed 1 in 6.

1907.8.1.2 Portions of bar above and below an offset shall be parallel to axis of column.

1907.8.1.3 Horizontal support at offset bends shall be provided by lateral ties, spirals or parts of the floor construction. Horizontal support provided shall be designed to resist one and one-half times the horizontal component of the computed force in the inclined portion of an offset bar. Lateral ties or spirals, if used, shall be placed not more than 6 inches (152 mm) from points of bend.

1907.8.1.4 Offset bars shall be bent before placement in the forms. See Section 1907.3.

1907.8.1.5 Where a column face is offset 3 inches (76 mm) or greater, longitudinal bars shall not be offset bent. Separate dowels, lap spliced with the longitudinal bars adjacent to the offset column faces, shall be provided. Lap splices shall conform to Section 1912.17.

1907.8.2 Steel cores. Load transfer in structural steel cores of composite compression members shall be provided by the following:

1907.8.2.1 Ends of structural steel cores shall be accurately finished to bear at end-bearing splices, with positive provision for alignment of one core above the other in concentric contact.

1907.8.2.2 At end-bearing splices, bearing shall be considered effective to transfer not more than 50 percent of the total compressive stress in the steel core.

1907.8.2.3 Transfer of stress between column base and footing shall be designed in accordance with Section 1915.8.

1907.8.2.4 Base of structural steel section shall be designed to transfer the total load from the entire composite member to the footing; or, the base may be designed to transfer the load from the steel core only, provided ample concrete section is available for transfer of the portion of the total load carried by the reinforced concrete section to the footing by compression in the concrete and by reinforcement.

1907.9 Connections.

1907.9.1 At connections of principal framing elements (such as beams and columns), enclosure shall be provided for splices of continuing reinforcement and for anchorage of reinforcement terminating in such connections.

1907.9.2 Enclosure at connections may consist of external concrete or internal closed ties, spirals or stirrups.

1907.10 Lateral Reinforcement for Compression Members.

1907.10.1 Lateral reinforcement for compression members shall conform to the provisions of Sections 1907.10.4 and 1907.10.5

and, where shear or torsion reinforcement is required, shall also conform to provisions of Section 1911.

1907.10.2 Lateral reinforcement requirements for composite compression members shall conform to Section 1910.16. Lateral reinforcement requirements for prestressing tendons shall conform to Section 1918.11.

1907.10.3 It shall be permitted to waive the lateral reinforcement requirements of Sections 1907.10, 1910.16 and 1918.11 where tests and structural analyses show adequate strength and feasibility of construction.

1907.10.4 Spirals. Spiral reinforcement for compression members shall conform to Section 1910.9.3 and to the following:

1907.10.4.1 Spirals shall consist of evenly spaced continuous bar or wire of such size and so assembled as to permit handling and placing without distortion from designed dimensions.

1907.10.4.2 For cast-in-place construction, size of spirals shall not be less than $3/8$-inch (9.5 mm) diameter.

1907.10.4.3 Clear spacing between spirals shall not exceed 3 inches (76 mm) or be less than 1 inch (25 mm). See also Section 1903.3.2.

1907.10.4.4 Anchorage of spiral reinforcement shall be provided by one and one-half extra turns of spiral bar or wire at each end of a spiral unit.

1907.10.4.5 Splices in spiral reinforcement shall be lap splices of $48d_b$, but not less than 12 inches (305 mm) or welded.

1907.10.4.6 Spirals shall extend from top of footing or slab in any story to level of lowest horizontal reinforcement in members supported above.

1907.10.4.7 Where beams or brackets do not frame into all sides of a column, ties shall extend above termination of spiral to bottom of slab or drop panel.

1907.10.4.8 In columns with capitals, spirals shall extend to a level at which the diameter or width of capital is two times that of the column.

1907.10.4.9 Spirals shall be held firmly in place and true to line.

1907.10.5 Ties. Tie reinforcement for compression members shall conform to the following:

1907.10.5.1 All nonprestressed bars shall be enclosed by lateral ties, at least No. 3 in size for longitudinal bars No. 10 or smaller, and at least No. 4 in size for Nos. 11, 14 and 18 and bundled longitudinal bars. Deformed wire or welded wire fabric of equivalent area shall be permitted.

1907.10.5.2 Vertical spacing of ties shall not exceed 16 longitudinal bar diameters, 48 tie bar or wire diameters, or least dimension of the compression member.

1907.10.5.3 Ties shall be arranged such that every corner and alternate longitudinal bar shall have lateral support provided by the corner of a tie with an included angle of not more than 135 degrees and a bar shall be not farther than 6 inches (152 mm) clear on each side along the tie from such a laterally supported bar. Where longitudinal bars are located around the perimeter of a circle, a complete circular tie shall be permitted.

1907.10.5.4 Ties shall be located vertically not more than one half a tie spacing above the top of footing or slab in any story and shall be spaced as provided herein to not more than one half a tie spacing below the lowest horizontal reinforcement *in members supported above.*

1907.10.5.5 Where beams or brackets frame from four directions into a column, termination of ties not more than 3 inches (76 mm) below reinforcement in shallowest of such beams or brackets shall be permitted.

1907.10.5.6 Column ties shall have hooks as specified in Section 1907.1.3.

1907.11 Lateral Reinforcement for Flexural Members.

1907.11.1 Compression reinforcement in beams shall be enclosed by ties or stirrups satisfying the size and spacing limitations in Section 1907.10.5 or by welded wire fabric of equivalent area. Such ties or stirrups shall be provided throughout the distance where compression reinforcement is required.

1907.11.2 Lateral reinforcement for flexural framing members subject to stress reversals or to torsion at supports shall consist of closed ties, closed stirrups, or spirals extending around the flexural reinforcement.

1907.11.3 Closed ties or stirrups may be formed in one piece by overlapping standard stirrup or tie end hooks around a longitudinal bar, or formed in one or two pieces lap spliced with a Class B splice (lap of 1.3 l_d), or anchored in accordance with Section 1912.13.

1907.12 Shrinkage and Temperature Reinforcement.

1907.12.1 Reinforcement for shrinkage and temperature stresses normal to flexural reinforcement shall be provided in structural slabs where the flexural reinforcement extends in one direction only.

1907.12.1.1 Shrinkage and temperature reinforcement shall be provided in accordance with either Section 1907.12.2 or 1907.12.3 below.

1907.12.1.2 Where shrinkage and temperature movements are significantly restrained, the requirements of Sections 1908.2.4 and 1909.2.7 shall be considered.

1907.12.2 Deformed reinforcement conforming to Section 1903.5.3 used for shrinkage and temperature reinforcement shall be provided in accordance with the following:

1907.12.2.1 Area of shrinkage and temperature reinforcement shall provide at least the following ratios of reinforcement area to gross concrete area, but not less than 0.0014:

1. Slabs where Grade 40 or 50 deformed bars are used   0.0020

2. Slabs where Grade 60 deformed bars or welded wire fabric (smooth or deformed) are used   0.0018

3. Slabs where reinforcement with yield stress exceeding 60,000 psi (413.7 MPa) measured at a yield strain of 0.35 percent is used

$$\frac{0.0018 \times 60,000}{f_y}$$

For **SI:** $\dfrac{0.0018 \times 413.7}{f_y}$

1907.12.2.2 Shrinkage and temperature reinforcement shall be spaced not farther apart than five times the slab thickness, or 18 inches (457 mm).

1907.12.2.3 At all sections where required, reinforcement for shrinkage and temperature stresses shall develop the specified yield strength f_y in tension in accordance with Section 1912.

1907.12.3 Prestressing tendons conforming to Section 1903.5.5 used for shrinkage and temperature reinforcement shall be provided in accordance with the following:

1907.12.3.1 Tendons shall be proportioned to provide a minimum average compressive stress of 100 psi (0.69 MPa) on gross concrete area using effective prestress, after losses, in accordance with Section 1918.6.

1907.12.3.2 Spacing of prestressed tendons shall not exceed 6 feet (1829 mm).

1907.12.3.3 When the spacing of prestressed tendons exceeds 54 inches (1372 mm), additional bonded shrinkage and temperature reinforcement conforming with Section 1907.12.2 shall be provided between the tendons at slab edges extending from the slab edge for a distance equal to the tendon spacing.

1907.13 Requirements for Structural Integrity.

1907.13.1 In the detailing of reinforcement and connections, members of a structure shall be effectively tied together to improve integrity of the overall structure.

1907.13.2 For cast-in-place construction, the following shall constitute minimum requirements:

1907.13.2.1 In joist construction, at least one bottom bar shall be continuous or shall be spliced over the support with a Class A tension splice and at noncontinuous supports be terminated with a standard hook.

1907.13.2.2 Beams at the perimeter of the structure shall have at least one sixth of the tension reinforcement required for negative moment at the support and one-quarter of the positive moment reinforcement required at midspan made continuous around the perimeter and tied with closed stirrups or stirrups anchored around the negative moment reinforcement with a hook having a bend of at least 135 degrees. Stirrups need not be extended through any joints. When splices are needed, the required continuity shall be provided with top reinforcement spliced at midspan and bottom reinforcement spliced at or near the support with Class A tension splices.

1907.13.2.3 In other than perimeter beams, when closed stirrups are not provided, at least one-quarter of the positive moment reinforcement required at midspan shall be continuous or shall be spliced over the support with a Class A tension splice and at noncontinuous supports be terminated with a standard hook.

1907.13.2.4 For two-way slab construction, see Section 1913.3.8.5.

1907.13.3 For precast concrete construction, tension ties shall be provided in the transverse, longitudinal, and vertical directions and around the perimeter of the structure to effectively tie elements together. The provisions of Section 1916.5 shall apply.

1907.13.4 For lift-slab construction, see Sections 1913.3.8.6 and 1918.12.6.

SECTION 1908 — ANALYSIS AND DESIGN

1908.0 Notations.

A_s = area of nonprestressed tension reinforcement, square inches (mm^2).

A'_s = area of compression reinforcement, square inches (mm^2).

b = width of compression face of member, inches (mm).

d = distance from extreme compression fiber to centroid of tension reinforcement, inches (mm).

E_c = modulus of elasticity of concrete, pounds per square inch (MPa). See Section 1908.5.1.

E_s = modulus of elasticity of reinforcement, pounds per square inch (MPa). See Sections 1908.2 and 1908.5.3.

f'_c = specified compressive strength of concrete, pounds per square inch (MPa).

f_y = specified yield strength of nonprestressed reinforcement, pounds per square inch (MPa).

l_n = clear span for positive moment or shear and average of adjacent clear spans for negative moment.

V_c = nominal shear strength provided by concrete.

w_c = unit weight of concrete, pounds per cubic foot (kg/m^3).

w_u = factored load per unit length of beam or per unit area of slab.

β_1 = factor defined in Section 1910.2.7.3.

ε_t = net tensile strain in extreme tension steel at nominal strength.

ρ = ratio of nonprestressed tension reinforcement.

   = A_s/bd.

ρ' = ratio of nonprestressed compression reinforcement.

   = A'_s/bd.

ρ_b = reinforcement ratio producing balanced strain conditions. See Section 1910.3.2.

ϕ = strength-reduction factor. See Section 1909.3.

1908.1 Design Methods.

1908.1.1 In design of structural concrete, members shall be proportioned for adequate strength in accordance with provisions of this code, using load factors and strength-reduction factors ϕ specified in Section 1909.

1908.1.2 Nonprestressed reinforced concrete members shall be permitted to be designed using the provisions of Section 1926.

1908.1.3 Design of reinforced concrete using Section 1927 shall be permitted.

1908.2 Loading.

1908.2.1 Design provisions of this code are based on the assumption that structures shall be designed to resist all applicable loads.

1908.2.2 Service loads shall be in accordance with Chapter 16 with appropriate live load reductions as permitted therein.

1908.2.3 In design for wind and earthquake loads, integral structural parts shall be designed to resist the total lateral loads.

1908.2.4 Consideration shall be given to effects of forces due to prestressing, crane loads, vibration, impact, shrinkage, temperature changes, creep, expansion of shrinkage-compensating concrete and unequal settlement of supports.

1908.3 Methods of Analysis.

1908.3.1 All members of frames or continuous construction shall be designed for the maximum effects of factored loads as determined by the theory of elastic analysis, except as modified by Section 1908.4. It is permitted to simplify the design by using the assumptions specified in Sections 1908.6 through 1908.9.

1908.3.2 Except for prestressed concrete, approximate methods of frame analysis may be used for buildings of usual types of construction, spans and story heights.

1908.3.3 As an alternate to frame analysis, the following approximate moments and shears shall be permitted to be used in design of continuous beams and one-way slabs (slabs reinforced to resist flexural stresses in only one direction), provided:

1. There are two or more spans,

2. Spans are approximately equal, with the larger of two adjacent spans not greater than the shorter by more than 20 percent,

3. Loads are uniformly distributed, and

4. Unit live load does not exceed three times unit dead load, and

5. Members are prismatic.

Positive moment:

End spans

Discontinuous end unrestrained . . . . $w_u l_n{}^2/11$

Discontinuous end integral with support . . . . . $w_u l_n{}^2/14$

Interior spans . . . . $w_u l_n{}^2/16$

Negative moment at exterior face of first interior support

Two spans . . . . $w_u l_n{}^2/9$

More than two spans . . . . $w_u l_n{}^2/10$

Negative moment at other faces of
interior supports . . . . $w_u l_n{}^2/11$

Negative moment at face of all supports for:

Slabs with spans not exceeding 10 feet (3048 mm),
and beams where ratio of sum of column stiffnesses
to beam stiffness exceeds eight at each end of
the span . . . . $w_u l_n{}^2/12$

Negative moment at interior face of exterior support for
members built integrally with supports:

Where support is a spandrel beam . . . . $w_u l_n{}^2/24$

Where support is a column . . . . $w_u l_n{}^2/16$

Shear in end members at face of first
interior support . . . . $1.15\, w_u l_n/2$

Shear at face of all other supports . . . . $w_u l_n/2$

1908.4 Redistribution of Negative Moments in Continuous Nonprestressed Flexural Members.

1908.4.1 Except where approximate values for moments are used, it is permitted to increase or decrease negative moments calculated by elastic theory at supports of continuous flexural members for any assumed loading arrangement by not more than

$$20 \left(1 - \frac{\rho - \rho'}{\rho_b} \right) \text{percent}$$

1908.4.2 The modified negative moments shall be used for calculating moments at sections within the spans.

1908.4.3 Redistribution of negative moments shall be made only when the section, at which moment is reduced, is so designed that ρ or $\rho - \rho'$ is not greater than $0.50\, \rho_b$, where

$$\rho_b = \frac{0.85\, \beta_1\, f'_c}{f_y} \frac{87,000}{87,000 + f_y} \qquad (8\text{-}1)$$

For **SI:**
$$\rho_b = \frac{0.85\, \beta_1\, f'_c}{f_y} \frac{600}{600 + f_y}$$

1908.4.4 *For criteria on moment redistribution for prestressed concrete members, see Section 1918.*

1908.5 Modulus of Elasticity.

1908.5.1 Modulus of elasticity E_c for concrete shall be permitted to be taken as $w_c^{1.5} 33 \sqrt{f'_c}$ (in psi) [For **SI:** $w_c^{1.5} 0.043 \sqrt{f'_c}$ (in MPA)] for values of w_c between 90 pcf and 155 pcf (1440 kg/m^3

and 2420 kg/m^3). For normal-weight concrete, E_c shall be permitted to be taken as $57,000 \sqrt{f'_c}$ (For **SI:** $4730 \sqrt{f'_c}$).

1908.5.2 Modulus of elasticity E_s for nonprestressed reinforcement shall be permitted to be taken as 29,000,000 psi (200 000 MPa).

1908.5.3 Modulus of elasticity E_s for prestressing tendons shall be determined by tests or supplied by the manufacturer.

1908.6 Stiffness.

1908.6.1 Use of any set of reasonable assumptions shall be permitted for computing relative flexural and torsional stiffnesses of columns, walls, floors and roof systems. The assumptions adopted shall be consistent throughout analysis.

1908.6.2 Effect of haunches shall be considered both in determining moments and in design of members.

1908.7 Span Length.

1908.7.1 Span length of members not built integrally with supports shall be considered the clear span plus depth of member, but need not exceed distance between centers of supports.

1908.7.2 In analysis of frames or continuous construction for determination of moments, span length shall be taken as the distance center to center of supports.

1908.7.3 For beams built integrally with supports, design on the basis of moments at faces of support shall be permitted.

1908.7.4 It shall be permitted to analyze solid or ribbed slabs built integrally with supports, with clear spans not more than 10 feet (3048 mm), as continuous slabs on knife edge supports with spans equal to the clear spans of the slab and width of beams otherwise neglected.

1908.8 Columns.

1908.8.1 Columns shall be designed to resist the axial forces from factored loads on all floors or roof and the maximum moment from factored loads on a single adjacent span of the floor or roof under consideration. Loading condition giving the maximum ratio of moment to axial load shall also be considered.

1908.8.2 In frames or continuous construction, consideration shall be given to the effect of unbalanced floor or roof loads on both exterior and interior columns and of eccentric loading due to other causes.

1908.8.3 In computing gravity load moments in columns, it shall be permitted to assume far ends of columns built integrally with the structure to be fixed.

1908.8.4 Resistance to moments at any floor or roof level shall be provided by distributing the moment between columns immediately above and below the given floor in proportion to the relative column stiffnesses and conditions of restraint.

1908.9 Arrangement of Live Load.

1908.9.1 It is permissible to assume that:

1. the live load is applied only to the floor or roof under consideration, and

2. the far ends of columns built integrally with the structure are considered to be fixed.

1908.9.2 It is permitted to assume that the arrangement of live load is limited to combinations of:

1. Factored dead load on all spans with full-factored live load on two adjacent spans, and

2. Factored dead load on all spans with full-factored live load on alternate spans.

1908.10 T-beam Construction.

1908.10.1 In T-beam construction, the flange and web shall be built integrally or otherwise effectively bonded together.

1908.10.2 Width of slab effective as a T-beam flange shall not exceed one fourth the span length of the beam, and the effective overhanging slab width on each side of the web shall not exceed:

1. Eight times the slab thickness, or

2. One half the clear distance to the next web.

1908.10.3 For beams with a slab on one side only, the effective overhanging flange width shall not exceed:

1. One twelfth the span length of the beam,

2. Six times the slab thickness, or

3. One half the clear distance to the next web.

1908.10.4 Isolated beams, in which the T-shape is used to provide a flange for additional compression area, shall have a flange thickness not less than one half the width of web and an effective flange width not more than four times the width of web.

1908.10.5 Where primary flexural reinforcement in a slab that is considered as a T-beam flange (excluding joist construction) is parallel to the beam, reinforcement perpendicular to the beam shall be provided in the top of the slab in accordance with the following:

1908.10.5.1 Transverse reinforcement shall be designed to carry the factored load on the overhanging slab width assumed to act as a cantilever. For isolated beams, the full width of overhanging flange shall be considered. For other T-beams, only the effective overhanging slab width need be considered.

1908.10.5.2 Transverse reinforcement shall be spaced not farther apart than five times the slab thickness or 18 inches (457 mm).

1908.11 Joist Construction.

1908.11.1 Joist construction consists of a monolithic combination of regularly spaced ribs and a top slab arranged to span in one direction or two orthogonal directions.

1908.11.2 Ribs shall not be less than 4 inches (102 mm) in width and shall have a depth of not more than three and one-half times the minimum width of rib.

1908.11.3 Clear spacing between ribs shall not exceed 30 inches (762 mm).

1908.11.4 Joist construction not meeting the limitations of the preceding two paragraphs shall be designed as slabs and beams.

1908.11.5 When permanent burned clay or concrete tile fillers of material having a unit compressive strength at least equal to that of the specified strength of concrete in the joists are used:

1908.11.5.1 For shear and negative-moment strength computations, it shall be permitted to include the vertical shells of fillers in contact with ribs. Other portions of fillers shall not be included in strength computations.

1908.11.5.2 Slab thickness over permanent fillers shall not be less than one twelfth the clear distance between ribs nor less than $1^1/_2$ inches (38 mm).

1908.11.5.3 In one-way joists, reinforcement normal to the ribs shall be provided in the slab as required by Section 1907.12.

1908.11.6 When removable forms or fillers not complying with Section 1908.11.5 are used:

1908.11.6.1 Slab thickness shall not be less than one twelfth the clear distance between ribs, or less than 2 inches (51 mm).

1908.11.6.2 Reinforcement normal to the ribs shall be provided in the slab as required for flexure, considering load concentrations, if any, but not less than required by Section 1907.12.

1908.11.7 Where conduits or pipes as permitted by Section 1906.3 are embedded within the slab, slab thickness shall be at least 1 inch (25 mm) greater than the total overall depth of the conduits or pipes at any point. Conduits or pipes shall not impair significantly the strength of the construction.

1908.11.8 For joist construction, contribution of concrete to shear strength V_c is permitted to be 10 percent more than that specified in Section 1911. It shall be permitted to increase shear strength using shear reinforcement or by widening the ends of the ribs.

1908.12 Separate Floor Finish.

1908.12.1 A floor finish shall not be included as part of a structural member unless placed monolithically with the floor slab or designed in accordance with requirements of Section 1917.

1908.12.2 It shall be permitted to consider all concrete floor finishes may be considered as part of required cover or total thickness for nonstructural considerations.

SECTION 1909 — STRENGTH AND SERVICEABILITY REQUIREMENTS

1909.0 Notations.

A_g = gross area of section, square inches (mm^2).

A'_s = area of compression reinforcement, square inches (mm^2).

b = width of compression face of member, inches (mm).

c = distance from extreme compression fiber to neutral axis in inches (mm).

D = dead loads, or related internal moments and forces.

d = distance from extreme compression fiber to centroid of tension reinforcement, inches (mm).

d' = distance from extreme compression fiber to centroid of compression reinforcement, inches (mm).

d_s = distance from extreme tension fiber to centroid of tension reinforcement, inches (mm).

d_t = distance from extreme compression fiber to extreme tension steel, inches (mm).

E = load effects of earthquake, or related internal moments and forces.

E_c = modulus of elasticity of concrete, pounds per square inch (MPa). See Section 1908.1.

F = loads due to weight and pressures of fluids with well-defined densities and controllable maximum heights, or related internal moments and forces.

f'_c = specified compressive strength of concrete, pounds per square inch (MPa).

$\sqrt{f'_c}$ = square root of specified compressive strength of concrete, pounds per square inch (MPa).

f_{ct} = average splitting tensile strength of lightweight aggregate concrete, pounds per square inch (MPa).

f_r = modulus of rupture of concrete, pounds per square inch (MPa).

f_y = specified yield strength of nonprestressed reinforcement, pounds per square inch (MPa).

H = loads due to weight and pressure of soil, water in soil, or other materials, or related internal moments and forces.

h = overall thickness of member, inches (mm).

I_{cr} = moment of inertia of cracked section transformed to concrete.

I_e = effective moment of inertia for computation of deflection.

I_g = moment of inertia of gross concrete section about centroidal axis, neglecting reinforcement.

L = live loads, or related internal moments and forces.

l = span length of beam or one-way slab, as defined in Section 1908.7; clear projection of cantilever, inches (mm).

l_n = length of clear span in long direction of two-way construction, measured face to face of supports in slabs without beams and face to face of beams or other supports in other cases.

M_a = maximum moment in member at stage deflection is computed.

M_{cr} = cracking moment. See Formula (9-8).

P_b = nominal axial load strength at balanced strain conditions. See Section 1910.3.2.

P_n = nominal axial load strength at given eccentricity.

T = cumulative effects of temperature, creep, shrinkage, differential settlement and shrinkage compensating concrete.

U = required strength to resist factored loads or related internal moments and forces.

W = wind load, or related internal moments and forces.

w_c = weight of concrete, pounds per cubic foot (kg/m^3).

y_t = distance from centroidal axis of gross section, neglecting reinforcement, to extreme fiber in tension.

α = ratio of flexural stiffness of beam section to flexural stiffness of a width of slab bounded laterally by center line of adjacent panel (if any) on each side of beam. See Section 1913.

α_m = average value of α for all beams on edges of a panel.

β = ratio of clear spans in long-to-short direction of two-way slabs.

ξ = time-dependent factor for sustained load. See Section 1909.5.2.5.

ε_t = net tensile strain in extreme tension steel at nominal strength.

λ = multiplier for additional long-time deflection as defined in Section 1909.5.2.5.

ρ = ratio of nonprestressed tension reinforcement, A_s/bd.

ρ' = reinforcement ratio for nonprestressed compression reinforcement, A'_g/bd.

ρ_b = reinforcement ratio producing balanced strain conditions. See Section B1910.3.2.

ϕ = strength-reduction factor. See Section 1909.3.

1909.1 General.

1909.1.1 Structures and structural members shall be designed to have design strengths at all sections at least equal to the required strengths calculated for the factored loads and forces in such combinations as are stipulated in this code.

1909.1.2 Members also shall meet all other requirements of this code to ensure adequate performance at service load levels.

1909.2 Required Strength.

1909.2.1 Required strength U to resist dead load D and live load L shall be at least equal to

$$U = 1.4D + 1.7L \qquad (9\text{-}1)$$

1909.2.2 If resistance to structural effects of a specified wind load W are included in design, the following combinations of D, L and W shall be investigated to determine the greatest required strength U

$$U = 0.75 (1.4D + 1.7L + 1.7W) \qquad (9\text{-}2)$$

where load combinations shall include both full value and zero value of L to determine the more severe condition, and

$$U = 0.9D + 1.3W \qquad (9\text{-}3)$$

but for any combination of D, L and W, required strength U shall not be less than Formula (9-1).

1909.2.3 If resistance to specified earthquake loads or forces E are included in design, load combinations of Section *1612.2.1* shall apply.

1909.2.4 If resistance to earth pressure H is included in design, required strength U shall be at least equal to

$$U = 1.4D + 1.7L + 1.7H \qquad (9\text{-}4)$$

except that where D or L reduces the effect of H, 0.9D shall be substituted for 1.4D and zero value of L shall be used to determine the greatest required strength U. For any combination of D, L and H, required strength U shall not be less than Formula (9-1).

1909.2.5 If resistance to loadings due to weight and pressure of fluids with well-defined densities and controllable maximum heights F is included in design, such loading shall have a load factor of 1.4 and be added to all loading combinations that include live load.

1909.2.6 If resistance to impact effects is taken into account in design, such effects shall be included with live load L.

1909.2.7 Where structural effects T of differential settlement, creep, shrinkage, expansion of shrinkage-compensating concrete or temperature change may be significant in design, required strength U shall be at least equal to

$$U = 0.75 (1.4D + 1.4T + 1.7L) \qquad (9\text{-}5)$$

but required strength U shall not be less than

$$U = 1.4 (D + T) \qquad (9\text{-}6)$$

Estimations of differential settlement, creep, shrinkage, expansion of shrinkage-compensating concrete or temperature change shall be based on a realistic assessment of such effects occurring in service.

1909.3 Design Strength.

1909.3.1 Design strength provided by a member, its connection to other members and its cross sections, in terms of flexure, axial load, shear and tension, shall be taken as the nominal strength calculated in accordance with requirements and assumptions of this code, multiplied by a strength-reduction factor ϕ in Sections 1909.3.2 and 1909.3.4.

1909.3.1.1 If the structural framing includes primary members of other materials proportioned to satisfy the load-factor combinations of Section 1928.1.2, it shall be permitted to proportion the concrete members using the set of strength-reduction factors, ϕ, listed in Section 1928.1.1 and the load-factor combinations in Section 1928.1.2.

1909.3.2 Strength-reduction factor ϕ shall be as follows:

1909.3.2.1 Flexure, without axial load 0.90

1909.3.2.2 Axial load and axial load with flexure. (For axial load with flexure, both axial load and moment nominal strength shall be multiplied by appropriate single value of ϕ.)

Axial tension and axial tension with flexure 0.90

Axial compression and axial compression with flexure:

Members with spiral reinforcement conforming to Section 1910.9.3 . . . . 0.75

Other reinforced members 0.70

except that for low values of axial compression, ϕ shall be permitted to be increased in accordance with the following:

For members in which f_y does not exceed 60,000 psi (413.7 MPa), with symmetric reinforcement, and with $(h - d' - d_s)/h$ not less than 0.70, ϕ shall be permitted to be increased linearly to 0.90 as ϕP_n decreases from $0.10 f'_c A_g$ to zero.

For other reinforced members, ϕ shall be permitted to be increased linearly to 0.90 as ϕP_n decreases from $0.10 f'_c A_g$ or ϕP_b, whichever is smaller, to zero.

1909.3.2.3 Shear and torsion *(See also Section 1909.3.4 for shear walls and frames in Seismic Zones 3 and 4)* 0.85

1909.3.2.4 Bearing on concrete (See also Section 1918.13) 0.70

1909.3.3 Development lengths specified in Section 1912 do not require a ϕ factor.

1909.3.4 *In Seismic Zones 3 and 4,* strength-reduction factors ϕ shall be as given above except for the following:

1909.3.4.1 *The shear strength-reduction factor shall be 0.6 for the design of walls, topping slabs used as diaphragms over precast concrete members and structural framing members, with the exception of joints, if their nominal shear strength is less than the shear corresponding to development of their nominal flexural strength.* The nominal flexural strength shall be determined *corresponding to* the most critical factored axial loads including earthquake effects. *The shear strength reduction factor for joints shall be 0.85.*

1909.3.4.2 *Reinforcement used for diaphragm chords or collectors placed in topping slabs over precast concrete members shall be designed using a strength-reduction factor of 0.6.*

1909.3.5 Strength reduction factor ϕ for flexure compression, shear and bearing of structural plain concrete in Section 1922 shall be 0.65.

1909.4 Design Strength for Reinforcement. Designs shall not be based on a yield strength of reinforcement f_y in excess of 80,000 psi (551.6 MPa), except for prestressing tendons.

1909.5 Control of Deflections.

1909.5.1 Reinforced concrete members subject to flexure shall be designed to have adequate stiffness to limit deflections or any deformations that affect strength or serviceability of a structure adversely.

1909.5.2 One-way construction (nonprestressed).

1909.5.2.1 Minimum thickness stipulated in Table 19-C-1 shall apply for one-way construction not supporting or attached to partitions or other construction likely to be damaged by large deflec-

tions, unless computation of deflection indicates a lesser thickness may be used without adverse effects.

1909.5.2.2 Where deflections are to be computed, deflections that occur immediately on application of load shall be computed by usual methods or formulas for elastic deflections, considering effects of cracking and reinforcement on member stiffness.

1909.5.2.3 Unless stiffness values are obtained by a more comprehensive analysis, immediate deflection shall be computed with the modulus of elasticity E_c for concrete as specified in Section 1908.5.1 (normal-weight or lightweight concrete) and with the effective moment of inertia as follows, but not greater than I_g.

$$I_e = \left(\frac{M_{cr}}{M_a}\right)^3 I_g + \left[1 - \left(\frac{M_{cr}}{M_a}\right)^3\right]I_{cr} \qquad (9\text{-}7)$$

WHERE:

$$M_{cr} = \frac{f_r I_g}{y_t} \qquad (9\text{-}8)$$

and for normal-weight concrete

$$f_r = 7.5 \sqrt{f'_c} \qquad (9\text{-}9)$$

For **SI:** $\qquad f_r = 0.62 \sqrt{f'_c}$

When lightweight aggregate concrete is used, one of the following modifications shall apply:

1. When f_{ct} is specified and concrete is proportioned in accordance with Section 1905.2, f_r shall be modified by substituting $f_{ct}/6.7$ (For **SI:** $1.8f_{ct}$) for $\sqrt{f'_c}$, but the value of $f_{ct}/6.7$ (For **SI:** $1.8f_{ct}$) shall not exceed $\sqrt{f'_c}$.

2. When f_{ct} is not specified, f_r shall be multiplied by 0.75 for "all-lightweight" concrete, and 0.85 for "sand-lightweight" concrete. Linear interpolation shall be permitted to be used when partial sand replacement is used.

1909.5.2.4 For continuous members, effective moment of inertia shall be permitted to be taken as the average of values obtained from Formula (9-7) for the critical positive and negative moment sections. For prismatic members, effective moment of inertia shall be permitted to be taken as the value obtained from Formula (9-7) at midspan for simple and continuous spans, and at support for cantilevers.

1909.5.2.5 Unless values are obtained by a more comprehensive analysis, additional longtime deflection resulting from creep and shrinkage of flexural members (normal-weight or lightweight concrete) shall be determined by multiplying the immediate deflection caused by the sustained load considered, by the factor

$$\lambda = \frac{\xi}{1 + 50\rho'} \qquad (9\text{-}10)$$

where ρ' shall be the value at midspan for simple and continuous spans, and at support for cantilevers. It is permitted to assume the time-dependent factor for sustained loads to be equal to

Five years or more	2.0
12 months	1.4
Six months	1.2
Three months	1.0

1909.5.2.6 Deflection computed in accordance with this section shall not exceed limits stipulated in Table 19-C-2.

1909.5.3 Two-way construction (nonprestressed).

1909.5.3.1 This section shall govern the minimum thickness of slabs or other two-way construction designed in accordance with the provisions of Section 1913 and conforming with the requirements of Section 1913.6.1.2. The thickness of slabs without interior beams spanning between the supports on all sides shall satisfy the requirements of Section 1909.5.3.2 or 1909.5.3.4. Thickness of slabs with beams spanning between the supports on all sides shall satisfy the requirements of Section 1909.5.3.3 or 1909.5.3.4.

1909.5.3.2 For slabs without interior beams spanning between the supports and having a ratio of long to short span not greater than 2, the minimum thickness shall be in accordance with the provisions of Table 19-C-3 and shall not be less than the following values:

1. Slabs without drop panels as defined in
   Sections 1913.3.7.1 and 1913.3.7.2 . . 5 inches (127 mm)

2. Slabs with drop panels as defined in
   Sections 1913.3.7.1 and 1913.3.7.2 . . 4 inches (102 mm)

1909.5.3.3 For slabs with beams spanning between the supports on all sides, the minimum thickness shall be as follows:

1. For α_m equal to or less than 0.2, the provisions of Section 1909.5.3.2 shall apply.

2. For α_m greater than 0.2 but not greater than 2.0, the thickness shall not be less than

$$h = \frac{l_n\left(0.8 + \frac{f_y}{200,000}\right)}{36 + 5\beta(\alpha_m - 0.2)} \qquad (9\text{-}11)$$

For **SI:** $\qquad h = \dfrac{l_n\left(0.8 + \frac{f_y}{1370}\right)}{36 + 5\beta(\alpha_m - 0.2)}$

but not less than 5 inches (127 mm).

3. For α_m greater than 2.0, the thickness shall not be less than

$$h = \frac{l_n\left(0.8 + \frac{f_y}{200,000}\right)}{36 + 9\beta} \qquad (9\text{-}12)$$

For **SI:** $\qquad h = \dfrac{l_n\left(0.8 + \frac{f_y}{1370}\right)}{36 + 9\beta}$

but not less than 3.5 inches (89 mm).

4. At discontinuous edges, an edge beam shall be provided with a stiffness ratio α not less than 0.80; or the minimum thickness required by Formula (9-11) or (9-12) shall be increased by at least 10 percent in the panel with a discontinuous edge.

1909.5.3.4 Slab thickness less than the minimum thickness required by Sections 1909.5.3.1, 1909.5.3.2 and 1909.5.3.3 shall be permitted to be used if shown by computation that the deflection will not exceed the limits stipulated in Table 19-C-2. Deflections shall be computed taking into account size and shape of the panel, conditions of support, and nature of restraints at the panel edges. The modulus of elasticity of concrete E_c shall be as specified in Section 1908.5.1. The effective moment of inertia shall be that given by Formula (9-7); other values shall be permitted to be used if they result in computed deflections in reasonable agreement with the results of comprehensive tests. Additional long-term deflection shall be computed in accordance with Section 1909.5.2.5.

1909.5.4 Prestressed concrete construction.

1909.5.4.1 For flexural members designed in accordance with provisions of Section 1918, immediate deflection shall be computed by usual methods or formulas for elastic deflections, and the moment of inertia of the gross concrete section shall be permitted to be used for uncracked sections.

1909.5.4.2 Additional long-time deflection of prestressed concrete members shall be computed taking into account stresses in concrete and steel under sustained load and including effects of creep and shrinkage of concrete and relaxation of steel.

1909.5.4.3 Deflection computed in accordance with this section shall not exceed limits stipulated in Table *19-C-2*.

1909.5.5 Composite construction.

1909.5.5.1 Shored construction. If composite flexural members are supported during construction so that, after removal of temporary supports, dead load is resisted by the full composite section, it shall be permitted to consider the composite member equivalent to a monolithically cast member for computation of deflection. For nonprestressed members, the portion of the member in compression shall determine whether values in Table 19-C-1 for normal-weight or lightweight concrete shall apply. If deflection is computed, account shall be taken of curvatures resulting from differential shrinkage of precast and cast-in-place components, and of axial creep effects in a prestressed concrete member.

1909.5.5.2 Unshored construction. If the thickness of a nonprestressed precast flexural member meets the requirements of Table 19-C-1, deflection need not be computed. If the thickness of a nonprestressed composite member meets the requirements of Table *19-C-1*, it is not required to compute deflection occurring after the member becomes composite, but the long-time deflection of the precast member shall be investigated for magnitude and duration of load prior to beginning of effective composite action.

1909.5.5.3 Deflection computed in accordance with this section shall not exceed limits stipulated in *Table 19-C-2*.

SECTION 1910 — FLEXURE AND AXIAL LOADS

1910.0 Notations.

A = effective tension area of concrete surrounding the flexural tension reinforcement and having the same centroid as that reinforcement, divided by the number of bars or wires, square inches (mm²). When the flexural reinforcement consists of different bar or wire sizes, the number of bars or wires shall be computed as the total area of reinforcement divided by the area of the largest bar or wire used.

A_c = area of core of spirally reinforced compression member measured to outside diameter of spiral, square inches (mm²).

A_g = gross area of section, square inches (mm²).

A_s = area of nonprestressed tension reinforcement, square inches (mm²).

$A_{s,min}$ = minimum amount of flexural reinforcement, inches squared (mm²). See Section 1910.5.

A_{sk} = area of skin reinforcement per unit height in one side face, square inches per foot (mm²/m).

A_{st} = total area of longitudinal reinforcement (bars or steel shapes), square inches (mm²).

A_t = area of structural steel shape, pipe or tubing in a composite section, square inches (mm²).

A_1 = loaded area.

A_2 = the area of the lower base of the largest frustum of a pyramid, cone, or tapered wedge contained wholly within the support and having for its upper base the loaded area, and having side slopes of 1 unit vertical in 2 units horizontal (50% slope).

a = depth of equivalent rectangular stress block as defined in Section 1910.2.7.1.

b = width of compression face of member, inches (mm).

b_w = web width, inches (mm).

C_m = a factor relating actual moment diagram to an equivalent uniform moment diagram.

c = distance from extreme compression fiber to neutral axis, inches (mm).

d = distance from extreme compression fiber to centroid of tension reinforcement, inches (mm).

d_c = thickness of concrete cover measured from extreme tension fiber to center of bar or wire located closest thereto, inches (mm).

d_t = distance from extreme compression fiber to extreme tension steel, inches (mm).

E_c = modulus of elasticity of concrete, pounds per square inch (MPa). See Section 1908.5.1.

E_s = modulus of elasticity of reinforcement, pounds per square inch (MPa). See Sections 1908.5.2 and 1908.5.3.

EI = flexural stiffness of compression member. See Formula (10-9).

f'_c = specified compressive strength of concrete, pounds per square inch (MPa).

f_s = calculated stress in reinforcement at service loads, kips per square inch (MPa).

f_y = specified yield strength of nonprestressed reinforcement, pounds per square inch (MPa).

h = overall dimension of member in direction of action considered, inches (mm).

I_g = moment of inertia of gross concrete section about centroidal axis, neglecting reinforcement.

I_{se} = moment of inertia of reinforcement about centroidal axis of member cross section.

I_t = moment of inertia of structural steel shape, pipe or tubing about centroidal axis of composite member cross section.

k = effective length factor for compression members.

l_c = length of a compression member in a frame, measured from center to center of the joints in the frame.

l_u = unsupported length of compression member.

M_c = factored moment to be used for design of compression member.

M_s = moment due to loads causing appreciable sway.

M_u = factored moment at section.

M_1 = smaller factored end moment on a compression member, positive if member is bent in single curvature, negative if bent in double curvature.

M_{1ns} = factored end moment on a compression member at the end at which M_1 acts, due to loads that cause no appreciable sidesway, calculated using a first-order elastic frame analysis.

M_{1s} = factored end moment on compression members at the end at which M_1 acts, due to loads that cause appreciable sidesway, calculated using a first-order elastic frame analysis.

M_2 = larger factored end moment on compression member, always positive.

$M_{2,min}$ = minimum value of M_2.

M_{2ns} = factored end moment on compression member at the end at which M_2 acts, due to loads that cause no appreciable sidesway, calculated using a first-order elastic frame analysis.

M_{2s} = factored end moment on compression member at the end at which M_2 acts, due to loads that cause appreciable sidesway, calculated using a first-order elastic frame analysis.

P_b = nominal axial load strength at balanced strain conditions. See Section 1910.3.2.

P_c = critical load. See Formula (10-9).

P_n = nominal axial load strength at given eccentricity.

P_o = nominal axial load strength at zero eccentricity.

P_u = factored axial load at given eccentricity $\leq \phi P_n$.

Q = stability index for a story. See Section 1910.11.4.

r = radius of gyration of cross section of a compression member.

V_u = factored horizontal shear in a story.

z = quantity limiting distribution of flexural reinforcement. See Section 1910.6.

β_1 = factor defined in Section 1910.2.7.3.

β_d = (a) for nonsway frames, β_d is the ratio of the maximum factored axial dead load to the total factored axial load.

= (b) for sway frames, except as required in Item 3, β_d is the ratio of the maximum factored sustained shear within a story to the total factored shear in that story.

= (c) for stability checks of sway frames carried out in accordance with Section 1910.13.6, β_d is the ratio of the maximum factored sustained axial load to the total factored axial load.

Δ_o = relative lateral deflection between the top and bottom of a story due to V_u, computed using a first-order elastic frame analysis and stiffness values satisfying Section 1910.11.1.

δ_{ns} = moment magnification factor for frames braced against sidesway to reflect effects of member curvature between ends of compression members.

δ_s = moment magnification factor for frames not braced against sidesway to reflect lateral drift resulting from lateral and gravity loads.

ε_t = net tensile strain in extreme tension steel at nominal strength.

ρ = ratio of nonprestressed tension reinforcement.

= A_s/bd.

ρ_b = reinforcement ratio producing balanced strain conditions. See Section 1910.3.2.

ρ_s = ratio of volume of spiral reinforcement to total volume of core (out-to-out of spirals) of a spirally reinforced compression member.

ϕ = strength-reduction factor. See Section 1909.3.

ϕ_k = stiffness reduction factor.

1910.1 Scope. Provisions of Section 1910 shall apply for design of members subject to flexure or axial loads or to combined flexure and axial loads.

1910.2 Design Assumptions.

1910.2.1 Strength design of members for flexure and axial loads shall be based on assumptions given in *the following items* and on satisfaction of applicable conditions of equilibrium and compatibility of strains.

1910.2.2 Strain in reinforcement and concrete shall be assumed directly proportional to the distance from the neutral axis, except, for deep flexural members with overall depth-to-clear-span ratios greater than two fifths for continuous spans and four fifths for simple spans, a nonlinear distribution of strain shall be considered. See Section 1910.7.

1910.2.3 Maximum usable strain at extreme concrete compression fiber shall be assumed equal to 0.003.

1910.2.4 Stress in reinforcement below specified yield strength f_y for grade of reinforcement used shall be taken as E_s times steel strain. For strains greater than that corresponding to f_y, stress in reinforcement shall be considered independent of strain and equal to f_y.

1910.2.5 Tensile strength of concrete shall be neglected in axial and flexural calculations of reinforced concrete, except where meeting requirements of Section 1918.4.

1910.2.6 Relationship between concrete compressive stress distribution and concrete strain shall be assumed to be rectangular, trapezoidal, parabolic or any other shape that results in prediction of strength in substantial agreement with results of comprehensive tests.

1910.2.7 Requirements of Section 1910.2.6 may be considered satisfied by an equivalent rectangular concrete stress distribution defined by the following:

1910.2.7.1 Concrete stress of $0.85f'_c$ shall be assumed uniformly distributed over an equivalent compression zone bounded by edges of the cross section and a straight line located parallel to the neutral axis at a distance $a = \beta_1 c$ from the fiber of maximum compressive strain.

1910.2.7.2 Distance c from fiber of maximum strain to the neutral axis shall be measured in a direction perpendicular to the axis.

1910.2.7.3 Factor β_1 shall be taken as 0.85 for concrete strengths f'_c up to and including 4,000 psi (27.58 MPa). For strengths above 4,000 psi (27.58 MPa), β_1 shall be reduced continuously at a rate of 0.05 for each 1,000 psi (6.89 MPa) of strength in excess of 4,000 psi (27.58 MPa), but β_1 shall not be taken less than 0.65.

1910.3 General Principles and Requirements.

1910.3.1 Design of cross section subject to flexure or axial loads or to combined flexure and axial loads shall be based on stress and strain compatibility using assumptions in Section 1910.2.

1910.3.2 Balanced strain conditions exist at a cross section when tension reinforcement reaches the strain corresponding to its specified yield strength f_y just as concrete in compression reaches its assumed ultimate strain of 0.003.

1910.3.3 For flexural members, and for members subject to combined flexure and compressive axial load when the design axial load strength ϕP_n is less than the smaller of $0.10f'_c A_g$ or ϕP_b, the ratio of reinforcement ρ provided shall not exceed 0.75 of the ratio ρ_b that would produce balanced strain conditions for the section

under flexure without axial load. For members with compression reinforcement, the portion of ρ_b equalized by compression reinforcement need not be reduced by the 0.75 factor.

1910.3.4 Use of compression reinforcement shall be permitted in conjunction with additional tension reinforcement to increase the strength of flexural members.

1910.3.5 Design axial load strength ϕP_n of compression members shall not be taken greater than the following:

1910.3.5.1 For nonprestressed members with spiral reinforcement conforming to Section 1907.10.4 or composite members conforming to Section 1910.16:

$$\phi P_{n(max.)} = 0.85\phi[0.85f'_c(A_g - A_{st}) + f_yA_{st}] \quad (10\text{-}1)$$

1910.3.5.2 For nonprestressed members with tie reinforcement conforming to Section 1907.10.5:

$$\phi P_{n(max.)} = 0.80\phi[0.85f'_c(A_g - A_{st}) + f_yA_{st}] \quad (10\text{-}2)$$

1910.3.5.3 For prestressed members, design axial load strength ϕP_n shall not be taken greater than 0.85 (for members with spiral reinforcement) or 0.80 (for members with tie reinforcement) of the design axial load strength at zero eccentricity ϕP_o.

1910.3.6 Members subject to compressive axial load shall be designed for the maximum moment that can accompany the axial load. The factored axial load P_u at given eccentricity shall not exceed that given in Section 1910.3.5. The maximum factored moment M_u shall be magnified for slenderness effects in accordance with Section 1910.10.

1910.4 Distance between Lateral Supports of Flexural Members.

1910.4.1 Spacing of lateral supports for a beam shall not exceed 50 times the least width b of compression flange or face.

1910.4.2 Effects of lateral eccentricity of load shall be taken into account in determining spacing of lateral supports.

1910.5 Minimum Reinforcement of Flexural Members.

1910.5.1 At every section of a flexural member where tensile reinforcement is required by analysis, except as provided in Sections 1910.5.2, 1910.5.3 and 1910.5.4, the area A_s provided shall not be less than that given by:

$$A_{s,min} = \frac{3\sqrt{f'_c}}{f_y} b_wd \text{ and not less than } \frac{200b_wd}{f_y} \quad (10\text{-}3)$$

1910.5.2 For a statically determinate T-section with flange in tension, the area $A_{s,min}$ shall be equal to or greater than the smaller value given either by:

$$A_{s,min} = \frac{6\sqrt{f'_c}}{f_y} b_wd \quad (10\text{-}4)$$

or Formula (10-3) with b_w set equal to the width of the flange.

1910.5.3 The requirements of Sections 1910.5.1 and 1910.5.2 need not be applied if at every section the area of tensile reinforcement provided is at least one-third greater than that required by analysis.

1910.5.4 For structural slabs and footings of uniform thickness, the minimum area of tensile reinforcement in the direction of span shall be the same as that required by Section 1907.12. Maximum spacing of this reinforcement shall not exceed the lesser of three times the thickness and 18 inches (457 mm).

1910.6 Distribution of Flexural Reinforcement in Beams and One-way Slabs.

1910.6.1 The rules for distribution of flexural reinforcement to control flexural cracking in beams and in one-way slabs (slabs reinforced to resist flexural stresses in only one direction) are as follows:

1910.6.2 Distribution of flexural reinforcement in two-way slabs shall be as required by Section 1913.3.

1910.6.3 Flexural tension reinforcement shall be well distributed within maximum flexural tension zones of a member cross section as required by Section 1910.6.4.

1910.6.4 When design yield strength f_y for tension reinforcement exceeds 40,000 psi (275.8 MPa), cross sections of maximum positive and negative moment shall be so proportioned that the quantity z given by:

$$z = f_s \sqrt[3]{d_c A} \qquad (10\text{-}5)$$

does not exceed 175 kips per inch (30.6 MN/m) for interior exposure and 145 kips per inch (25.4 MN/m) for exterior exposure. Calculated stress in reinforcement at service load f_s (kips per square inch) (MPa) shall be computed as the moment divided by the product of steel area and internal moment arm. Alternatively, it shall be permitted to take f_s as 60 percent of specified yield strength f_y.

1910.6.5 Provisions of Section 1910.6.4 may not be sufficient for structures subject to very aggressive exposure or designed to be watertight. For such structures, special investigations and precautions are required.

1910.6.6 Where flanges of T-beam construction are in tension, part of the flexural tension reinforcement shall be distributed over an effective flange width as defined in Section 1908.10, or a width equal to one tenth the span, whichever is smaller. If the effective flange width exceeds one tenth the span, some longitudinal reinforcement shall be provided in the outer portions of the flange.

1910.6.7 If the effective depth d of a beam or joist exceeds 36 inches (914 mm), longitudinal skin reinforcement shall be uniformly distributed along both side faces of the member for a distance $d/2$ nearest the flexural tension reinforcement. The area of skin reinforcement, A_{sk}, per foot (per mm) of height on each side face shall be $\geq 0.012 (d-30)$ [For **SI:** $\geq 0.012 (d-762)$]. The maximum spacing of the skin reinforcement shall not exceed the lesser of $d/6$ and 12 inches (305 mm). It shall be permitted to include such reinforcement in strength computations if a strain compatibility analysis is made to determine stresses in the individual bars or wires. The total area of longitudinal skin reinforcement in both faces need not exceed one half of the required flexural tensile reinforcement.

1910.7 Deep Flexural Members.

1910.7.1 Flexural members with overall depth-to-clear-span ratios greater than two fifths for continuous spans, or four fifths for simple spans, shall be designed as deep flexural members, taking into account nonlinear distribution of strain and lateral buckling.

1910.7.2 Shear strength of deep flexural members shall be in accordance with Section 1911.8.

1910.7.3 Minimum flexural tension reinforcement shall conform to Section 1910.5.

1910.7.4 Minimum horizontal and vertical reinforcement in the side faces of deep flexural members shall be the greater of the requirements of Sections 1911.8.8 and 1911.8.9 or Sections 1914.3.2 and 1914.3.3.

1910.8 Design Dimensions for Compression Members.

1910.8.1 Isolated compression member with multiple spirals. Outer limits of the effective cross section of a compression member with two or more interlocking spirals shall be taken at a distance outside the extreme limits of the spirals equal to the minimum concrete cover required by Section 1907.7.

1910.8.2 Compression member built monolithically with wall. Outer limits of the effective cross section of a spirally reinforced or tied reinforced compression member built monolithically with a concrete wall or pier shall be taken not greater than $1^1/_2$ inches (38 mm) outside the spiral or tie reinforcement.

1910.8.3 As an alternate to using the full gross area for design of a compressive member with a square, octagonal or other shaped cross section, it shall be permitted to use a circular section with a diameter equal to the least lateral dimension of the actual shape.

1910.8.4 Limits of section. For a compression member with a cross section larger than required by considerations of loading, it shall be permitted to base the minimum reinforcement and *design* strength on a reduced effective area A_g not less than one half the total area. This provision shall not apply in Seismic Zones 3 and 4.

1910.9 Limits for Reinforcement of Compression Members.

1910.9.1 Area of longitudinal reinforcement for noncomposite compression members shall not be less than 0.01 or more than 0.08 times gross area A_g of section.

1910.9.2 Minimum number of longitudinal bars in compression members shall be four for bars within rectangular or circular ties, three for bars within triangular ties, and six for bars enclosed by spirals conforming to *the following ratio:*

1910.9.3 Ratio of spiral reinforcement ρ_s shall not be less than the value given by

$$\rho_s = 0.45 \left(\frac{A_g}{A_c} - 1 \right) \frac{f'_c}{f_y} \qquad (10\text{-}6)$$

where f_y is the specified yield strength of spiral reinforcement but not more than 60,000 psi (413.7 MPa).

1910.10 Slenderness Effects in Compression Members.

1910.10.1 Except as allowed in Section 1910.10.2, the design of compression members, restraining beams and other supporting members shall be based on the factored forces and moments from a second order analysis considering materials nonlinearity and cracking, as well as the effects of member curvature and lateral drift, duration of loads, shrinkage and creep, and interaction with the supporting foundation. The dimensions of each member cross section used in the analysis shall be within 10 percent of the dimensions of the members shown on the design drawings and the analysis shall be repeated. The analysis procedure shall have been shown to result in prediction of strength in substantial agreement with the results of comprehensive tests of columns in statically indeterminate reinforced concrete structures.

1910.10.2 As an alternate of the procedure prescribed in Section 1910.10.1, it shall be permitted to base the design of compression members, restraining beams, and other supporting members on axial forces and moments from the analyses described in Section 1910.11.

1910.11 Magnified Moments—General.

1910.11.1 The factored axial forces, P_u, the factored moments, M_1 and M_2, at the ends of the column and, where required, the relative lateral story deflections, Δ_o, shall be computed using an elastic first-order frame analysis with the section properties determined taking into account the influence of axial loads, the presence of cracked regions along the length of the member and effects of duration of loads. Alternatively, it shall be permitted to use the following properties for the members in the structure:

1. Modulus of elasticity = E_c from Section 1908.5.1.

2. Moment of inertia:

Beams	$0.35\,I_g$
Columns	$0.70\,I_g$
Walls—Uncracked	$0.70\,I_g$
—Cracked	$0.35\,I_g$
Flat plates and flat slabs	$0.25\,I_g$

3. Area $1.0\,A_g$

The moments of inertia shall be divided by $(1 + ß_d)$ when:

1. sustained lateral loads act, or for

2. stability checks made in accordance with Section 1910.13.6.

1910.11.2 It shall be permitted to take the radius of gyration, r, equal to 0.30 times the overall dimension of the direction stability is being considered for rectangular compression members and 0.25 times the diameter for circular compression members. For other shapes, it shall be permitted to compute the radius of gyration for the gross concrete section.

1910.11.3 Unsupported length of compression members.

1910.11.3.1 The unsupported length l_u of a compression member shall be taken as the clear distance between floor slabs, beams or other members capable of providing lateral support in the direction being considered.

1910.11.3.2 Where column capitals or haunches are present, the unsupported length shall be measured to the lower extremity of the capital or haunch in the plane considered.

1910.11.4 Columns and stories in structures shall be designated as nonsway or sway columns or stories. The design of columns in nonsway frames or stories shall be based on Section 1910.12. The design of columns in sway frames or stories shall be based on Section 1910.13.

1910.11.4.1 It shall be permitted to assume a column in a structure is nonsway if the increase in column end moments due to second-order effects does not exceed 5 percent of the first-order end moments.

1910.11.4.2 It also shall be permitted to assume a story within a structure is nonsway if:

$$Q = \frac{\sum P_u \Delta_o}{V_u l_c} \text{ is less than or equal to } 0.05, \quad (10\text{-}7)$$

where $\sum P_u$ and V_u are the total vertical load and the story shear, respectively, in the story in question and Δ_o is the first-order relative deflection between the top and bottom of that story due to V_u.

1910.11.5 Where an individual compression member in the frame has a slenderness, kl_u/r, of more than 100, Section 1910.10.1 shall be used to compute the forces and moments in the frame.

1910.11.6 For compression members subject to bending about both principal axes, the moment about each axis shall be magnified separately based on the conditions of restraint corresponding to that axis.

1910.12 Magnified Moments—Nonsway Frames.

1910.12.1 For compression members in nonsway frames, the effective length factor k shall be taken as 1.0, unless analysis shows that a lower value is justified. The calculation of k shall be based on the E and I values used in Section 1910.11.1.

1910.12.2 In nonsway frames, it shall be permitted to ignore slenderness effect for compression members which satisfy:

$$\frac{kl_u}{r} \le 34 - 12\,(M_1/M_2) \quad (10\text{-}8)$$

where M_1/M_2 is not taken less than –0.5. The term M_1/M_2 is positive if the column is bent in single curvature.

1910.12.3 Compression members shall be designed for the factored axial load, P_u, and the moment amplified for the effects of member curvature, M_c, as follows:

$$M_c = \delta_{ns}\,M_2 \quad (10\text{-}9)$$

WHERE:

$$\delta_{ns} = \frac{C_m}{1 - \dfrac{P_u}{0.75P_c}} \ge 1.0 \quad (10\text{-}10)$$

$$P_c = \frac{\pi^2 EI}{(kl_u)^2} \quad (10\text{-}11)$$

EI shall be taken as

$$EI = \frac{(0.2E_c I_g + E_s I_{se})}{1 + \beta_d} \quad (10\text{-}12)$$

or

$$EI = \frac{0.40E_c\,I_g}{1 + \beta_d} \quad (10\text{-}13)$$

1910.12.3.1 For members without transverse loads between supports, C_m shall be taken as

$$C_m = 0.6 + 0.4\,\frac{M_1}{M_2} \ge 0.4 \quad (10\text{-}14)$$

where M_1/M_2 is positive if the column is bent in single curvature. For members with transverse loads between supports, C_m shall be taken as 1.0.

1910.12.3.2 The factored moment M_2 in Formula (10-9) shall not be taken less than

$$M_{2,\min} = P_u\,(0.6 + 0.03h) \quad (10\text{-}15)$$

about each axis separately, where 0.6 and h are in inches. For members for which $M_{2,\min}$ exceeds M_2, the value of C_m in Formula (10-14) shall either be taken equal to 1.0, or shall be based on the ratio of the computed end moments M_1 and M_2.

1910.13 Magnified Moments—Sway Frames.

1910.13.1 For compression members not braced against sidesway, the effective length factor k shall be determined using E and I values in accordance with Section 1910.11.1 and shall be greater than 1.0.

1910.13.2 For compression members not braced against sidesway, effects of slenderness may be neglected when kl_u/r is less than 22.

1910.13.3 The moments M_1 and M_2 at the ends of an individual compression member shall be taken as

$$M_1 = M_{1ns} + \delta_s M_{1s} \quad (10\text{-}16)$$

$$M_2 = M_{2ns} + \delta_s M_{2s} \qquad (10\text{-}17)$$

where $\delta_s M_s$ and $\delta_s M_{2s}$ shall be computed according to Section 1910.13.4.

1910.13.4 Calculation of $\delta_s M_s$.

1910.13.4.1 The magnified sway moments $\delta_s M_s$ shall be taken as the column end moments calculated using a second-order elastic analysis based on the member stiffnesses given in Section 1910.11.1.

1910.13.4.2 Alternatively, it shall be permitted to calculate $\delta_s M_s$ as

$$\delta_s M_s = \frac{M_s}{1 - Q} \geq M_s \qquad (10\text{-}18)$$

If δ_s calculated in this way exceeds 1.5, $\delta_s M_s$ shall be calculated using Section 1910.13.4.1 or 1910.13.4.3.

1910.13.4.3 Alternatively, it shall be permitted to calculate the magnified sway moment $\delta_s M_s$ as

$$\delta_s M_s = \frac{M_s}{1 - \dfrac{\Sigma P_u}{0.75 \Sigma P_c}} \geq M_s \qquad (10\text{-}19)$$

where ΣP_u is the summation for all the vertical loads in a story and ΣP_c is the summation for all sway resisting columns in a story, P_c is calculated using Formula (10-11) using k from Section 1910.13.1 and EI from Formula (10-12) or (10-13).

1910.13.5 If an individual compression member has

$$\frac{l_u}{r} > \frac{35}{\sqrt{\dfrac{P_u}{f'_c A_g}}} \qquad (10\text{-}20)$$

it shall be designed for the factored axial load, P_u, and the moment, M_c, calculated using Section 1910.12.3 in which M_1 and M_2 are computed in accordance with Section 1910.13.3, β_d as defined for the load combination under consideration and k as defined in Section 1910.12.1.

1910.13.6 In addition to load cases involving lateral loads, the strength and stability of the structure as a whole under factored gravity loads shall be considered.

1. When $\delta_s M_s$ is computed from Section 1910.13.4.1, the ratio of second-order lateral deflections to first-order lateral deflections for 1.4 dead load and 1.7 live load plus lateral load applied to the structure shall not exceed 2.5.

2. When $\delta_s M_s$ is computed according to Section 1910.13.4.2, the value of Q computed using ΣP_u for 1.4 dead load plus 1.7 live load shall not exceed 0.60.

3. When $\delta_s M_s$ is computed from Section 1910.13.4.3, δ_s computed using ΣP_u and ΣP_c corresponding to the factored dead and live loads shall be positive and shall not exceed 2.5.

In cases 1, 2 and 3 above, β_d shall be taken as the ratio of the maximum factored sustained axial load to the total factored axial load.

1910.13.7 In sway frames, flexural members shall be designed for the total magnified end moments of the compression members at the joint.

1910.14 Axially Loaded Members Supporting Slab System.
Axially loaded members supporting slab system included within the scope of Section 1913.1 shall be designed as provided in Section 1910 and in accordance with the additional requirements of Section 1913.

1910.15 Transmission of Column Loads through Floor System.
When the specified compressive strength of concrete in a column is greater than 1.4 times that specified for a floor system, transmission of load through the floor system shall be provided by one of the following:

1910.15.1 Concrete of strength specified for the column shall be placed in the floor at the column location. Top surface of the column concrete shall extend 2 feet (610 mm) into the slab from face of column. Column concrete shall be well integrated with floor concrete, and shall be placed in accordance with Sections 1906.4.5 and 1906.4.6.

1910.15.2 Strength of a column through a floor system shall be based on the lower value of concrete strength with vertical dowels and spirals as required.

1910.15.3 For columns laterally supported on four sides by beams of approximately equal depth or by slabs, strength of the column may be based on an assumed concrete strength in the column joint equal to 75 percent of column concrete strength plus 35 percent of floor concrete strength.

1910.16 Composite Compression Members.

1910.16.1 Composite compression members shall include all such members reinforced longitudinally with structural steel shapes, pipe or tubing with or without longitudinal bars.

1910.16.2 Strength of a composite member shall be computed for the same limiting conditions applicable to ordinary reinforced concrete members.

1910.16.3 Any axial load strength assigned to concrete of a composite member shall be transferred to the concrete by members or brackets in direct bearing on the composite member concrete.

1910.16.4 All axial load strength not assigned to concrete of a composite member shall be developed by direct connection to the structural steel shape, pipe or tube.

1910.16.5 For evaluation of slenderness effects, radius of gyration of a composite section shall not be greater than the value given by:

$$r = \sqrt{\frac{(E_c I_g / 5) + E_s I_t}{(E_c A_g / 5) + E_s A_t}} \qquad (10\text{-}21)$$

and, as an alternative to a more accurate calculation, EI in Formula (10-11) shall be taken either as Formula (10-12) or

$$EI = \frac{E_c I_g / 5}{1 + \beta_d} + E_s I_t \qquad (10\text{-}22)$$

1910.16.6 Structural steel-encased concrete core.

1910.16.6.1 For a composite member with concrete core encased by structural steel, thickness of the steel encasement shall not be less than

$$b \sqrt{\frac{f_y}{3E_s}}, \text{ for each face of width } b$$

nor

$$h \sqrt{\frac{f_y}{8E_s}}, \text{ for circular sections of diameter } h$$

1910.16.6.2 Longitudinal bars located within the encased concrete core shall be permitted to be used in computing A_t and I_t.

1910.16.7 Spiral reinforcement around structural steel core.
A composite member with spirally reinforced concrete around a structural steel core shall conform to the following:

1910.16.7.1 Specified compressive strength of concrete f'_c shall not be less than 2,500 psi (17.24 MPa).

1910.16.7.2 Design yield strength of structural steel core shall be the specified minimum yield strength for grade of structural steel used but not to exceed 50,000 psi (344.7 MPa).

1910.16.7.3 Spiral reinforcement shall conform to Section 1910.9.3.

1910.16.7.4 Longitudinal bars located within the spiral shall not be less than 0.01 or more than 0.08 times net area of concrete section.

1910.16.7.5 Longitudinal bars located within the spiral shall be permitted to be used in computing A_t and I_t.

1910.16.8 Tie reinforcement around structural steel core. A composite member with laterally tied concrete around a structural steel core shall conform to the following:

1910.16.8.1 Specified compressive strength of concrete f'_c shall not be less than 2,500 psi (17.24 MPa).

1910.16.8.2 Design yield strength of structural steel core shall be the specified minimum yield strength for grade of structural steel used but not to exceed 50,000 psi (344.7 MPa).

1910.16.8.3 Lateral ties shall extend completely around the structural steel core.

1910.16.8.4 Lateral ties shall have a diameter not less than $^1/_{50}$ times the greatest side dimension of composite member, except that ties shall not be smaller than No. 3 and are not required to be larger than No. 5. Welded wire fabric of equivalent area shall be permitted.

1910.16.8.5 Vertical spacing of lateral ties shall not exceed 16 longitudinal bar diameters, 48 tie bar diameters, or one half times the least side dimension of the composite member.

1910.16.8.6 Longitudinal bars located within the ties shall not be less than 0.01 or more than 0.08 times net area of concrete section.

1910.16.8.7 A longitudinal bar shall be located at every corner of a rectangular cross section, with other longitudinal bars spaced not farther apart than one half the least side dimension of the composite member.

1910.16.8.8 Longitudinal bars located within the ties shall be permitted to be used in computing A_t for strength but not in computing I_t for evaluation of slenderness effects.

1910.17 Bearing Strength.

1910.17.1 Design bearing strength on concrete shall not exceed ϕ $(0.85f'_c A_1)$, except when the supporting surface is wider on all sides than the loaded area, design bearing strength on the loaded area shall be permitted to be multiplied by $\sqrt{A_2/A_1}$, but not more than 2.

1910.17.2 Section 1910.17 does not apply to posttensioning anchorages.

SECTION 1911 — SHEAR AND TORSION

1911.0 Notations.

A_c = area of concrete section resisting shear transfer, square inches (mm²).

A_{cp} = area enclosed by outside perimeter of concrete cross section, inches squared (mm²). See Section 1911.6.1.

A_f = area of reinforcement in bracket or corbel resisting factored moment $[V_u a + N_{uc} (h-d)]$, square inches (mm²).

A_g = gross area of section, square inches (mm²).

A_h = area of shear reinforcement parallel to flexural tension reinforcement, square inches (mm²).

A_l = total area of longitudinal reinforcement to resist torsion, square inches (mm²).

A_n = area of reinforcement in bracket or corbel resisting tensile force N_{uc}, square inches (mm²).

A_o = gross area enclosed by shear flow, inches squared (mm²).

A_{oh} = area enclosed by centerline of the outermost closed transverse torsional reinforcement, inches squared (mm²).

A_{ps} = area of prestressed reinforcement in tension zone, square inches (mm²).

A_s = area of nonprestressed tension reinforcement, square inches (mm²).

A_t = area of one leg of a closed stirrup resisting torsion within a distance s, square inches (mm²).

A_v = area of shear reinforcement within a distance s, or area of shear reinforcement perpendicular to flexural tension reinforcement within a distance s for deep flexural members, square inches (mm²).

A_{vf} = area of shear-friction reinforcement, square inches (mm²).

A_{vh} = area of shear reinforcement parallel to flexural tension reinforcement within a distance s_2, square inches (mm²).

a = shear span, distance between concentrated load and face of supports.

b = width of compression face of member, inches (mm).

b_o = perimeter of critical section for slabs and footings, inches (mm).

b_t = width of that part of cross section containing the closed stirrups resisting torsion.

b_w = web width, or diameter of circular section, inches (mm).

b_1 = width of the critical section defined in Section 1911.12.6.1 measured in the direction of the span for which moments are determined, inches (mm).

b_2 = width of the critical section defined in Section 1911.12.6.1 measured in the direction perpendicular to b_1, inches (mm).

c_1 = size of rectangular or equivalent rectangular column, capital or bracket measured in the direction of the span for which moments are being determined, inches (mm).

c_2 = size of rectangular or equivalent rectangular column, capital or bracket measured transverse to the direction of the span for which moments are being determined, inches (mm).

d = distance from extreme compression fiber to centroid of longitudinal tension reinforcement, but need not be less than $0.80h$ for prestressed members, inches (mm). (For circular sections, d need not be less than the distance from extreme compression fiber to centroid of tension reinforcement in opposite half of member.)

f'_c = specified compressive strength of concrete, pounds per square inch (MPa).

$\sqrt{f'_c}$ = square root of specified compressive strength of concrete, pounds per square inch (MPa).

f_{ct} = average splitting tensile strength of lightweight aggregate concrete, pounds per square inch (MPa).

f_d = stress due to unfactored dead load, at extreme fiber of section where tensile stress is caused by externally applied loads, pounds per square inch (MPa).

f_{pc} = compressive stress in concrete (after allowance for all prestress losses) at centroid of cross section resisting externally applied loads or at junction of web and flange when the centroid lies within the flange, pounds per square inch (MPa). (In a composite member, f_{pc} is resultant compressive stress at centroid of composite section, or at junction of web and flange when the centroid lies within the flange, due to both prestress and moments resisted by precast member acting alone.)

f_{pe} = compressive stress in concrete due to effective prestress forces only (after allowance for all prestress losses) at extreme fiber of section where tensile stress is caused by externally applied loads, pounds per square inch (MPa).

f_{pu} = specified tensile strength of prestressing tendons, pounds per square inch (MPa).

f_y = specified yield strength of nonprestressed reinforcement, pounds per square inch (MPa).

f_{yl} = yield strength of longitudinal torsional reinforcement.

f_{yv} = yield strength of closed transverse torsional reinforcement.

h = overall thickness of member, inches (mm).

h_v = total depth of shearhead cross section, inches (mm).

h_w = total height of wall from base to top, inches (mm).

I = moment of inertia of section resisting externally applied factored loads.

l_n = clear span measured face to face of supports.

l_v = length of shearhead arm from centroid of concentrated load or reaction, inches (mm).

l_w = horizontal length of wall, inches (mm).

M_{cr} = moment causing flexural cracking at section due to externally applied loads. See Section 1911.4.2.1.

M_m = modified moment.

M_{max} = maximum factored moment at section due to externally applied loads.

M_p = required plastic moment strength of shearhead cross section.

M_u = factored moment at section.

M_v = moment resistance contributed by shearhead reinforcement.

N_u = factored axial load normal to cross section occurring simultaneously with V_u; to be taken as positive for compression, negative for tension, and to include effects of tension due to creep and shrinkage.

N_{uc} = factored tensile force applied at top of bracket or corbel acting simultaneously with V_u to be taken as positive for tension.

P_{cp} = outside perimeter of the concrete cross section, inches (mm).

Ph = perimeter of centerline of outermost closed transverse torsional reinforcement, inches (mm).

s = spacing of shear or torsion reinforcement in direction parallel to longitudinal reinforcement, inches (mm).

s_1 = spacing of vertical reinforcement in wall, inches (mm).

s_2 = spacing of shear or torsion reinforcement in direction perpendicular to longitudinal reinforcement—or spacing of horizontal reinforcement in wall, inches (mm).

T_n = nominal torsional moment strength.

T_u = factored torsional moment at section.

t = thickness of a wall of a hollow section, inches (mm).

V_c = nominal shear strength provided by concrete.

V_{ci} = nominal shear strength provided by concrete when diagonal cracking results from combined shear and moment.

V_{cw} = nominal shear strength provided by concrete when diagonal cracking results from excessive principal tensile stress in web.

V_d = shear force at section due to unfactored dead load.

V_i = factored shear force at section due to externally applied loads occurring simultaneously with M_{max}.

V_n = nominal shear strength.

V_p = vertical component of effective prestress force at section.

V_s = nominal shear strength provided by shear reinforcement.

V_u = factored shear force at section.

v_n = nominal shear stress, pounds per square inch (MPa). See Section 1911.12.6.2.

y_t = distance from centroidal axis of gross section, neglecting reinforcement, to extreme fiber in tension.

α = angle between included stirrups and longitudinal axis of member.

α_f = angle between shear-friction reinforcement and shear plane.

α_s = constant used to compute V_c in slabs and footings.

α_v = ratio of stiffness of shearhead arm to surrounding composite slab section. See Section 1911.12.4.5.

β_c = ratio of long side to short side of concentrated load or reaction area.

β_d = constant used to compute V_c in prestressed slabs.

γ_f = fraction of unbalanced moment transferred by flexure at slab-column connection. See Section 1913.5.3.2.

γ_v = fraction of unbalanced moment transferred by eccentricity of shear at slab-column connections. See Section 1911.12.6.1.

   = $1 - \gamma_f$.

η = number of identical arms of shearhead.

μ = coefficient of friction. See Section 1911.7.4.3.

λ = correction factor related to unit weight of concrete.

ρ = ratio of nonprestressed tension reinforcement.

   = A_s/bd.

ρ_h = ratio of horizontal shear reinforcement area to gross concrete area of vertical section.

ρ_n = ratio of vertical shear reinforcement area to gross concrete area of horizontal section.

ρ_w = $A_s/b_w d$.

θ = angle of compression diagonals in truss analogy for torsion.

ϕ = strength-reduction factor. See Section 1909.3.

1911.1 Shear Strength.

1911.1.1 Design of cross sections subject to shear shall be based on

$$\phi V_n \geq V_u \qquad (11\text{-}1)$$

where V_u is factored shear force at section considered and V_n is nominal shear strength computed by

$$V_n = V_c + V_s \qquad (11\text{-}2)$$

where V_c is nominal shear strength provided by concrete in accordance with Section 1911.3 or Section 1911.4, and V_s is nominal shear strength provided by shear reinforcement in accordance with Section 1911.5.6.

1911.1.1.1 In determining shear strength V_n, the effect of any openings in members shall be considered.

1911.1.1.2 In determining shear strength V_c, whenever applicable, effects of axial tension due to creep and shrinkage in restrained members shall be considered and effects of inclined flexural compression in variable-depth members shall be permitted to be included.

1911.1.2 The values of $\sqrt{f'_c}$ used in Section 1911 shall not exceed 100 psi (0.69 MPa).

> **EXCEPTION:** Values of $\sqrt{f'_c}$ greater than 100 psi (0.69 MPa) is allowed in computing V_c, V_{ci} and V_{cw} for reinforced or prestressed concrete beams and concrete joist construction having minimum web reinforcement equal to $f'_c/5{,}000$ ($f'_c/34.47$) times, but not more than three times the amounts required by Sections 1911.5.5.3, 1911.5.5.4 and 1911.6.5.2.

1911.1.3 Computations of maximum factored shear force V_u at supports in accordance with Section 1911.1.3.1 or 1911.1.3.2 shall be permitted when both of the following two conditions are satisfied:

1. Support reaction, in direction of applied shear, introduces compression into the end regions of member, and

2. No concentrated load occurs between face of support and location of critical section defined in *this section*.

1911.1.3.1 For nonprestressed members, sections located less than a distance d from face of support shall be permitted to be designed for the same shear V_u as that computed at a distance d.

1911.1.3.2 For prestressed members, sections located less than a distance $h/2$ from face of support shall be permitted to be designed for the same shear V_u as that computed at a distance $h/2$.

1911.1.4 For deep flexural members, brackets and corbels, walls and slabs and footings, the special provisions of Sections 1911.8 through 1911.12 shall apply.

1911.2 Lightweight Concrete.

1911.2.1 Provisions for shear strength V_c apply to normal-weight concrete. When lightweight aggregate concrete is used, one of the following modifications shall apply:

1911.2.1.1 When f_{ct} is specified and concrete is proportioned in accordance with Section 1905.2, provisions for V_c shall be modified by substituting $f_{ct}/6.7$ (For **SI:** $1.8\sqrt{f'_c}$) for $\sqrt{f'_c}$, but the value of $f_{ct}/6.7$ (For **SI:** $1.8\sqrt{f'_c}$) shall not exceed $\sqrt{f'_c}$.

1911.2.1.2 When f_{ct} is not specified, all values of $\sqrt{f'_c}$ affecting V_c, T_c and M_{cr} shall be multiplied by 0.75 for all-lightweight concrete and 0.85 for sand-lightweight concrete. Linear interpolation shall be permitted when partial sand replacement is used.

1911.3 Shear Strength Provided by Concrete for Nonprestressed Members.

1911.3.1 Shear strength V_c shall be computed by provisions of Sections 1911.3.1.1 through 1911.3.1.3 unless a more detailed calculation is made in accordance with Section 1911.3.2.

1911.3.1.1 For members subject to shear and flexure only,

$$V_c = 2\sqrt{f'_c}\, b_w d \qquad (11\text{-}3)$$

For **SI:** $\qquad V_c = 0.166\sqrt{f'_c}\, b_w d$

1911.3.1.2 For members subject to axial compression,

$$V_c = 2\left(1 + \frac{N_u}{2{,}000 A_g}\right)\sqrt{f'_c}\, b_w d \qquad (11\text{-}4)$$

For **SI:** $\quad V_c = 0.166\left(1 + 0.073\,\frac{N_u}{A_g}\right)\sqrt{f'_c}\, b_w d$

Quantity N_u/A_g shall be expressed in psi (MPa).

1911.3.1.3 For members subject to significant axial tension, shear reinforcement shall be designed to carry total shear, unless a more detailed analysis is made using Section 1911.3.2.3.

1911.3.2 Shear strength V_c shall be permitted to be computed by the more detailed calculation of Sections 1911.3.2.1 through 1911.3.2.3.

1911.3.2.1 For members subject to shear and flexure only,

$$V_c = \left(1.9\sqrt{f'_c} + 2{,}500\rho_w\frac{V_u d}{M_u}\right)b_w d \qquad (11\text{-}5)$$

For **SI:** $\quad V_c = \left(0.158\sqrt{f'_c} + 17.1\rho_w\frac{V_u d}{M_u}\right)b_w d$

but not greater than $3.5\sqrt{f'_c}\, b_w d$ (For **SI:** $0.29\sqrt{f'_c}\, b_w d$). Quantity $V_u d/M_u$ shall not be taken greater than 1.0 in computing V_c by Formula (11-5), where M_u is factored moment occurring simultaneously with V_u at section considered.

1911.3.2.2 For members subject to axial compression, it shall be permitted to compute V_c using Formula (11-5) with M_m substituted for M_u and $V_u d/M_u$ not then limited to 1.0, where

$$M_m = M_u - N_u\frac{(4h - d)}{8} \qquad (11\text{-}6)$$

However, V_c shall not be taken greater than

$$V_c = 3.5\sqrt{f'_c}\, b_w d\sqrt{1 + \frac{N_u}{500 A_g}} \qquad (11\text{-}7)$$

For **SI:** $\quad V_c = 0.29\sqrt{f'_c}\, b_w d\sqrt{1 + 0.29\frac{N_u}{A_g}}$

Quantity N_u/A_g shall be expressed in psi (MPa). When M_m as computed by Formula (11-6) is negative, V_c shall be computed by Formula (11-7).

1911.3.2.3 For members subject to significant axial tension,

$$V_c = 2\left(1 + \frac{N_u}{500 A_g}\right)\sqrt{f'_c}\, b_w \qquad (11\text{-}8)$$

For **SI:** $\quad V_c = 0.166\left(1 + 0.29\frac{N_u}{A_g}\right)\sqrt{f'_c}\, b_w d$

but not less than zero, where N_u is negative for tension. Quantity N_u/A_g shall be expressed in psi (MPa).

1911.4 Shear Strength Provided by Concrete for Prestressed Members.

1911.4.1 For members with effective prestress force not less than 40 percent of the tensile strength of flexural reinforcement, unless a more detailed calculation is made in accordance with Section 1911.4.2.

$$V_c = \left(0.6\sqrt{f'_c} + 700\frac{V_u d}{M_u}\right) b_w d \qquad (11\text{-}9)$$

For **SI:** $\quad V_c = \left(0.05\sqrt{f'_c} + 4.8\frac{V_u d}{M_u}\right) b_w d$

but V_c need not be taken less than $2\sqrt{f'_c}\,b_w d$ (For **SI:** $0.166\sqrt{f'_c}\,b_w d$) nor shall V_c be taken greater than $5\sqrt{f'_c}\,b_w d$ (For **SI:** $0.42\sqrt{f'_c}\,b_w d$) or the value given in Section 1911.4.3 or 1911.4.4. The quantity $V_u d/M_u$ shall not be taken greater than 1.0, where M_u is factored moment occurring simultaneously with V_u at section considered. When applying Formula (11-9), d in the term $V_u d/M_u$ shall be the distance from extreme compression fiber to centroid of prestressed reinforcement.

1911.4.2 Shear strength V_c shall be permitted to be computed in accordance with Sections 1911.4.2.1 and 1911.4.2.2 where V_c shall be the lesser of V_{ci} or V_{cw}.

1911.4.2.1 Shear strength V_{ci} shall be computed by

$$V_{ci} = 0.6\sqrt{f'_c}\,b_w d + V_d + \frac{V_i M_{cr}}{M_{max}} \qquad (11\text{-}10)$$

For **SI:** $\quad V_{ci} = 0.05\sqrt{f'_c}\,b_w d + V_d + \frac{V_i M_{cr}}{M_{max}}$

but V_{ci} need not be taken less than $1.7\sqrt{f'_c}\,b_w d$ $(0.14\sqrt{f'_c}\,b_w d)$, where

$$M_{cr} = (l/y_t)\,(6\sqrt{f'_c} + f_{pe} - f_d) \qquad (11\text{-}11)$$

For **SI:** $\quad M_{cr} = (l/y_t)\,(0.5\sqrt{f'_c} + f_{pe} - f_d)$

and values of M_{max} and V_i shall be computed from the load combination causing maximum moment to occur at the section.

1911.4.2.2 Shear strength V_{cw} shall be computed by

$$V_{cw} = (3.5\sqrt{f'_c} + 0.3f_{pc})\,b_w d + V_P \qquad (11\text{-}12)$$

For **SI:** $\quad V_{cw} = (0.29\sqrt{f'_c} + 0.3f_{pc})\,b_w d + V_P$

Alternatively, V_{cw} may be computed as the shear force corresponding to dead load plus live load that results in a principal tensile stress of $4\sqrt{f'_c}$ (For **SI:** $0.33\sqrt{f'_c}$) at centroidal axis of member, or at intersection of flange and web when centroidal axis is in the flange. In composite members, principal tensile stress shall be computed using the cross section that resists live load.

1911.4.2.3 In Formulas (11-10) and (11-12), d shall be the distance from extreme compression fiber to centroid of prestressed reinforcement or $0.8h$, whichever is greater.

1911.4.3 In a pretensioned member in which the section at a distance $h/2$ from face of support is closer to end of member than the transfer length of the prestressing tendons, the reduced prestress shall be considered when computing V_{cw}. This value of V_{cw} shall also be taken as the maximum limit for Formula (11-9). Prestress force may be assumed to vary linearly from zero at end of tendon to a maximum at a distance from end of tendon equal to the transfer length, assumed to be 50 diameters for strand and 100 diameters for single wire.

1911.4.4 In a pretensioned member where bonding of some tendons does not extend to end of member, a reduced prestress shall be considered when computing V_c in accordance with Section 1911.4.1 or 1911.4.2. Value of V_{cw} calculated using the reduced prestress shall also be taken as the maximum limit for Formula

(11-9). Prestress force due to tendons for which bonding does not extend to end of member may be assumed to vary linearly from zero at the point at which bonding commences to a maximum at a distance from this point equal to the transfer length, assumed to be 50 diameters for strand and 100 diameters for single wire.

1911.5 Shear Strength Provided by Shear Reinforcement.

1911.5.1 Types of shear reinforcement.

1911.5.1.1 Shear reinforcement consisting of the following shall be permitted:

1. Stirrups perpendicular to axis of member.

2. Welded wire fabric with wires located perpendicular to axis of member.

1911.5.1.2 For nonprestressed members, shear reinforcement shall be permitted to also consist of:

1. Stirrups making an angle of 45 degrees or more with longitudinal tension reinforcement.

2. Longitudinal reinforcement with bent portion making an angle of 30 degrees or more with the longitudinal tension reinforcement.

3. Combination of stirrups and bent longitudinal reinforcement.

4. Spirals.

1911.5.2 Design yield strength of shear reinforcement shall not exceed 60,000 psi (413.7 MPa), except that the design yield strength of welded deformed wire fabric shall not exceed 80,000 psi (551.6 MPa).

1911.5.3 Stirrups and other bars or wires used as shear reinforcement shall extend to a distance d from extreme compression fiber and shall be anchored at both ends according to Section 1912.13 to develop the design yield strength of reinforcement.

1911.5.4 Spacing limits for shear reinforcement.

1911.5.4.1 Spacing of shear reinforcement placed perpendicular to axis of member shall not exceed $d/2$ in nonprestressed members and $(^3/_4)h$ in prestressed members or 24 inches (610 mm).

1911.5.4.2 Inclined stirrups and bent longitudinal reinforcement shall be so spaced that every 45-degree line, extending toward the reaction from middepth of member $d/2$ to longitudinal tension reinforcement, shall be crossed by at least one line of shear reinforcement.

1911.5.4.3 When V_s exceeds $4\sqrt{f'_c}\,b_w d$ (For **SI:** $0.33\sqrt{f'_c}\,b_w d$), maximum spacings given in the paragraphs above shall be reduced by one half.

1911.5.5 Minimum shear reinforcement.

1911.5.5.1 A minimum area of shear reinforcement shall be provided in all reinforced concrete flexural members (prestressed and nonprestressed) where factored shear force V_u exceeds one half the shear strength provided by concrete ϕV_c, except:

1. Slabs and footings.

2. Concrete joist construction defined by Section 1908.11.

3. Beams with total depth not greater than 10 inches (254 mm), two and one half times thickness of flange or one half the width of web, whichever is greater.

1911.5.5.2 Minimum shear reinforcement requirements of Section 1911.5.5.1 shall be waived if shown by test that required nominal flexural and shear strengths can be developed when shear reinforcement is omitted. Such tests shall simulate effects of differential settlement, creep, shrinkage and temperature change, based on a realistic assessment of such effects occurring in service.

1911.5.5.3 Where shear reinforcement is required by Section 1911.5.5.1 or for strength and where Section 1911.6.1 allows torsion to be neglected, the minimum area of shear reinforcement for prestressed (except as provided in Section 1911.5.5.4) and nonprestressed members shall be computed by:

$$A_v = 50 \frac{b_w s}{f_y} \tag{11-13}$$

For **SI:**
$$A_v = 0.34 \frac{b_w s}{f_y}$$

where b_w and s are in inches.

1911.5.5.4 For prestressed members with effective prestress force not less than 40 percent of the tensile strength of flexural reinforcement, the area of shear reinforcement shall not be less than the smaller A_v, computed by Formula (11-13) or (11-14).

$$A_v = \frac{A_{ps}}{80} \frac{f_{pu}}{f_y} \frac{s}{d} \sqrt{\frac{d}{b_w}} \tag{11-14}$$

1911.5.6 Design of shear reinforcement.

1911.5.6.1 Where factored shear force V_u exceeds shear strength ϕV_c, shear reinforcement shall be provided to satisfy Formulas (11-1) and (11-2), where shear strength V_s shall be computed in accordance with Sections 1911.5.6.2 through 1911.5.6.8.

1911.5.6.2 When shear reinforcement perpendicular to axis of member is used,

$$V_s = \frac{A_v f_y d}{s} \tag{11-15-1}$$

where A_v is the area of shear reinforcement within a distance s.

For circular columns, the area used to compute V_c shall be 0.8 A_g. The shear strength V_s provided by the circular transverse reinforcing shall be computed by

$$V_s = \frac{\pi A_b f_{yh} D'}{2 s} \tag{11-15-2}$$

where A_b is the area of the hoop or spiral bar of yield strength f_{yh} with pitch s and hoop diameter D'.

1911.5.6.3 When inclined stirrups are used as shear reinforcement,

$$V_s = \frac{A_v f_y (\sin \alpha + \cos \alpha) d}{s} \tag{11-16}$$

1911.5.6.4 When shear reinforcement consists of a single bar or a single group of parallel bars, all bent up at the same distance from the support,

$$V_s = A_v f_y \sin \alpha \tag{11-17}$$

but not greater than $3 \sqrt{f'_c} b_w d$ (For **SI:** $0.25 \sqrt{f'_c} b_w d$).

1911.5.6.5 When shear reinforcement consists of a series of parallel bent-up bars or groups of parallel bent-up bars at different distances from the support, shear strength V_s shall be computed by Formula (11-16).

1911.5.6.6 Only the center three fourths of the inclined portion of any longitudinal bent bar shall be considered effective for shear reinforcement.

1911.5.6.7 Where more than one type of shear reinforcement is used to reinforce the same portion of a member, shear strength V_s shall be computed as the sum of the V_s values computed for the various types.

1911.5.6.8 Shear strength V_s shall not be taken greater than $8 \sqrt{f'_c} b_w d$ (For **SI:** $0.66 \sqrt{f'_c} b_w d$).

1911.6 Design for Torsion.

1911.6.1 It shall be permitted to neglect torsion effects when the factored torsional moment T_u is less than:

1. for nonprestressed members:

$$\phi \sqrt{f'_c} \left(\frac{A_{cp}^2}{P_{cp}} \right)$$

2. for prestressed members:

$$\phi \sqrt{f'_c} \left(\frac{A_{cp}^2}{P_{cp}} \right) \sqrt{1 + \frac{f_{pc}}{4 \sqrt{f'_c}}}$$

For members cast monolithically with a slab, the overhanging flange width used in computing A_{cp} and P_{cp} shall conform to Section 1913.2.4.

1911.6.2 Calculation of factored torsional moment T_u.

1911.6.2.1 If the factored torsional moment T_u in a member is required to maintain equilibrium and exceeds the minimum value given in Section 1911.6.1, the member shall be designed to carry that torsional moment in accordance with Sections 1911.6.3 through 1911.6.6.

1911.6.2.2 In a statically indeterminate structure where reduction of the torsional moment in a member can occur due to redistribution of internal forces upon cracking, the maximum factored torsional moment T_u shall be permitted to be reduced to

1. for nonprestressed members, at the sections described in Section 1911.6.2.4:

$$\phi_4 \sqrt{f'_c} \left(\frac{A_{cp}^2}{P_{cp}} \right)$$

2. for prestressed members, at the sections described in Section 1911.6.2.5:

$$\phi_4 \sqrt{f'_c} \left(\frac{A_{cp}^2}{P_{cp}} \right) \sqrt{1 + \frac{f_{pc}}{4 \sqrt{f'_c}}}$$

In such a case, the correspondingly redistributed bending moments and shears in the adjoining members shall be used in the design of those members.

1911.6.2.3 Unless determined by a more exact analysis, it shall be permitted to take the torsional loading from a slab as uniformly distributed along the member.

1911.6.2.4 In nonprestressed members, sections located less than a distance d from the face of a support shall be designed for not less than the torsion T_u computed at a distance d. If a concentrated torque occurs within this distance, the critical section for design shall be at the face of the support.

1911.6.2.5 In prestressed members, sections located less than a distance $h/2$ from the face of a support shall be designed for not less than the torsion T_u computed at a distance $h/2$. If a concentrated torque occurs within this distance, the critical section for design shall be at the face of the support.

1911.6.3 Torsional moment strength.

1911.6.3.1 The cross-sectional dimensions shall be such that:

1. for solid sections:

$$\sqrt{\left(\frac{V_u}{b_w d}\right)^2 + \left(\frac{T_u p_h}{1.7 A_{oh}^2}\right)^2} \leq \phi \left(\frac{V_c}{b_w d} + 8\sqrt{f'_c}\right) \quad (11\text{-}18)$$

2. for hollow sections:

$$\sqrt{\left(\frac{V_u}{b_w d}\right) + \left(\frac{T_u p_h}{1.7 A_{oh}^2}\right)} \leq \phi \left(\frac{V_c}{b_w d} + 8\sqrt{f'_c}\right) \quad (11\text{-}19)$$

1911.6.3.2 If the wall thickness varies around the perimeter of a hollow section, Formula (11-19) shall be evaluated at the location where the left-hand side of Formula (11-19) is a maximum.

1911.6.3.3 If the wall thickness is less than A_{oh}/p_h, the second term in Formula (11-19) shall be taken as:

$$\left(\frac{T_u}{1.7 A_{oh} t}\right)$$

where t is the thickness of the wall of the hollow section at the location where the stresses are being checked.

1911.6.3.4 Design yield strength of nonprestressed torsion reinforcement shall not exceed 60,000 psi (413.7 MPa).

1911.6.3.5 The reinforcement required for torsion shall be determined from:

$$\phi T_n \geq T_u \quad (11\text{-}20)$$

1911.6.3.6 The transverse reinforcement for torsion shall be designed using:

$$T_n = \frac{2 A_o A_t f_{vv}}{s} \cot\theta \quad (11\text{-}21)$$

where A_o shall be determined by analysis except that it shall be permitted to take A_o equal to $0.85 A_{oh}$; θ shall not be taken smaller than 30 degrees nor larger than 60 degrees. It shall be permitted to take θ equal to:

1. 45 degrees for nonprestressed members or members with less prestress than in Item 2 below,

2. 37.5 degrees for prestressed members with an effective prestress force not less than 40 percent of the tensile strength of the longitudinal reinforcement.

1911.6.3.7 The additional longitudinal reinforcement required for torsion shall not be less than:

$$A_l = \frac{A_t}{s} p_h \left(\frac{f_{yv}}{f_{yl}}\right) \cot^2\theta \quad (11\text{-}22)$$

where θ shall be the same value used in Formula (11-21) and A_t/s shall be taken as the amount computed from Formula (11-21) not modified in accordance with Section 1911.6.5.2 or 1911.6.5.3.

1911.6.3.8 Reinforcement required for torsion shall be added to that required for the shear, moment and axial force that act in combination with the torsion. The most restrictive requirements for reinforcement spacing and placement must be met.

1911.6.3.9 It shall be permitted to reduce the area of longitudinal torsion reinforcement in the flexural compression zone by an amount equal to $M_u/(0.9 d f_{yl})$, where M_u is the factored moment acting at the section in combination with T_u, except that the reinforcement provided shall not be less than that required by Section 1911.6.5.3 or 1911.6.6.2.

1911.6.3.10 In prestressed beams:

1. The total longitudinal reinforcement including tendons at each section shall resist the factored bending moment at that section plus an additional concentric longitudinal tensile force equal to $A_l f_{yl}$, based on the factored torsion at that section, and

2. The spacing of the longitudinal reinforcement including tendons shall satisfy the requirements in Section 1911.6.6.2.

1911.6.3.11 In prestressed beams, it shall be permitted to reduce the area of longitudinal torsional reinforcement on the side of the member in compression due to flexure below that required by Section 1911.6.3.10 in accordance with Section 1911.6.3.9.

1911.6.4 Details of torsional reinforcement.

1911.6.4.1 Torsion reinforcement shall consist of longitudinal bars or tendons and one or more of the following:

1. Closed stirrups or closed ties, perpendicular to the axis of the member, or

2. A closed cage of welded wire fabric with transverse wires perpendicular to the axis of the member, or

3. In nonprestressed beams, spiral reinforcement.

1911.6.4.2 Transverse torsional reinforcement shall be anchored by one of the following:

1. A 135-degree standard hook around a longitudinal bar, or

2. According to Section 1912.13.2.1, 1912.13.2.2 or 1912.13.2.3 in regions where the concrete surrounding the anchorage is restrained against spalling by a flange or slab or similar member.

1911.6.4.3 Longitudinal torsion reinforcement shall be developed at both ends.

1911.6.4.4 For hollow sections in torsion, the distance measured from the centerline of the transverse torsional reinforcement to the inside face of the wall of a hollow section shall not be less than $0.5 A_{oh}/p_h$.

1911.6.5 Minimum torsion reinforcement.

1911.6.5.1 A minimum area of torsion reinforcement shall be provided in all regions where the factored torsional moment T_u exceeds the values specified in Section 1911.6.1.

1911.6.5.2 Where torsional reinforcement is required by Section 1911.6.5.1, the minimum area of transverse closed stirrups shall be computed by:

$$(A_v + 2A_t) \geq \frac{50 b_w s}{f_{yv}} \quad (11\text{-}23)$$

1911.6.5.3 Where torsional reinforcement is required by Section 1911.6.5.1, the minimum total area of longitudinal torsional reinforcement shall be computed by:

$$A_{l,min} = \frac{5\sqrt{f_c} A_{cp}}{f_{vl}} - \left(\frac{A_t}{s}\right) p_h \frac{f_{yv}}{f_{yl}} \quad (11\text{-}24)$$

where A_t/s shall not be taken less than $25 b_w/f_{yv}$.

1911.6.6 Spacing of torsion reinforcement.

1911.6.6.1 The spacing of transverse torsion reinforcement shall not exceed the smaller of $p_h/8$ or 12 inches (305 mm).

1911.6.6.2 The longitudinal reinforcement required for torsion shall be distributed around the perimeter of the closed stirrups with a maximum spacing of 12 inches (305 mm). The longitudinal bars or tendons shall be inside the stirrups. There shall be at least one longitudinal bar or tendon in each corner of the stirrups. Bars shall have a diameter at least $^1/_{24}$ of the stirrup spacing but not less than a No. 3 bar.

1911.6.6.3 Torsion reinforcement shall be provided for a distance of at least $(b_t + d)$ beyond the point theoretically required.

1911.7 Shear-friction.

1911.7.1 The following provisions shall be applied where it is appropriate to consider shear transfer across a given plane, such as an existing or potential crack, an interface between dissimilar materials, or an interface between two concretes cast at different times.

1911.7.2 Design of cross sections subject to shear transfer as described in Section 1911.7 shall be based on Formula (11-1) where V_n is calculated in accordance with provisions of Section 1911.7.3 or 1911.7.4.

1911.7.3 A crack shall be assumed to occur along the shear plane considered. Required area of shear-friction reinforcement A_{vf} across the shear plane may be designed using either Section 1911.7.4 or any other shear transfer design methods that result in prediction of strength in substantial agreement with results of comprehensive tests.

1911.7.3.1 Provisions of Sections 1911.7.5 through 1911.7.10 shall apply for all calculations of shear transfer strength.

1911.7.4 Shear-friction design methods.

1911.7.4.1 When shear-friction reinforcement is perpendicular to shear plane, shear strength V_n shall be computed by

$$V_n = A_{vf} f_y \mu \qquad (11\text{-}25)$$

where μ is coefficient of friction in accordance with Section 1911.7.4.3.

1911.7.4.2 When shear-friction reinforcement is inclined to shear plane such that the shear force produces tension in shear-friction reinforcement, shear strength V_n shall be computed by

$$V_n = A_{vf} f_y (\mu \sin \alpha_1 + \cos \alpha_1) \qquad (11\text{-}26)$$

where α_1 is angle between shear-friction reinforcement and shear plane.

1911.7.4.3 Coefficient of friction μ in Formula (11-25) and Formula (11-26) shall be

Concrete placed monolithically	1.4λ
Concrete placed against hardened concrete with surface intentionally roughened as specified in Section 1911.7.9	1.0λ
Concrete placed against hardened concrete not intentionally roughened	0.6λ
Concrete anchored to as-rolled structural steel by headed studs or by reinforcing bars (see Section 1911.7.10)	0.7λ

where $\lambda = 1.0$ for normal-weight concrete, 0.85 for sand-lightweight concrete and 0.75 for all-lightweight concrete. Linear interpolation shall be permitted when partial sand replacement is used.

1911.7.5 Shear strength V_n shall not be taken greater than $0.2f'_c A_c$ or $800 A_c$ in pounds ($5.5 A_c$ in newtons), where A_c is area of concrete section resisting shear transfer.

1911.7.6 Design yield strength of shear-friction reinforcement shall not exceed 60,000 psi (413.7 MPa).

1911.7.7 Net tension across shear plane shall be resisted by additional reinforcement. Permanent net compression across shear plane shall be permitted to be taken as additive to the force in the shear-friction reinforcement $A_{vf} f_y$ when calculating required A_{vf}.

1911.7.8 Shear-friction reinforcement shall be appropriately placed along the shear plane and shall be anchored to develop the specified yield strength on both sides by embedment, hooks or welding to special devices.

1911.7.9 For the purpose of Section 1911.7, when concrete is placed against previously hardened concrete, the interface for shear transfer shall be clean and free of laitance. If μ is assumed equal to 1.0λ, interface shall be roughened to a full amplitude of approximately $^1/_4$ inch (6.4 mm).

1911.7.10 When shear is transferred between as-rolled steel and concrete using headed studs or welded reinforcing bars, steel shall be clean and free of paint.

1911.8 Special Provisions for Deep Flexural Members.

1911.8.1 Provisions of this section shall apply for members with l_n/d less than 5 that are loaded on one face and supported on the opposite face so that the compression struts can develop between the loads and the supports. See also Section 1912.10.6.

1911.8.2 The design of simple supported deep flexural members for shear shall be based on Formulas (11-1) and (11-2), where shear strength V_c shall be in accordance with Section 1911.8.6 or 1911.8.7, and shear strength V_s shall be in accordance with Section 1911.8.8.

1911.8.3 The design of continuous deep flexural members for shear shall be based on Sections 1911.1 through 1911.5 with Section 1911.8.5 substituted for Section 1911.1.3, or on methods satisfying equilibrium and strength requirements. In either case, the design shall also satisfy Sections 1911.8.4, 1911.8.9 and 1911.8.10.

1911.8.4 Shear strength V_n for deep flexural members shall not be taken greater than $8\sqrt{f'_c}\, b_w d$ (For **SI**: $0.66\sqrt{f'_c}\, b_w d$) when l_n/d is less than 2. When l_n/d is between 2 and 5,

$$V_n = \frac{2}{3}\left(10 + \frac{l_n}{d}\right)\sqrt{f'_c}\, b_w d \qquad (11\text{-}27)$$

For **SI**:
$$V_n = 0.055\left(10 + \frac{l_n}{d}\right)\sqrt{f'_c}\, b_w d$$

1911.8.5 Critical section for shear measured from face of support shall be taken at a distance $0.15l_n$ for uniformly loaded beams and $0.50a$ for beams with concentrated loads, but not greater than d.

1911.8.6 Unless a more detailed calculation is made in accordance with Section 1911.8.7.

$$V_c = 2\sqrt{f'_c}\, b_w d \qquad (11\text{-}28)$$

For **SI**:
$$V_c = 0.166\sqrt{f'_c}\, b_w d$$

1911.8.7 Shear strength V_c shall be permitted to be computed by

$$V_c = \left(3.5 - 2.5\,\frac{M_u}{V_u d}\right)$$

$$\left(1.9\sqrt{f'_c} + 2{,}500\rho_w\frac{V_u d}{M_u}\right)b_w d \qquad (11\text{-}29)$$

For **SI:**

$$V_c = \left(3.5 - 2.5\,\frac{M_u}{V_u d}\right)\left(0.16\sqrt{f'_c} + 17.2\rho_w\frac{V_u d}{M_u}\right)b_w d$$

except that the term

$$\left(3.5 - 2.5\,\frac{M_u}{V_u d}\right)$$

shall not exceed 2.5, and V_c shall not be taken greater than $6\sqrt{f'_c}\,b_w d$ (For **SI:** $0.5\sqrt{f'_c}\,b_w d$). M_u is factored moment occurring simultaneously with V_u at the critical section defined in Section 1911.8.5.

1911.8.8 Where factored shear force V_u exceeds shear strength ϕV_c, shear reinforcement shall be provided to satisfy Formulas (11-1) and (11-2), where shear strength V_s shall be computed by

$$V_s = \left[\frac{A_v}{s}\left(\frac{1+\frac{l_n}{d}}{12}\right) + \frac{A_{vh}}{s_2}\left(\frac{11-\frac{l_n}{d}}{12}\right)\right]f_y d \quad (11\text{-}30)$$

where A_v is area of shear reinforcement perpendicular to flexural tension reinforcement within a distance s, and A_{vh} is area of shear reinforcement parallel to flexural reinforcement within a distance s_2.

1911.8.9 Area of shear reinforcement A_v shall not be less than $0.0015\,b_w s$, and s shall not exceed $d/5$ or 18 inches (457 mm).

1911.8.10 Area of horizontal shear reinforcement A_{vh} shall not be less than $0.0025\,b_w s_2$, and s_2 shall not exceed $d/3$ or 18 inches (457 mm).

1911.8.11 Shear reinforcement required at the critical section defined in Section 1911.8.5 shall be used throughout the span.

1911.9 Special Provisions for Brackets and Corbels.

1911.9.1 The following provisions apply to brackets and corbels with a shear span-to-depth ratio a/d not greater than unity, and subject to a horizontal tensile force N_{uc} not larger than V_u. Distance d shall be measured at face of support.

1911.9.2 Depth at outside edge of bearing area shall not be less than $0.5d$.

1911.9.3 Section at face of support shall be designed to resist simultaneously a shear V_u, a moment $[V_u a + N_{uc}(h-d)]$, and a horizontal tensile force N_{uc}.

1911.9.3.1 In all design calculations in accordance with Section 1911.9, strength-reduction factor ϕ shall be taken equal to 0.85.

1911.9.3.2 Design of shear-friction reinforcement A_{vf} to resist shear V_u shall be in accordance with Section 1911.7.

1911.9.3.2.1 For normal-weight concrete, shear strength V_n shall not be taken greater than $0.2f'_c b_w d$ nor $800\,b_w d$ in pounds ($5.5\,b_w d$ in newtons)

1911.9.3.2.2 For all lightweight or sand-lightweight concrete, shear strength V_n shall not be taken greater than $(0.2 - 0.07\,a/d)$ $f'_c b_w d$ or $(800 - 280\,a/d)\,b_w d$ in pounds $[(5.5 - 1.9\,a/d)\,b_w d$ in newtons].

1911.9.3.3 Reinforcement A_f to resist moment $[V_u a + N_{uc}(h-d)]$ shall be computed in accordance with Sections 1910.2 and 1910.3.

1911.9.3.4 Reinforcement A_n to resist tensile force N_{uc} shall be determined from $N_{uc} \le \phi A_n f_y$. Tensile force N_{uc} shall not be taken less than $0.2\,V_u$ unless special provisions are made to avoid tensile forces. Tensile force N_{uc} shall be regarded as a live load even when tension results from creep, shrinkage or temperature change.

1911.9.3.5 Area of primary tension reinforcement A_s shall be made equal to the greater of $(A_f + A_n)$ or $(2A_{vf}/3 + A_n)$.

1911.9.4 Closed stirrups or ties parallel to A_s, with a total area A_n not less than $0.5\,(A_s - A_n)$, shall be uniformly distributed within two thirds of the effective depth adjacent to A_s.

1911.9.5 Ratio $\rho = A_s/bd$ shall not be less than $0.04\,(f'_c/f_y)$.

1911.9.6 At front face of bracket or corbel, primary tension reinforcement A_s shall be anchored by one of the following: (1) by a structural weld to a transverse bar of at least equal size; weld to be designed to develop specified yield strength f_y of A_s bars; (2) by bending primary tension bars A_s back to form a horizontal loop; or (3) by some other means of positive anchorage.

1911.9.7 Bearing area of load on bracket or corbel shall not project beyond straight portion of primary tension bar A_s, or project beyond interior face of transverse anchor bar (if one is provided).

1911.10 . Special Provisions for Walls.

1911.10.1 Design for shear forces perpendicular to face of wall shall be in accordance with provisions for slabs in Section 1911.12. Design for horizontal shear forces in plane of wall shall be in accordance with Section 1911.10.2 through 1911.10.8.

1911.10.2 Design of horizontal section for shear in plane of wall shall be based on Formulas (11-1) and (11-2), where shear strength V_c shall be in accordance with Section 1911.10.5 or 1911.10.6 and shear strength V_s shall be in accordance with Section 1911.10.9.

1911.10.3 Shear strength V_n at any horizontal section for shear in plane of wall shall not be taken greater than $10\sqrt{f'_c}\,hd$ (For **SI:** $0.83\sqrt{f'_c}\,hd$).

1911.10.4 For design for horizontal shear forces in plane of wall, d shall be taken equal to $0.8\,l_w$. A larger value of d, equal to the distance from extreme compression fiber to center of force of all reinforcement in tension shall be permitted to be used when determined by a strain compatibility analysis.

1911.10.5 Unless a more detailed calculation is made in accordance with Section 1911.10.6, shear strength V_c shall not be taken greater than $2\sqrt{f'_c}\,hd$ (For **SI:** $0.166\sqrt{f'_c}\,hd$) for walls subject to N_u in compression, or V_c shall not be taken greater than the value given in Section 1911.3.2.3 for walls subject to N_u in tension.

1911.10.6 Shear strength V_c shall be permitted to be computed by *Formulas* (11-31) and (11-32), where V_c shall be the lesser of *Formula* (11-31) or (11-32).

$$V_c = 3.3\sqrt{f'_c}\,hd + \frac{N_u d}{4l_w} \qquad (11\text{-}31)$$

For **SI:**

$$V_c = 0.27\sqrt{f'_c}\,hd + \frac{N_u d}{4l_w}$$

or

$$V_c = \left[0.6\sqrt{f'_c} + \frac{l_w\left(1.25\sqrt{f'_c} + 0.2\frac{N_u}{l_w h}\right)}{\frac{M_u}{V_u} - \frac{l_w}{2}}\right]hd \quad (11\text{-}32)$$

For **SI:**
$$V_c = \left[0.05 \sqrt{f'_c} + \frac{l_w\left(0.10\sqrt{f'_c} + 0.2\frac{N_u}{l_w h}\right)}{\frac{M_u}{V_u} - \frac{l_w}{2}} \right] hd$$

where N_u is negative for tension. When $(M_u/V_u - l_w/2)$ is negative, Formula (11-32) shall not apply.

1911.10.7 Sections located closer to wall base than a distance $l_w/2$ or one half the wall height, whichever is less, shall be permitted to be designed for the same V_c as that computed at a distance $l_w/2$ or one half the height.

1911.10.8 When factored shear force V_u is less than $\phi V_c/2$, reinforcement shall be provided in accordance with Section 1911.10.9 or in accordance with Section 1914. When V_u exceeds $\phi V_c/2$, wall reinforcement for resisting shear shall be provided in accordance with Section 1911.10.9.

1911.10.9 Design of shear reinforcement for walls.

1911.10.9.1 Where factored shear force V_u exceeds shear strength ϕV_c, horizontal shear reinforcement shall be provided to satisfy Formulas (11-1) and (11-2), where shear strength V_s shall be computed by

$$V_s = \frac{A_v f_y d}{s_2} \qquad (11\text{-}33)$$

where A_v is area of horizontal shear reinforcement within a distance s_2 and distance d is in accordance with Section 1911.10.4. Vertical shear reinforcement shall be provided in accordance with Section 1911.10.9.4.

1911.10.9.2 Ratio ρ_h of horizontal shear reinforcement area to gross concrete area of vertical section shall not be less than 0.0025.

1911.10.9.3 Spacing of horizontal shear reinforcement s_2 shall not exceed $l_w/5$, $3h$ or 18 inches (457 mm).

1911.10.9.4 Ratio ρ_n of vertical shear reinforcement area to gross concrete area of horizontal section shall not be less than

$$\rho_n = 0.0025 + 0.5\left(2.5 - \frac{h_w}{l_w}\right)(\rho_h - 0.0025) \qquad (11\text{-}34)$$

or 0.0025, but need not be greater than the required horizontal shear reinforcement.

1911.10.9.5 Spacing of vertical shear reinforcement s_1 shall not exceed $l_w/3$, $3h$ or 18 inches (457 mm).

1911.11 Transfer of Moments to Columns.

1911.11.1 When gravity load, wind, earthquake or other lateral forces cause transfer of moment at connections of framing elements to columns, the shear resulting from moment transfer shall be considered in the design of lateral reinforcement in the columns.

1911.11.2 Except for connections not part of a primary seismic load-resisting system that are restrained on four sides by beams or slabs of approximately equal depth, connections shall have lateral reinforcement not less than that required by Formula (11-13) within the column for a depth not less than that of the deepest connection of framing elements to the columns. See also Section 1907.9.

1911.12 Special Provisions for Slabs and Footings.

1911.12.1 The shear strength of slabs and footings in the vicinity of columns, concentrated loads or reactions is governed by the more severe of two conditions:

1911.12.1.1 Beam action where each critical section to be investigated extends in a plane across the entire width. For beam action the slab or footing shall be designed in accordance with Sections 1911.1 through 1911.5.

1911.12.1.2 Two-way action where each of the critical sections to be investigated shall be located so that its perimeter, b_o, is a minimum, but need not approach closer than $d/2$ to:

1. Edges or corners of columns, concentrated loads or reaction areas, or

2. Changes in slab thickness such as edges of capitals or drop panels.

For two-way action, the slab of footing shall be designed in accordance with Sections 1911.12.2 through 1911.12.6.

1911.12.1.3 For square or rectangular columns, concentrated loads or reactions areas, the critical sections with four straight sides shall be permitted.

1911.12.2 The design of a slab or footing for two-way action is based on Formulas (11-1) and (11-2). V_c shall be computed in accordance with Section 1911.12.2.1, 1911.12.2.2 or 1911.12.3.1. V_s shall be computed in accordance with Section 1911.12.3. For slabs with shear heads, V_n shall be in accordance with Section 1911.12.4. When moment is transferred between a slab and a column, Section 1911.12.6 shall apply.

1911.12.2.1 For nonprestressed slabs and footings, V_c shall be the smallest of:

1.
$$V_c = \left(2 + \frac{4}{\beta_c}\right)\sqrt{f'_c}\, b_o d \qquad (11\text{-}35)$$

For **SI:**
$$V_c = 0.083\left(2 + \frac{4}{\beta_c}\right)\sqrt{f'_c}\, b_o d$$

where β_c is the ratio of long side to short side of the column, concentrated load or reaction area

2.
$$V_c = \left(\frac{\alpha_s d}{b_o} + 2\right)\sqrt{f'_c}\, b_o d \qquad (11\text{-}36)$$

For **SI:**
$$V_c = 0.083\left(\frac{\alpha_s d}{b_o} + 2\right)\sqrt{f'_c}\, b_o d$$

where α_s is 40 for interior columns, 30 for edge columns and 20 for corner columns, and

3.
$$V_c = 4\sqrt{f'_c}\, b_o d \qquad (11\text{-}37)$$

For **SI:**
$$V_c = 0.33\sqrt{f'_c}\, b_o d$$

1911.12.2.2 At columns of two-way prestressed slabs and footings that meet the requirements of Section 1918.9.3:

$$V_c = (\beta_p \sqrt{f'_c} + 0.3f_{pc})\, b_o d + V_p \qquad (11\text{-}38)$$

For **SI:**
$$V_c = (0.083\,\beta_p\sqrt{f'_c} + 0.3f_{pc})\, b_o d + V_p$$

where β_p is the smaller of 3.5 or $(\alpha_s d/b_o + 1.5)$, α_s is 40 for interior columns, 30 for edge columns and 20 for corner columns, b_o is perimeter of critical section defined in Section 1911.12.1.2, f_{pc} is the average value of f_{pc} for the two directions, and V_p is the vertical component of all effective prestress forces crossing the critical section. V_c shall be permitted to be computed by Formula (11-38) if the following are satisfied; otherwise, Section 1911.12.2.1 shall apply:

1. No portion of the column cross section shall be closer to the discontinuous edge than four times the slab thickness, and

2. f'_c in Formula (11-38) shall not be taken greater than 5,000 psi (34.47 MPa), and

3. f_{pc} in each direction shall not be less than 125 psi (0.86 MPa), or be taken greater than 500 psi (3.45 MPa).

1911.12.3 Shear reinforcement consisting of bars or wires shall be permitted in slabs and footings in accordance with the following:

1911.12.3.1 V_n shall be computed by Formula (11-2), where V_c shall not be taken greater than $2\sqrt{f'_c}\,b_od$ (For **SI:** $0.166\sqrt{f'_c}\,b_od$), and the required area of shear reinforcement A_v and V_s shall be calculated in accordance with Section 1911.5 and anchored in accordance with Section 1912.13.

1911.12.3.2 V_n shall not be taken greater than $6\sqrt{f'_c}\,b_od$ (For **SI:** $0.50\sqrt{f'_c}\,b_od$).

1911.12.4 Shear reinforcement consisting of steel I- or channel-shaped sections (shearheads) shall be permitted in slabs. The provisions of Sections 1911.12.4.1 through 1911.12.4.9 shall apply where shear due to gravity load is transferred at interior column supports. Where moment is transferred to columns, Section 1911.12.6.3 shall apply.

1911.12.4.1 Each shearhead shall consist of steel shapes fabricated by welding with a full penetration weld into identical arms at right angles. Shearhead arms shall not be interrupted within the column section.

1911.12.4.2 A shearhead shall not be deeper than 70 times the web thickness of the steel shape.

1911.12.4.3 The ends of each shearhead arm shall be permitted to be cut at angles not less than 30 degrees with the horizontal, provided the plastic moment strength of the remaining tapered section is adequate to resist the shear force attributed to the arm of the shearhead.

1911.12.4.4 All compression flanges of steel shapes shall be located within $0.3d$ of compression surface of slab.

1911.12.4.5 The ratio α_v between the stiffness of each shearhead arm and that of the surrounding composite cracked slab section of width $(c_2 + d)$ shall not be less than 0.15.

1911.12.4.6 The plastic moment strength M_p required for each arm of the shearhead shall be computed by

$$\phi M_p = \frac{V_u}{2\eta}\left[h_v + \alpha_v\left(l_v - \frac{c_1}{2}\right)\right] \qquad (11\text{-}39)$$

where ϕ is the strength-reduction factor for flexure, η is the number of arms, and l_v is the minimum length of each shearhead arm required to comply with requirements of Section 1911.12.4.7 and 1911.12.4.8.

1911.12.4.7 The critical slab section for shear shall be perpendicular to the plane of the slab and shall cross each shearhead arm at three fourths the distance $[l_v - (c_1/2)]$ from the column face to the end of the shearhead arm. The critical section shall be located so that its perimeter b_o is a minimum, but need not be closer than the perimeter defined in Section 1911.12.1.2, Item 1.

1911.12.4.8 V_n shall not be taken greater than $4\sqrt{f'_c}\,b_od$ (For **SI:** $0.33\sqrt{f'_c}\,b_od$), on the critical section defined in Section 1911.12.4.7. When shearhead reinforcement is provided, V_n shall not be taken greater than $7\sqrt{f'_c}\,b_od$ (For **SI:** $0.58\sqrt{f'_c}\,b_od$), on the critical section defined in Section 1911.12.1.2, Item 1.

1911.12.4.9 The moment resistance M_v contributed to each slab column strip computed by a shearhead shall not be taken greater than

$$M_v = \frac{\phi\alpha_v V_u}{2\eta}\left(l_v - \frac{c_1}{2}\right) \qquad (11\text{-}40)$$

where ϕ is the strength-reduction factor for flexure, η is the number of arms, and l_v is the length of each shearhead arm actually provided. However, M_v shall not be taken larger than the smaller of:

1. Thirty percent of the total factored moment required for each slab column strip,

2. The change in column strip moment over the length l_v,

3. The value of M_p computed by Formula (11-39).

1911.12.4.10 When unbalanced moments are considered, the shearhead must have adequate anchorage to transmit M_p to column.

1911.12.5 Opening in slabs. When openings in slabs are located at a distance less than 10 times the slab thickness from a concentrated load or reaction area, or when openings in flat slabs are located within column strips as defined in Section 1913, the critical slab sections for shear defined in Section 1911.12.1.2 and Section 1911.12.4.7 shall be modified as follows:

1911.12.5.1 For slabs without shearheads, that part of the perimeter of the critical section that is enclosed by straight lines projecting from the centroid of the column, concentrated load or reaction area and tangent to the boundaries of the openings shall be considered ineffective.

1911.12.5.2 For slabs with shearheads, the ineffective portion of the perimeter shall be one half of that defined in Section 1911.12.5.1.

1911.12.6 Transfer of moment in slab-column connections.

1911.12.6.1 When gravity load, wind, earthquake or other lateral forces cause transfer of unbalanced moment, M_u, between a slab and a column, a fraction $\gamma_f M_u$ of the unbalanced moment shall be transferred by flexure in accordance with Section 1913.5.3. The remainder of the unbalanced moment given by $\gamma_v M_u$ shall be considered to be transferred by eccentricity of shear about the centroid of the critical section defined in Section 1911.12.1.2 where:

$$\gamma_v = (1 - \gamma_f) \qquad (11\text{-}41)$$

1911.12.6.2 The shear stress resulting from moment transfer by eccentricity of shear shall be assumed to vary linearly about the centroid of the critical sections defined in Section 1911.12.1.2. The maximum shear stress due to the factored shear force and moment shall not exceed ϕv_n:

For members without shear reinforcement:

$$\phi v_n = \phi V_c/(b_od) \qquad (11\text{-}42)$$

where V_c is as defined in Section 1911.12.2.1 and 1911.12.2.2.

For members with shear reinforcement other than shearheads:

$$\phi v_n = \phi(V_c + V_s)/(b_od) \qquad (11\text{-}43)$$

where V_c and V_s are defined in Section 1911.12.3. If shear reinforcement is provided, the design shall take into account the variation of shear stress around the column.

1911.12.6.3 When shear reinforcement consisting of steel I- or channel-shaped sections (shearheads) is provided, the sum of the shear stresses due to vertical load acting on the critical section defined by Section 1911.12.4.7 and the shear stresses resulting from moment transferred by eccentricity of shear about the centroid of the critical section defined in Section 1911.12.1.2 shall not exceed $\phi 4 \sqrt{f'_c}$ (For **SI:** $\phi 0.33 \sqrt{f'_c}$).

SECTION 1912 — DEVELOPMENT AND SPLICES OF REINFORCEMENT

1912.0 Notations.

A_b = area of an individual bar, square inches (mm^2).

A_s = area of nonprestressed tension reinforcement, square inches (mm^2).

A_{tr} = total cross-sectional area of all transverse reinforcement which is within the spacing s and which crosses the potential plane of splitting through the reinforcement being developed, inches squared (mm^2).

A_v = area of shear reinforcement within a distance s, square inches (mm^2).

A_w = area of an individual wire to be developed or spliced, square inches (mm^2).

a = depth of equivalent rectangular stress block as defined in Section 1910.2.7.1.

b_w = web width, or diameter of circular section, inches (mm).

c = spacing or cover dimension, inches (mm). See Section 1912.2.4.

d = distance from extreme compression fiber to centroid of tension reinforcement, inches (mm).

d_b = nominal diameter of bar, wire or prestressing strand, inches (mm).

f'_c = specified compressive strength of concrete, pounds per square inch (MPa).

$\sqrt{f'_c}$ = square root of specified compressive strength of concrete, pounds per square inch (MPa).

f_{ct} = average splitting tensile strength of lightweight aggregate concrete, pounds per square inch (MPa).

f_{ps} = stress in prestressed reinforcement at nominal strength, kips per square inch (MPa).

f_{se} = effective stress in prestressed reinforcement (after allowance for all prestress losses), kips per square inch (MPa).

f_y = specified yield strength of nonprestressed reinforcement, pounds per square inch (MPa).

f_{yt} = specified yield strength of transverse reinforcement, psi (MPa).

h = overall thickness of member, inches (mm).

K_{tr} = transverse reinforcement index.

= $\dfrac{A_{tr} f_{yt}}{1,500 s n}$ (constant 1,500 carries the unit lb./in.).

l_a = additional embedment length at support or at point of inflection, inches (mm).

l_d = development length, inches (mm).

= $l_{db} \times$ applicable modification factors.

l_{db} = basic development length, inches (mm).

l_{dh} = development length of standard hook in tension, measured from critical section to outside end of hook [straight embedment length between critical section and start of hook (point of tangency) plus radius of bend and one bar diameter], inches (mm).

= $l_{hb} \times$ applicable modification factors.

l_{hb} = basic development length of standard hook in tension, inches (mm).

M_n = nominal moment strength at section, inch-pounds (N·m).

= $A_s f_y (d - a/2)$.

N = number of bars in a layer being spliced or developed at a critical section.

n = number of bars or wires being spliced or developed along the plane of splitting.

s = maximum center to center spacing of transverse reinforcement within l_d, inches (mm).

s_w = spacing of wire to be developed or spliced, inches (mm).

V_u = factored shear force at section.

α = reinforcement location factor. See Section 1912.2.4.

β = coating factor. See Section 1912.2.4.

β_b = ratio of area of reinforcement cut off to total area of tension reinforcement at section.

γ = reinforcement size factor. See Section 1912.2.4.

λ = lightweight aggregate concrete factor. See Section 1912.2.4.

1912.1 Development of Reinforcement—General.

1912.1.1 Calculated tension or compression in reinforcement at each section of structural concrete members shall be developed on each side of that section by embedment length, hook or mechanical device, or a combination thereof. Hooks shall not be used to develop bars in compression.

1912.1.2 The values of $\sqrt{f'_c}$ used in Section 1912 shall not exceed 100 psi (0.69 MPa).

1912.2 Development of Deformed Bars and Deformed Wire in Tension.

1912.2.1 Development length, l_d, in terms of diameter, d_b, for deformed bars and deformed wire in tension shall be determined from either Section 1912.2.2 or 1912.2.3, but l_d shall not be less than 12 inches (305 mm).

1912.2.2 For deformed bars or deformed wire, l_d/d_b shall be as follows:

	NO. 6 AND SMALLER BARS AND DEFORMED WIRES	NO. 7 AND LARGER BARS
Clear spacing of bars being developed or spliced not less than d_b, clear cover not less than d_b, and stirrups or ties throughout l_d not less than the prescribed minimum or Clear spacing of bars being developed or spliced not less than $2d_b$ and clear cover not less than d_b	$\dfrac{l_d}{d_b} = \dfrac{f_y \alpha \beta \lambda}{25 \sqrt{f'_c}}$	$\dfrac{l_d}{d_b} = \dfrac{f_y \alpha \beta \lambda}{20 \sqrt{f'_c}}$
Other cases	$\dfrac{l_d}{d_b} = \dfrac{3 f_y \alpha \beta \lambda}{50 \sqrt{f'_c}}$	$\dfrac{l_d}{d_b} = \dfrac{3 f_y \alpha \beta \lambda}{40 \sqrt{f'_c}}$

1912.2.3 For deformed bars or deformed wire, l_d/d_b shall be:

$$\frac{l_d}{d_b} = \frac{3}{40} \frac{f_y}{\sqrt{f'_c}} \frac{\alpha \beta \gamma \lambda}{\left(\dfrac{c + K_{tr}}{d_b} \right)} \tag{12-1}$$

in which the term $\dfrac{c + K_{tr}}{d_b}$ shall not be taken greater than 2.5.

1912.2.4 The factors for use in the expressions for development of deformed bars and deformed wires in tension in Sections 1912.0 through 1912.19 are as follows:

α = reinforcement location factor

Horizontal reinforcement so placed that more than 12 inches (305 mm) of fresh concrete is cast in the member below the development length or splice 1.3

Other reinforcement 1.0

β = coating factor

Epoxy-coated bars or wires with cover less than $3d_b$, or clear spacing less than $6d_b$ 1.5

All other epoxy-coated bars or wires 1.2

Uncoated reinforcement 1.0

However, the product $\alpha\beta$ need not be taken greater than 1.7.

γ = reinforcement size factor

No. 6 and smaller bars and deformed wires 0.8

No. 7 and larger bars 1.0

λ = lightweight aggregate concrete factor.

When lightweight aggregate concrete is used 1.3

However, when f_{ct} is specified, λ shall be permitted to be taken as $6.7\sqrt{f'_c/f_{ct}}$ but not less than 1.0

when normal weight concrete is used 1.0

c = spacing or cover dimension, inches (mm).

Use the smaller of either the distance from the center of the bar to the nearest concrete surface or one-half the center-to-center spacing of the bars being developed.

K_{tr} = transverse reinforcement index

$$= \frac{A_{tr} f_{yt}}{1,500sn}$$

WHERE:

A_{tr} = total cross-sectional area of all transverse reinforcement which is within the spacing s and which crosses the potential plane of splitting through the reinforcement being developed, inches squared (mm^2).

f_{yt} = specified yield strength of transverse reinforcement, inches squared (mm^2).

s = maximum spacing of transverse reinforcement within l_d, center-to-center, inches (mm).

n = number of bars or wires being developed along the plane of splitting.

It shall be permitted to use $K_{tr} = 0$ as a design simplification even if transverse reinforcement is present.

1912.2.5 Excess reinforcement. Reduction in development length shall be permitted where reinforcement in a flexural member is in excess of that required by analysis except where anchorage or development for f_y is specifically required or the reinforcement is designed under provisions of Section 1921.2.1.4 . . . . [(A_s required)/(A_s provided)]

1912.3 Development of Deformed Bars in Compression.

1912.3.1 Development length l_d, in inches, for deformed bars in compression shall be computed as the product of the basic development length l_{db} and applicable modification factors as defined in this section, but l_d shall not be less than 8 inches (203 mm).

1912.3.2 Basic development length l_{db} shall be . $0.02d_b f_y/\sqrt{f'_c}$
(For **SI:** $0.24d_b f_y/\sqrt{f'_c}$)

but not less than . . . . $0.0003d_b f_y$
(For **SI:** $0.044\,d_b f_y$)

1912.3.3 Basic development length l_{db} shall be permitted to be multiplied by applicable factors for:

1912.3.3.1 Excess reinforcement. Reinforcement in excess of that required by analysis . . . . (A_s required)/(A_s provided)

1912.3.3.2 Spirals and ties. Reinforcement enclosed within spiral reinforcement not less than $1/4$-inch (6.4 mm) diameter and not more than 4-inch (102 mm) pitch or within No. 4 ties in conformance with Section 1907.10.5 and spaced not more than 4 inches (102 mm) on center 0.75

1912.4 Development of Bundled Bars.

1912.4.1 Development length of individual bars within a bundle, in tension or compression, shall be that for the individual bar, increased 20 percent for 3-bar bundle, and 33 percent for 4-bar bundle.

1912.4.2 For determining the appropriate factors in Section 1912.2, a unit of bundled bars shall be treated as a single bar of a diameter derived from the equivalent total area.

1912.5 Development of Standard Hooks in Tension.

1912.5.1 Development length l_{dh} in inches (mm) for deformed bars in tension terminating in a standard hook shall be computed as the product of the basic development length l_{hb} of Section 1912.5.2 and the applicable modification factor or factors of Section 1912.5.3, but l_{dh} shall not be less than $8d_b$ or less than 6 inches (152 mm).

1912.5.2 Basic development length l_{hb} for a hooked bar with f_y equal to 60,000 psi (413.7 MPa) shall be . . . . . $1,200\,d_b/\sqrt{f'_c}$
(For **SI:** $99.7\,d_b/\sqrt{f'_c}$)

1912.5.3 Basic development length l_{hb} shall be multiplied by applicable factor or factors for:

1912.5.3.1 Bar yield strength. Bars with f_y other than 60,000 psi (413.7 MPa) . . . . $f_y/60,000$
(For **SI:** $f_y/413.7$)

1912.5.3.2 Concrete cover. For No. 11 bar and smaller, side cover (normal to plane of the hook) not less than $2^1/2$ inches (64 mm), and for 90-degree hook, cover on bar extension beyond hook not less than 2 inches (51 mm) 0.7

1912.5.3.3 Ties or stirrups. For No. 11 bar and smaller, hook enclosed vertically or horizontally within ties or stirrup ties spaced along the full development length l_{dh} not greater than $3d_b$, where d_b is diameter of hooked bar 0.8

1912.5.3.4 Excessive reinforcement. Where anchorage or development for f_y is not specifically required, reinforcement in excess of that required by analysis . [(A_s required)/(A_s provided)]

1912.5.3.5 Lightweight aggregate concrete 1.3

1912.5.3.6 Epoxy-coated reinforcement. Hooked bars with epoxy coating 1.2

1912.5.4 For bars being developed by a standard hook at discontinuous ends of members with side cover and top (or bottom) cover over hook less than $2^1/2$ inches (64 mm), hooked bar shall be enclosed within ties or stirrups spaced along the full development length l_{dh} not greater than $3d_b$, where d_b is diameter of hooked bar. For this case, factor of Section 1912.5.3.3 shall not apply.

1912.5.5 Hooks shall not be considered effective in developing bars in compression.

1912.6 Mechanical Anchorage.

1912.6.1 Any mechanical device capable of developing the strength of reinforcement without damage to concrete may be used as anchorage.

1912.6.2 Test results showing adequacy of such mechanical devices shall be presented to the building official.

1912.6.3 Development of reinforcement shall be permitted to consist of a combination of mechanical anchorage plus additional embedment length of reinforcement between the point of maximum bar stress and the mechanical anchorage.

1912.7 Development of Welded Deformed Wire Fabric in Tension.

1912.7.1 Development length l_d, in inches (mm), of welded deformed wire fabric measured from the point of critical section to the end of wire shall be computed as the product of the development length l_d, from Section 1912.2.2 or 1912.2.3 times a wire fabric factor from Section 1912.7.2 or 1912.7.3. It shall be permitted to reduce the development length in accordance with 1912.2.5 when applicable, but l_d shall not be less than 8 inches (203 mm) except in computation of lap splices by Section 1912.18. When using the wire fabric factor from Section 1912.7.2, it shall be permitted to use an epoxy-coating factor β of 1.0 for epoxy-coated welded wire fabric in Sections 1912.2.2 and 1912.2.3.

1912.7.2 For welded deformed wire fabric with at least one cross wire within the development length and not less than 2 inches (51 mm) from the point of the critical section, the wire fabric factor shall be the greater of:

$$\left(\frac{f_y \ - \ 35,000}{f_y} \right)$$

or

$$\left(\frac{5d_b}{s_w} \right)$$

but need not be taken greater than 1.

1912.7.3 For welded deformed wire fabric with no cross wires within the development length or with a single cross wire less than 2 inches (51 mm) from the point of the critical section, the wire fabric factor shall be taken as 1, and the development length shall be determined as for deformed wire.

1912.7.4 When any plain wires are present in the deformed wire fabric in the direction of the development length, the fabric shall be developed in accordance with Section 1912.8.

1912.8 Development of Welded Plain Wire Fabric in Tension. Yield strength of welded plain wire fabric shall be considered developed by embedment of two cross wires with the closer cross wire not less than 2 inches (51 mm) from the point of the critical section. However, the development length l_d, in inches (mm), measured from the point of the critical section to the outermost cross wire shall not be less than

$$0.27 \ \frac{A_w}{s_w} \left(\frac{f_y}{\sqrt{f'_c}} \right) \lambda$$

For SI:

$$3.3 \ \frac{A_w}{s_w} \left(\frac{f_y}{\sqrt{f'_c}} \right) \lambda$$

except that when reinforcement provided is in excess of that required, this length may be reduced in accordance with Section 1912.2.5. l_d shall not be less than 6 inches (152 mm) except in computation of lap splices by Section 1912.19.

1912.9 Development of Prestressing Strand.

1912.9.1 Three- or seven-wire pretensioning strand shall be bonded beyond the critical section for a development length, in inches (mm), not less than

$$\left(f_{ps} \ - \ \frac{2}{3} f_{se} \right) d_b \dagger$$

For SI:

$$0.145 \left(f_{ps} \ - \ \frac{2}{3} f_{se} \right) d_b$$

†Expression in parentheses used as a constant without units.

where d_b is strand diameter in inches (mm), and f_{ps} and f_{se} are expressed in kips per square inch (MPa).

1912.9.2 Limiting the investigation to cross sections nearest each end of the member that are required to develop full design strength under specified factored loads shall be permitted.

1912.9.3 Where bonding of a strand does not extend to end of member, and design includes tension at service load in precompressed tensile zone as permitted by Section 1918.4.2, development length specified in Section 1912.9.1 shall be doubled.

1912.10 Development of Flexural Reinforcement—General.

1912.10.1 Development of tension reinforcement by bending across the web to be anchored or made continuous with reinforcement on the opposite face of member shall be permitted.

1912.10.2 Critical sections for development of reinforcement in flexural members are at points of maximum stress and at points within the span where adjacent reinforcement terminates or is bent. Provisions of Section 1912.11.3 must be satisfied.

1912.10.3 Reinforcement shall extend beyond the point at which it is no longer required to resist flexure for a distance equal to the effective depth of member or $12d_b$, whichever is greater, except at supports of simple spans and at free end of cantilevers.

1912.10.4 Continuing reinforcement shall have an embedment length not less than the development length l_d beyond the point where bent or terminated tension reinforcement is no longer required to resist flexure.

1912.10.5 Flexural reinforcement shall not be terminated in a tension zone unless one of the following conditions is satisfied:

1912.10.5.1 Shear at the cutoff point does not exceed two thirds that permitted, including shear strength of shear reinforcement provided.

1912.10.5.2 Stirrup area in excess of that required for shear and torsion is provided along each terminated bar or wire over a distance from the termination point equal to three fourths the effective depth of member. Excess stirrup area A_v shall not be less than $60b_w s/f_y$ (For SI: $0.41b_w s/f_y$). Spacing s shall not exceed $d/8\beta_b$ where β_b is the ratio of area of reinforcement cut off to total area of tension reinforcement at the section.

1912.10.5.3 For No. 11 bar and smaller, continuing reinforcement provides double the area required for flexure at the cutoff point and shear does not exceed three fourths that permitted.

1912.10.6 Adequate anchorage shall be provided for tension reinforcement in flexural members where reinforcement stress is not directly proportional to moment, such as sloped, stepped or tapered footings, brackets, deep flexural members, or members in which tension reinforcement is not parallel to compression face. See Sections 1912.11.4 and 1912.12.4 for deep flexural members.

1912.11 Development of Positive Moment Reinforcement.

1912.11.1 At least one third the positive moment reinforcement in simple members and one fourth the positive moment reinforcement in continuous members shall extend along the same face of member into the support. In beams, such reinforcement shall extend into the support at least 6 inches (152 mm).

1912.11.2 When a flexural member is part of a primary lateral-load-resisting system, positive moment reinforcement required to be extended into the support by Section 1912.11.1 shall be anchored to develop the specified yield strength f_y in tension at the face of support.

1912.11.3 At simple supports and at points of inflection, positive moment tension reinforcement shall be limited to a diameter such that l_d computed for f_y by Section 1912.2 satisfies Formula (12-2), except Formula (12-2) need not be satisfied for reinforcement terminating beyond center line of simple supports by a standard hook or a mechanical anchorage at least equivalent to a standard hook.

$$l_d \leq \frac{M_n}{V_u} + l_a \qquad (12\text{-}2)$$

WHERE:

l_a = at a support shall be the embedment length beyond center of support.

l_a = at a point of inflection shall be limited to the effective depth of member or $12d_b$, whichever is greater.

M_n = nominal strength assuming all reinforcement at the section to be stressed to the specified yield strength f_y.

V_u = factored shear force at the section.

An increase of 30 percent in the value of M_n/V_u shall be permitted when the ends of reinforcement are confined by a compressive reaction.

1912.11.4 At simple supports of deep flexural members, positive moment tension reinforcement shall be anchored to develop the specified yield strength f_y in tension at the face of support. At interior supports of deep flexural members, positive moment tension reinforcement shall be continuous or be spliced with that of the adjacent spans.

1912.12 Development of Negative Moment Reinforcement.

1912.12.1 Negative moment reinforcement in a continuous, restrained or cantilever member, or in any member of a rigid frame, shall be anchored in or through the supporting member by embedment length, hooks or mechanical anchorage.

1912.12.2 Negative moment reinforcement shall have an embedment length into the span as required by Sections 1912.1 and 1912.10, Item 3.

1912.12.3 At least one third the total tension reinforcement provided for negative moment at a support shall have an embedment length beyond the point of inflection not less than effective depth of member, $12d_b$, or $^1/_{16}$ the clear span, whichever is greater.

1912.12.4 At interior supports of deep flexural members, negative moment tension reinforcement shall be continuous with that of the adjacent spans.

1912.13 Development of Web Reinforcement.

1912.13.1 Web reinforcement shall be carried as close to compression and tension surfaces of member as cover requirements and proximity of other reinforcement will permit.

1912.13.2 Ends of single leg, simple U- or multiple U-stirrups shall be anchored by one of the following means:

1912.13.2.1 For No. 5 bar and D31 wire, and smaller, and for Nos. 6, 7 and 8 bars with f_y of 40,000 psi (275.8 MPa) or less, a standard stirrup hook around longitudinal reinforcement.

1912.13.2.2 For Nos. 6, 7 and 8 stirrups with f_y greater than 40,000 psi (275.8 MPa), a standard stirrup hook around a longitudinal bar plus an embedment between midheight of the member and the outside end of the hook equal to or greater than 0.014 $d_b f_y / \sqrt{f'_c}$ (For **SI:** 0.169 $d_b f_y / \sqrt{f'_c}$).

1912.13.2.3 For each leg of welded smooth wire fabric forming simple U-stirrups, either:

1. Two longitudinal wires spaced at a 2-inch (51 mm) spacing along the member at the top of the U.

2. One longitudinal wire located not more than $d/4$ from the compression face and a second wire closer to the compression face and spaced not less than 2 inches (51 mm) from the first wire. The second wire shall be permitted to be located on the stirrup leg beyond a bend, or on a bend with an inside diameter of bend not less than $8d_b$.

1912.13.2.4 For each end of a single-leg stirrup of welded plain or deformed wire fabric, two longitudinal wires at a minimum spacing of 2 inches (51 mm) and with the inner wire at least the greater of $d/4$ or 2 inches (51 mm) from middepth of member $d/2$. Outer longitudinal wire at tension face shall not be farther from the face than the portion of primary flexural reinforcement closest to the face.

1912.13.2.5 In joist construction as defined in Section 1908.11, for No. 4 bar and D20 wire and smaller, a standard hook.

1912.13.3 Between anchored ends, each bend in the continuous portion of a simple U-stirrup or multiple U-stirrups shall enclose a longitudinal bar.

1912.13.4 Longitudinal bars bent to act as shear reinforcement, if extended into a region of tension, shall be continuous with longitudinal reinforcement and, if extended into a region of compression, shall be anchored beyond middepth $d/2$ as specified for development length in Section 1912.2 for that part of f_y required to satisfy Formula (11-19).

1912.13.5 Pairs of U-stirrups or ties so placed as to form a closed unit shall be considered properly spliced when lengths of laps are $1.3l_d$. In members at least 18 inches (457 mm) deep, such splices with $A_b f_y$ not more than 9,000 pounds (40 000 N) per leg may be considered adequate if stirrup legs extend the full available depth of member.

1912.14 Splices of Reinforcement.

1912.14.1 Splices of reinforcement shall be made only as required or permitted on design drawings or in specifications, or as authorized by the *building official*.

1912.14.2 Lap splices.

1912.14.2.1 Lap splices shall not be used for bars larger than No. 11, except as provided in Sections 1912.16.2 and 1915.8.2.3.

1912.14.2.2 Lap splices of bars in a bundle shall be based on the lap splice length required for individual bars within the bundle, increased in accordance with Section 1912.4. Individual bar splices within a bundle shall not overlap. Entire bundles shall not be lap spliced.

1912.14.2.3 Bars spliced by noncontact lap splices in flexural members shall not be spaced transversely farther apart than one fifth the required lap splice length, or 6 inches (152 mm).

1912.14.3 Welded splices and mechanical connections.

1912.14.3.1 Welded splices and other mechanical connections may be used.

1912.14.3.2 Except as provided in this code, all welding shall conform to UBC Standard 19-1.

1912.14.3.3 A full-welded splice shall develop at least 125 percent of specified yield strength, f_y, of the bar.

1912.14.3.4 A full mechanical connection shall develop in tension or compression, as required, at least 125 percent of specified yield strength f_y of the bar.

1912.14.3.5 Welded splices and mechanical connections not meeting requirements of Section 1912.14.3.3 or 1912.14.3.4 are allowed only for No. 5 bars and smaller and in accordance with Section 1912.15.4.

1912.14.3.6 *Welded splices and mechanical connections shall maintain the clearance and coverage requirements of Sections 1907.6 and 1907.7.*

1912.15 Splices of Deformed Bars and Deformed Wire in Tension.

1912.15.1 Minimum length of lap for tension lap splices shall be as required for Class A or B splice, but not less than 12 inches (305 mm), where:

Class A splice . . . . $1.0l_d$

Class B splice . . . . $1.3l_d$

where l_d is the tensile development length for the specified yield strength f_y in accordance with Section 1912.2 without the modification factor of Section 1912.2.5.

1912.15.2 Lap splices of deformed bars and deformed wire in tension shall be Class B splices except that Class A splices may be used when (1) the area of reinforcement provided is at least twice that required by analysis over the entire length of the splice, and (2) one half or less of the total reinforcement is spliced within the required lap length.

1912.15.3 Welded splices or mechanical connections used where area of reinforcement provided is less than twice that required by analysis shall meet requirements of Section 1912.14.3.3 and 1912.14.3.4.

1912.15.4 Welded splices or mechanical connections not meeting the requirements of Section 1912.14.3.3 or 1912.14.3.4 are allowed for No. 5 bars and smaller when the area of reinforcement provided is at least twice that required by analysis, and the following requirements are met:

1912.15.4.1 Splices shall be staggered at least 24 inches (610 mm) and in such manner as to develop at every section at least twice the calculated tensile force at that section but not less than 20,000 psi (137.9 MPa) for total area of reinforcement provided.

1912.15.4.2 In computing tensile forces developed at each section, rate the spliced reinforcement at the specified splice strength. Unspliced reinforcement shall be rated at that fraction of f_y defined by the ratio of the shorter actual development length to l_d required to develop the specified yield strength f_y.

1912.15.4.3 *Mechanical connections need not be staggered as required by Section 1912.15.4.1 or 1912.15.5 provided the clearance and coverage requirements of Sections 1907.6 and 1907.7 are maintained and, at 90 percent of the yield stress, the strain measured over the full length of the connector does not exceed 50 percent of the strain of an unspliced bar when the maximum computed design load stress does not exceed 50 percent of the yield stress.*

1912.15.5 Splices in "tension tie members" shall be made with a full-welded splice or full mechanical connection in accordance with Section 1912.14.3.3 and 1912.14.3.4, and splices in adjacent bar shall be staggered at least 30 inches (762 mm).

1912.16 Splices of Deformed Bars in Compression.

1912.16.1 Compression lap splice length shall be 0.0005 $f_y d_b$ (For **SI**: 0.073 $f_y d_b$) for f_y of 60,000 psi (413.7 MPa) or less, or $(0.0009 f_y - 24) d_b$ [For **SI**: $(0.13 f_y - 24) d_b$] for f_y greater than 60,000 psi (413.7 MPa), but not less than 12 inches (305 mm). For f'_c less than 3,000 psi (20.68 MPa), length of lap shall be increased by one third.

1912.16.2 When bars of different size are lap spliced in compression, splice length shall be the larger of: development length of larger bar, or splice length of smaller bar. Lap splices of No. 14 and No. 18 bars to No. 11 and smaller bars shall be permitted.

1912.16.3 Welded splices or mechanical connections used in compression shall meet requirements of Sections 1912.14.3.3 and 1912.14.3.4.

1912.16.4 End-bearing splices.

1912.16.4.1 In bars required for compression only, transmission of compressive stress by bearing of square cut ends held in concentric contact by a suitable device shall be permitted.

1912.16.4.2 Bar ends shall terminate in flat surfaces within $1^1/2$ degrees of a right angle to the axis of the bars and shall be fitted within 3 degrees of full bearing after assembly.

1912.16.4.3 End-bearing splices shall be used only in members containing closed ties, closed stirrups or spirals.

1912.17 Special Splice Requirements for Columns.

1912.17.1 Lap splices, butt welded splices, mechanical connections or end-bearing splices shall be used with the limitations of Sections 1912.17.2 through 1912.17.4. A splice shall satisfy requirements for all load combinations for the column.

1912.17.2 Lap splices in columns.

1912.17.2.1 Where the bar stress due to factored loads is compressive, lap splices shall conform to Sections 1912.16.1 and 1912.16.2, and where applicable, to Section 1912.17.2.4 or 1912.17.2.5.

1912.17.2.2 Where the bar stress due to factored loads is tensile and does not exceed $0.5f_y$ in tension, lap splices shall be Class B tension lap splices if more than one half of the bars are spliced at any section, or Class A tension lap splices if one half or fewer of the bars are spliced at any section and alternate lap splices are staggered by l_d.

1912.17.2.3 Where the bar stress due to factored loads is greater than 0.5 f_y in tension, lap splices shall be Class B tension lap splices.

1912.17.2.4 In tied reinforced compression members, where ties throughout the lap splice length have an effective area not less than $0.0015hs$, lap splice length shall be permitted to be multiplied by 0.83, but lap length shall not be less than 12 inches (305 mm).

Tie legs perpendicular to dimension h shall be used in determining effective area.

1912.17.2.5 In spirally reinforced compression members, lap splice length of bars within a spiral shall be permitted to be multiplied by 0.75, but lap length shall not be less than 12 inches (305 mm).

1912.17.3 Welded splices or mechanical connectors in columns. Welded splices or mechanical connectors in columns shall meet the requirements of Section 1912.14.3.3 or 1912.14.3.4.

1912.17.4 End-bearing splices in columns. End-bearing splices complying with Section 1912.16.4 shall be permitted to be used for column bars stressed in compression provided the splices are staggered or additional bars are provided at splice locations. The continuing bars in each face of the column shall have a tensile strength, based on the specified yield strength f_y, not less than $0.25f_y$ times the area of the vertical reinforcement in that face.

1912.18 Splices of Welded Deformed Wire Fabric in Tension.

1912.18.1 Minimum length of lap for lap splices of welded deformed wire fabric measured between the ends of each fabric sheet shall not be less than $1.3l_d$ or 8 inches (203 mm), and the overlap measured between outermost cross wires of each fabric sheet shall not be less than 2 inches (51 mm), l_d shall be the development length for the specified yield strength f_y in accordance with Section 1912.7.

1912.18.2 Lap splices of welded deformed wire fabric, with no cross wires within the lap splice length, shall be determined as for deformed wire.

1912.18.3 When any plain wires are present in the deformed wire fabric in the direction of the lap splice or when deformed wire fabric is lap spliced to plain wire fabric, the fabric shall be lap spliced in accordance with Section 1912.19.

1912.19 Splices of Welded Plain Wire Fabric in Tension. Minimum length of lap for lap splices of welded smooth wire fabric shall be in accordance with the following:

1912.19.1 When area of reinforcement provided is less than twice that required by analysis at splice location, length of overlap measured between outermost cross wires of each fabric sheet shall not be less than one spacing of cross wires plus 2 inches (51 mm), or less than 1.5 l_d, or 6 inches (152 mm), l_d shall be the development length for the specified yield strength f_y in accordance with Section 1912.8.

1912.19.2 When area of reinforcement provided is at least twice that required by analysis at splice location, length of overlap measured between outermost cross wires of each fabric sheet shall not be less than 1.5 l_d, or 2 inches (51 mm), l_d shall be the development length for the specified yield strength f_y in accordance with Section 1912.8.

SECTION 1913 — TWO-WAY SLAB SYSTEMS

1913.0 Notations.

b_1 = width of the critical section defined in Section 1911.12.1.2 measured in the direction of the span for which moments are determined, inches (mm).

b_2 = width of the critical section defined in Section 1911.12.1.2 measured in the direction perpendicular to b_1, inches (mm).

C = cross-sectional constant to define torsional properties.

$$\sum \left(1 - 0.63 \, \frac{x}{y} \right) \frac{x^3 y}{3}$$

The constant C for T- or L-sections shall be permitted to be evaluated by dividing the section into separate rectangular parts and summing the values of C for each part.

c_1 = size of rectangular or equivalent rectangular column, capital, or bracket measured in the direction of the span for which moments are being determined, inches (mm).

c_2 = size of rectangular or equivalent rectangular column, capital or bracket measured transverse to the direction of the span for which moments are being determined, inches (mm).

E_{cb} = modulus of elasticity of beam concrete.

E_{cs} = modulus of elasticity of slab concrete.

h = overall thickness of member, inches (mm).

I_b = moment of inertia about centroidal axis of gross section of beam as defined in Section 1913.3.

I_s = moment of inertia about centroidal axis of gross section of slab.

= $h^3/12$ times width of slab defined in notations α and β_t.

K_t = torsional stiffness of torsional member; moment per unit rotation.

l_n = length of clear span in direction that moments are being determined, measured face to face of supports.

l_1 = length of span in direction that moments are being determined, measured center to center of supports.

l_2 = length of span transverse to l_1, measured center to center of supports. See also Sections 1913.6.2.3 and 1913.6.2.4.

M_o = total factored static moment.

M_u = factored moment at section.

V_c = nominal shear strength provided by concrete. See Section 1911.12.2.1.

V_u = factored shear force at section.

w_d = factored dead load per unit area.

w_l = factored live load per unit area.

w_u = factored load per unit area.

x = shorter overall dimension of rectangular part of cross section.

y = longer overall dimension of rectangular part of cross section.

α = ratio of flexural stiffness of beam section to flexural stiffness of a width of slab bounded laterally by center lines of adjacent panels (if any) on each side of the beam.

= $\dfrac{E_{cb}I_b}{E_{cs}I_s}$

α_1 = α in direction of l_1.

α_2 = α in direction of l_2.

β_t = ratio of torsional stiffness of edge beam section to flexural stiffness of a width of slab equal to span length of beam, center to center of supports.

= $\dfrac{E_{cb}C}{2E_{cs}I_s}$

γ_f = fraction of unbalanced moment transferred by flexure at slab-column connections. See Section 1913.5.3.2.

γ_v = fraction of unbalanced moment transferred by eccentricity of shear at slab-column connections.

= $1 - \gamma_f$

ρ = ratio of nonprestressed tension reinforcement.

ρ_b = reinforcement ratio producing balanced strain conditions.

ϕ = strength reduction factor.

1913.1 Scope.

1913.1.1 The provisions of this section shall apply for design of slab systems reinforced for flexure in more than one direction, with or without beams between supports.

1913.1.2 For a slab system supported by columns or walls, the dimensions c_1 and c_2 and the clear span l_n shall be based on an effective support area defined by the intersection of the bottom surface of the slab, or the drop panel if there is one, with the largest right circular cone, right pyramid, or tapered wedge whose surfaces are located within the column and capital or bracket and are oriented no greater than 45 degrees to the axis of the column.

1913.1.3 Solid slabs and slabs with recesses or pockets made by permanent or removable fillers between ribs or joists in two directions are included within the scope of this section.

1913.1.4 Minimum thickness of slabs designed in accordance with this section shall be as required by Section 1909.5.3.

1913.2 Definitions.

1913.2.1 Column strip is a design strip with a width on each side of a column center line equal to $0.25l_2$ or $0.25l_1$, whichever is less. Column strip includes beams, if any.

1913.2.2 Middle strip is a design strip bounded by two column strips.

1913.2.3 A panel is bounded by column, beam or wall center lines on all sides.

1913.2.4 For monolithic or fully composite construction, a beam includes that portion of slab on each side of the beam extending a distance equal to the projection of the beam above or below the slab, whichever is greater, but not greater than four times the slab thickness.

1913.3 Slab Reinforcement.

1913.3.1 Area of reinforcement in each direction for two-way slab systems shall be determined from moments at critical sections, but shall not be less than required by Section 1907.12.

1913.3.2 Spacing of reinforcement at critical sections shall not exceed two times the slab thickness, except for portions of slab area of cellular or ribbed construction. In the slab over cellular spaces, reinforcement shall be provided as required by Section 1907.12.

1913.3.3 Positive moment reinforcement perpendicular to a discontinuous edge shall extend to the edge of slab and have embedment, straight or hooked, at least 6 inches (152 mm) in spandrel beams, columns or walls.

1913.3.4 Negative moment reinforcement perpendicular to a discontinuous edge shall be bent, hooked or otherwise anchored, in spandrel beams, columns or walls, to be developed at face of support according to provisions of Section 1912.

1913.3.5 Where a slab is not supported by a spandrel beam or wall at a discontinuous edge or where a slab cantilevers beyond the support, anchorage of reinforcement shall be permitted within the slab.

1913.3.6 In slabs with beams between supports with a value of α greater than 1.0, special top and bottom slab reinforcement shall be provided at exterior corners in accordance with the following:

1913.3.6.1 The special reinforcement in both top and bottom of slab shall be sufficient to resist a moment equal to the maximum positive moment (per foot of width) (per meter of width) in the slab.

1913.3.6.2 The moment shall be assumed to be about an axis perpendicular to the diagonal from the corner in the top of the slab and perpendicular to the diagonal in the bottom of the slab.

1913.3.6.3 The special reinforcement shall be provided for a distance in each direction from the corner equal to one fifth the longer span.

1913.3.6.4 The special reinforcement shall be placed in a band parallel to the diagonal in the top of the slab and a band perpendicular to the diagonal in the bottom of the slab. Alternatively, the special reinforcement shall be placed in two layers parallel to the sides of the slab in either the top or bottom of the slab.

1913.3.7 Where a drop panel is used to reduce amount of negative moment reinforcement over the column of a flat slab, size of drop panel shall be in accordance with the following:

1913.3.7.1 Drop panel shall extend in each direction from center line of support a distance not less than one sixth the span length measured from center to center of supports in that direction.

1913.3.7.2 Projection of drop panel below the slab shall be at least one fourth the slab thickness beyond the drop.

1913.3.7.3 In computing required slab reinforcement, thickness of drop panel below the slab shall not be assumed greater than one fourth the distance from edge of drop panel to edge of column or column capital.

1913.3.8 Details of reinforcement in slabs without beams.

1913.3.8.1 In addition to the other requirements of Section 1913.3, reinforcement in slabs without beams shall have minimum extensions as prescribed in Figure 19-1.

1913.3.8.2 Where adjacent spans are unequal, extension of negative moment reinforcement beyond the face of support as prescribed in Figure 19-1 shall be based on requirements of longer span.

1913.3.8.3 Bent bars shall be permitted only when depth-span ratio permits use of bends 45 degrees or less.

1913.3.8.4 For slabs in frames not braced against sidesway, lengths of reinforcement shall be determined by analysis but shall not be less than those prescribed in Figure 19-1.

1913.3.8.5 All bottom bars or wires within the column strip, in each direction, shall be continuous or spliced with Class A splices located as shown in Figure 19-1. At least two of the column strip bottom bars or wires in each direction shall pass within the column core and shall be anchored at exterior supports.

1913.3.8.6 In slabs with shearheads and in lift-slab construction, at least two bonded bottom bars or wires in each direction shall pass through the shearhead or lifting collar as close to the column as practicable and be continuous or spliced with a Class A splice. At exterior columns, the reinforcement shall be anchored at the shearhead or lifting collar.

1913.4 Openings in Slab Systems.

1913.4.1 Openings of any size shall be permitted in slab systems if shown by analysis that the design strength is at least equal to the required strength considering Sections 1909.2 and 1909.3, and that all serviceability conditions, including the specified limits on deflections, are met.

1913.4.2 In lieu of special analysis as required by Section 1913.4.1, openings shall be permitted in slab systems without beams only in accordance with the following:

1913.4.2.1 Openings of any size shall be permitted in the area common to intersecting middle strips, provided total amount of reinforcement required for the panel without the opening is maintained.

1913.4.2.2 In the area common to intersecting column strips, not more than one eighth the width of column strip in either span shall be interrupted by openings. An amount of reinforcement equivalent to that interrupted by an opening shall be added on the sides of the opening.

1913.4.2.3 In the area common to one column strip and one middle strip, not more than one fourth the reinforcement in either strip shall be interrupted by openings. An amount of reinforcement equivalent to that interrupted by an opening shall be added on the sides of the opening.

1913.4.2.4 Shear requirements of Section 1911.12.5 shall be satisfied.

1913.5 Design Procedures.

1913.5.1 A slab system shall be designed by any procedure satisfying conditions of equilibrium and geometric compatibility if shown that the design strength at every section is at least equal to the required strength considering Sections 1909.2 and 1909.3 and that all serviceability conditions, including specified limits on deflections, are met.

1913.5.1.1 Design of a slab system for gravity loads including the slab and beams (if any) between supports and supporting columns or walls forming orthogonal frames, by either the Direct Design Method of Section 1913.6 or the Equivalent Frame Method of Section 1913.7, shall be permitted.

1913.5.1.2 For lateral loads, analysis of unbraced frames shall take into account effects of cracking and reinforcement on stiffness of frame members.

1913.5.1.3 Combining the results of the gravity load analysis with the results of the lateral load analysis shall be permitted.

1913.5.2 The slab and beams (if any) between supports shall be proportioned for factored moments prevailing at every section.

1913.5.3 When gravity load, wind, earthquake or other lateral forces cause transfer of moment between slab and column, a fraction of the unbalanced moment shall be transferred by flexure in accordance with Sections 1913.5.3.2 and 1913.5.3.3.

1913.5.3.1 Fraction of unbalanced moment not transferred by flexure shall be transferred by eccentricity of shear in accordance with Section 1911.12.6.

1913.5.3.2 A fraction of the unbalanced moment given by $\gamma_f M_u$ shall be considered to be transferred by flexure within an effective slab width between lines that are one and one-half slab or drop panel thickness ($1.5h$) outside opposite faces of the column or capital, where M_u is the moment to be transferred and

$$\gamma_f = \frac{1}{1 + \frac{2}{3}\sqrt{b_1/b_2}} \qquad (13\text{-}1)$$

1913.5.3.3 For unbalanced moments about an axis parallel to the edge at exterior supports, the value of γ_f by Formula (13-1) shall

be permitted to be increased up to 1.0 provided that V_u at an edge support does not exceed $0.75\phi V_c$ or at a corner support does not exceed $0.5\phi V_c$. For unbalanced moments at interior supports, and for unbalanced moments about an axis transverse to the edge at exterior supports, the value of γ_f in Formula (13-1) shall be permitted to be increased by up to 25 percent provided that V_u at the support does not exceed $0.4\phi V_c$. The reinforcement ratio ρ, within the effective slab width defined in Section 1913.5.3.2, shall not exceed $0.375\,\rho_b$. No adjustments to γ_f shall be permitted for prestressed slab systems.

1913.5.3.4 Concentration of reinforcement over the column by closer spacing or additional reinforcement shall be used to resist moment on the effective slab width defined in Section 1913.5.3.2.

1913.5.4 Design for transfer of load from slab to supporting columns or walls through shear and torsion shall be in accordance with Sections 1911.0 through 1911.12.

1913.6 Direct Design Method.

1913.6.1 Limitations. Design of slab systems within the following limitations by the Direct Design Method shall be permitted:

1913.6.1.1 There shall be a minimum of three continuous spans in each direction.

1913.6.1.2 Panels shall be rectangular, with a ratio of longer to shorter span center-to-center supports within a panel not greater than 2.

1913.6.1.3 Successive span lengths center-to-center supports in each direction shall not differ by more than one third the longer span.

1913.6.1.4 Offset of columns by a maximum of 10 percent of the span (in direction of offset) from either axis between center lines of successive columns shall be permitted.

1913.6.1.5 All loads shall be due to gravity only and uniformly distributed over an entire panel. Live load shall not exceed two times dead load.

1913.6.1.6 For a panel with beams between supports on all sides, the relative stiffness of beams in two perpendicular directions

$$\frac{\alpha_1 l_2^{\,2}}{\alpha_2 l_1^{\,2}} \qquad (13\text{-}2)$$

shall not be less than 0.2 or greater than 5.0.

1913.6.1.7 Moment redistribution as permitted by Section 1908.4 shall not be applied for slab systems designed by the direct design method. See Section 1913.6.7.

1913.6.1.8 Variations from the limitations of Section 1913.6.1 shall be permitted if demonstrated by analysis that requirements of Section 1913.5.1 are satisfied.

1913.6.2 Total factored static moment for a span.

1913.6.2.1 Total factored static moment for a span shall be determined in a strip bounded laterally by center line of panel on each side of center line of supports.

1913.6.2.2 Absolute sum of positive and average negative factored moments in each direction shall not be less than

$$M_o = \frac{w_u l_2 l_n^{\,2}}{8} \qquad (13\text{-}3)$$

1913.6.2.3 Where the transverse span of panels on either side of the center line of supports varies, l_2 in Formula (13-3) shall be taken as the average of adjacent transverse spans.

1913.6.2.4 When the span adjacent and parallel to an edge is being considered, the distance from edge to panel center line shall be substituted for l_2 in Formula (13-3).

1913.6.2.5 Clear span l_n shall extend from face to face of columns, capitals, brackets or walls. Value of l_n used in Formula (13-3) shall not be less than $0.65l_1$. Circular or regular polygon-shaped supports shall be treated as square supports with the same area.

1913.6.3 Negative and positive factored moments.

1913.6.3.1 Negative factored moments shall be located at face of rectangular supports. Circular or regular polygon-shaped supports shall be treated as square supports with the same area.

1913.6.3.2 In an interior span, total static moment M_o shall be distributed as follows:

Negative factored moment 0.65

Positive factored moment 0.35

1913.6.3.3 In an end span, total factored static moment M_o shall be distributed as follows:

	(1)	(2)	(3)	(4)	(5)
		Slab with Beams between All Supports	Slab without Beams between Interior Supports		
	Exterior Edge Unrestrained		Without Edge Beam	With Edge Beam	Exterior Edge Fully Restrained
Interior negative factored moment	0.75	0.70	0.70	0.70	0.65
Positive factored moment	0.63	0.57	0.52	0.50	0.35
Exterior negative factored moment	0	0.16	0.26	0.30	0.65

1913.6.3.4 Negative moment sections shall be designed to resist the larger of the two interior negative factored moments determined for spans framing into a common support unless an analysis is made to distribute the unbalanced moment in accordance with stiffness of adjoining elements.

1913.6.3.5 Edge beams or edges of slab shall be proportioned to resist in torsion their share of exterior negative factored moments.

1913.6.3.6 The gravity load moment to be transferred between slab and edge column in accordance with Section 1913.5.3.1 shall be $0.3M$.

1913.6.4 Factored moments in column strips.

1913.6.4.1 Column strips shall be proportioned to resist the following percentage of interior negative factored moments:

l_2/l_1	0.5	1.0	2.0
$(\alpha_1 l_2/l_1) = 0$	75	75	75
$(\alpha_1 l_2/l_1) \geq 1.0$	90	75	45

Linear interpolations shall be made between values shown.

1913.6.4.2 Column strips shall be proportioned to resist the following percentage of exterior negative factored moments:

l_2/l_1		0.5	1.0	2.0
$(\alpha_1 l_2/l_1) = 0$	$\beta_t = 0$	100	100	100
	$\beta_t \geq 2.5$	75	75	75
$(\alpha_1 l_2/l_1) \geq 1.0$	$\beta_t = 0$	100	100	100
	$\beta_t \geq 2.5$	90	75	45

Linear interpolations shall be made between values shown.

1913.6.4.3 Where supports consist of columns or walls extending for a distance equal to or greater than three fourths the span length l_2 used to compute M_o, negative moments shall be considered to be uniformly distributed across l_2.

1913.6.4.4 Column strips shall be proportioned to resist the following percentage of positive factored moments:

l_2/l_1	0.5	1.0	2.0
$(\alpha_1 l_2/l_1) = 0$	60	60	60
$(\alpha_1 l_2/l_1) \geq 1.0$	90	75	45

Linear interpolations shall be made between values shown:

1913.6.4.5 For slabs with beams between supports, the slab portion of column strips shall be proportioned to resist that portion of column strip moments not resisted by beams.

1913.6.5 Factored moments in beams.

1913.6.5.1 Beams between supports shall be proportioned to resist 85 percent of column strip moments if $(\alpha_1 l_2/l_1)$ is equal to or greater than 1.0.

1913.6.5.2 For values of $(\alpha_1 l_2/l_1)$ between 1.0 and zero, proportion of column strip moments resisted by beams shall be obtained by linear interpolation between 85 and zero percent.

1913.6.5.3 In addition to moments calculated for uniform loads according to Sections 1913.6.2.2, 1913.6.5.1 and 1913.6.5.2, beams shall be proportioned to resist all moments caused by concentrated or linear loads applied directly to beams, including weight of projecting beam stem above or below the slab.

1913.6.6 Factored moments in middle strips.

1913.6.6.1 That portion of negative and positive factored moments not resisted by column strips shall be proportionately assigned to corresponding half middle strips.

1913.6.6.2 Each middle strip shall be proportioned to resist the sum of the moments assigned to its two half middle strips.

1913.6.6.3 A middle strip adjacent to and parallel with an edge supported by a wall shall be proportioned to resist twice the moment assigned to the half middle strip corresponding to the first row of interior supports.

1913.6.7 Modification of factored moments.
Modification of negative and positive factored moments by 10 percent shall be permitted provided the total static moment for a panel in the direction considered is not less than that required by Formula (13-3).

1913.6.8 Factored shear in slab systems with beams.

1913.6.8.1 Beams with $(\alpha_1 l_2/l_1)$ equal to or greater than 1.0 shall be proportioned to resist shear caused by factored loads on tributary areas bounded by 45-degree lines drawn from the corners of the panels and the center lines of the adjacent panels parallel to the long sides.

1913.6.8.2 In proportioning of beams with $(\alpha_1 l_2/l_1)$ less than 1.0 to resist shear, linear interpolation, assuming beams carry no load at $\alpha_1 = 0$, shall be permitted.

1913.6.8.3 In addition to shears calculated according to this section, beams shall be proportioned to resist shears caused by factored loads applied directly on beams.

1913.6.8.4 Computations of slab shear strength on the assumption that load is distributed to supporting beams in accordance with Section 1913.6.8.1 or 1913.6.8.2 shall be permitted. Resistance to total shear occurring on a panel shall be provided.

1913.6.8.5 Shear strength shall satisfy requirements of Section 1911.

1913.6.9 Factored moments in columns and walls.

1913.6.9.1 Columns and walls built integrally with a slab system shall resist moments caused by factored loads on the slab system.

1913.6.9.2 At an interior support, supporting elements above and below the slab shall resist the moment specified by Formula (13-4) in direct proportion to their stiffnesses unless a general analysis is made.

$$M = 0.07 \left[(w_d + 0.5wl)l_2l_n{}^2 - w'_d l'_2(l'_n)^2 \right] \quad (13\text{-}4)$$

where w'_d, l'_2 and l'_n refer to shorter span.

1913.7 Equivalent Frame Method.

1913.7.1 Design of slab systems by the equivalent frame method shall be based on assumptions given in Sections 1913.7.2 through 1913.7.6, and all sections of slabs and supporting members shall be proportioned for moments and shears thus obtained.

1913.7.1.1 Where metal column capitals are used, it shall be permitted to take account of their contributions to stiffness and resistance to moment and to shear.

1913.7.1.2 Neglecting the change in length of columns and slabs due to direct stress, and deflections due to shear, shall be permitted.

1913.7.2 Equivalent frame.

1913.7.2.1 The structure shall be considered to be made up of equivalent frames on column lines taken longitudinally and transversely through the building.

1913.7.2.2 Each frame shall consist of a row of columns or supports and slab-beam strips, bounded laterally by the center line of panel on each side of the center line of columns or supports.

1913.7.2.3 Columns or supports shall be assumed to be attached to slab-beam strips by torsional members (Section 1913.7.5) transverse to the direction of the span for which moments are being determined and extending to bounding lateral panel center lines on each side of a column.

1913.7.2.4 Frames adjacent and parallel to an edge shall be bounded by that edge and the center line of adjacent panel.

1913.7.2.5 Analysis of each equivalent frame in its entirety shall be permitted. Alternatively for gravity loading, a separate analysis of each floor or roof with far ends of columns considered fixed shall be permitted.

1913.7.2.6 Where slab-beams are analyzed separately, determination of moment at a given support assuming that the slab-beam is fixed at any support two panel distance therefrom shall be permitted, provided the slab continues beyond that point.

1913.7.3 Slab-beams.

1913.7.3.1 Determination of the moment of inertia of slab-beams at any cross section outside of joints or column capitals using the gross area of concrete shall be permitted.

1913.7.3.2 Variation in moment of inertia along axis of slab-beams shall be taken into account.

1913.7.3.3 Moment of inertia of slab-beams from center of column to face of column, bracket or capital shall be assumed equal to the moment of inertia of the slab-beam at face of column, bracket or capital divided by the quantity $(1 - c_2/l_2)^2$ where c_2 and l_2 are

measured transverse to the direction of the span for which moments are being determined.

1913.7.4 Columns.

1913.7.4.1 Determination of the moment of inertia of columns at any cross section outside of joints or column capitals using the gross area of concrete shall be permitted.

1913.7.4.2 Variation in moment of inertia along axis of columns shall be taken into account.

1913.7.4.3 Moment of inertia of columns from top to bottom of the slab-beam at a joint shall be assumed infinite.

1913.7.5 Torsional members.

1913.7.5.1 Torsional members shall be assumed to have a constant cross section throughout their length consisting of the largest of:

1. A portion of slab having a width equal to that of the column, bracket or capital in the direction of the span for which moments are being determined, or

2. For monolithic or fully composite construction, the portion of slab specified in A above plus that part of the transverse beam above and below the slab, and

3. The transverse beam as defined in Section 1913.2.4.

1913.7.5.2 Where beams frame into columns in the direction of the span for which moments are being determined, the torsional stiffness shall be multiplied by the ratio of moment of inertia of slab with such beam to moment of inertia of slab without such beam.

1913.7.6 Arrangement of live load.

1913.7.6.1 When loading pattern is known, the equivalent frame shall be analyzed for that load.

1913.7.6.2 When live load is variable but does not exceed three-quarters of the dead load, or the nature of live load is such that all panels will be loaded simultaneously, it shall be permitted to assume that maximum factored moments occur at all sections with full factored live load on entire slab system.

1913.7.6.3 For loading conditions other than those defined in Section 1913.7.6.2, it shall be permitted to assume that maximum positive factored moment near midspan of a panel occurs with three-quarters of the full factored live load on the panel and on alternate panels; and it shall be permitted to assume that maximum negative factored moment in the slab at a support occurs with three-quarters of the full live load on adjacent panels only.

1913.7.6.4 Factored moments shall not be taken less than those occurring with full factored live load on all panels.

1913.7.7 Factored moments.

1913.7.7.1 At interior supports, critical section for negative factored moment (in both column and middle strips) shall be taken at face of rectilinear supports, but not greater than $0.175l_1$ from center of a column.

1913.7.7.2 At exterior supports provided with brackets or capitals, critical section for negative factored moment in the span perpendicular to an edge shall be taken at a distance from face of supporting element not greater than one half the projection of bracket or capital beyond face of supporting element.

1913.7.7.3 Circular or regular polygon-shaped supports shall be treated as square supports with the same area for location of critical section for negative design moment.

1913.7.7.4 When slab systems within limitations of Section 1913.6.1 are analyzed by the Equivalent Frame Method, it shall be permitted to reduce the resulting computed moments in such proportion that the absolute sum of the positive and average negative moments used in the design need not exceed the value obtained from Formula (13-3).

1913.7.7.5 Distribution of moments at critical sections across the slab-beam strip of each frame to column strips, beams and middle strips as provided in Sections 1913.6.4, 1913.6.5 and 1913.6.6 shall be permitted if the requirement of Section 1913.6.1.6 is satisfied.

SECTION 1914 — WALLS

1914.0 Notations.

A_g = gross area of section, square inches (mm^2).

f'_c = specified compressive strength of concrete, pounds per square inch (MPa).

h = overall thickness of member, inches (mm).

k = effective length factor.

l_c = vertical distance between supports, inches (mm).

M_{cr} = *cracking moment* $5\sqrt{f'_c}Ig/y_t$ *(For* **SI:** *$0.42\sqrt{f'_c}Ig/y_t$) for regular concrete.*

M_n = *nominal moment strength at section, inch-pound (N·m).*

M_u = *factored moment at section, inch-pound (N·m). See Section 1914.8.3.*

P_{nw} = nominal axial load strength of wall designed by Section 1914.4.

P_u = *factored axial load at midheight of wall, including tributary wall weight.*

ρ = *ratio of nonprestressed tension reinforcement.*

ρ_b = *reinforcement ratio producing balanced strain conditions. See Formula (8-1).*

ϕ = strength-reduction factor. See Section 1909.3.

1914.1 Scope.

1914.1.1 Provisions of Section 1914 shall apply for design of walls subjected to axial load, with or without flexure.

1914.1.2 Cantilever retaining walls are designed according to flexural design provisions of Section 1910 with minimum horizontal reinforcement according to Section 1914.3.3.

1914.2 General.

1914.2.1 Walls shall be designed for eccentric loads and any lateral or other loads to which they are subjected.

1914.2.2 Walls subject to axial loads shall be designed in accordance with Sections 1914.2, 1914.3 and either Section 1914.4 or 1914.5.

1914.2.3 Design for shear shall be in accordance with Section 1911.10.

1914.2.4 Unless demonstrated by a detailed analysis, horizontal length of wall to be considered as effective for each concentrated load shall not exceed center-to-center distance between loads, or width of bearing plus four times the wall thickness.

1914.2.5 Compression members built integrally with walls shall conform to Section 1910.8.2.

1914.2.6 Walls shall be anchored to intersecting elements such as floors or roofs or to columns, pilasters, buttresses, and intersecting walls and footings.

1914.2.7 Quantity of reinforcement and limits of thickness required by Sections 1914.3 and 1914.5 shall be permitted to be waived where structural analysis shows adequate strength and stability.

1914.2.8 Transfer of force to footing at base of wall shall be in accordance with Section 1915.8.

1914.3 Minimum Reinforcement.

1914.3.1 Minimum vertical and horizontal reinforcement shall be in accordance with Sections 1914.3.2 and 1914.3.3 unless a greater amount is required for shear by Sections 1911.10.8 and 1911.10.9.

1914.3.2 Minimum ratio of vertical reinforcement area to gross concrete area shall be:

1. 0.0012 for deformed bars not larger than No. 5 with a specified yield strength not less than 60,000 psi (413.7 MPa), or

2. 0.0015 for other deformed bars, or

3. 0.0012 for welded wire fabric (plain or deformed) not larger than W31 or D31.

1914.3.3 Minimum ratio of horizontal reinforcement area to gross concrete area shall be:

1. 0.0020 for deformed bars not larger than No. 5 with a specified yield strength not less than 60,000 psi (413.7 MPa), or

2. 0.0025 for other deformed bars, or

3. 0.0020 for welded wire fabric (plain or deformed) not larger than W31 or D31.

1914.3.4 Walls more than 10 inches (254 mm) thick, except basement walls, shall have reinforcement for each direction placed in two layers parallel with faces of wall in accordance with the following:

1. One layer consisting of not less than one half and not more than two thirds of total reinforcement required for each direction shall be placed not less than 2 inches (51 mm) or more than one third the thickness of wall from exterior surface.

2. The other layer, consisting of the balance of required reinforcement in that direction, shall be placed not less than $3/4$ inch (19 mm) or more than one third the thickness of wall from interior surface.

1914.3.5 Vertical and horizontal reinforcement shall not be spaced farther apart than three times the wall thickness, nor 18 inches (457 mm). *Unless otherwise required by the engineer, the upper- and lowermost horizontal reinforcement shall be placed within one half of the specified spacing at the top and bottom of the wall.*

1914.3.6 Vertical reinforcement need not be enclosed by lateral ties if vertical reinforcement area is not greater than 0.01 times gross concrete area, or where vertical reinforcement is not required as compression reinforcement.

1914.3.7 In addition to the minimum reinforcement required by Section 1914.3.1, not less than two No. 5 bars shall be provided around all window and door openings. Such bars shall be extended to develop the bar beyond the corners of the openings but not less than 24 inches (610 mm).

1914.3.8 The minimum requirements for horizontal and vertical steel of Sections 1914.3.2 and 1914.3.3 may be interchanged for precast panels which are not restrained along vertical edges to inhibit temperature expansion or contraction.

1914.4 Walls Designed as Compression Members. Except as provided in Section 1914.5, walls subject to axial load or com-

bined flexure and axial load shall be designed as compression members in accordance with provisions of Sections 1910.2, 1910.3, 1910.10, 1910.11, 1910.12, 1910.13, 1910.14, 1910.17, 1914.2 and 1914.3.

1914.5 Empirical Design Method.

1914.5.1 Walls of solid rectangular cross section shall be permitted to be designed by the empirical provisions of Section 1914.5 if resultant of all factored loads is located within the middle third of the overall thickness of wall and all limits of Sections 1914.2, 1914.3 and 1914.5 are satisfied.

1914.5.2 Design axial load strength ϕP_{nw} of a wall satisfying limitations of Section 1914.5.1 shall be computed by Formula (14-1) unless designed in accordance with Section 1914.4.

$$\phi P_{nw} = 0.55 \, \phi f'_c A_g \left[1 - \left(\frac{k l_c}{32 h} \right)^2 \right] \quad (14\text{-}1)$$

where $\phi = 0.70$ and effective length factor k shall be:

For walls braced top and bottom against lateral translation and

1. Restrained against rotation at one or both ends (top and/or bottom)   0.8

2. Unrestrained against rotation at both ends   1.0

For walls not braced against lateral translation   2.0

1914.5.3 Minimum thickness of walls designed by empirical design method.

1914.5.3.1 Thickness of bearing walls shall not be less than $^1/_{25}$ the supported height or length, whichever is shorter, or not less than 4 inches (102 mm).

1914.5.3.2 Thickness of exterior basement walls and foundation walls shall not be less than $7^1/_2$ inches (191 mm).

1914.6 Nonbearing Walls.

1914.6.1 Thickness of nonbearing walls shall not be less than 4 inches (102 mm), or not less than $^1/_{30}$ the least distance between members that provide lateral support.

1914.7 Walls as Grade Beams.

1914.7.1 Walls designed as grade beams shall have top and bottom reinforcement as required for moment in accordance with provisions of Sections 1910.2 through 1910.7. Design for shear shall be in accordance with provisions of Section 1911.

1914.7.2 Portions of grade beam walls exposed above grade shall also meet requirements of Section 1914.3.

1914.8 Alternate Design Slender Walls.

1914.8.1 When flexural tension controls design of walls, the requirements of Section 1910.10 may be satisfied by complying with the limitations and procedures set forth in this section.

1914.8.2 The following limitations apply when this section is employed.

1. Vertical service load stress at the location of maximum moment does not exceed 0.04 f'_c.

2. The reinforcement ratio ρ does not exceed 0.6 ρ_b.

3. Sufficient reinforcement is provided so that the nominal moment capacity times the ϕ factor is greater than M_{cr}.

4. Distribution of concentrated load does not exceed the width of bearing plus a width increasing at a slope of 2 vertical to 1 horizontal down to the design flexural section.

1914.8.3 The required factored moment, M_u at the midheight cross section for combined axial and lateral factored loads, including the P Δ moments, shall be as set forth in Formula (14-2).

$$M_u \leq \phi M_n \quad (14\text{-}2)$$

Unless a more comprehensive analysis is used, the P Δ moment shall be calculated using the maximum potential deflection, Δ_n, as defined in Section 1914.8.4.

1914.8.4 The midheight deflection Δ_s, under service lateral and vertical loads (without load factors), shall be limited by the relation

$$\Delta_s = \frac{l_c}{150} \quad (14\text{-}3)$$

Unless a more comprehensive analysis is used, the midheight deflection shall be computed with the following formulas:

$$\Delta_s = \Delta_{cr} + \left(\frac{M_s - M_{cr}}{M_n - M_{cr}} \right)(\Delta_n - \Delta_{cr}); \text{ for } M_s > M_{cr}$$
$$(14\text{-}4)$$

$$\Delta_s = \frac{5 M_s l_c^{\,2}}{48 E_c I_g}; \text{ for } M_s < M_{cr} \quad (14\text{-}5)$$

WHERE:

$$A_{se} = \frac{P_u + A_s f_y}{f_y}$$

$$I_{cr} = n A_{se}(d - c)^2 + \frac{bc^3}{3}$$

M_s = *the maximum moment in the wall resulting from the application of the unfactored load combinations.*

$$\Delta_{cr} = \frac{5 M_{cr} l_c^{\,2}}{48 E_c I_g}$$

$$\Delta_n = \frac{5 M_n l_c^{\,2}}{48 E_c I_{cr}}$$

SECTION 1915 — FOOTINGS

1915.0 Notations.

A_g = gross area of section, square inches (mm²).

d_p = diameter of pile at footing base.

β = ratio of long side to short side of footing.

1915.1 Scope.

1915.1.1 Provisions of this section shall apply for design of isolated footings and, where applicable, to combined footings and mats.

1915.1.2 Additional requirements for design of combined footings and mats are given in Section 1915.10.

1915.2 Loads and Reactions.

1915.2.1 Footings shall be proportioned to resist the factored loads and induced reactions, in accordance with the appropriate design requirements of this code and as provided in this section.

1915.2.2 Base area of footing or number and arrangement of piles shall be determined from the external forces and moments (transmitted by footing to soil or piles) and permissible soil pressure or permissible pile capacity selected through principles of soil mechanics. *External forces and moments are those resulting from unfactored loads (D, L, W and E) specified in Chapter 16.*

1915.2.3 For footings on piles, computations for moments and shears may be based on the assumption that the reaction from any pile is concentrated at pile center.

1915.2.4 External forces and moments applied to footings shall be transferred to supporting soil without exceeding permissible soil pressures.

1915.3 Footings Supporting Circular or Regular Polygon-shaped Columns or Pedestals. For location of critical sections for moment, shear and development of reinforcement in footings, it shall be permitted to treat circular or regular polygon-shaped concrete columns or pedestals as square members with the same area.

1915.4 Moment in Footings.

1915.4.1 External moment on any section of a footing shall be determined by passing a vertical plane through the footing and computing the moment of the forces acting over entire area of footing on one side of that vertical plane.

1915.4.2 Maximum factored moment for an isolated footing shall be computed as prescribed in Section 1915.4.1 at critical sections located as follows:

1. At face of column, pedestal or wall, for footings supporting a concrete column, pedestal or wall.

2. Halfway between middle and edge of wall, for footings supporting a masonry wall.

3. Halfway between face of column and edge of steel base, for footings supporting a column with steel base plates.

1915.4.3 In one-way footings, and two-way square footings, reinforcement shall be distributed uniformly across entire width of footing.

1915.4.4 In two-way rectangular footings, reinforcement shall be distributed as follows:

1915.4.4.1 Reinforcement in long direction shall be distributed uniformly across entire width of footing.

1915.4.4.2 For reinforcement in short direction, a portion of the total reinforcement given by Formula (15-1) shall be distributed uniformly over a band width (centered on center line of column or pedestal) equal to the length of short side of footing. Remainder of reinforcement required in short direction shall be distributed uniformly outside center band width of footing.

$$\frac{\text{Reinforcement in band width}}{\text{Total reinforcement in short direction}} = \frac{2}{(\beta + 1)} \quad (15\text{-}1)$$

1915.5 Shear in Footings.

1915.5.1 Shear strength in footings shall be in accordance with Section 1911.12.

1915.5.2 Location of critical section for shear in accordance with Section 1911 shall be measured from face of column, pedestal or wall, for footings supporting a column, pedestal or wall. For footings supporting a column or pedestal with steel base plates, the critical section shall be measured from location defined in Section 1915.4.2, Item 3.

1915.5.3 Computation of shear on any section through a footing supported on piles shall be in accordance with the following:

1915.5.3.1 Entire reaction from any pile whose center is located $d_p/2$ or more outside the section shall be considered as producing shear on that section.

1915.5.3.2 Reaction from any pile whose center is located $d_p/2$ or more inside the section shall be considered as producing no shear in that section.

1915.5.3.3 For intermediate positions of pile center, the portion of the pile reaction to be considered as producing shear on the section shall be based on straight-line interpolation between full value at $d_p/2$ outside the section and zero value at $d_p/2$ inside the section.

1915.6 Development of Reinforcement in Footings.

1915.6.1 Development of reinforcement in footings shall be in accordance with Section 1912.

1915.6.2 Calculated tension or compression in reinforcement at each section shall be developed on each side of that section by embedment length, hooks (tension only), mechanical device or combinations thereof.

1915.6.3 Critical sections for development of reinforcement shall be assumed at the same locations as defined in Section 1915.4.2 for maximum factored moment, and at all other vertical planes where changes of section or reinforcement occur. See also Section 1912.10.6.

1915.7 Minimum Footing Depth. Depth of footing above bottom reinforcement shall not be less than 6 inches (152 mm) for footings on soil, or not less than 12 inches (305 mm) for footings on piles.

1915.8 Transfer of Force at Base of Column, Wall or Reinforced Pedestal.

1915.8.1 Forces and moments at base of column, wall, or pedestal shall be transferred to supporting pedestal or footing by bearing on concrete and by reinforcement, dowels and mechanical connectors.

1915.8.1.1 Bearing on concrete at contact surface between supported and supporting member shall not exceed concrete bearing strength for either surface as given by Section 1910.17.

1915.8.1.2 Reinforcement, dowels or mechanical connectors between supported and supporting members shall be adequate to transfer:

1. All compressive force that exceeds concrete bearing strength of either member.

2. Any computed tensile force across interface.

In addition, reinforcement, dowels or mechanical connectors shall satisfy Section 1915.8.2 or 1915.8.3.

1915.8.1.3 If calculated moments are transferred to supporting pedestal or footing, reinforcement, dowels or mechanical connectors shall be adequate to satisfy Section 1912.17.

1915.8.1.4 Lateral forces shall be transferred to supporting pedestal or footing in accordance with shear-friction provisions of Section 1911.7 or by other appropriate means.

1915.8.2 In cast-in-place construction, reinforcement required to satisfy Section 1915.8.1 shall be provided either by extending longitudinal bars into supporting pedestal or footing, or by dowels.

1915.8.2.1 For cast-in-place columns and pedestals, area of reinforcement across interface shall not be less than 0.005 times gross area of supported member.

1915.8.2.2 For cast-in-place walls, area of reinforcement across interface shall not be less than minimum vertical reinforcement given in Section 1914.3.2.

1915.8.2.3 At footings, No. 14 and No. 18 longitudinal bars, in compression only, may be lap spliced with dowels to provide reinforcement required to satisfy Section 1915.8.1. Dowels shall not be larger than No. 11 bar and shall extend into supported member a distance not less than the development length of No. 14 or No. 18 bars or the splice length of the dowels, whichever is greater, and into the footing a distance not less than the development length of the dowels.

1915.8.2.4 If a pinned or rocker connection is provided in cast-in-place construction, connection shall conform to Sections 1915.8.1 and 1915.8.3.

1915.8.3 In precast construction, reinforcement required to satisfy Section 1915.8.1 may be provided by anchor bolts or suitable mechanical connectors.

1915.8.3.1 Connection between precast columns or pedestals and supporting members shall meet the requirements of Section 1916.5.1.3, Item 1.

1915.8.3.2 Connection between precast walls and supporting members shall meet the requirements of Section 1916.5.1.3, Items 2 and 3.

> **EXCEPTION:** *In tilt-up construction, this connection may be to an adjacent floor slab. In no case shall the connection provided be less than that required by Section 1611.*

1915.8.3.3 Anchor bolts and mechanical connectors shall be designed to reach their design strength prior to anchorage failure or failure of surrounding concrete.

1915.9 Sloped or Stepped Footings.

1915.9.1 In sloped or stepped footings, angle of slope or depth and location of steps shall be such that design requirements are satisfied at every section.

1915.9.2 Sloped or stepped footings designed as a unit shall be constructed to assure action as a unit.

1915.10 Combined Footings and Mats.

1915.10.1 Footings supporting more than one column, pedestal or wall (combined footings or mats) shall be proportioned to resist the factored loads and induced reactions in accordance with appropriate design requirements of this code.

1915.10.2 The direct design method of Section 1913 shall not be used for design of combined footings and mats.

1915.10.3 Distribution of soil pressure under combined footings and mats shall be consistent with properties of the soil and the structure and with established principles of soil mechanics.

1915.11 Plain Concrete Pedestals and Footings. See Section 1922.

SECTION 1916 — PRECAST CONCRETE

1916.0 Notations.

A_g = gross area of column, inches squared (mm²).

l = clear span, inches (mm).

1916.1 Scope.

1916.1.1 All provisions of this code, not specifically excluded and not in conflict with the provisions of Section 1916, shall apply to structures incorporating precast concrete structural members.

1916.2 General.

1916.2.1 Design of precast members and connections shall include loading and restraint conditions from initial fabrication to end use in the structure, including form removal, storage, transportation and erection.

1916.2.2 When precast members are incorporated into a structural system, the forces and deformations occurring in and adjacent to connections shall be included in the design.

1916.2.3 Tolerances for both precast members and interfacing members shall be specified. Design of precast members and connections shall include the effects of these tolerances.

1916.2.4 In addition to the requirements for drawings and specifications in Section *106.3.2*, the following shall be included in either the contract documents or shop drawings:

1. Details of reinforcement, inserts and lifting devices required to resist temporary loads from handling, storage, transportation and erection.

2. Required concrete strength at stated ages or stages of construction.

1916.3 Distribution of Forces among Members.

1916.3.1 Distribution of forces that are perpendicular to the plane of members shall be established by analysis or by test.

1916.3.2 Where the system behavior requires in-plane forces to be transferred between the members of a precast floor or wall system, the following shall apply:

1916.3.2.1 In-plane force paths shall be continuous through both connections and members.

1916.3.2.2 Where tension forces occur, a continuous path of steel or steel reinforcement shall be provided.

1916.4 Member Design.

1916.4.1 In one-way precast floor and roof slabs and in one-way precast, prestressed wall panels, all not wider than 12 feet (4 m), and where members are not mechanically connected to cause restraint in the transverse direction, the shrinkage and temperature reinforcement requirements of Section 1907.12 in the direction normal to the flexural reinforcement shall be permitted to be waived. This waiver shall not apply to members which require reinforcement to resist transverse flexural stresses.

1916.4.2 For precast, nonprestressed walls the reinforcement shall be designed in accordance with the provisions of Section 1910 or 1914 except that the area of horizontal and vertical reinforcement shall each be not less than 0.001 times the gross cross-sectional area of the wall panel. Spacing of reinforcement shall not exceed five times the wall thickness or 30 inches (762 mm) for interior walls or 18 inches (457 mm) for exterior walls.

1916.5 Structural Integrity.

1916.5.1 Except where the provisions of Section 1916.5.2 govern, the following minimum provisions for structural integrity shall apply to all precast concrete structures:

1916.5.1.1 Longitudinal and transverse ties required by Section 1907.13.3 shall connect members to a lateral-load-resisting system.

1916.5.1.2 Where precast elements form floor or roof diaphragms, the connections between diaphragm and those members being laterally supported shall have a nominal tensile strength capable of resisting not less than 300 pounds per linear foot (630 N/mm).

1916.5.1.3 Vertical tension tie requirements of Section 1907.13.3 shall apply to all vertical structural members, except cladding, and shall be achieved by providing connections at horizontal joints in accordance with the following:

1. Precast columns shall have a nominal strength in tension not less than $200 A_g$ in pounds (N). For columns with a larger cross section than required by consideration of loading, a reduced effective area A_g, based on cross section required but not less than one-half the total area, shall be permitted.

2. Precast wall panels shall have a minimum of two ties per panel, with a nominal tensile strength not less than 10,000 pounds (44 500 N) per tie.

3. When design forces result in no tension at the base, the ties required by Section 1916.5.1.3, Item 2, shall be permitted to be anchored into an appropriately reinforced concrete floor slab on grade.

1916.5.1.4 Connection details that rely solely on friction caused by gravity loads shall not be used.

1916.5.2 For precast concrete bearing wall structures three or more stories in height, the following minimum provisions shall apply:

1916.5.2.1 Longitudinal and transverse ties shall be provided in floor and roof systems to provide a nominal strength of 1,500 pounds per foot (315 N/mm) of width or length. Ties shall be provided over interior wall supports and between members and exterior walls. Ties shall be positioned in or within 2 feet (610 mm) of the plane of the floor or roof system.

1916.5.2.2 Longitudinal ties parallel to floor or roof slab spans shall be spaced not more than 10 feet (3048 mm) on centers. Provisions shall be made to transfer forces around openings.

1916.5.2.3 Transverse ties perpendicular to floor or roof slab spans shall be spaced not greater than the bearing wall spacing.

1916.5.2.4 Ties around the perimeter of each floor and roof, within 4 feet (1219 mm) of the edge, shall provide a nominal strength in tension not less than 16,000 pounds (71 200 N).

1916.5.2.5 Vertical tension ties shall be provided in all walls and shall be continuous over the height of the building. They shall provide a nominal tensile strength not less than 3,000 pounds per horizontal foot (6300 N/mm) of wall. Not less than two ties shall be provided for each precast panel.

1916.6 Connection and Bearing Design.

1916.6.1 Forces shall be permitted to be transferred between members by grouted joints, shear keys, mechanical connectors, reinforcing steel connections, reinforced topping or a combination of these means.

1916.6.1.1 The adequacy of connections to transfer forces between members shall be determined by analysis or by test. Where shear is the primary imposed loading, it shall be permitted to use the provisions of Section 1911.7 as applicable.

1916.6.1.2 When designing a connection using materials with different structural properties, their relative stiffnesses, strengths and ductilities shall be considered.

1916.6.2 Bearing for precast floor and roof members on simple supports shall satisfy the following:

1916.6.2.1 The allowable bearing stress at the contact surface between supported and supporting members and between any intermediate bearing elements shall not exceed the bearing strength for either surface and the bearing element. Concrete bearing strength shall be as given in Section 1910.17.

1916.6.2.2 Unless shown by test or analysis that performance will not be impaired, the following minimum requirements shall be met:

1. Each member and its supporting system shall have design dimensions selected so that, after consideration of tolerances, the distance from the edge of the support to the end of the precast member in the direction of the span is at least $1/180$ of the clear span, l, but not less than:

For solid or hollow-core slabs . . . . 2 inches (51 mm).

For beams or stemmed members . . . . 3 inches (76 mm).

2. Bearing pads at unarmored edges shall be set back a minimum of $1/2$ inch (12.7 mm) from the face of the support, or at least the chamfer dimension at chamfered edges.

1916.6.2.3 The requirements of Section 1912.11.1 shall not apply to the positive bending moment reinforcement for statically determinate precast members, but at least one-third of such reinforcement shall extend to the center of the bearing length.

1916.7 Items Embedded after Concrete Placement.

1916.7.1 When approved by the engineer, embedded items (such as dowels or inserts) that either protrude from the concrete or remain exposed for inspection shall be permitted to be embedded while the concrete is in a plastic state provided that:

1916.7.1.1 Embedded items are not required to be hooked or tied to reinforcement within the concrete.

1916.7.1.2 Embedded items are maintained in the correct position while the concrete remains plastic.

1916.7.1.3 The concrete is properly consolidated around the embedded item.

1916.8 Marking and Identification.

1916.8.1 Each precast member shall be marked to indicate its location and orientation in the structure and date of manufacture.

1916.8.2 Identification marks shall correspond to placing drawings.

1916.9 Handling.

1916.9.1 Member design shall consider forces and distortions during curing, stripping, storage, transportation and erection so that precast members are not overstressed or otherwise damaged.

1916.9.2 Precast members and structures shall be adequately supported and braced during erection to ensure proper alignment and structural integrity until permanent connections are completed.

1916.10 Strength Evaluation of Precast Construction.

1916.10.1 A precast element to be made composite with cast-in-place concrete shall be permitted to be tested in flexure as a precast element alone in accordance with the following:

1916.10.1.1 Test loads shall be applied only when calculations indicate the isolated precast element will not be critical in compression or buckling.

1916.10.1.2 The test load shall be that load which, when applied to the precast member alone, induces the same total force in the tension reinforcement as would be induced by loading the composite member with the test load required by Section 1920.3.2.

1916.10.2 The provisions of Sections 1920.5 shall be the basis for acceptance or rejection of the precast element.

SECTION 1917 — COMPOSITE CONCRETE FLEXURAL MEMBERS

1917.0 Notations.

A_c = area of contact surface being investigated for horizontal shear, square inches (mm^2).

A_v = area of ties within a distance s, square inches (mm^2).

b_v = width of cross section at contact surface being investigated for horizontal shear.

d = distance from extreme compression fiber to centroid of tension reinforcement for entire composite section, inches (mm).

h = overall thickness of composite members, inches (mm).

s = spacing of ties measured along the longitudinal axis of the member, inches (mm).

V_{nh} = nominal horizontal shear strength.

V_u = factored shear force at section.

λ = correction factor related to unit weight of concrete.

ρ_v = ratio of tie reinforcement area to area of contact surface.

= $A_v/b_v s$

ϕ = strength-reduction factor. See Section 1909.3.

1917.1 Scope.

1917.1.1 Provisions of this section shall apply for design of composite concrete flexural members defined as precast or cast-in-place concrete elements or both constructed in separate placements but so interconnected that all elements respond to loads as a unit.

1917.1.2 All provisions of this code shall apply to composite concrete flexural members, except as specifically modified in this section.

1917.2 General.

1917.2.1 The use of an entire composite member or portions thereof for resisting shear and moment shall be permitted.

1917.2.2 Individual elements shall be investigated for all critical stages of loading.

1917.2.3 If the specified strength, unit weight or other properties of the various elements are different, properties of the individual elements or the most critical values shall be used in design.

1917.2.4 In strength computations of composite members, no distinction shall be made between shored and unshored members.

1917.2.5 All elements shall be designed to support all loads introduced prior to full development of design strength of composite members.

1917.2.6 Reinforcement shall be provided as required to control cracking and to prevent separation of individual elements of composite members.

1917.2.7 Composite members shall meet requirements for control of deflections in accordance with Section 1909.5.5.

1917.3 Shoring. When used, shoring shall not be removed until supported elements have developed design properties required to support all loads and limit deflections and cracking at time of shoring removal.

1917.4 Vertical Shear Strength.

1917.4.1 When an entire composite member is assumed to resist vertical shear, design shall be in accordance with requirements of Section 1911 as for a monolithically cast member of the same cross-sectional shape.

1917.4.2 Shear reinforcement shall be fully anchored into interconnected elements in accordance with Section 1912.13.

1917.4.3 Extended and anchored shear reinforcement shall be permitted to be included as ties for horizontal shear.

1917.5 Horizontal Shear Strength.

1917.5.1 In a composite member, full transfer of horizontal shear forces shall be assured at contact surfaces of interconnected elements.

1917.5.1.1 Full transfer of horizontal shear forces may be assumed when all of the following are satisfied:

1. *Contact surfaces are clean, free of laitance, and intentionally roughened to a full amplitude of approximately $^1/_4$ inch (6.4 mm),*

2. *Minimum ties are provided in accordance with Section 1917.6,*

3. *Web members are designed to resist total vertical shear, and*

4. *All shear reinforcement is fully anchored into all interconnected elements.*

1917.5.1.2 If all requirements of Section 1917.5.1.1 are not satisfied, horizontal shear shall be investigated in accordance with Section 1917.5.2 or 1917.5.3.

1917.5.2 Unless calculated in accordance with Section 1917.5.3, design of cross sections subject to horizontal shear shall be based on

$$V_u \leq \phi V_{nh} \tag{17-1}$$

where V_u is factored shear force at section considered and V_{nh} is nominal horizontal shear strength in accordance with the following:

1917.5.2.1 When contact surfaces are clear, free of laitance and intentionally roughened, shear strength V_{nh} shall not be taken greater than $80b_v d$, in pounds ($0.55\,b_v d$, in newtons).

1917.5.2.2 When minimum ties are provided in accordance with Section 1917.6 and contact surfaces are clean and free of laitance, but not intentionally roughened, shear strength V_{nh} shall not be taken greater than $80b_v d$, in pounds ($0.55\,b_v d$, in newtons).

1917.5.2.3 When minimum ties are provided in accordance with Section 1917.6 and contact surfaces are clean, free of laitance, and intentionally roughened to a full amplitude of approximately $^1/_4$ inch (6.4 mm), shear strength V_{nh} shall be taken equal to $(260 + 0.6\rho_v f_y)b_v d$ in pounds $[(1.79 + 0.6\rho_v f_y)b_v d$, in newtons], but not greater than $500\,b_v d$ in pounds ($3.5\,b_v d$, in newtons). Values for λ in Section 1911.7.4.3 shall apply.

1917.5.2.4 When factored shear force V_u at section considered exceeds $\phi(500b_v d)$ [For **SI:** $\phi(3.5b_v d)$], design for horizontal shear shall be in accordance with Section 1911.7.4.

1917.5.2.5 When determining nominal horizontal shear strength over prestressed concrete elements, d shall be as defined or $0.8h$, whichever is greater.

1917.5.3 As an alternative to Section 1917.5.2, horizontal shear shall be determined by computing the actual compressive or tensile force in any segment, and provisions shall be made to transfer that force as horizontal shear to the supporting element. The factored horizontal shear force shall not exceed horizontal shear strength ϕV_{nh} as given in Sections 1917.5.2.1 through 1917.5.2.4 where area of contact surface A_c shall be substituted for $b_v d$.

1917.5.3.1 When ties provided to resist horizontal shear are designed to satisfy Section 1917.5.3, the tie-area-to-tie-spacing ratio along the member shall approximately reflect the distribution of shear forces in the member.

1917.5.4 When tension exists across any contact surface between interconnected elements, shear transfer by contact may be assumed only when minimum ties are provided in accordance with Section 1917.6.

1917.6 Ties for Horizontal Shear.

1917.6.1 When ties are provided to transfer horizontal shear, tie area shall not be less than that required by Section 1911.5.5.3 and tie spacing shall not exceed four times the least dimension of supported element, or 24 inches (610 mm).

1917.6.2 Ties for horizontal shear may consist of single bars or wire, multiple leg stirrups or vertical legs of welded wire fabric (plain or deformed).

1917.6.3 All ties shall be fully anchored into interconnected elements in accordance with Section 1912.13.

SECTION 1918 — PRESTRESSED CONCRETE

1918.0 Notations.

A = area of that part of cross section between flexural tension face and center of gravity of gross section, square inches (mm^2).

A_{ps} = area of prestressed reinforcement in tension zone, square inches (mm^2).

A_s = area of nonprestressed tension reinforcement, square inches (mm^2).

A'_s = area of compression reinforcement, square inches (mm^2).

b = width of compression face of member, inches (mm).

D = dead loads or related internal moments and forces.

d = distance from extreme compression fiber to centroid of nonprestressed tension reinforcement, inches (mm).

d' = distance from extreme compression fiber to centroid of compression reinforcement, inches (mm).

d_p = distance from extreme compression fiber to centroid of prestressed reinforcement.

e = base of Napierian logarithms.

f'_c = specified compressive strength of concrete, pounds per square inch (MPa).

$\sqrt{f'_c}$ = square root of specified compressive strength of concrete, pounds per square inch; or *square root of compressive strength of concrete at time of initial prestress, pounds per square inch (MPa)*.

f'_{ci} = compressive strength of concrete at time of initial prestress, pounds per square inch (MPa).

f_{pc} = average compressive stress in concrete due to effective prestress force only (after allowance for all prestress losses), pounds per square inch (MPa).

f_{ps} = stress in prestressed reinforcement at nominal strength, pounds per square inch (MPa).

f_{pu} = specified tensile strength of prestressing tendons, pounds per square inch (MPa).

f_{py} = specified yield strength of prestressing tendons, pounds per square inch (MPa).

f_r = modulus of rupture of concrete, pounds per square inch (MPa).

f_{se} = effective stress in prestressed reinforcement (after allowance for all prestress losses), pounds per square inch (MPa).

f_y = specified yield strength of nonprestressed reinforcement, pounds per square inch (MPa).

h = overall dimension of member in direction of action considered, inches (mm).

K = wobble friction coefficient per foot (per mm) of prestressing tendon.

L = live loads or related internal moments and forces.

l = length of span of two-way flat plates in direction parallel to that of the reinforcement being determined, inches (mm). See Formula (18-8).

l_x = length of prestressing tendon element from jacking end to any point x, feet (mm). See Formulas (18-1) and (18-2).

N_c = tensile force in concrete due to unfactored dead load plus live load ($D + L$).

P_s = prestressing tendon force at jacking end.

P_x = prestressing tendon force at any point x.

α = total angular change of prestressing tendon profile in radians from tendon jacking end to any point x.

β_1 = factor defined in Section 1910.2.7.1.

γ_p = factor for type of prestressing tendon.

= 0.55 for f_{py}/f_{pu} not less than 0.80.

= 0.40 for f_{py}/f_{pu} not less than 0.85.

= 0.28 for f_{py}/f_{pu} not less than 0.90.

μ = curvature friction coefficient.

ρ = ratio of nonprestressed tension reinforcement.

= A_s/bd.

ρ' = ratio of compression reinforcement.

= A'_s/bd.

ρ_p = ratio of prestressed reinforcement.

= A_{ps}/bd_p.

ϕ = strength-reduction factor. See Section 1909.3.

ω = $\rho f_y/f'_c$.

ω' = $\rho' f_y/f'_c$.

ω_p = $\rho_p f_{ps}/f'_c$.

$\omega_w, \omega_{pw}, \omega'_w$

= reinforcement indices for flanged sections computed as for ω, ω_p, and ω' except that b shall be the web width, and reinforcement area shall be that required to develop compressive strength of web only.

1918.1 Scope.

1918.1.1 Provisions of this section shall apply to members prestressed with wire, strands or bars conforming to provisions for prestressing tendons.

1918.1.2 All provisions of this code not specifically excluded, and not in conflict with provisions of this section, shall apply to prestressed concrete.

1918.1.3 The following provisions of this code shall not apply to prestressed concrete, except as specifically noted: Sections

1907.6.5, 1908.4, 1908.10.2 through 1908.10.4, 1908.11, 1910.3.2 and 1910.3.3, 1910.5, 1910.6, 1910.9.1, 1910.9.2, 1913, 1914.3, 1914.5 and 1914.6.

1918.2 General.

1918.2.1 Prestressed members shall meet the strength requirements specified in this code.

1918.2.2 Design of prestressed members shall be based on strength and on behavior at service conditions at all load stages that may be critical during the life of the structure from the time prestress is first applied.

1918.2.3 Stress concentrations due to prestressing shall be considered in design.

1918.2.4 Provisions shall be made for effects on adjoining construction of elastic and plastic deformations, deflections, changes in length and rotations due to prestressing. Effects of temperature and shrinkage shall also be included.

1918.2.5 Possibility of buckling in a member between points where concrete and prestressing tendons are in contact and of buckling in thin webs and flanges shall be considered.

1918.2.6 In computing section properties prior to bonding of prestressing tendons, effect of loss of area due to open ducts shall be considered.

1918.3 Design Assumptions.

1918.3.1 Strength design of prestressed members for flexure and axial loads shall be based on assumptions given in Section 1910.2, except Section 1910.2.4, shall apply only to reinforcement conforming to Section 1903.5.3.

1918.3.2 For investigation of stresses at transfer of prestress, at service loads and at cracking loads, straight-line theory may be used with the following assumptions:

1918.3.2.1 Strains vary linearly with depth through entire load range.

1918.3.2.2 At cracked sections, concrete resists no tension.

1918.4 Permissible Stresses in Concrete—Flexural Members.

1918.4.1 Stresses in concrete immediately after prestress transfer (before time-dependent prestress losses) shall not exceed the following:

1. Extreme fiber stress in compression . . . . $0.60f'_c$

2. Extreme fiber stress in tension except as permitted in . . . . $3\sqrt{f'_{ci}}$

(For **SI**: $0.25\sqrt{f'_{ci}}$)

3. Extreme fiber stress in tension at ends of simply supported members . . . . $6\sqrt{f'_{ci}}$

(For **SI**: $0.50\sqrt{f'_{ci}}$)

Where computed tensile stresses exceed these values, bonded auxiliary reinforcement (nonprestressed or prestressed) shall be provided in the tensile zone to resist the total tensile force in concrete computed with the assumption of an uncracked section.

1918.4.2 Stresses in concrete at service loads (after allowance for all prestress losses) shall not exceed the following:

1. Extreme fiber stress in compression due to prestress plus sustained loads . . . . $0.45f'_c$

2. Extreme fiber stress in compression due to prestress plus total load . . . . $0.60f'_c$

3. Extreme fiber stress in tension in precompressed tensile zone . . . . $6\sqrt{f'_c}$

(For **SI**: $0.50\sqrt{f'_c}$)

4. Extreme fiber stress in tension in precompressed tensile zone of members (except two-way slab systems) where analysis based on transformed cracked sections and on bilinear moment-deflection relationships show that immediate and long-time deflections comply with requirements of Section 1909.5.4, and where cover requirements comply with Section 1907.7.3.2 . . . . $12\sqrt{f'_c}$

(For **SI**: $1.0\sqrt{f'_c}$)

1918.4.3 Permissible stresses in concrete of Sections 1918.4.1 and 1918.4.2 may be exceeded if shown by test or analysis that performance will not be impaired.

1918.5 Permissible Stress in Prestressing Tendons.

1918.5.1 Tensile stress in prestressing tendons shall not exceed the following:

1. Due to tendon jacking force . . . . $0.94f_{py}$
but not greater than the lesser of $0.80 f_{pu}$ and the maximum value recommended by manufacturer of prestressing tendons or anchorages.

2. Immediately after prestress transfer . . . . $0.82f_{py}$
but not greater than $0.74f_{pu}$.

3. Posttensioning tendons, at anchorages and couplers, immediately after tendon anchorage . . . . $0.70f_{pu}$

1918.6 Loss of Prestress.

1918.6.1 To determine effective prestress f_{se}, allowance for the following sources of loss of prestress shall be considered:

1. Anchorage seating loss.

2. Elastic shortening of concrete.

3. Creep of concrete.

4. Shrinkage of concrete.

5. Relaxation of tendon stress.

6. Friction loss due to intended or unintended curvature in posttensioning tendons.

1918.6.2 Friction loss in posttensioning tendons.

1918.6.2.1 Effect of friction loss in posttensioning tendons shall be computed by

$$P_s = P_x e^{(Kl_x + \mu\alpha)} \qquad (18\text{-}1)$$

When $(Kl \pm \mu\alpha)$ is not greater than 0.3, effect of friction loss shall be permitted to be computed by

$$P_s = P_x(1 + Kl_x + \mu\alpha) \qquad (18\text{-}2)$$

1918.6.2.2 Friction loss shall be based on experimentally determined wobble K and curvature μ friction coefficients and shall be verified during tendon stressing operations.

1918.6.2.3 Values of wobble and curvature coefficients used in design shall be shown on design drawings.

1918.6.3 Where loss of prestress in member may occur due to connection of member to adjoining construction, such loss of prestress shall be allowed for in design.

1918.7 Flexural Strength.

1918.7.1 Design moment strength of flexural members shall be computed by the strength design methods of this chapter. For prestressing tendons, f_{ps} shall be substituted for f_y in strength computations.

1918.7.2 As an alternative to a more accurate determination of f_{ps} based on strain compatibility, the following approximate values of f_{ps} shall be used if f_{se} is not less than $0.5 f_{pu}$.

1. For members with bonded prestressing tendons:

$$f_{ps} = f_{pu}\left(1 - \frac{\gamma_p}{\beta_1}\left[\rho_p\frac{f_{pu}}{f'_c} + \frac{d}{d_p}(\omega - \omega')\right]\right) \quad (18\text{-}3)$$

If any compression reinforcement is taken into account when calculating f_{ps} by Formula (18-3), the term

$$\left[\rho_p\frac{f_{pu}}{f'_c} + \frac{d}{d_p}(\omega - \omega')\right]$$

shall be taken not less than 0.17 and d' shall be no greater than $0.15 d_p$.

2. For members with unbonded prestressing tendons and with a span-to-depth ratio of 35 or less:

$$f_{ps} = f_{se} + 10,000 + \frac{f'_c}{100\rho_p} \quad (18\text{-}4)$$

For **SI:** $\qquad f_{ps} = f_{se} + 69 + \dfrac{f'_c}{100\rho_p}$

but f_{ps} in Formula (18-4) shall not be taken greater than f_{py}, or (f_{se} + 60,000) (For **SI:** f_{se} + 413.7).

3. For members with unbonded prestressing tendons and with a span-to-depth ratio greater than 35:

$$f_{ps} = f_{se} + 10,000 + \frac{f'_c}{300\rho_p} \quad (18\text{-}5)$$

For **SI:** $\qquad f_{ps} = f_{se} + 69 + \dfrac{f'_c}{300\rho_p}$

but f_{ps} in Formula (18-5) shall not be taken greater than f_{py}, or (f_{se} + 30,000) (For **SI:** f_{se} + 206.9).

1918.7.3 Nonprestressed reinforcement conforming to Section 1903.5.3, if used with prestressing tendons, shall be permitted to be considered to contribute to the tensile force and to be included in moment strength computations at a stress equal to the specified yield strength f_y. Other nonprestressed reinforcement shall be permitted to be included in strength computations only if a strain compatibility analysis is made to determine stresses in such reinforcement.

1918.8 Limits for Reinforcement of Flexural Members.

1918.8.1 Ratio of prestressed and nonprestressed reinforcement used for computation of moment strength of a member, except as provided in this section shall be such that ω_p, $[\omega_p + (d/d_p)$ $(\omega - \omega')]$, or $[\omega_{pw} + (d/d_p)(\omega_w - \omega'_w)]$ is not greater than $0.36 \beta_1$.

1918.8.2 When a reinforcement ratio in excess of that specified in this section is provided, design moment strength shall not exceed the moment strength based on the compression portion of the moment couple.

1918.8.3 Total amount of prestressed and nonprestressed reinforcement shall be adequate to develop a factored load at least 1.2 times the cracking load computed on the basis of the modulus of rupture f_r specified in Section 1909.5.2.3, except for flexural

members with shear and flexural strength at least twice that required by Section 1909.2.

1918.9 Minimum Bonded Reinforcement.

1918.9.1 A minimum area of bonded reinforcement shall be provided in all flexural members with unbonded prestressing tendons as required by Sections 1918.9.2 and 1918.9.3.

1918.9.2 Except as provided in Section 1918.9.3, minimum area of bonded reinforcement shall be computed by

$$A_s = 0.004A \quad (18\text{-}6)$$

1918.9.2.1 Bonded reinforcement required by Formula (18-6) shall be uniformly distributed over precompressed tensile zone as close as practicable to extreme tension fiber.

1918.9.2.2 Bonded reinforcement shall be required regardless of service load stress conditions.

One-way, unbonded, posttensioned slabs and beams shall be designed to carry the dead load of the slab or beam plus 25 percent of the unreduced superimposed live load by some method other than the primary unbonded posttensioned reinforcement. Design shall be based on the strength method of design with a load factor and capacity reduction factor of one. All reinforcement other than the primary unbonded reinforcement provided to meet other requirements of this section may be used in the design.

1918.9.2.3 *Maximum spacing limitations of Sections 1907.6.1 and 1908.10.5.2, for bonded reinforcement in slabs are not applicable to spacing of bonded reinforcement in members with unbonded tendons.*

1918.9.3 For two-way flat plates, defined as solid slabs of uniform thickness, minimum area and distribution of bonded reinforcement shall be as follows:

1918.9.3.1 Bonded reinforcement shall not be required in positive moment areas where computed tensile stress in concrete at service load (after allowance for prestress losses) does not exceed $2\sqrt{f'_c}$ (For **SI:** $0.166\sqrt{f'_c}$).

1918.9.3.2 In positive moment areas where computed tensile stress in concrete at service load exceeds $2\sqrt{f'_c}$ (For **SI:** $0.166\sqrt{f'_c}$) minimum area of bonded reinforcement shall be computed by

$$A_s = \frac{N_c}{0.5f_y} \quad (18\text{-}7)$$

where design yield strength f_y shall not exceed 60,000 pounds per square inch (413.7 MPa). Bonded reinforcement shall be uniformly distributed over precompressed tensile zone as close as practicable to extreme tension fiber.

1918.9.3.3 In negative moment areas at column supports, minimum area of bonded reinforcement in each direction shall be computed by

$$A_s = 0.00075hl \quad (18\text{-}8)$$

where l is length of span in direction parallel to that of the reinforcement being determined. Bonded reinforcement required by Formula (18-8) shall be distributed within a slab width between lines that are $1.5h$ outside opposite faces of the column support. At least four bars or wires shall be provided in each direction. Spacing of bonded reinforcement shall not exceed 12 inches (305 mm).

1918.9.4 Minimum length of bonded reinforcement required by Sections 1918.9.2 and 1918.9.3 shall be as follows:

1918.9.4.1 In positive moment areas, minimum length of bonded reinforcement shall be one third the clear span length and centered in positive moment area.

1918.9.4.2 In negative moment areas, bonded reinforcement shall extend one sixth the clear span on each side of support.

1918.9.4.3 Where bonded reinforcement is provided for design moment strength in accordance with Section 1918.7.3, or for tensile stress conditions in accordance with Section 1918.9.3.2, minimum length also shall conform to provisions of Section 1912.

1918.10 Statically Indeterminate Structures.

1918.10.1 Frames and continuous construction of prestressed concrete shall be designed for satisfactory performance at service load conditions and for adequate strength.

1918.10.2 Performance at service load conditions shall be determined by elastic analysis, considering reactions, moments, shears, and axial forces produced by prestressing, creep, shrinkage, temperature change, axial deformation, restraint of attached structural elements and foundation settlement.

1918.10.3 Moments to be used to compute required strength shall be the sum of the moments due to reactions induced by prestressing (with a load factor of 1.0) and the moments due to factored loads. Adjustment of the sum of these moments shall be permitted as allowed in Section 1918.10.4.

1918.10.4 Redistribution of negative moments in continuous prestressed flexural members.

1918.10.4.1 Where bonded reinforcement is provided at supports in accordance with Section 1918.9.2, negative moments calculated by elastic theory for any assumed loading arrangement shall be permitted to be increased or decreased by not more than

$$20\left[1 - \frac{\omega_p + (d/d_p)(\omega - \omega')}{0.36\beta_1}\right] \text{ percent}$$

1918.10.4.2 The modified negative moments shall be used for calculating moments at sections within spans for the same loading arrangement.

1918.10.4.3 Redistribution of negative moments shall be made only when the section at which moment is reduced is so designed that ω_p, $[\omega_p + (d/d_p)(\omega - \omega')]$, or $[\omega_{pw} + (d/d_p)(\omega_w - \omega'_w)]$, whichever is applicable, is not greater than 0.24 β_1.

1918.11 Compression Members—Combined Flexure and Axial Loads.

1918.11.1 Prestressed concrete members subject to combined flexure and axial load, with or without nonprestressed reinforcement, shall be proportioned by the strength design methods of this chapter for members without prestressing. Effects of prestress, creep, shrinkage and temperature change shall be included.

1918.11.2 Limits for reinforcement of prestressed compression members.

1918.11.2.1 Members with average prestress f_{pc} less than 225 psi (1.55 MPa) shall have minimum reinforcement in accordance with Sections 1907.10, 1910.9.1 and 1910.9.2 for columns, or Section 1914.3 for walls.

1918.11.2.2 Except for walls, members with average prestress f_{pc} equal to or greater than 225 psi (1.55 MPa) shall have all prestressing tendons enclosed by spirals or lateral ties in accordance with the following:

1. Spirals shall conform to Section 1907.10.4.

2. Lateral ties shall be at least No. 3 in size or welded wire fabric of equivalent area, and spaced vertically not to exceed 48 tie bar or wire diameters or least dimension of compression member.

3. Ties shall be located vertically not more than half a tie spacing above top of footing or slab in any story, and shall be spaced as provided herein to not more than half a tie spacing below lowest horizontal reinforcement in members supported above.

4. Where beams or brackets frame into all sides of a column, it shall be permitted to terminate ties not more than 3 inches (76 mm) below lowest reinforcement in such beams or brackets.

1918.11.2.3 For walls with average prestress f_{pc} equal to or greater than 225 pounds per square inch (1.55 MPa), minimum reinforcement required by Section 1914.3 may be waived where structural analysis shows adequate strength and stability.

1918.12 Slab Systems.

1918.12.1 Factored moments and shears in prestressed slab systems reinforced for flexure in more than one direction shall be determined in accordance with provisions of Section 1913.7, excluding Sections 1913.7.7.4 and 1913.7.7.5, or by more detailed design procedures.

1918.12.2 Moment strength of prestressed slabs at every section shall be at least equal to the required strength considering Sections 1909.2, 1909.3, 1918.10.3 and 1918.10.4. Shear strength of prestressed slabs at columns shall be at least equal to the required strength considering Sections 1909.2, 1909.3, 1911.1, 1911.12.2 and 1911.12.6.2.

1918.12.3 At service load conditions, all serviceability limitations, including specified limits on deflections, shall be met, with appropriate consideration of the factors listed in Section 1918.10.2.

1918.12.4 For normal live loads and load uniformly distributed, spacing of prestressing tendons or groups of tendons in one direction shall not exceed eight times the slab thickness, or 5 feet (1524 mm). Spacing of tendons also shall provide a minimum average prestress, after allowance for all prestress losses, of 125 psi (0.86 MPa) on the slab section tributary to the tendon or tendon group. A minimum of two tendons shall be provided in each direction through the critical shear section over columns. Special consideration of tendon spacing shall be provided for slabs with concentrated loads.

1918.12.5 In slabs with unbonded prestressing tendons, bonded reinforcement shall be provided in accordance with Sections 1918.9.3 and 1918.9.4.

1918.12.6 In lift slabs, bonded bottom reinforcement shall be detailed in accordance with the last paragraph of Section 1913.3.8.6.

1918.13 Tendon Anchorage Zones.

1918.13.1 Reinforcement shall be provided where required in tendon anchorage zones to resist bursting, splitting and spalling forces induced by tendon anchorages. Regions of abrupt change in section shall be adequately reinforced.

1918.13.2 End blocks shall be provided where required for support bearing or for distribution of concentrated prestressing forces.

1918.13.3 Posttensioning anchorages and supporting concrete shall be designed to resist maximum jacking force for strength of concrete at time of prestressing.

1918.13.4 Posttensioning anchorage zones shall be designed to develop the guaranteed ultimate tensile strength of prestressing tendons using a strength-reduction factor φ of 0.90 for concrete.

1918.14 Corrosion Protection for Unbonded Prestressing Tendons.

1918.14.1 Unbonded tendons shall be completely coated with suitable material to ensure corrosion protection.

1918.14.2 Tendon cover shall be continuous over entire length to be unbonded, and shall prevent intrusion of cement paste or loss of coating materials during concrete placement.

1918.14.3 Unbonded single strand tendons shall be protected against corrosion.

1918.15 Posttensioning Ducts.

1918.15.1 Ducts for grouted or unbonded tendons shall be mortar-tight and nonreactive with concrete, tendons or filler materials.

1918.15.2 Ducts for grouted single wire, strand or bar tendons shall have an inside diameter at least $^1/_4$ inch (6.4 mm) larger than tendon diameter.

1918.15.3 Ducts for grouted multiple wire, strand or bar tendons shall have an inside cross-sectional area at least two times the area of tendons.

1918.15.4 Ducts shall be maintained free of water if members to be grouted are exposed to temperatures below freezing prior to grouting.

1918.16 Grout for Bonded Prestressing Tendons.

1918.16.1 Grout shall consist of portland cement and water; or portland cement, sand and water.

1918.16.2 Materials for grout shall conform to the following:

1918.16.2.1 Portland cement shall conform to Section 1903.2.

1918.16.2.2 Water shall conform to Section 1903.4.

1918.16.2.3 The gradation of sand shall be permitted to be modified as necessary to obtain satisfactory workability.

1918.16.2.4 Admixtures conforming to Section 1903.6 and known to have no injurious effects on grout, steel or concrete may be used. Calcium chloride shall not be used.

1918.16.3 Selection of grout proportions.

1918.16.3.1 Proportions of materials for grout shall be based on either of the following:

1. Results of tests on fresh and hardened grout prior to beginning grouting operations, or

2. Prior documented experience with similar materials and equipment and under comparable field conditions.

1918.16.3.2 Cement used in the work shall correspond to that on which selection of grout proportions was based.

1918.16.3.3 Water content shall be the minimum necessary for proper pumping of grout; however, water-cementitious materials ratio shall not exceed 0.45 by weight.

1918.16.3.4 Water shall not be added to increase grout flowability that has been decreased by delayed use of grout.

1918.16.4 Mixing and pumping grout.

1918.16.4.1 Grout shall be mixed in equipment capable of continuous mechanical mixing and agitation that will produce uniform distribution of materials, passed through screens, and pumped in a manner that will completely fill tendon ducts.

1918.16.4.2 Temperature of members at time of grouting shall be above 35°F (1.7°C) and shall be maintained above 35°F (1.7°C) until field-cured 2-inch (51 mm) cubes of grout reach a minimum compressive strength of 800 pounds per square inch (5.52 MPa).

1918.16.4.3 Grout temperatures shall not be above 90°F (32.2°C) during mixing and pumping.

1918.17 Protection for Prestressing Tendons. Burning or welding operations in vicinity of prestressing tendons shall be carefully performed so that tendons are not subject to excessive temperatures, welding sparks or ground currents.

1918.18 Application and Measurement of Prestressing Force.

1918.18.1 Prestressing force shall be determined by both of the following methods:

1. Measurement of tendon elongation. Required elongation shall be determined from average load-elongation curves for the prestressing tendons used.

2. Observation of jacking force on a calibrated gauge or load cell or by use of a calibrated dynamometer.

Cause of any difference in force determination between 1 and 2 that exceeds 5 percent for pretensioned elements or 7 percent for post-tensioned construction shall be ascertained and corrected.

1918.18.2 Where transfer of force from bulkheads of pretensioning bed to concrete is accomplished by flame cutting prestressing tendons, cutting points and cutting sequence shall be predetermined to avoid undesired temporary stresses.

1918.18.3 Long lengths of exposed pretensioned strand shall be cut near the member to minimize shock to concrete.

1918.18.4 Total loss of prestress due to unreplaced broken tendons shall not exceed 2 percent of total prestress.

1918.19 Posttensioning Anchorages and Couplers.

1918.19.1 Anchorages and couplers for bonded and unbonded prestressed tendons shall develop at least 95 percent of the specified breaking strength of the tendons, when tested in an unbonded condition, without exceeding anticipated set. For bonded tendons, anchorages and couplers shall be located so that 100 percent of the specified breaking strength of the tendons shall be developed at critical sections after tendons are bonded in the member.

1918.19.2 Couplers shall be placed in areas approved by the *building official* and enclosed in housing long enough to permit necessary movements.

1918.19.3 In unbonded construction subject to repetitive loads, special attention shall be given to the possibility of fatigue in anchorages and couplers.

1918.19.4 Anchorages, couplers and end fittings shall be permanently protected against corrosion.

SECTION 1919 — SHELLS AND FOLDED PLATES

1919.0 Notations.

E_c = modulus of elasticity of concrete, pounds per square inch (MPa). See Section 1908.5.1.

f'_c = specified compressive strength of concrete, pounds per square inch (MPa).

$\sqrt{f'_c}$ = square root of specified compressive strength of concrete, pounds per square inch (MPa).

f_y = specified yield strength of nonprestressed reinforcement, pounds per square inch (MPa).

h = thickness of shell or folded plate, inches (mm).

l_d = development length, inches (mm).

ϕ = strength-reduction factor. See Section 1909.3.

1919.1 Scope and Definitions.

1919.1.1 Provisions of Section 1919 shall apply to thin-shell and folded-plate concrete structures, including ribs and edge members.

1919.1.2 All provisions of Chapter 19 not specifically excluded, and not in conflict with provisions of Section 1919, shall apply to thin-shell structures.

AUXILIARY MEMBERS are ribs or edge beams which serve to strengthen, stiffen and/or support the shell; usually, auxiliary members act jointly with the shell.

ELASTIC ANALYSIS is an analysis of deformations and internal forces based on equilibrium, compatibility of strains and assumed elastic behavior and representing to suitable approximation the three-dimensional action of the shell together with its auxiliary members.

EXPERIMENTAL ANALYSIS is an analysis procedure based on the measurement of deformations and/or strains of the structure or its model; experimental analysis may be based on either elastic or inelastic behavior.

FOLDED PLATES are a special class of shell structures formed by joining flat, thin slabs along their edges to create a three-dimensional spatial structure.

INELASTIC ANALYSIS is an analysis of deformations and internal forces based on equilibrium, nonlinear stress-strain relations for concrete and reinforcement, consideration of cracking and time-dependent effects and compatibility of strains. The analysis shall represent a suitable approximation of the three-dimensional action of the shell together with its auxiliary members.

RIBBED SHELLS are spatial structures with material placed primarily along certain preferred rib lines, with the area between the ribs filled with thin slabs or left open.

THIN SHELLS are three-dimensional spatial structures made up of one or more curved slabs or folded plates whose thicknesses are small compared to their other dimensions. Thin shells are characterized by their three-dimensional load-carrying behavior which is determined by the geometry of their forms, by the manner in which they are supported and by the nature of the applied load.

1919.2 Analysis and Design.

1919.2.1 Elastic behavior shall be an accepted basis for determining internal forces and displacements of thin shells. This behavior shall be permitted to be established by computations based on an analysis of the uncracked concrete structure in which the material is assumed linearly elastic, homogeneous and isotropic. Poisson's ratio of concrete shall be permitted to be taken equal to zero.

1919.2.2 Inelastic analysis shall be permitted to be used where it can be shown that such methods provide a safe basis for design.

1919.2.3 Equilibrium checks of internal resistances and external loads shall be made to ensure consistency of results.

1919.2.4 Experimental or numerical analysis procedures shall be permitted where it can be shown that such procedures provide a safe basis for design.

1919.2.5 Approximate methods of analysis shall be permitted where it can be shown that such methods provide a safe basis for design.

1919.2.6 In prestressed shells, the analysis shall also consider behavior under loads induced during prestressing, at cracking load and at factored load. Where prestressing tendons are draped within a shell, design shall take into account force components on the shell resulting from the tendon profile not lying in one plane.

1919.2.7 The thickness of a shell and its reinforcement shall be proportioned for the required strength and serviceability, using either the strength design method of Section 1908.1.1 or the alternate design method of Section *1926*.

1919.2.8 Shell instability shall be investigated and shown by design to be precluded.

1919.2.9 Auxiliary members shall be designed according to the applicable provisions of this code. It shall be permitted to assume that a portion of the shell equal to the flange width, as specified in Section 1908.10, acts with the auxiliary member. In such portions of the shell, the reinforcement perpendicular to the auxiliary member shall be at least equal to that required for the flange of a T-beam by Section 1908.10.5.

1919.2.10 Strength design of shell slabs for membrane and bending forces shall be based on the distribution of stresses and strains as determined from either an elastic or an inelastic analysis.

1919.2.11 In a region where membrane cracking is predicted, the nominal compressive strength parallel to the cracks shall be taken as $0.4f'_c$.

1919.3 Design Strength of Materials.

1919.3.1 Specified compressive strength of concrete f'_c at 28 days shall not be less than 3,000 psi (20.69 MPa).

1919.3.2 Specified yield strength of nonprestressed reinforcement f_y shall not exceed 60,000 psi (413.7 MPa).

1919.4 Shell Reinforcement.

1919.4.1 Shell reinforcement shall be provided to resist tensile stresses from internal membrane forces, to resist tension from bending and twisting moments, to control shrinkage and temperature cracking and as special reinforcement as shell boundaries, load attachments and shell openings.

1919.4.2 Tensile reinforcement shall be provided in two or more directions and shall be proportioned such that its resistance in any direction equals or exceeds the component of internal forces in that direction.

Alternatively, reinforcement for the membrane forces in the slab shall be calculated as the reinforcement required to resist axial tensile forces plus the tensile force due to shear friction required to transfer shear across any cross section of the membrane. The assumed coefficient of friction shall not exceed 1.0 where $\lambda = 1.0$ for normal-weight concrete, 0.85 for sand-lightweight concrete, and 0.75 for all lightweight concrete. Linear interpolation shall be permitted when partial sand replacement is used.

1919.4.3 The area of shell reinforcement at any section as measured in two orthogonal directions shall not be less than the slab shrinkage or temperature reinforcement required by Section 1907.12.

1919.4.4 Reinforcement for shear and bending moments about axes in the plane of the shell slab shall be calculated in accordance with Sections 1910, 1911 and 1913.

1919.4.5 The area of shell tension reinforcement shall be limited so that the reinforcement will yield before either crushing of concrete in compression or shell buckling can take place.

1919.4.6 In regions of high tension, membrane reinforcement shall, if practical, be placed in the general directions of the principal tensile membrane forces. Where this is not practical, it shall be permitted to place membrane reinforcement in two or more component directions.

1919.4.7 If the direction of reinforcement varies more than 10 degrees from the direction of principal tensile membrane force, the amount of reinforcement shall be reviewed in relation to cracking at service loads.

1919.4.8 Where the magnitude of the principal tensile membrane stress within the shell varies greatly over the area of the shell surface, reinforcement resisting the total tension may be concentrated in the regions of largest tensile stress where it can be shown that this provides a safe basis for design. However, the ratio of shell reinforcement in any portion of the tensile zone shall not be less than 0.0035 based on the overall thickness of the shell.

1919.4.9 Reinforcement required to resist shell bending moments shall be proportioned with due regard to the simultaneous action of membrane axial forces at the same location. Where shell reinforcement is required in only one face to resist bending moments, equal amounts shall be placed near both surfaces of the shell even though a reversal of bending moments is not indicated by the analysis.

1919.4.10 Shell reinforcement in any direction shall not be spaced farther apart than 18 inches (457 mm), or five times the shell thickness. Where the principal membrane tensile stress on the gross concrete area due to factored loads exceeds $4 \phi \sqrt{f'_c}$ (For **SI:** $0.33 \phi \sqrt{f'_c}$), reinforcement shall not be spaced farther apart than three times the shell thickness.

1919.4.11 Shell reinforcement at the junction of the shell and supporting members or edge members shall be anchored in or extended through such members in accordance with the requirements of Section 1912, except that the minimum development length shall be 1.2 d but not less than 18 inches (457 mm).

1919.4.12 Splice development lengths of shell reinforcement shall be governed by the provisions of Section 1912, except that the minimum splice length of tension bars shall be 1.2 times the value required by Section 1912 but not less than 18 inches (457 mm). The number of splices in principal tensile reinforcement shall be kept to a practical minimum. Where splices are necessary, they shall be staggered at least l_d with not more than one third of the reinforcement spliced at any section.

1919.5 Construction.

1919.5.1 When removal of formwork is based on a specific modulus of elasticity of concrete because of stability or deflection considerations, the value of the modulus of elasticity E_c shall be determined from flexural tests of field-cured beam specimens. The number of test specimens, the dimensions of test beam specimens and test procedures shall be specified.

1919.5.2 The tolerances for the shape of the shell shall be specified. If construction results in deviations from the shape greater than the specified tolerances, an analysis of the effect of the devi-

ations shall be made and any required remedial actions shall be taken to ensure safe behavior.

SECTION 1920 — STRENGTH EVALUATION OF EXISTING STRUCTURES

1920.0 Notations.

D = dead loads, or related internal moments and forces.

f'_c = specified compressive strength concrete, psi (MPa).

h = overall thickness of member in direction of action considered, inches (mm).

L = live loads or related internal moments and forces.

l_t = span of member under load test, inches (mm). (The shorter span for two-way slab systems.) Span is the smaller of (1) distance between centers of supports and (2) clear distance between supports plus thickness, h, of member, inches (mm). In Formula (20-1), span for a cantilever shall be taken as twice the distance from support to cantilever end, inches (mm).

Δ_{max} = measured maximum deflection, inches (mm), see Formula (20-1).

$\Delta_{r\,max}$ = measured residual deflection, inches (mm), see Formulas (20-2) and (20-3).

$\Delta_{f\,max}$ = maximum deflection measured during the second test relative to the position of the structure at the beginning of the second test, inches (mm). See Formula (20-3).

1920.1 Strength Evaluation—General.

1920.1.1 If there is a doubt that a part or all of a structure meets the safety requirements of this code, a strength evaluation shall be carried out as required by the engineer or building official.

1920.1.2 If the effect of the strength deficiency is well understood and if it is feasible to measure the dimensions and materials properties required for analysis, analytical evaluations of strength based on those measurements shall suffice. Required data shall be determined in accordance with Section 1920.2.

1920.1.3 If the effect of the strength deficiency is not well understood or if it is not feasible to establish the required dimensions and material properties by measurement, a load test shall be required if the structure is to remain in service.

1920.1.4 If the doubt about safety of a part or all of a structure involves deterioration and if the observed response during the load test satisfies the acceptance criteria, the structure or part of the structure shall be permitted to remain in service for a specified time period. If deemed necessary by the engineer, periodic reevaluations shall be conducted.

1920.2 Determination of Required Dimensions and Material Properties.

1920.2.1 Dimensions of the structural elements shall be established at critical sections.

1920.2.2 Locations and sizes of the reinforcing bars, welded wire fabric or tendons shall be determined by measurement. It shall be permitted to base reinforcement locations on available drawings if spot checks are made confirming the information on the drawings.

1920.2.3 If required, concrete strength shall be based on results of cylinder tests or tests of cores removed from the part of the structure where the strength is in doubt. Concrete strength shall be determined as specified in Section 1905.6.4.

1920.2.4 If required, reinforcement or tendon strength shall be based on tensile tests of representative samples of the material in the structure in question.

1920.2.5 If the required dimensions and material properties are determined through measurements and testing, and if calculations can be made in accordance with Section 1920.1.2, it shall be permitted to increase the strength-reduction factor in Section 1909.3 but the strength-reduction factor shall not be more than:

Flexure without axial load 1.0

Axial tension and axial tension with flexure 1.0

Axial compression and axial compression with flexure:

Members with spiral reinforcement conforming to
Section 1910.9.3 . . . . 0.9

Other members 0.85

Shear and/or torsion 0.9

Bearing on concrete 0.85

1920.3 Load Test Procedure.

1920.3.1 Load arrangement. The number and arrangement of spans or panels loaded shall be selected to maximize the deflection and stresses in the critical regions of the structural elements of which strength is in doubt. More than one test load arrangement shall be used if a single arrangement will not simultaneously result in maximum values of the effects (such as deflection, rotation or stress) necessary to demonstrate the adequacy of the structure.

1920.3.2 Load intensity. The total test load (including dead load already in place) shall not be less than $0.85 (1.4D + 1.7L)$. It shall be permitted to reduce L in accordance with the requirements of this code.

1920.3.3 A load test shall not be made until that portion of the structure to be subject to load is at least 56 days old. If the owner of the structure, the contractor and all involved parties agree, it shall be permitted to make the test at an earlier age.

1920.4 Loading Criteria.

1920.4.1 The initial value for all applicable response measurements (such as deflection, rotation, strain, slip, crack widths) shall be obtained not more than one hour before application of the first load increment. Measurements shall be made at locations where maximum response is expected. Additional measurements shall be made if required.

1920.4.2 Test load shall be applied in not less than four approximately equal increments.

1920.4.3 Uniform test load shall be applied in a manner to ensure uniform distribution of the load transmitted to the structure or portion of the structure being tested. Arching of the applied load shall be avoided.

1920.4.4 A set of response measurements shall be made after each load increment is applied and after the total load has been applied on the structure for at least 24 hours.

1920.4.5 Total test load shall be removed immediately after all response measurements defined in Section 1920.4.4 are made.

1920.4.6 A set of final response measurements shall be made 24 hours after the test load is removed.

1920.5 Acceptance Criteria.

1920.5.1 The portion of the structure tested shall show no evidence of failure. Spalling and crushing of compressed concrete shall be considered an indication of failure.

1920.5.2 Measured maximum deflections shall satisfy one of the following conditions:

$$\Delta_{max} \leq \frac{l_t^2}{20,000h} \qquad (20\text{-}1)$$

$$\Delta_{r\,max} \leq \frac{\Delta_{max}}{4} \qquad (20\text{-}2)$$

If the measured maximum and residual deflections do not satisfy Formula (20-1) or (20-2), it shall be permitted to repeat the load test.

The repeated test shall be conducted not earlier than 72 hours after removal of the first test load. The portion of the structure tested in the repeat test shall be considered acceptable if deflection recovery satisfied the condition:

$$\Delta_{r\,max} \leq \frac{\Delta_{f\,max}}{5} \qquad (20\text{-}3)$$

where $\Delta_{f\,max}$ is the maximum deflection measured during the second test relative to the position of the structure at the beginning of the second test.

1920.5.3 Structural members tested shall not have cracks indicating the imminence of shear failure.

1920.5.4 In regions of structural members without transverse reinforcement, appearance of structural cracks inclined to the longitudinal axis and having a horizontal projection longer than the depth of the member at mid-point of the crack shall be evaluated.

1920.5.5 In regions of anchorage and lap splices, the appearance along the line of reinforcement of a series of short inclined cracks or horizontal cracks shall be evaluated.

1920.6 Provisions for Lower Load Rating. If the structure under investigation does not satisfy conditions or criteria of Section 1920.1.2, 1920.5.2 or 1920.5.3, the structure may be permitted for use at a lower load rating based on the results of the load test or analysis, if approved by the building official.

1920.7 Safety.

1920.7.1 Load tests shall be conducted in such a manner as to provide for safety of life and structure during the test.

1920.7.2 No safety measures shall interfere with load test procedures or affect results.

SECTION 1921 — REINFORCED CONCRETE STRUCTURES RESISTING FORCES INDUCED BY EARTHQUAKE MOTIONS

1921.0 Notations.

A_{ch} = cross-sectional area of a structural member measured out-to-out of transverse reinforcement, square inches (mm²).

A_{cp} = area of concrete section, resisting shear, of an individual pier or horizontal wall segment, square inches (mm²).

A_{cv} = net area of concrete section bounded by web thickness and length of section in the direction of shear force considered, square inches (mm²).

A_g = gross area of section, square inches (mm²).

A_j = effective cross-sectional area within a joint (see Section 1921.5.3.1) in a plane parallel to plane of reinforcement generating shear in the joint. The joint depth shall be the overall depth of the column. Where a beam frames into a support of larger width, the effective width of the joint shall not exceed the smaller of:

    1. beam width plus the joint depth

    2. twice the smaller perpendicular distance from the

longitudinal axis of the beam to the column side. See Section 1921.5.3.1.

A_{sh} = total cross-sectional area of transverse reinforcement (including crossties) within spacing, s, and perpendicular to dimension, h_c.

b = effective compressive flange width of a structural member, inches (mm).

b_w = web width, or diameter of circular section, inches (mm).

d = effective depth of section.

d_b = bar diameter.

E = load effects of earthquake, or related internal moments and forces.

f'_c = specified compressive strength of concrete, psi (MPa).

f_y = specified yield strength of reinforcement, psi (MPa).

f_{yh} = specified yield strength of transverse reinforcement, psi (MPa).

h = overall dimension of member in the direction of action considered.

h_c = cross-sectional dimension of a column core or shear wall boundary zone measured center-to-center of confining reinforcement.

h_w = height of entire wall (diaphragm) or of the segment of wall (diaphragm) considered.

l_d = development length for a straight bar.

l_{dh} = development length for a bar with a standard hook as defined in Formula (21-5).

l_o = minimum length, measured from joint face along axis of structural member, over which transverse reinforcement must be provided, inches (mm).

l_u = unsupported length of compression member (see Section 1910.11.3.1).

l_w = length of entire wall (diaphragm) or of segment of wall (diaphragm) considered in direction of shear force.

M_{pr} = probable flexural strength of members, with or without axial load, determined using the properties of the member at the joint faces assuming a tensile strength in the longitudinal bars of at least 1.25 f_y and a strength-reduction factor ϕ of 1.0.

M_s = portion of slab moment balanced by support moment.

s = spacing of transverse reinforcement measured along the longitudinal axis of the structural member, inches (mm).

$S_{e\ CONNECTION}$
  = *moment, shear or axial force at connection cross section other than the nonlinear action location corresponding to probable strength at the nonlinear action location, taking gravity load effects into consideration, per Section 1921.2.7.3.*

$S_{n\ CONNECTION}$
  = *nominal strength of connection cross section in flexural, shear or axial action, per Section 1921.2.7.3.*

s_o = maximum spacing of transverse reinforcement, inches (mm).

V_c = nominal shear strength provided by concrete.

V_e = design shear force determined from Section 1921.3.4.1 or 1921.4.5.1.

V_n = nominal shear strength.

V_u = factored shear force at section.

α_c = coefficient defining the relative contribution of concrete strength to wall strength.

ρ = ratio of nonprestressed tension reinforcement
  = A_s/bd.

ρ_g = ratio of total reinforcement area to cross-sectional area of column.

ρ_n = ratio of distributed shear reinforcement on a plane perpendicular to plane of A_{cv}.

ρ_s = ratio of volume of spiral reinforcement to the core volume confined by the spiral reinforcement (measured out-to-out).

ρ_v = A_{sv}/A_{cv}; where A_{sv} is the projection on A_{cv} of area of distributed shear reinforcement crossing the plane of A_{cv}.

ϕ = strength-reduction factor.

Δ_m = $0.7\,R\Delta_s$.

Δ_s = *Design Level Response Displacement, which is the total drift or total story drift that occurs when the structure is subjected to the design seismic forces.*

ψ = *Dynamic Amplification Factor from Sections 1921.2.7.3 and 1921.2.7.4.*

1921.1 Definitions. For the purposes of this section, certain terms are defined as follows:

BASE OF STRUCTURE is the level at which earthquake motions are assumed to be imparted to a building. This level does not necessarily coincide with the ground level.

BOUNDARY ELEMENTS (*or ZONES*) are portions along wall and diaphragm edges strengthened by longitudinal and transverse reinforcement. Boundary elements do not necessarily require an increase in the thickness of the wall or diaphragm. Edges of openings within walls and diaphragms shall be provided with boundary elements if required by Sections 1921.6.6.1 and 1921.6.7.1.

COLLECTOR ELEMENTS are elements that serve to transmit the inertial forces with the diaphragms to members of the lateral-force-resisting systems.

CONFINED CORE is the area within the core defined by h_c.

CONNECTION *is an element that joins two precast members or a precast member and a cast-in-place member.*

COUPLING BEAMS are a horizontal element in plane with and connecting two shear walls.

CROSSTIE is a continuous reinforcing bar having a seismic hook at one end and a hook of not less than 90 degrees with at least six diameters at the other end. The hooks shall engage peripheral longitudinal bars. The 90-degree hooks of two successive crossties engaging the same longitudinal bar shall be alternated end for end.

DESIGN LOAD COMBINATIONS are combinations of factored loads and forces specified in Sections *1612.2.1 and* 1909.2.

DEVELOPMENT LENGTH FOR A BAR WITH A STANDARD HOOK is the shortest distance between the critical section (where the strength of the bar is to be developed) and a tangent to the outer edge of the 90-degree hook.

DRY CONNECTION *is a connection used between precast members, which does not qualify as a wet connection.*

FACTORED LOADS AND FORCES are the specified loads and forces modified by the factors in Sections *1612.2.1 and 1909.2.*

HOOP is a closed tie or continuously wound tie. A closed tie can be made up of several reinforcing elements, each having seismic hooks at both ends. A continuously wound tie shall have a seismic hook at both ends.

JOINT *is the geometric volume common to intersecting members.*

LATERAL-FORCE-RESISTING SYSTEM is that portion of the structure composed of members proportioned to resist forces related to earthquake effects.

LIGHTWEIGHT-AGGREGATE CONCRETE is all lightweight or sanded lightweight aggregate concrete made with lightweight aggregates conforming to Section 1903.3.

NONLINEAR ACTION LOCATION *is the center of the region of yielding in flexure, shear or axial action.*

NONLINEAR ACTION REGION *is the member length over which nonlinear action takes place. It shall be taken as extending a distance of no less than h/2 on either side of the nonlinear action location.*

SEISMIC HOOK is a hook on a stirrup, hoop or crosstie having a bend not less than 135 degrees with a six-bar-diameter [but not less than 3 inches (76 mm)], extension that engages the longitudinal reinforcement and projects into the interior of the stirrup or hoop.

SHELL CONCRETE is concrete outside the transverse reinforcement confining the concrete.

SPECIFIED LATERAL FORCES are lateral forces corresponding to the appropriate distribution of the design base shear force prescribed by the governing code for earthquake-resistant design.

STRONG CONNECTION *is a connection that remains elastic, while the designated nonlinear action regions undergo inelastic response under the Design Basis Ground Motion.*

STRUCTURAL DIAPHRAGMS are structural members, such as floor and roof slabs, which transmit inertial forces to lateral-force-resisting members.

STRUCTURAL TRUSSES are assemblages of reinforced concrete members subjected primarily to axial forces.

STRUT is an element of a structural diaphragm used to provide continuity around an opening in the diaphragm.

TIE ELEMENTS are elements which serve to transmit inertia forces and prevent separation of such building components as footings and walls.

WALL PIER *is a wall segment with a horizontal length-to-thickness ratio between 2.5 and 6, and whose clear height is at least two times its horizontal length.*

WET CONNECTION *uses any of the splicing methods, per Section 1921.2.6.1 or 1921.3.2.3, to connect precast members and uses cast-in-place concrete or grout to fill the splicing closure.*

1921.2 General Requirements.

1921.2.1 Scope.

1921.2.1.1 *Section 1921* contains special requirements for design and construction of reinforced concrete members of a structure for which the design forces, related to earthquake motions, have been determined on the basis of energy dissipation in the nonlinear range of response.

1921.2.1.2 The provisions of Sections 1901 through 1918 shall apply except as modified by the provisions of Section 1921.

1921.2.1.3 *In Seismic Zones 0 and 1, the provisions of Section 1921 shall not apply.*

In Seismic Zone 2, reinforced concrete frames resisting forces induced by earthquake motions shall be intermediate moment-resisting frames proportioned to satisfy only Section 1921.8 in addition to the requirements of Sections 1901 through 1918. In Seismic Zone 2, frame members which are not designated to be part of the lateral-force-resisting system shall conform to Section 1921.7.

1921.2.1.4 *In Seismic Zones 3 and 4,* all reinforced concrete structural members that are part of the lateral-force-resisting system shall satisfy the requirements of Sections 1921.2 through 1921.7, in addition to the requirements of Sections 1901 through 1917.

1921.2.1.5 A reinforced concrete structural system not satisfying the requirements of this section may be used if it is demonstrated by experimental evidence and analysis that the proposed system will have strength and toughness equal to or exceeding those provided by a comparable monolithic reinforced concrete structure satisfying this section.

1921.2.1.6 *Precast lateral-force-resisting systems shall satisfy either of the following criteria:*

1. Emulate the behavior of monolithic reinforced concrete construction and satisfy Section 1921.2.2.5, or

2. Rely on the unique properties of a structural system composed of interconnected precast elements and conform to Section 1629.9.2.

1921.2.1.7 *In structures having precast gravity systems, the lateral-force-resisting system shall be one of the systems listed in Table 16-N and shall be well distributed using one of the following methods:*

1. The lateral-force-resisting systems shall be spaced such that the span of the diaphragm or diaphragm segment between lateral-force-resisting systems shall be no more than three times the width of the diaphragm or diaphragm segment.

Where the lateral-force-resisting system consists of moment-resisting frames, at least $[(N_b/4) + 1]$ of the bays (rounded up to the nearest integer) along any frame line at any story shall be part of the lateral-force-resisting system, where N_b is the total number of bays along that line at that story. This requirement applies to only the lower two thirds of the stories of buildings three stories or taller.

2. All beam-to-column connections that are not part of the lateral-force-resisting system shall be designed in accordance with the following:

Connection design force. The connection shall be designed to develop strength M. M is the moment developed at the connection when the frame is displaced by Δ_s assuming fixity at the connection and a beam flexural stiffness of no more than one-half of the gross section stiffness. M shall be sustained through a deformation of Δ_m.

Connection characteristics. The connection shall be permitted to resist moment in one direction only, positive or negative. The connection at the opposite end of the member shall resist moment with same positive or negative sign. The connection shall be permitted to have zero flexural stiffness up to a frame displacement of Δ_s.

In addition, complete calculations for the deformation compatibility of the gravity load carrying system shall be made in accordance with Section 1633.2.4 using cracked section stiffnesses in the lateral-force-resisting system and the diaphragm.

Where gravity columns are not provided with lateral support on all sides, a positive connection shall be provided along each unsupported direction parallel to a principal plan axis of the structure. The connection shall be designed for a horizontal force equal to 4 percent of the axial load strength (P_0) of the column.

The bearing length shall be 2 inches (51 mm) more than that required for bearing strength.

1921.2.2 Analysis and proportioning of structural members.

1921.2.2.1 The interaction of all structural and nonstructural members which materially affect the linear and nonlinear response of the structure to earthquake motions shall be considered in the analysis.

1921.2.2.2 Rigid members assumed not to be a part of the lateral-force-resisting system shall be permitted, provided their effect on the response of the system is considered and accommodated in the structural design. Consequences of failure of structural and nonstructural members which are not a part of the lateral-force-resisting system shall also be considered.

1921.2.2.3 Structural members below base of structure required to transmit to the foundation forces resulting from earthquake effects shall also comply with the requirements of Section 1921.

1921.2.2.4 All structural members assumed not to be part of the lateral-force-resisting system shall conform to Section 1921.7.

1921.2.2.5 Precast structural systems using frames and emulating the behavior of monolithic reinforced concrete construction shall satisfy either Section 1921.2.2.6 or 1921.2.2.7.

1921.2.2.6 Precast structural systems, utilizing wet connections, shall comply with all the applicable requirements of monolithic concrete construction for resisting seismic forces.

1921.2.2.7 Precast structural systems not meeting Section 1921.2.2.6 shall utilize strong connections resulting in nonlinear response away from connections. Design shall satisfy the requirements of Section 1921.2.7 in addition to all the applicable requirements of monolithic concrete construction for resisting seismic forces, except that provisions of Section 1921.3.1.2 shall apply to the segments between nonlinear action locations.

1921.2.3 Strength-reduction factors. Strength-reduction factors shall be as given in Section 1909.3.4.

1921.2.4 Concrete in members resisting earthquake-induced forces.

1921.2.4.1 Compressive strength f'_c shall not be less than 3,000 psi (20.69 MPa).

> **EXCEPTION:** *Footings of buildings three stories or less may have concrete with f'_c of not less than 2,500 psi (17.24 MPa).*

1921.2.4.2 Compressive strength of lightweight-aggregate concrete used in design shall not exceed 4,000 psi (27.58 MPa). Lightweight aggregate concrete with higher design compressive strength shall be permitted if demonstrated by experimental evidence that structural members made with that lightweight aggregate concrete provide strength and toughness equal to or exceeding those of comparable members made with normal-weight aggregate concrete of the same strength. *In no case shall the compressive strength of lightweight concrete used in design exceed 6,000 psi (41.37 MPa).*

1921.2.5 Reinforcement in members resisting earthquake-induced forces.

1921.2.5.1 *Alloy A 706 reinforcement. Except as permitted in Sections 1921.2.5.2 through 1921.2.5.5, reinforcement resisting earthquake-induced flexural and axial forces in frame members and in wall boundary elements shall comply with low alloy A 706 except as allowed in Section 1921.2.5.2.*

1921.2.5.2 Billet steel A 615 reinforcement. Billet steel A 615 Grades 40 and 60 reinforcement shall be permitted to be used in frame members and wall boundary elements if (1) the actual yield strength based on mill tests does not exceed the specified yield strength by more than 18,000 psi (124.1 MPa) [retests shall not exceed this value by more than an additional 3,000 psi (20.69 MPa)], and (2) the ratio of the actual ultimate tensile stress to the actual yield strength is not less than 1.25.

1921.2.5.3 The average prestress f_{pc}, calculated for an area equal to the member's shortest cross-sectional dimension multiplied by the perpendicular dimension, shall be the lesser of 350 psi (2.41 MPa) or $f'_c/12$ at locations of nonlinear action, where prestressing tendons are used in members of frames.

1921.2.5.4 For members in which prestressing tendons are used together with mild reinforcement to resist earthquake-induced forces, prestressing tendons shall not provide more than one-quarter of the strength for both positive and negative moments at the joint face and shall extend through exterior joints and be anchored at the exterior face of the joint or beyond.

1921.2.5.5 Shear strength provided by prestressing tendons shall not be considered in design.

1921.2.6 Welded splices and mechanically connected reinforcement.

1921.2.6.1 Reinforcement resisting earthquake-induced flexural or axial forces in frame members or in wall boundary members shall be permitted to be spliced using welded splices or mechanical connectors conforming to Section 1912.14.3.3 or 1912.14.3.4.

Splice locations in frame members shall conform to Section 1921.2.6.1.1 or 1921.2.6.1.2.

1921.2.6.1.1 Welded splices. In Seismic Zones 2, 3 and 4, welded splices on billet steel A 615 or low allow A 706 reinforcement shall not be used within an anticipated plastic hinge region nor within a distance of one beam depth on either side of the plastic hinge region or within a joint.

1921.2.6.1.2 Mechanical connection splices. Splices with mechanical connections shall be classified according to strength capacity as follows:

Type 1 splice. Mechanical connections meeting the requirements of Sections 1912.14.3.4 and 1912.14.3.5.

Type 2 splice. Mechanical connections that develop in tension the lesser of 95 percent of the ultimate tensile strength or 160 percent of specified yield strength, f_y, of the bar.

Mechanical connection splices shall be permitted to be located as follows:

Type 1 splice. In Seismic Zone 1, a Type 1 splice shall be permitted in any location within a member. In Seismic Zones 2, 3 and 4, a Type 1 splice shall not be used within an anticipated plastic hinge region or within a distance of one beam depth on either side of the plastic hinge region or within a joint.

Type 2 splice. A Type 2 splice shall be permitted in any location within a member.

1921.2.6.2 Welding of stirrups, ties, inserts or other similar elements to longitudinal reinforcement required by design shall not be permitted.

1921.2.7 Emulation of monolithic construction using strong connections. Members resisting earthquake-induced forces in precast frames using strong connections shall satisfy the following:

1921.2.7.1 Location. Nonlinear action location shall be selected so that there is a strong column/weak beam deformation mecha-

nism under seismic effects. The nonlinear action location shall be no closer to the near face of strong connection than h/2. For column-to-footing connections, where nonlinear action may occur at the column base to complete the mechanism, the nonlinear action location shall be no closer to the near face of the connection than h/2.

1921.2.7.2 Anchorage and splices. *Reinforcement in the nonlinear action region shall be fully developed outside both the strong connection region and the nonlinear action region. Noncontinuous anchorage reinforcement of strong connection shall be fully developed between the connection and the beginning of nonlinear action region. Lap splices are prohibited within connections adjacent to a joint.*

1921.2.7.3 Design forces. *Design strength of strong connections shall be based on*

$$\phi \, S_{n \; CONNECTION} > \psi \, S_{e \; CONNECTION} \qquad (21\text{-}1)$$

Dynamic amplification factor ψ shall be taken as 1.0.

1921.2.7.4 Column-to-column connection. *The strength of such connections shall comply with Section 1921.2.7.3 with ψ taken as 1.4. Where column-to-column connections occur, the columns shall be provided with transverse reinforcement as specified in Sections 1921.4.4.1 through 1921.4.4.3 over their full height if the factored axial compressive force in these members, including seismic effects, exceeds $A_g f'_c/10$.*

> **EXCEPTION:** *Where column-to-column connection is located within the middle third of the column clear height, the following shall apply: (1) The design moment strength ϕM_n of the connection shall not be less than 0.4 times the maximum M_{pr} for the column within the story height, and (2) the design shear strength ϕV_n of the connection shall not be less than that determined per Section 1921.4.5.1.*

1921.2.7.5 Column-face connection. *Any strong connection located outside the middle half of a beam span shall be a wet connection, unless a dry connection can be substantiated by approved cyclic test results. Any mechanical connector located within such a column-face strong connection shall develop in tension or compression, as required, at least 140 percent of specified yield strength, f_y, of the bar.*

1921.3 Flexural Members of Frames.

1921.3.1 Scope. Requirements of this section apply to frame members (1) resisting earthquake-induced forces and (2) proportioned primarily to resist flexure. These frame members shall also satisfy the following conditions:

1921.3.1.1 Factored axial compressive force on the member shall not exceed $(A_g f'_c/10)$.

1921.3.1.2 Clear span for the members shall not be less than four times its effective depth.

1921.3.1.3 The width-to-depth ratio shall not be less than 0.3.

1921.3.1.4 The width shall not be (1) less than 10 inches (254 mm) and (2) more than the width of the supporting member (measured on a plane perpendicular to the longitudinal axis of the flexural member) plus distances on each side of the supporting member not exceeding three fourths of the depth of the flexural member.

1921.3.2 Longitudinal reinforcement.

1921.3.2.1 At any section of a flexural member, except as provided in Section 1910.5.3, for top as well as for bottom reinforcement, the amount of reinforcement shall not be less than that given by Formula (10-3) but not less than $200 \, b_w d/f_y$, (For **SI:** $1.38 \, b_w d/$

f_y) and the reinforcement ratio, ρ, shall not exceed 0.025. At least two bars shall be provided continuously, both top and bottom.

1921.3.2.2 Positive-moment strength at joint face shall not be less than one half of the negative-moment strength provided at that face of the joint. Neither the negative nor the positive-moment strength at any section along member length shall be less than one fourth the maximum moment strength provided at face of either joint.

1921.3.2.3 Lap splices of flexural reinforcement shall be permitted only if hoop or spiral reinforcement is provided over the lap length. Maximum spacing of the transverse reinforcement enclosing the lapped bars shall not exceed $d/4$ or 4 inches (102 mm). Lap splices shall not be used (1) within the joints, (2) within a distance of twice the member depth from the face of joint, and (3) at locations where analysis indicates flexural yielding caused by inelastic lateral displacements of the frame.

1921.3.2.4 Welded splices and mechanical connections shall conform to Section 1921.2.6.1.

1921.3.3 Transverse reinforcement.

1921.3.3.1 Hoops shall be provided in the following regions of frame members:

1. Over a length equal to twice the member depth measured from the face of the supporting member toward midspan, at both ends of the flexural members.

2. Over lengths equal to twice the member depth on both sides of a section where flexural yielding may occur in connection with inelastic lateral displacements of the frame.

1921.3.3.2 The first hoop shall be located not more than 2 inches (51 mm) from the face of a supporting member. Maximum spacing of the hoops shall not exceed (1) $d/4$, (2) eight times the diameter of the smallest longitudinal bars, (3) 24 times the diameter of the hoop bars, and (4) 12 inches (305 mm).

1921.3.3.3 Where hoops are required, longitudinal bars on the perimeter shall have lateral support conforming to Section 1907.10.5.3.

1921.3.3.4 Where hoops are not required, stirrups with seismic hooks at both ends shall be spaced at a distance not more than $d/2$ throughout the length of the member.

1921.3.3.5 Stirrups or ties required to resist shear shall be hoops over lengths of members as specified in Sections 1921.3.3, 1921.4.4 and 1921.5.2.

1921.3.3.6 Hoops in flexural members shall be permitted to be made up of two pieces of reinforcement: a stirrup having seismic hooks at both ends and closed by a crosstie. Consecutive crossties engaging the same longitudinal bar shall have their 90-degree hooks at opposite sides of the flexural member. If the longitudinal reinforcing bars secured by the crossties are confined by a slab on only one side of the flexural frame member, the 90-degree hooks of the crossties shall all be placed on that side.

1921.3.4 Shear strength.

1921.3.4.1 Design forces. The design shear forces V_e shall be determined from consideration of the static forces on the portion of the member between faces of the joint. It shall be assumed that moments of opposite sign corresponding to probable strength M_{pr} act at the joint faces and that the member is loaded with the tributary gravity load along its span.

1921.3.4.2 Transverse reinforcement. Transverse reinforcement over the lengths identified in Section 1921.3.3.1 shall be proportioned to resist shear assuming $V_c = 0$ when both of the following conditions occur:

1. The earthquake-induced shear force calculated in accordance with Section 1921.3.4.1 represents one-half or more of the maximum required shear strength within those lengths.

2. The factored axial compressive force including earthquake effects is less than $A_g f'_c/20$.

1921.4 Frame Members Subjected to Bending and Axial Load.

1921.4.1 Scope. The requirements of Section 1921.4 apply to frame members (1) resisting earthquake-induced forces and (2) having a factored axial force exceeding $A_g f'_c/10$. These frame members shall also satisfy the following conditions:

1921.4.1.1 The shortest cross-sectional dimension, measured on a straight line passing through the geometric centroid, shall not be less than 12 inches (305 mm).

1921.4.1.2 The ratio of the shortest cross-sectional dimension to the perpendicular dimension shall not be less than 0.4.

1921.4.2 Minimum flexural strength of columns.

1921.4.2.1 Flexural strength of any column proportioned to resist a factored axial compressive force exceeding $A_g f'_c/10$ shall satisfy Section 1921.4.2.2 or 1921.4.2.3.

Lateral strength and stiffness of columns not satisfying Section 1921.4.2.2 shall be ignored in determining the calculated strength and stiffness of the structure but shall conform to Section 1921.7.

1921.4.2.2 The flexural strengths of the columns shall satisfy Formula (21-1).

$$\Sigma M_e \geq (^6/_5)\Sigma M_g \qquad (21\text{-}1)$$

WHERE:

ΣM_e = sum of moments, at the center of the joint, corresponding to the design flexural strength of the columns framing into that joint. Column flexural strength shall be calculated for the factored axial force, consistent with the direction of the lateral forces considered, resulting in the lowest flexural strength.

ΣM_g = sum of moments, at the center of the joint, corresponding to the design flexural strengths of the girders framing into that joint.

Flexural strengths shall be summed such that the column moments oppose the beam moments. Formula (21-1) shall be satisfied for beam moments acting in both directions in the vertical plane of the frame considered.

1921.4.2.3 If Section 1921.4.2.2 is not satisfied at a joint, columns supporting reactions from that joint shall be provided with transverse reinforcement as specified in Section 1921.4.4 over their full height.

1921.4.3 Longitudinal reinforcement.

1921.4.3.1 The reinforcement ratio ρ_g shall not be less than 0.01 and shall not exceed 0.06.

1921.4.3.2 Welded splices and mechanical connections shall conform to Section 1921.2.6.1. Lap splices shall be permitted only within the center half of the member length and shall be proportioned as tension splices.

1921.4.4 Transverse reinforcement.

1921.4.4.1 Transverse reinforcement as specified below shall be provided unless a larger amount is required by Section 1921.4.5.

1. The volumetric ratio of spiral or circular hoop reinforcement, ρ_s, shall not be less than that indicated by Formula (21-2).

$$\rho_s = 0.12 f'_c/f_{yh} \qquad (21\text{-}2)$$

and shall not be less than that required by Formula (10-6).

2. The total cross-sectional area of rectangular hoop reinforcement shall not be less than that given by Formulas (21-3) and (21-4).

$$A_{sh} = 0.3 (sh_c f'_c/f_{yh})[(A_g/A_{ch}) - 1] \qquad (21\text{-}3)$$

$$A_{sh} = 0.09 (sh_c f'_c/f_{yh}) \qquad (21\text{-}4)$$

3. Transverse reinforcement shall be provided by either single or overlapping hoops. Crossties of the same bar size and spacing as the hoops shall be permitted to be used. Each end of the crosstie shall engage a peripheral longitudinal reinforcing bar. Consecutive crossties shall be alternated end for end along the longitudinal reinforcement.

4. If the design strength of member core satisfies the requirement of the specified loading combinations including earthquake effect, Formulas (21-3) and (10-6) need not be satisfied.

5. *Any area of a column which extends more than 4 inches (102 mm) beyond the confined core shall have minimum reinforcement as required for nonseismic columns as specified in Section 1921.7.*

6. *Where the calculated point of contraflexure is not within the middle half of the member clear height, provide transverse reinforcement as specified in Sections 1921.4.4.1, Items 1 through 3, over the full height of the member.*

1921.4.4.2 Transverse reinforcement shall be spaced at distances not exceeding (1) one-quarter minimum member dimension and (2) 4 inches (102 mm). *Anchor bolts set in the top of a column shall be enclosed with ties as specified in Section 1921.4.4.8.*

1921.4.4.3 Crossties or legs of overlapping hoops shall not be spaced more than 14 inches (356 mm) on center in the direction perpendicular to the longitudinal axis of the structural member.

1921.4.4.4 Transverse reinforcement in amount specified in Sections 1921.4.4.1 through 1921.4.4.3 shall be provided over a length l_o from each joint face and on both sides of any section where flexural yielding may occur in connection with inelastic lateral displacements of the frame. The length l_o shall not be less than (1) the depth of the member at the joint face or at the section where flexural yielding may occur, (2) one sixth of the clear span of the member, and (3) 18 inches (457 mm).

1921.4.4.5 Columns supporting reactions from discontinued stiff members, such as walls, shall be provided with transverse reinforcement as specified in Sections 1921.4.4.1 through 1921.4.4.3 over their full height beneath the level at which the discontinuity occurs if the factored axial compressive force in these members, including earthquake effect, exceeds $A_g f'_c/10$. Transverse reinforcement as specified in Sections 1921.4.4.1 through 1921.4.4.3 shall extend into the discontinued member for at least the development length of the largest longitudinal reinforcement in the column in accordance with Section 1921.5.4. If the lower end of the column terminates on a wall, transverse reinforcement as specified in Sections 1921.4.4.1 through 1921.4.4.3 shall extend into the wall for at least the development length of the largest longitudinal reinforcement in the column at the point of termination. If the column terminates on a footing or mat, transverse reinforcement as specified in Sections 1921.4.4.1 through 1921.4.4.3 shall extend at least 12 inches (305 mm) into the footing or mat.

1921.4.4.6 Where transverse reinforcement as specified in Sections 1921.4.4.1 through 1921.4.4.3 is not provided throughout the full length of the column, the remainder of the column length shall contain spiral or hoop reinforcement with center-to-center spacing not exceeding the smaller of six times the diameter of the longitudinal column bars or 6 inches (152 mm).

1921.4.4.7 *At any section where the design strength,* ϕP_n, *of the column is less than the sum of the shears* V_e *computed in accordance with Sections 1921.3.4.1 and 1921.4.5.1 for all the beams framing into the column above the level under consideration, transverse reinforcement as specified in Sections 1921.4.4.1 through 1921.4.4.3 shall be provided. For beams framing into opposite sides of the column, the moment components may be assumed to be of opposite sign. For the determination of the design strength,* ϕP_n, *of the column, these moments may be assumed to result from the deformation of the frame in any one principal axis.*

1921.4.4.8 **Ties at anchor bolts.** *Anchor bolts which are set in the top of a column shall be provided with ties which enclose at least four vertical column bars. Such ties shall be in accordance with Section 1907.1.3, Item 3, shall be within 5 inches (127 mm) of the top of the column, and shall consist of at least two No. 4 or three No. 3 bars.*

1921.4.5 Shear strength requirements.

1921.4.5.1 Design forces. The design shear force V_e shall be determined from the consideration of the maximum forces that can be generated at the faces of the joints at each end of the member. These joint forces shall be determined using the maximum probable moment strengths, M_{pr}, of the member associated with the range of factored axial loads on the member. The member shear need not exceed those determined from joint strengths based on the probable moment strength, M_{pr}, of the transverse members framing in the joint. In no case shall V_e be less than the factored shear determined by analysis of the structure.

1921.4.5.2 Transverse reinforcement over the lengths l_o, identified in Section 1921.4.4.4, shall be proportioned to resist shear assuming $V_c = 0$ when both of the following conditions occur:

1. The earthquake-induced shear force calculated in accordance with Section 1921.4.5.1 represents one-half or more of the maximum required shear strength within those lengths.

2. The factored axial compressive force including earthquake effects is less than $A_g f'_c/20$.

1921.5 Joints of Frames.

1921.5.1 General requirements.

1921.5.1.1 Forces in longitudinal beam reinforcement at the joint face shall be determined by assuming that the stress in the flexural tensile reinforcement is 1.25 f_y.

1921.5.1.2 Strength of joint shall be governed by the appropriate strength-reduction factors specified in Section 1909.3.

1921.5.1.3 Beam longitudinal reinforcement terminated in a column shall be extended to the far face of the confined column core and anchored in tension according to Section 1921.5.4, and in compression according to Section 1912.

1921.5.1.4 Where longitudinal beam reinforcement extends through a beam-column joint, the column dimension parallel to the beam reinforcement shall not be less than 20 times the diameter of the largest longitudinal bar for normal-weight concrete. For lightweight concrete, the dimension shall not be less than 26 times the bar diameter.

1921.5.2 Transverse reinforcement.

1921.5.2.1 Transverse hoop reinforcement as specified in Section 1921.4.4 shall be provided within the joint, unless the joint is confined by structural members as specified in Section 1921.5.2.2.

1921.5.2.2 Within the depth of the shallowest framing member, transverse reinforcement equal to at least one half the amount required by Section 1921.4.4.1 shall be provided where members frame into all four sides of the joint and where each member width is at least three fourths the column width. At these locations, the spacing specified in Section 1921.4.4.2 shall be permitted to be increased to 6 inches (152 mm).

1921.5.2.3 Transverse reinforcement as required by Section 1921.4.4 shall be provided through the joint to provide confinement for longitudinal beam reinforcement outside the column core if such confinement is not provided by a beam framing into the joint.

1921.5.3 Shear strength.

1921.5.3.1 The nominal shear strength of the joint shall not be taken greater than the forces specified below for normal-weight aggregate concrete.

For joints confined on all four faces .... $20\sqrt{f'_c}A_j$

(For **SI:** $1.66\sqrt{f'_c}A_j$)

For joints confined on three faces or on two

opposite faces .... $15\sqrt{f'_c}A_j$

(For **SI:** $1.25\sqrt{f'_c}A_j$)

For others .... $12\sqrt{f'_c}A_j$

(For **SI:** $1.00\sqrt{f'_c}A_j$)

A member that frames into a face is considered to provide confinement to the joint if at least three fourths of the face of the joint is covered by the framing member. A joint is considered to be confined is such confining members frame into all faces of the joint.

1921.5.3.2 For lightweight aggregate concrete, the nominal shear strength of the joint shall not exceed three fourths of the limits for normal-weight aggregate concrete.

1921.5.4 Development length for reinforcement in tension.

1921.5.4.1 The development length, l_{dh}, for a bar with a standard 90-degree hook in normal-weight aggregate concrete shall not be less than $8d_b$, 6 inches (152 mm), and the length required by Formula (21-5).

$$l_{dh} = f_y d_b/65\sqrt{f'_c} \qquad (21\text{-}5)$$

For **SI:** $\qquad\qquad l_{dh} = f_y d_b/5.4\sqrt{f'_c}$

for bar sizes No. 3 through No. 11.

For lightweight aggregate concrete, the development length for a bar with a standard 90-degree hook shall not be less than $10d_b$, 7.5 inches (191 mm), and 1.25 times that required by Formula (21-5).

The 90-degree hook shall be located within the confined core of a column or of a boundary member.

1921.5.4.2 For bar sizes No. 3 through No. 11, the development length, l_d, for a straight bar shall not be less than (1) 2.5 times the length required by Section 1921.5.4.1 if the depth of the concrete cast in one lift beneath the bar does not exceed 12 inches (305 mm), and (2) 3.5 times the length required by Section 1921.5.4.1 if the depth of the concrete cast in one lift beneath the bar exceeds 12 inches (305 mm).

1921.5.4.3 Straight bars terminated at a joint shall pass through the confined core of a column or of a boundary member. Any portion of the straight embedment length not within the confined core shall be increased by a factor of 1.6.

1921.5.4.4 If epoxy-coated reinforcement is used, the development lengths in Sections 1921.5.4.1 through Section 1921.5.4.3 shall be multiplied by the applicable factor specified in Section 1912.2.4 or 1912.5.3.6.

1921.6 Shear Walls, Diaphragms and Trusses.

1921.6.1 Scope. The requirements of this section apply to *shear* walls and trusses serving as parts of the earthquake-force-resisting systems as well as to diaphragms, struts, ties, chords and collector members which transmit forces induced by earthquake.

1921.6.2 Reinforcement.

1921.6.2.1 The reinforcement ratio, ρ_v, for *shear* walls shall not be less than 0.0025 along the longitudinal and transverse axes. If the design shear force does not exceed $A_{cv}\sqrt{f'_c}$ (For **SI:** $0.08A_{cv}\sqrt{f'_c}$), the minimum reinforcement for *shear* walls shall be in conformance with Section 1914.3. The minimum reinforcement ratio for structural diaphragms shall be in conformance with Section 1907.12. Reinforcement spacing each way in *shear* walls and diaphragms shall not exceed 18 inches (457 mm). Reinforcement provided for shear strength shall be continuous and shall be distributed across the shear plane.

1921.6.2.2 At least two curtains of reinforcement shall be used in a wall if the in-plane factored shear force assigned to the wall exceeds $2A_{cv}\sqrt{f'_c}$ (For **SI:** $0.166A_{cv}\sqrt{f'_c}$).

*When V_u in the plane of the wall exceeds $A_{cv}\sqrt{f'_c}$ (For **SI:** $0.08A_{cv}\sqrt{f'_c}$), horizontal reinforcement terminating at the edges of shear walls shall have a standard hook engaging the edge reinforcement, or the edge reinforcement shall be enclosed in "U" stirrups having the same size and spacing as, and spliced to, the horizontal reinforcement.*

1921.6.2.3 Structural-truss elements, struts, ties and collector elements with compressive stresses exceeding $0.2f'_c$ shall have special transverse reinforcement, as specified in Section 1921.4.4, over the total length of the element. The special transverse reinforcement may be discontinued at a section where the calculated compressive stress is less than $0.15f'_c$. Stresses shall be calculated for the factored forces using a linearly elastic model and gross-section properties of the elements considered.

1921.6.2.4 All continuous reinforcement in *shear* walls, diaphragms, trusses, struts, ties, chords and collector elements shall be anchored or spliced in accordance with the provisions for reinforcement in tension as specified in Section 1921.5.4.

1921.6.3 Design forces. The design shear force V_u shall be obtained from the lateral load analysis in accordance with the factored loads and combinations specified in Section 1909.2 *and as modified in Section 1612.2.1.*

1921.6.4 Diaphragms. *See Sections 1921.6.11 and 1921.6.12.*

1921.6.5 Shear strength.

1921.6.5.1 Nominal shear strength of *shear* walls and diaphragms shall be determined using either Section 1921.6.5.2 or 1921.6.5.3.

1921.6.5.2 Nominal shear strength, V_n, of *shear* walls and diaphragms shall be assumed not to exceed the shear force calculated from

$$V_n = A_{cv}(2\sqrt{f'_c} + \rho_n f_y) \qquad (21\text{-}6)$$

For **SI:** $\qquad V_n = A_{cv}(0.166\sqrt{f'_c} + \rho_n f_y)$

1921.6.5.3 For walls (diaphragms) and wall (diaphragm) segments having a ratio of (h_w/l_w) less than 2.0, nominal shear strength of wall (diaphragm) shall be determined from Formula (21-7)

$$V_n = A_{cv}(\alpha_c\sqrt{f'_c} + \rho_n f_y) \qquad (21\text{-}7)$$

For **SI:** $\qquad V_n = A_{cv}(0.08\alpha_c\sqrt{f'_c} + \rho_n f_y)$

Where the coefficient α_c varies linearly from 3.0 for $h_w/l_w = 1.5$ to 2.0 for $h_w/l_w = 2.0$.

1921.6.5.4 In Section 1921.6.5.3 above, the value of ratio (h_w/l_w) used for determining V_n for segments of a wall or diaphragm shall be the largest of the ratios for the entire wall (diaphragm) and the segment of wall (diaphragm) considered.

1921.6.5.5 Walls (diaphragms) shall have distributed shear reinforcement providing resistance in two orthogonal directions in the plane of the wall (diaphragm). If the ratio (h_w/l_w) does not exceed 2.0, reinforcement ratio ρ_v shall not be less than reinforcement ratio ρ_n.

1921.6.5.6 Nominal shear strength of all wall piers sharing a common lateral force shall not be assumed to exceed $8A_{cv}\sqrt{f'_c}$ (For **SI:** $0.66A_{cv}\sqrt{f'_c}$) where A_{cv} is the total cross-sectional area and the nominal shear strength of any one of the individual wall piers shall not be assumed to exceed $10A_{cp}\sqrt{f'_c}$ (For **SI:** $0.83A_{cp}\sqrt{f'_c}$) where A_{cp} represents the cross-sectional area of the pier considered.

1921.6.5.7 Nominal shear strength of horizontal wall segments shall not be assumed to exceed $10A_{cp}\sqrt{f'_c}$ (For **SI:** $0.83A_{cp}\sqrt{f'_c}$) where A_{cp} represents the cross-sectional area of a horizontal wall segment.

1921.6.6 Design of shear walls for flexural and axial loads.

1921.6.6.1 Shear walls and portions of shear walls subject to combined flexural and axial loads shall be designed in accordance with Sections 1910.2 and 1910.3, except Section 1910.3.6 and the nonlinear strain requirements of Section 1910.2.2 do not apply. The strength-reduction factor ϕ shall be in accordance with Section 1909.3.

1921.6.6.2 The effective flange widths to be used in the design of I-, L-, C- or T-shaped sections shall not be assumed to extend further from the face of the web than (1) one half the distance to an adjacent shear wall web, or (2) 15 percent of the total wall height for the flange in compression or 30 percent of the total wall height for the flange in tension, not to exceed the total projection of the flange.

1921.6.6.3 Walls and portions of walls with $P_u > 0.35P_o$ shall not be considered to contribute to the calculated strength of the structure for resisting earthquake-induced forces. Such walls shall conform to the requirements of Section 1631.2, Item 4.

1921.6.6.4 Shear wall boundary zone detail requirements as defined in Section 1921.6.6.6 need not be provided in shear walls or portions of shear walls meeting the following conditions:

   1. $P_u \le 0.10A_gf'_c$ for geometrically symmetrical wall sections
   $P_u \le 0.05A_gf'_c$ for geometrically unsymmetrical wall sections

and either

   2. $\dfrac{M_u}{V_u l_w} \le 1.0$

or

3. $V_u \leq 3A_{cv} \sqrt{f'_c}$ and $\dfrac{M_u}{V_u l_w} \leq 3$ (For **SI**: $V_u \leq 0.25A_{cv} \sqrt{f'_c}$)

Shear walls and portions of shear walls not meeting the conditions of Section 1921.6.6.4 and having $P_u < 0.35P_o$ shall have boundary zones at each end a distance varying linearly from $0.25 \, l_w$ to $0.15 \, l_w$ for P_u varying from $0.35 \, P_o$ to $0.15 \, P_o$. The boundary zone shall have minimum length of $0.15 \, l_w$ and shall be detailed in accordance with Section 1921.6.6.6.

1921.6.6.5 *Alternatively, the requirements for boundary zones in shear walls or portions of shear walls not meeting the conditions of Section 1921.6.6.4 may be based on determination of the compressive strain levels at edges when the wall or portion of wall is subjected to displacement levels resulting from the ground motions specified in Section 1629.2 using cracked section properties and considering the response modification effects of possible nonlinear behavior of the building.*

Boundary zone detail requirements as defined in Section 1921.6.6.6 shall be provided over those portions of the wall where compressive strains exceed 0.003. In no instance shall designs be permitted in which compressive strains exceed ε_{max}.

WHERE:

$$\epsilon_{max} = 0.015 \qquad (21\text{-}8)$$

1. Using the displacement of Section 1921.6.6.5, determine the curvature of the wall cross section at each location of potential flexural yielding assuming the possible nonlinear response of the wall and its elements. Using a strain compatibility analysis of the wall cross section, determine the compressive strains resulting from these curvatures.

2. For shear walls in which the flexural limit state response is governed by yielding at the base of the wall, compressive strains at wall edges may be approximated as follows:

Determine the total curvature demand (ϕ_t) as given in Formula (21-9):

$$\phi_t = \frac{\Delta_i}{(h_w - l_p/2)l_p} + \phi_y \qquad (21\text{-}9)$$

WHERE:

c'_u = neutral axis depth at P'_u and M'_n.

l_p = height of the plastic hinge above critical section and which shall be established on the basis of substantiated test data or may be alternatively taken at $0.5l_w$.

P'_u = $1.2D + 0.5L + E_h$.

Δ_E = elastic design displacement at the top of the wall using gross section properties and code-specified seismic forces.

Δ_i = inelastic deflection at top of wall.

= $\Delta_t - \Delta_y$

Δ_t = total deflection at the top of the wall equal to Δ_M, using cracked section properties, or may be taken as $2\Delta_M$, using gross section properties.

Δ_y = displacement at top of wall corresponding to yielding of the tension reinforcement at critical section, or may be taken as $(M'_n/M_E)\Delta_E$, where M_E equals unfactored moment at critical section when top of wall is displaced Δ_E. M'_n is nominal flexural strength of critical section at P'_u.

ϕ_y = yield curvature which may be estimated as $0.003/l_w$.

If ϕ_t is less than or equal to $0.003/c'_u$, boundary zone details as defined in Section 1921.6.6.6 are not required. If ϕ_t exceeds $0.003/c'_u$, the compressive strains may be assumed to vary linearly over the depth c'_u and have maximum value equal to the product of c'_u and ϕ_t.

1921.6.6.6 Shear wall boundary zone detail requirements. When required by Section 1921.6.6.1 through 1921.6.6.5, boundary zones shall meet the following:

1. Dimensional requirements.

   1.1 All portions of the boundary zones shall have a thickness of $l_u/16$ or greater.

   1.2 Boundary zones shall extend vertically a distance equal to the development length of the largest vertical bar within the boundary zone above the elevation where the requirements of Section 1921.6.6.4 or 1921.6.6.5 are met.

   Extensions below the base of the boundary zone shall conform to Section 1921.4.4.6.

   EXCEPTION: The boundary zone reinforcement need not extend above the base of the boundary zone a distance greater than the larger of l_w or $M_u/4V_u$.

   1.3 Boundary zones as determined by the requirements of Section 1921.6.6.5 shall have a minimum length of 18 inches (457 mm) at each end of the wall or portion of wall.

   1.4 In I-, L-, C- or T-shaped sections, the boundary zone at each end shall include the effective flange width and shall extend at least 12 inches (305 mm) into the web.

2. Confinement reinforcement.

   2.1 All vertical reinforcement within the boundary zone shall be confined by hoops or cross ties producing an area of steel not less than:

   $$A_{sh} = 0.09sh_c f'_c/f_{yh} \qquad (21\text{-}10)$$

   2.2 Hoops and cross ties shall have a vertical spacing not greater than the smaller of 6 inches (152 mm) or 6 diameters of the largest vertical bar within the boundary zone.

   2.3 The ratio of the length to the width of the hoops shall not exceed 3. All adjacent hoops shall be overlapping.

   2.4 Cross ties or legs of overlapping hoops shall not be spaced further apart than 12 inches (305 mm) along the wall.

   2.5 Alternate vertical bars shall be confined by the corner of a hoop or cross tie.

3. Horizontal reinforcement.

   3.1 All horizontal reinforcement terminating within a boundary zone shall be anchored in accordance with Section 1921.6.2.

   3.2 Horizontal reinforcement shall not be lap spliced within the boundary zone.

4. Vertical reinforcement.

   4.1 Vertical reinforcement shall be provided to satisfy all tension and compression requirements.

   4.2 Area of reinforcement shall not be less than 0.005 times the area of boundary zone or less than two No. 5 bars at each edge of boundary zone.

   4.3 Lap splices of vertical reinforcement within the boundary zone shall be confined by hoops or cross ties. Spacing of hoops and cross ties confining lap-spliced reinforcement shall not exceed 4 inches (102 mm).

1921.6.6.7 Welded splices and mechanical connections of longitudinal reinforcement in the boundary *zone* shall conform to Section 1921.2.6.1.

1921.6.7 Boundaries of structural diaphragms.

1921.6.7.1 Boundary elements of structural diaphragms shall be proportioned to resist the sum of the factored axial force acting in the plane of the diaphragm and the force obtained from dividing the factored moment at the section by the distance between the edges of the diaphragm at that section.

1921.6.7.2 Splices of tensile reinforcement in the boundaries and collector elements of all diaphragms shall develop the yield strength of the reinforcement. Welded splices and mechanical connections shall conform to Section 1921.2.7.1.

1921.6.7.3 *Reinforcement for chords and collectors at splices and anchorage zones shall have a minimum spacing of three bar diameters, but not less than $1^1/_2$ inches (38 mm), and a minimum concrete cover of two and one-half bar diameters, but not less than 2 inches (51 mm), and shall have transverse reinforcement as specified by Section 1911.5.5.3, except as required in Section 1921.6.2.3.*

1921.6.8 Construction joints.

1921.6.8.1 All construction joints in walls and diaphragms shall conform to Section 1906.4, and contact surfaces shall be roughened as specified in Section 1911.7.9.

1921.6.9 Discontinuous walls. Columns supporting discontinuous walls shall be reinforced in accordance with Section 1921.4.4.5.

1921.6.10 Coupling beams.

1921.6.10.1 *For coupling beams with $1_n/d \geq 4$, the design shall conform to the requirements of Sections 1921.2 and 1921.3. It shall be permitted to waive the requirements of Sections 1921.3.1.3 and 1921.3.1.4 if it can be shown by analysis that lateral stability is adequate or if alternative means of maintaining lateral stability is provided.*

1921.6.10.2 *Coupling beams with $1_n/d < 4$ shall be permitted to be reinforced with two intersecting groups of symmetrical diagonal bars. Coupling beams with $1_n/d < 4$ and with factored shear force V_u exceeding $4\sqrt{f'_c}\,b_w d$ (For SI: $0.33\sqrt{f'_c}\,b_w d$) shall be reinforced with two intersecting groups of symmetrical diagonal bars. Each group shall consist of a minimum of four bars assembled in a core with a lateral dimension of each side not less than $b_w/2$ or 4 inches (102 mm). The design shear strength, ϕV_n, of these coupling beams shall be determined by:*

$$\phi V_n = 2\phi\, f_y \sin \alpha\, A_d \leq 10\phi\sqrt{f'_c}\,b_w d \qquad (21\text{-}11)$$

For SI: $\qquad \phi V_n = 2\phi\, f_y \sin \alpha\, A_d \leq 0.83\phi\sqrt{f'_c}\,b_w d$

WHERE:

α = *the angle between the diagonal reinforcement and the longitudinal axis.*

A_{vd} = *the total area of reinforcement of each group of diagonal bars.*

ϕ = *0.85.*

> EXCEPTION: *The design of coupling beams need not comply with the requirements for diagonal reinforcement if it can be shown that failure of the coupling beams will not impair the vertical load carrying capacity of the structure, the egress from the structure, or the integrity of nonstructural components and connections. The analysis shall take into account the effects of the failure of the coupling beams on foundation rotation and overall system displacements. Design strength of cou-*

pling beams assumed to be part of the seismic force resisting system shall not be reduced below the values otherwise required.

1921.6.10.3 *Each group of diagonally placed bars shall be enclosed in transverse reinforcement conforming to Sections 1921.4.4.1 through 1921.4.4.3. For the purpose of computing A_g, as per Formulas 10-6 and 21-3, the minimum cover, as specified in Section 1907.7, shall be assumed over each group of diagonally placed reinforcing bars.*

1921.6.10.4 *Reinforcement parallel and transverse to the longitudinal axis shall be provided and, as a minimum, shall conform to Sections 1910.5, 1911.8.9 and 1911.8.10.*

1921.6.10.5 *Contribution of the diagonal reinforcement to nominal flexural strength of the coupling beam shall be considered.*

1921.6.11 Floor topping. *A cast-in-place topping on a precast floor system may serve as the diaphragm, provided the cast-in-place topping acting alone is proportioned and detailed to resist the design forces.*

1921.6.12 Diaphragms. *Diaphragms used to resist prescribed lateral forces shall comply with the following:*

> *1. Thickness shall not be less than 2 inches (51 mm).*

> *2. When mechanical connectors are used to transfer forces between the diaphragm and the lateral system, the anchorage shall be adequate to develop $1.4\, A_s\, f_y$, where A_s is the connector's cross-sectional area.*

> *3. Collector and boundary elements in topping slabs placed over precast floor and roof elements shall not be less than 3 inches (76 mm) or 6 d_b thick, where d_b is the diameter of the largest reinforcement in the topping slab.*

> *4. Prestressing tendons shall not be used as primary reinforcement in boundaries and collector elements of structural diaphragms. Precompression from unbonded tendons may be used to resist diaphragm forces.*

1921.6.13 Wall piers.

1921.6.13.1 *Wall piers not designed as part of a special moment-resisting frame shall have transverse reinforcement designed to satisfy the requirements in Section 1921.6.13.2.*

> **EXCEPTIONS:** *1. Wall piers that satisfy Section 1921.7.*
>
> *2. Wall piers along a wall line within a story where other shear wall segments provide lateral support to the wall piers, and such segments have a total stiffness of at least six times the sum of the stiffnesses of all the wall piers.*

1921.6.13.2 *Transverse reinforcement shall be designed to resist the shear forces determined from Sections 1921.4.5.1 and 1921.3.4.2. When the axial compressive force, including earthquake effects, is less than $A_g\, f'_c/20$, transverse reinforcement in wall piers may have standard hooks at each end in lieu of hoops. Spacing of transverse reinforcement shall not exceed 6 inches (152 mm). Transverse reinforcement shall be extended beyond the pier clear height for at least the development length of the largest longitudinal reinforcement in the wall pier.*

1921.6.13.3 *Wall segments with horizontal length-to-thickness ratio less than $2^1/_2$ shall be designed as columns.*

1921.7 Frame Members Not Part of the Lateral-force-resisting System.

1921.7.1 Frame members assumed not to contribute to lateral resistance shall be detailed according to Section 1921.7.2 or 1921.7.3, depending on the magnitude of moments induced in those members when subjected to Δ_M. When induced moments under lateral displacements are not calculated, Section 1921.7.3 shall apply.

1921.7.2 When the induced moments and shears under lateral displacements of Section 1921.7.1 combined with the factored gravity moments and shear loads do not exceed the design moment and shear strength of the frame member, the following conditions shall be satisfied. For this purpose, the load combinations (1.4D + 1.4L) and 0.9D shall be used.

1921.7.2.1 Members with factored gravity axial forces not exceeding $(A_g f'_c/10)$, shall satisfy Section 1921.3.2.1. Stirrups shall be placed at not more than $d/2$ throughout the length of the member.

1921.7.2.2 Members with factored gravity axial forces exceeding $(A_g f'_c/10)$, but not exceeding $0.3P_o$ shall satisfy Sections 1921.4.3, 1921.4.4.1, Item 3, and 1921.4.4.3. Design shear strength shall not be less than the shear associated with the development of nominal moment strengths of the member at each end of the clear span. The maximum longitudinal spacing of ties shall be s_o for the full column height. The spacing s_o shall not be more than (1) 6 diameters of the smallest longitudinal bar enclosed, (2) 16 tie-bar diameters, (3) one-half the least cross-sectional dimension of the column and (4) 6 inches (152 mm).

1921.7.2.3 Members with factored gravity axial forces exceeding $0.3P_o$ shall satisfy Sections 1921.4.4 and 1921.4.5.

1921.7.3 When the induced moments under lateral displacements of Section 1921.7.1 exceed the design moment strength of the frame member, or where induced moments are not calculated, the following conditions in Sections 1921.7.3.1 through 1921.7.3.3 shall be satisfied.

1921.7.3.1 Materials shall satisfy Sections 1921.2.4, 1921.2.5 and 1921.2.6.

1921.7.3.2 Members with factored gravity axial forces not exceeding $(A_g f'_c/10)$ shall satisfy Sections 1921.3.2.1 and 1921.3.4. Stirrups shall be placed at not more than $d/2$ throughout the length of the member.

1921.7.3.3 Members with factored gravity axial forces exceeding $(A_g f'_c/10)$ shall satisfy Sections 1921.4.4, 1921.4.5 and 1921.5.2.1.

1921.7.4 Ties at anchor bolts. Anchor bolts set in the top of a column shall be enclosed with ties as specified in Section 1921.4.4.8.

1921.8 Requirements for Frames *in Seismic Zone 2.*

1921.8.1 *In Seismic Zone 2,* structural frames proportioned to resist forces induced by earthquake motions shall satisfy the requirements of Section 1921.8 in addition to those of Sections 1901 through 1918.

1921.8.2 Reinforcement details in a frame member shall satisfy Section 1921.8.4 if the factored compressive axial load for the member does not exceed $(A_g f'_c/10)$. If the factored compressive axial load is larger, frame reinforcement details shall satisfy Section 1921.8.5 unless the member has spiral reinforcement according to Formula (10-5). If a two-way slab system without beams is treated as part of a frame-resisting earthquake effect, reinforcement details in any span resisting moments caused by lateral force shall satisfy Section 1921.8.6.

1921.8.3 Design shear strength of beams, columns and two-way slabs resisting earthquake effect shall not be less than either (1) the sum of the shear associated with development of nominal moment strengths of the member at each restrained end of the clear span and the shear calculated for gravity loads, or (2) the maximum shear obtained from design load combinations which include

earthquake effect E, with E assumed to be twice that prescribed in Section 1626.

1921.8.4 Beams.

1921.8.4.1 The positive-moment strength at the face of the joint shall not be less than one third the negative-moment strength provided at that face of the joint. Neither the negative- nor the positive-moment strength at any section along the length of the member shall be less than one fifth the maximum moment strength provided at the face of either joint.

1921.8.4.2 At both ends of the member, stirrups shall be provided over lengths equal to twice the member depth measured from the face of the supporting member toward midspan. The first stirrup shall be located at not more than 2 inches (51 mm) from the face of the supporting member. Maximum stirrup spacing shall not exceed (1) $d/4$, (2) eight times the diameter of the smallest longitudinal bar enclosed, (3) 24 times the diameter of the stirrup bar, and (4) 12 inches (305 mm).

1921.8.4.3 Stirrups shall be placed at not more than $d/2$ throughout the length of the member.

1921.8.5 Columns.

1921.8.5.1 Maximum tie spacing shall not exceed s_o over a length l_o measured from the joint face. Spacing s_o shall not exceed (1) eight times the diameter of the smallest longitudinal bar enclosed, (2) 24 times the diameter of the tie bar, (3) one half of the smallest cross-sectional dimension of the frame member, and (4) 12 inches (305 mm). Length l_o shall not be less than (1) one sixth of the clear span of the member, (2) maximum cross-sectional dimension of the member, and (3) 18 inches (457 mm).

1921.8.5.2 The first tie shall be located at not more than $s_o/2$ from the joint face.

1921.8.5.3 Joint reinforcement shall conform to Section 1911.11.2.

1921.8.5.4 Tie spacing shall not exceed twice the spacings s_o.

1921.8.5.5 *Column lateral ties shall be as specified in Section 1907.1.3. Anchor bolts set in the top of a column shall be enclosed with ties as specified in Section 1921.4.4.8.*

1921.8.6 Two-way slabs without beams.

1921.8.6.1 Factored slab moment at support related to earthquake effect shall be determined for load combinations defined by Formulas (9-2) and (9-3). All reinforcement provided to resist M_s, the portion of slab moment balanced by support moment, shall be placed within the column strip defined in Section 1913.2.1.

1921.8.6.2 The fraction, defined by Formula (13-1), of moment M_s shall be resisted by reinforcement placed within the effective width specified in Section 1913.5.2.

1921.8.6.3 Not less than one half of the reinforcement in the column strip at support shall be placed within the effective slab width specified in Section 1913.5.2.

1921.8.6.4 Not less than one fourth of the top reinforcement at the support in the column strip shall be continuous throughout the span.

1921.8.6.5 Continuous bottom reinforcement in the column strip shall not be less than one third of the top reinforcement at the support in the column strip.

1921.8.6.6 Not less than one half of all bottom reinforcement at midspan shall be continuous and shall develop its yield strength at face of support as defined in Section 1913.6.2.5.

1921.8.6.7 At discontinuous edges of the slab, all top and bottom reinforcement at support shall be developed at the face of support as defined in Section 1913.6.2.5.

SECTION 1922 — STRUCTURAL PLAIN CONCRETE

1922.0 Notations.

A_g = gross area of section, inches squared (mm^2).

A_1 = loaded area, inches squared (mm^2).

A_2 = the area of the lower base of the largest frustum of a pyramid, cone or tapered wedge contained wholly within the support and having for its upper base the loaded area, and having side slopes of 1 unit vertical to 2 units horizontal, inches squared (mm^2).

b = width of member, inches (mm).

b_o = perimeter of critical section for shear in footings, inches (mm).

B_n = nominal bearing strength of loaded area.

f'_c = specified compressive strength of concrete, psi (MPa). See Section 1905.

$\sqrt{f'_c}$ = square root of specified compressive strength of concrete, psi (MPa).

f_{ct} = average splitting tensile strength of lightweight aggregate concrete, psi (MPa). See Sections 1905.1.4 and 1905.1.5.

h = overall thickness of member, inches (mm).

l_c = vertical distance between supports, inches (mm).

M_n = nominal moment strength at section.

M_u = factored moment at section.

P_n = nominal strength of cross section subject to compression.

P_{nw} = nominal axial load strength of wall designed by Section 1922.6.5.

P_u = factored axial load at given eccentricity.

S = elastic section modulus of section.

V_n = nominal shear strength at section.

v_u = shear stress due to factored shear force at section.

V_u = factored shear force at section.

β_c = ratio of long side to short side of concentrated load or reaction area.

ϕ = strength reduction factor. See Section 1909.3.5.

1922.1 Scope.

1922.1.1 This section provides minimum requirements for design and construction of structural plain concrete members (cast-in-place or precast) except as specified in Sections 1922.1.1.1 and 1922.1.1.2.

> **EXCEPTION:** *The design is not required when the minimum foundation for stud walls is in accordance with Table 18-I-C.*

1922.1.1.1 Structural plain concrete basement walls shall be exempted from the requirements for special exposure conditions of Section 1904.2.2.

1922.1.1.2 Design and construction of soil-supported slabs, such as sidewalks and slabs on grade shall not be regulated by this code unless they transmit vertical loads from other parts of the structure to the soil.

1922.1.2 For special structures, such as arches, underground utility structures, gravity walls, and shielding walls, provisions of this section shall govern where applicable.

1922.2 Limitations.

1922.2.1 Provisions of this section shall apply for design of structural plain concrete members defined as either unreinforced or containing less reinforcement than the minimum amount specified in this code for reinforced concrete.

1922.2.2 Use of structural plain concrete shall be limited to (1) members that are continuously supported by soil or supported by other structural members capable of providing continuous vertical support, (2) members for which arch action provides compression under all conditions of loading, or (3) walls and pedestals. See Sections 1922.6 and 1922.8. The use of structural plain concrete columns is not permitted.

1922.2.3 This section does not govern design and installation of cast-in-place concrete piles and piers embedded in ground.

1922.2.4 Minimum strength. Specified compressive strength of concrete, f'_c, used in structural plain concrete elements shall not be less than 2,500 psi (17.2 MPa).

1922.2.5 *Seismic Zones 2, 3 and 4. Plain concrete shall not be used in Seismic Zone 2, 3 or 4 except where specifically permitted by Section 1922.10.3.*

1922.3 Joints.

1922.3.1 Contraction or isolation joints shall be provided to divide structural plain concrete members into flexurally discontinuous elements. Size of each element shall be limited to control buildup of excessive internal stresses within each element caused by restraint to movements from creep, shrinkage and temperature effects.

1922.3.2 In determining the number and location of contraction or isolation joints, consideration shall be given to: influence of climatic conditions; selection and proportioning of materials; mixing, placing and curing of concrete; degree of restraint to movement; stresses due to loads to which an element is subject; and construction techniques.

1922.4 Design Method.

1922.4.1 Structural plain concrete members shall be designed for adequate strength in accordance with provisions of this chapter, using load factors and design strength.

1922.4.2 Factored loads and forces shall be in such combinations as specified in Section 1909.2.

1922.4.3 Where required strength exceeds design strength, reinforcement shall be provided and the member designed as a reinforced concrete member in accordance with appropriate design requirements of this chapter.

1922.4.4 Strength design of structural plain concrete members for flexure and axial loads shall be based on a linear stress-strain relationship in both tension and compression.

1922.4.5 Tensile strength of concrete shall be permitted to be considered in design of plain concrete members when provisions of Section 1922.3 have been followed.

1922.4.6 No strength shall be assigned to steel reinforcement that may be present.

1922.4.7 Tension shall not be transmitted through outside edges, construction joints, contraction joints or isolation joints of an individual plain concrete element. No flexural continuity due to tension shall be assumed between adjacent structural plain concrete elements.

1922.4.8 In computing strength in flexure, combined flexure and axial load, and shear, the gross cross section of a member shall be considered in design, except for concrete cast against soil, overall thickness h shall be taken as 2 inches (51 mm) less than actual thickness.

1922.5 Strength Design.

1922.5.1 Design of cross sections subject to flexure shall be based on

$$\phi M_n \geq M_u \qquad (22\text{-}1)$$

where M_u is factored moment and M_n is nominal moment strength* computed by

$$M_n = 5 \sqrt{f'_c} \, S \qquad (22\text{-}2)$$

where S is the elastic section modulus of the cross section.

1922.5.2 Design of cross sections subject to compression shall be based on

$$\phi P_N \geq P_u \qquad (22\text{-}3)$$

where P_u is factored load and P_n is nominal compression strength computed by

$$P_n = 0.60 f'_c \left[1 - \left(\frac{l_c}{32h} \right)^2 \right] A_1 \qquad (22\text{-}4)$$

where A_1 is the loaded area.

1922.5.3 Members subject to combined flexure and axial load in compression shall be proportioned such that on the compression face:

$$P_u / \phi P_n + M_u / \phi M_n \leq 1 \qquad (22\text{-}5)$$

and on the tension face:

$$M_u / S - P_u / A_g \leq 5\phi \sqrt{f'_c} \qquad (22\text{-}6)$$

1922.5.4 Design of rectangular cross sections subject to shear* shall be based on

$$\phi V_n \geq V_u \qquad (22\text{-}7)$$

where V_u is factored shear and V_n is nominal shear strength computed by

$$V_n = \frac{4}{3} \sqrt{f'_c} \, bh \qquad (22\text{-}8)$$

for beam action and by

$$V_n = \left[\frac{4}{3} + \frac{8}{3\beta_c} \right] \sqrt{f'_c} \, b_o h \leq 2.66 \sqrt{f'_c} \, b_o h \qquad (22\text{-}9)$$

for two-way action but not greater than $2.66 \sqrt{f'_c} \, b_o h$.

1922.5.5 Design of bearing areas subject to compression shall be based on

$$\phi B_n \geq P_u \qquad (22\text{-}10)$$

where P_u is factored bearing load and B_n is the nominal bearing strength of loaded area A_1 computed by

$$B_n = 0.85 f'_c A_1 \qquad (22\text{-}11)$$

except when the supporting surface is wider on all sides than the loaded area, design bearing strength on the loaded area shall be multiplied by $\sqrt{A_2 / A_1}$ but not more than 2.

1922.6 Walls.

1922.6.1 Structural plain concrete walls shall be continuously supported by soil, footings, foundation walls, grade beams or other structural members capable of providing continuous vertical support.

1922.6.2 Structural plain concrete walls shall be designed for vertical, lateral and other loads to which they are subjected.

1922.6.3 Structural plain concrete walls shall be designed for an eccentricity corresponding to the maximum moment that can accompany the axial load but not less than $0.10h$. If the resultant of all factored loads is located within the middle third of the overall wall thickness, the design shall be in accordance with Section 1922.5.3 or 1922.6.5. Otherwise, walls shall be designed in accordance with Section 1922.5.3.

1922.6.4 Design for shear shall be in accordance with Section 1922.5.4.

1922.6.5 Empirical design method.

1922.6.5.1 Structural plain concrete walls of solid rectangular cross section shall be permitted to be designed by Formula (22-13) if the resultant of all factored loads is located within the middle third of the overall thickness of wall.

1922.6.5.2 Design of walls subject to axial loads in compression shall be based on

$$\phi P_{nw} \geq P_u \qquad (22\text{-}12)$$

where P_u is the factored axial load and P_{nw} is nominal axial load strength computed by

$$P_{nw} = 0.45 f'_c A_g \left[1 - \left(\frac{l_c}{32h} \right)^2 \right] \qquad (22\text{-}13)$$

1922.6.6 Limitations.

1922.6.6.1 Unless demonstrated by a detailed analysis, horizontal length of wall to be considered effective for each vertical concentrated load shall not exceed center-to-center distance between loads, nor width of bearing plus four times the wall thickness.

1922.6.6.2 Except as provided for in Section 1922.6.6.3, thickness of bearing walls shall not be less than $1/24$ the unsupported height or length, whichever is shorter, nor less than $5^{1}/_{2}$ inches (140 mm).

1922.6.6.3 Thickness of exterior basement walls and foundation walls shall not be less than $7^{1}/_{2}$ inches (191 mm).

1922.6.6.4 Walls shall be braced against lateral translation. See Sections 1924.3 and 1922.4.7.

---

*Equations for nominal flexural and shear strengths apply for normal concrete; for lightweight aggregate concrete, one of the following modifications shall apply:

1. When f_{ct} is specified and concrete is proportioned in accordance with Section 1905.2, $f_{ct}/6.7$ shall be substituted for f'_c but the value of $f_{ct}/6.7$ shall not exceed.

2. When f_{ct} is not specified, the value for nominal flexural and shear shall be multiplied by 0.75 for "all-lightweight" concrete and by 0.85 for "sand-lightweight" concrete. Linear interpolation is permitted when partial sand replacement is used.

1922.6.6.5 Not less than two No. 5 bars shall be provided around all window and door openings. Such bars shall extend at least 24 inches (610 mm) beyond the corners of openings.

1922.7 Footings.

1922.7.1 Structural plain concrete footings shall be designed for factored loads and induced reactions in accordance with appropriate design requirements of this code and as provided in Sections 1922.7.2 through 1922.7.8.

1922.7.2 Base area of footing shall be determined from unfactored forces and moments transmitted by footing to soil and permissible soil pressure selected through principles of soil mechanics.

1922.7.3 Plain concrete shall not be used for footings on piles.

1922.7.4 Thickness of structural plain concrete footings shall not be less than 8 inches (203 mm). See Section 1922.4.8.

1922.7.5 Maximum factored moment shall be computed at critical sections located as follows:

1. At face of column, pedestal or wall for footing supporting a concrete column, pedestal or wall.

2. Halfway between middle and edge of wall, for footing supporting a masonry wall.

3. Halfway between face of column and edge of steel base plate, for footing supporting a column with steel base plate.

1922.7.6 Shear in plain concrete footing.

1922.7.6.1 Maximum factored shear shall be computed in accordance with Section 1922.7.6.2, with location of critical section measured from face of column, pedestal or wall for footing supporting a column, pedestal or wall. For footing supporting a column with steel base plates, the critical section shall be measured from location defined in Section 1922.7.5, Item 3.

1922.7.6.2 Shear strength of structural plain concrete footings in the vicinity of concentrated loads or reactions shall be governed by the more severe of two conditions:

1. Beam action for footing, with a critical section extending in a plane across the entire footing width and located at a distance h from face of concentrated load or reaction area. For this condition, the footing shall be designed in accordance with Formula (22-8).

2. Two-way action for footing, with a critical section perpendicular to plane of footing and located so that its perimeter b_o is a minimum, but need not approach closer than $h/2$ to perimeter of concentrated load or reaction area. For this condition, the footing shall be designed in accordance with Formula (22-9).

1922.7.7 Circular or regular polygon shaped concrete columns or pedestals shall be permitted to be treated as square members with the same area for location of critical sections for moment and shear.

1922.7.8 Factored bearing load on concrete at contact surface between supporting and supported member shall not exceed design bearing strength for either surface as given in Section 1922.5.5.

1922.8 Pedestals.

1922.8.1 Plain concrete pedestals shall be designed for vertical, lateral and other loads to which they are subjected.

1922.8.2 Ratio of unsupported height to average least lateral dimension of plain concrete pedestals shall not exceed 3.

1922.8.3 Maximum factored axial load applied to plain concrete pedestals shall not exceed design bearing strength given in Section 1922.5.5.

1922.9 Precast Members.

1922.9.1 Design of precast plain concrete members shall consider all loading conditions from initial fabrication to completion of the structure, including form removal, storage, transportation and erection.

1922.9.2 Limitations of Section 1922.2 apply to precast members of plain concrete not only to the final condition but also during fabrication, transportation and erection.

1922.9.3 Precast members shall be connected securely to transfer all lateral forces into a structural system capable of resisting such forces.

1922.9.4 Precast members shall be adequately braced and supported during erection to ensure proper alignment and structural integrity until permanent connections are completed.

1922.10 Seismic Requirements for Plain Concrete.

1922.10.1 General. *The design and construction of plain concrete components that resist seismic forces shall conform to the requirements of Section 1922, except as modified by this section.*

1922.10.2 Seismic Zones 0 and 1. *Structural plain concrete members located in Seismic Zones 0 and 1 shall be designed in accordance with the provisions of Sections 1922.1 through 1922.9.*

1922.10.3 Seismic Zones 2, 3 and 4. *Structural plain concrete members are not permitted in buildings located in Seismic Zones 2, 3 and 4.*

> **EXCEPTIONS:** *1. Footings for buildings of Group R, Division 3 or Group U, Division 1 Occupancy constructed in accordance with Table 18-1-C.*
>
> *2. Nonstructural slabs supported directly on the ground or by approved structural systems.*

Division III—DESIGN STANDARD FOR ANCHORAGE TO CONCRETE

SECTION 1923 — ANCHORAGE TO CONCRETE

1923.1 Service Load Design. *Bolts and headed stud anchors shall be solidly cast in concrete and the service load shear and tension shall not exceed the values set forth in Table 19-D.*

For combined tension and shear:

$$(P_s/P_t)^{5/3} + (V_s/V_t)^{5/3} \leq 1$$

WHERE:

P_s = *applied service tension load.*

P_t = *Table 19-D service tension load.*

V_s = *applied service shear load.*

V_t = *Table 19-D service shear load.*

1923.2 Strength Design. *The factored loads on embedded anchor bolts and headed studs shall not exceed the design strengths determined by Section 1923.3.*

In addition to the load factors in Section 1909.2, a multiplier of 2 shall be used if special inspection is not provided, or of 1.3 if it is provided. When anchors are embedded in the tension zone of a member, the load factors in Section 1909.2 shall have a multiplier of 3 if special inspection is not provided, or of 2 if it is provided.

1923.3 Strength of Anchors.

1923.3.1 General. *The strength of headed bolts and headed studs solidly cast in concrete shall be taken as the average of 10 tests approved by the building official for each concrete strength and anchor size. Alternatively, the strength of the anchor shall be calculated in accordance with Sections 1923.3.2 through 1923.3.4. The bearing area of headed anchors shall be at least one and one-half times the shank area.*

1923.3.2 Design strength in tension. *The design strength of anchors in tension shall be the minimum of P_{ss} or $\phi\, P_c$ where:*

$$P_{ss} = 0.9\, A_b\, f_{ut}$$

and for an anchor group where the distance between anchors is less than twice their embedment length or for a single anchor or anchor group where the distance between anchors is equal to or greater than twice their embedment length

$$\phi\, P_c = \phi\lambda\, 4\, A_p\, \sqrt{f'_c}$$

For **SI:** $\phi\, P_c = 0.32\, \phi\lambda\, A_p\, \sqrt{f'_c}$

WHERE:

A_b = *area [in square inches (mm²)] of anchor. Must be used with the corresponding steel properties to determine the weakest part of the assembly in tension.*

A_p = *the effective area [in square inches (mm²)] of the projection of an assumed concrete failure surface upon the surface from which the anchor protrudes. For a single anchor or for an anchor group where the distance between anchors is equal to or greater than twice their embedment length, the surface is assumed to be that of a truncated cone radiating at a 45-degree slope from the bearing edge of the anchor toward the surface from which the anchor protrudes. The effective area is the projection of the cone on this surface. For an anchor which is perpendicular to the surface from which it protrudes, the effective area is a circle.*

For an anchor group where the distance between anchors is less than twice their embedment length, the failure surface is assumed to be that of a truncated pyramid radiating at a 45-degree slope from the bearing

edge of the anchor group toward the surface from which the anchors protrudes. The effective area is the projection of this truncated pyramid on this surface. In addition, for thin sections with anchor groups, the failure surface shall be assumed to follow the extension of this slope through to the far side rather than be truncated, and the failure mode resulting in the lower value of ϕP_c shall control.

d_b = *anchor shank diameter.*

f'_c = *specified compression strength of concrete, which shall not be taken as greater than 6,000 psi (41.37 MPa) for design.*

f_{ut} = *minimum specified tensile strength [in psi (MPa)] of the anchor. May be assumed to be 60,000 psi (413.7 MPa) for A 307 bolts or A 108 studs.*

P_c = *design tensile strength [in pounds (MPa)].*

P_u = *required tensile strength from factored loads, pounds (N).*

V_c = *design shear strength [in pounds (MPa)].*

V_u = *required shear strength from factored loads, pounds (N).*

λ = *1 for normal-weight concrete, 0.75 for "all lightweight" concrete, and 0.85 for "sand-lightweight" concrete.*

ϕ = *strength reduction factor*

= *0.65.*

EXCEPTION: *When the anchor is attached to or hooked around reinforcing steel or otherwise terminated to effectively transfer forces to the reinforcing steel that is designed to distribute forces and avert sudden local failure, ϕ may be taken as 0.85.*

Where edge distance is less than embedment length, reduce ϕP_c proportionally. For multiple edge distances less than embedment length, use multiple reductions.

1923.3.3 Design strength in shear. *The design strength of anchors in shear shall be the minimum of V_{ss} or $\phi\, V_c$ where:*

$$V_{ss} = 0.75\, A_b\, f_{ut}$$

and where loaded toward an edge greater than 10 diameters away,

$$\phi\, V_c = \phi\, 800\, A_b\, \lambda\, \sqrt{f'_c}$$

For **SI:** $\phi V_c = 66.4\, \phi A_b\, \lambda\, \sqrt{f'_c}$

or where loaded toward an edge equal to or less than 10 diameters away,

$$\phi V_c = \phi\, 2\pi d_e{}^2\, \lambda\, \sqrt{f'_c}$$

For **SI:** $\phi\, V_c = 0.166\, \phi\pi d_e{}^2\, \lambda\, \sqrt{f'_c}$

where d_e equals the edge distance from the anchor axis to the free edge.

For groups of anchors, the concrete design shear strength shall be taken as the smallest of:

1. The design strength of the weakest anchor times the number of anchors,

2. The design strength of the row of anchors nearest the free edge in the direction of shear times the number of rows or

3. The design strength of the row farthest from the free edge in the direction of shear.

For shear loading toward an edge equal to or less than 10 diameters away, or tension or shear not toward an edge less than five diameters away, reinforcing sufficient to carry the load shall be provided to prevent failure of the concrete in tension. In no case shall the edge distance be less than four diameters.

1923.3.4 Combined tension and shear. *When tension and shear act simultaneously, all of the following shall be met:*

$$\frac{1}{\phi}\left(\frac{P_u}{P_c}\right) \leq 1 \qquad \frac{1}{\phi}\left(\frac{V_u}{V_c}\right) \leq 1$$

$$\frac{1}{\phi}\left[\left(\frac{P_u}{P_c}\right)^{5/3} + \left(\frac{V_u}{V_c}\right)^{5/3}\right] \leq 1 \qquad \left(\frac{P_u}{P_{ss}}\right)^2 + \left(\frac{V_u}{V_{ss}}\right)^2 \leq 1$$

Division IV—DESIGN AND CONSTRUCTION STANDARD FOR SHOTCRETE

SECTION 1924 — SHOTCRETE

1924.1 General. *Shotcrete shall be defined as mortar or concrete pneumatically projected at high velocity onto a surface. Except as specified in this section, shotcrete shall conform to the regulations of this chapter for plain concrete or reinforced concrete.*

1924.2 Proportions and Materials. *Shotcrete proportions shall be selected that allow suitable placement procedures using the delivery equipment selected and shall result in finished in-place hardened shotcrete meeting the strength requirements of this code.*

1924.3 Aggregate. *Coarse aggregate, if used, shall not exceed $^3/_4$ inch (19 mm).*

1924.4 Reinforcement. *The maximum size of reinforcement shall be No. 5 bars unless it can be demonstrated by preconstruction tests that adequate encasement of larger bars can be achieved. When No. 5 or smaller bars are used, there shall be a minimum clearance between parallel reinforcement bars of $2^1/_2$ inches (64 mm). When bars larger than No. 5 are permitted, there shall be a minimum clearance between parallel bars equal to six diameters of the bars used. When two curtains of steel are provided, the curtain nearest the nozzle shall have a minimum spacing equal to 12 bar diameters and the remaining curtain shall have a minimum spacing of six bar diameters.*

> **EXCEPTION:** *Subject to the approval of the building official, reduced clearances may be used where it can be demonstrated by preconstruction tests that adequate encasement of the bars used in the design can be achieved.*

Lap splices in reinforcing bars shall be by the noncontact lap splice method with at least 2 inches (51 mm) clearance between bars. The building official may permit the use of contact lap splices when necessary for the support of the reinforcing provided it can be demonstrated by means of preconstruction testing, that adequate encasement of the bars at the splice can be achieved, and provided that the splices are placed so that a line through the center of the two spliced bars is perpendicular to the surface of the shotcrete work.

Shotcrete shall not be applied to spirally tied columns.

1924.5 Preconstruction Tests. *When required by the building official a test panel shall be shot, cured, cored or sawn, examined and tested prior to commencement of the project. The sample panel shall be representative of the project and simulate job conditions as closely as possible. The panel thickness and reinforcing shall reproduce the thickest and most congested area specified in the structural design. It shall be shot at the same angle, using the same nozzleman and with the same concrete mix design that will be used on the project.*

1924.6 Rebound. *Any rebound or accumulated loose aggregate shall be removed from the surfaces to be covered prior to placing the initial or any succeeding layers of shotcrete. Rebound shall not be reused as aggregate.*

1924.7 Joints. *Except where permitted herein, unfinished work shall not be allowed to stand for more than 30 minutes unless all edges are sloped to a thin edge. Before placing additional material adjacent to previously applied work, sloping and square edges shall be cleaned and wetted.*

1924.8 Damage. *In-place shotcrete which exhibits sags or sloughs, segregation, honeycombing, sand pockets or other obvious defects shall be removed and replaced. Shotcrete above sags and sloughs shall be removed and replaced while still plastic.*

1924.9 Curing. *During the curing periods specified herein, shotcrete shall be maintained above 40°F (4.4°C) and in moist condition. In initial curing, shotcrete shall be kept continuously moist for 24 hours after placement is complete. Final curing shall continue for seven days after shotcreting, for three days if high-early-strength cement is used, or until the specified strength is obtained. Final curing shall consist of a fog spray or an approved moisture-retaining cover or membrane. In sections of a depth in excess of 12 inches (305 mm), final curing shall be the same as that for initial curing.*

1924.10 Strength Test. *Strength test for shotcrete shall be made by an approved agency on specimens which are representative of work and which have been water soaked for at least 24 hours prior to testing. When the maximum size aggregate is larger than $^3/_8$ inch (9.5 mm), specimens shall consist of not less than three 3-inch-diameter (76 mm) cores or 3-inch (76 mm) cubes. When the maximum size aggregate is $^3/_8$ inch (9.5 mm) or smaller, specimens shall consist of not less than three 2-inch-diameter (51 mm) cores or 2-inch (51 mm) cubes. Specimens shall be taken in accordance with one of the following:*

1. From the in-place work: taken at least once each shift or less than one for each 50 cubic yards (38.2 m^3) of shotcrete; or

2. From test panels: made not less than once each shift or not less than one for each 50 cubic yards (38.2 m^3) of shotcrete placed. When the maximum size aggregate is larger than $^3/_8$ inch (9.5 mm), the test panels shall have a minimum dimension of 18 inches by 18 inches (457 mm by 457 mm). When the maximum size aggregate is $^3/_8$ inch (9.5 mm) or smaller, the test panels shall have a minimum dimension of 12 inches by 12 inches (305 mm by 305 mm). Panels shall be gunned in the same position as the work, during the course of the work and by nozzlepersons doing the work. The condition under which the panels are cured shall be the same as the work.

The average of three cores from a single panel shall be equal to or exceed $0.85 f'_c$ with no single core less than $0.75 f'_c$. The average of three cubes taken from a single panel must equal or exceed f'_c with no individual cube less than $0.88 f'_c$. To check testing accuracy, locations represented by erratic core strengths may be retested.

1924.11 Inspections.

1924.11.1 During placement. *When shotcrete is used for structural members, a special inspector is required by Section 1701.5, Item 12. The special inspector shall provide continuous inspection of the placement of the reinforcement and shotcreting and shall submit a statement indicating compliance with the plans and specifications.*

1924.11.2 Visual examination for structural soundness of in-place shotcrete. *Completed shotcrete work shall be checked visually for reinforcing bar embedment, voids, rock pockets, sand streaks and similar deficiencies by examining a minimum of three 3-inch (76 mm) cores taken from three areas chosen by the design engineer which represent the worst congestion of reinforcing bars occurring in the project. Extra reinforcing bars may be added to noncongested areas and cores may be taken from these areas. The cores shall be examined by the special inspector and a report submitted to the building official prior to final approval of the shotcrete.*

> **EXCEPTION:** *Shotcrete work fully supported on earth, minor repairs, and when, in the opinion of the building official, no special hazard exists.*

1924.12 Equipment. *The equipment used in preconstruction testing shall be the same equipment used in the work requiring such testing, unless substitute equipment is approved by the building official.*

Division V—DESIGN STANDARD FOR REINFORCED GYPSUM CONCRETE

SECTION 1925 — REINFORCED GYPSUM CONCRETE

1925.1 General. *Reinforced gypsum concrete shall conform to UBC Standard 19-2.*

Reinforced gypsum concrete shall develop the minimum ultimate compressive strength in pounds per square inch (MPa) set forth in Table 19-E when dried to constant weight, with tests made on cylinders 2 inches (51 mm) in diameter and 4 inches (102 mm) long or on 2-inch (51 mm) cubes.

For special inspection, see Section 1701.

1925.2 Design. *The minimum thickness of reinforced gypsum concrete shall be 2 inches (51 mm) except the thickness may be reduced to $1^1/_2$ inches (38 mm), provided all of the following conditions are satisfied:*

1. The overall thickness, including the formboard, is not less than 2 inches (51 mm).

2. The clear span of the gypsum concrete between supports does not exceed 2 feet 9 inches (838 mm).

3. Diaphragm action is not required.

4. The design live load does not exceed 40 pounds per square foot (195 kg/m^2).

1925.3 Stresses. *The maximum allowable unit working stresses in reinforced gypsum concrete shall not exceed the values set forth in Table 19-F except as specified in Chapter 16. Bolt values shall not exceed those set forth in Table 19-G.*

Allowable shear in poured-in-place reinforced gypsum concrete diaphragms using standard hot-rolled bulb tee subpurlins shall be determined by UBC Standard 19-2. (See Table 19-2-A in the standard for values for commonly used roof systems.)

Division VI—ALTERNATE DESIGN METHOD

SECTION 1926 — ALTERNATE DESIGN METHOD

1926.0 Notations. The following symbols and notations apply only to the provisions of this section:

A_g = gross area of section, square inches (mm²).

A_1 = loaded area.

A_2 = maximum area of the portion of the supporting surface that is geometrically similar to and concentric with the loaded area.

A_v = area of shear reinforcement within a distance s, square inches (mm²).

b = width of compression face of member, inches (mm).

b_o = perimeter of critical section for slabs and footings, inches (mm).

b_w = web width, or diameter of circular section, inches (mm).

d = distance from extreme compression fiber to centroid of tension reinforcement, inches (mm).

E_c = modulus of elasticity of concrete, psi (MPa). See Section 1908.5.1.

E_s = modulus of elasticity of reinforcement, psi (MPa). See Section 1908.5.2.

f'_c = specified compressive strength of concrete, psi (MPa). See Section 1905.

$\sqrt{f'_c}$ = square root of specified compressive strength of concrete, psi (MPa).

f_{ct} = average splitting tensile strength of lightweight aggregate concrete, psi (MPa). See Section 1905.1.4.

f_s = permissible tensile stress in reinforcement, psi (MPa).

f_y = specified yield strength of reinforcement, psi (MPa).

M = design moment.

N = design axial load normal to cross section occurring simultaneously with V; to be taken as positive for compression, negative for tension and to include effects of tension due to creep and shrinkage.

n = modular ratio of elasticity.

   = E_s/E_c.

s = spacing of shear reinforcement in direction parallel to longitudinal reinforcement, inches (mm).

V = design shear force at section.

v = design shear stress.

v_c = permissible shear stress carried by concrete, psi (MPa).

v_h = permissible horizontal shear stress, psi (MPa).

α = angle between inclined stirrups and longitudinal axis of member.

β_c = ratio of long side to short side of concentrated load or reaction area.

ρ = ratio of tension reinforcement.

   = A_s/bd.

ϕ = strength-reduction factor. See Section 1926.2.1.

1926.1 Scope.

1926.1.1 Nonprestressed reinforced concrete members shall be permitted to be designed using service loads (without load factors) and permissible service load stresses in accordance with provisions of this section.

1926.1.2 For design of members not covered by this section, appropriate provisions of this code shall apply.

1926.1.3 All applicable provisions of this code for nonprestressed concrete, except Section 1908.4, shall apply to members designed by the alternate design method.

1926.1.4 Flexural members shall meet requirements for deflection control in Section 1909.5 and requirements of Sections 1910.4 through 1910.7 of this code.

1926.2 General.

1926.2.1 Load factors and strength-reduction factors ϕ shall be taken as unity for members designed by the alternate design method.

1926.2.2 It shall be permitted to proportion members for 75 percent of capacities required by other parts of the section when considering wind or earthquake forces combined with other loads, provided the resulting section is not less than that required for the combination of dead and live load.

1926.2.3 When dead load reduces effects of other loads, members shall be designed for 85 percent of dead load in combination with the other loads.

1926.3 Permissible Service Load Stresses.

1926.3.1 Stresses in concrete shall not exceed the following:

1. Flexure.
   Extreme fiber stress in compression .... $0.45 f'_c$
2. Shear.†
   Beams and one-way slabs and footings:
   Shear carried by concrete, v_c .... $1.1\sqrt{f'_c}$
   (For **SI**: $0.09\sqrt{f'_c}$)

   Maximum shear carried by
   concrete plus shear reinforcement .... $v_c + 4.4\sqrt{f'_c}$
   (For **SI**: $v_c + 0.37\sqrt{f'_c}$)

   Joists.*
   Shear carried by concrete, v_c .... $1.2\sqrt{f'_c}$
   (For **SI**: $0.10\sqrt{f'_c}$)

†For more detailed calculation of shear stress carried by concrete v_c and shear values for lightweight aggregate concrete, see Section 1926.7.4.
*Designed in accordance with Section 1908.11.

   Two-way slabs and footings:
   Shear carried by concrete, v_c‡ .... $(1 + 2/\beta_c)\sqrt{f'_c}$
   [For **SI**: $(1 + 2/\beta_c)0.08\sqrt{f'_c}$]
   .... but not greater than $2\sqrt{f'_c}$
   (For **SI**: $0.166\sqrt{f'_c}$)

3. Bearing on loaded area** .... $0.3 f'_c$

‡If shear reinforcement is provided, see Sections 1926.7.7.4 and 1926.7.7.5.
**When the supporting surface is wider on all sides than the loaded area, permissible bearing stress on the loaded area shall be permitted to be increased by $\sqrt{A_2/A_1}$ but not more than 2. When the supporting surface is sloped or stepped, A_2 shall be permitted to be taken as the area of the lower base of the largest frustum of a right pyramid or cone contained wholly within the support and having for its upper base the loaded area and having side slopes of 1 vertical to 2 horizontal.

1926.3.2 Tensile stress in reinforcement f_s shall not exceed the following:

1. Grade 40 or Grade 50 reinforcement .... 20,000 psi
   (137.9 MPa)

2. Grade 60 reinforcement or greater and
welded wire fabric (smoothed or
deformed) . . . . 24,000 psi
(165.5 MPa)

3. For flexural reinforcement, $^3/_8$ inch (9.5 mm) or
less in diameter, in one-way slabs of not more
than 12-foot (3658 mm) span, but not greater
than 30,000 psi (206.8 MPa) . . . . 0.50 f_y

1926.4 Development and Splices of Reinforcement.

1926.4.1 Development and splices of reinforcement shall be as
required in Section 1912.

1926.4.2 In satisfying requirements of Section 1912.11.3, M_n
shall be taken as computed moment capacity assuming all positive
moment tension reinforcement at the section to be stressed to the
permissible tensile stress f_s, and V_u shall be taken as unfactored
shear force at the section.

1926.5 Flexure. For investigation of stresses at service loads,
straight-line theory (for flexure) shall be used with the following
assumptions:

1926.5.1 Strains vary linearly as the distance from the neutral
axis, except for deep flexural members with overall depth-span ratios greater than 2:5 for continuous spans and 4:5 for simple spans,
a nonlinear distribution of strain shall be considered. (See Section
1910.7.)

1926.5.2 Stress-strain relationship of concrete is a straight line
under service loads within permissible service load stresses.

1926.5.3 In reinforced concrete members, concrete resists no tension.

1926.5.4 It shall be permitted to take the modular ratio, $n = E_s/E_c$,
as the nearest whole number (but not less than 6). Except in calculations for deflections, value of n for lightweight concrete shall be
assumed to be the same as for normal-weight concrete of the same
strength.

1926.5.5 In doubly reinforced flexural members, an effective
modular ratio of $2 E_s/E_c$ shall be used to transform compression
reinforcement for stress computations. Compressive stress in such
reinforcement shall not exceed permissible tensile stress.

1926.6 Compression Members with or without Flexure.

1926.6.1 Combined flexure and axial load capacity of compression members shall be taken as 40 percent of that computed in accordance with provisions in Section 1910.

1926.6.2 Slenderness effects shall be included according to requirements of Sections 1910.10 and 1910.11. In Formulas (10-7)
and (10-8), the term P_u shall be replaced by 2.5 times the design
axial load, and ϕ shall be taken equal to 1.0.

1926.6.3 Walls shall be designed in accordance with Section
1914 with flexure and axial load capacities taken as 40 percent of
that computed using Section 1914. In Formula (14-1), ϕ shall be
taken equal to 1.0.

1926.7 Shear and Torsion.

1926.7.1 Design shear stress v shall be computed by:

$$v = \frac{V}{b_w d} \qquad (26\text{-}1)$$

where V is design shear force at section considered.

1926.7.2 When the reaction, in direction of applied shear, introduces compression into the end regions of a member, sections

located less than a distance d from face of support shall be permitted to be designed for the same shear v as that computed at a
distance d.

1926.7.3 Whenever applicable, effects of torsion, in accordance
with provisions of Section 1911, shall be added. Shear and torsional moment strengths provided by concrete and limiting maximum strengths for torsion shall be taken as 55 percent of the values
given in Section 1911.

1926.7.4 Shear stress carried by concrete.

1926.7.4.1 For members subject to shear and flexure only, shear
stress carried by concrete v_c shall not exceed $1.1 \sqrt{f'_c}$ (For **SI:**
$0.09 \sqrt{f'_c}$) unless a more detailed calculation is made in accordance with Section 1926.7.4.4.

1926.7.4.2 For members subject to axial compression, shear
stress carried by concrete v_c shall not exceed $1.1 \sqrt{f'_c}$ (For **SI:**
$0.09 \sqrt{f'_c}$) unless a more detailed calculation is made in accordance with Section 1926.7.4.5.

1926.7.4.3 For members subject to significant axial tension,
shear reinforcement shall be designed to carry total shear, unless a
more detailed calculation is made using

$$v_c = 1.1\left(1 + 0.004\frac{N}{A_g}\right)\sqrt{f'_c} \qquad (26\text{-}2)$$

For **SI:** $\qquad v_c = 0.09\left(1 + 0.004\frac{N}{A_g}\right)\sqrt{f'_c}$

where N is negative for tension. Quantity N/A_g shall be expressed
in psi (MPa).

1926.7.4.4 For members subject to shear and flexure only, v_c it
shall be permitted to compute by

$$v_c = \sqrt{f'_c} + 1,300\,\rho_w\frac{Vd}{M} \qquad (26\text{-}3)$$

For **SI:** $\qquad v_c = 0.083\sqrt{f'_c} + 9\,\rho_w\frac{Vd}{M}$

but v_c shall not exceed $1.9\sqrt{f'_c}$ (For **SI:** $0.16\sqrt{f'_c}$). Quantity Vd/M
shall not be taken greater than 1.0 where M is design moment oc-
curring simultaneously with V at section considered.

1926.7.4.5 For members subject to axial compression, v_c may be
computed by

$$v_c = 1.1\left(1 + 0.0006\frac{N}{A_g}\right)\sqrt{f'_c} \qquad (26\text{-}4)$$

For **SI:** $\qquad v_c = 0.09\left(1 + 0.0006\frac{N}{A_g}\right)\sqrt{f'_c}$

Quantity N/A_g shall be expressed in psi (MPa).

1926.7.4.6 Shear stresses carried by concrete v_c apply to normal-weight concrete. When lightweight aggregate concrete is
used, one of the following modifications shall apply:

1. When f_{ct} is specified and concrete is proportioned in accordance with Section 1904.2, $f_{ct}/6.7$ (For **SI:** $1.8 f_{ct}$) shall be substituted for $\sqrt{f'_c}$, but the value of $f_{ct}/6.7$ (For **SI:** $1.8 f_{ct}$) shall not
exceed $\sqrt{f'_c}$.

2. When f_{ct} is not specified, the value of $\sqrt{f'_c}$ (For **SI:**
$0.083\sqrt{f'_c}$) shall be multiplied by 0.75 for all-lightweight concrete and by 0.85 for sand-lightweight concrete. Linear interpolation may be applied when partial sand replacement is used.

1926.7.4.7 In determining shear stress by concrete v_c, whenever applicable, effects of axial tension due to creep and shrinkage in restrained members shall be included and it shall be permitted to include effects of inclined flexural compression in variable-depth members.

1926.7.5 Shear stress carried by shear reinforcement.

1926.7.5.1 Types of shear reinforcement. Shear reinforcement shall consist of the following:

1. Stirrups perpendicular to axis of member.

2. Welded wire fabric with wires located perpendicular to axis of member making an angle of 45 degrees or more with longitudinal tension reinforcement.

3. Longitudinal reinforcement with bent portion making an angle of 30 degrees or more with longitudinal tension reinforcement.

4. Combinations of stirrups and bent longitudinal reinforcement.

5. Spirals.

1926.7.5.2 Design yield strength of shear reinforcement shall not exceed 60,000 psi (413.7 MPa).

1926.7.5.3 Stirrups and other bars or wires used as shear reinforcement shall extend to a distance d from extreme compression fiber and shall be anchored at both ends according to Section 1912.13 to develop design yield strength of reinforcement.

1926.7.5.4 Spacing limits for shear reinforcement.

1926.7.5.4.1 Spacing of shear reinforcement placed perpendicular to axis of member shall not exceed $d/2$ or 24 inches (610 mm).

1926.7.5.4.2 Inclined stirrups and bent longitudinal reinforcement shall be so spaced that every 45-degree line, extending toward the reaction from middepth of member $d/2$ to longitudinal tension reinforcement, shall be crossed by at least one line of shear reinforcement.

1926.7.5.4.3 When $(v - v_c)$ exceeds $2 \sqrt{f'_c}$ (For **SI:** $0.166 \sqrt{f'_c}$) maximum spacing given by this subsection shall be reduced by one half.

1926.7.5.5 Minimum shear reinforcement.

1926.7.5.5.1 A minimum area of shear reinforcement shall be provided in all reinforced concrete flexural members where design shear stress v is greater than one half the permissible shear stress v_c carried by concrete, except the following:

1. Slab and footings.

2. Concrete joist construction defined by Section 1908.11 of this code.

3. Beams with total depth not greater than 10 inches (254 mm), two and one-half times thickness of flange or one half the width of web, whichever is greater.

1926.7.5.5.2 Minimum shear reinforcement requirements of this section may be waived if shown by test that required ultimate flexural and shear strength can be developed when shear reinforcement is omitted.

1926.7.5.5.3 Where shear reinforcement is required by this subsection or by analysis, minimum area of shear reinforcement shall be computed by

$$A_v = 50 \frac{b_w s}{f_y} \qquad (26\text{-}5)$$

For **SI:**
$$A_v = 0.34 \frac{b_w s}{f_y}$$

where b_w and s are in inches (mm).

1926.7.5.6 Design of shear reinforcement.

1926.7.5.6.1 Where design shear stress v exceeds shear stress carried by concrete v_c, shear reinforcement shall be provided in accordance with this subsection.

1926.7.5.6.2 When shear reinforcement perpendicular to axis of member is used,

$$A_v = \frac{(v - v_c)b_w s}{f_s} \qquad (26\text{-}6)$$

1926.7.5.6.3 When inclined stirrups are used as shear reinforcement,

$$A_v = \frac{(v - v_c)b_w s}{f_s (\sin \alpha + \cos \alpha)} \qquad (26\text{-}7)$$

1926.7.5.6.4 When shear reinforcement consists of a single bar or a single group of parallel bars, all bent up at the same distance from the support,

$$A_v = \frac{(v - v_c)b_w d}{f_s \sin \alpha} \qquad (26\text{-}8)$$

where $(v - v_c)$ shall not exceed $1.6 \sqrt{f'_c}$ (For **SI:** $0.13 \sqrt{f'_c}$).

1926.7.5.6.5 When shear reinforcement consists of a series of parallel bent-up bars or groups of parallel bent-up bars at different distances from the support, required area shall be computed by Formula (26-7).

1926.7.5.6.6 Only the center three fourths of the inclined portion of any longitudinal bent bar shall be considered effective for shear reinforcement.

1926.7.5.6.7 When more than one type of shear reinforcement is used to reinforce the same portion of a member, required area shall be computed as the sum of the various types separately. In such computations, v_c shall be included only once.

1926.7.5.6.8 Value of $(v - v_c)$ shall not exceed $4.4 \sqrt{f'_c}$ (For **SI:** $0.37 \sqrt{f'_c}$).

1926.7.6 Shear friction. Where it is appropriate to consider shear transfer across a given plane such as an existing or potential crack, an interface between dissimilar materials, or an interface between two concretes cast at different times, shear friction provisions of Section 1911.7 shall be permitted to be applied with limiting maximum stress for shear taken as 55 percent of that given in Section 1911.7.5. Permissible stress in shear friction reinforcement shall be that given in Section 1926.3.2.

1926.7.7 Special provisions for slabs and footings.

1926.7.7.1 Shear capacity of slabs and footings in the vicinity of concentrated loads or reactions is governed by the more severe of the following two conditions:

1926.7.7.1.1 Beam action for slab or footing with a critical section extending in a plane across the entire width and located at a distance d from face of concentrated load or reaction area. For this condition, the slab or footing shall be designed in accordance with Sections 1926.7.1 through 1926.7.5.

1926.7.7.1.2 Two way action for slab or footing with a critical section perpendicular to plane of slab and located so that its perimeter is a minimum but need not approach closer than $d/2$ to perimeter of concentrated load or reaction area. For this condition, the slab or footing shall be designed in accordance with Sections 1926.7.7.2 and 1926.7.7.3.

1926.7.7.2 Design shear stress v shall be computed by

$$v = \frac{V}{b_o d} \qquad (26\text{-}9)$$

where V and b_o shall be taken at the critical section defined in Section 1926.7.7.1.2.

1926.7.7.3 Design shear stress v shall not exceed v_c given by Formula (26-10) unless shear reinforcement is provided.

$$v_c = \left(1 + \frac{2}{\beta_c}\right)\sqrt{f'_c} \qquad (26\text{-}10)$$

For **SI:** $\qquad v_c = 0.083\left(1 + \frac{2}{\beta_c}\right)\sqrt{f'_c}$

but v_c shall not exceed $2\sqrt{f'_c}$ (For **SI:** $0.166\sqrt{f'_c}$). β_c is the ratio of long side to short side of concentrated load or reaction area. When lightweight aggregate concrete is used, the modifications of Section 1926.7.4.6 shall apply.

1926.7.7.4 If shear reinforcement consisting of bars or wires is provided in accordance with Section 1911.12.3, v_c shall not exceed $\sqrt{f'_c}$ (For **SI:** $0.083\sqrt{f'_c}$), and v shall not exceed $3\sqrt{f'_c}$ (For **SI:** $0.25\sqrt{f'_c}$).

1926.7.7.5 If shear reinforcement consisting of steel I or channel shapes (shearheads) is provided in accordance with Section 1911.12.4 of this code, v on the critical section defined in Section 1926.7.7.1.2 shall not exceed $3.5\sqrt{f'_c}$ (For **SI:** $0.29\sqrt{f'_c}$) and v on the critical section defined in Section 1911.12.4.7 shall not exceed $2\sqrt{f'_c}$ (For **SI:** $0.166\sqrt{f'_c}$). In Formulas (11-38) and (11-39), design shear force V shall be multiplied by 2 and substituted for V_u.

1926.7.8 Special provisions for other members. For design of deep flexural members, brackets and corbels and walls, the special provisions of Section 1911 shall be used with shear strengths provided by concrete and limiting maximum strengths for shear taken as 55 percent of the values given in Section 1911. In Section 1911.10.6, the design axial load shall be multiplied by 1.2 if compression and 2.0 if tension and substituted for N_u.

1926.7.9 Composite concrete flexural members. For design of composite concrete flexural members, permissible horizontal shear stress v_h shall not exceed 55 percent of the horizontal shear strengths given in Section 1917.5.2.

Division VII—UNIFIED DESIGN PROVISIONS
NOTE: This is a new division.

SECTION 1927 — UNIFIED DESIGN PROVISIONS FOR REINFORCED AND PRESTRESSED CONCRETE FLEXURAL AND COMPRESSION MEMBERS

B.1927.0

B.1927.1 Scope. Design for flexure and axial load by provisions of Section 1927.0 shall be permitted. When Section 1927.0 is used in design, all numbered sections in Section 1927.0 shall be used in place of the corresponding numbered sections in Sections 1908, 1909, 1910 and 1918. If any section in Section 1927.0 is used, all sections in Section 1927.0 shall be substituted for the corresponding sections in Chapter 19.*

B.1908.4 Redistribution of Negative Moments in Continuous Flexural Members.

B.1908.4.1 Except where approximate values for moments are used, it shall be permitted to increase or decrease negative moments calculated by elastic theory at supports of continuous flexural members for any assumed loading arrangement by not more than 1,000 percent ε_t, with a maximum of 20 percent.

B.1908.4.2 The modified negative moments shall be used for calculating moments at sections within the spans.

B.1908.4.3 Redistribution of negative moments shall be made only when ε_t is equal to or greater than 0.0075 at the section at which moment is reduced.

B.1909.2 Required Strength.

B.1909.2.1 Required strength U to resist dead load D and live load L shall be at least equal to

$$U = 1.4D + 1.7L \qquad \text{(B.9-1)}$$

B.1909.2.2 If resistance to structural effects of a specified wind load W are included in design, the following combinations of D, L and W shall be investigated to determine the greatest required strength U:

$$U = 0.75 (1.4D + 1.7L + 1.7W) \qquad \text{(B.9-2)}$$

where load combinations shall include both full value and zero value of L to determine the more severe condition, and

$$U = 0.9D + 1.3W \qquad \text{(B.9-3)}$$

but for any combination of D, L and W, required strength U shall not be less than Formula (B.9-1).

B.1909.2.3 If resistance to specified earthquake loads or forces E are included in design, load combinations of Section 1612.2.1 shall apply.

B.1909.2.4 If resistance to earth pressure H is included in design, required strength U shall be at least equal to

$$U = 1.4D + 1.7L + 1.7H \qquad \text{(B.9-4)}$$

except that where D or L reduce the effect of H, $0.9D$ shall be substituted for $1.4D$ and zero value of L shall be used to determine

the greatest required U. For any combination of D, L and H, required strength U shall not be less than Formula (B.9-1).

B.1909.2.5 If resistance to loadings due to weight and pressure of fluids with well-defined densities and controllable maximum heights F is included in design, such loading shall have a load factor of 1.4, and be added to all loading combinations that include live load.

B.1909.2.6 If resistance to impact effects is taken into account in design, such effects shall be included with live load L.

B.1909.2.7 Where structural effects T of differential settlement, creep, shrinkage, expansion and shrinkage-compensating concrete, or temperature change are significant in design, required strength U shall be at least equal to

$$U = 0.75 (1.4D + 1.4T + 1.7L) \qquad \text{(B.9-5)}$$

but required strength U shall not be less than

$$U = 1.4 (D + T) \qquad \text{(B.9-6)}$$

Estimations of differential settlement, creep, shrinkage, expansion of shrinkage-compensating concrete, or temperature change shall be based on a realistic assessment of such effects occurring in service.

B.1909.3 Design Strength.

B.1909.3.1 Design strength provided by a member, its connections to other members, and its cross sections, in terms of flexure, axial load, shear, and torsion, shall be taken as the nominal strength calculated in accordance with requirements and assumptions of this code, multiplied by a strength reduction factor ϕ.

B.1909.3.2 Strength reduction factor ϕ shall be as follows:

B.1909.3.2.1 Tension-controlled sections 0.90

B.1909.3.2.2 Compression-controlled sections:

1. Members with spiral reinforcement conforming to Section 1910.9.3 . . . . 0.75

2. Other reinforced members 0.70

For sections in which the net tensile strain in the extreme tension steel at nominal strength is between the limits for compression-controlled and tension-controlled sections, ϕ shall be linearly increased from that for compression-controlled sections to 0.90 as the net tensile strain in the extreme tension steel at nominal strength increases from the compression-controlled strain limit to 0.005. Alternatively, it shall be permitted to take ϕ as that for compression-controlled sections.

B.1909.3.2.3 Shear and torsion 0.85

B.1909.3.2.4 Bearing on concrete. See also Section 1918.13. . . . . 0.70

B.1910.3.2 Balanced strain conditions exist at a cross section when tension reinforcement reaches the strain corresponding to its specified yield strength f_y just as concrete in compression reaches its assumed strain limit of 0.003.

The compression-controlled strain limit is the net tensile strain in the reinforcement at balanced strain conditions. For prestressed

---

*When Section 1927.0 is used, each section of Section 1927.0 must be substituted for the corresponding section in Chapter 19. For instance, Section B.1908.4 is substituted for Section 1908.4, etc., through Section B.1918.10.4 being substituted for Section 1918.10.4. The corresponding commentary sections should also be substituted.

Section 1927.0 introduces substantial changes in design for flexure and axial loads to Chapter 19. Reinforcement limits, strength reduction factor and moment redistribution are affected. Designs using the provisions of Section 1927.0 satisfy Chapter 19, and are equally acceptable.

sections, it shall be permitted to use the same compression-controlled strain limit as that for reinforcement with a design yield strength f_y of 60,000 psi (413.7 MPa).

B.1910.3.3 Sections are compression-controlled when the net tensile strain in the extreme tension steel is equal to or less than the compression-controlled strain limit at the time the concrete in compression reaches its assumed strain limit of 0.003. Sections are tension-controlled when the net tensile strain in the extreme tension steel is equal to or greater than 0.005 just as the concrete in compression reaches its assumed strain limit of 0.003. Sections with net tensile strain in the extreme tension steel between the compression-controlled strain limit and 0.005 constitute a transition region between compression-controlled and tension-controlled sections.

B.1918.1.3 The following provisions of this code shall not apply to prestressed concrete, except as specifically noted: Sections 1907.6.5, 1908.10.2, 1908.10.3, 1908.10.4, 1908.11, 1910.5, 1910.6, 1910.9.1 and 1910.9.2; Section 1913; and Sections 1914.3, 1914.5 and 1914.6.

B.1918.8 Limits for Reinforcement of Flexural Members.

B.1918.8.1 Prestressed concrete sections shall be classified as tension-controlled and compression-controlled sections in accordance with Section B.1910.3.3. The appropriate ϕ-factors from Section B.1909.3.2 shall apply.

B.1918.8.2 Total amount of prestressed and nonprestressed reinforcement shall be adequate to develop a factored load at least 1.2 times the cracking load computed on the basis of the modulus of rupture specified in Section 1909.5.2.3, except for flexural members with shear and flexural strength at least twice that required by Section 1909.2.

B.1918.8.3 Part or all of the bonded reinforcement consisting of bars or tendons shall be provided as close as practicable to the extreme tension fiber in all prestressed flexural members, except that in members prestressed with unbonded tendons, the minimum bonded reinforcement consisting of bars or tendons shall be as required by Section 1918.9.

B.1918.10.4 Redistribution of negative moments in continuous prestressed flexural members.

B.1918.10.4.1 Where bonded reinforcement is provided at supports in accordance with Section 1918.9.2, it shall be permitted to increase or decrease negative moments calculated by elastic theory for any assumed loading, in accordance with Section B.1908.4.

B.1918.10.4.2 The modified negative moments shall be used for calculating moments at sections within spans for the same loading arrangement.

Division VIII—ALTERNATIVE LOAD-FACTOR COMBINATION AND STRENGTH REDUCTION FACTORS
NOTE: This is a new division.

SECTION 1928 — ALTERNATIVE LOAD-FACTOR COMBINATION AND STRENGTH REDUCTION FACTORS

1928.1 General. It shall be permitted to proportion concrete structural elements using the alternate load-factor combinations in Section 1928.1.2 in conjunction with the alternate strength reduction factors in Section 1928.1.1 if the structural framing includes primary members of other materials proportioned to satisfy the alternate load-factor combinations in Section 1928.1.2. Loads shall be determined in accordance with Chapter 16 of this code.

1928.1.1 Alternate strength reduction factors.

1928.1.1.1 Flexure, without axial load 0.80

1928.1.1.2 Axial tension and axial tension with flexure . 0.80

1928.1.1.3 Axial compression and axial compression with flexure

   1. Members with spiral reinforcement conforming to Section 1910.9.3 . . . . 0.70

   2. Other reinforced members 0.65

except that for low values of axial compression, it shall be permitted to increase ϕ toward the value for flexure, 0.80, using the linear interpolation provided in either Section 1909.3.2.2 or B.1909.3.2.2.

   3. In Seismic Zones 3 and 4, members resisting earthquake forces without transverse reinforcement conforming to 21.4.4 . . . . 0.50

1928.1.1.4 Shear and torsion 0.75

except that in Seismic Zones 3 and 4:

   1. Shear in members resisting earthquake forces if the nominal shear strength of the member is less than the nominal shear corresponding to the development of the nominal flexural strength of the member 0.55

   2. Shear in joints of building structures 0.80

1928.1.1.5 Bearing 0.65

1928.1.1.6 Plain concrete 0.55

1928.1.2 Alternate load-factor combinations.

1928.1.2.1 Symbols and notations.

D = dead load consisting of: (1) weight of the member, (2) weight of all materials of construction incorporated into the building to be permanently supported by the member, including built-in partitions, and (3) weight of permanent equipment.

E = earthquake load.

F = loads due to fluids with well-defined pressures and maximum heights.

H = loads due to the weight and lateral pressure of soil and water in soil.

L = live loads due to intended use and occupancy, including loads due to movable objects and movable partitions and loads temporarily supported by the structure during maintenance. L includes any permissible reduction. If resistance to impact loads is taken into account in design, such effects shall be included with the live load L.

L_r = roof live loads.

P = loads, forces and effects due to ponding.

R = rain loads, except ponding.

S = snow loads.

T = self-straining forces and effects arising from contraction or expansion resulting from temperature changes, shrinkage, moisture changes, creep in component materials, movement due to differential settlement or combinations thereof.

W = wind load.

1928.1.2.2 Combining loads using strength design.

1928.1.2.3 Basic combinations. When permitted by Section 1928.1, structures, components and foundations shall be designed so that their design strength exceeds the effects of the factored loads in the following combinations:

   1. $1.4D$

   2. $1.2D + 1.6L + 0.5(L_r$ or S or $R)$

   3. $1.2D + 1.6(L_r$ or S or $R) + (0.5L$ or $0.8W)$

   4. $1.2D + 1.3W + 0.5L + 0.5(L_r$ or S or $R)$

   5. $1.2D + 1.5E + (0.5L$ or $0.2S)$

   6. $0.9D - (1.3W$ or $1.5E)$

      EXCEPTIONS: 1. The load factor on L in combinations 3, 4 and 5 shall equal 1.0 for garages, areas occupied and places of public assembly, and all areas where the live load is greater than 100 lb./ft.^{2} (pounds-force per square foot) (4.79 kPa).

      2. Each relevant strength limit state shall be considered. The most unfavorable effect may occur when one or more of the contributing loads are not acting.

1928.1.2.4 Other combinations. The structural effects of F, H, P or T shall be considered in design as the following factored loads: $1.3F$, $1.6H$, $1.2P$ and $1.2T$.

TABLE 19-A-1—TOTAL AIR CONTENT FOR FROST-RESISTANT CONCRETE

NOMINAL MAXIMUM AGGREGATE SIZE (inches)	AIR CONTENT, PERCENTAGE	
× 25.4 for mm	Severe Exposure	Moderate Exposure
$3/8$	$7^1/_2$	6
$1/2$	7	$5^1/_2$
$3/4$	6	5
1	6	$4^1/_2$
$1^1/_2$	$5^1/_2$	$4^1/_4$
2^1	5	4
3^1	$4^1/_2$	$3^1/_2$

[1]These air contents apply to total mix, as for the preceding aggregate sizes. When testing this concrete, however, aggregate larger than $1^1/_2$ inches (38 mm) is removed by hand picking or sieving, and air content is determined on the minus $1^1/_2$-inch (38 mm) fraction.

TABLE 19-A-2—REQUIREMENTS FOR SPECIAL EXPOSURE CONDITIONS

EXPOSURE CONDITION	MAXIMUM WATER-CEMENTITIOUS MATERIALS RATIO, BY WEIGHT, NORMAL-WEIGHT AGGREGATE CONCRETE	MINIMUM f'_c, NORMAL-WEIGHT AND LIGHTWEIGHT AGGREGATE CONCRETE, psi
		× 0.00689 for MPa
Concrete intended to have low permeability when exposed to water	0.50	4,000
Concrete exposed to freezing and thawing in a moist condition or to deicing chemicals	0.45	4,500
For corrosion protection for reinforced concrete exposed to chlorides from deicing chemicals, salt, saltwater, brackish water, seawater or spray from these sources	0.40	5,000

TABLE 19-A-3—REQUIREMENTS FOR CONCRETE EXPOSED TO DEICING CHEMICALS

CEMENTITIOUS MATERIALS	MAXIMUM PERCENT OF TOTAL CEMENTITIOUS MATERIALS BY WEIGHT[1]
Fly ash or other pozzolans conforming to ASTM C 618	25
Slag conforming to ASTM C 989	50
Silica fume conforming to ASTM C 1240	10
Total of fly ash or other pozzolans, slag and silica fume	50[2]
Total of fly ash or other pozzolans and silica fume	35[2]

[1]The total cementitious materials also includes ASTM C 150, C 595 and C 845 cement.

[2]Fly ash or other pozzolans and silica fume shall constitute no more than 25 and 10 percent, respectively, of the total weight of the cementitious materials. The maximum percentages above shall include:
1. Fly ash or other pozzolans present in Type IP or I(PM) blended cement in accordance with ASTM C 595.
2. Slag used in the manufacture of a IS or I(SM) blended cement in accordance with ASTM C 595.
3. Silica fume, ASTM C 1240, present in a blended cement.

TABLE 19-A-4—REQUIREMENTS FOR CONCRETE EXPOSED TO SULFATE-CONTAINING SOLUTIONS

SULFATE EXPOSURE	WATER- SOLUBLE SULFATE (SO$_4$) IN SOIL, PERCENTAGE BY WEIGHT	SULFATE (SO$_4$) IN WATER, ppm	CEMENT TYPE	MAXIMUM WATER-CEMENTITIOUS MATERIALS RATIO, BY WEIGHT, NORMAL-WEIGHT AGGREGATE CONCRETE[1]	MINIMUM f'_c, NORMAL-WEIGHT AND LIGHTWEIGHT AGGREGATE CONCRETE, psi
					× 0.00689 for MPa
Negligible	0.00-0.10	0-150	—	—	—
Moderate[2]	0.10-0.20	150-1,500	II, IP(MS), IS (MS)	0.50	4,000
Severe	0.20-2.00	1,500-10,000	V	0.45	4,500
Very severe	Over 2.00	Over 10,000	V plus pozzolan[3]	0.45	4,500

[1]A lower water-cementitious materials ratio or higher strength may be required for low permeability or for protection against corrosion of embedded items or freezing and thawing (Table 19-A-2).

[2]Seawater.

[3]Pozzolan that has been determined by test or service record to improve sulfate resistance when used in concrete containing Type V cement.

TABLE 19-A-5
TABLE 19-C-1

1997 UNIFORM BUILDING CODE

TABLE 19-A-5—MAXIMUM CHLORIDE ION CONTENT FOR CORROSION PROTECTION REINFORCEMENT

TYPE OF MEMBER	MAXIMUM WATER-SOLUBLE CHLORIDE ION (Cl) IN CONCRETE, PERCENTAGE BY WEIGHT OF CEMENTITIOUS MATERIALS
Prestressed concrete	0.06
Reinforced concrete exposed to chloride in service	0.15
Reinforced concrete that will be dry or protected from moisture in service	1.00
Other reinforced concrete construction	0.30

TABLE 19-A-6—MODIFICATION FACTOR FOR STANDARD DEVIATION WHEN LESS THAN 30 TESTS ARE AVAILABLE

NUMBER OF TESTS[1]	MODIFICATION FACTOR FOR STANDARD DEVIATION[2]
Less than 15	Use Table 19-A-7
15	1.16
20	1.08
25	1.03
30 or more	1.00

[1]Interpolate for intermediate numbers of tests.
[2]Modified standard deviation to be used to determine required average strength f'_{cr} from Section 1905.3.2.1.

TABLE 19-A-7—REQUIRED AVERAGE COMPRESSIVE STRENGTH WHEN DATA ARE NOT AVAILABLE TO ESTABLISH A STANDARD DEVIATION

SPECIFIED COMPRESSIVE STRENGTH f'_c psi	REQUIRED AVERAGE COMPRESSIVE STRENGTH f'_{cr} psi
$\times$ 0.00689 for MPa	
Less than 3,000 psi	$f'_c + 1,000$
3,000 to 5,000	$f'_c + 1,200$
Over 5,000	$f'_c + 1,400$

TABLE 19-B—MINIMUM DIAMETERS OF BEND

BAR SIZE	MINIMUM DIAMETER
Nos. 3 through 8	$6d_b$
Nos. 9, 10 and 11	$8d_b$
Nos. 14 and 18	$10d_b$

TABLE 19-C-1—MINIMUM THICKNESS OF NONPRESTRESSED BEAMS OR ONE-WAY SLABS UNLESS DEFLECTIONS ARE COMPUTED[1]

MEMBER	MINIMUM THICKNESS, h			
	Simply Supported	One End Continuous	Both Ends Continuous	Cantilever
	Members not supporting or attached to partitions or other construction likely to be damaged by large deflections			
Solid one-way slabs	$l/20$	$l/24$	$l/28$	$l/10$
Beams or ribbed one-way slabs	$l/16$	$l/18.5$	$l/21$	$l/8$

[1]Span length l is in inches.
Values given shall be used directly for members with normal-weight concrete [w_c = 145 pcf (2323 kg/m³)] and Grade 60 reinforcement. For other conditions, the values shall be modified as follows:
   (a)   For structural lightweight concrete having unit weights in the range 90 to 120 pounds per cubic foot (1442 to 1922 kg/m³), the value shall be multiplied by (1.65 − 0.005 w_c) (For SI: 1.65 − 0.0003 w_c) but not less than 1.09, where w_c is the unit weight in pounds per cubic foot (kg/m³).
   (b)   For f_y other than 60,000 psi (413.7 MPa), the values shall be multiplied by (0.4 + f_y/100,000) (For SI: 0.4 + f_y/689.5).

TABLE 19-C-2—MAXIMUM PERMISSIBLE COMPUTED DEFLECTIONS

TYPE OF MEMBER	DEFLECTION TO BE CONSIDERED	DEFLECTION LIMITATION
Flat roofs not supporting or attached to nonstructural elements likely to be damaged by large deflections	Immediate deflection due to live load L	$\dfrac{\ell^1}{180}$
Floors not supporting or attached to nonstructural elements likely to be damaged by large deflections	Immediate deflection due to live load L	$\dfrac{\ell}{360}$
Roof or floor construction supporting or attached to nonstructural elements likely to be damaged by large deflections	That part of the total deflection occurring after attachment of nonstructural elements (sum of the long-time deflection due to all sustained loads and the immediate deflection due to any additional live loads)[3]	$\dfrac{\ell^2}{480}$
Roof or floor construction supporting or attached to nonstructural elements likely to not be damaged by large deflections		$\dfrac{\ell^4}{240}$

[1]The limit is not intended to safeguard against ponding. The member shall be checked for ponding by suitable calculations of deflection, including added deflections due to ponded water, and considering long-term effects of all sustained loads, camber, construction tolerances, and reliability of provisions for drainage.
[2]The limit may be exceeded if adequate measures are taken to prevent damage to supported or attached elements.
[3]Long-time deflection shall be determined in accordance with Section 1909.5.2.5 or 1909.5.4.2, but may be reduced by the amount of deflection calculated to occur before attachment of nonstructural elements. This amount shall be determined on basis of accepted engineering data relating to time-deflection characteristics of members similar to those being considered.
[4]But not greater than tolerance provided for nonstructural elements. The limits may be exceeded if camber is provided so that total deflection minus camber does not exceed limit.

TABLE 19-C-3—MINIMUM THICKNESS OF SLABS WITHOUT INTERIOR BEAMS

YIELD STRENGTH, f_y, psi*	WITHOUT DROP PANELS*			WITH DROP PANELS[1]		
	Exterior Panels			Exterior Panels		
× 0.00689 for MPa	Without edge beams	With edge beams[2]	Interior panels	Without edge beams	With edge beams[2]	Interior panels
40,000	$\dfrac{l_n}{33}$	$\dfrac{l_n}{36}$	$\dfrac{l_n}{36}$	$\dfrac{l_n}{36}$	$\dfrac{l_n}{40}$	$\dfrac{l_n}{40}$
60,000	$\dfrac{l_n}{30}$	$\dfrac{l_n}{33}$	$\dfrac{l_n}{33}$	$\dfrac{l_n}{33}$	$\dfrac{l_n}{36}$	$\dfrac{l_n}{36}$
75,000	$\dfrac{l_n}{28}$	$\dfrac{l_n}{31}$	$\dfrac{l_n}{31}$	$\dfrac{l_n}{31}$	$\dfrac{l_n}{34}$	$\dfrac{l_n}{34}$

*For values of reinforcement yield strength between the values given in the table, minimum thickness shall be determined by linear interpolation.
[1]Drop panel is defined in Section 1913.3.7.
[2]Slabs with beams between columns along exterior edges. The value of α for the edge beam shall not be less than 0.8.

TABLE 19-D—ALLOWABLE SERVICE LOAD ON EMBEDDED BOLTS (Pounds) (Newtons)[1,2,3]

BOLT DIAMETER (inches)	MINIMUM[4] EMBEDMENT (inches)	EDGE DISTANCE (inches)	SPACING (inches)	MINIMUM CONCRETE STRENGTH (psi)					
				× 0.00689 for MPa					
				$f'_c = 2,000$		$f'_c = 3,000$		$f'_c = 4,000$	
				Tension[5]	Shear[6]	Tension[5]	Shear[6]	Tension[5]	Shear[6]
	× 25.4 for mm			× 4.5 for newtons					
$1/4$	$2^1/2$	$1^1/2$	3	200	500	200	500	200	500
$3/8$	3	$2^1/4$	$4^1/2$	500	1,100	500	1,100	500	1,100
$1/2$	4 4	3 5	6 6	950 1,400	1,250 1,550	950 1,500	1,250 1,650	950 1,550	1,250 1,750
$5/8$	$4^1/2$ $4^1/2$	$3^3/4$ $6^1/4$	$7^1/2$ $7^1/2$	1,500 2,050	2,750 2,900	1,500 2,200	2,750 3,000	1,500 2,400	2,750 3,050
$3/4$	5 5	$4^1/2$ $7^1/2$	9 9	2,250 2,700	2,940 4,250	2,250 2,950	3,560 4,300	2,250 3,200	3,560 4,400
$7/8$	6	$5^1/4$	$10^1/2$	2,550	3,350	2,550	4,050	2,550	4,050
1	7	6	12	2,850	3,750	3,250	4,500	3,650	5,300
$1^1/8$	8	$6^3/4$	$13^1/2$	3,400	4,750	3,400	4,750	3,400	4,750
$1^1/4$	9	$7^1/2$	15	4,000	5,800	4,000	5,800	4,000	5,800

[1]Values are natural stone aggregate concrete and bolts of at least A 307 quality. Bolts shall have a standard head or an equal deformity in the embedded portion.
[2]The tabulated values are for anchors installed at the specified spacing and edge distances. Such spacing and edge distance may be reduced 50 percent with an equal reduction in value. Use linear interpolation for intermediate spacings and edge margins.
[3]The allowable values may be increased per Section 1612.3 for duration of loads such as wind or seismic forces.
[4]An additional 2 inches (51 mm) of embedment shall be provided for anchor bolts located in the top of columns located in Seismic Zones 2, 3 and 4.
[5]Values shown are for work without special inspection. Where special inspection is provided, values may be increased 100 percent.
[6]Values shown are for work with or without special inspection.

TABLE 19-E
TABLE 19-G

1997 UNIFORM BUILDING CODE

TABLE 19-E—MINIMUM COMPRESSIVE STRENGTH AND MODULUS OF ELASTICITY AND OF RIGIDITY OF REINFORCED GYPSUM CONCRETE

CLASS	COMPRESSIVE STRENGTH psi (f_g)	MODULUS OF ELASTICITY psi (E)	E_s/E_g (n)	MODULUS OF RIGIDITY (G)
	× 0.00689 for MPa			
A	500	200,000	150	.36E
B	1,000	600,000	50	.40E

TABLE 19-F—ALLOWABLE UNIT WORKING STRESS REINFORCED GYPSUM CONCRETE

TYPE OF STRESS	FACTOR	CLASS A	CLASS B
		(pounds per square inch)	
		× 0.00689 for MPa	
Flexural compression	.25f_g	125	250
Axial compression or bearing	.20f_g	100	200
Bond for plain bars and shear[1]	.02f_g	10	20
Bond for deformed bars and electrically welded wire mesh[1]	.03f_g	15	30

[1]Electrically welded wire mesh reinforcement shall be considered as meeting the bond and shear requirements of this section. In no case shall the area of principal reinforcement be less than 0.26 square inch per foot (550 mm²/m) of slab width.

TABLE 19-G—SHEAR ON ANCHOR BOLTS AND DOWELS—REINFORCED GYPSUM CONCRETE[1]

BOLT OR DOWEL SIZE (inches)	EMBEDMENT (inches)	SHEAR[2] (inches)
	× 25.4 for mm	
$^3/_8$ bolt	4	325
$^1/_2$ bolt	5	450
$^5/_8$ bolt	5	650
$^3/_8$ deformed dowel	6	325
$^1/_2$ deformed dowel	6	450

[1]The bolts or dowels shall be spaced not closer than 6 inches (152 mm) on center.
[2]The tabulated values may be increased one third for bolts or dowels resisting wind or seismic forces.

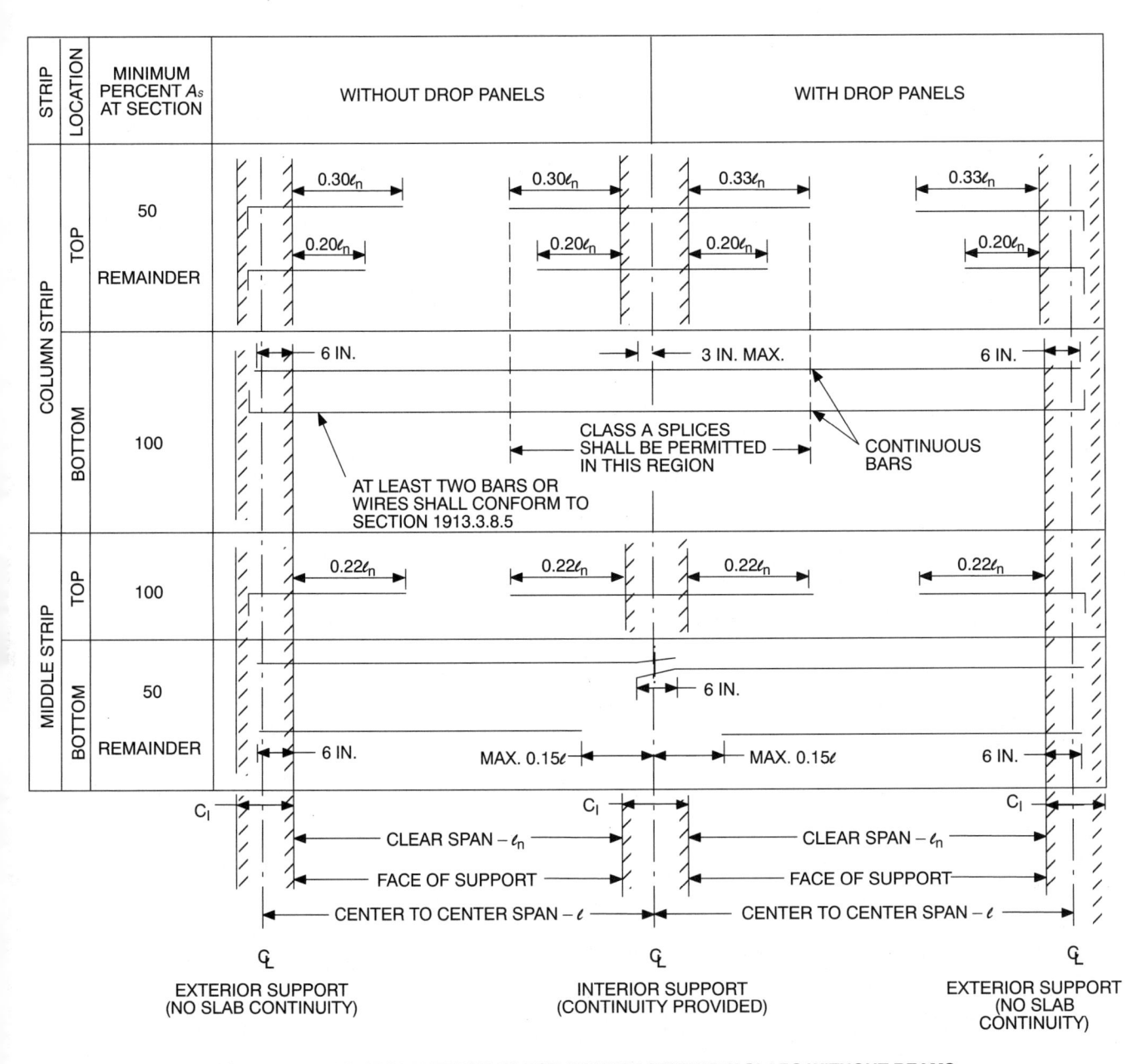

FIGURE 19-1—MINIMUM EXTENSIONS FOR REINFORCEMENT IN SLABS WITHOUT BEAMS
(See Section 1912.11.1 for reinforcement extension into supports.)

Chapter 20

LIGHTWEIGHT METALS

Division I—GENERAL

SECTION 2001 — MATERIAL STANDARDS AND SYMBOLS

2001.1 General. The quality, design, fabrication and erection of aluminum used structurally in buildings and structures shall conform to the requirements of this chapter, to other applicable requirements of this code and to Division II.

Allowable stresses and design formulas provided in this chapter shall be used with the allowable stress design load combinations specified in Section 1612.3.

2001.2 Alloys. The use of aluminum alloys and tempers other than those covered by this chapter and other lightweight metal alloys are allowed for structural members and assemblies, provided standards of performance not less than those required by this chapter are substantiated to the satisfaction of the building official. When required by the building official, certification that the alloys and tempers called for on the plans have been furnished shall be provided.

2001.3 Symbols and Notations. The symbols and notations used in this chapter are defined as follows:

A = area, square inches (mm^2).

A_c = area of compression element, square inches (mm^2) (compression flange plus one third of area of web between compression flange and neutral axis).

A_w = area of cross section lying within 1.0 inch of a weld, square inches (mm^2).

a_1 = shorter dimension of rectangular panel, inches (mm).

a_2 = longer dimension of rectangular panel, inches (mm).

a_e = equivalent width of rectangular panel, inches (mm).

B, D, C

= buckling formula constants, with following subscript:
  c—compression in columns
  p—compression in flat plates
  t—compression in round tubes
  tb—bending in round tubes
  b—bending in rectangular bars
  s—shear in flat plates

b = width of sections, inches (mm).

b/t = width-to-thickness ratio or rectangular element of a cross section.

c = distance from neutral axis to extreme fiber, inches (mm).

D = diameter, inches (mm).

d = depth of section or beam, inches (mm).

E = compressive modulus of elasticity, kips per square inch (ksi) (MPa).

F = allowable stress, ksi (MPa).

F_a = allowable compressive stress for member considered as an axially loaded column, ksi (MPa).

F_b = allowable compressive stress for member considered as a beam, ksi (MPa).

F_{bu} = bearing ultimate strength, ksi (MPa).

F_{buw} = bearing ultimate strength within 1.0 inch (25.4 mm) of a weld, ksi (MPa).

F_{by} = bearing yield strength, ksi (MPa).

F_{byw} = bearing yield strength within 1.0 inch (25.4 mm) of a weld, ksi (MPa).

F_c = allowable compressive stress, ksi (MPa).

F_{cy} = compressive yield strength, ksi (MPa).

F_{cyw} = compressive yield strength across a butt weld [0.2 percent offset in 10-inch (254 mm) gage length], ksi (MPa).

F_{ec} = $\pi^2 E / [n_u (l/r)^2]$, where l/r is slenderness ratio for member considered as a column tending to fail in the plane of the applied bending moments, ksi (MPa).

F_n = allowable stress for cross section 1.0 inch (25.4 mm) or more from weld, ksi (MPa).

F_{pw} = allowable stress on cross section, part of whose area lies within 1.0 (25.4 mm) inch of a weld, ksi (MPa).

F_s = allowable shear stress for members subjected only to torsion or shear, ksi (MPa).

F_{su} = shear ultimate strength, ksi (MPa).

F_{suw} = shear ultimate strength within 1.0 inch (25.4 mm) of a weld, ksi (MPa).

F_{sy} = shear yield strength, ksi (MPa).

F_{syw} = shear yield strength within 1.0 inch (25.4 mm) of a weld, ksi (MPa).

F_{tu} = tensile ultimate strength, ksi (MPa).

F_{tuw} = tensile ultimate strength across a butt weld, ksi (MPa).

F_{ty} = tensile yield strength, ksi (MPa).

F_{tyw} = tensile yield strength across a butt weld [0.2 percent offset in 10-inch (254 mm) gage length], ksi (MPa).

F_y = either F_{ty} or F_{cy}, whichever is smaller, ksi (MPa).

f = calculated stress, ksi (MPa).

f_a = average compressive stress on cross section of member produced by axial compressive load, ksi (MPa).

f_b = maximum bending stress (compressive) caused by transverse loads or end moments, ksi (MPa).

f_s = shear stress caused by torsion or transverse shear, ksi (MPa).

G = modulus of elasticity in shear, ksi (MPa).

g = spacing of rivet or bolt holes perpendicular to direction of load, inches (mm).

h = clear height of shear web, inches (mm).

I = moment of inertia, inches^4 (mm^4).

I_h = moment of inertia of horizontal stiffener, inches^4 (mm^4).

I_s = moment of inertia of transverse stiffener to resist shear buckling, inches^4 (mm^4).

I_x = moment of inertia of a beam about axis perpendicular to web, inches^4 (mm^4).

I_y = moment of inertia of a beam about axis parallel to web, inches^4 (mm^4).

I_{yc} = moment of inertia of compression element about axis parallel to vertical web, inches^4 (mm^4).

J = torsion constant, inches^4 (mm^4).

k_1 = coefficient for determining slenderness limit S_2 for sections for which the allowable compressive stress is based on crippling strength.

k_2 = coefficient for determining allowable compressive stress in sections with slenderness ratio above S_2 for which the allowable compressive stress is based on crippling strength.

k_c = coefficient for compression members.

k_t = coefficient for tension members.

L = length of compression member between points of lateral support, or twice the length of a cantilever column (except where analysis shows that a shorter length can be used), inches (mm).

L_b = length of beam between points at which the compression flange is supported against lateral movement, or length of cantilever beam from free end to point at which the compression flange is supported against lateral movement, inches (mm).

L_h = total length of portion of column lying within 1.0 inch (25.4 mm) of a weld (excluding welds at ends of columns that are supported at both ends), inches (mm).

L_w = increased length to be substituted in column formula to determine allowable stress for welded column, inches (mm).

l/r = slenderness ratio for columns.

M = bending moment, inch-kips (kN·m).

M_1, M_2 = bending moments at two ends of a beam, inch-kips (kN·m).

M_c = bending moment at center of span resulting from applied bending loads, inch-kips (kN·m).

M_m = maximum bending moment in span resulting from applied bending loads, inch-kips (kN·m).

N = length of bearing at reaction or concentrated load, inches (mm).

n_a = factor of safety on appearance of buckling.

n_u = factor of safety on ultimate strength.

n_y = factor of safety on yield strength.

P = local load concentration on bearing stiffener, kips (kN).

P_c = allowable reaction or concentrated load per web, kips (kN).

P_t = allowable tensile load per fastener, sheet to purlin or girt, kips (kN).

R = outside radius of round tube or maximum outside radius for an oval tube, inches (mm).

R_b = radius of curvature of tubular members, inches (mm).

R_t = transition radius, the radius of an attachment of the weld detail.

r = least radius of gyration of a column, inches (mm).

r_L = radius of gyration of lip or bulb about face of flange from which lip projects, inches (mm).

r_y = radius of gyration of a beam (about axis parallel to web), inches (mm). (For beams that are unsymmetrical about the horizontal axis, r_y should be calculated as though both flanges were the same as the compression flange.)

S_1, S_2 = slenderness limits.

S_c = section modulus of a beam, compression side, inches³ (mm³).

SR = stress ratio, the ratio of minimum stress to maximum stress.

S_t = section modulus of a beam, tension side, inches³ (mm³).

s = spacing of transverse stiffeners (clear distance between stiffeners for stiffeners consisting of a pair of members, one on each side of the web, center-to-center distance between stiffeners consisting of a member on one side of the web only), inches (mm); spacing of rivet or bolt holes parallel to direction of load, inches (mm).

t = thickness of flange, plate, web or tube, inches (mm). (For tapered flanges, t is the average thickness.)

V = shear force on web at stiffener location, kips (kN).

α = a factor equal to unity for a stiffener consisting of equal members on both sides of the web and equal to 3.5 for a stiffener consisting of a member on one side only.

θ = angle between plane of web and plane of bearing surface ($\theta \leq 90$), degrees.

2001.4 Identification. Aluminum for structural elements shall at all times be segregated or otherwise handled in the fabricator's plant so that the separate alloys and tempers are positively identified and, after completion of fabrication, shall be marked to identify the alloy and temper. Such markings shall be affixed to complete members and assemblies or to boxed or bundled shipments of multiple units prior to shipment from the fabricator's plant.

SECTION 2002 — ALLOWABLE STRESSES FOR MEMBERS AND FASTENERS

2002.1 Allowable Unit Stresses. Except as modified by Division II, allowable unit stresses in aluminum alloy structural members shall be determined in accordance with the formulas of Table 20-I-C utilizing the safety factors listed in Table 20-I-D, and the constants and coefficients listed in Tables 20-I-E, 20-I-F and 20-I-G. Where two formulas are given, the smaller of the resulting stresses shall be used.

2002.2 Welded Structural Members. Allowable unit stresses for structural members whose entire cross-sectional area lies within 1 inch (25.4 mm) of the center line of a butt weld of the heel of a fillet weld shall be determined by means of the formulas of Table 20-I-C utilizing the applicable minimum expected mechanical properties for welded aluminum alloys listed in Division II. The tensile ultimate strength, F_{tuw}, shall be 90 percent of the American Society of Mechanical Engineers weld qualification test value of ultimate strength. Except as modified by Division II, buckling constants determined in accordance with the formulas of Tables 20-I-E and 20-I-G shall be calculated using the nonwelded mechanical properties of the respective aluminum alloys.

If less than 15 percent of the area of a given cross section lies within 1 inch (25.4 mm) of the center line of a butt weld or the heel of a fillet weld, the effect of the weld may be neglected and allowable stresses for nonwelded structural members may be used.

If the area of a cross section that lies within 1 inch (25.4 mm) of a weld is between 15 percent and 100 percent of the total area of the cross section, the allowable stress shall be calculated by the following formula:

$$F_{pw} = F_n - \frac{A_w}{A}(F_n - F_w)$$

WHERE:

A = net area of cross section of a tension member or tension flange of a beam, or gross area of cross section of a compression member or compression flange of a beam, square inches (mm²). (A beam flange is considered to consist of that portion of the member further than $2c/3$ from the neutral axis, where c is the distance from the neutral axis to the extreme fiber.)

A_w = area of cross section lying within 1.0 inch (25.4 mm) of a weld.

F_n = allowable stress for cross section 1.0 inch (25.4 mm) or more from weld.

F_{pw} = allowable stress on cross section part of whose area lies within 1.0 inch (25.4 mm) of a weld.

F_w = allowable stress on cross section if entire area were to lie within 1.0 inch (25.4 mm) of a weld.

For columns and beams with welds at locations other than at their supported ends (not farther from the supports than 0.05 L from the ends), and for cantilever columns and single web beams with transverse welds at or near the supported end, the effect of welding on allowable stresses shall be determined in accordance with the provisions of Division II.

2002.3 Rivets and Bolts. Allowable stresses in aluminum rivets and bolts shall be as set forth in Table 20-I-A.

2002.4 Fillet Welds. Allowable shear stresses in fillet welds shall be as set forth in Table 20-I-B.

SECTION 2003 — DESIGN

2003.1 Combined Stresses. Members subjected to combinations of compression and bending or shear, compression and bending shall be proportioned in accordance with the provisions of Division II.

2003.2 Light Gage Members. Where the design of light gage structural members is involved, the special provisions of Division II shall be applied.

2003.3 Structural Roofing and Siding. The live load deflection of structural roofing and siding made of formed sheet shall not exceed $^1/_{60}$ of the span length.

2003.4 Connections. The design of mechanical and welded connections shall be in accordance with this chapter and the provisions of Division II.

SECTION 2004 — FABRICATION AND ERECTION

2004.1 Cutting. Oxygen cutting of aluminum alloys shall not be permitted.

2004.2 Fasteners. Bolts and other fasteners shall be aluminum, stainless steel or aluminized, hot-dip galvanized or electrogalvanized steel. Double cadmium-plated AN steel bolts may also be used. Steel rivets shall not be used except where aluminum is to be joined to steel or where corrosion resistance of the structure is not a requirement or where the structure is to be protected against corrosion.

2004.3 Dissimilar Materials. Where aluminum alloy parts are in contact with dissimilar metals, other than stainless, aluminized or galvanized steel or absorbent building materials likely to be continuously or intermittently wet, the faying surfaces shall be painted or otherwise separated in accordance with Division II.

2004.4 Painting. Except as prescribed in Section 2004.3, painting or coating of aluminum alloy parts shall be required only when called for on the plans.

2004.5 Welding. Aluminum parts shall be welded with an inert gas shielded arc or resistance welding process. No welding process that requires a welding flux shall be used. Filler alloys complying with the requirements of Division II shall be used.

2004.6 Welder Qualification. All welds of structural members shall be performed by welders qualified in accordance with the procedures of Division II.

2004.7 Erection. During erection, structural aluminum shall be adequately braced and fastened to resist dead, wind and erection loads.

TABLE 20-I-A
TABLE 20-I-B

1997 UNIFORM BUILDING CODE

TABLE 20-I-A—ALLOWABLE STRESSES FOR RIVETS

DESIGNATION BEFORE DRIVING	DRIVING PROCEDURE	DESIGNATION AFTER DRIVING	MINIMUM EXPECTED SHEAR STRENGTH (ksi)	ALLOWABLE SHEAR STRESS ON EFFECTIVE AREA (ksi)
			× 6.89 for MPa	
1100-H14	Cold, as received	1100-F	9.5	4
2017-T4	Cold, as received	2017-T3	34	14.5
2117-T4	Cold, as received	2117-T3	29	12
5056-H32	Cold, as received	5056-H321	26	11
6053-T61	Cold, as received	6053-T61	20	8.5
6061-T4	Hot, 990°F to 1050°F	6061-T43	21	9
6061-T6	Cold, as received	6061-T6	26	11[1]

ALLOWABLE STRESSES FOR BOLTS			
ALLOY AND TEMPER	MINIMUM EXPECTED SHEAR STRENGTH (ksi)	ALLOWABLE[2] SHEAR STRESS ON EFFECTIVE AREA (ksi)	ALLOWABLE TENSILE STRESS ON ROOT AREA (ksi)
	× 6.89 for MPa		
2024-T4	37	16	26
6061-T6	27	12	18
7075-T73	40	17	28

[1]Also applies to 6061-T6 pins.
[2]Values apply to either turned bolts or unfinished bolts in holes not more than $^1/_{16}$ inch (1.6 mm) oversized.

TABLE 20-I-B—ALLOWABLE SHEAR STRESSES IN FILLET WELDS (ksi)
(Shear stress is considered equal to the load divided by the throat area.)

FILLER ALLOY	1100	4043	5356 5554	5556
	× 6.89 for MPa			
Parent Alloy				
1100	3.2	4.8	*	*
3003	3.2	5	*	*
Alclad 3004	*	5	7	8
5052	*	5	7	*
5083	*	*	*	8.5
5086	*	*	7	8.5
5454	*	*	7	8.5
5456	*	*	*	8.5
6061	*	5	7	8.5
6063	*	5	6.5	6.5

* Not permitted.

TABLE 20-I-C—GENERAL FORMULAS FOR DETERMINING ALLOWABLE STRESSES

TYPE OF STRESS	TYPE OF MEMBER OR COMPONENT	SPEC. NO.	ALLOWABLE STRESS (ksi)		
Tension, axial, net section	Any tension member:	1	F_{ty}/n_y or $F_{tu}/(k_t n_u)$		
Tension in beams, extreme fiber, net section	Rectangular tubes, structural shapes bent about strong axis	2	F_{ty}/n_y or $F_{tu}/(k_t n_u)$	$\times$ 6890 for kN/m²	
	Round or oval tubes	3	$1.17F_{ty}/n_y$ or $1.24F_{tu}/(k_t n_u)$		
	Rectangular bars, plates, shapes bent about weak axis	4	$1.30F_{ty}/n_y$ or $1.42F_{tu}/(k_t n_u)$		
Bearing	On rivets and bolts	5	F_{by}/n_y or $F_{bu}/(1.2n_u)$		
	On flat surfaces and pins and on bolts in slotted holes	6	$F_{by}/(1.5n_y)$ or $F_{bu}/(1.8n_u)$		

TYPE OF STRESS	TYPE OF MEMBER OR COMPONENT	SPEC. NO.	ALLOWABLE STRESS, KSI, SLENDERNESS < S_1	SLENDERNESS LIMIT, S_1	ALLOWABLE STRESS, KSI SLENDERNESS BETWEEN S_1 AND S_2	SLENDERNESS LIMIT, S_2	ALLOWABLE STRESS, KSI SLENDERNESS ≥ S_2
Compression in columns, axial, gross section	All columns	7	$\dfrac{F_{cy}}{k_c n_y}$	$\dfrac{kL}{r} = \dfrac{B_c - \dfrac{n_u F_{cy}}{k_c n_y}}{D_c}$	$\dfrac{1}{n_u}\left(B_c - D_c\dfrac{kL}{r}\right)$	$\dfrac{kL}{r} = C_c$	$\dfrac{\pi^2 E}{n_u\,(kL/r)^2}$
Compression in components of columns, gross section	Outstanding flanges and legs	8	$\dfrac{F_{cy}}{k_c n_y}$	$\dfrac{b}{t} = \dfrac{B_p - \dfrac{n_u F_{cy}}{k_c n_y}}{5.1D_p}$	$\dfrac{1}{n_u}\left(B_p - 5.1D_p\dfrac{b}{t}\right)$	$\dfrac{b}{t} = \dfrac{C_p}{5.1}$	$\dfrac{\pi^2 E}{n_u(5.1b/t)^2}$
	Flat plates with both edges supported	9	$\dfrac{F_{cy}}{k_c n_y}$	$\dfrac{b}{t} = \dfrac{B_p - \dfrac{n_u F_{cy}}{k_c n_y}}{1.6D_p}$	$\dfrac{1}{n_u}\left(B_p - 1.6D_p\dfrac{b}{t}\right)$	$\dfrac{b}{t} = \dfrac{k_1 B_p}{1.6D_p}$	$\dfrac{k_2\sqrt{B_p E}}{n_u(1.6b/t)}$
	Curved plates supported on both edges, walls of round or oval tubes	10	$\dfrac{F_{cy}}{k_c n_y}$	$\dfrac{R}{t} = \left(\dfrac{B_t - \dfrac{n_u F_{cy}}{k_c n_y}}{D_t}\right)^2$	$\dfrac{1}{n_u}\left(B_t - D_t\sqrt{\dfrac{R}{t}}\right)$	$\dfrac{R}{t} = C_t$	$\dfrac{\pi^2 E}{16n_u\left(\dfrac{R}{t}\right)\left(1 + \dfrac{\sqrt{R/t}}{35}\right)^2}$
Compression in beams, extreme fiber, gross section	Single web beams bent about strong axis	11	$\dfrac{F_{cy}}{n_y}$	$\dfrac{L_b}{r_y} = \dfrac{1.2(B_c - F_{cy})}{D_c}$	$\dfrac{1}{n_y}\left(B_c - \dfrac{D_c L_b}{1.2r_y}\right)$	$\dfrac{L_b}{r_y} = 1.2C_c$	$\dfrac{\pi^2 E}{n_y(L_b/1.2r_y)^2}$
	Round or oval tubes	12	$\dfrac{1.17F_{cy}}{n_y}$	$\dfrac{R_b}{t} = \left(\dfrac{B_{tb} - 1.17F_{cy}}{D_{tb}}\right)^2$	$\dfrac{1}{n_y}\left(B_{tb} - D_{tb}\sqrt{\dfrac{R_b}{t}}\right)$	$\dfrac{R_b}{t} = \left(\dfrac{\dfrac{n_u}{n_y}B_{tb} - B_t}{\dfrac{n_u}{n_y}D_{tb} - D_t}\right)^2$	Same as Specification 10[1]
	Curved sections		$\dfrac{1.17F_{cy}}{n_y}$	$\dfrac{R}{t} = \left(\dfrac{B_t - 1.17F_{cy}}{D_t}\right)^2$	$\dfrac{1}{n_y}\left(B_t - D_t\sqrt{\dfrac{R}{t}}\right)$	$\dfrac{R}{t} = C_c$	$\dfrac{\pi^2 E}{16n_y\left(\dfrac{R}{t}\right)\left(1 + \dfrac{\sqrt{R/t}}{35}\right)^2}$
	Solid rectangular beams	13	$\dfrac{1.3F_{cy}}{n_y}$	$\dfrac{d}{t}\sqrt{\dfrac{L_b}{d}} = \dfrac{B_{br} - 1.3F_{cy}}{2.3D_{br}}$	$\dfrac{1}{n_y}\left(B_{br} - 2.3D_{br}\dfrac{d}{t}\sqrt{\dfrac{L_b}{d}}\right)$	$\dfrac{d}{t}\sqrt{\dfrac{L_b}{d}} = \dfrac{C_{br}}{2.3}$	$\dfrac{\pi^2 E}{5.29n_y(d/t)^2(L_b/d)}$
	Rectangular tubes and box sections	14	$\dfrac{F_{cy}}{n_y}$	$\dfrac{L_b S_c}{0.5\sqrt{I_y J}} = \left(\dfrac{B_c - F_{cy}}{1.6D_c}\right)^2$	$\dfrac{1}{n_y}\left(B_c - 1.6D_c\sqrt{\dfrac{L_b S_c}{0.5\sqrt{I_y J}}}\right)$	$\dfrac{L_b S_c}{0.5\sqrt{I_y J}} = \left(\dfrac{C_c}{1.6}\right)^2$	$\dfrac{\pi^2 E}{2.56n_y(L_b S_c/0.5\sqrt{I_y J})}$
Compression in components of beams (component under uniform compression), gross section	Outstanding flanges	15	$\dfrac{F_{cy}}{n_y}$	$\dfrac{b}{t} = \dfrac{B_p - F_{cy}}{5.1D_p}$	$\dfrac{1}{n_y}\left(B_p - 5.1D_p\dfrac{b}{t}\right)$	$\dfrac{b}{t} = \dfrac{k_1 B_p}{5.1D_p}$	$\dfrac{k_2\sqrt{B_p E}}{n_y(5.1b/t)}$
	Flat plates with both edges supported	16	$\dfrac{F_{cy}}{n_y}$	$\dfrac{b}{t} = \dfrac{B_p - F_{cy}}{1.6D_p}$	$\dfrac{1}{n_y}\left(B_p - 1.6D_p\dfrac{b}{t}\right)$	$\dfrac{b}{t} = \dfrac{k_1 B_p}{1.6D_p}$	$\dfrac{k_2\sqrt{B_p E}}{n_y(1.6b/t)}$
Compression in components of beams (component under bending in own plane), gross section	Flat plates with compressed edge free tension edge supported	17	$\dfrac{1.3F_{cy}}{n_y}$	$\dfrac{b}{t} = \dfrac{B_{br} - 1.3F_{cy}}{3.5D_{br}}$	$\dfrac{1}{n_y}\left(B_{br} - 3.5D_{br}\dfrac{b}{t}\right)$	$\dfrac{b}{t} = \dfrac{C_{br}}{3.5}$	$\dfrac{\pi^2 E}{n_y(3.5b/t)^2}$
	Flat plates with both edges supported	18	$\dfrac{1.3F_{cy}}{n_y}$	$\dfrac{h}{t} = \dfrac{B_{br} - 1.3F_{cy}}{0.67D_{br}}$	$\dfrac{1}{n_y}\left(B_{br} - 0.67D_{br}\dfrac{h}{t}\right)$	$\dfrac{h}{t} = \dfrac{k_1 B_{br}}{0.67D_{br}}$	$\dfrac{k_2\sqrt{B_{br}E}}{n_y(0.67h/t)}$
	Flat plates with horizontal stiffener, both edges supported	19	$\dfrac{1.3F_{cy}}{n_y}$	$\dfrac{h}{t} = \dfrac{B_{br} - 1.3F_{cy}}{0.29D_{br}}$	$\dfrac{1}{n_y}\left(B_{br} - 0.29D_{br}\dfrac{h}{t}\right)$	$\dfrac{h}{t} = \dfrac{k_1 B_{br}}{0.29D_{br}}$	$\dfrac{k_2\sqrt{B_{br}E}}{n_y(0.29h/t)}$

(Continued)

TABLE 20-I-C
TABLE 20-I-E

1997 UNIFORM BUILDING CODE

TABLE 20-I-C—GENERAL FORMULAS FOR DETERMINING ALLOWABLE STRESSES—(Continued)

TYPE OF STRESS	TYPE OF MEMBER OR COMPONENT	SPEC. NO.	ALLOWABLE STRESS, KSI, SLENDERNESS < S_1	SLENDERNESS LIMIT, S_1	ALLOWABLE STRESS, KSI SLENDERNESS BETWEEN S_1 AND S_2	SLENDERNESS LIMIT, S_2	ALLOWABLE STRESS, KSI SLENDERNESS ≥ S_2
Shear in webs, gross section	Unstiffened flat webs	20	$\dfrac{F_{sy}}{n_y}$	$\dfrac{h}{t} = \dfrac{B_s - F_{sy}}{1.25D_s}$	$\dfrac{1}{n_y}\left(B_s - 1.25D_s\dfrac{h}{t}\right)$	$\dfrac{h}{t} = \dfrac{C_s}{1.25}$	$\dfrac{\pi^2 E}{n_y(1.25h/t)^2}$
	Stiffened flat webs	21	$\dfrac{F_{sy}}{n_y}$	$\dfrac{a_e}{t} = \dfrac{B_s - \dfrac{n_a F_{sy}}{n_y}}{1.25D_s}$	$\dfrac{1}{n_a}\left(B_s - 1.25D_s\dfrac{a_e}{t}\right)$	$\dfrac{a_e}{t} = \dfrac{C_s}{1.25}$	$\dfrac{\pi^2 E}{n_a(1.25a_e/t)^2}$

[1]For R_b/t values greater than S_2, the allowable bending shall be determined from the formula for tubes in compression, Specification 10, using the formula that is appropriate for the particular value of R_b/t. Note that in this case, R_b/t may be either less than or greater than the value of S_2 for tubes in compression.

TABLE 20-I-D—FACTORS OF SAFETY FOR USE WITH ALUMINUM ALLOWABLE STRESS SPECIFICATIONS

	BUILDING AND SIMILAR TYPE STRUCTURES
1. Tension members	
F.S. on tensile strength, n_u	1.95
F.S. on yield strength, n_y	1.65
2. Columns	
F.S. on buckling strength, n_u	1.95
F.S. on crippling strength of thin sections, n_u	1.95
F.S. on yield strength for short columns, n_y	1.65
3. Beams	
F.S. on tensile strength, n_u	1.95
F.S. on tensile yield strength, n_y	1.65
F.S. on compressive yield strength for short beams, n_y	1.65
F.S. on buckling strength, n_y	1.65
F.S. on crippling strength of thin sections, n_y	1.65
F.S. on shear buckling of webs, n_a	1.20
4. Connections	
F.S. on bearing strength	$1.2 \times 1.95 = 2.34$
F.S. on bearing yield strength, n_y	1.65
F.S. on shear strength of rivets and bolts	$1.2 \times 1.95 = 2.34$
F.S. on shear strength of fillet welds	$1.2 \times 1.95 = 2.34$
F.S. on tensile strength of butt welds, n_u	1.95
F.S. on tensile yield strength of butt welds, n_y	1.65

TABLE 20-I-E—FORMULAS FOR BUCKLING CONSTANTS
For All Products Whose Temper Designation begins with -O, -H, -T1, -T2, -T3 or -T4

TYPE OF MEMBER AND STRESS	INTERCEPT (ksi)	SLOPE (ksi)	INTERSECTION
	× 6.89 for MPa		
1. Compression in columns and beam flanges	$B_c = F_{cy}\left[1 + \left(\dfrac{F_{cy}}{1000}\right)^{1/2}\right]$	$D_c = \dfrac{B_c}{20}\left(\dfrac{6B_c}{E}\right)^{1/2}$	$C_c = \dfrac{2B_c}{3D_c}$
2. Compression in flat plates	$B_p = F_{cy}\left[1 + \dfrac{(F_{cy})^{1/3}}{7.6}\right]$	$D_p = \dfrac{B_p}{20}\left(\dfrac{6B_p}{E}\right)^{1/2}$	$C_p = \dfrac{2B_p}{3D_p}$
3. Compression in round tubes under axial end load	$B_t = F_{cy}\left[1 + \dfrac{(F_{cy})^{1/5}}{5.8}\right]$	$D_t = \dfrac{B_t}{3.7}\left(\dfrac{B_t}{E}\right)^{1/3}$	C_t *
4. Compressive bending stress in solid rectangular bars	$B_b = 1.3F_{cy}\left[1 + \dfrac{(F_{cy})^{1/3}}{7}\right]$	$D_b = \dfrac{B_b}{20}\left(\dfrac{6B_b}{E}\right)^{1/2}$	$C_b = \dfrac{2B_b}{3D_b}$
5. Compressive bending stress in round tubes	$B_{tb} = 1.5F_y\left[1 + \dfrac{(F_y)^{1/5}}{5.8}\right]$	$D_{tb} = \dfrac{B_{tb}}{2.7}\left(\dfrac{B_{tb}}{E}\right)^{1/3}$	$C_{tb} = \left(\dfrac{B_{tb} - B_t}{D_{tb} - D_t}\right)^2$
6. Shear stress in flat plates	$B_s = F_{sy}\left[1 + \dfrac{(F_{sy})^{1/3}}{6.2}\right]$	$D_s = \dfrac{B_s}{20}\left(\dfrac{6B_s}{E}\right)^{1/2}$	$C_s = \dfrac{2B_s}{3D_s}$
7. Crippling of flat plates in compression or bending	$k_1 = 0.50$	$k_2 = 2.04$	

*C_t can be found from a plot of the curves of allowable stress based on elastic and inelastic buckling or by a trial-and-error solution.

TABLE 20-I-F—VALUES OF COEFFICIENTS k_t and k_c

ALLOY AND TEMPER	NONWELDED OR REGIONS FARTHER THAN 1.0 INCH (25.4 mm) FROM A WELD		REGIONS WITHIN 1.0 INCH (25.4 mm) OF A WELD	
	k_t	k_c	k_t	k_c[1]
2014-T6, -T651	1.25	1.12	—	—
Alclad 2014-T6, -T651	1.25	1.12	—	—
6061-T6, -T651	1.0	1.12	1.0	1.0
6063-T5, -T6, -T83	1.0	1.12	1.0	1.0
6351-T5	1.0	1.12	1.0	1.0
All others listed in Division II	1.0	1.10	1.0	1.0

[1]If the weld yield strength exceeds 0.9 of the parent metal yield strength, the allowable compressive stress within 1.0 inch (25.4 mm) of a weld should be taken equal to the allowable stress for nonwelded material.

TABLE 20-I-G—FORMULAS FOR BUCKLING CONSTANTS
For all products whose temper designation begins with -T5, -T6, -T7, -T8 or -T9

TYPE OF MEMBER AND STRESS	INTERCEPT (ksi) × 6.89 for MPa	SLOPE (ksi) × 6.89 for MPa	INTERSECTION
1. Compression in columns and beam flanges	$B_c = F_{cy}\left[1 + \left(\frac{F_{cy}}{2250}\right)^{1/2}\right]$	$D_c = \frac{B_c}{10}\left(\frac{B_c}{E}\right)^{1/2}$	$C_c = 0.41\frac{B_c}{D_c}$
2. Compression in flat plates	$B_p = F_{cy}\left[1 + \frac{(F_{cy})^{1/3}}{11.4}\right]$	$D_p = \frac{B_p}{10}\left(\frac{B_p}{E}\right)^{1/2}$	$C_p = 0.41\frac{B_p}{D_p}$
3. Compression in round tubes under axial end load	$B_t = F_{cy}\left[1 + \frac{(F_{cy})^{1/5}}{8.7}\right]$	$D_t = \frac{B_t}{4.5}\left(\frac{B_t}{E}\right)^{1/3}$	C_t *
4. Compressive bending stress in solid rectangular bars	$B_b = 1.3F_{cy}\left[1 + \frac{(F_{cy})^{1/3}}{7}\right]$	$D_b = \frac{B_b}{20}\left(\frac{6B_b}{E}\right)^{1/2}$	$C_b = \frac{2B_b}{3D_b}$
5. Compressive bending stress in round tubes	$B_{tb} = 1.5F_y\left[1 + \frac{(F_y)^{1/5}}{8.7}\right]$	$D_{tb} = \frac{B_{tb}}{2.7}\left(\frac{B_{tb}}{E}\right)^{1/3}$	$C_{tb} = \left(\frac{B_{tb} - B_t}{D_{tb} - D_t}\right)^2$
6. Shear stress in flat plates	$B_s = F_{sy}\left[1 + \frac{(F_{sy})^{1/3}}{9.3}\right]$	$D_s = \frac{B_s}{10}\left(\frac{B_s}{E}\right)^{1/2}$	$C_s = 0.41\frac{B_s}{D_s}$
7. Crippling of flat plates in compression	$k_1 = 0.35$	$k_2 = 2.27$	
8. Crippling of flat plates in bending	$k_1 = 0.50$	$k_2 = 2.04$	

*C_t can be found from a plot of the curves of allowable stress based on elastic and inelastic buckling or by a trial-and-error solution.

Division II—DESIGN STANDARD FOR ALUMINUM STRUCTURES
Based on Specifications for Aluminum Structures of The Aluminum Association (December, 1986)

SECTION 2005 — SCOPE

This standard covers design of aluminum alloy load-carrying members.

SECTION 2006 — MATERIALS

The principal materials to which this standard applies are aluminum alloys registered with The Aluminum Association. Those frequently used for structural members are listed in Table 20-II-A. Applicable American Society for Testing and Materials (ASTM) specifications are designations B 209, B 210, B 211, B 221, B 241, B 247, B 308 and B 429.

SECTION 2007 — DESIGN

Design shall be in accordance with Division I and other applicable provisions of this code.

Properties of section, such as cross-sectional area, moment of inertia, section modulus, radius of gyration, etc., shall be determined by accepted methods of engineering design. Computations of forces, moments, stresses and deflection shall be in accordance with accepted principles of elastic structural analyses.

SECTION 2008 — ALLOWABLE STRESSES

Allowable stresses shall be determined in accordance with the provisions of Division I.

SECTION 2009 — SPECIAL DESIGN RULES

2009.1 Combined Compression and Bending. A member subjected to axial compression and carrying a bending moment due to lateral or eccentric loads shall be proportioned in accordance with the following formulas:

1. Bending moment at center equal to or greater than 0.9 of maximum bending moment in span:

$$\frac{f_a}{F_a} + \frac{f_b}{F_b(1 - f_a/F_{ec})} \le 1$$

2. Bending moment at center equal to or less than 0.5 of maximum bending moment in span:

$$\frac{f_a}{F_a} + \frac{f_b}{F_b} \le 1$$

3. Bending moment at center between 0.5 and 0.9 maximum bending moment in span:

$$\frac{f_a}{F_a} + \frac{f_b}{F_b\left[1 - \left(\frac{2M_c}{M_m} - 1\right)\frac{f_a}{F_{ec}}\right]} \le 1$$

WHERE:

M_c = bending moment at center of span.

M_m = maximum bending moment in span.

2009.2 Torsion and Shear in Tubes. Allowable shear stresses in round or oval tubes due to torsion or transverse shear loads shall be determined from Specification 20 in Table 20-I-C with the ratio h/t replaced by an equivalent h/t given by the following:

$$\text{Equivalent } \frac{h}{t} = 2.9\left(\frac{R}{t}\right)^{5/8}\left(\frac{L_t}{R}\right)^{1/4}$$

WHERE:

L_t = length of tube between circumferential stiffeners, inches (mm). Equivalent (h/t) = value to be substituted for h/t in Specification 20 in Table 20-I-C.

2009.3 Combined Shear, Compression and Bending. Allowable combinations of shear, compression and bending, as in the web of a beam column or the wall of a tube, shall be determined from the following formula:

$$\frac{f_a}{F_a} + \frac{f_b}{F_b} + \left(\frac{f_s}{F_s}\right)^2 \le 1.0$$

2009.4 Stiffeners for Outstanding Flanges. Outstanding flanges stiffened by lips or bulbs at the free edge shall be considered as supported on both edges if the radius of gyration of the lip or bulb meets the following requirement:

$$r_L = \frac{b}{5}$$

For simple rectangular lips having the same thickness as the flange, as in the case of formed sheet construction, the preceding requirement can be expressed as:

$$b_L = b/3$$

WHERE:

b_L = clear width of lip, inches (mm).

Allowable stresses for flanges with lips or bulbs meeting the foregoing requirements shall be determined from Specifications 15 and 16 in Table 20-I-C. The area of stiffening lips or bulbs may be included with the area of the rest of the section in calculating the stresses caused by the loads.

2009.5 Horizontal Stiffeners for Shear Webs. If a horizontal stiffener is used on a beam web, it shall be located so that the distance from the toe of the compression flange to the centroid of the stiffener is 0.4 of the distance from the toe of the compression flange to the toe of the tension flange. The horizontal stiffener shall have a moment of inertia about the web of the beam not less than that given by the expression:

$$I_h = 2_\alpha f t h^3 \left[\left(1 + \frac{6A_h}{ht}\right)\left(\frac{s}{h}\right)^2 + 0.4\right]10^{-6}$$

For **SI:** 1 inch^4 = 416 231 mm^4.

WHERE:

A_h = gross area of cross section of horizontal stiffener, inches^2.

α = 1, for stiffener consisting of equal members on both sides of the web.

α = 3.5, for stiffener consisting of member on only one side of web.

For stiffener consisting of equal members on both sides of the web, the moment of inertia, I_h, shall be the sum of the moments of inertia about the center line of the web. For a stiffener consisting of a member on one side only, the moment of inertia shall be taken about the face of the web in contact with the stiffener.

2009.6 Vertical Stiffeners for Shear Webs. Stiffeners applied to beam webs to resist shear buckling shall have a moment of inertia not less than the value given by the following expressions:

$$\frac{s}{h} \le 0.4, \quad I_s = \frac{n_a V h^2}{22,400}\left(\frac{s}{h}\right)$$

For **SI:** 1 inch^4 = 416 231 mm^4.

$$\frac{s}{h} \geq 0.4, \quad I_s = \frac{n_a V h^2}{140,000}\left(\frac{h}{s}\right)$$

For **SI:** 1 inch^4 = 416 231 mm^4.

When a stiffener is composed of a pair of members, one on each side of the web, the stiffener spacing, *s*, shall be the clear distance between the pairs of stiffeners. When a stiffener is composed of a member on one side only of the web, the stiffener spacing, *s*, shall be the distance between rivet lines or other connecting lines.

For a stiffener composed of members of equal size on each side of the web, the moment of inertia of the stiffener shall be computed about the center line of the web. For a stiffener composed of a member on one side only of the web, the moment of inertia of the stiffener shall be computed about the face of the web in contact with the stiffener.

In the determination of the required moment of inertia of stiffeners, the distance, *h*, shall always be taken as the full clear height of the web regardless of whether or not a horizontal stiffener is present.

Stiffeners shall extend from flange to flange but need not be connected to either flange.

Unless the outer edge of a stiffener is continuously stiffened, its thickness shall not be less than one twelfth the clear width of the outstanding leg.

Vertical stiffeners shall, where possible, be placed in pairs at end bearings and at points of support of concentrated loads. They shall be connected to the web by enough rivets, or other means, to transmit the load. Such stiffeners shall be fitted to form a tight and uniform bearing against the loaded flanges unless welds, designed to transmit the full reaction or load, are provided between flange and stiffener.

Only that part of a stiffener cross section which lies outside the fillet of the flange angle shall be considered as effective in bearing. Bearing stiffeners shall not be joggled.

The moment of inertia of the bearing stiffener shall not be less than that given by the following expression:

$$I_b = I_s + \frac{P h^2 n_u}{\pi^2 E}$$

For **SI:** 1 inch^4 = 416 231 mm^4.

WHERE:

I_b = required moment of inertia of bearing stiffener, inches^4 (mm^4).

2009.7 Special Provisions for Thin Sections.

2009.7.1 Appearance of buckling. For very thin sections the allowable compressive stresses given in Specifications 9, 15, 16, 18 and 19 of Table 20-I-C may result in visible local buckling, even though an adequate margin of safety is provided against ultimate failure. In applications where any appearance of buckling must be avoided, the allowable stresses for thin sections shall not exceed the value of F_{ab} given by the following formulas:

SPECIFICATION	ALLOWABLE STRESS, F_{ab} ksi (MPa)
9, 16	$F_{ab} = \dfrac{\pi^2 E}{n_a (1.6 b/t)^2}$
15	$F_{ab} = \dfrac{\pi^2 E}{n_a (5.1 b/t)^2}$
18	$F_{ab} = \dfrac{\pi^2 E}{n_a (0.67 h/t)^2}$
19	$F_{ab} = \dfrac{\pi^2 E}{n_a (0.29 h/t)^2}$

2009.7.2 Weighted average allowable compressive stress. The cross section of a compression member may be composed of several thin elements, for which allowable stresses are determined by Specification 8, 9 or 10 of Table 20-I-C. The allowable compressive stress for the section as a whole may be considered to be the weighted average allowable stress for the individual elements, where the allowable stress for each element is weighted in accordance with the ratio of the area of the element to the total area of the section. The allowable compressive stress for the section as a whole used as a column must not exceed that given by Specification 7 of Table 20-I-C.

Weighted average allowable compressive stresses for beam flanges may be calculated in the same way, where the allowable stresses for individual elements are determined from Specifications 15 through 19 of Table 20-I-C. The beam flange may be considered to consist of the flange proper plus one sixth of the area of the web or webs.

2009.7.3 Trapezoidal-formed sheet beams. The weighted average allowable compressive stress for a trapezoidal-formed sheet beam, calculated according to paragraph 2, is:

$$F_{ba} = \frac{F_{bf} + F_{bh}\left(\frac{h}{3b}\right)}{1 + \frac{h}{3b}}$$

WHERE:

F_{ba} = weighted average allowable compressive stress for beam flange, ksi (MPa).

F_{bf} = allowable stress for flange proper based on Specification 16 of Table 20-I-C.

F_{bh} = allowable stress for webs based on Specification 18 or 19 of Table 20-I-C.

The foregoing formula may also be applied to the allowable tensile stress in trapezoidal-formed sheet beams, if the designer wishes to take full advantage of the strength of the section. In this case, F_{ba} is the weighted average allowable tensile stress, F_{bf} is determined from Specification 2 in Table 20-I-C, and F_{bh} is given by Specification 4 in Table 20-I-C.

In regions of negative bending moment (for example, at interior supports of multiple-span beams) the allowable tensile stress on the tension flange of a formed sheet beam shall not exceed the compressive stress that would be allowed on the same flange if it were in compression.

2009.7.4 Effect of local buckling on column strength. An additional limitation must be placed on the allowable stress for very thin-walled columns whose cross section is a rectangular tube or a formed sheet shape such that the flanges consist of flat elements supported on both edges. If the b/t for the flange of such a column is less than the value of S_2 in Specification 9 of Table 20-I-C, or less than 0.6 of the maximum slenderness ratio (L/r) for the column, no additional reduction in allowable stress is necessary. However, if the maximum b/t for the flange is greater than the value of S_2 from Specification 9 of Table 20-I-C, and also greater than 0.6 of the maximum slenderness ratio for the column, the allowable column stress shall not exceed the value given by

$$F_{rc} = \frac{\pi^2 E}{n_u (L/r)^{2/3} (1.6 b/t)^{4/3}}$$

WHERE:

F_{rc} = reduced allowable stress on column, ksi.

The allowable stress shall also not exceed the value given by Specification 9 of Table 20-I-C.

2009.7.5 Effect of local buckling on beam strength. The allowable compressive bending stress for single web beams whose flanges consist of thin, flat elements supported on one edge shall also be reduced in the case where the value of b/t for the flange is greater than the value of S_2 from Specification 15 of Table 20-I-C, and also greater than $0.16\,(L_b/r_y)$. In this case, the allowable beam stress shall not exceed

$$F_{rb} = \frac{\pi^2 E}{n_u\,(L_b/1.2r_y)^{2/3}\,(5.1b/t)^{4/3}}$$

WHERE:

F_{rb} = reduced allowable compressive bending stress in beam flange, ksi.

L_b/r_y = slenderness ratio for beam.

2009.7.6 Effective width for calculation of deflection of thin gage sections. As noted in paragraph 1, the allowable compressive stresses given in Specifications 9, 15, 16, 18 and 19 of Table 20-I-C may result in some local buckling at design loads for very thin sections even though an adequate margin of safety is provided against ultimate failure. This local buckling may result in increased deflections for sections containing thin elements with b/t value exceeding $1.65\,S_2$, where the value of S_2 is obtained for the element in question from Specifications 9, 15, 16, 18 and 19 of Table 20-I-C.

Where deflection at design loads is critical, the effective width concept may be used to determine an effective section to be used in deflection calculations. The effective width, b_e, of a thin element subjected to direct compression stresses is:

$$\text{If } f_a \leq n_a F_{ab}, \quad b_e = b$$

$$\text{If } f_a > n_a F_{ab}, \quad b_e = b\sqrt{n_a F_{ab}/f_a}$$

WHERE:

b_e = effective width of flat plate element to be used in deflection calculations, inches (mm).

F_{ab} = allowable stress for element from Subsection (g), ksi (MPa).

The same expression may be used to calculate the effective width on the compression side of a web in bending, with the compressive bending stress due to the applied loads, f_b, replacing f_a.

2009.7.7 Web crippling. For structural formed sheet roofing and siding, allowable interior reactions and concentrated loads for flat webs shall not exceed

$$P_c = 600\frac{F_{cy}\,dt^2}{w}\left(6 + 0.04\frac{N}{t}\right)$$

$$\left[1.1 - 0.1\left(\frac{\theta}{90}\right)\sqrt{\frac{r}{t}}\right] \text{ for } \frac{w}{t} \leq C_p$$

$$\text{and } P_c = 1,500\,Ed\,(N + w)\left(\frac{t}{w}\right)^3 \text{ for } \frac{w}{t} > C_p$$

$$\text{in which } C_p = \frac{2.5E\left(\frac{N}{w} + 1\right)}{F_{cy}\left(6 + 0.04\frac{N}{t}\right)\left[1.1 - 0.1\left(\frac{\theta}{90}\right)\sqrt{\frac{r}{t}}\right]}$$

Allowable end reactions shall not exceed

$$P_c = 600\frac{F_{cy}\,dt^2}{w}\left(3 + 0.04\frac{N}{t}\right)$$

$$\left[1.1 - 0.1\left(\frac{\theta}{90}\right)\sqrt{\frac{r}{t}}\right] \text{ for } \frac{w}{t} \leq C_p$$

$$\text{and } P_c = 1,500\,Ed\,\left(N + \frac{w}{2}\right)\left(\frac{t}{w}\right)^3 \text{ for } \frac{w}{t} > C_p$$

$$\text{in which } C_p = \frac{1.25E\left(\frac{2N}{w} + 1\right)}{F_{cy}\left(3 + 0.04\frac{N}{t}\right)\left[1.1 - 0.1\left(\frac{\theta}{90}\right)\sqrt{\frac{r}{t}}\right]}$$

WHERE:

d = depth (vertical projection), inches (mm).

F_{cy} = minimum compressive yield strength of sheet, in kips per square inch (MPa).

N = length of bearing at reaction or concentrated load, inches (mm).

P_c = allowable reaction or concentrated load per web, pounds (N).

r = bend radius at juncture of flange and web of trapezoidal section, measured to inside surface of bend, inches (mm).

t = sheet thickness, inches (mm).

w = slope width of web (shear element spanning between flats) of trapezoidal section, inches (mm).

θ = angle between plane of web of trapezoidal section and plane of bearing surface ($\theta \leq 90$), degrees.

2009.8 Fatigue. For up to 100,000 repetitions of maximum live load, if nonwelded, and 20,000 repetitions of maximum live load if welded, allowable stresses shall be determined in accordance with Table 20-I-C and Section 2010.1 provided that the structural members are free of re-entrant corners and other unusual stress raisers. For repetitions of loads in excess of these values, allowable stresses shall be determined by a special analysis.

2009.9 Compression in Single-web Beams. The formulas of Specification 11 of Table 20-I-C for single-web beams and girders, are based on an approximation in which the term L_b/r_y replaces a more complicated expression involving several different properties of the beam cross section. Because of this approximation, the formulas give very conservative results for certain conditions, namely for values of L_b/r_y exceeding about 50; for load distributions such that the bending moment near the center of the beam is appreciably less than the maximum bending moment in the beam; and for beams with transverse loads applied to the bottom flange. If the designer wishes to compute more precise values of allowable compressive stress for these cases, the value of r_y in Specification 11 of Table 20-I-C may be replaced by an "effective r_y" given by one of the following formulas:

Beam spans subjected to end moment only or to transverse loads applied at the neutral axis of the beam:

$$\text{Effective } r_y = \frac{k_b}{1.7}\sqrt{\frac{I_y d}{S_c}\sqrt{1 + 0.152\frac{J}{I_y}\left(\frac{L_b}{d}\right)^2}}$$

For **SI:** 1 inch = 25.4 mm.

Beams subjected to transverse loads applied on the top or bottom flange (where the load is free to move laterally with the beam if the beam should buckle):

Effective $r_y =$

$$\frac{k_b}{1.7}\sqrt{\frac{I_y d}{S_c}\left[\pm 0.5 + \sqrt{1.25 + 0.152\frac{J}{I_y}\left(\frac{L_b}{d}\right)^2}\right]}$$

For **SI:** 1 inch = 25.4 mm.

The plus sign in front of the term "0.5" applies if the load is on the bottom flange, the minus sign if the load is on the top flange.

Effective r_y = value to be substituted for r_y in Specification 11 of Table 20-I-C.

The terms appearing in the above formulas are defined in this code.

Values of the coefficient, k_b, follow:

BEAMS RESTRAINED AGAINST LATERAL DISPLACEMENT AT BOTH ENDS OF SPAN	VALUE OF COEFFCIENT k_b
Uniform bending moment, uniform transverse load, or two equal concentrated loads equidistant from the center of the span	1.00
Bending moment varying uniformly from a value of M_1 at one end to M_2 at the other end	
$M_1/M_2 =$ 0.5	1.14
$M_1/M_2 =$ 0	1.33
$M_1/M_2 = -0.5$	1.53
$M_1/M_2 = -1.0$	1.60
Concentrated load at center of span	1.16
CANTILEVER BEAMS	
Concentrated load at end of span	1.13
Uniform transverse load	1.43

2009.10 Compression in Elastically Supported Flanges. Allowable compressive stresses in elastically supported flanges, such as the compression flange of a standing seam roof or of a hat-shaped beam loaded with the two flanges in compression, shall be determined from Specification No. 11 with the following effective value of L_b/r_y substituted in the formulas for allowable stress.

$$\text{Effective } \frac{L_b}{r_y} = 2.7 \sqrt[4]{\frac{EA_c^2}{\beta I_{yc}}}$$

WHERE:

β = spring constant [transverse force in kips applied to a 1-inch (25.4 mm) length of the member at the compression flange to cause a 1-inch (25.4 mm) deflection of the flange], ksi (MPa).

SECTION 2010 — MECHANICAL CONNECTIONS

2010.1 Riveted and Bolted Connections. Aluminum alloys used for rivets and bolts shall be those listed in Table 20-I-A. Nuts for $1/4$-inch (6.4 mm) bolts and smaller shall be 2024-T4. Nuts for larger diameter bolts shall be alloy 6061-T6 or 6262-T9. Flat washers shall be Alclad 2024-T4. Spring lock washers shall be alloy 7075-T6. For improved corrosion resistance, a 0.0002-inch (0.005 mm) minimum thickness anodic coating may be applied to alloy 2024 bolts.

2010.1.1 Allowable loads. The allowable loads on rivets and bolts shall be calculated using the allowable bearing stresses in Table 20-I-C and the allowable shear stresses in Table 20-I-A. The allowable bearing stress depends on the ratio of edge distance to rivet or bolt diameter where the edge distance is the distance from the center of the rivet or bolt to the edge of the load-carrying member toward which the pressure of the rivet or bolt is directed.

Allowable bearing stresses on bolts apply to either threaded or unthreaded surfaces.

2010.1.2 Effective diameter. The effective diameter of rivets shall be taken as the hole diameter, but shall not exceed the nominal diameter of the rivet by 4 percent for cold driven rivets and 7 percent for hot driven rivets. The effective diameter of bolts shall be taken as the nominal diameter of the bolt.

2010.1.3 Shear area. The effective area of a rivet or bolt in any shear plane shall be based on the effective diameter except that for bolts with threads included in the shear plane, the effective shear area shall be based on the root diameter.

2010.1.4 Bearing area. The effective bearing area of rivets or bolts shall be the effective diameter multiplied by the length in bearing except that for countersunk rivets, half of the depth of the countersink shall be deducted from the length.

2010.1.5 Arrangements and strength of connections. Insofar as possible, connections shall be arranged so that the center of resistance of the connection shall coincide with the resultant line of action of the load. Where eccentricity exists, members and connections shall be proportioned to take into account any eccentricity of loading at the connections.

2010.1.6 Net section. The net section of a riveted or bolted tension member shall be determined as the sum of the net sections of its component parts. The net section of a part is the product of the thickness of the part multiplied by its least net width. The net width for a chain of holes extending across the part in any straight or broken line shall be obtained by deducting from the gross width the sum of the diameters of all the holes in the chain and adding $s^2/4g$ for each gage space in the chain. In the correction quantity $s^2/4g$, s denotes spacing parallel to the direction of the load (pitch) of any two successive holes in the chain, in inches, and g refers to gage, the spacing perpendicular to the direction of the load of the same holes, in inches (mm).

The net section of the part shall be obtained from that chain which gives the least net width. The hole diameter to be deducted shall be the actual hole diameter for drilled or reamed holes and the hole diameter plus $1/32$ inch (0.794 mm) for punched holes.

For angles, the gross width shall be the sum of the widths of the legs less the thickness. The gage for holes in opposite legs shall be the sum of the gages from the back of the angles less the thickness.

For splice members, the thickness shall be only that part of the thickness of the member that has been developed by rivets or bolts, beyond the section considered.

2010.1.7 Effective sections of angles. If a discontinuous angle (single or paired) in tension is connected to one side of a gusset plate, the effective net section shall be the net section of the connected leg plus one third of the section of the outstanding leg unless the outstanding leg is connected by a lug angle. In the latter case, the effective net section shall be the entire net section of the angle. The lug angle shall be designed to develop at least one half the total load in the member and shall be connected to the main member by at least two fasteners.

For double angles placed back to back and connected to both sides of a gusset plate, the effective net section shall be the net section of the connected legs plus two thirds of the section of the outstanding legs.

For intermediate joints of continuous angles, the effective net area shall be the gross sectional area less deductions for holes.

2010.1.8 Grip of rivets and bolts. If the grip (total thickness of metal being fastened) of rivets or bolts carrying calculated stress exceeds four and one-half times the diameter, the allowable load per rivet or bolt shall be reduced. The reduced allowable load shall be the normal allowable load divided by $[1/2 + G/(9\,D)]$ in which G is the grip and D is the nominal diameter of the rivet or bolt. If the grip of the rivet exceeds six times the diameter, special care shall be taken to ensure that holes will be filled completely.

2010.1.9 Spacing of rivets and bolts. Minimum distance of rivet centers shall be three times the nominal rivet diameter; minimum distance of bolt centers shall be two and one-half times the nominal bolt diameter. In built-up compression members, the pitch in the direction of stress shall be such that the allowable stress on the individual outside sheets and shapes treated as columns having a length equal to the rivet or bolt pitch exceeds the calculated stress. The gage at right angles to the direction of stress shall be such that the allowable stress in the outside sheets, calculated from Specification 9 of Table 20-I-C exceeds the calculated stress. In this case the width, b, may be taken as $0.8s$ where s is the gage in inches (mm).

2010.1.10 Stitch rivets and bolts. Where two or more web plates are in contact, there shall be stitch rivets or bolts to make them act in unison. In compression members, the pitch and gage of such rivets or bolts shall be determined as outlined in paragraph 9. In tension members, the maximum pitch or gage of such rivets or bolts shall not exceed a distance, in inches, equal to $(3 + 20t)$ in which t is the thickness of the outside plates, in inches (mm).

2010.1.11 Edge distance of rivets or bolts. The distance from the center of rivet or bolt under computed stress to the edge of the sheet or shape toward which the pressure is directed shall be twice the nominal diameter of the rivet or bolt. When a shorter edge distance is used, the allowable bearing stress as determined by Table 20-I-C shall be reduced by the ratio: actual edge distance/twice rivet or bolt diameter. The edge distance shall not be less than 1.5 times the rivet or bolt diameter to sheared, sawed, rolled or planed edges.

2010.1.12 Blind rivets. Blind rivets may be used only when the grip lengths and rivethole tolerances are as recommended by the respective manufacturers.

2010.1.13 Hollow-end rivets. If hollow-end rivets with solid cross sections for a portion of the length are used, the strength of these rivets may be taken equal to the strength of solid rivets of the same material, provided that the bottom of the cavity is at least 25 percent of the rivet diameter from the plane of shear as measured toward the hollow end; and, further, provided that they are used in locations where they will not be subjected to appreciable tensile stresses.

2010.1.14 Lock bolts. Lock bolts may be used when installed in conformance with the lock bolt manufacturer's recommended practices and provided the body diameter and bearing areas under the head and nut, or their equivalent, are not less than those of a conventional nut and bolt.

2010.2 Thread Forming (Tapping) Screws and Metal Stitching Staples. If joints carrying calculated loads are to be made with thread-forming screws or metal stitches, allowable strength values for these connections shall be established on the basis of specific acceptable tests.

2010.3 Fasteners for Structural Formed Sheet Roofing and Siding.

2010.3.1 General. Fasteners shall have tensile and tensile anchorage strengths in resisting back loads, or uplift, in excess of the strength of the connection between fastener and sheet.

2010.3.2 Allowable loads for fasteners. The allowable tensile load per fastener shall be:

$$P_t = (1/2.2) \times (\text{minimum strength of connection between fastener and sheet})$$

2010.3.3 Allowable loads for specific fasteners. The allowable loads for the specific fasteners listed, expressed in pounds (N), shall be used unless other allowable loads can be justified. Allowable loads for fasteners not listed shall be based on the results of

tests and shall comply with the provisions of Sections 2010.3.1 and 2010.3.2 above.

1. No. 14 stainless steel alloy self-tapping screws, hex head, cadmium plated, with composite aluminum-neoprene washer, the aluminum portion of which has minimum dimensions of 0.050-inch (1.27 mm) thickness and $5/8$-inch (16 mm) OD, or with a stainless steel neoprene washer, the stainless steel portion of which has minimum dimensions of 0.038-inch (0.965 mm) (No. 20 gage) thickness and $5/8$-inch (16 mm) OD. In crowns,

$$P_t = 140t \, F_{ty}$$
For **SI:**
$$P_t = 3.56t \, F_{ty}$$

and in valleys,

$$P_t = 170t \, F_{ty}$$
For **SI:**
$$P_t = 4.32t \, F_{ty}$$

For steel supporting members, screw holes should be made with a No. 8 drill for No. 14-gage through No. 11-gage material, a No. 4 drill for No. 10-gage up to $3/16$ inch (4.76 mm) and a No. 1 drill for $3/16$ inch (4.76 mm) and thicker.

2. Stainless steel alloy welded studs, $5/16$-inch-diameter (7.9 mm) base, $3/16$-inch-diameter (4.76 mm) serrated top, with field-installed swaged aluminum cap of $1/2$-inch (13 mm) diameter,

$$P_t = 230$$
For **SI:**
$$P_t = 1023 \, N$$

SECTION 2011 — FABRICATION

2011.1 Laying Out. Hole centers may be center punched and cutoff lines may be punched or scribed. Center punching and scribing shall not be used where such marks would remain on fabricated material.

A temperature correction shall be applied where necessary in the layout of critical dimensions. The coefficient of expansion shall be taken as 0.000013 per °F (0.0000072 per °C).

2011.2 Cutting. Material may be sheared, sawed, cut with a router or arc cut. All edges which have been cut by the arc process shall be planed to remove edge cracks.

Cut edges shall be true, smooth and free from excessive burrs or ragged breaks.

Re-entrant cuts shall be avoided wherever possible. If used, they shall be filleted by drilling prior to cutting.

Oxygen cutting of aluminum alloys shall not be permitted.

2011.3 Heating. Structural material shall not be heated.

> **EXCEPTION:** Material may be heated to a temperature not exceeding 400°F (204°C) for a period not exceeding 30 minutes in order to facilitate bending. Such heating shall be done only when proper temperature controls and supervision are provided to ensure that the limitations on temperature and time are carefully observed.

2011.4 Punching, Drilling and Reaming. The following rules for punching, drilling and reaming shall be observed:

1. Rivet or bolt holes may be either punched or drilled. Punching shall not be used if the metal thickness is greater than the diameter of the hole. The amount by which the diameter of a sub-punched hole is less than that of the finished hole shall be at least one fourth the thickness of the piece and in no case less than $1/32$ inch (0.8 mm).

2. The finished diameter of holes for cold-driven rivets shall not be more than 4 percent greater than the nominal diameter of the rivet.

3. The finished diameter of holes for hot-driven rivets shall not be more than 7 percent greater than the nominal diameter of the rivet.

4. The finished diameter of holes for bolts shall not be more than $1/16$ inch (1.6 mm) larger than the nominal bolt diameter.

5. If any holes must be enlarged to admit the rivets or bolts, they shall be reamed. Poor matching of holes shall be cause for rejection. Holes shall not be drifted in such a manner as to distort the metal. All chips lodged between contacting surfaces shall be removed before assembly.

2011.5 Riveting.

2011.5.1 Driven head. The driven head of aluminum alloy rivets shall be of the flat or the cone-point type with dimensions as follows:

1. Flat heads shall have a diameter not less than 1.4 times the nominal rivet diameter and a height not less than 0.4 times the nominal rivet diameter.

2. Cone-point heads shall have a diameter not less than 1.4 times the nominal rivet diameter and a height to the apex of the cone not less than 0.65 times the nominal rivet diameter. The included angle at the apex of the cone shall be approximately 127 degrees.

2011.5.2 Hole filling. Rivets shall fill holes completely. Rivet heads shall be concentric with the rivet holes and shall be in proper contact with the surface of the metal.

2011.5.3 Defective rivets. Defective rivets shall be removed by drilling.

2011.6 Painting.

2011.6.1 General. Structures of the alloys covered by these standards are not ordinarily painted (with the exception of 2014-T6 when exposed to corrosive environments). Surfaces shall be painted where:

1. The aluminum alloy parts are in contact with, or are fastened to, steel members or other dissimilar materials.

2. The structures are to be exposed to extremely corrosive conditions, or for reason of appearance. Painting procedure is covered in the following paragraphs and methods of cleaning and preparation are found in Section 2011.7. (Treatment and painting of the structure in accordance with United States Military Specification MIL-T-704 is also acceptable.)

2011.6.2 Contact with dissimilar materials. Where the aluminum alloy parts are in contact with, or are fastened to, steel members or other dissimilar materials, the aluminum shall be kept from direct contact with the steel or other dissimilar material by painting as follows:

1. Aluminum surfaces to be placed in contact with steel shall be given one coat of zinc chromate primer in accordance with Federal Specification TT-P-645 or the equivalent, or one coat of a suitable nonhardening joint compound capable of excluding moisture from the joint during prolonged service. Where severe corrosion conditions are expected, additional protection can be obtained by applying the joint compound in addition to the zinc chromate primer. Zinc chromate paint shall be allowed to dry hard (air dry 24 hours) before assembly of the parts. The steel surfaces to be placed in contact with aluminum shall be painted with good quality priming paint, such as zinc chromate primer in accordance with Federal Specification TT-P-645, followed by one coat of paint consisting of 2 pounds of aluminum paste pigment (ASTM Specification D 96266, Type 2, Class B) per gallon (0.24 kg/L) of varnish meeting Federal Specification TT-V-81d, Type II, or the

equivalent. Stainless steel, or aluminized, hot-dip galvanized or electrogalvanized steel placed in contact with aluminum need not be painted.

2. When aluminum is in direct contact with wood, fiberboard or other porous material that may absorb water, an insulating barrier shall be installed between the aluminum and the porous material. Such aluminum surfaces shall be given a heavy coat of alkali-resistant bituminous paint or other coating providing equivalent protection before installation. Aluminum in contact with concrete or masonry shall be similarly protected in cases where moisture is present and corrodents can be entrapped between the surfaces.

3. Aluminum surfaces to be embedded in concrete ordinarily need not be painted, unless corrosive components are added to the concrete or unless the concrete is subjected for extended periods to extremely corrosive conditions. In such cases, aluminum surfaces shall be given one coat of suitable quality paint, such as zinc chromate primer conforming to Federal Specification TT-P-645 or equivalent, or shall be wrapped with a suitable plastic tape applied in such a manner as to provide adequate protection at the overlap.

4. Water that comes in contact with aluminum after first running over a heavy metal such as copper may contain trace quantities of the dissimilar metal or its corrosion product, which will cause corrosion of the aluminum. Protection shall be obtained by painting or plastic coating the dissimilar metal or by designing the structure so that the drainage from the dissimilar metal is diverted away from the aluminum.

2011.6.3 Overall painting. Structures of the alloys covered by this standard are either not ordinarily painted for surface protection (with the exception of 2014-T6 when exposed to corrosive environments) or are made of prepainted aluminum components. There may be applications where the structures are to be exposed to extremely corrosive conditions. In these cases overall painting shall be specified.

2011.7 Cleaning and Treatment of Metal Surfaces. Prior to field painting of structures, all surfaces to be painted shall be cleaned immediately before painting by a method that will remove all dirt, oil, grease, chips and other foreign substances.

Exposed metal surfaces shall be cleaned with a suitable chemical cleaner such as a solution of phosphoric acid and organic solvents meeting United States Military Specification MIL-M-10578. If the metal is more than $1/8$ inch (3.2 mm) thick, sandblasting may be used.

SECTION 2012 — WELDED CONSTRUCTION

2012.1 Filler Wire. Verification shall be provided to show that the choice of filler metal for general purpose welding is appropriate.

2012.2 Columns and Single-web Beams with Welds at Locations Other than Ends and Cantilever Columns and Single-web Beams. The allowable stresses determined in accordance with the provisions of Division I apply to members supported at both ends with welds at the ends only (not farther from the supports than 0.05 L from the ends).

For columns with transverse welds at locations other than the supports, cantilever columns with transverse welds at or near the supported end and columns with longitudinal welds having A_w equal to or greater than 15 percent of A, the effect of welding on column strength shall be taken into account by using an increased slenderness ratio, L_w/r, in the column formula, as follows:

$$\text{If } \frac{L}{r} > \sqrt{\frac{250,000}{F_{cyw}}} \; ; \frac{L_w}{r} = \frac{L}{r}$$

For **SI:**   If $\dfrac{L}{r} > \sqrt{\dfrac{24.4E}{F_{cyw}}}$; $\dfrac{L_w}{r} = \dfrac{L}{r}$

                      If $\dfrac{L}{r} \leq \sqrt{\dfrac{250,000}{F_{cyw}}}$;

For **SI:**   If $\dfrac{L}{r} \leq \sqrt{\dfrac{24.4E}{F_{cyw}}}$;

$$\dfrac{L_w}{r} = \dfrac{L}{r} \sqrt{\dfrac{1 + 100\frac{L_h}{L}}{1 + \left(\frac{L_h}{L}\right)\left(\frac{L}{r}\right)^2 \left(\frac{F_{cyw}}{2500}\right)}}$$

For **SI:** $$\dfrac{L_w}{r} = \dfrac{L}{r} \sqrt{\dfrac{1 + 100\frac{L_h}{L}}{1 + \left(\frac{L_h}{L}\right)\left(\frac{L}{r}\right)^2 \left(\frac{4.1F_{cyw}}{E}\right)}}$$

The above formulas assume that the entire cross section within the length, L_h, is affected by the heat of welding. If only part of the cross section is so affected, the allowable stress based on L_w/r shall be substituted for F_w in the formula in Section 2002.2.

2012.3 Welding Fabrication. Welding of aluminum shall be in accordance with approved nationally recognized standards.

SECTION 2013 — TESTING

2013.1 General. Testing shall be considered an acceptable method for substantiating the design of aluminum alloy load-carrying members. Tests shall be conducted by an independent testing laboratory or by a manufacturer's testing laboratory.

2013.2 Test Loading and Behavior. In order to test a structure or load-carrying member adequately, the loading shall be applied in a fashion that reasonably approximates the application of the loading during service. Further, the structure or member shall be supported in a manner that is no more sustaining to the structure than the supports available will be when the structure is in service.

Determination of allowable load-carrying capacity shall be made on the basis that the member, assembly or connection shall be capable of sustaining during the test without failure a total load, including the weight of the test specimen, equal to twice the live load plus one and one-half the dead load. Furthermore, harmful local distortions shall not develop during the test at a total load, including the weight of the test specimen, equal to the dead load plus one and one-half times the live load.

The factors by which the design live and dead loads are multiplied to determine the test loads are reduced to three fourths of the values given in the preceding paragraph when wind or seismic forces represent all or a portion of the live load, provided the structure or member meets the test requirements with the full load factors applied to the dead load and to that portion of the live load not attributable to wind or seismic forces.

Differences that may exist between nominal section properties and those of tested sections shall be considered.

TABLE 20-II-A—MINIMUM MECHANICAL PROPERTIES FOR ALUMINUM ALLOYS
Values Are Given in Units of ksi (1,000 lb/in²)

ALLOY AND TEMPER	PRODUCT[1]	THICKNESS RANGE[1] (inch) × 25.4 for mm	TENSION[2] F_{tu} ksi	F_{ty} ksi	COMPRESSION F_{cy} ksi	SHEAR F_{su} ksi	F_{sy} ksi	BEARING F_{bu} ksi	F_{by} ksi	COMPRESSIVE MODULUS OF ELASTICITY[3] E ksi
						× 6.89 for MPa				
1100-H12	Sheet, plate	All	14	11	10	9	6.5	28	18	10,100
-H14	Rolled rod and bar Drawn tube	All	16	14	13	10	8	32	21	10,100
2014-T6	Sheet	0.040-0.249	66	58	59	40	33	125	93	10,900
-T651	Plate	0.250-2.000	67	59	58	40	34	127	94	10,900
-T6, -T6510[1]	Extrusions	All	60	53	55	35	31	114	85	10,900
-T6, -T651	Rolled rod and bar Drawn tube	All	65	55	53	38	32	124	88	10,900
Alclad 2014-T6	Sheet	0.020-0.039	63	55	56	38	32	120	88	10,800
-T6	Sheet	0.040-0.249	64	57	58	39	33	122	91	10,800
-T651	Plate	0.250-0.499	64	57	56	39	33	122	91	10,800
3003-H12	Sheet and plate	0.017-2.000	17	12	10	11	7	34	19	10,100
-H14	Sheet and plate	0.009-1.000	20	17	14	12	10	40	25	10,100
-H16	Sheet	0.006-0.162	24	21	18	14	12	46	31	10,100
-H18	Sheet	0.006-0.128	27	24	20	15	14	49	34	10,100
3003-H12	Drawn tube	All	17	12	11	11	7	34	19	10,100
-H14	Drawn tube	All	20	17	16	12	10	40	25	10,100
-H16	Drawn tube	All	24	21	19	14	12	46	31	10,100
-H18	Drawn tube	All	27	24	21	15	14	49	34	10,100
Alclad 3003-H12	Sheet and plate	0.017-2.000	16	11	9	10	6.5	32	18	10,100
-H14	Sheet and plate	0.009-1.000	19	16	13	12	9	38	24	10,100
-H16	Sheet	0.006-0.162	23	20	17	14	12	44	30	10,100
-H18	Sheet	0.006-0.128	26	23	19	15	13	47	32	10,100
Alclad 3003-H14	Drawn tube	0.010-0.500	19	16	15	12	9	38	24	10,100
-H18	Drawn Tube	0.010-0.500	26	23	20	15	13	47	32	10,100
3004-H32	Sheet and plate	0.017-2.000	28	21	18	17	12	56	36	10,100
-H34	Sheet and plate	0.009-1.000	32	25	22	19	14	64	40	10,100
-H36	Sheet	0.006-0.162	35	28	25	20	16	70	45	10,100
3004-H34	Drawn tube	0.018-0.450	32	25	24	19	14	64	40	10,100
-H36	Drawn tube	0.018-0.450	35	28	27	20	16	70	45	10,100
Alclad 3004-H32	Sheet	0.017-0.249	27	20	17	16	12	54	34	10,100
-H34	Sheet	0.009-0.249	31	24	21	18	14	62	38	10,100
-H36	Sheet	0.006-0.162	34	27	24	19	16	68	43	10,100
-H14	Sheet	0.009-0.249	32	26	22	19	15	64	39	10,100
-H16	Sheet	0.006-0.050	35	30	28	20	17	66	45	10,100
-H16	Sheet	0.051-0.162	35	30	26	20	17	66	45	10,100
-H131, -H241, -H341	Sheet	0.024-0.050	31	26	22	18	15	62	39	10,100
-H151, -H261, -H361	Sheet	0.024-0.050	34	30	28	19	17	66	45	10,100
3005-H25	Sheet	0.013-0.050	26	22	20	15	13	49	35	10,100
3006-H391	Sheet	0.010-0.050	31	27	27	20	16	60	44	10,100
3105-H25	Sheet	0.013-0.080	23	19	17	14	11	44	28	10,100
5005-H12	Sheet and plate	0.018-2.000	18	14	13	11	8	34	22	10,100
-H14	Sheet and plate	0.009-1.000	21	17	15	12	10	40	25	10,100
-H16	Sheet	0.006-0.162	24	20	18	14	12	48	30	10,100
-H32	Sheet and plate	0.017-2.000	17	12	11	11	7	34	20	10,100
-H34	Sheet and plate	0.009-1.000	20	15	14	12	8.5	40	24	10,100
-H36	Sheet	0.006-0.162	23	18	16	13	11	48	29	10,100
5050-H32	Sheet	0.017-0.249	22	16	14	14	9	44	27	10,100
-H34	Sheet	0.009-0.249	25	20	18	15	12	50	32	10,100
-H32	Rolled rod and bar Drawn tube	All	22	16	15	13	9	44	27	10,100
-H34	Rolled rod and bar Drawn tube	All	25	20	19	15	12	50	32	10,100
5052-H32	Sheet and plate	All	31	23	21	19	13	60	39	10,200
-H34	Rolled rod and bar Drawn tube	All	34	26	24	20	15	65	44	10,200
-H36	Sheet	0.006-0.162	37	29	26	22	17	70	46	10,200

(Continued)

TABLE 20-II-A

1997 UNIFORM BUILDING CODE

TABLE 20-II-A—MINIMUM MECHANICAL PROPERTIES FOR ALUMINUM ALLOYS—(Continued)
Values Are Given in Units of ksi (1,000 lb/in²)

ALLOY AND TEMPER	PRODUCT[1]	THICKNESS RANGE[1] (inch) ×25.4 for mm	TENSION[2] F_{tu} ksi	TENSION[2] F_{ty} ksi	COMPRES-SION F_{cy} ksi	SHEAR F_{su} ksi	SHEAR F_{sy} ksi	BEARING F_{bu} ksi	BEARING F_{by} ksi	COMPRESSIVE MODULUS OF ELASTICITY[3] E ksi
						×6.89 for MPa				
5083 -H111	Extrusions	up to 0.500	40	24	21	24	14	78	41	10,400
-H111	Extrusions	0.501 and over	40	24	21	23	14	78	38	10,400
-H321	Sheet and plate	0.188-1.500	44	31	26	26	18	84	53	10,400
-H323	Sheet	0.051-0.249	45	34	32	26	20	88	58	10,400
-H343	Sheet	0.051-0.249	50	39	37	29	23	95	66	10,400
-H321	Plate	1.501-3.000	41	29	24	24	17	78	49	10,400
5086 -H111	Extrusions	up to 0.500	36	21	18	21	12	70	36	10,400
-H111	Extrusions	0.501 and over	36	21	18	21	12	70	34	10,400
-H112	Plate	0.250-0.499	36	18	17	22	10	72	31	10,400
-H112	Plate	0.500-1.000	35	16	16	21	9	70	28	10,400
-H112	Plate	1.001-2.000	35	14	15	21	8	70	28	10,400
-H112	Plate	2.001-3.000	34	14	15	21	8	68	28	10,400
-H32	Sheet and plate	All	40	28	26	24	16	78	48	10,400
-H34	Drawn tube	All	44	34	32	26	20	84	58	10,400
5154-H38	Sheet	0.006-0.128	45	35	33	24	20	81	56	10,300
5454-H111	Extrusions	up to 0.500	33	19	16	20	11	64	32	10,400
-H111	Extrusions	0.501 and over	33	19	16	19	11	64	30	10,400
-H112	Extrusions	up to 5.000	31	12	13	19	7	62	24	10,400
-H32	Sheet and plate	0.020-2.000	36	26	24	21	15	70	44	10,400
-H34	Sheet and plate	0.020-1.000	39	29	27	23	17	74	49	10,400
5456-H111	Extrusions	up to 0.500	42	26	22	25	15	82	44	10,400
-H111	Extrusions	0.501 and over	42	26	22	24	15	82	42	10,400
-H112	Extrusions	up to 5.000	41	19	20	24	11	82	38	10,400
-H321	Sheet and plate	0.188-1.250	46	33	27	27	19	87	56	10,400
-H321	Plate	1.251-1.500	44	31	25	25	18	84	53	10,400
-H321	Plate	1.501-3.000	41	29	25	25	17	82	49	10,400
-H323	Sheet	0.051-0.249	48	36	34	28	21	94	61	10,400
-H343	Sheet	0.051-0.249	53	41	39	31	24	101	70	10,400
6005-T5	Extrusions	up to 0.500	38	35	35	24	20	80	56	10,100
6061-T6, -T651	Sheet and plate	0.010-4.000	42	35	35	27	20	88	58	10,100
-T6, -T6510[1]	Extrusions	up to 3.000	38	35	35	24	20	80	56	10,100
-T6, -T651	Rolled rod and bar	up to 8.000	42	35	35	27	20	88	56	10,100
-T6	Drawn tube	0.025-0.500	42	35	35	27	20	88	56	10,100
-T6	Pipe	up to 0.999	42	35	35	27	20	88	56	10,100
-T6	Pipe	over 0.999	38	35	35	24	20	80	56	10,100
6063-T5	Extrusions	up to 0.500	22	16	16	13	9	46	26	10,100
-T5	Extrusions	over 0.500	21	15	15	12	8.5	44	24	10,100
-T6	Extrusions Pipe	All	30	25	25	19	14	63	40	10,100
6351-T5	Extrusions	up to 1.00	38	35	35	24	20	80	56	10,100

[1]Values also apply to -T6511 temper.

[2]F_{tu} and F_{ty} are minimum specified values (except for Alclad 3004-H14, -H16 and F_{ty} for Alclad 3003-H18). Other strength properties are corresponding minimum expected values.

[3]For deflection calculations an average modulus of elasticity is used; numerically this is 100 ksi (689 MPa) lower than the values in this column.

TABLE 20-II-B—MINIMUM MECHANICAL PROPERTIES FOR WELDED ALUMINUM ALLOYS[1]
(Gas Tungsten Arc or Gas Metal Arc Welding with No Postweld Heat Treatment)

ALLOY AND TEMPER	PRODUCT AND THICKNESS RANGE (inch)	TENSION		COMPRESSION	SHEAR		BEARING	
		F_{tuw}[1] ksi	F_{tyw}[2] ksi	F_{cyw}[2] ksi	F_{suw} ksi	F_{syw} ksi	F_{buw} ksi	F_{byw} ksi
	× 25.4 for mm	× 6.89 for MPa						
1100-H12, -H14	All	11	4.5	4.5	8	2.5	23	8
3003-H12, -H14, -H16, -H18	All	14	7	7	10	4	30	12
Alclad 3003-H12, -H14, -H16, -H18	All	13	6	6	10	3.5	30	11
3004-H32, -H34, -H36	All	22	11	11	14	6.5	46	20
Alclad 3004-H32, -H34, -H14, -H16	All	21	11	11	13	6.5	44	19
3005-H25	Sheet 0.013-0.050	17	9	9	12	5	36	15
5005-H12, -H14, -H32, -H34	All	14	7	7	9	4	28	10
5050-H32, -H34	All	18	8	8	12	4.5	36	12
5052-H32, -H34	All	25	13	13	16	7.5	50	19
5083-H111 -H321 -H321 -H323, -H343	Extrusions / Sheet and plate 0.188-1.500 / Plate 1.501-3.000 / Sheet	39 / 40 / 39 / 40	21 / 24 / 23 / 24	20 / 24 / 23 / 24	23 / 24 / 24 / 24	12 / 14 / 13 / 14	78 / 80 / 78 / 80	32 / 36 / 34 / 36
5086-H111 -H112 -H112 -H112 -H32, -H34	Extrusions / Plate 0.250-0.499 / Plate 0.500-1.000 / Plate 1.001-2.000 / Sheet and plate	35 / 35 / 35 / 35 / 35	18 / 17 / 16 / 14 / 19	17 / 17 / 16 / 14 / 19	21 / 21 / 21 / 21 / 21	10 / 9.5 / 9 / 8 / 11	70 / 70 / 70 / 70 / 70	28 / 28 / 28 / 28 / 28
5086-H111 -H112 -H112 -H112 -H32, -H34	Extrusions / Plate 0.250-0.499 / Plate 0.500-1.000 / Plate 1.001-2.000 / Sheet and plate	35 / 35 / 35 / 35 / 35	18 / 17 / 16 / 14 / 19	17 / 17 / 16 / 14 / 19	21 / 21 / 21 / 21 / 21	10 / 9.5 / 9 / 8 / 11	70 / 70 / 70 / 70 / 70	28 / 28 / 28 / 28 / 28
5154-H38	Sheet	30	15	15	19	8.5	60	23
5454-H111 -H112 -H32, -H34	Extrusions / Extrusions / Sheet and plate	31 / 31 / 31	16 / 12 / 16	15 / 12 / 16	19 / 19 / 19	9.5 / 7 / 9.5	62 / 62 / 62	24 / 24 / 24
5456-H111 -H112 -H321 -H321 -H323, -H343	Extrusions / Extrusions / Sheet and plate 0.188-1.500 / Plate 1.501-3.000 / Sheet	41 / 41 / 42 / 41 / 42	24 / 19 / 26 / 24 / 26	22 / 19 / 24 / 23 / 26	24 / 24 / 25 / 25 / 25	14 / 11 / 15 / 14 / 15	82 / 82 / 84 / 82 / 84	38 / 38 / 38 / 36 / 38
6005-T5	Extrusions Up to 0.250	24	17	17	15	10	50	30
6061-T6, -T651[3] -T6, -T651[4]	All / Over 0.375	24 / 24	20 / 15	20 / 15	15 / 15	12 / 9	50 / 50	30 / 30
6063-T5, -T6	All	17	11	11	11	6.5	34	22
6351-T5[1] -T5[4]	Extrusions / Over 0.375	24 / 24	20 / 15	20 / 15	15 / 15	12 / 9	50 / 50	30 / 30

[1]Values of F_{tuw} are ASME weld qualification test values.
[2]0.2 percent offset in 10-inch (254 mm) gauge length across a butt weld.
[3]Values when welded with 5183, 5356 or 5556 alloy filler wire regardless of thickness. Values also apply to thicknesses less than 0.375 inch (9.5 mm) when welded with 4043, 5154, 5254 or 5554 alloy filler wire.
[4]Values when welded with 4043, 5154, 5254 or 5554 alloy filler wire.

Chapter 21
MASONRY

SECTION 2101 — GENERAL

2101.1 Scope. The materials, design, construction and quality assurance of masonry shall be in accordance with this chapter.

2101.2 Design Methods. Masonry shall comply with the provisions of one of the following design methods in this chapter as well as the requirements of Sections 2101 through 2105.

2101.2.1 Working stress design. Masonry designed by the working stress design method shall comply with the provisions of Sections 2106 and 2107.

2101.2.2 Strength design. Masonry designed by the strength design method shall comply with the provisions of Sections 2106 and 2108.

2101.2.3 Empirical design. Masonry designed by the empirical design method shall comply with the provisions of Sections 2106.1 and 2109.

2101.2.4 Glass masonry. Glass masonry shall comply with the provisions of Section 2110.

2101.3 Definitions. For the purpose of this chapter, certain terms are defined as follows:

AREAS:

Bedded Area is the area of the surface of a masonry unit which is in contact with mortar in the plane of the joint.

Effective Area of Reinforcement is the cross-sectional area of reinforcement multiplied by the cosine of the angle between the reinforcement and the direction for which effective area is to be determined.

Gross Area is the total cross-sectional area of a specified section.

Net Area is the gross cross-sectional area minus the area of ungrouted cores, notches, cells and unbedded areas. Net area is the actual surface area of a cross section of masonry.

Transformed Area is the equivalent area of one material to a second based on the ratio of moduli of elasticity of the first material to the second.

BOND:

Adhesion Bond is the adhesion between masonry units and mortar or grout.

Reinforcing Bond is the adhesion between steel reinforcement and mortar or grout.

BOND BEAM is a horizontal grouted element within masonry in which reinforcement is embedded.

CELL is a void space having a gross cross-sectional area greater than $1^1/_2$ square inches (967 mm^2).

CLEANOUT is an opening to the bottom of a grout space of sufficient size and spacing to allow the removal of debris.

COLLAR JOINT is the mortared or grouted space between wythes of masonry.

COLUMN, REINFORCED, is a vertical structural member in which both the reinforcement and masonry resist compression.

COLUMN, UNREINFORCED, is a vertical structural member whose horizontal dimension measured at right angles to the thickness does not exceed three times the thickness.

DIMENSIONS:

Actual Dimensions are the measured dimensions of a designated item. The actual dimension shall not vary from the specified dimension by more than the amount allowed in the appropriate standard of quality in Section 2102.

Nominal Dimensions of masonry units are equal to its specified dimensions plus the thickness of the joint with which the unit is laid.

Specified Dimensions are the dimensions specified for the manufacture or construction of masonry, masonry units, joints or any other component of a structure.

GROUT LIFT is an increment of grout height within the total grout pour.

GROUT POUR is the total height of masonry wall to be grouted prior to the erection of additional masonry. A grout pour will consist of one or more grout lifts.

GROUTED MASONRY:

Grouted Hollow-unit Masonry is that form of grouted masonry construction in which certain designated cells of hollow units are continuously filled with grout.

Grouted Multiwythe Masonry is that form of grouted masonry construction in which the space between the wythes is solidly or periodically filled with grout.

JOINTS:

Bed Joint is the mortar joint that is horizontal at the time the masonry units are placed.

Head Joint is the mortar joint having a vertical transverse plane.

MASONRY UNIT is brick, tile, stone, glass block or concrete block conforming to the requirements specified in Section 2102.

Hollow-masonry Unit is a masonry unit whose net cross-sectional areas (solid area) in any plane parallel to the surface containing cores, cells or deep frogs is less than 75 percent of its gross cross-sectional area measured in the same plane.

Solid-masonry Unit is a masonry unit whose net cross-sectional area in any plane parallel to the surface containing the cores or cells is at least 75 percent of the gross cross-sectional area measured in the same plane.

PRISM is an assemblage of masonry units and mortar with or without grout used as a test specimen for determining properties of the masonry.

REINFORCED MASONRY is that form of masonry construction in which reinforcement acting in conjunction with the masonry is used to resist forces.

SHELL is the outer portion of a hollow masonry unit as placed in masonry.

WALLS

Bonded Wall is a masonry wall in which two or more wythes are bonded to act as a structural unit.

Cavity Wall is a wall containing continuous air space with a minimum width of 2 inches (51 mm) and a maximum width of

$4^1/_2$ inches (114 mm) between wythes which are tied with metal ties.

WALL TIE is a mechanical metal fastener which connects wythes of masonry to each other or to other materials.

WEB is an interior solid portion of a hollow-masonry unit as placed in masonry.

WYTHE is the portion of a wall which is one masonry unit in thickness. A collar joint is not considered a wythe.

2101.4 Notations.

A_b = cross-sectional area of anchor bolt, square inches (mm^2).

A_e = effective area of masonry, square inches (mm^2).

A_g = gross area of wall, square inches (mm^2).

A_{jh} = total area of special horizontal reinforcement through wall frame joint, square inches (mm^2).

A_{mv} = net area of masonry section bounded by wall thickness and length of section in direction of shear force considered, square inches (mm^2).

A_p = area of tension (pullout) cone of embedded anchor bolt projected onto surface of masonry, square inches (mm^2).

A_s = effective cross-sectional area of reinforcement in column or flexural member, square inches (mm^2).

A_{se} = effective area of reinforcement, square inches (mm^2).

A_{sh} = total cross-sectional area of rectangular tie reinforcement for confined core, square inches (mm^2).

A_v = area of reinforcement required for shear reinforcement perpendicular to longitudinal reinforcement, square inches (mm^2).

A'_s = effective cross-sectional area of compression reinforcement in flexural member, square inches (mm^2).

a = depth of equivalent rectangular stress block, inches (mm).

B_{sn} = nominal shear strength of anchor bolt, pounds (N).

B_t = allowable tensile force on anchor bolt, pounds (N).

B_{tn} = nominal tensile strength of anchor bolt, pounds (N).

B_v = allowable shear force on anchor bolt, pounds (N).

b = effective width of rectangular member or width of flange for T and I sections, inches (mm).

b_{su} = factored shear force supported by anchor bolt, pounds (N).

b_t = computed tensile force on anchor bolt, pounds (N).

b_{tu} = factored tensile force supported by anchor bolt, pounds (N).

b_v = computed shear force on anchor bolt, pounds (N).

b' = width of web in T or I section, inches (mm).

C_d = nominal shear strength coefficient as obtained from Table 21-K.

c = distance from neutral axis to extreme fiber, inches (mm).

D = dead loads, or related internal moments and forces.

d = distance from compression face of flexural member to centroid of longitudinal tensile reinforcement, inches (mm).

d_b = diameter of reinforcing bar, inches (mm).

d_{bb} = diameter of largest beam longitudinal reinforcing bar passing through, or anchored in, a joint, inches (mm).

d_{bp} = diameter of largest pier longitudinal reinforcing bar passing through a joint, inches (mm).

E = load effects of earthquake, or related internal moments and forces.

E_m = modulus of elasticity of masonry, pounds per square inch (MPa).

e = eccentricity of P_{uf}, inches (mm).

e_{mu} = maximum usable compressive strain of masonry.

F = loads due to weight and pressure of fluids or related moments and forces.

F_a = allowable average axial compressive stress in columns for centroidally applied axial load only, pounds per square inch (MPa).

F_b = allowable flexural compressive stress in members subjected to bending load only, pounds per square inch (MPa).

F_{br} = allowable bearing stress in masonry, pounds per square inch (MPa).

F_s = allowable stress in reinforcement, pounds per square inch (MPa).

F_{sc} = allowable compressive stress in column reinforcement, pounds per square inch (MPa).

F_t = allowable flexural tensile stress in masonry, pounds per square inch (MPa).

F_v = allowable shear stress in masonry, pounds per square inch (MPa).

f_a = computed axial compressive stress due to design axial load, pounds per square inch (MPa).

f_b = computed flexural stress in extreme fiber due to design bending loads only, pounds per square inch (MPa).

f_{md} = computed compressive stress due to dead load only, pounds per square inch (MPa).

f_r = modulus of rupture, pounds per square inch (MPa).

f_s = computed stress in reinforcement due to design loads, pounds per square inch (MPa).

f_v = computed shear stress due to design load, pounds per square inch (MPa).

f_y = tensile yield stress of reinforcement, pounds per square inch (MPa).

f_{yh} = tensile yield stress of horizontal reinforcement, pounds per square inch (MPa).

f'_g = specified compressive strength of grout at age of 28 days, pounds per square inch (MPa).

f'_m = specified compressive strength of masonry at age of 28 days, pounds per square inch (MPa).

G = shear modulus of masonry, pounds per square inch (MPa).

H = loads due to weight and pressure of soil, water in soil or related internal moments and forces.

h = height of wall between points of support, inches (mm).

h_b = beam depth, inches (mm).

h_c = cross-sectional dimension of grouted core measured center to center of confining reinforcement, inches (mm).

h_p = pier depth in plane of wall frame, inches (mm).

h' = effective height of wall or column, inches (mm).

I = moment of inertia about neutral axis of cross-sectional area, inches^4 (mm^4).

I_e = effective moment of inertia, inches^4 (mm^4).

I_g, I_{cr} = gross, cracked moment of inertia of wall cross section, inches^4 (mm^4).

j = ratio or distance between centroid of flexural compressive forces and centroid of tensile forces of depth, d.

K = reinforcement cover or clear spacing, whichever is less, inches (mm).

k = ratio of depth of compressive stress in flexural member to depth, d.

L = live loads, or related internal moments and forces.

L_w = length of wall, inches (mm).

l = length of wall or segment, inches (mm).

l_b = embedment depth of anchor bolt, inches (mm).

l_{be} = anchor bolt edge distance, the least distance measured from edge of masonry to surface of anchor bolt, inches (mm).

l_d = required development length of reinforcement, inches (mm).

M = design moment, inch-pounds (N·mm).

M_a = maximum moment in member at stage deflection is computed, inch-pounds (N·mm).

M_c = moment capacity of compression reinforcement in flexural member about centroid of tensile force, inch-pounds (N·mm).

M_{cr} = nominal cracking moment strength in masonry, inch-pounds (N·mm).

M_m = moment of compressive force in masonry about centroid of tensile force in reinforcement, inch-pounds (N·mm).

M_n = nominal moment strength, inch-pounds (N·mm).

M_s = moment of tensile force in reinforcement about centroid of compressive force in masonry, inch-pounds (N·mm).

M_{ser} = service moment at midheight of panel, including $P\Delta$ effects, inch-pounds (N·mm).

M_u = factored moment, inch-pounds (N·mm).

n = modular ratio.

= E_s/E_m.

P = design axial load, pounds (N).

P_a = allowable centroidal axial load for reinforced masonry columns, pounds (N).

P_b = nominal balanced design axial strength, pounds (N).

P_f = load from tributary floor or roof area, pounds (N).

P_n = nominal axial strength in masonry, pounds (N).

P_o = nominal axial load strength in masonry without flexure, pounds (N).

P_u = factored axial load, pounds (N).

P_{uf} = factored load from tributary floor or roof loads, pounds (N).

P_{uw} = factored weight of wall tributary to section under consideration, pounds (N).

P_w = weight of wall tributary to section under consideration, pounds (N).

r = radius of gyration (based on specified unit dimensions or Tables 21-H-1, 21-H-2 and 21-H-3), inches (mm).

r_b = ratio of area of reinforcing bars cut off to total area of reinforcing bars at the section.

S = section modulus, inches^3 (mm^3).

s = spacing of stirrups or of bent bars in direction parallel to that of main reinforcement, inches (mm).

T = effects of temperature, creep, shrinkage and differential settlement.

t = effective thickness of wythe, wall or column, inches (mm).

U = required strength to resist factored loads, or related internal moments and forces.

u = bond stress per unit of surface area of reinforcing bar, pounds per square inch (MPa).

V = total design shear force, pounds (N).

V_{jh} = total horizontal joint shear, pounds (N).

V_m = nominal shear strength of masonry, pounds (N).

V_n = nominal shear strength, pounds (N).

V_s = nominal shear strength of shear reinforcement, pounds (N).

V_u = required shear strength in masonry, pounds (N).

W = wind load, or related internal moments in forces.

w_u = factored distributed lateral load.

Δ_s = horizontal deflection at midheight under factored load, inches (mm).

Δ_u = deflection due to factored loads, inches (mm).

ρ = ratio of area of flexural tensile reinforcement, A_s, to area bd.

ρ_b = reinforcement ratio producing balanced strain conditions.

ρ_n = ratio of distributed shear reinforcement on plane perpendicular to plane of A_{mv}.

Σ_o = sum of perimeters of all longitudinal reinforcement, inches (mm).

$\sqrt{f'_m}$ = square root of specified strength of masonry at the age of 28 days, pounds per square inch (MPa).

ϕ = strength-reduction factor.

SECTION 2102 — MATERIAL STANDARDS

2102.1 Quality. Materials used in masonry shall conform to the requirements stated herein. If no requirements are specified in this section for a material, quality shall be based on generally accepted good practice, subject to the approval of the building official.

Reclaimed or previously used masonry units shall meet the applicable requirements as for new masonry units of the same material for their intended use.

2102.2 Standards of Quality. The standards listed below labeled a "UBC Standard" are also listed in Chapter 35, Part II, and are part of this code. The other standards listed below are recognized standards. See Sections 3503 and 3504.

1. **Aggregates.**

    1.1 ASTM C 144, Aggregates for Masonry Mortar

    1.2 ASTM C 404, Aggregates for Grout

2. **Cement.**

    2.1 UBC Standard 21-11, Cement, Masonry. (Plastic cement conforming to the requirements of UBC Standard 25-1 may be used in lieu of masonry cement when it also conforms to UBC Standard 21-11.)

    2.2 ASTM C 150, Portland Cement

    2.3 UBC Standard 21-14, Mortar Cement

3. **Lime.**

    3.1 UBC Standard 21-12, Quicklime for Structural Purposes

    3.2 UBC Standard 21-13, Hydrated Lime for Masonry Purposes. When Types N and NA hydrated lime are used in masonry mortar, they shall comply with the provisions of UBC Standard 21-15, Section 21.1506.7, excluding the plasticity requirement.

4. **Masonry units of clay or shale.**

    4.1 ASTM C 34, Structural Clay Load-bearing Wall Tile

    4.2 ASTM C 56, Structural Clay Nonload-bearing Tile

    4.3 UBC Standard 21-1, Section 21.101, Building Brick (solid units)

    4.4 ASTM C 126, Ceramic Glazed Structural Clay Facing Tile, Facing Brick and Solid Masonry Units. Load-bearing glazed brick shall conform to the weathering and structural requirements of UBC Standard 21-1, Section 21.106, Facing Brick

    4.5 UBC Standard 21-1, Section 21.106, Facing Brick (solid units)

    4.6 UBC Standard 21-1, Section 21.107, Hollow Brick

    4.7 ASTM C 67, Sampling and Testing Brick and Structural Clay Tile

    4.8 ASTM C 212, Structural Clay Facing Tile

    4.9 ASTM C 530, Structural Clay Non-Loadbearing Screen Tile

5. **Masonry units of concrete.**

    5.1 UBC Standard 21-3, Concrete Building Brick

    5.2 UBC Standard 21-4, Hollow and Solid Load-bearing Concrete Masonry Units

    5.3 UBC Standard 21-5, Nonload-bearing Concrete Masonry Units

    5.4 ASTM C 140, Sampling and Testing Concrete Masonry Units

    5.5 ASTM C 426, Standard Test Method for Drying Shrinkage of Concrete Block

6. **Masonry units of other materials.**

    6.1 **Calcium silicate.**

    UBC Standard 21-2, Calcium Silicate Face Brick (Sand-lime Brick)

    6.2 UBC Standard 21-9, Unburned Clay Masonry Units and Standard Methods of Sampling and Testing Unburned Clay Masonry Units

    6.3 ACI-704, Cast Stone

    6.4 UBC Standard 21-17, Test Method for Compressive Strength of Masonry Prisms

7. **Connectors.**

    7.1 Wall ties and anchors made from steel wire shall conform to UBC Standard 21-10, Part II, and other steel wall ties and anchors shall conform to A 36 in accordance with UBC Standard 22-1. Wall ties and anchors made from copper, brass or other nonferrous metal shall have a minimum tensile yield strength of 30,000 psi (207 MPa).

    7.2 All such items not fully embedded in mortar or grout shall either be corrosion resistant or shall be coated after fabrication with copper, zinc or a metal having at least equivalent corrosion-resistant properties.

8. **Mortar.**

    8.1 UBC Standard 21-15, Mortar for Unit Masonry and Reinforced Masonry other than Gypsum

    8.2 UBC Standard 21-16, Field Tests Specimens for Mortar

    8.3 UBC Standard 21-20, Standard Test Method for Flexural Bond Strength of Mortar Cement

9. **Grout.**

    9.1 UBC Standard 21-18, Method of Sampling and Testing Grout

    9.2 UBC Standard 21-19, Grout for Masonry

10. **Reinforcement.**

    10.1 UBC Standard 21-10, Part I, Joint Reinforcement for Masonry

    10.2 ASTM A 615, A 616, A 617, A 706, A 767, and A 775, Deformed and Plain Billet-steel Bars, Rail-steel Deformed and Plain Bars, Axle-steel Deformed and Plain Bars, and Deformed Low-alloy Bars for Concrete Reinforcement

    10.3 UBC Standard 21-10, Part II, Cold-drawn Steel Wire for Concrete Reinforcement

SECTION 2103 — MORTAR AND GROUT

2103.1 General. Mortar and grout shall comply with the provisions of this section. Special mortars, grouts or bonding systems may be used, subject to satisfactory evidence of their capabilities when approved by the building official.

2103.2 Materials. Materials used as ingredients in mortar and grout shall conform to the applicable requirements in Section 2102. Cementitious materials for grout shall be one or both of the following: lime and portland cement. Cementitious materials for mortar shall be one or more of the following: lime, masonry cement, portland cement and mortar cement. Cementitious materials or additives shall not contain epoxy resins and derivatives, phenols, asbestos fibers or fireclays.

Water used in mortar or grout shall be clean and free of deleterious amounts of acid, alkalies or organic material or other harmful substances.

2103.3 Mortar.

2103.3.1 General. Mortar shall consist of a mixture of cementitious materials and aggregate to which sufficient water and approved additives, if any, have been added to achieve a workable, plastic consistency.

2103.3.2 Selecting proportions. Mortar with specified proportions of ingredients that differ from the mortar proportions of Table 21-A may be approved for use when it is demonstrated by laboratory or field experience that this mortar with the specified proportions of ingredients, when combined with the masonry units to be used in the structure, will achieve the specified compressive strength f'_m. Water content shall be adjusted to provide proper workability under existing field conditions. When the proportion of ingredients is not specified, the proportions by mortar type shall be used as given in Table 21-A.

2103.4 Grout.

2103.4.1 General. Grout shall consist of a mixture of cementitious materials and aggregate to which water has been added such that the mixture will flow without segregation of the constituents. The specified compressive strength of grout, f'_g, shall not be less than 2,000 psi (13.8 MPa).

2103.4.2 Selecting proportions. Water content shall be adjusted to provide proper workability and to enable proper placement under existing field conditions, without segregation. Grout shall be specified by one of the following methods:

1. Proportions of ingredients and any additives shall be based on laboratory or field experience with the grout ingredients and the masonry units to be used. The grout shall be specified by the proportion of its constituents in terms of parts by volume, or

2. Minimum compressive strength which will produce the required prism strength, or

3. Proportions by grout type shall be used as given in Table 21-B.

2103.5 Additives and Admixtures.

2103.5.1 General. Additives and admixtures to mortar or grout shall not be used unless approved by the building official.

2103.5.2 Antifreeze compounds. Antifreeze liquids, chloride salts or other such substances shall not be used in mortar or grout.

2103.5.3 Air entrainment. Air-entraining substances shall not be used in mortar or grout unless tests are conducted to determine compliance with the requirements of this code.

2103.5.4 Colors. Only pure mineral oxide, carbon black or synthetic colors may be used. Carbon black shall be limited to a maximum of 3 percent of the weight of the cement.

SECTION 2104 — CONSTRUCTION

2104.1 General. Masonry shall be constructed according to the provisions of this section.

2104.2 Materials: Handling, Storage and Preparation. All materials shall comply with applicable requirements of Section 2102. Storage, handling and preparation at the site shall conform also to the following:

1. Masonry materials shall be stored so that at the time of use the materials are clean and structurally suitable for the intended use.

2. All metal reinforcement shall be free from loose rust and other coatings that would inhibit reinforcing bond.

3. At the time of laying, burned clay units and sand lime units shall have an initial rate of absorption not exceeding 0.035 ounce per square inch (1.6 L/m^2) during a period of one minute. In the absorption test, the surface of the unit shall be held $^1/_8$ inch (3 mm) below the surface of the water.

4. Concrete masonry units shall not be wetted unless otherwise approved.

5. Materials shall be stored in a manner such that deterioration or intrusion of foreign materials is prevented and that the material will be capable of meeting applicable requirements at the time of mixing or placement.

6. The method of measuring materials for mortar and grout shall be such that proportions of the materials can be controlled.

7. Mortar or grout mixed at the jobsite shall be mixed for a period of time not less than three minutes or more than 10 minutes in a mechanical mixer with the amount of water required to provide the desired workability. Hand mixing of small amounts of mortar is permitted. Mortar may be retempered. Mortar or grout which has hardened or stiffened due to hydration of the cement shall not be used. In no case shall mortar be used two and one-half hours, nor grout used one and one-half hours, after the initial mixing water has been added to the dry ingredients at the jobsite.

EXCEPTION: Dry mixes for mortar and grout which are blended in the factory and mixed at the jobsite shall be mixed in mechanical mixers until workable, but not to exceed 10 minute

2104.3 Cold-weather Construction.

2104.3.1 General. All materials shall be delivered in a usable condition and stored to prevent wetting by capillary action, rain and snow.

The tops of all walls not enclosed or sheltered shall be covered with a strong weather-resistive material at the end of each day or shutdown.

Partially completed walls shall be covered at all times when work is not in progress. Covers shall be draped over the wall and extend a minimum of 2 feet (600 mm) down both sides and shall be securely held in place, except when additional protection is required in Section 2104.3.4.

2104.3.2 Preparation. If ice or snow has inadvertently formed on a masonry bed, it shall be thawed by application of heat carefully applied until top surface of the masonry is dry to the touch.

A section of masonry deemed frozen and damaged shall be removed before continuing construction of that section.

2104.3.3 Construction. Masonry units shall be dry at time of placement. Wet or frozen masonry units shall not be laid.

Special requirements for various temperature ranges are as follows:

1. Air temperature 40°F to 32°F (4.5°C to 0°C): Sand or mixing water shall be heated to produce mortar temperatures between 40°F and 120°F (4.5°C and 49°C).

2. Air temperature 32°F to 25°F (0°C to –4°C): Sand and mixing water shall be heated to produce mortar temperatures between 40°F and 120°F (4.5°C and 49°C). Maintain temperatures of mortar on boards above freezing.

3. Air temperature 25°F to 20°F (–4°C to –7°C): Sand and mixing water shall be heated to produce mortar temperatures between 40°F and 120°F (4.5°C and 49°C). Maintain mortar temperatures on boards above freezing. Salamanders or other sources of heat shall be used on both sides of walls under construction. Windbreaks shall be employed when wind is in excess of 15 miles per hour (24 km/h).

4. Air temperature 20°F (–7°C) and below: Sand and mixing water shall be heated to produce mortar temperatures between 40°F and 120°F (4.5°C and 49°C). Enclosure and auxiliary heat shall be provided to maintain air temperature above freezing. Temperature of units when laid shall not be less than 20°F (–7°C).

2104.3.4 Protection. When the mean daily air temperature is 40°F to 32°F (4.5°C to 0°C), masonry shall be protected from rain or snow for 24 hours by covering with a weather-resistive membrane.

When the mean daily air temperature is 32°F to 25°F (0°C to –4°C), masonry shall be completely covered with a weather-resistive membrane for 24 hours.

When the mean daily air temperature is 25°F to 20°F (–4°C to –7°C), masonry shall be completely covered with insulating blankets or equally protected for 24 hours.

When the mean daily air temperature is 20°F (–7°C) or below, masonry temperature shall be maintained above freezing for 24 hours by enclosure and supplementary heat, by electric heating blankets, infrared heat lamps or other approved methods.

2104.3.5 Placing grout and protection of grouted masonry. When air temperatures fall below 40°F (4.5°C), grout mixing water and aggregate shall be heated to produce grout temperatures between 40°F and 120°F (4.5°C and 49°C).

Masonry to be grouted shall be maintained above freezing during grout placement and for at least 24 hours after placement.

When atmospheric temperatures fall below 20°F (–7°C), enclosures shall be provided around the masonry during grout placement and for at least 24 hours after placement.

2104.4 Placing Masonry Units.

2104.4.1 Mortar. The mortar shall be sufficiently plastic and units shall be placed with sufficient pressure to extrude mortar from the joint and produce a tight joint. Deep furrowing which produces voids shall not be used.

The initial bed joint thickness shall not be less than $1/4$ inch (6 mm) or more than 1 inch (25 mm); subsequent bed joints shall not be less than $1/4$ inch (6 mm) or more than $5/8$ inch (16 mm) in thickness.

2104.4.2 Surfaces. Surfaces to be in contact with mortar or grout shall be clean and free of deleterious materials.

2104.4.3 Solid masonry units. Solid masonry units shall have full head and bed joints.

2104.4.4 Hollow-masonry units. All head and bed joints shall be filled solidly with mortar for a distance in from the face of the unit not less than the thickness of the shell.

Head joints of open-end units with beveled ends that are to be fully grouted need not be mortared. The beveled ends shall form a grout key which permits grout within $5/8$ inch (16 mm) of the face of the unit. The units shall be tightly butted to prevent leakage of grout.

2104.5 Reinforcement Placing. Reinforcement details shall conform to the requirements of this chapter. Metal reinforcement shall be located in accordance with the plans and specifications. Reinforcement shall be secured against displacement prior to grouting by wire positioners or other suitable devices at intervals not exceeding 200 bar diameters.

Tolerances for the placement of reinforcement in walls and flexural elements shall be plus or minus $1/2$ inch (12.7 mm) for d equal to 8 inches (200 mm) or less, $\pm$ 1 inch ($\pm$ 25 mm) for d equal to 24 inches (600 mm) or less but greater than 8 inches (200 mm), and $\pm$ $1^1/4$ inches (32 mm) for d greater than 24 inches (600 mm).

Tolerance for longitudinal location of reinforcement shall be $\pm$ 2 inches (51 mm).

2104.6 Grouted Masonry.

2104.6.1 General conditions. Grouted masonry shall be constructed in such a manner that all elements of the masonry act together as a structural element.

Prior to grouting, the grout space shall be clean so that all spaces to be filled with grout do not contain mortar projections greater than $1/2$ inch (12.7 mm), mortar droppings or other foreign material. Grout shall be placed so that all spaces designated to be grouted shall be filled with grout and the grout shall be confined to those specific spaces.

Grout materials and water content shall be controlled to provide adequate fluidity for placement without segregation of the constituents, and shall be mixed thoroughly.

The grouting of any section of wall shall be completed in one day with no interruptions greater than one hour.

Between grout pours, a horizontal construction joint shall be formed by stopping all wythes at the same elevation and with the grout stopping a minimum of $1^1/2$ inches (38 mm) below a mortar joint, except at the top of the wall. Where bond beams occur, the grout pour shall be stopped a minimum of $1/2$ inch (12.7 mm) below the top of the masonry.

Size and height limitations of the grout space or cell shall not be less than shown in Table 21-C. Higher grout pours or smaller cavity widths or cell size than shown in Table 21-C may be used when approved, if it is demonstrated that grout spaces will be properly filled.

Cleanouts shall be provided for all grout pours over 5 feet (1524 mm) in height.

Where required, cleanouts shall be provided in the bottom course at every vertical bar but shall not be spaced more than 32 inches (813 mm) on center for solidly grouted masonry. When cleanouts are required, they shall be sealed after inspection and before grouting.

Where cleanouts are not provided, special provisions must be made to keep the bottom and sides of the grout spaces, as well as the minimum total clear area as required by Table 21-C, clean and clear prior to grouting.

Units may be laid to the full height of the grout pour and grout shall be placed in a continuous pour in grout lifts not exceeding 6 feet (1830 mm). When approved, grout lifts may be greater than 6 feet (1830 mm) if it can be demonstrated the grout spaces can be properly filled.

All cells and spaces containing reinforcement shall be filled with grout.

2104.6.2 Construction requirements. Reinforcement shall be placed prior to grouting. Bolts shall be accurately set with templates or by approved equivalent means and held in place to prevent dislocation during grouting.

Segregation of the grout materials and damage to the masonry shall be avoided during the grouting process.

Grout shall be consolidated by mechanical vibration during placement before loss of plasticity in a manner to fill the grout space. Grout pours greater than 12 inches (300 mm) in height shall be reconsolidated by mechanical vibration to minimize voids due to water loss. Grout pours 12 inches (300 mm) or less in height shall be mechanically vibrated or puddled.

In one-story buildings having wood-frame exterior walls, foundations not over 24 inches (600 mm) high measured from the top of the footing may be constructed of hollow-masonry units laid in running bond without mortared head joints. Any standard shape unit may be used, provided the masonry units permit horizontal flow of grout to adjacent units. Grout shall be solidly poured to the full height in one lift and shall be puddled or mechanically vibrated.

In nonstructural elements which do not exceed 8 feet (2440 mm) in height above the highest point of lateral support, including fireplaces and residential chimneys, mortar of pouring consistency may be substituted for grout when the masonry is constructed and grouted in pours of 12 inches (300 mm) or less in height.

In multiwythe grouted masonry, vertical barriers of masonry shall be built across the grout space the entire height of the grout pour and spaced not more than 30 feet (9144 mm) horizontally. The grouting of any section of wall between barriers shall be completed in one day with no interruption longer than one hour.

2104.7 Aluminum Equipment. Grout shall not be handled nor pumped utilizing aluminum equipment unless it can be demonstrated with the materials and equipment to be used that there will be no deleterious effect on the strength of the grout.

2104.8 Joint Reinforcement. Wire joint reinforcement used in the design as principal reinforcement in hollow-unit construction shall be continuous between supports unless splices are made by lapping:

1. Fifty-four wire diameters in a grouted cell, or

2. Seventy-five wire diameters in the mortared bed joint, or

3. In alternate bed joints of running bond masonry a distance not less than 54 diameters plus twice the spacing of the bed joints, or

4. As required by calculation and specific location in areas of minimum stress, such as points of inflection.

Side wires shall be deformed and shall conform to UBC Standard 21-10, Part I, Joint Reinforcement for Masonry.

SECTION 2105 — QUALITY ASSURANCE

2105.1 General. Quality assurance shall be provided to ensure that materials, construction and workmanship are in compliance with the plans and specifications, and the applicable requirements of this chapter. When required, inspection records shall be maintained and made available to the building official.

2105.2 Scope. Quality assurance shall include, but is not limited to, assurance that:

1. Masonry units, reinforcement, cement, lime, aggregate and all other materials meet the requirements of the applicable standards of quality and that they are properly stored and prepared for use.

2. Mortar and grout are properly mixed using specified proportions of ingredients. The method of measuring materials for mortar and grout shall be such that proportions of materials are controlled.

3. Construction details, procedures and workmanship are in accordance with the plans and specifications.

4. Placement, splices and reinforcement sizes are in accordance with the provisions of this chapter and the plans and specifications.

2105.3 Compliance with f'_m.

2105.3.1 General. Compliance with the requirements for the specified compressive strength of masonry f'_m shall be in accordance with one of the sections in this subsection.

2105.3.2 Masonry prism testing. The compressive strength of masonry determined in accordance with UBC Standard 21-17 for each set of prisms shall equal or exceed f'_m. Compressive strength of prisms shall be based on tests at 28 days. Compressive strength at seven days or three days may be used provided a relationship between seven-day and three-day and 28-day strength has been established for the project prior to the start of construction. Verification by masonry prism testing shall meet the following:

1. A set of five masonry prisms shall be built and tested in accordance with UBC Standard 21-17 prior to the start of construction. Materials used for the construction of the prisms shall be taken from those specified to be used in the project. Prisms shall be constructed under the observation of the engineer or special inspector or an approved agency and tested by an approved agency.

2. When full allowable stresses are used in design, a set of three prisms shall be built and tested during construction in accordance with UBC Standard 21-17 for each 5,000 square feet (465 m^2) of wall area, but not less than one set of three masonry prisms for the project.

3. When one half the allowable masonry stresses are used in design, testing during construction is not required. A letter of certification from the supplier of the materials used to verify the f'_m in accordance with Section 2105.3.2, Item 1, shall be provided at the time of, or prior to, delivery of the materials to the jobsite to ensure the materials used in construction are representative of the materials used to construct the prisms prior to construction.

2105.3.3 Masonry prism test record. Compressive strength verification by masonry prism test records shall meet the following:

1. A masonry prism test record approved by the building official of at least 30 masonry prisms which were built and tested in accordance with UBC Standard 21-17. Prisms shall have been constructed under the observation of an engineer or special inspector or an approved agency and shall have been tested by an approved agency.

2. Masonry prisms shall be representative of the corresponding construction.

3. The average compressive strength of the test record shall equal or exceed 1.33 f'_m.

4. When full allowable stresses are used in design, a set of three masonry prisms shall be built during construction in accordance with UBC Standard 21-17 for each 5,000 square feet (465 m^2) of wall area, but not less than one set of three prisms for the project.

5. When one half the allowable masonry stresses are used in design, field testing during construction is not required. A letter of certification from the supplier of the materials to the jobsite shall be provided at the time of, or prior to, delivery of the materials to assure the materials used in construction are representative of the materials used to develop the prism test record in accordance with Section 2105.3.3, Item 1.

2105.3.4 Unit strength method. Verification by the unit strength method shall meet the following:

1. When full allowable stresses are used in design, units shall be tested prior to construction and test units during construction for each 5,000 square feet (465 m^2) of wall area for compressive strength to show compliance with the compressive strength required in Table 21-D; and

> **EXCEPTION:** Prior to the start of construction, prism testing may be used in lieu of testing the unit strength. During construction, prism testing may also be used in lieu of testing the unit strength and the grout as required by Section 2105.3.4, Item 4.

2. When one half the allowable masonry stresses are used in design, testing is not required for the units. A letter of certification from the manufacturer of the units shall be provided at the time of, or prior to, delivery of the units to the jobsite to assure the units comply with the compressive strength required in Table 21-D; and

3. Mortar shall comply with the mortar type required in Table 21-D; and

4. When full stresses are used in design for concrete masonry, grout shall be tested for each 5,000 square feet (465 m^2) of wall area, but not less than one test per project, to show compliance with the compressive strength required in Table 21-D, Footnote 4.

5. When one half the allowable stresses are used in design for concrete masonry, testing is not required for the grout. A letter of certification from the supplier of the grout shall be provided at the time of, or prior to, delivery of the grout to the jobsite to assure the grout complies with the compressive strength required in Table 21-D, Footnote 4; or

6. When full allowable stresses are used in design for clay masonry, grout proportions shall be verified by the engineer or special inspector or an approved agency to conform with Table 21-B.

7. When one half the allowable masonry stresses are used in design for clay masonry, a letter of certification from the supplier of the grout shall be provided at the time of, or prior to, delivery of the grout to the jobsite to assure the grout conforms to the proportions of Table 21-B.

2105.3.5 Testing prisms from constructed masonry. When approved by the building official, acceptance of masonry which does not meet the requirements of Section 2105.3.2, 2105.3.3 or 2105.3.4 shall be permitted to be based on tests of prisms cut from the masonry construction in accordance with the following:

1. A set of three masonry prisms that are at least 28 days old shall be saw cut from the masonry for each 5,000 square feet (465 m²) of the wall area that is in question but not less than one set of three masonry prisms for the project. The length, width and height dimensions of the prisms shall comply with the requirements of UBC Standard 21-17. Transporting, preparation and testing of prisms shall be in accordance with UBC Standard 21-17.

2. The compressive strength of prisms shall be the value calculated in accordance with UBC Standard 21-17, Section 21.1707.2, except that the net cross-sectional area of the prism shall be based on the net mortar bedded area.

3. Compliance with the requirement for the specified compressive strength of masonry, f'_m, shall be considered satisfied provided the modified compressive strength equals or exceeds the specified f'_m. Additional testing of specimens cut from locations in question shall be permitted.

2105.4 Mortar Testing. When required, mortar shall be tested in accordance with UBC Standard 21-16.

2105.5 Grout Testing. When required, grout shall be tested in accordance with UBC Standard 21-18.

SECTION 2106 — GENERAL DESIGN REQUIREMENTS

2106.1 General.

2106.1.1 Scope. The design of masonry structures shall comply with the working stress design provisions of Section 2107, or the strength design provisions of Section 2108 or the empirical design provisions of Section 2109, and with the provisions of this section. Unless otherwise stated, all calculations shall be made using or based on specified dimensions.

2106.1.2 Plans. Plans submitted for approval shall describe the required design strengths of masonry materials and inspection requirements for which all parts of the structure were designed, and any load test requirements.

2106.1.3 Design loads. See Chapter 16 for design loads.

2106.1.4 Stack bond. In bearing and nonbearing walls, except veneer walls, if less than 75 percent of the units in any transverse vertical plane lap the ends of the units below a distance less than one half the height of the unit, or less than one fourth the length of the unit, the wall shall be considered laid in stack bond.

2106.1.5 Multiwythe walls.

2106.1.5.1 General. All wythes of multiwythe walls shall be bonded by grout or tied together by corrosion-resistant wall ties or joint reinforcement conforming to the requirements of Section 2102, and as set forth in this section.

2106.1.5.2 Wall ties in cavity wall construction. Wall ties shall be of sufficient length to engage all wythes. The portion of the wall ties within the wythe shall be completely embedded in mortar or grout. The ends of the wall ties shall be bent to 90-degree angles with an extension not less than 2 inches (51 mm) long. Wall ties not completely embedded in mortar or grout between wythes shall be a single piece with each end engaged in each wythe.

There shall be at least one $^3/_{16}$-inch-diameter (9.5 mm) wall tie for each $4^1/_2$ square feet (0.42 m²) of wall area. For cavity walls in which the width of the cavity is greater than 3 inches (75 mm), but not more than $4^1/_2$ inches (115 mm), at least one $^3/_{16}$-inch-diameter (9.5 mm) wall tie for each 3 square feet (0.28 m²) of wall area shall be provided.

Ties in alternate courses shall be staggered. The maximum vertical distance between ties shall not exceed 24 inches (610 mm) and the maximum horizontal distance between ties shall not exceed 36 inches (914 mm).

Additional ties spaced not more than 36 inches (914 mm) apart shall be provided around openings within a distance of 12 inches (305 mm) from the edge of the opening.

Adjustable wall ties shall meet the following requirements:

1. One tie shall be provided for each 1.77 square feet (0.16 m²) of wall area. Horizontal and vertical spacing shall not exceed 16 inches (406 mm). Maximum misalignment of bed joints from one wythe to the other shall be $1^1/_4$ inches (32 mm).

2. Maximum clearance between the connecting parts of the tie shall be $^1/_{16}$ inch (1.6 mm). When used, pintle ties shall have at least two $^3/_{16}$-inch-diameter (4.8 mm) pintle legs.

Wall ties of different size and spacing that provide equivalent strength between wythes may be used.

2106.1.5.3 Wall ties for grouted multiwythe construction. Wythes of multiwythe walls shall be bonded together with at least $^3/_{16}$-inch-diameter (4.8 mm) steel wall tie for each 2 square feet (0.19 m²) of area. Wall ties of different size and spacing that provide equivalent strength between wythes may be used.

2106.1.5.4 Joint reinforcement. Prefabricated joint reinforcement for masonry walls shall have at least one cross wire of at least No. 9 gage steel for each 2 square feet (0.19 m²) of wall area. The vertical spacing of the joint reinforcement shall not exceed 16 inches (406 mm). The longitudinal wires shall be thoroughly embedded in the bed joint mortar. The joint reinforcement shall engage all wythes.

Where the space between tied wythes is solidly filled with grout or mortar, the allowable stresses and other provisions for masonry bonded walls shall apply. Where the space is not filled, tied walls shall conform to the allowable stress, lateral support, thickness (excluding cavity), height and tie requirements for cavity walls.

2106.1.6 Vertical support. Structural members providing vertical support of masonry shall provide a bearing surface on which the initial bed joint shall not be less than $^1/_4$ inch (6 mm) or more than 1 inch (25 mm) in thickness and shall be of noncombustible material, except where masonry is a nonstructural decorative feature or wearing surface.

2106.1.7 Lateral support. Lateral support of masonry may be provided by cross walls, columns, pilasters, counterforts or buttresses where spanning horizontally or by floors, beams, girts or roofs where spanning vertically.

The clear distance between lateral supports of a beam shall not exceed 32 times the least width of the compression area.

2106.1.8 Protection of ties and joint reinforcement. A minimum of $^5/_8$-inch (16 mm) mortar cover shall be provided between ties or joint reinforcement and any exposed face. The thickness of grout or mortar between masonry units and joint reinforcement shall not be less than $^1/_4$ inch (6 mm), except that $^1/_4$ inch (6 mm) or smaller diameter reinforcement or bolts may be placed in bed joints which are at least twice the thickness of the reinforcement or bolts.

2106.1.9 Pipes and conduits embedded in masonry. Pipes or conduit shall not be embedded in any masonry in a manner that will reduce the capacity of the masonry to less than that necessary for required strength or required fire protection.

Placement of pipes or conduits in unfilled cores of hollow-unit masonry shall not be considered as embedment.

> **EXCEPTIONS:** 1. Rigid electric conduits may be embedded in structural masonry when their locations have been detailed on the approved plan.
>
> 2. Any pipe or conduit may pass vertically or horizontally through any masonry by means of a sleeve at least large enough to pass any hub or coupling on the pipeline. Such sleeves shall not be placed closer than three diameters, center to center, nor shall they unduly impair the strength of construction.

2106.1.10 Load tests. When a load test is required, the member or portion of the structure under consideration shall be subjected to a superimposed load equal to twice the design live load plus one half of the dead load. This load shall be left in position for a period of 24 hours before removal. If, during the test or upon removal of the load, the member or portion of the structure shows evidence of failure, such changes or modifications as are necessary to make the structure adequate for the rated capacity shall be made; or where approved, a lower rating shall be established. A flexural member shall be considered to have passed the test if the maximum deflection D at the end of the 24-hour period does not exceed the value of Formulas (6-1) or (6-2) and the beams and slabs show a recovery of at least 75 percent of the observed deflection within 24 hours after removal of the load.

$$D = \frac{l}{200} \tag{6-1}$$

$$D = \frac{l^2}{4,000t} \tag{6-2}$$

2106.1.11 Reuse of masonry units. Masonry units may be reused when clean, whole and conforming to the other requirements of this section. All structural properties of masonry of reclaimed units shall be determined by approved test.

2106.1.12 Special provisions in areas of seismic risk.

2106.1.12.1 General. Masonry structures constructed in the seismic zones shown in Figure 16-2 shall be designed in accordance with the design requirements of this chapter and the special provisions for each seismic zone given in this section.

2106.1.12.2 Special provisions for Seismic Zones 0 and 1. There are no special design and construction provisions in this section for structures built in Seismic Zones 0 and 1.

2106.1.12.3 Special provisions for Seismic Zone 2. Masonry structures in Seismic Zone 2 shall comply with the following special provisions:

1. Columns shall be reinforced as specified in Sections 2106.3.6, 2106.3.7 and 2107.2.13.

2. Vertical wall reinforcement of at least 0.20 square inch (130 mm²) in cross-sectional area shall be provided continuously from support to support at each corner, at each side of each opening, at the ends of walls and at maximum spacing of 4 feet (1219 mm) apart horizontally throughout walls.

3. Horizontal wall reinforcement not less than 0.2 square inch (130 mm²) in cross-sectional area shall be provided (1) at the bot-

tom and top of wall openings and shall extend not less than 24 inches (610 mm) or less than 40 bar diameters past the opening, (2) continuously at structurally connected roof and floor levels and at the top of walls, (3) at the bottom of walls or in the top of foundations when doweled in walls, and (4) at maximum spacing of 10 feet (3048 mm) unless uniformly distributed joint reinforcement is provided. Reinforcement at the top and bottom of openings when continuous in walls may be used in determining the maximum spacing specified in Item 1 of this paragraph.

4. Where stack bond is used, the minimum horizontal reinforcement ratio shall be $0.0007bt$. This ratio shall be satisfied by uniformly distributed joint reinforcement or by horizontal reinforcement spaced not over 4 feet (1219 mm) and fully embedded in grout or mortar.

5. The following materials shall not be used as part of the vertical or lateral load-resisting systems: Type O mortar, masonry cement, plastic cement, nonloadbearing masonry units and glass block.

2106.1.12.4 Special provisions for Seismic Zones 3 and 4. All masonry structures built in Seismic Zones 3 and 4 shall be designed and constructed in accordance with requirements for Seismic Zone 2 and with the following additional requirements and limitations:

> **EXCEPTION:** One- and two-story masonry buildings of Group R, Division 3 and Group U Occupancies located in Seismic Zone 3 having masonry wall h'/t ratios not greater than 27 and using running bond construction when provisions of Section 2106.1.12.3 are met.

1. **Column reinforcement ties.** In columns that are stressed by tensile or compressive axial overturning forces from seismic loading, the spacing of column ties shall not exceed 8 inches (203 mm) for the full height of such columns. In all other columns, ties shall be spaced a maximum of 8 inches (203 mm) in the tops and bottoms of the columns for a distance of one sixth of the clear column height, 18 inches (457 mm), or the maximum column cross-sectional dimension, whichever is greater. Tie spacing for the remaining column height shall not exceed the lessor of 16 bar diameters, 48 tie diameters, the least column cross-sectional dimension, or 18 inches (457 mm).

Column ties shall terminate with a minimum 135-degree hook with extensions not less than six bar diameters or 4 inches (102 mm). Such extensions shall engage the longitudinal column reinforcement and project into the interior of the column. Hooks shall comply with Section 2107.2.2.5, Item 3.

> **EXCEPTION:** Where ties are placed in horizontal bed joints, hooks shall consist of a 90-degree bend having an inside radius of not less than four tie diameters plus an extension of 32 tie diameters.

2. **Shear Walls.**

   2.1 **Reinforcement.** The portion of the reinforcement required to resist shear shall be uniformly distributed and shall be joint reinforcement, deformed bars or a combination thereof. The spacing of reinforcement in each direction shall not exceed one half the length of the element, nor one half the height of the element, nor 48 inches (1219 mm).

   Joint reinforcement used in exterior walls and considered in the determination of the shear strength of the member shall be hot-dipped galvanized in accordance with UBC Standard 21-10.

   Reinforcement required to resist in-plane shear shall be terminated with a standard hook as defined in Section 2107.2.2.5 or with an extension of proper embedment length beyond the reinforcement at the end of the wall section. The hook or extension may be turned up, down or horizontally. Provisions shall be made not to obstruct grout placement. Wall reinforcement terminating in columns or beams shall be fully anchored into these elements.

2.2 **Bond.** Multiwythe grouted masonry shear walls shall be designed with consideration of the adhesion bond strength between the grout and masonry units. When bond strengths are not known from previous tests, the bond strength shall be determined by tests.

2.3 **Wall reinforcement.** All walls shall be reinforced with both vertical and horizontal reinforcement. The sum of the areas of horizontal and vertical reinforcement shall be at least 0.002 times the gross cross-sectional area of the wall, and the minimum area of reinforcement in either direction shall not be less than 0.0007 times the gross cross-sectional area of the wall. The minimum steel requirements for Seismic Zone 2 in Section 2106.1.12.3, Items 2 and 3, may be included in the sum. The spacing of reinforcement shall not exceed 4 feet (1219 mm). The diameter of reinforcement shall not be less than $^3/_8$ inch (9.5 mm) except that joint reinforcement may be considered as a part or all of the requirement for minimum reinforcement. Reinforcement shall be continuous around wall corners and through intersections. Only reinforcement which is continuous in the wall or element shall be considered in computing the minimum area of reinforcement. Reinforcement with splices conforming to Section 2107.2.2.6 shall be considered as continuous reinforcement.

2.4 **Stack bond.** Where stack bond is used, the minimum horizontal reinforcement ratio shall be 0.0015bt. Where open-end units are used and grouted solid, the minimum horizontal reinforcement ratio shall be 0.0007bt.

Reinforced hollow-unit stacked bond construction which is part of the seismic-resisting system shall use open-end units so that all head joints are made solid, shall use bond beam units to facilitate the flow of grout and shall be grouted solid.

3. **Type N mortar.** Type N mortar shall not be used as part of the vertical- or lateral-load-resisting system.

4. **Concrete abutting structural masonry.** Concrete abutting structural masonry, such as at starter courses or at wall intersections not designed as true separation joints, shall be roughened to a full amplitude of $^1/_{16}$ inch (1.6 mm) and shall be bonded to the masonry in accordance with the requirements of this chapter as if it were masonry. Unless keys or proper reinforcement is provided, vertical joints as specified in Section 2106.1.4 shall be considered to be stack bond and the reinforcement as required for stack bond shall extend through the joint and be anchored into the concrete.

2106.2 Working Stress Design and Strength Design Requirements for Unreinforced and Reinforced Masonry.

2106.2.1 General. In addition to the requirements of Section 2106.1, the design of masonry structures by the working stress design method and strength design method shall comply with the requirements of this section. Additionally, the design of reinforced masonry structures by these design methods shall comply with the requirements of Section 2106.3.

2106.2.2 Specified compressive strength of masonry. The allowable stresses for the design of masonry shall be based on a value of f'_m selected for the construction.

Verification of the value of f'_m shall be based on compliance with Section 2105.3. Unless otherwise specified, f'_m shall be based on 28-day tests. If other than a 28-day test age is used, the value of f'_m shall be as indicated in design drawings or specifications. Design drawings shall show the value of f'_m for which each part of the structure is designed.

2106.2.3 Effective thickness.

2106.2.3.1 Single-wythe walls. The effective thickness of single-wythe walls of either solid or hollow units is the specified thickness of the wall.

2106.2.3.2 Multiwythe walls. The effective thickness of multiwythe walls is the specified thickness of the wall if the space between wythes is filled with mortar or grout. For walls with an open space between wythes, the effective thickness shall be determined as for cavity walls.

2106.2.3.3 Cavity walls. Where both wythes of a cavity wall are axially loaded, each wythe shall be considered to act independently and the effective thickness of each wythe is as defined in Section 2106.2.3.1. Where only one wythe is axially loaded, the effective thickness of the cavity wall is taken as the square root of the sum of the squares of the specified thicknesses of the wythes.

Where a cavity wall is composed of a single wythe and a multiwythe, and both sides are axially loaded, each side of the cavity wall shall be considered to act independently and the effective thickness of each side is as defined in Sections 2106.2.3.1 and 2106.2.3.2. Where only one side is axially loaded, the effective thickness of the cavity wall is the square root of the sum of the squares of the specified thicknesses of the sides.

2106.2.3.4 Columns. The effective thickness for rectangular columns in the direction considered is the specified thickness. The effective thickness for nonrectangular columns is the thickness of the square column with the same moment of inertia about its axis as that about the axis considered in the actual column.

2106.2.4 Effective height. The effective height of columns and walls shall be taken as the clear height of members laterally supported at the top and bottom in a direction normal to the member axis considered. For members not supported at the top normal to the axis considered, the effective height is twice the height of the member above the support. Effective height less than clear height may be used if justified.

2106.2.5 Effective area. The effective cross-sectional area shall be based on the minimum bedded area of hollow units, or the gross area of solid units plus any grouted area. Where hollow units are used with cells perpendicular to the direction of stress, the effective area shall be the lesser of the minimum bedded area or the minimum cross-sectional area. Where bed joints are raked, the effective area shall be correspondingly reduced. Effective areas for cavity walls shall be that of the loaded wythes.

2106.2.6 Effective width of intersecting walls. Where a shear wall is anchored to an intersecting wall or walls, the width of the overhanging flange formed by the intersected wall on either side of the shear wall, which may be assumed working with the shear wall for purposes of flexural stiffness calculations, shall not exceed six times the thickness of the intersected wall. Limits of the effective flange may be waived if justified. Only the effective area of the wall parallel to the shear forces may be assumed to carry horizontal shear.

2106.2.7 Distribution of concentrated vertical loads in walls. The length of wall laid up in running bond which may be considered capable of working at the maximum allowable compressive stress to resist vertical concentrated loads shall not exceed the center-to-center distance between such loads, nor the width of bearing area plus four times the wall thickness. Concentrated vertical loads shall not be assumed to be distributed across continuous vertical mortar or control joints unless elements designed to distribute the concentrated vertical loads are employed.

2106.2.8 Loads on nonbearing walls. Masonry walls used as interior partitions or as exterior surfaces of a building which do not carry vertical loads imposed by other elements of the building shall be designed to carry their own weight plus any superimposed finish and lateral forces. Bonding or anchorage of nonbearing walls shall be adequate to support the walls and to transfer lateral forces to the supporting elements.

2106.2.9 Vertical deflection. Elements supporting masonry shall be designed so that their vertical deflection will not exceed $^1/_{600}$ of the clear span under total loads. Lintels shall bear on supporting masonry on each end such that allowable stresses in the supporting masonry are not exceeded. A minimum bearing length of 4 inches (102 mm) shall be provided for lintels bearing on masonry.

2106.2.10 Structural continuity. Intersecting structural elements intended to act as a unit shall be anchored together to resist the design forces.

2106.2.11 Walls intersecting with floors and roofs. Walls shall be anchored to all floors, roofs or other elements which provide lateral support for the wall. Where floors or roofs are designed to transmit horizontal forces to walls, the anchorage to such walls shall be designed to resist the horizontal force.

2106.2.12 Modulus of elasticity of materials.

2106.2.12.1 Modulus of elasticity of masonry. The moduli for masonry may be estimated as provided below. Actual values, where required, shall be established by test. The modulus of elasticity of masonry shall be determined by the secant method in which the slope of the line for the modulus of elasticity is taken from $0.05 f'_m$ to a point on the curve at $0.33 f'_m$. These values are not to be reduced by one half as set forth in Section 2107.1.2.

Modulus of elasticity of clay or shale unit masonry.

$$E_m = 750 f'_m, \text{ 3,000,000 psi (20.5 GPa) maximum} \qquad (6\text{-}3)$$

Modulus of elasticity of concrete unit masonry.

$$E_m = 750 f'_m, \text{ 3,000,000 psi (20.5 GPa) maximum} \qquad (6\text{-}4)$$

2106.2.12.2 Modulus of elasticity of steel.

$$E_s = 29,000,000 \text{ psi (200 GPa)} \qquad (6\text{-}5)$$

2106.2.13 Shear modulus of masonry.

$$G = 0.4 E_m \qquad (6\text{-}6)$$

2106.2.14 Placement of embedded anchor bolts.

2106.2.14.1 General. Placement requirements for plate anchor bolts, headed anchor bolts and bent bar anchor bolts shall be determined in accordance with this subsection. Bent bar anchor bolts shall have a hook with a 90-degree bend with an inside diameter of three bolt diameters, plus an extension of one and one half bolt diameters at the free end. Plate anchor bolts shall have a plate welded to the shank to provide anchorage equivalent to headed anchor bolts.

The effective embedment depth l_b for plate or headed anchor bolts shall be the length of embedment measured perpendicular from the surface of the masonry to the bearing surface of the plate

or head of the anchorage, and l_b for bent bar anchors shall be the length of embedment measured perpendicular from the surface of the masonry to the bearing surface of the bent end minus one anchor bolt diameter. All bolts shall be grouted in place with at least 1 inch (25 mm) of grout between the bolt and the masonry, except that $^1/_4$-inch-diameter (6.4 mm) bolts may be placed in bed joints which are at least $^1/_2$ inch (12.7 mm) in thickness.

2106.2.14.2 Minimum edge distance. The minimum anchor bolt edge distance l_{be} measured from the edge of the masonry parallel with the anchor bolt to the surface of the anchor bolt shall be $1^1/_2$ inches (38 mm).

2106.2.14.3 Minimum embedment depth. The minimum embedment depth of anchor bolts l_b shall be four bolt diameters but not less than 2 inches (51 mm).

2106.2.14.4 Minimum spacing between bolts. The minimum center-to-center distance between anchor bolts shall be four bolt diameters.

2106.2.15 Flexural resistance of cavity walls. For computing the flexural resistance of cavity walls, lateral loads perpendicular to the plane of the wall shall be distributed to the wythes according to their respective flexural rigidities.

2106.3 Working Stress Design and Strength Design Requirements for Reinforced Masonry.

2106.3.1 General. In addition to the requirements of Sections 2106.1 and 2106.2, the design of reinforced masonry structures by the working stress design method or the strength design method shall comply with the requirements of this section.

2106.3.2 Plain bars. The use of plain bars larger than $^1/_4$ inch (6.4 mm) in diameter is not permitted.

2106.3.3 Spacing of longitudinal reinforcement. The clear distance between parallel bars, except in columns, shall not be less than the nominal diameter of the bars or 1 inch (25 mm), except that bars in a splice may be in contact. This clear distance requirement applies to the clear distance between a contact splice and adjacent splices or bars.

The clear distance between the surface of a bar and any surface of a masonry unit shall not be less than $^1/_4$ inch (6.4 mm) for fine grout and $^1/_2$ inch (12.7 mm) for coarse grout. Cross webs of hollow units may be used as support for horizontal reinforcement.

2106.3.4 Anchorage of flexural reinforcement. The tension or compression in any bar at any section shall be developed on each side of that section by the required development length. The development length of the bar may be achieved by a combination of an embedment length, anchorage or, for tension only, hooks.

Except at supports or at the free end of cantilevers, every reinforcing bar shall be extended beyond the point at which it is no longer needed to resist tensile stress for a distance equal to 12 bar diameters or the depth of the beam, whichever is greater. No flexural bar shall be terminated in a tensile zone unless at least one of the following conditions is satisfied:

1. The shear is not over one half that permitted, including allowance for shear reinforcement where provided.

2. Additional shear reinforcement in excess of that required is provided each way from the cutoff a distance equal to the depth of the beam. The shear reinforcement spacing shall not exceed $d/8r_b$.

3. The continuing bars provide double the area required for flexure at that point or double the perimeter required for reinforcing bond.

At least one third of the total reinforcement provided for negative moment at the support shall be extended beyond the extreme position of the point of inflection a distance sufficient to develop one half the allowable stress in the bar, not less than $1/16$ of the clear span, or the depth d of the member, whichever is greater.

Tensile reinforcement for negative moment in any span of a continuous restrained or cantilever beam, or in any member of a rigid frame, shall be adequately anchored by reinforcement bond, hooks or mechanical anchors in or through the supporting member.

At least one third of the required positive moment reinforcement in simple beams or at the freely supported end of continuous beams shall extend along the same face of the beam into the support at least 6 inches (153 mm). At least one fourth of the required positive moment reinforcement at the continuous end of continuous beams shall extend along the same face of the beam into the support at least 6 inches (153 mm).

Compression reinforcement in flexural members shall be anchored by ties or stirrups not less than $1/4$ inch (6.4 mm) in diameter, spaced not farther apart than 16 bar diameters or 48 tie diameters, whichever is less. Such ties or stirrups shall be used throughout the distance where compression reinforcement is required.

2106.3.5 Anchorage of shear reinforcement. Single, separate bars used as shear reinforcement shall be anchored at each end by one of the following methods:

1. Hooking tightly around the longitudinal reinforcement through 180 degrees.

2. Embedment above or below the mid-depth of the beam on the compression side a distance sufficient to develop the stress in the bar for plain or deformed bars.

3. By a standard hook, as defined in Section 2107.2.2.5, considered as developing 7,500 psi (52 MPa), plus embedment sufficient to develop the remainder of the stress to which the bar is subjected. The effective embedded length shall not be assumed to exceed the distance between the mid-depth of the beam and the tangent of the hook.

The ends of bars forming a single U or multiple U stirrup shall be anchored by one of the methods set forth in Items 1 through 3 above or shall be bent through an angle of at least 90 degrees tightly around a longitudinal reinforcing bar not less in diameter than the stirrup bar, and shall project beyond the bend at least 12 stirrup diameters.

The loops or closed ends of simple U or multiple U stirrups shall be anchored by bending around the longitudinal reinforcement through an angle of at least 90 degrees and project beyond the end of the bend at least 12 stirrup diameters.

2106.3.6 Lateral ties. All longitudinal bars for columns shall be enclosed by lateral ties. Lateral support shall be provided to the longitudinal bars by the corner of a complete tie having an included angle of not more than 135 degrees or by a standard hook at the end of a tie. The corner bars shall have such support provided by a complete tie enclosing the longitudinal bars. Alternate longitudinal bars shall have such lateral support provided by ties and no bar shall be farther than 6 inches (152 mm) from such laterally supported bar.

Lateral ties and longitudinal bars shall be placed not less than $1^1/_2$ inches (38 mm) and not more than 5 inches (127 mm) from the surface of the column. Lateral ties may be placed against the longitudinal bars or placed in the horizontal bed joints where the requirements of Section 2106.1.8 are met. Spacing of ties shall not exceed 16 longitudinal bar diameters, 48 tie diameters or the least dimension of the column but not more than 18 inches (457 mm).

Ties shall be at least $1/4$ inch (6.4 mm) in diameter for No. 7 or smaller longitudinal bars and at least No. 3 for longitudinal bars larger than No. 7. Ties smaller than No. 3 may be used for longitudinal bars larger than No. 7, provided the total cross-sectional area of such smaller ties crossing a longitudinal plane is equal to that of the larger ties at their required spacing.

2106.3.7 Column anchor bolt ties. Additional ties shall be provided around anchor bolts which are set in the top of columns. Such ties shall engage at least four bolts or, alternately, at least four vertical column bars or a combination of bolts and bars totaling at least four. Such ties shall be located within the top 5 inches (127 mm) of the column and shall provide a total of 0.4 square inch (260 mm^2) or more in cross-sectional area. The uppermost tie shall be within 2 inches (51 mm) of the top of the column.

2106.3.8 Effective width b of compression area. In computing flexural stresses in walls where reinforcement occurs, the effective width assumed for running bond masonry shall not exceed six times the nominal wall thickness or the center-to-center distance between reinforcement. Where stack bond is used, the effective width shall not exceed three times the nominal wall thickness or the center-to-center distance between reinforcement or the length of one unit, unless solid grouted open-end units are used.

SECTION 2107 — WORKING STRESS DESIGN OF MASONRY

2107.1 General.

2107.1.1 Scope. The design of masonry structures using working stress design shall comply with the provisions of Section 2106 and this section. Stresses in clay or concrete masonry under service loads shall not exceed the values given in this section.

2107.1.2 Allowable masonry stresses. When quality assurance provisions do not include requirements for special inspection as prescribed in Section 1701, the allowable stresses for masonry in Section 2107 shall be reduced by one half.

When one half allowable masonry stresses are used in Seismic Zones 3 and 4, the value of f'_m from Table 21-D shall be limited to a maximum of 1,500 psi (10 MPa) for concrete masonry and 2,600 psi (18 MPa) for clay masonry unless the value of f'_m is verified by tests in accordance with Section 2105.3.4, Items 1 and 4 or 6. A letter of certification is not required.

When one half allowable masonry stresses are used for design in Seismic Zones 3 and 4, the value of f'_m shall be limited to 1,500 psi (10 MPa) for concrete masonry and 2,600 psi (18 MPa) for clay masonry for Section 2105.3.2, Item 3, and Section 2105.3.3, Item 5, unless the value of f'_m is verified during construction by the testing requirements of Section 2105.3.2, Item 2. A letter of certification is not required.

2107.1.3 Minimum dimensions for masonry structures located in Seismic Zones 3 and 4. Elements of masonry structures located in Seismic Zones 3 and 4 shall be in accordance with this section.

2107.1.3.1 Bearing walls. The nominal thickness of reinforced masonry bearing walls shall not be less than 6 inches (152 mm) except that nominal 4-inch-thick (102 mm) load-bearing reinforced hollow-clay unit masonry walls may be used, provided net area unit strength exceeds 8,000 psi (55 MPa), units are laid in running bond, bar sizes do not exceed $1/2$ inch (12.7 mm) with no more than two bars or one splice in a cell, and joints are flush cut, concave or a protruding V section.

2107.1.3.2 Columns. The least nominal dimension of a reinforced masonry column shall be 12 inches (305 mm) except that, for working stress design, if the allowable stresses are reduced by one half, the minimum nominal dimension shall be 8 inches (203 mm).

2107.1.4 Design assumptions. The working stress design procedure is based on working stresses and linear stress-strain distribution assumptions with all stresses in the elastic range as follows:

1. Plane sections before bending remain plane after bending.

2. Stress is proportional to strain.

3. Masonry elements combine to form a homogenous member.

2107.1.5 Embedded anchor bolts.

2107.1.5.1 General. Allowable loads for plate anchor bolts, headed anchor bolts and bent bar anchor bolts shall be determined in accordance with this section.

2107.1.5.2 Tension. Allowable loads in tension shall be the lesser value selected from Tables 21-E-1 and 21-E-2 or shall be determined from the lesser of Formula (7-1) or Formula (7-2).

$$B_t = 0.5 A_p \sqrt{f'_m} \qquad (7\text{-}1)$$

For **SI:** $\qquad B_t = 0.042 A_p \sqrt{f'_m}$

$$B_t = 0.2 A_b f_y \qquad (7\text{-}2)$$

The area A_p shall be the lesser of Formula (7-3) or Formula (7-4) and where the projected areas of adjacent anchor bolts overlap, A_p of each anchor bolt shall be reduced by one half of the overlapping area.

$$A_p = \pi l_b{}^2 \qquad (7\text{-}3)$$

$$A_p = \pi l_{b_e}{}^2 \qquad (7\text{-}4)$$

2107.1.5.3 Shear. Allowable loads in shear shall be the value selected from Table 21-F or shall be determined from the lesser of Formula (7-5) or Formula (7-6).

$$B_v = 350 \sqrt[4]{f'_m A_b} \qquad (7\text{-}5)$$

For **SI:** $\qquad B_v = 1070 \sqrt[4]{f'_m A_b}$

$$B_v = 0.12 A_b f_y \qquad (7\text{-}6)$$

Where the anchor bolt edge distance l_{be} in the direction of load is less than 12 bolt diameters, the value of B_v in Formula (7-5) shall be reduced by linear interpolation to zero at an l_{be} distance of $1\frac{1}{2}$ inches (38 mm). Where adjacent anchors are spaced closer than $8d_b$, the allowable shear of the adjacent anchors determined by Formula (7-5) shall be reduced by linear interpolation to 0.75 times the allowable shear value at a center-to-center spacing of four bolt diameters.

2107.1.5.4 Combined shear and tension. Anchor bolts subjected to combined shear and tension shall be designed in accordance with Formula (7-7).

$$\frac{b_t}{B_t} + \frac{b_v}{B_v} \leq 1.0 \qquad (7\text{-}7)$$

2107.1.6 Compression in walls and columns.

2107.1.6.1 Walls, axial loads. Stresses due to compressive forces applied at the centroid of wall may be computed by Formula (7-8) assuming uniform distribution over the effective area.

$$f_a = P / A_e \qquad (7\text{-}8)$$

2107.1.6.2 Columns, axial loads. Stresses due to compressive forces applied at the centroid of columns may be computed by

Formula (7-8) assuming uniform distribution over the effective area.

2107.1.6.3 Columns, bending or combined bending and axial loads. Stresses in columns due to combined bending and axial loads shall satisfy the requirements of Section 2107.2.7 where f_a/F_a is replaced by P/P_a. Columns subjected to bending shall meet all applicable requirements for flexural design.

2107.1.7 Shear walls, design loads. When calculating shear or diagonal tension stresses, shear walls which resist seismic forces in Seismic Zones 3 and 4 shall be designed to resist 1.5 times the forces required by Section 1630.

2107.1.8 Design, composite construction.

2107.1.8.1 General. The requirements of this section govern multiwythe masonry in which at least one wythe has strength or composition characteristics different from the other wythe or wythes and is adequately bonded to act as a single structural element.

The following assumptions shall apply to the design of composite masonry:

1. Analysis shall be based on elastic transformed section of the net area.

2. The maximum computed stress in any portion of composite masonry shall not exceed the allowable stress for the material of that portion.

2107.1.8.2 Determination of moduli of elasticity. The modulus of elasticity of each type of masonry in composite construction shall be measured by tests if the modular ratio of the respective types of masonry exceeds 2 to 1 as determined by Section 2106.2.12.

2107.1.8.3 Structural continuity.

2107.1.8.3.1 Bonding of wythes. All wythes of composite masonry elements shall be tied together as specified in Section 2106.1.5.2 as a minimum requirement. Additional ties or the combination of grout and metal ties shall be provided to transfer the calculated stress.

2107.1.8.3.2 Material properties. The effect of dimensional changes of the various materials and different boundary conditions of various wythes shall be included in the design.

2107.1.8.4 Design procedure, transformed sections. In the design of transformed sections, one material is chosen as the reference material, and the other materials are transformed to an equivalent area of the reference material by multiplying the areas of the other materials by the respective ratios of the moduli of elasticity of the other materials to that of the reference material. Thickness of the transformed area and its distance perpendicular to a given bending axis remain unchanged. Effective height or length of the element remains unchanged.

2107.1.9 Reuse of masonry units. The allowable working stresses for reused masonry units shall not exceed 50 percent of those permitted for new masonry units of the same properties.

2107.2 Design of Reinforced Masonry.

2107.2.1 Scope. The requirements of this section are in addition to the requirements of Sections 2106 and 2107.1, and govern masonry in which reinforcement is used to resist forces.

Walls with openings used to resist lateral loads whose pier and beam elements are within the dimensional limits of Section 2108.2.6.1.2 may be designed in accordance with Section 2108.2.6. Walls used to resist lateral loads not meeting the dimensional limits of Section 2108.2.6.1.2 may be designed as walls in accordance with this section or Section 2108.2.5.

2107.2.2 Reinforcement.

2107.2.2.1 Maximum reinforcement size. The maximum size of reinforcement shall be No. 11 bars. Maximum reinforcement area in cells shall be 6 percent of the cell area without splices and 12 percent of the cell area with splices.

2107.2.2.2 Cover. All reinforcing bars, except joint reinforcement, shall be completely embedded in mortar or grout and have a minimum cover, including the masonry unit, of at least $3/4$ inch (19 mm), $1^1/2$ inches (38 mm) of cover when the masonry is exposed to weather and 2 inches (51 mm) of cover when the masonry is exposed to soil.

2107.2.2.3 Development length. The required development length l_d for deformed bars or deformed wire shall be calculated by:

$$l_d = 0.002 \, d_b \, f_s \text{ for bars in tension} \qquad (7\text{-}9)$$

For **SI:** $\qquad l_d = 0.29 \, d_b \, f_s$ for bars in tension

$$l_d = 0.0015 \, d_b \, f_s \text{ for bars in compression} \qquad (7\text{-}10)$$

For **SI:** $\qquad l_d = 0.22 \, d_b \, f_s$ for bars in compression

Development length for smooth bars shall be twice the length determined by Formula (7-9).

2107.2.2.4 Reinforcement bond stress. Bond stress u in reinforcing bars shall not exceed the following:

Plain Bars	60 psi (413 kPa)
Deformed Bars	200 psi (1378 kPa)
Deformed Bars without Special Inspection	100 psi (689 kPa)

2107.2.2.5 Hooks.

1. The term "standard hook" shall mean one of the following:

    1.1 A 180-degree turn plus extension of at least four bar diameters, but not less than $2^1/2$ inches (63 mm) at free end of bar.

    1.2 A 90-degree turn plus extension of at least 12 bar diameters at free end of bar.

    1.3 For stirrup and tie anchorage only, either a 90-degree or a 135-degree turn, plus an extension of at least six bar diameters, but not less than $2^1/2$ inches (63 mm) at the free end of the bar.

2. Inside diameter of bend of the bars, other than for stirrups and ties, shall not be less than that set forth in Table 21-G.

3. Inside diameter of bend for No. 5 or smaller stirrups and ties shall not be less than four bar diameters. Inside diameter of bend for No. 5 or larger stirrups and ties shall not be less than that set forth in Table 21-G.

4. Hooks shall not be permitted in the tension portion of any beam, except at the ends of simple or cantilever beams or at the freely supported end of continuous or restrained beams.

5. Hooks shall not be assumed to carry a load which would produce a tensile stress in the bar greater than 7,500 psi (52 MPa).

6. Hooks shall not be considered effective in adding to the compressive resistance of bars.

7. Any mechanical device capable of developing the strength of the bar without damage to the masonry may be used in lieu of a hook. Data must be presented to show the adequacy of such devices.

2107.2.2.6 Splices. The amount of lap of lapped splices shall be sufficient to transfer the allowable stress of the reinforcement as specified in Sections 2106.3.4, 2107.2.2.3 and 2107.2.12. In no case shall the length of the lapped splice be less than 30 bar diameters for compression or 40 bar diameters for tension.

Welded or mechanical connections shall develop 125 percent of the specified yield strength of the bar in tension.

> **EXCEPTION:** For compression bars in columns that are not part of the seismic-resisting system and are not subject to flexure, only the compressive strength need be developed.

When adjacent splices in grouted masonry are separated by 3 inches (76 mm) or less, the required lap length shall be increased 30 percent.

> **EXCEPTION:** Where lap splices are staggered at least 24 bar diameters, no increase in lap length is required.

See Section 2107.2.12 for lap splice increases.

2107.2.3 Design assumptions. The following assumptions are in addition to those stated in Section 2107.1.4:

1. Masonry carries no tensile stress.

2. Reinforcement is completely surrounded by and bonded to masonry material so that they work together as a homogenous material within the range of allowable working stresses.

2107.2.4 Nonrectangular flexural elements. Flexural elements of nonrectangular cross section shall be designed in accordance with the assumptions given in Sections 2107.1.4 and 2107.2.3.

2107.2.5 Allowable axial compressive stress and force. For members other than reinforced masonry columns, the allowable axial compressive stress F_a shall be determined as follows:

$$F_a = 0.25 f'_m \left[1 - \left(\frac{h'}{140r} \right)^2 \right] \text{ for } h'/r \le 99 \quad (7\text{-}11)$$

$$F_a = 0.25 f'_m \left(\frac{70r}{h'} \right)^2 \text{ for } h'/r > 99 \qquad (7\text{-}12)$$

For reinforced masonry columns, the allowable axial compressive force P_a shall be determined as follows:

$$P_a = [0.25 f'_m A_e + 0.65 A_s F_{sc}] \left[1 - \left(\frac{h'}{140r} \right)^2 \right]$$

$$\text{for } h'/r \le 99 \qquad (7\text{-}13)$$

$$P_a = [0.25 f'_m A_e + 0.65 A_s F_{sc}] \left(\frac{70r}{h'} \right)^2$$

$$\text{for } h'/r > 99 \qquad (7\text{-}14)$$

2107.2.6 Allowable flexural compressive stress. The allowable flexural compressive stress F_b is:

$$F_b = 0.33 f'_m, \ 2,000 \text{ psi (13.8 MPa) maximum} \qquad (7\text{-}15)$$

2107.2.7 Combined compressive stresses, unity formula. Elements subjected to combined axial and flexural stresses shall be designed in accordance with accepted principles of mechanics or in accordance with Formula (7-16):

$$\frac{f_a}{F_a} + \frac{f_b}{F_b} \le 1 \qquad (7\text{-}16)$$

2107.2.8 Allowable shear stress in flexural members. Where no shear reinforcement is provided, the allowable shear stress F_v in flexural members is:

$$F_v = 1.0 \sqrt{f'_m}, \ 50 \text{ psi maximum} \qquad (7\text{-}17)$$

For **SI:** $F_v = 0.083 \sqrt{f'_m}$, 345 kPa maximum

> **EXCEPTION:** For a distance of $^1/_{16}$ the clear span beyond the point of inflection, the maximum stress shall be 20 psi (140 kPa).

Where shear reinforcement designed to take entire shear force is provided, the allowable shear stress F_v in flexural members is:

$$F_v = 3.0 \sqrt{f'_m}, \text{ 150 psi maximum} \qquad (7\text{-}18)$$

For **SI:** $F_v = 0.25 \sqrt{f'_m}$, 1.0 MPa maximum

2107.2.9 Allowable shear stress in shear walls. Where in-plane flexural reinforcement is provided and masonry is used to resist all shear, the allowable shear stress F_v in shear walls is:

For $M/Vd < 1$,

$$F_v = {}^1/_3 \left(4 - \frac{M}{Vd} \right) \sqrt{f'_m}, \; \left(80 - 45 \frac{M}{Vd} \right) \text{ maximum} \quad (7\text{-}19)$$

For **SI:** $F_v = {}^1/_{36} \left(4 - \frac{M}{Vd} \right) \sqrt{f'_m}, \; \left(80 - 45 \frac{M}{Vd} \right)$ maximum

For $M/Vd \geq 1$, $F_v = 1.0 \sqrt{f'_m}$, 35 psi maximum (7-20)

For **SI:** $F_v = {}^1/_{12} \sqrt{f'_m}$, 240 kPa maximum

Where shear reinforcement designed to take all the shear is provided, the allowable shear stress F_v in shear walls is:

For $M/Vd < 1$,

$$F_v = {}^1/_2 \left(4 - \frac{M}{Vd} \right) \sqrt{f'_m}, \; \left(120 - 45 \frac{M}{Vd} \right) \text{ maximum} \quad (7\text{-}21)$$

For **SI:** For $M/Vd < 1$,

$$F_v = {}^1/_{24} \left(4 - \frac{M}{Vd} \right) \sqrt{f'_m}, \; \left(120 - 45 \frac{M}{Vd} \right) \text{ maximum}$$

For $M/Vd \geq 1$, $F_v = 1.5 \sqrt{f'_m}$, 75 psi maximum (7-22)

For **SI:** For $M/Vd \geq 1$, $F_v = 0.12 \sqrt{f'_m}$, 520 kPa maximum

2107.2.10 Allowable bearing stress. When a member bears on the full area of a masonry element, the allowable bearing stress F_{br} is:

$$F_{br} = 0.26 f'_m \qquad (7\text{-}23)$$

When a member bears on one third or less of a masonry element, the allowable bearing stress F_{br} is:

$$F_{br} = 0.38 f'_m \qquad (7\text{-}24)$$

Formula (7-24) applies only when the least dimension between the edges of the loaded and unloaded areas is a minimum of one fourth of the parallel side dimension of the loaded area. The allowable bearing stress on a reasonably concentric area greater than one third but less than the full area shall be interpolated between the values of Formulas (7-23) and (7-24).

2107.2.11 Allowable stresses in reinforcement. The allowable stresses in reinforcement shall be as follows:

1. **Tensile stress.**

   1.1 Deformed bars,

   $$F_s = 0.5 f_y, \text{ 24,000 psi (165 MPa) maximum} \quad (7\text{-}25)$$

   1.2 Wire reinforcement,

   $$F_s = 0.5 f_y, \text{ 30,000 psi (207 MPa) maximum} \quad (7\text{-}26)$$

   1.3 Ties, anchors and smooth bars,

   $$F_s = 0.4 f_y, \text{ 20,000 psi (138 MPa) maximum} \quad (7\text{-}27)$$

2. **Compressive stress.**

   2.1 Deformed bars in columns,

$$F_{sc} = 0.4 f_y, \text{ 24,000 psi (165 MPa) maximum} \quad (7\text{-}28)$$

2.2 Deformed bars in flexural members,

$$F_s = 0.5 f_y, \text{ 24,000 psi (165 MPa) maximum} \quad (7\text{-}29)$$

2.3 Deformed bars in shear walls which are confined by lateral ties throughout the distance where compression reinforcement is required and where such lateral ties are not less than $^1/_4$ inch in diameter and spaced not farther apart than 16 bar diameters or 48 tie diameters,

$$F_{sc} = 0.4 f_y, \text{ 24,000 psi (165 MPa) maximum} \quad (7\text{-}30)$$

2107.2.12 Lap splice increases. In regions of moment where the design tensile stresses in the reinforcement are greater than 80 percent of the allowable steel tensile stress F_s, the lap length of splices shall be increased not less than 50 percent of the minimum required length. Other equivalent means of stress transfer to accomplish the same 50 percent increase may be used.

2107.2.13 Reinforcement for columns. Columns shall be provided with reinforcement as specified in this section.

2107.2.13.1 Vertical reinforcement. The area of vertical reinforcement shall not be less than $0.005 A_e$ and not more than $0.04 A_e$. At least four No. 3 bars shall be provided. The minimum clear distance between parallel bars in columns shall be two and one half times the bar diameter.

2107.2.14 Compression in walls and columns.

2107.2.14.1 General. Stresses due to compressive forces in walls and columns shall be calculated in accordance with Section 2107.2.5.

2107.2.14.2 Walls, bending or combined bending and axial loads. Stresses in walls due to combined bending and axial loads shall satisfy the requirements of Section 2107.2.7 where f_a is given by Formula (7-8). Walls subjected to bending with or without axial loads shall meet all applicable requirements for flexural design.

The design of walls with an h'/t ratio larger than 30 shall be based on forces and moments determined from an analysis of the structure. Such analysis shall consider the influence of axial loads and variable moment of inertia on member stiffness and fixed-end moments, effect of deflections on moments and forces and the effects of duration of loads.

2107.2.15 Flexural design, rectangular flexural elements. Rectangular flexural elements shall be designed in accordance with the following formulas or other methods based on the assumptions given in Sections 2107.1.4, 2107.2.3 and this section.

1. Compressive stress in the masonry:

$$f_b = \frac{M}{bd^2} \left(\frac{2}{jk} \right) \qquad (7\text{-}31)$$

2. Tensile stress in the longitudinal reinforcement:

$$f_s = \frac{M}{A_s jd} \qquad (7\text{-}32)$$

3. Design coefficients:

$$k = \sqrt{(n\rho)^2 + 2n\rho} - n\rho \qquad (7\text{-}33)$$

or

$$k = \frac{1}{1 + \dfrac{f_s}{n f_b}} \qquad (7\text{-}34)$$

$$j = 1 - \frac{k}{3} \qquad (7\text{-}35)$$

2107.2.16 Bond of flexural reinforcement. In flexural members in which tensile reinforcement is parallel to the compressive face, the bond stress shall be computed by the formula:

$$u = \frac{V}{\Sigma_o jd} \qquad (7\text{-}36)$$

2107.2.17 Shear in flexural members and shear walls. The shear stress in flexural members and shear walls shall be computed by:

$$f_v = \frac{V}{bjd} \qquad (7\text{-}37)$$

For members of T or I section, b' shall be substituted for b. Where f_v as computed by Formula (7-37) exceeds the allowable shear stress in masonry, F_v, web reinforcement shall be provided and designed to carry the total shear force. Both vertical and horizontal shear stresses shall be considered.

The area required for shear reinforcement placed perpendicular to the longitudinal reinforcement shall be computed by:

$$A_v = \frac{sV}{F_s d} \qquad (7\text{-}38)$$

Where web reinforcement is required, it shall be so spaced that every 45-degree line extending from a point at $d/2$ of the beam to the longitudinal tension bars shall be crossed by at least one line of web reinforcement.

2107.3 Design of Unreinforced Masonry.

2107.3.1 General. The requirements of this section govern masonry in which reinforcement is not used to resist design forces and are in addition to the requirements of Sections 2106 and 2107.1.

2107.3.2 Allowable axial compressive stress. The allowable axial compressive stress F_a is:

$$F_a = 0.25 f'_m \left[1 - \left(\frac{h'}{140r} \right)^2 \right] \text{ for } h'/r \leq 99 \qquad (7\text{-}39)$$

$$F_a = 0.25 f'_m \left(\frac{70r}{h'} \right)^2 \text{ for } h'/r > 99 \qquad (7\text{-}40)$$

2107.3.3 Allowable flexural compressive stress. The allowable flexural compressive stress F_b is:

$$F_b = 0.33 f'_m, \ 2{,}000 \text{ psi (13.8 MPa) maximum} \qquad (7\text{-}41)$$

2107.3.4 Combined compressive stresses, unity formula. Elements subjected to combined axial and flexural stresses shall be designed in accordance with accepted principles of mechanics or in accordance with the Formula (7-42):

$$\frac{f_a}{F_a} + \frac{f_b}{F_b} \leq 1 \qquad (7\text{-}42)$$

2107.3.5 Allowable tensile stress. Resultant tensile stress due to combined bending and axial load shall not exceed the allowable flexural tensile stress, F_t.

The allowable tensile stress for walls in flexure without tensile reinforcement using portland cement and hydrated lime, or using mortar cement Type M or S mortar, shall not exceed the values in Table 21-I.

Values in Table 21-I for tension normal to head joints are for running bond; no tension is allowed across head joints in stack bond masonry. These values shall not be used for horizontal flexural members.

2107.3.6 Allowable shear stress in flexural members. The allowable shear stress F_v in flexural members is:

$$F_v = 1.0 \sqrt{f'_m}, \ 50 \text{ psi maximum} \qquad (7\text{-}43)$$

For **SI:** $\qquad F_v = 0.083 \sqrt{f'_m}, \ 345 \text{ kPa maximum}$

EXCEPTION: For a distance of $1/16$th the clear span beyond the point of inflection, the maximum stress shall be 20 psi (138 kPa).

2107.3.7 Allowable shear stress in shear walls. The allowable shear stress F_v in shear walls is as follows:

1. Clay units $\qquad F_v = 0.3 \sqrt{f'_m}, \ 80 \text{ psi maximum} \qquad (7\text{-}44)$

For **SI:** $\qquad F_v = 0.025 \sqrt{f'_m}, \ 551 \text{ kPa maximum}$

2. Concrete units with Type M or S mortar, $F_v = 34$ psi (234 kPa) maximum.

3. Concrete units with Type N mortar, $F_v = 23$ psi (158 kPa) maximum.

4. The allowable shear stress in unreinforced masonry may be increased by $0.2 f_{md}$.

2107.3.8 Allowable bearing stress. When a member bears on the full area of a masonry element, the allowable bearing stress F_{br} shall be:

$$F_{br} = 0.26 f'_m \qquad (7\text{-}45)$$

When a member bears on one-third or less of a masonry element, the allowable bearing stress F_{br} shall be:

$$F_{br} = 0.38 f'_m \qquad (7\text{-}46)$$

Formula (7-46) applies only when the least dimension between the edges of the loaded and unloaded areas is a minimum of one fourth of the parallel side dimension of the loaded area. The allowable bearing stress on a reasonably concentric area greater than one third but less than the full area shall be interpolated between the values of Formulas (7-45) and (7-46).

2107.3.9 Combined bending and axial loads, compressive stresses. Compressive stresses due to combined bending and axial loads shall satisfy the requirements of Section 2107.3.4.

2107.3.10 Compression in walls and columns. Stresses due to compressive forces in walls and columns shall be calculated in accordance with Section 2107.2.5.

2107.3.11 Flexural design. Stresses due to flexure shall not exceed the values given in Sections 2107.1.2, 2107.3.3 and 2107.3.5, where:

$$f_b = Mc/I \qquad (7\text{-}47)$$

2107.3.12 Shear in flexural members and shear walls. Shear calculations for flexural members and shear walls shall be based on Formula (7-48).

$$f_v = V / A_e \qquad (7\text{-}48)$$

2107.3.13 Corbels. The slope of corbelling (angle measured from the horizontal to the face of the corbelled surface) of unreinforced masonry shall be not less than 60 degrees.

The maximum horizontal projection of corbelling from the plane of the wall shall be such that allowable stresses are not exceeded.

2107.3.14 Stack bond. Masonry units laid in stack bond shall have longitudinal reinforcement of at least 0.00027 times the vertical cross-sectional area of the wall placed horizontally in the bed joints or in bond beams spaced vertically not more than 48 inches (1219 mm) apart.

SECTION 2108 — STRENGTH DESIGN OF MASONRY

2108.1 General.

2108.1.1 General provisions. The design of hollow-unit clay and concrete masonry structures using strength design shall comply with the provisions of Section 2106 and this section.

> **EXCEPTION:** Two-wythe solid-unit masonry may be used under Sections 2108.2.1 and 2108.2.4.

2108.1.2 Quality assurance provisions. Special inspection during construction shall be provided as set forth in Section 1701.5, Item 7.

2108.1.3 Required strength. The required strength shall be determined in accordance with the factored load combinations of Section 1612.2.

2108.1.4 Design strength. Design strength is the nominal strength, multiplied by the strength-reduction factor, ϕ, as specified in this section. Masonry members shall be proportioned such that the design strength exceeds the required strength.

2108.1.4.1 Beams, piers and columns.

2108.1.4.1.1 Flexure. Flexure with or without axial load, the value of ϕ shall be determined from Formula (8-1):

$$\phi = 0.8 - \frac{P_u}{A_e f'_m} \qquad (8\text{-}1)$$

and $0.60 \leq \phi \leq 0.80$

2108.1.4.1.2 Shear. Shear: $\phi = 0.60$.

2108.1.4.2 Wall design for out-of-plane loads.

2108.1.4.2.1 Walls with unfactored axial load of 0.04 f'_m or less. Flexure: $\phi = 0.80$.

2108.1.4.2.2 Walls with unfactored axial load greater than 0.04 f'_m. Axial load and axial load with flexure: $\phi = 0.80$. Shear: $\phi = 0.60$.

2108.1.4.3 Wall design for in-plane loads.

2108.1.4.3.1 Axial load. Axial load and axial load with flexure: $\phi = 0.65$.

For walls with symmetrical reinforcement in which f_y does not exceed 60,000 psi (413 MPa), the value of ϕ may be increased linearly to 0.85 as the value of ϕP_n decreases from $0.10 f'_m A_e$ or 0.25 P_b to zero.

For solid grouted walls, the value of P_b may be calculated by Formula (8-2)

$$P_b = 0.85 f'_m b a_b \qquad (8\text{-}2)$$

WHERE:

$$a_b = 0.85d \left\{ e_{mu} / [e_{mu} + (f_y / E_s)] \right\} \qquad (8\text{-}3)$$

2108.1.4.3.2 Shear. Shear: $\phi = 0.60$.

The value of ϕ may be 0.80 for any shear wall when its nominal shear strength exceeds the shear corresponding to development of its nominal flexural strength for the factored-load combination.

2108.1.4.4 Moment-resisting wall frames.

2108.1.4.4.1 Flexure with or without axial load. The value of ϕ shall be as determined from Formula (8-4); however, the value of ϕ shall not be less than 0.65 nor greater than 0.85.

$$\phi = 0.85 - 2\left(\frac{P_u}{A_e f'_m}\right) \qquad (8\text{-}4)$$

2108.1.4.4.2 Shear. Shear: $\phi = 0.80$.

2108.1.4.5 Anchor. Anchor bolts: $\phi = 0.80$.

2108.1.4.6 Reinforcement.

2108.1.4.6.1 Development. Development: $\phi = 0.80$.

2108.1.4.6.2 Splices. Splices: $\phi = 0.80$.

2108.1.5 Anchor bolts.

2108.1.5.1 Required strength. The required strength of embedded anchor bolts shall be determined from factored loads as specified in Section 2108.1.3.

2108.1.5.2 Nominal anchor bolt strength. The nominal strength of anchor bolts times the strength-reduction factor shall equal or exceed the required strength.

The nominal tensile capacity of anchor bolts shall be determined from the lesser of Formula (8-5) or (8-6).

$$B_{tn} = 1.0 A_p \sqrt{f'_m} \qquad (8\text{-}5)$$

For **SI:** $\qquad B_{tn} = 0.084 A_p \sqrt{f'_m}$

$$B_{tn} = 0.4 A_b f_y \qquad (8\text{-}6)$$

The area A_p shall be the lesser of Formula (8-7) or (8-8) and where the projected areas of adjacent anchor bolts overlap, the value of A_p of each anchor bolt shall be reduced by one half of the overlapping area.

$$A_p = \pi l_b^2 \qquad (8\text{-}7)$$

$$A_p = \pi l_{be}^2 \qquad (8\text{-}8)$$

The nominal shear capacity of anchor bolts shall be determined from the lesser of Formula (8-9) or (8-10).

$$B_{sn} = 900 \sqrt[4]{f'_m A_b} \qquad (8\text{-}9)$$

For **SI:** $\qquad B_{sn} = 2750 \sqrt[4]{f'_m A_b}$

$$B_{sn} = 0.25 A_b f_y \qquad (8\text{-}10)$$

Where the anchor bolt edge distance, l_{be}, in the direction of load is less than 12 bolt diameters, the value of B_{tn} in Formula (8-9) shall be reduced by linear interpolation to zero at an l_{be} distance of $1\frac{1}{2}$ inches (38 mm). Where adjacent anchor bolts are spaced closer than $8d_b$, the nominal shear strength of the adjacent anchors determined by Formula (8-9) shall be reduced by linear interpolation to 0.75 times the nominal shear strength at a center-to-center spacing of four bolt diameters.

Anchor bolts subjected to combined shear and tension shall be designed in accordance with Formula (8-11).

$$\frac{b_{tu}}{\phi B_{tn}} + \frac{b_{su}}{\phi B_{sn}} \leq 1.0 \qquad (8\text{-}11)$$

2108.1.5.3 Anchor bolt placement. Anchor bolts shall be placed so as to meet the edge distance, embedment depth and spacing requirements of Sections 2106.2.14.2, 2106.2.14.3 and 2106.2.14.4.

2108.2 Reinforced Masonry.

2108.2.1 General.

2108.2.1.1 Scope. The requirements of this section are in addition to the requirements of Sections 2106 and 2108.1 and govern masonry in which reinforcement is used to resist forces.

2108.2.1.2 Design assumptions. The following assumptions apply:

Masonry carries no tensile stress greater than the modulus of rupture.

Reinforcement is completely surrounded by and bonded to masonry material so that they work together as a homogeneous material.

Nominal strength of singly reinforced masonry wall cross sections for combined flexure and axial load shall be based on applicable conditions of equilibrium and compatibility of strains. Strain in reinforcement and masonry walls shall be assumed to be directly proportional to the distance from the neutral axis.

Maximum usable strain, e_{mu}, at the extreme masonry compression fiber shall:

1. Be 0.003 for the design of beams, piers, columns and walls.

2. Not exceed 0.003 for moment-resisting wall frames, unless lateral reinforcement as defined in Section 2108.2.6.2.6 is utilized.

Strain in reinforcement and masonry shall be assumed to be directly proportional to the distance from the neutral axis.

Stress in reinforcement below specified yield strength f_y for grade of reinforcement used shall be taken as E_s times steel strain. For strains greater than that corresponding to f_y, stress in reinforcement shall be considered independent of strain and equal to f_y.

Tensile strength of masonry walls shall be neglected in flexural calculations of strength, except when computing requirements for deflection.

Relationship between masonry compressive stress and masonry strain may be assumed to be rectangular as defined by the following:

Masonry stress of $0.85 f'_m$ shall be assumed uniformly distributed over an equivalent compression zone bounded by edges of the cross section and a straight line located parallel to the neutral axis at a distance $a = 0.85c$ from the fiber of maximum compressive strain. Distance c from fiber of maximum strain to the neutral axis shall be measured in a direction perpendicular to that axis.

2108.2.2 Reinforcement requirements and details.

2108.2.2.1 Maximum reinforcement. The maximum size of reinforcement shall be No. 9. The diameter of a bar shall not exceed one fourth the least dimension of a cell. No more than two bars shall be placed in a cell of a wall or a wall frame.

2108.2.2.2 Placement. The placement of reinforcement shall comply with the following:

In columns and piers, the clear distance between vertical reinforcing bars shall not be less than one and one-half times the nominal bar diameter, nor less than $1^1/_2$ inches (38 mm).

2108.2.2.3 Cover. All reinforcing bars shall be completely embedded in mortar or grout and shall have a cover of not less than $1^1/_2$ inches (38 mm) nor less than $2.5 d_b$.

2108.2.2.4 Standard hooks. A standard hook shall be one of the following:

1. A 180-degree turn plus an extension of at least four bar diameters, but not less than $2^1/_2$ inches (63 mm) at the free end of the bar.

2. A 135-degree turn plus an extension of at least six bar diameters at the free end of the bar.

3. A 90-degree turn plus an extension of at least 12 bar diameters at the free end of the bar.

2108.2.2.5 Minimum bend diameter for reinforcing bars. Diameter of bend measured on the inside of a bar other than for stirrups and ties in sizes No. 3 through No. 5 shall not be less than the values in Table 21-G.

Inside diameter of bends for stirrups and ties shall not be less than $4d_b$ for No. 5 bars and smaller. For bars larger than No. 5, diameter of bend shall be in accordance with Table 21-G.

2108.2.2.6 Development. The calculated tension or compression reinforcement shall be developed in accordance with the following provisions:

The embedment length of reinforcement shall be determined by Formula (8-12).

$$l_d = l_{de} / \phi \qquad (8\text{-}12)$$

WHERE:

$$l_{de} = \frac{0.15 d_b^2 f_y}{K \sqrt{f'_m}} \le 52 d_b \qquad (8\text{-}13)$$

For **SI:**

$$l_{de} = \frac{1.8 d_b^2 f_y}{K \sqrt{f'_m}} \le 52 d_b$$

K shall not exceed $3d_b$.

The minimum embedment length of reinforcement shall be 12 inches (305 mm).

2108.2.2.7 Splices. Reinforcement splices shall comply with one of the following:

1. The minimum length of lap for bars shall be 12 inches (305 mm) or the length determined by Formula (8-14).

$$l_d = l_{de} / \phi \qquad (8\text{-}14)$$

Bars spliced by noncontact lap splices shall be spaced transversely a distance not greater than one fifth the required length of lap or more than 8 inches (203 mm).

2. A welded splice shall have the bars butted and welded to develop in tension 125 percent of the yield strength of the bar, f_y.

3. Mechanical splices shall have the bars connected to develop in tension or compression, as required, at least 125 percent of the yield strength of the bar, f_y.

2108.2.3 Design of beams, piers and columns.

2108.2.3.1 General. The requirements of this section are for the design of masonry beams, piers and columns.

The value of f'_m shall not be less than 1,500 psi (10.3 MPa). For computational purposes, the value of f'_m shall not exceed 4,000 psi (27.6 MPa).

2108.2.3.2 Design assumptions.

Member design forces shall be based on an analysis which considers the relative stiffness of structural members. The calculation of lateral stiffness shall include the contribution of all beams, piers and columns.

The effects of cracking on member stiffness shall be considered.

The drift ratio of piers and columns shall satisfy the limits specified in Chapter 16.

2108.2.3.3 Balanced reinforcement ratio for compression limit state. Calculation of the balanced reinforcement ratio, ρ_b, shall be based on the following assumptions:

1. The distribution of strain across the section shall be assumed to vary linearly from the maximum usable strain, e_{mu}, at the extreme compression fiber of the element, to a yield strain of f_y/E_s at the extreme tension fiber of the element.

2. Compression forces shall be in equilibrium with the sum of tension forces in the reinforcement and the maximum axial load associated with a loading combination $1.0D + 1.0L + (1.4E$ or $1.3W)$.

3. The reinforcement shall be assumed to be uniformly distributed over the depth of the element and the balanced reinforcement ratio shall be calculated as the area of this reinforcement divided by the net area of the element.

4. All longitudinal reinforcement shall be included in calculating the balanced reinforcement ratio except that the contribution of compression reinforcement to resistance of compressive loads shall not be considered.

2108.2.3.4 Required strength. Except as required by Sections 2108.2.3.6 through 2108.2.3.12, the required strength shall be determined in accordance with Section 2108.1.3.

2108.2.3.5 Design strength. Design strength provided by beam, pier or column cross sections in terms of axial force, sheer and moment shall be computed as the nominal strength multiplied by the applicable strength-reduction factor, ϕ, specified in Section 2108.1.4.

2108.2.3.6 Nominal strength.

2108.2.3.6.1 Nominal axial and flexural strength. The nominal axial strength, P_n, and the nominal flexural strength, M_n, of a cross section shall be determined in accordance with the design assumptions of Section 2108.2.1.2 and 2108.2.3.2.

The maximum nominal axial compressive strength shall be determined in accordance with Formula (8-15).

$$P_n = 0.80[0.85f'_m(A_e - A_s) + f_y A_s] \qquad (8\text{-}15)$$

2108.2.3.6.2 Nominal shear strength. The nominal shear strength shall be determined in accordance with Formula (8-16).

$$V_n = V_m + V_s \qquad (8\text{-}16)$$

WHERE:

$$V_m = C_d A_e \sqrt{f'_m}, \ 63C_d A_e \text{ maximum} \qquad (8\text{-}17)$$

For **SI:** $\qquad V_m = 0.083 \, C_d A_e \sqrt{f'_m}, \ 63C_d A_e \text{ maximum}$

and

$$V_s = A_e \rho_n f_y \qquad (8\text{-}18)$$

1. The nominal shear strength shall not exceed the value given in Table 21-J.

2. The value of V_m shall be assumed to be zero within any region subjected to net tension factored loads.

3. The value of V_m shall be assumed to be 25 psi (172 kPa) where M_u is greater than $0.7 M_n$. The required moment, M_u, for seismic design for comparison with the $0.7 M_n$ value of this section shall be based on an R of 2.

2108.2.3.7 Reinforcement.

1. Where transverse reinforcement is required, the maximum spacing shall not exceed one half the depth of the member nor 48 inches (1219 mm).

2. Flexural reinforcement shall be uniformly distributed throughout the depth of the element.

3. Flexural elements subjected to load reversals shall be symmetrically reinforced.

4. The nominal moment strength at any section along a member shall not be less than one fourth of the maximum moment strength.

5. The flexural reinforcement ratio, ρ, shall not exceed $0.5 \, \rho_b$.

6. Lap splices shall comply with the provisions of Section 2108.2.2.7.

7. Welded splices and mechanical splices which develop at least 125 percent of the specified yield strength of a bar may be used for splicing the reinforcement. Not more than two longitudinal bars shall be spliced at a section. The distance between splices of adjacent bars shall be at least 30 inches (762 mm) along the longitudinal axis.

8. Specified yield strength of reinforcement shall not exceed 60,000 psi (413 MPa). The actual yield strength based on mill tests shall not exceed 1.3 times the specified yield strength.

2108.2.3.8 Seismic design provisions. The lateral seismic load resistance in any line or story level shall be provided by shear walls or wall frames, or a combination of shear walls and wall frames. Shear walls and wall frames shall provide at least 80 percent of the lateral stiffness in any line or story level.

EXCEPTION: Where seismic loads are determined based on R not greater than 2 and where all joints satisfy the provisions of Section 2108.2.6.2.9, the piers may be used to provide seismic load resistance.

2108.2.3.9 Dimensional limits. Dimensions shall be in accordance with the following:

1. **Beams.**

1.1 The nominal width of a beam shall not be less than 6 inches (153 mm).

1.2 The clear distance between locations of lateral bracing of the compression side of the beam shall not exceed 32 times the least width of the compression area.

1.3 The nominal depth of a beam shall not be less than 8 inches (203 mm).

2. **Piers.**

2.1 The nominal width of a pier shall not be less than 6 inches (153 mm) and shall not exceed 16 inches (406 mm).

2.2 The distance between lateral supports of a pier shall not exceed 30 times the nominal width of the piers except as provided for in Section 2108.2.3.9, Item 2.3.

2.3 When the distance between lateral supports of a pier exceeds 30 times the nominal width of the pier, the provisions of Section 2108.2.4 shall be used for design.

2.4 The nominal length of a pier shall not be less than three times the nominal width of the pier. The nominal length of a pier shall not be greater than six times the nominal width of the pier. The clear height of a pier shall not exceed five times the nominal length of the pier.

EXCEPTION: The length of a pier may be equal to the width of the pier when the axial force at the location of maximum moment is less than $0.04 f'_m A_g$.

3. **Columns.**

3.1 The nominal width of a column shall not be less than 12 inches (305 mm).

3.2 The distance between lateral supports of a column shall not exceed 30 times the nominal width of the column.

3.3 The nominal length of a column shall not be less than 12 inches (305 mm) and not greater than three times the nominal width of the column.

2108.2.3.10 Beams.

2108.2.3.10.1 Scope. Members designed primarily to resist flexure shall comply with the requirements of this section. The factored axial compressive force on a beam shall not exceed 0.05 $A_e f'_m$.

2108.2.3.10.2 Longitudinal reinforcement.

1. The variation in the longitudinal reinforcing bars shall not be greater than one bar size. Not more than two bar sizes shall be used in a beam.

2. The nominal flexural strength of a beam shall not be less than 1.3 times the nominal cracking moment strength of the beam. The modulus of rupture, f_r, for this calculation shall be assumed to be 235 psi (1.6 MPa).

2108.2.3.10.3 Transverse reinforcement. Transverse reinforcement shall be provided where V_u exceeds V_m. Required shear, V_u, shall include the effects of drift. The value of V_u shall be based on Δ_M. When transverse shear reinforcement is required, the following provisions shall apply:

1. Shear reinforcement shall be a single bar with a 180-degree hook at each end.

2. Shear reinforcement shall be hooked around the longitudinal reinforcement.

3. The minimum transverse shear reinforcement ratio shall be 0.0007.

4. The first transverse bar shall not be more than one fourth of the beam depth from the end of the beam.

2108.2.3.10.4 Construction. Beams shall be solid grouted.

2108.2.3.11 Piers.

2108.2.3.11.1 Scope. Piers proportioned to resist flexure and shear in conjunction with axial load shall comply with the requirements of this section. The factored axial compression on the piers shall not exceed 0.3 $A_e f'_m$.

2108.2.3.11.2 Longitudinal reinforcement. A pier subjected to in-plane stress reversals shall be longitudinally reinforced symmetrically on both sides of the neutral axis of the pier.

1. One bar shall be provided in the end cells.

2. The minimum longitudinal reinforcement ratio shall be 0.0007.

2108.2.3.11.3 Transverse reinforcement. Transverse reinforcement shall be provided where V_u exceeds V_m. Required shear, V_u, shall include the effects of drift. The value of V_u shall be based on Δ_M. When transverse shear reinforcement is required, the following provisions shall apply:

1. Shear reinforcement shall be hooked around the extreme longitudinal bars with a 180-degree hook. Alternatively, at wall intersections, transverse reinforcement with a 90-degree standard hook around a vertical bar in the intersecting wall shall be permitted.

2. The minimum transverse reinforcement ratio shall be 0.0015.

2108.2.3.12 Columns.

2108.2.3.12.1 Scope. Columns shall comply with the requirements of this section.

2108.2.3.12.2 Longitudinal reinforcement. Longitudinal reinforcement shall be a minimum of four bars, one in each corner of the column.

1. Maximum reinforcement area shall be 0.03 A_e.

2. Minimum reinforcement area shall be 0.005 A_e.

2108.2.3.12.3 Lateral ties.

1. Lateral ties shall be provided in accordance with Section 2106.3.6.

2. Minimum lateral reinforcement area shall be 0.0018 A_g.

2108.2.3.12.4 Construction. Columns shall be solid grouted.

2108.2.4 Wall design for out-of-plane loads.

2108.2.4.1 General. The requirements of this section are for the design of walls for out-of-plane loads.

2108.2.4.2 Maximum reinforcement. The reinforcement ratio shall not exceed 0.5ρ_b.

2108.2.4.3 Moment and deflection calculations. All moment and deflection calculations in Section 2108.2.4 are based on simple support conditions top and bottom. Other support and fixity conditions, moments and deflections shall be calculated using established principles of mechanics.

2108.2.4.4 Walls with axial load of 0.04f'_m or less. The procedures set forth in this section, which consider the slenderness of walls by representing effects of axial forces and deflection in calculation of moments, shall be used when the vertical load stress at the location of maximum moment does not exceed 0.04f'_m as computed by Formula (8-19). The value of f'_m shall not exceed 6,000 psi (41.3 MPa).

$$\frac{P_w + P_f}{A_g} \leq 0.04 f'_m \qquad (8\text{-}19)$$

Walls shall have a minimum nominal thickness of 6 inches (153 mm).

Required moment and axial force shall be determined at the midheight of the wall and shall be used for design. The factored moment, M_u, at the midheight of the wall shall be determined by Formula (8-20).

$$M_u = \frac{w_u h^2}{8} + P_{uf}\frac{e}{2} + P_u \Delta_u \qquad (8\text{-}20)$$

WHERE:

Δ_u = deflection at midheight of wall due to factored loads

$$P_u = P_{uw} + P_{uf} \qquad (8\text{-}21)$$

The design strength for out-of-plane wall loading shall be determined by Formula (8-22).

$$M_u \leq \phi M_n \qquad (8\text{-}22)$$

WHERE:

$$M_n = A_{se} f_y (d - a/2) \qquad (8\text{-}23)$$

$$A_{se} = (A_s f_y + P_u) / f_y, \text{ effective area of steel} \qquad (8\text{-}24)$$

$$a = (P_u + A_s f_y) / 0.85 f'_m b, \text{ depth of stress block}$$
$$\text{due to factored loads} \qquad (8\text{-}25)$$

2108.2.4.5 Wall with axial load greater than 0.04f'_m. The procedures set forth in this section shall be used for the design of masonry walls when the vertical load stresses at the location of maximum moment exceed 0.04f'_m but are less than 0.2f'_m and the slenderness ratio h'/t does not exceed 30.

Design strength provided by the wall cross section in terms of axial force, shear and moment shall be computed as the nominal strength multiplied by the applicable strength-reduction factor, ϕ, specified in Section 2108.1.4. Walls shall be proportioned such that the design strength exceeds the required strength.

The nominal shear strength shall be determined by Formula (8-26).

$$V_n = 2A_{mv} \sqrt{f'_m} \qquad (8\text{-}26)$$

For **SI**:
$$V_n = 0.166 A_{mv} \sqrt{f'_m}$$

2108.2.4.6 Deflection design. The midheight deflection, Δ_s, under service lateral and vertical loads (without load factors) shall be limited by the relation:

$$\Delta_s = 0.007h \qquad (8\text{-}27)$$

$P\Delta$ effects shall be included in deflection calculation. The midheight deflection shall be computed with the following formula:

$$\Delta_s = \frac{5\,M_s h^2}{48\,E_m I_g} \text{ for } M_{ser} \leq M_{cr} \qquad (8\text{-}28)$$

$$\Delta_s = \frac{5\,M_{cr} h^2}{48\,E_m I_g} + \frac{5\,(M_{ser} - M_{cr})h^2}{48\,E_m I_{cr}} \text{ for } M_{cr} < M_{ser} < M_n \qquad (8\text{-}29)$$

The cracking moment strength of the wall shall be determined from the formula:

$$M_{cr} = S f_r \qquad (8\text{-}30)$$

The modulus of rupture, f_r, shall be as follows:

1. For fully grouted hollow-unit masonry,

$$f_r = 4.0 \sqrt{f'_m}, \text{ 235 psi maximum} \qquad (8\text{-}31)$$

For **SI**: $\quad f_r = 0.33 \sqrt{f'_m}, \text{ 1.6 MPa maximum}$

2. For partially grouted hollow-unit masonry,

$$f_r = 2.5 \sqrt{f'_m}, \text{ 125 psi maximum} \qquad (8\text{-}32)$$

For **SI**: $\quad f_r = 0.21 \sqrt{f'_m}, \text{ 861 kPa maximum}$

3. For two-wythe brick masonry,

$$f_r = 2.0 \sqrt{f'_m}, \text{ 125 psi maximum} \qquad (8\text{-}33)$$

For **SI**: $\quad f_r = 0.166 \sqrt{f'_m}, \text{ 861 kPa maximum}$

2108.2.5 Wall design for in-plane loads.

2108.2.5.1 General. The requirements of this section are for the design of walls for in-plane loads.

The value of f'_m shall not be less than 1,500 psi (10.3 MPa) nor greater than 4,000 psi (27.6 MPa).

2108.2.5.2 Reinforcement. Reinforcement shall be in accordance with the following:

1. Minimum reinforcement shall be provided in accordance with Section 2106.1.12.4, Item 2.3, for all seismic areas using this method of analysis.

2. When the shear wall failure mode is in flexure, the nominal flexural strength of the shear wall shall be at least 1.8 times the cracking moment strength of a fully grouted wall or 3.0 times the cracking moment strength of a partially grouted wall from Formula (8-30).

3. The amount of vertical reinforcement shall not be less than one half the horizontal reinforcement.

4. Spacing of horizontal reinforcement within the region defined in Section 2108.2.5.5, Item 3, shall not exceed three times the nominal wall thickness nor 24 inches (610 mm).

2108.2.5.3 Design strength. Design strength provided by the shear wall cross section in terms of axial force, shear and moment shall be computed as the nominal strength multiplied by the applicable strength-reduction factor, ϕ, specified in Section 2108.1.4.3.

2108.2.5.4 Axial strength. The nominal axial strength of the shear wall supporting axial loads only shall be calculated by Formula (8-34).

$$P_o = 0.85 f'_m (A_e - A_s) + f_y A_s \qquad (8\text{-}34)$$

Axial design strength provided by the shear wall cross section shall satisfy Formula (8-35).

$$P_u \leq 0.80\,\phi\,P_o \qquad (8\text{-}35)$$

2108.2.5.5 Shear strength. Shear strength shall be as follows:

1. The nominal shear strength shall be determined using either Item 2 or 3 below. Maximum nominal shear strength values are determined from Table 21-J.

2. The nominal shear strength of the shear wall shall be determined from Formula (8-36), except as provided in Item 3 below

$$V_n = V_m + V_s \qquad (8\text{-}36)$$

WHERE:

$$V_m = C_d A_{mv} \sqrt{f'_m} \qquad (8\text{-}37)$$

For **SI**: $\qquad V_m = 0.083\,C_d A_{mv} \sqrt{f'_m}$

and

$$V_s = A_{mv}\,\rho_n f_y \qquad (8\text{-}38)$$

3. For a shear wall whose nominal shear strength exceeds the shear corresponding to development of its nominal flexural strength, two shear regions exist.

For all cross sections within the region defined by the base of the shear wall and a plane at a distance L_w above the base of the shear wall, the nominal shear strength shall be determined from Formula (8-39).

$$V_n = A_{mv}\,\rho_n f_y \qquad (8\text{-}39)$$

The required shear strength for this region shall be calculated at a distance $L_w/2$ above the base of the shear wall, but not to exceed one half story height.

For the other region, the nominal shear strength of the shear wall shall be determined from Formula (8-36).

2108.2.5.6 Boundary members. Boundary members shall be as follows:

1. Boundary members shall be provided at the boundaries of shear walls when the compressive strains in the wall exceed 0.0015. The strain shall be determined using factored forces and R equal to 1.1.

2. The minimum length of the boundary member shall be three times the thickness of the wall, but shall include all areas where the compressive strain per Section 2108.2.6.2.7 is greater than 0.0015.

3. Lateral reinforcement shall be provided for the boundary elements. The lateral reinforcement shall be a minimum of No. 3 bars at a maximum of 8-inch (203 mm) spacing within the grouted core or equivalent confinement which can develop an ultimate compressive masonry strain of at least 0.006.

2108.2.6 Design of moment-resisting wall frames.

2108.2.6.1 General requirements.

2108.2.6.1.1 Scope. The requirements of this section are for the design of fully grouted moment-resisting wall frames constructed of reinforced open-end hollow-unit concrete or hollow-unit clay masonry.

2108.2.6.1.2 Dimensional limits. Dimensions shall be in accordance with the following.

Beams. Clear span for the beam shall not be less than two times its depth.

The nominal depth of the beam shall not be less than two units or 16 inches (406 mm), whichever is greater. The nominal beam depth to nominal beam width ratio shall not exceed 6.

The nominal width of the beam shall be the greater of 8 inches (203 mm) or $^1/_{26}$ of the clear span between pier faces.

Piers. The nominal depth of piers shall not exceed 96 inches (2438 mm). Nominal depth shall not be less than two full units or 32 inches (813 mm), whichever is greater.

The nominal width of piers shall not be less than the nominal width of the beam, nor less than 8 inches (203 mm) or $^1/_{14}$ of the clear height between beam faces, whichever is greater.

The clear height-to-depth ratio of piers shall not exceed 5.

2108.2.6.1.3 Analysis. Member design forces shall be based on an analysis which considers the relative stiffness of pier and beam members, including the stiffening influence of joints.

The calculation of beam moment capacity for the determination of pier design shall include any contribution of floor slab reinforcement.

The out-of-plane drift ratio of all piers shall satisfy the drift-ratio limits specified in Section 1630.10.2.

2108.2.6.2 Design procedure.

2108.2.6.2.1 Required strength. Except as required by Sections 2108.2.6.2.7 and 2108.2.6.2.8, the required strength shall be determined in accordance with Section 2108.1.3.

2108.2.6.2.2 Design strength. Design strength provided by frame member cross sections in terms of axial force, shear and moment shall be computed as the nominal strength multiplied by the applicable strength-reduction factor, ϕ, specified in Section 2108.1.4.4.

Members shall be proportioned such that the design strength exceeds the required strength.

2108.2.6.2.3 Design assumptions for nominal strength. The nominal strength of member cross sections shall be based on assumptions prescribed in Section 2108.2.1.2.

The value of f'_m shall not be less than 1,500 psi (10.3 MPa) or greater than 4,000 psi (27.6 MPa).

2108.2.6.2.4 Reinforcement. The nominal moment strength at any section along a member shall not be less than one fourth of the higher moment strength provided at the two ends of the member.

Lap splices shall be as defined in Section 2108.2.2.7. The center of the lap splice shall be at the center of the member clear length.

Welded splices and mechanical connections conforming to Section 1912.14.3, Items 1 through 4, may be used for splicing the reinforcement at any section provided not more than alternate longitudinal bars are spliced at a section, and the distance between splices of alternate bars is at least 24 inches (610 mm) along the longitudinal axis.

Reinforcement shall not have a specified yield strength greater than 60,000 psi (413 MPa). The actual yield strength based on mill tests shall not exceed the specified yield strength times 1.3.

2108.2.6.2.5 Flexural members (beams). Requirements of this section apply to beams proportioned primarily to resist flexure as follows:

The axial compressive force on beams due to factored loads shall not exceed $0.10 A_n f'_m$.

1. **Longitudinal reinforcement.** At any section of a beam, each masonry unit through the beam depth shall contain longitudinal reinforcement.

The variation in the longitudinal reinforcement area between units at any section shall not be greater than 50 percent, except multiple No. 4 bars shall not be greater than 100 percent of the minimum area of longitudinal reinforcement contained by any one unit, except where splices occur.

Minimum reinforcement ratio calculated over the gross cross section shall be 0.002.

Maximum reinforcement ratio calculated over the gross cross section shall be $0.15f'_m / f_y$.

2. **Transverse reinforcement.** Transverse reinforcement shall be hooked around top and bottom longitudinal bars with a standard 180-degree hook, as defined in Section 2108.2.2.4, and shall be single pieces.

Within an end region extending one beam depth from pier faces and at any region at which beam flexural yielding may occur during seismic or wind loading, maximum spacing of transverse reinforcement shall not exceed one fourth the nominal depth of the beam.

The maximum spacing of transverse reinforcement shall not exceed one half the nominal depth of the beam.

Minimum reinforcement ratio shall be 0.0015.

The first transverse bar shall not be more than 4 inches (102 mm) from the face of the pier.

2108.2.6.2.6 Members subjected to axial force and flexure.

The requirements set forth in this subsection apply to piers proportioned to resist flexure in conjunction with axial loads.

1. **Longitudinal reinforcement.** A minimum of four longitudinal bars shall be provided at all sections of every pier.

Flexural reinforcement shall be distributed across the member depth. Variation in reinforcement area between reinforced cells shall not exceed 50 percent.

Minimum reinforcement ratio calculated over the gross cross section shall be 0.002.

Maximum reinforcement ratio calculated over the gross cross section shall be $0.15f'_m / f_y$.

Maximum bar diameter shall be one eighth nominal width of the pier.

2. **Transverse reinforcement.** Transverse reinforcement shall be hooked around the extreme longitudinal bars with standard 180-degree hook as defined in Section 2108.2.2.4.

Within an end region extending one pier depth from the end of the beam, and at any region at which flexural yielding may occur during seismic or wind loading, the maximum spacing of transverse reinforcement shall not exceed one fourth the nominal depth of the pier.

The maximum spacing of transverse reinforcement shall not exceed one half the nominal depth of the pier.

The minimum transverse reinforcement ratio shall be 0.0015.

3. **Lateral reinforcement.** Lateral reinforcement shall be provided to confine the grouted core when compressive strains due to axial and bending forces exceed 0.0015, corresponding to factored forces with R_w equal to 1.5. The unconfined portion of the cross section with strain exceeding 0.0015 shall be neglected in computing the nominal strength of the section.

The total cross-sectional area of rectangular tie reinforcement for the confined core shall not be less than:

$$A_{sh} = 0.09 sh_c f'_m / f_{yh} \qquad (8\text{-}40)$$

Alternatively, equivalent confinement which can develop an ultimate compressive strain of at least 0.006 may be substituted for rectangular tie reinforcement.

2108.2.6.2.7 Pier design forces. Pier nominal moment strength shall not be less than 1.6 times the pier moment corresponding to the development of beam plastic hinges, except at the foundation level.

Pier axial load based on the development of beam plastic hinges in accordance with the paragraph above and including factored dead and live loads shall not exceed $0.15 A_n f'_m$.

The drift ratio of piers shall satisfy the limits specified in Chapter 16.

The effects of cracking on member stiffness shall be considered.

The base plastic hinge of the pier must form immediately adjacent to the level of lateral support provided at the base or foundation.

2108.2.6.2.8 Shear design.

1. **General.** Beam and pier nominal shear strength shall not be less than 1.4 times the shears corresponding to the development of beam flexural yielding.

It shall be assumed in the calculation of member shear force that moments of opposite sign act at the joint faces and that the member is loaded with the tributary gravity load along its span.

2. **Vertical member shear strength.** The nominal shear strength shall be determined from Formula (8-41):

$$V_n = V_m + V_s \qquad (8\text{-}41)$$

WHERE:

$$V_m = C_d A_{mv} \sqrt{f'_m} \qquad (8\text{-}42)$$

For **SI:** $\qquad V_m = 0.083 C_d A_{mv} \sqrt{f'_m}$

and

$$V_s = A_{mv} \rho_n f_y \qquad (8\text{-}43)$$

The value of V_m shall be zero within an end region extending one pier depth from beam faces and at any region where pier flexural yielding may occur during seismic loading, and at piers subjected to net tension factored loads.

The nominal pier shear strength, V_n, shall not exceed the value determined from Table 21-J.

3. **Beam shear strength.** The nominal shear strength shall be determined from Formula (8-44),

WHERE:

$$V_m = 1.2 A_{mv} \sqrt{f'_m} \qquad (8\text{-}44)$$

For **SI:** $\qquad V_m = 0.01 A_{mv} \sqrt{f'_m}$

The value of V_m shall be zero within an end region extending one beam depth from pier faces and at any region at which beam flexural yielding may occur during seismic loading.

The nominal beam shear strength, V_n, shall be determined from Formula (8-45).

$$V_n \leq 4 A_{mv} \sqrt{f'_m} \qquad (8\text{-}45)$$

For **SI:** $\qquad V_n \leq 0.33 A_{mv} \sqrt{f'_m}$

2108.2.6.2.9 Joints.

1. **General requirements.** Where reinforcing bars extend through a joint, the joint dimensions shall be proportioned such that

$$h_p > 4800 d_{bb} / \sqrt{f'_g} \qquad (8\text{-}46)$$

For **SI:** $\qquad h_p > 400 d_{bb} / \sqrt{f'_g}$

and

$$h_b > 1800 d_{bp} / \sqrt{f'_g} \qquad (8\text{-}47)$$

For **SI:** $\qquad h_b > 150 d_{bp} / \sqrt{f'_g}$

The grout strength shall not exceed 5,000 psi (34.4 MPa) for the purposes of Formulas (8-46) and (8-47).

Joint shear forces shall be calculated on the assumption that the stress in all flexural tension reinforcement of the beams at the pier faces is $1.4 f_y$.

Strength of joint shall be governed by the appropriate strength-reduction factors specified in Section 2108.1.4.4.

Beam longitudinal reinforcement terminating in a pier shall be extended to the far face of the pier and anchored by a standard 90- or 180-degree hook, as defined in Section 2108.2.2.4, bent back to the beam.

Pier longitudinal reinforcement terminating in a beam shall be extended to the far face of the beam and anchored by a standard 90- or 180-degree hook, as defined in Section 2108.2.2.4, bent back to the beam.

2. **Transverse reinforcement.** Special horizontal joint shear reinforcement crossing a potential corner-to-corner diagonal joint shear crack, and anchored by standard hooks, as defined in Section 2108.2.2.4, around the extreme pier reinforcing bars shall be provided such that

$$A_{jh} = 0.5 V_{jh} / f_y \qquad (8\text{-}48)$$

Vertical shear forces may be considered to be carried by a combination of masonry shear-resisting mechanisms and truss mechanisms involving intermediate pier reinforcing bars.

3. **Shear strength.** The nominal horizontal shear strength of the joint shall not exceed $7 \sqrt{f'_m}$ (For **SI:** $0.58 \sqrt{f'_m}$) or 350 psi (2.4 MPa), whichever is less.

SECTION 2109 — EMPIRICAL DESIGN OF MASONRY

2109.1 General. The design of masonry structures using empirical design located in those portions of Seismic Zones 0 and 1 as defined in Part III of Chapter 16 where the basic wind speed is less than 80 miles per hour as defined in Part II of Chapter 16 shall comply with the provisions of Section 2106 and this section, subject to approval of the building official.

2109.2 Height. Buildings relying on masonry walls for lateral load resistance shall not exceed 35 feet (10 668 mm) in height.

2109.3 Lateral Stability. Where the structure depends on masonry walls for lateral stability, shear walls shall be provided parallel to the direction of the lateral forces resisted.

Minimum nominal thickness of masonry shear walls shall be 8 inches (203 mm).

In each direction in which shear walls are required for lateral stability, the minimum cumulative length of shear walls provided shall be 0.4 times the long dimension of the building. The cumulative length of shear walls shall not include openings.

The maximum spacing of shear walls shall not exceed the ratio listed in Table 21-L.

2109.4 Compressive Stresses.

2109.4.1 General. Compressive stresses in masonry due to vertical dead loads plus live loads, excluding wind or seismic loads, shall be determined in accordance with Section 2109.4.3. Dead and live loads shall be in accordance with this code with permitted live load reductions.

2109.4.2 Allowable stresses. The compressive stresses in masonry shall not exceed the values set forth in Table 21-M. The allowable stresses given in Table 21-M for the weakest combination of the units and mortar used in any load wythe shall be used for all loaded wythes of multiwythe walls.

2109.4.3 Stress calculations. Stresses shall be calculated based on specified rather than nominal dimensions. Calculated compressive stresses shall be determined by dividing the design load by the gross cross-sectional area of the member. The area of openings, chases or recesses in walls shall not be included in the gross cross-sectional area of the wall.

2109.4.4 Anchor bolts. Bolt values shall not exceed those set forth in Table 21-N.

2109.5 Lateral Support. Masonry walls shall be laterally supported in either the horizontal or vertical direction not exceeding the intervals set forth in Table 21-O.

Lateral support shall be provided by cross walls, pilasters, buttresses or structural framing members horizontally or by floors, roof or structural framing members vertically.

Except for parapet walls, the ratio of height to nominal thickness for cantilever walls shall not exceed 6 for solid masonry or 4 for hollow masonry.

In computing the ratio for cavity walls, the value of thickness shall be the sums of the nominal thickness of the inner and outer wythes of the masonry. In walls composed of different classes of units and mortars, the ratio of height or length to thickness shall not exceed that allowed for the weakest of the combinations of units and mortar of which the member is composed.

2109.6 Minimum Thickness.

2109.6.1 General. The nominal thickness of masonry bearing walls in buildings more than one story in height shall not be less than 8 inches (203 mm). Solid masonry walls in one-story buildings may be of 6-inch nominal thickness when not over 9 feet (2743 mm) in height, provided that when gable construction is used, an additional 6 feet (1829 mm) is permitted to the peak of the gable.

> **EXCEPTION:** The thickness of unreinforced grouted brick masonry walls may be 2 inches (51 mm) less than required by this section, but in no case less than 6 inches (152 mm).

2109.6.2 Variation in thickness. Where a change in thickness due to minimum thickness occurs between floor levels, the greater thickness shall be carried up to the higher floor level.

2109.6.3 Decrease in thickness. Where walls of masonry of hollow units or masonry-bonded hollow walls are decreased in thickness, a course or courses of solid masonry shall be constructed between the walls below and the thinner wall above, or special units or construction shall be used to transmit the loads from face shells or wythes to the walls below.

2109.6.4 Parapets. Parapet walls shall be at least 8 inches (203 mm) in thickness and their height shall not exceed three times their thickness. The parapet wall shall not be thinner than the wall below.

2109.6.5 Foundation walls. Mortar used in masonry foundation walls shall be either Type M or S.

Where the height of unbalanced fill (height of finished grade above basement floor or inside grade) and the height of the wall between lateral support does not exceed 8 feet (2438 mm), and when the equivalent fluid weight of unbalanced fill does not exceed 30 pounds per cubic foot (480 kg/m^2), the minimum thickness of foundation walls shall be as set forth in Table 21-P. Maximum depths of unbalanced fill permitted in Table 21-P may be increased with the approval of the building official when local soil conditions warrant such an increase.

Where the height of unbalanced fill, height between lateral supports or equivalent fluid weight of unbalanced fill exceeds that set forth above, foundation walls shall be designed in accordance with Chapter 18.

2109.7 Bond.

2109.7.1 General. The facing and backing of multiwythe masonry walls shall be bonded in accordance with this section.

2109.7.2 Masonry headers. Where the facing and backing of solid masonry construction are bonded by masonry headers, not less than 4 percent of the wall surface of each face shall be composed of headers extending not less than 3 inches (76 mm) into the backing. The distance between adjacent full-length headers shall not exceed 24 inches (610 mm) either vertically or horizontally. In walls in which a single header does not extend through the wall, headers from opposite sides shall overlap at least 3 inches (76 mm), or headers from opposite sides shall be covered with another header course overlapping the header below at least 3 inches (76 mm).

Where two or more hollow units are used to make up the thickness of the wall, the stretcher courses shall be bonded at vertical intervals not exceeding 34 inches (864 mm) by lapping at least 3 inches (76 mm) over the unit below, or by lapping at vertical intervals not exceeding 17 inches (432 mm) with units which are at least 50 percent greater in thickness than the units below.

2109.7.3 Wall ties. Where the facing and backing of masonry walls are bonded with $^3/_{16}$-inch-diameter (4.8 mm) wall ties or metal ties of equivalent stiffness embedded in the horizontal mortar joints, there shall be at least one metal tie for each $4^1/_2$ square feet (0.42 m^2) of wall area. Ties in alternate courses shall be staggered, the maximum vertical distance between ties shall not exceed 24 inches (610 mm), and the maximum horizontal distance shall not exceed 36 inches (914 mm). Rods bent to rectangular shape shall be used with hollow-masonry units laid with the cells vertical. In other walls, the ends of ties shall be bent to 90-degree angles to provide hooks not less than 2 inches (51 mm) long. Additional ties shall be provided at all openings, spaced not more than 3 feet (914 mm) apart around the perimeter and within 12 inches (305 mm) of the opening.

The facing and backing of masonry walls may be bonded with prefabricated joint reinforcement. There shall be at least one cross wire serving as a tie for each $2^2/_3$ square feet (0.25 m^2) of wall area. The vertical spacing of the joint reinforcement shall not exceed 16 inches (406 mm). Cross wires of prefabricated joint reinforcement shall be at least No. 9 gage wire. The longitudinal wire shall be embedded in mortar.

2109.7.4 Longitudinal bond. In each wythe of masonry, head joints in successive courses shall be offset at least one fourth of the unit length or the walls shall be reinforced longitudinally as required in Section 2106.1.12.3, Item 4.

2109.8 Anchorage.

2109.8.1 Intersecting walls. Masonry walls depending on one another for lateral support shall be anchored or bonded at locations where they meet or intersect by one of the following methods:

1. Fifty percent of the units at the intersection shall be laid in an overlapping pattern, with alternating units having a bearing of not less than 3 inches (76 mm) on the unit below.

2. Walls shall be anchored by steel connectors having a minimum section of $^1/_4$ inch by $1^1/_2$ inches (6.4 mm by 38 mm) with ends bent up at least 2 inches (51 mm), or with cross pins to form anchorage. Such anchors shall be at least 24 inches (610 mm) long and the maximum spacing shall be 4 feet (1219 mm) vertically.

3. Walls shall be anchored by joint reinforcement spaced at a maximum distance of 8 inches (203 mm) vertically. Longitudinal rods of such reinforcement shall be at least No. 9 gage and shall extend at least 30 inches (762 mm) in each direction at the intersection.

4. Interior nonbearing walls may be anchored at their intersection, at vertical spacing of not more than 16 inches (406 mm) with joint reinforcement or $^1/_4$-inch (6.4 mm) mesh galvanized hardware cloth.

5. Other metal ties, joint reinforcement or anchors may be used, provided they are spaced to provide equivalent area of anchorage to that required by this section.

2109.8.2 Floor and roof anchorage. Floor and roof diaphragms providing lateral support to masonry walls shall be connected to the masonry walls by one of the following methods:

1. Wood floor joists bearing on masonry walls shall be anchored to the wall by approved metal strap anchors at intervals not exceeding 6 feet (1829 mm). Joists parallel to the wall shall be anchored with metal straps spaced not more than 6 feet (1829 mm) on center extending over and under and secured to at least three joists. Blocking shall be provided between joists at each strap anchor.

2. Steel floor joists shall be anchored to masonry walls with No. 3 bars, or their equivalent, spaced not more than 6 feet (1829 mm) on center. Where joists are parallel to the wall, anchors shall be located at joist cross bridging.

3. Roof structures shall be anchored to masonry walls with $^1/_2$-inch-diameter (12.7 mm) bolts at 6 feet (1829 mm) on center or their equivalent. Bolts shall extend and be embedded at least 15 inches (381 mm) into the masonry, or be hooked or welded to not less than 0.2 square inch (129 mm^2) of bond beam reinforcement placed not less than 6 inches (152 mm) from the top of the wall.

2109.8.3 Walls adjoining structural framing. Where walls are dependent on the structural frame for lateral support, they shall be anchored to the structural members with metal anchors or keyed to the structural members. Metal anchors shall consist of $^1/_2$-inch-diameter (12.7 mm) bolts spaced at a maximum of 4 feet (1219 mm) on center and embedded at least 4 inches (102 mm) into the masonry, or their equivalent area.

2109.9 Unburned Clay Masonry.

2109.9.1 General. Masonry of stabilized unburned clay units shall not be used in any building more than one story in height. The unsupported height of every wall of unburned clay units shall not be more than 10 times the thickness of such walls. Bearing walls shall in no case be less than 16 inches (406 mm) in thickness. All

footing walls which support masonry of unburned clay units shall extend to an elevation not less than 6 inches (152 mm) above the adjacent ground at all points.

2109.9.2 Bolts. Bolt values shall not exceed those set forth in Table 21-Q.

2109.10 Stone Masonry.

2109.10.1 General. Stone masonry is that form of construction made with natural or cast stone in which the units are laid and set in mortar with all joints filled.

2109.10.2 Construction. In ashlar masonry, bond stones uniformly distributed shall be provided to the extent of not less than 10 percent of the area of exposed facets. Rubble stone masonry 24 inches (610 mm) or less in thickness shall have bond stones with a maximum spacing of 3 feet (914 mm) vertically and 3 feet (914 mm) horizontally and, if the masonry is of greater thickness than 24 inches (610 mm), shall have one bond stone for each 6 square feet (0.56 m^2) of wall surface on both sides.

2109.10.3 Minimum thickness. The thickness of stone masonry bearing walls shall not be less than 16 inches (406 mm).

SECTION 2110 — GLASS MASONRY

2110.1 General. Masonry of glass blocks may be used in non-load-bearing exterior or interior walls and in openings which might otherwise be filled with windows, either isolated or in continuous bands, provided the glass block panels have a minimum thickness of 3 inches (76 mm) at the mortar joint and the mortared surfaces of the blocks are treated for mortar bonding. Glass block may be solid or hollow and may contain inserts.

2110.2 Mortar Joints. Glass block shall be laid in Type S or N mortar. Both vertical and horizontal mortar joints shall be at least $^1/_4$ inch (6 mm) and not more than $^3/_8$ inch (9.5 mm) thick and shall be completely filled. All mortar contact surfaces shall be treated to ensure adhesion between mortar and glass.

2110.3 Lateral Support. Glass panels shall be laterally supported along each end of the panel.

Lateral support shall be provided by panel anchors spaced not more than 16 inches (406 mm) on center or by channels. The lateral support shall be capable of resisting the horizontal design forces determined in Chapter 16 or a minimum of 200 pounds per lineal foot (2920 N per linear meter) of wall, whichever is greater. The connection shall accommodate movement requirements of Section 2110.6.

2110.4 Reinforcement. Glass block panels shall have joint reinforcement spaced not more than 16 inches (406 mm) on center and located in the mortar bed joint extending the entire length of the panel. A lapping of longitudinal wires for a minimum of 6 inches (152 mm) is required for joint reinforcement splices. Joint reinforcement shall also be placed in the bed joint immediately below and above openings in the panel. Joint reinforcement shall conform to UBC Standard 21-10, Part I. Joint reinforcement in exterior panels shall be hot-dip galvanized in accordance with UBC Standard 21-10, Part I.

2110.5 Size of Panels. Glass block panels for exterior walls shall not exceed 144 square feet (13.4 m^2) of unsupported wall surface or 15 feet (4572 mm) in any dimension. For interior walls, glass block panels shall not exceed 250 square feet (23.2 m^2) of unsupported area or 25 feet (7620 mm) in any dimension.

2110.6 Expansion Joints. Glass block shall be provided with expansion joints along the sides and top, and these joints shall have sufficient thickness to accommodate displacements of the supporting structure, but not less than $^3/_8$ inch (9.5 mm). Expansion joints shall be entirely free of mortar and shall be filled with resilient material.

2110.7 Reuse of Units. Glass block units shall not be reused after being removed from an existing panel.

SECTION 2111 — CHIMNEYS, FIREPLACES AND BARBECUES

Chimneys, flues, fireplaces and barbecues and their connections carrying products of combustion shall be designed, anchored, supported and reinforced as set forth in Chapter 31 and any applicable provisions of this chapter.

TABLE 21-A—MORTAR PROPORTIONS FOR UNIT MASONRY

MORTAR	TYPE	PROPORTIONS BY VOLUME (CEMENTITIOUS MATERIALS)								AGGREGATE MEASURED IN A DAMP, LOOSE CONDITION
		Portland Cement or Blended Cement	Masonry Cement[1]			Mortar Cement[2]			Hydrated Lime or Lime Putty	
			M	S	N	M	S	N		
Cement-lime	M	1	—	—	—	—	—	—	$1/4$	Not less than $2^1/4$ and not more than 3 times the sum of the separate volumes of cementitious materials.
	S	1	—	—	—	—	—	—	over $1/4$ to $1/2$	
	N	1	—	—	—	—	—	—	over $1/2$ to $1^1/4$	
	O	1	—	—	—	—	—	—	over $1^1/4$ to $2^1/2$	
Mortar cement	M	1	—	—	—	—	—	1	—	
	M	—	—	—	—	1	—	—	—	
	S	$1/2$	—	—	—	—	—	1	—	
	S	—	—	—	—	—	1	—	—	
	N	—	—	—	—	—	—	1	—	
Masonry cement	M	1	—	—	1	—	—	—	—	
	M	—	1	—	—	—	—	—	—	
	S	$1/2$	—	—	1	—	—	—	—	
	S	—	—	1	—	—	—	—	—	
	N	—	—	—	1	—	—	—	—	
	O	—	—	—	1	—	—	—	—	

[1]Masonry cement conforming to the requirements of UBC Standard 21-11.

[2]Mortar cement conforming to the requirements of UBC Standard 21-14.

TABLE 21-B—GROUT PROPORTIONS BY VOLUME[1]

TYPE	PARTS BY VOLUME OF PORTLAND CEMENT OR BLENDED CEMENT	PARTS BY VOLUME OF HYDRATED LIME OR LIME PUTTY	AGGREGATE MEASURED IN A DAMP, LOOSE CONDITION	
			Fine	Coarse
Fine grout	1	0 to $1/10$	$2^1/4$ to 3 times the sum of the volumes of the cementitious materials	
Coarse grout	1	0 to $1/10$	$2^1/4$ to 3 times the sum of the volumes of the cementitious materials	1 to 2 times the sum of the volumes of the cementitious materials

[1]Grout shall attain a minimum compressive strength at 28 days of 2,000 psi (13.8 MPa). The building official may require a compressive field strength test of grout made in accordance with UBC Standard 21-18.

TABLE 21-C—GROUTING LIMITATIONS

GROUT TYPE	GROUT POUR MAXIMUM HEIGHT (feet)[1]	MINIMUM DIMENSIONS OF THE TOTAL CLEAR AREAS WITHIN GROUT SPACES AND CELLS[2,3]	
		× 25.4 for mm	
	× 304.8 for mm	Multiwythe Masonry	Hollow-unit Masonry
Fine	1	$3/4$	$1^1/2 \times 2$
Fine	5	$1^1/2$	$1^1/2 \times 2$
Fine	8	$1^1/2$	$1^1/2 \times 3$
Fine	12	$1^1/2$	$1^3/4 \times 3$
Fine	24	2	3×3
Coarse	1	$1^1/2$	$1^1/2 \times 3$
Coarse	5	2	$2^1/2 \times 3$
Coarse	8	2	3×3
Coarse	12	$2^1/2$	3×3
Coarse	24	3	3×4

[1]See also Section 2104.6.

[2]The actual grout space or grout cell dimensions must be larger than the sum of the following items: (1) The required minimum dimensions of total clear areas in Table 21-C; (2) The width of any mortar projections within the space; and (3) The horizontal projections of the diameters of the horizontal reinforcing bars within a cross section of the grout space or cell.

[3]The minimum dimensions of the total clear areas shall be made up of one or more open areas, with at least one area being $3/4$ inch (19 mm) or greater in width.

TABLE 21-D
TABLE 21-E-1

1997 UNIFORM BUILDING CODE

TABLE 21-D—SPECIFIED COMPRESSIVE STRENGTH OF MASONRY, f'_m (psi) BASED ON SPECIFYING THE COMPRESSIVE STRENGTH OF MASONRY UNITS

COMPRESSIVE STRENGTH OF CLAY MASONRY UNITS[1, 2] (psi)	SPECIFIED COMPRESSIVE STRENGTH OF MASONRY, f'_m	
	Type M or S Mortar[3] (psi)	Type N Mortar[3] (psi)
	× 6.89 for kPa	
14,000 or more	5,300	4,400
12,000	4,700	3,800
10,000	4,000	3,300
8,000	3,350	2,700
6,000	2,700	2,200
4,000	2,000	1,600

COMPRESSIVE STRENGTH OF CONCRETE MASONRY UNITS[2, 4] (psi)	SPECIFIED COMPRESSIVE STRENGTH OF MASONRY, f'_m	
	Type M or S Mortar[3] (psi)	Type N Mortar[3] (psi)
	× 6.89 for kPa	
4,800 or more	3,000	2,800
3,750	2,500	2,350
2,800	2,000	1,850
1,900	1,500	1,350
1,250	1,000	950

[1]Compressive strength of solid clay masonry units is based on gross area. Compressive strength of hollow clay masonry units is based on minimum net area. Values may be interpolated. When hollow clay masonry units are grouted, the grout shall conform to the proportions in Table 21-B.

[2]Assumed assemblage. The specified compressive strength of masonry f'_m is based on gross area strength when using solid units or solid grouted masonry and net area strength when using ungrouted hollow units.

[3]Mortar for unit masonry, proportion specification, as specified in Table 21-A. These values apply to portland cement-lime mortars without added air-entraining materials.

[4]Values may be interpolated. In grouted concrete masonry, the compressive strength of grout shall be equal to or greater than the compressive strength of the concrete masonry units.

TABLE 21-E-1—ALLOWABLE TENSION, B_t, FOR EMBEDDED ANCHOR BOLTS FOR CLAY AND CONCRETE MASONRY, pounds[1,2,3]

f'_m (psi)	EMBEDMENT LENGTH, l_b, or EDGE DISTANCE, l_{be} (inches)						
	2	3	4	5	6	8	10
× 6.89 for kPa	× 25.4 for mm × 4.45 for N						
1,500	240	550	970	1,520	2,190	3,890	6,080
1,800	270	600	1,070	1,670	2,400	4,260	6,660
2,000	280	630	1,120	1,760	2,520	4,500	7,020
2,500	310	710	1,260	1,960	2,830	5,030	7,850
3,000	340	770	1,380	2,150	3,100	5,510	8,600
4,000	400	890	1,590	2,480	3,580	6,360	9,930
5,000	440	1,000	1,780	2,780	4,000	7,110	11,100
6,000	480	1,090	1,950	3,040	4,380	7,790	12,200

[1]The allowable tension values in Table 21-E-1 are based on compressive strength of masonry assemblages. Where yield strength of anchor bolt steel governs, the allowable tension in pounds is given in Table 21-E-2.

[2]Values are for bolts of at least A 307 quality. Bolts shall be those specified in Section 2106.2.14.1.

[3]Values shown are for work with or without special inspection.

TABLE 21-E-2—ALLOWABLE TENSION, B_t, FOR EMBEDDED ANCHOR BOLTS FOR CLAY AND CONCRETE MASONRY, pounds[1,2]

ANCHOR BOLT DIAMETER (inches)							
× 25.4 for mm							
$^1/_4$	$^3/_8$	$^1/_2$	$^5/_8$	$^3/_4$	$^7/_8$	1	$1^1/_8$
× 4.45 for N							
350	790	1,410	2,210	3,180	4,330	5,650	7,160

[1]Values are for bolts of at least A 307 quality. Bolts shall be those specified in Section 2106.2.14.1.
[2]Values shown are for work with or without special inspection.

TABLE 21-F—ALLOWABLE SHEAR, B_v, FOR EMBEDDED ANCHOR BOLTS FOR CLAY AND CONCRETE MASONRY, pounds[1,2]

f'_m (psi)	ANCHOR BOLT DIAMETER (inches)						
	× 25.4 for mm						
	$^3/_8$	$^1/_2$	$^5/_8$	$^3/_4$	$^7/_8$	1	$1^1/_8$
	× 4.45 for N						
1,500	480	850	1,330	1,780	1,920	2,050	2,170
1,800	480	850	1,330	1,860	2,010	2,150	2,280
2,000	480	850	1,330	1,900	2,060	2,200	2,340
2,500	480	850	1,330	1,900	2,180	2,330	2,470
3,000	480	850	1,330	1,900	2,280	2,440	2,590
4,000	480	850	1,330	1,900	2,450	2,620	2,780
5,000	480	850	1,330	1,900	2,590	2,770	2,940
6,000	480	850	1,330	1,900	2,600	2,900	3,080

[1]Values are for bolts of at least A 307 quality. Bolts shall be those specified in Section 2106.2.14.1.
[2]Values shown are for work with or without special inspection.

TABLE 21-G—MINIMUM DIAMETERS OF BEND

BAR SIZE	MINIMUM DIAMETER
No. 3 through No. 8	6 bar diameters
No. 9 through No. 11	8 bar diameters

TABLE 21-H-1—RADIUS OF GYRATION[1] FOR CONCRETE MASONRY UNITS[2]

GROUT SPACING (inches)	NOMINAL WIDTH OF WALL (inches)				
× 25.4 for mm	× 25.4 for mm				
	4	6	8	10	12
Solid grouted	1.04	1.62	2.19	2.77	3.34
16	1.16	1.79	2.43	3.04	3.67
24	1.21	1.87	2.53	3.17	3.82
32	1.24	1.91	2.59	3.25	3.91
40	1.26	1.94	2.63	3.30	3.97
48	1.27	1.96	2.66	3.33	4.02
56	1.28	1.98	2.68	3.36	4.05
64	1.29	1.99	2.70	3.38	4.08
72	1.30	2.00	2.71	3.40	4.10
No grout	1.35	2.08	2.84	3.55	4.29

[1]For single-wythe masonry or for an individual wythe of a cavity wall.

$$r = \sqrt{I/A_e}$$

[2]The radius of gyration shall be based on the specified dimensions of the masonry units or shall be in accordance with the values shown which are based on the minimum dimensions of hollow concrete masonry unit face shells and webs in accordance with UBC Standard 21-4 for two cell units.

TABLE 21-H-2
TABLE 21-I

1997 UNIFORM BUILDING CODE

TABLE 21-H-2—RADIUS OF GYRATION[1] FOR CLAY MASONRY UNIT LENGTH, 16 INCHES[2]

GROUP SPACING (inches)	NOMINAL WIDTH OF WALL (inches)				
× 25.4 for mm	× 25.4 for mm				
	4	6	8	10	12
Solid grouted	1.06	1.64	2.23	2.81	3.39
16	1.16	1.78	2.42	3.03	3.65
24	1.20	1.85	2.51	3.13	3.77
32	1.23	1.88	2.56	3.19	3.85
40	1.25	1.91	2.59	3.23	3.90
48	1.26	1.93	2.61	3.26	3.93
56	1.27	1.94	2.63	3.28	3.95
64	1.27	1.95	2.64	3.30	3.97
72	1.28	1.95	2.65	3.31	3.99
No grout	1.32	2.02	2.75	3.42	4.13

[1]For single-wythe masonry or for an individual wythe of a cavity wall.

$$r = \sqrt{I/A_e}$$

[2]The radius of gyration shall be based on the specified dimensions of the masonry units or shall be in accordance with the values shown which are based on the minimum dimensions of hollow clay masonry face shells and webs in accordance with UBC Standard 21-1 for two cell units.

TABLE 21-H-3—RADIUS OF GYRATION[1] FOR CLAY MASONRY UNIT LENGTH, 12 INCHES[2]

GROUT SPACING (inches)	NOMINAL WIDTH OF WALL (inches)				
× 25.4 for mm	× 25.4 for mm				
	4	6	8	10	12
Solid grouted	1.06	1.65	2.24	2.82	3.41
12	1.15	1.77	2.40	3.00	3.61
18	1.19	1.82	2.47	3.08	3.71
24	1.21	1.85	2.51	3.12	3.76
30	1.23	1.87	2.53	3.15	3.80
36	1.24	1.88	2.55	3.17	3.82
42	1.24	1.89	2.56	3.19	3.84
48	1.25	1.90	2.57	3.20	3.85
54	1.25	1.90	2.58	3.21	3.86
60	1.26	1.91	2.59	3.21	3.87
66	1.26	1.91	2.59	3.22	3.88
72	1.26	1.91	2.59	3.22	3.88
No grout	1.29	1.95	2.65	3.28	3.95

[1]For single-wythe masonry or for an individual wythe of a cavity wall.

$$r = \sqrt{I/A_e}$$

[2]The radius of gyration shall be based on the specified dimensions of the masonry units or shall be in accordance with the values shown which are based on the minimum dimensions of hollow clay masonry face shells and webs in accordance with UBC Standard 21-1 for two cell units.

TABLE 21-I—ALLOWABLE FLEXURAL TENSION (psi)

	MORTAR TYPE			
	Cement-lime and Mortar Cement		Masonry Cement	
	M or S	N	M or S	N
UNIT TYPE	× 6.89 for kPa			
Normal to bed joints				
Solid	40	30	24	15
Hollow	25	19	15	9
Normal to head joints				
Solid	80	60	48	30
Hollow	50	38	30	18

TABLE 21-J—MAXIMUM NOMINAL SHEAR STRENGTH VALUES[1,2]

M/Vd	V_n MAXIMUM
≤ 0.25	$6.0 A_e \sqrt{f'_m} \leq 380 A_e\ (322 A_e \sqrt{f'_m} \leq 1691 A_e)$
≥ 1.00	$4.0 A_e \sqrt{f'_m} \leq 250 A_e\ (214 A_e \sqrt{f'_m} \leq 1113 A_e)$

[1]M is the maximum bending moment that occurs simultaneously with the shear load V at the section under consideration. Interpolation may be by straight line for M/Vd values between 0.25 and 1.00.

[2]V_n is in pounds (N), and f'_m is in pounds per square inches (kPa).

TABLE 21-K—NOMINAL SHEAR STRENGTH COEFFICIENT

M/Vd[1]	C_d
≤ 0.25	2.4
≥ 1.00	1.2

[1]M is the maximum bending moment that occurs simultaneously with the shear load V at the section under consideration. Interpolation may be by straight line for M/Vd values between 0.25 and 1.00.

TABLE 21-L—SHEAR WALL SPACING REQUIREMENTS FOR EMPIRICAL DESIGN OF MASONRY

FLOOR OR ROOF CONSTRUCTION	MAXIMUM RATIO Shear Wall Spacing to Shear Wall Length
Cast-in-place concrete	5:1
Precast concrete	4:1
Metal deck with concrete fill	3:1
Metal deck with no fill	2:1
Wood diaphragm	2:1

TABLE 21-M
TABLE 21-N

1997 UNIFORM BUILDING CODE

TABLE 21-M—ALLOWABLE COMPRESSIVE STRESSES FOR EMPIRICAL DESIGN OF MASONRY

| CONSTRUCTION: COMPRESSIVE STRENGTH OF UNIT, GROSS AREA | ALLOWABLE COMPRESSIVE STRESSES[1] GROSS CROSS-SECTIONAL AREA (psi) | |
| | × 6.89 for kPa | |
× 6.89 for kPa	Type M or S Mortar	Type N Mortar
Solid masonry of brick and other solid units of clay or shale; sand-lime or concrete brick:		
8,000 plus, psi	350	300
4,500 psi	225	200
2,500 psi	160	140
1,500 psi	115	100
Grouted masonry, of clay or shale; sand-lime or concrete:		
4,500 plus, psi	275	200
2,500 psi	215	140
1,500 psi	175	100
Solid masonry of solid concrete masonry units:		
3,000 plus, psi	225	200
2,000 psi	160	140
1,200 psi	115	100
Masonry of hollow load-bearing units:		
2,000 plus, psi	140	120
1,500 psi	115	100
1,000 psi	75	70
700 psi	60	55
Hollow walls (cavity or masonry bonded)[2] solid units:		
2,500 plus, psi	160	140
1,500 psi	115	100
Hollow units	75	70
Stone ashlar masonry:		
Granite	720	640
Limestone or marble	450	400
Sandstone or cast stone	360	320
Rubble stone masonry Coarse, rough or random	120	100
Unburned clay masonry	30	—

[1]Linear interpolation may be used for determining allowable stresses for masonry units having compressive strengths which are intermediate between those given in the table.

[2]Where floor and roof loads are carried upon one wythe, the gross cross-sectional area is that of the wythe under load. If both wythes are loaded, the gross cross-sectional area is that of the wall minus the area of the cavity between the wythes.

TABLE 21-N—ALLOWABLE SHEAR ON BOLTS FOR EMPIRICALLY DESIGNED MASONRY EXCEPT UNBURNED CLAY UNITS

| DIAMETER BOLT (inches) | EMBEDMENT[1] (inches) | SOLID MASONRY (shear in pounds) | GROUTED MASONRY (shear in pounds) |
× 25.4 for mm		× 4.45 for N	
$^{1}/_{2}$	4	350	550
$^{5}/_{8}$	4	500	750
$^{3}/_{4}$	5	750	1,100
$^{7}/_{8}$	6	1,000	1,500
1	7	1,250	1,850[2]
$1^{1}/_{8}$	8	1,500	2,250[2]

[1]An additional 2 inches of embedment shall be provided for anchor bolts located in the top of columns for buildings located in Seismic Zones 2, 3 and 4.

[2]Permitted only with not less than 2,500 pounds per square inch (17.24 MPa) units.

TABLE 21-O—WALL LATERAL SUPPORT REQUIREMENTS
FOR EMPIRICAL DESIGN OF MASONRY

CONSTRUCTION	MAXIMUM *l/t* or *h/t*

Bearing walls
  Solid or solid grouted
  All other
	20
	18

Nonbearing walls
  Exterior
  Interior
	18
	36

TABLE 21-P—THICKNESS OF FOUNDATION WALLS FOR EMPIRICAL DESIGN OF MASONRY

FOUNDATION WALL CONSTRUCTION	NOMINAL THICKNESS (inches) × 25.4 for mm	MAXIMUM DEPTH OF UNBALANCED FILL (feet) × 304.8 for mm
Masonry of hollow units, ungrouted	8 10 12	4 5 6
Masonry of solid units	8 10 12	5 6 7
Masonry of hollow or solid units, fully grouted	8 10 12	7 8 8
Masonry of hollow units reinforced vertically with No. 4 bars and grout at 24″ o.c. Bars located not less than 4$^1/_2$″ from pressure side of wall.	8	7

TABLE 21-Q—ALLOWABLE SHEAR ON BOLTS FOR MASONRY OF UNBURNED CLAY UNITS

DIAMETER OF BOLTS (inches) × 25.4 for mm	EMBEDMENTS (inches)	SHEAR (pounds) × 4.45 for N
$^1/_2$	—	—
$^5/_8$	12	200
$^3/_4$	15	300
$^7/_8$	18	400
1	21	500
1$^1/_8$	24	600

Chapter 22
STEEL
Division I—GENERAL

SECTION 2201 — SCOPE

The quality, testing and design of steel used structurally in buildings or structures shall conform to the requirements specified in this chapter.

SECTION 2202 — STANDARDS OF QUALITY

The standards listed below labeled a "UBC Standard" are also listed in Chapter 35, Part II, and are part of this code. The other standards listed below are recognized standards. (See Sections 3503 and 3504.)

2202.1 Material Standards.

UBC Standard 22-1, Material Specifications for Structural Steel

2202.2 Design Standards.

ANSI/ASCE 8, Specification for the Design of Cold-formed Stainless Steel Structural Members, American Society of Civil Engineers

2202.3 Connectors.

ASTM A 502, Structural Rivet Steel

SECTION 2203 — MATERIAL IDENTIFICATION

2203.1 General. Steel furnished for structural load-carrying purposes shall be properly identified for conformity to the ordered grade in accordance with approved national standards, the provisions of this chapter and the appropriate UBC standards. Steel which is not readily identifiable as to grade from marking and test records shall be tested to determine conformity to such standards.

2203.2 Structural Steel. structural steel shall be identified by the mill in accordance with approved national standards. When such steel is furnished to a specified minimum yield point greater than 36,000 pounds per square inch (psi) (248 MPa), the American Society for Testing and Materials (ASTM) or other specification designation shall be so indicated.

The fabricator shall maintain identity of the material and shall maintain suitable procedures and records attesting that the specified grade has been furnished in conformity with the applicable standard. The fabricator's identification mark system shall be established and on record prior to fabrication.

When structural steel is furnished to a specified minimum yield point greater than 36,000 psi (248 MPa), the ASTM or other specification designation shall be included near the erection mark on each shipping assembly or important construction component over any shop coat of paint prior to shipment from the fabricator's plant. Pieces of such steel which are to be cut to smaller sizes shall, before cutting, be legibly marked with the fabricator's identification mark on each of the smaller-sized pieces to provide continuity of identification. When subject to fabrication operations, prior to assembling into members, which might obliterate paint marking, such as blast cleaning, galvanizing or heating for forming, such pieces of steel shall be marked by steel die stamping or by a substantial tag firmly attached.

Individual pieces of steel having a minimum specified yield point in excess of 36,000 psi (248 MPa), which are received by the fabricator in a tagged bundle or lift or which have only the top shape or plate in the bundle or lift marked by the mill shall be marked by the fabricator prior to use in accordance with the fabricator's established identification marking system.

2203.3 Cold-formed Carbon and Low-alloy Steel. Cold-formed carbon and low-alloy steel used for structural purposes shall be identified by the mill in accordance with approved national standards. When such steel is furnished to a specified minimum yield point greater than 33,000 psi (228 MPa), the fabricator shall indicate the ASTM or other specification designation, by painting, decal, tagging or other suitable means, on each lift or bundle of fabricated elements.

When cold-formed carbon and low-alloy steel used for structural purposes has a specified yield point equal to or greater than 33,000 psi (228 MPa), which was obtained through additional treatment, the resulting minimum yield point shall be identified in addition to the specification designation.

2203.4 Cold-formed Stainless Steel. Cold-formed stainless steel structural members designed in accordance with recognized standards shall be identified as to grade through mill test reports. (See reference to ANSI/ASCE 8 in Chapter 35.) A certification shall be furnished that the chemical and mechanical properties of the material supplied equals or exceeds that considered in the design. Each lift or bundle of fabricated elements shall be identified by painting, decal, tagging or other suitable means.

2203.5 Open-web Steel Joists. Open-web steel joists and similar fabricated light steel load-carrying members shall be identified in accordance with Division II as to type, size and manufacturer by tagging or other suitable means at the time of manufacture or fabrication. Such identification shall be maintained continuously to the point of their installation in a structure.

SECTION 2204 — DESIGN METHODS

Design shall be by one of the following methods.

2204.1 Load and Resistance Factor Design. Steel design based on load and resistance factor design methods shall resist the factored load combinations of Section 1612.2 in accordance with the applicable requirements of Section 2205. Seismic design of structures, where required, shall comply with Division IV for structures designed in accordance with Division II (LRFD).

2204.2 Allowable Stress Design. Steel design based on allowable stress design methods shall resist the load combinations of Section 1612.3 in accordance with the applicable requirements of Section 2205. Seismic design of structures, where required, shall comply with Division V for structures designed in accordance with Division III (ASD).

SECTION 2205 — DESIGN AND CONSTRUCTION PROVISIONS

2205.1 General. The following design standards shall apply.

2205.2 Structural Steel Construction. The design, fabrication and erection of structural steel shall be in accordance with the requirements of Division II for Load and Resistance Factor Design or Division III for Allowable Stress Design.

2205.3 Seismic Design Provisions for Structural Steel. Steel structural elements that resist seismic forces shall, in addition to the requirements of Section 2205.2, be designed in accordance with Division IV or V.

2205.4 Cold-formed Steel Construction. The design of cold-formed carbon or low-alloy steel structural members shall be in accordance with the requirements of Division VI for Load and Resistance Factor Design or Division VII for Allowable Stress Design.

2205.5 Cold-formed Stainless Steel Construction. The design of cold-formed stainless steel structural members shall be in accordance with approved national standards (see Section 2202).

2205.6 Design Provisions for Stud Wall Systems. Cold-formed steel stud wall systems that serve as part of the lateral-force-resisting system shall, in addition to the requirements of Section 2205.4 or 2205.5, be designed and constructed in accordance with Division VIII.

2205.7 Open-web Steel Joists and Joist Girders. The design, manufacture and use of steel joist, K, LH, and KLH series and joist girders shall be in accordance with Division IX.

2205.8 Steel Storage Racks. Steel storage racks may be designed in accordance with the provisions of Division X, except that in Seismic Zones 3 and 4 wholesale and retail sales areas, the W used in the design of racks over 8 feet (2438 mm) in height shall

be equal to the weight of the rack structure and contents with no reductions.

2205.9 Steel Cables. Structural applications of steel cables for buildings shall be in accordance with the provisions of Division XI.

2205.10 Welding. Welding procedures, welder qualification requirements and welding electrodes shall be in accordance with Division II, III, VI or VII and approved national standards.

2205.11 Bolts. The use of high-strength A 325 and A 490 bolts shall be in accordance with the requirements of Divisions II and III.

Anchor bolts shall be set accurately to the pattern and dimensions called for on the plans. The protrusion of the threaded ends through the connected material shall be sufficient to fully engage the threads of the nuts, but shall not be greater than the length of threads on the bolts. Base plate holes for anchor bolts may be oversized as follows:

Bolt Size, inches (mm)	Hole Size, inches (mm)
$^3/_4$ (19.1)	$^5/_{16}$ (7.9) oversized
$^7/_8$ (22.2)	$^5/_{16}$ (7.9) oversized
1 < 2 (25.4 < 50.8)	$^1/_2$ (12.7) oversized
> 2 (> 50.8)	1 (25.4) > bolt diameter

Division II—DESIGN STANDARD FOR LOAD AND RESISTANCE FACTOR DESIGN SPECIFICATION FOR STRUCTURAL STEEL BUILDINGS

American Institute of Steel Construction
(December 1, 1993)

See Section 1602, *Uniform Building Code*

SECTION 2206 — ADOPTION

Except for the modifications as set forth in Section 2207 of this division and the requirements of the *Uniform Building Code,* the design, fabrication, erection and quality control of structural steel shall be in accordance with the *Load and Resistance Factor Design Specifications for Structural Steel Buildings,* December 1, 1993, published by the American Institute of Steel Construction, 1 East Wacker Drive, Suite 3100, Chicago, Illinois 60601, as if set out at length herein.

SECTION 2207 — AMENDMENTS

The *Load and Resistance Factory Design Specification for Structural Steel Buildings* (hereinafter referred to as LRFD) adopted by this division applies to the design, fabrication, erection and quality control of structural steel, except as modified by this section. Where other codes, standards or specifications are referred to in LRFD they are considered as supplemental standards and only considered guidelines subject to the approval of the building official.

1. Appendices. Appendices Sections B Design Requirements; E Columns and Other Compression Members; F Beams and Other Flexural Members; G Plate Girders; H Members Under Combined Forces and Torsion; J Connections, Joints and Fasteners; and K Concentrated Forces, Ponding and Fatigue are specifically adopted and made a part of this division.

2. Glossary. The glossary is specifically adopted and made a part of this division.

3. Sec. A4 is amended as follows:

The nominal loads shall be the minimum design loads required by the code, and the load combinations shall be as specified in Section A4.1, as amended.

4. Sec. A4.1 is amended as follows:

E: earthquake load

Division III—DESIGN STANDARD FOR SPECIFICATION FOR STRUCTURAL STEEL BUILDINGS ALLOWABLE STRESS DESIGN AND PLASTIC DESIGN

SECTION 2208 — ADOPTION

Except for the modifications as set forth in Section 2209 of this division and the requirements of the building code, the design, fabrication and erection of structural steel shall be in accordance with the *Specification for Structural Steel Buildings Allowable Stress Design and Plastic Design,* June 1, 1989, published by the American Institute of Steel Construction, 1 East Wacker Drive, Suite 3100, Chicago, IL 60601, as if set out at length herein. The adoption of the Specification for Structural Steel Buildings Allowable Stress Design and Plastic Design hereinafter referred to as AISC-ASD shall include the following appendices: B5, F7 and K4.

Where other codes, standards or specifications are referred to in this specification, they are to be considered as only an indication of an acceptable method or material that can be used with the approval of the building official.

SECTION 2209 — AMENDMENTS

The following amendments shall be made to the AISC ASD specification, as adopted in Section 2208:

1. Sec. A4. Revise as follows:

The nominal loads shall be the minimum design loads required by the code.

2. Secs. A4.1, A4.4 and A4.5. Delete in their entirety without replacement.

3. Replace Sec. A5.2 as follows:

Wind and Seismic Stresses. Allowable stresses may be increased for load combinations, including wind and seismic, as permitted by Section 1612.3.2. No increase in allowable stress is permitted for Section 1612.3.1 load combinations.

Division IV—SEISMIC PROVISIONS FOR STRUCTURAL STEEL BUILDINGS

Based on Seismic Provisions for Structural Steel Buildings, of the American Institute of Steel Construction (June 15, 1992)

Section 2210 of this division contains the exceptions to the referenced specification. Section 2211 of this division, "Seismic Provisions for Structural Steel Buildings" is reproduced with permission of the publisher.

SECTION 2210 — AMENDMENTS

The American Institute of Steel Construction Specification transcribed in this division applies to the seismic design of structural steel members and is to be used in conjunction with Division II (LRFD). Where other codes, standards or specifications are referred to in this division, they are to be considered as only an indication of an acceptable method or material that can be used with the approval of the building official.

1. Part I, Sec. 1. Revise as follows:

1. SCOPE

These special seismic requirements are to be applied in conjunction with the building code and Chapter 22, Division II (AISC-LRFD) hereinafter referred to as the *Specification*. They are intended for the design and construction of structural steel members and connections in buildings for which the design forces resulting from earthquake motions have been determined on the basis of energy dissipation in the nonlinear range of response.

2. Part I, Sec. 2. Revise as follows:

2. REQUIREMENTS IN SEISMIC ZONES

2.1 Seismic Zone 0 and 1 and Zone 2 (I = 1.0).

Buildings in Seismic Zones 0 and 1 and buildings in Seismic Zone 2 having an importance factor equal to 1.0 shall be designed in accordance with solely the *Specification* or in accordance with the *Specification* and these provisions.

2.2 Seismic Zone 2 (I > 1.0).

Buildings in Seismic Zone 2 having an importance factor I greater than 1.0 shall be designed in accordance with the *Specification* as modified by the additional provisions of this section.

2.2.a. Steel used in seismic-resisting systems shall be limited by the provisions of Section 5.

2.2.b. Columns in seismic-resisting systems shall be designed in accordance with Section 6.

2.2.c. Ordinary Moment Frames (OMF) shall be designed in accordance with the provisions of Section 7.

2.2.d. Special Moment Frames (SMF) are required to conform only to the requirements of Sections 8.2, 8.7 and 8.8.

2.2.e. Braced framed systems shall conform to the requirements of Section 9 or 10 when used alone or in combination with the moment frames of the seismic-resisting system.

2.3 Seismic Zones 3 and 4.

Buildings in Seismic Zones 3 and 4 shall be designed in accordance with the *Specification* as modified by the additional provisions of this section.

2.3.a. Steel used in seismic-resisting systems shall be limited by the provisions of Section 5.

2.3.b. Columns in seismic-resisting systems shall be designed in accordance with Section 6.

2.3.c. Ordinary Moment Frames (OMF) shall be designed in accordance with the provisions of Section 7.

2.3.d. Special Moment Frames (SMF) shall be designed in accordance with the provisions of Section 8.

2.3.e. Braced framed systems shall conform to the requirements of Section 9 (CBF) or 10 (EBF) when used alone or in combination with the moment frames of the seismic-resisting system.

The use of K-bracing systems shall not be permitted as part of the seismic resisting system except as permitted by Section 9.5 (Low Buildings).

2.3.f. A quality assurance plan shall be submitted to the regulatory agency for the seismic-force-resisting system of the building.

3. Part I, Sec. 3. Revise as follows:

3. LOADS, LOAD COMBINATIONS AND NOMINAL STRENGTHS

3.1 Loads and Load Combinations

The required strength of the structure and its elements shall be determined from the appropriate critical combination of factored loads, as determined by the load combinations in accordance with Section 1612.2. Wherever load combinations 3-1 through 3-6 are used in these provisions, they shall be taken as combinations 12-1 through 12-6 in Section 1612.2.1 of this code.

Orthogonal earthquake effects shall be included in the analysis unless noted specifically otherwise in Chapter 16.

Where required by these provisions, an amplified horizontal earthquake load of Ω_o as defined in Section 1630.3.1 shall be applied in load combinations 3-7 through 3-8 below.

The additional load combinations using the amplified horizontal earthquake are:

$$1.2\ D\ +\ f_1 L\ +\ \Omega_o\ E$$

$$0.9\ D\ \pm\ \Omega_o\ E$$

WHERE:

f_1 = 1.0 for floors in places of public assembly, for live loads in excess of 100 pounds per square foot (4.79 kN/m^2), and for garage live load.

= 0.5 for other live loads.

Where the amplified load is required, orthogonal effects are not required to be included.

3.2 Nominal Strengths

The nominal strengths shall be as provided in the *Specification*.

4. Sec. 5. Modify the second sentence by adding A 913 at the end of the sentence.

5. Sec. 6.1. Revise as follows:

6.1 Column Strength

When $P_u/\phi P_n > 0.5$, columns in seismic resisting frames, in addition to complying with the *Specification*, shall be limited by the following requirements.

6.1.a. The required axial compression strength shall be determined from Load Combination 3-7.

6.1.b. The required axial tension strength shall be determined from Load Combination 3-8.

6.1.c. The axial load combinations 3-7 and 3-8 are not required to exceed either of the following:

1. The maximum loads transferred to the column, considering 1.25 times the design strength of the connecting beams or brace elements of the structure.

2. The limit as determined by the foundation capacity to resist overturning uplift.

6. Part I, Sec. 6.2.a. Delete.

7. Sec. 7.2.c.2. Revise as follows:

2. The connections have been demonstrated by cyclic tests to have adequate rotation capacity at a design story drift Δ_m, as defined in Section 1630.9.

8. Part I, Sec. 8. Revise as follows:

8.2.c. Connection Strength. Connection configurations utilizing welds or high-strength bolts shall demonstrate, by approved cyclic testing results or calculation, the ability to sustain inelastic rotation and to develop the strength criteria in Section 8.2.a considering the expected value of yield strength and strain hardening.

8.2.d. Delete.

Sec. 8.4.b. Add the following to the end of the section:

The outside wall width thickness ratio of rectangular tubes used for columns shall not exceed $110/\sqrt{F_y}$ (For **SI:** $0.65\sqrt{E/F_y}$), unless otherwise stiffened.

Sec. 8.7.b.1. Revise as follows:

1. The required column strength shall be determined as the lesser of:

a. The loads resulting from the application of Load Combination No. 12.5 in Section 1612.2.1 except $\Omega_0 E$ shall be substituted for E, or

b. 125 percent of the frame design strength based on either beam or panel zone design strengths.

8.9 Add section as follows:

8.9 Moment Frame Drift Calculations.

Moment frame drift calculations shall include bending and shear contributions from the clear girder and column spans, column axial deformation and the rotation and distortion of the panel zone.

8.9.a. Drift calculations may be based on column and girder center lines where either of the following conditions is met:

1. Where it can be demonstrated that drift so computed for frames of similar configuration is typically within 15 percent of that determined above, or

2. The nominal panel zone strength is equal to or greater than 0.8 ΣM_p of girders framing into the column flanges at the connection.

8.9.b. Column axial deformations may be neglected if they contribute less than 10 percent to the total drift.

9. Part I, Sec. 10.5. Add the following to the end of the section:

Intermediate bracing shall be provided at the top and bottom flanges of the link at intervals not exceeding $76/\sqrt{F_y}$ times the beam flange width. Such intermediate bracing shall have a design strength of 1.0 percent of the link flange nominal strength computed as $F_y b_f t_f$.

10. Part I. Add new requirements for Special Concentrically Braced Frames Requirements for Special Concentrically Braced Frames (SCBF) in Section 12:

Special CBFs shall be designed in accordance with the requirements of Section 9 except as modified herein. The following modifications shall apply to SCBFs and shall not modify the requirements for ordinary CBFs in Section 9.

a. Sec. 9.2.a. Revise as follows:

Slenderness: Bracing members shall have an $L/r < 1,000/\sqrt{F_y}$

(For **SI:** $5.87\sqrt{E/F_y}$).

b. Sec. 9.2.b. Revise as follows:

9.2.b. Compressive Design Strength. The design strength of a bracing member in axial compression shall not exceed $\phi_c P_n$.

c. Sec. 9.2.d. Revise as follows:

9.2.d. Width-thickness Ratio. Width-thickness ratios of stiffened and unstiffened compression elements of braces shall comply with Section B5 of the *Specification*. Braces shall be compact (i.e., $\lambda < \lambda p$). The width-thickness ratio of angle sections shall not exceed $52/\sqrt{F_y}$. Circular sections shall have an outside diameter to wall thickness ratio not exceeding $1,300/\sqrt{F_y}$; rectangular tubes shall have an outside wall width-thickness ratio not exceeding $100/\sqrt{F_y}$ unless the circular section or tube walls are stiffened.

d. Sec. 9.2.e. Revise as follows:

9.2.e. Built-up Member Stitches. For all built-up braces, the spacing of stitches shall be uniform and not less than two stitches shall be used:

1. For a brace in which stitches can be subjected to postbuckling shear, the spacing of the stitches shall be such that the slenderness ratio, L/r, of individual elements between the stitches does not exceed 0.4 times the governing slenderness ratio of the built-up member. The total shear strength of the stitches shall be at least equal to the tensile strength of each element. Bolted stitches shall not be located within the middle one-fourth of the clear brace length.

2. For braces that can buckle without causing shear in the stitches, the spacing of the stitches shall be such that the slenderness ratio, L/r, of the individual elements between the stitches does not exceed 0.75 times the governing slenderness ratio of the built-up member.

e. Sec. 9.4.a. Revise as follows:

9.4.a. V and Inverted V Type Bracing. V braced and inverted V braced frames shall comply with the following:

1. A beam intersected by braces shall be continuous between columns.

2. A beam intersected by braces shall be capable of supporting all tributary dead and live loads assuming the bracing is not present.

3. A beam intersected by braces shall be capable of resisting the combination of load effects caused by the application of the load combinations in Section 2213.5.1, Items 1 and 2,

except that the term Q_b shall be substituted for the term $3 (R_w/8)P_E$ where Q_b = the maximum unbalanced load effect applied to the beam by the braces. This load effect shall be permitted to be calculated using a minimum P_y for the brace in tension and a maximum $0.3 \phi_c P_n$ for the brace in compression.

4. The top and bottom flanges of the beam at the point of intersection of V braces shall be designed to support a lateral force equal to 1.5 percent of the nominal beam flange strength, $F_y b_f t_f$.

f. Sec. 9.4.b. Delete.

g. Sec. 9.5. Delete and substitute as follows:

9.5 Column.

9.5.a. Compactness. Columns used in SCBFs shall be compact according to Section B5 of the *Specification*. The outside wall width-thickness ratio of rectangular tube used for columns shall not exceed $110/\sqrt{F_y}$ unless otherwise stiffened.

9.5.b. Splices. In addition to meeting the requirements of Section 6.2, column splices in SBCFs also shall be designed to develop the nominal shear strength and 50 percent of the nominal moment strength of the section. Splices shall not be located in the middle one-third of the column clear height.

11. Sec. 10.2.g. Revise as follows:

The link rotation angle is the plastic angle between the link and the beam outside of the link determined at a design story drift, Δ_m, as defined in Section 1630.9.

12. Part II. Delete.

SECTION 2211 — ADOPTION

Reproduced with permission from American Institute of Steel Construction, Inc., One East Wacker Drive, Suite 3100, Chicago, IL 60601-2001. Persons desiring to reprint in whole or in part any portion of this specification must secure permission from the American Institute of Steel Construction.

Seismic Provisions for Structural Steel Buildings
June 15, 1992

2211.1 Table of Contents.

2211.2 Symbols.

The section numbers in parentheses after the definition of a symbol refers to the section where the symbol is first used.

A_e Effective net area, in.^{2} (9)

A_f Flange area of member, in.^{2} (6)

A_g Gross area, in.^{2} (8)

A_{st} Area of link stiffener, in.^{2} (10)

A_v Seismic coefficient representing the effective peak velocity-related acceleration. (2)

A_w Effective area of weld, in.^{2} (6)

A_w Link web area, in.^{2} (10)

C_s Response factor related to the fundamental period of the building. (3)

D Dead load due to the self-weight of the structure and the permanent elements on the structure, kips. (3)

E Earthquake load. (3)

F_{BM} Nominal strength of the base material to be welded, ksi. (6)

F_{EXX} Classification strength of weld metal, ksi. (6)

F_w Nominal strength of the weld electrode material, ksi. (6)

F_y Specified minimum yield strength of the type of steel being used, ksi. (8)

F_{yb} F_y of a beam, ksi. (8)

F_{yc} F_y of a column, ksi. (6)

H Average story height above and below a beam-to-column connection, in. (8)

L Live load due to occupancy and moveable equipment, kips. (3)

L Unbraced length of compression or bracing member, in. (8)

L_r Roof live load, kips. (3)

M_n Nominal moment strength of a member or joint, kip-in. (8)

M_p Plastic bending moment, kip-in. (8)

M_{pa} Plastic bending moment modified by axial load ratio, kip-in. (10)

M_u Required flexural strength on a member or joint, kip-in. (8)

P_D Required axial strength on a column resulting from application of dead load, D, kips. (6)

P_E Required axial strength on a column resulting from application of the specified earthquake load, E, kips. (6)

P_L Required axial strength on a column resulting from application of live load, L, kips. (6)

P_u Required axial strength on a column or a link, kips. (10)

P_n Nominal axial strength of a column, kips. (6)

$P_u{}^*$ Required axial strength on a brace, kips. (9)

P_{uc} Required axial strength on a column based on load combination with seismic loads, kips. (8)

P_y Nominal yield axial strength of a member = $F_y A_g$, kips. (10)

R Response modification factor. (3)

R' Load due to initial rainwater or ice exclusive of the ponding contribution, kips. (Symbol R is used in the *Specification*). (3)

R_n Nominal strength of a member. (8)

S Snow load, kips. (3)

V Base shear due to earthquake load, kips. (3)

V_n Nominal shear strength of a member, kips. (8)

V_u Required shear strength on a member, kips. (8)

V_p Nominal shear strength of an active link, kips. (10)

V_{pa} Nominal shear strength of an active link modified by the axial load magnitude, kips. (10)

W Wind load, kips. (3)

W_g Total weight of the building, kips. (3)

Z_b Plastic section modulus of a beam, in.^{3} (8)

Z_c Plastic section modulus of a column, in.^{3} (8)

b Width of compression element, in. (Table 8-1)

b_f Flange width, in. (8)

b_{cf} Column flange width, in. (8)

d_b Overall beam depth, in. (8)

d_c Overall column depth, in. (8)

d_z Overall panel zone depth between continuity plates, in. (8)

e EBF link length, in. (10)

h Assumed web depth for stability, in. (Table 8-1)

r Governing radius of gyration, in. (9)

r_y Radius of gyration about y axis, in. (8)

t_{bf} Thickness of beam flange, in. (8)

t_{cf}	Thickness of column flange, in. (8)
t_f	Thickness of flange, in. (8)
t_p	Thickness of panel zone including doubler plates, in. (8)
t_w	Thickness of web, in. (8)
t_z	Thickness of panel zone (doubler plates not necessarily included), in. (8)
w_z	Width of panel zone between column flanges, in. (8)
α	Fraction of member force transferred across a particular net section. (9)
ρ	Ratio of required axial force P_u to required shear strength V_u of a link. (10)
k	Slenderness parameter. (9)
k_p	Limiting slenderness parameter for compact element. (8)
k_r	Limiting slenderness parameter for non-compact element. (9)
ϕ	Resistance factor. (6,10)
ϕ_b	Resistance factor for beams. (6)
ϕ_c	Resistance factor for columns in compression. (6,10)
ϕ_t	Resistance factor for columns in tension. (6)
ϕ_v	Resistance factor for shear strength of panel zone of beam-to-column connections. (8)
ϕ_w	Resistance factor for welds. (6)

2211.3 Glossary.

Beam. A structural member whose primary function is to carry loads transverse to its longitudinal axis, usually a horizontal member in a seismic frame system.

Braced Frame. An essentially vertical truss system of concentric or eccentric type that resists lateral forces on the structural system.

Concentrically Braced Frame (CBF). A braced frame in which all members of the bracing system are subjected primarily to axial forces. The CBF shall meet the requirements of Sect. 9.

Connection. Combination of joints used to transmit forces between two or more members. Categorized by the type and amount of force transferred (moment, shear, end reaction).

Continuity Plates. Column stiffeners at top and bottom of the panel zone.

Design strength. Resistance (force, moment, stress, as appropriate) provided by element or connection; the product of the nominal strength and the resistance factor.

Diagonal Bracing. Inclined structural members carrying primarily axial load employed to enable a structural frame to act as a truss to resist horizontal loads.

Dual System. A dual system is a structural system with the following features:

- An essentially complete space frame which provides support for gravity loads.

- Resistance to lateral load is provided by moment resisting frames (SMF) or (OMF) which is capable of resisting at least 25 percent of the base shear and concrete or steel shear walls, steel eccentrically (EBF) or concentrically (CBF) braced frames.

- Each system shall be also designed to resist the total lateral load in proportion to its relative rigidity.

Eccentrically Braced Frame (EBF). A diagonal braced frame in which at least one end of each bracing member connects to a beam a short distance from a beam-to-column connection or from another beam-to-brace connection. The EBF shall meet the requirements of Sect. 10.

Essential Facilities. Those facilities defined as essential in the applicable code under which the structure is designed. In the absence of such a code, see ASCE 7-92.

Joint. Area where two or more ends, surfaces, or edges are attached. Categorized by type of fastener or weld used and method of force transfer.

K Braced Frame. A concentric braced frame (CBF) in which a pair of diagonal braces located on one side of a column is connected to a single point within the clear column height.

Lateral Support Member. Member designed to inhibit lateral buckling or lateral-torsional buckling of primary frame members.

Link. In EBF, the segment of a beam which extends from column to column, located between the end of a diagonal brace and a column or between the ends of two diagonal braces of the EBF. The length of the link is defined as the clear distance between the diagonal brace and the column face or between the ends of two diagonal braces.

Link Intermediate Web Stiffeners. Vertical web stiffeners placed within the link.

Link Rotation Angle. The link rotation angle is the plastic angle between the link and the beam outside of the link when the total story drift is E'/E times the drift derived using the specified base shear, V.

Link Shear Design Strength. The lesser of ϕV_p or $2\phi M_p/e$, where $\phi = 0.9$, $V_p = 0.55 F_y dt_w$ and e = the link length except as modified by Sect. S9.2.f.

LRFD. (Load and Resistance Factor Design). A method of proportioning structural components (members, connectors, connecting elements, and assemblages) such that no applicable limit state is exceeded when the structure is subjected to all design load combinations.

Moment Frame. A building frame system in which seismic shear forces are resisted by shear and flexure in members and joints of the frame.

Nominal loads. The magnitudes of the loads specified by the applicable code.

Nominal strength. The capacity of a structure or component to resist the effects of loads, as determined by computations using specified material strengths and dimensions and formulas derived from accepted principles of structural mechanics or by field tests or laboratory tests of scaled models, allowing for modeling effects, and differences between laboratory and field conditions.

Ordinary Moment Frame (OMF). A moment frame system which meets the requirements of Sect. 7.

P - Delta effect. Secondary effect of column axial loads and lateral deflection on the shears and moments in members.

Panel Zone. Area of beam-to column connection delineated by beam and column flanges.

Required Strength. Load effect (force, moment, stress, as appropriate) acting on element of connection determined by structural analysis from the factored loads (using most appropriate critical load combinations).

Resistance Factor. A factor that accounts for unavoidable deviations of the actual strength from the nominal value and the manner and consequences of failure.

Slip-Critical Joint. A bolted joint in which slip resistance of the connection is required.

Special Moment Frame (SMF). A moment frame system which meets the requirements of Sect. 8.

Structural System. An assemblage of load carrying components which are joined together to provide regular interaction or interdependence.

V Braced Frame. A concentrically braced frame (CBF) in which a pair of diagonal braces located either above or below a beam is connected to a single point within the clear beam span. Where the diagonal braces are below the beam, the system is also referred to as an Inverted V Braced Frame.

X Braced Frame. A concentrically braced frame (CBF) in which a pair of diagonal braces crosses near midlength of the braces.

Y Braced Frame. An eccentrically braced frame (EBF) in which the stem of the Y is the link of the EBF system.

2211.4 Part I—Load and Resistance Factor Design (LRFD).

1. SCOPE

These special seismic requirements are to be applied in conjunction with the AISC *Load and Resistance Factor Design Specification for Structural Steel Buildings* (LRFD), 1986; hereinafter referred to as the *Specification.* They are intended for the design and construction of structural steel members and connections in buildings for which the design forces resulting from earthquake motions have been determined on the basis of energy dissipation in the non-linear range of response.

Seismic provisions and the nominal loads for each Seismic Performance Category, Seismic Hazard Exposure Group, or Seismic Zone shall be as specified by the applicable code under which the structure is designed or where no code applies, as dictated by the conditions involved. In the absence of a code, the Performance Categories, Seismic Hazard Exposure Groups, loads and load combinations shall be as given herein.

2. SEISMIC PERFORMANCE CATEGORIES

Seismic Performance Categories vary with the Seismic Hazard Exposure Group shown in Table 2-1, the Effective Peak Velocity Related Acceleration, A_v, and the Seismic Hazard Exposure Group shown in Table 2-2.

In addition to the general requirements assigned to the various Seismic Performance Categories in the applicable building code for all types of construction, the following requirements apply to fabricated steel construction for buildings and structures with similar structural characteristics.

2.1. Seismic Performance Categories A, B, and C

Buildings assigned to Categories A, B, and C, except Category C in Seismic Hazard Exposure Group III where the value of $A_v \geq 0.10$, shall be designed either in accordance with solely the *Specification* or in accordance with the *Specification* and these provisions.

TABLE 2-1
Seismic Hazard Exposure Groups

Group III	Buildings having essential facilities that are necessary for post-earthquake recovery and requiring special requirements for access and functionality.
Group II	Buildings that constitute a substantial public hazard because of occupancy or use.
Group I	All buildings not classified in Groups II and III.

2.2. Seismic Performance Category C

Buildings assigned to Category C in Seismic Hazard Exposure Group III where the value of $A_v \geq 0.10$ shall be designed in accordance with the *Specification* as modified by the additional provisions of this section.

2.2.a. Steel used in seismic resisting systems shall be limited by the provisions of Sect. 5.

2.2.b. Columns in seismic resisting systems shall be designed in accordance with Sect. 6.

2.2.c. Ordinary Moment Frames (OMF) shall be designed in accordance with the provisions of Sect. 7.

2.2.d. Special Moment Frames (SMF) are required to conform only to the requirements of Sects. 8.2, 8.7, and 8.8.

2.2.e. Braced framed systems shall conform to the requirements of Sects. 9 or 10 when used alone or in combination with the moment frames of the seismic resisting system.

2.2.f. A quality assurance plan shall be submitted to the regulatory agency for the seismic force resisting system of the building.

2.3. Seismic Performance Categories D and E

Buildings assigned to Categories D and E shall be designed in accordance with the *Specification* as modified by the additional provisions of this section.

2.3.a. Steel used in seismic resisting systems shall be limited by the provisions of Sect. 5.

2.3.b. Columns in seismic resisting systems shall be designed in accordance with Sect. 6.

2.3.c. Ordinary Moment Frames (OMF) shall be designed in accordance with the provisions of Sect. 7.

2.3.d. Special Moment Frames (SMF) shall be designed in accordance with the provisions of Sect. 8.

2.3.e. Braced framed systems shall conform to the requirements of Sects. 9. (CBF) or 10. (EBF) when used alone or in combination with the moment frames of the seismic resisting system.

The use of K-bracing systems shall not be permitted as part of the seismic resisting system except as permitted by Sect. 9.5. (Low Buildings)

TABLE 2-2

Seismic Performance Categories

	Seismic Hazard Exposure Group		
Value of A_v	I	II	III
$0.20 \leq A_v$	D	D	E
$0.15 \leq A_v < 0.20$	C	D	D
$0.10 \leq A_v < 0.15$	C	C	C
$0.05 \leq A_v < 0.10$	B	B	C
$A_v < 0.05$	A	A	A

2.3.f. A quality assurance plan shall be submitted to the regulatory agency for the seismic force resisting system of the building.

3. LOADS, LOAD COMBINATIONS, AND NOMINAL STRENGTHS

3.1. Loads and Load Combinations

The following specified loads and their effects on the structure shall be taken into account:

D: dead load due to the weight of the structural elements and the permanent features on the structure.

L: live load due to occupancy and moveable equipment.

L_r: roof live load.

W: wind load.

S: snow load.

E: earthquake load (where the horizontal component is derived from base shear Formula $V = C_s W_g$).

R': load due to initial rainwater or ice exclusive of the ponding contribution.

In the Formula $V = C_s W_g$ for base shear:

C_s = Seismic design coefficient

W_g = Total weight of the building, see the applicable code.

For the nominal loads as defined above, see the applicable code.

The required strength of the structure and its elements shall be determined from the appropriate critical combination of factored loads. The following Load Combinations and corresponding load factors shall be investigated:

$$1.4D \tag{3-1}$$

$$1.2D + 1.6L + 0.5(L_r \text{ or } S \text{ or } R') \tag{3-2}$$

$$1.2D + 1.6(L_r \text{ or } S \text{ or } R') + (0.5L \text{ or } 0.8W) \tag{3-3}$$

$$1.2D + 1.3W + 0.5L + 0.5(L_r \text{ or } S \text{ or } R') \tag{3-4}$$

$$1.2D \pm 1.0E + 0.5L + 0.2S \tag{3-5}$$

$$0.9D \pm (1.0E \text{ or } 1.3W) \tag{3-6}$$

Exception: The load factor on L in Load Combinations 3-3, 3-4, and 3-5 shall equal 1.0 for garages, areas occupied as places of public assembly, and all areas where the live load is greater than 100 psf.

Other special load combinations are included with specific design requirements throughout these provisions.

Orthogonal earthquake effects shall be included in the analysis unless noted specifically otherwise in the governing building code.

Where required by these provisions, an amplified horizontal earthquake load of $0.4R \times E$ (where the term $0.4R$ is

greater or equal to 1.0) shall be applied in lieu of the horizontal component of earthquake load E in the load combinations above. The term R is the earthquake response modification coefficient contained in the applicable code. The additional load combinations using the amplified horizontal earthquake load are:

$$1.2D + 0.5L + 0.2S \pm 0.4R \times E \tag{3-7}$$

$$0.9D \pm 0.4R \times E \tag{3-8}$$

Exception: The load factor on L in Load Combinations 3-7 shall equal 1.0 for garages, areas occupied as places of public assembly and all areas where the live load is greater than 100 psf.

The term $0.4R$ in Load Combinations 3-7 and 3-8 shall be greater or equal to 1.0.

Where the amplified load is required, orthogonal effects are not required to be included.

3.2. Nominal Strengths

The nominal strengths shall be as provided in the *Specification*.

4. STORY DRIFT

Story drift shall be calculated using the appropriate load effects consistent with the structural system and the method of analysis. Limits on story drift shall be in accordance with the governing code and shall not impair the stability of the structure.

5. MATERIAL SPECIFICATIONS

Steel used in seismic force resisting systems shall be as listed in Sect. A3.1 of the *Specification*, except for buildings over one story in height. The steel used in seismic resisting systems described in Sections 8, 9, and 10 shall be limited to the following ASTM Specifications: A36, A500 (Grades B and C), A501, A572 (Grades 42 and 50), and A588. The steel used for base plates shall meet one of the preceding ASTM Specifications or ASTM A283 Grade D.

6. COLUMN REQUIREMENTS

6.1. Column Strength

When $P_u / \phi P_n > 0.5$, columns in seismic resisting frames, in addition to complying with the *Specification*, shall be limited by the following requirements:

6.1.a. Axial compression loads:

$$1.2P_D + 0.5P_L + 0.2P_S + 0.4R \times P_E \leq \phi_c P_n \tag{6-1}$$

where the term $0.4R$ is greater or equal to 1.0.

Exception: The load factor on P_L in Load Combination 6-1 shall equal 1.0 for garages, areas occupied as places of public assembly, and all areas where the live load is greater than 100 psf.

6.1.b. Axial tension loads:

$$0.9P_D - 0.4R \times P_E \leq \phi_t P_n \tag{6-2}$$

where the term $0.4R$ is greater or equal to 1.0.

6.1.c. The axial Load Combinations 6-1 and 6-2 are not required to exceed either of the following:

1. The maximum loads transferred to the column, considering 1.25 times the design strengths of the connecting beam or brace elements of the structure.

2. The limit as determined by the foundation capacity to resist overturning uplift.

6.2. Column Splices

Column splices shall have a design strength to develop the column axial loads given in Sect. 6.1.a, b, and c as well as the Load Combinations 3-1 to 3-6.

6.2.a. In column splices using either complete or partial penetration welded joints, beveled transitions are not required when changes in thickness and width of flanges and webs occur.

6.2.b. Splices using partial penetration welded joints shall not be within 3 ft of the beam-to-column connection. Column splices that are subject to net tension forces shall comply with the more critical of the following:

1. The design strength of partial penetration welded joints, the lesser of $\phi_w F_w A_w$ or $\phi_w F_{BM} A_w$ shall be at least 150 percent of the required strength, where $\phi_w = 0.8$ and $F_w = 0.6 F_{EXX}$.

2. The design strength of welds shall not be less than $0.5 F_{yc} A_f$, where F_{yc} is the yield strength of the column material and A_f is the flange area of the smaller column connected.

7. REQUIREMENTS FOR ORDINARY MOMENT FRAMES (OMF)

7.1. Scope

Ordinary Moment Frames (OMF) shall have a design strength as provided in the *Specification* to resist the Load Combinations 3-1 through 3-6 as modified by the following added provisions:

7.2. Joint Requirements

All beam-to-column and column to beam connections in OMF which resist seismic forces shall meet one of the following requirements:

7.2.a. FR (fully restrained) connections conforming with Sect. 8.2, except that the required flexural strength, M_u, of a column-to-beam joint is not required to exceed the nominal plastic flexural strength of the connection.

7.2.b. FR connections with design strengths of the connections meeting the requirements of Sect. 7.1 using the Load Combinations 3-7 and 3-8.

7.2.c. Either FR or PR (partially restrained) connections shall meet the following:

1. The design strengths of the members and connections meet the requirements of Sect. 7.1.

2. The connections have been demonstrated by cyclic tests to have adequate rotation capacity at a story drift calculated at a horizontal load of $0.4R \times E$, (where the term $0.4R$ is equal to or greater than 1.0).

3. The additional drift due to PR connections shall be considered in design.

FR and PR connections are described in detail in Sect. A2 of the *Specification*.

8. REQUIREMENTS FOR SPECIAL MOMENT FRAMES (SMF)

8.1. Scope

Special Moment Frames (SMF) shall have a design strength as provided in the *Specification* to resist the Load Combinations 3-1 through 3-6 as modified by the following added provisions:

8.2. Beam-to-Column Joints

8.2.a. The required flexural strength, M_u, of each beam-to-column joint shall be the lesser of the following quantities:

1. The plastic bending moment, M_p, of the beam.

2. The moment resulting from the panel zone nominal shear strength, V_n, as determined using Equation 8-1.

The joint is not required to develop either of the strengths defined above if it is shown that under an amplified frame deformation produced by Load Combinations 3-7 and 3-8, the design strength of the members at the connection is adequate to support the vertical loads, and the required lateral force resistance is provided by other means.

8.2.b. The required shear strength, V_u, of a beam-to-column joint shall be determined using the Load Combination $1.2D + 0.5L + 0.2S$ plus the shear resulting from M_u, as defined in Sect. 8.2.a., on each end of the beam. Alternatively, V_u shall be justified by a rational analysis. The required shear strength is not required to exceed the shear resulting from Load Combination 3-7.

8.2.c. The design strength, ϕR_n, of a beam-to-column joint shall be considered adequate to develop the required flexural strength, M_u, of the beam if it conforms to the following:

1. The beam flanges are welded to the column using complete penetration welded joints.

2. The beam web joint has a design shear strength ϕV_n greater than the required shear, V_u, and conforms to either:

a. Where the nominal flexural strength of the beam, M_n, considering only the flanges is greater than 70 percent of the nominal flexural strength of the entire beam section [i.e., $b_f t_f (d - t_f) F_{yf} \geq 0.7 M_p$]; the web joint shall be made by means of welding or slip-critical high strength bolting, or;

b. Where $b_f t_f (d - t_f) F_{yf} < 0.7 M_p$, the web joint shall be made by means of welding the web to the column directly or through shear tabs. That welding shall have a design strength of at least 20 percent of the nominal flexural strength of the beam web. The required beam shear, V_u, shall be resisted by further welding or by slip-critical high-strength bolting or both.

8.2.d. Alternate Joint Configurations: For joint configurations utilizing welds or high-strength bolts, but not conforming to Sect. 8.2.c, the design strength shall be determined by test or calculations to meet the criteria of Sect. 8.2.a. Where conformance is shown by calculation, the design strength of the joint shall be 125 percent of the design strengths of the connected elements.

8.3. Panel Zone of Beam-to-Column Connections
(Beam web parallel to column web)

8.3.a. Shear Strength: The required shear strength, V_u, of the panel zone shall be based on beam bending moments determined from the Load Combinations 3-5 and 3-6. However, V_u is not required to exceed the shear forces determined from $0.9\Sigma\phi_b M_p$ of the beams framing into the column flanges at the connection. The design shear strength, $\phi_v V_n$, of the panel zone shall be determined by the following formula:

$$\phi_v V_n = 0.6\phi_v F_y d_c t_p \left[1 + \frac{3b_{cf}t_{cf}^2}{d_b d_c t_p}\right] \text{ where for this}$$

case $\phi_v = 0.75$. (8-1)

where:

t_p = Total thickness of panel zone including doubler plates, in.

d_c = Overall column section depth, in.

b_{cf} = Width of the column flange, in.

t_{cf} = Thickness of the column flange, in.

d_b = Overall beam depth, in.

F_y = Specified yield strength of the panel zone steel, ksi.

8.3.b. Panel Zone Thickness: The panel zone thickness, t_z, shall conform to the following:

$$t_z \geq (d_z + w_z) / 90 \tag{8-2}$$

where:

d_z = the panel zone depth between continuity plates, in.

w_z = the panel zone width between column flanges, in.

For this purpose, t_z shall not include any doubler plate thickness unless the doubler plate is connected to the web with plug welds adequate to prevent local buckling of the plate.

Where a doubler plate is used without plug welds to the column web, the doubler plate shall conform to Eq. 8-2.

8.3.c. Panel Zone Doubler Plates: Doubler plates provided to increase the design strength of the panel zone or to reduce the web depth thickness ratio shall be placed next to the column web and welded across the plate width along the top and bottom with at least a minimum fillet weld. The doubler plates shall be fastened to the column flanges using either butt or fillet welded joints to develop the design shear strength of the doubler plate.

8.4. Beam and Column Limitations

8.4.a. Beam Flange Area: There shall be no abrupt changes in beam flange areas in plastic hinge regions.

8.4.b. Width-Thickness Ratios: Beams and columns shall comply with λ_p in Table 8-1 in lieu of those in Table B5. 1 of the *Specification*.

8.5. Continuity Plates

Continuity plates shall be provided if required by the provisions in the *Specification* for webs and flanges with concentrated forces and if the nominal column local flange bending strength R_n is less than $1.8F_{yb}b_{ft}t_{bf}$, where:

$R_n = 6.25(t_{cf})^2 F_{yf}$, and

F_{yb} = Specified minimum yield strength of beam, ksi.

F_{yf} = Specified minimum yield strength of column flange, ksi.

b_f = Beam flange width, in.

t_{bf} = Beam flange thickness, in.

t_{cf} = Column flange thickness, in.

Continuity plates shall be fastened by welds to both the column flanges and either the column webs or doubler plates.

8.6. Column-Beam Moment Ratio

At any beam-to-column connection, one of the following relationships shall be satisfied:

TABLE 8-1
Limiting Width Thickness Ratios λ_p for Compression Elements

Description of Element	Width-Thickness Ratio	Limiting Width-Thickness Ratios λ_p
Flanges of I-shaped nonhybrid sections and channels in flexure.	b/t	$52 / \sqrt{F_y}$
Flanges of I-shaped hybrid beams in flexure.		For $P_u/\phi_b P_y \leq 0.125$
Webs in combined flexural and axial compression	h/t_w	$\frac{520}{\sqrt{F_y}}\left[1 - \frac{1.54P_u}{\phi_b P_y}\right]$
		For $P_u/\phi_b P_y > 0.125$
		$\frac{191}{\sqrt{F_y}}\left[2.33 - \frac{P_u}{\phi_b P_y}\right] \geq \frac{253}{\sqrt{F_y}}$

$$\frac{\Sigma Z_c(F_{yc} - P_{uc}/A_g)}{\Sigma Z_b F_{yb}} \geq 1.0, \tag{8-3}$$

$$\frac{\Sigma Z_c(F_{yc} - P_{uc}/A_g)}{V_n d_b H/(H - d_b)} \geq 1.0, \tag{8-4}$$

where:

A_g = Gross area of a column, in.^{2}

F_{yb} = Specified minimum yield strength of a beam, ksi.

F_{yc} = Specified minimum yield strength of a column, ksi.

H = Average of the story heights above and below the joint, in.

P_{uc} = Required axial strength in the column (in compression) ≥ 0

V_n = Nominal strength of the panel zone as determined from Equation 8-1, ksi.

Z_b = Plastic section modulus of a beam, in.^{3}

Z_c = Plastic section modulus of a column, in.^{3}

d_b = Average overall depth of beams framing into the connection, in.

These requirements do not apply in any of the following cases, provided the columns conform to the requirements of Sect. 8.4:

8.6.a. Columns with $P_{uc} < 0.3F_{yc}A_g$.

8.6.b. Columns in any story that has a ratio of design shear strength to design force 50 percent greater than the story above.

8.6.c. Any column not included in the design to resist the required seismic shears, but included in the design to resist axial overturning forces.

8.7. Beam-to-Column Connection Restraint

8.7.a. Restrained Connection:

1. Column flanges at a beam-to-column connection require lateral support only at the level of the top flanges of the beams when a column is shown to remain elastic outside of the panel zone, using one of the following conditions:

   a. Ratios calculated using Eqs. 8-3 or 8A are greater than 1.25.

   b. Column remains elastic when loaded with Load Combination 3-7.

2. When a column cannot be shown to remain elastic outside of the panel zone, the following provisions apply:

   a. The column flanges shall be laterally supported at the levels of both top and bottom beam flanges.

   b. Each column flange lateral support shall be designed for a required strength equal to 2.0 percent of the nominal beam flange strength ($F_y b_f t_f$).

   c. Column flanges shall be laterally supported either directly, or indirectly, by means of the column web or beam flanges.

8.7.b. Unrestrained Connections: A column containing a beam-to-column connection with no lateral support transverse to the seismic frame at the connection shall be designed using the distance between adjacent lateral supports as the column height for buckling transverse to the seismic frame and conform to Sect. H of the *Specification* except that:

1. The required column strength shall be determined from the Load Combination 3-5 where E is the least of:

   a. The amplified earthquake force $0.4R \times E$ (where the term $0.4R$ shall be equal to or greater than 1.0).

   b. 125 percent of the frame design strength based on either beam or panel zone design strengths.

2. The L/r for these columns shall not exceed 60.

3. The required column moment transverse to the seismic frame shall include that caused by the beam flange force specified in Sect. 8.7.a.2.b plus the added second order moment due to the resulting column displacement in this direction.

8.8. Lateral Support of Beams

Both flanges of beams shall be laterally supported directly or indirectly. The unbraced length between lateral supports shall not exceed 2,500 r_y/F_y. In addition, lateral supports shall be placed at concentrated loads where an analysis indicates a hinge will be formed during inelastic deformations of the SMF.

9. REQUIREMENTS FOR CONCENTRICALLY BRACED (CBF) BUILDINGS

9.1. Scope

Concentrically Braced Frames (CBF) are braced systems whose worklines essentially intersect at points. Minor eccentricities, where the worklines intersect within the width of the bracing members, are acceptable if accounted for in the design. CBF shall have a design strength as provided in the *Specification* to resist the Load Combinations 3-1 through 3-6 as modified by the following added provisions:

9.2. Bracing Members

9.2.a. Slenderness: Bracing members shall have an

$$\frac{L}{r} \leq \frac{720}{\sqrt{F_y}} \quad \text{except as permitted in Sect. 9.5.}$$

9.2.b. Compressive Design Strength: The design strength of a bracing member in axial compression shall not exceed $0.8\phi_c P_n$.

9.2.c. Lateral Force Distribution: Along any line of bracing, braces shall be deployed in alternate directions such that, for either direction of force parallel to the bracing, at least 30 percent but no more than 70 percent of the total horizontal force shall be resisted by tension braces, unless the nominal strength, P_n, of each brace in compression is larger than the required strength, P_u, resulting from the application of the Load Combinations 3-7 or 3-8. A line of bracing, for the purpose of this provision, is defined as a single line or parallel lines whose plan offset is 10 percent or less of the building dimension perpendicular to the line of bracing.

9.2.d. Width-Thickness Ratios: Width-thickness ratios of stiffened and unstiffened compression elements in braces shall comply with Sect. B5 in the *Specification.* Braces shall be compact or non-compact, but not slender (i.e., $\lambda < \lambda_r$). Circular sections shall have an outside diameter to wall thickness ratio not exceeding $1,300/F_y$, rectangular tubes shall have a flat-width to wall thickness not exceeding $110/\sqrt{F_y}$, unless the circular section or tube walls are stiffened.

9.2.e. Built-up Member Stitches: For all built-up braces, the first bolted or welded stitch on each side of the midlength of a built up member shall be designed to transmit a force equal to 50 percent of the nominal strength of one element to the adjacent element. Not less than two stitches shall be equally spaced about the member centerline.

9.3. Bracing Connections

9.3.a. Forces: The required strength of bracing joints (including beam-to-column joints if part of the bracing system) shall be the least of the following:

1. The design axial tension strength of the bracing member.

2. The force in the brace resulting from the Load Combinations 3-7 or 3-8.

3. The maximum force, indicated by an analysis, that is transferred to the brace by the system.

9.3.b. Net Area: In bolted brace joints, the minimum ratio of effective net section area to gross section area shall be limited by:

$$\frac{A_e}{A_g} \geq \frac{1.2\alpha P_u{}^*}{\phi_t P_n} \qquad (9\text{--}1)$$

where:

A_e = Effective net area as defined in Equation B3-1 of the *Specification*.

$P_u{}^*$ = Required strength on the brace as determined in Sect. 9.3.a.

P_n = Nominal tension strength as specified in Chapter D of the *Specification*.

ϕ_t = Special resistance factor for tension = 0.75.

α = Fraction of the member force from Sect. 9.3.a that is transferred across a particular net section.

9.3.c. Gusset Plates:

1. Where analysis indicates that braces buckle in the plane of the gusset plates, the gusset and other parts of the connection shall have a design strength equal to or greater than the in-plane nominal bending strength of the brace.

2. Where the critical buckling strength is out-of-plane of the gusset plate, the brace shall terminate on the gusset a minimum of two times the gusset thickness from the theoretical line of bending which is unrestrained by the column or beam joints. The gusset plate shall have a required compressive strength to resist the compressive design strength of the brace member without local buckling of the gusset plate. For braces designed for axial load only, the bolts or welds shall be designed to transmit the brace forces along the centroids of the brace elements.

9.4. Special Bracing Configuration Requirements

9.4.a. V and Inverted V Type Bracing:

1. The design strength of the brace members shall be at least 1.5 times the required strength using Load Combinations 3-5 and 3-6.

2. The beam intersected by braces shall be continuous between columns.

3. A beam intersected by V braces shall be capable of supporting all tributary dead and live loads assuming the bracing is not present.

4. The top and bottom flanges of the beam at the point of intersection of V braces shall be designed to support a lateral force equal to 1.5 percent of the nominal beam flange strength ($F_y b_f t_f$).

9.4.b. K bracing, where permitted:

1. The design strength of K brace members shall be at least 1.5 times the required strength using Load Combinations 3-5 and 3-6.

2. A column intersected by K braces shall be continuous between beams.

3. A column intersected by K braces shall be capable of supporting all dead and live loads assuming the bracing is not present.

4. Both flanges of the column at the point of intersection of K braces shall be designed to support a lateral force equal to 1.5 percent of the nominal column flange strength ($F_y b_f t_f$).

9.5. Low Buildings

Braced frames not meeting the requirements of Sect. 9.2 through 9.4 shall only be used in buildings not over two stories and in roof structures if Load Combinations 3-7 and 3-8 are used for determining the required strength of the members and connections.

10. REQUIREMENTS FOR ECCENTRICALLY BRACED FRAMES (EBF)

10.1. Scope

Eccentrically braced frames shall be designed so that under inelastic earthquake deformations, yielding will occur in the links. The diagonal braces, the columns, and the beam segments outside of the links shall be designed to remain elastic under the maximum forces that will be generated by the fully yielded and strain hardened links, except where permitted by this section.

10.2. Links

10.2.a. Beams with links shall comply with the width-thickness ratios in Table 8-1.

10.2.b. The specified minimum yield stress of steel used for links shall not exceed $F_y = 50$ ksi.

10.2.c. The web of a link shall be single thickness without doubler plate reinforcement and without openings.

10.2.d. Except as limited by Sect. 10.2.f., the required shear strength of the link, V_u, shall not exceed the design shear strength of the link, ϕV_n, where:

ϕV_n = Link design shear strength of the link = the lesser of ϕV_p or $2\phi M_p / e$, kips.

$V_p = 0.6 F_y (d - 2t_f) t_w$, kips.

$\phi = 0.9$.

e = link length, in.

10.2.e. If the required axial strength, P_u, in a link is equal to or less than $0.15 P_y$, where $P_y = A_g F_y$, the effect of axial force on the link design shear strength need not be considered.

10.2.f. If the required axial strength, P_u, in a link exceeds $0.15P_y$, the following additional limitations shall be required:

1. The link design shear strength shall be the lesser of ϕV_{pa} or $2\phi M_{pa} / e$, where:

$$V_{pa} = V_p \sqrt{1 - (P_u/P_y)^2}$$

$$M_{pa} = 1.18 \, M_p \, [\, 1 - (P_u \, / \, P_y)]$$

$$\phi = 0.9$$

2. The length of the link shall not exceed:

$[1.15 - 0.5\rho(A_w/A_g)] \, 1.6 \, M_p \, / \, V_p$ for $\rho(A_w/A_g) \geq 0.3$ and

$1.6M_p \, / \, V_p$ for $\rho(A_w \, /A_g) <0.3$, where:

$$A_w = (d\text{-}2t_f)t_w$$

$$P = P_u \, / \, V_u$$

10.2.g. The link rotation angle is the plastic angle between the link and the beam outside of the link when the total story drift is $0.4R$ times the drift determined using the specified base shear V. The term $0.4R$ shall be equal to or greater than 1.0. Except as noted in Sect. 10.4.d, the link rotation angle shall not exceed the following values:

1. 0.09 radians for links of length $1.6M_p \, / \, V_p$ or less.

2. 0.03 radians for links of length $2.6M_p \, / \, Vp$ or greater.

3. Linear interpolation shall be used for links of length between $1.6Mp \, / \, V_p$ and $2.6M_p \, / \, V_p$.

10.2.h. Alternatively, the top story of an EBF building having over five stories shall be a CBF.

10.3. Link Stiffeners

10.3.a. Full depth web stiffeners shall be provided on both sides of the link web at the diagonal brace ends of the link. These stiffeners shall have a combined width not less than $(b_f - 2t_w)$ and a thickness not less than $0.75t_w$ or $^3/_8$-in., whichever is larger, where b_f and t_w are the link flange width and link web thickness, respectively.

10.3.b. Links shall be provided with intermediate web stiffeners as follows:

1. Links of lengths $1.6M_p \, / \, V_p$ or less shall be provided with intermediate web stiffeners spaced at intervals not exceeding $(30t_w - d/5)$ for a link rotation angle of 0.09 radians or $(52t_w - d/5)$ for link rotation angles of 0.03 radians or less. Linear interpolation shall be used for values between 0.03 and 0.09 radians.

2. Links of length greater than $2.6M_p \, / \, V_p$ and less than $5M_p \, / \, V_p$ shall be provided with intermediate web stiffeners placed at a distance of $1.5b_f$ from each end of the link.

3. Links of length between $1.6M_p \, / \, V_p$ and $2.6M_p/V_p$ shall be provided with intermediate web stiffeners meeting the requirements of 1 and 2 above.

4. No intermediate web stiffeners are required in links of lengths greater than $5M_p \, / \, V_p$.

5. Intermediate link web stiffeners shall be full depth. For links less than 25 inches in depth, stiffeners are required on oniy one side of the link web. The thickness of one-sided stiffeners shall be not less than t_w or $^3/_8$-in., whichever is larger, and the width shall be not less than $(b_f/2) - t_w$. For links 25 inches in depth or greater, similar intermediate stiffeners are required on both sides of the web.

10.3.c. Fillet welds connecting link stiffener to the link web shall have a design strength adequate to resist a force of $A_{st}F_y$, in which A_{st} = area of the stiffener. The design strength of fillet welds fastening the stiffener to the flanges shall be adequate to resist a force of $A_{st}F_y \, / \, 4$.

10.4. Link-to-Column Connections

Where a link is connected to a column, the following additional requirements shall be met:

10.4.a. The length of links connected to columns shall not exceed $1.6M_p \, / \, V_p$ unless it is demonstrated that the link-to-column connection is adequate to develop the required inelastic rotation of the link.

10.4.b. The link flanges shall have complete penetration welded joints to the column. The joint of the link web to the column shall be welded. The required strength of the welded joint shall be at least the nominal axial, shear, and flexural strengths of the link web.

10.4.c. The need for continuity plates shall be determined according to the requirements of Sect. 8.5.

10.4.d. Where the link is connected to the column web, the link flanges shall have complete penetration welded joints to plates and the web joint shall be welded. The required strength of the link web shall be at least the nominal axial, shear, and flexural strength of the link web. The link rotation angle shall not exceed 0.015 radians for any link length.

10.5. Lateral Support of Link

Lateral supports shall be provided at both the top and bottom flanges of link at the ends of the link. End lateral supports of links shall have a design strength of 6 percent of the link flange nominal strength computed as $F_y b_f t_f$.

10.6. Diagonal Brace and Beam Outside of Link

10.6.a. The required combined axial and moment strength of the diagonal brace shall be the axial forces and moments generated by 1.25 times the nominal shear strength of the link as defined in Sect. 10.2. The design strengths of the diagonal brace, as determined by Sect. H (including Appendix H) of the *Specification*, shall exceed the required strengths as defined above.

10.6.b. The required strength of the beam outside of the link shall be the forces generated by at least 1.25 times the nominal shear strength of the link and shall be provided with lateral support to maintain the stability of the beam. Lateral supports shall be provided at both top and bottom flanges of the beam and each shall have a design strength to resist at least 1.5 percent of the beam flange nominal strength computed as $F_y b_f t_f$.

10.6.c. At the connection between the diagonal brace and the beam at the link end of the brace, the intersection of the brace and beam centerlines shall be at the end of the link or in the link. The beam shall not be spliced within or adjacent to the connection between the beam and the brace.

10.6.d. The required strength of the diagonal brace-to-beam connection at the link end of the brace shall be at least the nominal strength of the brace. No part of this connection shall extend over the link length. If the brace resists a portion of the link end moment, the connection shall be designed as Type FR (Fully Restrained).

10.6.e. The width-thickness ratio of brace shall satisfy λ_p of Table B5.1 of the *Specification*.

10.7. Beam-to-Column Connections

Beam-to-column connections away from links are permitted to be designed as a pin in the plane of the web. The connection shall have a design strength to resist torsion about the longitudinal axis of the beam based on two equal and opposite forces of at least 1.5 percent of the beam flange nominal strength computed as $F_y b_f t_f$ acting laterally on the beam flanges.

10.8. Required Column Strength

The required strength of columns shall be determined by Load Combinations 3-5 and 3-6 except that the moments and axial loads introduced into the column at the connection of a link or brace shall not be less than those generated by 1.25 times the nominal strength of the link.

11. QUALITY ASSURANCE

The general requirements and responsibilities for performance of a quality assurance plan shall be in accordance with the requirements of the regulatory agency and specifications by the design engineer.

The special inspections and special tests needed to establish that the construction is in conformance with these provisions shall be included in a quality assurance plan.

The minimum special inspection and testing contained in the quality assurance plan beyond that required by the *Specification* shall be as follows:

Groove welded joints subjected to net tensile forces which are part of the seismic force resisting systems of Sects. 8, 9, and 10 shall be tested 100 percent either by ultrasonic testing or by other approved equivalent methods conforming to AWS D1.1.

Exception: The nondestructive testing rate for an individual welder shall be reduced to 25 percent with the concurrence of the person responsible for structural design, provided the reject rate is demonstrated to be 5 percent or less of the welds tested for the welder.

2211.5 Part II—Allowable Stress Design (ASD) Alternative.

As an alternative to the LRFD seismic design procedures for structural steel design given in PART I, the design procedures in the *Specification for Structural Steel Buildings—Allowable Stress Design and Plastic Design*, AISC 1989 are permitted as modified by PART II of these provisions. When using ASD, the provisions of PART I of these seismic provisions shall apply except the following sections shall be substituted for, or added to, the appropriate sections as indicated:

1. SCOPE

Revise the first paragraph of Part I, Sect. 1 to read as follows:

These special requirements are to be applied in conjunction with the AISC *Specification for Structural Steel Buildings—Allowable Stress Design and Plastic Design* hereinafter referred to as *Specification*. They are intended for the design and construction of structural steel members and connections in buildings for which the design forces resulting from earthquake motions have been determined on the basis of energy dissipation in the nonlinear range of response.

3. LOADS, LOAD COMBINATIONS AND NOMINAL STRENGTHS

Substitute the following for Section 3.2 in Part I:

3.2. Nominal Strengths

The nominal strengths of members shall be determined as follows:

3.2.a. Replace Sect. A5.2 of the *Specification* to read: "The nominal strength of structural steel members for resisting seismic forces acting alone or in combination with dead and live loads shall be determined by multiplying 1.7 times the allowable stresses in Sect. D, E, F, G, J, and K."

3.2.b. Amend the first paragraph of Sect. N1 of the *Specification* by deleting "or earthquake" and adding: "The nominal strength of members shall be determined by the requirements contained herein. Except as modified by these rules, all pertinent provisions of Chapters A through M shall govern."

3.2.c. In Sect H1 of the *Specification* the definition of F_e' shall read as follows:

$$F_e' = \frac{\pi^2 E}{(Kl_b/r_b)^2}$$

where:

l_b = the actual length in the plane of bending.

r_b = the corresponding radius of gyration.

K = the effective length factor in the plane of bending.

Add the following section to Part I:

3.3. Design Strengths

3.3.a. The design strengths of structural steel members and connections subjected to seismic forces in combination with other prescribed loads shall be determined by converting allowable stresses into nominal strengths and multiplying such nominal strengths by the resistance factors herein.

3.3.b. Resistance factors, ϕ, for use in Part II shall be as follows:

Flexure $\phi_b = 0.90$

Compression and axially loaded
composite members $\quad\phi_c = 0.85$

Eyebars and pin connected members:

Shear of the effective area $\quad\phi_{sf} = 0.75$

Tension on net effective area $\quad\phi_t = 0.75$

Bearing on the project area of pin $\quad\phi_t = 1.0$

Tension members:

Yielding on gross section $\quad\phi_t = 0.90$

Fracture in the net section $\quad\phi_t = 0.75$

Shear $\quad\phi_v = 0.90$

Connections:

Base plates that develop the strength
of the members or structural systems $\quad\phi = 0.90$

Welded connections that do not
develop the strength of the member
or structural system, including
connection of base plates and
anchor bolts $\quad\phi = 0.67$

Partial Penetration welds in columns
when subjected to tension stresses $\quad\phi = 0.80$

High strength bolts (A325 and A490)
and rivets:

Tensile strength $\quad\phi = 0.75$

Shear strength in hearing-type joints $\quad\phi = 0.65$

Slip-critical joints $\quad\phi = 1.0$

A307 bolts:

Tensile strength $\quad\phi = 0.75$

Shear strength in bearing-type joints $\quad\phi = 0.60$

Substitute the following for Section 7 in Part I in its entirety:

7. REQUIREMENTS FOR ORDINARY MOMENT FRAMES (OMF)

7.1. Scope

Ordinary Moment Frames (OMF) shall have a design strength as provided in the *Specification* to resist the Load Combinations 3-5 and 3-6 as modified by the following added provisions:

7.2. Joint Requirements

All beam-to-column and column to beam connections in OMF which resist seismic forces shall meet one of the following requirements:

7.2.a. Type 1 connections conforming with Sect. 8.2, except that the required flexural strength, M_u, of a column-to-beam joint are not required to exceed that required to develop the nominal plastic flexural strength of the connection.

7.2.b. Type 1 connections capable of inelastic deformation and the design strengths of the connections meeting the requirements of Sect. 7.1 using the Load Combinations 3-7 and 3-8.

7.2.c. Either Type 1 or Type 3 connections are permitted provided:

1. The design strengths of the members and connections meet the requirements of Sect. 7.1.

2. The connections have been demonstrated by cyclic tests to have adequate rotation capacity at a story drift calculated at a horizontal load of $0.4R \times E$ (where the term $0.4R$ is equal to or greater than 1.0).

3. The additional drift due to Type 3 connections shall be considered in design.

Type 1 and Type 3 connections are described in detail in Sect. A2 of the *Specification*.

Substitute the following in Sections 10.6.a and 10. 6.d in Part I:

10.6.a. Delete reference to Appendix H.

10.6.d. The last sentence shall read: "If the brace resists a portion of the link end moment as described above, the connection shall be designed as a Type 1 connection."

Division V—SEISMIC PROVISIONS FOR STRUCTURAL STEEL BUILDINGS FOR USE WITH ALLOWABLE STRESS DESIGN

SECTION 2212 — GENERAL

When the load combinations of Section 1612.3 for Allowable Stress Design are used, structural steel buildings shall be designed in accordance with the provisions of Chapter 22, Division III (AISC-ASD), and this division where applicable.

SECTION 2213 — SEISMIC PROVISIONS FOR STRUCTURAL STEEL BUILDINGS IN SEISMIC ZONES 3 AND 4

2213.1 General. Design and construction of steel framing in lateral-force-resisting systems in Seismic Zones 3 and 4 shall conform to the requirements of the code and to the requirements of this section.

2213.2 Definitions.

ALLOWABLE STRESSES are prescribed in Divisons III and VII.

CHEVRON BRACING is that form of bracing where a pair of braces located either above or below a beam terminates at a single point within the clear beam span.

CONNECTION is the group of elements that connect the member to the joint.

DIAGONAL BRACING is that form of bracing that diagonally connects joints at different levels.

ECCENTRICALLY BRACED FRAME (EBF) is a diagonal braced frame in which at least one end of each bracing member connects to a beam a short distance from a beam-to-column connection or from another beam-to-brace connection.

GIRDER is the horizontal member in a seismic frame. The words beam and girder may be used interchangeably.

JOINT is the entire assemblage at the intersections of the members.

K BRACING is that form of bracing where a pair of braces located on one side of a column terminates at a single point within the clear column height.

LINK BEAM is that part of a beam in an eccentrically braced frame which is designed to yield in shear and/or bending so that buckling of the bracing members is prevented.

STRENGTH is the strength as prescribed in Section 2213.4.2.

V BRACING is that form of chevron bracing that intersects a beam from above and inverted V bracing is that form of chevron bracing that intersects a beam from below.

X BRACING is that form of bracing where a pair of diagonal braces cross near midlength of the bracing members.

2213.3 Symbols and Notations. The symbols and notations unique to this section are as follows:

M_s = flexural strength.

P_{DL} = axial dead load.

P_E = axial load on member due to earthquake.

P_{LL} = axial live load.

P_{sc} = compressive axial strength of member.

P_{st} = tensile axial strength of member.

V_s = shear strength of member.

Z = plastic section modulus.

2213.4 Materials.

2213.4.1 Quality. Structural steel used in lateral-force-resisting systems shall conform to A 36, A 500, A 501, A 572 (Grades 42 and 50), A 913 (Grades 50 and 65) and A 588. Structural steel conforming to A 283 (Grade D) may be used for base plates and anchor bolts.

> **EXCEPTION:** Other steels permitted in this code may be used for the following:
> 1. One-story buildings.
> 2. Light-framed wall systems in accordance with Division VIII.

2213.4.2 Member strength. Where this section requires the strength of the member to be developed, the following shall be used:

	Strength
Moment	$M_s = ZF_y$
Shear	$V_s = 0.55\,F_y dt$
Axial compression	$P_{sc} = 1.7\,F_a A$
Axial tension	$P_{st} = F_y A$
Connectors	
Full-penetration welds	$F_y A$
Partial penetration welds	$1.7\,F_s$
Bolts and fillet welds	$1.7\,F_s$

Where F_s is the allowable stress value defined in the applicable chapter of Division III. For the purpose of determining member or connection strengths, the allowable stress values specified in Division III shall not be increased by the one-third allowable stress increase per Section 1612.3.2.

Members need not be compact unless otherwise required by this section.

2213.5 Column Requirements.

2213.5.1 Column strength. Columns shall satisfy the load combinations required by Section 1612.2 at load and resistance factor limits or Section 1612.3 at allowable stress limits with stress increases allowed by Section 1612.3.2. In addition, in Seismic Zones 3 and 4, columns in frames shall have the strength to resist the axial loads resulting from the load combinations in Items 1 and 2.

1. **Axial compression**

$$1.0\,P_{DL} + 0.7\,P_{LL} + \Omega_o P_E$$

2. **Axial tension**

$$0.85\,P_{DL} \pm \Omega_o P_E$$

> **EXCEPTION:** The axial load combination as outlined in Items 1 and 2 above:
> 1. Need not exceed either the maximum force that can be transferred to the column, by elements of the structure, or the limit as determined by the overturning uplift which the foundation is capable of resisting.
> 2. Need not apply to columns in moment-resisting frames complying with Formula (13-3-1) or (13-3-2) where f_a is equal to or less than $0.3\,F_y$ for all load combinations.

The load combinations from Items 1 and 2 need be used only when specifically referred to.

2213.5.2 Column splices. Column splices shall have sufficient strength to develop the column forces determined from Section 2213.5.1. Welded column splices subject to net tensile forces shall comply with the more critical of the following:

1. Partial penetration welds shall be designed to resist 150 percent of the force determined from Section 2213.5.1, Item 2.

2. Welding shall develop not less than 50 percent of the flange area strength of the smaller column.

Splices employing partial penetration welds shall be located at least three feet (914 mm) from girder flanges.

2213.5.3 Slenderness evaluation. This paragraph is applicable when the provisions are applied to the effective length determination of columns of moment frames resisting earthquake forces. In the plane of the earthquake forces the factor K may be taken as unity when all of the following conditions are met:

1. The column is either continuous or is fixed at each joint.

2. The maximum axial compressive stress, f_a, does not exceed $0.4 F_y$ under design loads.

3. The calculated drift ratios are less than the values given in Section 1630.8.

2213.6 Ordinary Moment Frame Requirements. Ordinary moment frames (OMF) shall be designed to resist the load combinations in Section 1612.3.

All beam-to-column connections in OMFs which resist earthquake forces shall meet one of the following requirements:

1. Fully restrained (Type F.R. or Type 1) conforming with Section 2213.7.1.

2. Fully restrained (Type F.R. or Type 1) connections with the design strengths of the connections capable of resisting a combination of gravity loads and Ω_o times the design seismic forces.

3. Partially restrained (Type P.R. or Type 3) connections are permitted provided:

3.1 The connections are designed to resist the load combinations in Section 1612.2 or 1612.3, and

3.2 The connections have been demonstrated by cyclic tests to have adequate rotation capacity to accommodate a story drift due to Ω_o times the design seismic forces.

3.3 The moment frame drift calculations shall include the contribution due to the rotation and distortion of the connection.

See Divisions II and III for definitions of fully restrained and partially restrained connections.

2213.7 Special Moment-resisting Frame (SMRF) Requirements.

2213.7.1 Girder-to-column connection.

2213.7.1.1 Required strength. The girder-to-column connection shall be adequate to develop the lesser of the following:

1. The strength of the girder in flexure.

2. The moment corresponding to development of the panel zone shear strength as determined from Formula (13-1).

> **EXCEPTION:** Where a connection is not designed to contribute flexural resistance at the joint, it need not develop the required strength if it can be shown to meet the deformation compatibility requirements of Section 1633.2.4.

2213.7.1.2 Connection strength. Connection configurations utilizing welds or high-strength bolts shall demonstrate, by approved cyclic test results or calculation, the ability to sustain inelastic rotation and develop the strength criteria in Section 2213.7.1.1 considering the effect of steel overstrength and strain hardening.

2213.7.1.3 Flange detail limitations. For steel whose specified ultimate strength is less than 1.5 times the specified yield strength, plastic hinges shall not form at locations in which the beam flange area has been reduced, such as for bolt holes. Bolted connections of flange plates of beam-column joints shall have the net-to-gross area ratio A_e/A_g equal to or greater than $1.2 F_y/F_u$.

2213.7.2 Panel zone.

2213.7.2.1 Strength. The panel zone of the joint shall be capable of resisting the shear induced by beam bending moments due to gravity loads plus 1.85 times the prescribed seismic forces, but the shear strength need not exceed that required to develop $0.8 \Sigma M_s$ of the girders framing into the column flanges at the joint. The joint panel zone shear strength may be obtained from the following formula:

$$V = 0.55 F_y d_c t \left[1 + \frac{3 b_c t_{cf}^2}{d_b d_c t} \right] \quad (13\text{-}1)$$

WHERE:

b_c = the width of the column flange.

d_b = the depth of the beam.

d_c = the column depth.

t = the total thickness of the joint panel zone including doubler plates.

t_{cf} = the thickness of the column flange.

2213.7.2.2 Thickness. The panel zone thickness, t_z, shall conform to the following formula:

$$t_z \geq (d_z + w_z)/90 \quad (13\text{-}2)$$

WHERE:

d_z = the panel zone depth between continuity plates.

w_z = the panel zone width between column flanges.

For this purpose, t_z, shall not include any double plate thickness unless the doubler plate is connected to the column web with plug welds adequate to prevent local buckling of the plate.

2213.7.2.3 Doubler plates. Doubler plates provided to reduce panel zone shear stress or to reduce the web depth thickness ratio shall be placed not more than $^1/_{16}$ inch (1.6 mm) from the column web and shall be welded across the plate width top and bottom with at least a $^3/_{16}$-inch (4.7 mm) fillet weld. They shall be either butt or fillet welded to the column flanges to develop the shear strength of the doubler plate. Weld strength shall be as given in Section 2213.4.2.

2213.7.3 Width-thickness ratio. Girders shall comply with Division III, except that the flange width-thickness ratio, $b_f/2t_f$, shall not exceed $52/\sqrt{F_y}$ (For **SI:** $0.31\sqrt{E/F_y}$). The width-thickness ratio of column sections shall meet the requirements of Division III, Section 2251N7. The outside wall width-thickness ratio of rectangular tubes used for columns shall not exceed $110/\sqrt{F_y}$ (For **SI:** $0.65\sqrt{E/F_y}$), unless otherwise stiffened.

2213.7.4 Continuity plates. When determining the need for girder tension flange continuity plates, the value of P_{bf} in Division III shall be taken as $1.8 (b t_f) F_{yb}$.

2213.7.5 Strength ratio. At any moment frame joint, the following relationships shall be satisfied:

$$\Sigma Z_c (F_{yc} - f_a) / \Sigma M_c > 1.0 \quad (13\text{-}3\text{-}1)$$

or

$$\Sigma Z_c \, (F_{yc} - f_a) \, / \, 1.25 \Sigma M_{pz} > 1.0 \qquad (13\text{-}3\text{-}2)$$

WHERE:

$f_a > 0$

M_c = the moment at column center line due to the development of plastic hinging in the beam accounting for over-strength and strain hardening.

M_{pz} = the sum of beam moments when panel zone shear strength reaches the value specified in Formula (13-1).

EXCEPTION: Columns meeting the compactness limitations for beams given in Section 2213.7.3 need not comply with this requirement provided they conform to one of the following conditions:

1. Columns with f_a less than $0.4 \, F_y$ for all load combinations other than loads specified in Section 2213.5.1, and

    1.1 Which are used in the top story of a multistory building with building period greater than 0.7 second

    1.2 Where the sum of their resistance is less than 20 percent of the shear in a story, and is less than 33 percent of the shear on each of the column lines within that story. A column line is defined for the purpose of this exception as a single line of columns, or parallel lines of columns located within 10 percent of the plan dimension perpendicular to the line of columns; or

    1.3 When the design for combined axial compression and bending is proportioned to satisfy Division III without the one-third permissible stress increase.

2. Columns in any story which have lateral shear strength 50 percent greater than that of the story above.

3. Columns which lateral shear strengths are not included in the design to resist code-required shears.

2213.7.6 Trusses in SMRF.
Trusses may be used as horizontal members in SMRF if the sum of the truss seismic force flexural strength exceeds the sum of the column seismic force flexural strength immediately above and below the truss by a factor of at least 1.25. For this determination the strengths of the members shall be reduced by the gravity load effects. In buildings of more than one story, the column axial stress shall not exceed $0.4F_y$ and the ratio of the unbraced column height to the least radius of gyration shall not exceed 60. Columns shall have allowable stresses reduced 25 percent when one end frames into a truss, and 50 percent when both ends frame into trusses. The connection of the truss chords to the column shall develop the lesser of the following:

1. The strength of the truss chord.

2. The chord force necessary to develop 125 percent of the flexural strength of the column.

2213.7.7 Girder-column joint restraint.

2213.7.7.1 Restrained joint. Where it can be shown that the columns of SMRF remain elastic, the flanges of the columns need be laterally supported only at the level of the girder top flange.

Columns may be assumed to remain elastic if one of the following conditions is satisfied:

1. The ratio in Formula (13-3-1) or (13-3-2) is greater than 1.25.

2. The flexural strength of the column is at least 1.25 times the moment that corresponds to the panel zone shear strength.

3. Girder flexural strength or panel zone strength will limit column stress $(f_a + f_{bx} + f_{by})$ to F_y of the column.

4. The column will remain elastic under gravity loads plus Ω_o times the design seismic forces.

Where the column cannot be shown to remain elastic, the column flanges shall be laterally supported at the levels of the girder top and bottom flanges. The column flange lateral support shall be capable of resisting a force equal to one percent of the girder flange capacity at allowable stresses and at a limiting displacement perpendicular to the frame of 0.2 inch (5.1 mm). Required bracing members may brace the column flanges directly or indirectly through the column web or the girder flanges.

2213.7.7.2 Unrestrained joint. Columns without lateral support transverse to a joint shall conform to the requirements of Division III, with the column considered as pin ended and the length taken as the distance between lateral supports conforming with Section 2213.7.7.1. The column stress, f_a, shall be determined from gravity loads plus the lesser of the following:

1. Ω_o times the design seismic forces.

2. The forces corresponding to either 125 percent of the girder flexural strength or the panel zone shear strength.

The stress, f_{by}, shall include the effects of the bracing force specified in Section 2213.7.7.1 and $P\Delta$ effects.

l/r for such columns shall not exceed 60.

At truss frames the column shall be braced at each truss chord for a lateral force equal to one percent of the compression yield strength of the chord.

2213.7.8 Beam bracing. Both flanges of beams shall be braced directly or indirectly. The beam bracing between column center lines shall not exceed $96r_y$. In addition, braces shall be placed at concentrated loads where a hinge may form.

2213.7.9 Changes in beam flange area. Abrupt changes in beam flange area are not permitted within possible plastic hinge regions of special moment-resistant frames.

2213.7.10 Moment frame drift calculations. Moment frame drift calculations shall include bending and shear contributions from the clear girder and column spans, column axial deformation and the rotation and distortion of the panel zone.

EXCEPTIONS: 1. Drift calculations may be based on column and girder center lines where either of the following conditions is met:

    1.1 It can be demonstrated that the drift so computed for frames of similar configuration is typically within 15 percent of that determined above.

    1.2 The column panel zone strength can develop $0.8 \Sigma M_s$ of girders framing to the column flanges at the joint.

2. Column axial deformations may be neglected if they contribute less than 10 percent to the total drift.

2213.8 Requirements for Braced Frames.

2213.8.1 General. The provisions of this section apply to all braced frames except special concentrically braced frames designed in accordance with Section 2213.9 or eccentrically braced frames (EBF) designed in accordance with Section 2213.10. Those members which resist seismic forces totally or partially by shear or flexure shall be designed in accordance with Section 2213.7 except Section 2213.7.3.

2213.8.2 Bracing members.

2213.8.2.1 Slenderness. In Seismic Zones 3 and 4, the l/r ratio for bracing members shall not exceed $720/\sqrt{F_y}$ (For **SI:** $4.23\sqrt{E/F_y}$), except as permitted in Sections 2213.8.5 and 2213.8.6.

2213.8.2.2 Stress reduction. The allowable stress, F_{as}, for bracing members resisting seismic forces in compression shall be determined from the following formula:

$$F_{as} = BF_a \qquad (13\text{-}4)$$

WHERE:

B = the stress-reduction factor determined from the following formula:

$$B = 1/\{1 + [(Kl/r)/2C_c]\} \qquad (13\text{-}5)$$

F_a = the allowable axial compressive stress allowed in Division III.

> **EXCEPTION:** Bracing members carrying gravity loads may be designed using the column strength requirement and load combinations of Section 2213.5.1, Item 1.

2213.8.2.3 Lateral-force distribution. The seismic lateral force along any line of bracing shall be distributed to the various members so that neither the sum of the horizontal components of the forces in members acting in tension nor the sum of the horizontal components of forces in members acting in compression exceed 70 percent of the total force.

> **EXCEPTION:** Where compression bracing acting alone has the strength, neglecting the stress-reduction factor B, to resist Ω_o times the design seismic force such distribution is not required.

A line of bracing is defined, for the purpose of this provision, as a single line or parallel lines within 10 percent of the dimension of the structure perpendicular to the line of bracing.

2213.8.2.4 Built-up members. The l/r of individual parts of built-up bracing members between stitches, when computed about a line perpendicular to the axis through the parts, shall not be greater than 75 percent of the l/r of the member as a whole.

2213.8.2.5 Compression elements in braces. The width-thickness ratio of stiffened and unstiffened compression elements used in braces shall be as shown in Division III, Table B5.1, for compact sections.

The width-thickness ratio of angle sections shall be limited to $52 / \sqrt{f_y}$ (For **SI:** $0.31 \sqrt{E/f_y}$). Circular sections shall have outside diameter-wall thickness ratio not exceeding $1,300/F_y$ (For **SI:** $7.63\ E/f_y$). Rectangular tubes shall have outside width-thickness ratio not exceeding $110 / \sqrt{F_y}$ (For **SI:** $0.65 \sqrt{E/F_y}$).

> **EXCEPTION:** Compression elements stiffened to resist local buckling.

2213.8.3 Bracing connection.

2213.8.3.1 Forces. Bracing connections shall have the strength to resist the least of the following:

1. The strength of the bracing in axial tension, P_{st}.

2. Ω_o times the force in the brace due to the design seismic forces, in combination with gravity loads.

3. The maximum force that can be transferred to the brace by the system.

Bracing connections shall, as a minimum, satisfy the load combinations required by Section 1612.2 at load and resistance factor design limits or Section 1612.3 at allowable stress design limits with stress increases allowed by Section 1612.3.2. These combinations shall include the provisions for Sections 2213.8.2.2 and 2213.8.4.1.

Beam-to-column connections for beams that are part of the bracing system shall have the capacity to transfer the force determined above. Where eccentricities in the frame geometry or connection load path exist, the affected members and connections shall have the strength to resist all secondary forces resulting from the eccentricities in combination with all primary forces using the lesser of the forces determined above.

2213.8.3.2 Net area. In bolted brace connections, the ratio of effective net section area to gross section area shall satisfy the formula:

$$\frac{A_e}{A_g} \geq \frac{1.2\,\alpha F^*}{F_u} \qquad (13\text{-}6)$$

WHERE:

A_e = effective net area as defined in Division III.

F_u = minimum tensile strength.

F^* = stress in brace as determined in Section 2213.8.3.1.

α = fraction of the member force from Section 2213.8.3.1 that is transferred across a particular net section.

2213.8.4 Bracing configuration.

2213.8.4.1 Chevron bracing. Chevron bracing shall conform with the following:

1. Bracing members shall be designed for 1.5 times the otherwise prescribed seismic forces, in addition to the requirements of Section 2213.8.2.2.

2. The beam intersected by chevron braces shall be continuous between columns.

3. Where chevron braces intersect a beam from below, i.e., inverted V brace, the beam shall be capable of supporting all tributary gravity loads presuming the bracing not to exist.

> **EXCEPTION:** This limitation need not apply to penthouses, one-story buildings or the top story of buildings.

2213.8.4.2 K bracing. K bracing is prohibited except as permitted in Section 2213.8.5.

2213.8.4.3 Nonconcentric bracing. Nonconcentric bracing shall conform with the following:

1. Any member intersected by the brace shall be continuous through the connection.

2. When the eccentricity of the brace is greater than the depth of the intersected member at the eccentric location, the affected member shall have the strength to resist the forces prescribed in Section 2213.8.3.1, including the effects of all secondary forces resulting from the eccentricities.

2213.8.5 One- and two-story buildings. Braced frames not meeting the requirements of Sections 2213.8.2 and 2213.8.4 may be used in buildings not over two stories in height and in roof structures as defined in Chapter 15 if the braces have the strength to resist Ω_o times the design seismic forces.

2213.8.6 Nonbuilding structures. Nonbuilding structures with R values defined by Table 16-P need comply only with the provisions of Section 2213.8.3.

2213.9 Requirements for Special Concentrically Braced Frames.

2213.9.1 General. The provisions of this section apply to special concentrically braced frame structures as defined in Section 1625. All members and connections in special braced frames shall be designed and detailed to resist shear and flexure caused by eccentricities in the geometry of the members comprising the frame in accordance with Section 2213.9. Any member intersected by a brace shall be continuous through the connection. Horizontal bracing that transfers forces between horizontally offset bracing in the vertical plane shall be subject to the requirements of Section 2213.9, except Sections 2213.9.2.3; 2213.9.4.1, Item 3; and 2213.9.4.2. Horizontal bracing other than the above is not subjected to the requirements of Section 2213.9.

2213.9.2 Bracing members.

2213.9.2.1 Slenderness. The kl/r ratio for bracing members shall not exceed $1,000/ \sqrt{F_y}$ (For **SI:** $5.87 \sqrt{E/F_y}$), except as permitted in Section 2213.9.6.

2213.9.2.2 Lateral-force distribution. The seismic lateral force along any line of bracing shall be distributed to the various members so that neither the sum of the horizontal components of forces in members acting in compression or tension exceed 70 percent of the total force.

> **EXCEPTION:** Where compression bracing acting alone has the strength to resist Ω_o times the design seismic force, such distribution is not required.

A line of bracing is defined, for the purposes of this provision, as a single line or parallel lines within 10 percent of the dimension of the structure perpendicular to the line of bracing.

2213.9.2.3 Built-up members. The spacing of stitches shall be such that the slenderness ratio (l/r) of individual elements between the stitches does not exceed 0.4 times the governing slenderness ratio of the built-up member. The total shear strength of the stitches shall be at least equal to the tensile strength of each element. The spacing of the stitches shall be uniform and not less than two stitches shall be used. Bolted stitches shall not be located within the middle one fourth of the clear brace length.

> **EXCEPTION:** Where it can be shown that braces can buckle without causing shear in the stitches, the spacing of the stitches shall be such that the slenderness ration (l/r) of the individual element between the stitches does not exceed 0.75 times the governing slenderness ratio of the built-up member.

2213.9.2.4 Compression elements in braces. The width-thickness ratio of compression elements used in braces shall meet the requirements of Division III, Table B5.1, for compact sections. The width-thickness ratio of angle section shall be limited to $52/\sqrt{F_y}$ (For **SI:** $0.31\sqrt{E/F_y}$). Circular sections shall have outside diameter-wall thickness ratio not exceeding $1{,}300/F_y$ (For **SI:** $7.63\,E/F_y$), rectangular tubes shall have outside wall width-thickness ratio not exceeding $110/\sqrt{F_y}$ (For **SI:** $0.65\sqrt{E/F_y}$).

> **EXCEPTION:** Compression elements stiffened to resist local buckling.

2213.9.3 Bracing connections.

2213.9.3.1 Forces. Bracing connections shall have the strength to resist the lesser of the following:

1. The strength of the brace in axial tension, P_{st}.

2. Ω_o times the force in the brace due to the design seismic forces, in combination with gravity loads.

3. The maximum force that can be transferred to the brace by the system.

Bracing connections shall, as a minimum, satisfy the load combinations required by Section 1612.3 at allowable stress limits with stress increases allowed by Section 1612.3.2. Beam-to-column connections for beams that are part of the bracing system shall have the capacity to transfer the force determined above. Where eccentricities in the frame geometry or connection load path exist, the affected members and connections shall have the strength to resist all secondary forces resulting from the eccentricities in combination with all primary forces using the lesser of the forces determined above.

2213.9.3.2 Net area. In bolted brace connections, the ratio of effective net section area to gross section shall satisfy Formula (13-6) of Section 2213.8.3.2.

2213.9.3.3 Gusset plates. End connections of braces shall provide a flexural strength in excess of that of the brace gross section about the critical buckling axis.

> **EXCEPTION:** Where the out-of-plane buckling strength of the brace is less than the in-plane buckling strength, the brace is permitted to terminate on a single gusset plate connection with a setback of two times the gusset thickness from a line about which the gusset plate may bend unrestrained by the column or beam joints. The gusset plate shall be designed to carry the compressive strength of the brace without buckling.

2213.9.4 Bracing configuration.

2213.9.4.1 Chevron bracing. Chevron bracing shall conform with the following:

1. The beam intersected by chevron braces shall be continuous between columns.

2. Where chevron braces intersect a beam from below, i.e., inverted V brace, the beam shall be capable of supporting all tributary gravity loads presuming the bracing not to exist.

3. A beam intersected by chevron braces shall have the strength to support the following tributary gravity loads and unbalanced brace force combinations:

$$1.2D + 0.5L + P_b$$
$$0.9D - P_b$$

WHERE:

D = tributary dead load.

L = tributary live load.

P_b = the maximum unbalanced post-buckling force that can be applied to the beam by the braces. For this purpose, the maximum unbalanced force may be computed using a minimum of P_{st} for the tension and a maximum of $0.3\,P_{sc}$ for the compression brace.

4. Both flanges of beams at the point of intersection of chevron braces shall be laterally supported directly or indirectly.

> **EXCEPTION:** Limitations 2 and 3 need not apply to penthouses, one-story buildings or the top story of buildings.

2213.9.4.2 K bracing. K bracing is prohibited.

2213.9.5 Columns. Columns in braced frames shall meet the requirements of Section 2213.7.3. In addition to meeting the requirements of Sections 2213.5.1 and 2213.5.2, column splices shall be designed to develop the full shear strength and 50 percent of the full moment strength of the section. Splices shall be located in the middle one third of the column clear height.

2213.9.6 Nonbuilding structures. Nonbuilding structures with R_w values defined by Table 16-P need comply only with the provisions of Sections 2213.9.3.1 and 2213.9.3.2.

2213.10 Eccentrically Braced Frame (EBF) Requirements.

2213.10.1 General. Eccentrically braced frames shall be designed in accordance with this section.

2213.10.2 Link beam. There shall be a link beam provided at least at one end of each brace. Beams in EBFs shall comply with the requirements of Division III, except that the flange width-thickness ratio, $b_f/2t_f$, shall not exceed $52/\sqrt{F_y}$. (For **SI:** $0.31\sqrt{E/F_y}$.)

2213.10.3 Link beam strength. Link beam shear strength, V_s, and flexural strength, M_s, are the strengths as defined in Section 2213.4.2. Where link beam strength is governed by shear, the flexural and axial capacities within the link shall be calculated using the beam flanges only.

A reduced flexural strength, M_{rs}, for use in Sections 2213.10.8 and 2213.10.13 is defined as $Z(F_y - f_a)$. Where f_a is less than $0.15F_y$, f_a may be neglected.

2213.10.4 Link beam rotation. The rotation of the link segment relative to the rest of the beam, at a total frame drift of Δ_M, shall not exceed the following:

1. 0.090 radian for link segments having clear lengths of 1.6 M_s/V_s or less.

2. 0.030 radian for link segments having clear lengths of 3.0 M_s/V_s or greater.

3. A value obtained by linear interpolation for clear lengths between the above limits.

2213.10.5 Link beam web. The web of the link beam shall be single thickness without doubler plate reinforcement. No openings shall be placed in the web of a link beam. The web shear shall not exceed $0.8V_s$ under prescribed lateral forces.

2213.10.6 Beam connection braces. Brace-to-beam connections shall develop the compression strength of the brace and transfer this force to the beam web. No part of the brace-to-beam connection shall extend into the web area of a link beam.

2213.10.7 Link beam stiffeners. Link beams shall have full-depth web stiffeners on both sides of the beam web at the brace end of the link beam. In addition, for link beams with clear lengths within the limits in Section 2213.10.4, Item 3, full-depth stiffeners shall be placed at a distance b_f from each end of the link. The stiffeners shall have a combined width not less than $b - 2t_w$ and a thickness not less than $0.75 \, t_w$ or less than $^3/_8$ inch (9.5 mm).

2213.10.8 Intermediate stiffeners. Intermediate full-depth web stiffeners shall be provided in either of the following conditions:

1. Where the link beam strength is controlled by V_s.

2. Where the link beam strength is controlled by flexure and the shear determined by applying the reduced flexural strength, M_{rs}, exceeds $0.45 \, F_y dt$.

2213.10.9 Web stiffener spacing. Where intermediate web stiffeners are required, the spacing shall conform to the requirements given below.

1. For link beams with rotation angle of 0.09 radian, the spacing shall not exceed $38 t_w - d/5$.

2. For link beams with a rotation angle of 0.03 radian or less, the spacing shall not exceed $56 t_w - d/5$. Interpolation may be used for rotation angles between 0.03 and 0.09 radian.

2213.10.10 Web stiffener location. For beams 24 inches (610 mm) in depth and greater, intermediate full-depth web stiffeners are required on both sides of the web. Such web stiffeners are required only on one side of the beam web for beams less than 24 inches (610 mm) in depth. The stiffener thickness, t_w, of one side stiffeners shall not be less than $^3/_8$ inch (9.5 mm) and the width shall not be less than $(b_f/2) - t_w$.

2213.10.11 Stiffener welds. Fillet welds connecting the stiffener to the beam web shall develop a stiffener force of $A_{st} F_y$. Fillet welds connecting the stiffener to the flanges shall develop a stiffener force of $A_{st} F_y/4$.

WHERE:

A_{st} = bt of stiffener.

b = width of stiffener plate.

2213.10.12 Link beam-column connections. Length of link beam connected to columns shall not exceed 1.6 M_s/V_s.

1. Where a link beam is connected to the column flange, the following requirements shall be met:

1.1 The beam flanges shall have full-penetration welds to the column.

1.2 Where the link beam strength is controlled by shear in conformance with Section 2213.10.8, the web connection shall be welded to develop the full link beam web shear strength.

2. Where the link beam is connected to the column web, the beam flanges shall have full-penetration welds to the connection plates and the web connection shall be welded to develop the link beam web shear strength. Rotation between the link beam and the column shall not exceed 0.015 radian at a total frame drift of Δ_M.

2213.10.13 Brace and beam strengths. The controlling link beam strength is either the shear strength, V_s, or the reduced flexural strength, M_{rs}, whichever results in the lesser axial force in the brace.

Each brace and beam outside the link shall have the axial strength or reduced flexural strength, M_{rs}, at least 1.5 times the forces corresponding to the controlling link beam strength. Each brace and beam assembly outside the link shall have combined reduced flexural strengths, M_{rs}, at least 1.3 times the forces corresponding to the controlling link beam strength.

2213.10.14 Column strength. Columns shall be designed to remain elastic at 1.25 times the strength of the EBF bay, as defined in Section 2213.10.13. Column strength need not exceed the requirements of Section 2213.5.

2213.10.15 Roof link beam. A link beam is not required in roof beams for EBF over five stories.

2213.10.16 Concentric brace in combination. The first story of an EBF bay over five stories in height may be concentrically braced if this story can be shown to have an elastic capacity 50 percent greater than the yield capacity of the story frames above the first story.

2213.10.17 Axial forces. Axial forces in beams of EBF frames due to braces and due to transfer of seismic force to the end of the frames shall be included in the frame calculations.

2213.10.18 Beam flanges. Top and bottom flanges of EBF beams shall be laterally braced at the ends of link beams and at intervals not exceeding $76/\sqrt{F_y}$ (For **SI:** $0.45\sqrt{E/F_y}$) times the beam flange width. End bracing shall be designed to resist 6.0 percent of the beam flange strength, defined as $F_y b_f t_f$. Intermediate bracing shall be designed to resist 1.0 percent of the beam flange force at the brace point using the link beam strength determined in Section 2213.10.13.

2213.10.19 Beam-column connection. Beam connections to columns may be designed as pins in the plane of the beam web if the link beam is not adjacent to the column. Such connection shall have the capacity to resist a torsional moment of $0.01 F_y \, b_f t_f d$.

2213.11 Requirements for Special Truss Moment Frames.

2213.11.1 General. Special truss moment frames of steel shall be designed in accordance with this section.

2213.11.2 Special segment. Each horizontal truss which is part of the moment frame shall have a special segment located within the middle one-half length of the truss. Such trusses shall be limited to span lengths between columns not to exceed 50 feet (15 240 mm) and overall depth not to exceed 6 feet (1829 mm). The length of the special segment shall range from 0.1 to 0.5 times the truss span length. The length-to-depth ratio of any panel in the special segment shall be limited to a maximum of 1.5 and a minimum of 0.67. All panels within the special segment shall be either Vierendeel or X braced, not a combination thereof. Where diagonal members are used in the special segment, they shall be arranged in an X pattern separated by vertical members. Such diagonal members shall be interconnected at points of crossing. The interconnection shall have the strength to resist a force at least equal to 0.25 times the diagonal member tension strength. Bolted connections shall not be used for web members within the special segment. Splicing of chord members shall not be permitted within the special segment or within a one-half panel length from the ends of the special segment. Axial stresses in diagonal web members due to concentrated dead plus live loads acting within the special segment shall not exceed $0.03F_y$.

2213.11.3 Special segment members strength. In the fully yielded state, the special segment shall develop vertical shear strength through flexural strength of the chord members and through axial tension and compression strength of diagonal web members. The top and bottom chord members in the special segment shall be made of identical sections and shall provide at least 25 percent of the required vertical shear strength. The maximum axial stress in the chord members shall not exceed $0.4F_y$. Diagonal members in any panel of the special segment shall be made of identical sections. The end connections of diagonal web members in the special segment shall have strength to resist a minimum force equal to P_{st} of the member.

2213.11.4 Nonspecial segment member strength. All members and connections of special truss moment frames, except those in Section 2213.11.3, shall have strength to resist the forces due to combination of specified gravity loads and lateral forces necessary to develop maximum amplified vertical shear force in all special segments, V_{ss}, given by the following formula:

$$V_{ss} = \frac{3.4 \, M_s}{L_s} + 0.11 \, EI \, \frac{(L - L_s)}{L_s^3} +$$
$$1.25 \, (P_{st} + 0.3 \, P_{sc}) \, \sin \, \alpha \qquad (13\text{-}7)$$

WHERE:

EI = flexural stiffness of the chord members.

L = span length of the truss.

L_s = 0.9 times the length of the special segment.

M_s = flexural strength of the chord members.

P_{sc} = axial compression strength of diagonal members.

P_{st} = axial tension strength of diagonal members.

α = angle that the diagonal members make with the horizontal.

2213.11.5 Connections. Connections of all elements in the truss frames, including those within the truss, shall conform to the requirements of Section 1633.2.3.

2213.11.6 Compactness. Diagonal web members of the special segment shall be made of flat bars. The width-thickness ratio of such flat bars shall not exceed 2.5. The width-thickness ratio of angles, and flanges and webs of T sections used for chord members in the special segment shall not exceed $52 / \sqrt{F_y}$ (For **SI:** $0.31 \sqrt{E/F_y}$).

2213.11.7 Lateral bracing. Top and bottom chords of the trusses shall be laterally braced at the ends of the special segment, and at intervals not to exceed L_p according to Chapter 22, Division II, Section F1.1, along the entire length of the truss. Each lateral brace at the ends of and within the special segment shall have strength to resist at least 5 percent of P_{st} of the chord member. Lateral braces outside of the special segment shall have strength to resist at least 2.5 percent of P_{st} of the chord member.

2213.11.8 Materials. The material specifications of Division III are superseded by the requirements of Section 2213.4.1 for all elements in the special trusses.

SECTION 2214 — SEISMIC PROVISIONS FOR STRUCTURAL STEEL BUILDINGS IN SEISMIC ZONES 1 AND 2

2214.1 General. Design and construction of steel framing in lateral-force-resisting systems in Seismic Zones 1 and 2 shall conform to the requirements of this code. Additionally, in Seismic Zone 2, such framing shall conform to the requirements of this section. Ordinary moment frames in Seismic Zone 1, meeting the requirements of Section 2214.4, may use an R value of 8.5. Other framing in Seismic Zone 1 need not comply with this section.

2214.2 Definitions. Definitions shall be as prescribed in Section 2213.2.

2214.3 Materials. Materials shall be as prescribed in Section 2213.4.

> **EXCEPTION:** Ordinary moment frames in accordance with Section 2214.4.

2214.4 Ordinary Moment Frame Requirements. Ordinary moment frames (OMF) shall be designed to resist the load combinations of Section 1612.2 for Load and Resistance Factor Design or Section 1612.3 for Allowable Stress Design. Ordinary moment frames in Seismic Zone 1 meeting these requirements may use an R value of 8.5.

All beam-to-column connections in OMFs that resist earthquake forces shall meet one of the following requirements:

1. Fully restrained (Type F.R. or Type 1) conforming to Section 2214.5.1.

2. Fully restrained (Type F.R. or Type 1) connections with the design strengths of the connections capable of resisting a combination of gravity loads and Ω_o times the design seismic forces.

3. Partially restrained (Type P.R. or Type 3) connections are permitted provided:

    3.1 The connections are designed to resist the load combinations in Section 1612.2 for Load and Resistance Factor Design or Section 1612.3 for Allowable Stress Design, and

    3.2 The connections have been demonstrated by cyclic tests to have adequate rotation capacity to accommodate a story drift due to Ω_o times the design seismic forces.

    3.3 The moment frame drift calculations shall include the contribution due to the rotation and distortion of the connection.

See Divisions II and III for definitions of fully restrained and partially restrained connections.

2214.5 Special Moment-resisting Frame (SMRF) Requirements.

2214.5.1 Girder-to-column connection.

2214.5.1.1 Required strength. The girder-to-column connection shall be adequate to develop the lesser of the following:

1. The strength of the girder in flexure.

2. The moment corresponding to development of the panel zone shear strength as determined from Formula (13-1).

> **EXCEPTION:** Where a connection is not designed to contribute flexural resistance at the joint, it need not develop the required strength if it can be shown to meet the deformation compatibility requirements of Section 1633.2.4.

2214.5.1.2 Connection strength. The girder-to-column connection may be considered to be adequate to develop the flexural strength of the girder if it conforms to the following:

1. The flanges have full-penetration butt welds to the columns.

2. The girder web-to-column connection shall be capable of resisting the girder shear determined for the combination of gravity loads and the seismic shear forces which result from compliance with Section 2214.5.1.1. This connection strength need not exceed that required to develop gravity loads plus Ω_o times the girder shear resulting from the design seismic forces.

Where the flexural strength of the girder flanges is greater than 70 percent of the flexural strength of the entire section [i.e., bt_f $(d-t_f)F_y>0.7Z_xF_y$] the web connection may be made by means of welding or high-strength bolting.

For girders not meeting the criteria in the paragraph above, the girder web-to-column connection shall be made by means of welding the web directly or through shear tabs to the column. That welding shall have a strength capable of development at least 20 percent of the flexural strength of the girder web. The girder shear shall be resisted by means of additional welds or friction-type high-strength bolts or both.

2214.5.1.3 Alternate connection. Connection configurations utilizing welds or high-strength bolts not conforming with Section 2214.5.1.2 may be used if they are shown by test or calculation to meet the criteria in Section 2214.5.1.1. Where conformance is shown by calculation, 125 percent of the strengths of the connecting elements may be used.

2214.5.1.4 Flange detail limitations. For steel whose specified strength is less than 1.5 times the specified yield strength, plastic hinges shall not form at locations in which the beam flange area has been reduced, such as for bolt holes. Bolted connections of flange plates of beam-column joints shall have the net-to-gross area ratio A_e/A_g equal to or greater than $1.2F_y/F_u$.

2214.5.2 Trusses in SMRF. Trusses may be used as horizontal members in SMRF if the sum of the truss seismic force flexural strength exceeds the sum of the column seismic force flexural strength immediately above and below the truss by a factor of at least 1.25. For this determination, the strengths of the members shall be reduced by the gravity load effects. In buildings of more than one story, the column axial stress shall not exceed $0.4F_y$ and the ratio of the unbraced column height to the least radius of gyration shall not exceed 60. The connection of the truss chords to the column shall develop the lesser of the following:

1. The strength of the truss chord.

2. The chord force necessary to develop 125 percent of the flexural strength of the column.

2214.5.3 Girder-column joint restraint.

2214.5.3.1 Restrained joint. Where it can be shown that the columns of SMRF remain elastic, the flanges of the columns need be laterally supported only at the level of the girder top flange.

Columns may be assumed to remain elastic if one of the following conditions is satisfied:

1. The ratio in Formula (13-3-1) or (13-3-2) is greater than 1.25.

2. The flexural strength of the column is at least 1.25 times the moment that corresponds to the panel zone shear strength.

3. Girder flexural strength or panel zone strength will limit column stress ($f_a + f_{bx} + f_{by}$) to F_y of the column.

4. The column will remain elastic under gravity loads plus Ω_o times the design seismic forces.

Where the column cannot be shown to remain elastic, the column flanges shall be laterally supported at the levels of the girder top and bottom flanges. The column flange lateral support shall be capable of resisting a force equal to one percent of the girder flange capacity at allowable stresses [and at a limiting displacement perpendicular to the frame of 0.2 inch (5.08 mm)]. Required bracing members may brace the column flanges directly or indirectly through the column web or the girder flanges.

2214.5.3.2 Unrestrained joint. Columns without lateral support transverse to a joint shall conform to the requirements of Division III, with the column considered as a pin ended and the length taken as the distance between lateral supports conforming with Section 2214.5.3.1. The column stress, F_a, shall be determined from gravity loads plus the lesser of the following:

1. Ω_o times the design seismic forces.

2. The forces corresponding to either 125 percent of the girder flexural strength or the panel zone shear strength.

The stress, f_{by}, shall include the effects of the bracing force specified in Section 2214.5.3.1 and $P\Delta$ effects.

l/r for such columns shall not exceed 60.

At truss frames, the column shall be braced at each truss chord for a lateral force equal to one percent of the compression yield strength of the chord.

2214.5.4 Changes in beam flange area. Abrupt changes in beam flange area are not permitted within possible plastic hinge regions of special moment-resistant frames.

2214.6 Requirements for Braced Frames.

2214.6.1 General. The provisions of this section apply to all braced frames, except special concentrically braced frames designed in accordance with Section 2213.9 and eccentrically braced frames designed in accordance with Section 2213.9. Those members which resist seismic forces totally or partially by shear or flexure shall be designed in accordance with Section 2214.5.

2214.6.2 Bracing members.

2214.6.2.1 Stress reduction. The allowable stress, F_{as}, for bracing members resisting seismic forces in compression shall be determined from the following formula:

$$F_{as} = BF_a \qquad (14\text{-}1)$$

WHERE:

B = the stress-reduction factor determined from the following formula:

$$B = 1/\{1 + [(Kl/r)/2C_c]\} \geq 0.8 \qquad (14\text{-}2)$$

F_a = the allowable axial compressive stress allowed in Division III.

> **EXCEPTION:** Bracing members carrying gravity loads may be designed using the column strength requirement and load combinations of Section 2213.5.1, Item 1.

2214.6.2.2 Built-up members. The l/r of individual parts of built-up bracing members between stitches, when computed about a line perpendicular to the axis through the parts, shall not be greater than 75 percent of the l/r of the member as a whole.

2214.6.2.3 Compression elements in braces. The width-to-thickness ratio of stiffened and unstiffened compression elements used in braces shall be shown in Division III.

2214.6.3 Bracing connections.

2214.6.3.1 Forces. Bracing connections shall be designed for the lesser of the following:

1. The tensile strength of the bracing.

2. Ω_o times the force in the brace due to design seismic forces.

3. The maximum force that can be transferred to the brace by the system.

Beam-to-column connections for beams that are part of the bracing system shall have the capacity to transfer the force determined above.

2214.6.3.2 Net area. In bolted brace connections, the ratio of effective net section area to gross section area shall satisfy the formula:

$$\frac{A_e}{A_g} = \frac{1.2\alpha F^*}{F_u} \qquad (14\text{-}3)$$

WHERE:

A_e = effective net area as defined in Division III.

A_g = gross area of the member.

F_u = minimum tensile strength.

F^* = stress in brace due to the forces determined in Section 2213.8.3.1.

α = fraction of the member force from Section 2213.8.3.1 that is transferred across a particular net section.

2214.6.4 Bracing configuration for chevron and K bracing. Bracing members shall be designed for 1.5 times the otherwise prescribed forces.

The beam intersected by chevron braces shall be continuous between columns.

Where chevron braces intersect a beam from below, i.e., inverted V brace, the beam shall be capable of supporting all tributary gravity loads presuming the bracing not to exist.

EXCEPTION: This limitation need not apply to penthouses, one-story buildings or the top story of buildings.

2214.6.5 One- and two-story buildings. Braced frames not meeting the requirements of Sections 2214.6.2 and 2214.6.4 may be used in buildings not over two stories in height and in roof structures as defined in Chapter 15 if the braces have the strength to resist Ω_o times the design seismic forces.

2214.6.6 Nonbuilding structures. Nonbuilding structures with R values defined by Table 16-P, need comply only with the provisions of Section 2214.6.3.

2214.7 Special Concentrically Braced Frames. Special concentrically braced frames shall comply with the requirements of Section 2213.9.

2214.8 Eccentrically Braced Frames. Eccentrically braced frames shall comply with the requirements of Section 2213.10.

2214.9 Nondestructive Testing. Nondestructive testing shall comply with the provisions of Section 1703.

2214.10 Special Truss Moment Frames. Special truss moment frames shall comply with the requirements of Section 2213.11.

Division VI—LOAD AND RESISTANCE FACTOR
DESIGN SPECIFICATION FOR COLD-FORMED STEEL STRUCTURAL MEMBERS

SECTION 2215 — ADOPTION

Except for the modifications as set forth in Section 2216 of this division and the requirements of the building code, the design of cold-formed steel structural members shall be in accordance with the *Load and Resistance Factor Design Specification for Cold-Formed Steel Structural Members,* March 16, 1991, published by the American Iron and Steel Institute, 1101 17th Street, NW, Suite 1300, Washington, DC, as if set out at length herein. The adoption of Load and Resistance Factor Design Specification for Cold-Formed Steel Structural Members, hereinafter referred to as the AISI-LRFD.

Where other codes, standards or specifications are referred to in this specification, they are to be considered as only an indication of an acceptable method or material that can be used with the approval of the building official.

SECTION 2216 — AMENDMENTS

The following amendments shall be made to the AISI-LRFD specifications, as adopted in Section 2215:

1. Sec. A4.1 is deleted in its entirety.

2. Sec. A4.2 is deleted in its entirety.

3. Sec. A4.4 is deleted in its entirety.

4. Sec. A5.1.4 is deleted and substituted as follows:

The nominal loads shall be the minimum design loads required by the code.

Division VII—SPECIFICATION FOR DESIGN OF COLD-FORMED STEEL STRUCTURAL MEMBERS

SECTION 2217 — ADOPTION

Except for the modifications as set forth in Section 2218 of this division and the requirements of the building code, the design of cold-formed steel structural members shall be in accordance with the *Specification for Design of Cold-Formed Steel Structural Members,* 1986 (with December 1989 Addendum), published by the American Iron and Steel Institute, 1101 17th Street, NW, Suite 1300, Washington, DC, as if set out at length herein. The Specification for Design of Cold-Formed Steel Structural Members shall hereinafter be referred to as the AISI-ASD.

Where other codes, standards or specifications are referred to in this specification, they are to be considered as only an indication of an acceptable method or material that can be used with the approval of the building official.

SECTION 2218 — AMENDMENTS

The following amendments shall be made to the AISI ASD specification, as adopted in Section 2217:

1. Secs. A4.1 and A4.2. are deleted in their entirety without replacement.

2. Sec. A4.4. Revise as follows:

Where load combinations specified by Section 1612.3 include wind or earthquake loads, the resulting forces may be multiplied by 0.75 for strength determination. Such reduction shall not be allowed in combination with stress increases in accordance with Section 1612.3.1.

3. Sec. E6. Revise as follows:

The following notations apply to this section:

d = nominal screw diameter.

F_{u1} = tensile strength of member in contact with the screw head.

F_{u2} = tensile strength of member not in contact with the screw head.

P_{as} = allowable shear force per screw.

P_{at} = allowable tension force per screw.

P_{not} = pull-out force per screw.

P_{nov} = pull-over force per screw.

P_{ns} = nominal shear strength per screw.

P_{nt} = nominal tension strength per screw.

t_1 = thickness of member in contact with the screw head.

t_2 = thickness of member not in contact with the screw head.

Ω = factor of safety = 3.0.

All E6 requirements shall apply to self-tapping screws with 0.08 inch (2.03 mm) < d < 0.25 inch (6.35 mm). The screws shall be thread-forming or thread-cutting, with or without a self-drilling point. Alternatively, design values for a particular application shall be permitted to be based on tests according to Section F. For diaphragm applications, Section D5 shall be used.

Screws shall be installed and tightened in accordance with the manufacturer's recommendations.

The tension force on the net section of each member joined by a screw connection shall not exceed T_a from Section C2 or P_a from Section E3.2.

E6.1 Minimum Spacing. The distance between the centers of fasteners shall be not less than $3d$.

E6.2 Minimum Edge and End Distance. The distance from the center of a fastener to the edge of any part shall not be less than $3d$. If the connection is subjected to shear force in one direction only, the minimum edge distance shall be reduced to $1.5d$ in the direction perpendicular to the force.

E6.3 Shear.

E6.3.1 Connection shear. The shear force per screw shall not exceed P_{as} calculated as follows:

$$P_{as} = P_{ns}/\Omega$$

For $t_2/t_1 \leq 1.0$, P_{ns} = shall be taken as the smallest of

$$P_{ns} = 4.2\ (t_2^3\ d)^{1/2}\ F_{u2} \qquad \text{(Eq. E6.3.1)}$$
$$P_{ns} = 2.7t_1\ d\ F_{u1} \qquad \text{(Eq. E6.3.2)}$$
$$P_{ns} = 2.7\ t_2\ d\ F_{u2} \qquad \text{(Eq. E6.3.3)}$$

For $t_2/t_1 \geq 2.5$, P_{ns} shall be taken as the smaller of

$$P_{ns} = 2.7t_1\ d\ F_{u1} \qquad \text{(Eq. E6.3.4)}$$
$$P_{ns} = 2.7\ t_2\ d\ F_{u2} \qquad \text{(Eq. E6.3.5)}$$

For $1.0 < t_2/t_1 < 2.5$, P_{ns} shall be determined by linear interpolation between the above two cases.

E6.3.2 Shear in screws. The shear capacity of the screw shall be determined by test according to Section F1(a). The shear capacity of the screw shall not be less than $1.25\ P_{ns}$.

E6.4 Tension. For screws that carry tensile loads, the head of the screw or washer, if a washer is provided, shall have a diameter d_w, not less than $5/16$ inch (7.95 mm). Washers shall be at least 0.050 inch (1.27 mm) thick.

The tension force per screw shall not exceed P_{at}, calculated as follows:

$$P_{at} = P_{nt}/\Omega \qquad \text{(Eq. E6.4.1)}$$

P_{nt} = shall be taken as the lesser of P_{not} and

P_{nov} as determined in Sections E4.4.1 and E4.4.2.

E6.4.1 Pull-out. The pull-out force, P_{not}, shall be calculated as follows:

$$P_{not} = 0.85t_c\ d\ F_{u2} \qquad \text{(Eq. E6.4.1.)}$$

where t_c is the lesser of the depth of the penetration and the thickness, t_2.

E6.4.2 Pull-over. The pull-over force, P_{nov}, shall be calculated as follows:

$$P_{nov} = 1.5t_1\ d_w\ F_{u1} \qquad \text{(Eq. E6.4.2.1)}$$

where d_w is the larger of the screw head diameter or the washer diameter and shall be taken not larger than $1/2$ inch (12.7 mm).

E6.4.3 Tension in screws. The tensile capacity of the screw shall be determined by test according to Section F1(a). The tensile capacity of the screw shall not be less than $1.25\ P_{nt}$.

Division VIII—LATERAL RESISTANCE FOR STEEL STUD WALL SYSTEMS

SECTION 2219 — GENERAL

Steel stud wall systems in which shear panels are used to resist lateral loads produced by wind or earthquake shall comply with the requirements of this section. The nominal shear value used to establish the allowable shear value or design shear value shall not exceed the values set forth in Table 22-VIII-A or Table 22-VIII-B for wind loads or Table 22-VIII-C for seismic loads. The allowable shear value (ASD) or design shear value (LRFD) shall be determined using the φ or Ω factors as set forth in Section 2219.3.

All boundary members and connections thereto shall be proportioned to transmit the induced forces. Framing members shall be of a minimum size, shape and of a minimum specified yield stress as listed in Table 22-VIII-A, 22-VIII-B or 22-VIII-C. Fasteners between framing members and between the panels and the framing members shall be as specified in Table 22-VIII-A, 22-VIII-B or 22-VIII-C. Fasteners along the edges in shear panels shall be placed not less than $^3/_8$ inch (9.5 mm) in from panel edges. Screws shall be of sufficient length to ensure penetration into the steel stud by at least two full diameter threads.

Panel thickness shown in Tables 22-VIII-A and 22-VIII-B shall be considered as minimum.

No panels less than 12 inches (305 mm) wide shall be used. All panel edges shall be fully blocked. Where horizontal strap blocking is used, it shall be a minimum $1^1/_2$ inches (38 mm) wide and of the same material and thickness as the track and studs. Studs shall be doubled (back to back) at shear wall ends.

The height to length ratio of wall systems listed in Tables 22-VIII-A, 22-VIII-B and 22-VIII-C shall not exceed 2:1.

2219.1 Wood Structural Panel Sheathing.

As an alternative to the provisions in Tables 22-VIII-A and 22-VIII-C, steel stud wall systems sheathed with wood structural panels may be used to resist horizontal forces from wind or seismic loads where allowable shear loads may be calculated by the principles of mechanics without limitation by using the wood structural panel shear values in the code and approved fastener values. Where $^7/_{16}$ inch (11 mm) OSB is specified, $^{15}/_{32}$-inch (12 mm) Structural 1 sheathing (plywood) may be substituted. Structural panels may be applied either parallel to or perpendicular to framing. No increase of the nominal loads shown in Tables 22-VIII-A and 22-VIII-C shall be permitted for duration of load nor shall an increase in nominal loads be permitted for installing sheathing on the opposite side unless indicated herein.

2219.2 Gypsum Board Panel Sheathing.

Stud wall systems sheathed with gypsum board may be used to resist horizontal forces produced by wind loads when the nominal load used to establish the allowable shear value or design shear value does not exceed the nominal value set forth in Table 22-VIII-B.

The values listed in Table 22-VIII-B shall not be cumulative with the shear values of other materials applied to the same wall; values shown shall not be increased when applied to both sides of the same panel.

End joints of adjacent courses of gypsum board sheets shall not occur over the same stud. Gypsum board shall be applied perpendicular to studs in accordance with Table 22-VIII-B.

2219.3 Design.

Where allowable stress design is used, the allowable shear value shall be determined by dividing the nominal shear value, shown in Tables 22-VIII-A and 22-VIII-B, by a factor of safety (Ω) which shall be taken as 3.0. The factor of safety (Ω) for the nominal loads shown in Table 22-VIII-C shall be taken as 2.5.

Where Load and Resistance Factor Design is used, the design shear value shall be determined by multiplying the nominal shear value, shown in Tables 22-VIII-A and 22-VIII-B, by a resistance factor (φ) which shall be taken as 0.45. The resistance factor (φ) for the nominal loads shown in Table 22-VIII-C shall be taken as 0.55.

SECTION 2220 — SPECIAL REQUIREMENTS IN SEISMIC ZONES 3 AND 4

2220.1 General. In Seismic Zones 3 and 4, in addition to the requirements of Section 2219, steel stud wall systems may be used to resist the specified seismic forces in buildings not over five stories in height. Such systems shall comply with the following:

1. The l/r of the brace may exceed 200 and is unlimited.

2. All boundary members, chords and collectors shall be designed and detailed to transmit the induced axial forces.

3. Connection of the diagonal bracing member, top chord splices, boundary members and collectors shall be designed to develop the full tensile strength of the member or Ω_o times the otherwise prescribed seismic forces.

4. Vertical and diagonal members of the braced bay shall be anchored so the bottom track is not required to resist uplift forces by bending of the track web.

5. Both flanges of studs in a bracing panel shall be braced to prevent lateral torsional buckling. Wire-tied bridging shall not be considered to provide such restraint.

6. Screws shall not be used to resist lateral forces by pullout resistance.

7. Provision shall be made for pretensioning or other methods of installation of tension-only bracing to guard against loose diagonal straps.

2220.2 Boundary Members and Anchorage. Boundary members and the uplift anchorage thereto shall have the strength to resist the forces determined by the load combinations in Section 2213.5.1.

2220.3 Wood Structural Panel Sheathing. Where wood structural panels provide lateral resistance, the design and construction of such walls shall be in accordance with the additional requirements of this section. Perimeter members at openings shall be provided and shall be detailed to distribute the shearing stresses. Wood sheathing shall not be used to splice these members.

Wood structural panels shall be manufactured using exterior glue.

Wall studs and track shall have a minimum uncoated base metal thickness of not less than 0.033 inch (0.84 mm) and shall not have an uncoated base metal thickness greater than 0.043 inch (1.10 mm).

**TABLE 22-VIII-A—NOMINAL SHEAR VALUES FOR WIND FORCES IN POUNDS PER FOOT
FOR SHEAR WALLS FRAMED WITH COLD-FORMED STEEL[1,2]**

ASSEMBLY DESCRIPTION	FASTENER SPACING AT PANEL EDGES[3] (inches)				FRAMING SPACING (inches o.c.)
	× 25.4 for mm × 0.0146 for N/mm				
× 25.4 for mm	6	4	3	2	× 25.4 for mm
$^{15}/_{32}$-inch Structural 1 sheathing (4-ply) one side	1,065[4]	—	—	—	24
$^{7}/_{16}$-inch rated sheathing (OSB) one side	910[4]	1,410	1,735	1,910	24

[1]Nominal shear values shall be multiplied by the appropriate strength reduction factor, Φ, to determine design strength or divided by the appropriate safety factor, Ω, to determine allowable shear values as set forth in Section 2219.3.

[2]Unless otherwise shown, studs shall be a minimum $1^{5}/_{8}$ inches (41 mm) by $3^{1}/_{2}$ inches (89 mm) with a $^{3}/_{8}$-inch (9.5 mm) return lip. Track shall be a minimum $1^{1}/_{4}$ inches (32 mm) by $3^{1}/_{2}$ inches (89 mm). Both studs and track shall have a minimum uncoated base metal thickness of 0.033 inch (0.84 mm) and shall be ASTM A 446 Grade A [or ASTM A 653, SQ, Grade 33 (new designation)]. Framing screws shall be No. 8 by $^{5}/_{8}$-inch (16 mm) wafer head self-drilling. Plywood and OSB screws shall be approved and shall be a minimum No. 8 by 1-inch (25 mm) flat head with a minimum head diameter of 0.292 inch (7.4 mm). Stud spacing shown are maximums.

[3]Screws in the field of the panel shall be installed 12 inches o.c. (305 mm) unless otherwise shown.

[4]Where fully blocked gypsum board is applied to the opposite side of this assembly, per Table 22-VIII-B, these nominal values may be increased by 30 percent.

**TABLE 22-VIII-B—NOMINAL SHEAR VALUES FOR WIND FORCES IN POUNDS PER FOOT FOR SHEAR WALLS FRAMED
WITH COLD-FORMED STEEL STUDS AND FACED WITH GYPSUM WALLBOARD[1,2]**

WALL CONSTRUCTION		SCREW SPACING (edge/field) (inches)	NOMINAL SHEAR VALUE (lbs/ft)
× 25.4 for mm	ORIENTATION	× 25.4 for mm	× 0.0146 for N/mm
$^{1}/_{2}$-inch gypsum board on both sides of wall with studs 24 inches o.c.	Gypsum board applied perpendicular to framing with strap blocking behind the horizontal joint and with solid blocking between the first two end studs.	7/7	585
		4/4	850

[1]Nominal shear values shall be multiplied by the appropriate strength reduction factor, Φ, to determine design strength or divided by the appropriate safety factor, Ω, to determine allowable shear values as set forth in Section 2219.3.

[2]Unless otherwise shown, studs shall be a minimum $1^{5}/_{8}$ inches (41 mm) by $3^{1}/_{2}$ inches (89 mm) with a $^{3}/_{8}$-inch (9.5 mm) return lip. Track shall be a minimum $1^{1}/_{4}$ inches (32 mm) by $3^{1}/_{2}$ inches (89 mm). Both studs and track shall have a minimum uncoated base metal thickness of 0.033 inch (0.84 mm) and shall be ASTM A 446 Grade A [or ASTM A 653, SQ, Grade 33 (new designation)]. Framing screws shall be No. 8 by $^{5}/_{8}$-inch (16 mm) wafer head self-drilling. Drywall screws shall be a minimum No. 6 by 1 inch (25 mm).

**TABLE 22-VIII-C—NOMINAL SHEAR VALUES FOR SEISMIC FORCES IN POUNDS PER FOOT
FOR SHEAR WALLS FRAMED WITH COLD-FORMED STEEL STUDS[1,2]**

ASSEMBLY DESCRIPTION	FASTENER SPACING AT PANEL EDGES[3] (inches)				FRAMING SPACING (inches o.c.)
	× 25.4 for mm × 0.0146 for N/mm				
× 25.4 for mm	6	4	3	2	× 25.4 for mm
$^{15}/_{32}$-inch Structural 1 sheathing (4-ply) one side	780	990	1,465	1,625	24
$^{7}/_{16}$-inch (OSB) one side	700	915	1,275	1,625	24

[1]Nominal shear values shall be multiplied by the appropriate strength reduction factor, Φ, to determine design strength or divided by the appropriate safety factor, Ω, to determine allowable shear values as set forth in Section 2219.3.

[2]Unless otherwise shown, studs shall be a minimum $1^{5}/_{8}$ inches (41 mm) by $3^{1}/_{2}$ inches (89 mm) with a $^{3}/_{8}$-inch (9.5 mm) return lip. Track shall be a minimum $1^{1}/_{4}$ inches (32 mm) by $3^{1}/_{2}$ inches (89 mm). Both studs and track shall have a minimum uncoated base metal thickness of 0.033 inch (0.084 mm) and shall not have a base metal thickness greater than 0.043 inch (1.10 mm) and shall be ASTM A 446 Grade A [or ASTM A 653, SQ, Grade 33 (new designation)]. Stud spacing shown are maximums. Framing screws shall be No. 8 by $^{5}/_{8}$-inch (16 mm) wafer head self-drilling. Plywood and OSB screws shall be approved and shall be a minimum No. 8 by 1-inch (25 mm) flat head with a minimum head diameter of 0.292 inch (7.4 mm).

[3]Screws in the field of the panel shall be installed 12 inches (305 mm) o.c. unless otherwise shown.

Division IX—OPEN WEB STEEL JOISTS

SECTION 2221 — ADOPTION

In addition to the requirements in the building code, the design, manufacture and use of open web steel joists shall be in accordance with the *Standard Specification for Steel Joists, K-Series, LH-Series, DLH-Series and Joist Girders,* 1994, published by the Steel Joist Institute, 1205 48th Avenue, Suite A, Myrtle Beach, SC 29577, as if set out at length herein.

Where other codes, standards or specifications are referred to in this specification, they are to be considered as only an indication of an acceptable method or material that can be used with the approval of the building official.

Division X—DESIGN STANDARD FOR STEEL STORAGE RACKS

Based on the Specification for the Design, Testing and Utilization of Industrial Steel Storage Racks, 1990 Edition, by the Rack Manufacturers Institute

SECTION 2222 — GENERAL PROVISIONS

2222.1 Scope. This division shall apply to pallet racks, movable shelf racks and stacker-racks made of cold-formed or hot-rolled steel structural members. It shall not apply to other types of racks, such as cantilever racks, drive-in and drive-through racks and rack buildings.

The design of racks not covered by this standard shall be in accordance with the provisions of Section 1632 using the loads as defined in this division.

> **EXCEPTION:** The building official may waive the design requirements for storage racks less than or equal to 8 feet (2438 mm) in height.

2222.2 Definitions. For the purpose of this division, certain terms are defined as follows:

FRAMES are the rigid-connection frames composed of posts and pallet beams in pallet racks, and the upright (trussed) frames.

MOVABLE-SHELF RACKS are a type of rack where the shelves are removable and replaceable in multiple locations in the section opening.

PALLET is a portable platform on which goods are placed for storage or transportation.

PALLET GUIDE is a device to prevent a pallet from sliding off its support beam or rail.

PALLET RACK (also **Standard Pallet Rack**) is a rack which utilizes horizontal beams connected to prefabricated upright frames to provide independent, multiple-level storage (usually on either side of an aisle). The horizontal beams support pallets.

PALLET RACK BEAMS (also **Shelf Beams**) are horizontal structural members used to support pallets in a rack.

STACKER RACKS (also **Stacker Crane Racks**) are a rack arrangement, usually higher than other types of racks, where floor-running storage and retrieval machines are used to shuttle loads. These machines are generally rail mounted and often fully automated and computer controlled.

STACKER-RACK BEAMS are horizontal structural members used to support pallets in a stacker rack.

STEEL STORAGE RACK is a single or multilevel storage system in single or multi-bays consisting of vertical columns or posts and horizontal beams. Diagonal and horizontal trussed bracing is often used to resist horizontal loads. The beams generally support pallets, and the system may be one of a number of different types (Standard Pallet Rack, Stacker Rack, etc.); see definitions of each type.

UNIT LOAD is a standardized load, meaning, for example, pallet load if pallets are used. Also refers to other standardized loads such as barrels which may be stored without pallets.

2222.3 Materials. This standard contemplates the use of steel of structural quality as listed in Division I.

Steels not listed in the above provisions are not excluded, provided they conform to the chemical and mechanical requirements of one of the listed specifications or other published specifications which establish their properties and structural suitability; and provided they are subjected by either the producer or the purchaser to analyses, tests and other controls to the extent and in the manner prescribed by one of the two listed specifications, as applicable.

2222.4 Applicable Design Specifications. Except as either modified or supplemented herein, Division I, shall apply to the design of steel storage racks.

2222.5 Integrity of Rack Installations. Individual rack components and assemblies thereof shall comply with this standard.

All rack installations and racks manufactured in conformity with this standard shall display in one or more conspicuous locations a permanent plaque each not less than 50 square inches (32 258 mm^2] in area and showing the maximum permissible unit load in clear, legible print.

> **EXCEPTION:** The building official may waive plaque installation for racks not exceeding 12 feet (3658 mm) in height to top shelf, covering a floor area less than 300 square feet (27.87 m^2), with a unit load not exceeding 2,500 pounds (11 121 N), and without double stacking on top level.

Load application and rack configuration drawings shall be furnished with each rack installation. The drawings shall present the permissible configurations or limitations as to the maximum number of shelves or rails, the maximum distance between them, and the maximum distance from the floor to the bottom shelf or rail.

The bottom of all posts shall be furnished with bearing plates, according to Section 2227.2. Drive-in, drive-through and stacker racks shall be anchored to the floor by anchor bolts capable of resisting the horizontal shear forces caused by the horizontal and vertical loads on the rack.

The stability of movable-shelf racks shall not be dependent upon movable shelves. Those components which provide stability, such as permanently bolted or welded top shelves and the longitudinal and transverse diagonal bracing, shall be clearly identified on the rack configuration drawings. In specific movable-shelf rack installations where rack height requires it, a conspicuous warning is to be placed in the owner's utilization instruction manual of any restrictions on shelf placement or shelf removal. Such restrictions also are to be permanently posted in locations clearly visible to forklift operators.

Lower portions of posts exposed to damage by forklift trucks or other moving equipment shall have protective devices. If not so protected the rack structure may, at the option of the building official: (1) be designed to maintain its full design load capacity at allowable stresses with the exposed post capacity reduced by one-half, or (2) be designed to maintain its full design load capacity at 50 percent increased allowable stresses with the exposed post assumed to have no carrying capacity.

Where racks are braced against the building structure, the building structure shall be designed for the horizontal and vertical forces listed in Section 2226.1 imposed on the building structure.

Racks shall be installed with a maximum tolerance from the vertical of 1 inch in 10 feet (25.4 mm in 3048 mm) of height. Special conditions may require more restrictive tolerances.

Support of racks by foundations, concrete floor slabs or other means shall be in conformance with Chapter 18.

SECTION 2223 — DESIGN PROCEDURES AND DIMENSIONAL LIMITATIONS

2223.1 General. All computations for allowable loads, stresses and deflections shall be in accordance with conventional methods of structural design, and as specified in Section 2222.4, except where modified or supplemented herein. Where adequate methods of design calculations are not available, justification by a testing program acceptable to the building official may be used.

2223.2 Dimensional Limitations. The limitations on flat-width ratios and slenderness ratios in Division I and Division V shall apply except for the following conditions:

1. Slenderness limitations shall not be imposed on tension members which do not resist compression forces under any loading condition.

2. The unbraced length of compression or tension members shall be the length between connections to other structural members disposed in the direction of the pertinent radius of gyration, or from such a connection to the nearest attachment to an external fixed structure, such as a floor.

SECTION 2224 — ALLOWABLE STRESSES AND EFFECTIVE WIDTHS

2224.1 General. Allowable stresses and effective design widths shall be as specified in Division I and Division IV except as provided herein. Allowable stresses for working stress design may be increased one-third when considering wind or earthquake forces either acting alone or when combined with vertical loads.

2224.2 Perforated Compression Members. The effect of perforations on the carrying capacity of compression members shall be recognized by modification of the Q-factor. Q-values for perforated compression members shall be determined by stub column tests acceptable to the building official. These members shall be designed in an approved manner. The effects of perforations on the capacity of members may be considered by using the section properties based on the minimum net area.

2224.3 Torsional-flexural Buckling. Sections subject to torsional-flexural buckling shall be designed according to Division VII.

SECTION 2225 — PALLET AND STACKER-RACK BEAMS

2225.1 Allowable Loads. Where the shape of the cross section permits, allowable loads of pallet-carrying beams shall be determined in accordance with Division I or Division IV.

2225.2 Deflections. At working load the deflections, including possible deformations in the end connections, shall not exceed $^1/_{180}$ of the span measured with respect to the beam ends.

2225.3 Determination by Test. Where the configuration of the cross section precludes calculation of allowable loads and deflections, determination may be made with a testing program acceptable to the building official.

SECTION 2226 — FRAME DESIGN

2226.1 General. Frames shall be designed for the critical combinations of vertical loads in the most unfavorable positions, horizontal loads as specified in Section 2228.3, and the additional effects of horizontal sway caused by looseness, if any, of the top tie beam to post connections.

2226.2 Effective Lengths. Effective lengths based on valid engineering principles shall be used in the design of posts and upright frames.

SECTION 2227 — CONNECTIONS AND BEARING PLATES

2227.1 Connections. Adequate strength of connections to withstand calculated resultant forces and moments, and adequate rigidity where such is required, shall be demonstrated by calculation or by testing in an approved manner.

Beams shall have support connections capable of withstanding an upward force of 1,000 pounds (4448 N) per connection within allowable design values of this code.

For movable-shelf racks, the top shelf and other fixed shelves are to include support connections capable of withstanding an upward force of 1,000 pounds (4448 N) per connection within the allowable design values of this code.

The movable shelves are generally constructed of a set of front and rear longitudinal beams connected to each other rigidly by transverse members. The movable shelves are to be connected in such a way to prevent forward displacement when lifting out the front beam of the shelf.

2227.2 Bearing Plates. Provision shall be made to transfer post loads and moments into the floor. Said forces and moments shall be consistent in magnitude, sense and direction with the rack analysis. Allowable bearing stresses on the bottom of base plates shall be determined in accordance with Chapter 19.

SECTION 2228 — LOADS

2228.1 Gravity Loads. Racks shall be designed for dead loads, live loads and unit loads as posted on the rack installation under Section 2222.5.

2228.2 Vertical Impact. Unit load-carrying beams, supporting arms, if any, and end connections shall be designed for an additional vertical impact load of 25 percent of one unit load located to produce maximum moments and shears. Impact stresses shall not exceed stresses referenced in Section 2224, nor shall they cause detrimental permanent deformations in connections. When allowable loads are determined by tests, due allowance shall be made for the additional impact load. Impact loads may be omitted when checking beam deflections and designing upright frames, posts and other vertical components.

2228.3 Horizontal Loads.

2228.3.1 General. All racks shall be designed for the horizontal forces and allowable stresses specified in this standard. These forces shall not cause permanent distortions of connections when subject to test, or permanent residual sway deflections (of the entire rack when subject to full-scale rack tests) larger than 20 percent of the sway deflections measured under the simultaneous action of horizontal and vertical loads.

2228.3.2 Horizontal stability. Horizontal stability shall be determined by applying horizontal forces simultaneously at all beam-to-post connections equal to 1.5 percent of the maximum live load plus dead load at the connection. The forces shall be applied separately in each of the two principal directions of the rack and in conjunction with full dead and live loads.

2228.3.3 Stacker racks or racks wholly or partially supporting moving equipment. Racks shall be designed for maximum forces and their locations, transmitted from moving equipment to racks, and applicable longitudinal and transverse impact factors due to moving equipment.

Devices acting as bumpers to stop moving equipment shall be considered in the design.

Forces described in this section need not act concurrently with those described in Sections 2228.3.2 and 2228.5.1.

2228.4 Wind Loads. Outdoor racks exposed to wind shall be designed for the wind loads prescribed by Chapter 16 acting on the horizontal projection of rack plus contents. For stability, consideration shall be given to loading conditions that produce large wind forces combined with small stabilizing gravity forces, such as racks fully loaded, but with unit loads of much smaller weight than the maximum posted unit load.

Forces described in Sections 2228.3.2 and 2228.5.1 need not act concurrently with wind loads. Forces described in Section 2228.3.3 shall act concurrently with wind forces for design purposes.

2228.5 Earthquake Loads.

2228.5.1 General. Steel storage racks which are not connected to buildings or other structures shall be designed to resist seismic forces in conformance with this standard.

2228.5.2 Minimum earthquake forces. The total minimum lateral force at strength design levels shall be determined in accordance with Section 1630.2.1:

WHERE:

R = 4.4 for racks or portions thereof where lateral stability is dependent on diagonal or x-bracing. Connections for bracing members shall be capable of developing the required strength of the members.

R = 5.6 for racks where the lateral stability is wholly dependent on moment-resisting frame action.

W = weight of rack structure plus contents. Where a number of storage rack units are interconnected so there are a minimum of four columns in any direction on each column line designed to resist horizontal forces, W may be equal to the total dead load plus 50 percent of the rack-rated capacity. In Seismic Zones 3 and 4, wholesale and retail areas, the 50 percent may only be used when combined with C_v/RT taken equal to $0.70C_a$ in Formula (30-4) and with $2.5/R$ taken equal to 0.70 in Formula (30-5).

Total seismic force shall be assumed to act nonconcurrently in the direction of each of the main axes of the rack. For racks having more than two storage levels, the total lateral force, V, shall be distributed over the height of the rack in accordance with

$$F_i = \frac{VW_ih_i}{\sum_{i=1}^{n} W_ih_i}$$

in which F_i is the lateral force applied at Level i; W_i is the portion of the total weight; W, which is assigned to Level i; h_i is the height of Level i above the base of the rack; and n is the total number of storage levels. The lateral force, V, shall be distributed in proportion to the total weight, W.

Other R values may be considered based on submission of substantiating data.

2228.6 Storage Racks in Buildings. Storage racks located in buildings at levels above the ground level, rack buildings or racks that depend on attachments to buildings or other structures at other than the floor level for their lateral stability, shall be designed to resist earthquake forces that consider the responses of the building and storage rack to earthquake ground motions as specified in Chapter 16, Division III.

2228.7 Other Considerations.

2228.7.1 Overturning. In determining overturning moments, the total weight shall be assumed to act at a height equal to 1.15 times the distance from the floor to the actual center of gravity of all the horizontal forces.

Equal safety against an overturning moment shall be provided when only the top level of the rack is loaded, in which case it is to be assumed that the force acts through the center of gravity of the top load.

2228.7.2 Torsional forces. Torsional forces shall be considered based on the critical combination of loaded and unloaded storage spaces.

2228.7.3 Concurrent forces. Forces described in Sections 2228.3.2, 2228.3.3 and 2228.4 need not be assumed to act concurrently with earthquake loads.

SECTION 2229 — SPECIAL RACK DESIGN PROVISIONS

2229.1 Stability of Truss-braced Upright Frames. The maximum allowable compression stress in the posts of truss-braced upright frames shall be determined from Divisions I and IV.

2229.2 Overturning and Height-to-depth Ratio. Overturning shall be based on the critical combination of vertical and horizontal loads. Stabilizing forces provided by anchor bolts to the floor shall not be considered to resist overturning unless the anchors are specifically designed and installed to resist the uplift forces. Unless all columns are so anchored, the minimum ratio between righting moment and overturning moment shall be 1.5. Sections 2228.4 and 2228.7 shall be considered in the design.

The height-to-depth ratio of a storage rack shall not exceed 6 to 1 measuring to the top of the topmost load unless the rack is anchored or braced externally.

2229.3 Connections to Buildings. Connections of racks to buildings, if any, shall be designed and installed to prevent reactions or displacements of the buildings or racks from damaging one another. Section 2228.5 shall be considered.

Division XI—DESIGN STANDARD FOR STRUCTURAL APPLICATIONS OF STEEL CABLES FOR BUILDINGS

This standard is based on the American Society of Civil Engineers Standard 17-95, *Structural Applications of Steel Cables for Buildings* (ASCE 17-95) available from the American Society of Civil Engineers, 1801 Alexander Bell Drive, Reston, Virginia 20191-4400.

SECTION 2230 — ADOPTION

The American Society of Civil Engineers standard adopted by this division provides requirements for the structural design, fabrication and the installation of steel cables for use as primary structural elements for the support of roofs and floors of buildings. Where other codes, standards or specifications are referred to in this division, they are to be considered as only an indication of an acceptable method of material that can be used with the approval of the building official.

Chapter 23

WOOD

NOTE: This chapter has been revised in its entirety.

Division I—GENERAL DESIGN REQUIREMENTS

SECTION 2301 — GENERAL

2301.1 Scope. The quality and design of wood members and their fastenings shall conform to the provisions of this chapter.

2301.2 Design Methods. Design shall be based on one of the following methods.

2301.2.1 Allowable stress design. Design using allowable stress design methods shall resist the load combinations of Section 1612.3, in accordance with the applicable requirements of Section 2305.

2301.2.2 Conventional light-frame construction. The design and construction of conventional light-frame wood structures shall be in accordance with the applicable requirements of Section 2305.

SECTION 2302 — DEFINITIONS

2302.1 Definitions. The following terms used in this chapter shall have the meanings indicated in this section:

AFPA is the American Forest and Paper Association, 1111 19th Street, N.W., Suite 800, Washington, D.C. 20036 (formerly NFoPA, National Forest Products Association).

AHA is the American Hardboard Association, Inc., 1210 W. Northwest Highway, Palatine, Illinois 60067.

AITC is the American Institute of Timber Construction, 7012 S. Revere Parkway, Suite 140, Englewood, Colorado 80112.

ALSC is the American Lumber Standard Committee, Post Office Box 210, Germantown, Maryland 20875-0210.

APA is the American Plywood Association, 7011 South 19th Street, Tacoma, Washington 98411.

AWPA is the American Wood Preservers Association, Post Office Box 286, Woodstock, Maryland 21163-0286.

BLOCKED DIAPHRAGM is a diaphragm in which all sheathing edges not occurring on framing members are supported on and connected to blocking.

BRACED WALL LINE is a series of braced wall panels in a single story that meets the requirements of Section 2320.11.3.

BRACED WALL PANEL is a section of wall braced in accordance with Section 2320.11.3.

CONVENTIONAL LIGHT-FRAME CONSTRUCTION is a type of construction whose primary structural elements are formed by a system of repetitive wood-framing members. Refer to Section 2320 for conventional light-frame construction provisions.

DIAPHRAGM is a horizontal or nearly horizontal system acting to transmit lateral forces to the vertical-resisting elements. When the term "diaphragm" is used, it includes horizontal bracing systems.

FIBERBOARD is a fibrous-felted, homogeneous panel made from lignocellulosic fibers (usually wood or cane) and having a density of less than 31 pounds per cubic foot (497 kg/m^3) but more than 10 pounds per cubic foot (160 kg/m^3).

GLUED BUILT-UP MEMBERS are structural elements, the sections of which are composed of built-up lumber, wood structural panels or wood structural panels in combination with lumber, all parts bonded together with adhesives.

GRADE (Lumber) is the classification of lumber in regard to strength and utility in accordance with UBC Standard 23-1 and the grading rules of an approved lumber grading agency.

HARDBOARD is a fibrous-felted, homogeneous panel made from lignocellulosic fibers consolidated under heat and pressure in a hot press to a density not less than 31 pounds per cubic foot (497 kg/m^3).

NELMA is the Northeastern Lumber Manufacturers Association, 272 Tuttle Road, Post Office Box 87 A, Cumberland Center, Maine 04021.

NLGA is the National Lumber Grades Authority, 103-4000 Dominion Street, Burnaby B.C., Canada V5G 4G3.

NSLB is the Northern Softwood Lumber Bureau (serviced by NELMA), 272 Tuttle Road, Post Office Box 87 A, Cumberland Center, Maine 04021.

NOMINAL LOADING is a design load that stresses a member of fastening to the full allowable stress tabulated in this chapter. This loading may be applied for approximately 10 years, either continuously or cumulatively, and 90 percent of this load may be applied for the remainder of the life of the member or fastening.

NOMINAL SIZE (Lumber) is the commercial size designation of width and depth, in standard sawn lumber and glued-laminated lumber grades; somewhat larger than the standard net size of dressed lumber, in accordance with UBC Standard 23-1 for sawn lumber.

PARTICLEBOARD is a manufactured panel product consisting of particles of wood or combinations of wood particles and wood fibers bonded together with synthetic resins or other suitable bonding system by a bonding process in accordance with approved nationally recognized standards.

PLYWOOD is a panel of laminated veneers conforming to UBC Standard 23-2 or 23-3.

RIS is the Redwood Inspection Service, 405 Enfrente Drive, Suite 200, Novato, California 94949.

ROTATION is the torsional movement of a diaphragm about a vertical axis.

SPIB is the Southern Pine Inspection Bureau, 4709 Scenic Highway, Pensacola, Florida 32504.

STRUCTURAL GLUED-LAMINATED TIMBER is any member comprising an assembly of laminations of lumber in which the grain of all laminations is approximately parallel longitudinally, in which the laminations are bonded with adhesives.

SUBDIAPHRAGM is a portion of a larger wood diaphragm designed to anchor and transfer local forces to primary diaphragm struts and the main diaphragm.

TREATED WOOD is wood treated with an approved preservative under treating and quality control procedures.

WCLIB is the West Coast Lumber Inspection Bureau, 6980 S.W. Varnes Road, Post Office Box 23145, Portland, Oregon 97223.

WOOD OF NATURAL RESISTANCE TO DECAY OR TERMITES is the heartwood of the species set forth below. Corner sapwood is permitted on 5 percent of the pieces provided 90 percent or more of the width of each side on which it occurs is heartwood. Recognized species are:

Decay resistant: Redwood, Cedars, Black Locust

Termite resistant: Redwood, Eastern Red Cedar

WOOD STRUCTURAL PANEL is a structural panel product composed primarily of wood and meeting the requirements of UBC Standard 23-2 or 23-3. Wood structural panels include all-veneer plywood, composite panels containing a combination of veneer and wood-based material, and matformed panels such as oriented strand board and waferboard.

WWPA is the Western Wood Products Association, Yeon Building, 522 S. W. Fifth Avenue, Portland, Oregon 97204-2122.

SECTION 2303 — STANDARDS OF QUALITY

The standards listed below labeled a "UBC Standard" are also listed in Chapter 35, Part II, and are part of this code. The other standards listed below are recognized standards. (See Sections 3503 and 3504.)

1. **Grading rules.**

    1.1 UBC Standard 23-1, Classification, Definition, Methods of Grading and Development of Design Values for All Species of Lumber

    1.2 Standard Grading Rules for Canadian Lumber, United States Edition, NLGA

    1.3 Standard Grading Rules No. 17, WCLIB

    1.4 Standard Grading Rules, WWPA

    1.5 Grading Rules, NHPMA

    1.6 Grading Rules, SPIB

    1.7 Standard Specifications for Grades of California Redwood Lumber, RIS

    1.8 Standard Grading Rules, NELMA

2. **Structural glued-laminated timber.**

    2.1 ANSI/AITC Standard A190.1 and ASTM D 3737, Design and Manufacture of Structural Glued-laminated Timber

    2.2 Standard Specifications for Structural Glued-laminated Timber of Softwood Species, AITC 117; Manufacturing, AITC 117; Design and Standard Specifications for Hardwood Glued-laminated Timber, AITC 119.

    2.3 Inspection Manual AITC 200 of the American Institute of Timber Construction, Tests for Structural Glued-laminated Timber.

    2.4 AITC 500, Determination of Design Values for Structural Glued-laminated Timber in accordance with ASTM D 3737, American Institute of Timber Construction.

3. **Preservative treatment by pressure process and quality control.**

    3.1 Standard Specifications C1, C2, C3, C4, C9, C14, C15, C16, C22, C23, C24, C28 and M4, AWPA

4. **Product standards.**

    4.1 UBC Standard 23-2, Construction and Industrial Plywood

    4.2 UBC Standard 23-3, Performance Standard for Wood-Based Structural-Use Panels

    4.3 ANSI A208.1, Particleboard

    4.4 ASTM D 1037, Evaluating the Properties of Wood-based Fiber and Particle Panel Materials

    4.5 ASTM D 1333, Determining Formaldehyde Levels from Wood-based Products Under Defined Test Conditions Using a Large Chamber

    4.6 ANSI 05.1, Wood Poles—Specifications and Dimensions

    4.7 ASTM D 25, Round Timber Piles

    4.8 ANSI/AHA A194.1, Cellulosic Fiber Insulating Board (Fiberboard)

    4.9 ANSI/AHA 135.6, Hardboard Siding

5. **Design standards.**

    5.1 ASTM D 5055, Structural Capacities of Prefabricated Wood I-Joists

    5.2 ANSI/TPI 1 National Design Standard for Metal Plate Connected Wood Truss Construction

    5.3 ANSI/TPI 2 Standard for Testing Performance for Metal Plate Connected Wood Trusses

    5.4 ASCE 16, Load and Resistance Factor Design Standard for Engineered Wood Construction

6. **Fire retardancy.**

    6.1 UBC Standard 23-4, Fire-retardant-treated Wood Tests on Durability and Hygroscopic Properties

    6.2 UBC Standard 23-5, Fire-retardant-treated Wood

7. **Adhesives and glues.**

    7.1 ASTM D 3024, Dry Use Adhesive with Protein Base, Casein Type

    7.2 ASTM D 2559, Wet Use Adhesives

    7.3 APA Specification AFG-01, Adhesives for Field Gluing Plywood to Wood Framing

    7.4 ASTM D 1101 and AITC 200 in Testing of Glue Joints in Laminated Wood Product

8. **Design values.**

    8.1 ASTM D 1990, Establishing Allowable Properties for Visually-Graded Dimension Lumber from In-Grade Tests of Full-Size Specimens

    8.2 ASTM D 245, Establishing Structural Grades and Related Allowable Properties for Visually Graded Lumber

    8.3 ASTM D 2555, Standard Test Methods for Establishing Clear Wood Strength Values

SECTION 2304 — MINIMUM QUALITY

2304.1 Quality and Identification. All lumber, wood structural panels, particleboard, structural glued-laminated timber, end-jointed lumber, fiberboard sheathing (when used structurally), hardboard siding (when used structurally), piles and poles regulated by this chapter shall conform to the applicable standards and grading rules specified in this code and shall be so identified by the grade mark or certificate of inspection issued by an approved agency.

All preservatively treated wood required to be treated under Section 2306 shall be identified by the quality mark of an inspection agency which has been accredited by an accreditation body which complies with the requirements of the American Lumber Standard Committee Treated Wood Program, or equivalent.

2304.2 Minimum Capacity or Grade. Minimum capacity of structural framing members may be established by performance tests. When tests are not made, capacity shall be based on allowable stresses and design criteria specified in this code.

Studs, joists, rafters, foundation plates or sills, planking 2 inches (51 mm) or more in depth, beams, stringers, posts, structural sheathing and similar load-bearing members shall be of at least the minimum grades set forth in the tables in this chapter.

Approved end-jointed lumber may be used interchangeably with solid-sawn members of the same species and grade. Such use shall include, but not be limited to, light-framing joists, planks and decking.

Wood structural panels shall be of the grades specified in UBC Standard 23-2 or 23-3.

2304.3 Timber Connectors and Fasteners. Safe loads and design practices for types of connectors and fasteners not mentioned or fully covered in Division III, Part III, may be determined in a manner approved by the building official.

The number and size of nails connecting wood members shall not be less than that set forth in Tables 23-II-B-1 and 23-II-B-2. Other connections shall be fastened to provide equivalent strength. End and edge distances and nail penetrations shall be in accordance with the applicable provisions of Division III, Part III.

Fasteners for pressure-preservative treated and fire-retardant treated wood shall be of hot-dipped zinc coated galvanized, stainless steel, silicon bronze or copper. Fasteners for wood foundations shall be as required in Chapter 18, Division II. Fasteners required to be corrosion resistant shall be either zinc-coated fasteners, aluminum alloy wire fasteners or stainless steel fasteners.

Connections depending on joist hangers or framing anchors, ties, and other mechanical fastenings not otherwise covered may be used where approved.

2304.4 Fabrication, Installation and Manufacture.

2304.4.1 General. Preparation, fabrication and installation of wood members and their fastenings shall conform to accepted engineering practices and to the requirements of this code. All members shall be framed, anchored, tied and braced to develop the strength and rigidity necessary for the purposes for which they are used.

2304.4.2 Timber connectors and fasteners. The installation of timber connectors and fasteners shall be in accordance with the provisions set forth in Division III, Part III.

2304.4.3 Structural glued-laminated timber. The manufacture and fabrication of structural glued-laminated timber shall be under the supervision of qualified personnel.

2304.4.4 Metal-plate-connected wood trusses. Metal-plate-connected wood trusses shall conform to the provisions of Division V. Each manufacturer of trusses using metal plate connectors shall retain an approved agency having no financial interest in the plant being inspected to make nonscheduled inspections of truss fabrication, delivery, and operations. The inspection shall cover all phases of truss operation, including lumber storage, handling, cutting, fixtures, presses or rollers, fabrication, bundling and banding, handling and delivery.

2304.5 Dried Fire-retardant-treated Wood. Approved fire-retardant-treated wood shall be dried, following treatment, to a maximum moisture content as follows: solid-sawn lumber 2 inches (51 mm) in thickness or less to 19 percent, and plywood to 15 percent.

2304.6 Size of Structural Members. Sizes of lumber and structural glued-laminated timber referred to in this code are nominal sizes. Computations to determine the required sizes of members shall be based on the net dimensions (actual sizes) and not the nominal sizes.

2304.7 Shrinkage. Consideration shall be given in design to the possible effect of cross-grain dimensional changes considered vertically which may occur in lumber fabricated in a green condition.

2304.8 Rejection. The building official may deny permission for the use of a wood member where permissible grade characteristics or defects are present in such a combination that they affect the serviceability of the member.

SECTION 2305 — DESIGN AND CONSTRUCTION REQUIREMENTS

2305.1 General. The following design requirements apply.

2305.2 All wood structures shall be designed and constructed in accordance with the requirements of Division I and Division II, Part I.

2305.3 Wind and earthquake load-resisting systems for all engineered wood structures shall be designed and constructed in accordance with the requirements of Division II, Part II.

2305.4 The design and construction of wood structures using allowable stress design methods shall be in accordance with Division III.

2305.5 The design and construction of conventional light-frame wood structures shall be in accordance with Division IV.

2305.6 The design and installation of timber connectors and fasteners shall be in accordance with Division III, Part III.

2305.7 Metal-plate-connected wood trusses shall conform to the provisions of Division V.

2305.8 Design of structural glued built-up members with plywood components shall be in accordance with Division VI.

2305.9 Design of joists and rafters shall be permitted to be in accordance with Division VII.

2305.10 Design of plank and beam flooring shall be permitted to be in accordance with Division VIII.

Division II—GENERAL REQUIREMENTS

Part I—REQUIREMENTS APPLICABLE TO ALL DESIGN METHODS

SECTION 2306 — DECAY AND TERMITE PROTECTION

2306.1 Preparation of Building Site. Site preparation shall be in accordance with Section 3302.

2306.2 Wood Support Embedded in Ground. Wood embedded in the ground or in direct contact with the earth and used for the support of permanent structures shall be treated wood unless continuously below the groundwater line or continuously submerged in fresh water. Round or rectangular posts, poles and sawn timber columns supporting permanent structures that are embedded in concrete or masonry in direct contact with earth or embedded in concrete or masonry exposed to the weather shall be treated wood. The wood shall be treated for ground contact.

2306.3 Under-floor Clearance. When wood joists or the bottom of wood structural floors without joists are located closer than 18 inches (457 mm) or wood girders are located closer than 12 inches (305 mm) to exposed ground in crawl spaces or unexcavated areas located within the periphery of the building foundation, the floor assembly, including posts, girders, joists and subfloor, shall be approved wood of natural resistance to decay as listed in Section 2306.4 or treated wood.

When the above under-floor clearances are required, the under-floor area shall be accessible. Accessible under-floor areas shall be provided with a minimum 18-inch-by-24-inch (457 mm by 610 mm) opening unobstructed by pipes, ducts and similar construction. All under-floor access openings shall be effectively screened or covered. Pipes, ducts and other construction shall not interfere with the accessibility to or within under-floor areas.

2306.4 Plates, Sills and Sleepers. All foundation plates or sills and sleepers on a concrete or masonry slab, which is in direct contact with earth, and sills that rest on concrete or masonry foundations, shall be treated wood or Foundation redwood, all marked or branded by an approved agency. Foundation cedar or No. 2 Foundation redwood marked or branded by an approved agency may be used for sills in territories subject to moderate hazard, where termite damage is not frequent and when specifically approved by the building official. In territories where hazard of termite damage is slight, any species of wood permitted by this code may be used for sills when specifically approved by the building official.

2306.5 Columns and Posts. Columns and posts located on concrete or masonry floors or decks exposed to the weather or to water splash or in basements and that support permanent structures shall be supported by concrete piers or metal pedestals projecting above floors unless approved wood of natural resistance to decay or treated wood is used. The pedestals shall project at least 6 inches (152 mm) above exposed earth and at least 1 inch (25 mm) above such floors.

Individual concrete or masonry piers shall project at least 8 inches (203 mm) above exposed ground unless the columns or posts that they support are of approved wood of natural resistance to decay or treated wood is used.

2306.6 Girders Entering Masonry or Concrete Walls. Ends of wood girders entering masonry or concrete walls shall be provided with a $^1/_2$-inch (12.7 mm) air space on tops, sides and ends unless approved wood of natural resistance to decay or treated wood is used.

2306.7 Under-floor Ventilation. Under-floor areas shall be ventilated by an approved mechanical means or by openings into the under-floor area walls. Such openings shall have a net area of not less than 1 square foot for each 150 square feet (0.067 m^2 for each 10 m^2) of under-floor area. Openings shall be located as close to corners as practical and shall provide cross ventilation. The required area of such openings shall be approximately equally distributed along the length of at least two opposite sides. They shall be covered with corrosion-resistant wire mesh with mesh openings of $^1/_4$ inch (6.4 mm) in dimension. Where moisture due to climate and groundwater conditions is not considered excessive, the building official may allow operable louvers and may allow the required net area of vent openings to be reduced to 10 percent of the above, provided the under-floor ground surface area is covered with an approved vapor retarder.

2306.8 Wood and Earth Separation. Protection of wood against deterioration as set forth in the previous sections for specified applications is required. In addition, wood used in construction of permanent structures and located nearer than 6 inches (152 mm) to earth shall be treated wood or wood of natural resistance to decay, as defined in Section 2302.1. Where located on concrete slabs placed on earth, wood shall be treated wood or wood of natural resistance to decay. Where not subject to water splash or to exterior moisture and located on concrete having a minimum thickness of 3 inches (76 mm) with an impervious membrane installed between concrete and earth, the wood may be untreated and of any species.

Where planter boxes are installed adjacent to wood frame walls, a 2-inch-wide (51 mm) air space shall be provided between the planter and the wall. Flashings shall be installed when the air space is less than 6 inches (152 mm) in width. Where flashing is used, provisions shall be made to permit circulation of air in the air space. The wood-frame wall shall be provided with an exterior wall covering conforming to the provisions of Section 2310.

2306.9 Wood Supporting Roofs and Floors. Wood structural members supporting moisture-permeable floors or roofs that are exposed to the weather, such as concrete or masonry slabs, shall be approved wood of natural resistance to decay or treated wood unless separated from such floors or roofs by an impervious moisture barrier.

2306.10 Moisture Content of Treated Wood. When wood pressure treated with a water-borne preservative is used in enclosed locations where drying in service cannot readily occur, such wood shall be at a moisture content of 19 percent or less before being covered with insulation, interior wall finish, floor covering or other material.

2306.11 Retaining Walls. Wood used in retaining or crib walls shall be treated wood.

2306.12 Weather Exposure. Those portions of glued-laminated timbers that form the structural supports of a building or other structure and are exposed to weather and not properly protected by a roof, eave overhangs of similar covering shall be pressure treated with an approved preservative or be manufactured from wood of natural resistance to decay.

All wood structural panels, when designed to be exposed in outdoor applications, shall be of exterior type, except as provided in Section 2306.2. In geographical areas where experience has demonstrated a specific need, approved wood of natural resistance to decay or treated wood shall be used for those portions of wood members which form the structural supports of buildings, balconies, porches or similar permanent building appurtenances when such members are exposed to the weather without adequate protection from a roof, eave, overhang or other covering to prevent moisture or water accumulation on the surface or at joints between members. Depending on local experience, such members may include horizontal members such as girders, joists and decking; or vertical members such as posts, poles and columns; or both horizontal and vertical members.

2306.13 Water Splash. Where wood-frame walls and partitions are covered on the interior with plaster, tile or similar materials and are subject to water splash, the framing shall be protected with approved waterproof paper conforming to Section 1402.1.

SECTION 2307 — WOOD SUPPORTING MASONRY OR CONCRETE

Wood members shall not be used to permanently support the dead load of any masonry or concrete.

> **EXCEPTIONS:** 1. Masonry or concrete nonstructural floor or roof surfacing not more than 4 inches (102 mm) thick may be supported by wood members.
>
> 2. Any structure may rest upon wood piles constructed in accordance with the requirements of Chapter 18.
>
> 3. Veneer of brick, concrete or stone applied as specified in Section 1403.6.2 may be supported by approved treated wood foundations when the maximum height of veneer does not exceed 30 feet (9144 mm) above the foundations. Such veneer used as an interior wall finish may also be supported on wood floors that are designed to support the additional load and designed to limit the deflection and shrinkage to $^1/_{600}$ of the span of the supporting members.
>
> 4. Glass block masonry having an installed weight of 20 pounds per square foot (97.6 kg/m^2) or less and installed with the provisions of Section 2109.5. When glass block is supported on wood floors, the floors shall be designed to limit deflection and shrinkage to $^1/_{600}$ of the span of the supporting members and the allowable stresses for the framing members shall be reduced in accordance with Division III, Part I.

See Division II, Part II for wood members resisting horizontal forces contributed by masonry or concrete.

SECTION 2308 — WALL FRAMING

The framing of exterior and interior walls shall be in accordance with provisions specified in Division IV unless a specific design is furnished.

Wood stud walls and bearing partitions shall not support more than two floors and a roof unless an analysis satisfactory to the building official shows that shrinkage of the wood framing will not have adverse effects on the structure or any plumbing, electrical or mechanical systems, or other equipment installed therein due to excessive shrinkage or differential movements caused by shrinkage. The analysis shall also show that the roof drainage system and the foregoing systems or equipment will not be adversely affected or, as an alternate, such systems shall be designed to accommodate the differential shrinkage or movements.

SECTION 2309 — FLOOR FRAMING

Wood-joisted floors shall be framed and constructed and anchored to supporting wood stud or masonry walls as specified in Chapter 16.

Fire block and draft stops shall be in accordance with Section 708.

SECTION 2310 — EXTERIOR WALL COVERINGS

2310.1 General. Exterior wood stud walls shall be covered on the outside with the materials and in the manner specified in this section or elsewhere in this code. Studs or sheathing shall be covered on the outside face with a weather-resistive barrier when required by Section 1402.1. Exterior wall coverings of the minimum thickness specified in this section are based on a maximum stud spacing of 16 inches (406 mm) unless otherwise specified.

2310.2 Siding. Solid wood siding shall have an average thickness of $^3/_8$ inch (9.5 mm) unless placed over sheathing permitted by this code.

Siding patterns known as rustic, drop siding or shiplap shall have an average thickness in place of not less than $^{19}/_{32}$ inch (15 mm) and shall have a minimum thickness of not less than $^3/_8$ inch (9.5 mm). Bevel siding shall have a minimum thickness measured at the butt section of not less than $^7/_{16}$ inch (11 mm) and a tip thickness of not less than $^3/_{16}$ inch (4.8 mm). Siding of lesser dimensions may be used, provided such wall covering is placed over sheathing which conforms to the provisions specified elsewhere in this code.

All weatherboarding or siding shall be securely nailed to each stud with not less than one nail, or to solid 1-inch (25 mm) nominal wood sheathing or $^{15}/_{32}$-inch (12 mm) wood structural panel sheathing or $^1/_2$-inch (13 mm) particleboard sheathing with not less than one line of nails spaced not more than 24 inches (610 mm) on center in each piece of the weatherboarding or siding.

Wood board sidings applied horizontally, diagonally or vertically shall be fastened to studs, nailing strips or blocking set at a maximum 24 inches (610 mm) on center. Fasteners shall be nails or screws with a penetration of not less than 1$^1/_2$ inches (38 mm) into studs, studs and wood sheathing combined, or blocking. Distance between such fastenings shall not exceed 24 inches (610 mm) for horizontally or vertically applied sidings and 32 inches (813 mm) for diagonally applied sidings.

2310.3 Plywood. When plywood is used for covering the exterior of outside walls, it shall be of the exterior type not less than $^3/_8$ inch (9.5 mm) thick. Plywood panel siding shall be installed in accordance with Table 23-II-A-1. Unless applied over 1-inch (25 mm) wood sheathing or $^{15}/_{32}$-inch (12 mm) wood structural panel sheathing or $^1/_2$-inch (13 mm) particleboard sheathing, joints shall occur over framing members and shall be protected with a continuous wood batten, approved caulking, flashing, vertical or horizontal shiplaps; or joints shall be lapped horizontally or otherwise made waterproof.

2310.4 Shingles or Shakes. Wood shingles or shakes and asbestos cement shingles may be used for exterior wall covering, provided the frame of the structure is covered with building paper as specified in Section 1402.1. All shingles or shakes attached to sheathing other than wood sheathing shall be secured with approved corrosion-resistant fasteners or on furring strips attached to the studs. Wood shingles or shakes may be applied over fiberboard shingle backer and sheathing with annular grooved nails. The thickness of wood shingles or shakes between wood nailing boards shall not be less than $^3/_8$ inch (9.5 mm). Wood shingles or shakes and asbestos shingles or siding may be nailed directly to approved fiberboard nailbase sheathing not less than $^1/_2$-inch (13 mm) nominal thickness with annular grooved nails.

The weather exposure of wood shingle or shake siding used on exterior walls shall not exceed maximums set forth in Table 23-II-K.

2310.5 Particleboard. When particleboard is used for covering the exterior of outside walls, it shall be of the M-1, M-S and M-2 Exterior Glue grades. Particleboard panel siding shall be installed in accordance with Tables 23-II-A-2 and 23-II-B-1. Panels shall be gapped $^1/_8$ inch (3.2 mm) and nails shall be spaced not less than $^3/_8$ inch (9.5 mm) from edges and ends of sheathing. Unless applied over $^5/_8$-inch (16 mm) net wood sheathing or $^1/_2$-inch (13 mm) plywood sheathing or $^1/_2$-inch (13 mm) particleboard sheathing, joints shall occur over framing members and shall be covered with a continuous wood batt; or joints shall be lapped horizontally or otherwise made waterproof to the satisfaction of the building official. Particleboard shall be sealed and protected with exterior quality finishes.

2310.6 Hardboard. When hardboard siding is used for covering the outside of exterior walls, it shall conform to Table 23-II-C. Lap siding shall be installed horizontally and applied to sheathed or unsheathed walls. Corner bracing shall be installed in conformance with Division IV. A weather-resistive barrier shall be installed under the lap siding as required by Section 1402.1.

Square-edged nongrooved panels and shiplap grooved or nongrooved siding shall be applied vertically to sheathed or unsheathed walls. Siding that is grooved shall not be less than $^1/_4$ inch (6.4 mm) thick in the groove.

Nail size and spacing shall follow Table 23-II-C and shall penetrate framing $1^1/_2$ inches (38 mm). Lap siding shall overlap 1 inch (25 mm) minimum and be nailed through both courses and into framing members with nails located $^1/_2$ inch (13 mm) from bottom of the overlapped course. Square-edged nongrooved panels shall be nailed $^3/_8$ inch (9.5 mm) from the perimeter of the panel and intermediately into studs. Shiplap edge panel siding with $^3/_8$-inch (9.5 mm) shiplap shall be nailed $^3/_8$ inch (9.5 mm) from the edges on both sides of the shiplap. The $^3/_4$-inch (19 mm) shiplap shall be nailed $^3/_8$ inch (9.5 mm) from the edge and penetrate through both the overlap and underlap. Top and bottom edges of the panel shall be nailed $^3/_8$ inch (9.5 mm) from the edge. Shiplap and lap siding shall not be force fit. Square-edged panels shall maintain a $^1/_{16}$-inch (1.6 mm) gap at joints. All joints and edges of siding shall be over framing members, and shall be made resistant to weather penetration with battens, horizontal overlaps or shiplaps to the satisfaction of the building official. A $^1/_8$-inch (3.2 mm) gap shall be provided around all openings.

2310.7 Nailing. All fasteners used for the attachment of siding shall be of a corrosion-resistant type.

SECTION 2311 — INTERIOR PANELING

All softwood wood structural panels shall conform with the provisions of Chapter 8 and shall be installed in accordance with Table 23-II-B-1. Panels shall comply with UBC Standard 23-3.

SECTION 2312 — SHEATHING

2312.1 Structural Floor Sheathing. Structural floor sheathing shall be designed in accordance with the general provisions of this code and the special provisions in this section.

Sheathing used as subflooring shall be designed to support all loads specified in this code and shall be capable of supporting concentrated loads of not less than 300 pounds (1334 N) without failure. The concentrated load shall be applied by a loaded disc, 3 inches (76 mm) or smaller in diameter.

Flooring, including the finish floor, underlayment and subfloor, where used, shall meet the following requirements:

1. Deflection under uniform design load limited to $^1/_{360}$ of the span between supporting joists or beams.

2. Deflection of flooring relative to joists under a 1-inch-diameter (25 mm) concentrated load of 200 pounds (890 N) limited to 0.125 inch (3.2 mm) or less when loaded midway between supporting joists or beams not over 24 inches (610 mm) on center and $^1/_{360}$ of the span for spans over 24 inches (610 mm).

Floor sheathing conforming to the provisions of Table 23-II-D-1, 23-II-D-2, 23-II-E-1, 23-II-F-1 or 23-II-F-2 shall be deemed to meet the requirements of this section.

2312.2 Structural Roof Sheathing. Structural roof sheathing shall be designed in accordance with the general provisions of this code and the special provisions in this section. Structural roof sheathing shall be designed to support all loads specified in this code and shall be capable of supporting concentrated loads of not less than 300 pounds (1334 N) without failure. The concentrated load shall be applied by a loaded disc, 3 inches (76 mm) or smaller in diameter. Structural roof sheathing shall meet the following requirement:

1. Deflection under uniform design live and dead load limited to $^1/_{180}$ of the span between supporting rafters or beams and $^1/_{240}$ under live load only.

Roof sheathing conforming to the provisions of Tables 23-II-D-1 and 23-II-D-2 or 23-II-E-1 and 23-II-E-2 shall be deemed to meet the requirements of this section.

Wood structural panel roof sheathing shall be bonded by intermediate or exterior glue. Wood structural panel roof sheathing exposed on the underside shall be bonded with exterior glue.

SECTION 2313 — MECHANICALLY LAMINATED FLOORS AND DECKS

A laminated lumber floor or deck built up of wood members set on edge, when meeting the following requirements, may be designed as a solid floor or roof deck of the same thickness, and continuous spans may be designed on the basis of the full cross section using the simple span moment coefficient.

Nail length shall not be less than two and one-half times the net thickness of each lamination. When deck supports are 4 feet (1219 mm) on center or less, side nails shall be spaced not more than 30 inches (762 mm) on center and staggered one third of the spacing in adjacent laminations. When supports are spaced more than 4 feet (1219 mm) on center, side nails shall be spaced not more than 18 inches (457 mm) on center alternately near top and bottom edges, and also staggered one third of the spacing in adjacent laminations. Two side nails shall be used at each end of butt-jointed pieces.

Laminations shall be toenailed to supports with 20d or larger common nails. When the supports are 4 feet (1219 mm) on center or less, alternate laminations shall be toenailed to alternate supports; when supports are spaced more than 4 feet (1219 mm) on center, alternate laminations shall be toenailed to every support.

A single-span deck shall have all laminations full length.

A continuous deck of two spans shall not have more than every fourth lamination spliced within quarter points adjoining supports.

Joints shall be closely butted over supports or staggered across the deck but within the adjoining quarter spans.

No lamination shall be spliced more than twice in any span.

SECTION 2314 — POST-BEAM CONNECTIONS

Where post and beam or girder construction is used, the design shall be in accordance with the provisions of this code. Positive connection shall be provided to ensure against uplift and lateral displacement.

Part II—REQUIREMENTS APPLICABLE TO ENGINEERED DESIGN OF WIND AND EARTHQUAKE LOAD-RESISTING SYSTEMS

SECTION 2315 — WOOD SHEAR WALLS AND DIAPHRAGMS

2315.1 General. Particleboard vertical diaphragms and lumber and wood structural panel horizontal and vertical diaphragms may be used to resist horizontal forces in horizontal and vertical distributing or resisting elements, provided the deflection in the plane of the diaphragm, as determined by calculations, tests or analogies drawn therefrom, does not exceed the permissible deflection of attached distributing or resisting elements. See UBC Standard 23-2 for a method of calculating the deflection of a blocked wood structural panel diaphragm.

Permissible deflection shall be that deflection up to which the diaphragm and any attached distributing or resisting element will maintain its structural integrity under assumed load conditions, i.e., continue to support assumed loads without danger to occupants of the structure.

Connections and anchorages capable of resisting the design forces shall be provided between the diaphragms and the resisting elements. Openings in diaphragms that materially affect their strength shall be fully detailed on the plans and shall have their edges adequately reinforced to transfer all shearing stresses.

Size and shape of each horizontal diaphragm and shear wall shall be limited as set forth in Table 23-II-G. The height of a shear wall shall be defined as:

1. The maximum clear height from foundation to bottom of diaphragm framing above, or

2. The maximum clear height from top of diaphragm to bottom of diaphragm framing above.

The width of a shear wall shall be defined as the width of sheathing. See Figure 23-II-1, Section (a).

Where shear walls with openings are designed for force transfer around the openings, the limitations of Table 23-II-G shall apply to the overall shear wall including openings and to each wall pier at the side of an opening. The height of a wall pier shall be defined as the clear height of the pier at the side of an opening. The width of a wall pier shall be defined as the sheathed width of the pier at the side of an opening. Design for force transfer shall be based on a rational analysis. Detailing of boundary members around the opening shall be provided in accordance with Section 2315. See Figure 23-II-1, Section (b).

In buildings of wood-frame construction where rotation is provided for, the depth of the diaphragm normal to the open side shall not exceed 25 feet (7620 mm) or two thirds the diaphragm width, whichever is the smaller depth. Straight sheathing shall not be permitted to resist shears in diaphragms acting in rotation.

> **EXCEPTIONS:** 1. One-story, wood-framed structures with the depth normal to the open side not greater than 25 feet (7620 mm) may have a depth equal to the width.
>
> 2. Where calculations show that diaphragm deflections can be tolerated, the depth normal to the open end may be increased to a depth-to-width ratio not greater than $1^1/_2$:1 for diagonal sheathing or 2:1 for special diagonal sheathed or wood structural panel or particle board diaphragms.

In masonry or concrete buildings, lumber and wood structural panel diaphragms shall not be considered as transmitting lateral forces by rotation.

Diaphragm sheathing nails or other approved sheathing connectors shall be driven so that their head or crown is flush with the surface of the sheathing.

2315.2 Wood Members Resisting Horizontal Forces Contributed by Masonry and Concrete. Wood members shall not be used to resist horizontal forces contributed by masonry or concrete construction in buildings over one story in height.

> **EXCEPTIONS:** 1. Wood floor and roof members may be used in horizontal trusses and diaphragms to resist horizontal forces imposed by wind, earthquake or earth pressure, provided such forces are not resisted by rotation of the truss or diaphragm. See Section 2315.1.
>
> 2. Vertical wood structural panel-sheathed shear walls may be used to provide resistance to wind or earthquake forces in two-story buildings of masonry or concrete construction, provided the following requirements are met:
>
> 2.1 Story-to-story wall heights shall not exceed 12 feet (3658 mm).
>
> 2.2 Horizontal diaphragms shall not be considered to transmit lateral forces by rotation or cantilever action.
>
> 2.3 Deflections of horizontal and vertical diaphragms shall not permit per-story deflections of supported masonry or concrete walls to exceed 0.005 times each story height.
>
> 2.4 Wood structural panel sheathing in horizontal diaphragms shall have all unsupported edges blocked. Wood structural panel sheathing for both stories of vertical diaphragms shall have all unsupported edges blocked and for the lower story walls shall have a minimum thickness of $^{15}/_{32}$ inch (12 mm).
>
> 2.5 There shall be no out-of-plane horizontal offsets between the first and second stories of wood structural panel shear walls.

2315.3 Wood Diaphragms.

2315.3.1 Conventional lumber diaphragm construction. Such lumber diaphragms shall be made up of 1-inch (25 mm) nominal sheathing boards laid at an angle of approximately 45 degrees to supports. Sheathing boards shall be directly nailed to each intermediate bearing member with not less than two 8d nails for 1-inch-by-6-inch (25 mm by 152 mm) nominal boards and three 8d nails for boards 8 inches (203 mm) or wider; and three 8d nails and four 8d nails shall be used for 6-inch and 8-inch (152 mm and 203 mm) boards, respectively, at the diaphragm boundaries. End joints in adjacent boards shall be separated by at least one joist or stud space, and there shall be at least two boards between joints on the same support. Boundary members at edges of diaphragms shall be designed to resist direct tensile or compressive chord stresses and adequately tied together at corners.

2315.3.2 Special lumber diaphragm construction. Special diagonally sheathed diaphragms shall conform to conventional lumber diaphragm construction and shall have all elements designed in conformance with the provisions of this code.

Each chord or portion thereof may be considered as a beam loaded with a uniform load per foot equal to 50 percent of the unit shear due to diaphragm action. The load shall be assumed as acting normal to the chord, in the plane of the diaphragm, and either toward or away from the diaphragm. The span of the chord, or portion thereof, shall be the distance between structural members of the diaphragm, such as the joists, studs and blocking, which serve to transfer the assumed load to the sheathing.

Special diagonally sheathed diaphragms shall include conventional diaphragms sheathed with two layers of diagonal sheathing at 90 degrees to each other and on the same face of the supporting members.

2315.3.3 Wood structural panel diaphragms. Horizontal and vertical diaphragms sheathed with wood structural panels may be used to resist horizontal forces not exceeding those set forth in Table 23-II-H for horizontal diaphragms and Table 23-II-I-1 for vertical diaphragms, or may be calculated by principles of mechanics without limitation by using values of nail strength and wood structural panel shear values as specified elsewhere in this code. Wood structural panels for horizontal diaphragms shall be as set forth in Tables 23-II-E-1 and 23-II-E-2 for corresponding joist spacing and loads. Wood structural panels in shear walls shall be at least $^5/_{16}$ inch (7.9 mm) thick for studs spaced 16 inches (406 mm) on center and $^3/_8$ inch (9.5 mm) thick where studs are spaced 24 inches (610 mm) on center.

Maximum spans for wood structural panel subfloor underlayment shall be as set forth in Table 23-II-F-1. Wood structural panels used for horizontal and vertical diaphragms shall conform to UBC Standard 23-2 or 23-3.

All boundary members shall be proportioned and spliced where necessary to transmit direct stresses. Framing members shall be at least 2-inch (51 mm) nominal in the dimension to which the wood structural panel is attached. In general, panel edges shall bear on the framing members and butt along their center lines. Nails shall be placed not less than $^3/_8$ inch (9.5 mm) in from the panel edge, shall be spaced not more than 6 inches (152 mm) on center along panel edge bearings, and shall be firmly driven into the framing members. No unblocked panels less than 12 inches (305 mm) wide shall be used.

Diaphragms with panel edges supported in accordance with Tables 23-II-E-1, 23-II-E-2 and 23-II-F-1 shall not be considered as blocked diaphragms unless blocking or other means of shear transfer is provided.

2315.4 Particleboard Diaphragms. Vertical diaphragms sheathed with particleboard may be used to resist horizontal forces not exceeding those set forth in Table 23-II-I-2.

All boundary members shall be proportioned and spliced where necessary to transmit direct stresses. Framing members shall be at least 2-inch (51 mm) nominal in the dimension to which the particleboard is attached. In general, panel edges shall bear on the framing members and butt along their center lines. Nails shall be placed not less than $^3/_8$ inch (9.5 mm) in from the panel edge, shall be spaced not more than 6 inches (152 mm) on center along panel edge bearings, and shall be firmly driven into the framing members. No unblocked panels less than 12 inches (305 mm) wide shall be used.

Diaphragms with panel edges supported in accordance with Table 23-II-F-2 shall not be considered as blocked diaphragms unless blocking or other means of shear transfer is provided.

2315.5 Wood Shear Walls and Diaphragms in Seismic Zones 3 and 4.

2315.5.1 Scope. Design and construction of wood shear walls and diaphragms in Seismic Zones 3 and 4 shall conform to the requirements of this section.

2315.5.2 Framing. Collector members shall be provided to transmit tension and compression forces. Perimeter members at openings shall be provided and shall be detailed to distribute the shearing stresses. Diaphragm sheathing shall not be used to splice these members.

Diaphragm chords and ties shall be placed in, or tangent to, the plane of the diaphragm framing unless it can be demonstrated that the moments, shears and deflections and deformations resulting from other arrangements can be tolerated.

2315.5.3 Wood structural panels. Wood structural panels shall be manufactured using exterior glue.

Wood structural panel diaphragms and shear walls shall be constructed with wood structural panel sheets not less than 4 feet by 8 feet (1219 mm by 2438 mm), except at boundaries and changes in framing where minimum sheet dimension shall be 24 inches (610 mm) unless all edges of the undersized sheets are supported by framing members or blocking.

Framing members or blocking shall be provided at the edges of all sheets in shear walls.

Wood structural panel sheathing may be used for splicing members, other than those noted in Section 2315.5.2, where the additional nailing required to develop the transfer of forces will not cause cross-grain bending or cross-grain tension in the nailed member.

2315.5.4 Heavy wood panels. Diagonally sheathed panels utilizing 2-inch (51 mm) nominal boards may be used to resist the same permissible shears as 1-inch (25 mm) nominal lumber, except that 16d nails shall be used instead of 8d nails.

Panels utilizing straight decking overlaid with wood structural panels may be used to resist shear forces using the same shear values as permitted for the wood structural panel alone. Wood structural panel joints parallel to the decking shall be located at least 1 inch (25 mm) offset from any parallel decking joint.

Heavy decking panels utilizing dowel pins, or vertically laminated panels connected by nailing units to each other, resist shear forces based on the permissible shear values of their connectors.

2315.5.5 Particleboard. Particleboard shall not be less than Type M "Exterior Glue."

Shear walls shall be sheathed with particleboard sheets not less than 4 feet by 8 feet (1219 mm by 2438 mm) except at boundaries and changes in framing. The required nail size and spacing in Table 23-II-B-1 apply to panel edges only. All panel edges shall be backed with 2-inch (51 mm) nominal or wider framing. Sheets are permitted to be installed either horizontally or vertically. For $^3/_8$-inch (9.5 mm) particleboard sheets installed with the long dimension parallel to the studs spaced 24 inches (610 mm) on center, nails shall be spaced at 6 inches (152 mm) on center along intermediate framing members. For all other conditions, nails of the same size shall be spaced at 12 inches (305 mm) on center along intermediate framing members.

2315.6 Fiberboard Sheathing Diaphragms. Wood stud walls sheathed with fiberboard sheathing may be used to resist horizontal forces not exceeding those set forth in Division III, Part IV. The fiberboard sheathing, 4 feet by 8 feet (1219 mm by 2438 mm), shall be applied vertically to wood studs not less than 2-inch (51 mm) nominal in thickness spaced 16 inches (406 mm) on center. Nailing shown in Table 23-II-J shall be provided at the perimeter of the sheathing board and at intermediate studs. Blocking not less than 2-inch (51 mm) nominal in thickness shall be provided at horizontal joints when wall height exceeds length of sheathing panel, and sheathing shall be fastened to the blocking with nails sized as shown in Table 23-II-J spaced 3 inches (76 mm) on centers each side of joint. Nails shall be spaced not less than $^3/_8$ inch (9.5 mm) from edges and ends of sheathing. Marginal studs of shear walls or shear-resisting elements shall be adequately anchored at top and bottom and designed to resist all forces. The maximum height width ratio shall be $1^1/_2$:1.

TABLE 23-II-A-1—EXPOSED PLYWOOD PANEL SIDING

MINIMUM THICKNESS[1] (inch) × 25.4 for mm	MINIMUM NUMBER OF PLIES	STUD SPACING (inches) PLYWOOD SIDING APPLIED DIRECTLY TO STUDS OR OVER SHEATHING × 25.4 for mm
$3/8$	3	16^2
$1/2$	4	24

[1]Thickness of grooved panels is measured at bottom of grooves.
[2]May be 24 inches (610 mm) if plywood siding applied with face grain perpendicular to studs or over one of the following: (1) 1-inch (25 mm) board sheathing, (2) $7/16$-inch (11 mm) wood structural panel sheathing or (3) $3/8$-inch (9.5 mm) wood structural panel sheathing with strength axis (which is the long direction of the panel unless otherwise marked) of sheathing perpendicular to studs.

TABLE 23-II-A-2—ALLOWABLE SPANS FOR EXPOSED PARTICLEBOARD PANEL SIDING

GRADE	STUD SPACING (inches) × 25.4 for mm	MINIMUM THICKNESS (inches) × 25.4 for mm Siding Direct to Studs	MINIMUM THICKNESS (inches) × 25.4 for mm Siding Continuous Support	Exterior Ceilings and Soffits Direct to Supports
M-1	16	$5/8$	$3/8$	$3/8$
M-S M-2 "Exterior Glue"	24	$5/8$	$3/8$	$3/8$

TABLE 23-II-B-1   **1997 UNIFORM BUILDING CODE**

TABLE 23-II-B-1—NAILING SCHEDULE

CONNECTION	NAILING[1]
1. Joist to sill or girder, toenail	3-8d
2. Bridging to joist, toenail each end	2-8d
3. 1″ × 6″ (25 mm × 152 mm) subfloor or less to each joist, face nail	2-8d
4. Wider than 1″ × 6″ (25 mm × 152 mm) subfloor to each joist, face nail	3-8d
5. 2″ (51 mm) subfloor to joist or girder, blind and face nail	2-16d
6. Sole plate to joist or blocking, typical face nail Sole plate to joist or blocking, at braced wall panels	16d at 16″ (406 mm) o.c. 3-16d per 16″ (406 mm)
7. Top plate to stud, end nail	2-16d
8. Stud to sole plate	4-8d, toenail or 2-16d, end nail
9. Double studs, face nail	16d at 24″ (610 mm) o.c.
10. Doubled top plates, typical face nail Double top plates, lap splice	16d at 16″ (406 mm) o.c. 8-16d
11. Blocking between joists or rafters to top plate, toenail	3-8d
12. Rim joist to top plate, toenail	8d at 6″ (152 mm) o.c.
13. Top plates, laps and intersections, face nail	2-16d
14. Continuous header, two pieces	16d at 16″ (406 mm) o.c. along each edge
15. Ceiling joists to plate, toenail	3-8d
16. Continuous header to stud, toenail	4-8d
17. Ceiling joists, laps over partitions, face nail	3-16d
18. Ceiling joists to parallel rafters, face nail	3-16d
19. Rafter to plate, toenail	3-8d
20. 1″ (25 mm) brace to each stud and plate, face nail	2-8d
21. 1″ × 8″ (25 mm × 203 mm) sheathing or less to each bearing, face nail	2-8d
22. Wider than 1″ × 8″ (25 mm × 203 mm) sheathing to each bearing, face nail	3-8d
23. Built-up corner studs	16d at 24″ (610 mm) o.c.
24. Built-up girder and beams	20d at 32″ (813 mm) o.c. at top and bottom and staggered 2-20d at ends and at each splice
25. 2″ (51 mm) planks	2-16d at each bearing

26. Wood structural panels and particleboard:[2]
   Subfloor and wall sheathing (to framing):
      $^1/_2$″ (12.7 mm) and less
      $^{19}/_{32}$″-$^3/_4$″ (15 mm-19 mm)
      $^7/_8$″-1″ (22 mm-25 mm)
      $1^1/_8$″-$1^1/_4$″ (29 mm-32 mm)
   Combination subfloor-underlayment (to framing):
      $^3/_4$″ (19 mm) and less
      $^7/_8$″-1″ (22 mm-25 mm)
      $1^1/_8$″-$1^1/_4$″ (29 mm-32 mm)
	
	6d[3]
	8d[4] or 6d[5]
	8d[3]
	10d[4] or 8d[5]
	
	6d[5]
	8d[5]
	10d[4] or 8d[5]

27. Panel siding (to framing)[2]:
    $^1/_2$″ (12.7 mm) or less
    $^5/_8$″ (16 mm)
	
	6d[6]
	8d[6]

28. Fiberboard sheathing:[7]
    $^1/_2$″ (12.7 mm)
 	
    $^{25}/_{32}$″ (20 mm)
	
	No. 11 ga.[8]
	6d[4]
	No. 16 ga.[9]
	No. 11 ga.[8]
	8d[4]
	No. 16 ga.[9]

29. Interior paneling
    $^1/_4$″ (6.4 mm)
    $^3/_8$″ (9.5 mm)
	
	4d[10]
	6d[11]

[1]Common or box nails may be used except where otherwise stated.

[2]Nails spaced at 6 inches (152 mm) on center at edges, 12 inches (305 mm) at intermediate supports except 6 inches (152 mm) at all supports where spans are 48 inches (1219 mm) or more. For nailing of wood structural panel and particleboard diaphragms and shear walls, refer to Sections 2315.3.3 and 2315.4. Nails for wall sheathing may be common, box or casing.

[3]Common or deformed shank.

[4]Common.

[5]Deformed shank.

[6]Corrosion-resistant siding or casing nails conforming to the requirements of Section 2304.3.

[7]Fasteners spaced 3 inches (76 mm) on center at exterior edges and 6 inches (152 mm) on center at intermediate supports.

[8]Corrosion-resistant roofing nails with $^7/_{16}$-inch-diameter (11 mm) head and $1^1/_2$-inch (38 mm) length for $^1/_2$-inch (12.7 mm) sheathing and $1^3/_4$-inch (44 mm) length for $^{25}/_{32}$-inch (20 mm) sheathing conforming to the requirements of Section 2304.3.

[9]Corrosion-resistant staples with nominal $^7/_{16}$-inch (11 mm) crown and $1^1/_8$-inch (29 mm) length for $^1/_2$-inch (12.7 mm) sheathing and $1^1/_2$-inch (38 mm) length for $^{25}/_{32}$-inch (20 mm) sheathing conforming to the requirements of Section 2304.3.

[10]Panel supports at 16 inches (406 mm) [20 inches (508 mm) if strength axis in the long direction of the panel, unless otherwise marked]. Casing or finish nails spaced 6 inches (152 mm) on panel edges, 12 inches (305 mm) at intermediate supports.

[11]Panel supports at 24 inches (610 mm). Casing or finish nails spaced 6 inches (152 mm) on panel edges, 12 inches (305 mm) at intermediate supports.

TABLE 23-II-B-2—WOOD STRUCTURAL PANEL ROOF SHEATHING NAILING SCHEDULE[1]

WIND REGION	NAILS	PANEL LOCATION	ROOF FASTENING ZONE[2]		
			1	2	3
			Fastening Schedule (inches on center)		
			× 25.4 for mm		
Greater than 90 mph (145 km/h)	8d common	Panel edges[3]	6	6	4[4]
		Panel field	6	6	6[4]
Greater than 80 mph (129 km/h) to 90 mph (145 km/h)	8d common	Panel edges[3]	6	6	4
		Panel field	12	6	6
80 mph (129 km/h) or less	8d common	Panel edges[3]	6	6	6
		Panel field	12	12	12

[1]Applies only to mean roof heights up to 35 feet (10 700 mm). For mean roof heights over 35 feet (10 700 mm), the nailing shall be designed.
[2]The roof fastening zones are shown below:

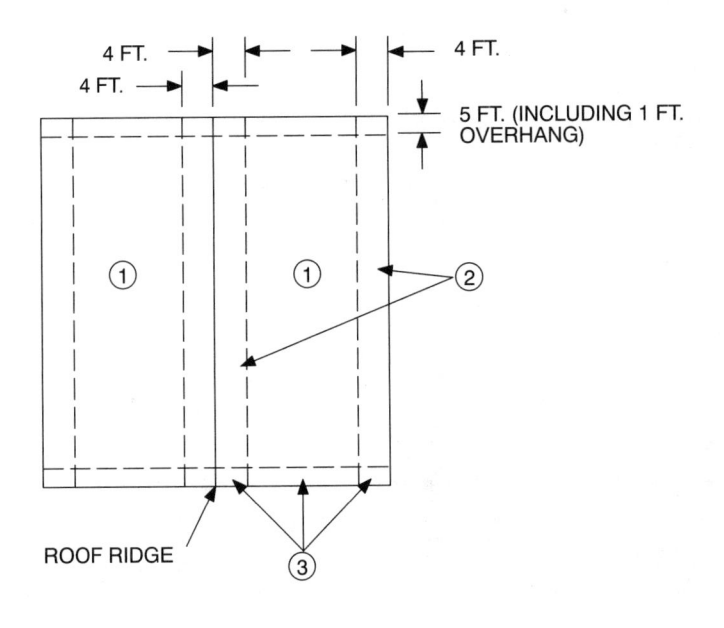

ROOF FASTENING ZONES

For **SI:** 1 foot = 304.8 mm.

[3]Edge spacing also applies over roof framing at gable-end walls.
[4]Use 8d ring-shank nails in this zone if mean roof height is greater than 25 feet (7600 mm).

TABLE 23-II-C
TABLE 23-II-D-2

1997 UNIFORM BUILDING CODE

TABLE 23-II-C—HARDBOARD SIDING

SIDING	MINIMAL NOMINAL THICKNESS (inch)	FRAMING (2″ x 4″) MAXIMUM SPACING	NAIL SIZE[1, 2]	NAIL SPACING	
				General	Bracing Panels[3]
				× 25.4 for mm	
1. LAP SIDING					
Direct to studs	$^3/_8$	16″ o.c.	8d	16″ o.c.	Not applicable
Over sheathing	$^3/_8$	16″ o.c.	10d	16″ o.c.	Not applicable
2. SQUARE EDGE PANEL SIDING					
Direct to studs	$^3/_8$	24″ o.c.	6d	6″ o.c. edges; 12″ o.c. at intermed. supports	4″ o.c. edges; 8″ o.c. intermed. supports
Over sheathing	$^3/_8$	24″ o.c.	8d	6″ o.c. edges; 12″ o.c. at intermed. supports	4″ o.c. edges; 8″ o.c. intermed. supports
3. SHIPLAP EDGE PANEL SIDING					
Direct to studs	$^3/_8$	16″ o.c.	6d	6″ o.c. edges; 12″ o.c. at intermed. supports	4″ o.c. edges; 8″ o.c. intermed. supports
Over sheathing	$^3/_8$	16″ o.c.	8d	6″ o.c. edges; 12″ o.c. at intermed. supports	4″ o.c. edges; 8″ o.c. intermed. supports

[1]Nails shall be corrosion resistant in accordance with Division III, Part III.
[2]Minimum acceptable nail dimensions (inches).

	Panel Siding (inch)	Lap Siding (inch)
	× 25.4 for mm	
Shank diameter	0.092	0.099
Head diameter	0.225	0.240

[3]When used to comply with Division IV, Section 2320.11.3.

TABLE 23-II-D-1—ALLOWABLE SPANS FOR LUMBER FLOOR AND ROOF SHEATHING[1, 2]

SPAN (inches)	MINIMUM NET THICKNESS (inches) OF LUMBER PLACED			
	Perpendicular to Supports		Diagonally to Supports	
	× 25.4 for mm			
× 25.4 for mm	Surfaced Dry[3]	Surfaced Unseasoned	Surfaced Dry[3]	Surfaced Unseasoned
	Floors			
1. 24	$^3/_4$	$^{25}/_{32}$	$^3/_4$	$^{25}/_{32}$
2. 16	$^5/_8$	$^{11}/_{16}$	$^5/_8$	$^{11}/_{16}$
	Roofs			
3. 24	$^5/_8$	$^{11}/_{16}$	$^3/_4$	$^{25}/_{32}$

[1]Installation details shall conform to Sections 2320.9.1 and 2320.12.8 for floor and roof sheathing, respectively.
[2]Floor or roof sheathing conforming with this table shall be deemed to meet the design criteria of Section 2312.
[3]Maximum 19 percent moisture content.

TABLE 23-II-D-2—SHEATHING LUMBER SHALL MEET THE FOLLOWING MINIMUM GRADE REQUIREMENTS: BOARD GRADE

SOLID FLOOR OR ROOF SHEATHING	SPACED ROOF SHEATHING	GRADING RULES
1. Utility	Standard	NLGA, WCLIB, WWPA
2. 4 common or utility	3 common or standard	NLGA, WCLIB, WWPA, NHPMA or NELMA
3. No. 3	No. 2	SPIB
4. Merchantable	Construction common	RIS

TABLE 23-II-E-1—ALLOWABLE SPANS AND LOADS FOR WOOD STRUCTURAL PANEL SHEATHING AND SINGLE-FLOOR GRADES CONTINUOUS OVER TWO OR MORE SPANS WITH STRENGTH AXIS PERPENDICULAR TO SUPPORTS[1,2]

SHEATHING GRADES		ROOF[3]				FLOOR[4]
		Maximum Span (inches)		Load[5] (pounds per square foot)		
		× 25.4 for mm		× 0.0479 for kN/m²		
Panel Span Rating	Panel Thickness (inches)					Maximum Span (inches)
Roof/Floor Span	× 25.4 for mm	With Edge Support[6]	Without Edge Support	Total Load	Live Load	× 25.4 for mm
12/0	5/16	12	12	40	30	0
16/0	5/16, 3/8	16	16	40	30	0
20/0	5/16, 3/8	20	20	40	30	0
24/0	3/8, 7/16, 1/2	24	20[7]	40	30	0
24/16	7/16, 1/2	24	24	50	40	16
32/16	15/32, 1/2, 5/8	32	28	40	30	16[8]
40/20	19/32, 5/8, 3/4, 7/8	40	32	40	30	20[8,9]
48/24	23/32, 3/4, 7/8	48	36	45	35	24
54/32	7/8, 1	54	40	45	35	32
60/48	7/8, 1, 1 1/8	60	48	45	35	48

SINGLE-FLOOR GRADES		ROOF[3]				FLOOR[4]
		Maximum Span (inches)		Load[5] (pounds per square foot)		
		× 25.4 for mm		× 0.0479 for kN/m²		
Panel Span Rating (inches)	Panel Thickness (inches)					Maximum Span (inches)
× 25.4 for mm		With Edge Support[6]	Without Edge Support	Total Load	Live Load	× 25.4 for mm
16 oc	1/2, 19/32, 5/8	24	24	50	40	16[8]
20 oc	19/32, 5/8, 3/4	32	32	40	30	20[8,9]
24 oc	23/32, 3/4	48	36	35	25	24
32 oc	7/8, 1	48	40	50	40	32
48 oc	1 3/32, 1 1/8	60	48	50	50	48

[1]Applies to panels 24 inches (610 mm) or wider.
[2]Floor and roof sheathing conforming with this table shall be deemed to meet the design criteria of Section 2312.
[3]Uniform load deflection limitations 1/180 of span under live load plus dead load, 1/240 under live load only.
[4]Panel edges shall have approved tongue-and-groove joints or shall be supported with blocking unless 1/4-inch (6.4 mm) minimum thickness underlayment or 1 1/2 inches (38 mm) of approved cellular or lightweight concrete is placed over the subfloor, or finish floor is 3/4-inch (19 mm) wood strip. Allowable uniform load based on deflection of 1/360 of span is 100 pounds per square foot (psf) (4.79 kN/m²) except the span rating of 48 inches on center is based on a total load of 65 psf (3.11 kN/m).
[5]Allowable load at maximum span.
[6]Tongue-and-groove edges, panel edge clips [one midway between each support, except two equally spaced between supports 48 inches (1219 mm) on center], lumber blocking, or other. Only lumber blocking shall satisfy blocked diaphragms requirements.
[7]For 1/2-inch (12.7 mm) panel, maximum span shall be 24 inches (610 mm).
[8]May be 24 inches (610 mm) on center where 3/4-inch (19 mm) wood strip flooring is installed at right angles to joist.
[9]May be 24 inches (610 mm) on center for floors where 1 1/2 inches (38 mm) of cellular or lightweight concrete is applied over the panels.

TABLE 23-II-E-2—ALLOWABLE LOAD (PSF) FOR WOOD STRUCTURAL PANEL ROOF SHEATHING CONTINUOUS OVER TWO OR MORE SPANS AND STRENGTH AXIS PARALLEL TO SUPPORTS
(Plywood structural panels are five-ply, five-layer unless otherwise noted.)[1,2]

PANEL GRADE	THICKNESS (inch)	MAXIMUM SPAN (inches)	LOAD AT MAXIMUM SPAN (psf)	
			× 0.0479 for kN/m²	
	× 25.4 for mm		Live	Total
Structural I	7/16	24	20	30
	15/32	24	35[3]	45[3]
	1/2	24	40[3]	50[3]
	19/32, 5/8	24	70	80
	23/32, 3/4	24	90	100
Other grades covered in UBC Standard 23-2 or 23-3	7/16	16	40	50
	15/32	24	20	25
	1/2	24	25	30
	19/32	24	40[3]	50[3]
	5/8	24	45[3]	55[3]
	23/32, 3/4	24	60[3]	65[3]

[1]Roof sheathing conforming with this table shall be deemed to meet the design criteria of Section 2312.
[2]Uniform load deflection limitations: 1/180 of span under live load plus dead load, 1/240 under live load only. Edges shall be blocked with lumber or other approved type of edge supports.
[3]For composite and four-ply plywood structural panel, load shall be reduced by 15 pounds per square foot (0.72 kN/m²).

TABLE 23-II-F-1
TABLE 23-II-G

1997 UNIFORM BUILDING CODE

TABLE 23-II-F-1—ALLOWABLE SPAN FOR WOOD STRUCTURAL PANEL COMBINATION SUBFLOOR-UNDERLAYMENT (SINGLE FLOOR)[1,2] Panels Continuous over Two or More Spans and Strength Axis Perpendicular to Supports

IDENTIFICATION	MAXIMUM SPACING OF JOISTS (inches)				
	× 25.4 for mm				
	16	20	24	32	48
Species Group[3]	Thickness (inches)				
	× 25.4 for mm				
1	$^1/_2$	$^5/_8$	$^3/_4$	—	—
2, 3	$^5/_8$	$^3/_4$	$^7/_8$	—	—
4	$^3/_4$	$^7/_8$	1	—	—
Span rating[4]	16 o.c.	20 o.c.	24 o.c.	32 o.c.	48 o.c.

[1]Spans limited to value shown because of possible effects of concentrated loads. Allowable uniform loads based on deflection of $^1/_{360}$ of span is 100 pounds per square foot (psf) (4.79 kN/m^2), except allowable total uniform load for $1^1/_8$-inch (29 mm) wood structural panels over joists spaced 48 inches (1219 mm) on center is 65 psf (3.11 kN/m^2). Panel edges shall have approved tongue-and-groove joints or shall be supported with blocking, unless $^1/_4$-inch (6.4 mm) minimum thickness underlayment or $1^1/_2$ inches (38 mm) of approved cellular or lightweight concrete is placed over the subfloor, or finish floor is $^3/_4$-inch (19 mm) wood strip.

[2]Floor panels conforming with this table shall be deemed to meet the design criteria of Section 2312.

[3]Applicable to all grades of sanded exterior-type plywood. See UBC Standard 23-2 for plywood species groups.

[4]Applicable to underlayment grade and C-C (plugged) plywood, and single floor grade wood structural panels.

TABLE 23-II-F-2—ALLOWABLE SPANS FOR PARTICLEBOARD SUBFLOOR AND COMBINED SUBFLOOR-UNDERLAYMENT[1,2]

GRADE	THICKNESS (inches) × 25.4 for mm	MAXIMUM SPACING OF SUPPORTS (inches)[3]	
		× 25.4 for mm	
		Subfloor	Combined Subfloor-Underlayment[4,5]
2-M-W	$^1/_2$	16	—
	$^5/_8$	20	16
	$^3/_4$	24	24
2-M-3	$^3/_4$	20	20

[1]All panels are continuous over two or more spans.

[2]Floor sheathing conforming with this table shall be deemed to meet the design criteria of Section 2312.

[3]Uniform deflection limitation: $^1/_{360}$ of the span under 100 pounds per square foot (4.79 kN/m^2) minimum load.

[4]Edges shall have tongue-and-groove joints or shall be supported with blocking. The tongue-and-groove panels are installed with the long dimension perpendicular to supports.

[5]A finish wearing surface is to be applied to the top of the panel.

TABLE 23-II-G—MAXIMUM DIAPHRAGM DIMENSION RATIOS

MATERIAL	HORIZONTAL DIAPHRAGMS	SHEAR WALLS
	Maximum Span-Width Ratios	Maximum Height-Width Ratios
1. Diagonal sheathing, conventional	3:1	1:1[1]
2. Diagonal sheathing, special	4:1	2:1[2]
3. Wood structural panels and particleboard, nailed all edges	4:1	2:1[2, 3]
4. Wood structural panels and particleboard, blocking omitted at intermediate joints.	4:1	4

[1]In Seismic Zones 0, 1, 2 and 3, the maximum ratio may be 2:1.

[2]In Seismic Zones 0, 1, 2 and 3, the maximum ratio may be $3^1/_2$:1.

[3]In Seismic Zone 4, the maximum ratio may be $3^1/_2$:1 for walls not exceeding 10 feet (3048 mm) in height on one side of the door to a one-story Group U Occupancy.

[4]Not permitted.

TABLE 23-II-H—ALLOWABLE SHEAR IN POUNDS PER FOOT FOR HORIZONTAL WOOD STRUCTURAL PANEL DIAPHRAGMS WITH FRAMING OF DOUGLAS FIR-LARCH OR SOUTHERN PINE[1]

PANEL GRADE	COMMON NAIL SIZE	MINIMUM NAIL PENETRATION IN FRAMING (inches) × 25.4 for mm	MINIMUM NOMINAL PANEL THICKNESS (inches)	MINIMUM NOMINAL WIDTH OF FRAMING MEMBER (inches)	BLOCKED DIAPHRAGMS — Nail spacing (in.) at diaphragm boundaries (all cases), at continuous panel edges parallel to load (Cases 3 and 4) and at all panel edges (Cases 5 and 6) — 6 (× 25.4 for mm); Nail spacing (in.) at other panel edges 6	4; other edges 6	2½[2]; other edges 4	2[2]; other edges 3	UNBLOCKED DIAPHRAGMS — Nails spaced 6″ (152 mm) max. at supported edges — Case 1 (No unblocked edges or continuous joints parallel to load)	All other configurations (Cases 2, 3, 4, 5 and 6)
					× 0.0146 for N/mm					
Structural 1	6d	1¼	5/16	2	185	250	375	420	165	125
				3	210	280	420	475	185	140
	8d	1½	3/8	2	270	360	530	600	240	180
				3	300	400	600	675	265	200
	10d[3]	1 5/8	15/32	2	320	425	640	730	285	215
				3	360	480	720	820	320	240
C-D, C-C, Sheathing, and other grades covered in UBC Standard 23-2 or 23-3	6d	1¼	5/16	2	170	225	335	380	150	110
				3	190	250	380	430	170	125
			3/8	2	185	250	375	420	165	125
				3	210	280	420	475	185	140
	8d	1½	3/8	2	240	320	480	545	215	160
				3	270	360	540	610	240	180
			7/16	2	255	340	505	575	230	170
				3	285	380	570	645	255	190
			15/32	2	270	360	530	600	240	180
				3	300	400	600	675	265	200
	10d[3]	1 5/8	15/32	2	290	385	575	655	255	190
				3	325	430	650	735	290	215
			19/32	2	320	425	640	730	285	215
				3	360	480	720	820	320	240

[1]These values are for short-time loads due to wind or earthquake and must be reduced 25 percent for normal loading. Space nails 12 inches (305 mm) on center along intermediate framing members.

Allowable shear values for nails in framing members of other species set forth in Division III, Part III, shall be calculated for all other grades by multiplying the shear capacities for nails in Structural I by the following factors: 0.82 for species with specific gravity greater than or equal to 0.42 but less than 0.49, and 0.65 for species with a specific gravity less than 0.42.

[2]Framing at adjoining panel edges shall be 3-inch (76 mm) nominal or wider and nails shall be staggered where nails are spaced 2 inches (51 mm) or 2½ inches (64 mm) on center.

[3]Framing at adjoining panel edges shall be 3-inch (76 mm) nominal or wider and nails shall be staggered where 10d nails having penetration into framing of more than 1 5/8 inches (41 mm) are spaced 3 inches (76 mm) or less on center.

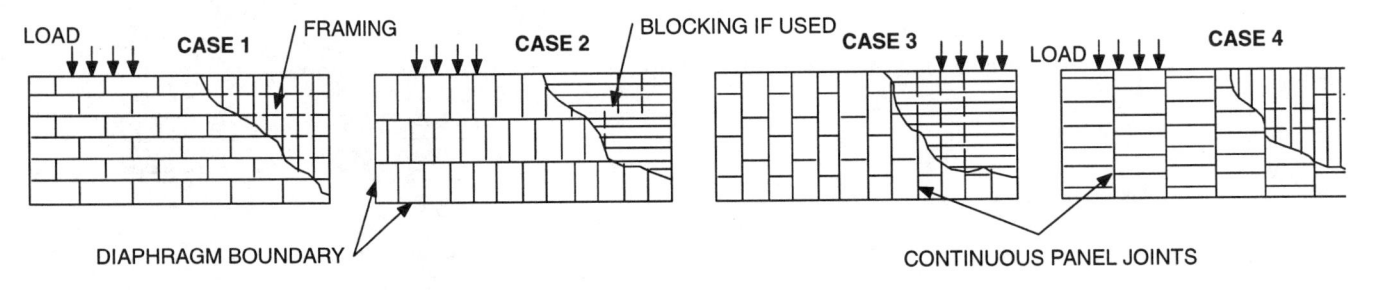

LOAD — CASE 1 — FRAMING | CASE 2 — BLOCKING IF USED | CASE 3 | LOAD — CASE 4

DIAPHRAGM BOUNDARY | CONTINUOUS PANEL JOINTS

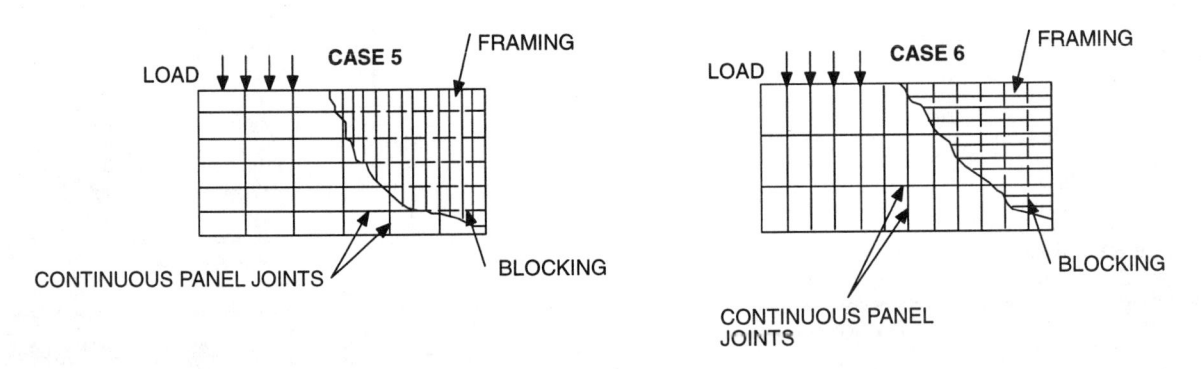

LOAD — CASE 5 — FRAMING | LOAD — CASE 6 — FRAMING

CONTINUOUS PANEL JOINTS — BLOCKING | CONTINUOUS PANEL JOINTS — BLOCKING

NOTE: Framing may be oriented in either direction for diaphragms, provided sheathing is properly designed for vertical loading.

TABLE 23-II-I-1
TABLE 23-II-I-2

1997 UNIFORM BUILDING CODE

TABLE 23-II-I-1—ALLOWABLE SHEAR FOR WIND OR SEISMIC FORCES IN POUNDS PER FOOT FOR WOOD STRUCTURAL PANEL SHEAR WALLS WITH FRAMING OF DOUGLAS FIR-LARCH OR SOUTHERN PINE[1,2,3]

PANEL GRADE	MINIMUM NOMINAL PANEL THICKNESS (inches) ×25.4 for mm	MINIMUM NAIL PENETRATION IN FRAMING (inches) ×25.4 for mm	PANELS APPLIED DIRECTLY TO FRAMING — Nail Size (Common or Galvanized Box)[5]	6	4	3	2	PANELS APPLIED OVER 1/2-INCH (13 mm) OR 5/8-INCH (16 mm) GYPSUM SHEATHING — Nail Size (Common or Galvanized Box)[5]	6	4	3	2
				×0.0146 for N/mm					×0.0146 for N/mm			
Structural I	5/16	1 1/4	6d	200	300	390	510	8d	200	300	390	510
	3/8	1 1/2	8d	230[4]	360[4]	460[4]	610[4]	10d	280	430	550	730
	7/16			255[4]	395[4]	505[4]	670[4]					
	15/32			280	430	550	730					
	15/32	1 5/8	10d	340	510	665	870	—	—	—	—	—
C-D, C-C Sheathing, plywood panel siding and other grades covered in UBC Standard 23-2 or 23-3	5/16	1 1/4	6d	180	270	350	450	8d	180	270	350	450
	3/8			200	300	390	510		200	300	390	510
	3/8	1 1/2	8d	220[4]	320[4]	410[4]	530[4]	10d	260	380	490	640
	7/16			240[4]	350[4]	450[4]	585[4]					
	15/32			260	380	490	640					
	15/32	1 5/8	10d	310	460	600	770	—	—	—	—	—
	19/32			340	510	665	870					
Plywood panel siding in grades covered in UBC Standard 23-2	5/16	1 1/4	Nail Size (Galvanized Casing) 6d	140	210	275	360	Nail Size (Galvanized Casing) 8d	140	210	275	360
	3/8	1 1/2	8d	160	240	310	410	10d	160	240	310	410

[1]All panel edges backed with 2-inch (51 mm) nominal or wider framing. Panels installed either horizontally or vertically. Space nails at 6 inches (152 mm) on center along intermediate framing members for 3/8-inch (9.5 mm) and 7/16-inch (11 mm) panels installed on studs spaced 24 inches (610 mm) on center and 12 inches (305 mm) on center for other conditions and panel thicknesses. These values are for short-time loads due to wind or earthquake and must be reduced 25 percent for normal loading.

Allowable shear values for nails in framing members of other species set forth in Division III, Part III, shall be calculated for all other grades by multiplying the shear capacities for nails in Structural I by the following factors: 0.82 for species with specific gravity greater than or equal to 0.42 but less than 0.49, and 0.65 for species with a specific gravity less than 0.42.

[2]Where panels are applied on both faces of a wall and nail spacing is less than 6 inches (152 mm) on center on either side, panel joints shall be offset to fall on different framing members or framing shall be 3-inch (76 mm) nominal or thicker and nails on each side shall be staggered.

[3]In Seismic Zones 3 and 4, where allowable shear values exceed 350 pounds per foot (5.11 N/mm), foundation sill plates and all framing members receiving edge nailing from abutting panels shall not be less than a single 3-inch (76 mm) nominal member and foundation sill plates shall not be less than a single 3-inch (76 mm) nominal member. In shear walls where total wall design shear does not exceed 600 pounds per foot (8.76 N/mm), a single 2-inch (51 mm) nominal sill plate may be used, provided anchor bolts are designed for a load capacity of 50 percent or less of the allowable capacity and bolts have a minimum of 2-inch-by-2-inch-by-3/16-inch (51 mm by 51 mm by 5 mm) thick plate washers. Plywood joint and sill plate nailing shall be staggered in all cases.

[4]The values for 3/8-inch (9.5 mm) and 7/16-inch (11 mm) panels applied direct to framing may be increased to values shown for 15/32-inch (12 mm) panels, provided studs are spaced a maximum of 16 inches (406 mm) on center or panels are applied with long dimension across studs.

[5]Galvanized nails shall be hot-dipped or tumbled.

TABLE 23-II-I-2—ALLOWABLE SHEAR IN POUNDS PER FOOT FOR PARTICLEBOARD SHEAR WALLS WITH FRAMING OF DOUGLAS FIR-LARCH OR SOUTHERN PINE[1,2,3]

PANEL GRADE	MINIMUM NOMINAL PANEL THICKNESS (Inches) ×25.4 for mm	MINIMUM NAIL PENETRATION IN FRAMING (Inches)	Nail size (Common or Galvanized Box)	PANELS APPLIED DIRECT TO FRAMING — Allowable Shear (pounds per foot)[1] Nail Spacing at Panel Edges (Inches) 6	4	3	2
				×0.0146 for N/mm			
M-S[4] and M-2[4]	3/8	1 1/2	6d	120	180	230	300
	3/8	1 1/2	8d	130	190	240	315
	1/2			140	210	270	350
	1/2	1 5/8	10d[5]	185	275	360	460
	5/8			200	305	395	520

[1]All panel edges backed with 2-inch (51 mm) nominal or wider framing. Space nails at 6 inches (152 mm) on center along intermediate framing members for 3/8-inch (9.5 mm) panel installed with the long dimension parallel to studs spaced 24 inches (610 mm) on center and 12 inches (305 mm) on center for other conditions and panel thicknesses. These values are for short-time loads due to wind or earthquake and must be reduced 25 percent for normal loading.

Allowable shear values for nails in framing members of other species set forth in Division III, Part III, shall be calculated for all grades by multiplying the values for common and galvanized box nails by the following factors: Group III, 0.82 and Group IV, 0.65.

[2]Where particleboard is applied on both faces of a wall and nail spacing is less than 6 inches (152 mm) on center on either side, panel joints shall be offset to fall on different framing members, or framing shall be 3-inch (76 mm) nominal or thicker and nails on each side shall be staggered.

[3]In Seismic Zones 3 and 4, where allowable shear values exceed 350 pounds per foot (5.11 N/mm), foundation sill plates and all framing members receiving edge nailing from abutting panels shall not be less than a single 3-inch (76 mm) nominal member and foundation sill plates shall not be less than a single 3-inch (76 mm) nominal member. In shear walls where total wall design shear does not exceed 600 pounds per foot (8.76 N/mm), a single 2-inch (51 mm) nominal sill plate may be used, provided anchor bolts are designed for a load capacity of 50 percent or less of the allowable capacity and bolts have a minimum of 2-inch-by-2-inch-by-3/16-inch (51 mm by 51 mm by 5 mm) thick plate washers. Plywood joint and sill plate nailing shall be staggered in all cases.

[4]Products shall be manufactured with exterior glue and shall be identified with the words "Exterior Glue" following the product grade designation.

[5]Framing at adjoining panel edges shall be 3-inch (76 mm) nominal or wider and nails shall be staggered where 10d nails having penetration into framing of more than 1 5/8 inches (41 mm) are spaced 3 inches (76 mm) or less on center.

TABLE 23-II-J—ALLOWABLE SHEARS FOR WIND OR SEISMIC LOADING ON
VERTICAL DIAPHRAGMS OF FIBERBOARD SHEATHING BOARD CONSTRUCTION
FOR TYPE V CONSTRUCTION ONLY[1]

SIZE AND APPLICATION	NAIL SIZE	SHEAR VALUE IN POUNDS PER FOOT (N/mm) 3-INCH (76 mm) NAIL SPACING AROUND PERIMETER AND 6-INCH (152 mm) AT INTERMEDIATE POINTS
		× 1.46 for N/mm
$1/2'' \times 4' \times 8'$ (13 × 1219 × 2438 mm)	No. 11 gage galvanized roofing nail $1^1/_2''$ (38 mm) long, $7/_{16}''$ (11 mm) head	125[2]
$25/_{32}'' \times 4' \times 8'$ (20 × 1219 × 2438 mm)	No. 11 gage galvanized roofing nail $1^3/_4''$ (44 mm) long, $7/_{16}''$ (11 mm) head	175

[1]Fiberboard sheathing diaphragms shall not be used to brace concrete or masonry walls.
[2]The shear value may be 175 (778 N) for $1/2$-inch-by-4-foot-by-8-foot (12.7 by 1219 by 2438 mm) fiberboard nail-base sheathing.

TABLE 23-II-K—WOOD SHINGLE AND SHAKE SIDE WALL EXPOSURES

SHINGLE OR SHAKE	MAXIMUM WEATHER EXPOSURES (inches)			
	× 25.4 for mm			
	Single-Coursing		Double-Coursing	
Length and Type	No. 1	No. 2	No. 1	No. 2
16-inch (405 mm) shingles	$7^1/_2$	$7^1/_2$	12	10
18-inch (455 mm) shingles	$8^1/_2$	$8^1/_2$	14	11
24-inch (610 mm) shingles	$11^1/_2$	$11^1/_2$	16	14
18-inch (455 mm) resawn shakes	$8^1/_2$	—	14	—
18-inch (455 mm) straight-split shakes	$8^1/_2$	—	16	—
24-inch (610 mm) resawn shakes	$11^1/_2$	—	20	—

FIGURE 23-II-1

1997 UNIFORM BUILDING CODE

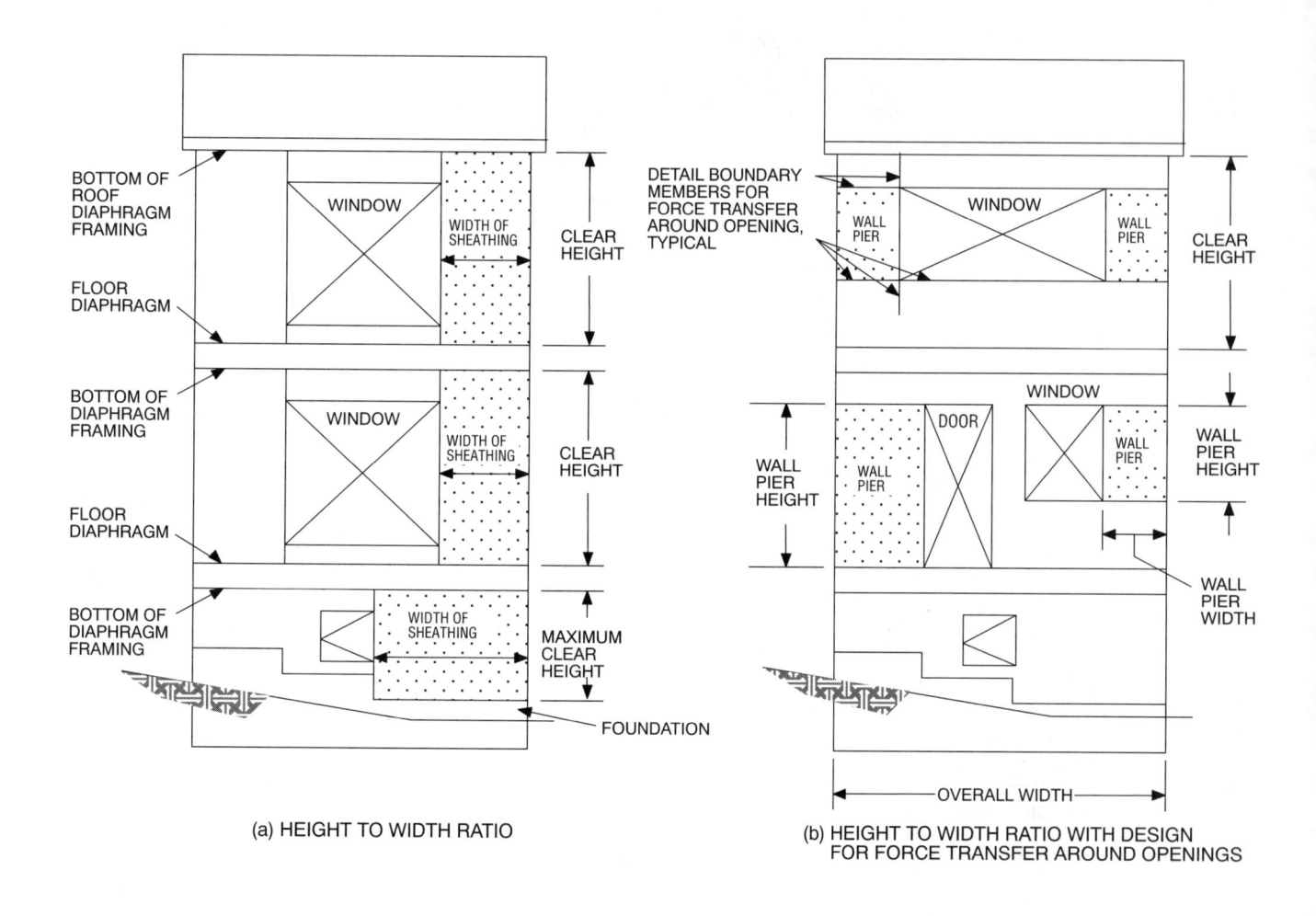

(a) HEIGHT TO WIDTH RATIO

(b) HEIGHT TO WIDTH RATIO WITH DESIGN
FOR FORCE TRANSFER AROUND OPENINGS

FIGURE 23-II-1—GENERAL DEFINITION OF SHEAR WALL HEIGHT TO WIDTH RATIO

Division III—DESIGN SPECIFICATIONS FOR ALLOWABLE STRESS DESIGN OF WOOD BUILDINGS

Part I—ALLOWABLE STRESS DESIGN OF WOOD

This standard, with certain exceptions, is the ANSI/NFoPA NDS-91 National Design Specification for Wood Construction of the American Forest and Paper Association, Revised 1991 Edition, and the Supplement to the 1991 Edition, National Design Specification, adopted by reference.

The National Design Specification for Wood Construction, Revised 1991 Edition, and supplement are available from the American Forest and Paper Association, 1111 19th Street, NW, Eighth Floor, Washington, DC, 20036.

SECTION 2316 — DESIGN SPECIFICATIONS

2316.1 Adoption and Scope. The National Design Specification for Wood Construction, Revised 1991 Edition (NDS), which is hereby adopted as a part of this code, shall apply to the design and construction of wood structures using visually graded lumber, mechanically graded lumber, structural glued laminated timber, and timber piles. National Design Specification Appendix Section F, Design for Creep and Critical Deflection Applications, Appendix Section G, Effective Column Length, and Appendix Section J, Solution of Hankinson Formula are specifically adopted and made a part of this standard. The Supplement to the 1991 Edition National Design Specification, Tables 2A, 4A, 4B, 4C, 4D, 4E, 5A, 5B and 5C are specifically adopted and made a part of this standard.

Other codes, standards or specifications referred to in this standard are to be considered as only an indication of an acceptable method or material that can be used with the approval of the building official, except where such other codes, standards or specifications are specifically adopted by this code as primary standards.

2316.2 Amendments.

1. Sec. 1.1. Delete and substitute the following:

The design of structures using visually graded lumber, mechanically graded lumber, structural glued laminated timber, timber piles, and design of their connections shall be in accordance with Chapter 23, Division III, Part 1.

2. Secs. 1.2 through 1.5. Delete.

3. Sec. 2.2. Delete first sentence and substitute the following:

Allowable stress design values for visually graded structural lumber, mechanically graded structural lumber and structural glued laminated timber shall be in accordance with NDS Supplement Tables 2A, 4A, 4B, 4C, 4D, 5A, 5B and 5C. Values for species and grades not tabulated shall be submitted to the building official for approval.

4. Sec. 2.3.2.1. In fourth sentence, delete "or Figure B1 (see Appendix B)."

5. Sec. 2.3.2.3. Delete and substitute the following:

2.3.2.3 When using Section 1612.3.1 basic load combinations, the Load Duration Factor, C_D, noted in Table 2.3.2 shall be permitted to be used. When using Section 1612.3.2 alternate load combinations, the one-third increase shall not be used concurrently with the Load Duration Factor, C_D.

6. Table 2.3.2. Delete and substitute as follows:

TABLE 2.3.2—LOAD DURATION FACTORS, C_D

DESIGN LOAD	LOAD DURATION	C_D
Dead Load	Permanent	0.9
Floor, Occupancy Live Load	Ten Years	1.0
Snow Load	Two Months	1.15
Roof Live Load	Seven Days	1.25
Earthquake Load[1]	—	1.33
Wind Load[2]	—	1.33
Impact	—	2.0

[1]1.60 may be used for nailed and bolted connections exhibiting Mode III or IV behavior, except that the increases for earthquake are not combined with the increase allowed in Section 1612.3. The 60-percent increase for nailed and bolted connections exhibiting Mode III or IV behavior for earthquake shall not be applicable to joist hangers, framing anchors, and other mechanical fastenings, including straps and hold-down anchors. The 60-percent increase shall not apply to the allowable shear values in Tables 23-II-H, 23-II-I-1, 23-II-I-2, 23-II-J or in Section 2315.3.

[2]1.60 may be used for members and nailed and bolted connections exhibiting Mode III or IV behavior, except that the increases for wind are not combined with the increase allowed in Section 1612.3. The 60-percent increase shall not apply to the allowable shear values in Tables 23-II-H, 23-II-I-1, 23-II-I-2, 23-II-J or in Section 2315.3.

7. Sec. 2.3.4. Add a second paragraph following Table 2.3.4:

The allowable unit stresses for fire-retardant-treated solid-sawn lumber and plywood, including fastener values, subject to prolonged elevated temperatures from manufacturing or equipment processes, but not exceeding 150°F (66°C), shall be developed from approved test methods that properly consider potential strength-reduction characteristics, including effects of heat and moisture.

8. Sec. 2.3.6. Add second, third and fourth paragraphs as follows:

The values for lumber and plywood impregnated with approved fire-retardant chemicals, including fastener values, shall be submitted to the building official for approval. Submittal to the building official shall include all substantiating data. Such values shall be developed from approved test methods and procedures that consider potential strength-reduction characteristics, including the effects of elevated temperatures and moisture. Other adjustments are applicable, except that the impact load-duration factor shall not apply.

Values for glued-laminated timber, including fastener design values, shall be recommended by the treater and submitted to the building official for approval. Submittal to the building official shall include all substantiating data.

In addition to the requirements specified in Section 207, fire-retardant lumber having structural applications shall be tested and identified by an approved inspection agency in accordance with UBC Standard 23-5.

9. Sec. 2.3.8. Add new second and third paragraphs following Table 2.3.8:

For lumber I beams and box beams, the form factor, C_f, shall be calculated as:

$$C_f = \left[1 + \left(\frac{d^2 + 143}{d^2 + 88} - 1 \right) C_g \right]$$

For **SI**:
$$C_f = \left[1 + \left(\frac{\left(\frac{d}{25.4}\right)^2 + 143}{\left(\frac{d}{25.4}\right)^2 + 88} - 1 \right) C_g \right]$$

WHERE:

C_f = form factor.

C_g = support factor = $p^2(6 - 8p + 3p^2)(1 - q) + q$.

d = depth of I or box beam.

p = ratio of depth of compression flange to full depth of beam.

q = ratio of thickness of web or webs to full width of beam.

10. Sec. 2.3.10. Add a paragraph at end of section as follows:

In joists supported on a ribbon or ledger board and spiked to the studding, the allowable stress in compression perpendicular to grain may be increased 50 percent.

11. Sec. 3.2.1. Add a second sentence as follows:

For continuous beams, the span shall be taken as the distance between centers of bearings on supports over which the beam is continuous.

12. Sec. 3.2.3.2. Add to end of paragraph as follows:

Cantilevered portions of beams less than 4 inches (102 mm) in nominal thickness shall not be notched unless the reduced section properties and lumber defects are considered in the design. For effects of notch on shear strength, see Section 3.4.4.

13. Sec. 3.3.2. Add a last paragraph as follows:

A beam of circular cross section may be assumed to have the same strength as a square beam having the same cross-sectional area. If a circular beam is tapered, it shall be considered a beam of variable cross section.

14. Sec. 3.4.4. Add a section as follows:

3.4.4.5 When girders, beams or joists are notched at points of support on the compression side, they shall meet design requirements for the net section in bending and in shear. The actual shear stress at such point shall be calculated as follows:

$$f_v = \frac{3V}{2b\left[d - \left(\frac{d-d'}{d'}\right)e\right]}$$

WHERE:

d = total depth of beam.

d' = actual depth of beam at notch.

e = distance notch extends inside the inner edge of support.

V = shear force.

Where e exceeds d', the actual shear stress for the notch on the compression side shall be calculated as follows:

$$f_v = \frac{3V}{2bd'}$$

15. Sec. 3.7.1.4. Delete and substitute as follows:

The slenderness ratio for solid columns, le/d shall not exceed 50.

16. Sec. 3.8.2. Delete and substitute as follows:

Where designs that induce tension stresses perpendicular to grain cannot be avoided, mechanical reinforcement sufficient to resist such forces shall be provided.

17. Sec. 4.2.5.5. Delete.

18. Sec. 4.4.1.1. Delete and substitute as follows:

Rectangular sawn lumber beams, rafters, joists or other bending members shall be supported laterally to prevent rotation or lateral displacement in accordance with Section 4.4.1.2, or shall be designed in accordance with the lateral stability provisions in Section 3.3.3.

19. Sec. 4.4.1.2. Delete first sentence.

20. Sec. 5.4.1. Delete second paragraph and substitute as follows:

For curved bending members having a varying cross section, the maximum actual radial stress induced, f_r, is given by:

$$f_r = K_r \frac{6M}{bd^2}$$

WHERE:

b = width of cross section, inches (mm).

d = depth of cross section at the apex in inches (mm).

K_r = radial stress factor determined from the following relationship:

$$K_r = A + B\left(\frac{d}{Rm}\right) + C\left(\frac{d}{Rm}\right)^2$$

M = bending moment at midspan in inch-pounds (N·mm).

WHERE:

Rm = radius of curvature at the center line of the member at midspan in inches (mm).

A, B

and C = constants as follows:

β (1)	A (2)	B (3)	C (4)
(0.0)	(0.0)	(0.2500)	(0.0)
2.5°	0.0079	0.1747	0.1284
5.0°	0.0174	0.1251	0.1939
7.5°	0.0279	0.0937	0.2162
10.0°	0.0391	0.0754	0.2119
15.0°	0.0629	0.0619	0.1722
20.0°	0.0893	0.0608	0.1393
25.0°	0.1214	0.0605	0.1238
30.0°	0.1649	0.0603	0.1115

and β = angle between the upper edge of the member and the horizontal in degrees. Values of K_r for intermediate values of β may be interpolated linearly.

When the beam is loaded with a uniform load, K_r may be modified by multiplying by the reduction factor C_r as calculated by the following formula:

$$C_r = A + B\left(\frac{L}{L_t}\right) + C\left(\frac{d_c}{R_m}\right) + D\left(\frac{L}{L_t}\right)^2$$
$$+ E\left(\frac{d_c}{R_m}\right)^2 + F\left(\frac{d_c}{R_m}\right)\left(\frac{L}{L_t}\right)$$
$$+ G\left(\frac{L}{L_t}\right)^3 + H\left(\frac{d_c}{R_m}\right)^3$$

WHERE:

C_r = reduction factor.

L = span of beam.

L_t = length of beam between tangent points.

A, B

$\ldots H$ = constants for a given β as follows:

β	A	B	C	D	E	F	G	H
2.3°	-0.142	0.418	-2.358	-0.053	—	—	0.002	—
9.7°	0.143	0.376	-0.541	-0.060	—	—	0.003	—
14.9°	0.406	0.293	-0.927	-0.041	—	—	0.002	—
20.0°	0.423	0.364	-1.022	-0.067	—	0.146	—	—
25.2°	0.540	0.360	-1.061	-0.070	—	0.156	—	—
29.8°	0.502	0.372	—	-0.076	-3.712	0.138	0.004	4.336

and β = angle between the upper edge of the member and the horizontal in degrees. Values of C_r for intermediate values may be interpolated linearly.

PITCHED AND TAPERED CURVED BEAM

21. Sec. 5.4.1.2. Add a second paragraph:

These values are subject to modification for duration of load. If these values are exceeded, mechanical reinforcing sufficient to resist all radial tension shall be provided, but in no case shall the calculated radial tension stress exceed one third the allowable unit stress in horizontal shear. When mechanical reinforcing is used, the maximum moisture content of the laminations at the time of manufacture shall not exceed 12 percent for dry conditions of use.

22. Sec. 5.4.4. Add a section as follows:

5.4.4 Ponding. Roof-framing members shall be designed for the deflection and drainage or ponding requirements specified in Section 1506 and Chapter 16. In glued-laminated timbers, the minimum slope for roof drainage required by Section 1506 shall be in addition to a camber of one and one-half times the calculated dead load deflection. The calculation of the required slope shall not include any vertical displacement created by short taper cuts. In no case shall the deflection of glued-laminated timber roof members exceed $1/2$-inch (13 mm) for a 5 pound-per-square-foot (239 Pa) uniform load.

23. Sec. 5.4.5. Add a new section as follows:

5.4.5 Tapered Faces. Sawn tapered cuts shall not be permitted on the tension face of any beam. Pitched or curved beams shall be so fabricated that the laminations are parallel to the tension face. Straight, pitched or curved beams may have sawn tapered cuts on the compression face.

For other members subject to bending, the slope of tapered faces, measured from the tangent to the lamination of the section under consideration, shall not be steeper than 1 unit vertical in 24 units horizontal (4% slope) on the tension side.

EXCEPTIONS: 1. This requirement does not apply to arches.

2. Taper may be steeper at sections increased in size beyond design requirements for architectural projections.

24. Sec. 8.3. Add a section as follows:

8.3 Allowable shear values for bolts used to connect a wood member to concrete or masonry are permitted to be determined as one half the tabulated double shear value for a wood member twice the thickness of the member attached to the concrete or masonry.

25. Sec. 12.4.1. Delete and substitute as follows:

12.4.1 For wood-to-wood joints, the spacing center to center of nails in the direction of stress shall not be less than the required penetration. Edge or end distances in the direction of stress shall not be less than one-half of the required penetration. All spacing and edge and end distances shall be such as to avoid splitting of the wood.

26. Sec. 13.2.1. Delete and substitute as follows:

13.2.1 Test for design values. Tests to determine design values for metal plate connectors in lateral withdrawal, net section shear and net section tension shall be conducted in accordance with the test and evaluation procedures in ANSI/TPI 1-1995. Design values determined in accordance with these test procedures shall be multiplied by all applicable adjustment factors (see Table 7.3.1) to obtain allowable design values.

27. NDS Supplement Table 5A. Add combinations and design values as follows:

COMBINATION SYMBOL[4]	SPECIES OUTER LAMINATIONS/ CORE LAMINATIONS[5]	DESIGN VALUES IN POUNDS PER SQUARE INCH (psi)												
		BENDING ABOUT X–X AXIS (Loaded Perpendicular to Wide Faces of Laminations)						BENDING ABOUT Y–Y AXIS (Loaded Parallel to Wide Faces of Laminations)					AXIALLY LOADED		
		Bending		Compression Perpendicular to Grain		Shear Parallel to Grain[10] F_{vxx}	Modulus of Elasticity E_{xx}	Bending F_{byy}	Compression Perpendicular to Grain (Side Faces) $F_{c\perp yy}$	Shear Parallel to Grain F_{vyy}	Shear Parallel to Grain (For Members With Multiple Piece Laminations Which are not Edge glued)[13] F_{vyy}	Modulus of Elasticity E_{yy}	Tension Parallel to Grain F_t	Compression Parallel to Grain F_c	Modulus of Elasticity E
		Tension Zone Stressed in Tension F_{bxx}	Compression Zone Stressed in Tension[6] F_{bxx}	Tension Face[9.1] $^0F_{c\perp xx}$	Compression Face[9,10] $F_{c\perp xx}$										
VISUALLY GRADED SOUTHERN PINE														
26F–V1	SP/SP	2600	1300	650	650	200	1,800,000	1900	560	175	90	1,600,000	1150	1600	1,600,000
26F–V2	SP/SP	2600	1300	650	650	200	1,900,000	2200	650	175	90	1,800,000	1200	1650	1,800,000
26F–V3	SP/SP	2600	1300	650	650	200	1,900,000	2100	560	175	90	1,800,000	1150	1600	1,800,000
26F–V4[8]	SP/SP	2600	2600	650	650	200	1,900,000	2100	560	175	90	1,800,000	1150	1600	1,800,000
E-RATED SOUTHERN PINE														
28F–E1	SP/SP	2800	1400	650	650	200	2,000,000	1600	560	175	90	1,700,000	1300	1850	1,700,000
28F–E2[8]	SP/SP	2800	2800	650	650	200	2,000,000	1600	560	175	90	1,700,000	1300	1850	1,700,000
30F–E1[15]	SP/SP	3000	1500	650	650	200	2,000,000	1750	560	175	90	1,700,000	1250	1750	1,700,000
30F–E2[8,15]	SP/SP	3000	3000	650	650	200	2,000,000	1750	560	175	90	1,700,000	1250	1750	1,700,000

[15]These combinations are only for nominal widths 6 inches and less, in accordance with AITC 117–93.

Part II—PLYWOOD STRUCTURAL PANELS

SECTION 2317 — PLYWOOD STRUCTURAL PANELS

Values for plywood structural panels shall be in accordance with Table 23-III-A.

Part III—FASTENINGS

SECTION 2318 — TIMBER CONNECTORS AND FASTENERS

2318.1 General. Timber connectors and fasteners may be used to transmit forces between wood members and between wood and metal members. Allowable design values, Z and W, shall be determined in accordance with Division III, Part I or this section. Modifications to allowable design values, and installation of timber connectors and fasteners shall be in accordance with the provisions set forth in Division III, Part I.

2318.2 Bolts. Allowable lateral design values, $Z_{\parallel}$, $Z_{m\perp}$ and $Z_{s\perp}$, in pounds for bolts in shear in seasoned lumber of Douglas fir-larch and Southern pine shall be as set forth in Tables 23-III-B-1 and 23-III-B-2.

2318.3 Nails and Spikes.

2318.3.1 Allowable lateral loads. Allowable lateral design values, Z, for common wire and box nails driven perpendicular to the grain of the wood, when used to fasten wood members together, shall be as set forth in Tables 23-III-C-1 and 23-III-C-2.

A wire nail driven parallel to the grain of the wood shall not be subjected to more than two thirds of the lateral load allowed when driven perpendicular to the grain. Toenails shall not be subjected to more than five sixths of the lateral load allowed for nails driven perpendicular to the grain.

In Seismic Zones 3 and 4, toenails shall not be used to transfer lateral forces in excess of 150 pounds per foot (2188 N/m) from diaphragms to shear walls, drag struts (collectors) or other elements, or from shear walls to other elements.

EXCEPTION: Structures built in accordance with Section 2320.

2318.3.2 Allowable withdrawal loads. Allowable withdrawal design values, W, for wire nails driven perpendicular to the grain of the wood shall be as set forth in Table 23-III-D.

Nails driven parallel to the grain of the wood shall not be allowed for resisting withdrawal forces.

2318.3.3 Spacing and penetration. Common wire nails shall have penetration into the piece receiving the point as set forth in Tables 23-III-C-1 and 23-III-C-2. Nails or spikes for which the gages or lengths are not set forth in Tables 23-III-C-1 and 23-III-C-2 shall have a required penetration of not less than 11 diameters, and allowable loads may be interpolated. Allowable loads shall not be increased when the penetration of nails into the member holding the point is larger than required by this section.

2318.4 Joist Hangers and Framing Anchors. Connections depending on joist hangers or framing anchors, ties and other mechanical fastenings not otherwise covered may be used where approved.

2318.5 Miscellaneous Fasteners.

2318.5.1 Drift Bolts and Drift Pins.

2318.5.1.1 Withdrawal design values. Drift bolt and drift pin connections loaded in withdrawal shall be designed in accordance with good engineering practice.

2318.5.1.2 Lateral design values. Allowable lateral design values for drift bolts and drift pins driven in the side grain of wood shall not exceed 75 percent of the allowable lateral design values for common bolts of the same diameter and length in main member. Additional penetration of pin into members should be provided in lieu of the washer, head and nut on a common bolt.

2318.5.2 Spike Grids. Wood-to-wood connections involving spike grids for load transfer shall be designed in accordance with good engineering practice.

Part IV—ALLOWABLE STRESS DESIGN FOR WIND AND EARTHQUAKE LOADS

SECTION 2319 — WOOD SHEAR WALLS AND DIAPHRAGMS

2319.1 Conventional Lumber Diaphragms. Conventional lumber diaphragms of Douglas fir-larch or Southern pine, constructed in accordance with Section 2315.3.1, may be used to resist shear due to wind or seismic forces not exceeding 300 pounds per lineal foot (4.37 kN/m) of width. Where nails are used with sheathing and framing members with a specific gravity less than 0.49, the allowable unit shear strength of the diaphragm shall be multiplied by the following factors: 0.82 for species with specific gravity greater than or equal to 0.42 but less than 0.49, and 0.65 for species with a specific gravity less than 0.42.

2319.2 Special Lumber Diaphragms. Special diagonally sheathed diaphragms of Douglas fir-larch or Southern pine, constructed in accordance with Section 2315.3.2, may be used to resist shears due to wind or seismic loads, provided such shear do not stress the nails beyond their allowable safe lateral strength and do not exceed 600 pounds per lineal foot (8.75 kN/m) of width. Where nails are used with sheathing and framing members with a specific gravity less than 0.49, the allowable unit shear strength of the diaphragm shall be multiplied by the following factors: 0.82 for species with specific gravity greater than or equal to 0.42 but less than 0.49, and 0.65 for species with a specific gravity less than 0.42.

2319.3 Wood Structural Panel Diaphragms. Horizontal and vertical diaphragms sheathed with wood structural panels may be used to resist horizontal forces not exceeding those set forth in Table 23-II-H for horizontal diaphragms and Table 23-II-I-1 for vertical diaphragms.

Where the wood structural panel is applied to both faces of a shear wall in accordance with Table 23-II-I-1, allowable shear for the wall may be taken as twice the tabulated shear for one side, except that where the shear capacities are not equal, the allowable shear shall be either the shear for the side with the higher capacity or twice the shear for the side with the lower capacity, whichever is greater.

2319.4 Particleboard Diaphragms. Vertical diaphragms sheathed with particleboard may be used to resist horizontal forces not exceeding those set forth in Table 23-II-I-2.

2319.5 Fiberboard Sheathing Diaphragms. Wood stud walls sheathed with fiberboard sheathing may be used to resist horizontal forces not exceeding those set forth in Table 23-II-J.

TABLE 23-III-A—ALLOWABLE UNIT STRESSES FOR CONSTRUCTION AND INDUSTRIAL SOFTWOOD PLYWOOD
(In pounds per square inch—normal loading)
(To be used with section properties in Plywood-design Specifications—See UBC Standard 23-2)

STRESS	SPECIES[1] GROUP OF FACE PLY	EXTERIOR A-A, A-C, C-C / Structural I A-C, C-C (Use Group I Stresses)		EXTERIOR A-B, B-B, B-C C-C (PLUGGED) / Structural I C-D (Use Group 1 Stresses) / Structural II C-D (Use Group 3 Stresses) / C-D Sheathing (Exterior Glue) / All Interior Grades with Exterior Glue		ALL OTHER GRADES OF INTERIOR INCLUDING C-D SHEATHING
		Wet[2]	Dry[3]	Wet[2]	Dry[3]	Dry[3]
		$\times$ 0.00689 for N/mm²				
1. Extreme fiber stress in bending (F_b) Tension in plane of plies (F_t) Face grain parallel or perpendicular to span (at 45° to face grain use $1/6\,F_t$)	1	1,430	2,000	1,190	1,650	1,650
	2, 3	980	1,400	820	1,200	1,200
	4	940	1,330	780	1,110	1,110
2. Compression in plane of plies (F_c) Parallel or perpendicular to face grain (at 45° to face grain use $1/3\,F_c$)	1	970	1,640	900	1,540	1,540
	2	730	1,200	680	1,100	1,100
	3	610	1,060	580	990	990
	4	610	1,000	580	950	950
3. Shear in plane perpendicular to plies F_v[4] Parallel or perpendicular to face grain (at 45° to face grain use $2\,F_v$)	1	155	190	155	190	160
	2, 3	120	140	120	140	120
	4	110	130	110	130	115
4. Shear, rolling, in the plane of plies Parallel or perpendicular to face grain (at 45° to face grain use $1^1/3\,F_s$)	Marine and Structural I	63	75	63	75	48
	Structural II	49	56	49	56	
	All others[5]	44	53	44	53	
5. Bearing (on face) Perpendicular to plane of plies	1	210	340	210	340	340
	2, 3	135	210	135	210	210
	4	105	160	105	160	160
6. Modulus of elasticity In bending in plane of plies Face grain parallel or perpendicular to span	1	1,500,000	1,800,000	1,500,000	1,800,000	1,800,000
	2	1,300,000	1,500,000	1,300,000	1,500,000	1,500,000
	3	1,100,000	1,200,000	1,100,000	1,200,000	1,200,000
	4	900,000	1,000,000	900,000	1,000,000	1,000,000

SPAN RATING

THICKNESS (inches) $\times$ 25.4 for mm	C-C AND C-D							UNDERLAYMENT AND C-C PLUGGED			
	12/0	16/0	20/0	24/0	32/16	40/20	48/24	16" o.c.	20" o.c.	24" o.c.	48" o.c.
								$\times$ 25.4 for mm			
$5/16$	4	3	1								
$3/8$			4[7]	1							
$15/32$[6], $1/2$				4[7]	1			1			
$19/32$, $5/8$					4[7]	1		4[7]	1		
$23/32$, $3/4$						4[7]	1		4[7]	1	
$7/8$							3[8]			3[8]	
$1^1/8$											1

[1]See UBC Standard 23-2 for plywood species groups. For C-C, C-D underlayment, and C-C plugged, the combination of span rating and panel thickness determines the species group and, therefore, the stress permitted, as in the above table.

[2]Wet condition of use corresponds to a moisture content of 16 percent or more.

[3]Dry condition of use corresponds to a moisture content of less than 16 percent.

[4]See UBC Standard 23-2, Section 23.221, for provisions under which F_v stresses may be increased.

[5]Reduce stresses 25 percent for three-layer (four-ply) panels over $5/8$ inch (16 mm) thick.

[6]Thickness not applicable to underlayment and C-C plugged.

[7]Use Group 3 stresses for Structural II.

[8]Use Group 4 stresses for underlayment and C-C plugged 24 inches (610 mm) on center.

TABLE 23-III-B-1
TABLE 23-III-B-2

1997 UNIFORM BUILDING CODE

TABLE 23-III-B-1—BOLT DESIGN VALUES (Z) FOR SINGLE SHEAR (Two Member) CONNECTIONS[1,2,3]
(For sawn lumber with both members of identical species)

THICKNESS			G=0.55 SOUTHERN PINE			G=0.50 DOUGLAS FIR-LARCH			G=0.42 SPRUCE-PINE-FIR		
MAIN MEMBER l_m (inches)	SIDE MEMBER l_s (inches)	BOLT DIAMETER D (inches)	$Z_\parallel$ lbs.	$Z_s \perp$ lbs.	$Z_m \perp$ lbs.	$Z_\parallel$ lbs.	$Z_s \perp$ lbs.	$Z_m \perp$ lbs.	$Z_\parallel$ lbs.	$Z_s \perp$ lbs.	$Z_m \perp$ lbs.
× 25.4 for mm			× 4.45 for N								
1 1/2	1 1/2	1/2	530	330	330	480	300	300	410	240	240
		5/8	660	400	400	600	360	360	510	290	290
		3/4	800	460	460	720	420	420	610	340	340
		7/8	930	520	520	850	470	470	710	380	380
		1	1,060	580	580	970	530	530	810	430	430
3 1/2	1 1/2	1/2	660	400	470	610	370	430	540	320	370
		5/8	940	560	620	880	520	540	780	410	430
		3/4	1,270	660	690	1,200	590	610	1,080	450	480
		7/8	1,680	720	770	1,590	630	680	1,340	490	540
		1	2,010	770	830	1,830	680	740	1,530	530	590
5 1/2	1 1/2	5/8	940	560	640	880	520	590	780	410	520
		3/4	1,270	660	850	1,200	590	790	1,080	450	690
		7/8	1,680	720	1,090	1,590	630	980	1,440	490	760
		1	2,150	770	1,190	2,050	680	1,060	1,760	530	830
7 1/2	1 1/2	5/8	940	560	640	880	520	590	780	410	520
		3/4	1,270	660	850	1,200	590	790	1,080	450	690
		7/8	1,680	720	1,090	1,590	630	1,010	1,440	490	890
		1	2,150	770	1,350	2,050	680	1,270	1,760	530	1,110

[1]Tabulated lateral design values (Z) for bolted connections shall be multiplied by all applicable adjustment factors (see Division III, Part I).
[2]Tabulated lateral design values (Z) are for "full diameter" bolts with a bending yield strength (F_{yb}) of 45,000 psi (310 N/mm^2).
[3]For other species and configurations, see Division III, Part I.

TABLE 23-III-B-2—BOLT DESIGN VALUES (Z) FOR DOUBLE SHEAR (Three Member) CONNECTIONS[1,2,3]
(For sawn lumber with all members of identical species)

THICKNESS			G=0.55 SOUTHERN PINE			G=0.50 DOUGLAS FIR-LARCH			G=0.42 SPRUCE-PINE-FIR		
MAIN MEMBER l_m (inches)	SIDE MEMBER l_s (inches)	BOLT DIAMETER D (inches)	$Z_\parallel$ lbs.	$Z_s \perp$ lbs.	$Z_m \perp$ lbs.	$Z_\parallel$ lbs.	$Z_s \perp$ lbs.	$Z_m \perp$ lbs.	$Z_\parallel$ lbs.	$Z_s \perp$ lbs.	$Z_m \perp$ lbs.
× 25.4 for mm			× 4.45 for N								
1 1/2	1 1/2	1/2	1,150	800	550	1,050	730	470	880	640	370
		5/8	1,440	1,130	610	1,310	1,040	530	1,100	830	410
		3/4	1,730	1,330	660	1,580	1,170	590	1,320	900	450
		7/8	2,020	1,440	720	1,840	1,260	630	1,540	970	490
		1	2,310	1,530	770	2,100	1,350	680	1,760	1,050	530
3 1/2	1 1/2	1/2	1,320	800	940	1,230	730	860	1,080	640	740
		5/8	1,870	1,130	1,290	1,760	1,040	1,190	1,570	830	960
		3/4	2,550	1,330	1,550	2,400	1,170	1,370	2,160	900	1,050
		7/8	3,360	1,440	1,680	3,280	1,260	1,470	2,880	970	1,130
		1	4,310	1,530	1,790	4,090	1,350	1,580	3,530	1,050	1,230
5 1/2	1 1/2	5/8	1,870	1,130	1,290	1,760	1,040	1,190	1,570	830	1,040
		3/4	2,550	1,330	1,690	2,400	1,170	1,580	2,160	900	1,380
		7/8	3,360	1,440	2,170	3,180	1,260	2,030	2,880	970	1,780
		1	4,310	1,530	2,700	4,090	1,350	2,480	3,530	1,050	1,930
7 1/2	1 1/2	5/8	1,870	1,130	1,290	1,760	1,040	1,190	1,570	830	1,040
		3/4	2,550	1,330	1,690	2,400	1,170	1,580	2,160	900	1,380
		7/8	3,360	1,440	2,170	3,180	1,260	2,030	2,880	970	1,780
		1	4,310	1,530	2,700	4,090	1,350	2,530	3,530	1,050	2,240

[1]Tabulated lateral design values (Z) for bolted connections shall be multiplied by all applicable adjustment factors (see Divison III, Part I).
[2]Tabulated lateral design values (Z) are for "full diameter" bolts with a bending yield strength (F_{yb}) of 45,000 psi (310 N/mm^2).
[3]For other species and configurations, see Division III, Part I.

TABLE 23-III-C-1—BOX NAIL DESIGN VALUES (Z) FOR SINGLE SHEAR (Two Member) CONNECTIONS[1,2,3]
(With both members of identical species)

SIDE MEMBER THICKNESS t_s (inches)	NAIL LENGTH L (inches)	NAIL DIAMETER D (inches)	PENNY-WEIGHT	G=0.55 SOUTHERN PINE Z lbs.	G=0.50 DOUGLAS-FIR LARCH Z lbs.	G=0.42 SPRUCE-PINE-FIR Z lbs.
	× 25.4 for mm			× 4.45 for N		
1/2	2	0.099	6d	55	48	38
	2 1/2	0.113	8d	67	59	47
	3	0.128	10d	82	73	59
	3 1/4	0.128	12d	82	73	59
	3 1/2	0.135	16d	89	79	65
	4	0.148	20d	101	90	73
	4 1/2	0.148	30d	101	90	73
	5	0.162	40d	117	105	87
3/4	2	0.099	6d	61	55	47
	2 1/2	0.113	8d	79	72	57
	3	0.128	10d	101	87	68
	3 1/4	0.128	12d	101	87	68
	3 1/2	0.135	16d	108	94	74
	4	0.148	20d	121	105	83
	4 1/2	0.148	30d	121	105	83
	5	0.162	40d	138	121	96
1	2 1/2	0.113	8d	79	72	61
	3	0.128	10d	101	93	79
	3 1/4	0.128	12d	101	93	79
	3 1/2	0.135	16d	113	103	86
	4	0.148	20d	128	118	96
	4 1/2	0.148	30d	128	118	96
	5	0.162	40d	154	141	109
1 1/2	3 1/4	0.128	12d	101	93	79
	3 1/2	0.135	16d	113	103	88
	4	0.148	20d	128	118	100
	4 1/2	0.148	30d	128	118	100
	5	0.162	40d	154	141	120

[1]Tabulated lateral design values (Z) for nailed connections shall be multiplied by all applicable adjustment factors (see Division III, Part I).

[2]Tabulated lateral design values (Z) are for box nails inserted in side grain with nail axis perpendicular to wood fibers and with the following nail bending yield strengths (F_{yb}):

F_{yb}=100,000 psi (690 N/mm^2) for 0.099- (2.5 mm), 0.113- (2.9 mm), 0.128- (3.3 mm) and 0.135-inch-diameter (3.4 mm) box nails.

F_{yb}=90,000 psi (621 N/mm^2) for 0.148- (3.8 mm) and 0.162-inch-diameter (4.1 mm) box nails.

[3]For other species and configurations, see Division III, Part I.

TABLE 23-III-C-2
TABLE 23-III-D

1997 UNIFORM BUILDING CODE

TABLE 23-III-C-2—COMMON WIRE NAIL DESIGN VALUES (Z) FOR SINGLE SHEAR (Two Member) CONNECTIONS[1,2,3]
(with both members of identical species)

SIDE MEMBER THICKNESS t_s (inches)	NAIL LENGTH L (inches)	NAIL DIAMETER D (inches)		G=0.55 SOUTHERN PINE Z lbs.	G=0.50 DOUGLAS-FIR LARCH Z lbs	G=0.42 SPRUCE-PINE-FIR Z lbs.
	× 25.4 for mm		PENNY-WEIGHT	× 4.45 for N		
$^1/_2$	2	0.113	6d	67	59	47
	$2^1/_2$	0.131	8d	85	76	61
	3	0.148	10d	101	90	73
	$3^1/_4$	0.148	12d	101	90	73
	$3^1/_2$	0.162	16d	117	105	87
	4	0.192	20d	137	124	103
	$4^1/_2$	0.207	30d	148	134	112
	5	0.225	40d	162	147	123
	$5^1/_2$	0.244	50d	166	151	127
	6	0.263	60d	188	171	144
$^3/_4$	$2^1/_2$	0.131	8d	104	90	70
	3	0.148	10d	121	105	83
	$3^1/_4$	0.148	12d	121	105	83
	$3^1/_2$	0.162	16d	138	121	96
	4	0.192	20d	157	138	111
	$4^1/_2$	0.207	30d	166	147	119
	5	0.225	40d	178	158	129
	$5^1/_2$	0.244	50d	182	162	132
	6	0.263	60d	203	181	149
1	3	0.148	10d	128	118	96
	$3^1/_4$	0.148	12d	128	118	96
	$3^1/_2$	0.162	16d	154	141	109
	4	0.192	20d	183	159	124
	$4^1/_2$	0.207	30d	192	167	131
	5	0.225	40d	202	177	140
	$5^1/_2$	0.244	50d	207	181	143
	6	0.263	60d	227	199	159
$1^1/_2$	$3^1/_2$	0.162	16d	154	141	120
	4	0.192	20d	185	170	144
	$4^1/_2$	0.207	30d	203	186	158
	5	0.225	40d	224	205	172
	$5^1/_2$	0.244	50d	230	211	175
	6	0.263	60d	262	240	191

[1]Tabulated lateral design values (Z) for nailed connections shall be multiplied by all applicable adjustment factors (see Division III, Part I).

[2]Tabulated lateral design values (Z) are for common wire nails inserted in side grain with nail axis perpendicular to wood fibers and with the following nail bending yield strengths (F_{yb}):

F_{yb}=100,000 psi (690 N/mm^2) for 0.113- (2.9 mm), 0.131- (3.3 mm) and 0.135-inch-diameter (3.4 mm) common wire nails.
F_{yb}=90,000 psi (621 N/mm^2) for 0.148- (3.8 mm) and 0.162-inch-diameter (4.1 mm) common wire nails.
F_{yb}=80,000 psi (552 N/mm^2) for 0.192- (4.9 mm), 0.207 (5.3 mm) and 0.225-inch-diameter (5.7 mm) common wire nails.
F_{yb}=70,000 psi (482 N/mm^2) for 0.244- (6.2 mm) and 0.263-inch-diameter (6.7 mm) common wire nails.

[3]For other species and configurations, see Division III, Part I.

TABLE 23-III-D—NAIL AND SPIKE WITHDRAWAL DESIGN VALUES (W)[1,2]
Tabulated Withdrawal Design Values (W) Are in Pounds per Inch of Penetration into Side Grain of Main Member

	SPECIFIC GRAVITY, G	COMMON WIRE NAILS, BOX NAILS AND COMMON WIRE SPIKES Diameter, D													
		0.099″	0.113″	0.128″	0.131″	0.135″	0.148″	0.162″	0.192″	0.207″	0.225″	0.244″	0.263″	0.283″	0.312″	0.375″
Southern Pine	0.55	31	35	40	41	42	46	50	59	64	70	76	81	88	97	116
Douglas-Fir Larch	0.50	24	28	31	32	33	36	40	47	50	55	60	64	69	76	91
Spruce-Pine-Fir	0.42	16	18	20	21	21	23	26	30	33	35	38	41	45	49	59

[1]Tabulated withdrawal design values (W) for nail or spike connections shall be multiplied by all applicable adjustment factors (see Division III, Part I).
[2]For other species and configurations, see Division III, Part I.

Division IV—CONVENTIONAL LIGHT-FRAME CONSTRUCTION

SECTION 2320 — CONVENTIONAL LIGHT-FRAME CONSTRUCTION DESIGN PROVISIONS

2320.1 General. The requirements in this section are intended for conventional light-frame construction. Other methods may be used provided a satisfactory design is submitted showing compliance with other provisions of this code.

Only the following occupancies may be constructed in accordance with this division:

1. One-, two- or three-story buildings housing Group R Occupancies.

2. One-story Occupancy Category 4 buildings, as defined in Table 16-K, when constructed on a slab-on-grade floor.

3. Group U Occupancies.

4. Top-story walls and roofs of Occupancy Category 4 buildings not exceeding two stories of wood framing.

5. Interior nonload-bearing partitions, ceilings and curtain walls in all occupancies.

When total loads exceed those specified in Tables 23-IV-J-1, 23-IV-J-3, and 23-IV-R-1, 23-IV-R-2, 23-IV-R-3, 23-IV-R-4, 23-IV-R-7, 23-IV-R-8, 23-IV-R-9, 23-IV-R-10, 23-IV-R-11 and 23-IV-R-12; 23-VII-R-1, 23-VII-R-3, 23-VII-R-7, 23-VII-R-9, 23-VIII-A, 23-VIII-B, 23-VIII-C, 23-VIII-D, an engineering design shall be provided for the gravity load system.

Other approved repetitive wood members may be used in lieu of solid-sawn lumber in conventional construction provided these members comply with the provisions of this code.

2320.2 Design of Portions. When a building of otherwise conventional construction contains nonconventional structural elements, those elements shall be designed in accordance with Section 1605.2.

2320.3 Additional Requirements for Conventional Construction in High-wind Areas. Appendix Chapter 23 provisions for conventional construction in high-wind areas shall apply when specifically adopted.

2320.4 Additional Requirements for Conventional Construction in Seismic Zones 0, 1, 2 and 3.

2320.4.1 Braced wall lines. Where the basic wind speed is not greater than 80 miles per hour (mph) (129 km/h), buildings shall be provided with exterior and interior braced wall lines. Spacing shall not exceed 34 feet (10 363 mm) on center in both the longitudinal and transverse directions in each story.

2320.4.2 Braced wall lines for high wind. Where the basic wind speed exceeds 80 mph (129 km/h), buildings shall be provided with exterior and interior braced wall lines. Spacing shall not exceed 25 feet (7620 mm) on center in both the longitudinal and transverse directions in each story.

> **EXCEPTION:** In one- and two-story Group R, Division 3 buildings, interior braced wall line spacing may be increased to not more than 34 feet (10 363 mm) on center in order to accommodate one single room per dwelling unit not exceeding 900 square feet (83.6 m²). The building official may require additional walls to contain braced panels when this exception is used.

2320.4.3 Veneer. Anchored masonry and stone wall veneer shall not exceed 5 inches (127 mm) in thickness and shall conform to the requirements of Chapter 14.

2320.4.4 Lateral force-resisting system. Buildings in Seismic Zone 3 that are not provided with braced wall lines in accordance

with Section 2320.4 or that are of unusual shape as described in Section 2320.5.4 shall have a lateral-force-resisting system designed to resist the forces specified in Chapter 16.

2320.5 Additional Requirements for Conventional Construction in Seismic Zone 4.

2320.5.1 Braced wall lines. Buildings shall be provided with exterior and interior braced wall lines. Spacing shall not exceed 25 feet (7620 mm) on center in both the longitudinal and transverse directions in each story.

> **EXCEPTION:** In one- and two-story Group R, Division 3 buildings, interior braced wall line spacing may be increased to not more than 34 feet (10 363 mm) on center in order to accommodate one single room per dwelling unit not exceeding 900 square feet (83.61 m²). The building official may require additional walls to contain braced panels when this exception is used.

2320.5.2 Lateral-force-resisting system. When total loads supported on wood framing exceed those specified in Tables 23-IV-J-1, 23-IV-J-3, 23-IV-R-1, 23-IV-R-2, 23-IV-R-3, 23-IV-R-4, 23-IV-R-7, 23-IV-R-8, 23-IV-R-9 and 23-IV-R-10, 23-VII-R-1. 23-VII-R-3, 23-VII-R-7, 23-VII-R-9, 23-VIII-A, 23-VIII-B, 23-VIII-C and 23-VIII-D, an engineering design shall be provided for the lateral-force-resisting system.

2320.5.3 Veneer. Anchored masonry and stone wall veneer shall not exceed 5 inches (127 mm) in thickness, shall conform to the requirements of Chapter 14 and shall not extend above the first story.

2320.5.4 Unusually shaped buildings. When of unusual shape, buildings of light-frame construction shall have a lateral-force-resisting system designed to resist the forces specified in Chapter 16. Buildings shall be considered to be of unusual shape when the building official determines that the structure has framing irregularities, offsets, split levels or any configuration that creates discontinuities in the seismic load path and may include one or more of the following.

2320.5.4.1 When exterior braced wall panels, as required by Section 2320.11.3, are not in one plane vertically from the foundation to the uppermost story in which they are required.

> **EXCEPTION:** Floors with cantilevers or setbacks not exceeding four times the nominal depth of the floor joists may support braced wall panels provided:
> 1. Floor joists are 2 inches by 10 inches (51 mm by 254 mm) or larger and spaced at not more than 16 inches (406 mm) on center.
> 2. The ratio of the back span to the cantilever is at least 2 to 1.
> 3. Floor joists at ends of braced wall panels are doubled.
> 4. A continuous rim joist is connected to ends of all cantilevered joists. The rim joist may be spliced using a metal tie not less than 0.058 inch (1.47 mm) (16 galvanized gage) and $1^1/_2$ inches (38 mm) wide fastened with six 16d nails.
> 5. Gravity loads carried at the end of cantilevered joists are limited to uniform wall and roof load and the reactions from headers having a span of 8 feet (2438 mm) or less.

2320.5.4.2 When a section of floor or roof is not laterally supported by braced wall lines on all edges.

> **EXCEPTION:** Portions of roofs or floors which do not support braced wall panels above may extend up to 6 feet (1829 mm) beyond a braced wall line.

2320.5.4.3 When the end of a required braced wall panel extends more than 1 foot (305 mm) over an opening in the wall below. This provision is applicable to braced wall panels offset in plane and to braced wall panels offset out of plane as permitted by Section 2320.5.4.1, exception.

EXCEPTION: Braced wall panels may extend over an opening not more than 8 feet (2438 mm) in width when the header is a 4-inch by 12-inch (102 mm by 305 mm) or larger member.

2320.5.4.4 When an opening in a floor or roof exceeds the lesser of 12 feet (3657 mm) or 50 percent of the least floor or roof dimension.

2320.5.4.5 Construction where portions of a floor level are vertically offset such that the framing members on either side of the offset cannot be lapped or tied together in an approved manner as required by Section 2320.8.3.

EXCEPTION: Framing supported directly by foundations.

2320.5.4.6 When braced wall lines do not occur in two perpendicular directions.

2320.5.5 Lumber roof decks. Lumber roof decks shall have solid sheathing.

2320.5.6 Interior braced wall support. In one-story buildings, interior braced wall lines shall be supported on continuous foundations at intervals not exceeding 50 feet (15 240 mm). In buildings more than one story in height, all interior braced wall panels shall be supported on continuous foundations.

EXCEPTION: Two-story buildings may have interior braced wall lines supported on continuous foundations at intervals not exceeding 50 feet (15 240 mm) provided:

1. Cripple wall height does not exceed 4 feet (1219 mm).

2. First-floor braced wall panels are supported on doubled floor joists, continuous blocking or floor beams.

3. Distance between bracing lines does not exceed twice the building width parallel to the braced wall line.

2320.6 Foundation Plates or Sills. Foundations and footings shall be as specified in Chapter 18. Foundation plates or sills resting on concrete or masonry foundations shall be bolted as required by Section 1806.6.

2320.7 Girders. Girders for single-story construction or girders supporting loads from a single floor shall not be less than 4 inches by 6 inches (102 mm by 153 mm) for spans 6 feet (1829 mm) or less, provided that girders are spaced not more than 8 feet (2438 mm) on center. Other girders shall be designed to support the loads specified in this code. Girder end joints shall occur over supports. When a girder is spliced over a support, an adequate tie shall be provided. The end of beams or girders supported on masonry or concrete shall not have less than 3 inches (76 mm) of bearing.

2320.8 Floor Joists.

2320.8.1 General. Spans for joists shall be in accordance with Tables 23-IV-J-1 and 23-IV-J-2.

2320.8.2 Bearing. Except where supported on a 1-inch by 4-inch (25 mm by 102 mm) ribbon strip and nailed to the adjoining stud, the ends of each joist shall not have less than 1 1/2 inches (38 mm) of bearing on wood or metal, or less than 3 inches (76 mm) on masonry.

2320.8.3 Framing details. Joists shall be supported laterally at the ends and at each support by solid blocking except where the ends of joists are nailed to a header, band or rim joist or to an adjoining stud or by other approved means. Solid blocking shall not be less than 2 inches (51 mm) in thickness and the full depth of joist.

Notches on the ends of joists shall not exceed one fourth the joist depth. Holes bored in joists shall not be within 2 inches (51 mm) of the top or bottom of the joist, and the diameter of any such hole shall not exceed one third the depth of the joist. Notches in the top

or bottom of joists shall not exceed one sixth the depth and shall not be located in the middle third of the span.

Joist framing from opposite sides of a beam, girder or partition shall be lapped at least 3 inches (76 mm) or the opposing joists shall be tied together in an approved manner.

Joists framing into the side of a wood girder shall be supported by framing anchors or on ledger strips not less than 2 inches by 2 inches (51 mm by 51 mm).

2320.8.4 Framing around openings. Trimmer and header joists shall be doubled, or of lumber of equivalent cross section, when the span of the header exceeds 4 feet (1219 mm). The ends of header joists more than 6 feet (1829 mm) long shall be supported by framing anchors or joist hangers unless bearing on a beam, partition or wall. Tail joists over 12 feet (3658 mm) long shall be supported at header by framing anchors or on ledger strips not less than 2 inches by 2 inches (51 mm by 51 mm).

2320.8.5 Supporting bearing partitions. Bearing partitions perpendicular to joists shall not be offset from supporting girders, walls or partitions more than the joist depth.

Joists under and parallel to bearing partitions shall be doubled.

2320.8.6 Blocking. Floor joists shall be blocked when required by the provisions of Division III, Part I or Section 2320.8.3.

2320.9 Subflooring.

2320.9.1 Lumber subfloor. Sheathing used as a structural subfloor shall conform to the limitations set forth in Tables 23-II-D-1 and 23-II-D-2.

Joints in subflooring shall occur over supports unless end-matched lumber is used, in which case each piece shall bear on at least two joists.

Subflooring may be omitted when joist spacing does not exceed 16 inches (406 mm) and 1-inch (25 mm) nominal tongue-and-groove wood strip flooring is applied perpendicular to the joists.

2320.9.2 Wood structural panels. Where used as structural subflooring, wood structural panels shall be as set forth in Tables 23-II-E-1 and 23-II-E-2. Wood structural panel combination subfloor underlayment shall have maximum spans as set forth in Table 23-II-F-1.

When wood structural panel floors are glued to joists with an adhesive in accordance with the adhesive manufacturer's directions, fasteners may be spaced a maximum of 12 inches (305 mm) on center at all supports.

2320.9.3 Plank flooring. Plank flooring shall be designed in accordance with the general provisions of this code.

In lieu of such design, 2-inch (51 mm) tongue-and-groove planking may be used in accordance with Table 23-IV-A. Joints in such planking may be randomly spaced, provided the system is applied to not less than three continuous spans, planks are center-matched and end-matched or splined, each plank bears on at least one support and joints are separated by at least 24 inches (610 mm) in adjacent pieces. One-inch (25 mm) nominal strip square-edged flooring, 1/2-inch (12.7 mm) tongue-and-groove flooring or 3/8-inch (9.5 mm) wood structural panel shall be applied over random-length decking used as a floor. The strip and tongue-and-groove flooring shall be applied at right angles to the span of the planks. The 3/8-inch (9.5 mm) plywood shall be applied with the face grain at right angles to the span of the planks.

2320.9.4 Particleboard. Where used as structural subflooring or as combined subfloor underlayment, particleboard shall be as set forth in Table 23-II-F-2.

2320.10 Particleboard Underlayment. In accordance with approved recognized standards, particleboard floor underlayment shall conform to Type PBU. Underlayment shall not be less than $^1/_4$ inch (6.4 mm) in thickness and shall be identified by the grade mark of an approved inspection agency. Underlayment shall be installed in accordance with this code and as recommended by the manufacturer.

2320.11 Wall Framing.

2320.11.1 Size, height and spacing. The size, height and spacing of studs shall be in accordance with Table 23-IV-B except that Utility grade studs shall not be spaced more than 16 inches (406 mm) on center, or support more than a roof and ceiling, or exceed 8 feet (2438 mm) in height for exterior walls and load-bearing walls or 10 feet (3048 mm) for interior nonload-bearing walls.

2320.11.2 Framing details. Studs shall be placed with their wide dimension perpendicular to the wall. Not less than three studs shall be installed at each corner of an exterior wall.

> **EXCEPTION:** At corners, a third stud may be omitted through the use of wood spacers or backup cleats of $^3/_8$-inch-thick (9.5 mm) wood structural panel, $^3/_8$-inch (9.5 mm) Type M "Exterior Glue" particleboard, 1-inch-thick (25 mm) lumber or other approved devices that will serve as an adequate backing for the attachment of facing materials. Where fire-resistance ratings or shear values are involved, wood spacers, backup cleats or other devices shall not be used unless specifically approved for such use.

Bearing and exterior wall studs shall be capped with double top plates installed to provide overlapping at corners and at intersections with other partitions. End joints in double top plates shall be offset at least 48 inches (2438 mm).

> **EXCEPTION:** A single top plate may be used, provided the plate is adequately tied at joints, corners and intersecting walls by at least the equivalent of 3-inch by 6-inch (76 mm by 152 mm) by 0.036-inch-thick (0.9 mm) galvanized steel that is nailed to each wall or segment of wall by six 8d nails or equivalent, provided the rafters, joists or trusses are centered over the studs with a tolerance of no more than 1 inch (25 mm).

When bearing studs are spaced at 24-inch (610 mm) intervals and top plates are less than two 2-inch by 6-inch (51 mm by 152 mm) or two 3-inch by 4-inch (76 mm by 102 mm) members and when the floor joists, floor trusses or roof trusses which they support are spaced at more than 16-inch (406 mm) intervals, such joists or trusses shall bear within 5 inches (127 mm) of the studs beneath or a third plate shall be installed.

Interior nonbearing partitions may be capped with a single top plate installed to provide overlapping at corners and at intersections with other walls and partitions. The plate shall be continuously tied at joints by solid blocking at least 16 inches (406 mm) in length and equal in size to the plate or by $^1/_8$-inch by $1^1/_2$-inch (3.2 mm by 38 mm) metal ties with spliced sections fastened with two 16d nails on each side of the joint.

Studs shall have full bearing on a plate or sill not less than 2 inches (51 mm) in thickness having a width not less than that of the wall studs.

2320.11.3 Bracing. Braced wall lines shall consist of braced wall panels which meet the requirements for location, type and amount of bracing specified in Table 23-IV-C-1 and are in line or offset from each other by not more than 4 feet (1219 mm). Braced wall panels shall start at not more than 8 feet (2438 mm) from each end of a braced wall line. All braced wall panels shall be clearly indicated on the plans. Construction of braced wall panels shall be by one of the following methods:

1. Nominal 1-inch by 4-inch (25 mm by 102 mm) continuous diagonal braces let into top and bottom plates and intervening studs, placed at an angle not more than 60 degrees or less than 45 degrees from the horizontal, and attached to the framing in conformance with Table 23-II-B-1.

2. Wood boards of $^5/_8$-inch (16 mm) net minimum thickness applied diagonally on studs spaced not over 24 inches (610 mm) on center.

3. Wood structural panel sheathing with a thickness not less than $^5/_{16}$ inch (7.9 mm) for 16-inch (406 mm) stud spacing and not less than $^3/_8$ inch (9.5 mm) for 24-inch (610 mm) stud spacing in accordance with Tables 23-II-A-1 and 23-IV-D-1.

4. Fiberboard sheathing 4-foot by 8-foot (1219 mm by 2438 mm) panels not less than $^1/_2$ inch (13 mm) thick applied vertically on studs spaced not over 16 inches (406 mm) on center when installed in accordance with Section 2315.6 and Table 23-II-J.

5. Gypsum board [sheathing $^1/_2$ inch (13 mm) thick by 4 feet (1219 mm) wide, wallboard or veneer base] on studs spaced not over 24 inches (610 mm) on center and nailed at 7 inches (178 mm) on center with nails as required by Table 25-I.

6. Particleboard wall sheathing panels where installed in accordance with Table 23-IV-D-2.

7. Portland cement plaster on studs spaced 16 inches (406 mm) on center installed in accordance with Table 25-I.

8. Hardboard panel siding when installed in accordance with Section 2310.6 and Table 23-II-C.

Method 1 is not permitted in Seismic Zones 2B, 3 and 4. For cripple wall bracing, see Section 2320.11.5. For Methods 2, 3, 4, 6, 7 and 8, each braced panel must be at least 48 inches (1219 mm) in length, covering three stud spaces where studs are spaced 16 inches (406 mm) apart and covering two stud spaces where studs are spaced 24 inches (610 mm) apart.

For Method 5, each braced wall panel must be at least 96 inches (2438 mm) in length when applied to one face of a braced wall panel and 48 inches (1219 mm) when applied to both faces.

All vertical joints of panel sheathing shall occur over studs. Horizontal joints shall occur over blocking equal in size to the studding except where waived by the installation requirements for the specific sheathing materials.

Braced wall panel sole plates shall be nailed to the floor framing and top plates shall be connected to the framing above in accordance with Table 23-II-B-1. Sills shall be bolted to the foundation or slab in accordance with Section 1806.6. Where joists are perpendicular to braced wall lines above, blocking shall be provided under and in line with the braced wall panels.

2320.11.4 Alternate braced wall panels. Any braced wall panel required by Section 2320.11.3 may be replaced by an alternate braced wall panel constructed in accordance with the following:

1. In one-story buildings, each panel shall have a length of not less than 2 feet 8 inches (813 mm) and a height of not more than 10 feet (3048 mm). Each panel shall be sheathed on one face with $^3/_8$-inch-minimum-thickness (9.5 mm) plywood sheathing nailed with 8d common or galvanized box nails in accordance with Table 23-II-B-1 and blocked at all plywood edges. Two anchor bolts installed in accordance with Section 1806.6, shall be provided in each panel. Anchor bolts shall be placed at panel quarter points. Each panel end stud shall have a tie-down device fastened to the foundation, capable of providing an approved uplift capacity of not less than 1,800 pounds (816.5 kg). The tie-down device shall be installed in accordance with the manufacturer's recommendations. The panels shall be supported directly on a foundation or on floor framing supported directly on a foundation which is continuous across the entire length of the braced wall line. This founda-

tion shall be reinforced with not less than one No. 4 bar top and bottom.

2. In the first story of two-story buildings, each braced wall panel shall be in accordance with Section 2320.11.4, Item 1, except that the plywood sheathing shall be provided on both faces, three anchor bolts shall be placed at one-fifth points, and tie-down device uplift capacity shall not be less than 3,000 pounds (1360.8 kg).

2320.11.5 Cripple walls. Foundation cripple walls shall be framed of studs not less in size than the studding above with a minimum length of 14 inches (356 mm), or shall be framed of solid blocking. When exceeding 4 feet (1219 mm) in height, such walls shall be framed of studs having the size required for an additional story.

Cripple walls having a stud height exceeding 14 inches (356 mm) shall be braced in accordance with Table 23-IV-C-2. Solid blocking or wood structural panel sheathing may be used to brace cripple walls having a stud height of 14 inches (356 mm) or less. In Seismic Zone 4, Method 7 is not permitted for bracing any cripple wall studs.

Spacing of boundary nailing for required wall bracing shall not exceed 6 inches (152 mm) on center along the foundation plate and the top plate of the cripple wall. Nail size, nail spacing for field nailing and more restrictive boundary nailing requirements shall be as required elsewhere in the code for the specific bracing material used.

2320.11.6 Headers. Headers and lintels shall conform to the requirements set forth in this paragraph and together with their supporting systems shall be designed to support the loads specified in this code. All openings 4 feet (1219 mm) wide or less in bearing walls shall be provided with headers consisting of either two pieces of 2-inch (51 mm) framing lumber placed on edge and securely fastened together or 4-inch (102 mm) lumber of equivalent cross section. All openings more than 4 feet (1219 mm) wide shall be provided with headers or lintels. Each end of a lintel or header shall have a length of bearing of not less than $1^1/_2$ inches (38 mm) for the full width of the lintel.

2320.11.7 Pipes in walls. Stud partitions containing plumbing, heating, or other pipes shall be so framed and the joists underneath so spaced as to give proper clearance for the piping. Where a partition containing such piping runs parallel to the floor joists, the joists underneath such partitions shall be doubled and spaced to permit the passage of such pipes and shall be bridged. Where plumbing, heating or other pipes are placed in or partly in a partition, necessitating the cutting of the soles or plates, a metal tie not less than 0.058 inch (1.47 mm) (16 galvanized gage) and $1^1/_2$ inches (38 mm) wide shall be fastened to each plate across and to each side of the opening with not less than six 16d nails.

2320.11.8 Bridging. Unless covered by interior or exterior wall coverings or sheathing meeting the minimum requirements of this code, all stud partitions or walls with studs having a height-to-least-thickness ratio exceeding 50 shall have bridging not less than 2 inches (51 mm) in thickness and of the same width as the studs fitted snugly and nailed thereto to provide adequate lateral support.

2320.11.9 Cutting and notching. In exterior walls and bearing partitions, any wood stud may be cut or notched to a depth not exceeding 25 percent of its width. Cutting or notching of studs to a depth not greater than 40 percent of the width of the stud is permitted in nonbearing partitions supporting no loads other than the weight of the partition.

2320.11.10 Bored holes. A hole not greater in diameter than 40 percent of the stud width may be bored in any wood stud. Bored holes not greater than 60 percent of the width of the stud are permitted in nonbearing partitions or in any wall where each bored stud is doubled, provided not more than two such successive doubled studs are so bored.

In no case shall the edge of the bored hole be nearer than $5/_8$ inch (16 mm) to the edge of the stud. Bored holes shall not be located at the same section of stud as a cut or notch.

2320.12 Roof and Ceiling Framing.

2320.12.1 General. The framing details required in this section apply to roofs having a minimum slope of 3 units vertical in 12 units horizontal (25% slope) or greater. When the roof slope is less than 3 units vertical in 12 units horizontal (25% slope), members supporting rafters and ceiling joists such as ridge board, hips and valleys shall be designed as beams.

2320.12.2 Spans. Allowable spans for ceiling joists shall be in accordance with Tables 23-IV-J-3 and 23-IV-J-4. Allowable spans for rafters shall be in accordance with Tables 23-IV-R-1 through 23-IV-R-12, where applicable.

2320.12.3 Framing. Rafters shall be framed directly opposite each other at the ridge. There shall be a ridge board at least 1-inch (25 mm) nominal thickness at all ridges and not less in depth than the cut end of the rafter. At all valleys and hips there shall be a single valley or hip rafter not less than 2-inch (51 mm) nominal thickness and not less in depth than the cut end of the rafter.

2320.12.4 Notches and holes. Notching at the ends of rafters or ceiling joists shall not exceed one fourth the depth. Notches in the top or bottom of the rafter or ceiling joist shall not exceed one sixth the depth and shall not be located in the middle one third of the span, except that a notch not exceeding one third of the depth is permitted in the top of the rafter or ceiling joist not further from the face of the support than the depth of the member.

Holes bored in rafters or ceiling joists shall not be within 2 inches (51 mm) of the top and bottom and their diameter shall not exceed one third the depth of the member.

2320.12.5 Framing around openings. Trimmer and header rafters shall be doubled, or of lumber of equivalent cross section, when the span of the header exceeds 4 feet (1219 mm). The ends of header rafters more than 6 feet (1829 mm) long shall be supported by framing anchors or rafter hangers unless bearing on a beam, partition or wall.

2320.12.6 Rafter ties. Rafters shall be nailed to adjacent ceiling joists to form a continuous tie between exterior walls when such joists are parallel to the rafters. Where not parallel, rafters shall be tied to 1-inch by 4-inch (25 mm by 102 mm) (nominal) minimum-size crossties. Rafter ties shall be spaced not more than 4 feet (1219 mm) on center.

2320.12.7 Purlins. Purlins to support roof loads may be installed to reduce the span of rafters within allowable limits and shall be supported by struts to bearing walls. The maximum span of 2-inch by 4-inch (51 mm by 102 mm) purlins shall be 4 feet (1219 mm). The maximum span of the 2-inch by 6-inch (51 mm by 152 mm) purlin shall be 6 feet (1829 mm) but in no case shall the purlin be smaller than the supported rafter. Struts shall not be smaller than 2-inch by 4-inch (51 mm by 102 mm) members. The unbraced length of struts shall not exceed 8 feet (2438 mm) and the minimum slope of the struts shall not be less than 45 degrees from the horizontal.

2320.12.8 Blocking. Roof rafters and ceiling joists shall be supported laterally to prevent rotation and lateral displacement when required by Division III, Part I, Section 4.4.1.2. Roof trusses shall be supported laterally at points of bearing by solid blocking to prevent rotation and lateral displacement.

2320.12.9 Roof sheathing. Roof sheathing shall be in accordance with Tables 23-II-E-1 and 23-II-E-2 for wood structural panels, and Tables 23-II-D-1 and 23-II-D-2 for lumber.

Joints in lumber sheathing shall occur over supports unless approved end-matched lumber is used, in which case each piece shall bear on at least two supports.

Wood structural panels used for roof sheathing shall be bonded by intermediate or exterior glue. Wood structural panel roof sheathing exposed on the underside shall be bonded with exterior glue.

2320.12.10 Roof planking. Planking shall be designed in accordance with the general provisions of this code.

In lieu of such design, 2-inch (51 mm) tongue-and-groove planking may be used in accordance with Table 23-IV-A. Joints in such planking may be randomly spaced, provided the system is applied to not less than three continuous spans, planks are center-matched and end-matched or splined, each plank bears on at least one support, and joints are separated by at least 24 inches (610 mm) in adjacent pieces.

2320.13 Exit Facilities. In Seismic Zones 3 and 4, exterior exit balconies, stairs and similar exit facilities shall be positively anchored to the primary structure at not over 8 feet (2438 mm) on center or shall be designed for lateral forces. Such attachment shall not be accomplished by use of toenails or nails subject to withdrawal.

TABLE 23-IV-A

1997 UNIFORM BUILDING CODE

TABLE 23-IV-A—ALLOWABLE SPANS FOR 2-INCH (51 mm) TONGUE-AND-GROOVE DECKING

SPAN[1] (feet) × 304.8 for mm	LIVE LOAD × 0.0479 for kN/m²	DEFLECTION LIMIT	f (psi) × 0.00689 for N/mm²	E (psi) × 0.00689 for N/mm²
		Roofs		
4	20	1/240 1/360	160	170,000 256,000
	30	1/240 1/360	210	256,000 384,000
	40	1/240 1/360	270	340,000 512,000
4.5	20	1/240 1/360	200	242,000 305,000
	30	1/240 1/360	270	363,000 405,000
	40	1/240 1/360	350	484,000 725,000
5.0	20	1/240 1/360	250	332,000 500,000
	30	1/240 1/360	330	495,000 742,000
	40	1/240 1/360	420	660,000 1,000,000
5.5	20	1/240 1/360	300	442,000 660,000
	30	1/240 1/360	400	662,000 998,000
	40	1/240 1/360	500	884,000 1,330,000
6.0	20	1/240 1/360	360	575,000 862,000
	30	1/240 1/360	480	862,000 1,295,000
	40	1/240 1/360	600	1,150,000 1,730,000
6.5	20	1/240 1/360	420	595,000 892,000
	30	1/240 1/360	560	892,000 1,340,000
	40	1/240 1/360	700	1,190,000 1,730,000
7.0	20	1/240 1/360	490	910,000 1,360,000
	30	1/240 1/360	650	1,370,000 2,000,000
	40	1/240 1/360	810	1,820,000 2,725,000
7.5	20	1/240 1/360	560	1,125,000 1,685,000
	30	1/240 1/360	750	1,685,000 2,530,000
	40	1/240 1/360	930	2,250,000 3,380,000
8.0	20	1/240 1/360	640	1,360,000 2,040,000
	30	1/240 1/360	850	2,040,000 3,060,000
		Floors		
4 4.5 5.0	40	1/360	840 950 1060	1,000,000 1,300,000 1,600,000

[1]Spans are based on simple beam action with 10 pounds per square foot (0.48 kN/m²) dead load and provisions for a 300-pound (1334 N) concentrated load on a 12-inch (305 mm) width of floor decking. Random lay-up permitted in accordance with the provisions of Section 2320.9.3 or 2320.12.9. Lumber thickness assumed at 1¹/₂ inches (38 mm), net.

TABLE 23-IV-B—SIZE, HEIGHT AND SPACING OF WOOD STUDS

STUD SIZE (inches)	BEARING WALLS				NONBEARING WALLS	
	Laterally Unsupported Stud Height[1] (feet)	Supporting Roof and Ceiling Only	Supporting One Floor, Roof and Ceiling	Supporting Two Floors, Roof and Ceiling	Laterally Unsupported Stud Height[1] (feet)	Spacing (inches)
		Spacing (inches)				
× 25.4 for mm	× 304.8 for mm	× 25.4 for mm			× 304.8 for mm	× 25.4 for mm
1. 2 × 3[2]	—	—	—	—	10	16
2. 2 × 4	10	24	16	—	14	24
3. 3 × 4	10	24	24	16	14	24
4. 2 × 5	10	24	24	—	16	24
5. 2 × 6	10	24	24	16	20	24

[1]Listed heights are distances between points of lateral support placed perpendicular to the plane of the wall. Increases in unsupported height are permitted where justified by an analysis.
[2]Shall not be used in exterior walls.

TABLE 23-IV-C-1—BRACED WALL PANELS[1]

SEISMIC ZONE	CONDITION	CONSTRUCTION METHOD[2,3]								BRACED PANEL LOCATION AND LENGTH[4]
		1	2	3	4	5	6	7	8	
0, 1 and 2A	One story, top of two or three story	X	X	X	X	X	X	X	X	Each end and not more than 25 feet (7620 mm) on center
	First story of two story or second story of three story	X	X	X	X	X	X	X	X	
	First story of three story		X	X	X	X[5]	X	X	X	
2B, 3 and 4	One story, top of two story or three story		X	X	X	X	X	X[6]	X	Each end and not more than 25 feet (7620 mm) on center
	First story of two story or second of three story		X	X	X	X[5]	X	X[6]	X	Each end and not more than 25 feet (7620 mm) on center but not less than 25% of building length[7]
	First story of three story		X	X	X	X[5]	X	X[6]	X	Each end and not more than 25 feet (7620 mm) on center but not less than 40% of building length[7]

[1]This table specifies minimum requirements for braced panels which form interior or exterior braced wall lines.
[2]See Section 2320.11.3 for full description.
[3]See Section 2320.11.4 for alternate braced panel requirement.
[4]Building length is the dimension parallel to the braced wall length.
[5]Gypsum wallboard applied to supports at 16 inches (406 mm) on center.
[6]Not permitted for bracing cripple walls in Seismic Zone 4. See Section 2320.11.5.
[7]The required lengths shall be doubled for gypsum board applied to only one face of a braced wall panel.

TABLE 23-IV-C-2—CRIPPLE WALL BRACING

SEISMIC ZONE	CONDITION	AMOUNT OF CRIPPLE WALL BRACING[1,2]
		× 25.4 for mm
4	One story above cripple wall	3/8″ wood structural panel with 8d at 6″/12″ nailing on 60 percent of wall length minimum
	Two story above cripple wall	3/8″ wood structural panel with 8d at 4″/12″ nailing on 50 percent of wall length minimum or 3/8″ wood structural panel with 8d at 6″/12″ nailing on 75 percent of wall length minimum
3	One story above cripple wall	3/8″ wood structural panel with 8d at 6″/12″ nailing on 40 percent of wall length minimum
0, 1 and 2	One story above cripple wall	3/8″ wood structural panel with 8d at 6″/12″ nailing on 30 percent of wall length minimum
0, 1, 2 and 3	Two story above cripple wall	3/8″ wood structural panel with 8d at 4″/12″ nailing on 40 percent of wall length minimum or 3/8″ wood structural panel with 8d at 6″/12″ nailing on 60 percent of wall length minimum

[1]Braced panel length shall be at least two times the height of the cripple wall, but not less than 48 inches (1219 mm).
[2]All panels along a wall shall be nearly equal in length and shall be nearly equally spaced along the length of the wall.

TABLE 23-IV-D-1
TABLE 23-IV-D-2

1997 UNIFORM BUILDING CODE

TABLE 23-IV-D-1—WOOD STRUCTURAL PANEL WALL SHEATHING[1]
(Not exposed to the weather, strength axis parallel or perpendicular to studs)

MINIMUM THICKNESS (inch) $\times$ 25.4 for mm	PANEL SPAN RATING	STUD SPACING (inches) $\times$ 25.4 for mm		
		Siding Nailed to Studs	Sheathing under Coverings Specified in Section 2310.4	
			Sheathing Parallel to Studs	Sheathing Perpendicular to Studs
$5/16$	12/0, 16/0, 20/0 Wall—16 o.c.	16	—	16
$3/8$, $15/32$, $1/2$	16/0, 20/0, 24/0, 32/16 Wall—24 o.c.	24	16	24
$7/16$, $15/32$, $1/2$	24/0, 24/16, 32/16 Wall—24 o.c.	24	24[2]	24

[1]In reference to Section 2320.11.3, blocking of horizontal joints is not required.
[2]Plywood shall consist of four or more plies.

TABLE 23-IV-D-2—ALLOWABLE SPANS FOR PARTICLEBOARD WALL SHEATHING[1]
(Not exposed to the weather, long dimension of the panel parallel or perpendicular to studs)

GRADE	THICKNESS (Inch)	STUD SPACING (inches) $\times$ 25.4 for mm	
		Siding Nailed to Studs	Sheathing under Coverings Specified in Section 2310.4 Parallel or Perpendicular to Studs
		$\times$ 25.4 for mm	
M-1 M-S	$3/8$	16	16
M-2 "Exterior Glue"	$1/2$	16	16

[1]In reference to Section 2320.11.3, blocking of horizontal joints is not required.

TABLE 23-IV-J-1—FLOOR JOISTS WITH _L_/360 DEFLECTION LIMITS
The allowable bending stress (_Fb_) and modulus of elasticity _(E)_ used in this table shall be from Tables 23-IV-V-1 and 23-IV-V-2 only.

DESIGN CRITERIA:
Deflection — For 40 psf (1.92 kN/m²) live load.
Limited to span in inches (mm) divided by 360.
Strength — Live load of 40 psf (1.92 kN/m²) plus dead load of 10 psf (0.48 kN/m²) determines the required bending design value.

Joist Size (in)	Spacing (in)	Modulus of Elasticity, E, in 1,000,000 psi × 0.00689 for N/mm²																
× 25.4 for mm		0.8	0.9	1.0	1.1	1.2	1.3	1.4	1.5	1.6	1.7	1.8	1.9	2.0	2.1	2.2	2.3	2.4
2 × 6	12.0	8-6	8-10	9-2	9-6	9-9	10-0	10-3	10-6	10-9	10-11	11-2	11-4	11-7	11-9	11-11	12-1	12-3
	16.0	7-9	8-0	8-4	8-7	8-10	9-1	9-4	9-6	9-9	9-11	10-2	10-4	10-6	10-8	10-10	11-0	11-2
	19.2	7-3	7-7	7-10	8-1	8-4	8-7	8-9	9-0	9-2	9-4	9-6	9-8	9-10	10-0	10-2	10-4	10-6
	24.0	6-9	7-0	7-3	7-6	7-9	7-11	8-2	8-4	8-6	8-8	8-10	9-0	9-2	9-4	9-6	9-7	9-9
2 × 8	12.0	11-3	11-8	12-1	12-6	12-10	13-2	13-6	13-10	14-2	14-5	14-8	15-0	15-3	15-6	15-9	15-11	16-2
	16.0	10-2	10-7	11-0	11-4	11-8	12-0	12-3	12-7	12-10	13-1	13-4	13-7	13-10	14-1	14-3	14-6	14-8
	19.2	9-7	10-0	10-4	10-8	11-0	11-3	11-7	11-10	12-1	12-4	12-7	12-10	13-0	13-3	13-5	13-8	13-10
	24.0	8-11	9-3	9-7	9-11	10-2	10-6	10-9	11-0	11-3	11-5	11-8	11-11	12-1	12-3	12-6	12-8	12-10
2 × 10	12.0	14-4	14-11	15-5	15-11	16-5	16-10	17-3	17-8	18-0	18-5	18-9	19-1	19-5	19-9	20-1	20-4	20-8
	16.0	13-0	13-6	14-0	14-6	14-11	15-3	15-8	16-0	16-5	16-9	17-0	17-4	17-8	17-11	18-3	18-6	18-9
	19.2	12-3	12-9	13-2	13-7	14-0	14-5	14-9	15-1	15-5	15-9	16-0	16-4	16-7	16-11	17-2	17-5	17-8
	24.0	11-4	11-10	12-3	12-8	13-0	13-4	13-8	14-0	14-4	14-7	14-11	15-2	15-5	15-8	15-11	16-2	16-5
2 × 12	12.0	17-5	18-1	18-9	19-4	19-11	20-6	21-0	21-6	21-11	22-5	22-10	23-3	23-7	24-0	24-5	24-9	25-1
	16.0	15-10	16-5	17-0	17-7	18-1	18-7	19-1	19-6	19-11	20-4	20-9	21-1	21-6	21-10	22-2	22-6	22-10
	19.2	14-11	15-6	16-0	16-7	17-0	17-6	17-11	18-4	18-9	19-2	19-6	19-10	20-2	20-6	20-10	21-2	21-6
	24.0	13-10	14-4	14-11	15-4	15-10	16-3	16-8	17-0	17-5	17-9	18-1	18-5	18-9	19-1	19-4	19-8	19-11
_Fb_	12.0	718	777	833	888	941	993	1,043	1,092	1,140	1,187	1,233	1,278	1,323	1,367	1,410	1,452	1,494
	16.0	790	855	917	977	1,036	1,093	1,148	1,202	1,255	1,306	1,357	1,407	1,456	1,504	1,551	1,598	1,644
	19.2	840	909	975	1,039	1,101	1,161	1,220	1,277	1,333	1,388	1,442	1,495	1,547	1,598	1,649	1,698	1,747
	24.0	905	979	1,050	1,119	1,186	1,251	1,314	1,376	1,436	1,496	1,554	1,611	1,667	1,722	1,776	1,829	1,882

NOTE: The required bending design value, _Fb_, in pounds per square inch (× 0.00689 for N/mm²) is shown at the bottom of this table and is applicable to all lumber sizes shown. Spans are shown in feet-inches (1 foot = 304.8 mm, 1 inch = 25.4 mm) and are limited to 26 feet (7925 mm) and less.

TABLE 23-IV-J-2

1997 UNIFORM BUILDING CODE

TABLE 23-IV-J-2—FLOOR JOISTS WITH $L/360$ DEFLECTION LIMITS
The allowable bending stress (F_b) and modulus of elasticity (E) used in this table shall be from Tables 23-IV-V-1 and 23-IV-V-2 only.

DESIGN CRITERIA:
Deflection — For 40 psf (1.92 kN/m²) live load.
Limited to span in inches (mm) divided by 360.
Strength — Live load of 40 psf (1.92 kN/m²) plus dead load of 20 psf (0.96 kN/m²) determines the required bending design value.

Joist Size (in) ×25.4 for mm	Spacing (in)	Modulus of Elasticity, E, in 1,000,000 psi × 0.00689 for N/mm²																
		0.8	**0.9**	**1.0**	**1.1**	**1.2**	**1.3**	**1.4**	**1.5**	**1.6**	**1.7**	**1.8**	**1.9**	**2.0**	**2.1**	**2.2**	**2.3**	**2.4**
2×6	12.0	8-6	8-10	9-2	9-6	9-9	10-0	10-3	10-6	10-9	10-11	11-2	11-4	11-7	11-9	11-11	12-1	12-3
	16.0	7-9	8-0	8-4	8-7	8-10	9-1	9-4	9-6	9-9	9-11	10-2	10-4	10-6	10-8	10-10	11-0	11-2
	19.2	7-3	7-7	7-10	8-1	8-4	8-7	8-9	9-0	9-2	9-4	9-6	9-8	9-10	10-0	10-2	10-4	10-6
	24.0	6-9	7-0	7-3	7-6	7-9	7-11	8-2	8-4	8-6	8-8	8-10	9-0	9-2	9-4	9-6	9-7	9-9
2×8	12.0	11-3	11-8	12-1	12-6	12-10	13-2	13-6	13-10	14-2	14-5	14-8	15-0	15-3	15-6	15-9	15-11	16-2
	16.0	10-2	10-7	11-0	11-4	11-8	12-0	12-3	12-7	12-10	13-1	13-4	13-7	13-10	14-1	14-3	14-6	14-8
	19.2	9-7	10-0	10-4	10-8	11-0	11-3	11-7	11-10	12-1	12-4	12-7	12-10	13-0	13-3	13-5	13-8	13-10
	24.0	8-11	9-3	9-7	9-11	10-2	10-6	10-9	11-0	11-3	11-5	11-8	11-11	12-1	12-3	12-6	12-8	12-10
2×10	12.0	14-4	14-11	15-5	15-11	16-5	16-10	17-3	17-8	18-0	18-5	18-9	19-1	19-5	19-9	20-1	20-4	20-8
	16.0	13-0	13-6	14-0	14-6	14-11	15-3	15-8	16-0	16-5	16-9	17-0	17-4	17-8	17-11	18-3	18-6	18-9
	19.2	12-3	12-9	13-2	13-7	14-0	14-5	14-9	15-1	15-5	15-9	16-0	16-4	16-7	16-11	17-2	17-5	17-8
	24.0	11-4	11-10	12-3	12-8	13-0	13-4	13-8	14-0	14-4	14-7	14-11	15-2	15-5	15-8	15-11	16-2	16-5
2×12	12.0	17-5	18-1	18-9	19-4	19-11	20-6	21-0	21-6	21-11	22-5	22-10	23-3	23-7	24-0	24-5	24-9	25-1
	16.0	15-10	16-5	17-0	17-7	18-1	18-7	19-1	19-6	19-11	20-4	20-9	21-1	21-6	21-10	22-2	22-6	22-10
	19.2	14-11	15-6	16-0	16-7	17-0	17-6	17-11	18-4	18-9	19-2	19-6	19-10	20-2	20-6	20-10	21-2	21-6
	24.0	13-10	14-4	14-11	15-4	15-10	16-3	16-8	17-0	17-5	17-9	18-1	18-5	18-9	19-1	19-4	19-8	19-11
F_b	12.0	862	932	1,000	1,066	1,129	1,191	1,251	1,310	1,368	1,424	1,480	1,534	1,587	1,640	1,692	1,742	1,793
	16.0	949	1,026	1,101	1,173	1,243	1,311	1,377	1,442	1,506	1,568	1,629	1,688	1,747	1,805	1,862	1,918	1,973
	19.2	1,008	1,090	1,170	1,246	1,321	1,393	1,464	1,533	1,600	1,666	1,731	1,794	1,857	1,918	1,978	2,038	2,097
	24.0	1,086	1,174	1,260	1,343	1,423	1,501	1,577	1,651	1,724	1,795	1,864	1,933	2,000	2,066	2,131	2,195	2,258

NOTE: The required bending design value, F_b, in pounds per square inch (× 0.00689 for N/mm²) is shown at the bottom of this table and is applicable to all lumber sizes shown. Spans are shown in feet-inches (1 foot = 304.8 mm, 1 inch = 25.4 mm) and are limited to 26 feet (7925 mm) and less.

TABLE 23-IV-J-3—CEILING JOISTS WITH L/240 DEFLECTION LIMITS
The allowable bending stress (F_b) and modulus of elasticity (E) used in this table shall be from Tables 23-IV-V-1 and 23-IV-V-2 only.

DESIGN CRITERIA:
Deflection — For 10 psf (0.48 kN/m²) live load.
Limited to span in inches (mm) divided by 240.
Strength — Live load of 10 psf (0.48 kN/mm²) plus dead load of 5 psf (0.24 kN/m²) determines the required fiber stress value.

Joist Size (in) × 25.4 for mm	Spacing (in)	Modulus of Elasticity, E, in 1,000,000 psi × 0.00689 for N/mm²																
		0.8	0.9	1.0	1.1	1.2	1.3	1.4	1.5	1.6	1.7	1.8	1.9	2.0	2.1	2.2	2.3	2.4
2 × 4	12.0	9-10	10-3	10-7	10-11	11-3	11-7	11-10	12-2	12-5	12-8	12-11	13-2	13-4	13-7	13-9	14-0	14-2
	16.0	8-11	9-4	9-8	9-11	10-3	10-6	10-9	11-0	11-3	11-6	11-9	11-11	12-2	12-4	12-6	12-9	12-11
	19.2	8-5	8-9	9-1	9-4	9-8	9-11	10-2	10-4	10-7	10-10	11-0	11-3	11-5	11-7	11-9	12-0	12-2
	24.0	7-10	8-1	8-5	8-8	8-11	9-2	9-5	9-8	9-10	10-0	10-3	10-5	10-7	10-9	10-11	11-1	11-3
2 × 6	12.0	15-6	16-1	16-8	17-2	17-8	18-2	18-8	19-1	19-6	19-11	20-3	20-8	21-0	21-4	21-8	22-0	22-4
	16.0	14-1	14-7	15-2	15-7	16-1	16-6	16-11	17-4	17-8	18-1	18-5	18-9	19-1	19-5	19-8	20-0	20-3
	19.2	13-3	13-9	14-3	14-8	15-2	15-7	15-11	16-4	16-8	17-0	17-4	17-8	17-11	18-3	18-6	18-10	19-1
	24.0	12-3	12-9	13-3	13-8	14-1	14-5	14-9	15-2	15-6	15-9	16-1	16-4	16-8	16-11	17-2	17-5	17-8
2 × 8	12.0	20-5	21-2	21-11	22-8	23-4	24-0	24-7	25-2	25-8								
	16.0	18-6	19-3	19-11	20-7	21-2	21-9	22-4	22-10	23-4	23-10	24-3	24-8	25-2	25-7	25-11		
	19.2	17-5	18-1	18-9	19-5	19-11	20-6	21-0	21-6	21-11	22-5	22-10	23-3	23-8	24-0	24-5	24-9	25-2
	24.0	16-2	16-10	17-5	18-0	18-6	19-0	19-6	19-11	20-5	20-10	21-2	21-7	21-11	22-4	22-8	23-0	23-4
2 × 10	12.0	26-0																
	16.0	23-8	24-7	25-5														
	19.2	22-3	23-1	23-11	24-9	25-5												
	24.0	20-8	21-6	22-3	22-11	23-8	24-3	24-10	25-5	26-0								
F_b	12.0	711	769	825	880	932	983	1,033	1,082	1,129	1,176	1,221	1,266	1,310	1,354	1,396	1,438	1,480
	16.0	783	847	909	968	1,026	1,082	1,137	1,191	1,243	1,294	1,344	1,394	1,442	1,490	1,537	1,583	1,629
	19.2	832	900	965	1,029	1,090	1,150	1,208	1,265	1,321	1,375	1,429	1,481	1,533	1,583	1,633	1,682	1,731
	24.0	896	969	1,040	1,108	1,174	1,239	1,302	1,363	1,423	1,481	1,539	1,595	1,651	1,706	1,759	1,812	1,864

NOTE: The required bending design value, F_b, in pounds per square inch (× 0.00689 for N/mm²) is shown at the bottom of this table and is applicable to all lumber sizes shown. Spans are shown in feet-inches (1 foot = 304.8 mm, 1 inch = 25.4 mm) and are limited to 26 feet (7925 mm) and less.

TABLE 23-IV-J-4

1997 UNIFORM BUILDING CODE

TABLE 23-IV-J-4—CEILING JOISTS WITH *L*/240 DEFLECTION LIMITS
The allowable bending stress (F_b) and modulus of elasticity *(E)* used in this table shall be from Tables 23-IV-V-1 and 23-IV-V-2 only.

DESIGN CRITERIA:
Deflection — For 20 psf (0.96 kN/m²) live load.
Limited to span in inches (mm) divided by 240.
Strength — Live load of 20 psf (0.96 kN/m²) plus dead load of 10 psf (0.48 kN/m²) determines the required bending design value.

Joist Size (in)	Spacing (in)	Modulus of Elasticity, *E*, in 1,000,000 psi × 0.00689 for N/mm²																
× 25.4 for mm		0.8	0.9	1.0	1.1	1.2	1.3	1.4	1.5	1.6	1.7	1.8	1.9	2.0	2.1	2.2	2.3	2.4
2 × 4	12.0	7-10	8-1	8-5	8-8	8-11	9-2	9-5	9-8	9-10	10-0	10-3	10-5	10-7	10-9	10-11	11-1	11-3
	16.0	7-1	7-5	7-8	7-11	8-1	8-4	8-7	8-9	8-11	9-1	9-4	9-6	9-8	9-9	9-11	10-1	10-3
	19.2	6-8	6-11	7-2	7-5	7-8	7-10	8-1	8-3	8-5	8-7	8-9	8-11	9-1	9-3	9-4	9-6	9-8
	24.0	6-2	6-5	6-8	6-11	7-1	7-3	7-6	7-8	7-10	8-0	8-1	8-3	8-5	8-7	8-8	8-10	8-11
2 × 6	12.0	12-3	12-9	13-3	13-8	14-1	14-5	14-9	15-2	15-6	15-9	16-1	16-4	16-8	16-11	17-2	17-5	17-8
	16.0	11-2	11-7	12-0	12-5	12-9	13-1	13-5	13-9	14-1	14-4	14-7	14-11	15-2	15-5	15-7	15-10	16-1
	19.2	10-6	10-11	11-4	11-8	12-0	12-4	12-8	12-11	13-3	13-6	13-9	14-0	14-3	14-6	14-8	14-11	15-2
	24.0	9-9	10-2	10-6	10-10	11-2	11-5	11-9	12-0	12-3	12-6	12-9	13-0	13-3	13-5	13-8	13-10	14-1
2 × 8	12.0	16-2	16-10	17-5	18-0	18-6	19-0	19-6	19-11	20-5	20-10	21-2	21-7	21-11	22-4	22-8	23-0	23-4
	16.0	14-8	15-3	15-10	16-4	16-10	17-3	17-9	18-1	18-6	18-11	19-3	19-7	19-11	20-3	20-7	20-11	21-2
	19.2	13-10	14-5	14-11	15-5	15-10	16-3	16-8	17-1	17-5	17-9	18-1	18-5	18-9	19-1	19-5	19-8	19-11
	24.0	12-10	13-4	13-10	14-3	14-8	15-1	15-6	15-10	16-2	16-6	16-10	17-2	17-5	17-9	18-0	18-3	18-6
2 × 10	12.0	20-8	21-6	22-3	22-11	23-8	24-3	24-10	25-5	26-0								
	16.0	18-9	19-6	20-2	20-10	21-6	22-1	22-7	23-1	23-8	24-1	24-7	25-0	25-5	25-10			
	19.2	17-8	18-4	19-0	19-7	20-2	20-9	21-3	21-9	22-3	22-8	23-1	23-7	23-11	24-4	24-9	25-1	25-5
	24.0	16-5	17-0	17-8	18-3	18-9	19-3	19-9	20-2	20-8	21-1	21-6	21-10	22-3	22-7	22-11	23-4	23-8
F_b	12.0	896	969	1,040	1,108	1,174	1,239	1,302	1,363	1,423	1,481	1,539	1,595	1,651	1,706	1,759	1,812	1,864
	16.0	986	1,067	1,145	1,220	1,293	1,364	1,433	1,500	1,566	1,631	1,694	1,756	1,817	1,877	1,936	1,995	2,052
	19.2	1,048	1,134	1,216	1,296	1,374	1,449	1,522	1,594	1,664	1,733	1,800	1,866	1,931	1,995	2,058	2,120	2,181
	24.0	1,129	1,221	1,310	1,396	1,480	1,561	1,640	1,717	1,793	1,866	1,939	2,010	2,080	2,149	2,217	2,283	2,349

NOTE: The required bending design value, F_b, in pounds per square inch (× 0.00689 for N/mm²) is shown at the bottom of this table and is applicable to all lumber sizes shown. Spans are shown in feet-inches (1 foot = 304.8 mm, 1 inch = 25.4 mm) and are limited to 26 feet (7925 mm) and less.

TABLE 23-IV-R-1—RAFTERS WITH L/240 DEFLECTION LIMITATION
The allowable bending stress (F_b) and modulus of elasticity (E) used in this table shall be from Tables 23-IV-V-1 and 23-IV-V-2 only.

DESIGN CRITERIA:
Strength — Live load of 20 psf (0.96 kN/m²) plus dead load of 10 psf (0.48 kN/m²) determines the required bending design value.
Deflection — For 20 psf (0.96 kN/m²) live load.
Limited to span in inches (mm) divided by 240.

Rafter Size (in) × 25.4 for mm	Spacing (in)	Bending Design Value, F_b (psi) × 0.00689 for N/mm²										
		300	400	500	600	700	800	900	1000	1100	1200	1300
2 × 6	12.0	7-1	8-2	9-2	10-0	10-10	11-7	12-4	13-0	13-7	14-2	14-9
	16.0	6-2	7-1	7-11	8-8	9-5	10-0	10-8	11-3	11-9	12-4	12-10
	19.2	5-7	6-6	7-3	7-11	8-7	9-2	9-9	10-3	10-9	11-3	11-8
	24.0	5-0	5-10	6-6	7-1	7-8	8-2	8-8	9-2	9-7	10-0	10-5
2 × 8	12.0	9-4	10-10	12-1	13-3	14-4	15-3	16-3	17-1	17-11	18-9	19-6
	16.0	8-1	9-4	10-6	11-6	12-5	13-3	14-0	14-10	15-6	16-3	16-10
	19.2	7-5	8-7	9-7	10-6	11-4	12-1	12-10	13-6	14-2	14-10	15-5
	24.0	6-7	7-8	8-7	9-4	10-1	10-10	11-6	12-1	12-8	13-3	13-9
2 × 10	12.0	11-11	13-9	15-5	16-11	18-3	19-6	20-8	21-10	22-10	23-11	24-10
	16.0	10-4	11-11	13-4	14-8	15-10	16-11	17-11	18-11	19-10	20-8	21-6
	19.2	9-5	10-11	12-2	13-4	14-5	15-5	16-4	17-3	18-1	18-11	19-8
	24.0	8-5	9-9	10-11	11-11	12-11	13-9	14-8	15-5	16-2	16-11	17-7
2 × 12	12.0	14-6	16-9	18-9	20-6	22-2	23-9	25-2				
	16.0	12-7	14-6	16-3	17-9	19-3	20-6	21-9	23-0	24-1	25-2	
	19.2	11-6	13-3	14-10	16-3	17-6	18-9	19-11	21-0	22-0	23-0	23-11
	24.0	10-3	11-10	13-3	14-6	15-8	16-9	17-9	18-9	19-8	20-6	21-5
E	12.0	0.15	0.24	0.33	0.44	0.55	0.67	0.80	0.94	1.09	1.24	1.40
	16.0	0.13	0.21	0.29	0.38	0.48	0.58	0.70	0.82	0.94	1.07	1.21
	19.2	0.12	0.19	0.26	0.35	0.44	0.53	0.64	0.75	0.86	0.98	1.10
	24.0	0.11	0.17	0.24	0.31	0.39	0.48	0.57	0.67	0.77	0.88	0.99

Rafter Size (in) × 25.4 for mm	Spacing (in)	Bending Design Value, F_b (psi) × 0.00689 for N/mm²										
		1400	1500	1600	1700	1800	1900	2000	2100	2200	2300	2400
2 × 6	12.0	15-4	15-11	16-5	16-11	17-5	17-10					
	16.0	13-3	13-9	14-2	14-8	15-1	15-6	15-11	16-3			
	19.2	12-2	12-7	13-0	13-4	13-9	14-2	14-6	14-10	15-2	15-7	
	24.0	10-10	11-3	11-7	11-11	12-4	12-8	13-0	13-3	13-7	13-11	14-2
2 × 8	12.0	20-3	20-11	21-7	22-3	22-11	23-7					
	16.0	17-6	18-1	18-9	19-4	19-10	20-5	20-11	21-5			
	19.2	16-0	16-7	17-1	17-7	18-1	18-7	19-1	19-7	20-0	20-6	
	24.0	14-4	14-10	15-3	15-9	16-3	16-8	17-1	17-6	17-11	18-4	18-9
2 × 10	12.0	25-10										
	16.0	22-4	23-1	23-11	24-7	25-4	26-0					
	19.2	20-5	21-1	21-10	22-6	23-1	23-9	24-5	25-0	25-7		
	24.0	18-3	18-11	19-6	20-1	20-8	21-3	21-10	22-4	22-10	23-5	23-11
2 × 12	12.0											
	16.0											
	19.2	24-10	25-8									
	24.0	22-2	23-0	23-9	24-5	25-2	25-10					
E	12.0	1.56	1.73	1.91	2.09	2.28	2.47					
	16.0	1.35	1.50	1.65	1.81	1.97	2.14	2.31	2.48			
	19.2	1.23	1.37	1.51	1.65	1.80	1.95	2.11	2.27	2.43	2.60	
	24.0	1.10	1.22	1.35	1.48	1.61	1.75	1.89	2.03	2.18	2.33	2.48

NOTE: The required modulus of elasticity, E, in 1,000,000 pounds per square inch (psi) (× 0.00689 for N/mm²) is shown at the bottom of this table, is limited to 2.6 million psi (17 914 N/mm²) and less, and is applicable to all lumber sizes shown. Spans are shown in feet-inches (1 foot = 304.8 mm, 1 inch = 25.4 mm) and are limited to 26 feet (7925 mm) and less.

TABLE 23-IV-R-2

1997 UNIFORM BUILDING CODE

TABLE 23-IV-R-2—RAFTERS WITH L/240 DEFLECTION LIMITATION
The allowable bending stress (F_b) and modulus of elasticity (E) used in this table shall be from Tables 23-IV-V-1 and 23-IV-V-2 only.

DESIGN CRITERIA:
Strength — Live load of 30 psf (1.44 kN/m²) plus dead load of 10 psf (0.48 kN/m²) determines the required bending design value.
Deflection — For 30 psf (1.44 kN/m²) live load.
Limited to span in inches (mm) divided by 240.

Rafter Size (in)	Spacing (in)	Bending Design Value, F_b (psi)										
		× 0.00689 for N/mm²										
× 25.4 for mm		300	400	500	600	700	800	900	1000	1100	1200	1300
2 × 6	12.0	6-2	7-1	7-11	8-8	9-5	10-0	10-8	11-3	11-9	12-4	12-10
	16.0	5-4	6-2	6-10	7-6	8-2	8-8	9-3	9-9	10-2	10-8	11-1
	19.2	4-10	5-7	6-3	6-10	7-5	7-11	8-5	8-11	9-4	9-9	10-1
	24.0	4-4	5-0	5-7	6-2	6-8	7-1	7-6	7-11	8-4	8-8	9-1
2 × 8	12.0	8-1	9-4	10-6	11-6	12-5	13-3	14-0	14-10	15-6	16-3	16-10
	16.0	7-0	8-1	9-1	9-11	10-9	11-6	12-2	12-10	13-5	14-0	14-7
	19.2	6-5	7-5	8-3	9-1	9-9	10-6	11-1	11-8	12-3	12-10	13-4
	24.0	5-9	6-7	7-5	8-1	8-9	9-4	9-11	10-6	11-0	11-6	11-11
2 × 10	12.0	10-4	11-11	13-4	14-8	15-10	16-11	17-11	18-11	19-10	20-8	21-6
	16.0	8-11	10-4	11-7	12-8	13-8	14-8	15-6	16-4	17-2	17-11	18-8
	19.2	8-2	9-5	10-7	11-7	12-6	13-4	14-2	14-11	15-8	16-4	17-0
	24.0	7-4	8-5	9-5	10-4	11-2	11-11	12-8	13-4	14-0	14-8	15-3
2 × 12	12.0	12-7	14-6	16-3	17-9	19-3	20-6	21-9	23-0	24-1	25-2	
	16.0	10-11	12-7	14-1	15-5	16-8	17-9	18-10	19-11	20-10	21-9	22-8
	19.2	9-11	11-6	12-10	14-1	15-2	16-3	17-3	18-2	19-0	19-11	20-8
	24.0	8-11	10-3	11-6	12-7	13-7	14-6	15-5	16-3	17-0	17-9	18-6
E	12.0	0.15	0.23	0.32	0.43	0.54	0.66	0.78	0.92	1.06	1.21	1.36
	16.0	0.13	0.20	0.28	0.37	0.47	0.57	0.68	0.80	0.92	1.05	1.18
	19.2	0.12	0.18	0.26	0.34	0.43	0.52	0.62	0.73	0.84	0.95	1.08
	24.0	0.11	0.16	0.23	0.30	0.38	0.46	0.55	0.65	0.75	0.85	0.96

Rafter Size (in)	Spacing (in)	Bending Design Value, F_b (psi)										
		× 0.00689 for N/mm²										
× 25.4 for mm		1400	1500	1600	1700	1800	1900	2000	2100	2200	2300	2400
2 × 6	12.0	13-3	13-9	14-2	14-8	15-1	15-6	15-11				
	16.0	11-6	11-11	12-4	12-8	13-1	13-5	13-9	14-1	14-5		
	19.2	10-6	10-10	11-3	11-7	11-11	12-3	12-7	12-10	13-2	13-6	
	24.0	9-5	9-9	10-0	10-4	10-8	10-11	11-3	11-6	11-9	12-0	12-4
2 × 8	12.0	17-6	18-1	18-9	19-4	19-10	20-5	20-11				
	16.0	15-2	15-8	16-3	16-9	17-2	17-8	18-1	18-7	19-0		
	19.2	13-10	14-4	14-10	15-3	15-8	16-2	16-7	16-11	17-4	17-9	
	24.0	12-5	12-10	13-3	13-8	14-0	14-5	14-10	15-2	15-6	15-10	16-3
2 × 10	12.0	22-4	23-1	23-11	24-7	25-4	26-0					
	16.0	19-4	20-0	20-8	21-4	21-11	22-6	23-1	23-8	24-3		
	19.2	17-8	18-3	18-11	19-6	20-0	20-7	21-1	21-8	22-2	22-8	
	24.0	15-10	16-4	16-11	17-5	17-11	18-5	18-11	19-4	19-10	20-3	20-8
2 × 12	12.0											
	16.0	23-6	24-4	25-2	25-11							
	19.2	21-6	22-3	23-0	23-8	24-4	25-0	25-8				
	24.0	19-3	19-11	20-6	21-2	21-9	22-5	23-0	23-6	24-1	24-8	25-2
E	12.0	1.52	1.69	1.86	2.04	2.22	2.41	2.60				
	16.0	1.32	1.46	1.61	1.76	1.92	2.08	2.25	2.42	2.60		
	19.2	1.20	1.33	1.47	1.61	1.75	1.90	2.05	2.21	2.37	2.53	
	24.0	1.08	1.19	1.31	1.44	1.57	1.70	1.84	1.98	2.12	2.27	2.41

NOTE: The required modulus of elasticity, E, in 1,000,000 pounds per square inch (psi) (× 0.00689 for N/mm²) is shown at the bottom of this table, is limited to 2.6 million psi (17 914 N/mm²) and less, and is applicable to all lumber sizes shown. Spans are shown in feet-inches (1 foot = 304.8 mm, 1 inch = 25.4 mm) and are limited to 26 feet (7925 mm) and less.

TABLE 23-IV-R-3—RAFTERS WITH L/240 DEFLECTION LIMITATION
The allowable bending stress (F_b) and modulus of elasticity (E) used in this table shall be from Tables 23-IV-V-1 and 23-IV-V-2 only.

DESIGN CRITERIA:
Strength — Live load of 20 psf (0.96 kN/m²) plus dead load of 15 psf (0.72 kN/m²) determines the required bending design value.
Deflection — For 20 psf (0.96 kN/m²) live load.
Limited to span in inches (mm) divided by 240.

Rafter Size (in)	Spacing (in)	Bending Design Value, F_b (psi) × 0.00689 for N/mm²												
× 25.4 for mm		300	400	500	600	700	800	900	1000	1100	1200	1300	1400	1500
2 × 6	12.0	6-7	7-7	8-6	9-4	10-0	10-9	11-5	12-0	12-7	13-2	13-8	14-2	14-8
	16.0	5-8	6-7	7-4	8-1	8-8	9-4	9-10	10-5	10-11	11-5	11-10	12-4	12-9
	19.2	5-2	6-0	6-9	7-4	7-11	8-6	9-0	9-6	9-11	10-5	10-10	11-3	11-7
	24.0	4-8	5-4	6-0	6-7	7-1	7-7	8-1	8-6	8-11	9-4	9-8	10-0	10-5
2 × 8	12.0	8-8	10-0	11-2	12-3	13-3	14-2	15-0	15-10	16-7	17-4	18-0	18-9	19-5
	16.0	7-6	8-8	9-8	10-7	11-6	12-3	13-0	13-8	14-4	15-0	15-7	16-3	16-9
	19.2	6-10	7-11	8-10	9-8	10-6	11-2	11-10	12-6	13-1	13-8	14-3	14-10	15-4
	24.0	6-2	7-1	7-11	8-8	9-4	10-0	10-7	11-2	11-9	12-3	12-9	13-3	13-8
2 × 10	12.0	11-1	12-9	14-3	15-8	16-11	18-1	19-2	20-2	21-2	22-1	23-0	23-11	24-9
	16.0	9-7	11-1	12-4	13-6	14-8	15-8	16-7	17-6	18-4	19-2	19-11	20-8	21-5
	19.2	8-9	10-1	11-3	12-4	13-4	14-3	15-2	15-11	16-9	17-6	18-2	18-11	19-7
	24.0	7-10	9-0	10-1	11-1	11-11	12-9	13-6	14-3	15-0	15-8	16-3	16-11	17-6
2 × 12	12.0	13-5	15-6	17-4	19-0	20-6	21-11	23-3	24-7	25-9				
	16.0	11-8	13-5	15-0	16-6	17-9	19-0	20-2	21-3	22-4	23-3	24-3	25-2	26-0
	19.2	10-8	12-3	13-9	15-0	16-3	17-4	18-5	19-5	20-4	21-3	22-2	23-0	23-9
	24.0	9-6	11-0	12-3	13-5	14-6	15-6	16-6	17-4	18-2	19-0	19-10	20-6	21-3
E	12.0	0.12	0.19	0.26	0.35	0.44	0.54	0.64	0.75	0.86	0.98	1.11	1.24	1.37
	16.0	0.11	0.16	0.23	0.30	0.38	0.46	0.55	0.65	0.75	0.85	0.96	1.07	1.19
	19.2	0.10	0.15	0.21	0.27	0.35	0.42	0.51	0.59	0.68	0.78	0.88	0.98	1.09
	24.0	0.09	0.13	0.19	0.25	0.31	0.38	0.45	0.53	0.61	0.70	0.78	0.88	0.97

Rafter Size (in)	Spacing (in)	Bending Design Value, F_b (psi) × 0.00689 for N/mm²												
× 25.4 for mm		1600	1700	1800	1900	2000	2100	2200	2300	2400	2500	2600	2700
2 × 6	12.0	15-2	15-8	16-1	16-7	17-0	17-5	17-10					
	16.0	13-2	13-7	13-11	14-4	14-8	15-1	15-5	15-9	16-1	16-5		
	19.2	12-0	12-4	12-9	13-1	13-5	13-9	14-1	14-5	14-8	15-0	15-4	
	24.0	10-9	11-1	11-5	11-8	12-0	12-4	12-7	12-10	13-2	13-5	13-8	13-11
2 × 8	12.0	20-0	20-8	21-3	21-10	22-4	22-11	23-6					
	16.0	17-4	17-10	18-5	18-11	19-5	19-10	20-4	20-9	21-3	21-8		
	19.2	15-10	16-4	16-9	17-3	17-8	18-1	18-7	19-0	19-5	19-9	20-2	
	24.0	14-2	14-7	15-0	15-5	15-10	16-3	16-7	17-0	17-4	17-8	18-0	18-5
2 × 10	12.0	25-6											
	16.0	22-1	22-10	23-5	24-1	24-9	25-4	25-11					
	19.2	20-2	20-10	21-5	22-0	22-7	23-1	23-8	24-2	24-9	25-3	25-9	
	24.0	18-1	18-7	19-2	19-8	20-2	20-8	21-2	21-8	22-1	22-7	23-0	23-5
2 × 12	12.0												
	16.0												
	19.2	24-7	25-4	26-0									
	24.0	21-11	22-8	23-3	23-11	24-7	25-2	25-9					
E	12.0	1.51	1.66	1.81	1.96	2.12	2.28	2.44					
	16.0	1.31	1.44	1.56	1.70	1.83	1.97	2.11	2.26	2.41	2.56		
	19.2	1.20	1.31	1.43	1.55	1.67	1.80	1.93	2.06	2.20	2.34	2.48	
	24.0	1.07	1.17	1.28	1.39	1.50	1.61	1.73	1.85	1.97	2.09	2.22	2.35

NOTE: The required modulus of elasticity, E, in 1,000,000 pounds per square inch (psi) (× 0.00689 for N/mm²) is shown at the bottom of this table, is limited to 2.6 million psi (17 914 N/mm²) and less, and is applicable to all lumber sizes shown. Spans are shown in feet-inches (1 foot = 304.8 mm, 1 inch = 25.4 mm) and are limited to 26 feet (7925 mm) and less.

TABLE 23-IV-R-4

1997 UNIFORM BUILDING CODE

TABLE 23-IV-R-4—RAFTERS WITH L/240 DEFLECTION LIMITATION
The allowable bending stress (F_b) and modulus of elasticity (E) used in this table shall be from Tables 23-IV-V-1 and 23-IV-V-2 only.

DESIGN CRITERIA:
Strength — Live load of 30 psf (1.44 kN/m^2) plus dead load of 15 psf (0.72 kN/m^2) determines the required bending design value.
Deflection — For 30 psf (1.44 kN/m^2) live load.
Limited to span in inches (mm) divided by 240.

Rafter Size (in)	Spacing (in)	Bending Design Value, F_b (psi)												
× 25.4 for mm		× 0.00689 for N/mm^2												
		300	400	500	600	700	800	900	1000	1100	1200	1300	1400	1500
2 × 6	12.0	5-10	6-8	7-6	8-2	8-10	9-6	10-0	10-7	11-1	11-7	12-1	12-6	13-0
	16.0	5-0	5-10	6-6	7-1	7-8	8-2	8-8	9-2	9-7	10-0	10-5	10-10	11-3
	19.2	4-7	5-4	5-11	6-6	7-0	7-6	7-11	8-4	8-9	9-2	9-6	9-11	10-3
	24.0	4-1	4-9	5-4	5-10	6-3	6-8	7-1	7-6	7-10	8-2	8-6	8-10	9-2
2 × 8	12.0	7-8	8-10	9-10	10-10	11-8	12-6	13-3	13-11	14-8	15-3	15-11	16-6	17-1
	16.0	6-7	7-8	8-7	9-4	10-1	10-10	11-6	12-1	12-8	13-3	13-9	14-4	14-10
	19.2	6-0	7-0	7-10	8-7	9-3	9-10	10-6	11-0	11-7	12-1	12-7	13-1	13-6
	24.0	5-5	6-3	7-0	7-8	8-3	8-10	9-4	9-10	10-4	10-10	11-3	11-8	12-1
2 × 10	12.0	9-9	11-3	12-7	13-9	14-11	15-11	16-11	17-10	18-8	19-6	20-4	21-1	21-10
	16.0	8-5	9-9	10-11	11-11	12-11	13-9	14-8	15-5	16-2	16-11	17-7	18-3	18-11
	19.2	7-8	8-11	9-11	10-11	11-9	12-7	13-4	14-1	14-9	15-5	16-1	16-8	17-3
	24.0	6-11	8-0	8-11	9-9	10-6	11-3	11-11	12-7	13-2	13-9	14-4	14-11	15-5
2 × 12	12.0	11-10	13-8	15-4	16-9	18-1	19-4	20-6	21-8	22-8	23-9	24-8	25-7	
	16.0	10-3	11-10	13-3	14-6	15-8	16-9	17-9	18-9	19-8	20-6	21-5	22-2	23-0
	19.2	9-4	10-10	12-1	13-3	14-4	15-4	16-3	17-1	17-11	18-9	19-6	20-3	21-0
	24.0	8-5	9-8	10-10	11-10	12-10	13-8	14-6	15-4	16-1	16-9	17-5	18-1	18-9
E	12.0	0.13	0.19	0.27	0.36	0.45	0.55	0.66	0.77	0.89	1.01	1.14	1.28	1.41
	16.0	0.11	0.17	0.24	0.31	0.39	0.48	0.57	0.67	0.77	0.88	0.99	1.10	1.22
	19.2	0.10	0.15	0.22	0.28	0.36	0.44	0.52	0.61	0.70	0.80	0.90	1.01	1.12
	24.0	0.09	0.14	0.19	0.25	0.32	0.39	0.46	0.54	0.63	0.72	0.81	0.90	1.00

Rafter Size (in)	Spacing (in)	Bending Design Value, F_b (psi)												
× 25.4 for mm		× 0.00689 for N/mm^2												
		1600	1700	1800	1900	2000	2100	2200	2300	2400	2500	2600	2700
2 × 6	12.0	13-5	13-10	14-2	14-7	15-0	15-4	15-8					
	16.0	11-7	11-11	12-4	12-8	13-0	13-3	13-7	13-11	14-2			
	19.2	10-7	10-11	11-3	11-6	11-10	12-2	12-5	12-8	13-0	13-3	13-6	
	24.0	9-6	9-9	10-0	10-4	10-7	10-10	11-1	11-4	11-7	11-10	12-1	12-4
2 × 8	12.0	17-8	18-2	18-9	19-3	19-9	20-3	20-8					
	16.0	15-3	15-9	16-3	16-8	17-1	17-6	17-11	18-4	18-9			
	19.2	13-11	14-5	14-10	15-2	15-7	16-0	16-4	16-9	17-1	17-5	17-9	
	24.0	12-6	12-10	13-3	13-7	13-11	14-4	14-8	15-0	15-3	15-7	15-11	16-3
2 × 10	12.0	22-6	23-3	23-11	24-6	25-2	25-10						
	16.0	19-6	20-1	20-8	21-3	21-10	22-4	22-10	23-5	23-11			
	19.2	17-10	18-4	18-11	19-5	19-11	20-5	20-10	21-4	21-10	22-3	22-8	
	24.0	15-11	16-5	16-11	17-4	17-10	18-3	18-8	19-1	19-6	19-11	20-4	20-8
2 × 12	12.0												
	16.0	23-9	24-5	25-2	25-10								
	19.2	21-8	22-4	23-0	23-7	24-2	24-10	25-5	25-11				
	24.0	19-4	20-0	20-6	21-1	21-8	22-2	22-8	23-3	23-9	24-2	24-8	25-2
E	12.0	1.56	1.71	1.86	2.02	2.18	2.34	2.51					
	16.0	1.35	1.48	1.61	1.75	1.89	2.03	2.18	2.33	2.48			
	19.2	1.23	1.35	1.47	1.59	1.72	1.85	1.99	2.12	2.26	2.41	2.55	
	24.0	1.10	1.21	1.31	1.43	1.54	1.66	1.78	1.90	2.02	2.15	2.28	2.41

NOTE: The required modulus of elasticity, E, in 1,000,000 pounds per square inch (psi) (× 0.00689 for N/mm^2) is shown at the bottom of this table, is limited to 2.6 million psi (17 914 N/mm^2) and less, and is applicable to all lumber sizes shown. Spans are shown in feet-inches (1 foot = 304.8 mm, 1 inch = 25.4 mm) and are limited to 26 feet (7925 mm) and less.

TABLE 23-IV-R-5—RAFTERS WITH L/240 DEFLECTION LIMITATION
The allowable bending stress (Fb) and modulus of elasticity (E) used in this table shall be from Tables 23-IV-V-1 and 23-IV-V-2 only.

DESIGN CRITERIA:
Strength — Live load of 20 psf (0.96 kN/m²) plus dead load of 20 psf (0.96 kN/m²) determines the required bending design value.
Deflection — For 20 psf (0.96 kN/m²) live load.
Limited to span in inches (mm) divided by 240.

Rafter Size (in)	Spacing (in)	Bending Design Value, F_b (psi) × 0.00689 for N/mm²												
× 25.4 for mm		300	400	500	600	700	800	900	1000	1100	1200	1300	1400	1500
2 × 6	12.0	6-2	7-1	7-11	8-8	9-5	10-0	10-8	11-3	11-9	12-4	12-10	13-3	13-9
	16.0	5-4	6-2	6-10	7-6	8-2	8-8	9-3	9-9	10-2	10-8	11-1	11-6	11-11
	19.2	4-10	5-7	6-3	6-10	7-5	7-11	8-5	8-11	9-4	9-9	10-1	10-6	10-10
	24.0	4-4	5-0	5-7	6-2	6-8	7-1	7-6	7-11	8-4	8-8	9-1	9-5	9-9
2 × 8	12.0	8-1	9-4	10-6	11-6	12-5	13-3	14-0	14-10	15-6	16-3	16-10	17-6	18-1
	16.0	7-0	8-1	9-1	9-11	10-9	11-6	12-2	12-10	13-5	14-0	14-7	15-2	15-8
	19.2	6-5	7-5	8-3	9-1	9-9	10-6	11-1	11-8	12-3	12-10	13-4	13-10	14-4
	24.0	5-9	6-7	7-5	8-1	8-9	9-4	9-11	10-6	11-0	11-6	11-11	12-5	12-10
2 × 10	12.0	10-4	11-11	13-4	14-8	15-10	16-11	17-11	18-11	19-10	20-8	21-6	22-4	23-1
	16.0	8-11	10-4	11-7	12-8	13-8	14-8	15-6	16-4	17-2	17-11	18-8	19-4	20-0
	19.2	8-2	9-5	10-7	11-7	12-6	13-4	14-2	14-11	15-8	16-4	17-0	17-8	18-3
	24.0	7-4	8-5	9-5	10-4	11-2	11-11	12-8	13-4	14-0	14-8	15-3	15-10	16-4
2 × 12	12.0	12-7	14-6	16-3	17-9	19-3	20-6	21-9	23-0	24-1	25-2			
	16.0	10-11	12-7	14-1	15-5	16-8	17-9	18-10	19-11	20-10	21-9	22-8	23-6	24-4
	19.2	9-11	11-6	12-10	14-1	15-2	16-3	17-3	18-2	19-0	19-11	20-8	21-6	22-3
	24.0	8-11	10-3	11-6	12-7	13-7	14-6	15-5	16-3	17-0	17-9	18-6	19-3	19-11
E	12.0	0.10	0.15	0.22	0.28	0.36	0.44	0.52	0.61	0.71	0.80	0.91	1.01	1.13
	16.0	0.09	0.13	0.19	0.25	0.31	0.38	0.45	0.53	0.61	0.70	0.79	0.88	0.97
	19.2	0.08	0.12	0.17	0.23	0.28	0.35	0.41	0.48	0.56	0.64	0.72	0.80	0.89
	24.0	0.07	0.11	0.15	0.20	0.25	0.31	0.37	0.43	0.50	0.57	0.64	0.72	0.80

Rafter Size (in)	Spacing (in)	Bending Design Value, F_b (psi) × 0.00689 for N/mm²												
× 25.4 for mm		1600	1700	1800	1900	2000	2100	2200	2300	2400	2500	2600	2700	
2 × 6	12.0	14-2	14-8	15-1	15-6	15-11	16-3	16-8	17-0	17-5	17-9	18-1		
	16.0	12-4	12-8	13-1	13-5	13-9	14-1	14-5	14-9	15-1	15-4	15-8	16-0	
	19.2	11-3	11-7	11-11	12-3	12-7	12-10	13-2	13-6	13-9	14-0	14-4	14-7	
	24.0	10-0	10-4	10-8	10-11	11-3	11-6	11-9	12-0	12-4	12-7	12-10	13-1	
2 × 8	12.0	18-9	19-4	19-10	20-5	20-11	21-5	21-11	22-5	22-11	23-5	23-10		
	16.0	16-3	16-9	17-2	17-8	18-1	18-7	19-0	19-5	19-10	20-3	20-8	21-1	
	19.2	14-10	15-3	15-8	16-2	16-7	16-11	17-4	17-9	18-1	18-6	18-10	19-3	
	24.0	13-3	13-8	14-0	14-5	14-10	15-2	15-6	15-10	16-3	16-7	16-10	17-2	
2 × 10	12.0	23-11	24-7	25-4	26-0									
	16.0	20-8	21-4	21-11	22-6	23-1	23-8	24-3	24-10	25-4	25-10			
	19.2	18-11	19-6	20-0	20-7	21-1	21-8	22-2	22-8	23-1	23-7	24-1	24-6	
	24.0	16-11	17-5	17-11	18-5	18-11	19-4	19-10	20-3	20-8	21-1	21-6	21-11	
2 × 12	12.0													
	16.0	25-2	25-11											
	19.2	23-0	23-8	24-4	25-0	25-8								
	24.0	20-6	21-2	21-9	22-5	23-0	23-6	24-1	24-8	25-2	25-8			
E	12.0	1.24	1.36	1.48	1.60	1.73	1.86	2.00	2.14	2.28	2.42	2.57		
	16.0	1.07	1.18	1.28	1.39	1.50	1.61	1.73	1.85	1.97	2.10	2.22	2.35	
	19.2	0.98	1.07	1.17	1.27	1.37	1.47	1.58	1.69	1.80	1.91	2.03	2.15	
	24.0	0.88	0.96	1.05	1.13	1.22	1.32	1.41	1.51	1.61	1.71	1.82	1.92	

NOTE: The required modulus of elasticity, E, in 1,000,000 pounds per square inch (psi) (× 0.00689 for N/mm²) is shown at the bottom of this table, is limited to 2.6 million psi (17 914 N/mm²) and less, and is applicable to all lumber sizes shown. Spans are shown in feet-inches (1 foot = 304.8 mm, 1 inch = 25.4 mm) and are limited to 26 feet (7925 mm) and less.

TABLE 23-IV-R-6

1997 UNIFORM BUILDING CODE

TABLE 23-IV-R-6—RAFTERS WITH L/240 DEFLECTION LIMITATION
The allowable bending stress (F_b) and modulus of elasticity (E) used in this table shall be from Tables 23-IV-V-1 and 23-IV-V-2 only.

DESIGN CRITERIA:
Strength — Live load of 30 psf (1.44 kN/m^2) plus dead load of 20 psf (0.96 kN/m^2) determines the required bending design value.
Deflection — For 30 psf (1.44 kN/m^2) live load.
Limited to span in inches (mm) divided by 240.

Rafter Size (in)	Spacing (in)	Bending Design Value, F_b (psi) $\times$ 0.00689 for N/mm^2												
$\times$ 25.4 for mm		300	400	500	600	700	800	900	1000	1100	1200	1300	1400	1500
2 × 6	12.0	5-6	6-4	7-1	7-9	8-5	9-0	9-6	10-0	10-6	11-0	11-5	11-11	12-4
	16.0	4-9	5-6	6-2	6-9	7-3	7-9	8-3	8-8	9-1	9-6	9-11	10-3	10-8
	19.2	4-4	5-0	5-7	6-2	6-8	7-1	7-6	7-11	8-4	8-8	9-1	9-5	9-9
	24.0	3-11	4-6	5-0	5-6	5-11	6-4	6-9	7-1	7-5	7-9	8-1	8-5	8-8
2 × 8	12.0	7-3	8-4	9-4	10-3	11-1	11-10	12-7	13-3	13-11	14-6	15-1	15-8	16-3
	16.0	6-3	7-3	8-1	8-11	9-7	10-3	10-10	11-6	12-0	12-7	13-1	13-7	14-0
	19.2	5-9	6-7	7-5	8-1	8-9	9-4	9-11	10-6	11-0	11-6	11-11	12-5	12-10
	24.0	5-2	5-11	6-7	7-3	7-10	8-4	8-11	9-4	9-10	10-3	10-8	11-1	11-6
2 × 10	12.0	9-3	10-8	11-11	13-1	14-2	15-1	16-0	16-11	17-9	18-6	19-3	20-0	20-8
	16.0	8-0	9-3	10-4	11-4	12-3	13-1	13-10	14-8	15-4	16-0	16-8	17-4	17-11
	19.2	7-4	8-5	9-5	10-4	11-2	11-11	12-8	13-4	14-0	14-8	15-3	15-10	16-4
	24.0	6-6	7-7	8-5	9-3	10-0	10-8	11-4	11-11	12-6	13-1	13-7	14-2	14-8
2 × 12	12.0	11-3	13-0	14-6	15-11	17-2	18-4	19-6	20-6	21-7	22-6	23-5	24-4	25-2
	16.0	9-9	11-3	12-7	13-9	14-11	15-11	16-10	17-9	18-8	19-6	20-3	21-1	21-9
	19.2	8-11	10-3	11-6	12-7	13-7	14-6	15-5	16-3	17-0	17-9	18-6	19-3	19-11
	24.0	7-11	9-2	10-3	11-3	12-2	13-0	13-9	14-6	15-3	15-11	16-7	17-2	17-9
E	12.0	0.11	0.17	0.23	0.31	0.38	0.47	0.56	0.66	0.76	0.86	0.97	1.09	1.21
	16.0	0.09	0.14	0.20	0.26	0.33	0.41	0.49	0.57	0.66	0.75	0.84	0.94	1.05
	19.2	0.09	0.13	0.18	0.24	0.30	0.37	0.44	0.52	0.60	0.68	0.77	0.86	0.95
	24.0	0.08	0.12	0.16	0.22	0.27	0.33	0.40	0.46	0.54	0.61	0.69	0.77	0.85

Rafter Size (in)	Spacing (in)	Bending Design Value, F_b (psi) $\times$ 0.00689 for N/mm^2												
$\times$ 25.4 for mm		1600	1700	1800	1900	2000	2100	2200	2300	2400	2500	2600	2700
2 × 6	12.0	12-8	13-1	13-6	13-10	14-2	14-7	14-11	15-3	15-7	15-11		
	16.0	11-0	11-4	11-8	12-0	12-4	12-7	12-11	13-2	13-6	13-9	14-0	14-3
	19.2	10-0	10-4	10-8	10-11	11-3	11-6	11-9	12-0	12-4	12-7	12-10	13-1
	24.0	9-0	9-3	9-6	9-9	10-0	10-3	10-6	10-9	11-0	11-3	11-5	11-8
2 × 8	12.0	16-9	17-3	17-9	18-3	18-9	19-2	19-8	20-1	20-6	20-11		
	16.0	14-6	14-11	15-5	15-10	16-3	16-7	17-0	17-5	17-9	18-1	18-6	18-10
	19.2	13-3	13-8	14-0	14-5	14-10	15-2	15-6	15-10	16-3	16-7	16-10	17-2
	24.0	11-10	12-2	12-7	12-11	13-3	13-7	13-11	14-2	14-6	14-10	15-1	15-5
2 × 10	12.0	21-4	22-0	22-8	23-3	23-11	24-6	25-1	25-7				
	16.0	18-6	19-1	19-7	20-2	20-8	21-2	21-8	22-2	22-8	23-1	23-7	24-0
	19.2	16-11	17-5	17-11	18-5	18-11	19-4	19-10	20-3	20-8	21-1	21-6	21-11
	24.0	15-1	15-7	16-0	16-6	16-11	17-4	17-9	18-1	18-6	18-11	19-3	19-7
2 × 12	12.0	26-0											
	16.0	22-6	23-2	23-10	24-6	25-2	25-9						
	19.2	20-6	21-2	21-9	22-5	23-0	23-6	24-1	24-8	25-2	25-8		
	24.0	18-4	18-11	19-6	20-0	20-6	21-1	21-7	22-0	22-6	23-0	23-5	23-10
E	12.0	1.33	1.46	1.59	1.72	1.86	2.00	2.14	2.29	2.44	2.60		
	16.0	1.15	1.26	1.37	1.49	1.61	1.73	1.86	1.99	2.12	2.25	2.39	2.53
	19.2	1.05	1.15	1.25	1.36	1.47	1.58	1.70	1.81	1.93	2.05	2.18	2.31
	24.0	0.94	1.03	1.12	1.22	1.31	1.41	1.52	1.62	1.73	1.84	1.95	2.06

NOTE: The required modulus of elasticity, E, in 1,000,000 pounds per square inch (psi) ($\times$ 0.00689 for N/mm^2) is shown at the bottom of this table, is limited to 2.6 million psi (17 914 N/mm^2) and less, and is applicable to all lumber sizes shown. Spans are shown in feet-inches (1 foot = 304.8 mm, 1 inch = 25.4 mm) and are limited to 26 foot (7925 mm) and less.

TABLE 23-IV-R-7—RAFTERS WITH L/180 DEFLECTION LIMITATION
The allowable bending stress (F_b) and modulus of elasticity (E) used in this table shall be from Tables 23-IV-V-1 and 23-IV-V-2 only.

DESIGN CRITERIA:
Strength — Live load of 20 psf (0.96 kN/m^2) plus dead load of 10 psf (0.48 kN/m^2) determines the required bending design value.
Deflection — For 20 psf (0.96 kN/m^2) live load.
Limited to span in inches (mm) divided by 180.

Rafter Size (in) × 25.4 for mm	Spacing (in)	Bending Design Value, F_b (psi) × 0.00689 for N/mm^2 200	300	400	500	600	700	800	900	1000	1100	1200	1300	1400	1500	1600
2×4	12.0	3-8	4-6	5-3	5-10	6-5	6-11	7-5	7-10	8-3	8-8	9-0	9-5	9-9	10-1	10-5
	16.0	3-2	3-11	4-6	5-1	5-6	6-0	6-5	6-9	7-2	7-6	7-10	8-2	8-5	8-9	9-0
	19.2	2-11	3-7	4-1	4-7	5-1	5-5	5-10	6-2	6-6	6-10	7-2	7-5	7-9	8-0	8-3
	24.0	2-7	3-2	3-8	4-1	4-6	4-11	5-3	5-6	5-10	6-1	6-5	6-8	6-11	7-2	7-5
2×6	12.0	5-10	7-1	8-2	9-2	10-0	10-10	11-7	12-4	13-0	13-7	14-2	14-9	15-4	15-11	16-5
	16.0	5-0	6-2	7-1	7-11	8-8	9-5	10-0	10-8	11-3	11-9	12-4	12-10	13-3	13-9	14-2
	19.2	4-7	5-7	6-6	7-3	7-11	8-7	9-2	9-9	10-3	10-9	11-3	11-8	12-2	12-7	13-0
	24.0	4-1	5-0	5-10	6-6	7-1	7-8	8-2	8-8	9-2	9-7	10-0	10-5	10-10	11-3	11-7
2×8	12.0	7-8	9-4	10-10	12-1	13-3	14-4	15-3	16-3	17-1	17-11	18-9	19-6	20-3	20-11	21-7
	16.0	6-7	8-1	9-4	10-6	11-6	12-5	13-3	14-0	14-10	15-6	16-3	16-10	17-6	18-1	18-9
	19.2	6-0	7-5	8-7	9-7	10-6	11-4	12-1	12-10	13-6	14-2	14-10	15-5	16-0	16-7	17-1
	24.0	5-5	6-7	7-8	8-7	9-4	10-1	10-10	11-6	12-1	12-8	13-3	13-9	14-4	14-10	15-3
2×10	12.0	9-9	11-11	13-9	15-5	16-11	18-3	19-6	20-8	21-10	22-10	23-11	24-10	25-10		
	16.0	8-5	10-4	11-11	13-4	14-8	15-10	16-11	17-11	18-11	19-10	20-8	21-6	22-4	23-1	23-11
	19.2	7-8	9-5	10-11	12-2	13-4	14-5	15-5	16-4	17-3	18-1	18-11	19-8	20-5	21-1	21-10
	24.0	6-11	8-5	9-9	10-11	11-11	12-11	13-9	14-8	15-5	16-2	16-11	17-7	18-3	18-11	19-6
E	12.0	0.06	0.12	0.18	0.25	0.33	0.41	0.51	0.60	0.71	0.82	0.93	1.05	1.17	1.30	1.43
	16.0	0.05	0.10	0.15	0.22	0.28	0.36	0.44	0.52	0.61	0.71	0.80	0.91	1.01	1.13	1.24
	19.2	0.05	0.09	0.14	0.20	0.26	0.33	0.40	0.48	0.56	0.64	0.73	0.83	0.93	1.03	1.13
	24.0	0.04	0.08	0.13	0.18	0.23	0.29	0.36	0.43	0.50	0.58	0.66	0.74	0.83	0.92	1.01

Rafter Size (in) × 25.4 for mm	Spacing (in)	Bending Design Value, F_b (psi) × 0.00689 for N/mm^2 1700	1800	1900	2000	2100	2200	2300	2400	2500	2600	2700	2800	2900	3000
2×4	12.0	10-9	11-1	11-4	11-8	11-11	12-3	12-6							
	16.0	9-4	9-7	9-10	10-1	10-4	10-7	10-10	11-1	11-4	11-6				
	19.2	8-6	8-9	9-0	9-3	9-5	9-8	9-11	10-1	10-4	10-6	10-9			
	24.0	7-7	7-10	8-0	8-3	8-5	8-8	8-10	9-0	9-3	9-5	9-7	9-9	9-11	10-1
2×6	12.0	16-11	17-5	17-10	18-4	18-9	19-3	19-8							
	16.0	14-8	15-1	15-6	15-11	16-3	16-8	17-0	17-5	17-9	18-1				
	19.2	13-4	13-9	14-2	14-6	14-10	15-2	15-7	15-11	16-2	16-6	16-10			
	24.0	11-11	12-4	12-8	13-0	13-3	13-7	13-11	14-2	14-6	14-9	15-1	15-4	15-7	15-11
2×8	12.0	22-3	22-11	23-7	24-2	24-9	25-4	25-11							
	16.0	19-4	19-10	20-5	20-11	21-5	21-11	22-5	22-11	23-5	23-10				
	19.2	17-7	18-1	18-7	19-1	19-7	20-0	20-6	20-11	21-4	21-9	22-2			
	24.0	15-9	16-3	16-8	17-1	17-6	17-11	18-4	18-9	19-1	19-6	19-10	20-3	20-7	20-11
2×10	12.0														
	16.0	24-7	25-4	26-0											
	19.2	22-6	23-1	23-9	24-5	25-0	25-7								
	24.0	20-1	20-8	21-3	21-10	22-4	22-10	23-5	23-11	24-5	24-10	25-4	25-10		
E	12.0	1.57	1.71	1.85	2.00	2.15	2.31	2.47							
	16.0	1.36	1.48	1.60	1.73	1.86	2.00	2.14	2.28	2.42	2.57				
	19.2	1.24	1.35	1.46	1.58	1.70	1.82	1.95	2.08	2.21	2.34	2.48			
	24.0	1.11	1.21	1.31	1.41	1.52	1.63	1.74	1.86	1.98	2.10	2.22	2.34	2.47	2.60

NOTE: The required modulus of elasticity, E, in 1,000,000 pounds per square inch (psi) (× 0.00689 for N/mm^2) is shown at the bottom of this table, is limited to 2.6 million psi (17 914 N/mm^2) and less, and is applicable to all lumber sizes shown. Spans are shown in feet-inches (1 foot = 304.8 mm, 1 inch = 25.4 mm) and are limited to 26 feet (7925 mm) and less.

TABLE 23-IV-R-8

1997 UNIFORM BUILDING CODE

TABLE 23-IV-R-8—RAFTERS WITH L/180 DEFLECTION LIMITATION
The allowable bending stress (Fb) and modulus of elasticity (E) used in this table shall be from Tables 23-IV-V-1 and 23-IV-V-2 only.

DESIGN CRITERIA:
Strength — Live load of 30 psf (1.44 kN/m²) plus dead load of 10 psf (0.48 kN/m²) determines the required bending design value.
Deflection — For 30 psf (1.44 kN/m²) live load.
Limited to span in inches (mm) divided by 180.

Rafter Size (in) × 25.4 for mm	Spacing (in)	Bending Design Value, F_b (psi) × 0.00689 for N/mm²														
		200	300	400	500	600	700	800	900	1000	1100	1200	1300	1400	1500	1600
2 × 4	12.0	3-2	3-11	4-6	5-1	5-6	6-0	6-5	6-9	7-2	7-6	7-10	8-2	8-5	8-9	9-0
	16.0	2-9	3-5	3-11	4-4	4-10	5-2	5-6	5-10	6-2	6-6	6-9	7-1	7-4	7-7	7-10
	19.2	2-6	3-1	3-7	4-0	4-4	4-9	5-1	5-4	5-8	5-11	6-2	6-5	6-8	6-11	7-2
	24.0	2-3	2-9	3-2	3-7	3-11	4-3	4-6	4-10	5-1	5-4	5-6	5-9	6-0	6-2	6-5
2 × 6	12.0	5-0	6-2	7-1	7-11	8-8	9-5	10-0	10-8	11-3	11-9	12-4	12-10	13-3	13-9	14-2
	16.0	4-4	5-4	6-2	6-10	7-6	8-2	8-8	9-3	9-9	10-2	10-8	11-1	11-6	11-11	12-4
	19.2	4-0	4-10	5-7	6-3	6-10	7-5	7-11	8-5	8-11	9-4	9-9	10-1	10-6	10-10	11-3
	24.0	3-7	4-4	5-0	5-7	6-2	6-8	7-1	7-6	7-11	8-4	8-8	9-1	9-5	9-9	10-0
2 × 8	12.0	6-7	8-1	9-4	10-6	11-6	12-5	13-3	14-0	14-10	15-6	16-3	16-10	17-6	18-1	18-9
	16.0	5-9	7-0	8-1	9-1	9-11	10-9	11-6	12-2	12-10	13-5	14-0	14-7	15-2	15-8	16-3
	19.2	5-3	6-5	7-5	8-3	9-1	9-9	10-6	11-1	11-8	12-3	12-10	13-4	13-10	14-4	14-10
	24.0	4-8	5-9	6-7	7-5	8-1	8-9	9-4	9-11	10-6	11-0	11-6	11-11	12-5	12-10	13-3
2 × 10	12.0	8-5	10-4	11-11	13-4	14-8	15-10	16-11	17-11	18-11	19-10	20-8	21-6	22-4	23-1	23-11
	16.0	7-4	8-11	10-4	11-7	12-8	13-8	14-8	15-6	16-4	17-2	17-11	18-8	19-4	20-0	20-8
	19.2	6-8	8-2	9-5	10-7	11-7	12-6	13-4	14-2	14-11	15-8	16-4	17-0	17-8	18-3	18-11
	24.0	6-0	7-4	8-5	9-5	10-4	11-2	11-11	12-8	13-4	14-0	14-8	15-3	15-10	16-4	16-11
E	12.0	0.06	0.11	0.17	0.24	0.32	0.40	0.49	0.59	0.69	0.79	0.91	1.02	1.14	1.27	1.39
	16.0	0.05	0.10	0.15	0.21	0.28	0.35	0.43	0.51	0.60	0.69	0.78	0.88	0.99	1.10	1.21
	19.2	0.05	0.09	0.14	0.19	0.25	0.32	0.39	0.47	0.54	0.63	0.72	0.81	0.90	1.00	1.10
	24.0	0.04	0.08	0.12	0.17	0.23	0.29	0.35	0.42	0.49	0.56	0.64	0.72	0.81	0.89	0.99

Rafter Size (in) × 25.4 for mm	Spacing (in)	Bending Design Value, F_b (psi) × 0.00689 for N/mm²													
		1700	1800	1900	2000	2100	2200	2300	2400	2500	2600	2700	2800	2900	3000
2 × 4	12.0	9-4	9-7	9-10	10-1	10-4	10-7	10-10	11-1						
	16.0	8-1	8-4	8-6	8-9	9-0	9-2	9-5	9-7	9-9	10-0				
	19.2	7-4	7-7	7-9	8-0	8-2	8-5	8-7	8-9	8-11	9-1	9-3	9-5		
	24.0	6-7	6-9	7-0	7-2	7-4	7-6	7-8	7-10	8-0	8-2	8-4	8-5	8-7	8-9
2 × 6	12.0	14-8	15-1	15-6	15-11	16-3	16-8	17-0	17-5						
	16.0	12-8	13-1	13-5	13-9	14-1	14-5	14-9	15-1	15-4	15-8				
	19.2	11-7	11-11	12-3	12-7	12-10	13-2	13-6	13-9	14-0	14-4	14-7	14-10		
	24.0	10-4	10-8	10-11	11-3	11-6	11-9	12-0	12-4	12-7	12-10	13-1	13-3	13-6	13-9
2 × 8	12.0	19-4	19-10	20-5	20-11	21-5	21-11	22-5	22-11						
	16.0	16-9	17-2	17-8	18-1	18-7	19-0	19-5	19-10	20-3	20-8				
	19.2	15-3	15-8	16-2	16-7	16-11	17-4	17-9	18-1	18-6	18-10	19-3	19-7		
	24.0	13-8	14-0	14-5	14-10	15-2	15-6	15-10	16-3	16-7	16-10	17-2	17-6	17-10	18-1
2 × 10	12.0	24-7	25-4	26-0											
	16.0	21-4	21-11	22-6	23-1	23-8	24-3	24-10	25-4	25-10					
	19.2	19-6	20-0	20-7	21-1	21-8	22-2	22-8	23-1	23-7	24-1	24-6	25-0		
	24.0	17-5	17-11	18-5	18-11	19-4	19-10	20-3	20-8	21-1	21-6	21-11	22-4	22-9	23-1
E	12.0	1.53	1.66	1.80	1.95	2.10	2.25	2.40	2.56						
	16.0	1.32	1.44	1.56	1.69	1.82	1.95	2.08	2.22	2.36	2.50				
	19.2	1.21	1.32	1.43	1.54	1.66	1.78	1.90	2.03	2.15	2.28	2.42	2.55		
	24.0	1.08	1.18	1.28	1.38	1.48	1.59	1.70	1.81	1.93	2.04	2.16	2.28	2.41	2.53

NOTE: The required modulus of elasticity, E, in 1,000,000 pounds per square inch (psi) (× 0.00689 for N/mm²) is shown at the bottom of this table, is limited to 2.6 million psi (17 914 N/mm²) and less, and is applicable to all lumber sizes shown. Spans are shown in feet-inches (1 foot = 304.8 mm, 1 inch = 25.4 mm) and are limited to 26 feet (7925 mm) and less.

TABLE 23-IV-R-9—RAFTERS WITH L/180 DEFLECTION LIMITATION
The allowable bending stress (F_b) and modulus of elasticity (E) used in this table shall be from Tables 23-IV-V-1 and 23-IV-V-2 only.

DESIGN CRITERIA:
Strength — Live load of 20 psf (0.96 kN/m²) plus dead load of 15 psf (0.72 kN/m²) determines the required bending design value.
Deflection — For 20 psf (0.96 kN/m²) live load.
Limited to span in inches (mm) divided by 180.

Rafter Size (in) × 25.4 for mm	Spacing (in)	Bending Design Value, F_b (psi) × 0.00689 for N/mm²														
		200	300	400	500	600	700	800	900	1000	1100	1200	1300	1400	1500	1600
2×4	12.0	3-5	4-2	4-10	5-5	5-11	6-5	6-10	7-3	7-8	8-0	8-4	8-8	9-0	9-4	9-8
	16.0	2-11	3-7	4-2	4-8	5-1	5-6	5-11	6-3	6-7	6-11	7-3	7-6	7-10	8-1	8-4
	19.2	2-8	3-4	3-10	4-3	4-8	5-1	5-5	5-9	6-0	6-4	6-7	6-11	7-2	7-5	7-8
	24.0	2-5	2-11	3-5	3-10	4-2	4-6	4-10	5-1	5-5	5-8	5-11	6-2	6-5	6-7	6-10
2×6	12.0	5-4	6-7	7-7	8-6	9-4	10-0	10-9	11-5	12-0	12-7	13-2	13-8	14-2	14-8	15-2
	16.0	4-8	5-8	6-7	7-4	8-1	8-8	9-4	9-10	10-5	10-11	11-5	11-10	12-4	12-9	13-2
	19.2	4-3	5-2	6-0	6-9	7-4	7-11	8-6	9-0	9-6	9-11	10-5	10-10	11-3	11-7	12-0
	24.0	3-10	4-8	5-4	6-0	6-7	7-1	7-7	8-1	8-6	8-11	9-4	9-8	10-0	10-5	10-9
2×8	12.0	7-1	8-8	10-0	11-2	12-3	13-3	14-2	15-0	15-10	16-7	17-4	18-0	18-9	19-5	20-0
	16.0	6-2	7-6	8-8	9-8	10-7	11-6	12-3	13-0	13-8	14-4	15-0	15-7	16-3	16-9	17-4
	19.2	5-7	6-10	7-11	8-10	9-8	10-6	11-2	11-10	12-6	13-1	13-8	14-3	14-10	15-4	15-10
	24.0	5-0	6-2	7-1	7-11	8-8	9-4	10-0	10-7	11-2	11-9	12-3	12-9	13-3	13-8	14-2
2×10	12.0	9-0	11-1	12-9	14-3	15-8	16-11	18-1	19-2	20-2	21-2	22-1	23-0	23-11	24-9	25-6
	16.0	7-10	9-7	11-1	12-4	13-6	14-8	15-8	16-7	17-6	18-4	19-2	19-11	20-8	21-5	22-1
	19.2	7-2	8-9	10-1	11-3	12-4	13-4	14-3	15-2	15-11	16-9	17-6	18-2	18-11	19-7	20-2
	24.0	6-5	7-10	9-0	10-1	11-1	11-11	12-9	13-6	14-3	15-0	15-8	16-3	16-11	17-6	18-1
E	12.0	0.05	0.09	0.14	0.20	0.26	0.33	0.40	0.48	0.56	0.65	0.74	0.83	0.93	1.03	1.14
	16.0	0.04	0.08	0.12	0.17	0.23	0.28	0.35	0.41	0.49	0.56	0.64	0.72	0.80	0.89	0.98
	19.2	0.04	0.07	0.11	0.16	0.21	0.26	0.32	0.38	0.44	0.51	0.58	0.66	0.73	0.81	0.90
	24.0	0.04	0.07	0.10	0.14	0.18	0.23	0.28	0.34	0.40	0.46	0.52	0.59	0.66	0.73	0.80

Rafter Size (in) × 25.4 for mm	Spacing (in)	Bending Design Value, F_b (psi) × 0.00689 for N/mm²													
		1700	1800	1900	2000	2100	2200	2300	2400	2500	2600	2700	2800	2900	3000
2×4	12.0	9-11	10-3	10-6	10-10	11-1	11-4	11-7	11-10	12-1	12-4	12-7			
	16.0	8-7	8-10	9-1	9-4	9-7	9-10	10-0	10-3	10-5	10-8	10-10	11-1	11-3	11-5
	19.2	7-10	8-1	8-4	8-6	8-9	8-11	9-2	9-4	9-7	9-9	9-11	10-1	10-3	10-5
	24.0	7-0	7-3	7-5	7-8	7-10	8-0	8-2	8-4	8-6	8-8	8-10	9-0	9-2	9-4
2×6	12.0	15-8	16-1	16-7	17-0	17-5	17-10	18-2	18-7	19-0	19-4	19-9			
	16.0	13-7	13-11	14-4	14-8	15-1	15-5	15-9	16-1	16-5	16-9	17-1	17-5	17-8	18-0
	19.2	12-4	12-9	13-1	13-5	13-9	14-1	14-5	14-8	15-0	15-4	15-7	15-11	16-2	16-5
	24.0	11-1	11-5	11-8	12-0	12-4	12-7	12-10	13-2	13-5	13-8	13-11	14-2	14-5	14-8
2×8	12.0	20-8	21-3	21-10	22-4	22-11	23-6	24-0	24-6	25-0	25-6	26-0			
	16.0	17-10	18-5	18-11	19-5	19-10	20-4	20-9	21-3	21-8	22-1	22-6	22-11	23-4	23-9
	19.2	16-4	16-9	17-3	17-8	18-1	18-7	19-0	19-5	19-9	20-2	20-7	20-11	21-4	21-8
	24.0	14-7	15-0	15-5	15-10	16-3	16-7	17-0	17-4	17-8	18-0	18-5	18-9	19-1	19-5
2×10	12.0														
	16.0	22-10	23-5	24-1	24-9	25-4	25-11								
	19.2	20-10	21-5	22-0	22-7	23-1	23-8	24-2	24-9	25-3	25-9				
	24.0	18-7	19-2	19-8	20-2	20-8	21-2	21-8	22-1	22-7	23-0	23-5	23-11	24-4	24-9
E	12.0	1.24	1.36	1.47	1.59	1.71	1.83	1.96	2.09	2.22	2.35	2.49			
	16.0	1.08	1.17	1.27	1.37	1.48	1.59	1.70	1.81	1.92	2.04	2.16	2.28	2.40	2.53
	19.2	0.98	1.07	1.16	1.25	1.35	1.45	1.55	1.65	1.75	1.86	1.97	2.08	2.19	2.31
	24.0	0.88	0.96	1.04	1.12	1.21	1.29	1.38	1.48	1.57	1.66	1.76	1.86	1.96	2.06

NOTE: The required modulus of elasticity, E, in 1,000,000 pounds per square inch (psi) (× 0.00689 for N/mm²) is shown at the bottom of this table, is limited to 2.6 million psi (17 914 N/mm²) and less, and is applicable to all lumber sizes shown. Spans are shown in feet-inches (1 foot = 304.8 mm, 1 inch = 25.4 mm) and are limited to 26 feet (7925 mm) and less.

TABLE 23-IV-R-10

1997 UNIFORM BUILDING CODE

TABLE 23-IV-R-10—RAFTERS WITH L/180 DEFLECTION LIMITATION
The allowable bending stress (F_b) and modulus of elasticity (E) used in this table shall be from Tables 23-IV-V-1 and 23-IV-V-2 only.

DESIGN CRITERIA:
Strength — Live load of 30 psf (1.44 kN/m²) plus dead load of 15 psf (0.72 kN/m²) determines the required bending design value.
Deflection — For 30 psf (1.44 kN/m²) live load.
Limited to span in inches (mm) divided by 180.

Rafter Size (in) ×25.4 for mm	Spacing (in)	Bending Design Value, F_b (psi) ×0.00689 for N/mm²														
		200	300	400	500	600	700	800	900	1000	1100	1200	1300	1400	1500	1600
2×4	12.0	3-0	3-8	4-3	4-9	5-3	5-8	6-0	6-5	6-9	7-1	7-5	7-8	8-0	8-3	8-6
2×4	16.0	2-7	3-2	3-8	4-1	4-6	4-11	5-3	5-6	5-10	6-1	6-5	6-8	6-11	7-2	7-5
2×4	19.2	2-5	2-11	3-4	3-9	4-1	4-5	4-9	5-1	5-4	5-7	5-10	6-1	6-4	6-6	6-9
2×4	24.0	2-2	2-7	3-0	3-4	3-8	4-0	4-3	4-6	4-9	5-0	5-3	5-5	5-8	5-10	6-0
2×6	12.0	4-9	5-10	6-8	7-6	8-2	8-10	9-6	10-0	10-7	11-1	11-7	12-1	12-6	13-0	13-5
2×6	16.0	4-1	5-0	5-10	6-6	7-1	7-8	8-2	8-8	9-2	9-7	10-0	10-5	10-10	11-3	11-7
2×6	19.2	3-9	4-7	5-4	5-11	6-6	7-0	7-6	7-11	8-4	8-9	9-2	9-6	9-11	10-3	10-7
2×6	24.0	3-4	4-1	4-9	5-4	5-10	6-3	6-8	7-1	7-6	7-10	8-2	8-6	8-10	9-2	9-6
2×8	12.0	6-3	7-8	8-10	9-10	10-10	11-8	12-6	13-3	13-11	14-8	15-3	15-11	16-6	17-1	17-8
2×8	16.0	5-5	6-7	7-8	8-7	9-4	10-1	10-10	11-6	12-1	12-8	13-3	13-9	14-4	14-10	15-3
2×8	19.2	4-11	6-0	7-0	7-10	8-7	9-3	9-10	10-6	11-0	11-7	12-1	12-7	13-1	13-6	13-11
2×8	24.0	4-5	5-5	6-3	7-0	7-8	8-3	8-10	9-4	9-10	10-4	10-10	11-3	11-8	12-1	12-6
2×10	12.0	8-0	9-9	11-3	12-7	13-9	14-11	15-11	16-11	17-10	18-8	19-6	20-4	21-1	21-10	22-6
2×10	16.0	6-11	8-5	9-9	10-11	11-11	12-11	13-9	14-8	15-5	16-2	16-11	17-7	18-3	18-11	19-6
2×10	19.2	6-4	7-8	8-11	9-11	10-11	11-9	12-7	13-4	14-1	14-9	15-5	16-1	16-8	17-3	17-10
2×10	24.0	5-8	6-11	8-0	8-11	9-9	10-6	11-3	11-11	12-7	13-2	13-9	14-4	14-11	15-5	15-11
E	12.0	0.05	0.09	0.15	0.20	0.27	0.34	0.41	0.49	0.58	0.67	0.76	0.86	0.96	1.06	1.17
E	16.0	0.04	0.08	0.13	0.18	0.23	0.29	0.36	0.43	0.50	0.58	0.66	0.74	0.83	0.92	1.01
E	19.2	0.04	0.08	0.12	0.16	0.21	0.27	0.33	0.39	0.46	0.53	0.60	0.68	0.76	0.84	0.92
E	24.0	0.04	0.07	0.10	0.14	0.19	0.24	0.29	0.35	0.41	0.47	0.54	0.61	0.68	0.75	0.83

Rafter Size (in) ×25.4 for mm	Spacing (in)	Bending Design Value, F_b (psi) ×0.00689 for N/mm²													
		1700	1800	1900	2000	2100	2200	2300	2400	2500	2600	2700	2800	2900	3000
2×4	12.0	8-9	9-0	9-3	9-6	9-9	10-0	10-3	10-5	10-8	10-10	11-1			
2×4	16.0	7-7	7-10	8-0	8-3	8-5	8-8	8-10	9-0	9-3	9-5	9-7	9-9	9-11	10-1
2×4	19.2	6-11	7-2	7-4	7-6	7-9	7-11	8-1	8-3	8-5	8-7	8-9	8-11	9-1	9-3
2×4	24.0	6-3	6-5	6-7	6-9	6-11	7-1	7-3	7-5	7-6	7-8	7-10	8-0	8-1	8-3
2×6	12.0	13-10	14-2	14-7	15-0	15-4	15-8	16-1	16-5	16-9	17-1	17-5			
2×6	16.0	11-11	12-4	12-8	13-0	13-3	13-7	13-11	14-2	14-6	14-9	15-1	15-4	15-7	15-11
2×6	19.2	10-11	11-3	11-6	11-10	12-2	12-5	12-8	13-0	13-3	13-6	13-9	14-0	14-3	14-6
2×6	24.0	9-9	10-0	10-4	10-7	10-10	11-1	11-4	11-7	11-10	12-1	12-4	12-6	12-9	13-0
2×8	12.0	18-2	18-9	19-3	19-9	20-3	20-8	21-2	21-7	22-1	22-6	22-11			
2×8	16.0	15-9	16-3	16-8	17-1	17-6	17-11	18-4	18-9	19-1	19-6	19-10	20-3	20-7	20-11
2×8	19.2	14-5	14-10	15-2	15-7	16-0	16-4	16-9	17-1	17-5	17-9	18-1	18-5	18-9	19-1
2×8	24.0	12-10	13-3	13-7	13-11	14-4	14-8	15-0	15-3	15-7	15-11	16-3	16-6	16-10	17-1
2×10	12.0	23-3	23-11	24-6	25-2	25-10									
2×10	16.0	20-1	20-8	21-3	21-10	22-4	22-10	23-5	23-11	24-5	24-10	25-4	25-10		
2×10	19.2	18-4	18-11	19-5	19-11	20-5	20-10	21-4	21-10	22-3	22-8	23-1	23-7	24-0	24-5
2×10	24.0	16-5	16-11	17-4	17-10	18-3	18-8	19-1	19-6	19-11	20-4	20-8	21-1	21-5	21-10
E	12.0	1.28	1.39	1.51	1.63	1.76	1.88	2.01	2.15	2.28	2.42	2.56			
E	16.0	1.11	1.21	1.31	1.41	1.52	1.63	1.74	1.86	1.98	2.10	2.22	2.34	2.47	2.60
E	19.2	1.01	1.10	1.20	1.29	1.39	1.49	1.59	1.70	1.80	1.91	2.03	2.14	2.25	2.37
E	24.0	0.90	0.99	1.07	1.15	1.24	1.33	1.42	1.52	1.61	1.71	1.81	1.91	2.02	2.12

NOTE: The required modulus of elasticity, E, in 1,000,000 pounds per square inch (psi) (× 0.00689 for N/mm²) is shown at the bottom of this table, is limited to 2.6 million psi (17 914 N/mm²) and less, and is applicable to all lumber sizes shown. Spans are shown in feet-inches (1 foot = 304.8 mm, 1 inch = 25.4 mm) and are limited to 26 feet (7925 mm) and less.

TABLE 23-IV-R-11—RAFTERS WITH *L*/180 DEFLECTION LIMITATION
The allowable bending stress (*F$_b$*) and modulus of elasticity (*E*) used in this table shall be from Tables 23-IV-V-1 and 23-IV-V-2 only.

DESIGN CRITERIA:
Strength — Live load of 20 psf (0.96 kN/m^2) plus dead load of 20 psf (0.96 kN/m^2) determines the required bending design value.
Deflection — For 20 psf (0.96 kN/m^2) live load.
Limited to span in inches (mm) divided by 180.

Rafter Size (in) ×25.4 for mm	Spacing (in) ×25.4 for mm	Bending Design Value, F_b (psi) ×0.00689 for N/mm^2														
		200	300	400	500	600	700	800	900	1000	1100	1200	1300	1400	1500	1600
2×4	12.0	3-2	3-11	4-6	5-1	5-6	6-0	6-5	6-9	7-2	7-6	7-10	8-2	8-5	8-9	9-0
	16.0	2-9	3-5	3-11	4-4	4-10	5-2	5-6	5-10	6-2	6-6	6-9	7-1	7-4	7-7	7-10
	19.2	2-6	3-1	3-7	4-0	4-4	4-9	5-1	5-4	5-8	5-11	6-2	6-5	6-8	6-11	7-2
	24.0	2-3	2-9	3-2	3-7	3-11	4-3	4-6	4-10	5-1	5-4	5-6	5-9	6-0	6-2	6-5
2×6	12.0	5-0	6-2	7-1	7-11	8-8	9-5	10-0	10-8	11-3	11-9	12-4	12-10	13-3	13-9	14-2
	16.0	4-4	5-4	6-2	6-10	7-6	8-2	8-8	9-3	9-9	10-2	10-8	11-1	11-6	11-11	12-4
	19.2	4-0	4-10	5-7	6-3	6-10	7-5	7-11	8-5	8-11	9-4	9-9	10-1	10-6	10-10	11-3
	24.0	3-7	4-4	5-0	5-7	6-2	6-8	7-1	7-6	7-11	8-4	8-8	9-1	9-5	9-9	10-0
2×8	12.0	6-7	8-1	9-4	10-6	11-6	12-5	13-3	14-0	14-10	15-6	16-3	16-10	17-6	18-1	18-9
	16.0	5-9	7-0	8-1	9-1	9-11	10-9	11-6	12-2	12-10	13-5	14-0	14-7	15-2	15-8	16-3
	19.2	5-3	6-5	7-5	8-3	9-1	9-9	10-6	11-1	11-8	12-3	12-10	13-4	13-10	14-4	14-10
	24.0	4-8	5-9	6-7	7-5	8-1	8-9	9-4	9-11	10-6	11-0	11-6	11-11	12-5	12-10	13-3
2×10	12.0	8-5	10-4	11-11	13-4	14-8	15-10	16-11	17-11	18-11	19-10	20-8	21-6	22-4	23-1	23-11
	16.0	7-4	8-11	10-4	11-7	12-8	13-8	14-8	15-6	16-4	17-2	17-11	18-8	19-4	20-0	20-8
	19.2	6-8	8-2	9-5	10-7	11-7	12-6	13-4	14-2	14-11	15-8	16-4	17-0	17-8	18-3	18-11
	24.0	6-0	7-4	8-5	9-5	10-4	11-2	11-11	12-8	13-4	14-0	14-8	15-3	15-10	16-4	16-11
E	12.0	0.04	0.08	0.12	0.16	0.21	0.27	0.33	0.39	0.46	0.53	0.60	0.68	0.76	0.84	0.93
	16.0	0.04	0.07	0.10	0.14	0.18	0.23	0.28	0.34	0.40	0.46	0.52	0.59	0.66	0.73	0.80
	19.2	0.03	0.06	0.09	0.13	0.17	0.21	0.26	0.31	0.36	0.42	0.48	0.54	0.60	0.67	0.73
	24.0	0.03	0.05	0.08	0.11	0.15	0.19	0.23	0.28	0.32	0.37	0.43	0.48	0.54	0.60	0.66

Rafter Size (in) ×25.4 for mm	Spacing (in) ×25.4 for mm	Bending Design Value, F_b (psi) ×0.00689 for N/mm^2													
		1700	1800	1900	2000	2100	2200	2300	2400	2500	2600	2700	2800	2900	3000
2×4	12.0	9-4	9-7	9-10	10-1	10-4	10-7	10-10	11-1	11-4	11-6	11-9	11-11	12-2	12-4
	16.0	8-1	8-4	8-6	8-9	9-0	9-2	9-5	9-7	9-9	10-0	10-2	10-4	10-6	10-9
	19.2	7-4	7-7	7-9	8-0	8-2	8-5	8-7	8-9	8-11	9-1	9-3	9-5	9-7	9-9
	24.0	6-7	6-9	7-0	7-2	7-4	7-6	7-8	7-10	8-0	8-2	8-4	8-5	8-7	8-9
2×6	12.0	14-8	15-1	15-6	15-11	16-3	16-8	17-0	17-5	17-9	18-1	18-5	18-9	19-1	19-5
	16.0	12-8	13-1	13-5	13-9	14-1	14-5	14-9	15-1	15-4	15-8	16-0	16-3	16-7	16-10
	19.2	11-7	11-11	12-3	12-7	12-10	13-2	13-6	13-9	14-0	14-4	14-7	14-10	15-1	15-4
	24.0	10-4	10-8	10-11	11-3	11-6	11-9	12-0	12-4	12-7	12-10	13-1	13-3	13-6	13-9
2×8	12.0	19-4	19-10	20-5	20-11	21-5	21-11	22-5	22-11	23-5	23-10	24-4	24-9	25-2	25-8
	16.0	16-9	17-2	17-8	18-1	18-7	19-0	19-5	19-10	20-3	20-8	21-1	21-5	21-10	22-2
	19.2	15-3	15-8	16-2	16-7	16-11	17-4	17-9	18-1	18-6	18-10	19-3	19-7	19-11	20-3
	24.0	13-8	14-0	14-5	14-10	15-2	15-6	15-10	16-3	16-7	16-10	17-2	17-6	17-10	18-1
2×10	12.0	24-7	25-4	26-0											
	16.0	21-4	21-11	22-6	23-1	23-8	24-3	24-10	25-4	25-10					
	19.2	19-6	20-0	20-7	21-1	21-8	22-2	22-8	23-1	23-7	24-1	24-6	25-0	25-5	25-10
	24.0	17-5	17-11	18-5	18-11	19-4	19-10	20-3	20-8	21-1	21-6	21-11	22-4	22-9	23-1
E	12.0	1.02	1.11	1.20	1.30	1.40	1.50	1.60	1.71	1.82	1.93	2.04	2.15	2.27	2.39
	16.0	0.88	0.96	1.04	1.13	1.21	1.30	1.39	1.48	1.57	1.67	1.76	1.86	1.96	2.07
	19.2	0.80	0.88	0.95	1.03	1.10	1.18	1.27	1.35	1.44	1.52	1.61	1.70	1.79	1.89
	24.0	0.72	0.78	0.85	0.92	0.99	1.06	1.13	1.21	1.28	1.36	1.44	1.52	1.60	1.69

NOTE: The required modulus of elasticity, *E*, in 1,000,000 pounds per square inch (psi) (× 0.00689 for N/mm^2) is shown at the bottom of this table, is limited to 2.6 million psi (17 914 N/mm^2) and less, and is applicable to all lumber sizes shown. Spans are shown in feet-inches (1 foot = 304.8 mm, 1 inch = 25.4 mm) and are limited to 26 feet (7925 mm) and less.

TABLE 23-IV-R-12

1997 UNIFORM BUILDING CODE

TABLE 23-IV-R-12—RAFTERS WITH $L/180$ DEFLECTION LIMITATION
The allowable bending stress (F_b) and modulus of elasticity (E) used in this table shall be from Tables 23-IV-V-1 and 23-IV-V-2 only.

DESIGN CRITERIA:
Strength — Live load of 30 psf (1.44 kN/m^2) plus dead load of 20 psf (0.96 kN/m^2) determines the required bending design value.
Deflection — For 30 psf (1.44 kN/m^2) live load.
Limited to span in inches (mm) divided by 180.

Rafter Size (in) ×25.4 for mm	Spacing (in)	Bending Design Value, F_b (psi) ×0.00689 for N/mm^2														
		200	300	400	500	600	700	800	900	1000	1100	1200	1300	1400	1500	1600
2×4	12.0	2-10	3-6	4-0	4-6	4-11	5-4	5-9	6-1	6-5	6-8	7-0	7-3	7-7	7-10	8-1
	16.0	2-6	3-0	3-6	3-11	4-3	4-8	4-11	5-3	5-6	5-10	6-1	6-4	6-7	6-9	7-0
	19.2	2-3	2-9	3-2	3-7	3-11	4-3	4-6	4-10	5-1	5-4	5-6	5-9	6-0	6-2	6-5
	24.0	2-0	2-6	2-10	3-2	3-6	3-9	4-0	4-3	4-6	4-9	4-11	5-2	5-4	5-6	5-9
2×6	12.0	4-6	5-6	6-4	7-1	7-9	8-5	9-0	9-6	10-0	10-6	11-0	11-5	11-11	12-4	12-8
	16.0	3-11	4-9	5-6	6-2	6-9	7-3	7-9	8-3	8-8	9-1	9-6	9-11	10-3	10-8	11-0
	19.2	3-7	4-4	5-0	5-7	6-2	6-8	7-1	7-6	7-11	8-4	8-8	9-1	9-5	9-9	10-0
	24.0	3-2	3-11	4-6	5-0	5-6	5-11	6-4	6-9	7-1	7-5	7-9	8-1	8-5	8-8	9-0
2×8	12.0	5-11	7-3	8-4	9-4	10-3	11-1	11-10	12-7	13-3	13-11	14-6	15-1	15-8	16-3	16-9
	16.0	5-2	6-3	7-3	8-1	8-11	9-7	10-3	10-10	11-6	12-0	12-7	13-1	13-7	14-0	14-6
	19.2	4-8	5-9	6-7	7-5	8-1	8-9	9-4	9-11	10-6	11-0	11-6	11-11	12-5	12-10	13-3
	24.0	4-2	5-2	5-11	6-7	7-3	7-10	8-4	8-11	9-4	9-10	10-3	10-8	11-1	11-6	11-10
2×10	12.0	7-7	9-3	10-8	11-11	13-1	14-2	15-1	16-0	16-11	17-9	18-6	19-3	20-0	20-8	21-4
	16.0	6-6	8-0	9-3	10-4	11-4	12-3	13-1	13-10	14-8	15-4	16-0	16-8	17-4	17-11	18-6
	19.2	6-0	7-4	8-5	9-5	10-4	11-2	11-11	12-8	13-4	14-0	14-8	15-3	15-10	16-4	16-11
	24.0	5-4	6-6	7-7	8-5	9-3	10-0	10-8	11-4	11-11	12-6	13-1	13-7	14-2	14-8	15-1
E	12.0	0.04	0.08	0.12	0.17	0.23	0.29	0.35	0.42	0.49	0.57	0.65	0.73	0.82	0.91	1.00
	16.0	0.04	0.07	0.11	0.15	0.20	0.25	0.31	0.36	0.43	0.49	0.56	0.63	0.71	0.78	0.86
	19.2	0.03	0.06	0.10	0.14	0.18	0.23	0.28	0.33	0.39	0.45	0.51	0.58	0.65	0.72	0.79
	24.0	0.03	0.06	0.09	0.12	0.16	0.20	0.25	0.30	0.35	0.40	0.46	0.52	0.58	0.64	0.71

Rafter Size (in) ×25.4 for mm	Spacing (in)	Bending Design Value, F_b (psi) ×0.00689 for N/mm^2													
		1700	1800	1900	2000	2100	2200	2300	2400	2500	2600	2700	2800	2900	3000
2×4	12.0	8-4	8-7	8-10	9-0	9-3	9-6	9-8	9-11	10-1	10-4	10-6	10-8	10-11	11-1
	16.0	7-3	7-5	7-8	7-10	8-0	8-2	8-5	8-7	8-9	8-11	9-1	9-3	9-5	9-7
	19.2	6-7	6-9	7-0	7-2	7-4	7-6	7-8	7-10	8-0	8-2	8-4	8-5	8-7	8-9
	24.0	5-11	6-1	6-3	6-5	6-7	6-8	6-10	7-0	7-2	7-3	7-5	7-7	7-8	7-10
2×6	12.0	13-1	13-6	13-10	14-2	14-7	14-11	15-3	15-7	15-11	16-2	16-6	16-10	17-1	17-5
	16.0	11-4	11-8	12-0	12-4	12-7	12-11	13-2	13-6	13-9	14-0	14-3	14-7	14-10	15-1
	19.2	10-4	10-8	10-11	11-3	11-6	11-9	12-0	12-4	12-7	12-10	13-1	13-3	13-6	13-9
	24.0	9-3	9-6	9-9	10-0	10-3	10-6	10-9	11-0	11-3	11-5	11-8	11-11	12-1	12-4
2×8	12.0	17-3	17-9	18-3	18-9	19-2	19-8	20-1	20-6	20-11	21-4	21-9	22-2	22-6	22-11
	16.0	14-11	15-5	15-10	16-3	16-7	17-0	17-5	17-9	18-1	18-6	18-10	19-2	19-6	19-10
	19.2	13-8	14-0	14-5	14-10	15-2	15-6	15-10	16-3	16-7	16-10	17-2	17-6	17-10	18-1
	24.0	12-2	12-7	12-11	13-3	13-7	13-11	14-2	14-6	14-10	15-1	15-5	15-8	15-11	16-3
2×10	12.0	22-0	22-8	23-3	23-11	24-6	25-1	25-7							
	16.0	19-1	19-7	20-2	20-8	21-2	21-8	22-2	22-8	23-1	23-7	24-0	24-6	24-11	25-4
	19.2	17-5	17-11	18-5	18-11	19-4	19-10	20-3	20-8	21-1	21-6	21-11	22-4	22-9	23-1
	24.0	15-7	16-0	16-6	16-11	17-4	17-9	18-1	18-6	18-11	19-3	19-7	20-0	20-4	20-8
E	12.0	1.09	1.19	1.29	1.39	1.50	1.61	1.72	1.83	1.95	2.07	2.19	2.31	2.43	2.56
	16.0	0.95	1.03	1.12	1.21	1.30	1.39	1.49	1.59	1.69	1.79	1.89	2.00	2.11	2.22
	19.2	0.86	0.94	1.02	1.10	1.19	1.27	1.36	1.45	1.54	1.63	1.73	1.83	1.92	2.03
	24.0	0.77	0.84	0.91	0.99	1.06	1.14	1.22	1.30	1.38	1.46	1.55	1.63	1.72	1.81

NOTE: The required modulus of elasticity, E, in 1,000,000 pounds per square inch (psi) (×0.00689 for N/mm^2) is shown at the bottom of this table, is limited to 2.6 million psi (17 914 N/mm^2) and less, and is applicable to all lumber sizes shown. Spans are shown in feet-inches (1 foot = 304.8 mm, 1 inch = 25.4 mm) and are limited to 26 feet (7925 mm) and less.

TABLE 23-IV-V-1—VALUES FOR JOISTS AND RAFTERS—VISUALLY GRADED LUMBER
For Use in Tables 23-IV-J-1 through 23-IV-R-12 and Chapter 23, Division VII only.

These "F_b" values are for use where repetitive members are spaced not more than 24 inches (610 mm). For wider spacing, the "F_b" values shall be reduced 13 percent.

Values for surfaced dry or surfaced green lumber apply at 19 percent maximum moisture content in use.

SPECIES AND GRADE	SIZE (inches) ×25.4 for mm	DESIGN VALUE IN BENDING "F_b" psi			MODULUS OF ELASTICITY "E" psi	GRADING RULES AGENCY
		Normal Duration	Snow Loading	7-day Loading		
		×0.00689 for N/mm²				
ASPEN						
Select Structural	2 × 4	1,510	1,735	1,885	1,100,000	
No. 1		1,080	1,240	1,350	1,100,000	
No. 2		1,035	1,190	1,295	1,000,000	
No. 3		605	695	755	900,000	
Stud		600	690	750	900,000	
Construction		805	925	1,005	900,000	
Standard		430	495	540	900,000	
Utility		200	230	250	800,000	
Select Structural	2 × 6	1,310	1,505	1,635	1,100,000	
No. 1		935	1,075	1,170	1,100,000	
No. 2		895	1,030	1,120	1,000,000	
No. 3		525	600	655	900,000	
Stud		545	630	685	900,000	NELMA NSLB WWPA
Select Structural	2 × 8	1,210	1,390	1,510	1,100,000	
No. 1		865	990	1,080	1,100,000	
No. 2		830	950	1,035	1,000,000	
No. 3		485	555	605	900,000	
Select Structural	2 × 10	1,105	1,275	1,385	1,100,000	
No. 1		790	910	990	1,100,000	
No. 2		760	875	950	1,000,000	
No. 3		445	510	555	900,000	
Select Structural	2 × 12	1,005	1,155	1,260	1,100,000	
No. 1		720	825	900	1,100,000	
No. 2		690	795	865	1,000,000	
No. 3		405	465	505	900,000	
BEECH-BIRCH-HICKORY						
Select Structural	2 × 4	2,500	2,875	3,125	1,700,000	
No. 1		1,810	2,085	2,265	1,600,000	
No. 2		1,725	1,985	2,155	1,500,000	
No. 3		990	1,140	1,240	1,300,000	
Stud		980	1,125	1,225	1,300,000	
Construction		1,325	1,520	1,655	1,400,000	
Standard		750	860	935	1,300,000	
Utility		345	395	430	1,200,000	
Select Structural	2 × 6	2,170	2,495	2,710	1,700,000	
No. 1		1,570	1,805	1,960	1,600,000	
No. 2		1,495	1,720	1,870	1,500,000	
No. 3		860	990	1,075	1,300,000	
Stud		890	1,025	1,115	1,300,000	NELMA
Select Structural	2 × 8	2,000	2,300	2,500	1,700,000	
No. 1		1,450	1,665	1,810	1,600,000	
No. 2		1,380	1,585	1,725	1,500,000	
No. 3		795	915	990	1,300,000	
Select Structural	2 × 10	1,835	2,110	2,295	1,700,000	
No. 1		1,330	1,525	1,660	1,600,000	
No. 2		1,265	1,455	1,580	1,500,000	
No. 3		725	835	910	1,300,000	
Select Structural	2 × 12	1,670	1,920	2,085	1,700,000	
No. 1		1,210	1,390	1,510	1,600,000	
No. 2		1,150	1,325	1,440	1,500,000	
No. 3		660	760	825	1,300,000	

(Continued)

TABLE 23-IV-V-1

1997 UNIFORM BUILDING CODE

TABLE 23-IV-V-1—VALUES FOR JOISTS AND RAFTERS—VISUALLY GRADED LUMBER—(Continued)

SPECIES AND GRADE	SIZE (inches) ×25.4 for mm	DESIGN VALUE IN BENDING "F_b" psi			MODULUS OF ELASTICITY "E" psi	GRADING RULES AGENCY
		Normal Duration	Snow Loading	7-day Loading		
		×0.00689 for N/mm²				
COTTONWOOD						
Select Structural		1,510	1,735	1,885	1,200,000	
No. 1		1,080	1,240	1,350	1,200,000	
No. 2		1,080	1,240	1,350	1,100,000	
No. 3	2 × 4	605	695	755	1,000,000	
Stud		600	690	750	1,000,000	
Construction		805	925	1,005	1,000,000	
Standard		460	530	575	900,000	
Utility		200	230	250	900,000	
Select Structural		1,310	1,505	1,635	1,200,000	
No. 1		935	1,075	1,170	1,200,000	
No. 2	2 × 6	935	1,075	1,170	1,100,000	
No. 3		525	600	655	1,100,000	
Stud		545	630	685	1,000,000	
Select Structural		1,210	1,390	1,510	1,200,000	NSLB
No. 1		865	990	1,080	1,200,000	
No. 2	2 × 8	865	990	1,080	1,100,000	
No. 3		485	555	605	1,000,000	
Select Structural		1,105	1,275	1,385	1,200,000	
No. 1		790	910	990	1,200,000	
No. 2	2 × 10	790	910	990	1,100,000	
No. 3		445	510	555	1,000,000	
Select Structural		1,005	1,155	1,260	1,200,000	
No. 1		720	825	900	1,200,000	
No. 2	2 × 12	720	825	900	1,100,000	
No. 3		405	465	505	1,000,000	
DOUGLAS FIR-LARCH						
Select Structural		2,500	2,875	3,125	1,900,000	
No. 1 and better		1,985	2,280	2,480	1,800,000	
No. 1		1,725	1,985	2,155	1,700,000	
No. 2		1,510	1,735	1,885	1,600,000	
No. 3	2 × 4	865	990	1,080	1,400,000	
Stud		855	980	1,065	1,400,000	
Construction		1,150	1,325	1,440	1,500,000	
Standard		635	725	790	1,400,000	
Utility		315	365	395	1,300,000	
Select Structural		2,170	2,495	2,710	1,900,000	
No. 1 and better		1,720	1,975	2,150	1,800,000	
No. 1		1,495	1,720	1,870	1,700,000	
No. 2	2 × 6	1,310	1,505	1,635	1,600,000	
No. 3		750	860	935	1,400,000	
Stud		775	895	970	1,400,000	
Select Structural		2,000	2,300	2,500	1,900,000	WCLIB
No. 1 and better		1,585	1,825	1,985	1,800,000	WWPA
No. 1	2 × 8	1,380	1,585	1,725	1,700,000	
No. 2		1,210	1,390	1,510	1,600,000	
No. 3		690	795	865	1,400,000	
Select Structural		1,835	2,110	2,295	1,900,000	
No. 1 and better		1,455	1,675	1,820	1,800,000	
No. 1	2 × 10	1,265	1,455	1,580	1,700,000	
No. 2		1,105	1,275	1,385	1,600,000	
No. 3		635	725	790	1,400,000	
Select Structural		1,670	1,920	2,085	1,900,000	
No. 1 and better		1,325	1,520	1,655	1,800,000	
No. 1	2 × 12	1,150	1,325	1,440	1,700,000	
No. 2		1,005	1,155	1,260	1,600,000	
No. 3		575	660	720	1,400,000	

(Continued)

TABLE 23-IV-V-1—VALUES FOR JOISTS AND RAFTERS—VISUALLY GRADED LUMBER—(Continued)

SPECIES AND GRADE	SIZE (inches)	DESIGN VALUE IN BENDING "F_b" psi			MODULUS OF ELASTICITY "E" psi	GRADING RULES AGENCY
		Normal Duration	Snow Loading	7-day Loading		
	× 25.4 for mm	× 0.00689 for N/mm²				
DOUGLAS FIR-LARCH (North)						
Select Structural		2,245	2,580	2,805	1,900,000	
No. 1/No. 2		1,425	1,635	1,780	1,600,000	
No. 3		820	940	1,025	1,400,000	
Stud	2 × 4	820	945	1,030	1,400,000	
Construction		1,095	1,255	1,365	1,500,000	
Standard		605	695	755	1,400,000	
Utility		290	330	360	1,300,000	
Select Structural		1,945	2,235	2,430	1,900,000	
No. 1/No. 2	2 × 6	1,235	1,420	1,540	1,600,000	
No. 3		710	815	890	1,400,000	NLGA
Stud		750	860	935	1,400,000	
Select Structural		1,795	2,065	2,245	1,900,000	
No. 1/No. 2	2 × 8	1,140	1,310	1,425	1,600,000	
No. 3		655	755	820	1,400,000	
Select Structural		1,645	1,890	2,055	1,900,000	
No. 1/No. 2	2 × 10	1,045	1,200	1,305	1,600,000	
No. 3		600	690	750	1,400,000	
Select Structural		1,495	1,720	1,870	1,900,000	
No. 1/No. 2	2 × 12	950	1,090	1,185	1,600,000	
No. 3		545	630	685	1,400,000	
DOUGLAS FIR (South)						
Select Structural		2,245	2,580	2,805	1,400,000	
No. 1		1,555	1,785	1,940	1,300,000	
No. 2		1,425	1,635	1,780	1,200,000	
No. 3		820	940	1,025	1,100,000	
Stud	2 × 4	820	945	1,030	1,100,000	
Construction		1,065	1,225	1,330	1,200,000	
Standard		605	695	755	1,100,000	
Utility		290	330	360	1,000,000	
Select Structural		1,945	2,235	2,430	1,400,000	
No. 1		1,345	1,545	1,680	1,300,000	
No. 2	2 × 6	1,235	1,420	1,540	1,200,000	
No. 3		710	815	890	1,100,000	
Stud		750	860	935	1,100,000	WWPA
Select Structural		1,795	2,065	2,245	1,400,000	
No. 1		1,240	1,430	1,555	1,300,000	
No. 2	2 × 8	1,140	1,310	1,425	1,200,000	
No. 3		655	755	820	1,100,000	
Select Structural		1,645	1,890	2,055	1,400,000	
No. 1		1,140	1,310	1,425	1,300,000	
No. 2	2 × 10	1,045	1,200	1,305	1,200,000	
No. 3		600	690	750	1,100,000	
Select Structural		1,495	1,720	1,870	1,400,000	
No. 1		1,035	1,190	1,295	1,300,000	
No. 2	2 × 12	950	1,090	1,185	1,200,000	
No. 3		545	630	685	1,100,000	

(Continued)

TABLE 23-IV-V-1

1997 UNIFORM BUILDING CODE

TABLE 23-IV-V-1—VALUES FOR JOISTS AND RAFTERS—VISUALLY GRADED LUMBER—(Continued)

SPECIES AND GRADE	SIZE (inches) × 25.4 for mm	DESIGN VALUE IN BENDING "F_b" psi			MODULUS OF ELASTICITY "E" psi	GRADING RULES AGENCY
		Normal Duration	Snow Loading	7-day Loading		
		× 0.00689 for N/mm²				
EASTERN HEMLOCK—TAMARACK						
Select Structural	2 × 4	2,155	2,480	2,695	1,200,000	NELMA NSLB
No. 1		1,335	1,535	1,670	1,100,000	
No. 2		990	1,140	1,240	1,100,000	
No. 3		605	695	755	900,000	
Stud		570	655	710	900,000	
Construction		775	895	970	1,000,000	
Standard		430	495	540	900,000	
Utility		200	230	250	800,000	
Select Structural	2 × 6	1,870	2,150	2,335	1,200,000	
No. 1		1,160	1,330	1,450	1,100,000	
No. 2		860	990	1,075	1,100,000	
No. 3		525	600	655	900,000	
Stud		520	595	645	900,000	
Select Structural	2 × 8	1,725	1,985	2,155	1,200,000	
No. 1		1,070	1,230	1,335	1,100,000	
No. 2		795	915	990	1,100,000	
No. 3		485	555	605	900,000	
Select Structural	2 × 10	1,580	1,820	1,975	1,200,000	
No. 1		980	1,125	1,225	1,100,000	
No. 2		725	835	910	1,100,000	
No. 3		445	510	555	900,000	
Select Structural	2 × 12	1,440	1,655	1,795	1,200,000	
No. 1		890	1,025	1,115	1,100,000	
No. 2		660	760	825	1,100,000	
No. 3		405	465	505	900,000	
EASTERN SOFTWOODS						
Select Structural	2 × 4	2,155	2,480	2,695	1,200,000	NELMA NSLB
No. 1		1,335	1,535	1,670	1,100,000	
No. 2		990	1,140	1,240	1,100,000	
No. 3		605	695	755	900,000	
Stud		570	655	710	900,000	
Construction		775	895	970	1,000,000	
Standard		430	495	540	900,000	
Utility		200	230	250	800,000	
Select Structural	2 × 6	1,870	2,150	2,335	1,200,000	
No. 1		1,160	1,330	1,450	1,100,000	
No. 2		860	990	1,075	1,100,000	
No. 3		525	600	655	900,000	
Stud		520	595	645	900,000	
Select Structural	2 × 8	1,725	1,985	2,155	1,200,000	
No. 1		1,070	1,230	1,335	1,100,000	
No. 2		795	915	990	1,100,000	
No. 3		485	555	605	900,000	
Select Structural	2 × 10	1,580	1,820	1,975	1,200,000	
No. 1		980	1,125	1,225	1,100,000	
No. 2		725	835	910	1,100,000	
No. 3		445	510	555	900,000	
Select Structural	2 × 12	1,440	1,655	1,795	1,200,000	
No. 1		890	1,025	1,115	1,100,000	
No. 2		660	760	825	1,100,000	
No. 3		405	465	505	900,000	

(Continued)

TABLE 23-IV-V-1—VALUES FOR JOISTS AND RAFTERS—VISUALLY GRADED LUMBER—(Continued)

SPECIES AND GRADE	SIZE (inches) × 25.4 for mm	DESIGN VALUE IN BENDING "F_b" psi			MODULUS OF ELASTICITY "E" psi	GRADING RULES AGENCY
		Normal Duration	Snow Loading	7-day Loading		
		× 0.00689 for N/mm²				
EASTERN WHITE PINE						
Select Structural	2 × 4	2,155	2,480	2,695	1,200,000	NELMA NSLB
No. 1		1,335	1,535	1,670	1,100,000	
No. 2		990	1,140	1,240	1,100,000	
No. 3		605	695	755	900,000	
Stud		570	655	710	900,000	
Construction		775	895	970	1,000,000	
Standard		430	495	540	900,000	
Utility		200	230	250	800,000	
Select Structural	2 × 6	1,870	2,150	2,335	1,200,000	
No. 1		1,160	1,330	1,450	1,100,000	
No. 2		860	990	1,075	1,100,000	
No. 3		525	600	655	900,000	
Stud		520	595	645	900,000	
Select Structural	2 × 8	1,725	1,985	2,155	1,200,000	
No. 1		1,070	1,230	1,335	1,100,000	
No. 2		795	915	990	1,100,000	
No. 3		485	555	605	900,000	
Select Structural	2 × 10	1,580	1,820	1,975	1,200,000	
No. 1		980	1,125	1,225	1,100,000	
No. 2		725	835	910	1,100,000	
No. 3		445	510	555	900,000	
Select Structural	2 × 12	1,440	1,655	1,795	1,200,000	
No. 1		890	1,025	1,115	1,100,000	
No. 2		660	760	825	1,100,000	
No. 3		405	465	505	900,000	
HEM-FIR						
Select Structural	2 × 4	2,415	2,775	3,020	1,600,000	WCLIB WWPA
No. 1 and better		1,810	2,085	2,265	1,500,000	
No. 1		1,640	1,885	2,050	1,500,000	
No. 2		1,465	1,685	1,835	1,300,000	
No. 3		865	990	1,080	1,200,000	
Stud		855	980	1,065	1,200,000	
Construction		1,120	1,290	1,400	1,300,000	
Standard		635	725	790	1,200,000	
Utility		290	330	360	1,100,000	
Select Structural	2 × 6	2,095	2,405	2,615	1,600,000	
No. 1 and better		1,570	1,805	1,960	1,500,000	
No. 1		1,420	1,635	1,775	1,500,000	
No. 2		1,270	1,460	1,590	1,300,000	
No. 3		750	860	935	1,200,000	
Stud		775	895	970	1,200,000	
Select Structural	2 × 8	1,930	2,220	2,415	1,600,000	
No. 1 and better		1,450	1,665	1,810	1,500,000	
No. 1		1,310	1,510	1,640	1,500,000	
No. 2		1,175	1,350	1,465	1,300,000	
No. 3		690	795	865	1,200,000	
Select Structural	2 × 10	1,770	2,035	2,215	1,600,000	
No. 1 and better		1,330	1,525	1,660	1,500,000	
No. 1		1,200	1,380	1,500	1,500,000	
No. 2		1,075	1,235	1,345	1,300,000	
No. 3		635	725	790	1,200,000	
Select Structural	2 × 12	1,610	1,850	2,015	1,600,000	
No. 1 and better		1,210	1,390	1,510	1,500,000	
No. 1		1,095	1,255	1,365	1,500,000	
No. 2		980	1,125	1,220	1,300,000	
No. 3		575	660	720	1,200,000	

(Continued)

TABLE 23-IV-V-1

1997 UNIFORM BUILDING CODE

TABLE 23-IV-V-1—VALUES FOR JOISTS AND RAFTERS—VISUALLY GRADED LUMBER—(Continued)

SPECIES AND GRADE	SIZE (inches) × 25.4 for mm	DESIGN VALUE IN BENDING "F_b" psi			MODULUS OF ELASTICITY "E" psi	GRADING RULES AGENCY
		Normal Duration	Snow Loading	7-day Loading		
		× 0.00689 for N/mm²				
HEM-FIR (North)						
Select Structural	2 × 4	2,245	2,580	2,805	1,700,000	
No. 1/No. 2		1,725	1,985	2,155	1,600,000	
No. 3		990	1,140	1,240	1,400,000	
Stud		980	1,125	1,225	1,400,000	
Construction		1,325	1,520	1,655	1,500,000	
Standard		720	825	900	1,400,000	
Utility		345	395	430	1,300,000	
Select Structural	2 × 6	1,945	2,235	2,430	1,700,000	
No. 1/No. 2		1,495	1,720	1,870	1,600,000	
No. 3		860	990	1,075	1,400,000	NLGA
Stud		890	1,025	1,115	1,400,000	
Select Structural	2 × 8	1,795	2,065	2,245	1,700,000	
No. 1/No. 2		1,380	1,585	1,725	1,600,000	
No. 3		795	915	990	1,400,000	
Select Structural	2 × 10	1,645	1,890	2,055	1,700,000	
No. 1/No. 2		1,265	1,455	1,580	1,600,000	
No. 3		725	835	910	1,400,000	
Select Structural	2 × 12	1,495	1,720	1,870	1,700,000	
No. 1/No. 2		1,150	1,325	1,440	1,600,000	
No. 3		660	760	825	1,400,000	
MIXED MAPLE						
Select Structural	2 × 4	1,725	1,985	2,155	1,300,000	
No. 1		1,250	1,440	1,565	1,200,000	
No. 2		1,210	1,390	1,510	1,100,000	
No. 3		690	795	865	1,000,000	
Stud		695	800	870	1,000,000	
Construction		920	1,060	1,150	1,100,000	
Standard		520	595	645	1,000,000	
Utility		260	300	325	900,000	
Select Structural	2 × 6	1,495	1,720	1,870	1,300,000	
No. 1		1,085	1,245	1,355	1,200,000	
No. 2		1,045	1,205	1,310	1,100,000	
No. 3		600	690	750	1,000,000	
Stud		635	725	790	1,000,000	NELMA
Select Structural	2 × 8	1,380	1,585	1,725	1,300,000	
No. 1		1,000	1,150	1,250	1,200,000	
No. 2		965	1,110	1,210	1,100,000	
No. 3		550	635	690	1,000,000	
Select Structural	2 × 10	1,265	1,455	1,580	1,300,000	
No. 1		915	1,055	1,145	1,200,000	
No. 2		885	1,020	1,105	1,100,000	
No. 3		505	580	635	1,000,000	
Select Structural	2 × 12	1,150	1,325	1,440	1,300,000	
No. 1		835	960	1,040	1,200,000	
No. 2		805	925	1,005	1,100,000	
No. 3		460	530	575	1,000,000	
MIXED OAK						
Select Structural	2 × 4	1,985	2,280	2,480	1,100,000	
No. 1		1,425	1,635	1,780	1,000,000	
No. 2		1,380	1,585	1,725	900,000	
No. 3		820	940	1,025	800,000	
Stud		790	910	990	800,000	NELMA
Construction		1,065	1,225	1,330	900,000	
Standard		605	695	755	800,000	
Utility		290	330	360	800,000	

(Continued)

TABLE 23-IV-V-1—VALUES FOR JOISTS AND RAFTERS—VISUALLY GRADED LUMBER—(Continued)

SPECIES AND GRADE	SIZE (inches) × 25.4 for mm	DESIGN VALUE IN BENDING "F_b" psi			MODULUS OF ELASTICITY "E" psi	GRADING RULES AGENCY
		Normal Duration	Snow Loading	7-day Loading		
		× 0.00689 for N/mm²				
MIXED OAK—(continued)						
Select Structural		1,720	1,975	2,150	1,100,000	
No. 1		1,235	1,420	1,540	1,000,000	
No. 2	2 × 6	1,195	1,375	1,495	900,000	
No. 3		710	815	890	800,000	
Stud		720	825	900	800,000	
Select Structural		1,585	1,825	1,985	1,100,000	
No. 1		1,140	1,310	1,425	1,000,000	
No. 2	2 × 8	1,105	1,270	1,380	900,000	
No. 3		655	755	820	800,000	NELMA
Select Structural		1,455	1,675	1,820	1,100,000	
No. 1		1,045	1,200	1,305	1,000,000	
No. 2	2 × 10	1,010	1,165	1,265	900,000	
No. 3		600	690	750	800,000	
Select Structural		1,325	1,520	1,655	1,100,000	
No. 1		950	1,090	1,185	1,000,000	
No. 2	2 × 12	920	1,060	1,150	900,000	
No. 3		545	630	685	800,000	
MIXED SOUTHERN PINE						
Select Structural		2,360	2,710	2,950	1,600,000	
No. 1		1,670	1,920	2,080	1,500,000	
No. 2		1,500	1,720	1,870	1,400,000	
No. 3		865	990	1,080	1,200,000	
Stud	2 × 4	890	1,020	1,110	1,200,000	
Construction		1,150	1,320	1,440	1,300,000	
Standard		635	725	790	1,200,000	
Utility		315	365	395	1,100,000	
Select Structural		2,130	2,450	2,660	1,600,000	
No. 1		1,490	1,720	1,870	1,500,000	
No. 2	2 × 6	1,320	1,520	1,650	1,400,000	
No. 3		775	895	970	1,200,000	
Stud		775	895	970	1,200,000	SPIB
Select Structural		2,010	2,310	2,520	1,600,000	
No. 1		1,380	1,590	1,720	1,500,000	
No. 2	2 × 8	1,210	1,390	1,510	1,400,000	
No. 3		720	825	900	1,200,000	
Select Structural		1,730	1,980	2,160	1,600,000	
No. 1		1,210	1,390	1,510	1,500,000	
No. 2	2 × 10	1,060	1,220	1,330	1,400,000	
No. 3		605	695	755	1,200,000	
Select Structural		1,610	1,850	2,010	1,600,000	
No. 1		1,120	1,290	1,400	1,500,000	
No. 2	2 × 12	1,010	1,160	1,260	1,400,000	
No. 3		575	660	720	1,200,000	

(Continued)

TABLE 23-IV-V-1—VALUES FOR JOISTS AND RAFTERS—VISUALLY GRADED LUMBER—(Continued)

SPECIES AND GRADE	SIZE (inches) × 25.4 for mm	DESIGN VALUE IN BENDING "F_b" psi			MODULUS OF ELASTICITY "E" psi	GRADING RULES AGENCY
		Normal Duration	Snow Loading	7-day Loading		
		× 0.00689 for N/mm²				
NORTHERN RED OAK						
Select Structural	2 × 4	2,415	2,775	3,020	1,400,000	
No. 1		1,725	1,985	2,155	1,400,000	
No. 2		1,680	1,935	2,100	1,300,000	
No. 3		950	1,090	1,185	1,200,000	
Stud		950	1,090	1,185	1,200,000	
Construction		1,265	1,455	1,580	1,200,000	
Standard		720	825	900	1,100,000	
Utility		345	395	430	1,000,000	
Select Structural	2 × 6	2,095	2,405	2,615	1,400,000	
No. 1		1,495	1,720	1,870	1,400,000	
No. 2		1,460	1,675	1,820	1,300,000	
No. 3		820	945	1,030	1,200,000	
Stud		865	990	1,080	1,200,000	
Select Structural	2 × 8	1,930	2,220	2,415	1,400,000	NELMA
No. 1		1,380	1,585	1,725	1,400,000	
No. 2		1,345	1,545	1,680	1,300,000	
No. 3		760	875	950	1,200,000	
Select Structural	2 × 10	1,770	2,035	2,215	1,400,000	
No. 1		1,265	1,455	1,580	1,400,000	
No. 2		1,235	1,420	1,540	1,300,000	
No. 3		695	800	870	1,200,000	
Select Structural	2 × 12	1,610	1,850	2,015	1,400,000	
No. 1		1,150	1,325	1,440	1,400,000	
No. 2		1,120	1,290	1,400	1,300,000	
No. 3		635	725	790	1,200,000	
NORTHERN SPECIES						
Select Structural	2 × 4	1,640	1,885	2,050	1,100,000	
No. 1/No. 2		990	1,140	1,240	1,100,000	
No. 3		605	695	755	1,000,000	
Stud		570	655	710	1,000,000	
Construction		775	895	970	1,000,000	
Standard		430	495	540	900,000	
Utility		200	230	250	900,000	
Select Structural	2 × 6	1,420	1,635	1,775	1,100,000	
No. 1/No. 2		860	990	1,075	1,100,000	
No. 3		525	600	655	1,000,000	
Stud		520	595	645	1,000,000	NLGA
Select Structural	2 × 8	1,310	1,510	1,640	1,100,000	
No. 1/No. 2		795	915	990	1,100,000	
No. 3		485	555	605	1,000,000	
Select Structural	2 × 10	1,200	1,380	1,500	1,100,000	
No. 1/No. 2		725	835	910	1,100,000	
No. 3		445	510	555	1,000,000	
Select Structural	2 × 12	1,095	1,255	1,365	1,100,000	
No. 1/No. 2		660	760	825	1,100,000	
No. 3		405	465	505	1,000,000	

(Continued)

TABLE 23-IV-V-1—VALUES FOR JOISTS AND RAFTERS—VISUALLY GRADED LUMBER—(Continued)

SPECIES AND GRADE	SIZE (inches) × 25.4 for mm	DESIGN VALUE IN BENDING "F_b" psi			MODULUS OF ELASTICITY "E" psi	GRADING RULES AGENCY
		Normal Duration	Snow Loading	7-day Loading		
			× 0.00689 for N/mm²			
NORTHERN WHITE CEDAR						
Select Structural		1,335	1,535	1,670	800,000	
No. 1		990	1,140	1,240	700,000	
No. 2		950	1,090	1,185	700,000	
No. 3	2 × 4	560	645	700	600,000	
Stud		540	620	670	600,000	
Construction		720	825	900	700,000	
Standard		405	465	505	600,000	
Utility		200	230	250	600,000	
Select Structural		1,160	1,330	1,450	800,000	
No. 1		860	990	1,075	700,000	
No. 2	2 × 6	820	945	1,030	700,000	
No. 3		485	560	605	600,000	
Stud		490	560	610	600,000	NELMA
Select Structural		1,070	1,230	1,335	800,000	
No. 1		795	915	990	700,000	
No. 2	2 × 8	760	875	950	700,000	
No. 3		450	515	560	600,000	
Select Structural		980	1,125	1,225	800,000	
No. 1		725	835	910	700,000	
No. 2	2 × 10	695	800	870	700,000	
No. 3		410	475	515	600,000	
Select Structural		890	1,025	1,115	800,000	
No. 1		660	760	825	700,000	
No. 2	2 × 12	635	725	790	700,000	
No. 3		375	430	465	600,000	
RED MAPLE						
Select Structural		2,245	2,580	2,805	1,700,000	
No. 1		1,595	1,835	1,995	1,600,000	
No. 2		1,555	1,785	1,940	1,500,000	
No. 3		905	1,040	1,130	1,300,000	
Stud	2 × 4	885	1,020	1,105	1,300,000	
Construction		1,210	1,390	1,510	1,400,000	
Standard		660	760	825	1,300,000	
Utility		315	365	395	1,200,000	
Select Structural		1,945	2,235	2,430	1,700,000	
No. 1		1,385	1,590	1,730	1,600,000	
No. 2	2 × 6	1,345	1,545	1,680	1,500,000	
No. 3		785	905	980	1,300,000	
Stud		805	925	1,005	1,300,000	NELMA
Select Structural		1,795	2,065	2,245	1,700,000	
No. 1		1,275	1,470	1,595	1,600,000	
No. 2	2 × 8	1,240	1,430	1,555	1,500,000	
No. 3		725	835	905	1,300,000	
Select Structural		1,645	1,890	2,055	1,700,000	
No. 1		1,170	1,345	1,465	1,600,000	
No. 2	2 × 10	1,140	1,310	1,425	1,500,000	
No. 3		665	765	830	1,300,000	
Select Structural		1,495	1,720	1,870	1,700,000	
No. 1		1,065	1,225	1,330	1,600,000	
No. 2	2 × 12	1,035	1,190	1,295	1,500,000	
No. 3		605	695	755	1,300,000	

(Continued)

TABLE 23-IV-V-1

1997 UNIFORM BUILDING CODE

TABLE 23-IV-V-1—VALUES FOR JOISTS AND RAFTERS—VISUALLY GRADED LUMBER—(Continued)

SPECIES AND GRADE	SIZE (inches) × 25.4 for mm	DESIGN VALUE IN BENDING "F_b" psi			MODULUS OF ELASTICITY "E" psi	GRADING RULES AGENCY
		Normal Duration	Snow Loading	7-day Loading		
		× 0.00689 for N/mm²				
RED OAK						
Select Structural	2 × 4	1,985	2,280	2,480	1,400,000	
No. 1		1,425	1,635	1,780	1,300,000	
No. 2		1,380	1,585	1,725	1,200,000	
No. 3		820	940	1,025	1,100,000	
Stud		790	910	990	1,100,006	
Construction		1,065	1,225	1,330	1,200,000	
Standard		605	695	755	1,100,000	
Utility		290	330	360	1,000,000	
Select Structural	2 × 6	1,720	1,975	2,150	1,400,000	
No. 1		1,235	1,420	1,540	1,300,000	
No. 2		1,195	1,375	1,495	1,200,000	
No. 3		710	815	890	1,100,000	
Stud		720	825	900	1,100,000	NELMA
Select Structural	2 × 8	1,585	1,825	1,985	1,400,000	
No. 1		1,140	1,310	1,425	1,300,000	
No. 2		1,105	1,270	1,380	1,200,000	
No. 3		655	755	820	1,100,000	
Select Structural	2 × 10	1,455	1,675	1,820	1,400,000	
No. 1		1,045	1,200	1,305	1,300,000	
No. 2		1,010	1,165	1,265	1,200,000	
No. 3		600	690	750	1,100,000	
Select Structural	2 × 12	1,325	1,520	1,655	1,400,000	
No. 1		950	1,090	1,185	1,300,000	
No. 2		920	1,060	1,150	1,200,000	
No. 3		545	630	685	1,100,000	
REDWOOD						
Clear Structural	2 × 4	3,020	3,470	3,775	1,400,000	
Select Structural		2,330	2,680	2,910	1,400,000	
Select Structural, open grain		1,900	2,180	2,370	1,100,000	
No. 1		1,680	1,935	2,100	1,300,000	
No. 1, open grain		1,335	1,535	1,670	1,100,000	
No. 2		1,595	1,835	1,995	1,200,000	
No. 2, open grain		1,250	1,440	1,565	1,000,000	
No. 3		905	1,040	1,130	1,100,000	
No. 3, open grain		735	845	915	900,000	
Stud		725	835	910	900,000	
Construction		950	1,090	1,185	900,000	
Standard		520	595	645	900,000	RIS
Utility		260	300	325	800,000	
Clear Structural	2 × 6	2,615	3,010	3,270	1,400,000	
Select Structural		2,020	2,320	2,525	1,400,000	
Select Structural, open grain		1,645	1,890	2,055	1,100,000	
No. 1		1,460	1,675	1,820	1,300,000	
No. 1, open grain		1,160	1,330	1,450	1,100,000	
No. 2		1,385	1,590	1,730	1,200,000	
No. 2, open grain		1,085	1,245	1,355	1,000,000	
No. 3		785	905	980	1,100,000	
No. 3, open grain		635	730	795	900,000	
Stud		660	760	825	900,000	

(Continued)

TABLE 23-IV-V-1—VALUES FOR JOISTS AND RAFTERS—VISUALLY GRADED LUMBER—(Continued)

SPECIES AND GRADE	SIZE (inches) × 25.4 for mm	DESIGN VALUE IN BENDING "F_b" psi			MODULUS OF ELASTICITY "E" psi	GRADING RULES AGENCY
		Normal Duration	Snow Loading	7-day Loading		
		× 0.00689 for N/mm²				
REDWOOD—(continued)						
Clear Structural		2,415	2,775	3,020	1,400,000	
Select Structural		1,865	2,140	2,330	1,400,000	
Select Structural, open grain		1,520	1,745	1,900	1,100,000	
No. 1		1,345	1,545	1,680	1,300,000	
No. 1, open grain	2 × 8	1,070	1,230	1,335	1,100,000	
No. 2		1,275	1,470	1,595	1,200,000	
No. 2, open grain		1,000	1,150	1,250	1,000,000	
No. 3		725	835	905	1,100,000	
No. 3, open grain		585	675	735	900,000	
Clear Structural		2,215	2,545	2,765	1,400,000	
Select Structural		1,710	1,965	2,135	1,400,000	
Select Structural, open grain		1,390	1,600	1,740	1,100,000	
No. 1		1,235	1,420	1,540	1,300,000	
No. 1, open grain	2 × 10	980	1,125	1,225	1,100,000	RIS
No. 2		1,170	1,345	1,465	1,200,000	
No. 2, open grain		915	1,055	1,145	1,000,000	
No. 3		665	765	830	1,100,000	
No. 3, open grain		540	620	670	900,000	
Clear Structural		2,015	2,315	2,515	1,400,000	
Select Structural		1,555	1,785	1,940	1,400,000	
Select Structural, open grain		1,265	1,455	1,580	1,100,000	
No. 1		1,120	1,290	1,400	1,300,000	
No. 1, open grain	2 × 12	890	1,025	1,115	1,100,000	
No. 2		1,065	1,225	1,330	1,200,000	
No. 2, open grain		835	960	1,040	1,000,000	
No. 3		605	695	755	1,100,000	
No. 3, open grain		490	560	610	900,000	
SOUTHERN PINE						
Dense Select Structural		3,510	4,030	4,380	1,900,000	
Select Structural		3,280	3,770	4,100	1,800,000	
Non-Dense Select Structural		3,050	3,500	3,810	1,700,000	
No. 1 Dense		2,300	2,650	2,880	1,800,000	
No. 1		2,130	2,450	2,660	1,700,000	
No. 1 Non-Dense		1,950	2,250	2,440	1,600,000	
No. 2 Dense		1,960	2,250	2,440	1,700,000	
No. 2	2 × 4	1,720	1,980	2,160	1,600,000	SPIB
No. 2 Non-Dense		1,550	1,790	1,940	1,400,000	
No. 3		980	1,120	1,220	1,400,000	
Stud		1,010	1,160	1,260	1,400,000	
Construction		1,270	1,450	1,580	1,500,000	
Standard		720	825	900	1,300,000	
Utility		345	395	430	1,300,000	

(Continued)

TABLE 23-IV-V-1

1997 UNIFORM BUILDING CODE

TABLE 23-IV-V-1—VALUES FOR JOISTS AND RAFTERS—VISUALLY GRADED LUMBER—(Continued)

SPECIES AND GRADE	SIZE (inches) × 25.4 for mm	DESIGN VALUE IN BENDING "F_b" psi			MODULUS OF ELASTICITY "E" psi	GRADING RULES AGENCY
		Normal Duration	Snow Loading	7-day Loading		
		× 0.00689 for N/mm²				
SOUTHERN PINE—(continued)						
Dense Select Structural	2 × 6	3,100	3,570	3,880	1,900,000	
Select Structural		2,930	3,370	3,670	1,800,000	
Non-Dense Select Structural		2,700	3,110	3,380	1,700,000	
No. 1 Dense		2,010	2,310	2,520	1,800,000	
No. 1		1,900	2,180	2,370	1,700,000	
No. 1 Non-Dense		1,720	1,980	2,160	1,600,000	
No. 2 Dense		1,670	1,920	2,080	1,700,000	
No. 2		1,440	1,650	1,800	1,600,000	
No. 2 Non-Dense		1,320	1,520	1,650	1,400,000	
No. 3		865	990	1,080	1,400,000	
Stud		890	1,020	1,110	1,400,000	
Dense Select Structural	2 × 8	2,820	3,240	3,520	1,900,000	
Select Structural		2,650	3,040	3,310	1,800,000	
Non-Dense Select Structural		2,420	2,780	3,020	1,700,000	
No. 1 Dense		1,900	2,180	2,370	1,800,000	
No. 1		1,730	1,980	2,160	1,700,000	
No. 1 Non-Dense		1,550	1,790	1,940	1,600,000	
No. 2 Dense		1,610	1,850	2,010	1,700,000	
No. 2		1,380	1,590	1,720	1,600,000	
No. 2 Non-Dense		1,260	1,450	1,580	1,400,000	
No. 3		805	925	1,010	1,400,000	
Dense Select Structural	2 × 10	2,470	2,840	3,090	1,900,000	SPIB
Select Structural		2,360	2,710	2,950	1,800,000	
Non-Dense Select Structural		2,130	2,450	2,660	1,700,000	
No. 1 Dense		1,670	1,920	2,080	1,800,000	
No. 1		1,500	1,720	1,870	1,700,000	
No. 1 Non-Dense		1,380	1,590	1,730	1,600,000	
No. 2 Dense		1,380	1,590	1,730	1,700,000	
No. 2		1,210	1,390	1,510	1,600,000	
No. 2 Non-Dense		1,090	1,260	1,370	1,400,000	
No. 3		690	795	865	1,400,000	
Dense Select Structural	2 × 12	2,360	2,710	2,950	1,900,000	
Select Structural		2,190	2,510	2,730	1,800,000	
Non-Dense Select Structural		2,010	2,310	2,520	1,700,000	
No. 1 Dense		1,550	1,790	1,940	1,800,000	
No. 1		1,440	1,650	1,800	1,700,000	
No. 1 Non-Dense		1,320	1,520	1,650	1,600,000	
No. 2 Dense		1,320	1,520	1,650	1,700,000	
No. 2		1,120	1,290	1,400	1,600,000	
No. 2 Non-Dense		1,040	1,190	1,290	1,400,000	
No. 3		660	760	825	1,400,000	

(Continued)

TABLE 23-IV-V-1—VALUES FOR JOISTS AND RAFTERS—VISUALLY GRADED LUMBER—(Continued)

SPECIES AND GRADE	SIZE (inches)	DESIGN VALUE IN BENDING "F_b" psi			MODULUS OF ELASTICITY "E" psi	GRADING RULES AGENCY
		Normal Duration	Snow Loading	7-day Loading		
	× 25.4 for mm	× 0.00689 for N/mm²				
SPRUCE-PINE-FIR						
Select Structural		2,155	2,480	2,695	1,500,000	
No. 1/No. 2		1,510	1,735	1,885	1,400,000	
No. 3		865	990	1,080	1,200,000	
Stud	2 × 4	855	980	1,065	1,200,000	
Construction		1,120	1,290	1,400	1,300,000	
Standard		635	725	790	1,200,000	
Utility		290	330	360	1,100,000	
Select Structural		1,870	2,150	2,335	1,500,000	
No. 1/No. 2	2 × 6	1,310	1,505	1,635	1,400,000	
No. 3		750	860	935	1,200,000	NLGA
Stud		775	895	970	1,200,000	
Select Structural		1,725	1,985	2,155	1,500,000	
No. 1/No. 2	2 × 8	1,210	1,390	1,510	1,400,000	
No. 3		690	795	865	1,200,000	
Select Structural		1,580	1,820	1,975	1,500,000	
No. 1/No. 2	2 × 10	1,105	1,275	1,385	1,400,000	
No. 3		635	725	790	1,200,000	
Select Structural		1,440	1,655	1,795	1,500,000	
No. 1/No. 2	2 × 12	1,005	1,155	1,260	1,400,000	
No. 3		575	660	720	1,200,000	
SPRUCE-PINE-FIR (South)						
Select Structural		2,245	2,580	2,805	1,300,000	
No. 1		1,465	1,685	1,835	1,200,000	
No. 2		1,295	1,490	1,615	1,100,000	
No. 3		735	845	915	1,000,000	
Stud	2 × 4	725	835	910	1,000,000	
Construction		980	1,125	1,220	1,000,000	
Standard		545	630	685	900,000	
Utility		260	300	325	900,000	
Select Structural		1,945	2,235	2,430	1,300,000	
No. 1		1,270	1,460	1,590	1,200,000	
No. 2	2 × 6	1,120	1,290	1,400	1,100,000	
No. 3		635	730	795	1,000,000	NELMA
Stud		660	760	825	1,000,000	NSLB
Select Structural		1,795	2,065	2,245	1,300,000	WCLIB
No. 1		1,175	1,350	1,465	1,200,000	WWPA
No. 2	2 × 8	1,035	1,190	1,295	1,100,000	
No. 3		585	675	735	1,000,000	
Select Structural		1,645	1,890	2,055	1,300,000	
No. 1		1,075	1,235	1,345	1,200,000	
No. 2	2 × 10	950	1,090	1,185	1,100,000	
No. 3		540	620	670	1,000,000	
Select Structural		1,495	1,720	1,870	1,300,000	
No. 1		980	1,125	1,220	1,200,000	
No. 2	2 × 12	865	990	1,080	1,100,000	
No. 3		490	560	610	1,000,000	
WESTERN CEDARS						
Select Structural		1,725	1,985	2,155	1,100,000	
No. 1		1,250	1,440	1,565	1,000,000	
No. 2		1,210	1,390	1,510	1,000,000	
No. 3		690	795	865	900,000	WCLIB
Stud	2 × 4	695	800	870	900,000	WWPA
Construction		920	1,060	1,150	900,000	
Standard		520	595	645	800,000	
Utility		260	300	325	800,000	

(Continued)

TABLE 23-IV-V-1

1997 UNIFORM BUILDING CODE

TABLE 23-IV-V-1—VALUES FOR JOISTS AND RAFTERS—VISUALLY GRADED LUMBER—(Continued)

SPECIES AND GRADE	SIZE (inches) × 25.4 for mm	DESIGN VALUE IN BENDING "F_b" psi			MODULUS OF ELASTICITY "E" psi	GRADING RULES AGENCY
		Normal Duration	Snow Loading	7-day Loading		
		× 0.00689 for N/mm²				
WESTERN CEDARS—(continued)						
Select Structural	2 × 6	1,495	1,720	1,870	1,100,000	
No. 1		1,085	1,245	1,355	1,000,000	
No. 2		1,045	1,205	1,310	1,000,000	
No. 3		600	690	750	900,000	
Stud		635	725	790	900,000	
Select Structural	2 × 8	1,380	1,585	1,725	1,100,000	
No. 1		1,000	1,150	1,250	1,000,000	
No. 2		965	1,110	1,210	1,000,000	
No. 3		550	635	690	900,000	WCLIB WWPA
Select Structural	2 × 10	1,265	1,455	1,580	1,100,000	
No. 1		915	1,055	1,145	1,000,000	
No. 2		885	1,020	1,105	1,000,000	
No. 3		505	580	635	900,000	
Select Structural	2 × 12	1,150	1,325	1,440	1,100,000	
No. 1		835	960	1,040	1,000,000	
No. 2		805	925	1,005	1,000,000	
No. 3		460	530	575	900,000	
WESTERN WOODS						
Select Structural	2 × 4	1,510	1,735	1,885	1,200,000	
No. 1		1,120	1,290	1,400	1,100,000	
No. 2		1,120	1,290	1,400	1,000,000	
No. 3		645	745	810	900,000	
Stud		635	725	790	900,000	
Construction		835	960	1,040	1,000,000	
Standard		460	530	575	900,000	
Utility		230	265	290	800,000	
Select Structural	2 × 6	1,310	1,505	1,635	1,200,000	
No. 1		970	1,120	1,215	1,100,000	
No. 2		970	1,120	1,215	1,000,000	
No. 3		560	645	700	900,000	
Stud		575	660	720	900,000	WCLIB WWPA
Select Structural	2 × 8	1,210	1,390	1,510	1,200,000	
No. 1		895	1,030	1,120	1,100,000	
No. 2		895	1,030	1,120	1,000,000	
No. 3		520	595	645	900,000	
Select Structural	2 × 10	1,105	1,275	1,385	1,200,000	
No. 1		820	945	1,030	1,100,000	
No. 2		820	945	1,030	1,000,000	
No. 3		475	545	595	900,000	
Select Structural	2 × 12	1,005	1,155	1,260	1,200,000	
No. 1		750	860	935	1,100,000	
No. 2		750	860	935	1,000,000	
No. 3		430	495	540	900,000	
WHITE OAK						
Select Structural	2 × 4	2,070	2,380	2,590	1,100,000	
No. 1		1,510	1,735	1,885	1,000,000	
No. 2		1,465	1,685	1,835	900,000	
No. 3		820	940	1,025	800,000	
Stud		820	945	1,030	800,000	NELMA
Construction		1,095	1,255	1,365	900,000	
Standard		605	695	755	800,000	
Utility		290	330	360	800,000	

(Continued)

TABLE 23-IV-V-1—VALUES FOR JOISTS AND RAFTERS—VISUALLY GRADED LUMBER—(Continued)

SPECIES AND GRADE	SIZE (inches) × 25.4 for mm	DESIGN VALUE IN BENDING "F_b" psi			MODULUS OF ELASTICITY "E" psi	GRADING RULES AGENCY
		Normal Duration	Snow Loading	7-day Loading		
		× 0.00689 for N/mm^2				
WHITE OAK—(continued)						
Select Structural		1,795	2,065	2,245	1,100,000	
No. 1		1,310	1,505	1,635	1,000,000	
No. 2	2 × 6	1,270	1,460	1,590	900,000	
No. 3		710	815	890	800,000	
Stud		750	860	935	800,000	
Select Structural		1,655	1,905	2,070	1,100,000	
No. 1		1,210	1,390	1,510	1,000,000	
No. 2	2 × 8	1,175	1,350	1,465	900,000	
No. 3		655	755	820	800,000	NELMA
Select Structural		1,520	1,745	1,900	1,100,000	
No. 1		1,105	1,275	1,385	1,000,000	
No. 2	2 × 10	1,075	1,235	1,345	900,000	
No. 3		600	690	750	800,000	
Select Structural		1,380	1,585	1,725	1,100,000	
No. 1		1,005	1,155	1,260	1,000,000	
No. 2	2 × 12	980	1,125	1,220	900,000	
No. 3		545	630	685	800,000	
YELLOW POPLAR						
Select Structural		1,725	1,985	2,155	1,500,000	
No. 1		1,250	1,440	1,565	1,400,000	
No. 2		1,210	1,390	1,510	1,300,000	
No. 3		690	795	865	1,200,000	
Stud	2 × 4	695	800	870	1,200,000	
Construction		920	1,060	1,150	1,300,000	
Standard		520	595	645	1,100,000	
Utility		230	265	290	1,100,000	
Select Structural		1,495	1,720	1,870	1,500,000	
No. 1		1,085	1,245	1,355	1,400,000	
No. 2	2 × 6	1,045	1,205	1,310	1,300,000	
No. 3		600	690	750	1,200,000	
Stud		635	725	790	1,200,000	NSLB
Select Structural		1,380	1,585	1,725	1,500,000	
No. 1		1,000	1,150	1,250	1,400,000	
No. 2	2 × 8	965	1,110	1,210	1,300,000	
No. 3		550	635	690	1,200,000	
Select Structural		1,265	1,455	1,580	1,500,000	
No. 1		915	1,055	1,145	1,400,000	
No. 2	2 × 10	885	1,020	1,105	1,300,000	
No. 3		505	580	635	1,200,000	
Select Structural		1,150	1,325	1,440	1,500,000	
No. 1		835	960	1,040	1,400,000	
No. 2	2 × 12	805	925	1,005	1,300,000	
No. 3		460	530	575	1,200,000	

TABLE 23-IV-V-2

1997 UNIFORM BUILDING CODE

TABLE 23-IV-V-2—VALUES FOR JOISTS AND RAFTERS—MECHANICALLY GRADED LUMBER
For use in Tables 23-V-J-1 through 23-V-R-12 and Division V only.

GRADE DESIGNATION	SIZE (inches) × 25.4 for mm	DESIGN VALUE IN BENDING "F_b" psi			MODULUS OF ELASTICITY "E" psi	GRADING RULES AGENCIES
		Normal Duration	Snow Loading	7-day Loading		
		× 0.00689 for N/mm²				
MACHINE STRESS RATED (MSR) LUMBER						
900f-1.0E		1,040	1,190	1,290	1,000,000	WCLIB,WWPA
1200f-1.2E		1,380	1,590	1,730	1,200,000	NLGA,SPIB,WCLIB,WWPA
1350f-1.3E		1,550	1,790	1,940	1,300,000	SPIB,WCLIB,WWPA
1450f-1.3E		1,670	1,920	2,080	1,300,000	NLGA,WCLIB,WWPA
1500f-1.3E		1,730	1,980	2,160	1,300,000	SPIB
1500f-1.4E		1,730	1,980	2,160	1,400,000	NLGA,SPIB,WCLIB,WWPA
1650f-1.4E		1,900	2,180	2,370	1,400,000	SPIB
1650f-1.5E		1,900	2,180	2,370	1,500,000	NLGA,SPIB,WCLIB,WWPA
1800f-1.6E		2,070	2,380	2,590	1,600,000	NLGA,SPIB,WCLIB,WWPA
1950f-1.5E		2,240	2,580	2,800	1,500,000	SPIB
1950f-1.7E	2 × 4 and wider	2,240	2,580	2,800	1,700,000	NLGA,SPIB,WWPA
2100f-1.8E		2,420	2,780	3,020	1,800,000	NLGA,SPIB,WCLIB,WWPA
2250f-1.6E		2,590	2,980	3,230	1,600,000	SPIB
2250f-1.9E		2,590	2,980	3,230	1,900,000	NLGA,SPIB,WWPA
2400f-1.7E		2,760	3,170	3,450	1,700,000	SPIB
2400f-2.0E		2,760	3,170	3,450	2,000,000	NLGA,SPIB,WCLIB,WWPA
2550f-2.1E		2,930	3,370	3,670	2,100,000	NLGA,SPIB,WWPA
2700f-2.2E		3,110	3,570	3,880	2,200,000	NLGA,SPIB,WCLIB,WWPA
2850f-2.3E		3,280	3,770	4,100	2,300,000	SPIB,WWPA
3000f-2.4E		3,450	3,970	4,310	2,400,000	NLGA,SPIB
3150f-2.5E		3,620	4,170	4,530	2,500,000	SPIB
3300f-2.6E		3,800	4,360	4,740	2,600,000	SPIB
900f-1.2E		1,040	1,190	1,290	1,200,000	NLGA,WCLIB
1200f-1.5E		1,380	1,590	1,730	1,500,000	NLGA,WCLIB
1350f-1.8E	2 × 6 and wider	1,550	1,790	1,940	1,800,000	NLGA
1500f-1.8E		1,730	1,980	2,160	1,800,000	WCLIB
1800f-2.1E		2,070	2,380	2,590	2,100,000	NLGA,WCLIB
MACHINE EVALUATED LUMBER (MEL)						
M-10		1,610	1,850	2,010	1,200,000	
M-11		1,780	2,050	2,230	1,500,000	
M-12		1,840	2,120	2,300	1,600,000	
M-13		1,840	2,120	2,300	1,400,000	
M-14		2,070	2,380	2,590	1,700,000	
M-15		2,070	2,380	2,590	1,500,000	
M-16		2,070	2,380	2,590	1,500,000	
M-17	2 × 4 and wider	2,240	2,580	2,800	1,700,000	SPIB
M-18		2,300	2,650	2,880	1,800,000	
M-19		2,300	2,650	2,880	1,600,000	
M-20		2,300	2,650	2,880	1,900,000	
M-21		2,650	3,040	3,310	1,900,000	
M-22		2,700	3,110	3,380	1,700,000	
M-23		2,760	3,170	3,450	1,800,000	
M-24		3,110	3,570	3,880	1,900,000	
M-25		3,160	3,640	3,950	2,200,000	
M-26	2 × 4 and wider	3,220	3,700	4,030	2,000,000	SPIB
M-27		3,450	3,970	4,310	2,100,000	

The note to Table 23-IV-V-1 applies also to mechanically graded lumber.

Division V—DESIGN STANDARD FOR METAL PLATE CONNECTED WOOD TRUSS
Based on ANSI/TPI 1-1995, National Design Standard for Metal Plate Connected
Wood Truss Construction, of the Truss Plate Institute

SECTION 2321 — METAL PLATE CONNECTED WOOD TRUSS DESIGN

2321.1 Design and Fabrication. The design and fabrication of metal plate connected wood trusses shall be in accordance with ANSI/TPI 1-1995.

2321.2 Performance. Full-scale load tests in accordance with ANSI/TPI 2 (see Section 2303.1, Item 5) may be required at the option of the building official to provide a means of demonstrating that minimum adequate performance is obtainable from specific metal connector plates, various lumber types and grades, a particular truss design and a particular fabrication procedure. ANSI/TPI 2 provides procedures for testing and evaluating wood trusses designed in accordance with ANSI/TPI 1.

2321.3 In-plant Inspection. Each truss manufacturer shall retain an approved agency having no financial interest in the plant being inspected to make nonscheduled inspections of truss fabrication and delivery operations. The inspections shall cover all phases of the truss operation, including lumber storage, handling, cutting, fixtures, presses or rollers, fabrication bundling and banding, handling, and delivery.

2321.4 Marking. Each truss shall be legibly branded, marked or otherwise have permanently affixed thereto the following information located within 2 feet (610 mm) of the center of the span on the face of the bottom chord:

1. Identity of the company manufacturing the truss.

2. The design load.

3. The spacing of trusses.

Division VI—DESIGN STANDARD FOR STRUCTURAL GLUED BUILT-UP MEMBERS—PLYWOOD COMPONENTS
Based on Design and Fabrication Specifications of the American Plywood Association

SECTION 2322 — PLYWOOD STRESSED SKIN PANELS

2322.1 Scope. This standard covers requirements for the design of plywood stressed skin panels as referred to in Section 2305.8.

2322.2 Definition. A panel with stressed covers, or a stressed skin panel, is one in which the covering acts integrally with the framing members to resist bending loads in proportion to its effective moment of inertia. It consists of several longitudinal framing members or ribs spaced by headers and covered on top alone or top and bottom with plywood panels.

2322.3 Design.

2322.3.1 Spacing of ribs. The clear distance between ribs shall not exceed twice the basic spacing set forth in Table 23-VI-A.

2322.3.2 Skin bending. The stressed skin shall be capable of resisting bending stresses due to loads normal to the skin with allowance for direction of the face grain.

2322.3.3 Determination of section properties. For determination of bending stresses, the section properties shall be calculated considering that only the ribs or plies of the ribs and the plies in the stressed skin with grain parallel to the span of the ribs are effective. If the clear distance between ribs is less than the basic spacing "b" as set forth in Table 23-VI-A, the effective width of the skin is equal to the full panel width. If the clear distance between ribs is greater than the basic spacing "b," the effective width of the skins equals the sum of the widths of the ribs in contact with the stressed skin plus a portion of the skin extending a distance equal to 0.5 "b" each side of each rib.

For determination of horizontal shear, rolling shear, deflection and splice stresses, the section properties shall be based on the gross section of all material having its grain parallel with the direction of the principal stress. The modulus of elasticity used to calculate deflection shall be based on the modulus of elasticity of the plywood stressed skins.

2322.3.4 Method of design. Stressed skin panels shall be designed in accordance with accepted engineering formulas, without exceeding the allowable stresses specified in Section 2322.4.

2322.3.5 Splices. The stressed skin panels and ribs shall be continuous in the longitudinal direction, except where adequately spliced and designed in accordance with Section 2322.4.

2322.4 Allowable Stresses.

2322.4.1 Panel stresses.

2322.4.1.1 Direct stress from bending. The allowable working stresses in the stressed skin flanges shall not exceed the values set forth in Table 23-III-A for tension or compression when multiplied by the factor "F":

b = the clear distance between ribs in inches (mm).

"b" = the basic spacing shown in Table 23-VI-A in inches (mm).

For $b/$"b" equal or less than 0.5, F equals 1.

For $b/$"b" greater than 0.5 and less than 1.0

$$F = 0.67 (2 - b/\text{"}b\text{"}).$$

For $b/$"b" equal or greater than 1.0 and not greater than 2.0

$$F = 0.67.$$

2322.4.1.2 Rolling shear. The shear between interior ribs and plywood skin shall not exceed the values set forth in Table 23-III-A for rolling shear. The allowable working stresses shall be reduced 50 percent at exterior ribs. This reduction applies to exterior stringers whose clear distance to the panel edge is less than half the clear distance between stringers.

2322.4.1.3 Stress in plywood skin splices.

2322.4.1.3.1 Scarf-jointed splice in compression. The allowable stress in the skin at the splice shall not exceed the values specified in Section 2322.4.1.1. The slope of the scarf shall not be steeper than 1 unit vertical in 5 units horizontal (20% slope).

2322.4.1.3.2 Butt-jointed splice in compression. Where a splice plate is installed on one side of the joint, the allowable stress in the skin at the splice shall not exceed the values specified in Section 2322.4.1.1, provided said values are reduced in proportion to the ratio of the width of the splice plate to the width of the stressed skin. The splice plate shall be at least equal in thickness and grade to the skin being spliced. The splice plate shall be centered over the joint with the face grain parallel to the face grain of the skin. The minimum length of the splice plate, measured parallel to the span, shall not be less than the values set forth in Table 23-VI-B. Splice plates shall be preglued to the skins prior to assembly of the panel. No percentage reduction is taken for compressive stress.

2322.4.1.3.3 Scarf-jointed splice in tension. The allowable stress in the skin at the splice shall not exceed the values specified in Section 2322.4.1.1. The slope of scarf shall not be steeper than 1 unit vertical in 8 units horizontal (12.5% slope).

2322.4.1.3.4 Butt-jointed splice in tension. Where a splice plate is installed, the allowable stress in the skin at the splice shall not exceed the values set forth in Table 23-VI-C. The splice plate thickness, position, orientation to face grain, installation and length shall conform to Section 2322.4.1.3.2.

2322.4.2 Rib or longitudinal stresses.

2322.4.2.1 Bending. Bending stresses shall not exceed the values set forth in Chapter 23, Division III.

2322.4.2.2 Horizontal shear. Horizontal shear shall not exceed the values set forth in Chapter 23, Division III.

2322.4.2.3 Splices in ribs or longitudinal members. Accurately fitted and well-glued scarf or finger joints are permitted provided the maximum stress in bending does not exceed the percentages of the basic flexural stress shown in ANSI/AITC A190.1 and ASTM D 3737. In the finger joint, the portion of the joint occupied by the tips of the fingers at that cross section shall not be considered as effective. Finger joints and other nonstandard configurations of scarf joints may be prequalified in accordance with ANSI/AITC A190.1 and ASTM D 3737. The effective slope of the joint shall also be determined.

2322.5 Fabrication. The plywood stressed skins shall be fabricated in accordance with the procedures specified in Section 2326.

SECTION 2323 — PLYWOOD CURVED PANELS

2323.1 Scope. This division covers requirements for the design of plywood curved panels as referred to in Section 2305.8. Fabrication procedure shall be in accordance with the requirements of Section 2326.

2323.2 General Design. Curved panels may be designed as either curved flexural panels or arch panels. The curved flexural panels act as a simple beam, without developing horizontal thrust. Bearing details are designed accordingly, with provision for horizontal deflection. Tie rods are not required. Arch panels are stressed both in compression and in flexure. They exert horizontal thrust at the supports and therefore require tie rods or abutments.

2323.3 Definition.

2323.3.1 Core types. Curved plywood panels consist of full-length plywood faces top and bottom spaced by and joined with glue to a structural core capable of resisting the shearing forces. The core may consist of one or more layers of plywood, butt-jointed or full-length, glued over the full areas to the faces or spaced ribs constructed of single-piece or laminated plywood or lumber strips, spaced as required, either preglued or glued during panel assembly.

2323.4 Allowable Stresses.

2323.4.1 Plywood core. The allowable working stresses in plywood core type of construction in flexure, compression and tension shall not exceed the values set forth in Table 23-III-A and reduced for curvature by multiplication by the following curvature factor:

$$1 - 2000\left[\frac{t}{R}\right]^2$$

WHERE:

R = radius of curvature at center line of member, inches (mm).

t = thickness, out-to-out of plies parallel with the stress, inches (mm).

$\frac{t}{R}$ = shall not exceed $^1/_{125}$.

2323.4.2 Spaced ribs. The allowable working stresses in compression tension shall be established as specified in Section 2323.4.1. Additionally, the allowable stresses shall be further reduced as for stressed skin panels as specified in Section 2322.

2323.4.3 Radial stresses. The radial stresses induced by bending moment in a curved spaced rib or plywood core panel shall be computed by the formula:

$$S_R = \frac{3M}{2Rbh}$$

WHERE:

b = width of stringer, inches (mm).

h = overall height of panel, inches (mm).

M = bending moment, in inch-pounds (N·mm), on a width of panel equal to the rib spacing.

R = radius of curvature at center line of member, inches (mm).

S_R = actual stress in a radial direction, pounds per square inch (N/mm^2).

When M is in the direction tending to decrease curvature (increase the radius) the stress is in tension and S_R for plywood shall be limited to one half the values shown for rolling shear in the plane of the plies in Table 23-III-A. S_R for lumber ribs shall be limited to one third the values shown for horizontal shear in Chapter 23, Division III. When M is in the direction tending to increase curvature (decrease the radius) the stress is in compression and S_R for plywood shall be limited to the values shown for compression perpendicular to the grain in Table 23-III-A. S_R for lumber ribs shall be limited to the values shown for compression perpendicular to the grain in Chapter 23, Division III.

2323.5 Effective Cross Section.

2323.5.1 Direction of grain. All plywood and lumber having its grain parallel with the direction of stress shall be considered effective in resisting the stress, except for spaced-rib panels which should be as outlined in Section 2322.

2323.5.2 Scarfed plywood joints. The slope of scarfed joints shall not be steeper than one in eight in order to fully develop the strength of the plywood elements.

2323.5.3 Glued-plywood butt joints with splice plates. Butt joints with splice plates shall be designed in accordance with this standard. Splice plates shall be preglued to the skins prior to assembly of the panel.

2323.5.4 Butt joints without splice plates. The effective strength of solid plywood core panels and of laminated ribs in a spaced rib panel at a section containing a butt joint in any lamination shall be determined by ignoring the butted lamination and any other lamination containing a butt joint closer than 50 times the lamination thickness.

2323.5.5 Deflection. The deflection of curved panels containing butt joints in either the core or the faces may be based on the gross section of all material having its grain parallel with the direction of principal stress, provided all butt joints are staggered at least 10 times the lamination thickness.

2323.6 Design of Curved Flexural Panels. Curved flexural panels shall be designed as simple beams in accordance with the applicable parts of this standard, with provisions to permit horizontal displacement at supports. The extent of this displacement shall be determined in accordance with accepted engineering practice.

2323.7 Design of Arch Panels.

2323.7.1 General. Arch panels shall be designed in accordance with this standard as arches which are subjected to combined bending and direct stress. Determination of vertical and horizontal reactions and maximum axial loads, moments and shears shall be based on accepted engineering practice.

2323.7.2 Bending and direct stress. Combined bending and direct stress shall be calculated by the formula:

$$\frac{P}{Ac} + \frac{M}{Sf} = 1$$

$$c = \frac{0.3E}{(l/d)^2}$$

WHERE:

A = area in square inches (mm^2) of arch cross section per foot (mm) width of arch, of plies with their grain parallel to the direction of the stress.

c = allowable unit compressive stress in pounds per square inch (N/mm^2), as indicated by the above formula but not to exceed the values in Chapter 23, Division III.

d = thickness, out-to-out, of plies with their grain parallel to the direction of the stress in inches (mm).

E = modulus of elasticity of skins in pounds per square inch (N/mm^2).

f = allowable stress in extreme fiber, pounds per square inch (N/mm^2), tension or compression, modified by Sections 2322.4.1 and 2323.4.1 as applicable.

l = chord length in inches (mm) between points of zero moment.

M = moment, in inch-pounds per foot (N·mm/mm) width of arch.

P = direct force, pounds per foot (N/mm) width of arch.

S = section modulus, inches cubed per foot (mm³/mm) width of arch, of plies with their grain parallel to the direction of the stress.

2323.7.3 Shear stresses. Shear stresses shall be calculated by the following formula:

$$S = \frac{VQ}{Ib}$$

WHERE:

b = width of rib, inches (mm). [b = 12 inches (305 mm) for solid plywood core panel.]

I = moment of inertia of arch panel, inches to the fourth power, per foot (mm⁴/mm) width of arch panel, or per rib section.

Q = first moment about neutral axis of area of parallel grain material from panel face into plane at which shear stress is to be calculated, inches cubed per foot (mm³/mm) width of arch panel, or per rib section.

S = shear stress, pounds per square inch (N/mm²), either rolling shear in plywood, or horizontal shear in rib.

V = shear acting normal to the slope of the arch, pounds per foot (N/mm) of arch width, or per rib section.

2323.7.4 Connections. Connections of panels to supporting members shall be with nails, lag screws, bolts, or other means adequate to resist the maximum horizontal thrust and uplift, as well as any shear developed by assumed diaphragm action.

2323.7.5 Supports. The horizontal members supporting arch panels shall be adequate to resist vertical and horizontal loads between supports without deflection sufficient to alter the design basis of the arch panels. Horizontal thrust shall be resisted by properly designed tie rods, struts or abutments.

Where tie rods are used to resist thrust, such rods or other effective means of resisting thrust shall be placed in all bays.

Where abutments are used at the outer edges of exterior bays, struts or other means of resisting thrust, such as shear walls, shall be placed in all bays. Spacing shall be as required by the interior panel supporting members. Such struts are required to provide for unbalanced loads when all bays are not equally loaded.

2323.7.6 Buckling. Adequate design provisions shall be provided to ensure stability against buckling due to axial and/or shear forces.

SECTION 2324 — PLYWOOD BEAMS

2324.1 Scope. This standard covers requirements for design of plywood beams as referred to in Section 2305.8.

2324.2 Definition. Plywood beams are structural units consisting of one or more vertical plywood webs that are attached to lumber flanges. The lumber flanges carry the bending forces while the plywood webs transmit the shear. At intervals along the beam, stiffeners, inserted between the flanges and attached to the webs, serve to distribute concentrated loads and resist web buckling.

2324.3 Flanges.

2324.3.1 Design.

2324.3.1.1 Bending in symmetrical sections. The allowable resisting moment for a symmetrical section shall not exceed the values established by the following formula:

$$M = \frac{2t\,I}{h}$$

WHERE:

h = depth of the beam in inches (mm).

I = net moment of inertia in inches to the fourth power (mm⁴) of all the continuous parallel grain material in the flange and web section. Location of the neutral axis shall be calculated without considering butt joints.

M = resisting moment in inch-pounds (N·mm).

t = allowable working stress in pounds per square inch (N/mm²) in tension parallel to the grain of the flange lumber. Also see Sections 2324.3.1.3 and 2324.3.1.5.

2324.3.1.2 Bending in unsymmetrical sections. The allowable resisting moment for an unsymmetrical beam section shall not exceed the value established by the following formula:

$$M = \frac{f\,I}{Z}$$

WHERE:

f = allowable working stress in pounds per square inch (N/mm²). Note that the allowable working stress for the tension flange and compression flange corresponds to allowable values for t and c, respectively. Also see Sections 2324.3.1.3 and 2324.3.1.5.

I = net moment of inertia in inches to the fourth power (mm⁴) of all continuous parallel grain material in the flange and web section.

M = resisting moment in inch-pounds (N·mm).

Z = distance from the neutral axis to an outer flange fiber in inches (mm). Location of the neutral axis shall be calculated without considering butt joints.

2324.3.1.3 Lateral support of compression flange. The actual stress in the compression flange shall not exceed the allowable stress established by the following formula and in no case shall exceed the allowable unit stress for compression parallel to the grain:

$$\frac{P}{A} = \frac{0.3E}{\left(\frac{L}{d}\right)^2}$$

WHERE:

d = width of the upper flange in inches (mm).

E = modulus of elasticity in pounds per square inch (N/mm²).

L = distance between lateral supports of the compression flange in inches (mm).

P/A = allowable compression stress parallel to the grain in pounds per square inch (N/mm²).

Bracing elements and their connections providing restraint for the compression flange shall be capable of resisting a force F applied in a horizontal direction of not less than that established by the following formula:

$$F = 0.02\,A\,(c')\,\frac{L}{Lm}$$

WHERE:

A = area of the compression flange in square inches (mm²).

c' = the allowable compressive stress or the actual stress in compression in pounds per square inch (N/mm²) as governed by Section 2324.3.1

F = force applied in a horizontal direction in pounds (N).

Lm = maximum permissible bracing for the actual stress.

When the member is not symmetrical about the vertical neutral axis, due consideration shall be taken in the design for adequate lateral restraint.

2324.3.1.4 Lateral stability of deep narrow beam. The ratio I_x/I_y of the gross moment of inertia of all parallel grain material about the horizontal neutral axis to that about the vertical axis shall determine the type of bracing required as set forth in Table 23-VI-D.

2324.3.1.5 Flange splices. Scarf joints in tension and compression flanges may be assumed fully effective for the determination of moment of inertia. Scarf joints in adjacent laminations shall be spaced no closer than 16 times the lamination thickness, measured center to center, except that the spacing may be reduced to zero where the design indicates no bending stress.

Butt joints shall not be permitted in tension flanges except where the butted lamination is omitted in determination of the moment of inertia at that section. In addition, the moment of inertia shall be reduced at the above section where butted joints occur in adjacent laminations based upon the area reduction factors set forth in Table 23-VI-E.

The allowable stress in tension flanges having butt joints shall not exceed 80 percent of the allowable code values. Compression flanges having butt joints shall be designed as required for tension flanges except that a reduction in stress is not required.

The allowable stress in flanges having finger joints shall not exceed 2,200 pounds per square inch (0.015 16 N/mm²). The portion of the joint occupied by the tips of the fingers at the cross section shall not be considered as effective.

Approved finger joints based on the performance tests set forth in ANSI/AITC A190.1 and ASTM D 3737, with the exception that the low test value be limited to twice the design stress, may be used in lieu of the preceding requirement.

2324.3.2 Fabrication. Flanges shall consist of one or more laminations of Douglas fir (Coast Region), West Coast hemlock or southern pine dry lumber not more than 2 inches (51 mm) thick that is stress-graded in accordance with WCLIB Standard Grading Rules No. 16 and SPIB Southern Pine Grading Rules. Finger joints may be used in lieu of scarf joints. All provisions of WCLIB Standard Grading Rules No. 16 are applicable except as further limited by Section 2324.3.1.5.

2324.4 Plywood Webs.

2324.4.1 Design.

2324.4.1.1 Horizontal shear. The allowable shear on the plywood web shall not exceed the value established by the following formula:

$$V = \frac{vI\,\Sigma\,t}{Q}$$

WHERE:

I = total moment of inertia in inches to the fourth power (mm⁴) about the neutral axis of all parallel grain material regardless of any butt joints.

Q = statical moment about the neutral axis of all parallel grain material regardless of any butt joints lying above (or below) the neutral axis in inches to the third power (mm³).

V = allowable total shear on the section in pounds (N).

v = allowable plywood shear stress through the panel thickness in pounds per square inch (N/mm²). See Table 23-III-A and Section 2324.4.1.4.

Σt = total shear thickness of all webs at the section in inches (mm).

2324.4.1.2 Flange-web shear (rolling shear). For beams having one or two webs, the allowable flange-web shear on pressure-glued or nail-glued systems shall not exceed the value established by the following formula:

$$V = \left(\frac{nsdI}{Q_{fl}}\right)$$

WHERE:

d = depth of contact area between flange and plywood web in inches (mm).

I = total moment of inertia about the neutral axis of all parallel grain material, regardless of any butt joints in inches to the fourth power (mm⁴).

n = number of contact surfaces between web and flange.

Q_{fl} = statical moment about the neutral axis of all parallel grain material, regardless of any butt joints, in the upper (or lower) flanges in inches to the third power (mm³).

s = one half of the allowable plywood rolling shear stress in pounds per square inch (N/mm²) as given in Table 23-III-A. Shear values shall not be assigned to the nails in the nail-glued systems.

V = allowable total shear on the section in pounds (N).

2324.4.1.3 Stiffeners for concentrated loads. Stiffeners shall be placed over reactions and where other heavy concentrated loads occur. They shall fit tightly against the flanges and their attachment to the webs shall be capable of transmitting the concentrated load or reaction. The width of the stiffeners shall be equal to the lumber flange width at the section. Their dimension parallel to the beam span shall not be less than w.

$$w = \frac{P}{c_{\perp}\,b}$$

WHERE:

b = flange width in inches (mm).

$c_{\perp}$ = allowable stress in compression perpendicular to grain in pounds per square inch (N/mm²) for the flange lumber as set forth in Division III.

P = concentrated load or reaction in pounds (N).

2324.4.1.4 Intermediate stiffeners to prevent web buckling. Intermediate stiffeners shall be spaced not to exceed 48 inches (1219 mm) on center. The width of the stiffeners shall be equal to the lumber flange width at the section and shall be of not less than 2-inch (51 mm) nominal dimensioned lumber.

In regions of high shear, where shear stress is 100 percent of values set forth in Chapter 23, Division III, the spacing of the stiffeners specified above shall be reduced in accordance with Table 23-VI-F.

Where the webs are stressed to less than 100 percent of the shear strength, the stiffener spacing b shall be calculated by the following formula, except that where the shear stress is 90 percent or less of the shear strength of the webs, the intermediate stiffeners may be spaced at 48 inches (1219 mm) on center. In no case shall the spacing exceed 48 inches (1219 mm).

$$b' = b\left(1 + \frac{100 - p}{25}\right)$$

WHERE:

b = stiffener spacing in inches (mm) from Table 23-VI-F.

b' = actual stiffener spacing in inches (mm).

p = actual percentage of allowable plywood shear stress existing at the section.

2324.4.1.5 Tapered beams. Where the depth of the beam is tapered, the net vertical component of the direct forces in the flanges shall be added to or subtracted from the external shear to obtain the net shear acting at a section of web. This vertical component shall be calculated as follows:

$$P_v = \frac{M}{L_l}$$

WHERE:

L_l = distance from section to the intersection of the flange center lines in inches (mm).

M = bending moment acting on section in inch-pounds (N·mm).

P_v = vertical component of flange in pounds (N).

2324.4.1.6 Web splices. Scarfed and butt end joints in plywood webs shall be designed in accordance with Part I of this standard. Joints subject to more than one type of stress or to stress reversal shall be designed for the most severe case. The slope of scarf joints, 1 unit vertical in 8 units horizontal (12.5% slope) or flatter, and splice plate lengths in Table 23-VI-B of this standard are adequate to transmit 100 percent of the shear strength of the plywood webs being spliced, provided that the provisions of Section 2322.4.1.3.2 regarding splice plate thickness, grade and orientation, are adhered to.

All end joints in plywood webs shall be staggered at least 24 inches (610 mm) with intermediate stiffeners at each joint.

SECTION 2325 — PLYWOOD SANDWICH PANELS

2325.1 Scope. This standard covers requirements for the design of plywood sandwich panels as referred to in Section 2305.8.

2325.2 Definition. A structural sandwich panel is an assembly consisting of a lightweight core securely laminated between two relatively thin, strong facings. Axial compression forces are carried by compression in the facings, stabilized by the core material against buckling; bending moments are resisted by an internal couple composed of forces in the facings; shearing forces are resisted by the core.

2325.3 General Design.

2325.3.1 Material. For purposes of this standard, facings are assumed to be plywood meeting the requirements of UBC Standard 23-2. Cores may be a variety of material, including polystyrene foams, polyurethane foams and paper honeycombs.

2325.3.2 Bond between faces and core. Core may be glued to faces, or in the case of some foam materials, may adhere directly to the faces during expansion. In exterior wall panels, the bond shall be waterproof. The combination of core material and bond shall be such as not to creep excessively under the long-term loads and temperatures.

2325.3.3 Trial section. A trial section of an exterior wall sandwich panel shall be determined as described below. It shall then be investigated for all possible modes of failure.

2325.4 Faces and Cores.

2325.4.1 Plywood faces. The required parallel-grain plywood area shall be determined by the following formula:

$$A_1 + A_2 = \frac{P}{F_c}$$

WHERE:

A_1 = parallel-grain area of outside (top) skin in in.^{2}/ft. (mm^2/mm) of width.

A_2 = parallel-grain area of inside (bottom) skin in in.^{2}/ft. (mm^2/mm).

F_c = allowable compressive stress in parallel plies of plywood in pounds per square inch (N/mm^2).

P = axial load in pounds per foot (N/mm) of panel width.

2325.4.2 Core thickness. The minimum core thickness for structural purposes shall satisfy the following formula. In practice, core thickness is usually chosen on the basis of required insulating value.

$$S = \frac{M}{F_c} = \frac{A(h+c)^2}{4h}$$

WHERE:

A = parallel-grain area of skin in in.^{2}/ft. (mm^2/mm) of panel width, $= A_1 = A_2$.

c = core thickness in inches (mm).

h = total panel thickness in inches (mm).

L = panel span in feet (mm).

M = maximum bending moment applied to panel, in lb.-in. per foot (N·m/mm) of panel width (for simply supported end conditions, $M = 1.5wL^2$).

S = section modulus of panel in in.^{3}/ft. (mm^3/mm) of width.

w = normal loading in pounds per square foot (N/mm^2).

2325.5 Analysis.

2325.5.1 Neutral axis. The location of the neutral axis for the panel shall be calculated from the following formula:

$$\bar{y} = \frac{A_1\left(h - \frac{t_1}{2}\right) + A_2\left(\frac{t_2}{2}\right)}{A_1 + A_2}$$

WHERE:

t_1 = thickness of outside (top) skin in inches (mm).

t_2 = thickness of inside (bottom) skin in inches (mm).

$\bar{y}$ = distance from bottom (or inside) of panel to neutral axis in inches (mm).

2325.5.2 Moment of inertia and section modulus. Moment of inertia and section modulus shall be calculated from the following formulas:

$$I = \frac{A_1 A_2(h+c)^2}{4(A_1 + A_2)}$$

$$S_1 = \frac{I}{h - \bar{y}}, \quad S_2 = \frac{I}{\bar{y}}$$

WHERE:

I = panel moment of inertia in inches^4 per foot (mm^4/mm) of width.

S_1 = section modulus calculated from the compression side of the panel, in inches^3 (mm^3).

S_2 = section modulus calculated from the tension side of the panel. in inches^3 (mm^3).

2325.5.3 Column buckling. Allowable axial load shall be no larger than the value calculated from the following formula:

$$P_{cr} = \frac{\pi^2 EI}{(12L)^2\left[1 + \dfrac{\pi^2 EI}{(12L)^2 \times 6(h+c)G_c}\right]}$$

For **SI:**
$$P_{cr} = \frac{\pi^2 EI}{L^2 \left[1 + \dfrac{2\pi^2 EI}{L^2(h+c)G_c} \right]}$$

WHERE:

E = modulus of elasticity of plywood in pounds per square inch (N/mm²). This value should include a 10 percent increase over published data to restore an allowance made when shear deflection is not computed separately.

G_c = modulus of rigidity of core in direction of span, in pounds per square inch (N/mm²).

P_{cr} = theoretical column buckling load in pounds per foot (N/mm) of panel width.

2325.5.4 Skin buckling. Stress tending to cause buckling of skin shall be no larger than that given in the following formula:

$$C_{cr} = 0.5 \sqrt[3]{EE_c G_c}$$

WHERE:

C_{cr} = theoretical skin buckling stress in pounds per square inch (N/mm²).

E_c = modulus of elasticity of the core perpendicular to the skin, in pounds per square inch (N/mm²).

2325.5.5 Deflection. The maximum deflection shall conform to Table 16-D. Deflection shall be calculated according to the following formulas.

Deflection due to transverse loading only is equal to:

$$\Delta = \Delta_b + \Delta_s = \frac{5wL^4 \times 1728}{384EI} + \frac{wL^2}{4(h+c)G_c}$$

For **SI:**
$$\Delta = \Delta_b + \Delta_s = \frac{5wL^4}{384EI} + \frac{wL^2}{4(h+c)G_c}$$

WHERE:

Δ = total deflection in inches (mm).

Δ_b = deflection due to loading.

Δ_s = deflection due to shear.

Total deflection including effects of axial load is approximately equal to:

$$\Delta_{max} = \frac{\Delta}{1 - P/P_{cr}}$$

WHERE:

Δ_{max} = maximum deflection in inches (mm).

2325.5.6 Bending stress. Bending stress shall be calculated using the following formula. This stress includes the bending due to the axial load through the initial transverse load deflection.

$$f_{b\,max} = \frac{1.5wL^2 + P\Delta_{max}}{S_1}$$

WHERE:

$f_{b\,max}$ = applied bending stress in the facings.

2325.5.7 Combined stress. Maximum combined stress will occur at midlength or midheight of the panel. It is the sum of the axial stress and the compressive bending stress in the concave side of the panel. It shall be calculated in accordance with the following formula:

$$f_{c\,max} = \frac{P}{A_1 + A_2} + f_{b\,max}$$

WHERE:

$f_{c\,max}$ = maximum combined stress in pounds per square inch (N/mm²) (compression).

This stress shall be less than F_c and less than $\frac{1}{3}C_{cr}$.

2325.5.8 Shear stress. Shear stress shall be computed by the formula:

$$f_v = \frac{wL}{(h+c)12} \le F_v$$

For **SI:**
$$f_v = \frac{wL}{(h+c)} \le F_v$$

WHERE:

f_v = applied shear stress in the core, in pounds per square inch (N/mm²).

F_v = allowable shear stress in the core, in pounds per square inch (N/mm²).

SECTION 2326 — FABRICATION OF PLYWOOD COMPONENTS

2326.1 Scope. This standard applies to the fabrication of glued plywood-lumber structural assemblies such as stressed skin panels, curved panels, beams, etc., that have been designed in accordance with accepted engineering principles, including Division VI, Sections 2322, 2323 and 2324, for stressed skin panels, curved panels and beams.

2326.2 Materials.

2326.2.1 Plywood.

2326.2.1.1 General. Plywood shall be as specified in the design and conform to UBC Standard 23-2. Each original panel shall bear the grade trademark of an approved independent inspection and testing agency.

2326.2.1.2 Type. When the equilibrium moisture content of the member in use exceeds 16 percent or if any edge or surface of the plywood is permanently exposed to the weather, the plywood affected shall be exterior type. Otherwise, it may be interior type with exterior glue.

2326.2.1.3 Moisture content. At time of gluing the plywood shall be conditioned to a moisture content of approximately that which it will attain in service, but shall be between 7 percent and 16 percent. Difference in average moisture content between panels glued to each other in any member shall not exceed 5 percent.

2326.2.1.4 Surface requirements. Surfaces of plywood to be glued shall be clean and free from oil, dust, paper tape and other material which would be detrimental to satisfactory gluing.

Medium density overlaid surfaces shall not be relied on for a structural glue bond.

2326.2.1.5 Dimensional tolerances. Scarfed panels of plywood shall be square, as measured on the diagonals, within $\frac{1}{8}$ inch (3.8 mm) for a 4-foot-wide (1219 mm) panel, and proportionately for other widths.

2326.2.2 Lumber.

2326.2.2.1 Grading. Lumber shall be uniformly manufactured and shall be of the grade required by the design. Lumber shall be graded and grade marked in accordance with applicable UBC standards for the species to be used, except as modified herein.

When lumber is resawn, it shall be regraded and grade marked on the basis of the new size.

2326.2.2.2 Knotholes. Knotholes to the same size as the sound and tight knots specified for the grade by applicable UBC standards may be permitted.

2326.2.2.3 Moisture content. At time of gluing, the lumber shall be conditioned to a moisture content of approximately that which it will attain in service, but shall be between 7 percent and 16 percent.

The range of moisture content of the various pieces assembled into a single panel or flange of a beam shall not exceed 5 percent.

2326.2.2.4 Surface requirements. Surfaces of lumber to be glued shall be clean and free from oil, dust and other foreign matter which would be detrimental to satisfactory gluing.

Each piece of lumber shall be machine finished, but not sanded, to a smooth surface with a maximum allowable variation of $^1/_{64}$ inch (0.40 mm) in the surface to be glued.

Warp, twist, cup or other characteristics which would prevent intimate contact of adjacent glued surfaces shall not be permitted.

2326.2.2.5 Flanges. Lumber for laminated flanges of beams and ribs for curved panels shall conform with the applicable requirements of ANSI/AITC Standard A190.1 and ASTM D 3737 regardless of the number of laminations.

2326.2.2.6 Transverse members. Lumber for headers, blocking and stiffeners shall be of minimum 2-inch (51 mm) nominal thickness and Standard Grade or higher Douglas fir or West Coast hemlock or equal.

2326.2.3 Glue.

2326.2.3.1 General. Mixing, spreading, storage life, pot life and working life, and assembly time and temperature shall be in accordance with the manufacturer's recommendations.

2326.2.3.2 Interior type. When the equilibrium moisture content of the member in use does not exceed 16 percent, glue may be casein type, containing a mold inhibitor, and conforming with ASTM D 3024.

2326.2.3.3 Exterior type. When the equilibrium moisture content of the member in use exceeds 16 percent, glue shall be of a phenol, resorcinol or melamine base and conform to ASTM D 3024 and D 2559 and APA AFG-01.

2326.3 Fabrication.

2326.3.1 General. Units may be assembled in a one-step process using either nail-gluing or mechanical pressure including pressure from clamps, presses, or other reasonably uniform measurable pressure, externally applied, to attach the plywood to the framing. If the unit is longer than available lengths of plywood, the skins and webs shall be spliced to full length, either with scarf joints, or butt joints with glued splice plates. Flange laminations of beams having glue lines parallel to the web may be glued at the same time as webs or preassembled. In either case, the laminating shall conform to the applicable requirements of ANSI/AITC Standard A1901.1 and ASTM D 3737.

If more than one unit is pressed at a time, care shall be taken to prevent distortion of any core material used in the assembly.

All cutouts for openings shall be reinforced as required in the design.

2326.3.2 Surfacing. The edges of framing members to which plywood skins or webs are to be glued shall be surfaced prior to assembly so that the members have a maximum variation in surface of $^1/_{64}$ inch (0.4 mm) and $^1/_{32}$ inch (0.8 mm) in depth for all framing members (allowing for actual thickness of any splice plates superimposed on blocking). This variation in depth shall apply to each framing member, as well as to the entire group of framing members within a unit.

Stiffeners for glued beams shall be surfaced prior to gluing so that their surfaces are flush with those of the flanges within $^1/_{32}$ inch (0.8 mm), allowing for any superimposed plywood splice plates.

For glued beams, flanges at all stiffener locations shall have a maximum deviation from square of $^1/_{16}$ inch (1.6 mm) in 6 inches (152 mm), measured perpendicular to an accurate square gage. Twist or bow which would prevent intimate contact of webs, or would cause the beam to deform, shall not be permitted.

Surfaces of high density overlaid plywood to be glued shall be roughened, as by a light sanding, before gluing.

2326.3.3 Glued joints. Plywood skins and webs shall be glued to all framing members over their full contact area, except that solid core curved panels may have the width and spacing of the contact area specified in the design.

All splice plates at butt joints in plywood skins and webs and all scarfed end joints shall be glued over their full contact area. Scarf joints shall be glued, under pressure, and those plywood surfaces in contact with the framing members shall be sanded smooth with tape removed prior to assembly.

2326.3.4 End joints in lumber.

2326.3.4.1 General. End joints in stringers, ribs and flange material, including shims, shall be as specified in the design for the grade and stress used.

2326.3.4.2 Scarf joints. Scarf joints in the lumber flanges shall be well scattered throughout. Unless otherwise specified, in adjoining laminations they shall be spaced not closer than 16 times the lamination thickness, measured from center to center. If shown on the approved plans, this spacing may be decreased to zero where the design indicates no bending stress. In flanges of three or less laminations, only one scarf joint shall be allowed at any one cross section; in flanges of four or more laminations, two scarf joints shall be allowed at the same cross section. Scarf slopes shall not be steeper than 1 unit vertical in 10 units horizontal (10% slope) unless otherwise permitted by the design.

2326.3.4.3 Butt joints. Stringer, rib and flange laminations shall not be butt jointed unless specified in the design. If permitted and not otherwise stipulated in the design, they shall be spaced at least 30 times the lamination thickness in adjoining (actually touching) laminations, and at least 10 times the lamination thickness in nonadjoining laminations.

2326.3.5 End joints in plywood.

2326.3.5.1 Scarf joints. When a skin or web is composed of panels less than full length, end joints shall be scarfed and glued, unless butt joints with plywood splice plates are permitted by the design. Slope of scarf joints shall not be steeper than 1 unit vertical in 8 units horizontal (12.5% slope).

2326.3.5.2 Butt joints in skins of panels. Butt joints shall be backed with plywood splice plates glued to one side of the skin. For ribbed panels, when glued during panel assembly the splice plate shall be backed with one or more pieces of lumber blocking, accurately machined in width so as to obtain adequate pressure. Lumber blocking by itself shall not be used for splicing skins unless shown on the approved plans.

Splice plates shall be centered over the butt joint, and shall have their grain parallel with that of the skin. Plates shall extend to within $^1/_4$ inch (6.4 mm) of the ribs; the latter shall not be notched to receive them, unless permitted by the design. Splice plates shall be at least equal in thickness to the skin, except that minimum thickness shall be $^1/_2$ inch (12.7 mm) if nail- or staple-glued. Minimum splice plate lengths shall be as shown in Table 23-VI-B.

Solid plywood core panels shall be scarf jointed to length if so specified by the design. If designated on the approved plans, joints in solid plywood core panels may be tightly butted, provided they will conform readily with the curvature of the surface. Spacing of butt joints shall not be less than 10 times the lamination thickness.

2326.3.5.3 Butt joints in webs of beams. Butt joints shall be staggered at least 24 inches (610 mm).

Splices at butt joints in webs shall be in accordance with the design. Unless otherwise noted in the design, at all web butt joints a plywood shear splice plate shall be centered over the joint and preglued prior to assembly. The plate shall extend to within $^1/_4$ inch (6.4 mm) of each flange on the inside of the beam, shall be at least equal in thickness to the web being spliced, of a length as specified on the approved plans, and shall have its grain parallel to that of the web. No splice plate shall consist of more than two pieces of plywood. Where the design provides for the stiffeners to act as the web splices, web butt joints shall be located over the center of the stiffener, within $^1/_{16}$ inch (1.6 mm), and webs shall be glued to the stiffener. The stiffener alone may be used as a splice plate when the web is 24 inches (610 mm) deep or less, and is no thicker than $^3/_8$ inch (9.5 mm), or carries no more shear than would be allowed on a $^3/_8$-inch (9.5 mm) panel.

2326.3.6 Stiffener location for beams. Stiffeners shall be placed as shown on the approved plans, but in any case they shall be spaced not to exceed 4 feet (1219 mm) on centers, and at reactions and other concentrated load joints.

Stiffeners shall be held in tight contact with the flanges by positive lateral pressure during fabrication.

2326.3.7 Assembly. Ribs, stringers, flanges and other framing members shall be assembled accurately and shall be square. Joints shall be made tight, and all framing members shall be of uniform thickness within $^1/_{32}$ inch (0.8 mm).

All gluing surfaces shall be flush within $^1/_{32}$ inch (0.8 mm).

Plywood may be lightly tacked to framing members so as to maintain alignment during pressing.

2326.3.8 Tongue-and-groove edge joint. Where a tongue-and-groove-type edge joint for stressed skin and curved panels is specified (and not otherwise detailed), the longitudinal framing member forming the tongue shall be of at least 2-inch (51 mm) nominal thickness, set out $^3/_4$ inch (19 mm) from the plywood edge plus or minus $^1/_{16}$ inch (1.6 mm). Edges of the tongue shall be eased so as to leave a flat shoulder at least $^3/_8$ inch (9.5 mm) wide. Any corresponding framing member forming the base of the groove shall be set back $^1/_4$ inch to 1 inch (6.4 mm to 25 mm) more than the amount by which the tongue protrudes. Abutting plywood edges may be chamfered. One face may be cut back $^1/_{16}$ inch (1.6 mm) to provide a tight fit for the opposite face.

2326.3.9 Nail- and pressure-gluing. Curved units shall be laminated over a form. Means shall be used that will provide close contact and substantially uniform pressure on panels and beams. Application of pressure or nailing may start at any point, but shall progress to an end or ends. Movement of the units shall not be permitted while the glue is setting.

Where clamping or other positive mechanical means are used, the pressure on the net framing area shall be sufficient to provide adequate contact and ensure good glue bond. Pressure shall be uniformly distributed by caul plates, beams or other effective means.

In place of mechanical pressure methods, nail-gluing may be used for bonding plywood to lumber and plywood to plywood, but shall not be used for bonding lumber to lumber. Regardless of the method used, the resulting glue line must meet the required standards. Nails shall be at least 4d for plywood up to $^3/_8$ inch (9.5 mm) thick, and 6d for $^1/_2$-inch to $^7/_8$ inch (12.7 mm to 22 mm) plywood, and 8d for 1-inch to $1^1/_8$-inch (25 mm to 29 mm) plywood. Nails shall be spaced not to exceed 3 inches (76 mm) along the ribs for plywood through $^3/_8$ inch (9.5 mm), or 4 inches (102 mm) for plywood $^1/_2$ inch (12.7 mm) and thicker, using one line for glue lines 2 inches (51 mm) wide or less, two lines for glue lines more than 2 inches (51 mm) wide and up to 4 inches (102 mm) wide, three lines for glue lines up to 10 inches (254 mm) wide and four lines for 12-inch-wide (305 mm) glue lines. (If shorter nail lengths are used in order to avoid penetrating the opposite face, spacing shall be decreased in proportion to actual penetration.)

Panels having solid plywood cores may be glued using staples or nails, following a schedule of demonstrated effectiveness, but in no case spaced farther apart than 6 inches (152 mm) both ways. Length of fasteners is limited by the available thickness of the laminations. Additional nailing, into temporary solid framing, may be required to provide close contact for adequate glue bonds.

In any case, it shall be the responsibility of the fabricator to produce a glue bond which meets or exceeds the requirements of this division.

2326.3.10 Insulation and ventilation. In hollow panels, insulating and vapor-barrier materials and ventilation provided as specified in the design shall be securely fastened in the assembly in such a way that they cannot interfere with the process of gluing the plywood skins to the framing members or with the ventilation pattern.

When specified, longitudinal sections shall be vented through blocking and headers on the cool side of the insulation. Provision shall be made to line up the ventholes in assembling panels, with a definite ventilation pattern leading to the outdoors. Framing members shall not be notched for ventilation, unless shown on the approved plans.

2326.4 Identification. Each member shall be identified by the appropriate trademark of the approved independent inspection and testing agency, legibly applied so as to be clearly visible. If the strength of one surface of a beam or panel is different from the other, the top surface shall be identified.

2326.5 Test Samples. When glue bond test samples are taken from a member and from trim, they shall be taken as cores approximately 2 inches (51 mm) in diameter, drilled perpendicular to the plane of the skins or webs, and no deeper than $^3/_4$ inch (19 mm) into any framing member. Samples at beam flanges shall be taken from the ends of the beams only. Centers of cores shall normally be located in the top corner of the beam at points within 2 inches (51 mm) from one edge and up to 6 inches (152 mm) from the other.

No samples shall be taken from the same skin or web closer together than 12 inches (305 mm), except as detailed in Exception 2. Samples shall be taken at a distance from the panel ends not greater than the panel depth.

EXCEPTIONS: 1. One sample may be taken from one of the outside longitudinal framing members, within the outer fourth points of the panel length, provided the framing member is notched no deeper than $^1/_2$ inch (12.7 mm) below its edge. The other outside longitudinal framing member may be sampled similarly, but at the opposite end of the panel.

2. Two samples per butt joint at any one cross section may be taken from skin or web splice plates. They shall be located midway between longitudinal framing members, and shall be aligned longitudinally, one on each side of the butt joint.

Samples through beam web splice plates shall be taken within the middle half of the span, at the center line of the beam depth.

Where the core of curved panels consists of solid plywood or a solid core material, only one core shall be taken at any one cross section. Cross sections shall be no closer than 12 inches (305 mm) along the arc. Not more than four cores shall be taken from a 4-foot-wide (1219 mm) panel, with a proportionate number limited to other widths. Ribs in such panels shall be sampled as required by this section, except that at ribs the core shall not extend more than $^3/_4$ inch (19 mm) into the rib.

Samples may be taken from other locations when approved by the building official.

2326.6 Testing of Glued Joints.

2326.6.1 General. This section shall govern all glued joints which have not been subjected to prior inspection and testing such as the glue line between plywood and lumber as contained in ASTM D 1101 and AITC 200. These joints shall be referred to as "secondary" glue lines. Glued joints subjected to prior testing such as those in plywood shall be referred to as "primary" glue lines. Before testing, specimens shall be allowed to cure.

All secondary glue line testing shall be performed by approved testing agencies. Test specimens containing localized defects permitted by the grade of the material involved shall be discarded.

2326.6.2 Lumber-plywood side grain combinations.

2326.6.2.1 Selection of test samples. Lumber-plywood side grain samples shall be taken from members selected at random. They shall consist of 2-inch-diameter (51 mm) core of adequate depth and location to yield sufficient material on both sides of the secondary glue line.

Members to be sampled shall be selected to include production variables such as different lots of material, glue batches, shifts and crews.

Trim may be acceptable in lieu of cores for sampling purposes if it is truly representative of the member.

2326.6.2.2 Dry shear strength testing. Lumber and plywood side grain specimens shall be tested in a block shear tool with the load applied through a self-aligning seat to ensure uniform lateral distribution of the load. Rate of load application shall be 0.2 inch (5.1 mm) per minute. Ultimate load shall be read to the nearest 5 pounds (22.2 N) and wood failure shall be determined to the nearest 5 percent.

2326.6.2.3 Durability testing.

2326.6.2.3.1 Interior. This test method shall be used in all cases where the durability requirement for the completed component is Interior.

Lumber and plywood side grain specimens shall be submerged in water at room temperature for a period of four hours and then dried at a temperature between 100°F and 105°F (37.8°C. and 40.6°C) for a period of 19 hours with sufficient air circulation in drying cabinet to lower moisture content of specimen to a maximum of 8 percent based on oven-dry weight. This test procedure shall be conducted through three cycles, unless all specimens have failed. All specimens shall be inspected after the first and third cycles and all failures recorded. Total continuous delamination 1 inch (25 mm) along the edge of the test specimen and $^1/_4$ inch (6.4 mm) deep shall be considered as failure.

2326.6.2.3.2 Exterior. This test method shall be used where the test specimen contains exterior-type adhesives throughout and where the durability requirement for the completed component is Exterior.

Lumber and plywood specimens shall be cycled by one of the following methods:

1. **Cold soak.** Test specimens shall be submerged in water at room temperature for 48 hours and then dried for eight hours at a temperature of 145°F (62.8°C) [± 5°F (2.8°C)], followed by two cycles of soaking for 16 hours and drying for eight hours under conditions described above. The specimens shall be soaked again for a period of 16 hours.

2. **Vacuum pressure.** Place the specimens in the pressure vessel and weight them down. Admit water at a temperature of 65°F to 80°F (18.3°C. to 26.7°C) in sufficient quantity so that the specimens are completely submerged. Separate the specimens by stickers, wire screens or other means in such a manner that all end grain surfaces are freely exposed to the water. Draw a vacuum of 20 inches to 25 inches (67.5 kPa to 84.4 kPa) of mercury (at sea level) and hold for 30 minutes. Release the vacuum and apply a pressure of 40 plus or minus 5 pounds per square inch (275.8 kPa ± 34.5 kPa) for two hours.

While still wet, the specimens shall be tested in accordance with the method specified above in Section 2326.6.2.2.

2326.6.2.4 Performance requirements.

2326.6.2.4.1 General. The minimum performance requirements for laboratory testing of all side grain glue joint specimens shall be as specified in this section. If two or more grain orientations are present in any one test run or combinations of test runs, an adjusted average shear stress performance requirement shall be computed by the weighted average method.

2326.6.2.4.2 Dry shear strength requirements. The specimens shall have an average dry strength shear stress as set forth in Table 23-VI-G.

2326.6.2.4.3 Interior durability requirement. Ninety-five percent of all specimens shall pass the first cycle and 85 percent shall pass three cycles.

2326.6.2.4.4 Exterior durability requirement. Eighty-five percent of more wood failure shall be required for the overall average. Ninety percent of all specimens tested shall show 60 percent or more wood failure, 80 percent of all specimens tested shall show 80 percent or more wood failure. Failure in any part of test specimen, except glue failure in the secondary glue bond tested, is wood failure.

2326.6.3 Lumber end joints.

2326.6.3.1 General. Plain scarf joints and finger joints shall be tested in accordance with the provisions of ASTM D 1101 and AITC 200.

2326.6.3.2 Selection of test samples. Test samples shall be obtained from full-size joints, or from scarf joints selected at random from routine production. The latter shall be taken from the center line of the lamination containing the scarf being sampled. No more than one sample shall be taken from any component cross section and scarf samples shall not be closer than 4 feet (1219 mm).

2326.6.3.3 Dry tension testing. Specimens shall be tested in accordance with AITC 200.

2326.6.3.4 Durability testing.

2326.6.3.4.1 Interior. Test specimens shall be submerged in water at room temperature for a period of four hours and then dried at a temperature between 100°F and 105°F (37.8°C. and 40.6°C) for a period of 19 hours with sufficient air circulation in drying cabinet to lower moisture content of specimen to a maximum of 8 percent based on oven-dry weight. This test procedure shall be conducted through three cycles, unless all specimens have failed.

The specimens then shall be tested dry in accordance with AITC 200.

2326.6.3.4.2 Exterior. Test specimens shall be cycled by one of the following methods:

1. **Cold soak.** Test specimens shall be submerged in water at room temperature for 48 hours and then dried for eight hours at a temperature of 145°F (62.8°C) [± 5°F (2.8°C)], followed by two cycles of soaking for 16 hours and drying for eight hours under conditions described above. The specimens shall be soaked again for a period of 16 hours.

2. **Vacuum pressure.** Place the specimens in the pressure vessel and weight them down. Admit water at a temperature of 65°F to 80°F (18.3°C. to 26.7°C) in sufficient quantity so that the specimens are completely submerged. Separate the specimens by stickers, wire screens, or other means in such a manner that all end grain surfaces are freely exposed to the water. Draw a vacuum of 20 inches to 25 inches (67.5 kPa to 84.4 kPa) of mercury (at sea level) and hold for 30 minutes. Release the vacuum and apply a pressure of 40 plus or minus 5 pounds per square inch (275.8 kPa ± 34.5 kPa) for two hours.

The specimens shall then be tested wet in accordance with AITC 200.

2326.6.3.5 Performance requirements.

2326.6.3.5.1 General. The minimum performance requirements for laboratory testing of all lumber end joint specimens shall be as specified in this section.

2326.6.3.5.2 Dry tension stress requirements. The test specimens shall have the tension stresses set forth in Table 23-VI-H. No test value shall be less than twice the normal allowable stress.

Dry tension wood failure requirements regardless of species shall average not less than 80 percent except tests averaging at least 10,000 pounds per square inch (69 N/mm^2) in tension shall not have less than 70 percent. The minimum requirement for 90 percent of the individual specimens shall not be less than 40 percent wood failure.

2326.6.3.5.3 Interior durability requirement. Average wood failure shall not be less than 70 percent.

2326.6.3.5.4 Exterior durability requirement. Average wood failure shall not be less than 85 percent and the minimum requirement for 90 percent of the individual specimens shall not be less than 50 percent.

2326.6.4 Plywood end joints.

2326.6.4.1 General. Plain plywood scarf joints shall be tested in accordance with the provisions of this section. Splice butt joints shall be tested as specified in Section 2326.6.2.

2326.6.4.2 Selection of test samples. Full width scarfs shall be selected at random from routine production. A minimum of 2 percent of the scarfs produced shall be sampled. At least 20 specimens suitable for testing shall be obtained from each 4 feet (1219 mm) of scarf. They shall be equally distributed across the width.

2326.6.4.3 Dry tension. Specimens shall be tested in a tension machine equipped with wedge-type grips. Dry tension specimens shall be loaded at a rate of 0.4 inch (10 mm) per minute. Ultimate load shall be read to the nearest 5 pounds (22.2 N) and wood failure on the joined surface estimated to the nearest 5 percent. Wood failure shall be estimated on the scarf surface of the plies parallel to the direction of the tension load.

2326.6.4.4 Durability.

2326.6.4.4.1 Interior. Interior durability test specimens shall be sub-merged in water at room temperature for a period of four hours and then dried at a temperature between 100°F and 105°F (37.8°C. and 40.6°C) for a period of 19 hours with sufficient air circulation in drying cabinet to lower moisture content of specimen to a maximum of 8 percent based on oven-dry weight. This test procedure shall be conducted through three cycles, unless all specimens have failed. Test specimens showing delamination on the face or end in excess of $^1/_{16}$ inch (1.6 mm) deep and $^1/_2$ inch (12.7 mm) long at the scarf glue line shall be considered as failing.

2326.6.4.4.2 Exterior. Exterior durability test specimens shall be cycled by one of the following methods:

1. **Cold soak.** Test specimens shall be submerged in water at room temperature for 48 hours and then dried for eight hours at a temperature of 145°F (62.8°C) [± 5°F (2.8°C)], followed by two cycles of soaking for 16 hours and drying for eight hours under conditions described above. The specimens shall be soaked again for a period of 16 hours.

2. **Vacuum pressure.** Place the specimens in the pressure vessel and weight them down. Admit water at a temperature of 65°F to 80°F (18.3°C. to 26.7°C) in sufficient quantity so that the specimens are completely submerged. Separate the specimens by stickers, wire screens, or other means in such a manner that all end grain surfaces are freely exposed to the water. Draw a vacuum of 20 inches to 25 inches of mercury (67.5 kPa to 84.4 kPa) (at sea level) and hold for 30 minutes. Release the vacuum and apply a pressure of 40 ± 5 pounds per square inch (275.8 kPa ± 34.5 kPa) for two hours.

While still wet, the specimens shall be tested in a tension machine equipped with wedge-type grips. Dry tension specimens shall be loaded at a rate of 0.4 inch (10 mm) per minute. Ultimate load shall be read to the nearest 5 pounds (22.2 N) and wood failure on the joined surface estimated to the nearest 5 percent.

2326.6.4.5 Performance requirements.

2326.6.4.5.1 General. The minimum performance requirements for laboratory testing of all lumber end joint specimens shall be as specified in this section.

2326.6.4.5.2 Dry tension stress. The average tension stress for Group 1 shall be 4,000 pounds per square inch (28 N/mm^2); for Groups 2 and 3, 2,800 pounds per square inch (19 N/mm^2); and 2,400 pounds per square inch (17 N/mm^2) for Group 4, all adjusted as required for inner ply species. Regardless of species, wood failure shall average not less than 80 percent and the minimum requirement for 90 percent of the individual test specimens shall not be less than 50 percent.

2326.6.4.5.3 Durability requirements.

1. **Interior.** Ninety-five percent of all test specimens shall pass one cycle and 85 percent shall pass three cycles.

2. **Exterior.** Average wood failure shall not be less than 85 percent.

SECTION 2327 — ALL-PLYWOOD BEAMS

2327.1 Scope. This part covers requirements for the design of all-plywood beams as referred to in Section 2305.8.

2327.2 Definition. All-plywood beams are staple-glued structural units consisting of one or more vertical layers of butt-jointed plywood for webs which resist both bending and shear forces. Plywood web splice plates located at web joints, and/or one or more vertical layers of butt-jointed plywood for flanges, may also be staple glued to the webs for added resistance to shear and bending forces, if necessary. Plywood web stiffeners are staple-glued to single-layer webs to distribute concentrated loads and resist web buckling at end supports and interior concentrated load points. (For design of plywood beams consisting of plywood webs attached to lumber flanges, see Section 2324 of this division.)

2327.3 Allowable Stresses. Allowable stresses for plywood used in all-plywood beams shall not exceed the values set forth in Table 23-III-A, except that allowable bending stress may be multiplied by 2 for five-ply, five-layer plywood, and by 1.7 for three-, four- or five-ply, three-layer plywood. Five-ply, five-layer plywood with butt-jointed center ply shall be considered as three-layer plywood for design purposes.

2327.4 Design.

2327.4.1 Bending. The allowable resisting moment for a beam section shall not exceed the value established by the following formula:

$$M = \frac{fI}{Z}$$

WHERE:

f = allowable working stress in bending for plywood, in pounds per square inch (N/mm^2). Also, see Section 2327.3.

I = the minimum net moment of inertia, in inches to the fourth power (mm^4), of all continuous parallel grain material at any beam section containing a web or flange butt joint. Location of the neutral axis shall be calculated without considering butt joints in the web of flanges (if used). Also, see Section 2327.6.2. A web splice plate (if used) shall be included in the net amount of inertia only if directly glued to the web layer containing the butt joint.

M = resisting moment in inch-pounds (N·mm).

Z = distance in inches (mm) from the neutral axis to the extreme fiber in bending.

2327.4.2 Horizontal shear. The allowable shear on the plywood web shall not exceed the value established by the following formula:

$$V = \frac{vI \Sigma t}{Q}$$

WHERE:

I = total moment of inertia, in inches to the fourth power (mm^4), of all parallel grain material in the web and flanges regardless of butt joints. Stiffeners or web splice plates shall be disregarded.

Q = statical moment about the neutral axis, in inches to the third power (mm^3), of all parallel grain material in the web and flanges, regardless of butt joints, lying above (or below) the neutral axis. Stiffeners or web splice plates shall be disregarded.

V = allowable total shear on the section, in pounds (N). Loads may be disregarded that are located within a distance from the support equal to the beam depth.

v = allowable plywood shear stress through the panel thickness, in pounds per square inch (N/mm^2).

Σt = total shear thickness of continuous webs (and/or web splice plates and stiffeners, if applicable) at the neutral

axis of the section, in inches (mm). Also, see Section 2327.6.2.

2327.4.3 Flange-web shear (rolling shear). For beams containing plywood flanges, the allowable flange-web shear on glued joints shall not exceed the value established by the following formula:

$$V = \frac{nsdI}{Q_{fl}}$$

WHERE:

d = depth of contact area between flange and web, in inches (mm).

I = total moment of inertia, in inches to the fourth power (mm^4), of all parallel grain material in the web and flanges regardless of butt joint. Stiffeners or web splice plates shall be disregarded.

n = number of contact surfaces between web and flange.

Q_{fl} = statical moment about the neutral axis, in inches to the third power (mm^3), of all parallel grain material in the upper (or lower) flanges, regardless of butt joints.

s = one half of the allowable plywood rolling shear stress, in pounds per square inch (N/mm^2).

V = allowable total shear on the section, in pounds (N). Loads may be disregarded that are located within a distance from the support equal to the beam depth.

2327.4.4 Lateral stability. The compression edges of the beam shall be positively restrained from lateral buckling, as by fastening to structural panel sheathing, framing spaced no further than 24 inches (610 mm) on center, and/or by ceiling material.

2327.4.5 Connections. Bolts, lag screws or other similar large-diameter connectors shall not be used in web or flange tension areas where bending stresses exceed the allowable values set forth in Table 23-III-A.

2327.5 Materials.

2327.5.1 Plywood. Plywood shall be as required in Section 2326.2.1. Plywood pieces cut 4^1/$_2$ inches (114 mm) or less in width for flanges shall be visually inspected after cutting for size of knots. Pieces containing knots in face or back plies larger than two thirds of the flange width shall not be used in fabricating flanges for all-plywood beams.

2327.5.2 Glue. Glue shall be as specified in Section 2326.2.3.

2327.5.3 Staples. Staples shall be 16 gage by ^{7}/$_{16}$-inch (1.6 mm by 11 mm) crown, made from galvanized steel wire. Staple length shall be ^{1}/$_8$ inch (3.2 mm) less than the total thickness of materials joined.

2327.6 Fabrication.

2327.6.1 General. Plywood beams shall be fabricated with glue and staples. Plywood face grain for webs, flanges, web splice plates and stiffeners shall be oriented parallel to the span (i.e., horizontally).

2327.6.2 Web and flange end joints. End joints in plywood webs and flanges shall be located as specified in the design. Joints in any web or flange lamination shall be spaced at least 24 inches (610 mm) from the nearest joint in any other lamination. In single-web beams, joints in webs shall be located at least 24 inches (610 mm) from any end or interior support.

Butt joints at ends of plywood web or flange pieces shall be trimmed square and tightly butted [maximum gap ^{1}/$_{32}$ inch (0.8 mm)].

2327.6.3 Adhesive application. Adhesive shall be spread uniformly over the full contact area of mating web and/or flange surfaces.

If web or flange laminations are glued under pressure, such pressure shall be applied by clamping or other mechanical means. Pressure shall be sufficient to provide adequate contact and ensure good glue bonds [100 to 150 psi (689 to 1034 kPa) is suggested, unless otherwise specified by the adhesive manufacturer]. Movement of the members shall be prevented until the adhesive develops sufficient handling strength as recommended by the adhesive manufacturer.

In any case, it shall be the responsibility of the fabricator to produce a glue bond which meets or exceeds applicable specifications.

2327.6.4 Staple installation. Staples shall be installed with their crowns parallel to the plywood face-grain direction. Staple spacing shall be as shown in Figure 23-VI-1.

Installation of staples may start at any point but shall progress to the end or ends of each piece. Pressure may be needed during stapling to flatten webs and flanges to ensure uniform contact between mating glued surfaces.

2327.6.5 Web splices. Splices at butt joints in webs shall be as specified in the design. Butt joints shall be spliced with a plywood plate centered over the joint. The splice plate shall be glued over its full contact area and stapled to the web in accordance with Figure 23-VI-1. The plate shall extend to at least $^1/_4$ inch (6.4 mm) of each flange (if applicable); shall be equal in thickness to the web; shall be of a length as specified in the design or in Table 23-VI-B; and shall have its face-grain direction parallel with that of the web. In multiple-layer webs, the splice plate shall be directly glued to the web containing the butt joint.

2327.6.6 Web stiffeners. Stiffeners for single-layer webs shall be located as shown in the design, but in any case they shall be placed at end supports and at interior concentrated-load points.

The stiffeners shall consist of a plywood plate glued over its full contact area and stapled to the web in accordance with Figure 23-VI-1. The plate shall extend to at least $^1/_4$ inch (6.4 mm) of each flange (if applicable); shall be at least equal in thickness to the web; shall be of a length as specified in the design, but in no case less than 10 inches (254 mm); and shall have its face-grain direction parallel with that of the web. The end or ends of the stiffener shall extend at least 6 inches (152 mm) beyond the edges of the support at the end and any interior supports.

Single-layer webs shall be reinforced at cutouts with a plywood web stiffener.

2327.7 Identification. Each member shall be identified as specified in Section 2326.4.

2327.8 Test Samples. Test samples shall be as specified in Section 2326.5.

2327.9 Testing of Glued Joints. Plywood splice butt joints shall be tested as specified in Section 2326.6.2.

TABLE 23-VI-A
TABLE 23-VI-D

1997 UNIFORM BUILDING CODE

TABLE 23-VI-A—BASIC SPACING

PLYWOOD THICKNESS (inches)	PLYWOOD FINISH	BASIC SPACING "b" (inches)	
× 25.4 for mm		× 25.4 for mm	
		Face Grain Parallel to Longitudinal Members	Face Grain Perpendicular to Longitudinal Members
1/4	Sanded	10.3	11.6
5/16	Rough	11.9	16.8
3/8	Rough (3-ply)	14.2	20.1
3/8	Sanded (3-ply)	16.4	16.4
3/8	Sanded (5-ply)	18.1	20.2
1/2	Rough and sanded (5-ply)	23.2	28.5
5/8	Rough and sanded (5-ply)	29.1	35.6
3/4	Rough and sanded (5-ply)	38.2	38.2
7/8	Sanded (7-ply)	41.6	48.1
1	Rough (7-ply)	45.5	58.9
1	Sanded (7-ply)	54.5	47.9

TABLE 23-VI-B—LENGTH OF SPLICE PLATES

THICKNESS OF SKIN (inches)	LENGTH OF SPLICE PLATE (inches)
× 25.4 for mm	
1/4	6
5/16	8
3/8 (sanded)	10
3/8 (unsanded)	12
1/2	14
5/8	16
3/4	16

TABLE 23-VI-C—ALLOWABLE PLYWOOD TENSION STRESS FOR BUTT JOINT SPLICE[1]

PLYWOOD THICKNESS (inches)	ALL STRUCT. I GRADES	GROUP 1	GROUP 2 AND GROUP 3	GROUP 4
× 25.4 for mm	× 0.00689 for N/mm²			
1/4, 5/16, 3/8 sanded, 3/8 unsanded	1,200	960	800	720
1/2	1,200	800	760	720
5/8 and 3/4	960	640	600	560

[1]These values are based on stress acting parallel to the face grain and must be reduced in accordance with Table 23-III-A for stresses perpendicular or at 45 degrees to face grain. In addition, stresses are based on a ratio of width of splice plate to the width of the stressed skin of 80 percent. Where other ratios are used, the allowable stress shall be adjusted accordingly. For applicable species groups for unsanded panels, refer to Footnote 1 of Table 23-III-A.

TABLE 23-VI-D—BRACING FOR DEEP NARROW BEAM

$\dfrac{I_x}{I_y}$	TYPE OF BRACING REQUIRED
Not more than 5	None required
More than 5 to not more than 10	Ends held in position at bottom flanges at supports
More than 10 to not more than 20	Ends held in position at top and bottom flanges at supports
More than 20 to not more than 30	Top or bottom flange continuously supported in accordance with Section 2324.3.1.3
More than 30 to not more than 40	Beam restrained by bridging or other bracing at not more than 8 feet (2438 mm) on center
More than 40	Bridging at 8 feet (2438 mm) on center and compression flanges continuously supported in accordance with Section 2324.3.1.3

TABLE 23-VI-E—AREA REDUCTION FACTORS

BUTT JOINT SPACING *T* = LAMINATION THICKNESS	PERCENT OF GROSS AREA OF ADJACENT LAMINATION EFFECTIVE
30*T* 20*T*	90 80
10*T*	60
Less than 10*T*	0

TABLE 23-VI-F—WEB STIFFENER SPACING

PLYWOOD WEB THICKNESS (inches)	CLEAR DISTANCE BETWEEN FLANGES (inches)					
	10	20	30	40	50	60 or More
	× 25.4 for mm					
$^3/_8$	15	15	15	15	15	15
$^1/_2$	27	22	22	22	22	22
$^5/_8$	48	29	28	28	28	28
$^3/_4$	48	38	35	34	34	34
$^7/_8$	48	48	41	39	39	39
1	48	48	48	48	46	45
$1^3/_{16}$	48	48	48	48	48	48

TABLE 23-VI-G—DRY SHEAR STRENGTH STRESS REQUIREMENTS[1]

GRAIN ORIENTATION OF LUMBER OR PLYWOOD WITH RESPECT TO ADJACENT PLY OR LUMBER LAMINATION	AVERAGE SHEAR STRESS (pounds per square inch)
	× 0.00689 for N/mm²
(Douglas Fir and Southern Pine)	
Parallel 45° Perpendicular	650 500 250
(Other Species)	
Parallel 45° Perpendicular	520 400 200

[1]The minimum shear test value shall not be less than twice the normal allowable rolling shear value.

TABLE 23-VI-H—DRY TENSION STRESS

AVERAGE TENSION STRESS (pounds per square inch)	MINIMUM TENSION STRESS FOR AT LEAST 80 PERCENT OF SPECIMENS (pounds per square inch)
× 0.00689 for N/mm²	
[Douglas Fir (Coast type) and Southern Pine]	
7,000	6,000
(Other Species)	
6,000	5,000

FIGURE 23-VI-1

1997 UNIFORM BUILDING CODE

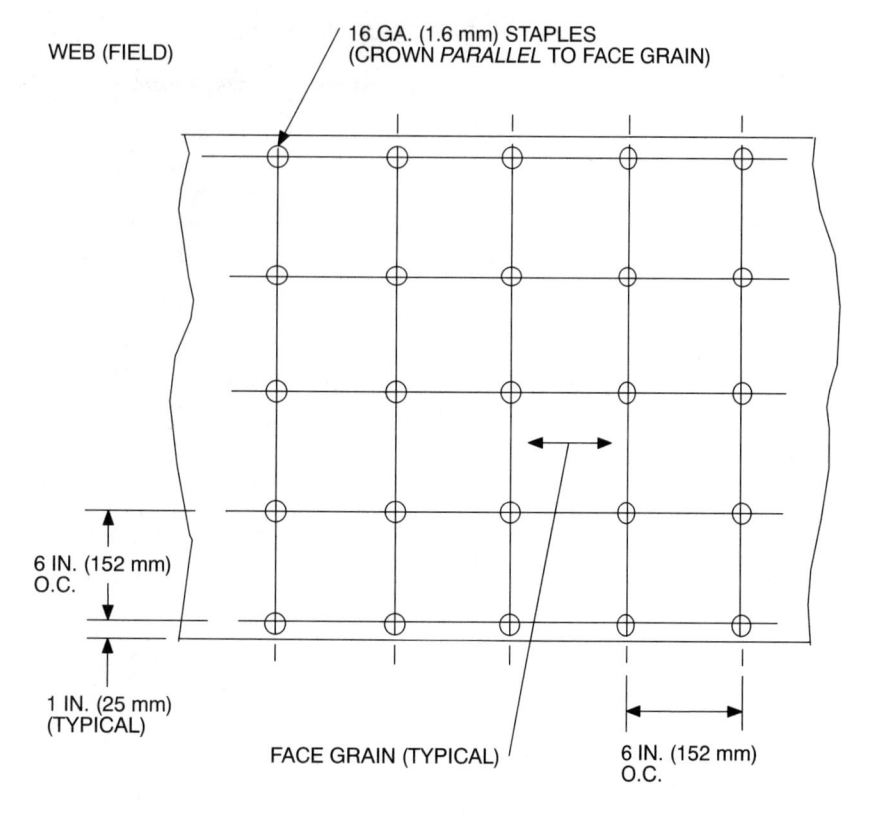

FIGURE 23-VI-1—STAPLE SPACING FOR PLYWOOD WEBS, FLANGES, SPLICE PLATES AND STIFFENERS

WEB (SPLICE)

16 GA. (1.6 mm) STAPLES
(CROWN *PARALLEL* TO FACE GRAIN)

4 IN.
O.C.
(102 mm)

$^{1}/_{4}$ IN. (6.4 mm) (MAXIMUM)

3 IN. (76 mm) O.C.

1 IN. (25 mm) (TYPICAL)

WEB (JOINT)

16 GA. (1.6 mm) STAPLES
(CROWN *PARALLEL* TO FACE GRAIN)

3 IN. O.C.
(76 mm)

3 IN. O.C.
(76 mm)

1 IN. (25 mm) (TYPICAL)

12 IN.
(305 mm)

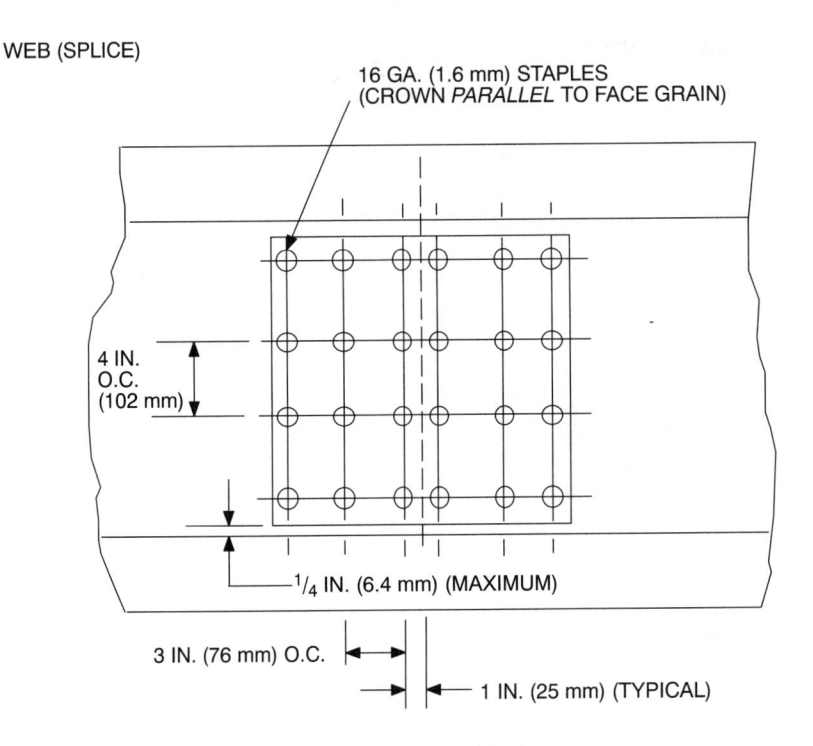

FIGURE 23-VI-1—STAPLE SPACING FOR PLYWOOD WEBS, FLANGES, SPLICE PLATES AND STIFFENERS—(Continued)

FIGURE 23-VI-1

1997 UNIFORM BUILDING CODE

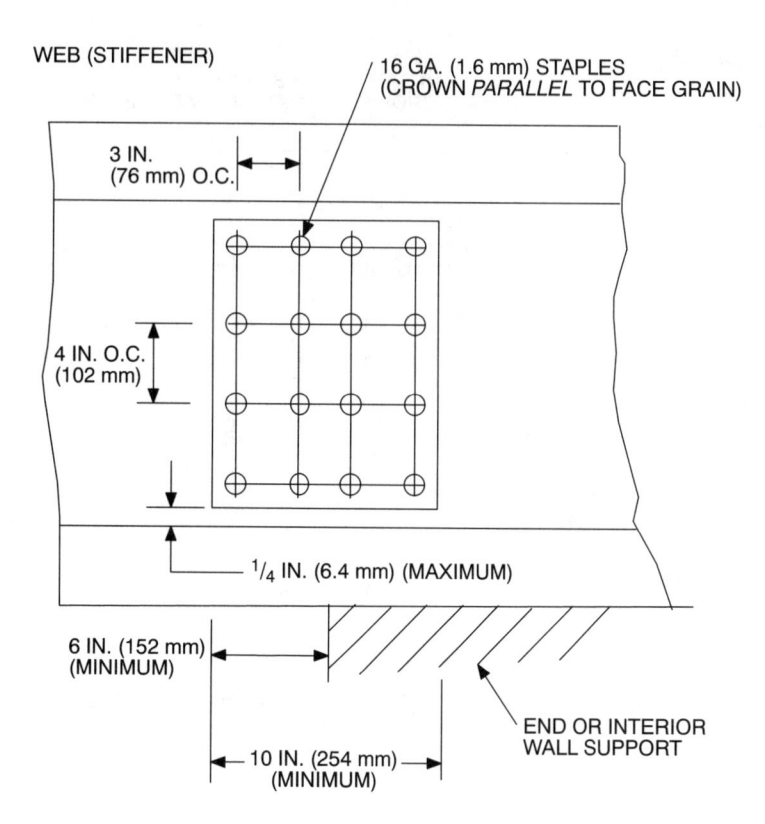

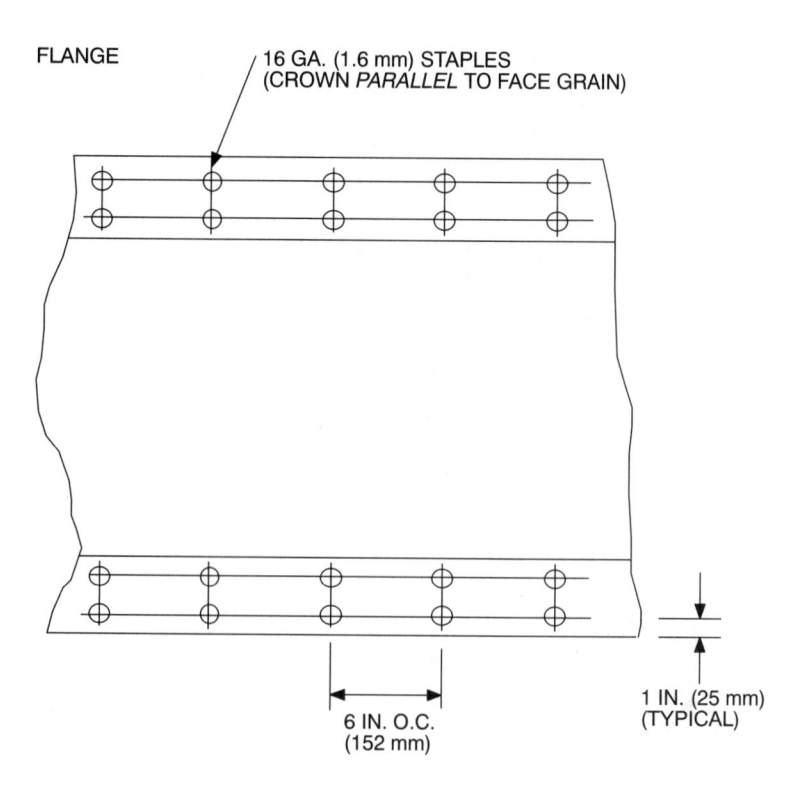

FIGURE 23-VI-1—STAPLE SPACING FOR PLYWOOD WEBS, FLANGES, SPLICE PLATES AND STIFFENERS   (Continued)

Division VII—DESIGN STANDARD FOR SPAN TABLES FOR JOISTS AND RAFTERS

SECTION 2328 — SPAN TABLES FOR JOISTS AND RAFTERS

2328.1 Scope. This standard covers the loading (DL and LL) and deflection criteria used to establish the allowable spans in Tables 23-IV-J-1 through 23-IV-R-12 and the tables of this division. The tables in this standard are intended for light-frame construction.

SECTION 2329 — DESIGN CRITERIA FOR JOISTS AND RAFTERS

The allowable spans of tables of this standard and Tables 23-IV-J-1 through 23-IV-R-12 are calculated on the basis of a series of modulus of elasticity (E), and fiber stress (F_b) values. The range of values in the tables provides allowable spans for all species and grades of nominal 2-inch lumber customarily used in construction. The allowable span is the clear distance between supports. For sloping rafters the span is measured along the horizontal projection.

SECTION 2330 — LUMBER STRESSES

The use of the span tables requires reference to the list of single or repetitive member bending stress values (F_b) and modulus of elasticity values (E) from Tables 23-IV-V-1 and 23-IV-V-2.

SECTION 2331 — MOISTURE CONTENT

Tabulated spans are calculated on the basis of the dry sizes and are also applicable to the corresponding green sizes. The spans in these tables are intended for use in covered structures or where moisture content in use does not exceed 19 percent.

SECTION 2332 — LUMBER SIZE

Tabulated spans apply to surfaced (S4S) lumber having dimensions which conform to UBC Standard 23-1.

SECTION 2333 — SPAN TABLES FOR JOISTS AND RAFTERS

The following tables are based on the design criteria set forth in each table and of this standard.

TABLE 23-VII-J-1

1997 UNIFORM BUILDING CODE

TABLE 23-VII-J-1—FLOOR JOISTS WITH *L*/360 DEFLECTION LIMITS
The allowable bending stress (*F$_b$*) and modulus of elasticity *(E)* used in this table shall be from Tables 23-IV-V-1 and 23-IV-V-2 only.

DESIGN CRITERIA:
Deflection — For 50 psf (2.4 kN/m^2) live load.
Limited to span in inches (mm) divided by 360.
Strength — Live load of 50 psf (2.4 kN/m^2) plus dead load of 10 psf (0.48 kN/m^2) determines the required bending design value.

Joist Size (in)	Spacing (in)	Modulus of Elasticity, *E*, in 1,000,000 psi (× 0.00689 for N/mm^2)																
× 25.4 for mm		0.8	0.9	1.0	1.1	1.2	1.3	1.4	1.5	1.6	1.7	1.8	1.9	2.0	2.1	2.2	2.3	2.4
2 × 6	12.0	7-11	8-3	8-6	8-9	9-1	9-3	9-6	9-9	9-11	10-2	10-4	10-6	10-9	10-11	11-1	11-3	11-5
	16.0	7-2	7-6	7-9	8-0	8-3	8-5	8-8	8-10	9-1	9-3	9-5	9-7	9-9	9-11	10-1	10-2	10-4
	19.2	6-9	7-0	7-3	7-6	7-9	7-11	8-2	8-4	8-6	8-8	8-10	9-0	9-2	9-4	9-6	9-7	9-9
	24.0	6-3	6-6	6-9	7-0	7-2	7-4	7-7	7-9	7-11	8-1	8-3	8-4	8-6	8-8	8-9	8-11	9-1
2 × 8	12.0	10-5	10-10	11-3	11-7	11-11	12-3	12-7	12-10	13-1	13-5	13-8	13-11	14-2	14-4	14-7	14-10	15-0
	16.0	9-6	9-10	10-2	10-6	10-10	11-1	11-5	11-8	11-11	12-2	12-5	12-7	12-10	13-1	13-3	13-5	13-8
	19.2	8-11	9-3	9-7	9-11	10-2	10-6	10-9	11-0	11-3	11-5	11-8	11-11	12-1	12-3	12-6	12-8	12-10
	24.0	8-3	8-7	8-11	9-2	9-6	9-9	10-0	10-2	10-5	10-8	10-10	11-0	11-3	11-5	11-7	11-9	11-11
2 × 10	12.0	13-3	13-10	14-4	14-9	15-2	15-7	16-0	16-5	16-9	17-1	17-5	17-9	18-0	18-4	18-7	18-11	19-2
	16.0	12-1	12-7	13-0	13-5	13-10	14-2	14-7	14-11	15-2	15-6	15-10	16-1	16-5	16-8	16-11	17-2	17-5
	19.2	11-4	11-10	12-3	12-8	13-0	13-4	13-8	14-0	14-4	14-7	14-11	15-2	15-5	15-8	15-11	16-2	16-5
	24.0	10-7	11-0	11-4	11-9	12-1	12-5	12-8	13-0	13-3	13-7	13-10	14-1	14-4	14-7	14-9	15-0	15-2
2 × 12	12.0	16-2	16-10	17-5	18-0	18-6	19-0	19-6	19-11	20-4	20-9	21-2	21-7	21-11	22-3	22-8	23-0	23-4
	16.0	14-8	15-3	15-10	16-4	16-10	17-3	17-8	18-1	18-6	18-10	19-3	19-7	19-11	20-3	20-7	20-10	21-2
	19.2	13-10	14-4	14-11	15-4	15-10	16-3	16-8	17-0	17-5	17-9	18-1	18-5	18-9	19-1	19-4	19-8	19-11
	24.0	12-10	13-4	13-10	14-3	14-8	15-1	15-5	15-10	16-2	16-6	16-10	17-1	17-5	17-8	18-0	18-3	18-6
F$_b$	12.0	743	803	862	918	973	1,026	1,078	1,129	1,179	1,228	1,275	1,322	1,368	1,413	1,458	1,502	1,545
	16.0	817	884	949	1,011	1,071	1,130	1,187	1,243	1,298	1,351	1,404	1,455	1,506	1,555	1,604	1,653	1,700
	19.2	869	940	1,008	1,074	1,138	1,201	1,261	1,321	1,379	1,436	1,491	1,546	1,600	1,653	1,705	1,756	1,807
	24.0	936	1,012	1,086	1,157	1,226	1,293	1,359	1,423	1,485	1,547	1,607	1,666	1,724	1,781	1,837	1,892	1,946

NOTE: The required bending design value, *F$_b$*, in pounds per square inch (N/mm^2) is shown at the bottom of this table and is applicable to all lumber sizes shown. Spans are shown in feet-inches (1 foot = 304.8 mm, 1 inch = 25.4 mm) and are limited to 26 feet (7925 mm) and less. Check sources of supply for availability of lumber in lengths greater than 20 feet (6096 mm).

TABLE 23-VII-J-2—FLOOR JOISTS WITH *L*/360 DEFLECTION LIMITS
The allowable bending stress *(Fb)* and modulus of elasticity *(E)* used in this table shall be from Tables 23-IV-V-1 and 23-IV-V-2 only.

DESIGN CRITERIA:
Deflection — For 60 psf (2.87 kN/m^2) live load.
Limited to span in inches (mm) divided by 360.
Strength — Live load of 60 psf (2.87 kN/m^2) plus dead load of 10 psf (0.48 kN/m^2) determines the required bending design value.

Joist Size (in)	Spacing (in)	Modulus of Elasticity, *E*, in 1,000,000 psi (× 0.00689 for N/mm^2)																
× 25.4 for mm		0.8	0.9	1.0	1.1	1.2	1.3	1.4	1.5	1.6	1.7	1.8	1.9	2.0	2.1	2.2	2.3	2.4
2 × 6	12.0	7-5	7-9	8-0	8-3	8-6	8-9	8-11	9-2	9-4	9-7	9-9	9-11	10-1	10-3	10-5	10-7	10-9
	16.0	6-9	7-0	7-3	7-6	7-9	7-11	8-2	8-4	8-6	8-8	8-10	9-0	9-2	9-4	9-6	9-7	9-9
	19.2	6-4	6-7	6-10	7-1	7-3	7-6	7-8	7-10	8-0	8-2	8-4	8-6	8-8	8-9	8-11	9-0	9-2
	24.0	5-11	6-2	6-4	6-7	6-9	6-11	7-1	7-3	7-5	7-7	7-9	7-10	8-0	8-2	8-3	8-5	8-6
2 × 8	12.0	9-10	10-2	10-7	10-11	11-3	11-6	11-10	12-1	12-4	12-7	12-10	13-1	13-4	13-6	13-9	13-11	14-2
	16.0	8-11	9-3	9-7	9-11	10-2	10-6	10-9	11-0	11-3	11-5	11-8	11-11	12-1	12-3	12-6	12-8	12-10
	19.2	8-5	8-9	9-0	9-4	9-7	9-10	10-1	10-4	10-7	10-9	11-0	11-2	11-4	11-7	11-9	11-11	12-1
	24.0	7-9	8-1	8-5	8-8	8-11	9-2	9-4	9-7	9-10	10-0	10-2	10-5	10-7	10-9	10-11	11-1	11-3
2 × 10	12.0	12-6	13-0	13-6	13-11	14-4	14-8	15-1	15-5	15-9	16-1	16-5	16-8	17-0	17-3	17-6	17-9	18-0
	16.0	11-4	11-10	12-3	12-8	13-0	13-4	13-8	14-0	14-4	14-7	14-11	15-2	15-5	15-8	15-11	16-2	16-5
	19.2	10-8	11-1	11-6	11-11	12-3	12-7	12-11	13-2	13-6	13-9	14-0	14-3	14-6	14-9	15-0	15-2	15-5
	24.0	9-11	10-4	10-8	11-0	11-4	11-8	11-11	12-3	12-6	12-9	13-0	13-3	13-6	13-8	13-11	14-1	14-4
2 × 12	12.0	15-2	15-10	16-5	16-11	17-5	17-11	18-4	18-9	19-2	19-7	19-11	20-3	20-8	21-0	21-4	21-7	21-11
	16.0	13-10	14-4	14-11	15-4	15-10	16-3	16-8	17-0	17-5	17-9	18-1	18-5	18-9	19-1	19-4	19-8	19-11
	19.2	13-0	13-6	14-0	14-5	14-11	15-3	15-8	16-0	16-5	16-9	17-0	17-4	17-8	17-11	18-3	18-6	18-9
	24.0	12-1	12-7	13-0	13-5	13-10	14-2	14-7	14-11	15-2	15-6	15-10	16-1	16-5	16-8	16-11	17-2	17-5
Fb	12.0	767	830	890	949	1,005	1,061	1,114	1,167	1,218	1,268	1,317	1,366	1,413	1,460	1,506	1,551	1,596
	16.0	844	913	980	1,044	1,107	1,167	1,226	1,284	1,341	1,396	1,450	1,503	1,556	1,607	1,658	1,707	1,757
	19.2	897	971	1,041	1,110	1,176	1,240	1,303	1,365	1,425	1,483	1,541	1,597	1,653	1,708	1,761	1,814	1,867
	24.0	967	1,046	1,122	1,195	1,267	1,336	1,404	1,470	1,535	1,598	1,660	1,721	1,781	1,840	1,897	1,955	2,011

NOTE: The required bending design value, *Fb*, in pounds per square inch (N/mm^2) is shown at the bottom of this table and is applicable to all lumber sizes shown. Spans are shown in feet-inches (1 foot = 304.8 mm, 1 inch = 25.4 mm) and are limited to 26 feet (7925 mm) and less. Check sources of supply for availability of lumber in lengths greater than 20 feet (6096 mm).

TABLE 23-VII-J-3

1997 UNIFORM BUILDING CODE

TABLE 23-VII-J-3—FLOOR JOISTS WITH L/360 DEFLECTION LIMITS
The allowable bending stress (F_b) and modulus of elasticity (E) used in this table shall be from Tables 23-IV-V-1 and 23-IV-V-2 only.

DESIGN CRITERIA:
Deflection — For 50 psf (2.4 kN/m^2) live load.
Limited to span in inches (mm) divided by 360.
Strength — Live load of 50 psf (2.4 kN/m^2) plus dead load of 20 psf (0.96 kN/m^2) determines the required bending design value.

Joist Size (in)	Spacing (in)	Modulus of Elasticity, E, in 1,000,000 psi ($\times$ 0.00689 for N/mm^2)																
$\times$ 25.4 for mm		0.8	0.9	1.0	1.1	1.2	1.3	1.4	1.5	1.6	1.7	1.8	1.9	2.0	2.1	2.2	2.3	2.4
2 × 6	12.0	7-11	8-3	8-6	8-9	9-1	9-3	9-6	9-9	9-11	10-2	10-4	10-6	10-9	10-11	11-1	11-3	11-5
	16.0	7-2	7-6	7-9	8-0	8-3	8-5	8-8	8-10	9-1	9-3	9-5	9-7	9-9	9-11	10-1	10-2	10-4
	19.2	6-9	7-0	7-3	7-6	7-9	7-11	8-2	8-4	8-6	8-8	8-10	9-0	9-2	9-4	9-6	9-7	9-9
	24.0	6-3	6-6	6-9	7-0	7-2	7-4	7-7	7-9	7-11	8-1	8-3	8-4	8-6	8-8	8-9	8-11	9-1
2 × 8	12.0	10-5	10-10	11-3	11-7	11-11	12-3	12-7	12-10	13-1	13-5	13-8	13-11	14-2	14-4	14-7	14-10	15-0
	16.0	9-6	9-10	10-2	10-6	10-10	11-1	11-5	11-8	11-11	12-2	12-5	12-7	12-10	13-1	13-3	13-5	13-8
	19.2	8-11	9-3	9-7	9-11	10-2	10-6	10-9	11-0	11-3	11-5	11-8	11-11	12-1	12-3	12-6	12-8	12-10
	24.0	8-3	8-7	8-11	9-2	9-6	9-9	10-0	10-2	10-5	10-8	10-10	11-0	11-3	11-5	11-7	11-9	11-11
2 × 10	12.0	13-3	13-10	14-4	14-9	15-2	15-7	16-0	16-5	16-9	17-1	17-5	17-9	18-0	18-4	18-7	18-11	19-2
	16.0	12-1	12-7	13-0	13-5	13-10	14-2	14-7	14-11	15-2	15-6	15-10	16-1	16-5	16-8	16-11	17-2	17-5
	19.2	11-4	11-10	12-3	12-8	13-0	13-4	13-8	14-0	14-4	14-7	14-11	15-2	15-5	15-8	15-11	16-2	16-5
	24.0	10-7	11-0	11-4	11-9	12-1	12-5	12-8	13-0	13-3	13-7	13-10	14-1	14-4	14-7	14-9	15-0	15-2
2 × 12	12.0	16-2	16-10	17-5	18-0	18-6	19-0	19-6	19-11	20-4	20-9	21-2	21-7	21-11	22-3	22-8	23-0	23-4
	16.0	14-8	15-3	15-10	16-4	16-10	17-3	17-8	18-1	18-6	18-10	19-3	19-7	19-11	20-3	20-7	20-10	21-2
	19.2	13-10	14-4	14-11	15-4	15-10	16-3	16-8	17-0	17-5	17-9	18-1	18-5	18-9	19-1	19-4	19-8	19-11
	24.0	12-10	13-4	13-10	14-3	14-8	15-1	15-5	15-10	16-2	16-6	16-10	17-1	17-5	17-8	18-0	18-3	18-6
F_b	12.0	866	937	1,005	1,071	1,135	1,198	1,258	1,317	1,375	1,432	1,488	1,542	1,596	1,649	1,701	1,752	1,802
	16.0	954	1,032	1,107	1,179	1,250	1,318	1,385	1,450	1,514	1,576	1,637	1,698	1,757	1,815	1,872	1,928	1,984
	19.2	1,013	1,096	1,176	1,253	1,328	1,401	1,472	1,541	1,609	1,675	1,740	1,804	1,867	1,928	1,989	2,049	2,108
	24.0	1,092	1,181	1,267	1,350	1,430	1,509	1,585	1,660	1,733	1,804	1,874	1,943	2,011	2,077	2,143	2,207	2,271

NOTE: The required bending design value, F_b, in pounds per square inch (N/mm^2) is shown at the bottom of this table and is applicable to all lumber sizes shown. Spans are shown in feet-inches (1 foot = 304.8 mm, 1 inch = 25.4 mm) and are limited to 26 feet (7925 mm) and less. Check sources of supply for availability of lumber in lengths greater than 20 feet (6096 mm).

TABLE 23-VII-J-4—FLOOR JOISTS WITH *L*/360 DEFLECTION LIMITS
The allowable bending stress *(Fb)* and modulus of elasticity *(E)* used in this table shall be from Tables 23-IV-V-1 and 23-IV-V-2 only.

DESIGN CRITERIA:
Deflection — For 60 psf (2.87 kN/m²) live load.
Limited to span in inches (mm) divided by 360.
Strength — Live load of 60 psf (2.87 kN/m²) plus dead load of 20 psf (0.96 kN/m²) determines the required bending design value.

Joist Size (in)	Spacing (in)	Modulus of Elasticity, *E*, in 1,000,000 psi (× 0.00689 for N/mm²)																
× 25.4 for mm		0.8	0.9	1.0	1.1	1.2	1.3	1.4	1.5	1.6	1.7	1.8	1.9	2.0	2.1	2.2	2.3	2.4
2 × 6	12.0	7-5	7-9	8-0	8-3	8-6	8-9	8-11	9-2	9-4	9-7	9-9	9-11	10-1	10-3	10-5	10-7	10-9
	16.0	6-9	7-0	7-3	7-6	7-9	7-11	8-2	8-4	8-6	8-8	8-10	9-0	9-2	9-4	9-6	9-7	9-9
	19.2	6-4	6-7	6-10	7-1	7-3	7-6	7-8	7-10	8-0	8-2	8-4	8-6	8-8	8-9	8-11	9-0	9-2
	24.0	5-11	6-2	6-4	6-7	6-9	6-11	7-1	7-3	7-5	7-7	7-9	7-10	8-0	8-2	8-3	8-5	8-6
2 × 8	12.0	9-10	10-2	10-7	10-11	11-3	11-6	11-10	12-1	12-4	12-7	12-10	13-1	13-4	13-6	13-9	13-11	14-2
	16.0	8-11	9-3	9-7	9-11	10-2	10-6	10-9	11-0	11-3	11-5	11-8	11-11	12-1	12-3	12-6	12-8	12-10
	19.2	8-5	8-9	9-0	9-4	9-7	9-10	10-1	10-4	10-7	10-9	11-0	11-2	11-4	11-7	11-9	11-11	12-1
	24.0	7-9	8-1	8-5	8-8	8-11	9-2	9-4	9-7	9-10	10-0	10-2	10-5	10-7	10-9	10-11	11-1	11-3
2 × 10	12.0	12-6	13-0	13-6	13-11	14-4	14-8	15-1	15-5	15-9	16-1	16-5	16-8	17-0	17-3	17-6	17-9	18-0
	16.0	11-4	11-10	12-3	12-8	13-0	13-4	13-8	14-0	14-4	14-7	14-11	15-2	15-5	15-8	15-11	16-2	16-5
	19.2	10-8	11-1	11-6	11-11	12-3	12-7	12-11	13-2	13-6	13-9	14-0	14-3	14-6	14-9	15-0	15-2	15-5
	24.0	9-11	10-4	10-8	11-0	11-4	11-8	11-11	12-3	12-6	12-9	13-0	13-3	13-6	13-8	13-11	14-1	14-4
2 × 12	12.0	15-2	15-10	16-5	16-11	17-5	17-11	18-4	18-9	19-2	19-7	19-11	20-3	20-8	21-0	21-4	21-7	21-11
	16.0	13-10	14-4	14-11	15-4	15-10	16-3	16-8	17-0	17-5	17-9	18-1	18-5	18-9	19-1	19-4	19-8	19-11
	19.2	13-0	13-6	14-0	14-5	14-11	15-3	15-8	16-0	16-5	16-9	17-0	17-4	17-8	17-11	18-3	18-6	18-9
	24.0	12-1	12-7	13-0	13-5	13-10	14-2	14-7	14-11	15-2	15-6	15-10	16-1	16-5	16-8	16-11	17-2	17-5
Fb	12.0	877	949	1,018	1,084	1,149	1,212	1,273	1,333	1,392	1,449	1,506	1,561	1,615	1,669	1,721	1,773	1,824
	16.0	965	1,044	1,120	1,193	1,265	1,334	1,402	1,468	1,532	1,595	1,657	1,718	1,778	1,837	1,894	1,951	2,008
	19.2	1,026	1,109	1,190	1,268	1,344	1,418	1,489	1,559	1,628	1,695	1,761	1,826	1,889	1,952	2,013	2,074	2,133
	24.0	1,105	1,195	1,282	1,366	1,448	1,527	1,604	1,680	1,754	1,826	1,897	1,967	2,035	2,102	2,169	2,234	2,298

NOTE: The required bending design value, *Fb*, in pounds per square inch (N/mm²) is shown at the bottom of this table and is applicable to all lumber sizes shown. Spans are shown in feet-inches (1 foot = 304.8 mm, 1 inch = 25.4 mm) and are limited to 26 feet (7925 mm) and less. Check sources of supply for availability of lumber in lengths greater than 20 feet (6096 mm).

TABLE 23-VII-R-1   1997 UNIFORM BUILDING CODE

TABLE 23-VII-R-1—RAFTERS WITH *L*/240 DEFLECTION LIMITS
The allowable bending stress *(F_b)* and modulus of elasticity *(E)* used in this table shall be from Tables 23-IV-V-1 and 23-IV-V-2 only.

DESIGN CRITERIA:
Strength — Live load of 40 psf (1.92 kN/m²) plus dead load of 10 psf (0.48 kN/m²) determines the required bending design value.
Deflection — For 40 psf live (1.02 kN/m²) load.
Limited to span in inches (mm) divided by 240.

Rafter Size (in)	Spacing (in)	Bending Design Value, F_b (psi) (× 0.00689 for N/mm²)										
× 25.4 for mm		300	400	500	600	700	800	900	1000	1100	1200	1300
2 × 6	12.0	5-6	6-4	7-1	7-9	8-5	9-0	9-6	10-0	10-6	11-0	11-5
	16.0	4-9	5-6	6-2	6-9	7-3	7-9	8-3	8-8	9-1	9-6	9-11
	19.2	4-4	5-0	5-7	6-2	6-8	7-1	7-6	7-11	8-4	8-8	9-1
	24.0	3-11	4-6	5-0	5-6	5-11	6-4	6-9	7-1	7-5	7-9	8-1
2 × 8	12.0	7-3	8-4	9-4	10-3	11-1	11-10	12-7	13-3	13-11	14-6	15-1
	16.0	6-3	7-3	8-1	8-11	9-7	10-3	10-10	11-6	12-0	12-7	13-1
	19.2	5-9	6-7	7-5	8-1	8-9	9-4	9-11	10-6	11-0	11-6	11-11
	24.0	5-2	5-11	6-7	7-3	7-10	8-4	8-11	9-4	9-10	10-3	10-8
2 × 10	12.0	9-3	10-8	11-11	13-1	14-2	15-1	16-0	16-11	17-9	18-6	19-3
	16.0	8-0	9-3	10-4	11-4	12-3	13-1	13-10	14-8	15-4	16-0	16-8
	19.2	7-4	8-5	9-5	10-4	11-2	11-11	12-8	13-4	14-0	14-8	15-3
	24.0	6-6	7-7	8-5	9-3	10-0	10-8	11-4	11-11	12-6	13-1	13-7
2 ×12	12.0	11-3	13-0	14-6	15-11	17-2	18-4	19-6	20-6	21-7	22-6	23-5
	16.0	9-9	11-3	12-7	13-9	14-11	15-11	16-10	17-9	18-8	19-6	20-3
	19.2	8-11	10-3	11-6	12-7	13-7	14-6	15-5	16-3	17-0	17-9	18-6
	24.0	7-11	9-2	10-3	11-3	12-2	13-0	13-9	14-6	15-3	15-11	16-7
E	12.0	0.14	0.22	0.31	0.41	0.51	0.63	0.75	0.88	1.01	1.15	1.30
	16.0	0.12	0.19	0.27	0.35	0.44	0.54	0.65	0.76	0.88	1.00	1.12
	19.2	0.11	0.18	0.24	0.32	0.41	0.50	0.59	0.69	0.80	0.91	1.03
	24.0	0.10	0.16	0.22	0.29	0.36	0.44	0.53	0.62	0.71	0.81	0.92

Rafter Size (in)	Spacing (in)	Bending Design Value, F_b (psi) (× 0.00689 for N/mm²)										
× 25.4 for mm		1400	1500	1600	1700	1800	1900	2000	2100	2200	2300	2400
2 × 6	12.0	11-11	12-4	12-8	13-1	13-6	13-10	14-2				
	16.0	10-3	10-8	11-0	11-4	11-8	12-0	12-4	12-7	12-11		
	19.2	9-5	9-9	10-0	10-4	10-8	10-11	11-3	11-6	11-9	12-0	12-4
	24.0	8-5	8-8	9-0	9-3	9-6	9-9	10-0	10-3	10-6	10-9	11-0
2 × 8	12.0	15-8	16-3	16-9	17-3	17-9	18-3	18-9				
	16.0	13-7	14-0	14-6	14-11	15-5	15-10	16-3	16-7	17-0		
	19.2	12-5	12-10	13-3	13-8	14-0	14-5	14-10	15-2	15-6	15-10	16-3
	24.0	11-1	11-6	11-10	12-2	12-7	12-11	13-3	13-7	13-11	14-2	14-6
2 × 10	12.0	20-0	20-8	21-4	22-0	22-8	23-3	23-11				
	16.0	17-4	17-11	18-6	19-1	19-7	20-2	20-8	21-2	21-8		
	19.2	15-10	16-4	16-11	17-5	17-11	18-5	18-11	19-4	19-10	20-3	20-8
	24.0	14-2	14-8	15-1	15-7	16-0	16-6	16-11	17-4	17-9	18-1	18-6
2 ×12	12.0	24-4	25-2	26-0								
	16.0	21-1	21-9	22-6	23-2	23-10	24-6	25-2	25-9			
	19.2	19-3	19-11	20-6	21-2	21-9	22-5	23-0	23-6	24-1	24-8	25-2
	24.0	17-2	17-9	18-4	18-11	19-6	20-0	20-6	21-1	21-7	22-0	22-6
E	12.0	1.45	1.61	1.77	1.94	2.12	2.30	2.48				
	16.0	1.26	1.39	1.54	1.68	1.83	1.99	2.15	2.31	2.48		
	19.2	1.15	1.27	1.40	1.54	1.67	1.81	1.96	2.11	2.26	2.42	2.58
	24.0	1.03	1.14	1.25	1.37	1.50	1.62	1.75	1.89	2.02	2.16	2.30

NOTE: The required modulus of elasticity, *E*, in 1,000,000 pounds per square inch is shown at the bottom of this table, is limited to 2.6 million psi (17.92 × 10⁶ kN/m²) and less, and is applicable to all lumber sizes shown. Spans are shown in feet-inches (1 foot = 304.8 mm, 1 inch = 25.4 mm) and are limited to 26 feet (7925 mm) and less.

TABLE 23-VII-R-2—RAFTERS WITH *L*/240 DEFLECTION LIMITS
The allowable bending stress (*F*ₒ) and modulus of elasticity *(E)* used in this table shall be from Tables 23-IV-V-1 and 23-IV-V-2 only.

DESIGN CRITERIA:
Strength — Live load of 50 psf (2.40 kN/m²) plus dead load of 10 psf (0.48 kN/m²) determines the required bending design value.
Deflection — For 50 psf (2.4 kN/m²) live load.
Limited to span in inches (mm) divided by 240.

Rafter Size (in)	Spacing (in)	Bending Design Value, F_b (psi) ($\times$ 0.00689 for N/mm²)										
× 25.4 for mm		300	400	500	600	700	800	900	1000	1100	1200	1300
2 × 6	12.0	5-0	5-10	6-6	7-1	7-8	8-2	8-8	9-2	9-7	10-0	10-5
	16.0	4-4	5-0	5-7	6-2	6-8	7-1	7-6	7-11	8-4	8-8	9-1
	19.2	4-0	4-7	5-1	5-7	6-1	6-6	6-10	7-3	7-7	7-11	8-3
	24.0	3-7	4-1	4-7	5-0	5-5	5-10	6-2	6-6	6-10	7-1	7-5
2 × 8	12.0	6-7	7-8	8-7	9-4	10-1	10-10	11-6	12-1	12-8	13-3	13-9
	16.0	5-9	6-7	7-5	8-1	8-9	9-4	9-11	10-6	11-0	11-6	11-11
	19.2	5-3	6-0	6-9	7-5	8-0	8-7	9-1	9-7	10-0	10-6	10-11
	24.0	4-8	5-5	6-0	6-7	7-2	7-8	8-1	8-7	9-0	9-4	9-9
2 × 10	12.0	8-5	9-9	10-11	11-11	12-11	13-9	14-8	15-5	16-2	16-11	17-7
	16.0	7-4	8-5	9-5	10-4	11-2	11-11	12-8	13-4	14-0	14-8	15-3
	19.2	6-8	7-8	8-7	9-5	10-2	10-11	11-7	12-2	12-9	13-4	13-11
	24.0	6-0	6-11	7-8	8-5	9-1	9-9	10-4	10-11	11-5	11-11	12-5
2 × 12	12.0	10-3	11-10	13-3	14-6	15-8	16-9	17-9	18-9	19-8	20-6	21-5
	16.0	8-11	10-3	11-6	12-7	13-7	14-6	15-5	16-3	17-0	17-9	18-6
	19.2	8-1	9-4	10-6	11-6	12-5	13-3	14-1	14-10	15-7	16-3	16-11
	24.0	7-3	8-5	9-4	10-3	11-1	11-10	12-7	13-3	13-11	14-6	15-1
E	12.0	0.14	0.21	0.29	0.39	0.49	0.60	0.71	0.83	0.96	1.10	1.24
	16.0	0.12	0.18	0.26	0.34	0.42	0.52	0.62	0.72	0.83	0.95	1.07
	19.2	0.11	0.17	0.23	0.31	0.39	0.47	0.56	0.66	0.76	0.87	0.98
	24.0	0.10	0.15	0.21	0.27	0.35	0.42	0.50	0.59	0.68	0.77	0.87

Rafter Size (in)	Spacing (in)	Bending Design Value, F_b (psi) ($\times$ 0.00689 for N/mm²)										
× 25.4 for mm		1400	1500	1600	1700	1800	1900	2000	2100	2200	2300	2400
2 × 6	12.0	10-10	11-3	11-7	11-11	12-4	12-8	13-0	13-3			
	16.0	9-5	9-9	10-0	10-4	10-8	10-11	11-3	11-6	11-9	12-0	
	19.2	8-7	8-11	9-2	9-5	9-9	10-0	10-3	10-6	10-9	11-0	11-3
	24.0	7-8	7-11	8-2	8-5	8-8	8-11	9-2	9-5	9-7	9-10	10-0
2 × 8	12.0	14-4	14-10	15-3	15-9	16-3	16-8	17-1	17-6			
	16.0	12-5	12-10	13-3	13-8	14-0	14-5	14-10	15-2	15-6	15-10	
	19.2	11-4	11-8	12-1	12-5	12-10	13-2	13-6	13-10	14-2	14-6	14-10
	24.0	10-1	10-6	10-10	11-2	11-6	11-9	12-1	12-5	12-8	12-11	13-3
2 × 10	12.0	18-3	18-11	19-6	20-1	20-8	21-3	21-10	22-4			
	16.0	15-10	16-4	16-11	17-5	17-11	18-5	18-11	19-4	19-10	20-3	
	19.2	14-5	14-11	15-5	15-11	16-4	16-10	17-3	17-8	18-1	18-6	18-11
	24.0	12-11	13-4	13-9	14-3	14-8	15-0	15-5	15-10	16-2	16-6	16-11
2 × 12	12.0	22-2	23-0	23-9	24-5	25-2	25-10					
	16.0	19-3	19-11	20-6	21-2	21-9	22-5	23-0	23-6	24-1	24-8	
	19.2	17-6	18-2	18-9	19-4	19-11	20-5	21-0	21-6	22-0	22-6	23-0
	24.0	15-8	16-3	16-9	17-3	17-9	18-3	18-9	19-3	19-8	20-1	20-6
E	12.0	1.38	1.53	1.69	1.85	2.01	2.18	2.36	2.54			
	16.0	1.20	1.33	1.46	1.60	1.74	1.89	2.04	2.20	2.35	2.52	
	19.2	1.09	1.21	1.33	1.46	1.59	1.73	1.86	2.00	2.15	2.30	2.45
	24.0	0.98	1.08	1.19	1.31	1.42	1.54	1.67	1.79	1.92	2.06	2.19

NOTE: The required modulus of elasticity, *E*, in 1,000,000 pounds per square inch is shown at the bottom of this table, is limited to 2.6 million psi (17.92 × 10⁶ kN/m²) and less, and is applicable to all lumber sizes shown. Spans are shown in feet-inches (1 foot = 304.8 mm, 1 inch = 25.4 mm) and are limited to 26 feet (7925 mm) and less.

TABLE 23-VII-R-3   1997 UNIFORM BUILDING CODE

TABLE 23-VII-R-3—RAFTERS WITH L/240 DEFLECTION LIMITS
The allowable bending stress (Fb) and modulus of elasticity (E) used in this table shall be from Tables 23-IV-V-1 and 23-IV-V-2 only.

DESIGN CRITERIA:
Strength — Live load of 40 psf (1.92 kN/m²) plus dead load of 15 psf (0.72 kN/m²) determines the required bending design value.
Deflection — For 40 psf (1.92 kN/m²) live load.
Limited to span in inches (mm) divided by 240.

Rafter Size (in)	Spacing (in)	Bending Design Value, F_b (psi) (× 0.00689 for kN/m²)												
× 25.4 for mm		300	400	500	600	700	800	900	1000	1100	1200	1300	1400	1500
2 × 6	12.0	5-3	6-1	6-9	7-5	8-0	8-7	9-1	9-7	10-0	10-6	10-11	11-4	11-9
	16.0	4-6	5-3	5-10	6-5	6-11	7-5	7-10	8-3	8-8	9-1	9-5	9-10	10-2
	19.2	4-2	4-9	5-4	5-10	6-4	6-9	7-2	7-7	7-11	8-3	8-8	8-11	9-3
	24.0	3-8	4-3	4-9	5-3	5-8	6-1	6-5	6-9	7-1	7-5	7-9	8-0	8-3
2 × 8	12.0	6-11	8-0	8-11	9-9	10-7	11-3	12-0	12-7	13-3	13-10	14-5	14-11	15-5
	16.0	6-0	6-11	7-9	8-6	9-2	9-9	10-4	10-11	11-6	12-0	12-6	12-11	13-5
	19.2	5-6	6-4	7-1	7-9	8-4	8-11	9-6	10-0	10-6	10-11	11-5	11-10	12-3
	24.0	4-11	5-8	6-4	6-11	7-6	8-0	8-6	8-11	9-4	9-9	10-2	10-7	10-11
2 × 10	12.0	8-10	10-2	11-5	12-6	13-6	14-5	15-3	16-1	16-11	17-8	18-4	19-1	19-9
	16.0	7-8	8-10	9-10	10-10	11-8	12-6	13-3	13-11	14-8	15-3	15-11	16-6	17-1
	19.2	7-0	8-1	9-0	9-10	10-8	11-5	12-1	12-9	13-4	13-11	14-6	15-1	15-7
	24.0	6-3	7-2	8-1	8-10	9-6	10-2	10-10	11-5	11-11	12-6	13-0	13-6	13-11
2 × 12	12.0	10-9	12-5	13-10	15-2	16-5	17-6	18-7	19-7	20-6	21-5	22-4	23-2	24-0
	16.0	9-3	10-9	12-0	13-2	14-2	15-2	16-1	17-0	17-9	18-7	19-4	20-1	20-9
	19.2	8-6	9-10	10-11	12-0	12-11	13-10	14-8	15-6	16-3	17-0	17-8	18-4	19-0
	24.0	7-7	8-9	9-10	10-9	11-7	12-5	13-2	13-10	14-6	15-2	15-9	16-5	17-0
E	12.0	0.12	0.19	0.27	0.35	0.44	0.54	0.65	0.76	0.88	1.00	1.13	1.26	1.40
	16.0	0.11	0.17	0.23	0.31	0.39	0.47	0.56	0.66	0.76	0.86	0.98	1.09	1.21
	19.2	0.10	0.15	0.21	0.28	0.35	0.43	0.51	0.60	0.69	0.79	0.89	0.99	1.10
	24.0	0.09	0.14	0.19	0.25	0.31	0.38	0.46	0.54	0.62	0.71	0.80	0.89	0.99

Rafter Size (in)	Spacing (in)	Bending Design Value, F_b (psi) (× 0.00689 for kN/m²)											
× 25.4 for mm		1600	1700	1800	1900	2000	2100	2200	2300	2400	2500	2600	2700
2 × 6	12.0	12-1	12-6	12-10	13-2	13-6	13-10	14-2					
	16.0	10-6	10-10	11-1	11-5	11-9	12-0	12-4	12-7	12-10			
	19.2	9-7	9-10	10-2	10-5	10-8	11-0	11-3	11-6	11-9	12-0	12-2	
	24.0	8-7	8-10	9-1	9-4	9-7	9-10	10-0	10-3	10-6	10-8	10-11	11-1
2 × 8	12.0	16-0	16-5	16-11	17-5	17-10	18-3	18-9					
	16.0	13-10	14-3	14-8	15-1	15-5	15-10	16-3	16-7	16-11			
	19.2	12-7	13-0	13-5	13-9	14-1	14-6	14-10	15-2	15-5	15-9	16-1	
	24.0	11-3	11-8	12-0	12-4	12-7	12-11	13-3	13-6	13-10	14-1	14-5	14-8
2 × 10	12.0	20-4	21-0	21-7	22-2	22-9	23-4	23-11					
	16.0	17-8	18-2	18-9	19-3	19-9	20-2	20-8	21-2	21-7			
	19.2	16-1	16-7	17-1	17-7	18-0	18-5	18-11	19-4	19-9	20-2	20-6	
	24.0	14-5	14-10	15-3	15-8	16-1	16-6	16-11	17-3	17-8	18-0	18-4	18-9
2 × 12	12.0	24-9	25-6										
	16.0	21-5	22-1	22-9	23-5	24-0	24-7	25-2	25-9				
	19.2	19-7	20-2	20-9	21-4	21-11	22-5	23-0	23-6	24-0	24-6	25-0	
	24.0	17-6	18-1	18-7	19-1	19-7	20-1	20-6	21-0	21-5	21-11	22-4	22-9
E	12.0	1.54	1.68	1.83	1.99	2.15	2.31	2.48					
	16.0	1.33	1.46	1.59	1.72	1.86	2.00	2.15	2.29	2.45			
	19.2	1.22	1.33	1.45	1.57	1.70	1.83	1.96	2.09	2.23	2.37	2.52	
	24.0	1.09	1.19	1.30	1.41	1.52	1.63	1.75	1.87	2.00	2.12	2.25	2.38

NOTE: The required modulus of elasticity, E, in 1,000,000 pounds per square inch is shown at the bottom of this table, is limited to 2.6 million psi (17.92 × 10⁶ kN/m²) and less, and is applicable to all lumber sizes shown. Spans are shown in feet-inches (1 foot = 304.8 mm, 1 inch = 25.4 mm) and are limited to 26 feet (7925 mm) and less.

TABLE 23-VII-R-4—RAFTERS WITH *L*/240 DEFLECTION LIMITS
The allowable bending stress (F_b) and modulus of elasticity (E) used in this table shall be from Tables 23-IV-V-1 and 23-IV-V-2 only.

DESIGN CRITERIA:
Strength — Live load of 50 psf (2.40 kN/m^2) plus dead load of 15 psf (0.72 kN/m^2) determines the required bending design value.
Deflection — For 50 psf (2.40 kN/m^2) live load.
Limited to span in inches (mm) divided by 240.

Rafter Size (in)	Spacing (in)	Bending Design Value, F_b (psi) (× 0.00689 for kN/m^2)												
× 25.4 for mm		300	400	500	600	700	800	900	1000	1100	1200	1300	1400	1500
2 × 6	12.0	4-10	5-7	6-3	6-10	7-4	7-11	8-4	8-10	9-3	9-8	10-0	10-5	10-9
	16.0	4-2	4-10	5-5	5-11	6-5	6-10	7-3	7-8	8-0	8-4	8-8	9-0	9-4
	19.2	3-10	4-5	4-11	5-5	5-10	6-3	6-7	7-0	7-4	7-8	7-11	8-3	8-6
	24.0	3-5	3-11	4-5	4-10	5-3	5-7	5-11	6-3	6-6	6-10	7-1	7-4	7-8
2 × 8	12.0	6-4	7-4	8-3	9-0	9-9	10-5	11-0	11-7	12-2	12-9	13-3	13-9	14-3
	16.0	5-6	6-4	7-1	7-9	8-5	9-0	9-6	10-1	10-7	11-0	11-6	11-11	12-4
	19.2	5-0	5-10	6-6	7-1	7-8	8-3	8-8	9-2	9-8	10-1	10-6	10-10	11-3
	24.0	4-6	5-2	5-10	6-4	6-10	7-4	7-9	8-3	8-7	9-0	9-4	9-9	10-1
2 × 10	12.0	8-1	9-4	10-6	11-6	12-5	13-3	14-1	14-10	15-6	16-3	16-11	17-6	18-2
	16.0	7-0	8-1	9-1	9-11	10-9	11-6	12-2	12-10	13-5	14-1	14-8	15-2	15-9
	19.2	6-5	7-5	8-3	9-1	9-10	10-6	11-1	11-9	12-3	12-10	13-4	13-10	14-4
	24.0	5-9	6-7	7-5	8-1	8-9	9-4	9-11	10-6	11-0	11-6	11-11	12-5	12-10
2 × 12	12.0	9-10	11-5	12-9	13-11	15-1	16-1	17-1	18-0	18-11	19-9	20-6	21-4	22-1
	16.0	8-7	9-10	11-0	12-1	13-1	13-11	14-10	15-7	16-4	17-1	17-9	18-6	19-1
	19.2	7-10	9-0	10-1	11-0	11-11	12-9	13-6	14-3	14-11	15-7	16-3	16-10	17-5
	24.0	7-0	8-1	9-0	9-10	10-8	11-5	12-1	12-9	13-4	13-11	14-6	15-1	15-7
E	12.0	0.12	0.19	0.26	0.34	0.43	0.53	0.63	0.74	0.85	0.97	1.10	1.22	1.36
	16.0	0.11	0.16	0.23	0.30	0.37	0.46	0.55	0.64	0.74	0.84	0.95	1.06	1.18
	19.2	0.10	0.15	0.21	0.27	0.34	0.42	0.50	0.58	0.67	0.77	0.87	0.97	1.07
	24.0	0.09	0.13	0.18	0.24	0.31	0.37	0.45	0.52	0.60	0.69	0.77	0.87	0.96

Rafter Size (in)	Spacing (in)	Bending Design Value, F_b (psi) (× 0.00689 for kN/m^2)												
× 25.4 for mm		1600	1700	1800	1900	2000	2100	2200	2300	2400	2500	2600	2700
2 × 6	12.0	11-2	11-6	11-10	12-2	12-5	12-9	13-1	13-4				
	16.0	9-8	9-11	10-3	10-6	10-9	11-1	11-4	11-7	11-10	12-1		
	19.2	8-10	9-1	9-4	9-7	9-10	10-1	10-4	10-7	10-9	11-0	11-3	11-5
	24.0	7-11	8-1	8-4	8-7	8-10	9-0	9-3	9-5	9-8	9-10	10-0	10-3
2 × 8	12.0	14-8	15-2	15-7	16-0	16-5	16-10	17-3	17-7				
	16.0	12-9	13-1	13-6	13-10	14-3	14-7	14-11	15-3	15-7	15-11		
	19.2	11-7	12-0	12-4	12-8	13-0	13-4	13-7	13-11	14-3	14-6	14-10	15-1
	24.0	10-5	10-8	11-0	11-4	11-7	11-11	12-2	12-5	12-9	13-0	13-3	13-6
2 × 10	12.0	18-9	19-4	19-10	20-5	20-11	21-6	22-0	22-6				
	16.0	16-3	16-9	17-3	17-8	18-2	18-7	19-0	19-5	19-10	20-3		
	19.2	14-10	15-3	15-9	16-2	16-7	17-0	17-4	17-9	18-2	18-6	18-11	19-3
	24.0	13-3	13-8	14-1	14-5	14-10	15-2	15-6	15-11	16-3	16-7	16-11	17-3
2 × 12	12.0	22-9	23-6	24-2	24-10	25-6							
	16.0	19-9	20-4	20-11	21-6	22-1	22-7	23-2	23-8	24-2	24-8		
	19.2	18-0	18-7	19-1	19-8	20-2	20-8	21-1	21-7	22-1	22-6	23-0	23-5
	24.0	16-1	16-7	17-1	17-7	18-0	18-6	18-11	19-4	19-9	20-2	20-6	20-11
E	12.0	1.50	1.64	1.78	1.94	2.09	2.25	2.41	2.58				
	16.0	1.30	1.42	1.55	1.68	1.81	1.95	2.09	2.23	2.38	2.53		
	19.2	1.18	1.30	1.41	1.53	1.65	1.78	1.91	2.04	2.17	2.31	2.45	2.59
	24.0	1.06	1.16	1.26	1.37	1.48	1.59	1.71	1.82	1.94	2.07	2.19	2.32

NOTE: The required modulus of elasticity, E, in 1,000,000 pounds per square inch is shown at the bottom of this table, is limited to 2.6 million psi (17.92 × 10^6 kN/m^2) and less, and is applicable to all lumber sizes shown. Spans are shown in feet-inches (1 foot = 304.8 mm, 1 inch = 25.4 mm) and are limited to 26 feet (7925 mm) and less.

TABLE 23-VII-R-5

1997 UNIFORM BUILDING CODE

TABLE 23-VII-R-5—RAFTERS WITH *L*/240 DEFLECTION LIMITS
The allowable bending stress (F_b) and modulus of elasticity *(E)* used in this table shall be from Tables 23-IV-V-1 and 23-IV-V-2 only.

DESIGN CRITERIA:
Strength — Live load of 40 psf (1.92 kN/m^2) plus dead load of 20 psf (0.96 kN/m^2) determines the required bending design value.
Deflection — For 40 psf (1.92 kN/m^2) live load.
Limited to span in inches (mm) divided by 240.

Rafter Size (in)	Spacing (in)	Bending Design Value, F_b (psi) (× 0.00689 for kN/m^2)												
× 25.4 for mm		300	400	500	600	700	800	900	1000	1100	1200	1300	1400	1500
2 × 6	12.0	5-0	5-10	6-6	7-1	7-8	8-2	8-8	9-2	9-7	10-0	10-5	10-10	11-3
	16.0	4-4	5-0	5-7	6-2	6-8	7-1	7-6	7-11	8-4	8-8	9-1	9-5	9-9
	19.2	4-0	4-7	5-1	5-7	6-1	6-6	6-10	7-3	7-7	7-11	8-3	8-7	8-11
	24.0	3-7	4-1	4-7	5-0	5-5	5-10	6-2	6-6	6-10	7-1	7-5	7-8	7-11
2 × 8	12.0	6-7	7-8	8-7	9-4	10-1	10-10	11-6	12-1	12-8	13-3	13-9	14-4	14-10
	16.0	5-9	6-7	7-5	8-1	8-9	9-4	9-11	10-6	11-0	11-6	11-11	12-5	12-10
	19.2	5-3	6-0	6-9	7-5	8-0	8-7	9-1	9-7	10-0	10-6	10-11	11-4	11-8
	24.0	4-8	5-5	6-0	6-7	7-2	7-8	8-1	8-7	9-0	9-4	9-9	10-1	10-6
2 × 10	12.0	8-5	9-9	10-11	11-11	12-11	13-9	14-8	15-5	16-2	16-11	17-7	18-3	18-11
	16.0	7-4	8-5	9-5	10-4	11-2	11-11	12-8	13-4	14-0	14-8	15-3	15-10	16-4
	19.2	6-8	7-8	8-7	9-5	10-2	10-11	11-7	12-2	12-9	13-4	13-11	14-5	14-11
	24.0	6-0	6-11	7-8	8-5	9-1	9-9	10-4	10-11	11-5	11-11	12-5	12-11	13-4
2 × 12	12.0	10-3	11-10	13-3	14-6	15-8	16-9	17-9	18-9	19-8	20-6	21-5	22-2	23-0
	16.0	8-11	10-3	11-6	12-7	13-7	14-6	15-5	16-3	17-0	17-9	18-6	19-3	19-11
	19.2	8-1	9-4	10-6	11-6	12-5	13-3	14-1	14-10	15-7	16-3	16-11	17-6	18-2
	24.0	7-3	8-5	9-4	10-3	11-1	11-10	12-7	13-3	13-11	14-6	15-1	15-8	16-3
E	12.0	0.11	0.17	0.24	0.31	0.39	0.48	0.57	0.67	0.77	0.88	0.99	1.10	1.22
	16.0	0.09	0.15	0.20	0.27	0.34	0.41	0.49	0.58	0.67	0.76	0.86	0.96	1.06
	19.2	0.09	0.13	0.19	0.24	0.31	0.38	0.45	0.53	0.61	0.69	0.78	0.87	0.97
	24.0	0.08	0.12	0.17	0.22	0.28	0.34	0.40	0.47	0.54	0.62	0.70	0.78	0.87

Rafter Size (in)	Spacing (in)	Bending Design Value, F_b (psi) (× 0.00689 for kN/m^2)											
× 25.4 for mm		1600	1700	1800	1900	2000	2100	2200	2300	2400	2500	2600	2700
2 × 6	12.0	11-7	11-11	12-4	12-8	13-0	13-3	13-7	13-11	14-2			
	16.0	10-0	10-4	10-8	10-11	11-3	11-6	11-9	12-0	12-4	12-7	12-10	13-1
	19.2	9-2	9-5	9-9	10-0	10-3	10-6	10-9	11-0	11-3	11-5	11-8	11-11
	24.0	8-2	8-5	8-8	8-11	9-2	9-5	9-7	9-10	10-0	10-3	10-5	10-8
2 × 8	12.0	15-3	15-9	16-3	16-8	17-1	17-6	17-11	18-4	18-9			
	16.0	13-3	13-8	14-0	14-5	14-10	15-2	15-6	15-10	16-3	16-7	16-10	17-2
	19.2	12-1	12-5	12-10	13-2	13-6	13-10	14-2	14-6	14-10	15-1	15-5	15-8
	24.0	10-10	11-2	11-6	11-9	12-1	12-5	12-8	12-11	13-3	13-6	13-9	14-0
2 × 10	12.0	19-6	20-1	20-8	21-3	21-10	22-4	22-10	23-5	23-11			
	16.0	16-11	17-5	17-11	18-5	18-11	19-4	19-10	20-3	20-8	21-1	21-6	21-11
	19.2	15-5	15-11	16-4	16-10	17-3	17-8	18-1	18-6	18-11	19-3	19-8	20-0
	24.0	13-9	14-3	14-8	15-0	15-5	15-10	16-2	16-6	16-11	17-3	17-7	17-11
2 × 12	12.0	23-9	24-5	25-2	25-10								
	16.0	20-6	21-2	21-9	22-5	23-0	23-6	24-1	24-8	25-2	25-8		
	19.2	18-9	19-4	19-11	20-5	21-0	21-6	22-0	22-6	23-0	23-5	23-11	24-4
	24.0	16-9	17-3	17-9	18-3	18-9	19-3	19-8	20-1	20-6	21-0	21-5	21-9
E	12.0	1.35	1.48	1.61	1.75	1.89	2.03	2.18	2.33	2.48			
	16.0	1.17	1.28	1.39	1.51	1.63	1.76	1.88	2.01	2.15	2.28	2.42	2.56
	19.2	1.07	1.17	1.27	1.38	1.49	1.60	1.72	1.84	1.96	2.08	2.21	2.34
	24.0	0.95	1.04	1.14	1.23	1.33	1.43	1.54	1.64	1.75	1.86	1.98	2.09

NOTE: The required modulus of elasticity, *E*, in 1,000,000 pounds per square inch is shown at the bottom of this table, is limited to 2.6 million psi (17.92 × 10^6 kN/m^2) and less, and is applicable to all lumber sizes shown. Spans are shown in feet-inches (1 foot = 304.8 mm, 1 inch = 25.4 mm) and are limited to 26 feet (7925 mm) and less.

TABLE 23-VII-R-6—RAFTERS WITH *L*/240 DEFLECTION LIMITS
The allowable bending stress (F_b) and modulus of elasticity (*E*) used in this table shall be from Tables 23-IV-V-1 and 23-IV-V-2 only.

DESIGN CRITERIA:
Strength — Live load of 50 psf (2.40 kN/m^2) plus dead load of 20 psf (0.96 kN/m^2) determines the required bending design value.
Deflection — For 50 psf (2.40 kN/m^2) live load.
Limited to span in inches (mm) divided by 240.

Rafter Size (in)	Spacing (in)	Bending Design Value, F_b (psi) ($\times$ 0.00689 for kN/m^2)												
$\times$ 25.4 for mm		300	400	500	600	700	800	900	1000	1100	1200	1300	1400	1500
2 × 6	12.0	4-8	5-4	6-0	6-7	7-1	7-7	8-1	8-6	8-11	9-4	9-8	10-0	10-5
	16.0	4-0	4-8	5-2	5-8	6-2	6-7	7-0	7-4	7-8	8-1	8-5	8-8	9-0
	19.2	3-8	4-3	4-9	5-2	5-7	6-0	6-4	6-9	7-0	7-4	7-8	7-11	8-3
	24.0	3-3	3-10	4-3	4-8	5-0	5-4	5-8	6-0	6-4	6-7	6-10	7-1	7-4
2 × 8	12.0	6-2	7-1	7-11	8-8	9-4	10-0	10-7	11-2	11-9	12-3	12-9	13-3	13-8
	16.0	5-4	6-2	6-10	7-6	8-1	8-8	9-2	9-8	10-2	10-7	11-1	11-6	11-10
	19.2	4-10	5-7	6-3	6-10	7-5	7-11	8-5	8-10	9-3	9-8	10-1	10-6	10-10
	24.0	4-4	5-0	5-7	6-2	6-7	7-1	7-6	7-11	8-4	8-8	9-0	9-4	9-8
2 × 10	12.0	7-10	9-0	10-1	11-1	11-11	12-9	13-6	14-3	15-0	15-8	16-3	16-11	17-6
	16.0	6-9	7-10	8-9	9-7	10-4	11-1	11-9	12-4	13-0	13-6	14-1	14-8	15-2
	19.2	6-2	7-2	8-0	8-9	9-5	10-1	10-8	11-3	11-10	12-4	12-10	13-4	13-10
	24.0	5-6	6-5	7-2	7-10	8-5	9-0	9-7	10-1	10-7	11-1	11-6	11-11	12-4
2 × 12	12.0	9-6	11-0	12-3	13-5	14-6	15-6	16-6	17-4	18-2	19-0	19-10	20-6	21-3
	16.0	8-3	9-6	10-8	11-8	12-7	13-5	14-3	15-0	15-9	16-6	17-2	17-9	18-5
	19.2	7-6	8-8	9-8	10-8	11-6	12-3	13-0	13-9	14-5	15-0	15-8	16-3	16-10
	24.0	6-9	7-9	8-8	9-6	10-3	11-0	11-8	12-3	12-10	13-5	14-0	14-6	15-0
E	12.0	0.11	0.17	0.23	0.31	0.39	0.47	0.56	0.66	0.76	0.87	0.98	1.10	1.21
	16.0	0.09	0.14	0.20	0.27	0.34	0.41	0.49	0.57	0.66	0.75	0.85	0.95	1.05
	19.2	0.09	0.13	0.18	0.24	0.31	0.37	0.45	0.52	0.60	0.69	0.77	0.87	0.96
	24.0	0.08	0.12	0.17	0.22	0.27	0.33	0.40	0.47	0.54	0.61	0.69	0.77	0.86

Rafter Size (in)	Spacing (in)	Bending Design Value, F_b (psi) ($\times$ 0.00689 for kN/m^2)												
$\times$ 25.4 for mm		1600	1700	1800	1900	2000	2100	2200	2300	2400	2500	2600	2700
2 × 6	12.0	10-9	11-1	11-5	11-8	12-0	12-4	12-7	12-10	13-2			
	16.0	9-4	9-7	9-10	10-2	10-5	10-8	10-11	11-2	11-5	11-7	11-10	12-1
	19.2	8-6	8-9	9-0	9-3	9-6	9-9	9-11	10-2	10-5	10-7	10-10	11-0
	24.0	7-7	7-10	8-1	8-3	8-6	8-8	8-11	9-1	9-4	9-6	9-8	9-10
2 × 8	12.0	14-2	14-7	15-0	15-5	15-10	16-3	16-7	17-0	17-4			
	16.0	12-3	12-8	13-0	13-4	13-8	14-0	14-4	14-8	15-0	15-4	15-7	15-11
	19.2	11-2	11-6	11-10	12-2	12-6	12-10	13-1	13-5	13-8	14-0	14-3	14-6
	24.0	10-0	10-4	10-7	10-11	11-2	11-6	11-9	12-0	12-3	12-6	12-9	13-0
2 × 10	12.0	18-1	18-7	19-2	19-8	20-2	20-8	21-2	21-8	22-1			
	16.0	15-8	16-1	16-7	17-0	17-6	17-11	18-4	18-9	19-2	19-7	19-11	20-4
	19.2	14-3	14-9	15-2	15-7	15-11	16-4	16-9	17-1	17-6	17-10	18-2	18-6
	24.0	12-9	13-2	13-6	13-11	14-3	14-8	15-0	15-4	15-8	15-11	16-3	16-7
2 × 12	12.0	21-11	22-8	23-3	23-11	24-7	25-2	25-9					
	16.0	19-0	19-7	20-2	20-9	21-3	21-9	22-4	22-10	23-3	23-9	24-3	24-8
	19.2	17-4	17-11	18-5	18-11	19-5	19-11	20-4	20-10	21-3	21-8	22-2	22-7
	24.0	15-6	16-0	16-6	16-11	17-4	17-9	18-2	18-7	19-0	19-5	19-10	20-2
E	12.0	1.34	1.47	1.60	1.73	1.87	2.01	2.16	2.31	2.46			
	16.0	1.16	1.27	1.38	1.50	1.62	1.74	1.87	2.00	2.13	2.26	2.40	2.54
	19.2	1.06	1.16	1.26	1.37	1.48	1.59	1.71	1.82	1.94	2.07	2.19	2.32
	24.0	0.95	1.04	1.13	1.22	1.32	1.42	1.53	1.63	1.74	1.85	1.96	2.07

NOTE: The required modulus of elasticity, *E*, in 1,000,000 pounds per square inch is shown at the bottom of this table, is limited to 2.6 million psi (17.92 $\times$ 10^6 kN/m^2) and less, and is applicable to all lumber sizes shown. Spans are shown in feet-inches (1 foot = 304.8 mm, 1 inch = 25.4 mm) and are limited to 26 feet (7925 mm) and less.

TABLE 23-VII-R-7

1997 UNIFORM BUILDING CODE

TABLE 23-VII-R-7—RAFTERS WITH L/180 DEFLECTION LIMITS
The allowable bending stress (Fb) and modulus of elasticity (E) used in this table shall be from Tables 23-IV-V-1 and 23-IV-V-2 only.

DESIGN CRITERIA:
Strength — Live load of 40 psf (1.92 kN/m²) plus dead load of 10 psf (0.48 kN/m²) determines the required bending design value.
Deflection — For 40 psf (1.92 kN/m²) live load.
Limited to span in inches (mm) divided by 180.

Rafter Size (in) × 25.4 for mm	Spacing (in)	Bending Design Value, Fb (psi) (× 0.00689 for kN/m²) 200	300	400	500	600	700	800	900	1000	1100	1200	1300	1400	1500	1600
2 × 4	12.0	2-10	3-6	4-0	4-6	4-11	5-4	5-9	6-1	6-5	6-8	7-0	7-3	7-7	7-10	8-1
	16.0	2-6	3-0	3-6	3-11	4-3	4-8	4-11	5-3	5-6	5-10	6-1	6-4	6-7	6-9	7-0
	19.2	2-3	2-9	3-2	3-7	3-11	4-3	4-6	4-10	5-1	5-4	5-6	5-9	6-0	6-2	6-5
	24.0	2-0	2-6	2-10	3-2	3-6	3-9	4-0	4-3	4-6	4-9	4-11	5-2	5-4	5-6	5-9
2 × 6	12.0	4-6	5-6	6-4	7-1	7-9	8-5	9-0	9-6	10-0	10-6	11-0	11-5	11-11	12-4	12-8
	16.0	3-11	4-9	5-6	6-2	6-9	7-3	7-9	8-3	8-8	9-1	9-6	9-11	10-3	10-8	11-0
	19.2	3-7	4-4	5-0	5-7	6-2	6-8	7-1	7-6	7-11	8-4	8-8	9-1	9-5	9-9	10-0
	24.0	3-2	3-11	4-6	5-0	5-6	5-11	6-4	6-9	7-1	7-5	7-9	8-1	8-5	8-8	9-0
2 × 8	12.0	5-11	7-3	8-4	9-4	10-3	11-1	11-10	12-7	13-3	13-11	14-6	15-1	15-8	16-3	16-9
	16.0	5-2	6-3	7-3	8-1	8-11	9-7	10-3	10-10	11-6	12-0	12-7	13-1	13-7	14-0	14-6
	19.2	4-8	5-9	6-7	7-5	8-1	8-9	9-4	9-11	10-6	11-0	11-6	11-11	12-5	12-10	13-3
	24.0	4-2	5-2	5-11	6-7	7-3	7-10	8-4	8-11	9-4	9-10	10-3	10-8	11-1	11-6	11-10
2 × 10	12.0	7-7	9-3	10-8	11-11	13-1	14-2	15-1	16-0	16-11	17-9	18-6	19-3	20-0	20-8	21-4
	16.0	6-6	8-0	9-3	10-4	11-4	12-3	13-1	13-10	14-8	15-4	16-0	16-8	17-4	17-11	18-6
	19.2	6-0	7-4	8-5	9-5	10-4	11-2	11-11	12-8	13-4	14-0	14-8	15-3	15-10	16-4	16-11
	24.0	5-4	6-6	7-7	8-5	9-3	10-0	10-8	11-4	11-11	12-6	13-1	13-7	14-2	14-8	15-1
E	12.0	0.06	0.11	0.17	0.23	0.31	0.38	0.47	0.56	0.66	0.76	0.86	0.97	1.09	1.21	1.33
	16.0	0.05	0.09	0.14	0.20	0.26	0.33	0.41	0.49	0.57	0.66	0.75	0.84	0.94	1.05	1.15
	19.2	0.05	0.09	0.13	0.18	0.24	0.30	0.37	0.44	0.52	0.60	0.68	0.77	0.86	0.95	1.05
	24.0	0.04	0.08	0.12	0.16	0.22	0.27	0.33	0.40	0.46	0.54	0.61	0.69	0.77	0.85	0.94

Rafter Size (in) × 25.4 for mm	Spacing (in)	Bending Design Value, Fb (psi) (× 0.00689 for kN/m²) 1700	1800	1900	2000	2100	2200	2300	2400	2500	2600	2700	2800	2900	3000
2 × 4	12.0	8-4	8-7	8-10	9-0	9-3	9-6	9-8	9-11	10-1					
	16.0	7-3	7-5	7-8	7-10	8-0	8-2	8-5	8-7	8-9	8-11	9-1			
	19.2	6-7	6-9	7-0	7-2	7-4	7-6	7-8	7-10	8-0	8-2	8-4	8-5	8-7	
	24.0	5-11	6-1	6-3	6-5	6-7	6-8	6-10	7-0	7-2	7-3	7-5	7-7	7-8	7-10
2 × 6	12.0	13-1	13-6	13-10	14-2	14-7	14-11	15-3	15-7	15-11					
	16.0	11-4	11-8	12-0	12-4	12-7	12-11	13-2	13-6	13-9	14-0	14-3			
	19.2	10-4	10-8	10-11	11-3	11-6	11-9	12-0	12-4	12-7	12-10	13-1	13-3	13-6	
	24.0	9-3	9-6	9-9	10-0	10-3	10-6	10-9	11-0	11-3	11-5	11-8	11-11	12-1	12-4
2 × 8	12.0	17-3	17-9	18-3	18-9	19-2	19-8	20-1	20-6	20-11					
	16.0	14-11	15-5	15-10	16-3	16-7	17-0	17-5	17-9	18-1	18-6	18-10			
	19.2	13-8	14-0	14-5	14-10	15-2	15-6	15-10	16-3	16-7	16-10	17-2	17-6	17-10	
	24.0	12-2	12-7	12-11	13-3	13-7	13-11	14-2	14-6	14-10	15-1	15-5	15-8	15-11	16-3
2 × 10	12.0	22-0	22-8	23-3	23-11	24-6	25-1	25-7							
	16.0	19-1	19-7	20-2	20-8	21-2	21-8	22-2	22-8	23-1	23-7	24-0			
	19.2	17-5	17-11	18-5	18-11	19-4	19-10	20-3	20-8	21-1	21-6	21-11	22-4	22-9	
	24.0	15-7	16-0	16-6	16-11	17-4	17-9	18-1	18-6	18-11	19-3	19-7	20-0	20-4	20-8
E	12.0	1.46	1.59	1.72	1.86	2.00	2.14	2.29	2.44	2.60					
	16.0	1.26	1.37	1.49	1.61	1.73	1.86	1.99	2.12	2.25	2.39	2.53			
	19.2	1.15	1.25	1.36	1.47	1.58	1.70	1.81	1.93	2.05	2.18	2.31	2.43	2.57	
	24.0	1.03	1.12	1.22	1.31	1.41	1.52	1.62	1.73	1.84	1.95	2.06	2.18	2.30	2.41

NOTE: The required modulus of elasticity, E, in 1,000,000 pounds per square inch is shown at the bottom of this table, is limited to 2.6 million psi (17.92 × 10⁶ kN/m²) and less, and is applicable to all lumber sizes shown. Spans are shown in feet-inches (1 foot = 304.8 mm, 1 inch = 25.4 mm) and are limited to 26 feet (7925 mm) and less.

TABLE 23-VII-R-8—RAFTERS WITH *L*/180 DEFLECTION LIMITS
The allowable bending stress (*F_b*) and modulus of elasticity *(E)* used in this table shall be from Tables 23-IV-V-1 and 23-IV-V-2 only.

DESIGN CRITERIA:
Strength — Live load of 50 psf (2.40 kN/m²) plus dead load of 10 psf (0.48 kN/m²) determines the required bending design value.
Deflection — For 50 psf (2.40 kN/m²) live load.
Limited to span in inches (mm) divided by 180.

Rafter Size (in) ×25.4 for mm	Spacing (in)	_ Bending Design Value, *F_b* (psi) (×0.00689 for kN/m²)													
		200	300	400	500	600	700	800	900	1000	1100	1200	1300	1400	1500	1600
2×4	12.0	2-7	3-2	3-8	4-1	4-6	4-11	5-3	5-6	5-10	6-1	6-5	6-8	6-11	7-2	7-5
	16.0	2-3	2-9	3-2	3-7	3-11	4-3	4-6	4-10	5-1	5-4	5-6	5-9	6-0	6-2	6-5
	19.2	2-1	2-6	2-11	3-3	3-7	3-10	4-1	4-4	4-7	4-10	5-1	5-3	5-5	5-8	5-10
	24.0	1-10	2-3	2-7	2-11	3-2	3-5	3-8	3-11	4-1	4-4	4-6	4-8	4-11	5-1	5-3
2×6	12.0	4-1	5-0	5-10	6-6	7-1	7-8	8-2	8-8	9-2	9-7	10-0	10-5	10-10	11-3	11-7
	16.0	3-7	4-4	5-0	5-7	6-2	6-8	7-1	7-6	7-11	8-4	8-8	9-1	9-5	9-9	10-0
	19.2	3-3	4-0	4-7	5-1	5-7	6-1	6-6	6-10	7-3	7-7	7-11	8-3	8-7	8-11	9-2
	24.0	2-11	3-7	4-1	4-7	5-0	5-5	5-10	6-2	6-6	6-10	7-1	7-5	7-8	7-11	8-2
2×8	12.0	5-5	6-7	7-8	8-7	9-4	10-1	10-10	11-6	12-1	12-8	13-3	13-9	14-4	14-10	15-3
	16.0	4-8	5-9	6-7	7-5	8-1	8-9	9-4	9-11	10-6	11-0	11-6	11-11	12-5	12-10	13-3
	19.2	4-3	5-3	6-0	6-9	7-5	8-0	8-7	9-1	9-7	10-0	10-6	10-11	11-4	11-8	12-1
	24.0	3-10	4-8	5-5	6-0	6-7	7-2	7-8	8-1	8-7	9-0	9-4	9-9	10-1	10-6	10-10
2×10	12.0	6-11	8-5	9-9	10-11	11-11	12-11	13-9	14-8	15-5	16-2	16-11	17-7	18-3	18-11	19-6
	16.0	6-0	7-4	8-5	9-5	10-4	11-2	11-11	12-8	13-4	14-0	14-8	15-3	15-10	16-4	16-11
	19.2	5-5	6-8	7-8	8-7	9-5	10-2	10-11	11-7	12-2	12-9	13-4	13-11	14-5	14-11	15-5
	24.0	4-11	6-0	6-11	7-8	8-5	9-1	9-9	10-4	10-11	11-5	11-11	12-5	12-11	13-4	13-9
E	12.0	0.06	0.10	0.16	0.22	0.29	0.37	0.45	0.53	0.63	0.72	0.82	0.93	1.04	1.15	1.26
	16.0	0.05	0.09	0.14	0.19	0.25	0.32	0.39	0.46	0.54	0.62	0.71	0.80	0.90	0.99	1.10
	19.2	0.04	0.08	0.13	0.17	0.23	0.29	0.35	0.42	0.49	0.57	0.65	0.73	0.82	0.91	1.00
	24.0	0.04	0.07	0.11	0.16	0.21	0.26	0.32	0.38	0.44	0.51	0.58	0.66	0.73	0.81	0.89

Rafter Size (in) ×25.4 for mm	Spacing (in)	_ Bending Design Value, *F_b* (psi) (×0.00689 for kN/m²)													
		1700	1800	1900	2000	2100	2200	2300	2400	2500	2600	2700	2800	2900	3000
2×4	12.0	7-7	7-10	8-0	8-3	8-5	8-8	8-10	9-0	9-3					
	16.0	6-7	6-9	7-0	7-2	7-4	7-6	7-8	7-10	8-0	8-2	8-4	8-5		
	19.2	6-0	6-2	6-4	6-6	6-8	6-10	7-0	7-2	7-3	7-5	7-7	7-9	7-10	8-0
	24.0	5-5	5-6	5-8	5-10	6-0	6-1	6-3	6-5	6-6	6-8	6-9	6-11	7-0	7-2
2×6	12.0	11-11	12-4	12-8	13-0	13-3	13-7	13-11	14-2	14-6					
	16.0	10-4	10-8	10-11	11-3	11-6	11-9	12-0	12-4	12-7	12-10	13-1	13-3		
	19.2	9-5	9-9	10-0	10-3	10-6	10-9	11-0	11-3	11-5	11-8	11-11	12-2	12-4	12-7
	24.0	8-5	8-8	8-11	9-2	9-5	9-7	9-10	10-0	10-3	10-5	10-8	10-10	11-0	11-3
2×8	12.0	15-9	16-3	16-8	17-1	17-6	17-11	18-4	18-9	19-1					
	16.0	13-8	14-0	14-5	14-10	15-2	15-6	15-10	16-3	16-7	16-10	17-2	17-6		
	19.2	12-5	12-10	13-2	13-6	13-10	14-2	14-6	14-10	15-1	15-5	15-8	16-0	16-3	16-7
	24.0	11-2	11-6	11-9	12-1	12-5	12-8	12-11	13-3	13-6	13-9	14-0	14-4	14-7	14-10
2×10	12.0	20-1	20-8	21-3	21-10	22-4	22-10	23-5	23-11	24-5					
	16.0	17-5	17-11	18-5	18-11	19-4	19-10	20-3	20-8	21-1	21-6	21-11	22-4		
	19.2	15-11	16-4	16-10	17-3	17-8	18-1	18-6	18-11	19-3	19-8	20-0	20-5	20-9	21-1
	24.0	14-3	14-8	15-0	15-5	15-10	16-2	16-6	16-11	17-3	17-7	17-11	18-3	18-7	18-11
E	12.0	1.39	1.51	1.64	1.77	1.90	2.04	2.18	2.32	2.47					
	16.0	1.20	1.31	1.42	1.53	1.65	1.77	1.89	2.01	2.14	2.27	2.40	2.54		
	19.2	1.10	1.19	1.29	1.40	1.50	1.61	1.72	1.84	1.95	2.07	2.19	2.32	2.44	2.57
	24.0	0.98	1.07	1.16	1.25	1.34	1.44	1.54	1.64	1.75	1.85	1.96	2.07	2.18	2.30

NOTE: The required modulus of elasticity, *E*, in 1,000,000 pounds per square inch is shown at the bottom of this table, is limited to 2.6 million psi (17.92 × 10⁶ kN/m²) and less, and is applicable to all lumber sizes shown. Spans are shown in feet-inches (1 foot = 304.8 mm, 1 inch = 25.4 mm) and are limited to 26 feet (7925 mm) and less.

TABLE 23-VII-R-9

1997 UNIFORM BUILDING CODE

TABLE 23-VII-R-9—RAFTERS WITH L/180 DEFLECTION LIMITS
The allowable bending stress (F_b) and modulus of elasticity (E) used in this table shall be from Tables 23-IV-V-1 and 23-IV-V-2 only.

DESIGN CRITERIA:
Strength — Live load of 40 psf (1.92 kN/m^2) plus dead load of 15 psf (0.72 kN/m^2) determines the required bending design value.
Deflection — For 40 psf (1.92 kN/m^2) live load.
Limited to span in inches (mm) divided by 180.

Rafter Size (in) × 25.4 for mm	Spacing (in)	200	300	400	500	600	700	800	900	1000	1100	1200	1300	1400	1500	1600
2×4	12.0	2-9	3-4	3-10	4-4	4-9	5-1	5-5	5-9	6-1	6-5	6-8	6-11	7-3	7-6	7-8
	16.0	2-4	2-11	3-4	3-9	4-1	4-5	4-9	5-0	5-3	5-6	5-9	6-0	6-3	6-6	6-8
	19.2	2-2	2-8	3-1	3-5	3-9	4-0	4-4	4-7	4-10	5-1	5-3	5-6	5-8	5-11	6-1
	24.0	1-11	2-4	2-9	3-1	3-4	3-7	3-10	4-1	4-4	4-6	4-9	4-11	5-1	5-3	5-5
2×6	12.0	4-3	5-3	6-1	6-9	7-5	8-0	8-7	9-1	9-7	10-0	10-6	10-11	11-4	11-9	12-1
	16.0	3-8	4-6	5-3	5-10	6-5	6-11	7-5	7-10	8-3	8-8	9-1	9-5	9-10	10-2	10-6
	19.2	3-5	4-2	4-9	5-4	5-10	6-4	6-9	7-2	7-7	7-11	8-3	8-8	8-11	9-3	9-7
	24.0	3-0	3-8	4-3	4-9	5-3	5-8	6-1	6-5	6-9	7-1	7-5	7-9	8-0	8-3	8-7
2×8	12.0	5-8	6-11	8-0	8-11	9-9	10-7	11-3	12-0	12-7	13-3	13-10	14-5	14-11	15-5	16-0
	16.0	4-11	6-0	6-11	7-9	8-6	9-2	9-9	10-4	10-11	11-6	12-0	12-6	12-11	13-5	13-10
	19.2	4-6	5-6	6-4	7-1	7-9	8-4	8-11	9-6	10-0	10-6	10-11	11-5	11-10	12-3	12-7
	24.0	4-0	4-11	5-8	6-4	6-11	7-6	8-0	8-6	8-11	9-4	9-9	10-2	10-7	10-11	11-3
2×10	12.0	7-2	8-10	10-2	11-5	12-6	13-6	14-5	15-3	16-1	16-11	17-8	18-4	19-1	19-9	20-4
	16.0	6-3	7-8	8-10	9-10	10-10	11-8	12-6	13-3	13-11	14-8	15-3	15-11	16-6	17-1	17-8
	19.2	5-8	7-0	8-1	9-0	9-10	10-8	11-5	12-1	12-9	13-4	13-11	14-6	15-1	15-7	16-1
	24.0	5-1	6-3	7-2	8-1	8-10	9-6	10-2	10-10	11-5	11-11	12-6	13-0	13-6	13-11	14-5
E	12.0	0.05	0.09	0.14	0.20	0.26	0.33	0.41	0.49	0.57	0.66	0.75	0.84	0.94	1.05	1.15
	16.0	0.04	0.08	0.12	0.17	0.23	0.29	0.35	0.42	0.49	0.57	0.65	0.73	0.82	0.91	1.00
	19.2	0.04	0.07	0.11	0.16	0.21	0.26	0.32	0.38	0.45	0.52	0.59	0.67	0.75	0.83	0.91
	24.0	0.04	0.07	0.10	0.14	0.19	0.24	0.29	0.34	0.40	0.46	0.53	0.60	0.67	0.74	0.82

Bending Design Value, F_b (psi) (× 0.00689 kN/m^2)

Rafter Size (in) × 25.4 for mm	Spacing (in)	1700	1800	1900	2000	2100	2200	2300	2400	2500	2600	2700	2800	2900	3000
2×4	12.0	7-11	8-2	8-5	8-7	8-10	9-0	9-3	9-5	9-8	9-10	10-0			
	16.0	6-11	7-1	7-3	7-6	7-8	7-10	8-0	8-2	8-4	8-6	8-8	8-10	9-0	9-2
	19.2	6-3	6-6	6-8	6-10	7-0	7-2	7-4	7-6	7-7	7-9	7-11	8-1	8-2	8-4
	24.0	5-7	5-9	5-11	6-1	6-3	6-5	6-6	6-8	6-10	6-11	7-1	7-3	7-4	7-6
2×6	12.0	12-6	12-10	13-2	13-6	13-10	14-2	14-6	14-10	15-2	15-5	15-9			
	16.0	10-10	11-1	11-5	11-9	12-0	12-4	12-7	12-10	13-1	13-4	13-7	13-10	14-1	14-4
	19.2	9-10	10-2	10-5	10-8	11-0	11-3	11-6	11-9	12-0	12-2	12-5	12-8	12-11	13-1
	24.0	8-10	9-1	9-4	9-7	9-10	10-0	10-3	10-6	10-8	10-11	11-1	11-4	11-6	11-9
2×8	12.0	16-5	16-11	17-5	17-10	18-3	18-9	19-2	19-7	19-11	20-4	20-9			
	16.0	14-3	14-8	15-1	15-5	15-10	16-3	16-7	16-11	17-3	17-7	18-0	18-3	18-7	18-11
	19.2	13-0	13-5	13-9	14-1	14-6	14-10	15-2	15-5	15-9	16-1	16-5	16-8	17-0	17-3
	24.0	11-8	12-0	12-4	12-7	12-11	13-3	13-6	13-10	14-1	14-5	14-8	14-11	15-2	15-5
2×10	12.0	21-0	21-7	22-2	22-9	23-4	23-11	24-5	24-11	25-6	26-0				
	16.0	18-2	18-9	19-3	19-9	20-2	20-8	21-2	21-7	22-1	22-6	22-11	23-4	23-9	24-2
	19.2	16-7	17-1	17-7	18-0	18-5	18-11	19-4	19-9	20-2	20-6	20-11	21-4	21-8	22-1
	24.0	14-10	15-3	15-8	16-1	16-6	16-11	17-3	17-8	18-0	18-4	18-9	19-1	19-5	19-9
E	12.0	1.26	1.38	1.49	1.61	1.73	1.86	1.99	2.12	2.25	2.39	2.53			
	16.0	1.09	1.19	1.29	1.40	1.50	1.61	1.72	1.83	1.95	2.07	2.19	2.31	2.44	2.56
	19.2	1.00	1.09	1.18	1.27	1.37	1.47	1.57	1.67	1.78	1.89	2.00	2.11	2.22	2.34
	24.0	0.89	0.97	1.06	1.14	1.23	1.31	1.41	1.50	1.59	1.69	1.79	1.89	1.99	2.09

Bending Design Value, F_b (psi) (× 0.00689 kN/m^2)

NOTE: The required modulus of elasticity, E, in 1,000,000 pounds per square inch is shown at the bottom of this table, is limited to 2.6 million psi (17.92 × 10^6 kN/m^2) and less, and is applicable to all lumber sizes shown. Spans are shown in feet-inches (1 foot = 304.8 mm, 1 inch = 25.4 mm) and are limited to 26 feet (7925 mm) and less.

TABLE 23-VII-R-10—RAFTERS WITH L/180 DEFLECTION LIMITS
The allowable bending stress (Fb) and modulus of elasticity (E) used in this table shall be from Tables 23-IV-V-1 and 23-IV-V-2 only.

DESIGN CRITERIA:
Strength — Live load of 50 psf (2.40 kN/m²) plus dead load of 15 psf (0.72 kN/m²) determines the required bending design value.
Deflection — For 50 psf (2.40 kN/m²) live load.
Limited to span in inches (mm) divided by 180.

Rafter Size (in)	Spacing (in)	Bending Design Value, F_b (psi) (× 0.00689 kN/m²)													
× 25.4 for mm		200	300	400	500	600	700	800	900	1000	1100	1200	1300	1400	1500	1600
2 × 4	12.0	2-6	3-1	3-7	4-0	4-4	4-8	5-0	5-4	5-7	5-11	6-2	6-5	6-8	6-10	7-1
	16.0	2-2	2-8	3-1	3-5	3-9	4-1	4-4	4-7	4-10	5-1	5-4	5-6	5-9	5-11	6-2
	19.2	2-0	2-5	2-10	3-2	3-5	3-8	4-0	4-2	4-5	4-8	4-10	5-1	5-3	5-5	5-7
	24.0	1-9	2-2	2-6	2-10	3-1	3-4	3-7	3-9	4-0	4-2	4-4	4-6	4-8	4-10	5-0
2 × 6	12.0	3-11	4-10	5-7	6-3	6-10	7-4	7-11	8-4	8-10	9-3	9-8	10-0	10-5	10-9	11-2
	16.0	3-5	4-2	4-10	5-5	5-11	6-5	6-10	7-3	7-8	8-0	8-4	8-8	9-0	9-4	9-8
	19.2	3-1	3-10	4-5	4-11	5-5	5-10	6-3	6-7	7-0	7-4	7-8	7-11	8-3	8-6	8-10
	24.0	2-9	3-5	3-11	4-5	4-10	5-3	5-7	5-11	6-3	6-6	6-10	7-1	7-4	7-8	7-11
2 × 8	12.0	5-2	6-4	7-4	8-3	9-0	9-9	10-5	11-0	11-7	12-2	12-9	13-3	13-9	14-3	14-8
	16.0	4-6	5-6	6-4	7-1	7-9	8-5	9-0	9-6	10-1	10-7	11-0	11-6	11-11	12-4	12-9
	19.2	4-1	5-0	5-10	6-6	7-1	7-8	8-3	8-8	9-2	9-8	10-1	10-6	10-10	11-3	11-7
	24.0	3-8	4-6	5-2	5-10	6-4	6-10	7-4	7-9	8-3	8-7	9-0	9-4	9-9	10-1	10-5
2 × 10	12.0	6-7	8-1	9-4	10-6	11-6	12-5	13-3	14-1	14-10	15-6	16-3	16-11	17-6	18-2	18-9
	16.0	5-9	7-0	8-1	9-1	9-11	10-9	11-6	12-2	12-10	13-5	14-1	14-8	15-2	15-9	16-3
	19.2	5-3	6-5	7-5	8-3	9-1	9-10	10-6	11-1	11-9	12-3	12-10	13-4	13-10	14-4	14-10
	24.0	4-8	5-9	6-7	7-5	8-1	8-9	9-4	9-11	10-6	11-0	11-6	11-11	12-5	12-10	13-3
E	12.0	0.05	0.09	0.14	0.20	0.26	0.32	0.40	0.47	0.55	0.64	0.73	0.82	0.92	1.02	1.12
	16.0	0.04	0.08	0.12	0.17	0.22	0.28	0.34	0.41	0.48	0.55	0.63	0.71	0.80	0.88	0.97
	19.2	0.04	0.07	0.11	0.15	0.20	0.26	0.31	0.37	0.44	0.51	0.58	0.65	0.73	0.81	0.89
	24.0	0.04	0.06	0.10	0.14	0.18	0.23	0.28	0.33	0.39	0.45	0.52	0.58	0.65	0.72	0.79

Rafter Size (in)	Spacing (in)	Bending Design Value, F_b (psi) (× 0.00689 kN/m²)													
× 25.4 for mm		1700	1800	1900	2000	2100	2200	2300	2400	2500	2600	2700	2800	2900	3000
2 × 4	12.0	7-4	7-6	7-9	7-11	8-1	8-4	8-6	8-8	8-10	9-0	9-3	9-5		
	16.0	6-4	6-6	6-8	6-10	7-0	7-2	7-4	7-6	7-8	7-10	8-0	8-1	8-3	8-5
	19.2	5-9	5-11	6-1	6-3	6-5	6-7	6-9	6-10	7-0	7-2	7-3	7-5	7-7	7-8
	24.0	5-2	5-4	5-6	5-7	5-9	5-11	6-0	6-2	6-3	6-5	6-6	6-8	6-9	6-10
2 × 6	12.0	11-6	11-10	12-2	12-5	12-9	13-1	13-4	13-8	13-11	14-2	14-6	14-9		
	16.0	9-11	10-3	10-6	10-9	11-1	11-4	11-7	11-10	12-1	12-4	12-6	12-9	13-0	13-3
	19.2	9-1	9-4	9-7	9-10	10-1	10-4	10-7	10-9	11-0	11-3	11-5	11-8	11-10	12-1
	24.0	8-1	8-4	8-7	8-10	9-0	9-3	9-5	9-8	9-10	10-0	10-3	10-5	10-7	10-9
2 × 8	12.0	15-2	15-7	16-0	16-5	16-10	17-3	17-7	18-0	18-4	18-9	19-1	19-5		
	16.0	13-1	13-6	13-10	14-3	14-7	14-11	15-3	15-7	15-11	16-3	16-6	16-10	17-1	17-5
	19.2	12-0	12-4	12-8	13-0	13-4	13-7	13-11	14-3	14-6	14-10	15-1	15-4	15-8	15-11
	24.0	10-8	11-0	11-4	11-7	11-11	12-2	12-5	12-9	13-0	13-3	13-6	13-9	14-0	14-3
2 × 10	12.0	19-4	19-10	20-5	20-11	21-6	22-0	22-6	22-11	23-5	23-11	24-4	24-9		
	16.0	16-9	17-3	17-8	18-2	18-7	19-0	19-5	19-10	20-3	20-8	21-1	21-6	21-10	22-3
	19.2	15-3	15-9	16-2	16-7	17-0	17-4	17-9	18-2	18-6	18-11	19-3	19-7	19-11	20-3
	24.0	13-8	14-1	14-5	14-10	15-2	15-6	15-11	16-3	16-7	16-11	17-3	17-6	17-10	18-2
E	12.0	1.23	1.34	1.45	1.57	1.69	1.81	1.93	2.06	2.19	2.32	2.46	2.60		
	16.0	1.06	1.16	1.26	1.36	1.46	1.57	1.67	1.78	1.90	2.01	2.13	2.25	2.37	2.49
	19.2	0.97	1.06	1.15	1.24	1.33	1.43	1.53	1.63	1.73	1.84	1.94	2.05	2.16	2.28
	24.0	0.87	0.95	1.03	1.11	1.19	1.28	1.37	1.46	1.55	1.64	1.74	1.84	1.94	2.04

NOTE: The required modulus of elasticity, E, in 1,000,000 pounds per square inch is shown at the bottom of this table, is limited to 2.6 million psi (17.92 × 10⁶ kN/m²) and less, and is applicable to all lumber sizes shown. Spans are shown in feet-inches (1 foot = 304.8 mm, 1 inch = 25.4 mm) and are limited to 26 feet (7925 mm) and less.

TABLE 23-VII-R-11

1997 UNIFORM BUILDING CODE

TABLE 23-VII-R-11—RAFTERS WITH L/180 DEFLECTION LIMITS
The allowable bending stress (F_b) and modulus of elasticity (E) used in this table shall be from Tables 23-IV-V-1 and 23-IV-V-2 only.

DESIGN CRITERIA:
Strength — Live load of 40 psf (1.92 kN/m^2) plus dead load of 20 psf (0.96 kN/m^2) determines the required bending design value.
Deflection — For 40 psf (1.92 kN/m^2) live load.
Limited to span in inches (mm) divided by 180.

Rafter Size (in) ×25.4 for mm	Spacing (in)	Bending Design Value, F_b (psi) (× 0.00689 kN/m^2)														
		200	300	400	500	600	700	800	900	1000	1100	1200	1300	1400	1500	1600
2×4	12.0	2-7	3-2	3-8	4-1	4-6	4-11	5-3	5-6	5-10	6-1	6-5	6-8	6-11	7-2	7-5
	16.0	2-3	2-9	3-2	3-7	3-11	4-3	4-6	4-10	5-1	5-4	5-6	5-9	6-0	6-2	6-5
	19.2	2-1	2-6	2-11	3-3	3-7	3-10	4-1	4-4	4-7	4-10	5-1	5-3	5-5	5-8	5-10
	24.0	1-10	2-3	2-7	2-11	3-2	3-5	3-8	3-11	4-1	4-4	4-6	4-8	4-11	5-1	5-3
2×6	12.0	4-1	5-0	5-10	6-6	7-1	7-8	8-2	8-8	9-2	9-7	10-0	10-5	10-10	11-3	11-7
	16.0	3-7	4-4	5-0	5-7	6-2	6-8	7-1	7-6	7-11	8-4	8-8	9-1	9-5	9-9	10-0
	19.2	3-3	4-0	4-7	5-1	5-7	6-1	6-6	6-10	7-3	7-7	7-11	8-3	8-7	8-11	9-2
	24.0	2-11	3-7	4-1	4-7	5-0	5-5	5-10	6-2	6-6	6-10	7-1	7-5	7-8	7-11	8-2
2×8	12.0	5-5	6-7	7-8	8-7	9-4	10-1	10-10	11-6	12-1	12-8	13-3	13-9	14-4	14-10	15-3
	16.0	4-8	5-9	6-7	7-5	8-1	8-9	9-4	9-11	10-6	11-0	11-6	11-11	12-5	12-10	13-3
	19.2	4-3	5-3	6-0	6-9	7-5	8-0	8-7	9-1	9-7	10-0	10-6	10-11	11-4	11-8	12-1
	24.0	3-10	4-8	5-5	6-0	6-7	7-2	7-8	8-1	8-7	9-0	9-4	9-9	10-1	10-6	10-10
2×10	12.0	6-11	8-5	9-9	10-11	11-11	12-11	13-9	14-8	15-5	16-2	16-11	17-7	18-3	18-11	19-6
	16.0	6-0	7-4	8-5	9-5	10-4	11-2	11-11	12-8	13-4	14-0	14-8	15-3	15-10	16-4	16-11
	19.2	5-5	6-8	7-8	8-7	9-5	10-2	10-11	11-7	12-2	12-9	13-4	13-11	14-5	14-11	15-5
	24.0	4-11	6-0	6-11	7-8	8-5	9-1	9-9	10-4	10-11	11-5	11-11	12-5	12-11	13-4	13-9
E	12.0	0.04	0.08	0.13	0.18	0.23	0.29	0.36	0.43	0.50	0.58	0.66	0.74	0.83	0.92	1.01
	16.0	0.04	0.07	0.11	0.15	0.20	0.25	0.31	0.37	0.43	0.50	0.57	0.64	0.72	0.80	0.88
	19.2	0.04	0.06	0.10	0.14	0.18	0.23	0.28	0.34	0.40	0.46	0.52	0.59	0.65	0.73	0.80
	24.0	0.03	0.06	0.09	0.13	0.16	0.21	0.25	0.30	0.35	0.41	0.46	0.52	0.59	0.65	0.72

Rafter Size (in) ×25.4 for mm	Spacing (in)	Bending Design Value, F_b (psi) (× 0.00689 kN/m^2)													
		1700	1800	1900	2000	2100	2200	2300	2400	2500	2600	2700	2800	2900	3000
2×4	12.0	7-7	7-10	8-0	8-3	8-5	8-8	8-10	9-0	9-3	9-5	9-7	9-9	9-11	10-1
	16.0	6-7	6-9	7-0	7-2	7-4	7-6	7-8	7-10	8-0	8-2	8-4	8-5	8-7	8-9
	19.2	6-0	6-2	6-4	6-6	6-8	6-10	7-0	7-2	7-3	7-5	7-7	7-9	7-10	8-0
	24.0	5-5	5-6	5-8	5-10	6-0	6-1	6-3	6-5	6-6	6-8	6-9	6-11	7-0	7-2
2×6	12.0	11-11	12-4	12-8	13-0	13-3	13-7	13-11	14-2	14-6	14-9	15-1	15-4	15-7	15-11
	16.0	10-4	10-8	10-11	11-3	11-6	11-9	12-0	12-4	12-7	12-10	13-1	13-3	13-6	13-9
	19.2	9-5	9-9	10-0	10-3	10-6	10-9	11-0	11-3	11-5	11-8	11-11	12-2	12-4	12-7
	24.0	8-5	8-8	8-11	9-2	9-5	9-7	9-10	10-0	10-3	10-5	10-8	10-10	11-0	11-3
2×8	12.0	15-9	16-3	16-8	17-1	17-6	17-11	18-4	18-9	19-1	19-6	19-10	20-3	20-7	20-11
	16.0	13-8	14-0	14-5	14-10	15-2	15-6	15-10	16-3	16-7	16-10	17-2	17-6	17-10	18-1
	19.2	12-5	12-10	13-2	13-6	13-10	14-2	14-6	14-10	15-1	15-5	15-8	16-0	16-3	16-7
	24.0	11-2	11-6	11-9	12-1	12-5	12-8	12-11	13-3	13-6	13-9	14-0	14-4	14-7	14-10
2×10	12.0	20-1	20-8	21-3	21-10	22-4	22-10	23-5	23-11	24-5	24-10	25-4	25-10		
	16.0	17-5	17-11	18-5	18-11	19-4	19-10	20-3	20-8	21-1	21-6	21-11	22-4	22-9	23-1
	19.2	15-11	16-4	16-10	17-3	17-8	18-1	18-6	18-11	19-3	19-8	20-0	20-5	20-9	21-1
	24.0	14-3	14-8	15-0	15-5	15-10	16-2	16-6	16-11	17-3	17-7	17-11	18-3	18-7	18-11
E	12.0	1.11	1.21	1.31	1.41	1.52	1.63	1.74	1.86	1.98	2.10	2.22	2.34	2.47	2.60
	16.0	0.96	1.05	1.13	1.22	1.32	1.41	1.51	1.61	1.71	1.82	1.92	2.03	2.14	2.25
	19.2	0.88	0.95	1.04	1.12	1.20	1.29	1.38	1.47	1.56	1.66	1.75	1.85	1.95	2.05
	24.0	0.78	0.85	0.93	1.00	1.08	1.15	1.23	1.31	1.40	1.48	1.57	1.66	1.75	1.84

NOTE: The required modulus of elasticity, E, in 1,000,000 pounds per square inch is shown at the bottom of this table, is limited to 2.6 million psi (17.92 × 10^6 kN/m^2) and less, and is applicable to all lumber sizes shown. Spans are shown in feet-inches (1 foot = 304.8 mm, 1 inch = 25.4 mm) and are limited to 26 feet (7925 mm) and less.

TABLE 23-VII-R-12—RAFTERS WITH L/180 DEFLECTION LIMITS
The allowable bending stress (Fb) and modulus of elasticity (E) used in this table shall be from Tables 23-IV-V-1 and 23-IV-V-2 only.

DESIGN CRITERIA:
Strength — Live load of 50 psf (2.40 kN/m^2) plus dead load of 20 psf (0.96 kN/m^2) determines the required bending design value.
Deflection — For 50 psf (2.40 kN/m^2) live load.
Limited to span in inches (mm) divided by 180.

Rafter Size (in) × 25.4 for mm	Spacing (in)	200	300	400	500	600	700	800	900	1000	1100	1200	1300	1400	1500	1600
		Bending Design Value, F_b (psi) (× 0.00689 kN/m^2)														
2×4	12.0	2-5	2-11	3-5	3-10	4-2	4-6	4-10	5-1	5-5	5-8	5-11	6-2	6-5	6-7	6-10
	16.0	2-1	2-7	2-11	3-4	3-7	3-11	4-2	4-5	4-8	4-11	5-1	5-4	5-6	5-9	5-11
	19.2	1-11	2-4	2-8	3-0	3-4	3-7	3-10	4-1	4-3	4-6	4-8	4-10	5-1	5-3	5-5
	24.0	1-8	2-1	2-5	2-8	2-11	3-2	3-5	3-7	3-10	4-0	4-2	4-4	4-6	4-8	4-10
2×6	12.0	3-10	4-8	5-4	6-0	6-7	7-1	7-7	8-1	8-6	8-11	9-4	9-8	10-0	10-5	10-9
	16.0	3-3	4-0	4-8	5-2	5-8	6-2	6-7	7-0	7-4	7-8	8-1	8-5	8-8	9-0	9-4
	19.2	3-0	3-8	4-3	4-9	5-2	5-7	6-0	6-4	6-9	7-0	7-4	7-8	7-11	8-3	8-6
	24.0	2-8	3-3	3-10	4-3	4-8	5-0	5-4	5-8	6-0	6-4	6-7	6-10	7-1	7-4	7-7
2×8	12.0	5-0	6-2	7-1	7-11	8-8	9-4	10-0	10-7	11-2	11-9	12-3	12-9	13-3	13-8	14-2
	16.0	4-4	5-4	6-2	6-10	7-6	8-1	8-8	9-2	9-8	10-2	10-7	11-1	11-6	11-10	12-3
	19.2	3-11	4-10	5-7	6-3	6-10	7-5	7-11	8-5	8-10	9-3	9-8	10-1	10-6	10-10	11-2
	24.0	3-6	4-4	5-0	5-7	6-2	6-7	7-1	7-6	7-11	8-4	8-8	9-0	9-4	9-8	10-0
2×10	12.0	6-5	7-10	9-0	10-1	11-1	11-11	12-9	13-6	14-3	15-0	15-8	16-3	16-11	17-6	18-1
	16.0	5-6	6-9	7-10	8-9	9-7	10-4	11-1	11-9	12-4	13-0	13-6	14-1	14-8	15-2	15-8
	19.2	5-1	6-2	7-2	8-0	8-9	9-5	10-1	10-8	11-3	11-10	12-4	12-10	13-4	13-10	14-3
	24.0	4-6	5-6	6-5	7-2	7-10	8-5	9-0	9-7	10-1	10-7	11-1	11-6	11-11	12-4	12-9
E	12.0	0.04	0.08	0.13	0.18	0.23	0.29	0.35	0.42	0.50	0.57	0.65	0.74	0.82	0.91	1.00
	16.0	0.04	0.07	0.11	0.15	0.20	0.25	0.31	0.37	0.43	0.50	0.56	0.64	0.71	0.79	0.87
	19.2	0.04	0.06	0.10	0.14	0.18	0.23	0.28	0.33	0.39	0.45	0.52	0.58	0.65	0.72	0.79
	24.0	0.03	0.06	0.09	0.12	0.16	0.21	0.25	0.30	0.35	0.40	0.46	0.52	0.58	0.64	0.71

Rafter Size (in) × 25.4 for mm	Spacing (in)	1700	1800	1900	2000	2100	2200	2300	2400	2500	2600	2700	2800	2900	3000
		Bending Design Value, F_b (psi) (× 0.00689 kN/m^2)													
2×4	12.0	7-0	7-3	7-5	7-8	7-10	8-0	8-2	8-4	8-6	8-8	8-10	9-0	9-2	9-4
	16.0	6-1	6-3	6-5	6-7	6-9	6-11	7-1	7-3	7-5	7-6	7-8	7-10	8-0	8-1
	19.2	5-7	5-9	5-11	6-0	6-2	6-4	6-6	6-7	6-9	6-11	7-0	7-2	7-3	7-5
	24.0	5-0	5-1	5-3	5-5	5-6	5-8	5-9	5-11	6-0	6-2	6-3	6-5	6-6	6-7
2×6	12.0	11-1	11-5	11-8	12-0	12-4	12-7	12-10	13-2	13-5	13-8	13-11	14-2	14-5	14-8
	16.0	9-7	9-10	10-2	10-5	10-8	10-11	11-2	11-5	11-7	11-10	12-1	12-4	12-6	12-9
	19.2	8-9	9-0	9-3	9-6	9-9	9-11	10-2	10-5	10-7	10-10	11-0	11-3	11-5	11-7
	24.0	7-10	8-1	8-3	8-6	8-8	8-11	9-1	9-4	9-6	9-8	9-10	10-0	10-3	10-5
2×8	12.0	14-7	15-0	15-5	15-10	16-3	16-7	17-0	17-4	17-8	18-0	18-5	18-9	19-1	19-5
	16.0	12-8	13-0	13-4	13-8	14-0	14-4	14-8	15-0	15-4	15-7	15-11	16-3	16-6	16-9
	19.2	11-6	11-10	12-2	12-6	12-10	13-1	13-5	13-8	14-0	14-3	14-6	14-10	15-1	15-4
	24.0	10-4	10-7	10-11	11-2	11-6	11-9	12-0	12-3	12-6	12-9	13-0	13-3	13-6	13-8
2×10	12.0	18-7	19-2	19-8	20-2	20-8	21-2	21-8	22-1	22-7	23-0	23-5	23-11	24-4	24-9
	16.0	16-1	16-7	17-0	17-6	17-11	18-4	18-9	19-2	19-7	19-11	20-4	20-8	21-1	21-5
	19.2	14-9	15-2	15-7	15-11	16-4	16-9	17-1	17-6	17-10	18-2	18-6	18-11	19-3	19-7
	24.0	13-2	13-6	13-11	14-3	14-8	15-0	15-4	15-8	15-11	16-3	16-7	16-11	17-2	17-6
E	12.0	1.10	1.20	1.30	1.40	1.51	1.62	1.73	1.84	1.96	2.08	2.20	2.32	2.45	2.58
	16.0	0.95	1.04	1.12	1.21	1.31	1.40	1.50	1.60	1.70	1.80	1.91	2.01	2.12	2.23
	19.2	0.87	0.95	1.03	1.11	1.19	1.28	1.37	1.46	1.55	1.64	1.74	1.84	1.94	2.04
	24.0	0.78	0.85	0.92	0.99	1.07	1.14	1.22	1.30	1.39	1.47	1.56	1.64	1.73	1.82

NOTE: The required modulus of elasticity, E, in 1,000,000 pounds per square inch is shown at the bottom of this table, is limited to 2.6 million psi (17.92 × 10^6 kN/m^2) and less, and is applicable to all lumber sizes shown. Spans are shown in feet-inches (1 foot = 304.8 mm, 1 inch = 25.4 mm) and are limited to 26 feet (7925 mm) and less.

Division VIII—DESIGN STANDARD FOR PLANK-AND-BEAM FRAMING
Based on Wood Construction Data No. 4 (1970) of the American Forest and Paper Association

SECTION 2334 — SCOPE

This division covers plank-and-beam construction under the provisions of this code and is subject to the regulations of this chapter. Tables in this division are for use where moderate uniform loads occur and are not applicable for concentrated loads such as partitions, bathtubs, refrigerators, etc. See also Table 23-IV-A.

SECTION 2335 — DEFINITION

The plank-and-beam structural floor or roof system consists of a plank subfloor or roof decking with supporting beams spaced a maximum of 8 feet (2438 mm) on center.

SECTION 2336 — DESIGN

The load strength and deflection requirements for 2-inch (51 mm) planks and beams are set forth in Tables 23-VIII-A, 23-VIII-B, 23-VIII-C and 23-VIII-D.

The allowable unit stress f appropriate for each grade and species of lumber is set forth in Chapter 23, Division III.

TABLE 23-VIII-A—2-INCH (51 mm) PLANK—REQUIRED MINIMUM *f* AND *E* LIVE LOAD: 20, 30 AND 40 POUNDS PER SQUARE FOOT (0.96, 1.44 and 1.92 kN/m²) WITHOUT PLASTERED CEILING BELOW

PLANK SPAN (feet)	LIVE LOAD (pounds per square foot)	DEFLECTION LIMITATION	TYPE A		TYPE B		TYPE C		TYPE D	
× 304.8 for mm	× 0.0479 for kN/m²		*f* psi	*E* psi	*f* psi	*E* psi	*f* psi	*E* psi	*f* psi	*E* psi
							× 0.00689 for N/mm²			
6	20	$\frac{l}{240}$	360	576,000	360	239,000	288	305,000	360	408,000
		$\frac{l}{360}$	360	864,000	360	359,000	288	457,000	360	611,000
	30	$\frac{l}{240}$	480	864,000	480	359,000	384	457,000	480	611,000
		$\frac{l}{360}$	480	1,296,000	480	538,000	384	685,000	480	917,000
	40	$\frac{l}{240}$	600	1,152,000	600	478,000	480	609,000	600	815,000
		$\frac{l}{360}$	600	1,728,000	600	717,000	480	914,000	600	1,223,000
7	20	$\frac{l}{240}$	490	915,000	490	380,000	392	484,000	490	647,000
		$\frac{l}{360}$	490	1,372,000	490	570,000	392	726,000	490	971,000
	30	$\frac{l}{240}$	653	1,372,000	653	570,000	522	726,000	653	971,000
		$\frac{l}{360}$	653	2,058,000	653	854,000	522	1,088,000	653	1,456,000
	40	$\frac{l}{240}$	817	1,829,000	817	759,000	653	968,000	817	1,294,000
		$\frac{l}{360}$	817	2,744,000	817	1,139,000	653	1,451,000	817	1,941,000
8	20	$\frac{l}{240}$	640	1,365,000	640	567,000	512	722,000	640	966,000
		$\frac{l}{360}$	640	2,048,000	640	850,000	512	1,083,000	640	1,449,000
	30	$\frac{l}{240}$	853	2,048,000	853	850,000	682	1,083,000	853	1,449,000
		$\frac{l}{360}$	853	3,072,000	853	1,275,000	682	1,625,000	853	2,174,000
	40	$\frac{l}{240}$	1,067	2,731,000	1067	1,134,000	853	1,444,000	1,067	1,932,000
		$\frac{l}{360}$	1,067	4,096,000	1067	1,700,000	853	2,166,000	1,067	2,898,000

TABLE 23-VIII-B

1997 UNIFORM BUILDING CODE

**TABLE 23-VIII-B—ROOF BEAMS—LIVE LOAD 20 POUNDS PER SQUARE FOOT (0.96 kN/m²)—
DEFLECTION LIMITATION L/240 (not for support of plaster)**

SPAN OF BEAM (feet)	NOMINAL SIZE OF BEAM	MINIMUM f AND E IN PSI FOR BEAMS SPACED:					
		6'-0" (1829 mm)		7'-0" (2134 mm)		8'-0" (2438 mm)	
		f	E	f	E	f	E
× 304.8 for mm	× 25.4 for mm	× 0.00689 for N/mm²					
10	2–3 × 6	1,070	780,000	1,250	910,000	1,430	1,040,000
	1–3 × 8	1,235	680,000	1,440	794,000	1,645	906,000
	2–2 × 8	1,030	570,000	1,200	665,000	1,370	760,000
	1–4 × 8	880	485,000	1,030	566,000	1,175	646,000
	3–2 × 8	685	380,000	800	443,000	915	506,000
	2–3 × 8	615	340,000	720	397,000	820	453,000
	2–2 × 10	630	273,000	735	219,000	840	364,000
11	2–3 × 6	1,295	1,037,000	1,510	121,000	1,730	1,382,000
	1–3 × 8	1,490	905,000	1,740	1,056,000	1,990	1,206,000
	2–2 × 8	1,245	754,000	1,450	880,000	1,660	1,005,000
	1–4 × 8	1,065	647,000	1,245	755,000	1,420	862,000
	3–2 × 8	830	503,000	970	587,000	1,105	670,000
	2–3 × 8	745	453,000	870	529,000	995	604,000
	2–2 × 10	765	363,000	890	424,000	1,020	484,000
12	2–3 × 6	1,545	1,346,000	1,800	1,571,000	2,060	1,794,000
	1–3 × 8	1,775	1,175,000	2,070	1,371,000	2,370	1,566,000
	2–2 × 8	1,480	980,000	1,725	1,144,000	1,970	1,306,000
	1–4 × 8	1,270	840,000	1,480	980,000	1,690	1,120,000
	3–2 × 8	985	653,000	1,150	762,000	1,315	870,000
	2–3 × 8	890	588,000	1,035	686,000	1,185	784,000
	1–6 × 8	755	483,000	880	564,000	1,005	644,000
	2–2 × 10	910	472,000	1,060	551,000	1,210	629,000
	1–3 × 10	1,090	566,000	1,275	660,000	1,455	754,000
13	2–3 × 6	1,815	1,711,000	2,110	1,997,000	2,415	2,281,000
	1–3 × 8	2,085	1,494,000	2,430	1,743,000	2,780	1,991,000
	2–2 × 8	1,740	1,245,000	2,025	1,453,000	2,315	1,660,000
	1–4 × 8	1,490	1,067,000	1,735	1,245,000	1,985	1,422,000
	3–2 × 8	1,160	830,000	1,350	969,000	1,545	1,106,000
	2–3 × 8	1,045	747,000	1,215	872,000	1,390	996,000
	1–6 × 8	885	614,000	1,040	716,000	1,185	818,000
	2–2 × 10	1,070	600,000	1,245	700,000	1,420	800,000
	1–3 × 10	1,280	719,000	1,495	839,000	1,710	958,000
14	2–2 × 8	2,015	1,555,000	2,350	1,815,000	2,685	2,073,000
	3–2 × 8	1,340	1,037,000	1,570	1,210,000	1,790	1,382,000
	2–3 × 8	1,210	933,000	1,410	1,089,000	1,610	1,244,000
	1–6 × 8	1,025	766,000	1,200	894,000	1,370	1,021,000
	1–3 × 10	1,485	899,000	1,730	1,049,000	1,980	1,198,000
	2–2 × 10	1,235	749,000	1,445	874,000	1,650	998,000
	1–4 × 10	1,060	642,000	1,240	749,000	1,415	856,000
	3–2 × 10	825	499,000	965	582,000	1,100	665,000
	2–3 × 10	740	449,000	865	524,000	990	598,000
15	3–2 × 8	1,540	1,275,000	1,800	1,488,000	2,055	1,699,000
	2–3 × 8	1,390	1,148,000	1,620	1,340,000	1,850	1,530,000
	1–6 × 8	1,180	943,000	1,375	1,100,000	1,570	1,257,000
	1–3 × 10	1,705	1,105,000	1,990	1,289,000	2,270	1,473,000
	2–2 × 10	1,420	921,000	1,660	1,075,000	1,895	1,228,000
	1–4 × 10	1,220	789,000	1,420	921,000	1,625	1,052,000
	3–2 × 10	950	614,000	1,105	717,000	1,265	818,000
	2–3 × 10	850	553,000	995	645,000	1,135	737,000
	1–6 × 10	735	464,000	855	541,000	980	618,000
	4–2 × 10	710	461,000	830	538,000	945	614,000
	2–2 × 10	960	512,000	1,120	597,000	1,280	682,000

TABLE 23-VIII-C—ROOF BEAMS—LIVE LOAD 30 POUNDS PER SQUARE FOOT (1.44 kN/m²)—DEFLECTION LIMITATION L/240 (not for support of plaster)

SPAN OF BEAM (feet)	NOMINAL SIZE OF BEAM	MINIMUM I AND E IN PSI FOR BEAMS SPACED:					
		6'-0" (1829 mm)		7'-0" (2134 mm)		8'-0" (2438 mm)	
		f	E	f	E	f	E
× 304.8 for mm	× 25.4 for mm	× 0.00689 for N/mm²					
10	2–3 × 6	1,430	1,170,000	1,670	1,365,000	1,905	1,560,000
	1–3 × 8	1,645	1,020,000	1,920	1,190,000	2,195	1,360,000
	1–4 × 8	1,175	727,000	1,370	848,000	1,565	969,000
	3–2 × 8	915	570,000	1,070	665,000	1,220	760,000
	2–3 × 8	820	510,000	955	595,000	1,095	680,000
	2–4 × 8	590	364,000	690	425,000	785	485,000
	2–2 × 10	840	409,000	980	477,000	1,120	545,000
11	2–3 × 6	1,725	1,555,000	2,015	1,815,000	2,300	2,073,000
	1–3 × 8	1,990	1,357,000	2,320	1,584,000	2,655	1,809,000
	1–4 × 8	1,420	970,000	1,660	1,132,000	1,895	1,293,000
	3–2 × 8	1,105	754,000	1,290	880,000	1,475	1,005,000
	2–3 × 8	995	679,000	1,160	792,000	1,325	905,000
	2–4 × 8	710	485,000	830	566,000	945	646,000
	2–2 × 10	1,020	544,000	1,190	635,000	1,360	725,000
12	1–4 × 8	1,690	1,260,000	1,970	1,470,000	2,255	1,679,000
	3–2 × 8	1,315	979,000	1,535	1,142,000	1,755	1,305,000
	2–3 × 8	1,185	882,000	1,385	1,029,000	1,580	1,176,000
	2–4 × 8	845	630,000	985	735,000	1,125	840,000
	1–6 × 8	1,005	724,000	1,175	845,000	1,340	965,000
	2–2 × 10	1,210	708,000	1,410	826,000	1,615	944,000
	3–2 × 10	810	472,000	945	551,000	1,080	629,000
	2–3 × 10	725	424,000	845	495,000	965	565,000
13	1–4 × 8	1,985	1,600,000	2,315	1,867,000	2,645	2,133,000
	3–2 × 8	1,545	1,245,000	1,805	1,453,000	2,060	1,659,000
	2–3 × 8	1,390	1,120,000	1,620	1,307,000	1,855	1,493,000
	2–4 × 8	990	801,000	1,155	935,000	1,320	1,068,000
	1–6 × 8	1,180	921,000	1,375	1,075,000	1,575	1,228,000
	2–2 × 10	1,425	900,000	1,665	1,050,000	1,900	1,200,000
	3–2 × 10	950	600,000	1,110	700,000	1,265	800,000
	2–3 × 10	855	540,000	1,000	630,000	1,140	720,000
	1–4 × 10	1,220	923,000	1,425	1,079,000	1,625	1,230,000
14	3–2 × 8	1,790	1,555,000	2,090	1,815,000	2,385	2,073,000
	2–3 × 8	1,610	1,400,000	1,880	1,634,000	2,145	1,866,000
	2–4 × 8	1,150	1,000,000	1,340	1,167,000	1,535	1,333,000
	1–6 × 8	1,370	1,149,000	1,600	1,341,000	1,825	1,532,000
	2–2 × 10	1,650	1,123,000	1,925	1,310,000	2,200	1,497,000
	3–2 × 10	1,100	748,000	1,285	873,000	1,465	997,000
	2–3 × 10	990	673,000	1,155	785,000	1,320	897,000
	1–4 × 10	1,415	963,000	1,650	1,124,000	1,885	1,283,000
	1–6 × 10	915	943,000	1,070	1,100,000	1,220	1,257,000
	2–4 × 10	705	481,000	825	561,000	940	641,000
15	2–4 × 8	1,320	1,230,000	1,540	1,435,000	1,760	1,640,000
	1–6 × 8	1,570	1,414,000	1,830	1,650,000	2,095	1,885,000
	2–2 × 10	1,895	1,381,000	2,210	1,612,000	2,525	1,841,000
	3–2 × 10	1,260	921,000	1,470	1,075,000	1,680	1,228,000
	2–3 × 10	1,135	829,000	1,325	967,000	1,515	1,105,000
	1–4 × 10	1,620	1,183,000	1,890	1,380,000	2,160	1,577,000
	1–6 × 10	980	696,000	1,145	812,000	1,305	928,000
	2–4 × 10	810	592,000	945	691,000	1,080	789,000
	4–2 × 10	945	691,000	1,105	806,000	1,260	921,000
	1–8 × 10	720	510,000	840	595,000	960	680,000
	2–2 × 12	1,280	768,000	1,495	896,000	1,705	1,024,000
	1–4 × 12	1,095	658,000	1,280	768,000	1,460	877,000

TABLE 23-VIII-D

1997 UNIFORM BUILDING CODE

TABLE 23-VIII-D—ROOF AND FLOOR BEAMS—LIVE LOAD 40 POUNDS PER SQUARE FOOT (1.92 kN/m²)—DEFLECTION LIMITATION L/360 (not for support of plaster)

SPAN OF BEAM (feet)	NOMINAL SIZE OF BEAM	MINIMUM I AND E IN PSI FOR BEAMS SPACED:					
		6'-0" (1829 mm)		7'-0" (2134 mm)		8'-0" (2438 mm)	
		f	E	f	E	f	E
× 304.8 for mm	× 25.4 for mm	× 0.00689 for N/mm²					
10	1–3 × 8	2,055	1,700,000	2,400	1,984,000	2,740	2,266,000
	2–2 × 8	1,710	1,417,000	1,995	1,654,000	2,280	1,889,000
	1–4 × 8	1,470	1,211,000	1,715	1,413,000	1,960	1,614,000
	1–6 × 8	875	697,000	1,020	813,000	1,165	929,000
	2–2 × 10	1,050	681,000	1,225	795,000	1,400	908,000
	1–3 × 10	1,260	819,000	1,470	956,000	1,680	1,092,000
	1–4 × 10	900	585,000	1,050	683,000	1,200	780,000
11	2–2 × 8	2,070	1,886,000	2,415	2,201,000	2,760	2,514,000
	1–4 × 8	1,775	1,616,000	2,070	1,886,000	2,365	2,154,000
	1–6 × 8	1,055	929,000	1,230	1,084,000	1,405	1,238,000
	2–2 × 10	1,275	906,000	1,490	1,057,000	1,700	1,208,000
	1–3 × 10	1,575	1,090,000	1,780	1,272,000	2,030	1,453,000
	1–4 × 10	1,090	779,000	1,270	909,000	1,455	1,038,000
	3–2 × 10	850	605,000	990	706,000	1,135	806,000
12	1–6 × 8	1,255	1,206,000	1,465	1,407,000	1,670	1,607,000
	3–2 × 8	1,645	1,631,000	1,920	1,903,000	2,190	2,174,000
	2–2 × 10	1,510	1,180,000	1,760	1,377,000	2,010	1,573,000
	1–3 × 10	1,820	1,415,000	2,125	1,651,000	2,425	1,886,000
	1–4 × 10	1,300	1,010,000	1,515	1,179,000	1,735	1,346,000
	3–2 × 10	1,010	786,000	1,180	917,000	1,345	1,048,000
	2–3 × 10	905	706,000	1,055	824,000	1,205	941,000
	1–6 × 10	785	594,000	915	693,000	1,045	792,000
	2–4 × 10	650	505,000	760	589,000	865	673,000
13	1–6 × 8	1,475	1,535,000	1,720	1,791,000	1,965	2,046,000
	2–3 × 8	1,735	1,866,000	2,025	2,178,000	2,315	2,487,000
	2–4 × 8	1,235	1,335,000	1,440	1,558,000	1,645	1,779,000
	3–2 × 10	1,185	1,000,000	1,380	1,167,000	1,580	1,333,000
	2–2 × 10	1,780	1,500,000	2,075	1,750,000	2,370	2,000,000
	1–3 × 10	2,130	1,799,000	2,485	2,099,000	2,840	2,398,000
	2–3 × 10	1,070	900,000	1,250	1,050,000	1,425	1,200,000
	1–4 × 10	1,525	1,537,000	1,780	1,794,000	2,035	2,049,000
	2–4 × 10	760	642,000	890	749,000	1,015	856,000
14	2–4 × 8	1,435	1,666,000	1,675	1,944,000	1,915	2,221,000
	3–2 × 10	1,375	1,246,000	1,605	1,454,000	1,830	1,661,000
	2–3 × 10	1,235	1,121,000	1,440	1,308,000	1,645	1,494,000
	1–4 × 10	1,770	1,605,000	2,065	1,873,000	2,360	2,139,000
	2–4 × 10	880	801,000	1,025	935,000	1,175	1,068,000
	3–3 × 10	825	749,000	960	874,000	1,100	998,000
	1–6 × 10	1,145	1,571,000	1,335	1,833,000	1,525	2,094,000
	1–8 × 10	780	691,000	910	806,000	1,040	921,000
	4–2 × 10	1,030	936,000	1,200	1,092,000	1,375	1,248,000
	2–2 × 12	1,395	1,040,000	1,630	1,214,000	1,860	1,386,000
15	3–2 × 10	1,575	1,535,000	1,840	1,791,000	2,100	2,046,000
	2–3 × 10	1,420	1,381,000	1,655	1,612,000	1,890	1,841,000
	2–4 × 10	1,010	986,000	1,175	1,151,000	1,345	1,314,000
	3–3 × 10	945	921,000	1,100	1,075,000	1,260	1,228,000
	1–6 × 10	1,225	1,160,000	1,430	1,354,000	1,635	1,546,000
	1–8 × 10	900	850,000	1,050	992,000	1,200	1,133,000
	4–2 × 10	1,180	1,151,000	1,375	1,343,000	1,575	1,534,000
	2–2 × 12	1,600	1,280,000	1,865	1,494,000	2,130	1,706,000
	3–2 × 12	1,065	854,000	1,240	997,000	1,420	1,138,000
	1–3 × 12	1,920	1,536,000	2,240	1,792,000	2,560	2,047,000
	4–2 × 12	800	640,000	935	747,000	1,065	853,000
	2–3 × 12	960	767,000	1,120	895,000	1,280	1,022,000

Chapter 24 is printed in its entirety in Volume 1 of the *Uniform Building Code*. Excerpts from Chapter 24 are reprinted herein.

Excerpts from Chapter 24
GLASS AND GLAZING

SECTION 2409 — SLOPED GLAZING AND SKYLIGHTS

2409.5 Design Loads. Sloped glazing and skylights shall be designed to withstand the tributary loads specified in Section 1605. Sizing limitations specified within Graph 24-1 and Table 24-A may be utilized for glazing materials set forth in Section 2409.2, provided the design loads are increased by a factor of 2.67.

TABLE 24-A—ADJUSTMENT FACTORS—RELATIVE RESISTANCE TO WIND LOADS

GLASS TYPE	ADJUSTMENT FACTOR[1]
Laminated[2]	0.75
Fully tempered	4.00
Heat strengthened	2.00
Wired	0.50
Insulating glass[3]—2 panes —3 panes	1.70 2.55
Patterned[4]	1.00
Regular (annealed)	1.00
Sandblasted	0.40[5]

[1]Loads determined from Chapter 16, Division III, shall be divided by this adjustment factor for use with Graph 24-1.
[2]Applies when two plies are identical in thickness and type; use total glass thickness, not thickness of one ply.
[3]Applies when each glass panel is the same thickness and type; use thickness of one panel.
[4]Use minimum glass thickness, i.e., measured at the thinnest part of the pattern; if necessary, interpolation of curves in Graph 24-1 may be required.
[5]Factor varies depending on depth and severity of sandblasting; value shown is minimum.

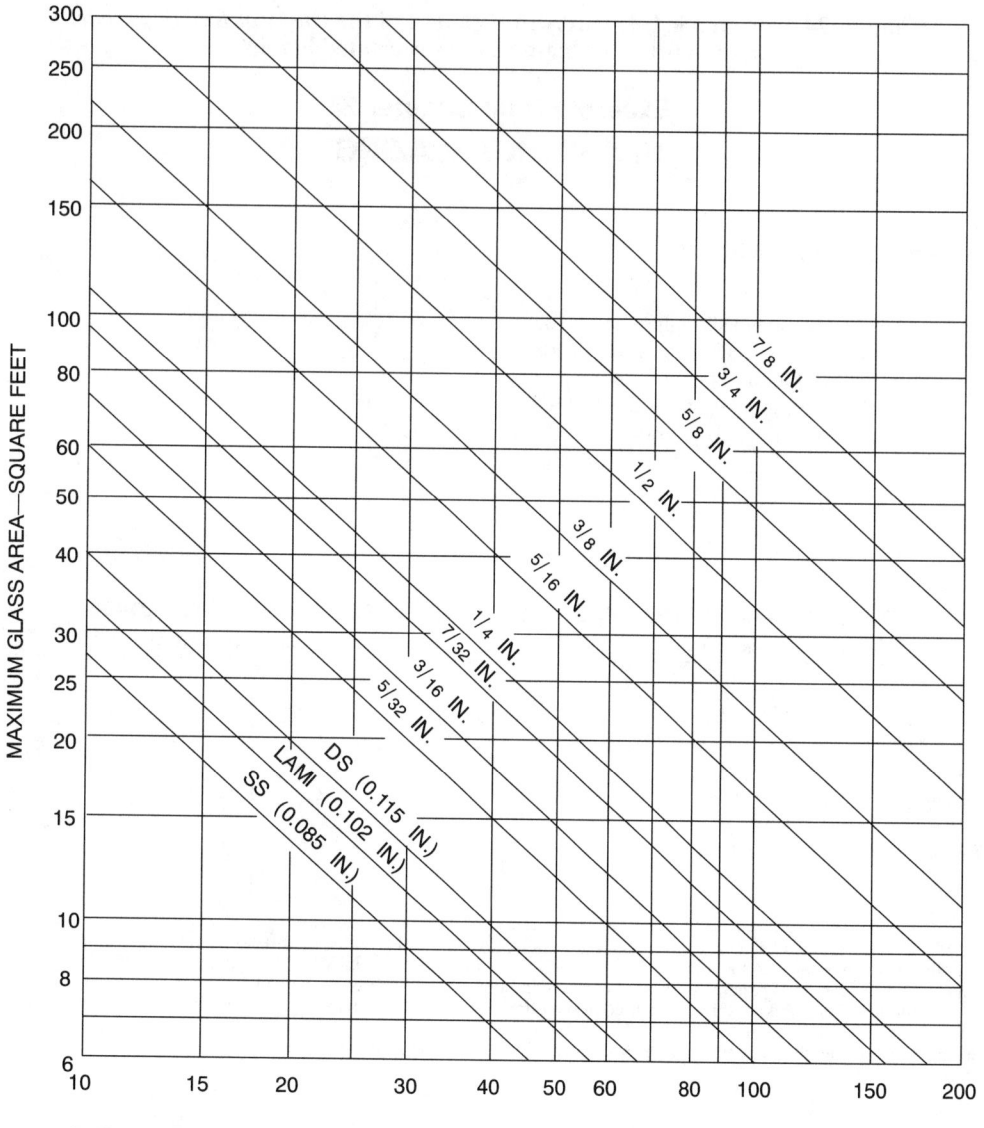

DESIGN WIND PRESSURE FROM CHAPTER 16, DIVISION III—POUNDS PER SQUARE FOOT

For **SI:** 1 inch = 25.4 mm, 1 square foot = 0.0929 m², 1 pound per square foot = 0.479 kN/m².

GRAPH 24-1—MAXIMUM ALLOWABLE AREA OF GLASS[1]

[1]Applicable for ratios of width to length of 1:1 to 5:1. Design safety factor = 2.5.

Chapter 25 is printed in its entirety in Volume 1 of the *Uniform Building Code*.
Excerpts from Chapter 25 are reprinted herein.

Excerpts from Chapter 25
GYPSUM BOARD AND PLASTER

SECTION 2513 — SHEAR-RESISTING CONSTRUCTION WITH WOOD FRAME

2513.1 General. Cement plaster, gypsum lath and plaster, gypsum veneer base, gypsum sheathing board, and gypsum wallboard may be used on wood studs for vertical diaphragms if applied in accordance with this section. Shear-resisting values shall not exceed those set forth in Table 25-I. The effects of overturning on vertical diaphragms shall be investigated in accordance with Section 1605.2.2.

The shear values tabulated shall not be cumulative with the shear value of other materials applied to the same wall. The shear values may be additive when the identical materials applied as specified in this section are applied to both sides of the wall.

2513.2 Masonry and Concrete Construction. Cement plaster, gypsum lath and plaster, gypsum veneer base, gypsum sheathing board, and gypsum wallboard shall not be used in vertical diaphragms to resist forces imposed by masonry or concrete construction.

2513.3 Wall Framing. Framing for vertical diaphragms shall comply with Section 2320.11 for bearing walls, and studs shall not be spaced farther apart than 16 inches (406 mm) center to center. Sills, plates and marginal studs shall be adequately connected to framing elements located above and below to resist all design forces.

2513.4 Height-to-length Ratio. The maximum allowable height-to-length ratio for the construction in this section shall be 2 to 1. Wall sections having height-to-length ratios in excess of $1^1/_2$ to 1 shall be blocked.

2513.5 Application. End joints of adjacent courses of gypsum lath, gypsum veneer base, gypsum sheathing board or gypsum wallboard sheets shall not occur over the same stud.

Where required in Table 25-I, blocking having the same cross-sectional dimensions as the studs shall be provided at all joints that are perpendicular to the studs.

The size and spacing of nails shall be as set forth in Table 25-I. Nails shall not be spaced less than $3/_8$ inch (9.5 mm) from edges and ends of gypsum lath, gypsum veneer base, gypsum sheathing board and gypsum wallboard, or from sides of studs, blocking, and top and bottom plates.

2513.5.1 Gypsum lath. Gypsum lath shall be applied perpendicular to the studs. Maximum allowable shear values shall be as set forth in Table 25-I.

2513.5.2 Gypsum sheathing board. Four-foot-wide (1219 mm) pieces may be applied parallel or perpendicular to studs. Two-foot-wide (610 mm) pieces shall be applied perpendicular to the studs. Maximum allowable shear values shall be as set forth in Table 25-I.

2513.5.3 Gypsum wallboard or veneer base. Gypsum wallboard or veneer base may be applied parallel or perpendicular to studs. Maximum allowable shear values shall be as set forth in Table 25-I.

TABLE 25-I—ALLOWABLE SHEAR FOR WIND OR SEISMIC FORCES IN POUNDS PER FOOT FOR VERTICAL DIAPHRAGMS OF LATH AND PLASTER OR GYPSUM BOARD FRAME WALL ASSEMBLIES[1]

TYPE OF MATERIAL	THICKNESS OF MATERIAL × 25.4 for mm × 304.8 for mm	WALL CONSTRUCTION	NAIL SPACING[2] MAXIMUM (inches) × 25.4 for mm	SHEAR VALUE × 14.6 for N/m	MINIMUM NAIL SIZE[3] × 25.4 for mm
1. Expanded metal, or woven wire lath and portland cement plaster	$7/8''$	Unblocked	6	180	No. 11 gage, $1^1/2''$ long, $7/16''$ head No. 16 gage staple, $7/8''$ legs
2. Gypsum lath	$3/8''$ lath and $1/2''$ plaster	Unblocked	5	100	No. 13 gage, $1^1/8''$ long, $19/64''$ head, plasterboard blued nail
3. Gypsum sheathing board	$1/2'' \times 2' \times 8'$	Unblocked	4	75	No. 11 gage, $1^3/4''$ long, $7/16''$ head, diamond-point, galvanized
	$1/2'' \times 4'$	Blocked	4	175	
	$1/2'' \times 4'$	Unblocked	7	100	
4. Gypsum wallboard or veneer base	$1/2''$	Unblocked	7	100	5d cooler (0.086″ dia., $1^5/8''$ long, $15/64''$ head) or wallboard (0.086″ dia., $1^5/8''$ long, $9/32''$ head)
			4	125	
		Blocked	7	125	
			4	150	
	$5/8''$	Unblocked	7	115	6d cooler (0.092″ dia., $1^7/8''$ long, $1/4''$ head) or wallboard (0.0915″ dia., $1^7/8''$ long, $19/64''$ head)
			4	145	
		Blocked	7	145	
			4	175	
		Blocked Two ply	Base ply: 9 Face ply: 7	250	Base ply—6d cooler (0.092″ dia., $1^7/8''$ long, $1/4''$ head) or wallboard (0.0915″ dia., $1^7/8''$ long, $19/64''$ head) Face ply—8d cooler (0.113″ dia., $2^3/8''$ long, $9/32''$ head) or wallboard (0.113″ dia., $2^3/8''$ long, $3/8''$ head)

[1]These vertical diaphragms shall not be used to resist loads imposed by masonry or concrete construction. See Section 2513.2. Values shown are for short-term loading due to wind or due to seismic loading. Values shown must be reduced 25 percent for normal loading. The values shown in Items 2, 3 and 4 shall be reduced 50 percent for loading due to earthquake in Seismic Zones 3 and 4.
[2]Applies to nailing at all studs, top and bottom plates, and blocking.
[3]Alternate nails may be used if their dimensions are not less than the specified dimensions.

Chapters 26-34

Chapters 26 through 34 are printed in Volume 1 of the *Uniform Building Code*.

**Chapter 35 is printed in its entirety in Volume 1 of the *Uniform Building Code*.
Excerpts from Chapter 35 are reprinted herein.**

Excerpts from Chapter 35
UNIFORM BUILDING CODE STANDARDS

Part I—General

SECTION 3501 — UBC STANDARDS

The Uniform Building Code standards referred to in various parts of this code, which are also listed in Part II of this chapter, are hereby declared to be part of this code and are referred to in this code as a "UBC standard."

SECTION 3502 — ADOPTED STANDARDS

The standards referred to in various parts of the code, which are listed in Part III of this chapter, are hereby declared to be part of this code.

SECTION 3503 — STANDARD OF DUTY

The standard of duty established for the recognized standards listed in Part IV of this chapter is that the design, construction and quality of materials of buildings and structures be reasonably safe for life, limb, health, property and public welfare.

SECTION 3504 — RECOGNIZED STANDARDS

The standards listed in Part IV of this chapter are recognized standards. Compliance with these recognized standards shall be prima facie evidence of compliance with the standard of duty set forth in Section 3503.

Part II—UBC Standards

UBC STD. AND SEC.	TITLE AND SOURCE

CHAPTER 18

18-1; 1801.2, 1803.1
　　Soils Classification. Standard Method D 2487-69 of the ASTM.

18-2; 1801.2, 1803.2
　　Expansion Index Test. Recommendation of the Los Angeles Section of the ASCE Soil Committee.

CHAPTER 19

19-1; 1903.5.2, 1912.14.3
　　Welding Reinforcing Steel, Metal Inserts and Connections in Reinforced Concrete Construction. Structural Welding Code—Reinforcing Steel ANSI/AWS D1.4-92 of the American Welding Society, Inc.

19-2; 1903.9, 1925.1, 1925.3
　　Mill-Mixed Gypsum Concrete and Poured Gypsum Roof Diaphragms. Standard Specification C 317-70 of the ASTM. Poured Gypsum Roof Diaphragm, based on reports of test programs by S. B. Barnes and Associates, dated February 1955, November 1956, January 1958 and February 1962.

CHAPTER 21

21-1; 2102.2, Item 4
　　Building Brick, Facing Brick and Hollow Brick. (Made from Clay or Shale.) Standard Specifications C 62-92c, C 216-94a and C 652-94a of the ASTM.

21-2; 2102.2, Item 6
　　Calcium Silicate Face Brick (Sand-lime Brick). Standard Specification C 73-95 of the ASTM.

21-3; 2102.2, Item 5
　　Concrete Building Brick. Standard Specification C 55-95 of the ASTM.

21-4; 2102.2, Item 5
　　Hollow and Solid Load-bearing Concrete Masonry Units. Standard Specification C 90-95 of the ASTM.

21-5; 2102.2, Item 5
　　Nonload-bearing Concrete Masonry Units. Standard Specification C 129-95 of the ASTM.

21-6; See *Uniform Code for Building Conservation.*
　　In-Place Masonry Shear Tests. Test Standard of the International Conference of Building Officials.

21-7; See *Uniform Code for Building Conservation.*
　　Tests of Anchors in Unreinforced Masonry Walls. Test Standard of the International Conference of Building Officials.

21-8; See *Uniform Code for Building Conservation.*
　　Pointing of Unreinforced Masonry Walls. Construction Specification of the International Conference of Building Officials.

21-9; 2102.2, Item 6
　　Unburned Clay Masonry Units and Standard Methods of Sampling and Testing Unburned Clay Masonry Units. Test Standard of the International Conference of Building Officials.

21-10; 2102.2, 2104.8
　　Part I—Joint Reinforcement for Masonry. Specification Standard of the International Conference of Building Officials. Part II—Cold-drawn Steel Wire for Concrete Reinforcement. Standard Specification A 82-90a of the ASTM.

21-11; 2102.2, Item 2; Table 21-A
　　Cement, Masonry. Standard Specification C 91-93a of the ASTM.

21-12; 2102.2, Item 3
　　Quicklime for Structural Purposes. Standard Specification C 5-79 (Reapproved 1992) of the ASTM.

21-13; 2102.2, Item 3
　　Hydrated Lime for Masonry Purposes. Standard Specification C 207-91 (Reapproved 1992) of the ASTM.

21-14; 2102.2, Item 2; Table 21-A
　　Mortar Cement. Test Standard of the International Conference of Building Officials.

21-15; 2102.2, Item 8
　　Mortar for Unit Masonry and Reinforced Masonry Other Than Gypsum. Standard Specification C 270-95T of the ASTM.

21-16; 2102.2, Item 8
　　Field Tests Specimens for Mortar. Test Standard of the International Conference of Building Officials.

21-17; 2102.2, Item 6; 2105.3.2, 2105.3.3
　　Test Method for Compressive Strength of Masonry Prisms. Standard Test Method E 447-80 of the ASTM.

21-18; 2102.2, Item 9; Table 21-B
　　Method of Sampling and Testing Grout. Standard Method C 1019-89a (93) of the ASTM.

21-19; 2102.2, Item 9
　　Grout for Masonry. Standard Specification C 476-91 of the ASTM.

21-20; 2102.2, Item 8

Standard Test Method for Flexural Bond Strength of Mortar Cement. Test Standard of the International Conference of Building Officials.

CHAPTER 22

22-1; 1808.6.1, 1808.7.1, 2202.2

Material Specifications for Structural Steel. Standard Specifications A 27, A 36, A 48, A 53, A 148, A 242, A 252, A 283, A 307, A 325, A 336, A 441, A 446, A 449, A 490, A 500, A 501, A 514, A 529, A 563, A 569, A 570, A 572, A 588, A 606, A 607, A 611, A 618, A 666, A 668, A 690, A 715 and A 852 of the ASTM.

CHAPTER 23

23-1; 2302.1, 2303

Classification, Definition and Methods of Grading for All Species of Lumber. Standard Methods D 245-88 and D 2555-88 of the ASTM, Handbook No. 72 of the United States Department of Agriculture, American Softwood Lumber Standard PS20-70 and National Grading Rule for Dimension Lumber of the National Grading Rule Committee.

23-2; 2302.1, 2303, 2304.2

Construction and Industrial Plywood. Product Standard PS 1-83 of the United States Department of Commerce, and National Bureau of Standards Calculation Diaphragm Action, an Engineering Standard of the International Conference of Building Officials.

23-3; 2302.1, 2303, 2304.2

Wood-Based Structural-Use Panels. Product Standard PS 2-92 of the United States Department of Commerce and the American Plywood Association.

23-4; 201.2, 207, 2303

Fire-retardant-treated Wood Tests on Durability and Hygroscopic Properties. Standard Test Methods D 2898-81 and D 3201-79 of the ASTM and Standards C 20-83 and C 27-83 of the American Wood Preservers Association.

23-5; 2303,

Fire-retardant-treated Wood. Design Values for Fire-retardant-treated Lumber.

Part III—Standards Adopted by Reference

TITLE AND SOURCE	SECTION REFERENCE
CHAPTER 22	
Load and Resistance Factor Design Specifications for Structural Steel Buildings American Institute of Steel Construction, December 1, 1993	Chapter 22, Div. II
Specification for Structural Steel Buildings Allowable Stress Design and Plastic Design, American Institute of Steel Construction, June 1, 1989	Chapter 22, Div. III
Load and Resistance Factor Design Specification for Cold-formed Steel Structural Members American Iron and Steel Institute, March 16, 1991	Chapter 22, Div. VI
Specification for Design of Cold-formed Steel Structural Members American Iron and Steel Institute, 1986 (with December 1989 Addendum)	Chapter 22, Div. VII
Standard Specification for Steel Joists, K-Series, LH-Series, DLH-Series and Joist Girders Steel Joist Institute, 1994	Chapter 22, Div. IX

Structural Applications of Steel Cables for Buildings American Society of Civil Engineers (ASCE 17-95) — Chapter 22, Div. XI

CHAPTER 23

National Design Specification for Wood Construction — Chapter 23, Div. III, Part I American Forest and Paper Association, Revised 1991 Edition

Part IV—Recognized Standards

TITLE AND SOURCE	SECTION REFERENCE
CHAPTER 16	
Guide Specifications for the Design Loads of Metal Flagpoles. ANSI/NAAMM FP1001, 1990.	1604
Minimum Design Loads for Buildings and Other Structures. ASCE 7-95—Chapter 6.	1604
Structural Standards for Steel Antenna Towers and Antenna Supporting Structures. ANSI EIA/TIA 222-2E, 1991.	1604
CHAPTER 19	
Concrete Aggregates. C 33-93 of the ASTM.	1903.3
Lightweight Aggregates for Structural and Insulating Concrete. C 330-89 and C 332-83 of the ASTM.	703.4, 1903.3
Reinforcing Bars for Concrete. A 615-94, A 616-93, A 617-93, A 706-92b, A 767-M-90 and A 775-M-94d of the ASTM.	1903.5
Smooth Steel Wire for Spiral Reinforcement A 82-94 of the ASTM.	1903.5
Fabricated Deformed Steel Bar Mats. A 184-90 of the ASTM.	1903.5
Welded Steel Wire Fabric and Deformed Steel Wire. A 185-94, A 496-94 and A 497-94a of the ASTM.	1903.5
Steel Wire, Strand and Bar for Prestressing. A 416-94, A 421-91 and A 722-90 of the ASTM.	1903.5
Air-entraining Admixtures for Concrete. C 260-94 of the ASTM.	1903.6
Chemical Admixtures for Concrete. C 494-92 and C 1017-92 of the ASTM.	1903.6
Fly Ash and Raw or Calcined Natural Pozzolans for Use as Admixtures in Portland Cement Concrete. C 618-94a of the ASTM.	1903.6
Concrete Tests. C 31-85, C 39-86, C 42-84a, C 172-82 and C 192-81 of the ASTM.	1903.8
Splitting Tensile Strength. C 496-85 of the ASTM.	1903.8
Specification for Expansive Hydraulic Cement. C 845-90 of the ASTM.	1903.6
Structural Steel. A 36-94 of the ASTM.	1903.5.6.1
Pipe, Steel, Black and Hot-Dipped, Zinc-Coated Welded and Seamless A 53-93a of the ASTM.	1903.5.6.2

Appendix

**Appendix Chapters 3 through 15 are printed in Volume 1
of the *Uniform Building Code***

Appendix Chapter 16
STRUCTURAL FORCES

Division I—SNOW LOAD DESIGN

SECTION 1637 — GENERAL

1637.1 Scope. Buildings, structures and portions thereof shall be designed and constructed to sustain all loads required by Chapter 16, combined in accordance with Section 1612.2 for load and resistance factor design or Section 1612.3 for allowable stress design.

1637.2 Definitions. For the purpose of this appendix, certain terms are defined as follows:

HEATED GREENHOUSE is a production greenhouse provided with heating facilities capable of maintaining an interior temperature of no less than $50°F$ ($10°C$) at any point 3 feet (914 mm) above the floor, and provided with roof-ceiling assembly with a thermal resistance (R) less than 2.0.

PRODUCTION GREENHOUSE is a greenhouse used exclusively for the growing of flowers or plants on a production basis or for research, with no public access.

SECTION 1638 — NOTATIONS

- a = roof slope expressed in degrees.
- B = width of projection measured parallel to ridge, feet (m). Minimum assumed width shall be 1 foot (305 mm).
- C_e = snow exposure factor (see Table A-16-A).
- C_s = slope-reduction factor.
- C_v = valley design coefficient (see Figure A-16-11).
- D = density of snow, pounds per cubic foot (pcf) (N/m^3) (refer to Formula 44-2).
- F_s = ice splitter horizontal load, pounds (N).
- F_v = ice splitter snow weight, pounds (N).
- h_b = height of balanced snow load on lower roof or deck, feet (m).
- h_d = maximum height of drift surcharge, feet (m) (refer to Formula 44-1).
- h_g = depth of ground snow (as determined by the building official), feet (m).
- h_r = difference in height between the upper and lower roof or deck, feet (m).
- I = importance factor (see Table A-16-B).
- L = horizontal distance between projection and ridge, feet (m).
- P_f = minimum roof snow load, pounds per square foot (psf) (N/m^2).
- P_g = basic ground snow load, psf (N/m^2).
- P_m = maximum intensity of the load at the height change, psf (N/m^2).
- S = horizontal separation between adjacent structures, feet (m) (see Figure A-16-5).
- W_b = horizontal dimension in feet of upper roof normal to the line of change in roof level, but not less than 50 feet (15.2 m), or greater than 500 feet (152.4 m).
- W_d = width of drift, base of triangular drift load, feet (m).
- X = vertical component of roof slope (rise), feet (m).
- Y = horizontal component of roof slope (run), feet (m).

SECTION 1639 — GROUND SNOW LOADS

The ground snow load, P_g, to be used in the determination of design snow loads for buildings and other structures shall be as shown in Figures A-16-1, A-16-2 and A-16-3. For the hatched areas in Figure A-16-1, the basic ground snow loads shall be determined by the building official. The ground snow load, P_g, may be adjusted by the building official when a registered engineer or architect submits data substantiating the adjustments.

SECTION 1640 — ROOF SNOW LOADS

The value of roof (or other member) snow load, P_f, shall be determined by the following formula:

$$P_f = C_e I P_g \qquad (40\text{-}1\text{-}1)$$

where C_e is given in Table A-16-A and I is given in Table A-16-B.

> **EXCEPTION:** The value of roof snow load, P_f, for heated greenhouses shall be determined by the following formula:
>
> $$P_f = 0.83 \, C_e \, IP_g \qquad (40\text{-}1\text{-}2)$$

The roof snow load shall be assumed to act vertically upon the area projected upon a horizontal plane. Portions of curved roofs or inclined walls having a slope exceeding 70 degrees shall be considered free from snow load. The point at which the slope exceeds 70 degrees shall be considered the eave for such roofs. For curved roofs, the roof slope factor, C_s, shall be determined from the appropriate slope reduction formula by basing the slope on the vertical angle from the eave to the crown.

Where roof snow loads are in excess of 20 psf (958 N/m^2) and a is greater than 30 degrees, Formula (40-1-1) or (40-1-2) may be multiplied by C_s given by the formula:

$$C_s = 1 - \frac{a - 30}{40} \qquad (40\text{-}2\text{-}1)$$

for unobstructed slippery surfaces, and

$$C_s = 1 - \frac{a - 45}{25} \qquad (40\text{-}2\text{-}2)$$

for all other surfaces where a is greater than 45 degrees.

> **EXCEPTION:** For heated greenhouses C_s is given in the formulas:
>
> $$C_s = 1 - \frac{(a - 15)}{55} \text{ for unobstructed slippery surfaces} \qquad (40\text{-}2\text{-}3)$$

$$C_s = 1 - \frac{(a - 30)}{40} \text{ for other surfaces} \qquad (40\text{-}2\text{-}4)$$

Where the ground snow load P_g is greater than 100 psf (4788 N/m^2) and a is greater than 20 degrees,

$$C_s = 1 - \left(\frac{P_f - 20}{P_f}\right)\left(\frac{a - 20}{40}\right) \qquad (40\text{-}2\text{-}5)$$

For **SI:** $\qquad C_s = 1 - \left(\frac{P_f - 958}{P_f}\right)\left(\frac{a - 20}{40}\right)$

The following conditions must be met before using Formulas (40-2-1), (40-2-2), (40-2-3), (40-2-4) and (40-2-5):

1. The height of all eaves exceeds h_g, and

2. There are no obstructions adjacent to the structure for a distance h_g measured from the eave normal to the ridge line.

Where the eave height is less than h_g but greater than $h_g/2$, and condition 2 above is met, the roof snow load reduction represented by the last term in Formulas (40-2-1), (40-2-2) and (40-2-3) shall be divided by 2.

3. If P_g is 20 psf (958 N/m^2) or less, design roof snow load must not be less than P_g. If P_g exceeds 20 psf (958 N/m^2), design roof snow load must not be less than 20 psf (958 N/m^2).

4. Reduced roof loads where P_g exceeds 70 psf (3352 N/m^2) shall not be less than those obtained through use of Formula (40-1-1) for P_g equal to 70 psf (3352 N/m^2).

SECTION 1641 — UNBALANCED SNOW LOADS, GABLE ROOFS

1641.1 General. In addition to the balanced load condition, unbalanced loading shall be considered for gable roofs in accordance with this section.

1641.2 Single-gable Roofs. Single-gable roofs with slopes greater than $1/2$ unit vertical in 12 units horizontal (4.2% slope) and less than 3 units vertical in 12 units horizontal (25% slope) shall be designed to sustain a uniformly distributed load of 0.5 P_f acting on one slope and 1.0 P_f on the opposite slope. Roofs with slopes greater than 3 units vertical in 12 units horizontal (25% slope) shall be designed to sustain a uniformly distributed load equal to 1.25 P_f applied to one slope only.

1641.3 Multiple-gable Roofs.

1641.3.1 With parallel ridge lines. For multiple-gable roofs with parallel ridge lines having slopes exceeding 2 units vertical in 12 units horizontal (16.7% slope), the roof snow load shall be increased from one half the applicable uniform roof load at the ridge ($0.5P_f$) to three times the uniform load at the valley (3.0 P_f).

1641.3.2 With nonparallel ridge lines. Structural members at roof valleys for multiple-gable roofs having intersecting ridge lines in areas where P_g is greater than 70 psf (3352 N/m^2) and where the slope is 3 units vertical in 12 units horizontal (16.7% slope) or greater shall be designed for P_f times C_v and the distribution of loads is as shown in Figures A-16-12 and A-16-13 where C_v shall be determined from Figure A-16-11.

SECTION 1642 — UNBALANCED SNOW LOAD FOR CURVED ROOFS

Portions of curved roofs having a slope exceeding 70 degrees shall be considered free of snow load. The equivalent slope of a curved roof for calculating C_s is equal to the slope of a line from the point at which the slope exceeds 70 degrees to the crown. If the equivalent slope is less than 10 degrees or greater than 60 degrees, unbalanced snow loads need not be considered.

In all cases the windward side shall be considered free of snow. The unbalanced load varies linearly from $0.5C_s P_f$ to $2C_s P_f/C_e$ where the slope is 30 degrees and decreasing to

$$\frac{2C_s P_f}{C_e}\left(1 - \frac{a - 30}{40}\right) \text{ at the eave.}$$

If the ground or another roof abuts an arched roof structure with a slope exceeding 30 degrees at or within 3 feet (914 mm) of its eave, the snow load shall not be decreased between the 30-degree point and the eave, but shall remain constant at $2C_s P_f/C_e$.

SECTION 1643 — SPECIAL EAVE REQUIREMENTS

Eave overhanging roof structures shall be designed to sustain a uniformly distributed load of 2.0 P_f, or as determined by the building official, to account for ice dams and snow accumulation as shown in Figure A-16-10. Heat strips or other exposed heat methods may not be used in lieu of this design criterion. For shingle or shake roofs, hot or cold underlayment roofing is required on all roofs from the building edge for a distance of 5 feet (1524 mm) or to the ridge, whichever is less. All building exits under down-slope eaves shall be protected from sliding snow and ice.

SECTION 1644 — DRIFT LOADS ON LOWER ROOFS, DECKS AND ROOF PROJECTIONS

1644.1 General. Multilevel roofs, lower roofs and decks of adjacent structures and roofs adjacent to projections shall be designed in accordance with this section.

1644.2 Drift Loads for Lower Roofs. The geometry of the surcharge load produced by snow drifting shall be taken as the triangular load distributions shown in Figures A-16-4, A-16-5, A-16-6, A-16-7 and A-16-8. The height of the drift shall be calculated according to the formula:

$$h_d = 0.43 \sqrt[3]{W_b} \sqrt[4]{P_g + 10} - 1.5 \qquad (44\text{-}1)$$

For **SI:** 1 foot = 304.8 mm.

The value of h_d can be taken from Figure A-16-9 for $W_b \leq 600$ feet (182 880 mm).

WHERE:

$$D = 0.13P_g + 14.0 \leq 35 \text{ pcf} \qquad (44\text{-}2)$$

For SI: $D = 0.43 P_g + 2198 \leq 5495$.

The width of the drift, W_d, in feet (mm), shall be taken as the smaller of $4h_d$ or $4 (h_r - h_b)$. If W_d exceeds the width of the lower roof, the drift shall be truncated at the far edge of the roof, not reduced to zero at this location. Drift loads need be considered only when:

$$(h_r - h_b)/h_b > 0.2 \qquad (44\text{-}3)$$

WHERE:

$$h_b = P_f /D$$

and where P_f is evaluated on the basis of the lower roof.

The maximum intensity, P_m, of the snow load at the high point of the drift in pounds per square foot (N/m^2) is given by:

$$P_m = D (h_d + h_b) \leq D\, h_r \qquad (44\text{-}4)$$

1644.3 Roof of an Adjacent Lower Structure. Drifts may occur on lower roofs of structures sited within 20 feet (6096 mm) of a higher structure as depicted in Figure A-16-5. The height of the surcharge on the lower structure shall be taken as h_d multiplied by $(20 - S)/20$ [For **SI:** $(6.1 - S)/6.1$]to account for the horizontal separation between structures, S, in feet (mm).

1644.4 Sliding Snow. Lower roofs that are located below roofs having a slope greater than 2 units vertical in 12 units horizontal (16.7% slope) shall be designed for an increase in drift height of $0.4h_d$, except that the total drift surcharge ($1.4\,h_d$) shall not exceed the height of the roof above the uniform snow depth ($h_r - h_b$). Sliding snow need not be considered if the lower roof is separated a distance, S, greater than h_r or 20 feet (6096 mm) as shown in Figures A-16-5 and A-16-6.

1644.5 Roof Projections. Mechanical equipment, penthouses, parapets and other projections above the roof can produce drifting as depicted in Figure A-16-7. Such drift loads shall be calculated on all sides of projections having horizontal dimensions exceeding 15 feet (4572 mm). Drifts created at the perimeter of the roof by projections shall be computed using half the drift height from Formula (44-1) (i.e., $0.5\,h_d$) with W_b taken equal to the length of the roof associated with the projections. The value of W_b shall be taken as the maximum distance from the projection to the edges of the roof, or 50 feet (15 240 mm), whichever is less.

1644.6 Intersecting Drifts. When one snow drift intersects another at an angle as depicted in Figure A-16-8, the maximum unit pressure of the drift shall be taken as the greater of the two individual drifts, but not the sum of the two. The total load on the area of intersection is increased, however, simply because of the assumed geometry of the intersecting drifts.

SECTION 1645 — RAIN ON SNOW

In geographic areas where intense rains may add to the roof snow load, the building official may require the use of an additional rain on snow surcharge of 5 psf (239 N/m²). This surcharge may be disregarded where roof slopes exceed $^1/_2$ unit vertical in 12 units horizontal (4.2% slope) or where the basic ground snow load, P_g, exceeds 50 psf (2394 N/m²). See Section 1611.7 for ponding.

SECTION 1646 — DEFLECTIONS

For roof slopes less than $^1/_2$ unit vertical in 12 units horizontal (4.2% slope), the deflection of any structural member shall not exceed L/180 evaluated on the basis of roof snow loads plus K times dead load, where K is defined in Table 16-E.

SECTION 1647 — IMPACT LOADS

Whenever P_g exceeds 70 psf (3352 N/m²), structures which could be subjected to impact loads (snow unloading from a higher roof) shall be designed for impact loading.

SECTION 1648 — VERTICAL OBSTRUCTIONS

Whenever P_g exceeds 70 psf (3352 N/m²), roof projections which could be subjected to sliding ice or snow shall be protected with ice splitters or crickets, or shall be designed for these forces. These conditions apply whenever the roof slope is 3 units vertical in 12 units horizontal (25% slope) or greater [except those projections within 36 inches (914 mm) of the ridge]. All ice splitters shall be constructed the full width of the projection base (see Figure A-16-14).

Ice splitters shall be designed for a horizontal force F_s and the resultant moment produced from F_s being applied at midheight of the splitter given by:

$$F_s = \frac{F_v X}{\sqrt{X^2 + Y^2}} \qquad (48\text{-}1)$$

WHERE:

$$F_v = L\,(0.5L + B)\,P_f$$

The projection width, B, shall not exceed 6 feet (1829 mm) where the roof slope is greater than 2 units vertical in 12 units horizontal (16.7% slope) unless approved by the building official. Chimneys and similar projections at or near the eave of a roof shall have footings and roof/wall ties designed to resist the force of the sliding snow. Cross-grain bending of wood ledgers and edge nailing of plywood shall not be considered to resist such forces. Snow melting equipment shall not be considered to reduce the required design loads.

TABLE A-16-A—SNOW EXPOSURE COEFFICIENT (C_e)[1,2]

1. Roofs located in generally open terrain extending one-half mile (804.7 m) or more from the structure.	0.6
2. Structures located in densely forested or sheltered areas.	0.9
3. All other structures.	0.7

[1]The building official may determine this coefficient for specific structures with special local conditions. For Alaska, Arizona and Hawaii, the coefficient shall be determined by the building official.

[2]For roofs at or near grade with slopes less than 3 units vertical in 12 units horizontal (25% slope) or decks at or near grade, C_e equals 1.0.

TABLE A-16-B—VALUES FOR OCCUPANCY IMPORTANCE FACTOR I

TYPE OF OCCUPANCY	I Snow
1. Essential facilities.	1.15
2. Any building where the primary occupancy is for assembly use for more than 300 persons (in one room).	1.15
3. Agricultural buildings, production greenhouses and other miscellaneous structures.	0.9
4. All others.	1.0

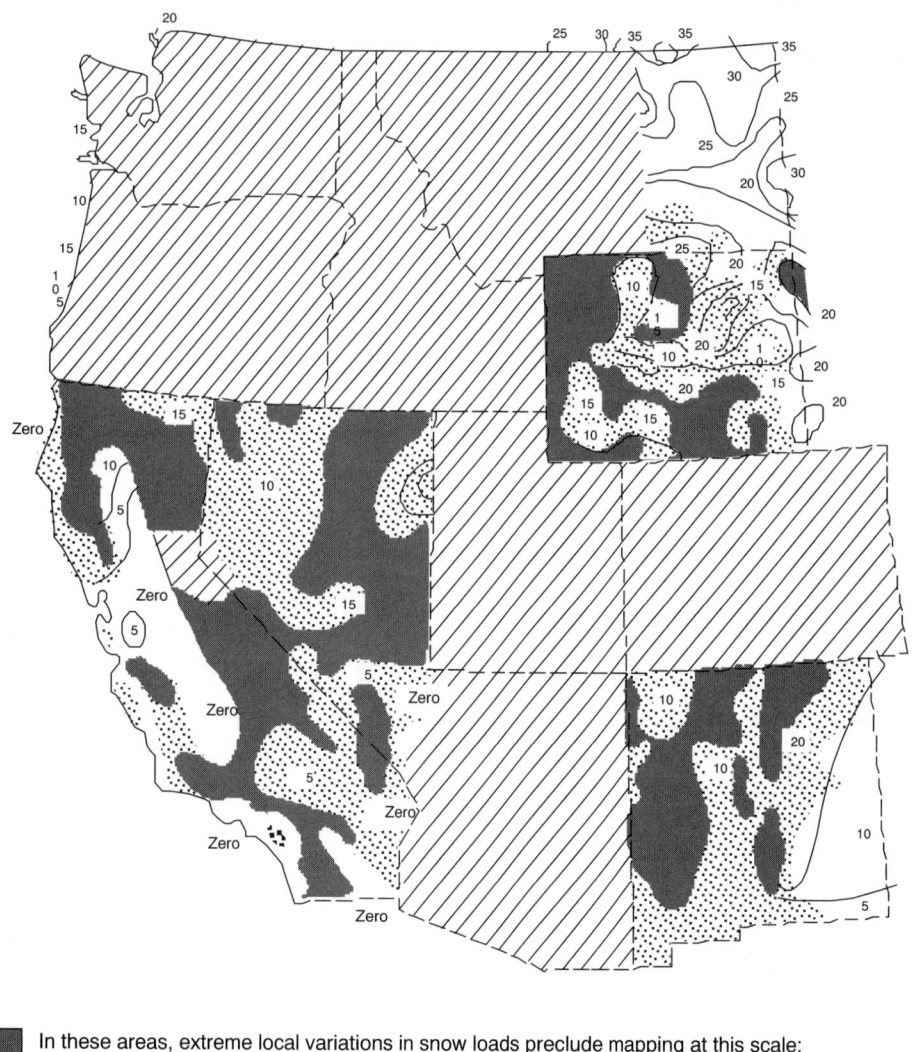

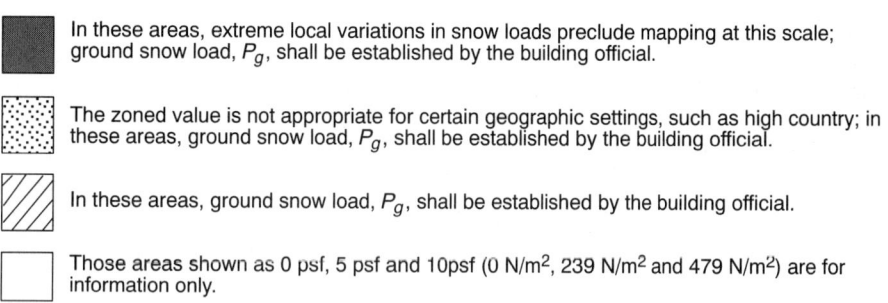

In these areas, extreme local variations in snow loads preclude mapping at this scale; ground snow load, P_g, shall be established by the building official.

The zoned value is not appropriate for certain geographic settings, such as high country; in these areas, ground snow load, P_g, shall be established by the building official.

In these areas, ground snow load, P_g, shall be established by the building official.

Those areas shown as 0 psf, 5 psf and 10psf (0 N/m², 239 N/m² and 479 N/m²) are for information only.

For **SI:** 1 psf = 47.8 N/m².

FIGURE A-16-1—GROUND SNOW LOAD, P_g, FOR 50-YEAR MEAN RECURRENCE INTERVAL FOR THE WESTERN UNITED STATES

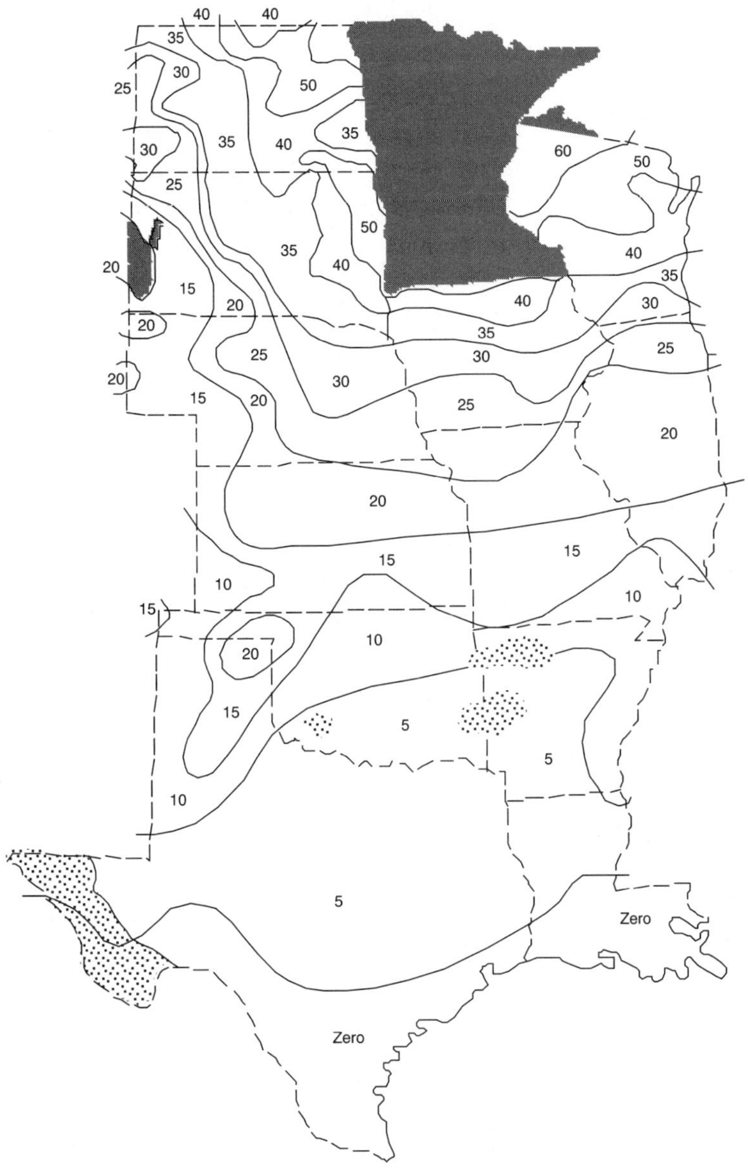

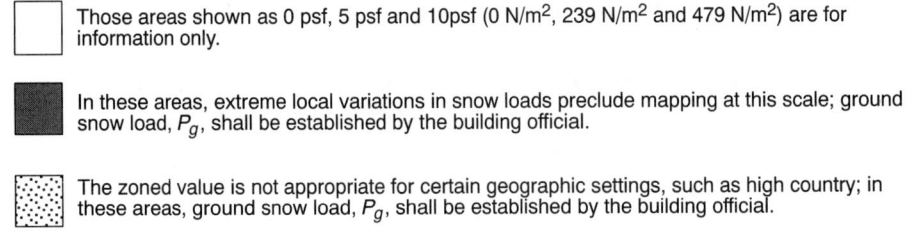

Those areas shown as 0 psf, 5 psf and 10psf (0 N/m², 239 N/m² and 479 N/m²) are for information only.

In these areas, extreme local variations in snow loads preclude mapping at this scale; ground snow load, P_g, shall be established by the building official.

The zoned value is not appropriate for certain geographic settings, such as high country; in these areas, ground snow load, P_g, shall be established by the building official.

For **SI:** 1 psf = 47.8 N/m².

FIGURE A-16-2—GROUND SNOW LOAD, P_g, FOR 50-YEAR MEAN RECURRENCE INTERVAL FOR THE CENTRAL UNITED STATES

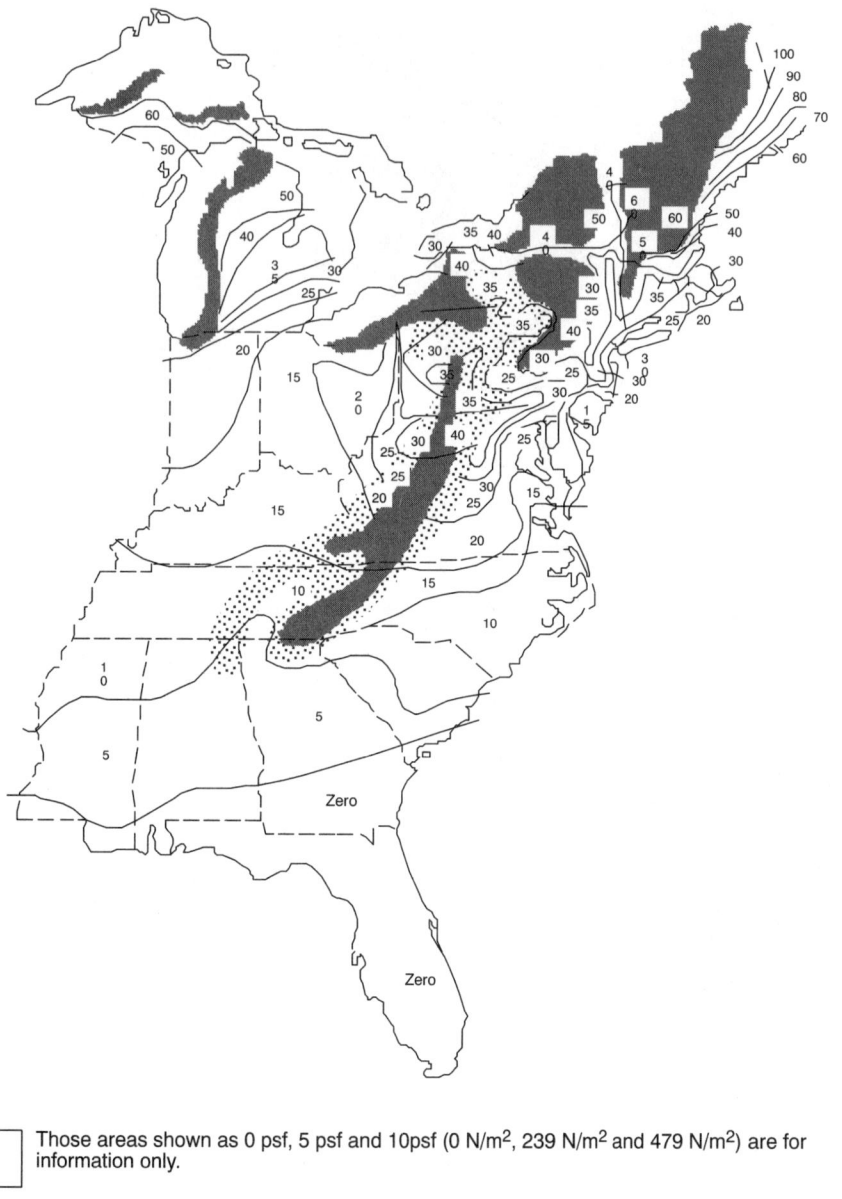

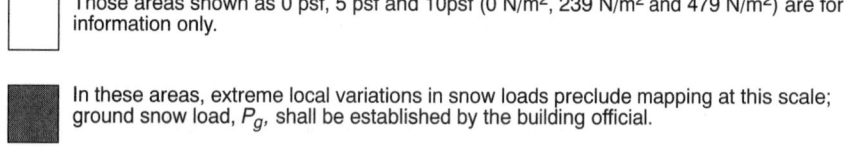

Those areas shown as 0 psf, 5 psf and 10psf (0 N/m², 239 N/m² and 479 N/m²) are for information only.

In these areas, extreme local variations in snow loads preclude mapping at this scale; ground snow load, P_g, shall be established by the building official.

The zoned value is not appropriate for certain geographic settings, such as high country; in these areas, ground snow load, P_g, shall be established by the building official.

For **SI:** 1 psf = 47.8 N/m².

FIGURE A-16-3—GROUND SNOW LOAD, P_g, FOR 50-YEAR MEAN RECURRENCE INTERVAL FOR THE EASTERN UNITED STATES

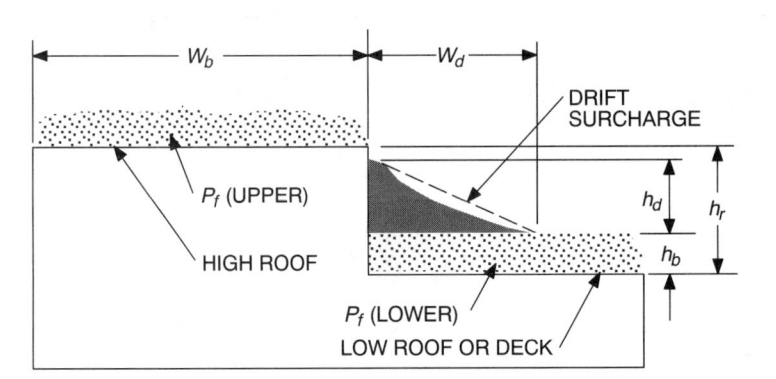

FIGURE A-16-4—DRIFTING SNOW ON LOW ROOFS AND DECKS

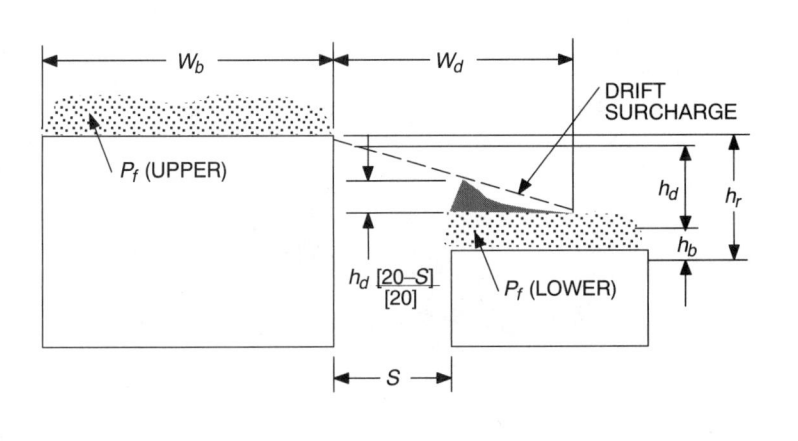

For **SI:** $h_d \left[\dfrac{6.1 - S}{6.1} \right]$.

FIGURE A-16-5—DRIFTING SNOW ONTO ADJACENT LOW STRUCTURES

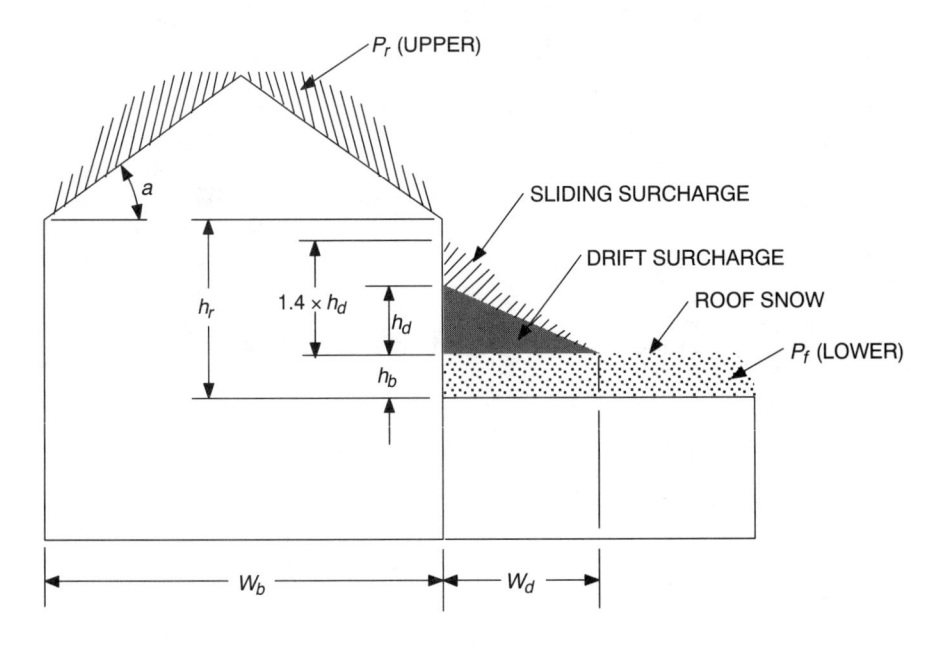

FIGURE A-16-6—ADDITIONAL SURCHARGE DUE TO SLIDING SNOW

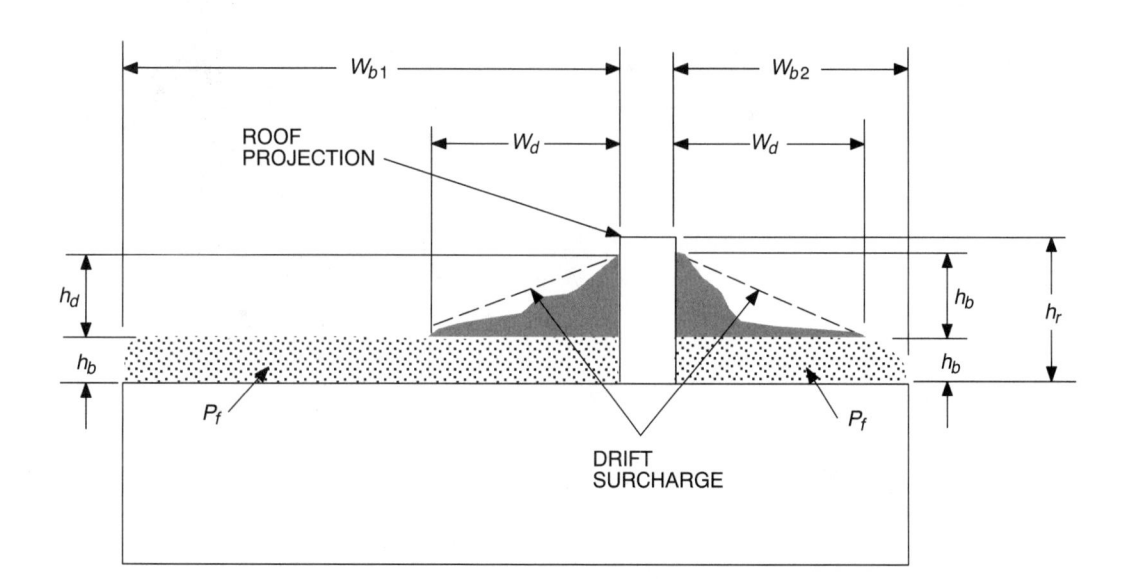

FIGURE A-16-7—SNOW DRIFTING AT ROOF PROJECTIONS

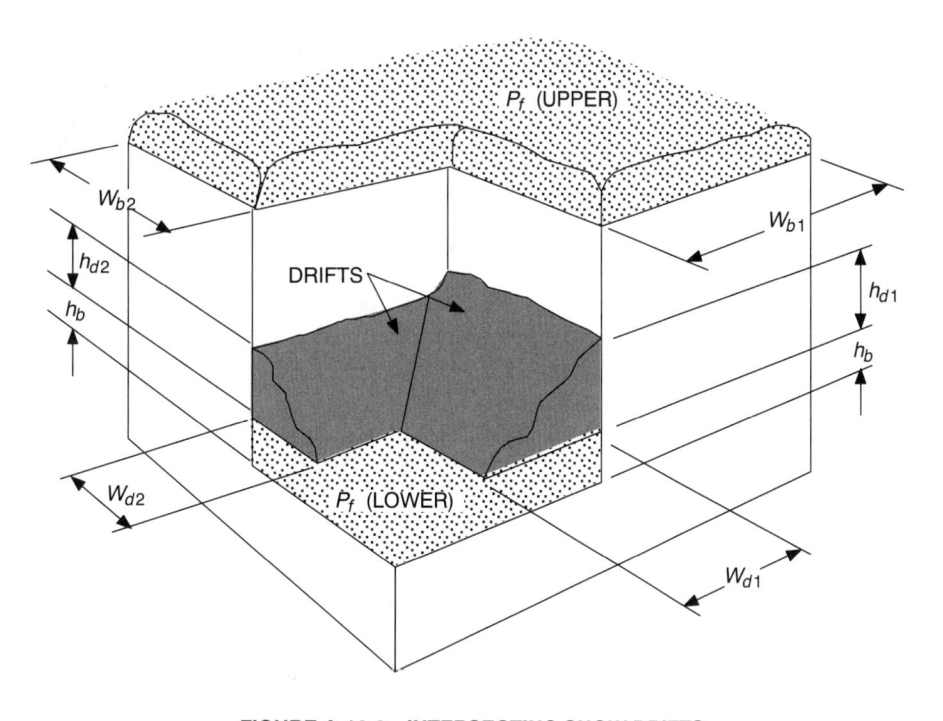

FIGURE A-16-8—INTERSECTING SNOW DRIFTS

NOTE:

$$h_{d1} = 0.43 \sqrt[3]{W_{b1}} \sqrt[4]{P_g + 10} - 1.5$$

$$h_{d2} = 0.43 \sqrt[3]{W_{b2}} \sqrt[4]{P_g + 10} - 1.5$$

For **SI:** 1 foot = 304.8 mm.

P_f is evaluated on the basis of upper roof.

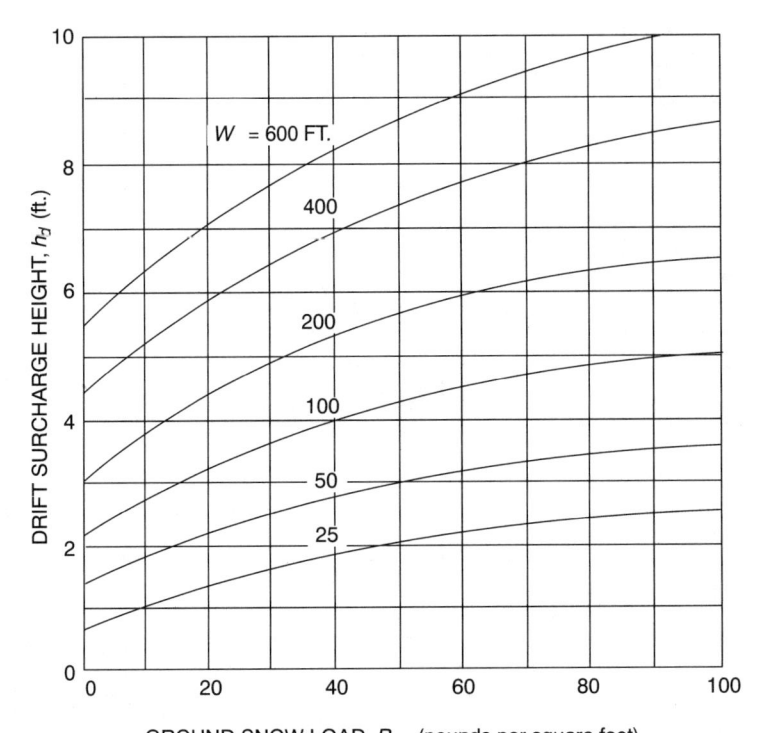

For **SI:** 1 foot = 304.8 mm, 1 psf = 47.8 N/m^2.

FIGURE A-16-9—DETERMINATION OF h_d

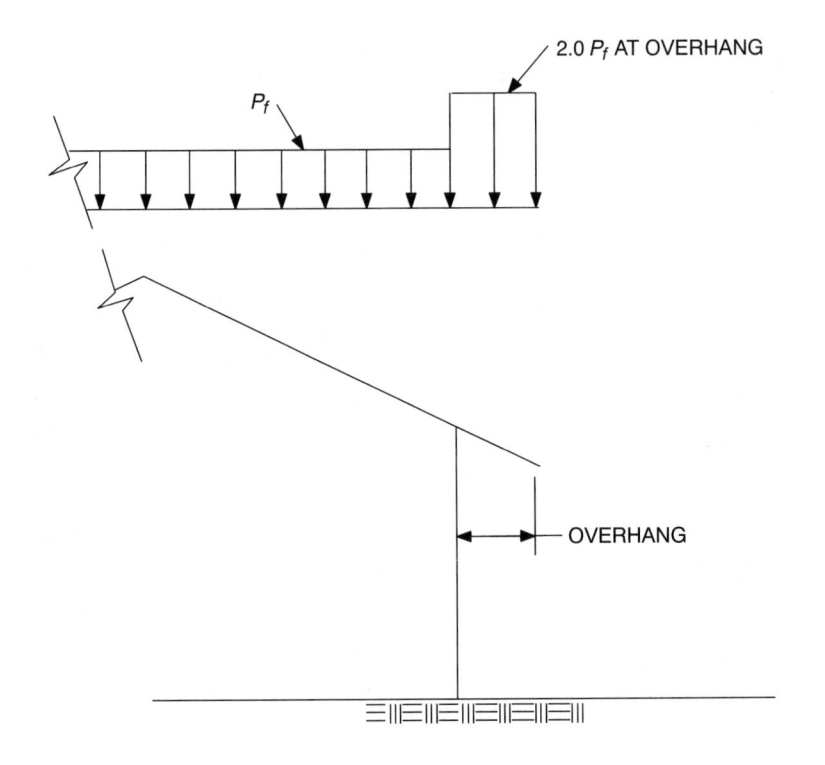

FIGURE A-16-10—OVERHANG LOADS

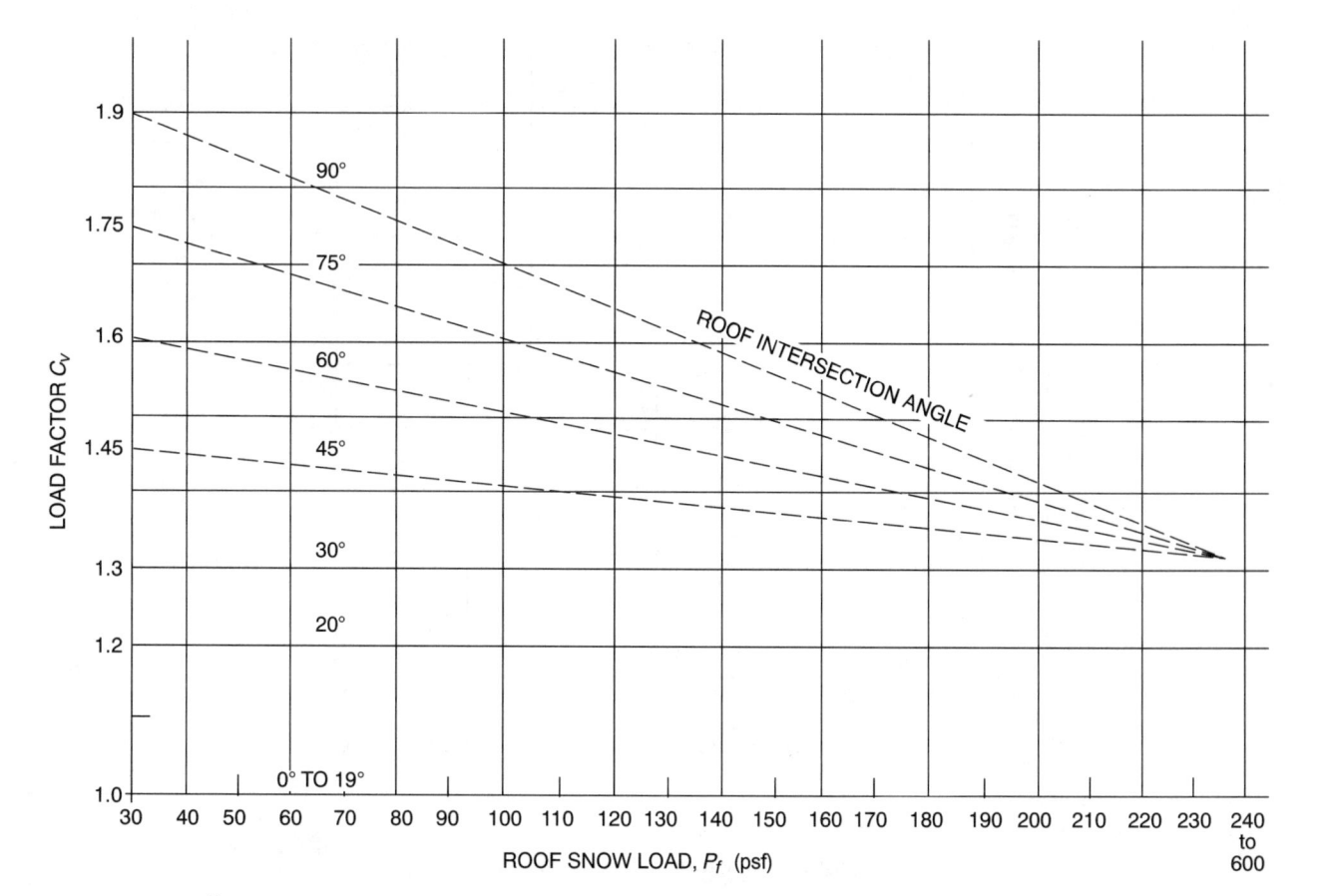

For **SI:** 1 psf = 47.8 N/m^2.

FIGURE A-16-11—VALLEY COEFFICIENT, C_v

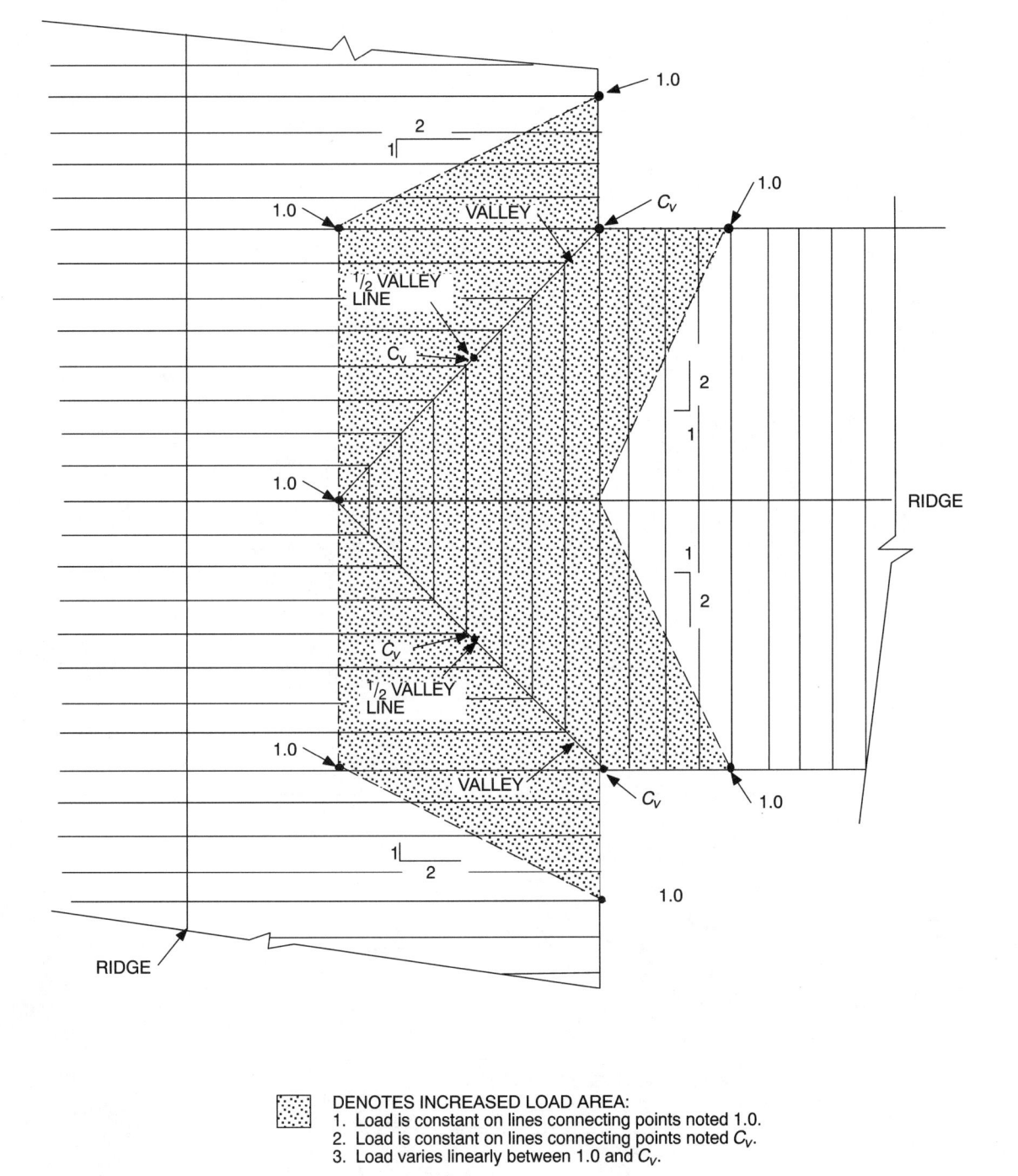

DENOTES INCREASED LOAD AREA:
1. Load is constant on lines connecting points noted 1.0.
2. Load is constant on lines connecting points noted C_v.
3. Load varies linearly between 1.0 and C_v.

FIGURE A-16-12—VALLEY DESIGN COEFFICIENTS, C_v

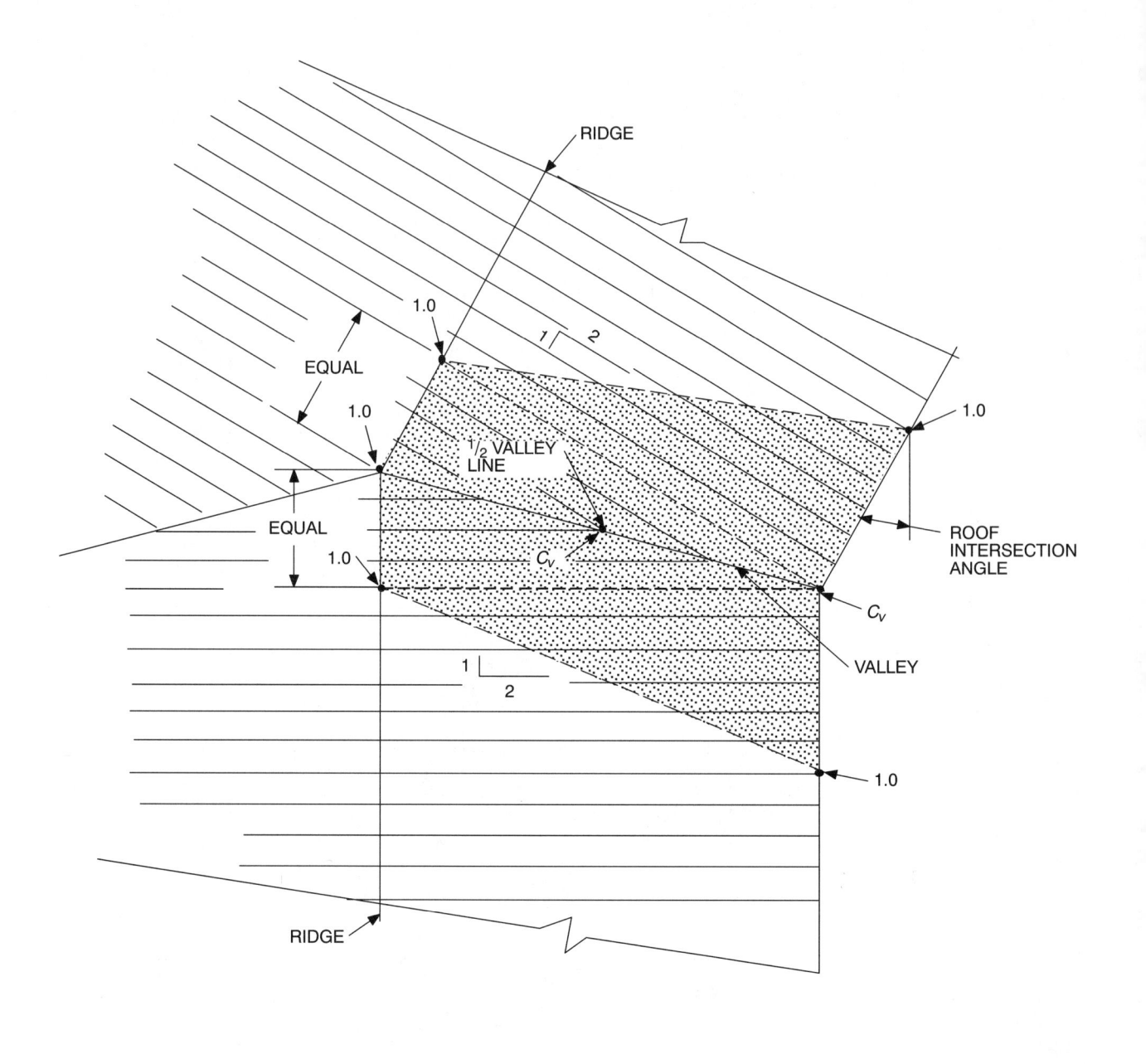

DENOTES INCREASED LOAD AREA:
1. Load is constant on lines connecting points noted 1.0.
2. Load is constant on lines connecting points noted C_v.
3. Load varies linearly between 1.0 and C_v.

FIGURE A-16-13—VALLEY DESIGN COEFFICIENTS, C_v

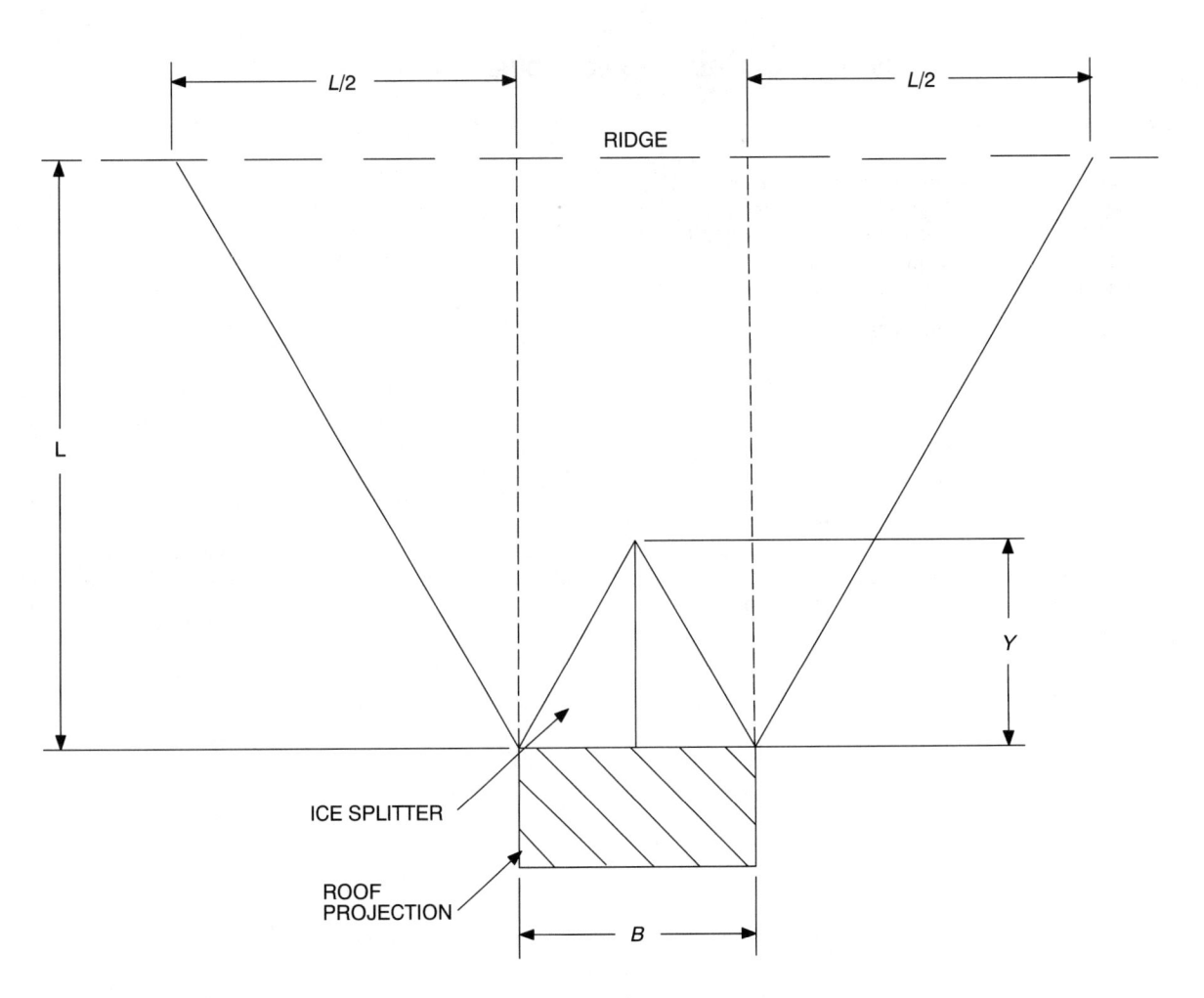

FIGURE A-16-14—ICE SPLITTER—PLAN VIEW

Division II—EARTHQUAKE RECORDING INSTRUMENTATION

SECTION 1649 — GENERAL

In Seismic Zones 3 and 4 every building over six stories in height with an aggregate floor area of 60,000 square feet (5574 m^2) or more, and every building over 10 stories in height regardless of floor area, shall be provided with not less than three approved recording accelerographs.

The accelerographs shall be interconnected for common start and common timing.

SECTION 1650 — LOCATION

The instruments shall be located in the basement, midportion, and near the top of the building. Each instrument shall be located so that access is maintained at all times and is unobstructed by room contents. A sign stating MAINTAIN CLEAR ACCESS TO THIS INSTRUMENT shall be posted in a conspicuous location.

SECTION 1651 — MAINTENANCE

Maintenance and service of the instruments shall be provided by the owner of the building, subject to the approval of the building official. Data produced by the instruments shall be made available to the building official on request.

SECTION 1652 — INSTRUMENTATION OF EXISTING BUILDINGS

All owners of existing structures selected by the jurisdiction authorities shall provide accessible space for the installation of appropriate earthquake-recording instruments. Location of said instruments shall be determined by the jurisdiction authorities. The jurisdiction authorities shall make arrangements to provide, maintain and service the instruments. Data shall be the property of the jurisdiction, but copies of individual records shall be made available to the public on request and the payment of an appropriate fee.

Division III—SEISMIC ZONE TABULATION

NOTE: This division has been revised in its entirety.

SECTION 1653 — FOR AREAS OUTSIDE THE UNITED STATES

Location	Seismic Zone	Location	Seismic Zone
AFRICA		Mali	
Algeria		Bamako	0
Alger	3	Mauritania	
Oran	3	Nouakchott	0
Angola		Mauritius	
Luanda	0	Port Louis	0
Benin		Morocco	
Cotonou	0	Casablanca	2A
Botswana		Port Lyautcy	1
Gaborone	0	Rabat	2A
Burundi		Tangier	3
Bujumbura	3	Mozambique	
Cameroon		Maputo	2A
Douala	0	Niger	
Yaounde	0	Niamey	0
Cape Verde		Nigeria	
Praia	0	Ibadan	0
Central African Republic		Kaduna	0
Bangui	0	Lagos	0
Chad		Republic of Rwanda	
Ndjamena	0	Kigali	3
Congo		Senegal	
Brazzaville	0	Dakar	0
Djibouti	3	Seychelles	
Egypt		Victoria	0
Alexandria	2A	Sierra Leone	
Cairo	2A	Freetown	0
Port Said	2A	Somalia	
Equatorial Guinea		Mogadishu	0
Malabo	0	South Africa	
Ethiopia		Cape Town	3
Addis Ababa	3	Durban	2A
Asmara	3	Johannesburg	2A
Gabon		Natal	1
Libreville	0	Pretoria	2A
Gambia		Swaziland	
Banjul	0	Mbabane	2A
Ghana		Tanzania	
Accra	3	Dar es Salaam	2A
Guinea		Zanzibar	2A
Bissau	1	Togo	
Conakry	0	Lome	1
Ivory Coast		Tunisia	
Abidjan	0	Tunis	3
Kenya		Uganda	
Nairobi	2A	Kampala	2A
Lesotho		Upper Volta	
Maseru	2A	Ougadougou	0
Liberia		Zaire	
Monrovia	1	Bukavu	3
Libya		Kinshasa	0
Tripoli	2A	Lubumbashi	2A
Wheelus AFB	2A	Zambia	
Malagasy Republic		Lukasa	2A
Tananarive	0	Zimbabwe	
Malawi		Harare (Salisbury)	3
Blantyre	3	ASIA	
Lilongwe	3	Afghanistan	
Zomba	3	Kabul	4

Location	Seismic Zone
Bahrain	
Manama	0
Bangladesh	
Dacca	3
Brunei	
Bandar Seri Begawan	1
Burma	
Mandalay	3
Rangoon	3
China	
Beijing	4
Chengdu	3
Guangzhou	2A
Nanjing	2A
Qingdao	3
Shanghai	2A
Shengyang	4
Taiwan	
All	4
Tihwa	4
Wuhan	2A
Xianggang	2A
Cyprus	
Nicosia	3
India	
Bombay	3
Calcutta	2A
Madras	1
New Delhi	3
Indonesia	
Bandung	4
Jakarta	4
Medan	3
Surabaya	4
Iran	
Isfahan	3
Shiraz	3
Tabriz	4
Tehran	4
Iraq	
Baghdad	3
Basra	1
Israel	
Haifa	3
Jerusalem	3
Tel Aviv	3
Japan	
Fukuoka	3
Itazuke AFB	3
Misawa AFB	3
Naha, Okinawa	4
Osaka/Kobe	4
Sapporo	3
Tokyo	4
Wakkami	3
Yokohama	4
Yokota	4
Jordan	
Amman	3
Korea	
Kimhae	1
Kwangju	1
Pusan	1
Seoul	0
Kuwait	
Kuwait	1

Location	Seismic Zone
Laos	
Vientiane	1
Lebanon	
Beirut	3
Malaysia	
Kuala Lumpur	1
Nepal	
Kathmandu	4
Oman	
Muscat	2A
Pakistan	
Islamabad	4
Karachi	4
Lahore	2A
Peshawar	4
Qatar	
Doha	0
Saudi Arabia	
Al Batin	1
Dharan	1
Jiddah	2A
Khamis Mushayf	1
Riyadh	0
Singapore	
All	1
South Yemen	
Aden City	3
Sri Lanka	
Colombo	0
Syria	
Aleppo	3
Damascus	3
Thailand	
Bangkok	1
Chiang Mai	2A
Songkhla	0
Udorn	1
Turkey	
Adana	2A
Ankara	2A
Ismir	4
Istanbul	4
Karamursel	3
United Arab Emirates	
Abu Dhabi	0
Dubai	0
Vict Nam	
Ho Chi Minh (Saigon)	0
Yemen Aran Republic	
Sanaa	3
ATLANTIC OCEAN AREA	
Azores	
All	2A
Bermuda	
All	1
CARIBBEAN SEA	
Bahama Islands	
All	1
Cuba	
All	2A
Dominican Republic	
Santo Domingo	3
French West Indies	
Martinique	3
Grenada	
Saint Georges	3

Location	Seismic Zone	Location	Seismic Zone
Haiti		Paris	0
Port au Prince	3	Strasbourg	2A
Jamaica		Germany, Federal Republic	
Kingston	3	Berlin	0
Leeward Islands		Bonn	2A
All	3	Bremen	0
Trinidad & Tobago		Dusseldorf	1
All	3	Frankfurt	2A
CENTRAL AMERICA		Hamburg	0
Belize		Munich	1
Belmopan	2A	Stuttgart	2A
Canal Zone		Vaihigen	2A
All	2A	Greece	
Costa Rica		Athens	3
San Jose	3	Kavalla	4
El Salvador		Makri	4
San Salvador	4	Rhodes	3
Guatemala		Sauda Bay	4
Guatemala	4	Thessaloniki	4
Honduras		Hungary	
Tegucigalpa	3	Budapest	2A
Mexico		Iceland	
Ciudad Juarez	2A	Keflavick	3
Guadalajara	3	Reykjavik	4
Hermosillo	3	Ireland	
Matamoros	0	Dublin	0
Mazatlan	2A	Italy	
Merida	0	Aviano AFB	3
Mexico City	3	Brindisi	0
Monterrey	0	Florence	3
Nuevo Laredo	0	Genoa	3
Tijuana	3	Milan	2A
Nicaragua		Naples	3
Managua	4	Palermo	3
Panama		Rome	2A
Colon	3	Sicily	3
Galeta	2B	Trieste	3
Panama	3	Turin	2A
EUROPE		Luxembourg	
Albania		Luxembourg	1
Tirana	3	Malta	
Austria		Valleta	2A
Salzburg	2A	Netherlands	
Vienna	2A	All	0
Belgium		Norway	
Antwerp	1	Oslo	2A
Brussels	2A	Poland	
Bosnia-Herzegovina		Krakow	2A
Belgrade	2A	Poznan	1
Bulgaria		Warszawa	1
Sofia	3	Portugal	
Croatia		Lisbon	4
Zagreb	3	Opporto	3
Czechoslovakia		Romania	
Bratislava	2A	Bucharest	3
Prague	1	Russia	
Denmark		Moscow	0
Copenhagen	1	St. Petersburg	0
Finland		Spain	
Helsinki	1	Barcelona	2A
France		Bilbao	2A
Bordeaux	2A	Madrid	0
Lyon	1	Rota	2A
Marseille	3	Seville	2A
Nice	3	Sweden	
		Goteborg	2A
		Stockholm	1

Location	Seismic Zone	Location	Seismic Zone
Switzerland		Peru	
Bern	2A	Lima	4
Geneva	1	Piura	4
Zurich	2A	Uruguay	
Ukraine		Montevideo	0
Kiev	0	Venezuela	
United Kingdom		Caracas	4
Belfast	0	Maracaibo	2A
Edinburgh	1	**PACIFIC OCEAN AREA**	
Edzell	1	Australia	
Glasgow/Renfrew	1	Brisbane	1
Hamilton	1	Canberra	1
Liverpool	1	Melbourne	1
London	2A	Perth	1
Londonderry	1	Sydney	1
Thurso	1	Caroline Islands	
NORTH AMERICA		Koror, Palau Is.	2A
Greenland		Ponape	0
All	1	Fiji	
Canada		Suva	3
Argentia NAS	2A	Johnson Island	
Calgary, Alb	1	All	1
Churchill, Man	0	Mariana Islands	
Cold Lake, Alb	1	Guam	3
Edmonton, Alb	1	Saipan	3
E. Harmon AFB	2A	Tinian	3
Fort Williams, Ont	0	Marshall Islands	
Frobisher N.W. Ter.	0	All	1
Goose Airport	1	New Zealand	
Halifax	1	Auckland	3
Montreal, Quebec	3	Wellington	4
Ottawa, Ont	2A	Papau New Guinea	
St. John's Nfd	3	Port Moresby	3
Toronto, Ont	1	Phillipine Islands	
Vancouver	3	Baguio	3
Winnepeg, Man	1	Cebu	4
SOUTH AMERICA		Manila	4
Argentina		Samoa	
Buenos Aires	0	All	3
Bolivia		Wake Island	
La Paz	3	All	0
Santa Cruz	1		
Brazil			
Belem	0		
Belo Horizonte	0		
Brasilia	0		
Manaus	0		
Porto Allegre	0		
Recife	0		
Rio de Janeiro	0		
Salvador	0		
Sao Paulo	1		
Chile			
Santiago	4		
Valparaiso	4		
Colombia			
Bogota	3		
Ecuador			
Guayaquil	3		
Quito	4		
Paraguay			
Asuncion	0		

The above compilation is a partial listing of seismic zones for cities and countries outside of the United States. It has been provided in this code primarily as a source of information, and may not, in all cases, reflect local ordinances or current scientific information.

When an authority having jurisdiction requires seismic design forces that are higher than would be indicated by the above zones, the local requirements shall govern. When an authority having jurisdiction requires seismic design forces that are lower than would be indicated by the above zones, and these forces have been developed with consideration of regional tectonics and up-to-date geologic and seismologic information, the local requirements may be used.

When no local seismic design requirements exist, properly determined information on site-specific ground motions may be used to justify a lower seismic zone. Such site-specific ground motions shall have been developed with proper consideration of regional tectonics and local geologic and seismologic information, and shall have no more than a 10 percent chance of being exceeded in a 50-year period.

Division IV—EARTHQUAKE REGULATIONS FOR SEISMIC-ISOLATED STRUCTURES

SECTION 1654 — GENERAL

Every seismic-isolated structure and every portion thereof shall be designed and constructed in accordance with the requirements of this division and the applicable requirements of Chapter 16, Part IV.

The lateral-force-resisting system and the isolation system shall be designed to resist the deformations and stresses produced by the effects of seismic ground motions as provided in this division.

Where wind forces prescribed by Chapter 16, Part III, produce greater deformations or stresses, such loads shall be used for design in lieu of the deformations and stresses resulting from earthquake forces.

SECTION 1655 — DEFINITIONS

The definitions of Section 1627 and the following apply to the provisions of this division:

DESIGN DISPLACEMENT is the design-basis earthquake lateral displacement, excluding additional displacement due to actual and accidental torsion, required for design of the isolation system.

DESIGN-BASIS EARTHQUAKE is defined in Section 1631.2.

EFFECTIVE DAMPING is the value of equivalent viscous damping corresponding to energy dissipated during cyclic response of the isolation system.

EFFECTIVE STIFFNESS is the value of the lateral force in the isolation system, or an element thereof, divided by the corresponding lateral displacement.

ISOLATION INTERFACE is the boundary between the upper portion of the structure, which is isolated, and the lower portion of the structure, which moves rigidly with the ground.

ISOLATION SYSTEM is the collection of structural elements that includes all individual isolator units, all structural elements that transfer force between elements of the isolation system, and all connections to other structural elements. The isolation system also includes the wind-restraint system if such a system is used to meet the design requirements of this section.

ISOLATOR UNIT is a horizontally flexible and vertically stiff structural element of the isolation system that permits large lateral deformations under design seismic load. An isolator unit may be used either as part of or in addition to the weight-supporting system of the building.

MAXIMUM CAPABLE EARTHQUAKE is the maximum level of earthquake ground shaking that may ever be expected at the building site within the known geological framework. In Seismic Zones 3 and 4, this intensity may be taken as the level of earthquake ground motion that has a 10 percent probability of being exceeded in a 100-year time period.

MAXIMUM DISPLACEMENT is the maximum capable earthquake lateral displacement, excluding additional displacement due to actual and accidental torsion, required for design of the isolation system.

TOTAL DESIGN DISPLACEMENT is the design-basis earthquake lateral displacement, including additional displacement due to actual and accidental torsion, required for design of the isolation system, or an element thereof.

TOTAL MAXIMUM DISPLACEMENT is the maximum capable earthquake lateral displacement, including additional displacement due to actual and accidental torsion, required for verification of the stability of the isolation system, or elements thereof, design of building separations, and vertical load testing of isolator unit prototypes.

WIND-RESTRAINT SYSTEM is the collection of structural elements that provide restraint of the seismic-isolated structure for wind loads. The wind-restraint system may be either an integral part of isolator units or may be a separate device.

SECTION 1656 — SYMBOLS AND NOTATIONS

The symbols and notations of Section 1628 and the following provisions apply to the provisions of this division:

B_D = numerical coefficient related to the effective damping of the isolation system at the design displacement, β_D, as set forth in Table A-16-C.

B_M = numerical coefficient related to the effective damping of the isolation system at the maximum displacement, β_M, as set forth in Table A-16-C.

b = the shortest plan dimension of the structure, in feet (mm), measured perpendicular to d.

C_{AD} = the seismic coefficient, C_a, as set forth in Table 16-Q.

C_{AM} = the seismic coefficient, C_a, as set forth in Table A-16-F for shaking intensity, $M_M Z N_a$.

C_{VD} = seismic coefficient, C_v, as set forth in Table 16-R.

C_{VM} = seismic coefficient, C_v, as set forth in Table A-16-G for shaking intensity, $M_M Z N_v$.

D_D = design displacement, in inches (mm), at the center of rigidity of the isolation system in the direction under consideration, as prescribed by Formula (58-1).

$D_D{}'$ = design displacement, in inches (mm), at the center of rigidity of the isolation system in the direction under consideration, as prescribed by Formula (59-1).

D_M = maximum displacement, in inches (mm), at the center of rigidity of the isolation system in the direction under consideration, as prescribed by Formula (58-3).

$D_M{}'$ = maximum displacement, in inches (mm), at the center of rigidity of the isolation system in the direction under consideration, as prescribed by Formula (59-2).

D_{TD} = total design displacement, in inches (mm), of an element of the isolation system including both translational displacement at the center of rigidity, D_D, and the component of torsional displacement in the direction under consideration, as specified in Section 1658.3.5.

D_{TM} = total maximum displacement, in inches (mm), of an element of the isolation system, including both translational displacement at the center of rigidity, D_M, and the component of torsional displacement in the direction under consideration, as specified by Section 1658.3.3.

d = the longest plan dimension of the structure, in feet (mm).

E_{LOOP} = energy dissipated in kip-inches (kN-mm), in an isolator unit during a full cycle of reversible load over a test displacement range from Δ^+ to Δ^-, as measured by the area enclosed by the loop of the force-deflection curve.

ΣE_D = total energy dissipated, in kip-inches (kN-mm), of all units of the isolation system during a full cycle of response at the design displacement, D_D.

ΣE_M = total energy dissipated, in kip-inches (kN-mm), of all units of the isolation system during a full cycle of response at the maximum displacement, D_M.

e = the actual eccentricity, in feet (mm), measured in plan between the center of mass of the structure above the isolation interface and the center of rigidity of the isolation system, plus accidental eccentricity, in feet (mm), taken as 5 percent of the maximum building dimension perpendicular to the direction of force under consideration.

$F-$ = negative force, in kips (kN), in an isolator unit during a single cycle of prototype testing at a displacement amplitude of $\Delta-$.

$F+$ = positive force, in kips (kN), in an isolator unit during a single cycle of prototype testing at a displacement amplitude of $\Delta+$.

$\Sigma|F_D{}^+|_{max}$
= sum, for all isolator units, of the absolute values of the individual isolator unit's maximum positive force in kips (kN) at positive displacement D_D. For a given isolator unit, the maximum positive force at positive displacement, D_D, is determined by comparing each of the maximum positive forces that occurred during each cycle of the prototype test sequence associated with displacement increment D_D, and selecting the maximum positive value at positive displacement, D_D.

$\Sigma|F_D{}^+|_{min}$
= sum, for all isolator units, of the absolute values of the individual isolator unit's minimum positive force in kips (kN) at positive displacement D_D. For a given isolator unit, the minimum positive force at positive displacement, D_D, is determined by comparing each of the minimum positive forces that occurred during each cycle of the prototype test sequence associated with displacement increment D_D, and selecting the minimum positive value at positive displacement, D_D.

$\Sigma|F_D{}^-|_{max}$
= sum, for all isolator units, of the absolute values of the individual isolator unit's maximum negative force in kips (kN) at negative displacement D_D. For a given isolator unit, the maximum negative force at negative displacement, D_D, is determined by comparing each of the maximum negative forces that occurred during each cycle of the prototype test sequence associated with displacement increment D_D, and selecting the maximum negative value at negative displacement, D_D.

$\Sigma|F_D{}^-|_{min}$
= sum, for all isolator units, of the absolute values of the individual isolator unit's minimum negative force in kips (kN) at negative displacement D_D. For a given isolator unit, the minimum negative force at negative displacement, D_D, is determined by comparing each of the minimum negative forces that occurred during each cycle of the prototype test sequence associated with displacement increment D_D, and selecting the minimum negative value at negative displacement, D_D.

$\Sigma|F_M{}^+|_{max}$
= sum, for all isolator units, of the absolute values of the individual isolator unit's maximum positive force in kips (kN) at positive displacement D_M. For a given isolator unit, the maximum positive force at positive dis-

placement, D_M, is determined by comparing each of the maximum positive forces that occurred during each cycle of the prototype test sequence associated with displacement increment D_M, and selecting the maximum positive value at positive displacement, D_M.

$\Sigma|F_M{}^+|_{min}$
= sum, for all isolator units, of the absolute values of the individual isolator unit's minimum positive force in kips (kN) at positive displacement, D_M. For a given isolator unit, the minimum positive force at positive displacement, D_M, is determined by comparing each of the minimum positive forces that occurred during each cycle of the prototype test sequence associated with displacement increment D_M and selecting the minimum positive value at positive displacement, D_M.

$\Sigma|F_M{}^-|_{max}$
= sum, for all isolator units, of the absolute values of the individual isolator unit's maximum negative force in kips (kN) at negative displacement D_M. For a given isolator unit, the maximum negative force at negative displacement, D_M, is determined by comparing each of the maximum negative forces that occurred during each cycle of the prototype test sequence associated with displacement increment D_M and selecting the maximum negative value at negative displacement, D_M.

$\Sigma|F_M{}^-|_{min}$
= sum, for all isolator units, of the absolute values of the individual isolator unit's minimum negative force in kips (kN) at negative displacement D_M. For a given isolator unit, the minimum negative force at negative displacement, D_M, is determined by comparing each of the minimum negative forces that occurred during each cycle of the prototype test sequence associated with displacement increment D_M and selecting the minimum negative value at negative displacement, D_M.

g = gravity constant (386.4 in/sec.^{2}, or 9,810 mm/sec.^{2}, for **SI**).

k_{eff} = effective stiffness of an isolator unit, in kips/inch as prescribed by Formula (65-1).

k_{Dmax} = maximum effective stiffness, in kips/inch (kN/mm), of the isolation system at the design displacement in the horizontal direction under consideration.

k_{Mmax} = maximum effective stiffness, in kips/inch (kN/mm), of the isolation system at the maximum displacement in the horizontal direction under consideration.

k_{Dmin} = minimum effective stiffness, in kips/inch (kN/mm), of the isolation system at the design displacement in the horizontal direction under consideration.

k_{Mmin} = minimum effective stiffness, in kips/inch (kN/mm), of the isolation system at the maximum displacement in the horizontal direction under consideration.

M_M = numerical coefficient related to maximum capable earthquake response as set forth in Table A-16-D.

N_a = near-source factor used in the determination of C_{AD} and C_{AM} related to both the proximity of the building or structure to known faults with magnitudes and slip rates as set forth in Tables 16-S and 16-U.

N_v = near-source factor used in the determination of C_{VD} and C_{VM} related to both the proximity of the building or structure to known faults with magnitudes and slip rates as set forth in Tables 16-T and 16-U.

R_l = numerical coefficient related to the type of lateral-force-resisting system above the isolation system as set forth in Table A-16-E for seismic-isolated structures.

T_D = effective period, in seconds, of seismic-isolated structure at the design displacement in the direction under consideration, as prescribed by Formula (58-2).

T_M = effective period, in seconds, of seismic-isolated structure at the maximum displacement in the direction under consideration, as prescribed by Formula (58-4).

V_b = the total lateral seismic design force or shear on elements of the isolation system or elements below the isolation system as prescribed by Formula (58-5).

V_s = the total lateral seismic design force or shear on elements above the isolation system as prescribed by Formula (58-8) and the limits specified in Section 1658.

W = the total seismic dead load defined in Section 1630.1. For design of the isolation system, W is the total seismic dead load weight of the structure above the isolation interface.

y = the distance, in feet (mm), between the center of rigidity of the isolation system rigidity and the element of interest, measured perpendicular to the direction of seismic loading under consideration.

β_{eff} = effective damping of the isolation system and isolator unit, as prescribed by Formula (65-2).

β_D = effective damping of the isolation system at the design displacement, as prescribed by Formula (65-3).

β_M = effective damping of the isolation system at the maximum displacement, as prescribed by Formula (65-4).

$\Delta+$ = maximum positive displacement of an isolator unit during each cycle of prototype testing.

$\Delta-$ = maximum negative displacement of an isolator unit during each cycle of prototype testing.

SECTION 1657 — CRITERIA SELECTION

1657.1 Basis for Design. The procedures and limitations for the design of seismic-isolated structures shall be determined considering zoning, site characteristics, vertical acceleration, cracked section properties of concrete and masonry members, occupancy, configuration, structural system and height in accordance with Section 1629, except as noted below.

1657.2 Stability of the Isolation System. The stability of the vertical load-carrying elements of the isolation system shall be verified by analysis and test, as required, for lateral seismic displacement equal to the total maximum displacement.

1657.3 Occupancy Categories. The importance factor, I, for a seismic-isolated building shall be taken as 1.0 regardless of occupancy category.

1657.4 Configuration Requirements. Each structure shall be designated as being regular or irregular on the basis of the structural configuration above the isolation system, in accordance with Section 1629.5.

1657.5 Selection of Lateral Response Procedure.

1657.5.1 General. Any seismic-isolated structure may be, and certain seismic-isolated structures defined below shall be, designed using the dynamic lateral response procedure of Section 1659.

1657.5.2 Static analysis. The static lateral response procedure of Section 1658 may be used for design of a seismic-isolated structure, provided:

1. The structure is located at least 10 kilometers (km) from all active faults.

2. The structure is located on Soil Profile Type S_A, S_B, S_C or S_D.

3. The structure above the isolation interface is equal to or less than four stories, or 65 feet (19.8 m), in height.

4. The effective period of the isolated structure, T_M, is equal to or less than 3.0 seconds.

5. The effective period of the isolated structure, T_D, is greater than three times the elastic, fixed-base period of the structure above the isolation system, as determined by Formula (30-8) of Section 1630.

6. The structure above the isolation system is of regular configuration.

7. The isolation system is defined by all of the following attributes:

   7.1 The effective stiffness of the isolation system at the design displacement is greater than one third of the effective stiffness at 20 percent of the design displacement.

   7.2 The isolation system is capable of producing a restoring force, as specified in Section 1661.2.4.

   7.3 The isolation system has force-deflection properties which are independent of the rate of loading.

   7.4 The isolation system has force-deflection properties which are independent of vertical load and bilateral load.

   7.5 The isolation system does not limit maximum capable earthquake displacement to less than C_{VM}/C_{VD} times the total design displacement.

1657.5.3 Dynamic analysis. The dynamic lateral response procedure of Section 1659 shall be used for design of seismic-isolated structures as specified below:

1. **Response spectrum analysis.** Response spectrum analysis may be used for design of a seismic-isolated structure, provided:

   1.1 The structure is located on Soil Profile Type S_A, S_B, S_C or S_D.

   1.2 The isolation system is defined by all of the attributes specified in Section 1657.5.2, Item 7.

2. **Time-history analysis.** Time-history analysis may be used for design of any seismic-isolated structure and shall be used for design of all seismic-isolated structures not meeting the criteria of Section 1657.5.3, Item 1.

3. **Site-specific design spectra.** Site-specific ground motion spectra of the design-basis earthquake and the maximum capable earthquake, developed in accordance with Section 1631.2, shall be used for design and analysis of all seismic-isolated structures as specified below:

1. The structure is located on Soil Profile Type S_E or S_F.

2. The structure is located within 10 km of an active fault.

SECTION 1658 — STATIC LATERAL RESPONSE PROCEDURE

1658.1 General. Except as provided in Section 1659, every seismic-isolated structure, or portion thereof, shall be designed and constructed to resist minimum earthquake displacements and forces as specified by this section and the applicable requirements of Section 1630.

1658.2 Deformation Characteristics of the Isolation System. Minimum lateral earthquake design displacements and forces on seismic-isolated structures shall be based on the deformation characteristics of the isolation system.

The deformation characteristics of the isolation system shall explicitly include the effects of the wind-restraint system if such a system is used to meet the design requirements of this document.

The deformation characteristics of the isolation system shall be based on properly substantiated tests performed in accordance with Section 1665.

1658.3 Minimum Lateral Displacements.

1658.3.1 Design displacement. The isolation system shall be designed and constructed to withstand minimum lateral earthquake displacements which act in the direction of each of the main horizontal axes of the structure in accordance with the formula:

$$D_D = \frac{\left(\frac{g}{4\pi^2}\right)C_{VD}T_D}{B_D} \qquad (58\text{-}1)$$

1658.3.2 Effective period at the design displacement. The effective period of the isolated structure at the design displacement, T_D, shall be determined using the deformational characteristics of the isolation system in accordance with the formula:

$$T_D = 2\pi\sqrt{\frac{W}{k_{Dmin}g}} \qquad (58\text{-}2)$$

1658.3.3 Maximum displacement. The maximum displacement of the isolation system, D_M, in the most critical direction of horizontal response shall be calculated in accordance with the formula:

$$D_M = \frac{\left(\frac{g}{4\pi^2}\right)C_{VM}T_M}{B_M} \qquad (58\text{-}3)$$

1658.3.4 Effective period at the maximum displacement. The effective period of the isolated structure at the maximum displacement, T_M, shall be determined using the deformational characteristics of the isolation system in accordance with the formula:

$$T_M = 2\pi\sqrt{\frac{W}{k_{Mmin}g}} \qquad (58\text{-}4)$$

1658.3.5 Total displacement. The total design displacement, D_{TD}, and the total maximum displacement, D_{TM}, of elements of the isolation system shall include additional displacement due to actual and accidental torsion calculated considering the spatial distribution of the lateral stiffness of the isolation system and the most disadvantageous location of mass eccentricity.

The total design displacement, D_{TD}, and the of total maximum displacement D_{TM}, of elements of an isolation system with uniform spatial distribution of lateral stiffness shall not be taken as less than that prescribed by the formulas:

$$D_{TD} = D_D\left[1 + y\frac{12e}{b^2 + d^2}\right] \qquad (58\text{-}5)$$

$$D_{TM} = D_M\left[1 + y\frac{12e}{b^2 + d^2}\right] \qquad (58\text{-}6)$$

The total design displacement, D_{TD}, and the total maximum displacement, D_{TM}, may be taken as less than the value prescribed by Formulas (58-5) and (58-6), but not less than 1.1 times D_D and 1.1 times D_M, respectively, provided the isolation system is shown by calculation to be configured to resist torsion accordingly.

1658.4 Minimum Lateral Forces.

1658.4.1 Isolation system and structural elements at or below the isolation system. The isolation system, the foundation, and all structural elements below the isolation system shall be designed and constructed to withstand a minimum lateral seismic force, V_b, using all of the appropriate provisions for a nonisolated structure where:

$$V_b = k_{Dmax}D_D \qquad (58\text{-}7)$$

1658.4.2 Structural elements above the isolation system. The structure above the isolation system shall be designed and constructed to withstand a minimum shear force, V_s, using all of the appropriate provisions for a nonisolated structure where:

$$V_s = \frac{k_{Dmax}D_D}{R_I} \qquad (58\text{-}8)$$

The R_I factor shall be based on the type of lateral-force-resisting system used for the structure above the isolation system.

1658.4.3 Limits on V_s. The value of V_s shall not be taken as less than the following:

1. The lateral seismic force required by Chapter 16, Division III, for a fixed-base structure of the same weight, W, and a period equal to the isolated period, T_D.

2. The base shear corresponding to the design wind load.

3. The lateral seismic force required to fully activate the isolation system factored by 1.5 (e.g., one and one-half times the yield level of a softening system, the ultimate capacity of a sacrificial wind-restraint system or the static friction level of a sliding system).

1658.5 Vertical Distribution of Force. The total force shall be distributed over the height of the structure above the isolation interface in accordance with the formula:

$$F_x = \frac{V_s w_x h_x}{\sum_{i=1}^{n} w_i h_i} \qquad (58\text{-}9)$$

At each level designated as x, the force F_x shall be applied over the area of the building in accordance with the mass distribution at the level. Stresses in each structural element shall be calculated as the effect of force, F_x, applied at the appropriate levels above the base.

1658.6 Drift Limits. The maximum interstory drift ratio of the structure above the isolation system shall not exceed $0.010/R_I$.

SECTION 1659 — DYNAMIC LATERAL-RESPONSE PROCEDURE

1659.1 General. As required by Section 1657, every seismic-isolated structure, or portion thereof, shall be designed and constructed to resist earthquake displacements and forces as specified in this section and the applicable requirements of Section 1631.

1659.2 Isolation System and Structural Elements below the Isolation System. The total design displacement of the isolation system shall not be taken as less than 90 percent of D_{TD} as specified by Section 1658.3.3.

The total maximum displacement of the isolation system shall not be taken as less than 80 percent of D_{TM} as prescribed by Formula (58-6).

The design lateral shear force on the isolation system and structural elements below the isolation system shall not be taken as less than 90 percent of V_b as prescribed by Formula (58-7).

The limits of the first and second paragraphs shall be evaluated using values of D_{TD} and D_{TM} determined in accordance with Section 1658.3, except that $D_D{}'$ may be used in lieu of D_D and $D_M{}'$ may be used in lieu of D_M, where $D_D{}'$ and $D_M{}'$ are prescribed by the formulas:

$$D_{D'} = \frac{D_D}{\sqrt{1 + \left(\frac{T}{T_D}\right)^2}} \qquad (59\text{-}1)$$

$$D_{M'} = \frac{D_M}{\sqrt{1 + \left(\frac{T}{T_M}\right)^2}} \qquad (59\text{-}2)$$

and T is the elastic, fixed-base period of the structure above the isolation system, as determined only by Formula (30-4) of Section 1630.

1659.3 Structural Elements above the Isolation System. The design lateral shear force on the structure above the isolation system, if regular in configuration, shall not be taken as less than 80 percent of V_S as prescribed by Formula (58-8) or less than the limits specified by Section 1658.4.3.

> **EXCEPTION:** The design lateral shear force on the structure above the isolation system, if regular in configuration, may be taken as less than 80 percent, but not less than 60 percent, of V_S provided time-history analysis is used for design of the structure.

The design lateral shear force on the structure above the isolation system, if irregular in configuration, shall not be taken as less than V_S as prescribed by Formula (58-8) or less than the limits specified by Section 1658.4.3.

> **EXCEPTION:** The design lateral shear force on the structure above the isolation system, if irregular in configuration, may be taken as less than 100 percent, but not less than 80 percent, of V_S, provided time-history analysis is used for design of the structure.

1659.4 Ground Motion.

1659.4.1 Design spectra. Properly substantiated, site-specific spectra are required for design of all structures with an isolated period, T_M, greater than 3.0 seconds, or located on Soil Profile Type S_E or S_F or located within 10 km of an active fault or located

in Seismic Zone 1, 2A or 2B. Structures that do not require site-specific spectra and for which site-specific spectra have not been calculated shall be designed using spectra based on Figure 16-3 of Chapter 16, Division III.

A design spectrum shall be constructed for the design-basis earthquake. This design spectrum shall not be taken as less than the response spectrum given in Figure 16-3 of Chapter 16, Division III, where the values of C_a shall be taken as equal to C_{AD} and C_v shall be taken as equal to C_{VD}.

> **EXCEPTION:** If a site-specific spectrum is calculated for the design-basis earthquake, then the design spectrum may be taken as less than 100 percent, but not less than 80 percent of the response spectrum given in Figure 16-3 of Chapter 16, Division III, where the values of C_a shall be taken as equal to C_{AD} and C_v shall be taken as equal to C_{VD}.

A design spectrum shall be constructed for the maximum capable earthquake. This spectrum shall not be taken as less than the spectrum given in Figure 16-3 of Chapter 16, Division III where the values of C_a shall be taken as equal to C_{AM} and C_v shall be taken as equal to C_{VM}. This spectrum shall be used to determine the total maximum displacement and overturning forces for design and testing of the isolation system.

> **EXCEPTION:** If a site-specific spectrum is calculated for the maximum capable earthquake, then the design spectrum may be taken as less than 100 percent, but not less than 80 percent of the response spectrum given in Figure 16-3 of Chapter 16, Division III, where the values of C_a shall be taken as equal to C_{AM} and C_v shall be taken as equal to C_{VM}.

1659.4.2 Time histories. Pairs of appropriate horizontal ground-motion time-history components shall be selected and scaled from not less than three recorded events. Appropriate time histories shall have magnitudes, fault distances and source mechanisms that are consistent with those that control the design-basis earthquake (or maximum capable earthquake). Where three appropriate recorded ground motion time history pairs are not available, appropriate simulated ground motion time history pairs may be used to make up the total number required. For each pair of horizontal ground-motion components, the square root sum of the squares (SRSS) of the 5 percent-damped spectrum of the scaled horizontal components shall be constructed. The motions shall be scaled such that the average value of the SRSS spectra does not fall below 1.3 times the 5 percent-damped spectrum of the design-basis earthquake (or maximum capable earthquake) by more than 10 percent for periods from $0.5T_D$ seconds to $1.25T_M$ seconds.

1659.5 Mathematical Model.

1659.5.1 General. The mathematical models of the isolated structure, including the isolation system, the lateral-force-resisting system and other structural elements, shall conform to Section 1631.3 and to the requirements of Sections 1659.5.2 and 1659.5.3 below.

1659.5.2 Isolation system. The isolation system shall be modeled using deformational characteristics developed and verified by test in accordance with the requirements of Section 1658.2.

The isolation system shall be modeled with sufficient detail to:

1. Account for the spatial distribution of isolator units,

2. Calculate translation, in both horizontal directions, and torsion of the structure above the isolation interface, considering the most disadvantageous location of mass eccentricity,

3. Assess overturning/uplift forces on individual isolator units; and

4. Account for the effects of vertical load, bilateral load and/or the rate of loading if the force deflection properties of the isolation system are dependent on one or more of these attributes.

1659.5.3 Isolated structure.

1659.5.3.1 Displacement. The maximum displacement of each floor and the total design displacement and total maximum displacement across the isolation system shall be calculated using a model of the isolated structure that incorporates the force-deflection characteristics of nonlinear elements of the isolation system and the lateral-force-resisting system.

Lateral-force-resisting systems with nonlinear elements include, but are not limited to, irregular structural systems designed for a lateral force less than V_s as prescribed by Formula (58-8) and the limits specified by Section 1658.4.3, and regular structural systems designed for a lateral force less than 80 percent of V_s.

1659.5.3.2 Forces and displacements in key elements. Design forces and displacements in key elements of the lateral-force-resisting system may be calculated using a linear elastic model of the isolated structure, provided:

1. Pseudo-elastic properties assumed for nonlinear isolation system components are based on the maximum effective stiffness of the isolation system.

2. All key elements of the lateral-force-resisting system are linear.

1659.6 Description of Analysis Procedures.

1659.6.1 General. A response spectrum analysis or a time-history analysis, or both, shall be performed in accordance with Sections 1631.4 and 1631.5 and the requirements of this section.

1659.6.2 Input earthquake. The design-basis earthquake shall be used to calculate the total design displacement of the isolation system and the lateral forces and displacements of the isolated structure. The maximum capable earthquake shall be used to calculate the total maximum displacement of the isolation system.

1659.6.3 Response spectrum analysis. Response spectrum analysis shall be performed using a modal damping value for the fundamental mode in the direction of interest not greater than the effective damping of the isolation system or 30 percent of critical, whichever is less. Modal damping values for higher modes shall be selected consistent with those appropriate for response spectrum analysis of the structure above the isolation system on a fixed base.

Response spectrum analysis used to determine the total design displacement and the total maximum displacement shall include simultaneous excitation of the model by 100 percent of the most critical direction of ground motion and 30 percent of the ground motion on the orthogonal axis. The maximum displacement of the isolation system shall be calculated as the vectorial sum of the two orthogonal displacements.

1659.6.4 Time-history analysis. Time-history analysis shall be performed with at least three appropriate pairs of horizontal time-history components, as defined in Section 1659.4.2.

Each pair of time histories shall be applied simultaneously to the model, considering the most disadvantageous location of mass eccentricity. The maximum displacement of the isolation system shall be calculated from the vectorial sum of the two orthogonal displacements at each time step.

The parameter of interest shall be calculated for each time-history analysis. If three time-history analyses are performed, then the maximum response of the parameter of interest shall be used for design. If seven or more time-history analyses are performed, then the average value of the response parameter of interest may be used for design.

1659.7 Design Lateral Force.

1659.7.1 Isolation system and structural elements at or below the isolation system. The isolation system, foundation and all structural elements below the isolation system shall be designed using all of the appropriate provisions for a nonisolated structure and the forces obtained from the dynamic analysis.

1659.7.2 Structural elements above the isolation system. Structural elements above the isolation system shall be designed using the appropriate provisions for a nonisolated structure and the forces obtained from the dynamic analysis divided by a factor of R_I. The R_I factor shall be based on the type of lateral-force-resisting system used for the structure above the isolation system.

1659.7.3 Scaling of results. When the factored lateral shear force on structural elements, determined using either response spectrum or time-history analysis, is less than minimum level prescribed by Sections 1659.1 and 1659.2, then all response parameters, including member forces and moments shall be adjusted upward proportionally.

1659.8 Drift Limits. Maximum interstory drift corresponding to the design lateral force, including displacement due to vertical deformation of the isolation system, shall not exceed the following limits:

1. The maximum interstory drift ratio of the structure above the isolation system, calculated by response spectrum analysis, shall not exceed $0.015/R_I$.

2. The maximum interstory drift ratio of the structure above the isolation system, calculated by time-history analysis considering the force-deflection characteristics of nonlinear elements of the lateral-force-resisting system, shall not exceed $0.020/R_I$.

The secondary effects of the maximum capable earthquake lateral displacement, Δ, of the structure above the isolation system combined with gravity forces shall be investigated if the interstory drift ratio exceeds $0.010/R_I$.

SECTION 1660 — LATERAL LOAD ON ELEMENTS OF STRUCTURES AND NONSTRUCTURAL COMPONENTS SUPPORTED BY STRUCTURES

1660.1 General. Parts or portions of an isolated structure, permanent nonstructural components and the attachments to them, and the attachments for permanent equipment supported by a structure shall be designed to resist seismic forces and displacements as prescribed by this section and the applicable requirements of Section 1632.

1660.2 Forces and Displacements.

1660.2.1 Components at or above the isolation interface. Elements of seismic-isolated structures and nonstructural components, or portions thereof, which are at or above the isolation interface, shall be designed to resist a total lateral seismic force equal to the maximum dynamic response of the element or component under consideration.

> **EXCEPTION:** Elements of seismic-isolated structures and nonstructural components, or portions thereof, may be designed to resist total lateral seismic force as prescribed by Formula (32-1) or (32-2) of Section 1632.

1660.2.2 Components that cross the isolation interface. Elements of seismic-isolated structures and nonstructural components, or portions thereof, that cross the isolation interface shall be designed to withstand the total maximum displacement.

1660.2.3 Components below the isolation interface. Elements of seismic-isolated structures and nonstructural components, or portions thereof, which are below the isolation interface shall be designed and constructed in accordance with the requirements of Section 1632.

SECTION 1661 — DETAILED SYSTEMS REQUIREMENTS

1661.1 General. The isolation system and the structural system shall comply with the requirements of Section 1633 and the material requirements of Chapters 19 through 23. In addition, the isolation system shall comply with the detailed system requirements of this section and the structural system shall comply with the detailed system requirements of this section and the applicable portions of Section 1633.

1661.2 Isolation System.

1661.2.1 Environmental conditions. In addition to the requirements for vertical and lateral loads induced by wind and earthquake, the isolation system shall be designed with consideration given to other environmental conditions including aging effects, creep, fatigue, operating temperature and exposure to moisture or damaging substances.

1661.2.2 Wind forces. Isolated structures shall resist design wind loads at all levels above the isolation interface in accordance with the general wind design provisions. At the isolation interface, a wind restraint system shall be provided to limit lateral displacement in the isolation system to a value equal to that required between floors of the structure above the isolation interface.

1661.2.3 Fire resistance. Fire resistance for the isolation system shall meet that required for the building columns, walls or other structural elements in which it is installed.

Isolator systems required to have a fire-resistive rating shall be protected with approved materials or construction assemblies designed to provide the same degree of fire resistance as the structural element in which it is installed when tested in accordance with UBC Standard 7-1. See Section 703.2.

Such isolation system protection applied to isolator units shall be capable of retarding the transfer of heat to the isolator unit in such a manner that the required gravity load-carrying capacity of the isolator unit will not be impaired after exposure to the standard time-temperature curve fire test prescribed in UBC Standard 7-1 for a duration not less than that required for the fire-resistive rating of the structural element in which it is installed.

Such isolation system protection applied to isolator units shall be suitably designed and securely installed so as not to dislodge, loosen, sustain damage, or otherwise impair its ability to accommodate the seismic movements for which the isolator unit is designed and to maintain its integrity for the purpose of providing the required fire-resistive protection.

1661.2.4 Lateral restoring force. The isolation system shall be configured to produce a restoring force such that the lateral force at the total design displacement is at least $0.025W$ greater than the lateral force at 50 percent of the total design displacement.

> **EXCEPTION:** The isolation system need not be configured to produce a restoring force, as required above, provided the isolation system is capable of remaining stable under full vertical load and accommodating a total maximum displacement equal to the greater of either 3.0 times the total design displacement 36 C_{VM}, inches (For **SI:** 914.4 C_{VM}, mm).

1661.2.5 Displacement restraint. The isolation system may be configured to include a displacement restraint that limits lateral displacement due to the maximum capable earthquake to less than C_{VM}/C_{VD} times the total design displacement, provided that the seismic-isolated structure is designed in accordance with the following criteria when more stringent than the requirements of Section 1629.

1. Maximum capable earthquake response is calculated in accordance with the dynamic analysis requirements of Sections 1631 and 1659, explicitly considering the nonlinear characteristics of the isolation system and the structure above the isolation system.

2. The ultimate capacity of the isolation system and structural elements below the isolation system shall exceed the strength and displacement demands of the maximum capable earthquake.

3. The structure above the isolation system is checked for stability and ductility demand of the maximum capable earthquake.

4. The displacement restraint does not become effective at a displacement less than 0.75 times the total design displacement unless it is demonstrated by analysis that earlier engagement does not result in unsatisfactory performance.

1661.2.6 Vertical load stability. Each element of the isolation system shall be designed to be stable under the maximum vertical load, $1.2D + 1.0L + |E|_{max}$ and the minimum vertical load, $0.80 \cdot |E|_{min}$, at a horizontal displacement equal to the total maximum displacement. The vertical earthquake load on an individual isolation unit due to overturning, $|E|_{max}$ and $|E|_{min}$, shall be based on peak response due to the maximum capable earthquake.

1661.2.7 Overturning. The factor of safety against global structural overturning at the isolation interface shall not be less than 1.0 for required load combinations. All gravity and seismic loading conditions shall be investigated. Seismic forces for overturning calculations shall be based on the maximum capable earthquake and W shall be used for the vertical restoring force.

Local uplift of individual elements is permitted provided the resulting deflections do not cause overstress or instability of the isolator units or other building elements.

1661.2.8 Inspection and replacement.

1. Access for inspection and replacement of all components of the isolation system shall be provided.

2. The architect or engineer of record or a person designated by the architect or engineer of record shall complete a final series of inspections or observations of building separation areas and of components that cross the isolation interface prior to the issuance of the certificate of occupancy for the seismic-isolated building. Such inspections and observations shall indicate that as-built conditions allow for free and unhindered displacement of the structure to maximum design levels and that all components that cross the isolation interface as installed, are able to accommodate the stipulated displacements.

3. Seismic-isolated buildings shall have a periodic monitoring, inspection and maintenance program for the isolation system established by the architect or engineer responsible for the design of the system. The objective of such a program shall be to ensure that all elements of the isolation system are able to perform to minimum design levels at all times.

4. Remodeling, repair or retrofitting at the isolation system interface, including that of components that cross the isolation interface, shall be performed under the direction of an architect or engineer licensed in the appropriate disciplines and experienced in the design and construction of seismic-isolated structures.

5. Horizontal displacement recording devices shall be installed at the isolation interface in seismic-isolated buildings.

1661.2.9 Quality control. A quality control testing program for isolator units shall be established by the engineer responsible for the structural design.

1661.3 Structural System.

1661.3.1 Horizontal distribution of force. A horizontal diaphragm or other structural elements shall provide continuity above the isolation interface and shall have adequate strength and ductility to transmit forces (due to nonuniform ground motion) from one part of the building to another.

1661.3.2 Building separations. Minimum separations between the isolated building and surrounding retaining walls or other fixed obstructions shall not be less than the total maximum displacement.

SECTION 1662 — NONBUILDING STRUCTURES

Nonbuilding structures shall be designed in accordance with the requirements of Section 1634 using design displacements and forces calculated in accordance with Section 1658 or 1659.

SECTION 1663 — FOUNDATIONS

Foundations shall be designed and constructed in accordance with the requirements of Chapter 18 using design forces calculated in accordance with Section 1658 or 1659.

SECTION 1664 — DESIGN AND CONSTRUCTION REVIEW

1664.1 General. A design review of the isolation system and related test programs shall be performed by an independent engineering team including persons licensed in the appropriate disciplines, experienced in seismic analysis methods and the theory and application of seismic isolation.

1664.2 Isolation System. Isolation system design review shall include, but not be limited to, the following:

1. Review of site-specific seismic criteria, including the development of site-specific spectra and ground motion time histories, and all other design criteria developed specifically for the project.

2. Review of the preliminary design, including the determination of the total design displacement of the isolation system design displacement and lateral force design level.

3. Overview and observation of prototype testing (Section 1665).

4. Review of the final design of the entire structural system and all supporting analyses.

5. Review of the isolation system quality control testing program (Section 1661.2.9).

The engineer of record shall submit with the plans and calculations a statement by all members of the independent engineering team stating that the above has been completed.

SECTION 1665 — REQUIRED TESTS OF ISOLATION SYSTEM

1665.1 General. The deformation characteristics and damping values of the isolation system used in the design and analysis of seismic-isolated structures shall be based on the following tests of a selected sample of the components prior to construction.

The isolation system components to be tested shall include the wind restraint system if such systems are used in the design.

The tests specified in this section are for establishing and validating the design properties of the isolation system, and shall not be considered as satisfying the manufacturing quality control tests of Section 1661.2.9.

1665.2 Prototype Tests.

1665.2.1 General. Prototype tests shall be performed separately on two full-size specimens or sets of specimens, as appropriate, of each type and size of isolator unit of the isolation system. The test specimens shall include the wind restraint system, as well as individual isolator units, if such systems are used in the design. Specimens tested shall not be used for construction.

1665.2.2 Record. For each cycle of tests the force-deflection behavior of the test specimen shall be recorded.

1665.2.3 Sequence and cycles. The following sequence of tests shall be performed for the prescribed number of cycles at a vertical load equal to the average $D + 0.5L$ on all isolator units of a common type and size:

1. Twenty fully reversed cycles of loading at a lateral force corresponding to the wind design force.

2. Three fully reversed cycles of loading at each of the following increments of displacement: $0.2\,D_D$, $0.5\,D_D$ and $1.0\,D_D$, $1.0\,D_M$.

3. Three fully reversed cycles at the total maximum displacement, $1.0D_{TM}$.

4. $(15C_{VD}/C_{VA}B_D)$, but not less than 10, fully reversed cycles of loading at 1.0 times the total design displacement, $1.0D_{TD}$.

If an isolator unit is also a vertical load-carrying element, then Item 2 of the sequence of cyclic tests specified above shall be performed for two additional vertical load cases:

$$(1)\quad 1.2D + 0.5L + |E|$$

$$(2)\quad 0.8D - |E|$$

where D and L are defined in Chapter 16, Division III. The vertical test load on an individual isolator unit shall include the load increment due to earthquake overturning, $|E|$, and shall be equal to or greater than the peak earthquake vertical force response corresponding to the test displacement being evaluated. In these tests, the combined vertical load shall be taken as the typical or average downward force on all isolator units of a common type and size.

1665.2.4 Units dependent on loading rates. If the force-deflection properties of the isolator units are dependent on the rate of loading, then each set of tests specified in Section 1665.2.3 shall be performed dynamically at a frequency equal to the inverse of the effective period, T_D, of the isolated structure.

If reduced-scale prototype specimens are used to quantify rate-dependent properties of isolators, the reduced-scale prototype specimens shall be of the same type and material and be manufactured with the same processes and quality as full-scale prototypes, and shall be tested at a frequency that represents full-scale prototype loading rates.

The force-deflection properties of an isolator unit shall be considered to be dependent on the rate of loading if there is greater than a plus or minus 10 percent difference in the effective stiffness at the design displacement when tested at a frequency equal to the inverse of the effective period, T_D, of the isolated structure and when tested at any frequency in the range of 0.1 to 2.0 times the inverse of the effective period, T_D, of the isolated structure.

1665.2.5 Units dependent on bilateral load. If the force-deflection properties of the isolator units are dependent on bilateral load, then the tests specified in Sections 1665.2.3 and 1665.2.4 shall be augmented to include bilateral load at increments of the total design displacement 0.25 and 1.0, 0.50 and 1.0, 0.75 and 1.0, and 1.0 and 1.0.

> **EXCEPTION:** If reduced-scale prototype specimens are used to quantify bilateral-load-dependent properties, then such scaled specimens shall be of the same type and material, and manufactured with the same processes and quality as full-scale prototypes.

The force-deflection properties of an isolator unit shall be considered to be dependent on bilateral load, if the bilateral and unilateral force-deflection properties have greater than a plus or minus 10 percent difference in effective stiffness at the design displacement.

1665.2.6 Maximum and minimum vertical load. Isolator units that carry vertical load shall be statically tested for the maximum and minimum vertical load, at the total maximum displacement. In these tests, the combined vertical loads of $1.2D + 1.0L + |E|_{max}$ shall be taken as the maximum vertical force, and the combined vertical load of $0.8D - |E|_{min}$ shall be taken as the minimum vertical force, on any one isolator unit of a common type and size. The vertical load on an individual isolator unit shall include the load increment due to earthquake overturning, $|E|_{max}$ and $|E|_{min}$, and shall be based on peak response due to the maximum capable earthquake.

1665.2.7 Sacrificial wind-restraint systems. If a sacrificial wind-restraint system is to be utilized, then the ultimate capacity shall be established by test.

1665.2.8 Testing similar units. The prototype tests are not required if an isolator unit is of similar dimensional characteristics and of the same type and material as the prototype isolator unit that has been previously tested using the specified sequence of tests.

1665.3 Determination of Force-deflection Characteristics. The force-deflection characteristics of the isolation system shall be based on the cyclic load tests of isolator prototypes specified in Section 1665.2.3.

As required, the effective stiffness of an isolator unit, k_{eff}, shall be calculated for each cycle of loading by the formula:

$$k_{eff} = \frac{F^+ - F^-}{\Delta^+ - \Delta^-} \qquad (65\text{-}1)$$

where $F+$ and $F-$ are the positive and negative forces at Δ^+ and Δ^-, respectively.

As required, the effective damping (β_{eff}) of an isolator unit shall be calculated for each cycle of loading by the formula:

$$\beta_{eff} = \frac{2}{\pi}\left[\frac{E_{Loop}}{k_{eff}(|\Delta^+| + |\Delta^-|)^2}\right] \qquad (65\text{-}2)$$

where the energy dissipated per cycle of loading, E_{Loop}, and the effective stiffness, k_{eff}, shall be based on test displacements of Δ^+ and Δ^-.

1665.4 System Adequacy. The performance of the test specimens shall be assessed as adequate if the following conditions are satisfied:

1. The force-deflection plots of all tests specified in Section 1665.2 have a positive incremental force-carrying capacity.

2. For each increment of test displacement specified in Section 1665.2.3, Item 2, and for each vertical load case specified in Section 1665.2.3:

 2.1 There is no greater than a plus or minus 10 percent difference between the effective stiffness at each of the three cycles of test and the average value of effective stiffness for each test specimen.

 2.2 There is no greater than a 10 percent difference in the average value of effective stiffness of the two test specimens of a common type and size of the isolator unit over the required three cycles of test.

3. For each specimen there is no greater than a plus or minus 20 percent change in the initial effective stiffness of each test specimen over the $(15C_{VD}/C_{VA}B_D)$, but not less than 10, cycles of the test specified in Section 1665.2.3, Item 4.

4. For each specimen there is no greater than a 20 percent decrease in the initial effective damping over for the $(15C_{VD}/C_{VA}B_D)$, but not less than 10, cycles of the test specified in Section 1665.2.3, Item 4.

5. All specimens of vertical load-carrying elements of the isolation system remain stable at the total maximum displacement for static load as prescribed in Section 1665.2.6.

1665.5 Design Properties of the Isolation System.

1665.5.1 Maximum and minimum effective stiffness. At the design displacement, the maximum and minimum effective stiffnesses of the isolation system, k_{Dmax} and k_{Dmin}, shall be based on the cyclic tests of Section 1665.2.3 and calculated by the formulas:

$$k_{Dmax} = \frac{\sum |F_D^+|_{max} + \sum |F_D^-|_{max}}{2D_D} \qquad (65\text{-}3)$$

$$k_{Dmin} = \frac{\sum |F_D^+|_{min} + \sum |F_D^-|_{min}}{2D_D} \qquad (65\text{-}4)$$

At the maximum displacement, the maximum and minimum effective stiffness of the isolation system, k_{Mmax} and k_{Mmin}, shall be based on the cyclic tests of Section 1665.2.3 and calculated by the formulas:

$$k_{Mmax} = \frac{\sum |F_M^+|_{max} + \sum |F_M^-|_{max}}{2D_M} \qquad (65\text{-}5)$$

$$k_{Mmin} = \frac{\sum |F_M^+|_{min} + \sum |F_M^-|_{min}}{2D_M} \qquad (65\text{-}6)$$

For isolator units that are found by the tests of Sections 1665.2.3, 1665.2.4 and 1665.2.5 to have force-deflection characteristics which vary with vertical load, rate of loading or bilateral load, respectively, the values of k_{Dmax} and k_{Mmax} shall be increased and the values of k_{Dmin} and k_{Mmin} shall be decreased, as necessary, to bound the effects of measured variation in effective stiffness.

1665.5.2 Effective damping. At the design displacement, the effective damping of the isolation system, β_D, shall be based on the cyclic tests of Section 1665.2.3 and calculated by the formula:

$$\beta_D = \frac{1}{2\pi}\left[\frac{\sum E_D}{k_{Dmax}D_D^2}\right] \qquad (65\text{-}7)$$

In Formula (65-7), the total energy dissipated in the isolation system per cycle of design displacement response, $\sum E_D$, shall be taken as the sum of the energy dissipated per cycle in all isolator units measured at test displacements, Δ^+ and Δ^-, that are equal in magnitude to the design displacement, D_D.

At the maximum displacement, the effective damping of the isolation system, β_M, shall be based on the cyclic tests of Section 1665.2.3 and calculated by the formula:

$$\beta_M = \frac{1}{2\pi}\left[\frac{\sum E_M}{k_{Mmax}D_M^2}\right] \qquad (65\text{-}8)$$

In Formula (65-8), the total energy dissipated in the isolation system per cycle of response, E_M, shall be taken as the sum of the energy dissipated per cycle in all isolator units measured at test displacements, Δ^+ and Δ^-, that are equal in magnitude to the maximum displacement, D_M.

TABLE A-16-C—DAMPING COEFFICIENTS, B_D AND B_M

EFFECTIVE DAMPING, β_D or β_M (percentage of critical)[1,2]	B_D or B_M FACTOR
≤ 2	0.8
5	1.0
10	1.2
20	1.5
30	1.7
40	1.9
≥ 50	2.0

[1]The damping coefficient shall be based on the effective damping of the isolation system determined in accordance with the requirements of Section 1665.5.

[2]The damping coefficient shall be based on linear interpolation for effective damping values other than those given.

TABLE A-16-D—MAXIMUM CAPABLE EARTHQUAKE RESPONSE COEFFICIENT, M_M

DESIGN BASIS EARTHQUAKE SHAKING INTENSITY, ZN_v	MAXIMUM CAPABLE EARTHQUAKE RESPONSE COEFFICIENT, M_M
0.075	2.67
0.15	2.0
0.20	1.75
0.30	1.50
0.40	1.25
≥ 0.50	1.20

TABLE A-16-E—STRUCTURAL SYSTEMS ABOVE THE ISOLATION INTERFACE[1]

BASIC STRUCTURAL SYSTEM[2]	LATERAL-FORCE-RESISTING SYSTEM DESCRIPTION	R_l	HEIGHT LIMIT FOR SEISMIC ZONES 3 AND 4 × 304.8 for mm
1. Bearing wall system	1. Light-framed walls with shear panels		
	a. Wood structural panel walls for structures three stories or less	2.0	65
	b. All other light-framed walls	2.0	65
	2. Shear walls		
	a. Concrete	2.0	160
	b. Masonry	2.0	160
	3. Light steel-framed bearing walls with tension-only bracing	1.6	65
	4. Braced frames where bracing carries gravity load		
	a. Steel	1.6	160
	b. Concrete[3]	1.6	—
	c. Heavy timber	1.6	65
2. Building frame system	1. Steel eccentrically braced frame (EBF)	2.0	240
	2. Light-framed walls with shear panels		
	a. Wood structural panel walls for structures three stories or less	2.0	65
	b. All other light-framed walls	2.0	65
	3. Shear walls		
	a. Concrete	2.0	240
	b. Masonry	2.0	160
	4. Ordinary braced frames		
	a. Steel	1.6	160
	b. Concrete[3]	1.6	—
	c. Heavy timber	1.6	65
	5. Special concentrically braced frames		
	a. Steel	2.0	240

(Continued)

TABLE A-16-E—STRUCTURAL SYSTEMS ABOVE THE ISOLATION INTERFACE[1]—(Continued)

BASIC STRUCTURAL SYSTEM[2]	LATERAL-FORCE-RESISTING SYSTEM DESCRIPTION	R_I	HEIGHT LIMIT FOR SEISMIC ZONES 3 AND 4 × 304.8 for mm
3. Moment-resisting frame system	1. Special moment-resisting frame (SMRF)		
	a. Steel	2.0	N.L.
	b. Concrete	2.0	N.L.
	2. Masonry moment-resisting wall frame (MMRWF)	2.0	160
	3. Concrete intermediate moment-resisting frame (IMRF)[4]	2.0	—
	4. Ordinary moment-resisting frame (OMRF)		
	a. Steel[5]	2.0	160
	b. Concrete[6]	2.0	—
	5. Special truss moment frames of steel (STMF)	2.0	240
4. Dual systems	1. Shear walls		
	a. Concrete with SMRF	2.0	N.L.
	b. Concrete with steel OMRF	2.0	160
	c. Concrete with IMRF[4]	2.0	160
	d. Masonry with SMRF	2.0	160
	e. Masonry with steel OMRF	2.0	160
	f. Masonry with concrete IMRF[3]	2.0	—
	g. Masonry with masonry MMRWF	2.0	160
	2. Steel EBF		
	a. With steel SMRF	2.0	N.L.
	b. With steel OMRF	2.0	160
	3. Ordinary braced frames		
	a. Steel with steel SMRF	2.0	N.L.
	b. Steel with steel OMRF	2.0	160
	c. Concrete with concrete SMRF[3]	2.0	—
	d. Concrete with concrete IMRF[3]	2.0	—
	4. Specially concentrically braced frames		
	a. Steel with steel SMRF	2.0	N.L.
	b. Steel with steel OMRF	2.0	160
5. Cantilevered column building systems	1. Cantilevered column elements	1.4	35[7]
6. Shear wall-frame interaction systems	1. Concrete[6]	2.0	—
7. Undefined systems	See Sections 1629.6.7 and 1629.9.2		—

N.L.—no limit.

[1]See Section 1630.4 for combination of structural systems.

[2]Basic structural systems are defined in Section 1629.6.

[3]Prohibited in Seismic Zones 3 and 4.

[4]Prohibited in Seismic Zones 3 and 4, except as permitted in Section 1633.2.

[5]Ordinary moment-resisting frames in Seismic Zone 1 meeting the requirements of Section 2213.6 may use an R_I value of 2.0.

[6]Prohibited in Seismic Zones 2A, 2B, 3 and 4. See Section 1633.2.7.

[7]Total height of the building including cantilevered columns.

TABLE A-16-F—SEISMIC COEFFICIENT, C_{AM}[1]

SOIL PROFILE TYPE	MAXIMUM CAPABLE EARTHQUAKE SHAKING INTENSITY $M_M Z N_a$				
	$M_M Z N_a = 0.075$	$M_M Z N_a = 0.15$	$M_M Z N_a = 0.2$	$M_M Z N_a = 0.3$	$M_M Z N_a \geq 0.4$
S_A	0.06	0.12	0.16	0.24	$0.8 M_M Z N_a$
S_B	0.08	0.15	0.20	0.30	$1.0 M_M Z N_a$
S_C	0.09	0.18	0.24	0.33	$1.0 M_M Z N_a$
S_D	0.12	0.22	0.28	0.36	$1.1 M_M Z N_a$
S_E	0.19	0.30	0.34	0.36	$0.9 M_M Z N_a$
S_F	See Footnote 2				

[1]Linear interpolation may be used to determine the value of C_{AM} for values of $M_M Z N_a$ for other than those shown in the table.

[2]Site-specific geotechnical investigation and dynamic site response analysis shall be performed to determine seismic coefficients for soil.

TABLE A-16-G—SEISMIC COEFFICIENT, C_{VM}[1]

SOIL PROFILE TYPE	MAXIMUM CAPABLE EARTHQUAKE SHAKING INTENSITY $M_M Z N_v$				
	$M_M Z N_v = 0.075$	$M_M Z N_v = 0.15$	$M_M Z N_v = 0.20$	$M_M Z N_v = 0.30$	$M_M Z N_v \geq 0.40$
S_A	0.06	0.12	0.16	0.24	$0.8 M_M Z N_v$
S_B	0.08	0.15	0.20	0.30	$1.0 M_M Z N_v$
S_C	0.13	0.25	0.32	0.45	$1.4 M_M Z N_v$
S_D	0.18	0.32	0.40	0.54	$1.6 M_M Z N_v$
S_E	0.26	0.50	0.64	0.84	$2.4 M_M Z N_v$
S_F	See Footnote 2				

[1]Linear Interpolation may be used to determine the value of C_{VM} for values of $M_M Z N_v$ for other than those shown in the table.

[2]Site-specific geotechnical investigation and dynamic site response analysis shall be performed to determine seismic coefficients for soil.

Appendix Chapter 18
WATERPROOFING AND DAMPPROOFING FOUNDATIONS

SECTION 1820 — SCOPE

Walls, or portions thereof, retaining earth and enclosing interior spaces and floors below grade shall be waterproofed or dampproofed according to this appendix chapter.

> **EXCEPTION:** Walls enclosing crawl spaces.

SECTION 1821 — GROUNDWATER TABLE INVESTIGATION

A subsurface soils investigation shall be made in accordance with Section 1804.3, Item 3, to determine the possibility of the groundwater table rising above the proposed elevation of the floor or floors below grade. The building official may require that this determination be made by an engineer or architect licensed by the state to practice as such.

> **EXCEPTIONS:** 1. When foundation waterproofing is provided.
>
> 2. When dampproofing is provided and the building official finds that there is satisfactory data from adjacent areas to demonstrate that groundwater has not been a problem.

SECTION 1822 — DAMPPROOFING REQUIRED

Where the groundwater investigation required by Section 1821 indicates that a hydrostatic pressure caused by the water table will not occur, floors and walls shall be dampproofed and a subsoil drainage system shall be installed in accordance with this appendix chapter.

> **EXCEPTION:** Wood foundation systems shall be constructed in accordance with Chapter 18, Division II.

SECTION 1823 — FLOOR DAMPPROOFING

1823.1 General. Dampproofing materials shall be installed between the floor and base materials required by Section 1825.2.

> **EXCEPTION:** Where a separate floor is provided above a concrete slab, the dampproofing may be installed on top of the slab.

1823.2 Dampproofing Materials. Dampproofing installed beneath the slab shall consist of not less than 6-mil (0.152 mm) polyethylene, or other approved methods or materials. When permitted to be installed on top of the slab, dampproofing shall consist of not less than 4-mil (0.1 mm) polyethylene, mopped-on bitumen or other approved methods or materials. Joints in membranes shall be lapped and sealed in an approved manner.

SECTION 1824 — WALL DAMPPROOFING

1824.1 General. Dampproofing materials shall be installed on the exterior surface of walls, and shall extend from a point 6 inches (152 mm) above grade, down to the top of the spread portion of the footing.

1824.2 Surface Preparation. Prior to application of dampproofing materials on concrete walls, fins or sharp projections that may pierce the membrane shall be removed and all holes and recesses resulting from the removal of form ties shall be sealed with a dry-pack mortar, bituminous material, or other approved methods or materials.

1824.3 Dampproofing Materials. Wall dampproofing shall consist of a bituminous material, acrylic modified cement base coating, any of the materials permitted for waterproofing in Section 1828.4, or other approved methods or materials. When such materials are not approved for direct application to unit masonry, the wall shall be parged on the exterior surface below grade with not less than $^3/_8$ inch (9.5 mm) of portland cement mortar.

SECTION 1825 — OTHER DAMPPROOFING REQUIREMENTS

1825.1 Subsoil Drainage System. When dampproofing is required, a base material shall be installed under the floor and a drain shall be installed around the foundation perimeter in accordance with this subsection.

> **EXCEPTION:** When the finished ground level is below the floor level for more than 25 percent of the perimeter of the building, the base material required by Section 1825.2 need not be provided and the foundation drain required by Section 1825.3 need be provided only around that portion of the building where the ground level is above the floor level.

1825.2 Base Material. Floors shall be placed over base material not less than 4 inches (102 mm) in thickness consisting of gravel or crushed stone containing not more than 10 percent material that passes a No. 4 sieve (4.75 mm).

1825.3 Foundation Drain. The drain shall consist of gravel, crushed stone or drain tile.

Gravel or crushed stone drains shall contain not more than 10 percent material that passes a No. 4 sieve (4.75 mm). The drain shall extend a minimum of 12 inches (305 mm) beyond the outside edge of the footing. The depth shall be such that the bottom of the drain is not higher than the bottom of the base material under the floor, and the top of the drain is not less than 6 inches (152 mm) above the spread portion of the footing. The top of the drain shall be covered with an approved filter membrane material.

When drain tile or perforated pipe is used, the invert of the pipe or tile shall be not higher than the floor elevation. The top of joints or the top of perforations shall be protected with an approved filter membrane material. The pipe or tile shall be placed on not less than 2 inches (51 mm) of gravel or crushed stone complying with this section and covered with not less than 6 inches (152 mm) of the same material.

1825.4 Drainage Disposal. The floor base and foundation perimeter drain shall discharge by gravity or mechanical means into an approved drainage system.

> **EXCEPTION:** Where a site is located in well-drained gravel or sand-gravel mixture soils, a dedicated drainage system need not be provided.

SECTION 1826 — WATERPROOFING REQUIRED

Where the groundwater investigation required by Section 1821 indicates that a hydrostatic pressure caused by the water table does exist, walls and floors shall be waterproofed in accordance with this appendix chapter.

> **EXCEPTIONS:** 1. When the groundwater table can be lowered and maintained at an elevation not less than 6 inches (152 mm) below the bottom of the lowest floor, dampproofing provisions in accordance with Section 1822 may be used in lieu of waterproofing.
>
> The design of the system to lower the groundwater table shall be based on accepted principles of engineering which shall consider, but not necessarily be limited to, the permeability of the soil, the rate at

which water enters the drainage system, the rated capacity of pumps, the head against which pumps are to pump, and the rated capacity of the disposal area of the system.

2. Wood foundation systems constructed in accordance with Chapter 18, Division II, are to be provided with additional moisture-control measures as specified in Section 1812.

SECTION 1827 — FLOOR WATERPROOFING

1827.1 General. Floors required to be waterproofed shall be of concrete designed to withstand anticipated hydrostatic pressure.

1827.2 Waterproofing Materials. Waterproofing of floors shall be accomplished by placing under the slab a membrane of rubberized asphalt, polymer-modified asphalt, butyl rubber, neoprene, or not less than 6-mil (0.15 mm) polyvinyl chloride or polyethylene, or other approved materials capable of bridging nonstructural cracks. Joints in the membrane shall be lapped not less than 6 inches (152 mm) and sealed in an approved manner.

SECTION 1828 — WALL WATERPROOFING

1828.1 General. Walls required to be waterproofed shall be of concrete or masonry designed to withstand the anticipated hydrostatic pressure and other lateral loads.

1828.2 Wall Preparation. Prior to the application of waterproofing materials on concrete or masonry walls, the wall surfaces shall be prepared in accordance with Section 1824.2.

1828.3 Where Required. Waterproofing shall be applied from a point 12 inches (305 mm) above the maximum elevation of the groundwater table down to the top of the spread portion of the footing. The remainder of the wall located below grade shall be dampproofed with materials in accordance with Section 1824.3.

1828.4 Waterproofing Materials. Waterproofing shall consist of rubberized asphalt, polymer-modified asphalt, butyl rubber, or other approved materials capable of bridging nonstructural cracks. Joints in the membrane shall be lapped and sealed in an approved manner.

1828.5 Joints. Joints in walls and floors, and between the wall and floor, and penetrations of the wall and floor shall be made watertight using approved methods and materials.

SECTION 1829 — OTHER DAMPPROOFING AND WATERPROOFING REQUIREMENTS

1829.1 Placement of Backfill. The excavation outside the foundation shall be backfilled with soil which is free of organic material, construction debris and large rocks. The backfill shall be placed in lifts and compacted in a manner which does not damage the waterproofing or dampproofing material or structurally damage the wall.

1829.2 Site Grading. The ground immediately adjacent to the foundation shall be sloped away from the building at not less than 1 unit vertical in 12 units horizontal (8.3% slope) for a minimum distance of 6 feet (1829 mm) measured perpendicular to the face of the wall or an approved alternate method of diverting water away from the foundation shall be used. Consideration shall be given to possible additional settlement of the backfill when establishing final ground level adjacent to the foundation.

1829.3 Erosion Protection. Where water impacts the ground from the edge of the roof, downspout, scupper, valley, or other rainwater collection or diversion device, provisions shall be used to prevent soil erosion and direct the water away from the foundation.

Appendix Chapter 19
PROTECTION OF RESIDENTIAL CONCRETE EXPOSED TO FREEZING AND THAWING

SECTION 1928 — GENERAL

1928.1 Purpose. The purpose of this appendix is to provide minimum standards for the protection of residential concrete exposed to freezing and thawing conditions.

1928.2 Scope. The provisions of this appendix apply to concrete used in buildings of Groups R and U Occupancies that are three stories or less in height.

1928.3 Special Provisions. Normal-weight aggregate concrete used in buildings of Groups R and U Occupancies three stories or less in height which are subject to de-icer chemicals or freezing and thawing conditions as determined from Figure A-19-1 shall comply with the requirements of Table A-19-A.

TABLE A-19-A—MINIMUM SPECIFIED COMPRESSIVE STRENGTH OF CONCRETE[1]

TYPE OR LOCATION OF CONCRETE CONSTRUCTION	MINIMUM SPECIFIED COMPRESSIVE STRENGTH[2] (f'_c)		
	× 6.89 for kPa		
	Weathering Potential[3]		
	Negligible	Moderate	Severe
Basement walls and foundations not exposed to the weather	2,500	2,500	2,500[4]
Basement slabs and interior slabs on grade, except garage floor slabs	2,500	2,500	2,500[4]
Basement walls, foundation walls, exterior walls and other vertical concrete work exposed to the weather	2,500	3,000[5]	3,000[5]
Porches, carport slabs and steps exposed to the weather, and garage floor slabs	2,500	3,000[5]	3,500[5]

[1]Increases in compressive strength above those used in the design shall not cause implementation of the special inspection provisions of Section 1701.5, Item 1.

[2]At 28 days, pounds per square inch (kPa).

[3]See Figure A-19-1 for weathering potential.

[4]Concrete in these locations which may be subject to freezing and thawing during construction shall be air-entrained concrete in accordance with Footnote 5.

[5]Concrete shall be air entrained. Total air content (percentage by volume of concrete) shall not be less than 5 percent or more than 7 percent.

SEVERE

MODERATE

MILD

WEATHERING REGIONS (WEATHERING INDEX)

FIGURE A-19-1—WEATHERING REGIONS FOR RESIDENTIAL CONCRETE

NOTES:

[1]The three exposures are:

A. Severe—Outdoor exposure in a cold climate where concrete may be exposed to the use of de-icing salts or where there may be a continuous presence of moisture during frequent cycles of freezing and thawing. Examples are pavements, driveways, walks, curbs, steps, porches and slabs in unheated garages. Destructive action from de-icing salts may occur either from direct application or from being carried onto an unsalted area from a salted area, such as on the undercarriage of a car traveling on a salted street but parked on an unsalted driveway or garage slab.

B. Moderate—Outdoor exposure in a climate where concrete will not be exposed to the application of de-icing salts but will occasionally be exposed to freezing and thawing.

C. Mild—Any exposure where freezing and thawing in the presence of moisture is rare or totally absent.

[2]Data needed to determine the weathering index for any locality may be found or estimated from the tables of Local Climatological Data, published by the Weather Bureau, U.S. Department of Commerce.

[3]The weathering regions map provides the location of severe, moderate and mild winter weathering areas as they occur in the United States (Alaska and Hawaii are classified as severe and mild, respectively). The map cannot be precise. This is especially true in mountainous areas where conditions change dramatically within very short distances. It is intended to classify as severe any area in which weathering conditions may cause de-icing salt to be used, either by individuals or for street or highway maintenance. These conditions are significant snowfall combined with extended periods during which there is little or no natural thawing. If there is any doubt about which of two regions is applicable, the more severe exposure should be selected.

[4]The Weathering Index:

Severe—As a guideline, the number of days during which the temperature does not rise above 32°F (0°C) is multiplied by the inches of snowfall. An index of 150 or more is classified as severe. Cold, humid climates may be more severe than cold, dry climates for a given index.

Moderate, Mild—Multiply the inches of precipitation times the number of days the temperature registers below 32°F (0°C) Use the occurrence between the first day in the fall and the last day in the spring that the temperature registers below 32°F (0°C) An index above 200 is moderate. An index below 200 is mild.

Appendix Chapter 21
PRESCRIPTIVE MASONRY CONSTRUCTION IN HIGH-WIND AREAS

SECTION 2112 — GENERAL

2112.1 Purpose. The provisions of this chapter are intended to promote public safety and welfare by reducing the risk of wind-induced damages to masonry construction.

2112.2 Scope. The requirements of this chapter shall apply to masonry construction in buildings when all of the following conditions are met:

1. The building is located in an area with a basic wind speed from 80 through 110 miles per hour (mph) (129 km/h through 177 km/h).

2. The building is located in Seismic Zone 0, 1 or 2.

3. The building does not exceed two stories.

4. Floor and roof joists shall be wood or steel or of precast hollowcore concrete planks with a maximum span of 32 feet (9754 mm) between bearing walls. Masonry walls shall be provided for the support of steel joists or concrete planks.

5. The building is of regular shape.

2112.3 General. The requirements of Chapter 21 are applicable except as specifically modified by this chapter. Other methods may be used provided a satisfactory design is submitted showing compliance with the provisions of Chapter 16, Part II, and other applicable provisions of this code.

Wood floor, roof and interior walls shall be constructed as specified in Appendix Chapter 23 and as further regulated in this section.

In areas where the wind speed exceeds 110 mph (177 km/h), masonry buildings shall be designed in accordance with Chapter 16, Part II, and other applicable provisions of this code.

Buildings of unusual shape or size, or split-level construction, shall be designed in accordance with Chapter 16, Part II, and other applicable provisions of this code.

In addition to the other provisions of this chapter, foundations for buildings in areas subject to wave action or tidal surge shall be designed in accordance with approved national standards.

All metal connectors and fasteners used in exposed locations or in areas otherwise subject to corrosion shall be of corrosion-resistant or noncorrosive material. When the terms "corrosion resistant" or "noncorrosive" are used in this chapter, they shall mean having a corrosion resistance equal to or greater than a hot-dipped galvanized coating of 1.5 ounces of zinc per square foot (3.95 g/m^2) of surface area. When an element is required to be corrosion resistant or noncorrosive, all of its parts, such as screws, nails, wire, dowels, bolts, nuts, washers, shims, anchors, ties and attachments, shall be corrosion resistant.

2112.4 Materials.

2112.4.1 General. All masonry materials shall comply with Section 2102.2 as applicable for standards of quality.

2112.4.2 Hollow-unit masonry.

1. Exterior concrete block shall be a minimum of Grade N-II with a compressive strength of not less than 1,900 pounds per square inch (psi) (13 091 kPa) on the net area.

2. Interior concrete block shall be a minimum of Grade S-II with a compressive strength of not less than 700 psi (4823 kPa) on the gross area.

3. Exterior clay or shale hollow brick shall have a compressive strength of not less than 2,500 psi (17 225 kPa) on the net area. Such hollow brick shall be at least Grade MW except that where subject to severe freezing it shall be Grade SW.

4. Interior clay or shale hollow brick shall be Grade MW with a compressive strength of 2,000 psi (13 780 kPa) on the net area.

2112.4.3 Solid masonry.

1. Exterior clay or shale bricks shall have a compressive strength of not less than 2,500 psi (17 225 kPa) on the net area.

2. Exterior clay or shale bricks shall be Grade MW, except that where subject to severe freezing they shall be Grade SW.

3. Interior clay or shale bricks shall have a compressive strength of not less than 2,000 psi (13 780 kPa).

2112.4.4 Grout. Grout shall achieve a compressive strength of not less than 2,000 psi (13 780 kPa).

2112.4.5 Mortar. Mortar for exterior walls and for interior shear walls shall be Type M or Type S.

2112.5 Construction Requirements. Grouted cavity wall and block wall construction shall comply with Section 2104.

Unburned clay masonry and stone masonry shall not be used.

2112.6 Foundations. Footings shall have a thickness of not less than 8 inches (203 mm) and shall comply with Tables A-21-A-1 and A-21-A-2 for width. See Figure A-21-1 for other applicable details.

Footings shall extend 18 inches (457 mm) below the undisturbed ground surface or the frost depth, whichever is deeper.

Foundation stem walls shall be as wide as the wall they support. They shall be reinforced with reinforcing bar sizes and spacing to match the reinforcement of the walls they support.

Basement and other below-grade walls shall comply with Table A-21-B.

2112.7 Drainage. Basement walls and other walls or portions thereof retaining more than 3 feet (914 mm) of earth and enclosing interior spaces or floors below grade shall have a minimum 4-inch-diameter (102 mm) footing drain as illustrated in Table A-21-B and Figure A-21-3.

The finish elevations around the building shall be graded to provide a slope away from the building of not less than $^1/_4$ unit vertical in 12 units horizontal (2% slope).

2112.8 Wall Construction.

2112.8.1 Minimum thickness. Reinforced exterior bearing walls shall have a minimum 8-inch (203 mm) nominal thickness. Interior masonry nonbearing walls shall have a minimum 6-inch (152 mm) nominal thickness. Unreinforced grouted brick walls shall have a minimum 10-inch (254 mm) thickness. Unreinforced hollow-unit and solid masonry shall have a minimum 8-inch (203 mm) nominal thickness.

EXCEPTION: In buildings not more than two stories or 26 feet (7924.8 mm) in height, masonry walls may be of 8-inch (203 mm) nominal thickness. Solid masonry walls in one-story buildings may be of 6-inch (152 mm) nominal thickness when not over 9 feet (2743 mm) in height, provided that when gable construction is used an additional 6 feet (1829 mm) are permitted to the peak of the gable.

2112.8.2 Lateral support and height. All walls shall be laterally supported at the top and bottom. The maximum unsupported height of bearing walls or other masonry walls shall be 12 feet (3658 mm). Gable-end walls may be 15 feet (4572 mm) at their peak.

Wood-framed gable-end walls on buildings shall comply with Table A-21-I and Figure A-21-17 or A-21-18.

2112.8.3 Walls in Seismic Zone 2 and use of stack bond. In Seismic Zone 2, walls shall comply with Figure A-21-2 as a minimum. Walls with stack bond shall be designed.

2112.8.4 Lintels. The span of lintels over openings shall not exceed 12 feet (3658 mm), and lintels shall be reinforced. The reinforcement bars shall extend not less than 2 feet (610 mm) beyond the edge of opening and into lintel supports.

Lintel reinforcement shall be within fully grouted cells in accordance with Table A-21-E.

2112.8.5 Reinforcement. Walls shall be reinforced as shown in Tables A-21-C-1 through A-21-C-5 and Figure A-21-2.

2112.8.6 Anchorage of walls to floors and roofs. Anchors between walls and floors or roofs shall be embedded in grouted cells or cavities and shall conform to Section 2112.9.

2112.9 Floor and Roof Systems. The anchorage of wood roof systems which are supported by masonry walls shall comply with Appendix Sections 2337.5.1 and 2337.5.8, Table A-21-D and Figure A-21-7.

Wood roof and floor systems which are supported by ledgers at the inside face of masonry walls shall comply with Table A-21-D, Part I.

The ends of joist girders shall extend a distance of not less than 6 inches (152 mm) over masonry or concrete supports and be attached to a steel bearing plate. This plate is to be located not more than $^1/_2$ inch (12.7 mm) from the face of the wall and is to be not less than 9 inches (229 mm) wide perpendicular to the length of the joist girder. Ends of joist girders resting on steel bearing plates on masonry or structural concrete shall be attached thereto with a minimum of two $^1/_4$-inch (6.4 mm) fillet welds 2 inches (51 mm) long, or with two $^3/_4$-inch (19 mm) bolts.

Ends of joist girders resting on steel supports shall be connected thereto with a minimum of two $^1/_4$-inch (6.4 mm) fillet welds 2 inches (51 mm) long, or with two $^3/_4$-inch (19 mm) bolts. In steel frames, joist girders at column lines shall be field bolted to the columns to provide lateral stability during construction.

Steel joist roof and floor systems shall be anchored in accordance with Table A-21-H.

Wall ties spaced as shown in Table A-21-D, Part II, shall connect to framing or blocking at roofs and walls. Wall ties shall enter grouted cells or cavities and shall be $1^1/_8$-inch (29 mm) minimum width by 0.036 inch (0.91 mm) (No. 20 galvanized sheet gage) sheet steel.

Roof and floor hollow-core precast plank systems shall be anchored in accordance with Table A-21-G.

Roof uplift anchorage shall enter a grouted bond beam reinforced with horizontal bars as shown in Tables A-21-C-1 through A-21-C-5 and Figure A-21-7.

2112.10 Lateral Force Resistance.

2112.10.1 Complete load path and uplift resistance. Strapping, approved framing anchors, and mechanical fasteners, bond beams, and vertical reinforcement shall be installed to provide a continuous tie from the roof to the foundation system. (See Figure A-21-8.) In addition, roof and floor systems, masonry shear walls, or masonry or wood cross walls shall provide lateral stability.

2112.10.2 Floor and roof diaphragms. Floor and roof diaphragms shall be connected to masonry walls as shown in Table A-21-F, Part II.

Gabled and sloped roof members not supported at the ridge shall be tied by ceiling joists or equivalent lateral ties located as close to where the roof member bears on the wall as is practically possible, at not more than 48 inches (1219 mm) on center. Collar ties shall not be used for these lateral ties. (See Figure A-21-17 and Table A-21-I.)

2112.10.3 Walls. Masonry walls shall be provided around all sides of floor and roof systems in accordance with Figure A-21-9 and Table A-21-F.

The cumulative length of exterior masonry walls along each side of the floor or roof systems shall be at least 20 percent of the parallel dimension. Required elements shall be without openings and shall not be less than 48 inches (1219 mm) in width.

Interior cross walls (nonbearing) at right angles to bearing walls shall be provided when the length of the building perpendicular to the span of the floor or roof framing exceeds twice the distance between shear walls or 32 feet (9754 mm), whichever is greater. Cross walls, when required, shall conform to Section 2112.10.4.

2112.10.4 Interior cross walls. When required by Table A-21-F, Part I, masonry walls shall be at least 6 feet (1829 mm) long and reinforced with 9 gage wire joint reinforcement spaced not more than 16 inches (406 mm) on center. Cross walls shall comply with Footnote 3 of Table A-21-F, Part I.

Interior wood stud walls may be used to resist the wind load from one-story masonry buildings in areas where the basic wind speed is 100 mph (161 km/h), Exposure C or less, and 110 mph (177 km/h), Exposure B. When wood stud walls are so used, they shall:

1. Be perpendicular to exterior masonry walls at 15 feet (4572 mm) or less on center.

2. Be at least 8 feet (2438 mm) long without openings and be sheathed on at least one side with $^{15}/_{32}$-inch (12 mm) wood structural panel nailed with 8d common or galvanized box nails at 6 inches (152 mm) on center edge and field nailing. All unsupported edges of wood structural panels shall be blocked.

3. Be connected to wood blocking or wood joists below with two 16d nails at 16 inches (406 mm) on center through their sill plates. They shall be connected to footings with $^1/_2$-inch-diameter (12.7 mm) bolts at 3 feet 6 inches (1067 mm) on center.

4. Connect to wood roof systems as outlined in Table A-21-F, Part II, as a cross wall. Wood structural panel roof sheathing shall have all unsupported edges blocked.

**TABLE A-21-A-1—EXTERIOR FOUNDATION REQUIREMENTS FOR
MASONRY BUILDINGS WITH 6- AND 8-INCH-THICK WALLS
(Wood or Steel Framing)
(Width of Footings in Inches)[1,2,3]
See Figure A-21-1 for typical details.**

		ONE-STORY BUILDINGS			TWO-STORY BUILDINGS					
		Roof Live Load[4]			Roof Live Load[4] (psf)					
		× 0.0479 for kN/m²			× 0.0479 for kN/m²					
					20		30		40	
					Plus Floor Live Load[5] (psf)					
					× 0.0479 for kN/m²					
WALL HEIGHT (feet)	SPAN TO BEARING WALLS (feet)	20 psf (inches)	30 psf (inches)	40 psf (inches)	50	100	50	100	50	100
× 304.8 for mm					Minimum Width of Footing (inches)					
					× 25.4 for mm					
8	8	12			12	12	12	12	12	12
	16				12	14	12	14	12	14
	24				14	18	14	18	16	18
	32				16	20	18	20	18	20
10	8	12			12	12	12	12	12	12
	16				14	16	14	16	14	16
	24				16	20	16	18	16	20
	32				20	24	20	22	20	24
12	8	12	12	12	12	14	12	14	12	14
	16	12	12	12	16	18	16	16	14	16
	24	12	12	14	18	20	18	20	18	20
	32	12	14	16	20	22	22	22	22	24

[1]For buildings with under-floor space or basements, footing thickness is to be a minimum of 12 inches (305 mm). It shall be reinforced with No. 4 bars at 24 inches (610 mm) on center when its width is required to be 18 inches (457 mm) or larger and it supports more than the roof and one floor.

[2]Soil to be at least Class 4 as shown in Table 18-I-A.

[3]Footings are a minimum of 10 inches (254 mm) thick for a one-story building and 12 inches (305 mm) thick for a two-story building. Bottom of footing to be 18 inches (457 mm) below grade or the frost depth, whichever is deeper. Footing to be reinforced with No. 4 bars at 24 inches (610 mm) on center when supporting more than the roof and one floor.

[4]From Table 21-C or local snow load tables. For areas without snow loads use 20 pounds per square foot (0.96 kN/m²).

[5]From Table 21-A. For intermediate floor loads go to next higher value.

TABLE A-21-A-2—INTERIOR FOUNDATION REQUIREMENTS FOR MASONRY BUILDINGS WITH 6- AND 8-INCH-THICK WALLS
(Wood or Steel Framing)
(Width of Footings in Inches)[1,2,3,4]
See Figure A-21-1 for typical details.

WALL HEIGHT (feet)	SPAN TO BEARING WALLS (feet)	ONE-STORY BUILDINGS Roof Live Load[5] × 0.0479 for kN/m²			TWO-STORY BUILDINGS Roof Live Load[5] (psf) × 0.0479 for kN/m² Plus Floor Live Load[6] (psf) × 0.0479 for kN/m²					
					20		30		40	
		20 psf (inches)	30 psf (inches)	40 psf (inches)	50	100	50	100	50	100
× 304.8 for mm		Minimum Width of Footing (inches) × 25.4 for mm								
8	8	12	12	12	12	14	12	14	12	14
	16	12	12	12	16	20	18	20	18	22
	24	12	12	14	20	26	22	28	22	28
	32	14	14	16	24	28	26	32	28	34
10	8	12	12	12	14	16	14	16	14	16
	16	12	12	12	20	24	20	22	20	22
	24	12	14	14	22	28	22	28	22	28
	32	14	14	16	26	34	26	32	28	34
12	8	12	12	12	14	16	16	18	16	18
	16	12	14	16	20	24	20	22	20	22
	24	14	14	16	24	28	22	28	24	28
	32	16	16	18	28	30	28	32	28	34

[1]For buildings with under-floor space or basements, footing thickness is to be a minimum of 12 inches (305 mm). It shall be reinforced with No. 4 bars at 24 inches (610 mm) on center when its width is required to be 18 inches (457 mm) or larger and it supports more than the roof and one floor.
[2]Soil to be at least Class 4 as shown in Table 18-I-A.
[3]Footings are 10 inches (254 mm) thick for up to 24 inches (610 mm) wide and 12 inches (305 mm) thick for up to 34 inches (864 mm) wide. Footings shall be reinforced with No. 4 bars at 24 inches (610 mm) on center when supporting more than the roof and one floor.
[4]These interior footings support roof-ceiling or floors or both for a distance on each side equal to the span length shown. A tributary width equal to the span length may be used.
[5]From Table 16-C or local snow load tables. For areas without snow loads use 20 pounds per square foot (0.96 kN/m²).
[6]From Table 16-A. For intermediate floor loads go to next higher value.

TABLE A-21-B—VERTICAL REINFORCEMENT AND TOP RESTRAINT FOR VARIOUS HEIGHTS OF BASEMENT AND OTHER BELOW-GRADE WALLS

DESIGN ASSUMPTIONS
A. Materials:
   1. **Concrete Masonry Units**—Grade hollow load-bearing units conforming to Section 2112.4.2 for strength of units should not be less than that required for applicable f'_m.
   2. **Mortar**—Type M, 2,500 psi (17 240 kPa) strength.
   3. **Corefill**—Fine or coarse grout (UBC Standard 21-19) with an ultimate strength (28 days) of at least 2,500 psi. (17 240 kPa)
   4. **Reinforcement**—Deformed billet-steel bars.
   5. 1,500 psf (71.8 kPa) soil bearing required.[1]
B. Allowable stresses in accordance with Section 2106 and Table 21-M.

Soil Equiv.-fluid wt. = 30 pcf[1] (4.71 kN/m³)			Vertical Reinforcement with Axial Compressive Load (P) Equal to or Less than 5,000 lb./lin. ft. (72.92 kN/m)		
	Floor Connection[2,3]		f'_m = 1,500 psi (10 335 kPa)		
Wall Depth below Grade h (feet)	Wood Floor		Spacing of Reinforcement (inches)[4]		
× 304.8 for mm	Bolt and Spacing	Angle Clip Spacing	× 25.4 for mm		
8-Inch Walls			No. 3	No. 4	No. 5
× 25.4 for mm					
4	1/2″ at 60″	48″ o.c.	24	40	56
5	1/2″ at 40″	32″ o.c.	16	24	40
6	5/8″ at 32″	20″ o.c.	—	16	24
			Spacing of Reinforcement (inches)		
10-Inch Walls			× 25.4 for mm		
× 25.4 for mm			No. 4	No. 5	No. 6	No. 7
6	5/8″ at 32″	20″ o.c.	40	56	64	72
7	5/8″ at 24″	16″ o.c.	24	40	48	56
9	3/4″ at 20″	2 at 24″ o.c.	16	24	32	40
			Spacing of Reinforcement (inches)		
12-Inch Walls			× 25.4 for mm		
× 25.4 for mm			No. 4	No. 5	No. 6	No. 7
7	5/8″ at 24″	16″ o.c.	40	56	80	80
8	3/4″ at 20″	2 at 24″ o.c.	32	48	56	64
9	7/8″ at 18″	2 at 18″ o.c.	24	40	48	48
10	1″ at 16″	2 at 16″ o.c.	16	32	40	40

[1]Soil type is at least Class 4 as shown in Table 18-I-A.
[2]There shall be no backfill placed until after the wall is anchored to the floor and seven days have passed after grouting.
[3]For Figure A-21-4 only.
[4]See Figure A-21-5 for placement of reinforcement.

TABLE A-21-C-1—VERTICAL REINFORCING STEEL REQUIREMENTS FOR 6-INCH-THICK (153 mm) MASONRY WALLS[1] IN AREAS WHERE BASIC WIND SPEEDS ARE 80 MILES PER HOUR (129 km/h) OR GREATER[2,3,4,5]
(Wood or Steel Roof and Floor Framing)

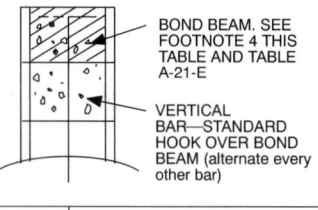

BOND BEAM. SEE FOOTNOTE 4 THIS TABLE AND TABLE A-21-E

VERTICAL BAR—STANDARD HOOK OVER BOND BEAM (alternate every other bar)

Criteria: Roof Live Load = 20 psf to 40 psf (0.96 kN/m² to 1.9 kN/m²);
Floor Live Load = 50 psf (2.4 kN/m²); enclosed building[6]

EXPO-SURE	STORIES	UNSUP-PORTED HEIGHT (feet) ×304.8 for mm	80 MPH 8	16	24	32	90 MPH 8	16	24	32	100 MPH 8	16	24	32	110 MPH 8	16	24	32
B	One-story building	8	NR*								No. 4 80	No. 4 80	No. 4 80	No. 4 80	No. 4 64	No. 4 64	No. 4 72	No. 4 88
		10	No. 4 80	No. 4 88	No. 4 96	No. 4 96	No. 4 64	No. 4 64	No. 4 72	No. 4 80	No. 4 48	No. 4 48	No. 4 48	No. 4 56	No. 4 40	No. 4 40	No. 4 40	No. 4 48
		12	No. 4 48	No. 4 48	No. 4 56	No. 4 64	No. 4 40	No. 4 40	No. 4 48	No. 4 48	No. 5 48	No. 5 48	No. 5 56	No. 5 56	No. 5 40	No. 5 40	No. 5 40	No. 5 40
	Two-story building		Design required or use 8-inch or larger units for two-story condition.															
C	One-story building	8	No. 4 72	No. 4 72	No. 4 72	No. 4 96	No. 4 56	No. 4 56	No. 4 56	No. 4 56	No. 4 40	No. 4 40	No. 4 48	No. 4 48	No. 4 32	No. 4 32	No. 4 32	No. 4 40
		10	No. 4 40	No. 4 40	No. 4 40	No. 4 48	No. 4 32	No. 4 32	No. 4 32	No. 4 32	No. 5 40	No. 5 40	No. 5 40	No. 5 48	No. 5 32	No. 5 32	No. 5 32	No. 5 40
		12	No. 5 40	No. 5 48	No. 5 48	No. 5 48	No. 5 32	No. 5 32	No. 5 32	No. 5 40	Use 8-inch or larger units							
	Two-story building		Design required or use 8-inch or larger units for two-story condition.															
D	One-story building	8	No. 4 56	No. 4 56	No. 4 64	No. 4 80	No. 4 48	No. 4 48	No. 4 48	No. 4 48	No. 4 32	No. 4 40	No. 4 40	No. 4 40	No. 4 32	No. 4 32	No. 4 32	No. 4 32
		10	No. 4 32	No. 4 32	No. 4 32	No. 4 40	No. 5 40	No. 5 40	No. 5 48	No. 5 48	No. 5 48	No. 5 32	No. 5 32	No. 5 40				
		12	No. 5 32	No. 5 40	No. 5 40	No. 5 40	Use 8-inch or larger units											
	Two-story building		Design required or use 8-inch or larger units for two-story condition.															

Column header notes: Span between Bearing Walls (feet) × 304.8 for mm; Size of Rebar and Spacing (inches) × 25.4 for mm; × 1.61 for km/h.

*NR — No vertical reinforcement required. However, see Table A-21-F for shear wall reinforcement.

[1]These values are for walls with running bond. For stack bond see Section 2112.8.3.

[2]The figure on top of the listed data is the bar size; the figure below it is the maximum spacing in inches (mm). Reinforcing bar strength shall be A 615 Grade 60. The vertical bars are centered in the middle of the wall.

[3]Roof load is assumed to be concentrically loaded on the wall. For roofs which hang on ledgers, a design is required.

[4]Minimum horizontal reinforcement shall be one No. 4 at the ledger and foundation. Also, see Table A-21-E for lintels and Table A-21-F for shear wall reinforcing where applicable.

[5]Hook vertical bars over bond beam bars as shown. Extend bars into footing using lap splices where necessary.

[6]Design required for open buildings of 6-inch-thick (153 mm) masonry.
    To use this table, check criteria by the following method:
   [6.1]Choose proper roof live load from Table 16-C or snow load criteria for the locality in which the building is located.
   [6.2]Check if building is enclosed or partially enclosed by the procedure in Chapter 16, Part III.
   [6.3]Choose proper floor load from Table 16-A. [For loads less than 50 pounds per square foot (psf) (2.4 kN/m²), use 50 psf (2.4 kN/m²), and for loads between 50 psf (2.4 kN/m²) and 100 psf (4.8 kN/m²), use 100 psf (4.8 kN/m²).]
   [6.4]Find proper wind speed and exposure for the site—see Figure 16-1, Chapter 16, Sections 1619 and 1620.
   [6.5]Within the proper vertical column, choose appropriate span-to-bearing wall and appropriate height and story.
   [6.6]Read proper size and spacing of reinforcement for the thickness of the wall mentioned in the title of the table. (Equivalent area of steel, taking spacing into account, may be substituted.)
   [6.7]For buildings in Seismic Zone 2 (see Figure 16-2 in Chapter 16), use minimum reinforcement in Figure A-21-2 if it is more restrictive than the table values.

TABLE A-21-C-2—VERTICAL REINFORCING STEEL REQUIREMENTS FOR 8-INCH-THICK (203 mm) MASONRY WALLS[1] IN AREAS WHERE BASIC WIND SPEEDS ARE 80 MILES PER HOUR (129 km/h) OR GREATER[2,3,4,5]
(Wood or Steel Roof and Floor Framing)

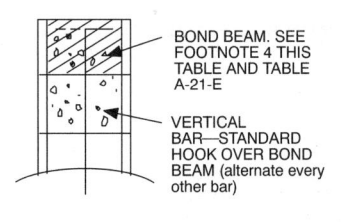

BOND BEAM. SEE FOOTNOTE 4 THIS TABLE AND TABLE A-21-E

VERTICAL BAR—STANDARD HOOK OVER BOND BEAM (alternate every other bar)

Criteria: Roof Live Load = 20 psf to 40 psf (0.96 kN/m^2 to 1.9 kN/m^2);
Floor Live Load = 50 psf (2.4 kN/m^2); enclosed building

× 1.61 for km/h — Span between Bearing Walls (feet) — × 304.8 for mm — Size of Rebar and Spacing (inches) — × 25.4 for mm

EXPO-SURE	STORIES	UNSUP-PORTED HEIGHT (feet) × 304.8 for mm	80 MPH 8	16	24	32	90 MPH 8	16	24	32	100 MPH 8	16	24	32	110 MPH 8	16	24	32
B	One-story building or top story of two-story building	8	NR*	NR*	NR*	NR*	NR*	NR*	NR*	NR*	NR*	NR*	NR*	NR*	No. 3 / 56	No. 3 / 56	No. 3 / 64	No. 3 / 64
		10	NR*	NR*	NR*	NR*	No. 4 / 80	No. 4 / 80	No. 4 / 88	No. 4 / 88	No. 4 / 64	No. 4 / 64	No. 4 / 64	No. 4 / 72	No. 4 / 48	No. 4 / 48	No. 4 / 56	No. 4 / 56
		12	No. 4 / 64	No. 4 / 72	No. 4 / 72	No. 4 / 72	No. 4 / 56	No. 4 / 56	No. 4 / 56	No. 4 / 56	No. 4 / 40	No. 4 / 40	No. 5 / 64	No. 5 / 64	No. 5 / 56	No. 5 / 56	No. 5 / 56	No. 5 / 56
	First story of a two-story building	8	No. 3 / 96	No. 3 / 96	No. 3 / 96	No. 3 / 96	No. 3 / 96	No. 3 / 96	No. 3 / 96	No. 3 / 96	No. 3 / 96	No. 3 / 88	No. 3 / 80	No. 3 / 72	No. 3 / 72	No. 3 / 72	No. 3 / 64	No. 3 / 64
		10	No. 3 / 88	No. 3 / 80	No. 3 / 72	No. 3 / 64	No. 3 / 64	No. 3 / 64	No. 3 / 56	No. 3 / 56	No. 4 / 72	No. 4 / 72	No. 4 / 64	No. 4 / 64	No. 4 / 64	No. 4 / 56	No. 4 / 56	No. 4 / 56
		12	No. 4 / 80	No. 4 / 72	No. 4 / 64	No. 4 / 64	No. 4 / 64	No. 4 / 56	No. 4 / 56	No. 4 / 48	No. 4 / 48	No. 4 / 48	No. 5 / 64	No. 5 / 56	No. 5 / 56	No. 5 / 56	No. 5 / 48	No. 5 / 48
C	One-story building or top story of two-story building	8	NR*	NR*	NR*	NR*	No. 3 / 48	No. 3 / 48	No. 3 / 48	No. 3 / 56	No. 4 / 64	No. 4 / 64	No. 4 / 72	No. 4 / 72	No. 4 / 48	No. 4 / 56	No. 4 / 56	No. 4 / 56
		10	No. 4 / 56	No. 4 / 56	No. 4 / 64	No. 4 / 64	No. 4 / 48	No. 4 / 48	No. 4 / 48	No. 4 / 48	No. 5 / 56	No. 5 / 56	No. 5 / 56	No. 5 / 56	No. 5 / 48	No. 5 / 48	No. 5 / 48	No. 5 / 48
		12	No. 5 / 56	No. 5 / 64	No. 5 / 64	No. 5 / 64	No. 5 / 48	No. 5 / 48	No. 5 / 48	No. 5 / 48	No. 6 / 56	No. 6 / 56	No. 6 / 56	No. 6 / 56	No. 6 / 40	No. 6 / 40	No. 6 / 40	No. 6 / 48
	First story of a two-story building	8	No. 3 / 80	No. 3 / 80	No. 3 / 56	No. 3 / 72	No. 3 / 56	No. 3 / 56	No. 3 / 56	No. 3 / 56	No. 4 / 72	No. 4 / 72	No. 4 / 72	No. 4 / 64	No. 4 / 56	No. 4 / 56	No. 4 / 56	No. 4 / 56
		10	No. 4 / 72	No. 4 / 64	No. 4 / 64	No. 4 / 56	No. 4 / 56	No. 4 / 48	No. 4 / 48	No. 4 / 48	No. 5 / 64	No. 5 / 56	No. 5 / 56	No. 5 / 56	No. 5 / 48	No. 5 / 48	No. 5 / 48	No. 5 / 48
		12	No. 5 / 64	No. 5 / 64	No. 5 / 56	No. 5 / 56	No. 5 / 48	No. 5 / 48	No. 5 / 48	No. 5 / 48	No. 6 / 56	No. 6 / 56	No. 6 / 48	No. 6 / 48	No. 6 / 48	No. 6 / 40	No. 6 / 40	No. 6 / 40
D	One-story building or top story of two-story building	8	No. 3 / 48	No. 3 / 48	No. 3 / 56	No. 3 / 56	No. 4 / 64	No. 4 / 72	No. 4 / 72	No. 4 / 80	No. 4 / 56	No. 4 / 56	No. 4 / 56	No. 4 / 56	No. 4 / 40	No. 4 / 48	No. 4 / 48	No. 4 / 48
		10	No. 4 / 48	No. 4 / 48	No 4 / 48	No. 4 / 56	No. 5 / 56	No. 5 / 64	No. 5 / 64	No. 5 / 64	No. 5 / 48	No. 5 / 48	No. 5 / 48	No. 5 / 48	No. 5 / 40	No. 5 / 40	No. 5 / 40	No. 5 / 40
		12	No. 5 / 48	No. 5 / 48	No. 5 / 56	No. 5 / 56	No. 6 / 56	No. 6 / 56	No. 6 / 56	No. 6 / 56	No. 6 / 48	No. 6 / 48	No .6 / 48	No. 6 / 48	No. 6 / 40	No. 6 / 40	No. 6 / 40	No. 6 / 40
	First story of a two-story building	8	No. 3 / 64	No. 3 / 64	No. 3 / 64	No. 3 / 56	No. 4 / 80	No. 4 / 80	No. 4 / 72	No. 4 / 72	No. 4 / 64	No. 4 / 56	No. 4 / 56	No. 4 / 56	No. 4 / 48	No. 4 / 48	No. 4 / 48	No. 4 / 48
		10	No. 4 / 56	No. 4 / 56	No. 4 / 56	No. 4 / 48	No. 4 / 48	No. 5 / 64	No. 5 / 64	No. 5 / 56	No. 5 / 48	No. 5 / 48	No. 5 / 48	No. 5 / 48	No. 5 / 40	No. 5 / 40	No. 5 / 40	No. 5 / 32
		12	No. 5 / 56	No. 5 / 56	No. 5 / 48	No. 5 / 48	No. 6 / 56	No. 6 / 56	No. 6 / 56	No. 6 / 48	No. 6 / 48	No. 6 / 48	No. 6 / 40	No. 6 / 40	No. 6 / 40	No. 6 / 32	No. 6 / 32	No. 6 / 32

*NR — No vertical reinforcement required. However, see Table A-21-F for shear wall reinforcement.

[1]These values are for walls with running bond. For stack bond see Section 2112.8.3.

[2]The figure on top of the listed data is the bar size; the figure below it is the maximum spacing in inches (mm). Reinforcing bar strength shall be A 615 Grade 60.

[3]Roof load is assumed to be concentrically loaded on the wall. For roofs which hang on ledgers, a design is required.

[4]Minimum horizontal reinforcement shall be one No. 4 at the ledger and foundation. Also, see Table A-21-E for lintels and Table A-21-F for shear wall reinforcing where applicable.

[5]Hook vertical bars over bond beam as shown. Extend bars into footing using lap splices where necessary. Where second-story bar spacing does not match those on the first story, hook bars around floor bond beam also.

To use this table, check criteria by the following method:

5.1 Choose proper roof live load from Table 16-C or snow load criteria for the locality in which the building is located.

5.2 Check if building is enclosed or partially enclosed by the procedure in Chapter 16, Part III.

5.3 Choose proper floor load from Table 16-A. [For loads less that 50 psf (2.4 kN/m^2), use 50 psf (2.4 kN/m^2), and for loads between 50 psf (2.4 kN/m^2) and 100 psf (4.8 kN/m^2), use 100 psf (4.8 kN/m^2).]

5.4 Find proper wind speed and exposure for the site—see Figure 16-1, Chapter 16, Sections 1619 and 1620.

5.5 Within the proper vertical column, choose appropriate span-to-bearing wall and appropriate height and story.

5.6 Read proper size and spacing of reinforcement for the thickness of the wall mentioned in the title of the table. (Equivalent area of steel, taking spacing into account, may be substituted.)

5.7 For buildings in Seismic Zone 2 (see Figure 16-2 in Chapter 16), use minimum reinforcement in Figure A-21-2 if it is more restrictive than the table values.

TABLE A-21-C-3—VERTICAL REINFORCING STEEL REQUIREMENTS FOR 8-INCH-THICK (203 mm) MASONRY WALLS[1] IN AREAS WHERE BASIC WIND SPEEDS ARE 80 MILES PER HOUR (129 km/h) OR GREATER[2,3,4,5]
(Wood or Steel Roof and Floor Framing)

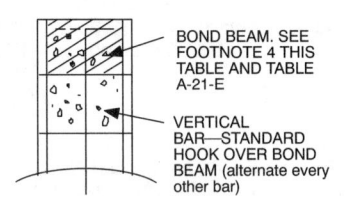

BOND BEAM. SEE FOOTNOTE 4 THIS TABLE AND TABLE A-21-E

VERTICAL BAR—STANDARD HOOK OVER BOND BEAM (alternate every other bar)

Criteria: Roof Live Load = 20 psf to 40 psf (0.96 kN/m² to 1.9 kN/m²);
Floor Live Load = 100 psf (4.8 kN/m²); enclosed building

			80 MPH				90 MPH				100 MPH				110 MPH			
		UNSUP-PORTED HEIGHT (feet) × 304.8 for mm	8	16	24	32	8	16	24	32	8	16	24	32	8	16	24	32
EXPO-SURE	STORIES		Size of Rebar and Spacing (inches) × 25.4 for mm															
B	One-story building or top story of two-story building	8	NR*	NR*	NR*	NR*	NR*	NR*	NR*	NR*	NR*	NR*	NR*	NR*	No. 3 56	No. 3 56	No. 3 64	No. 3 64
		10	NR*	NR*	NR*	NR*	No. 4 64	No. 4 64	No. 4 64	No. 4 72	No. 4 64	No. 4 64	No. 4 64	No. 4 72	No. 4 48	No. 4 48	No. 4 56	No. 4 56
		12	No. 4 64	No. 4 72	No. 4 72	No. 4 72	No. 4 56	No. 4 56	No. 4 56	No. 4 56	No. 4 40	No. 4 40	No. 4 40	No. 4 40	No. 5 56	No. 5 56	No. 5 56	No. 5 56
	First story of a two-story building	8	No. 3 96	No. 3 96	No. 3 80	No. 3 64	No. 3 96	No. 3 88	No. 3 72	No. 3 56	No. 3 80	No. 3 64	No. 3 56	No. 3 48	No. 3 64	No. 3 56	No. 3 48	No. 4 64
		10	No. 3 72	No. 3 64	No. 3 56	No. 3 48	No. 3 56	No. 3 48	No. 4 64	No. 4 56	No. 4 72	No. 4 64	No. 4 56	No. 4 48	No. 4 56	No. 4 48	No. 4 48	No. 5 56
		12	No. 4 72	No. 4 64	No. 4 56	No. 4 48	No. 4 56	No. 4 48	No. 4 48	No. 4 40	No. 4 48	No. 4 40	No. 5 48	No. 5 48	No. 5 56	No. 5 48	No. 5 48	No. 5 40
C	One-story building or top story of two-story building	8	NR*	NR*	NR*	NR*	No. 3 48	No. 3 48	No. 3 48	No. 3 56	No. 4 64	No. 4 64	No. 4 72	No. 4 72	No. 4 48	No. 4 56	No. 4 56	No. 4 56
		10	No. 4 56	No. 4 56	No. 4 64	No. 4 64	No. 4 48	No. 4 48	No. 4 48	No. 4 48	No. 5 56	No. 5 56	No. 5 56	No. 5 56	No. 5 48	No. 5 48	No. 5 48	No. 5 48
		12	No. 4 40	No. 5 64	No. 5 64	No. 5 64	No. 5 48	No. 5 48	No. 5 48	No. 5 48	No. 5 40	No. 6 56	No. 6 56	No. 6 56	No. 6 40	No. 6 40	No. 6 40	No. 5 32
	First story of a two-story building	8	No. 3 72	No. 3 64	No. 3 56	No. 3 48	No. 3 56	No. 3 48	No. 4 64	No. 4 56	No. 4 72	No. 4 64	No. 4 56	No. 4 48	No. 4 56	No. 4 48	No. 4 48	No. 5 56
		10	No. 4 64	No. 4 56	No. 4 48	No. 4 48	No. 4 48	No. 4 48	No. 5 56	No. 5 56	No. 5 56	No. 5 56	No. 5 48	No. 5 48	No. 5 48	No. 5 48	No. 5 40	No. 5 40
		12	No. 4 40	No. 5 56	No. 5 48	No. 5 48	No. 5 48	No. 5 40	No. 5 40	No. 6 48	No. 5 40	No. 6 48	No. 6 48	No. 6 40	No. 6 40	No. 6 40	No. 6 40	No. 6 32
D	One-story building or top story of two-story building	8	No. 3 48	No. 3 48	No. 3 56	No. 3 56	No. 3 64	No. 4 72	No. 4 72	No. 4 80	No. 4 56	No. 4 56	No. 4 56	No. 4 56	No. 4 40	No. 4 48	No. 4 48	No. 4 48
		10	No. 4 48	No. 4 48	No 4 48	No. 4 56	No. 5 56	No. 5 64	No. 5 64	No. 5 64	No. 5 48	No. 5 48	No. 5 48	No. 5 48	No. 5 40	No. 5 40	No. 5 40	No. 5 40
		12	No. 5 48	No. 5 48	No. 5 56	No. 5 56	No. 5 40	No. 5 40	No. 6 56	No. 6 56	No. 6 48	No .6 48	No. 6 48	No. 5 32	No. 6 40	No. 6 40	No. 6 40	No. 6 40
	First story of a two-story building	8	No. 3 56	No. 3 56	No. 3 48	No. 4 64	No. 3 48	No. 4 64	No. 4 56	No. 4 56	No. 4 56	No. 4 56	No. 4 48	No. 5 64	No. 5 48	No. 5 64	No. 5 56	No. 5 56
		10	No. 4 56	No. 4 48	No. 4 48	No. 5 56	No. 5 64	No. 5 56	No. 5 48	No. 5 48	No. 5 48	No. 5 48	No. 5 40	No. 5 40	No. 6 40	No. 6 40	No. 6 48	No. 6 48
		12	No. 5 48	No. 5 48	No. 5 40	No. 5 40	No. 5 40	No. 5 40	No. 6 48	No. 6 40	No. 6 48	No. 6 40	No. 6 40	No. 6 32	No. 6 32	No. 6 32	No. 6 32	No. 6 24

*NR — No vertical reinforcement required. However, see Table A-21-F for shear wall reinforcement.

[1]These values are for walls with running bond. For stack bond see Section 2112.8.3.
[2]The figure on top of the listed data is the bar size; the figure below it is the maximum spacing in inches (mm). Reinforcing bar strength shall be A 615 Grade 60.
[3]Roof load is assumed to be concentrically loaded on the wall. For roofs which hang on ledgers, a design is required.
[4]Minimum horizontal reinforcement shall be one No. 4 at the ledger and foundation. Also, see Table A-21-E for lintels and Table A-21-F for shear wall reinforcing where applicable.
[5]Hook vertical bars over bond beam as shown. Extend bars into footing using lap splices where necessary. Where second-story bar spacing does not match those on the first story, hook bars around floor bond beam also.

To use this table, check criteria by the following method:
5.1 Choose proper roof live load from Table 16-C or snow load criteria for the locality in which the building is located.
5.2 Check if building is enclosed or partially enclosed by the procedure in Chapter 16, Part III.
5.3 Choose proper floor load from Table 16-A. [For loads less than 50 psf (2.4 kN/m²), use 50 psf (2.4 kN/m²), and for loads between 50 psf (2.4 kN/m²) and 100 psf (4.8 kN/m²), use 100 psf (4.8 kN/m²).]
5.4 Find proper wind speed and exposure for the site—see Figure 16-1, Chapter 16, Sections 1619 and 1620.
5.5 Within the proper vertical column, choose appropriate span-to-bearing wall and appropriate height and story.
5.6 Read proper size and spacing of reinforcement for the thickness of the wall mentioned in the title of the table. (Equivalent area of steel, taking spacing into account, may be substituted.)
5.7 For buildings in Seismic Zone 2 (see Figure 16-2 in Chapter 16), use minimum reinforcement in Figure A-21-2 if it is more restrictive than the table values

TABLE A-21-C-4—VERTICAL REINFORCING STEEL REQUIREMENTS FOR 8-INCH-THICK (203 mm) MASONRY WALLS[1] IN AREAS WHERE BASIC WIND SPEEDS ARE 80 MILES PER HOUR (129 km/h) OR GREATER[2,3,4,5] (Wood or Steel Roof and Floor Framing)

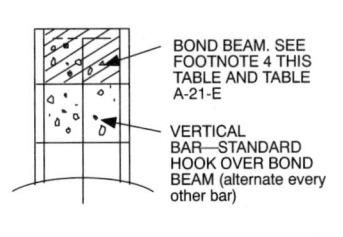

BOND BEAM. SEE FOOTNOTE 4 THIS TABLE AND TABLE A-21-E

VERTICAL BAR—STANDARD HOOK OVER BOND BEAM (alternate every other bar)

Criteria: Roof Live Load = 20 psf to 40 psf ($0.96 \ kN/m^2$ to $1.9 \ kN/m^2$); Floor Live Load = 50 psf ($2.4 \ kN/m^2$); partially enclosed building

EXPO-SURE	STORIES	UNSUP-PORTED HEIGHT (feet) $\times$ 304.8 for mm	80 MPH $\times$ 304.8 for mm 8	16	24	32	90 MPH 8	16	24	32	100 MPH 8	16	24	32	110 MPH 8	16	24	32
			Span between Bearing Walls (feet) — $\times$ 1.61 for km/h — Size of Rebar and Spacing (inches) — $\times$ 25.4 for mm															
B	One-story building or top story of two-story building	8	No. 4 / 96	No. 4 / 96	No. 3 / 80	No. 3 / 88	No. 3 / 56	No. 3 / 56	No. 3 / 64	No. 3 / 64	No. 3 / 40	No. 3 / 48	No. 3 / 48	No. 3 / 48	No. 4 / 64	No. 4 / 64	No. 4 / 64	No. 4 / 72
		10	No. 4 / 64	No. 4 / 64	No. 4 / 72	No. 4 / 72	No. 4 / 48	No. 4 / 56	No. 4 / 56	No. 4 / 56	No. 4 / 40	No. 4 / 40	No. 4 / 40	No. 4 / 40	No. 5 / 48	No. 5 / 56	No. 5 / 56	No. 5 / 56
		12	No. 4 / 40	No. 4 / 48	No. 4 / 48	No. 4 / 48	No. 5 / 56	No. 5 / 56	No. 5 / 56	No. 5 / 56	No. 5 / 40	No. 5 / 40	No. 5 / 40	No. 5 / 40	No. 6 / 48	No. 6 / 48	No. 6 / 48	No. 6 / 48
	First story of a two-story building	8	No. 3 / 96	No. 3 / 96	No. 3 / 88	No. 3 / 80	No. 3 / 72	No. 3 / 72	No. 3 / 64	No. 3 / 64	No. 3 / 48	No. 3 / 48	No. 3 / 48	No. 3 / 48	No. 4 / 64	No. 4 / 64	No. 4 / 64	No. 4 / 64
		10	No. 3 / 48	No. 3 / 48	No. 3 / 48	No. 3 / 48	No. 4 / 64	No. 4 / 56	No. 4 / 56	No. 4 / 56	No. 4 / 48	No. 4 / 48	No. 4 / 48	No. 4 / 40	No. 4 / 40	No. 5 / 56	No. 5 / 56	No. 5 / 48
		12	No. 4 / 48	No. 4 / 48	No. 4 / 48	No. 4 / 40	No. 4 / 40	No. 5 / 56	No. 5 / 56	No. 5 / 48	No. 5 / 48	No. 5 / 40	No. 5 / 40	No. 5 / 40	No. 6 / 48	No. 6 / 48	No. 6 / 48	No. 6 / 48
C	One-story building or top story of two-story building	8	No. 3 / 40	No. 3 / 40	No. 3 / 40	No. 3 / 40	No. 4 / 56	No. 4 / 56	No. 4 / 56	No. 4 / 56	No. 4 / 40	No. 4 / 40	No. 4 / 40	No. 4 / 48	No. 5 / 56	No. 5 / 56	No. 5 / 56	No. 5 / 56
		10	No. 4 / 40	No. 4 / 40	No. 4 / 40	No. 4 / 40	No. 5 / 48	No. 5 / 48	No. 5 / 48	No. 5 / 48	No. 5 / 32	No. 5 / 32	No. 5 / 40	No. 5 / 40	No. 6 / 40	No. 6 / 40	No. 6 / 40	No. 6 / 40
		12	No. 5 / 40	No. 5 / 40	No. 5 / 40	No. 5 / 40	No. 6 / 40	No. 6 / 48	No. 6 / 48	No. 6 / 48	No. 6 / 32	No. 6 / 32	No. 6 / 32	No. 6 / 32	Use 10-inch or larger units			
	First story of a two-story building	8	No. 3 / 48	No. 3 / 48	No. 3 / 48	No. 3 / 48	No. 4 / 56	No. 4 / 56	No. 4 / 56	No. 4 / 56	No. 4 / 48	No. 4 / 48	No. 4 / 40	No. 4 / 40	No. 5 / 56	No. 5 / 56	No. 5 / 56	No. 5 / 48
		10	No. 4 / 40	No. 4 / 40	No. 4 / 40	No. 4 / 40	No. 5 / 48	No. 5 / 48	No. 5 / 48	No. 5 / 48	No. 5 / 40	No. 5 / 40	No. 5 / 40	No. 6 / 48	No. 6 / 48	No. 6 / 48	No. 6 / 40	No. 6 / 40
		12	No. 5 / 40	No. 5 / 40	No. 5 / 40	No. 6 / 48	No. 6 / 48	No. 6 / 48	No. 6 / 40	No. 6 / 40	No. 6 / 32	No. 6 / 32	No. 6 / 32	No. 6 / 32	Use 10-inch or larger units			
D	One-story building or top story of two-story building	8	No. 4 / 56	No. 4 / 56	No. 4 / 56	No. 4 / 64	No. 4 / 40	No. 4 / 48	No. 4 / 48	No. 4 / 48	No. 5 / 56	No. 5 / 56	No. 5 / 56	No. 5 / 56	No. 5 / 48	No. 5 / 48	No. 5 / 48	No. 5 / 48
		10	No. 5 / 48	No. 5 / 48	No. 5 / 48	No. 5 / 56	No. 5 / 40	No. 5 / 40	No. 5 / 40	No. 5 / 40	No. 6 / 56	No. 6 / 48	No. 6 / 48	No. 6 / 48	No. 6 / 32	No. 6 / 32	No. 6 / 40	No. 6 / 40
		12	No. 6 / 48	No. 6 / 48	No. 6 / 48	No. 6 / 48	No. 6 / 40	No. 6 / 40	No. 6 / 40	No. 6 / 40	No. 6 / 24	No. 6 / 32	No. 6 / 32	No. 6 / 32	Use 10-inch or larger units			
	First story of a two-story building	8	No. 4 / 64	No. 4 / 64	No. 4 / 64	No. 4 / 56	No. 4 / 48	No. 4 / 48	No. 4 / 48	No. 4 / 48	No. 4 / 40	No. 4 / 40	No. 5 / 56	No. 5 / 56	No. 5 / 48	No. 5 / 48	No. 5 / 48	No. 5 / 40
		10	No. 5 / 56	No. 5 / 56	No. 5 / 48	No. 5 / 48	No. 5 / 40	No. 5 / 40	No. 5 / 40	No. 5 / 40	No. 6 / 48	No. 6 / 48	No. 6 / 48	No. 6 / 40	No. 6 / 40	No. 6 / 40	No. 6 / 32	No. 6 / 32
		12	No. 6 / 48	No. 6 / 48	No. 6 / 48	No. 6 / 48	No. 6 / 40	No. 6 / 40	Use 10-inch or larger units									

*NR — No vertical reinforcement required. However, see Table A-21-F for shear wall reinforcement.

[1] These values are for walls with running bond. For stack bond see Section 2112.8.3.

[2] The figure on top of the listed data is the bar size; the figure below it is the maximum spacing in inches (mm). Reinforcing bar strength shall be A 615 Grade 60.

[3] Roof load is assumed to be concentrically loaded on the wall. For roofs which hang on ledgers, a design is required.

[4] Minimum horizontal reinforcement shall be one No. 4 at the ledger and foundation. Also, see Table A-21-E for lintels and Table A-21-F for shear wall reinforcing where applicable.

[5] Hook vertical bars over bond beam as shown. Extend bars into footing using lap splices where necessary.

To use this table, check criteria by the following method:

[5.1] Choose proper roof live load from Table 16-C or snow load criteria for the locality in which the building is located.

[5.2] Check if building is enclosed or partially enclosed by the procedure in Chapter 16, Part III.

[5.3] Choose proper floor load from Table 16-A. [For loads less than 50 psf ($2.4 \ kN/m^2$), use 50 psf ($2.4 \ kN/m^2$), and for loads between 50 psf ($2.4 \ kN/m^2$) and 100 psf ($4.8 \ kN/m^2$) , use 100 psf ($4.8 \ kN/m^2$).]

[5.4] Find proper wind speed and exposure for the site—see Figure 16-1, Chapter 16, Sections 1619 and 1620.

[5.5] Within the proper vertical column, choose appropriate span-to-bearing wall and appropriate height and story.

[5.6] Read proper size and spacing of reinforcement for the thickness of the wall mentioned in the title of the table. (Equivalent area of steel, taking spacing into account, may be substituted.)

[5.7] For buildings in Seismic Zone 2 (see Figure 16-2 in Chapter 16), use minimum reinforcement in Figure A-21-2 if it is more restrictive than the table values.

TABLE A-21-C-5—VERTICAL REINFORCING STEEL REQUIREMENTS FOR 8-INCH-THICK (203 mm) MASONRY WALLS[1] IN AREAS WHERE BASIC WIND SPEEDS ARE 80 MILES PER HOUR (129 km/h) OR GREATER[2,3,4,5]
(Wood or Steel Roof and Floor Framing)

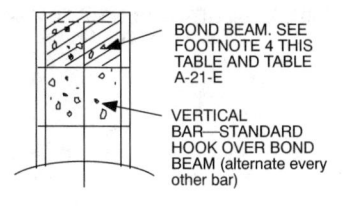

BOND BEAM. SEE FOOTNOTE 4 THIS TABLE AND TABLE A-21-E

VERTICAL BAR—STANDARD HOOK OVER BOND BEAM (alternate every other bar)

Criteria: Roof Live Load = 20 psf to 40 psf (0.96 kN/m² to 1.9 kN/m²); Floor Live Load = 100 psf (4.8 kN/m²); partially enclosed building

		UNSUPPORTED HEIGHT (feet)	80 MPH				90 MPH				100 MPH				110 MPH			
			colspan span between bearing walls (feet) × 304.8 for mm															
			8	16	24	32	8	16	24	32	8	16	24	32	8	16	24	32
EXPOSURE	STORIES	× 304.8 for mm	Size of Rebar and Spacing (inches) × 25.4 for mm															
B	One-story building or top story of two-story building	8	No. 3 72	No. 4 96	No. 3 80	No. 3 88	No. 3 56	No. 3 56	No. 3 64	No. 3 64	No. 4 80	No. 4 80	No. 4 80	No. 4 88	No. 4 64	No. 4 64	No. 4 64	No. 4 72
		10	No. 4 64	No. 4 64	No. 4 72	No. 4 72	No. 4 48	No. 4 56	No. 4 56	No. 4 56	No. 4 40	No. 4 40	No. 4 40	No. 4 40	No. 5 48	No. 5 56	No. 5 56	No. 5 56
		12	No. 4 40	No. 4 48	No. 4 48	No. 4 48	No. 5 56	No. 5 56	No. 5 56	No. 5 56	No. 5 40	No. 5 40	No. 5 40	No. 5 40	No. 6 48	No. 6 48	No. 6 48	No. 6 48
	First story of a two-story building	8	No. 3 88	No. 3 96	No. 3 56	No. 4 72	No. 3 64	No. 3 56	No. 3 64	No. 4 64	No. 4 80	No. 4 72	No. 4 64	No. 4 56	No. 4 64	No. 4 56	No. 4 48	No. 4 48
		10	No. 4 72	No. 4 64	No. 4 56	No. 4 48	No. 4 56	No. 4 48	No. 4 48	No. 4 40	No. 4 48	No. 4 40	No. 5 56	No. 5 48	No. 5 56	No. 5 48	No. 5 48	No. 5 40
		12	No. 4 48	No. 4 40	No. 4 40	No. 5 48	No. 5 56	No. 5 48	No. 5 48	No. 5 40	No. 5 40	No. 5 40	No. 6 48	No. 6 48	No. 6 48	No. 6 48	No. 6 40	No. 6 40
C	One-story building or top story of two-story building	8	No. 4 64	No. 4 64	No. 4 72	No. 4 72	No. 4 56	No. 4 56	No. 4 56	No. 4 56	No. 4 40	No. 4 40	No. 4 40	No. 4 48	No. 5 56	No. 5 56	No. 5 56	No. 5 56
		10	No. 5 56	No. 5 56	No. 4 40	No. 4 40	No. 5 48	No. 5 48	No. 5 48	No. 5 48	No. 5 32	No. 6 56	No. 6 56	No. 5 40	No. 6 40	No. 6 40	No. 6 40	No. 6 40
		12	No. 5 40	No. 5 40	No. 5 40	No. 5 40	No. 6 48	No. 6 48	No. 6 48	No. 6 48	No. 6 32	No. 6 32	No. 6 32	No. 6 32	Use 10-inch or larger units			
	First story of a two-story building	8	No. 4 72	No. 4 64	No. 4 56	No. 4 48	No. 4 56	No. 4 48	No. 4 48	No. 4 40	No. 4 40	No. 4 40	No. 4 40	No. 4 48	No. 5 56	No. 5 48	No. 5 48	No. 5 40
		10	No. 5 64	No. 5 56	No. 5 48	No. 5 48	No. 5 48	No. 5 48	No. 5 40	No. 5 40	No. 5 40	No. 6 48	No. 6 48	No. 6 40	No. 6 40	No. 6 40	No. 6 40	No. 6 32
		12	No. 5 40	No. 6 48	No. 6 48	No. 6 40	No. 6 40	No. 6 40	No. 6 40	No. 6 32	No. 6 32	No. 6 32	No. 6 32		Use 10-inch or larger units			
D	One-story building or top story of two-story building	8	No. 4 56	No. 4 56	No. 4 56	No. 4 64	No. 4 40	No. 5 72	No. 5 72	No. 5 72	No. 5 56	No. 5 56	No. 5 56	No. 5 56	No. 5 48	No. 5 48	No. 5 48	No. 5 48
		10	No. 5 48	No. 5 48	No. 5 48	No. 5 56	No. 5 40	No. 5 40	No. 5 40	No. 5 40	No. 6 56	No. 6 48	No. 6 48	No. 6 48	No. 6 32	No. 6 32	No. 6 40	No. 6 40
		12	No. 6 48	No. 6 48	No. 6 48	No. 6 48	No. 6 40	No. 6 40	No. 6 40	No. 6 40	No. 6 24	No. 6 32	No. 6 32	No. 6 32	Use 10-inch or larger units			
	First story of a two-story building	8	No. 4 64	No. 4 56	No. 4 48	No. 4 48	No. 4 48	No. 5 64	No. 5 56	No. 5 56	No. 5 56	No. 5 48	No. 5 48	No. 5 48	No. 5 48	No. 5 40	No. 5 40	No. 5 40
		10	No. 5 56	No. 5 48	No. 5 48	No. 5 40	No. 5 40	No. 5 40	No. 6 48	No. 6 48	No. 6 48	No. 6 40	No. 6 40	No. 6 40	No. 6 40	No. 6 32	No. 6 32	No. 6 32
		12	No. 6 48	No. 6 48	No. 6 40	No. 6 40	No. 6 40	No. 6 32	No. 6 32	No. 6 32	Use 10-inch or larger units							

*NR — No vertical reinforcement required. However, see Table A-21-F for shear wall reinforcement.

[1]These values are for walls with running bond. For stack bond see Section 2112.8.3.

[2]The figure on top of the listed data is the bar size; the figure below it is the maximum spacing in inches (mm). Reinforcing bar strength shall be A 615 Grade 60.

[3]Roof load is assumed to be concentrically loaded on the wall. For roofs which hang on ledgers, a design is required.

[4]Minimum horizontal reinforcement shall be one No. 4 at the ledger and foundation. Also, see Table A-21-E for lintels and Table A-21-F for shear wall reinforcing where applicable.

[5]Hook vertical bars over bond beam as shown. Extend bars into footing using lap splices where necessary.

To use this table, check criteria by the following method:

[5.1]Choose proper roof live load from Table 16-C or snow load criteria for the locality in which the building is located.

[5.2]Check if building is enclosed or partially enclosed by the procedure in Chapter 16, Part III.

[5.3]Choose proper floor load from Table 16-A. [For loads less than 50 psf (2.4 kN/m²), use 50 psf (2.4 kN/m²), and for loads between 50 psf (2.4 kN/m²) and 100 psf (4.8 kN/m²), use 100 psf (4.8 kN/m²).]

[5.4]Find proper wind speed and exposure for the site—see Figure 16-1, Chapter 16, Sections 1619 and 1620.

[5.5]Within the proper vertical column, choose appropriate span-to-bearing wall and appropriate height and story.

[5.6]Read proper size and spacing of reinforcement for the thickness of the wall mentioned in the title of the table. (Equivalent area of steel, taking spacing into account, may be substituted.)

[5.7]For buildings in Seismic Zone 2 (see Figure 16-2 in Chapter 16), use minimum reinforcement in Figure A-21-2 if it is more restrictive than the table values.

TABLE A-21-D—ANCHORAGE OF WOOD MEMBERS TO EXTERIOR WALLS FOR VERTICAL AND UPLIFT FORCES
[In areas where basic wind speeds are 80 miles per hour (129 km/h) or greater]
See Figure A-21-7 for details

**Part I—Anchor bolt size and spacing [in inches (mm)][1,2,3] on wood ledgers carrying vertical loads from roofs and floors[4,5]
Douglas fir-larch, California redwood (close grain) and southern pine[6,7]**

TYPE OF LOADING	LIVE LOAD[8,9] psf ($\times$ 0.0479 for kN/m²)	2-INCH (51 mm) $\times$ LEDGER				3-INCH (76 mm) $\times$ LEDGER				4-INCH (102 mm) $\times$ LEDGER			
		Span between Bearing Walls (feet) $\times$ 304.8 for mm ($\times$ 25.4 for mm)											
		8	16	24	32	8	16	24	32	8	16	24	32
Roof	20	$1/2$ 32	$(2)1/2$ 16	$5/8$ 16	$7/8$ 16	$1/2$ 32	$1/2$ 16	$(2)1/2$ 32	$7/8$ 16	—	$5/8$ 32	$7/8$ 32	$(2)5/8$ 32
	30	$(2)1/2$ 32	$1/2$ 16	$3/4$ 16	$7/8$ 16	$1/2$ 16	$(2)7/8$ 32	$7/8$ 16	$7/8$ 16	—	$(2)1/2$ 32	$5/8$ 16	$3/4$ 16
	40	$1/2$ 16	$5/8$ 16	$3/4$ 8	—	$5/8$ 16	$(2)5/8$ 32	$7/8$ 16	1 16	$5/8$ 32	$5/8$ 16	$3/4$ 16	$7/8$ 16
Floor[10]	50	$1/2$ 16	1 12	—	—	$5/8$ 24	$3/4$ 32	$3/4$ 12	$1\,1/4$ 12	$5/8$ 24	$7/8$ 24	$7/8$ 16	$7/8$ 12
	100	1 16	$(2)3/4$ 12	—	—	$5/8$ 16	1 12	$(2)3/4$ 12	$(2)1$ 12	$7/8$ 16	$3/4$ 12	1 12	$(2)3/4$ 12

[1]Closer spacing may be used.

[2]Use two bolts, one above the other, at splices and locate them away from the splice end by $3^1/2$ inches (89 mm) for $^1/2$-inch (13 mm) diameter, $4^1/2$ inches (114.3 mm) for $^5/8$-inch (15.9 mm) diameter, $5^1/4$ inches (133 mm) for $^3/4$-inch (19 mm) diameter, $6^1/4$ inches (158 mm) for $^7/8$-inch (22.2 mm) diameter and 7 inches (178 mm) for 1-inch (25.4 mm) diameter.

[3]See Table A-21-F for lateral force requirements (when applicable).

[4]Tabulated values are based on short-term loading due to roof loads (25 percent) or snow loads (15 percent), whichever controls. No increase is allowed for floor loads.

[5]See details in Figure A-21-7 for location relative to other construction. Note that roofs are concentrically loaded.

[6]See Chapter 23, Division III, Part I, for other species. Adjust spacing in direct proportion to the perpendicular-to-grain values for the applicable ledger and bolt sizes shown using the procedure described in Chapter 23, Division III, Part I. No increase is allowed for special inspection.

[7]Values on top are bolt sizes and underneath are spacing. Multiple bolts are shown in parenthesis: example (2) = two.

[8]See Table 16-C or Appendix Chapter 16, Division I, for values.

[9]Joist spacing is limited to 30 inches (762 mm) on center maximum.

[10]Where two bolts are required they shall be staggered at half the spacing shown or be placed one above the other.

Part II—Uplift anchors[1] for wood roof members [number of common nails in a 0.036 inch (0.91 mm) (No. 20 galvanized sheet gage) by $1^1/8$-inch (28.6 mm) tie strap embedded 5 inches (127 mm) into a masonry bond beam[2]]

ENCLOSURE[3]	EXPOSURE[4]	80 MPH				90 MPH				100 MPH				110 MPH			
		Span between Bearing Walls (feet)[5] $\times$ 1.61 for km/h $\times$ 304.8 for mm															
		8	16	24	32	8	16	24	32	8	16	24	32	8	16	24	32
Enclosed	B	NR	NR	NR	NR	NR	NR	NR	NR	NR	NR	NR	NR	NR	NR	2-8d	2-8d
	C	NR	NR	NR	NR	NR	2-8d	3-8d	4-8d	2-8d	4-8d	5-10d	5-10d	2-10d	4-10d	3-10d 24″	4-10d 24″
	D	NR	2-8d	3-8d	4-8d	2-8d	4-8d	4-10d	5-10d	3-8d	5-8d	5-10d	4-10d 24″	3-10d	5-10d	4-10d 24″	5-10d 24″
Open	B	NR	NR	NR	NR	NR	NR	2-8d	2-8d	NR	2-8d	4-8d	5-10d	2-8d	4-8d	5-8d	6-10d
	C	2-8d	4-8d	5-8d	5-10d	3-8d	5-8d	3-10d 24″	4-10d 24″	3-10d	5-10d	5-10d 24″	5-10d 16″	5-8d	4-10d 24″	5-10d 16″	6-10d 16″
	D	2-8d	5-8d	5-10d	5-10d 24″	4-8d	5-10d	4-10d 24″	5-10d 24″	5-8d	4-10d 24″	6-10d 24″	6-10d 16″	4-8d	5-10d 24″	6-10d 16″	6-10d 12″

NR — No requirements; use Table 23-II-B-1 minimum.

[1]Tie straps are at 48 inches (1219 mm) on center unless otherwise stated. See Figure A-21-7 for illustration of tie straps.

[2]Bond beam to be at least 48 inches (1219 mm) deep nominal and shall be reinforced as shown in Table A-21-E for lintels, or Tables A-21-C-1 through A-21-C-5 for walls in general where they are more restrictive.

[3]See Chapter 21, Part II, for definitions.

[4]See Section 1616 for definitions.

[5]For flat roofs connected to interior walls, the span shall be one half the larger distance on either side of the wall.

TABLE A-21-E—LINTEL REINFORCEMENT OVER EXTERIOR OPENINGS[1,2]—WOOD AND STEEL FRAMING[3]
[Lintels larger than 12 feet 0 inch (3658 mm) shall be designed.][4]
8-INCH (203 mm) MASONRY UNITS[5]

Part I—Roof Loads[5]

ANY WALL HEIGHT (feet)	SPAN TO BEARING WALLS (feet)[9]	SECOND STORY OF A TWO-STORY OR ONE-STORY BUILDINGS ROOF LIVE LOAD[6,7,8]					
		20-30 psf			40 psf		
		× 0.0479 for kN/m²					
		Width of Opening[9] (feet)					
		× 304.8 for mm					
		4	8	12	4	8	12
		Lintel depth (inches) number and size of rebar					
× 304.8 for mm		× 25.4 for mm					
Any (up to 12')	8	8 1 No. 3	8 1 No. 3	16 1 No. 4 (B)	8 1 No. 3	8 1 No. 4	16 1 No. 4 (B)
	16	8 1 No. 3	8 1 No. 3	16 1 No. 4 (B)	8 1 No. 3	8 2 No. 4 (A)	16 2 No. 5 (B)
	24	8 1 No. 3	8 1 No. 4	16 1 No. 4 (B)	8 1 No. 3	16 1 No. 4 (B)	24 2 No. 5 (B)
	32	8 1 No. 3	16 1 No. 4 (B)	16 1 No. 5 (B)	8 1 No. 3	16 1 No. 5 (B)	24 2 No. 5 (C)

Part II—Floor and Roof Loads[5]

WALL HEIGHT	SPAN TO BEARING[9,11] WALLS (feet)	FIRST STORY OF TWO-STORY BUILDINGS FLOOR LIVE LOAD[10]					
		50 psf			100 psf		
		× 0.0479 for kN/m²					
		Width of Opening[9] (feet)					
		× 304.8 for mm					
		4	8	12	4	8	12
		Lintel depth (inches) number and size of rebar					
× 304.8 for mm		× 25.4 for mm					
Any (up to 12')	8	8 1 No. 3	8 1 No. 4	16 1 No. 4 (B)	8 1 No. 3	8 2 No. 4 (A)	16 2 No. 5 (B)
	16	8 1 No. 3	8 2 No. 4 (A)	16 2 No. 5 (B)	8 1 No. 3	16 1 No. 4 (B)	24 2 No. 4 (C)
	24	8 1 No. 3	16 1 No. 4 (B)	24 2 No. 5 (B)	8 1 No. 3 (A)	16 1 No. 5 (B)	24 3 No. 5 (C)
	32	8 1 No. 3	16 1 No. 5 (B)	24 2 No. 5 (C)	8 1 No. 4 (B)	24 2 No. 5 (C)	Design Required

[1]The values shown are number and size of A 615, 60 grade steel reinforcement bars: Example—2 No. 4 is two $1/2$-inch-diameter (13 mm) deformed reinforcing bars. See also Figure A-21-8 for continuous load path.

[2]Stirrup spacing requirements: A = No. 3 at 8 inches (203 mm) on center, B = No. 3 at 4 inches (102 mm) on center, C = No. 4 at 8 inches (203 mm) on center. None are required unless specifically mentioned in the table.

[3]Design required for lintels supporting precast planks.

[4]Lintels are 8-inch (203 mm) nominal depth where supporting roof loads only and 16-inch (406 mm) nominal depth where supporting floor and roofs unless otherwise stated. All lintels are solidly grouted.

[5]Wall weight is included.

[6]The stirrup size and spacing, where required, as indicated in parenthesis below the reinforcing bar requirements.

[7]All exposure categories are included for wind uplift on the lintel. See Footnote 4 of Tables A-21-C-1 through A-21-C-5 as a minimum bond beam. Table A-21-F may also control.

[8]Two No. 5 vertical bars minimum are required on each side of the lintel for 100 and 110 miles per hour (161 and 177 km/h), Exposure D. Bar to extend 25 inches (635 mm) beyond opening or hook over top bars.

[9]For spans between the figures shown, go to next higher span width.

[10]From Table 21-A. For other floor loads go to next higher value. Where required floor load exceeds 100 pounds per square foot (4.8 kN/m²), a design is required.

[11]When interior walls support floors from each side, these values may be used if the spans on each side are less than 16 feet 0 inch (4877 mm) each. Enter the table with the total of both span widths.

**TABLE A-21-F—MASONRY SHEAR WALL[1,2,3] AND DIAPHRAGM
REQUIREMENTS IN HIGH-WIND AREAS[4]**

Part I—Minimum wall length and horizontal bar reinforcement required for exterior shear walls and cross walls[5] (all wall heights). [Design criteria: 20 psf to 40 psf (0.96 kN/m^2 to 1.9 kN/m^2) roof load; 50 psf or 100 psf (2.4 kN/m^2 or 4.8 kN/m^2) floor load; open or enclosed buildings.]

Wind Speed	Exposure	Distance between Shear Resisting Walls[7] "L" or "b" (feet)	One-story Building or Second Story of a Two-story Building	First Story of a Two-story Building
× 1.61 for km/h		× 304.8 for mm	inch × 25.4 for mm foot × 304.8 for mm	
			8-INCH (203 mm) WALLS[6]	
80 mph	B	32	NSR	9'-4"
		48	NSR	5'-4" DBL (D)
		64	10'-0"	7'-6" DBL (C)
	C	32	NSR	5'-4" DBL (C)
		48	11'-0"	8'-8" DBL (C)
		64	13'-4"	15'-0" (D)
	D	32	8'-8"	7'-0" (C)
		48	9'-4" (C)	10'-8" (D)
		64	10'-0" (D)	13'-8" (D)
90 mph	B	32	NSR	7'-8" DBL (C)
		48	NSR	8'-0" (D)
		64	12'-8"	12'-0" (D)
	C	32	NSR	14'-8"
		48	13'-8"	10'-0" (D)
		64	10'-8" (C)	15'-6" DBL (B)
	D	32	7'-8" (C)	11'-8" (D)
		48	12'-0" (C)	12'-8" DBL (B)
		64	11'-8" (D)	18'-4" DBL (C)
100 mph	B	32	NSR	5'-4" DBL (C)
		48	10'-0"	10'-0" (D)
		64	15'-4"	64'-8" DBL (C)

(Continued)

**TABLE A-21-F—MASONRY SHEAR WALL[1,2,3] AND DIAPHRAGM
REQUIREMENTS IN HIGH-WIND AREAS[4]—(Part I Continued)**

Wind Speed	Exposure	Distance between Shear Resisting Walls[7] "L" or "b" (feet)	One-story Building or Second Story of a Two-story Building	First Story of a Two-story Building
× 1.61 for km/h		× 304.8 for mm	inch × 25.4 for mm foot × 304.8 mm	
100 mph (cont.)	C	32	5'-4" (D)	11'-8" (D)
		48	12'-8" (C)	12'-8" DBL (C)
		64	12'-4" (D)	19'-8" DBL (C)
	D	32	5'-4" DBL (B)	9'-4" DBL (C)
		48	9'-4" (D)	14'-8" DBL (C)
		64	17'-4" (D)	21'-0" DBL (C)
110 mph	B	32	NSR	6'-0" DBL (C)
		48	12'-0"	10'-0" DBL (C)
		64	12'-8" (C)	14'-0" (D)
	C	32	5'-4" DBL (B)	9'-8" (D)
		48	12'-0" (C)	15'-4" (D)
		64	16'-8" (C)	18'-8" DBL (C)
	D	32	8'-8" (C)	11'-4" (D)
		48	12'-4" (C)	18'-0" (D)
		64	18'-8" (C)	20'-8" DBL (C)

*NSR—No special horizontal reinforcement required for shear resistance if 5 feet 4 inches (1626 mm) long minimum.

[1]Cumulative shear wall length is to be at least as long as is shown in this table. However, see Figure A-21-9. The top figure is the minimum length. When required, the figure below it in parenthesis is the spacing of steel reinforcing wire installed as shown in Figure A-21-10, below. (A) = two 0.148 inch (3.76 mm) (No. 9 B.W. gage) at 16 inches (406 mm) on center, (B) = two $^{3}/_{16}$ inch (4.76 mm) at 16 inches (406 mm) on center, (C) = two 0.148 inches (3.76 mm) (No. 9 B.W. gage) at 8 inches (203 mm) on center, (D) = two $^{3}/_{16}$ inch (4.76 mm) at 8 inches (203 mm) on center. The symbol DBL means double these amounts. Equivalent areas of reinforcing bars spaced not over 4 feet 0 inch (1219 mm) on center may be used.

[2]All bearing and shear walls are to be in-plane with vertical reinforcement, when required, extending from one floor to the other as dictated in Tables A-21-C-1 through A-21-C-5.

[3]Minimum bond beam shall be 100 miles per hour (mph) (161 km/h), Exposure B; 90 mph (145 km/h), Exposure B, and 80 mph (129 km/h), Exposures B and C, one No. 4; 100 mph (161 km/h), Exposure C; 80 and 90 mph (129 and 145 km/h); Exposures C and D, two No. 4; all others two No. 5.

[4]Table is adjusted to include provisions for Seismic Zones 0, 1 and 2.

[5]Cross walls are to be at least twice as long as shown in the table for shear walls. The tributary width ($L/2$) shall be the distance used in the third column above to find minimum reinforcement and length.

[6]For walls which width is equal to or less than half its height, add an extra No. 5 vertical bar at each end.

[7]Use 32-foot (9753 mm) requirements for distances less than 32 feet (9754 mm). Also use it for bearing walls used as shear walls.

(Continued)

TABLE A-21-F—MASONRY SHEAR WALL[1,2,3] AND DIAPHRAGM REQUIREMENTS IN HIGH-WIND AREAS[4]—(Continued)

Part II—Wood floor and roof diaphragms and connections[8,9]
[All wall heights 8 feet to 12 feet (2438 mm to 3657 mm).]

			FRAMING OF DOUGLAS FIR-LARCH OR SOUTHERN PINE		
				Minimum Wood Structural Panel/Particleboard Size[9] and Nailing[11,12]	
Wind Speed		**Distance between Shear Walls[10] "L" or "b" (feet)**	**Thickness (inches)**		**Nail Spacing (inches)**
× 1.61 for km/h	**Exposure**	× 304.8 for mm	× 25.4 for mm	**Common Nail Size (penny)**	× 25.4 for mm
80 mph	B	16	$5/16$	6	6 o.c.
		32	$3/8$	6	6 o.c.
		48	$3/8$	8	6 o.c.
		64	$3/8$	8	6 o.c.
	C	16	$3/8$	8	6 o.c.
		32	$1/2$ or $15/32$	8	6 o.c.
		48	$1/2$ or $15/32$	10	6 o.c.
		64	$5/8$ or $19/32$	10	6 o.c.
	D	16	$1/2$ or $15/32$	8	6 o.c.
		32	$5/8$ or $19/32$	10	6 o.c.
		48	$1/2$ or $15/32$ blocked	8	4/6 o.c.
		64	$1/2$ or $15/32$ blocked	8	4/6 o.c.
90 mph	B	16	$5/16$	6	6 o.c.
		32	$3/8$	8	6 o.c.
		48	$3/8$	8	6 o.c.
		64	$3/8$	8	6 o.c.
	C	16	$1/2$ or $15/32$	10	6 o.c.
		32	$3/8$ blocked	8	4/6 o.c.
		48	$3/8$ blocked	8	4/6 o.c.
		64	$5/8$ or $19/32$ blocked	10	6 o.c.
	D	16	$5/8$ or $19/32$	10	6 o.c.
		32	$1/2$ or $15/32$ blocked	10	4/6 o.c.
		48	$1/2$ or $15/32$ blocked	10	4/6 o.c.
		64	Design required or provide extra cross walls		
100 mph	B	16	$3/8$	8	6 o.c.
		32	$1/2$ or $15/32$	8	6 o.c.
		48	$1/2$ or $15/32$	8	6 o.c.
		64	$5/8$ or $19/32$	10	6 o.c.

(Continued)

TABLE A-21-F—MASONRY SHEAR WALL[1,2,3] AND DIAPHRAGM REQUIREMENTS IN HIGH-WIND AREAS[4]—(Part II Continued)

Wind Speed × 1.61 for km/h	Exposure	Distance between Shear Walls[10] "L" or "b" (feet) × 304.8 for mm	Minimum Wood Structural Panel/Particleboard Size[9] and Nailing[11,12]		
			Thickness (inches) × 25.4 for mm	Common Nail Size (penny)	Nail Spacing (inches) × 25.4 for mm
100 mph (cont.)	C	16	$3/8$ blocked	8	4/6 o.c.
		32	$5/8$ or $19/32$ blocked	10	4/6 o.c.
		48	$5/8$ or $19/32$ blocked	10	4/6 o.c.
		64	Design required or provide extra cross walls		
	D	16	$1/2$ or $15/32$ blocked	10	4/6 o.c.
		32	$5/8$ or $19/32$ blocked	10	4/6 o.c.
		48	Design required or provide extra cross walls		
		64	Design required or provide extra cross walls		
110 mph	B	16	$1/2$ or $15/32$	8	6 o.c.
		32	$1/2$ or $15/32$	10	6 o.c.
		48	$5/8$ or $19/32$	10	6 o.c.
		64	$1/2$ or $15/32$ blocked	8	4/6 o.c.
	C	16	$1/2$ or $15/32$ blocked	8	4/6 o.c.
		32	$5/8$ or $19/32$ blocked	10	4/6 o.c.
		48	Design required or provide extra cross walls		
		64	Design required or provide extra cross walls		
	D	16	$5/8$ or $19/32$ blocked	10	4/6 o.c.
		32	Design required or provide extra cross walls		
		48	Design required or provide extra cross walls		
		64	Design required or provide extra cross walls		

Note at top of table body: FRAMING OF DOUGLAS FIR-LARCH OR SOUTHERN PINE

[8]These requirements represent the maximum values for a diaphragm which is within a maximum 32-foot-by-64-foot (9.75 m by 19.5 m) module surrounded by shear walls, cross walls or bearing walls. (See Figure A-21-9.)

[9]See Tables 23-II-E-1 and 23-II-E-2 for minimum sizes depending on span between joists.

[10]See Figure A-21-9 for "L" and "b."

[11]The wood structural panel/particleboard (all grades) thickness is given first. The nailing size and boundary/supported edge spacing is shown next. Blocking of unsupported edges is stated where required. Twelve-inch (305 mm) spacing required in the field of the roof/floor. Boundary nailing is required over interior walls [see Figure A-21-12 (b)].

[12]Use Case 1 for unblocked diaphragms and any case for blocked diaphragms.

TABLE A-21-G—MINIMUM WALL CONNECTION REQUIREMENTS IN HIGH-WIND AREAS
Precast Hollow-core Plank Floors and Roofs

Spacing of No. 4 bent reinforcing bar in block or brick walls connected to precast concrete planks[1,2]

WIND SPEED AND EXPOSURE	EXTERIOR WALLS	INTERIOR WALLS
	× 25.4 for mm	
90 mph (145 km/h) Exposure C and less 100 mph (161 km/h) Exposure B	32″ o.c.	16″ o.c.
90 mph (145 km/h) Exposure D 100 mph (161 km/h) Exposure C 110 mph (177 km/h) Exposure B	24″ o.c.	12″ o.c.
100 mph (161 km/h) Exposure D 110 mph (177 km/h) Exposures C and D	16″ o.c.	12″ o.c.

[1]This table assumes maximum wall height of 12 feet (3.7 m) and a width-to-length ratio of diaphragm between shear walls of 3:1 or less.
[2]The precast planks shall be designed as shall the walls and footings supporting them.

TABLE A-21-H—MINIMUM HOLD-DOWN REQUIREMENTS IN HIGH-WIND AREAS
Steel Floors and Roofs

WIND SPEED AND EXPOSURE	MAXIMUM SPACING OF ROOF JOISTS WITH CONNECTION SHOWN[1,2,3]
	× 25.4 for mm
100 mph (161 km/h) Exposure B 90 mph (145 km/h) Exposures B and C 80 mph (129 km/h) Exposures B, C and D	48″
110 mph (177 km/h) Exposure B 100 mph (161 km/h) Exposure C	30″
110 mph (177 km/h) Exposures C and D 100 mph (161 km/h) Exposure D	Design required

[1]Maximum span is 32 feet (9.75 m) to bearing walls.
[2]Joists and decking to be designed.
[3]Bottom chord of joists to be braced for reversal of stresses caused by wind uplift.

TABLE A-21-I—DIAGONAL BRACING REQUIREMENTS
FOR GABLE-END WALL[1,2] ROOF PITCH 3:12 to 5:12

	BASIC WIND SPEED (mph)							
	× 1.61 for km/h							
	80		90		100		110	
	3:12 (25%)	4:12 (33%) and 5:12 (42%)	3:12 (25%)	4:12 (33%) and 5:12 (42%)	3:12 (25%)	4:12 (33%) and 5:12 (42%)	3:12 (25%)	4:12 (33%) and 5:12 (42%)
EXPOSURE	× 25.4 for mm							
B	I at 48″ o.c.	III at 48″ o.c.	I at 48″ o.c.	III at 48″ o.c.	I at 24″ o.c.	III at 24″ o.c.	I at 24″ o.c.	III at 24″ o.c.
C	I at 24″ o.c.	III at 48″ o.c.	I at 24″ o.c.	III at 24″ o.c.	II at 24″ o.c.	IV at 24″ o.c.	II at 24″ o.c.	IV at 24″ o.c.
D	I at 24″ o.c.	III at 48″ o.c.	II at 24″ o.c.	IV at 24″ o.c.	II at 24″ o.c.	IV at 24″ o.c.	Two-II at 24″ o.c.	Two-III at 24″ o.c.

[1]   I = 2-inch-by-4-inch brace, one clip angle (51 mm × 102 mm).

   II = 2-inch-by-4-inch brace, two clip angles (one each side) (51 mm × 102 mm).

   III = 3-inch-by-4-inch brace, one clip angle (76 mm × 102 mm).

   IV = 3-inch-by-4-inch brace, two clip angles (one each side) (76 mm × 102 mm).

The spacing requirements of the brace are shown below the symbol.
[2]See Figures A-21-17 and A-21-18 for details and size of clip angles.

NOTE: Horizontal and vertical reinforcement to be determined
by Tables A-21-C-1 through A-21-C-5 and A-21-F.

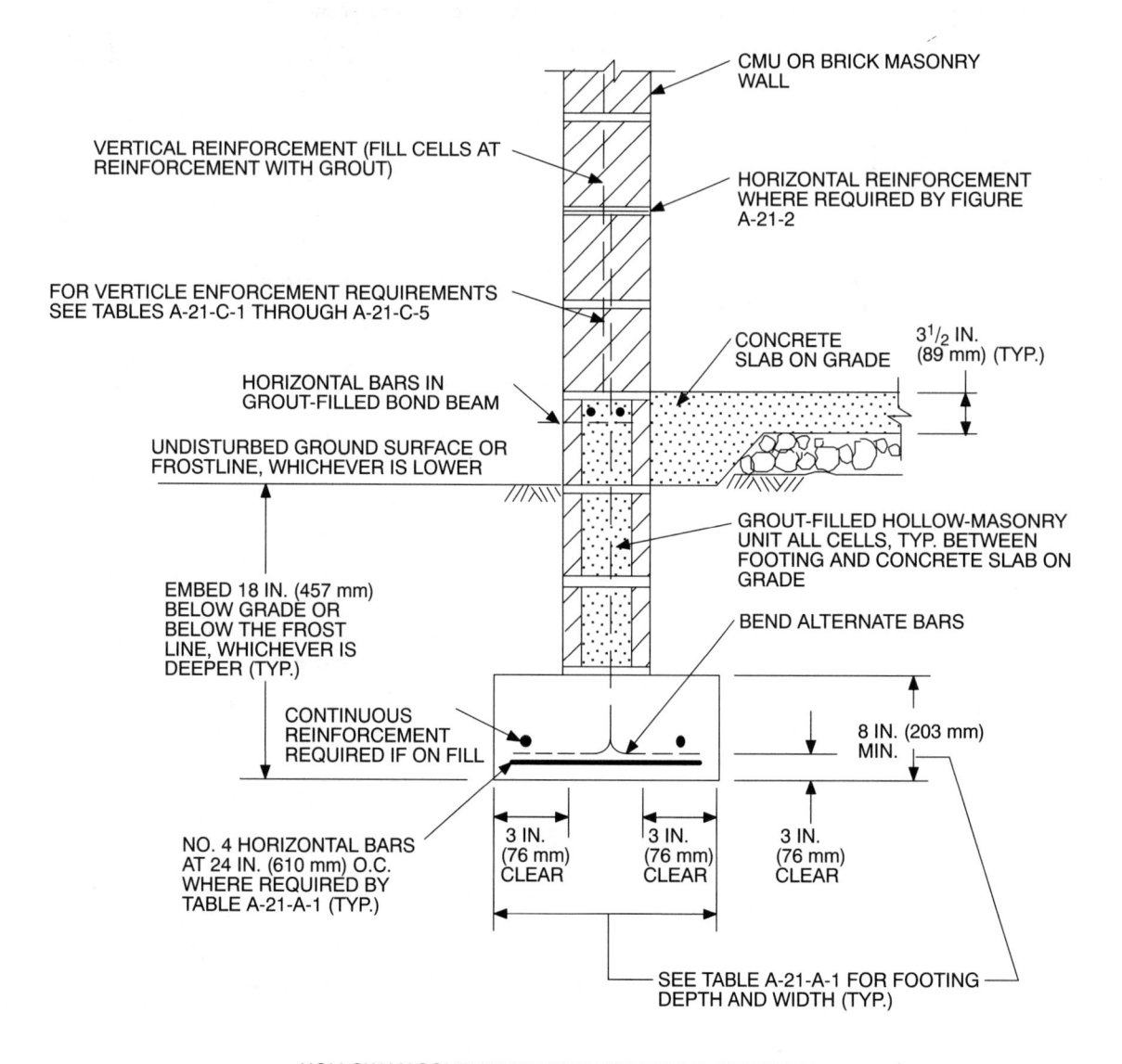

HOLLOW-MASONRY UNIT EXTERIOR FOUNDATION WALL

FIGURE A-21-1—VARIOUS DETAILS OF FOOTINGS
(See Tables A-21-A-1 and A-21-A-2 for widths.)

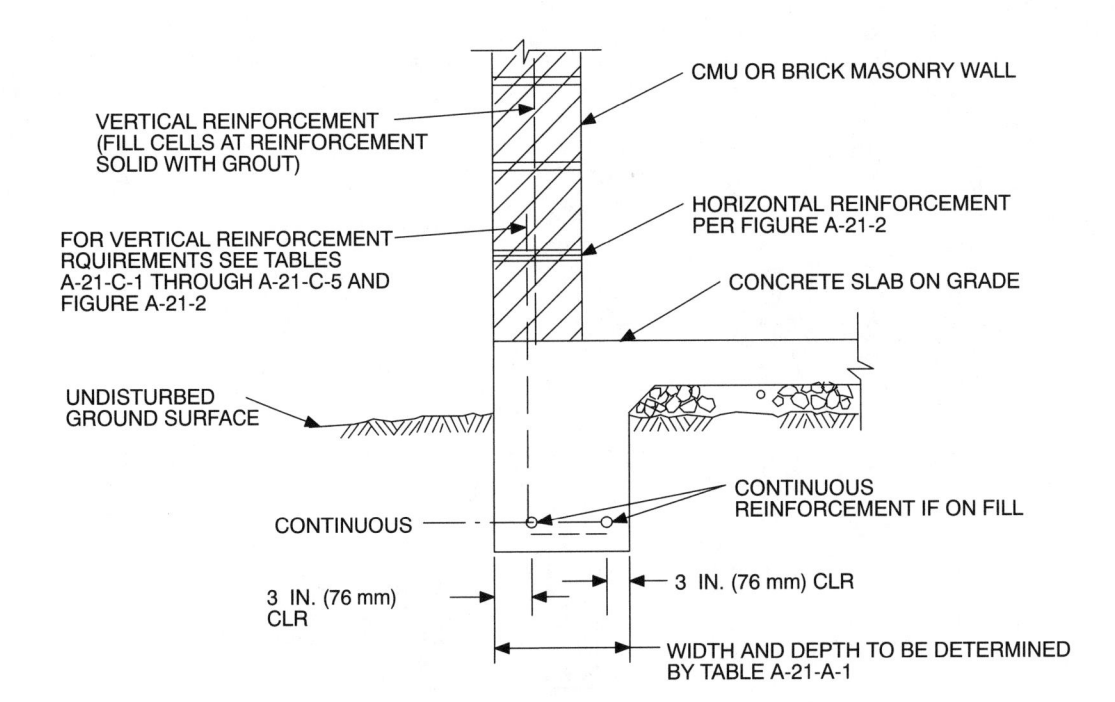

VERTICAL REINFORCEMENT (FILL CELLS AT REINFORCEMENT SOLID WITH GROUT)

FOR VERTICAL REINFORCEMENT RQUIREMENTS SEE TABLES A-21-C-1 THROUGH A-21-C-5 AND FIGURE A-21-2

UNDISTURBED GROUND SURFACE

CONTINUOUS

3 IN. (76 mm) CLR

CMU OR BRICK MASONRY WALL

HORIZONTAL REINFORCEMENT PER FIGURE A-21-2

CONCRETE SLAB ON GRADE

CONTINUOUS REINFORCEMENT IF ON FILL

3 IN. (76 mm) CLR

WIDTH AND DEPTH TO BE DETERMINED BY TABLE A-21-A-1

GRADE BEAM OR CONTINUOUS CONCRETE SLAB—TURN DOWN

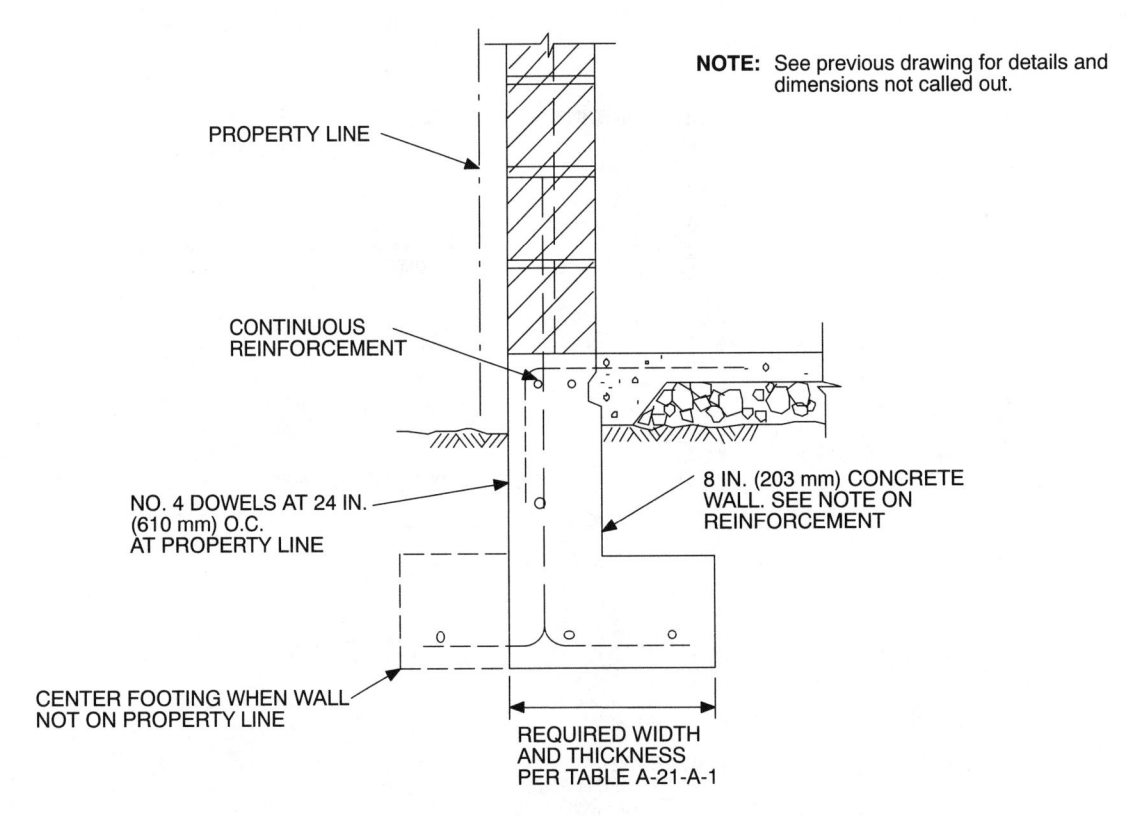

PROPERTY LINE

CONTINUOUS REINFORCEMENT

NO. 4 DOWELS AT 24 IN. (610 mm) O.C. AT PROPERTY LINE

CENTER FOOTING WHEN WALL NOT ON PROPERTY LINE

NOTE: See previous drawing for details and dimensions not called out.

8 IN. (203 mm) CONCRETE WALL. SEE NOTE ON REINFORCEMENT

REQUIRED WIDTH AND THICKNESS PER TABLE A-21-A-1

HOLLOW-MASONRY UNIT CONCRETE EXTERIOR FOUNDATION WALL

FIGURE A-21-1—VARIOUS DETAILS OF FOOTINGS—(Continued)
(See Tables A-21-A-1 and A-21-A-2 for widths.)

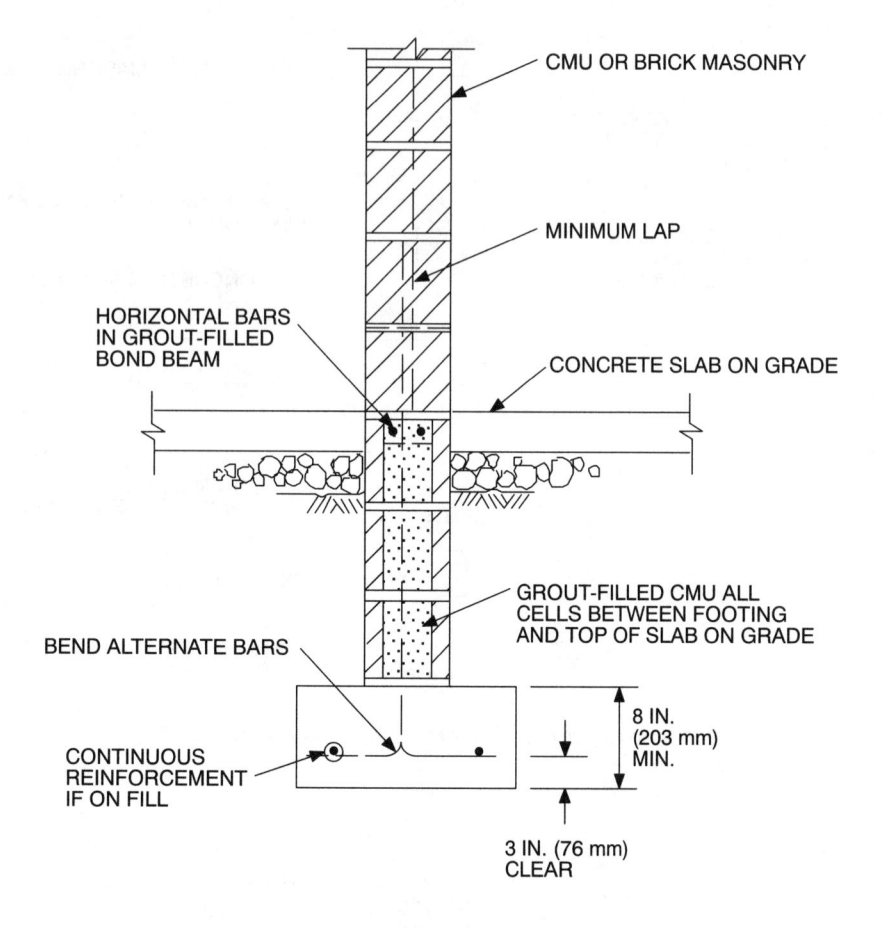

HOLLOW–MASONRY UNIT INTERIOR FOUNDATION WALL

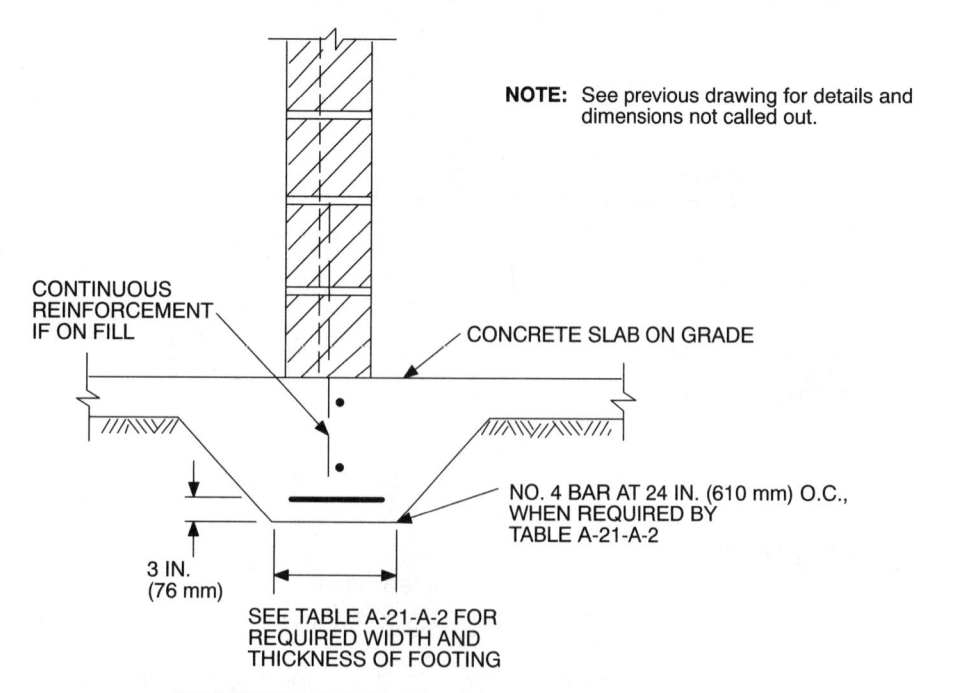

CONCRETE INTERIOR NONBEARING WALL FOOTING

FIGURE A-21-1—VARIOUS DETAILS OF FOOTINGS—(Continued)
(See Tables A-21-A-1 and A-21-A-2 for widths.)

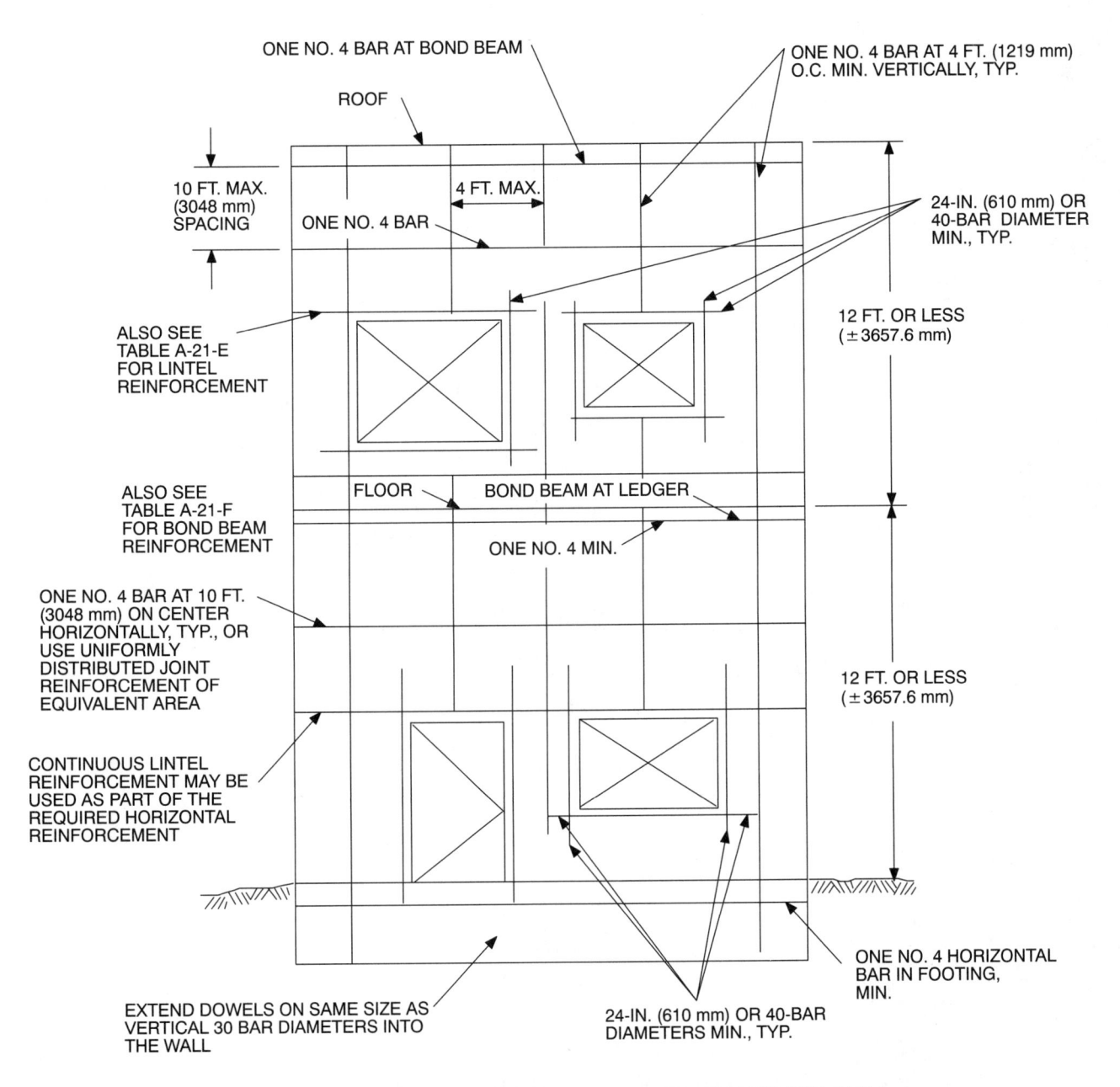

FIGURE A-21-2—MINIMUM MASONRY WALL REQUIREMENTS IN SEISMIC ZONE 2

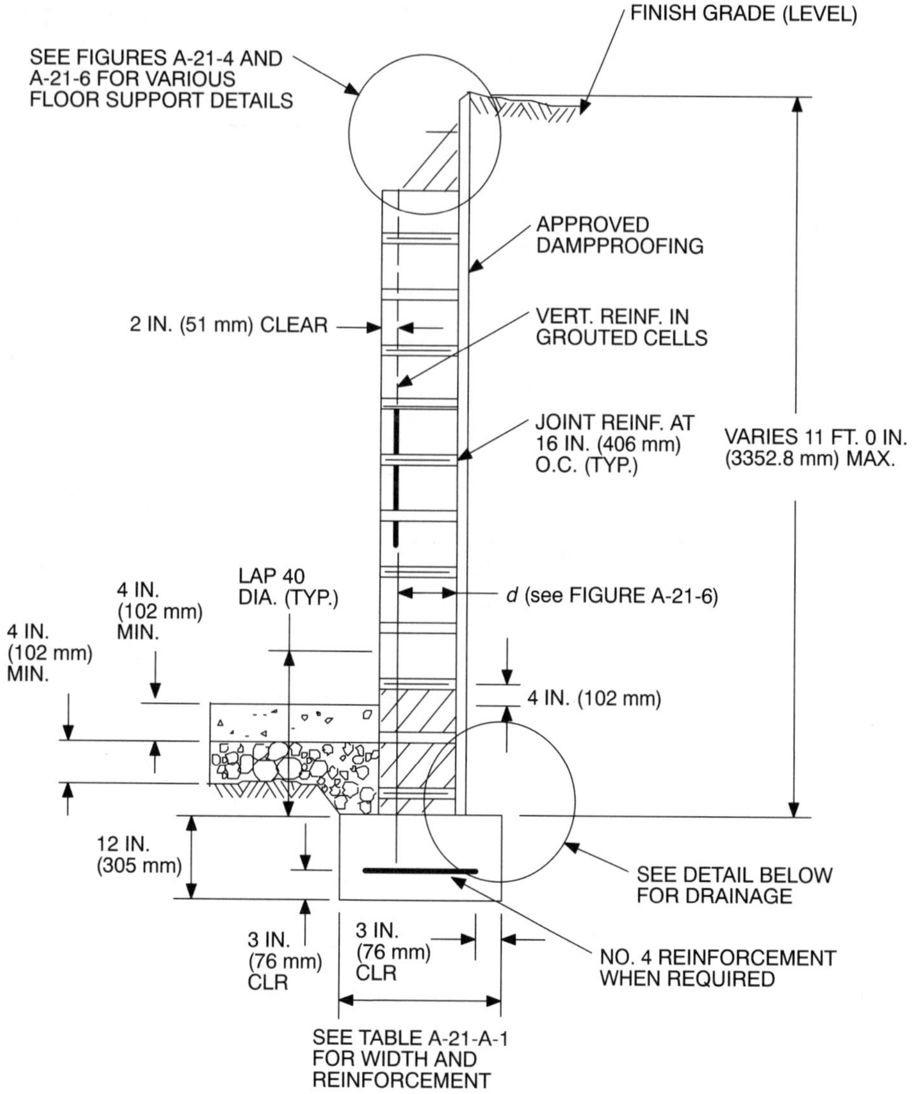

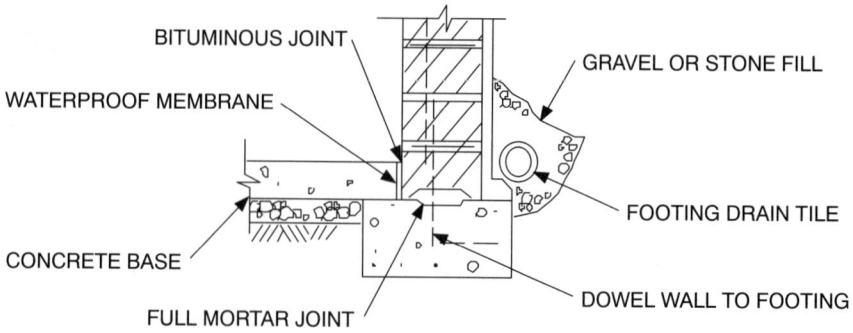

FIGURE A-21-3—BELOW-GRADE WALL AND DRAINAGE DETAILS

FINISH GRADE (LEVEL)

SEE FIGURES A-21-4 AND A-21-6 FOR VARIOUS FLOOR SUPPORT DETAILS

APPROVED DAMPPROOFING

2 IN. (51 mm) CLEAR

VERT. REINF. IN GROUTED CELLS

JOINT REINF. AT 16 IN. (406 mm) O.C. (TYP.)

VARIES 11 FT. 0 IN. (3352.8 mm) MAX.

4 IN. (102 mm) MIN.

4 IN. (102 mm) MIN.

LAP 40 DIA. (TYP.)

d (see FIGURE A-21-6)

4 IN. (102 mm)

12 IN. (305 mm)

3 IN. (76 mm) CLR

3 IN. (76 mm) CLR

SEE DETAIL BELOW FOR DRAINAGE

NO. 4 REINFORCEMENT WHEN REQUIRED

SEE TABLE A-21-A-1 FOR WIDTH AND REINFORCEMENT

BASEMENT WALL

BITUMINOUS JOINT

WATERPROOF MEMBRANE

CONCRETE BASE

FULL MORTAR JOINT

GRAVEL OR STONE FILL

FOOTING DRAIN TILE

DOWEL WALL TO FOOTING

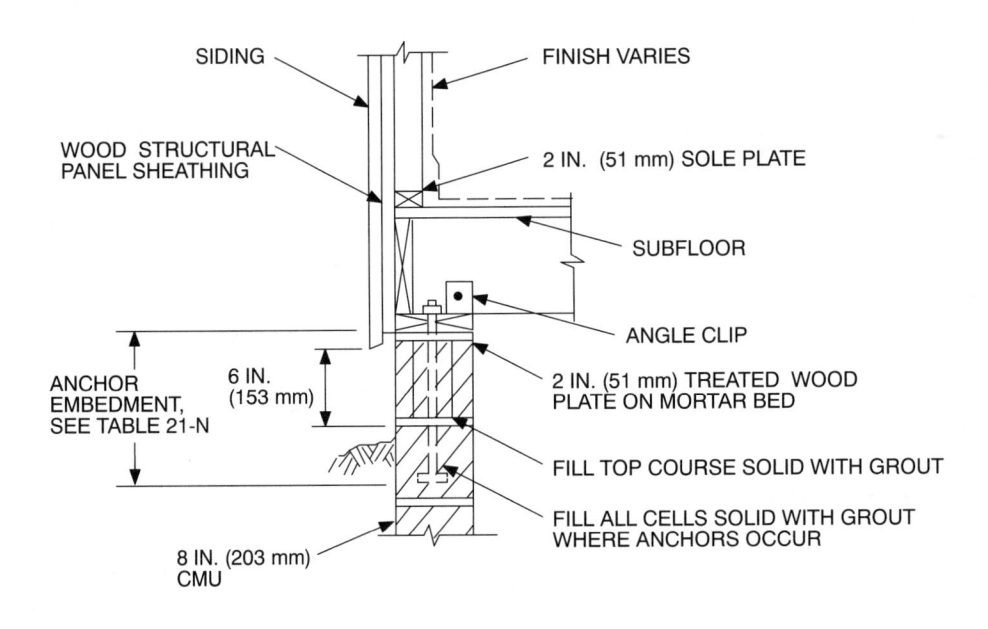

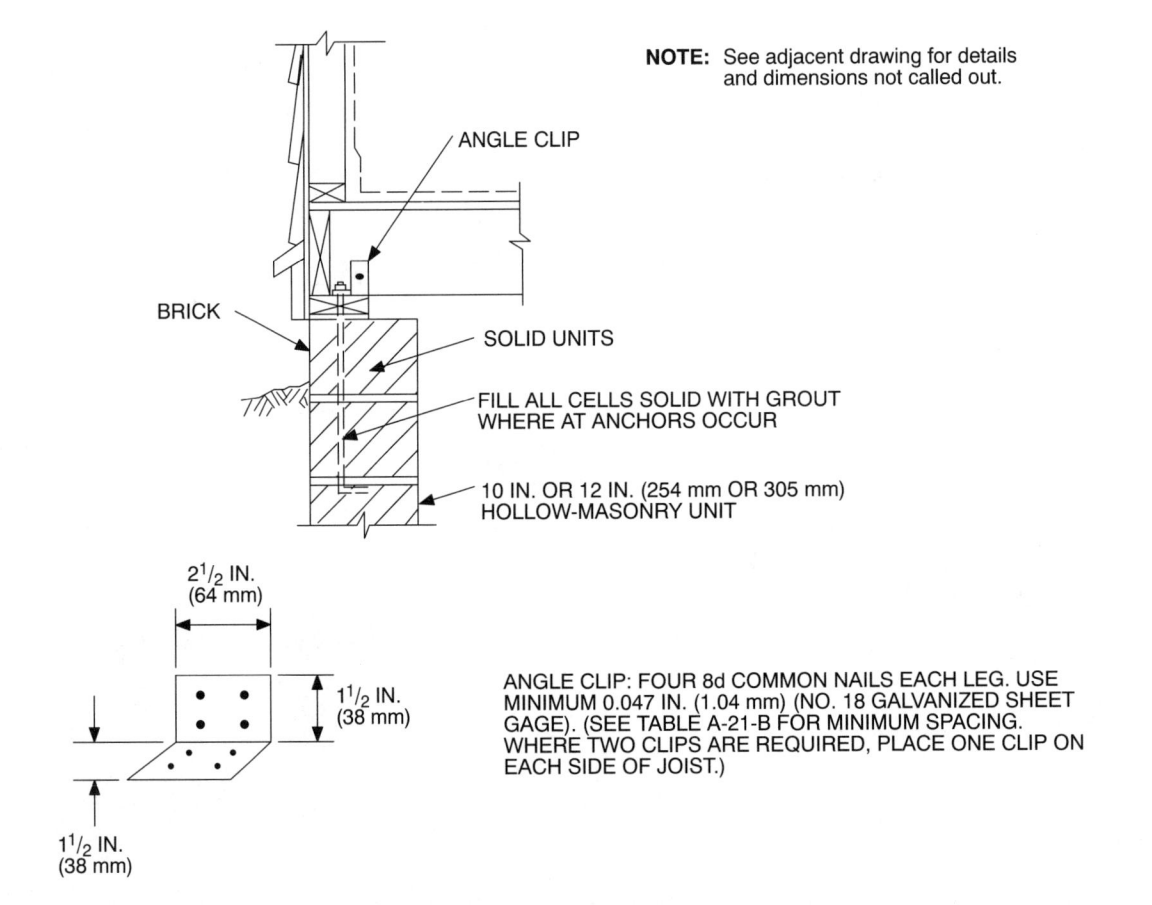

NOTE: See adjacent drawing for details and dimensions not called out.

ANGLE CLIP: FOUR 8d COMMON NAILS EACH LEG. USE MINIMUM 0.047 IN. (1.04 mm) (NO. 18 GALVANIZED SHEET GAGE). (SEE TABLE A-21-B FOR MINIMUM SPACING. WHERE TWO CLIPS ARE REQUIRED, PLACE ONE CLIP ON EACH SIDE OF JOIST.)

FIGURE A-21-4—HOLLOW-MASONRY UNIT FOUNDATION WALL—WOOD FLOOR

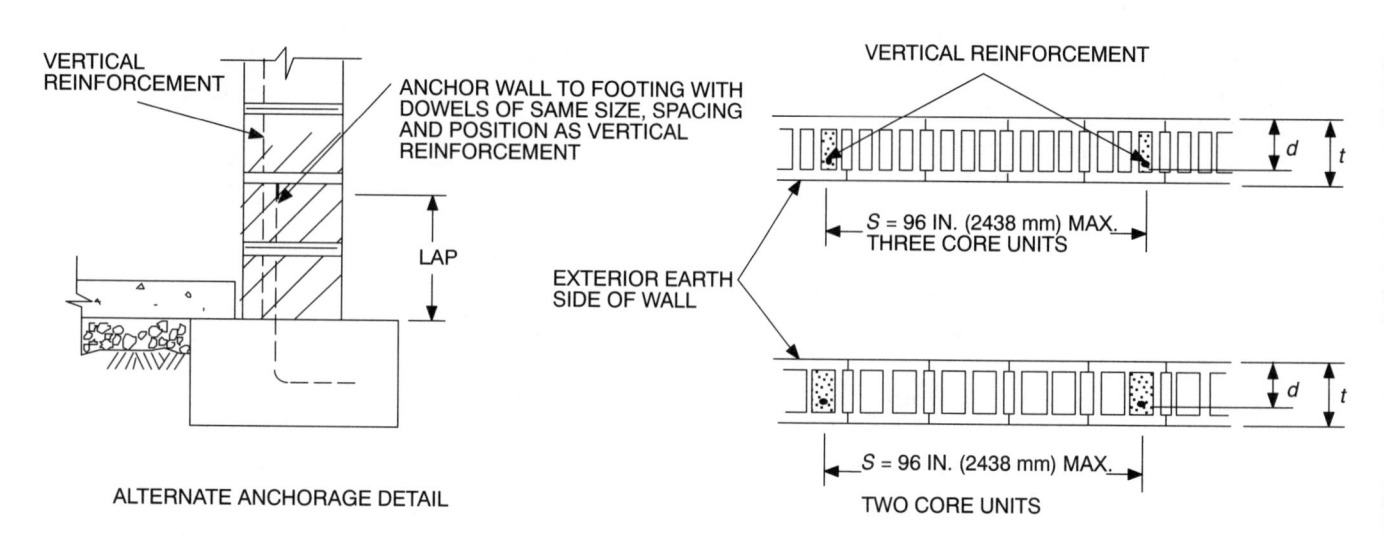

TYPICAL HORIZONTAL SECTION

8 IN. (203 mm) WALLS: $t = 7^5/_8$ IN. (194 mm) $d = 5$ IN. (127 mm)
10 IN. (254 mm) WALLS: $t = 9^5/_8$ IN. (245 mm) $d = 7$ IN. (153 mm)
12 IN. (305 mm) WALLS: $t = 11^5/_8$ IN. (295 mm) $d = 8^3/_4$ IN. (225 mm)

FIGURE A-21-5—PLACEMENT OF REINFORCEMENT

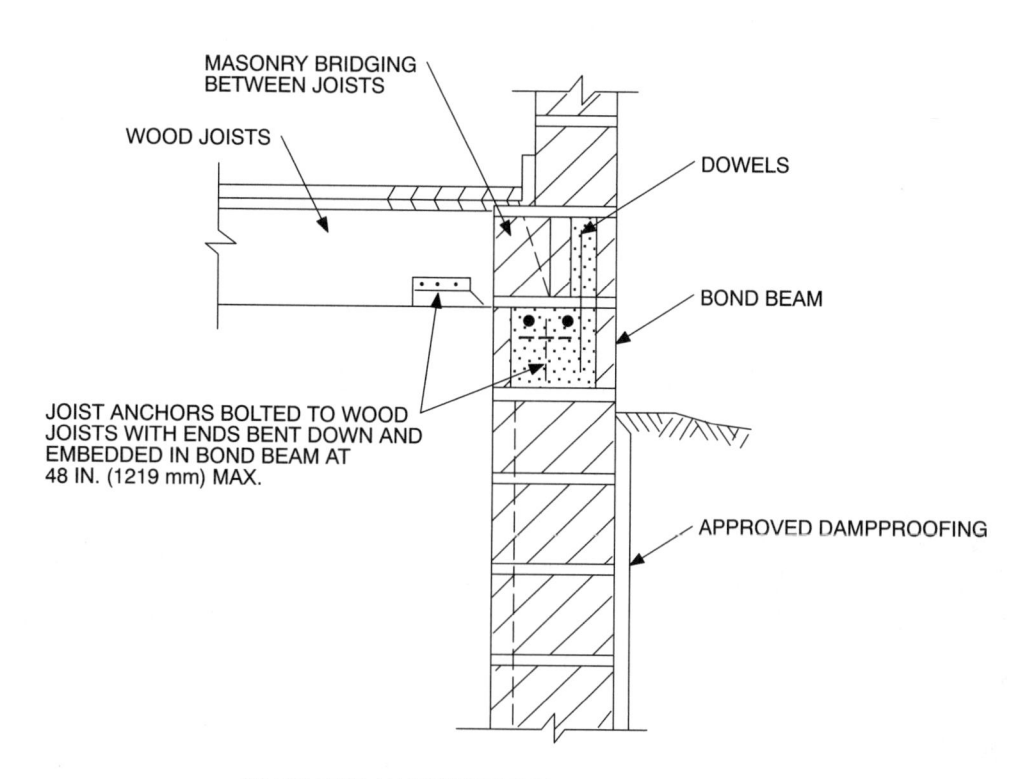

(A) HOLLOW—MASONRY UNIT WALL—WOOD FLOOR

FIGURE A-21-6—VARIOUS CONNECTIONS OF FLOORS TO BASEMENT WALLS

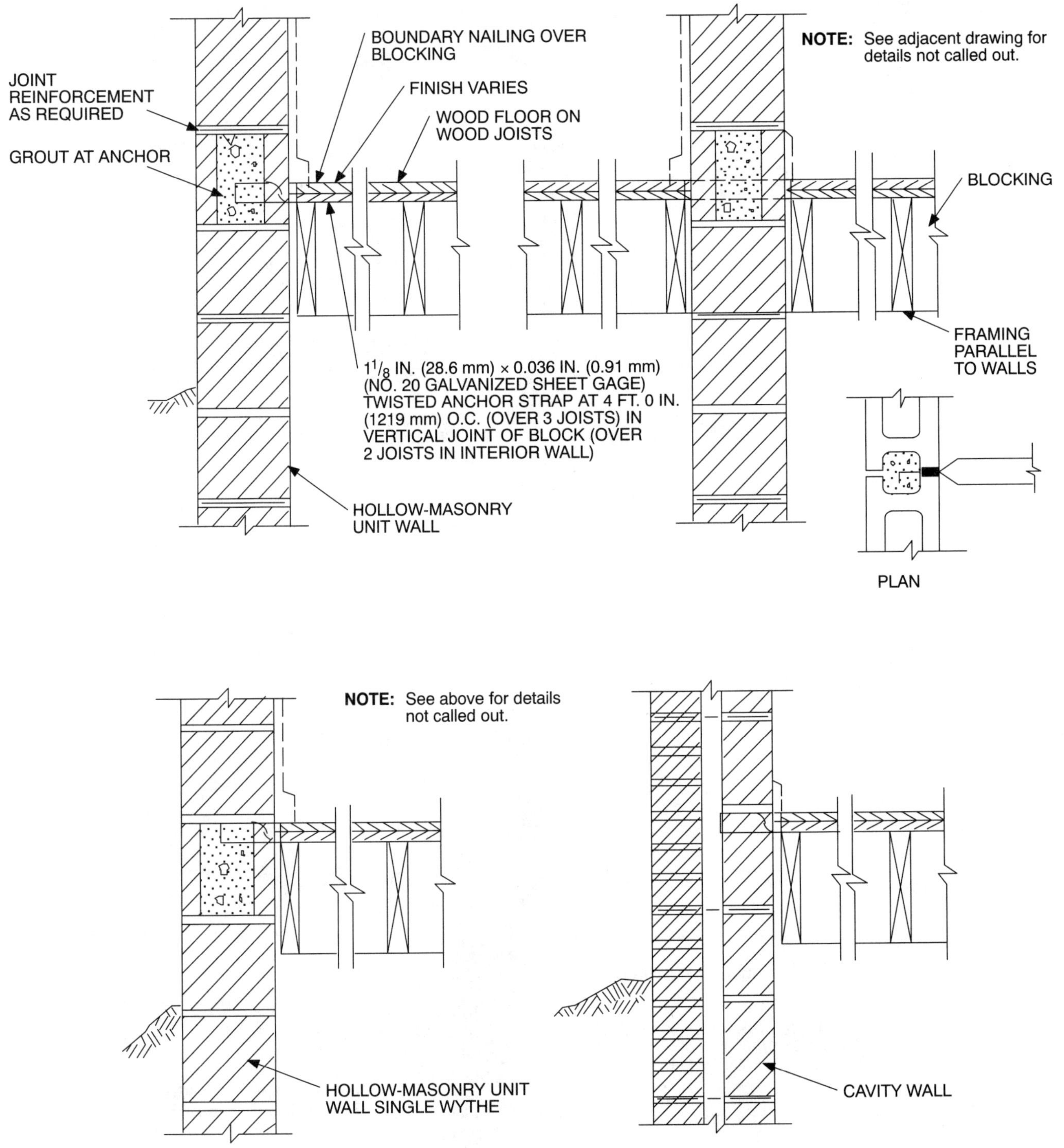

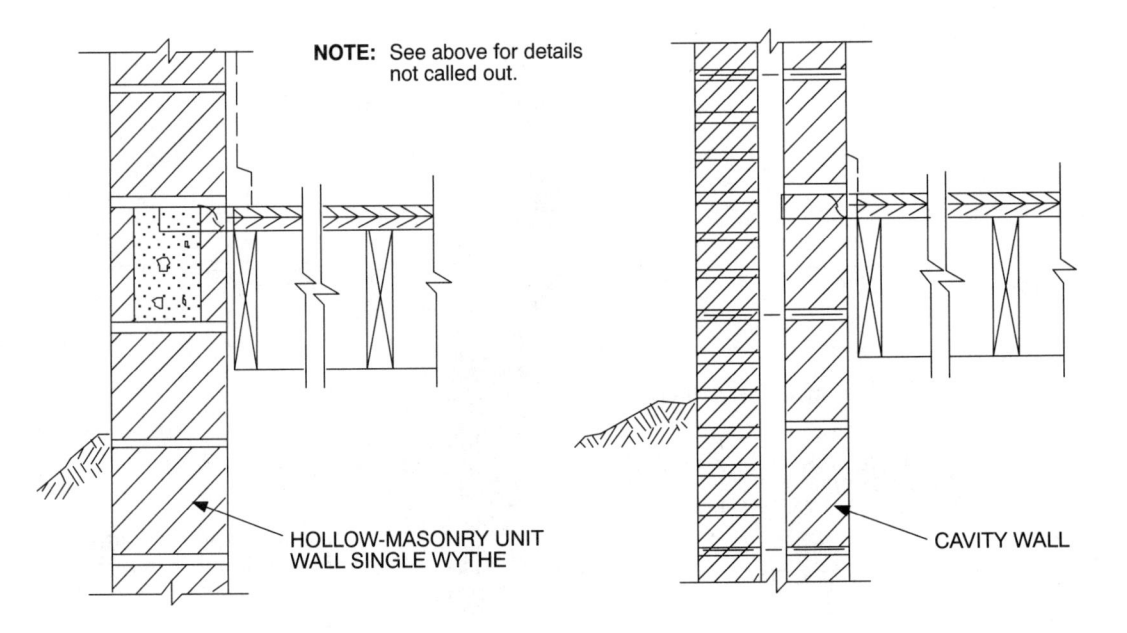

JOINT
REINFORCEMENT
AS REQUIRED

GROUT AT ANCHOR

BOUNDARY NAILING OVER
BLOCKING

FINISH VARIES

WOOD FLOOR ON
WOOD JOISTS

NOTE: See adjacent drawing for
details not called out.

BLOCKING

FRAMING
PARALLEL
TO WALLS

1¹/₈ IN. (28.6 mm) × 0.036 IN. (0.91 mm)
(NO. 20 GALVANIZED SHEET GAGE)
TWISTED ANCHOR STRAP AT 4 FT. 0 IN.
(1219 mm) O.C. (OVER 3 JOISTS) IN
VERTICAL JOINT OF BLOCK (OVER
2 JOISTS IN INTERIOR WALL)

HOLLOW-MASONRY
UNIT WALL

PLAN

NOTE: See above for details
not called out.

HOLLOW-MASONRY UNIT
WALL SINGLE WYTHE

CAVITY WALL

(B) WOOD FLOOR, JOISTS PARALLEL TO WALL

FIGURE A-21-6—VARIOUS CONNECTIONS OF FLOORS TO BASEMENT WALLS—(Continued)

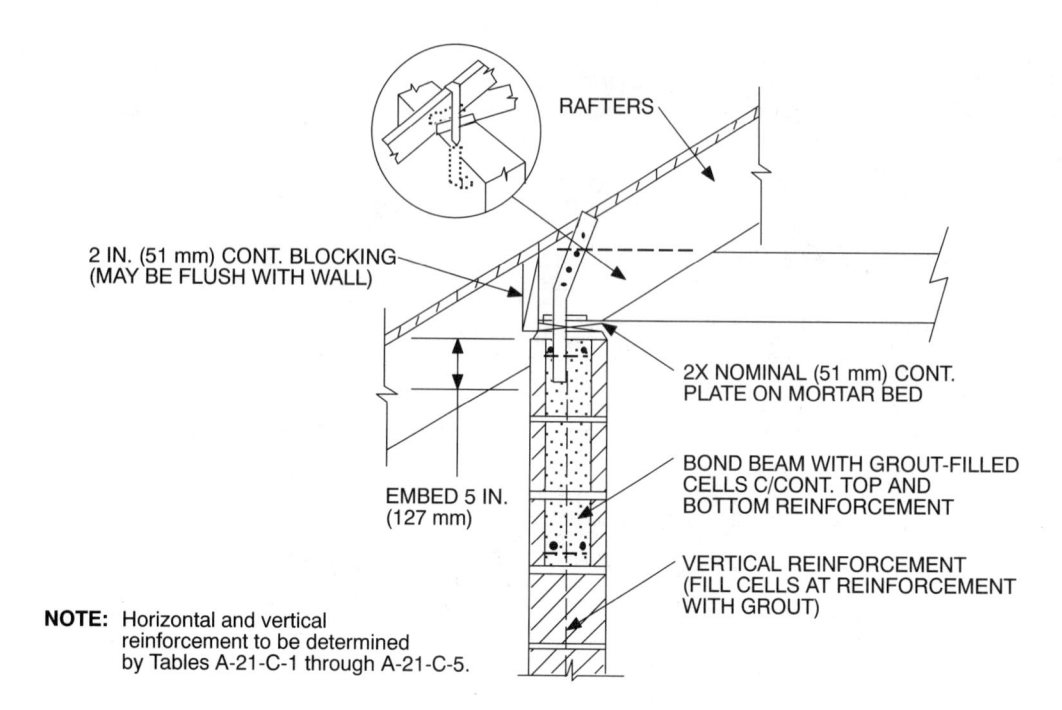

2 IN. (51 mm) CONT. BLOCKING
(MAY BE FLUSH WITH WALL)

RAFTERS

2X NOMINAL (51 mm) CONT.
PLATE ON MORTAR BED

BOND BEAM WITH GROUT-FILLED
CELLS C/CONT. TOP AND
BOTTOM REINFORCEMENT

VERTICAL REINFORCEMENT
(FILL CELLS AT REINFORCEMENT
WITH GROUT)

EMBED 5 IN.
(127 mm)

NOTE: Horizontal and vertical
reinforcement to be determined
by Tables A-21-C-1 through A-21-C-5.

ROOF WITH OVERHANG

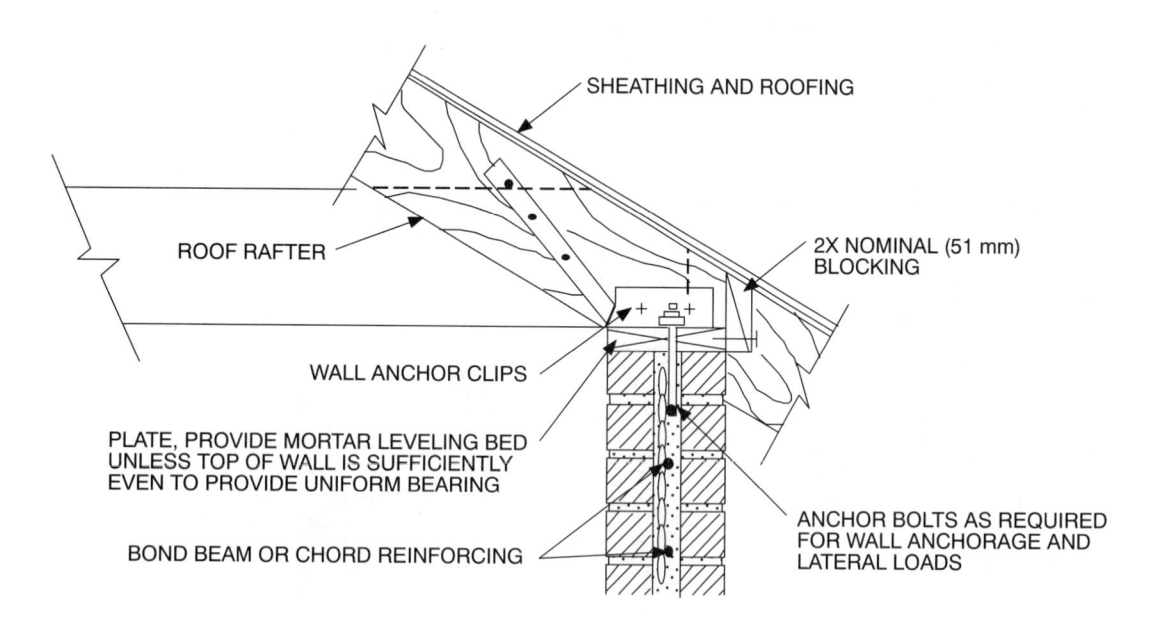

SHEATHING AND ROOFING

ROOF RAFTER

2X NOMINAL (51 mm)
BLOCKING

WALL ANCHOR CLIPS

PLATE, PROVIDE MORTAR LEVELING BED
UNLESS TOP OF WALL IS SUFFICIENTLY
EVEN TO PROVIDE UNIFORM BEARING

ANCHOR BOLTS AS REQUIRED
FOR WALL ANCHORAGE AND
LATERAL LOADS

BOND BEAM OR CHORD REINFORCING

**FIGURE A-21-7—VARIOUS DETAILS ASSOCIATED
WITH TABLE A-21-D (Uplift Resistance)**

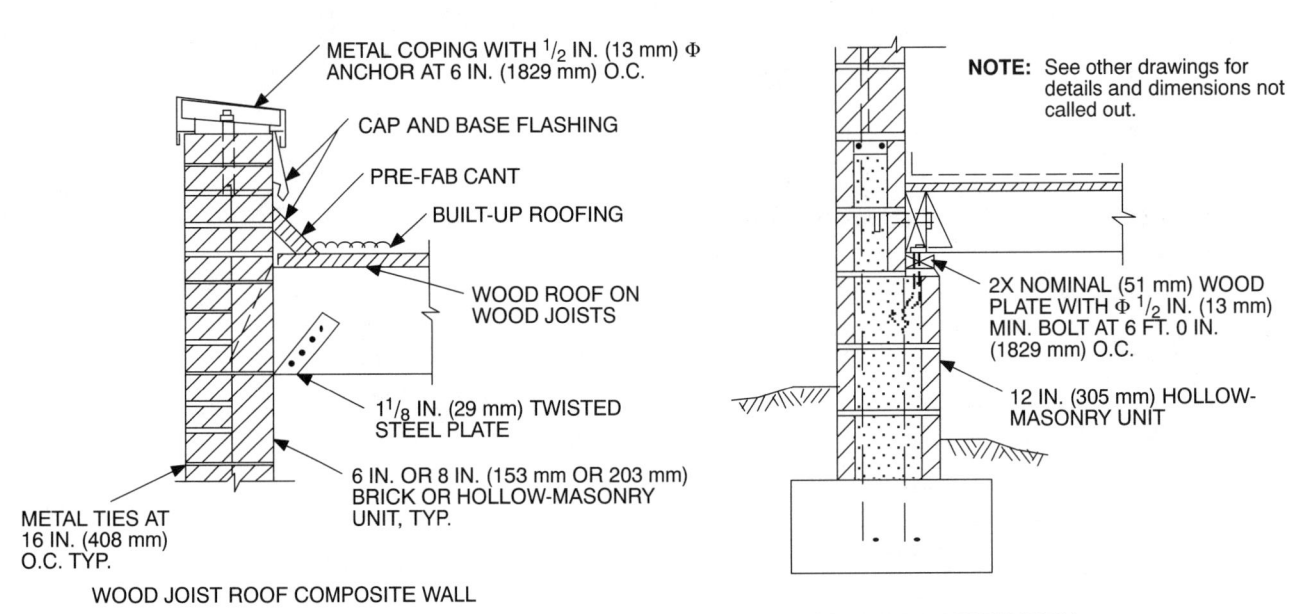

METAL COPING WITH $^1/_2$ IN. (13 mm) Φ ANCHOR AT 6 IN. (1829 mm) O.C.

CAP AND BASE FLASHING

PRE-FAB CANT

BUILT-UP ROOFING

WOOD ROOF ON WOOD JOISTS

$1^1/_8$ IN. (29 mm) TWISTED STEEL PLATE

6 IN. OR 8 IN. (153 mm OR 203 mm) BRICK OR HOLLOW-MASONRY UNIT, TYP.

METAL TIES AT 16 IN. (408 mm) O.C. TYP.

WOOD JOIST ROOF COMPOSITE WALL

NOTE: See other drawings for details and dimensions not called out.

2X NOMINAL (51 mm) WOOD PLATE WITH Φ $^1/_2$ IN. (13 mm) MIN. BOLT AT 6 FT. 0 IN. (1829 mm) O.C.

12 IN. (305 mm) HOLLOW-MASONRY UNIT

HOLLOW-MASONRY UNIT FOUNDATION WALL—JOIST PERPENDICULAR

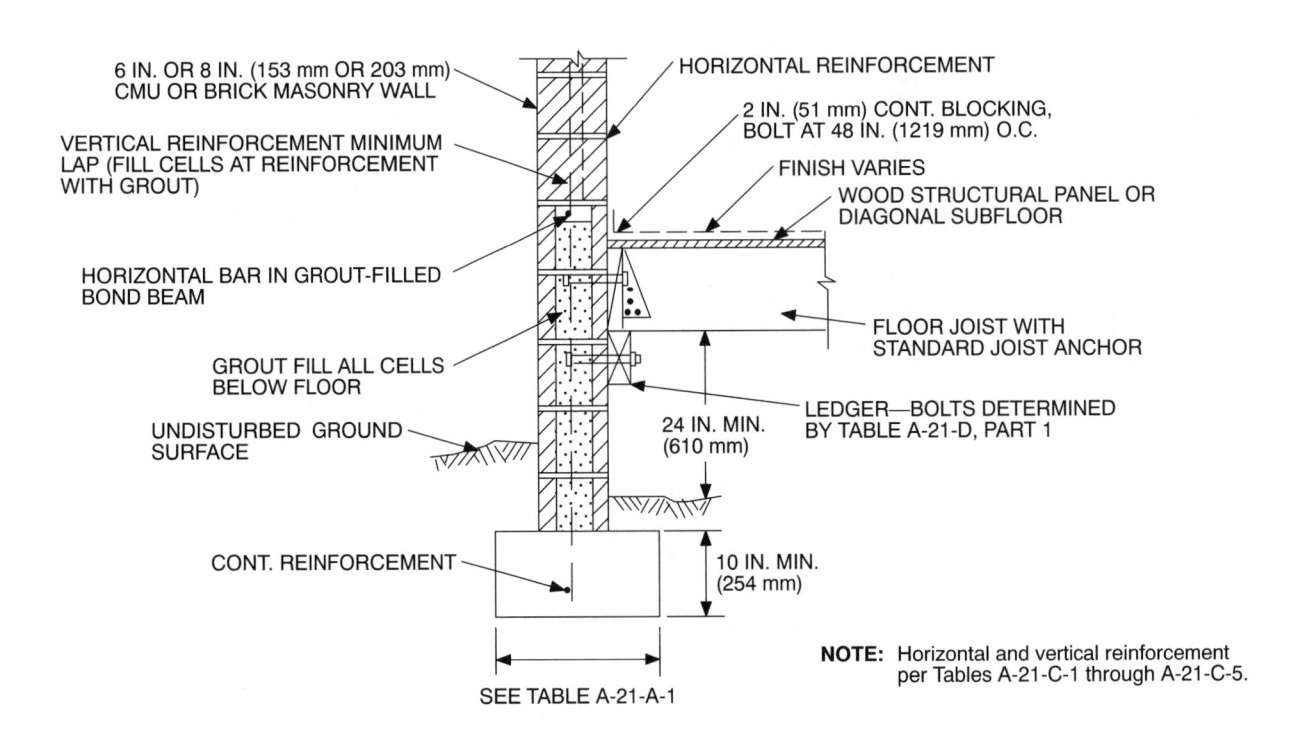

6 IN. OR 8 IN. (153 mm OR 203 mm) CMU OR BRICK MASONRY WALL

VERTICAL REINFORCEMENT MINIMUM LAP (FILL CELLS AT REINFORCEMENT WITH GROUT)

HORIZONTAL BAR IN GROUT-FILLED BOND BEAM

GROUT FILL ALL CELLS BELOW FLOOR

UNDISTURBED GROUND SURFACE

CONT. REINFORCEMENT

HORIZONTAL REINFORCEMENT

2 IN. (51 mm) CONT. BLOCKING, BOLT AT 48 IN. (1219 mm) O.C.

FINISH VARIES

WOOD STRUCTURAL PANEL OR DIAGONAL SUBFLOOR

FLOOR JOIST WITH STANDARD JOIST ANCHOR

LEDGER—BOLTS DETERMINED BY TABLE A-21-D, PART 1

24 IN. MIN. (610 mm)

10 IN. MIN. (254 mm)

SEE TABLE A-21-A-1

NOTE: Horizontal and vertical reinforcement per Tables A-21-C-1 through A-21-C-5.

HOLLOW-MASONRY UNIT FOUNDATION WALL—JOIST PERPENDICULAR—(Continued)

FIGURE A-21-7—VARIOUS DETAILS ASSOCIATED WITH TABLE A-21-D (Uplift Resistance)—(Continued)

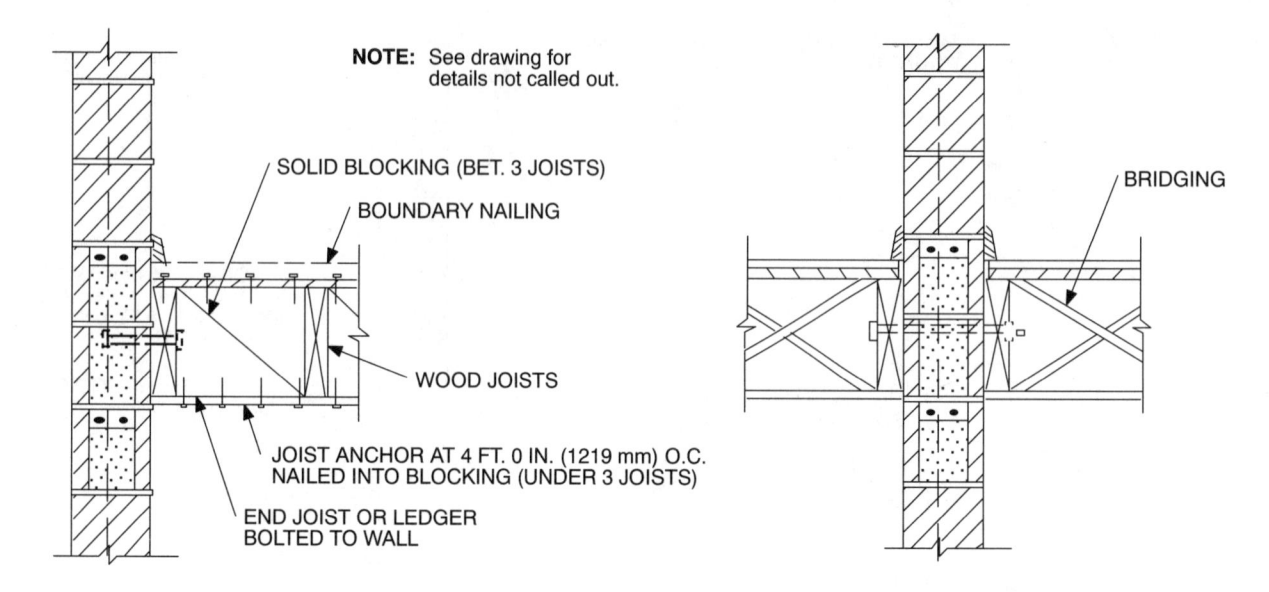

NOTE: See drawing for details not called out.

SOLID BLOCKING (BET. 3 JOISTS)

BOUNDARY NAILING

BRIDGING

WOOD JOISTS

JOIST ANCHOR AT 4 FT. 0 IN. (1219 mm) O.C.
NAILED INTO BLOCKING (UNDER 3 JOISTS)

END JOIST OR LEDGER
BOLTED TO WALL

EXTERIOR WALL—JOIST PARALLEL

INTERIOR WALL—JOIST PARALLEL

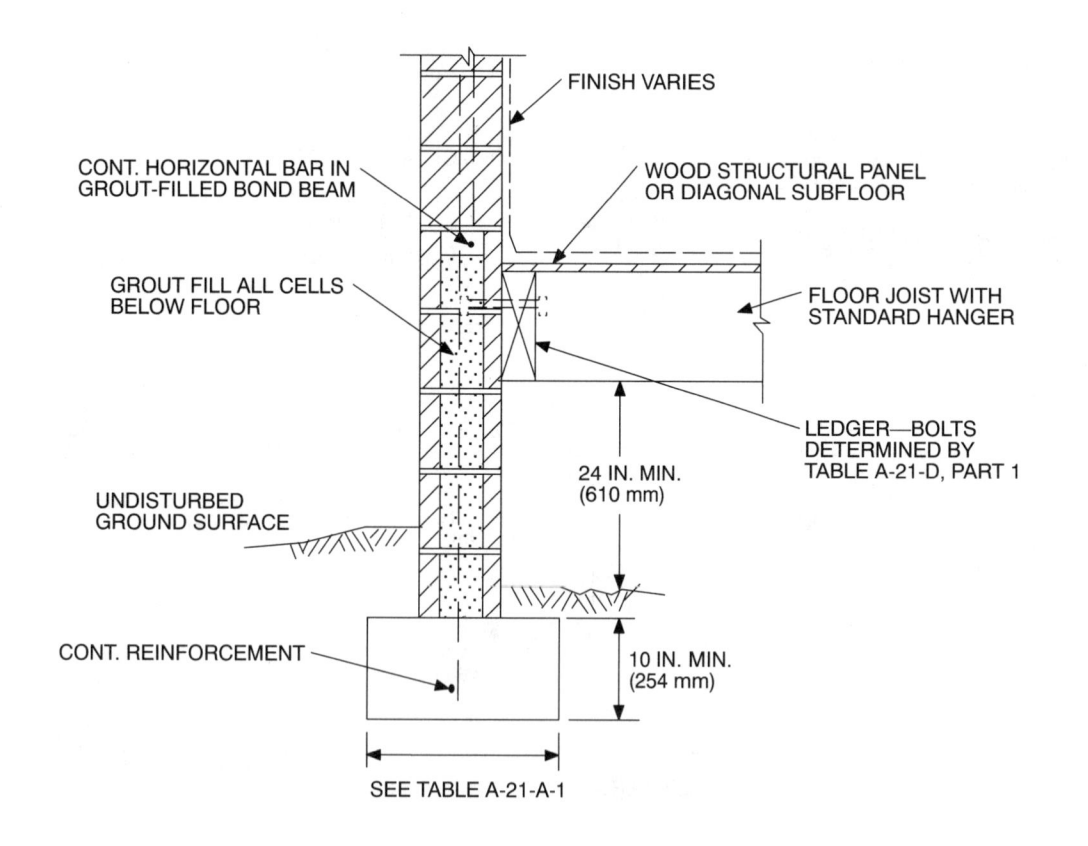

FINISH VARIES

CONT. HORIZONTAL BAR IN
GROUT-FILLED BOND BEAM

WOOD STRUCTURAL PANEL
OR DIAGONAL SUBFLOOR

GROUT FILL ALL CELLS
BELOW FLOOR

FLOOR JOIST WITH
STANDARD HANGER

LEDGER—BOLTS
DETERMINED BY
TABLE A-21-D, PART 1

24 IN. MIN.
(610 mm)

UNDISTURBED
GROUND SURFACE

CONT. REINFORCEMENT

10 IN. MIN.
(254 mm)

SEE TABLE A-21-A-1

HOLLOW-MASONRY UNIT FOUNDATION
WALL—JOIST PERPENDICULAR

FIGURE A-21-7—VARIOUS DETAILS ASSOCIATED WITH TABLE A-21-D (Uplift Resistance)—(Continued)

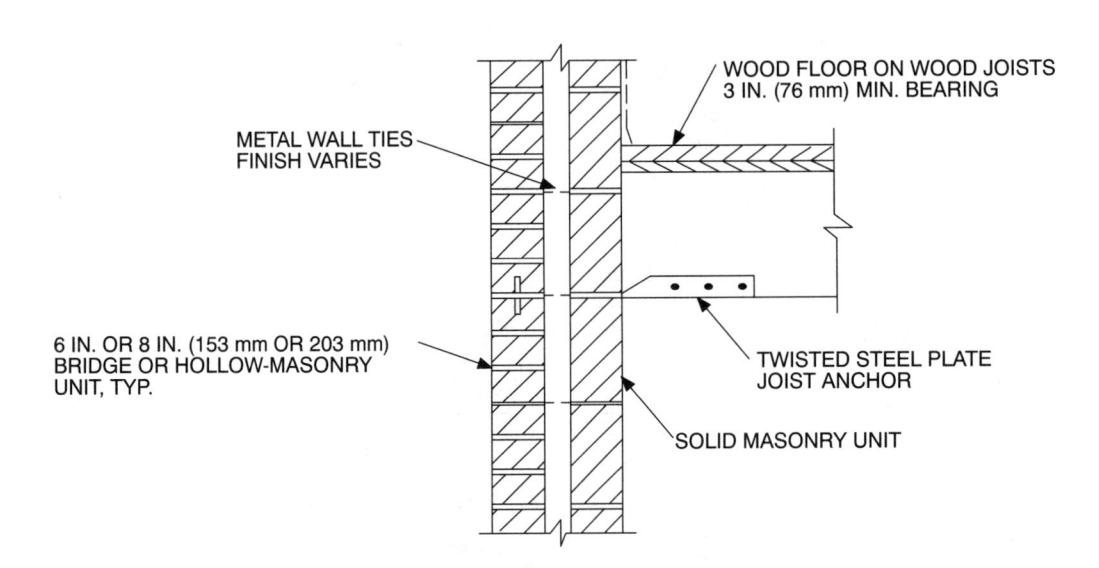

WOOD FLOOR

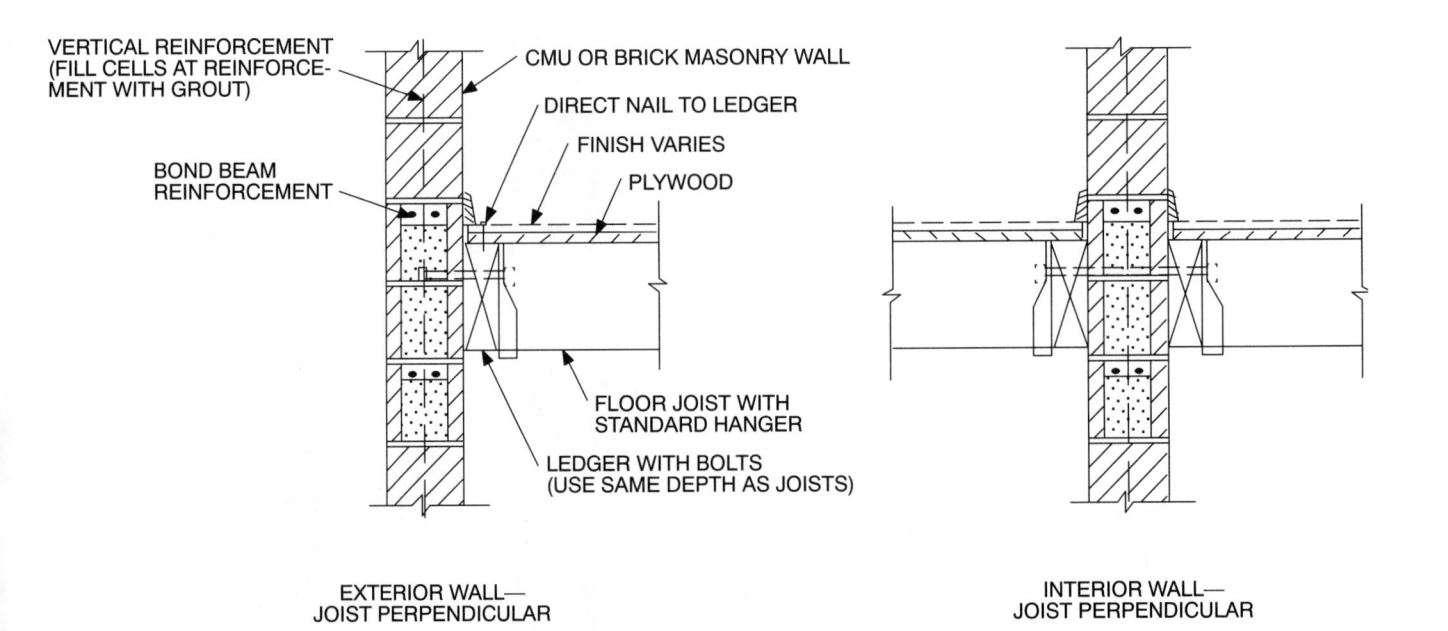

EXTERIOR WALL—
JOIST PERPENDICULAR

INTERIOR WALL—
JOIST PERPENDICULAR

FIGURE A-21-7—VARIOUS DETAILS ASSOCIATED WITH TABLE A-21-D (Uplift Resistance)—(Continued)

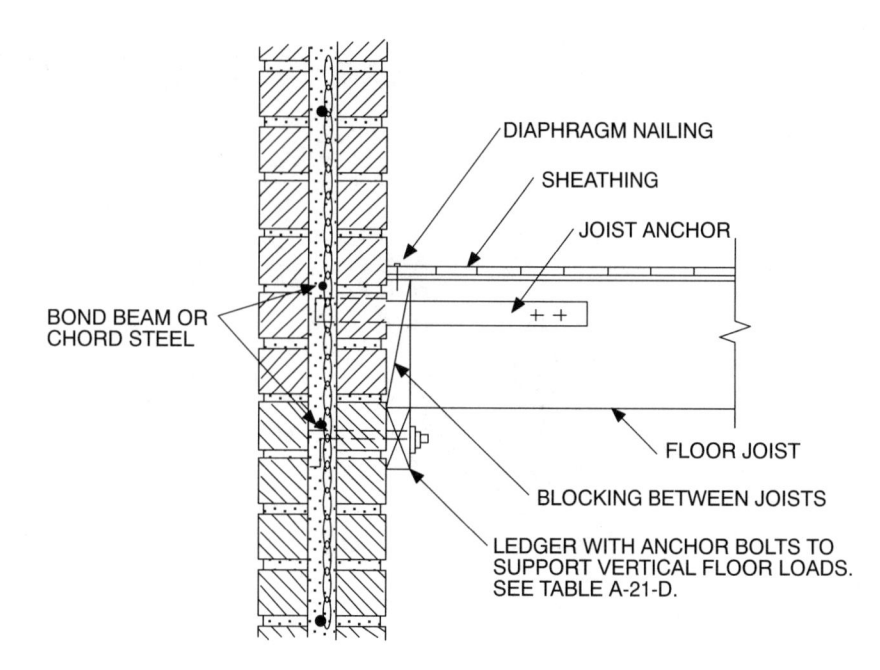

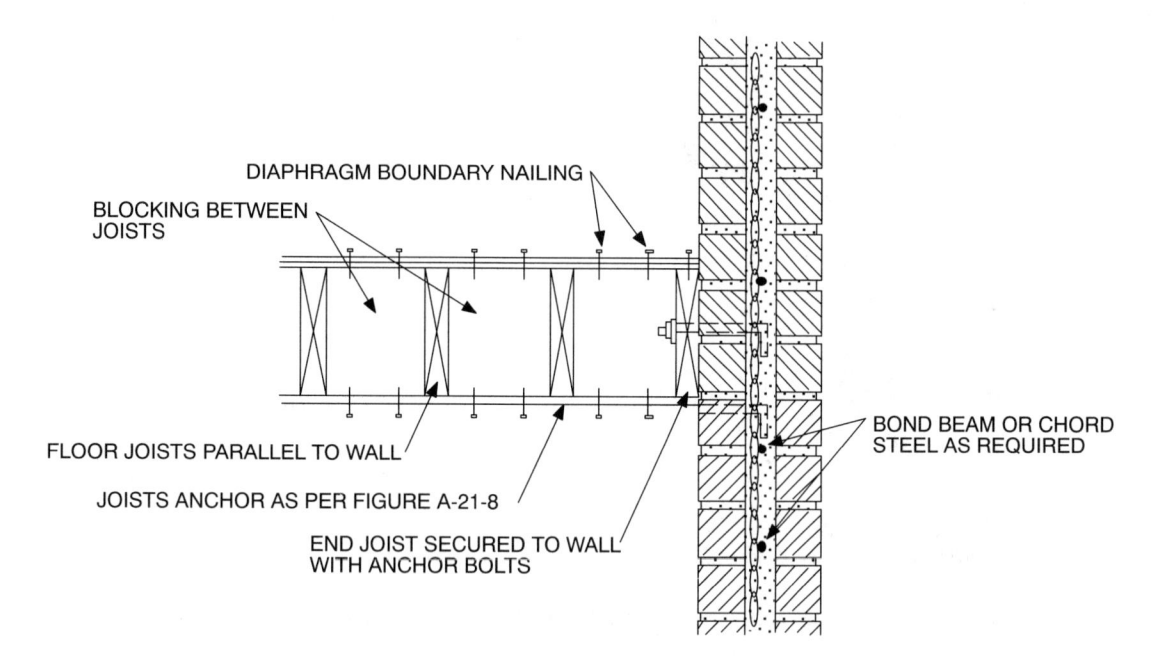

FIGURE A-21-7—VARIOUS DETAILS ASSOCIATED WITH TABLE A-21-D (Uplift Resistance)—(Continued)

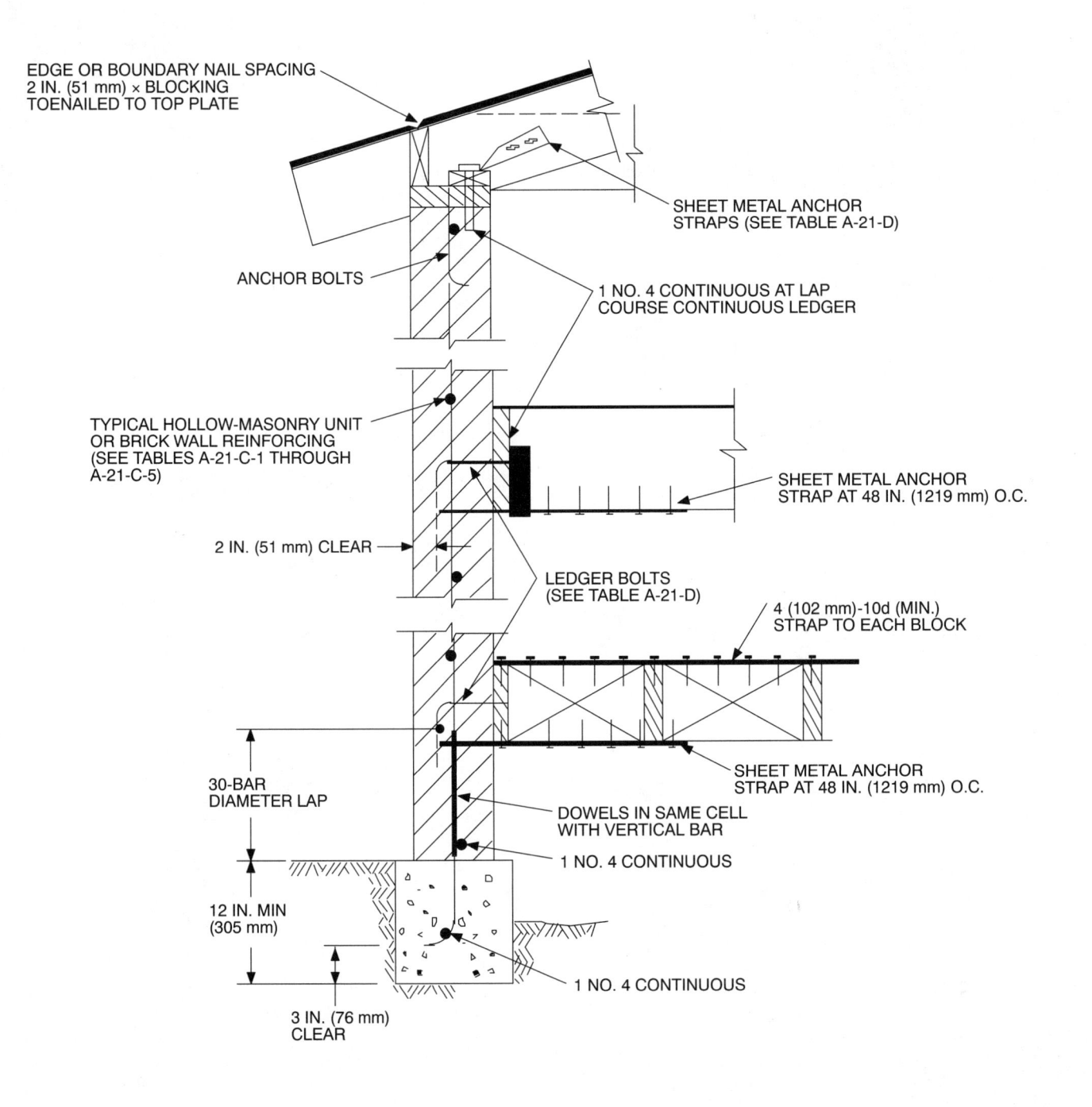

EDGE OR BOUNDARY NAIL SPACING
2 IN. (51 mm) × BLOCKING
TOENAILED TO TOP PLATE

SHEET METAL ANCHOR
STRAPS (SEE TABLE A-21-D)

ANCHOR BOLTS

1 NO. 4 CONTINUOUS AT LAP
COURSE CONTINUOUS LEDGER

TYPICAL HOLLOW-MASONRY UNIT
OR BRICK WALL REINFORCING
(SEE TABLES A-21-C-1 THROUGH
A-21-C-5)

SHEET METAL ANCHOR
STRAP AT 48 IN. (1219 mm) O.C.

2 IN. (51 mm) CLEAR

LEDGER BOLTS
(SEE TABLE A-21-D)

4 (102 mm)-10d (MIN.)
STRAP TO EACH BLOCK

SHEET METAL ANCHOR
STRAP AT 48 IN. (1219 mm) O.C.

30-BAR
DIAMETER LAP

DOWELS IN SAME CELL
WITH VERTICAL BAR

1 NO. 4 CONTINUOUS

12 IN. MIN
(305 mm)

1 NO. 4 CONTINUOUS

3 IN. (76 mm)
CLEAR

FIGURE A-21-8—CONTINUOUS LOAD PATH

FILL ALL CELLS WITH GROUT

REINFORCEMENT

NO. 3 TIES AT 16 IN. (406 mm) O.C.

24 IN. (610 mm)

STANDARD 8 IN. × 8 IN. × 16 IN.
(203 mm × 203 mm × 406 mm)
UNITS WITH WEB CUTOUTS

LINTEL OR BOND BEAM UNITS

PLACE METAL LATH OR HEAVY WATERPROOF PAPER
OVER CORES OF BEARING UNITS TO RETAIN CONCRETE

TWO NO. 4 EACH SIDE OF OPENING.
EXTEND 24 IN. (610 mm) BEYOND
OPENING OR HOOK OVER BOND BEAM BARS
(SEE TABLE A-21-E, FOOTNOTE 8.)

REINFORCING DETAILS

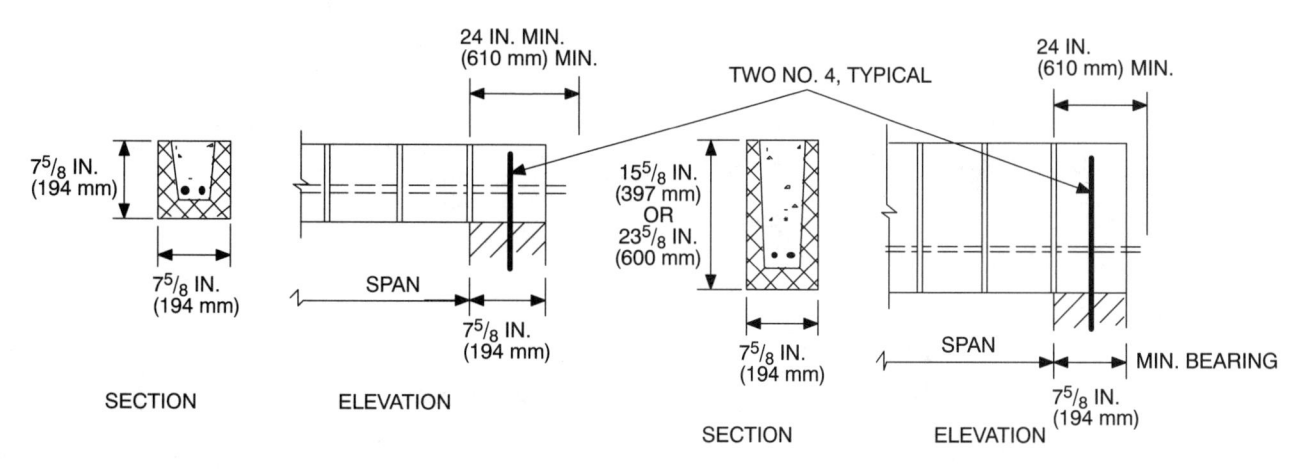

24 IN. MIN.
(610 mm) MIN.

TWO NO. 4, TYPICAL

24 IN.
(610 mm) MIN.

$7^5/_8$ IN.
(194 mm)

$7^5/_8$ IN.
(194 mm)

SPAN

$7^5/_8$ IN.
(194 mm)

SECTION

ELEVATION

$15^5/_8$ IN.
(397 mm)
OR
$23^5/_8$ IN.
(600 mm)

$7^5/_8$ IN.
(194 mm)

SECTION

SPAN

MIN. BEARING

ELEVATION

$7^5/_8$ IN.
(194 mm)

WITHOUT STIRRUPS

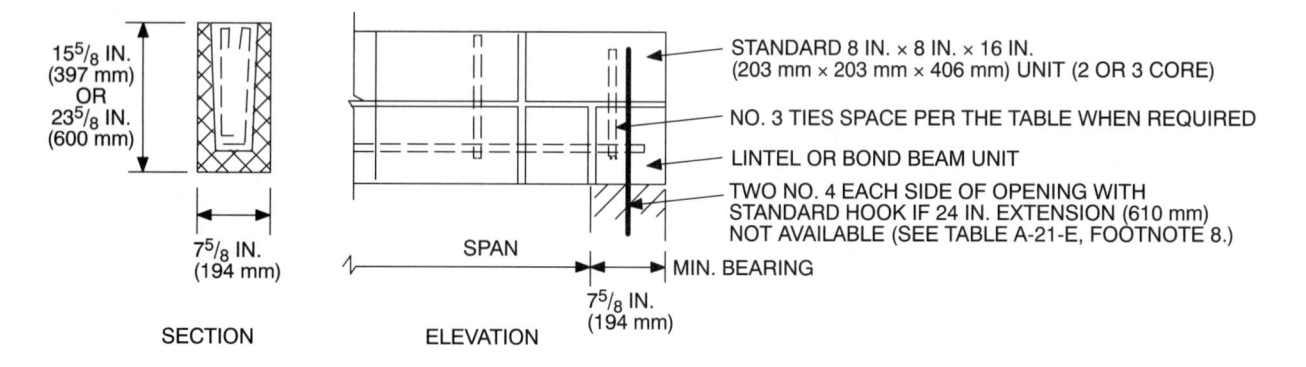

$15^5/_8$ IN.
(397 mm)
OR
$23^5/_8$ IN.
(600 mm)

STANDARD 8 IN. × 8 IN. × 16 IN.
(203 mm × 203 mm × 406 mm) UNIT (2 OR 3 CORE)

NO. 3 TIES SPACE PER THE TABLE WHEN REQUIRED

LINTEL OR BOND BEAM UNIT

TWO NO. 4 EACH SIDE OF OPENING WITH
STANDARD HOOK IF 24 IN. EXTENSION (610 mm)
NOT AVAILABLE (SEE TABLE A-21-E, FOOTNOTE 8.)

$7^5/_8$ IN.
(194 mm)

SECTION

SPAN

MIN. BEARING

ELEVATION

$7^5/_8$ IN.
(194 mm)

WITH STIRRUPS

FIGURE A-21-8—CONTINUOUS LOAD PATH—(Continued)

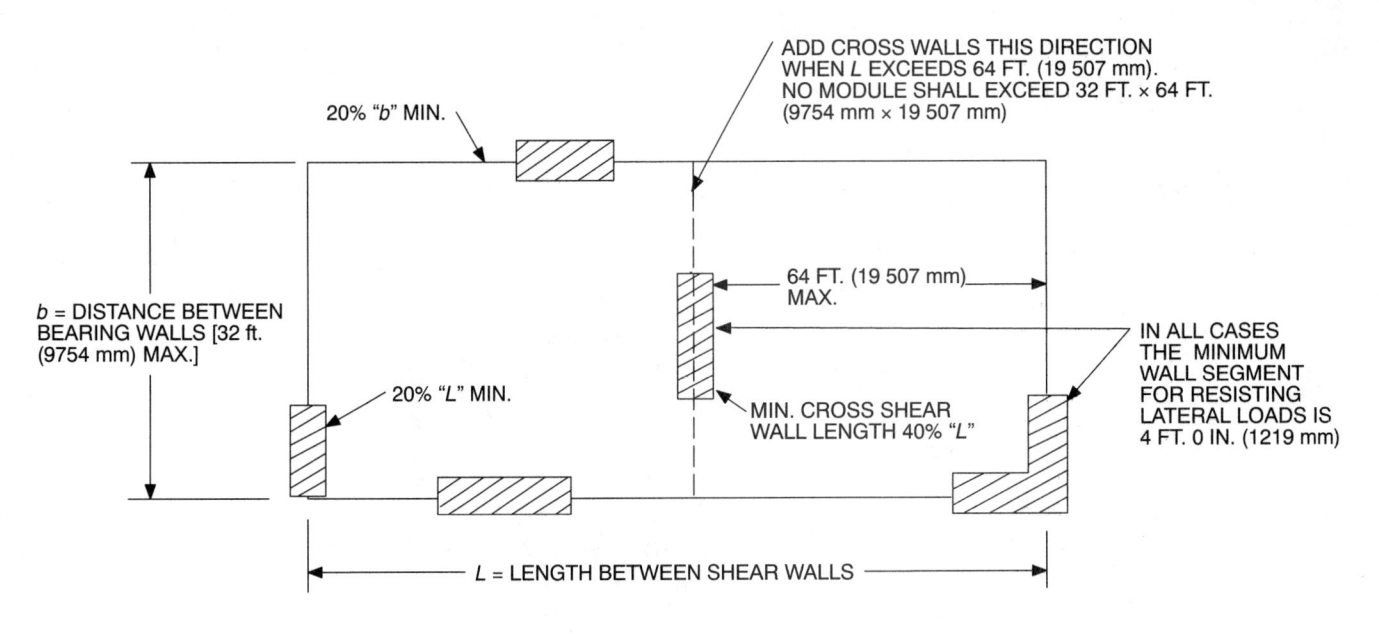

FIGURE A-21-9—SPACING AND LENGTHS OF SHEAR WALLS

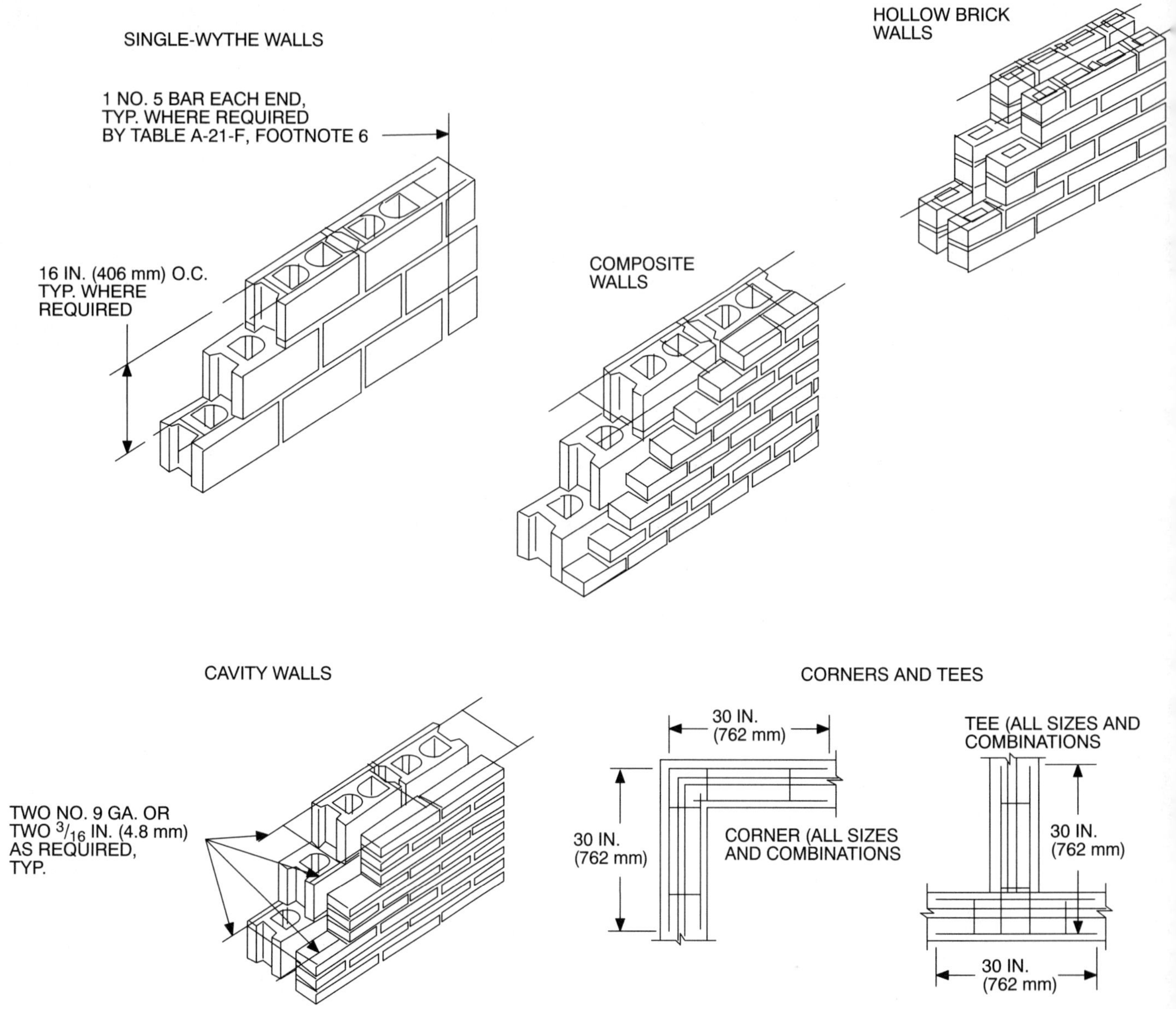

SINGLE-WYTHE WALLS

1 NO. 5 BAR EACH END,
TYP. WHERE REQUIRED
BY TABLE A-21-F, FOOTNOTE 6

16 IN. (406 mm) O.C.
TYP. WHERE
REQUIRED

HOLLOW BRICK
WALLS

COMPOSITE
WALLS

CAVITY WALLS

TWO NO. 9 GA. OR
TWO $^3/_{16}$ IN. (4.8 mm)
AS REQUIRED,
TYP.

CORNERS AND TEES

30 IN.
(762 mm)

30 IN.
(762 mm)

CORNER (ALL SIZES
AND COMBINATIONS

TEE (ALL SIZES AND
COMBINATIONS

30 IN.
(762 mm)

30 IN.
(762 mm)

FIGURE A-21-10—SPACING OF STEEL REINFORCING WIRE

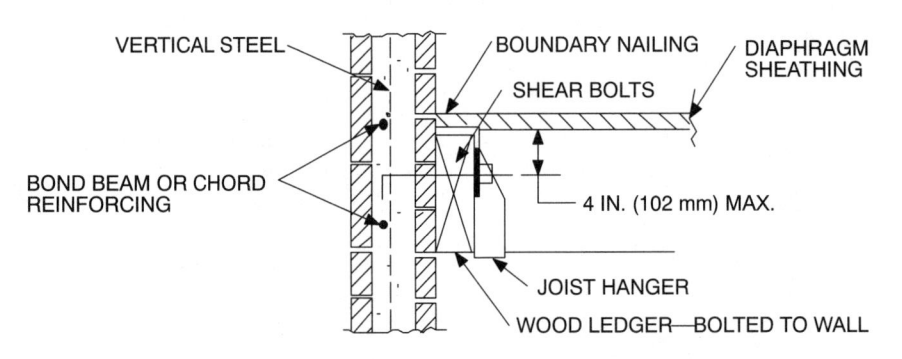

(a) FLOOR JOISTS PERPENDICULAR TO WALL JOIST HANGER SUPPORTS

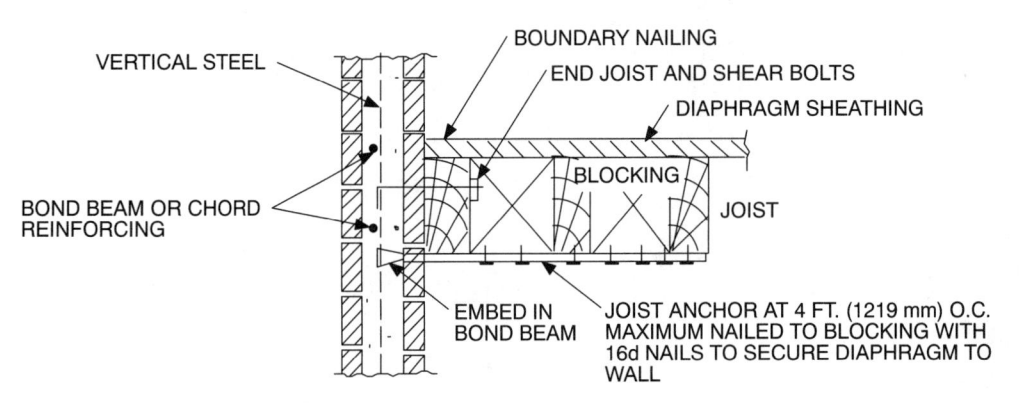

(b) FLOOR JOISTS PARALLEL TO WALL

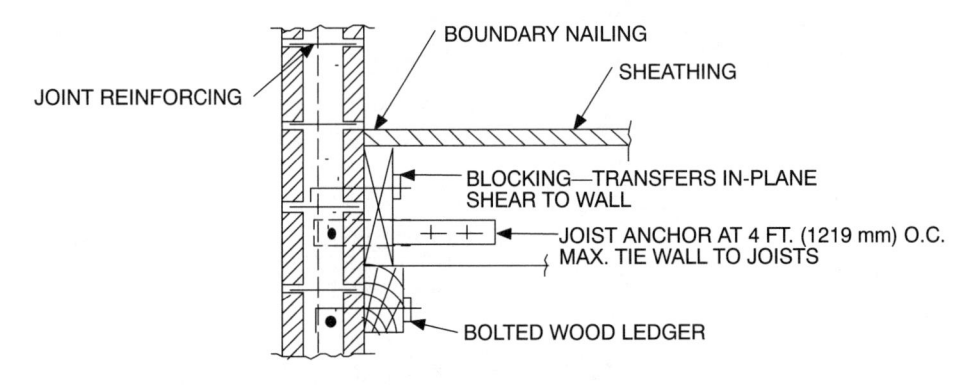

(c) WOOD LEDGER FLOOR JOIST SUPPORT

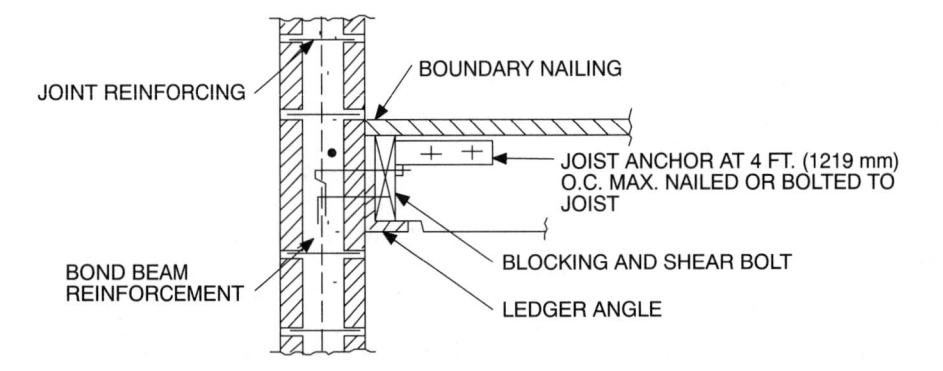

(d) STEEL LEDGER FLOOR JOIST SUPPORT

FIGURE A-21-11—FLOOR-TO-WALL CONNECTION DETAILS

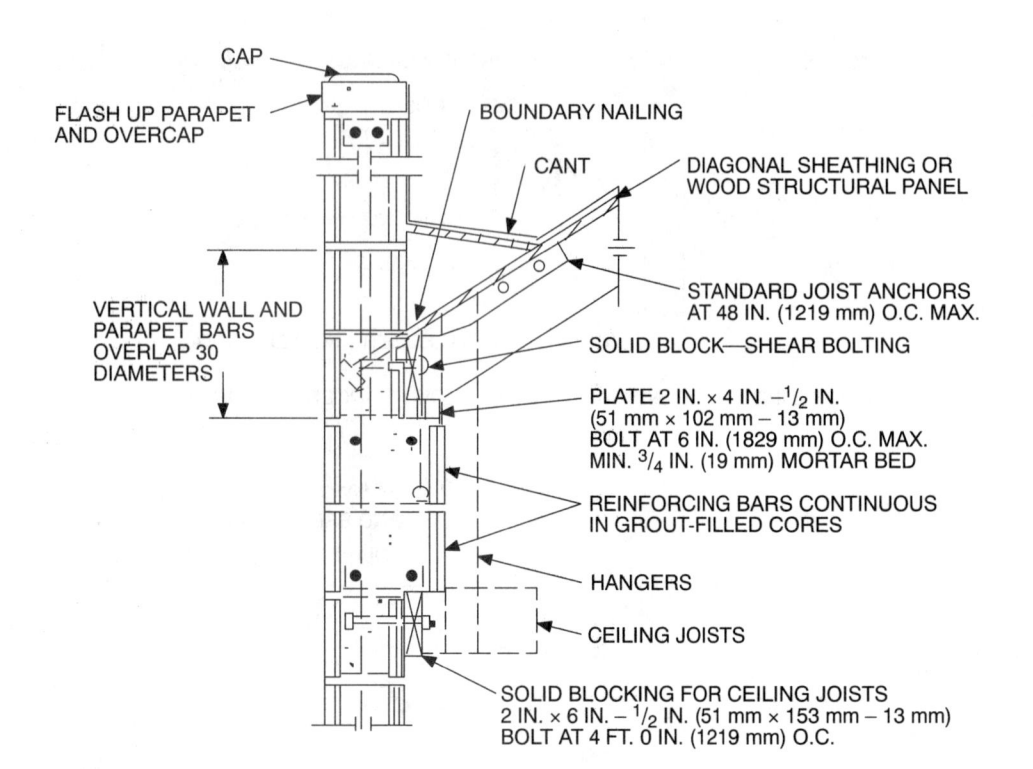

CAP

FLASH UP PARAPET
AND OVERCAP

BOUNDARY NAILING

CANT

DIAGONAL SHEATHING OR
WOOD STRUCTURAL PANEL

VERTICAL WALL AND
PARAPET BARS
OVERLAP 30
DIAMETERS

STANDARD JOIST ANCHORS
AT 48 IN. (1219 mm) O.C. MAX.

SOLID BLOCK—SHEAR BOLTING

PLATE 2 IN. × 4 IN. −$^1/_2$ IN.
(51 mm × 102 mm − 13 mm)
BOLT AT 6 IN. (1829 mm) O.C. MAX.
MIN. $^3/_4$ IN. (19 mm) MORTAR BED

REINFORCING BARS CONTINUOUS
IN GROUT-FILLED CORES

HANGERS

CEILING JOISTS

SOLID BLOCKING FOR CEILING JOISTS
2 IN. × 6 IN. − $^1/_2$ IN. (51 mm × 153 mm − 13 mm)
BOLT AT 4 FT. 0 IN. (1219 mm) O.C.

(a) EXTERIOR WALL SUPPORT

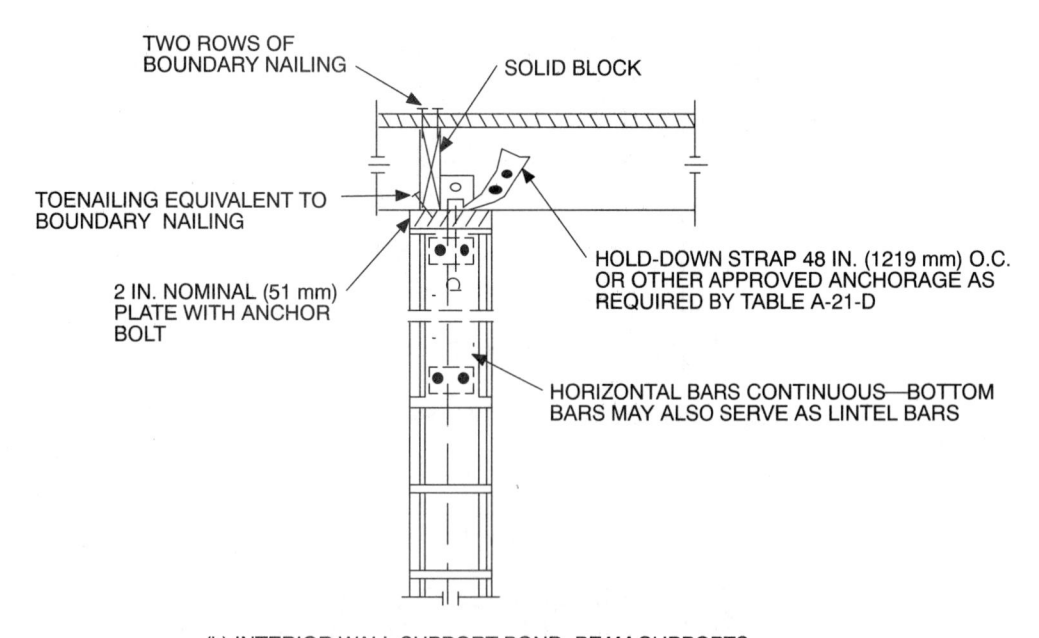

TWO ROWS OF
BOUNDARY NAILING

SOLID BLOCK

TOENAILING EQUIVALENT TO
BOUNDARY NAILING

2 IN. NOMINAL (51 mm)
PLATE WITH ANCHOR
BOLT

HOLD-DOWN STRAP 48 IN. (1219 mm) O.C.
OR OTHER APPROVED ANCHORAGE AS
REQUIRED BY TABLE A-21-D

HORIZONTAL BARS CONTINUOUS—BOTTOM
BARS MAY ALSO SERVE AS LINTEL BARS

(b) INTERIOR WALL SUPPORT BOND–BEAM SUPPORTS

FIGURE A-21-12—ROOF-TO-WALL CONNECTION DETAILS

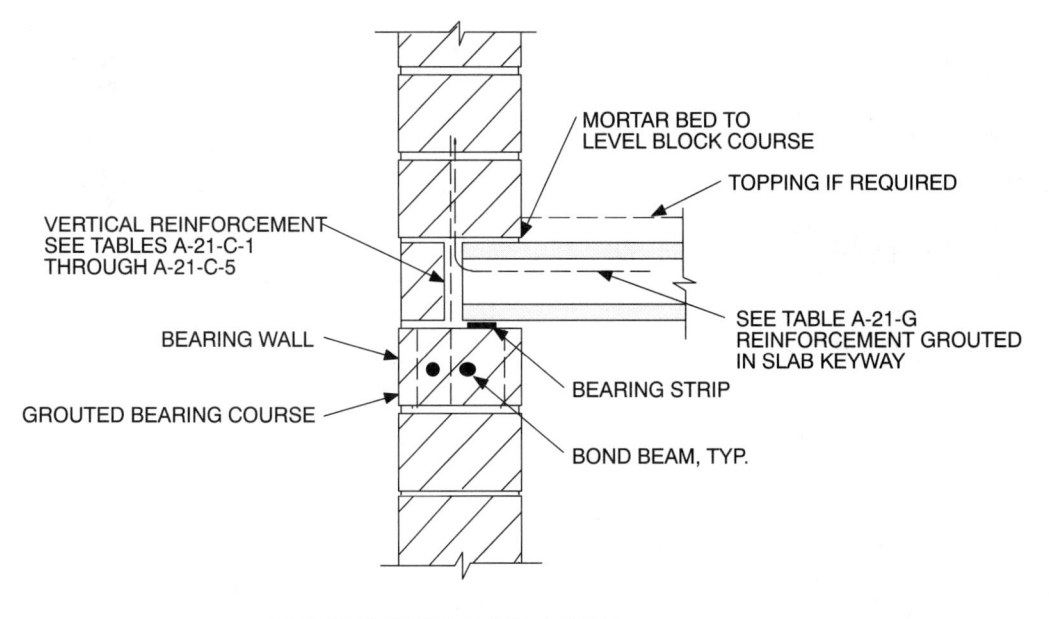

(a) SLAB PERPENDICULAR TO WALL

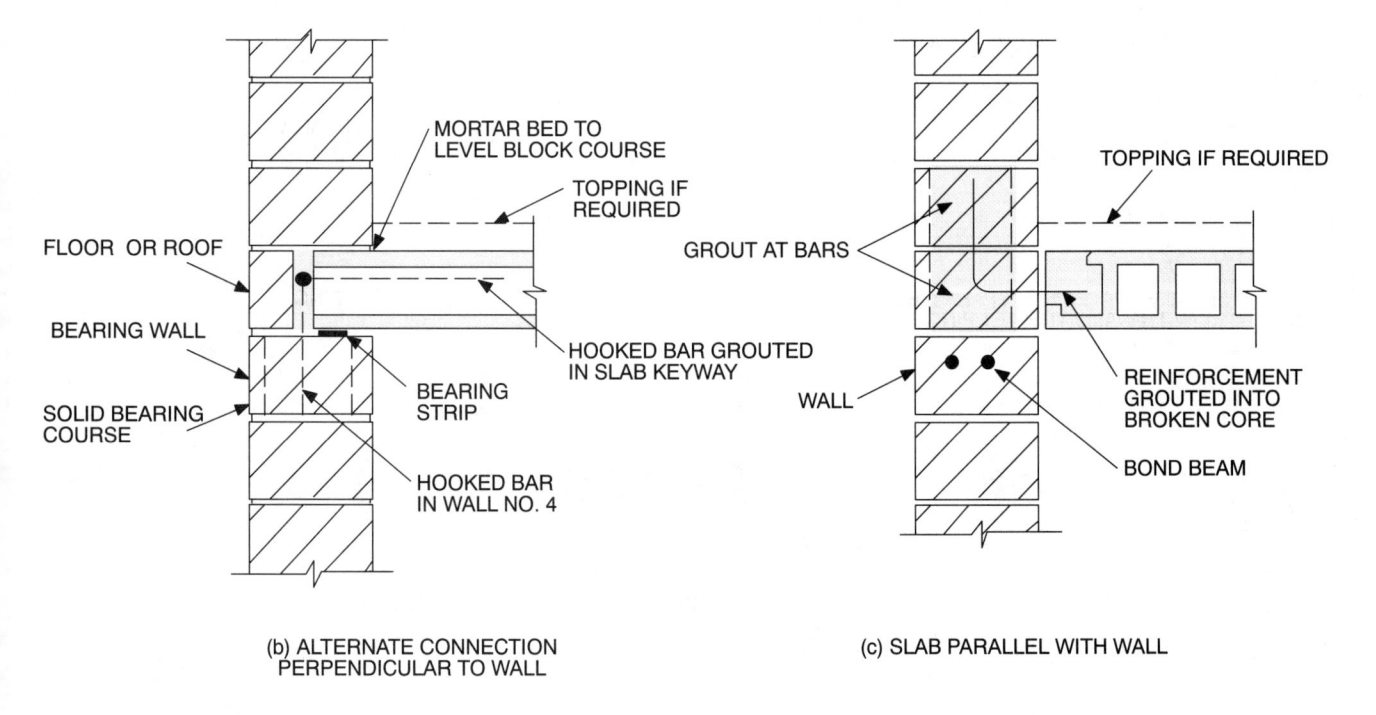

(b) ALTERNATE CONNECTION
PERPENDICULAR TO WALL

(c) SLAB PARALLEL WITH WALL

FIGURE A-21-13—VARIOUS TYPES OF WALL CONNECTIONS

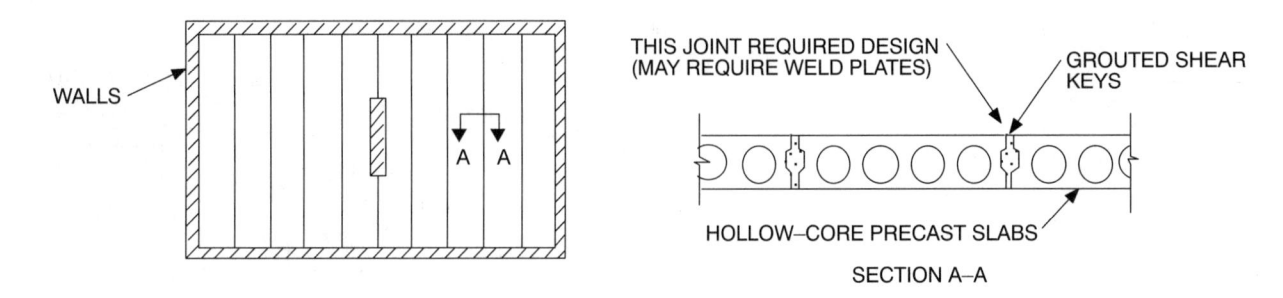

(d) PLAN VIEW OF FLOOR OR ROOF AND CROSS SECTION THROUGH PLANKS

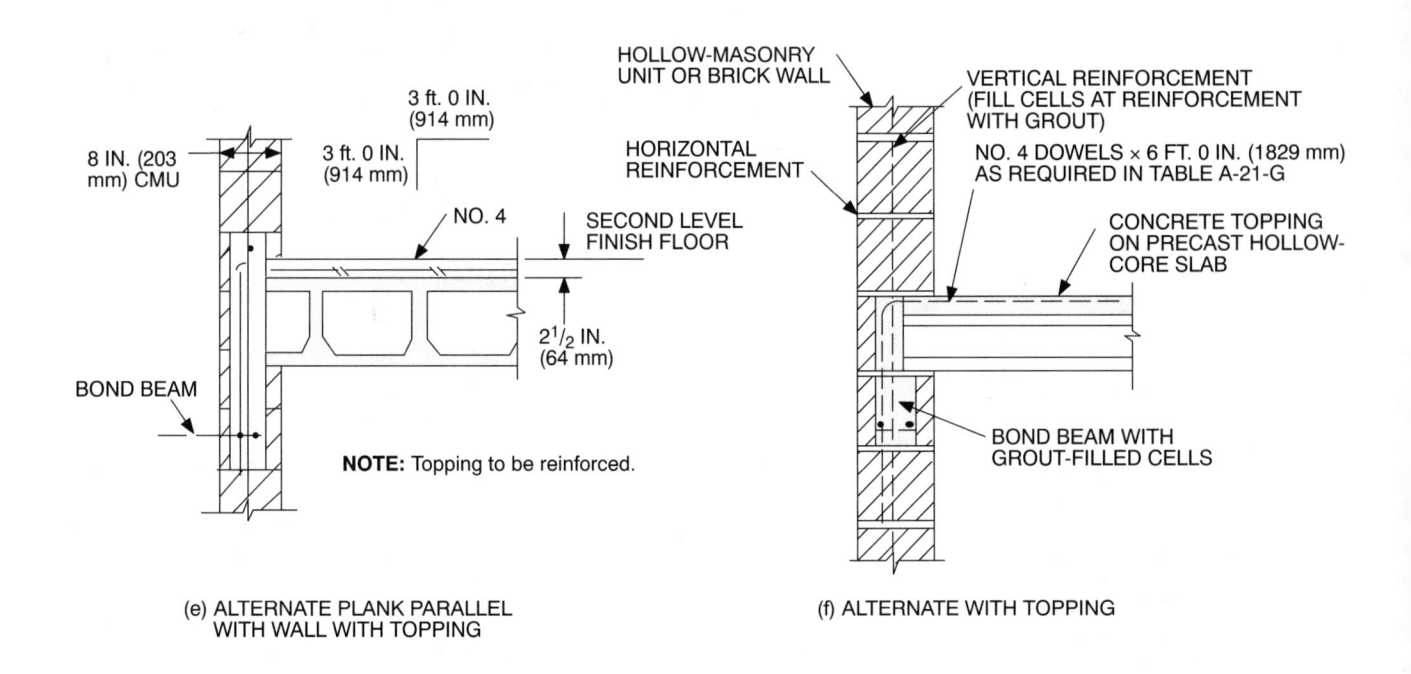

(e) ALTERNATE PLANK PARALLEL WITH WALL WITH TOPPING

(f) ALTERNATE WITH TOPPING

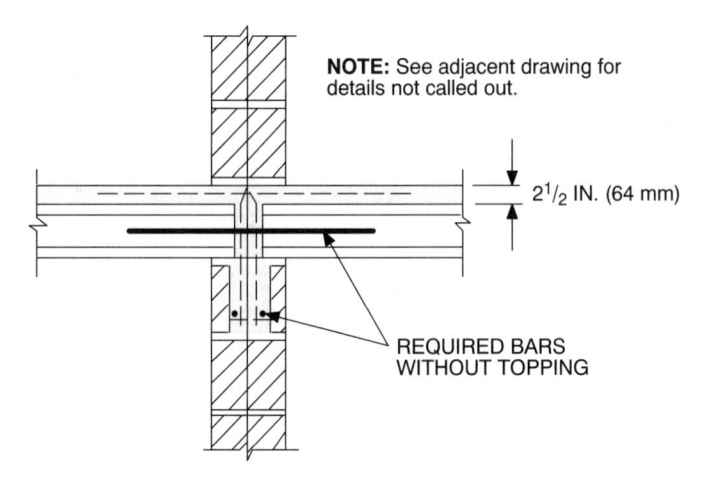

(g) INTERIOR WALL MINIMUM CONNECTION

FIGURE A-21-13—VARIOUS TYPES OF WALL CONNECTIONS—(Continued)

BENT ℝ 0.114 IN. (2.90 mm) (NO. 11 GALVANIZED SHEET GAGE) × ⌐ 4 IN. (102 mm) / 4 IN. (102 mm)

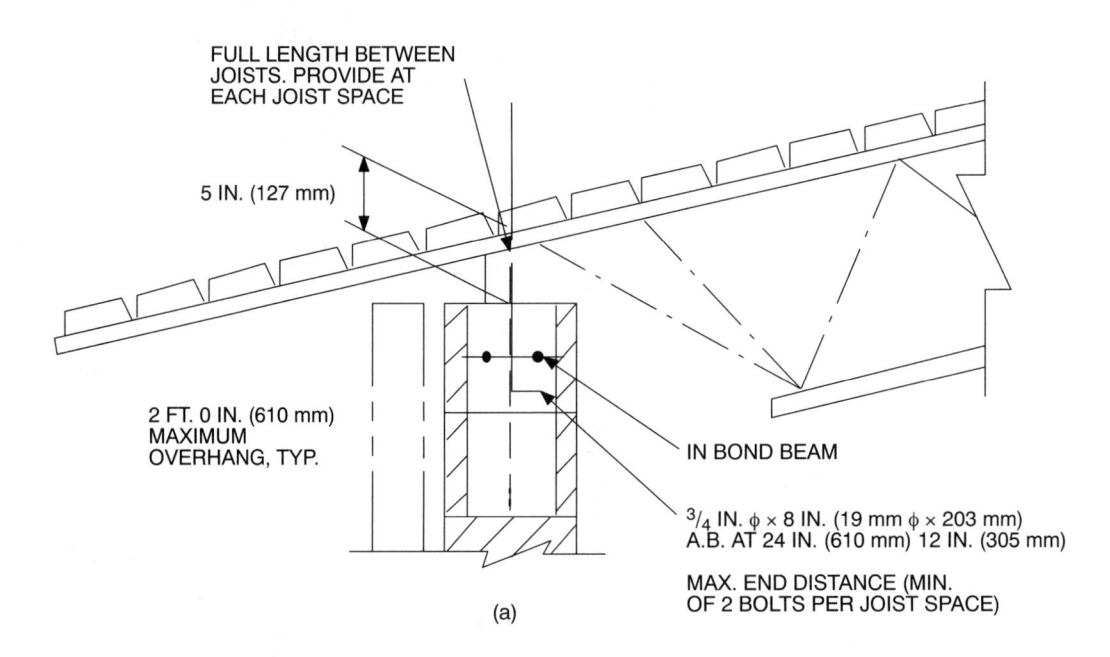

FULL LENGTH BETWEEN JOISTS. PROVIDE AT EACH JOIST SPACE

5 IN. (127 mm)

2 FT. 0 IN. (610 mm) MAXIMUM OVERHANG, TYP.

IN BOND BEAM

$3/_4$ IN. φ × 8 IN. (19 mm φ × 203 mm) A.B. AT 24 IN. (610 mm) 12 IN. (305 mm)

MAX. END DISTANCE (MIN. OF 2 BOLTS PER JOIST SPACE)

(a)

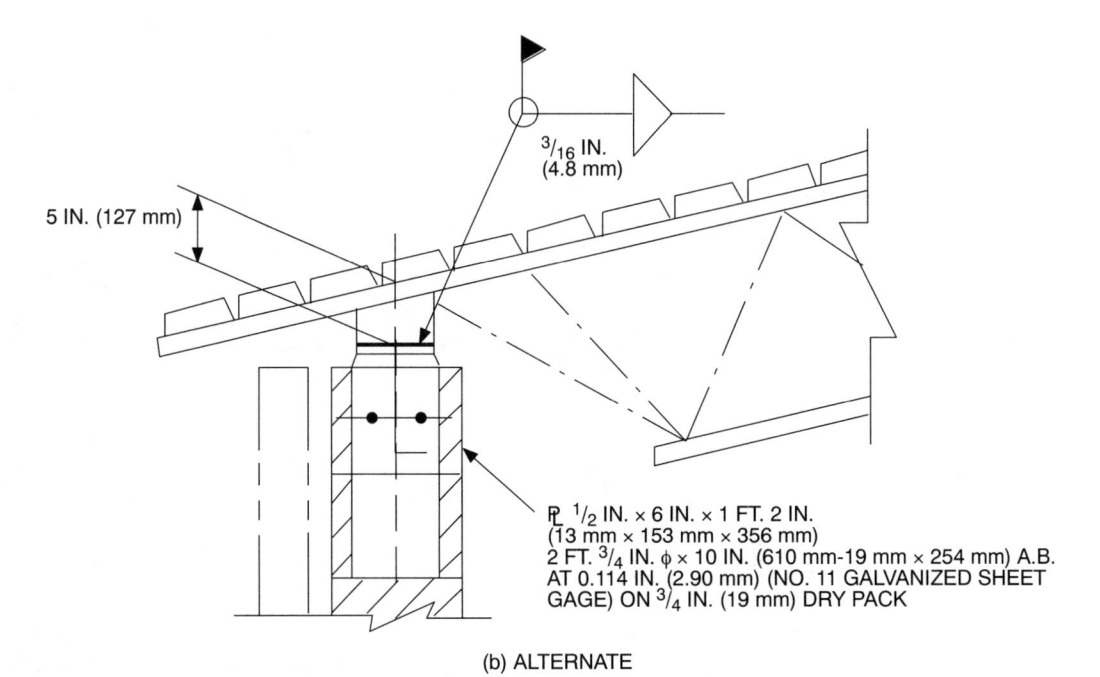

$3/_{16}$ IN. (4.8 mm)

5 IN. (127 mm)

ℝ $1/_2$ IN. × 6 IN. × 1 FT. 2 IN. (13 mm × 153 mm × 356 mm) 2 FT. $3/_4$ IN. φ × 10 IN. (610 mm-19 mm × 254 mm) A.B. AT 0.114 IN. (2.90 mm) (NO. 11 GALVANIZED SHEET GAGE) ON $3/_4$ IN. (19 mm) DRY PACK

(b) ALTERNATE

FIGURE A-21-14—EXTERIOR WALL DETAILS

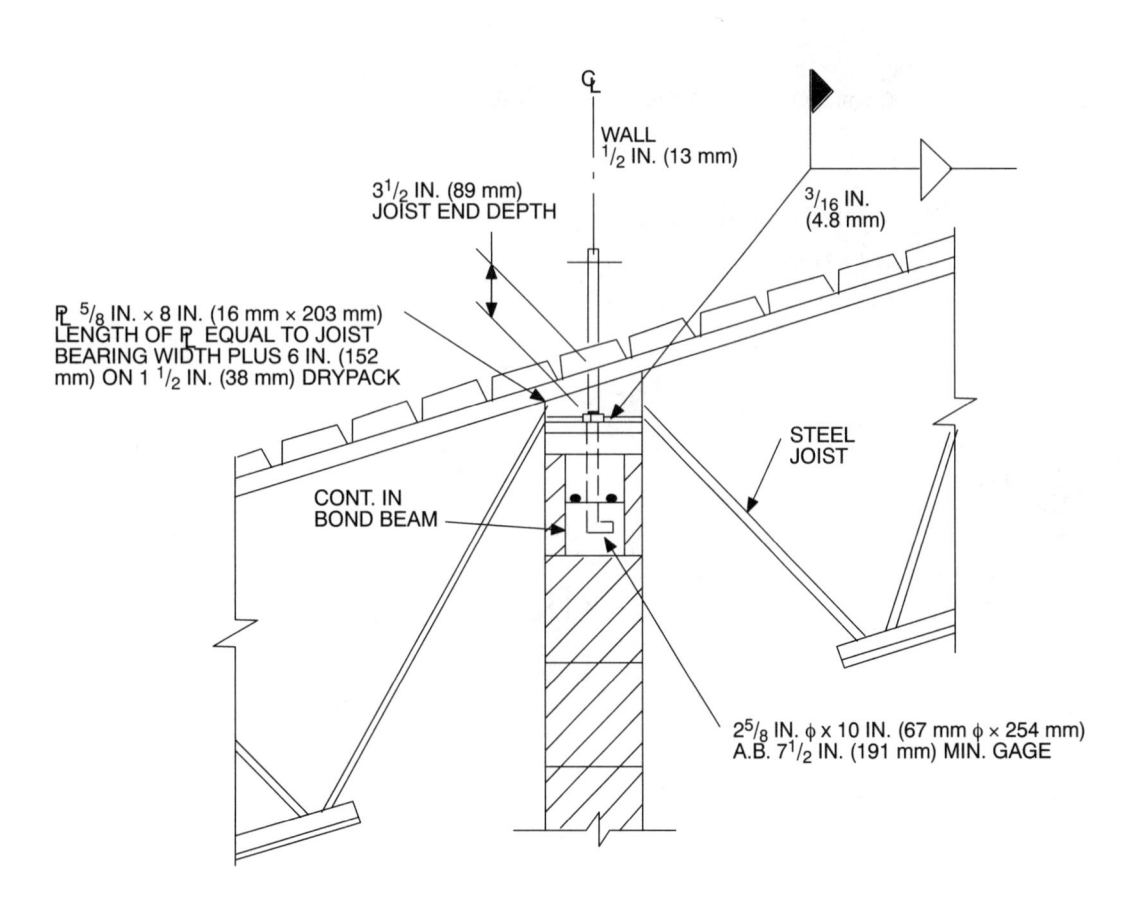

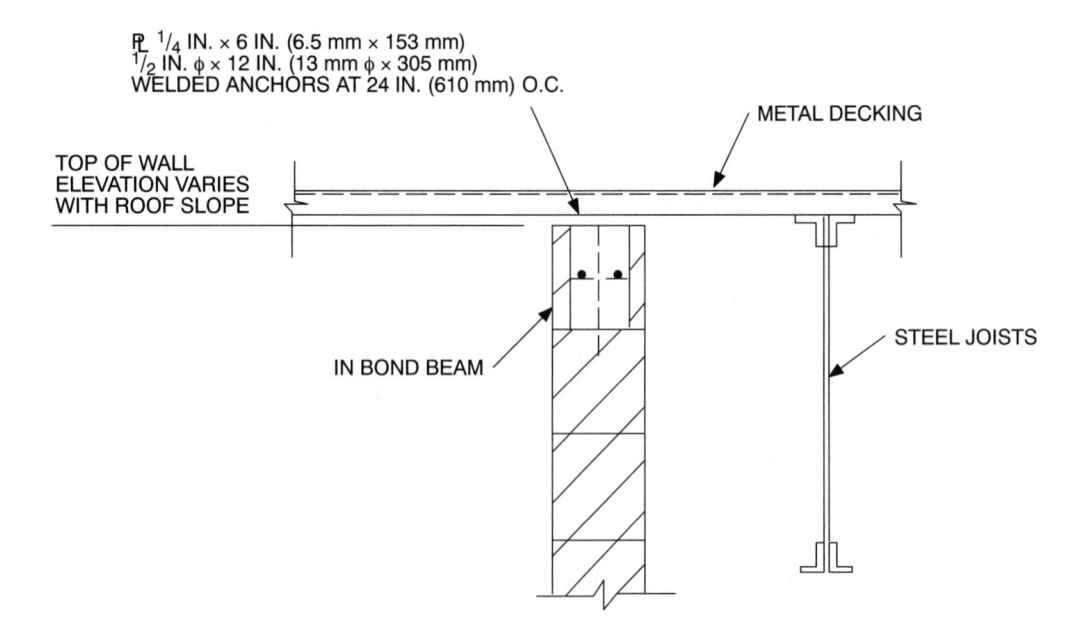

FIGURE A-21-15—INTERIOR WALL DETAILS

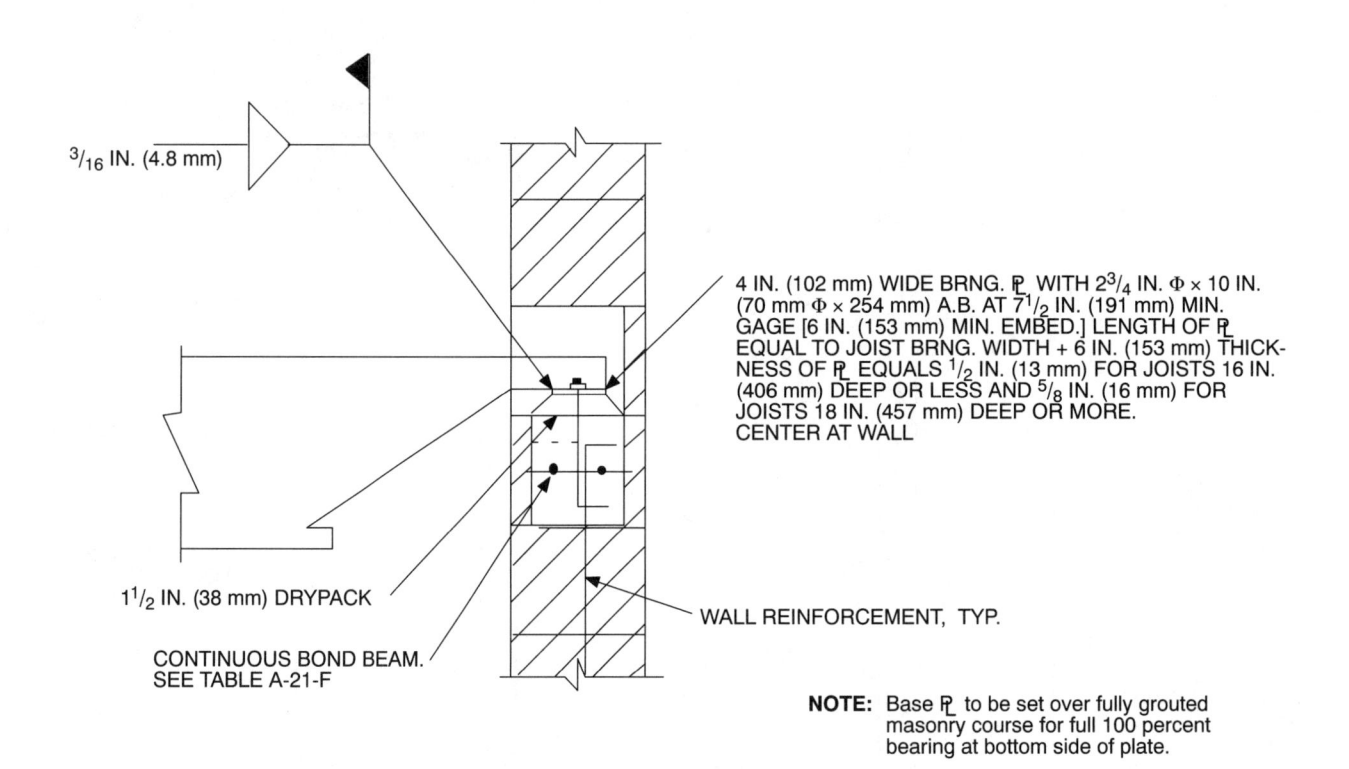

$^3/_{16}$ IN. (4.8 mm)

4 IN. (102 mm) WIDE BRNG. P̱ WITH 2$^3/_4$ IN. Φ × 10 IN. (70 mm Φ × 254 mm) A.B. AT 7$^1/_2$ IN. (191 mm) MIN. GAGE [6 IN. (153 mm) MIN. EMBED.] LENGTH OF P̱ EQUAL TO JOIST BRNG. WIDTH + 6 IN. (153 mm) THICK-NESS OF P̱ EQUALS $^1/_2$ IN. (13 mm) FOR JOISTS 16 IN. (406 mm) DEEP OR LESS AND $^5/_8$ IN. (16 mm) FOR JOISTS 18 IN. (457 mm) DEEP OR MORE. CENTER AT WALL

1$^1/_2$ IN. (38 mm) DRYPACK

CONTINUOUS BOND BEAM. SEE TABLE A-21-F

WALL REINFORCEMENT, TYP.

NOTE: Base P̱ to be set over fully grouted masonry course for full 100 percent bearing at bottom side of plate.

FIGURE A-21-16—FLOOR DETAILS
(Design Required for Joists and Wall)

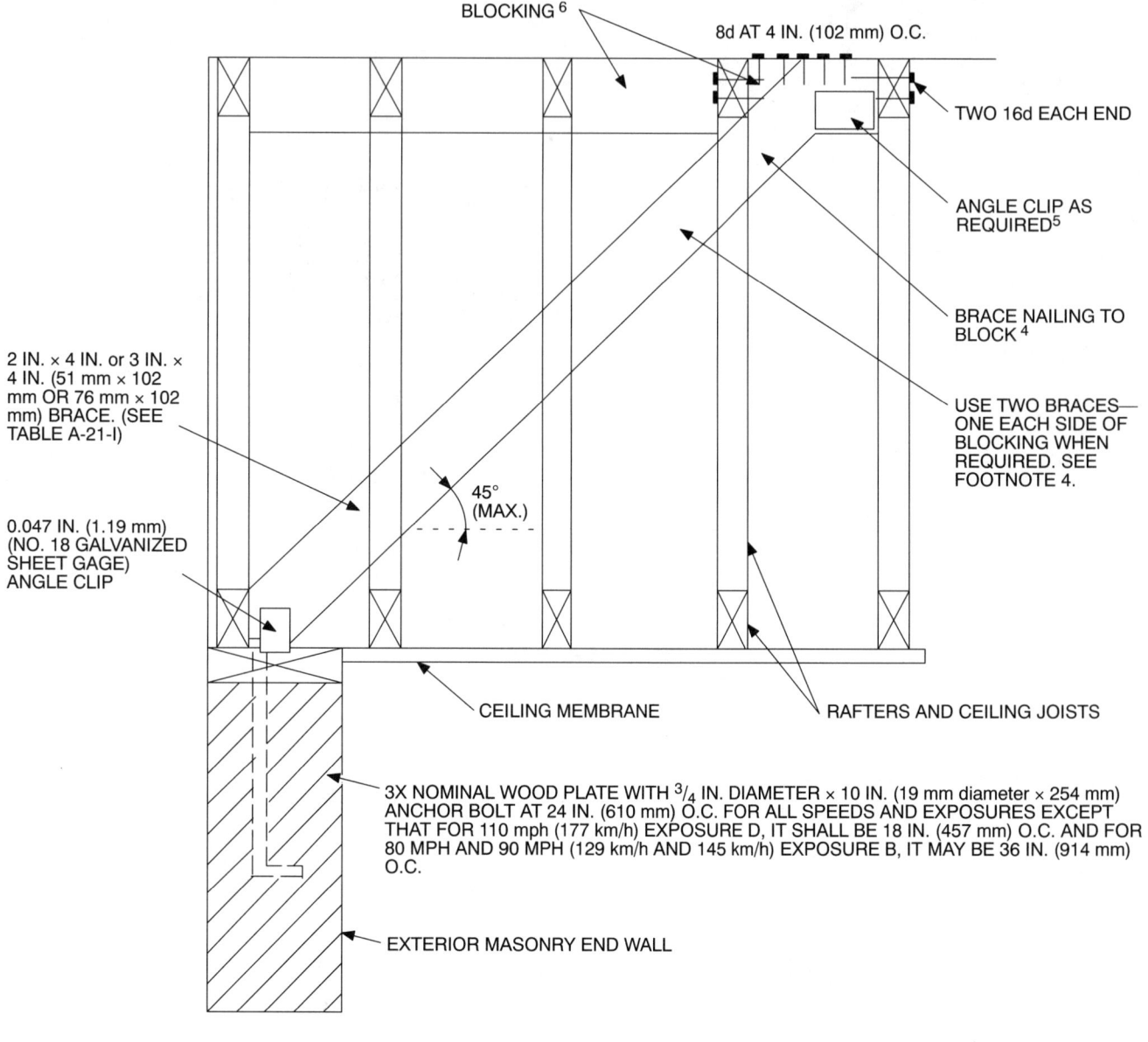

BLOCKING [6]

8d AT 4 IN. (102 mm) O.C.

TWO 16d EACH END

ANGLE CLIP AS REQUIRED[5]

BRACE NAILING TO BLOCK [4]

USE TWO BRACES— ONE EACH SIDE OF BLOCKING WHEN REQUIRED. SEE FOOTNOTE 4.

2 IN. × 4 IN. or 3 IN. × 4 IN. (51 mm × 102 mm OR 76 mm × 102 mm) BRACE. (SEE TABLE A-21-I)

45° (MAX.)

0.047 IN. (1.19 mm) (NO. 18 GALVANIZED SHEET GAGE) ANGLE CLIP

CEILING MEMBRANE

RAFTERS AND CEILING JOISTS

3X NOMINAL WOOD PLATE WITH 3/4 IN. DIAMETER × 10 IN. (19 mm diameter × 254 mm) ANCHOR BOLT AT 24 IN. (610 mm) O.C. FOR ALL SPEEDS AND EXPOSURES EXCEPT THAT FOR 110 mph (177 km/h) EXPOSURE D, IT SHALL BE 18 IN. (457 mm) O.C. AND FOR 80 MPH AND 90 MPH (129 km/h AND 145 km/h) EXPOSURE B, IT MAY BE 36 IN. (914 mm) O.C.

EXTERIOR MASONRY END WALL

[1] For roof slopes up to 5 units vertical in 12 units horizontal (42%); see Table A-21-I.
[2] See Detail 2, Table A-21-B, for size of angle clip.
[3] Angle clip one side or both sides as required by Table A-21-I.
[4] Use six 16d nails to fasten brace to block, except use two braces and six 16d nails each for 110 miles per hour (mph) (177 km/h), Exposure D. Place on brace on each side block.
[5] Add angle clip each end of block for 90 mph (145 km/h), Exposure D, and 100 and 110 mph (161 and 177 km/h) for Exposures C and D.
[6] Use 2 in. × 6 in. (51 mm × 153 mm) block with 2 in. (51 mm) × brace, 2 in. × 8 in. (51 mm × 203 mm) block with 3 in. (76 mm) × brace.

FIGURE A-21-17—DIAGONAL BRACING OF GABLE-END WALL[1]

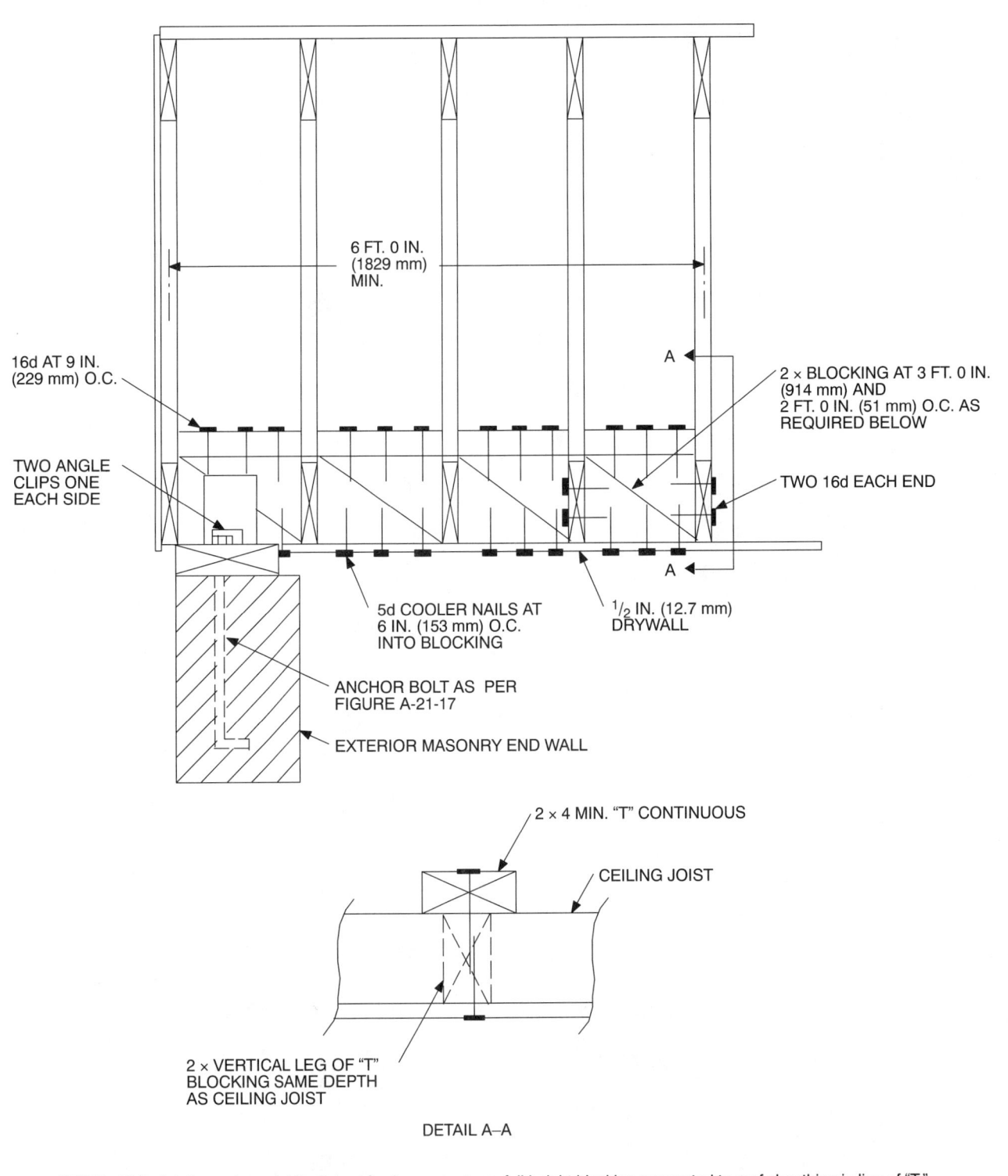

16d AT 9 IN.
(229 mm) O.C.

TWO ANGLE
CLIPS ONE
EACH SIDE

6 FT. 0 IN.
(1829 mm)
MIN.

A

2 × BLOCKING AT 3 FT. 0 IN.
(914 mm) AND
2 FT. 0 IN. (51 mm) O.C. AS
REQUIRED BELOW

TWO 16d EACH END

A

5d COOLER NAILS AT
6 IN. (153 mm) O.C.
INTO BLOCKING

¹/₂ IN. (12.7 mm)
DRYWALL

ANCHOR BOLT AS PER
FIGURE A-21-17

EXTERIOR MASONRY END WALL

2 × 4 MIN. "T" CONTINUOUS

CEILING JOIST

2 × VERTICAL LEG OF "T"
BLOCKING SAME DEPTH
AS CEILING JOIST

DETAIL A–A

NOTE: This detail may be used for flat roofs also, except use full height blocking connected to roof sheathing in lieu of "T."
2 × 4 "T" at 36 in. (914 mm) on center—90 miles per hour (mph) (145 km/h) Exposure C and less, and 100 mph and
110 mph (161 km/h and 177 km/h), Exposure B.
2 × 4 "T" at 24 in. (610 mm) on center—required for 90 mph (145 mm) exposure.
See Figure A-21-4 for details of clip angle and connections.

FIGURE A-21-18—ALTERNATE HORIZONTAL BRACING OF GABLE-END WALL

Appendix Chapter 23

CONVENTIONAL LIGHT-FRAME CONSTRUCTION IN HIGH-WIND AREAS

SECTION 2337 — GENERAL

2337.1 Purpose. The provisions of this chapter are intended to promote public safety and welfare by reducing the risk of wind-induced damages to conventional light-frame construction.

2337.2 Scope. This chapter applies to regular-shaped buildings which have roof structural members spanning 32 feet (9.75 m) or less, are not more than three stories in height, are of conventional light-frame construction and are located in areas with a basic wind speed from 80 through 110 miles per hour (mph) (129 km/h through 177 km/h).

> **EXCEPTION:** Detached carports and garages not exceeding 600 square feet (55.7 m^2) and accessory to Group R, Division 3 Occupancies need only comply with the roof-member-to-wall-tie requirements of Section 2337.5.8.

2337.3 Definitions. For the purpose of this chapter, certain terms are defined as follows:

CORROSION RESISTANT or **NONCORROSIVE** is material having a corrosion resistance equal to or greater than a hot-dipped galvanized coating of 1.5 ounces of zinc per square foot (4 g/m^2) of surface area.

2337.4 General. The requirements of Section 2320 are applicable except as specifically modified by this chapter. Other methods may be used, provided a satisfactory design is submitted showing compliance with the provisions of Section 1611.4 and other applicable portions of this code.

In addition to the other provisions of this chapter, foundations for buildings in areas subject to wave action or tidal surge shall be designed in accordance with approved national standards.

When an element is required to be corrosion resistant or noncorrosive, all of its parts, such as screws, nails, wire, dowels, bolts, nuts, washers, shims, anchors, ties and attachments, shall also be corrosion resistant or noncorrosive.

2337.5 Complete Load Path and Uplift Ties.

2337.5.1 General. Blocking, bridging, straps, approved framing anchors or mechanical fasteners shall be installed to provide continuous ties from the roof to the foundation system. (See Figure A-23-1.)

Tie straps shall be 1^1/$_8$-inch (28.6 mm) by 0.036-inch (0.91 mm) (No. 20 gage) sheet steel and shall be corrosion resistant as herein specified. All metal connectors and fasteners used in exposed locations or in areas otherwise subject to corrosion shall be of corrosion-resistant or noncorrosive material.

2337.5.2 Walls-to-foundation tie. Exterior walls shall be tied to a continuous foundation, or an elevated foundation system in accordance with Section 2337.10.

2337.5.3 Sills and foundation tie. Foundation plates resting on concrete or masonry foundations shall be bolted to the foundation with not less than ^{1}/$_2$-inch-diameter (13 mm) anchor bolts with 7-inch-minimum (178 mm) embedment into the foundation. In areas where the basic wind speed is 90 mph (145 km/h) or greater, the maximum spacing of anchor bolts shall be 4 feet (1219 mm) on center. Structures located where the basic wind speed is less than 90 mph (145 km/h) may have anchor bolts spaced not more than 6 feet (1829 mm) on center.

2337.5.4 Floor-to-foundation tie. The lowest-level exterior wall studs shall be connected to the foundation sill plate or an approved elevated foundation system with bent tie straps spaced not more than 48 inches (1219 mm) on center. Tie straps shall be nailed and installed in accordance with Table A-23-B and Figure A-23-1.

2337.5.5 Wall framing details. The spacing of 2-inch-by-4-inch studs (51 mm by 102 mm) in exterior walls shall not exceed 16 inches (406 mm) on center for areas with a basic wind speed of 90 mph (145 km/h) or greater.

Mechanical fasteners complying with this chapter shall be installed as required to connect studs to the sole plates, foundation sill plate and top plates of the wall.

Interior main cross-stud partitions shall be installed approximately perpendicular to the exterior wall when the length of the structure exceeds the width. The maximum distance between these partitions shall not exceed the width of the structure. Interior main cross-stud partition walls shall be securely fastened to exterior walls at the point of intersection with fasteners as required by Table 23-II-B-1. The main cross-stud partitions shall be covered on both sides by materials as described in Section 2337.5.6.

2337.5.6 Wall sheathing. All exterior walls and required interior main cross-stud partitions shall be sheathed in accordance with Table A-23-A. The total width of sheathed wall elements shall not be less than 50 percent of the exterior wall length or 60 percent of the width of the building for required interior main cross-stud partitions. The exterior wall sheathing or covering shall extend from the foundation sill plate or girder to the top plates at the roof level and shall be adequately attached thereto.

A sheathed wall element not less than 4 feet (1219 mm) in width shall be installed at each corner or as near thereto as possible. There shall not be less than one 4-foot (1219 mm) sheathed wall element for every 20 feet (6096 mm) or fraction thereof of wall length. The height-to-length ratio of required sheathed wall elements shall not exceed 3 for wood structural panel or particleboard and 1^1/$_2$ (38 mm) for other sheathing materials listed in Table A-23-A.

2337.5.7 Floor-to-floor tie. Upper-level exterior wall studs shall be aligned and connected to the wall studs below with a tie strap as required by Table A-23-B.

2337.5.8 Roof-members-to-wall tie. Tie straps shall be provided from the side of the roof-framing member to the exterior studs, posts or other supporting members below the roof. The wall studs to which the roof-framing members are tied shall be aligned with the roof-framing member and be connected in accordance with Table A-23-B.

The eave overhang shall not exceed 3 feet (914 mm) unless an analysis is provided showing that the required resistance is provided to prevent uplift.

Where openings exceed 6 feet (1829 mm) in width, the required tie straps shall be doubled at each edge of the opening and connected to a doubled full-height wall stud. When openings exceed 12 feet (3658 mm) in width, ties designed to prevent uplift shall be provided.

> **EXCEPTION:** The opening width may be increased to 16 feet (4877 mm) for garages and carports accessory to Group R, Division 3 Occupancies when constructed in accordance with the following:

1. Approved column bases shall be a minimum $^3/_{16}$-inch (4.8 mm) steel plate embedded not less than 8 inches (203 mm) into the concrete footing and connected to a minimum 4-inch-by-4-inch (102 mm by 102 mm) wood post with two $^5/_8$-inch-diameter (15.9 mm) through bolts.

2. Beams over openings shall be connected to minimum 4-inch by 4-inch (102 mm by 102 mm) wood posts below with an approved $^3/_{16}$-inch (4.8 mm) steel post cap with two $^5/_8$-inch-diameter (15.9 mm) through bolts to the posts and to the beams.

2337.5.9 Ridge ties. Opposing rafters shall be aligned at the ridge and be connected at the rafters with a tie strap in accordance with Table A-23-C.

2337.6 Masonry Veneer. Anchor ties shall be spaced so as to support not more than $1^1/_3$ square feet (860 mm^2) of wall area but not more than 12 inches (305 mm) on center vertically. The materials and connection details shall comply with Chapter 14.

2337.7 Roof Sheathing. Solid roof sheathing shall be applied and shall consist of a minimum 1-inch-thick (25 mm) nominal lumber applied diagonally or a minimum $^{15}/_{32}$-inch-thick (11.9 mm) wood structural panel or particleboard or other approved sheathing applied with the long dimension perpendicular to supporting rafters. Sheathing shall be nailed to roof framing in an approved manner. The end joints of wood structural panels or particleboard shall be staggered and shall occur over blocking, rafters or other supports.

2337.8 Gable-end Walls. The roof overhang at gabled ends shall not exceed 2 feet (610 mm) unless an analysis showing that the required resistance to prevent uplift is provided.

Gable-end wall studs shall be continuous between points of lateral support which are perpendicular to the plane of the wall.

Gable-end wall studs shall be attached with approved mechanical fasteners at the top and bottom.

2337.9 Roof Covering. Roof coverings shall be approved and shall be installed and fastened in accordance with Chapter 15 and with the manufacturer's instructions. In areas with basic wind speeds of 90 mph (145 km/h) or greater strip asphalt shingles shall be fastened with a minimum of six fasteners and hand sealed.

2337.10 Elevated Foundation.

2337.10.1 General. When approved, elevated foundations supporting not more than one story and meeting the provisions of this section may be used. A foundation investigation may be required by the building official.

2337.10.2 Material. All exposed wood-framing members shall be treated wood. All metal connectors and fasteners used in exposed locations shall be corrosion-resistant or noncorrosive steel.

2337.10.3 Wood piles. The spacing of wood piles shall not exceed 8 feet (2438 mm) on center. Square piles shall not be less than 10 inches (254 mm) and tapered piles shall have a tip of not less than 8 inches (203 mm). Ten-inch-square (64 516 mm^2) piles shall have a minimum embedment length of 10 feet (3048 mm) and shall project not more than 8 feet (2438 mm) above undisturbed ground surface. Eight-inch (203 mm) taper piles shall have a minimum embedment length of 14 feet (4267 mm) and shall project not more than 7 feet (2134 mm) above undisturbed ground surface.

2337.10.4 Girders. Floor girders shall be solid sawn timber, built-up 2-inch-thick (51 mm) lumber or trusses. Splices shall occur over wood piles. The floor girders shall span in the direction parallel to the potential floodwater and wave action.

2337.10.5 Connections. Wood piles may be notched to provide a shelf for supporting the floor girders. The total notching shall not exceed 50 percent of the pile cross section. Approved bolted connections with $^1/_4$-inch (6.4 mm) corrosion-resistant or noncorrosive steel plates and $^3/_4$-inch-diameter (19 mm) bolts shall be provided. Each end of the girder shall be connected to the piles using a minimum of two $^3/_4$-inch-diameter (19 mm) bolts.

TABLE A-23-A—WALL SHEATHING AT EXTERIOR WALLS AND INTERIOR MAIN CROSS-STUD PARTITIONS[1]

BASIC WIND SPEED (mph)			EXPOSURE		
× 1.61 for km/h	STORIES	LEVEL[2]	B	C	D
80	1		A	A	B
	2	2	A	A	B
		1	C	D	D
	3	3	A	A	B
		2	C	D	D
		1	C	D	E
90	1		A	B	B
	2	2	A	B	B
		1	C	D	D
	3	3	A	B	Not permitted
		2	C	D	
		1	D	E	
100	1		A	C	C
	2	2	A	C	C
		1	C	D	E
	3	3	A	Not permitted	Not permitted
		2	C		
		1	D		
110	1		B	C	C
	2	2	B	C	C
		1	D	E	E

[1]Sheathing types; exterior walls with sheathing at one face, interior main cross-stud partitions with sheathing at each face. The values for sheathing are listed in order of increased capacity. Sheathing with a capacity greater than required may be substituted for the sheathing listed. Particleboard sheathing in accordance with Table 23-IV-D-2 may be substituted for sheathing Types A and B.

A. One-half-inch (12.7 mm) gypsum board or gypsum sheathing with 5d cooler nails at 7 inches (178 mm) or $^3/_8$-inch (9.5 mm) gypsum lath and $^1/_2$-inch (12.7 mm) plaster.

B. One-half-inch (12.7 mm) gypsum board or gypsum sheathing with 5d cooler nails at 4 inches (102 mm).

C. Expanded metal lath and $^7/_8$-inch (22 mm) portland cement plaster.

D. Three-eighths-inch (9.5 mm) wood structural panel or particleboard sheathing with 8d nails at 6 inches (153 mm) all edges and 12 inches (305 mm) intermediate.

E. Three-eighths-inch (9.5 mm) plywood or particleboard sheathing with 8d nails at 4 inches (102 mm) all edges and 12 inches (305 mm) intermediate.
 The application of these sheathing materials shall comply with Section 2513.5 and Table 25-I for Types A, B and C, and Section 2315.1 and Table 23-II-I-1 or 23-II-I-2 for Types D and E. All panel edges of Types D and E shall be backed with 2-inch (51 mm) nominal or wider framing.

[2]Level refers to the space between the upper surface of any floor and upper surface of floor next above. The topmost level shall be the space between upper surface of the topmost floor and the ceiling or roof above. Wall sheathing at useable or unused under-floor space shall be provided as required for the level directly above.

TABLE A-23-B—ROOF AND FLOOR ANCHORAGE AT EXTERIOR WALLS

BASIC WIND SPEED (mph)		NUMBER OF NAILS[2]		
		Exposure		
× 1.61 for km/h	LOCATION[1]	B	C	D
80	roof to wall	6-8d	8-8d	8-10d
	floor to floor	—	4-10d	6-10d
	floor to foundation	—	4-10d	4-10d
90	roof to wall	8-8d	8-10d	10-10d
	floor to floor	—	6-10d	8-10d
	floor to foundation	—	4-10d	6-10d
100	roof to wall	8-10d	10-10d	12-10d
	floor to floor	6-10d	8-10d	10-10d
	floor to foundation	4-10d	6-10d	8-10d
110	roof to wall	10-10d	12-10d	12-10d
	floor to floor	8-10d	10-10d	10-10d
	floor to foundation	6-10d	8-10d	8-10d

[1]For floor-to-foundation anchorage, see Section 2337.5.4.

[2]Number of common nails listed is total required for each tie strap. The tie straps shall be spaced at 48 inches (1219 mm) on center along the length of the wall. The number of nails on each side of the roof or floor plate joints shall be equal. Nails shall be spaced to avoid splitting of the wood. See Figure A-23-1 for illustration of these tie straps.

TABLE A-23-C—RIDGE TIE-STRAP NAILING[1]

BASIC WIND SPEED (mph)	NUMBER OF NAILS[1]		
	Exposure		
× 1.61 for km/h	B	C	D
80	6-10d	8-10d	10-10d
90	8-10d	10-10d	12-10d
100	10-10d	12-10d	14-10d
110	12-10d	14-10d	16-10d

[1]Number of common nails listed is total required for each tie strap. The tie straps shall be spaced at 48 inches (1219 mm) on center along the length of the roof. The number of nails on each side of the rafter/ridge joint shall be equal. Nails shall be spaced to avoid splitting of the wood. See Figure A-23-1 for illustration of these tie straps.

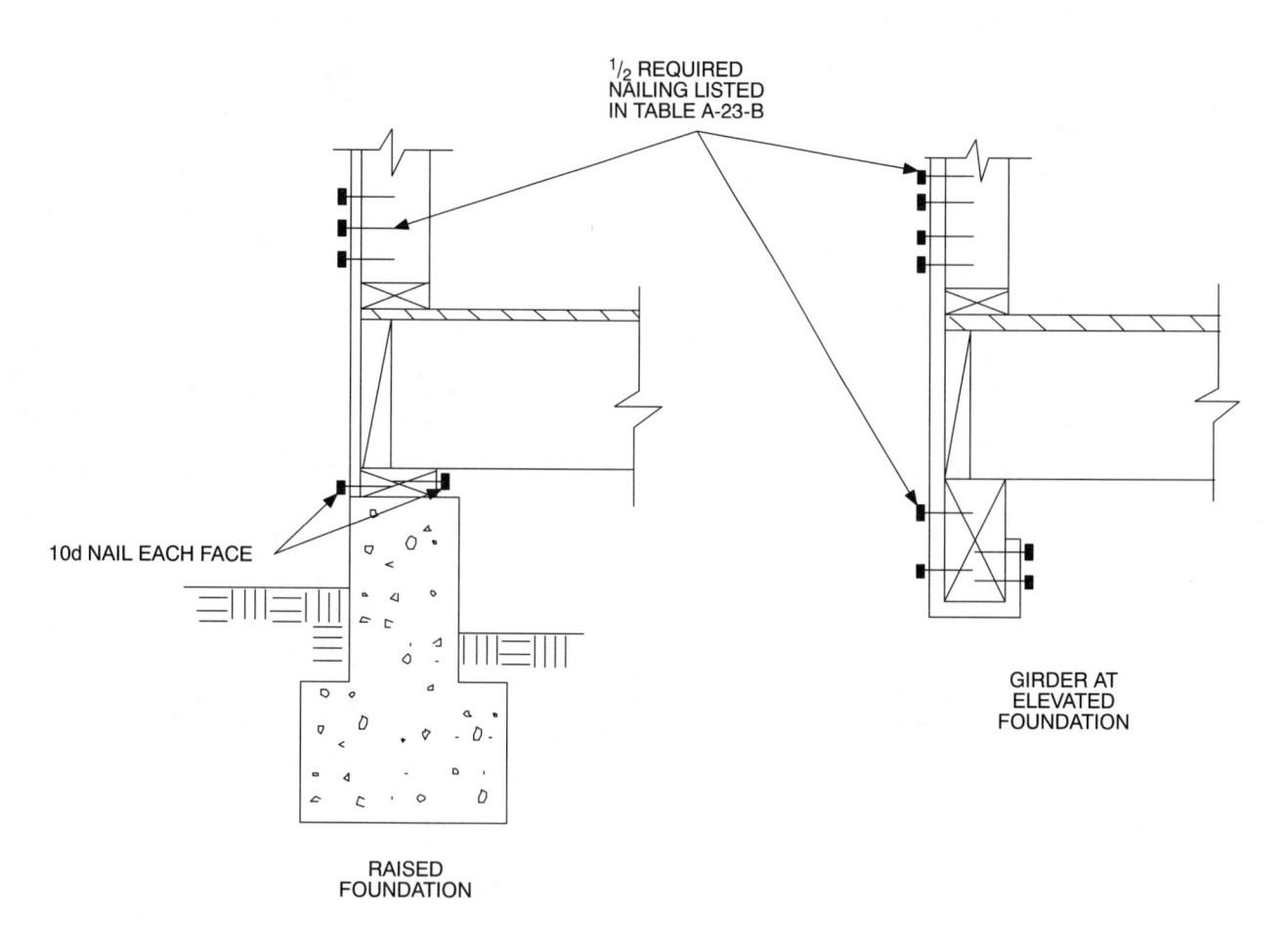

½ REQUIRED NAILING LISTED IN TABLE A-23-B

10d NAIL EACH FACE

RAISED FOUNDATION

GIRDER AT ELEVATED FOUNDATION

FIGURE A-23-1—COMPLETE LOAD PATH DETAILS

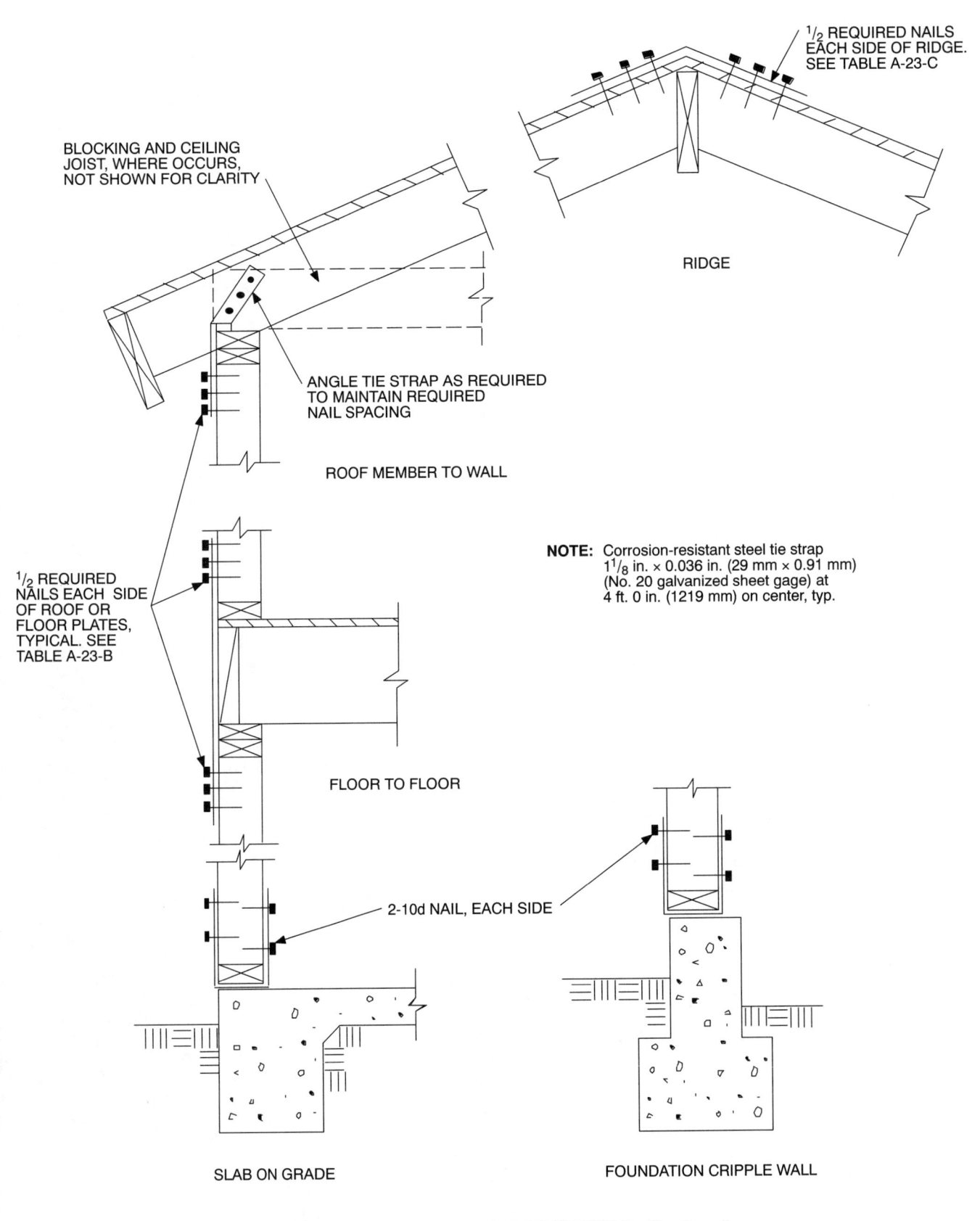

FIGURE A-23-1—COMPLETE LOAD PATH DETAILS—(Continued)

Appendix Chapters 30 through 34 are printed in Volume 1
of the *Uniform Building Code*.

UNIT CONVERSION TABLES

SI SYMBOLS AND PREFIXES

BASE UNITS		
Quantity	Unit	Symbol
Length	Meter	m
Mass	Kilogram	kg
Time	Second	s
Electric current	Ampere	A
Thermodynamic temperature	Kelvin	K
Amount of substance	Mole	mol
Luminous intensity	Candela	cd

SI SUPPLEMENTARY UNITS		
Quantity	Unit	Symbol
Plane angle	Radian	rad
Solid angle	Steradian	sr

SI PREFIXES		
Multiplication Factor	Prefix	Symbol
$1\ 000\ 000\ 000\ 000\ 000\ 000 = 10^{18}$	exa	E
$1\ 000\ 000\ 000\ 000\ 000 = 10^{15}$	peta	P
$1\ 000\ 000\ 000\ 000 = 10^{12}$	tera	T
$1\ 000\ 000\ 000 = 10^{9}$	giga	G
$1\ 000\ 000 = 10^{6}$	mega	M
$1\ 000 = 10^{3}$	kilo	k
$100 = 10^{2}$	hecto	h
$10 = 10^{1}$	deka	da
$0.1 = 10^{-1}$	deci	d
$0.01 = 10^{-2}$	centi	c
$0.001 = 10^{-3}$	milli	m
$0.000\ 001 = 10^{-6}$	micro	μ
$0.000\ 000\ 001 = 10^{-9}$	nano	n
$0.000\ 000\ 000\ 001 = 10^{-12}$	pico	p
$0.000\ 000\ 000\ 000\ 001 = 10^{-15}$	femto	f
$0.000\ 000\ 000\ 000\ 000\ 001 = 10^{-18}$	atto	a

SI DERIVED UNIT WITH SPECIAL NAMES			
Quantity	Unit	Symbol	Formula
Frequency (of a periodic phenomenon)	hertz	Hz	$1/s$
Force	newton	N	$kg \cdot m/s^2$
Pressure, stress	pascal	Pa	N/m^2
Energy, work, quantity of heat	joule	J	$N \cdot m$
Power, radiant flux	watt	W	J/s
Quantity of electricity, electric charge	coulomb	C	$A \cdot s$
Electric potential, potential difference, electromotive force	volt	V	W/A
Capacitance	farad	F	C/V
Electric resistance	ohm	Ω	V/A
Conductance	siemens	S	A/V
Magnetic flux	weber	Wb	$V \cdot s$
Magnetic flux density	tesla	T	Wb/m^2
Inductance	henry	H	Wb/A
Luminous flux	lumen	lm	$cd \cdot sr$
Illuminance	lux	lx	lm/m^2
Activity (of radionuclides)	becquerel	Bq	$1/s$
Absorbed dose	gray	Gy	J/kg

CONVERSION FACTORS

To convert	to	multiply by
LENGTH		
1 mile (U.S. statute)	km	1.609 344
1 yd	m	0.9144
1 ft	m	0.3048
	mm	304.8
1 in	mm	25.4
AREA		
1 mile^2 (U.S. statute)	km^2	2.589 998
1 acre (U.S. survey)	ha	0.404 6873
	m^2	4046.873
1 yd^2	m^2	0.836 1274
1 ft^2	m^2	0.092 903 04
1 in^2	mm^2	645.16
VOLUME, MODULUS OF SECTION		
l acre ft	m^3	1233.489
1 yd^3	m^3	0.764 5549
100 board ft	m^3	0.235 9737
1 ft^3	m^3	0.028 316 85
	L(dm^3)	28.3168
1 in^3	mm^3	16 387.06
	mL (cm^3)	16.3871
1 barrel (42 U.S. gallons)	m^3	0.158 9873
(FLUID) CAPACITY		
1 gal (U.S. liquid)*	L**	3.785 412
1 qt (U.S. liquid)	mL	946.3529
1 pt (U.S. liquid)	mL	473.1765
1 fl oz (U.S.)	mL	29.5735
1 gal (U.S. liquid)	m^3	0.003 785 412
*1 gallon (UK) approx. 1.2 gal (U.S.)	**1 liter approx. 0.001 cubic meter	
SECOND MOMENT OF AREA		
1 in^4	mm^4	416 231 4
	m^4	416 231 4 $\times$ 10^{-7}
PLANE ANGLE		
1° (degree)	rad	0.017 453 29
	mrad	17.453 29
1′ (minute)	urad	290.8882
1″ (second)	urad	4.848 137
VELOCITY, SPEED		
1 ft/s	m/s	0.3048
1 mile/h	km/h	1.609 344
	m/s	0.447 04
VOLUME RATE OF FLOW		
1 ft^3/s	m^3/s	0.028 316 85
1 ft^3/min	L/s	0.471 9474
1 gal/min	L/s	0.063 0902
1 gal/min	m^3/min	0.0038
1 gal/h	mL/s	1.051 50
1 million gal/d	L/s	43.8126
1 acre ft/s	m^3/s	1233.49
TEMPERATURE INTERVAL		
1°F	°C or K	0.555 556 $^5/_9$°C = $^5/_9$K
EQUIVALENT TEMPERATURE ($t_{°C} = T_K - 273.15$)		
$t_{°F}$	$t_{°C}$	$t_{°F} = {}^9/_5 t_{°C} + 32$

(Continued)

CONVERSION FACTORS—(Continued)

To convert	to	multiply by
MASS		
1 ton (short ***)	metric ton	0.907 185
	kg	907.1847
1 lb	kg	0.453 5924
1 oz	g	28.349 52
***1 long ton (2,240 lb)	kg	1016.047
MASS PER UNIT AREA		
1 lb/ft^2	kg/m^2	4.882 428
1 oz/yd^2	g/m^2	33.905 75
1 oz/ft^2	g/m^2	305.1517
DENSITY (MASS PER UNIT VOLUME)		
1 lb/ft^3	kg/m^3	16.01846
1 lb/yd^3	kg/m^3	0.593 2764
1 ton/yd^3	t/m^3	1.186 553
FORCE		
1 tonf (ton-force)	kN	8.896 44
1 kip (1,000 lbf)	kN	4.448 22
1 lbf (pound-force)	N	4.448 22
MOMENT OF FORCE, TORQUE		
1 lbf·ft	N·m	1.355 818
1 lbf·in	N·m	0.112 9848
1 tonf·ft	kN·m	2.711 64
1 kip·ft	kN·m	1.355 82
FORCE PER UNIT LENGTH		
1 lbf/ft	N/m	14.5939
1 lbf/in	N/m	175.1268
1 tonf/ft	kN/m	29.1878
PRESSURE, STRESS, MODULUS OF ELASTICITY (FORCE PER UNIT AREA) (1 Pa = 1 N/m^2)		
1 tonf/in^2	MPa	13.7895
1 tonf/ft^2	kPa	95.7605
1 kip/in^2	MPa	6.894 757
1 lbf/in^2	kPa	6.894 757
1 lbf/ft^2	Pa	47.8803
Atmosphere	kPa	101.3250
1 inch mercury	kPa	3.376 85
1 foot (water column at 32°F)	kPa	2.988 98
WORK, ENERGY, HEAT(1J = 1N·m = 1W·s)		
1 kWh (550 ft·lbf/s)	MJ	3.6
1 Btu (Int. Table)	kJ	1.055 056
	J	1055.056
1 ft·lbf	J	1.355 818
COEFFICIENT OF HEAT TRANSFER		
1 Btu/(ft^2·h·°F)	W/(m^2·K)	5.678 263
THERMAL CONDUCTIVITY		
1 Btu/(ft·h·°F)	W/(m·K)	1.730 735
ILLUMINANCE		
1 lm/ft^2 (footcandle)	lx (lux)	10.763 91
LUMINANCE		
1 cd/ft^2	cd/m^2	10.7639
1 foot lambert	cd/m^2	3.426 259
1 lambert	kcd/m^2	3.183 099

GAGE CONVERSION TABLE

APPROXIMATE MINIMUM THICKNESS (inch/mm) FOR CARBON SHEET STEEL CORRESPONDING TO MANUFACTURER'S STANDARD GAGE AND GALVANIZED SHEET GAGE NUMBERS

Manufacturer's Standard Gage No.	Decimal and Nominal Thickness Equivalent		Recommended Minimum Thickness Equivalent[1]		Galvanized Sheet Gage No.	Decimal and Nominal Thickness Equivalent		Recommended Minimum Thickness Equivalent[1]	
	(inch)	(mm)[2]	(inch)	(mm)[2]		(inch)	(mm)[2]	(inch)	(mm)[2]
8	0.1644	4.17	0.156	3.46	8	0.1681	4.27	0.159	4.04
9	0.1495	3.80	0.142	3.61	9	0.1532	3.89	0.144	3.66
10	0.1345	3.42	0.127	3.23	10	0.1382	3.51	0.129	3.23
11	0.1196	3.04	0.112	2.84	11	0.1233	3.13	0.114	2.90
12	0.1046	2.66	0.097	2.46	12	0.1084	2.75	0.099	2.51
13	0.0897	2.28	0.083	2.11	13	0.0934	2.37	0.084	2.13
14	0.0747	1.90	0.068	1.73	14	0.0785	1.97	0.070	1.78
15	0.0673	1.71	0.062	1.57	15	0.0710	1.80	0.065	1.65
16	0.0598	1.52	0.055	1.40	16	0.0635	1.61	0.058	1.47
17	0.0538	1.37	0.050	1.27	17	0.0575	1.46	0.053	1.35
18	0.0478	1.21	0.044	1.12	18	0.0516	1.31	0.047	1.19
19	0.0418	1.06	0.038	0.97	19	0.0456	1.16	0.041	1.04
20	0.0359	0.91	0.033	0.84	20	0.0396	1.01	0.036	0.91
21	0.0329	0.84	0.030	0.76	21	0.0366	0.93	0.033	0.84
22	0.0299	0.76	0.027	0.69	22	0.0336	0.85	0.030	0.76
23	0.0269	0.68	0.024	0.61	23	0.0306	0.78	0.027	0.69
24	0.0239	0.61	0.021	0.53	24	0.0276	0.70	0.024	0.61
25	0.0209	0.53	0.018	0.46	25	0.0247	0.63	0.021	0.53
26	0.0179	0.45	0.016	0.41	26	0.0217	0.55	0.019	0.48
27	0.0164	0.42	0.014	0.36	27	0.0202	0.51	0.017	0.43
28	0.0149	0.38	0.013	0.33	28	0.0187	0.47	0.016	0.41
					29	0.0172	0.44	0.014	0.36
					30	0.0157	0.40	0.013	0.33

[1]The thickness of the sheets set forth in the code correspond to the thickness shown under these columns. They are the approximate minimum thicknesses and are based on the following references:

Carbon sheet steel—Thickness 0.071 inch and over:
　ASTM A 568-74, Table 3, Thickness Tolerances of Hot-Rolled Sheet
　(Carbon Steel).

Carbon sheet steel—Thickness less than 0.071 inch:
　ASTM A 568-74, Table 23, Thickness Tolerances of Cold-Rolled Sheet
　(Carbon and High Strength Low Alloy).

Galvanized sheet steel—All thicknesses:
　ASTM A 525-79, Table 4, Thickness Tolerances of Hot-Dip Galvanized Sheet.

Minimum thickness is the difference between the thickness equivalent of each gage and the maximum negative tolerance for the widest rolled width.

[2]The SI equivalents are calculated and rounded to two significant figures following the decimal point.

INDEX

Index is not all inclusive of code items.